RSMeans
Light Commercial Cost Data
19th Annual Edition

2000

Senior Editor
Howard M. Chandler

Contributing Editors
Barbara Balboni
John H. Chiang, PE
Paul C. Crosscup
Jennifer L. Curran
Stephen E. Donnelly
J. Robert Lang
Robert C. McNichols
Robert W. Mewis
Melville J. Mossman, PE
John J. Moylan
Jeannene D. Murphy
Peter T. Nightingale
Stephen C. Plotner
Michael J. Regan
William R. Tennyson, II
Phillip R. Waier, PE

Manager, Engineering Operations
John H. Ferguson, PE

President
Durwood S. Snead

Vice President and General Manager
Roger J. Grant

Vice President, Sales and Marketing
John M. Shea

Production Manager
Michael Kokernak

Production Coordinator
Marion E. Schofield

Technical Support
Wayne D. Anderson
Thomas J. Dion
Michael H. Donelan
Jonathan Forgit
Gary L. Hoitt
Paula Reale-Camelio
Kathryn S. Rodriguez
Sheryl A. Rose
James N. Wills

Art Director
Helen A. Marcella

Book & Cover Design
Norman R. Forgit

R.S. Means Company, Inc. ("R.S. Means"), its authors, editors and engineers, apply diligence and judgment in locating and using reliable sources for the information published. **However, R.S. Means makes no express or implied warranty or guarantee in connection with the content of the information contained herein, including the accuracy, correctness, value, sufficiency, or completeness of the data, methods and other information contained herein. R.S. Means makes no express or implied warranty of merchantability or fitness for a particular purpose.** R.S. Means shall have no liability to any customer or third party for any loss, expense, or damage including consequential, incidental, special or punitive damages, including lost profits or lost revenue, caused directly or indirectly by any error or omission, or arising out of, or in connection with, the information contained herein.

No part of this publication may be reproduced, stored in a retrieval system, or transmitted in any form or by any means without prior written permission of R.S. Means Company, Inc.

Editorial Advisory Board

James E. Armstrong
Energy Consultant
EnergyVision

William R. Barry
Chief Estimator
Mitchell Construction Company

Robert F. Cox, PhD
Assistant Professor
ME Rinker Sr. School of Bldg. Constr.
University of Florida

Roy F. Gilley, AIA
Principal
Gilley-Hinkel Architects

Kenneth K. Humphreys, PhD,
   PE, CCE
Secretary-Treasurer
International Cost
Engineering Council

Patricia L. Jackson, PE
Vice President, Project Management
Aguirre Corporation

Martin F. Joyce
Vice President, Utility Division
Bond Brothers, Inc.

First Printing

Foreword

R.S. Means Co., Inc. is owned by CMD Group, a leading worldwide provider of proprietary construction information. CMD Group is comprised of three synergistic product groups crafted to be the complete resource for reliable, timely and actionable construction market data. In North America, CMD Group encompasses: Architects' First Source, an innovative product selection and specification solution in print and on the Internet; Construction Market Data (CMD), the source for construction activity information, as well as early planning reports for the design community; Associated Construction Publications, with 14 magazines, one of the largest editorial networks dedicated to U.S. highway and heavy construction coverage; Manufacturer's Survey Associates (MSA), a leading estimating and quantity survey firm in the U.S.; R.S. Means, the authority on construction cost data in North America; CMD Canada, the leading supplier of project information, industry news and forecasting data products for the Canadian construction industry; and BIMSA/Mexico, the dominant distributor of information on building projects and construction throughout Mexico. Worldwide, CMD Group includes Byggfakta Scandinavia, providing construction market data to Denmark, Estonia, Finland, Norway and Sweden; and Cordell Building Information Services, the market leader for construction and cost information in Australia.

Our Mission

Since 1942, R.S. Means Company, Inc. has been actively engaged in construction cost publishing and consulting throughout North America.

Today, over fifty years after the company began, our primary objective remains the same: to provide you, the construction and facilities professional, with the most current and comprehensive construction cost data possible.

Whether you are a contractor, an owner, an architect, an engineer, a facilities manager, or anyone else who needs a fast and reliable construction cost estimate, you'll find this publication to be a highly useful and necessary tool.

Today, with the constant flow of new construction methods and materials, it's difficult to find the time to look at and evaluate all the different construction cost possibilities. In addition, because labor and material costs keep changing, last year's cost information is not a reliable basis for today's estimate or budget.

That's why so many construction professionals turn to R.S. Means. We keep track of the costs for you, along with a wide range of other key information, from city cost indexes . . . to productivity rates . . . to crew composition . . . to contractor's overhead and profit rates.

R.S. Means performs these functions by collecting data from all facets of the industry, and organizing it in a format that is instantly accessible to you. From the preliminary budget to the detailed unit price estimate, you'll find the data in this book useful for all phases of construction cost determination.

The Staff, the Organization, and Our Services

When you purchase one of R.S. Means' publications, you are in effect hiring the services of a full-time staff of construction and engineering professionals.

Our thoroughly experienced and highly qualified staff works daily at collecting, analyzing, and disseminating comprehensive cost information for your needs. These staff members have years of practical construction experience and engineering training prior to joining the firm. As a result, you can count on them not only for the cost figures, but also for additional background reference information that will help you create a realistic estimate.

The Means organization is always prepared to help you solve construction problems through its five major divisions: Construction and Cost Data Publishing, Electronic Products and Services, Consulting Services, Insurance Division, and Educational Services.

Besides a full array of construction cost estimating books, Means also publishes a number of other reference works for the construction industry. Subjects include construction estimating and project and business management; special topics such as HVAC, roofing, plumbing, and hazardous waste remediation; and a library of facility management references.

In addition, you can access all of our construction cost data through your computer with Means CostWorks 2000 CD-ROM, an electronic tool that offers over 50,000 lines of Means construction cost data.

What's more, you can increase your knowledge and improve your construction estimating and management performance with a Means Construction Seminar or In-House Training Program. These two-day seminar programs offer unparalleled opportunities for everyone in your organization to get updated on a wide variety of construction-related issues.

Means also is a worldwide provider of construction cost management and analysis services for commercial and government owners and of claims and valuation services for insurers.

In short, R.S. Means can provide you with the tools and expertise for constructing accurate and dependable construction estimates and budgets in a variety of ways.

Robert Snow Means Established a Tradition of Quality That Continues Today

Robert Snow Means spent years building his company, making certain he always delivered a quality product.

Today, at R.S. Means, we do more than talk about the quality of our data and the usefulness of our books. We stand behind all of our data, from historical cost indexes... to construction materials and techniques... to current costs.

If you have any questions about our products or services, please call us toll-free at 1-800-334-3509. Our customer service representatives will be happy to assist you.

Table of Contents

Foreword	ii
How the Book Is Built: An Overview	iv
Quick Start	v
How To Use the Book: The Details	vi
Commercial/Industrial/Institutional Section	1
Assemblies Cost Section	92
Unit Price Section	267
Reference Section	537
Reference Numbers	538
Crew Listings	580
Historical Cost Indexes	603
Location Factors	611
Abbreviations	617
Index	620
Other Means Publications and Reference Works	Yellow Pages
Installing Contractor's Overhead & Profit	Inside Back Cover

COMMERCIAL/INDUSTRIAL/INSTITUTIONAL

BUILDING TYPES

UNIT PRICES
1. GENERAL REQUIREMENTS
2. SITE CONSTRUCTION
3. CONCRETE
4. MASONRY
5. METALS
6. WOOD & PLASTICS
7. THERMAL & MOISTURE PROTECTION
8. DOORS & WINDOWS
9. FINISHES
10. SPECIALTIES
11. EQUIPMENT
12. FURNISHINGS
13. SPECIAL CONSTRUCTION
14. CONVEYING SYSTEMS
15. MECHANICAL
16. ELECTRICAL

ASSEMBLIES
1. FOUNDATIONS
2. SUBSTRUCTURES
3. SUPERSTRUCTURES
4. EXTERIOR CLOSURE
5. ROOFING
6. INTERIOR CONSTRUCTION
7. CONVEYING SYSTEMS
8. MECHANICAL
9. ELECTRICAL
11. SPECIAL CONSTRUCTION
12. SITE WORK

REFERENCE INFORMATION
- REFERENCE NUMBERS
- CREWS
- COST INDEXES
- INDEX

How the Book Is Built: An Overview

A Powerful Construction Tool

You have in your hands one of the most powerful construction tools available today. A successful project is built on the foundation of an accurate and dependable estimate. This book will enable you to construct just such an estimate.

For the casual user the book is designed to be:

- quickly and easily understood so you can get right to your estimate
- filled with valuable information so you can understand the necessary factors that go into the cost estimate

For the regular user, the book is designed to be:

- a handy desk reference that can be quickly referred to for key costs
- a comprehensive, fully reliable source of current construction costs and productivity rates, so you'll be prepared to estimate any project
- a source book for preliminary project cost, product selections, and alternate materials and methods

To meet all of these requirements we have organized the book into the following clearly defined sections.

Quick Start

This one-page section (see following page) can quickly get you started on your assemblies or unit price estimate.

How To Use the Book: The Details

This section contains an in-depth explanation of how the book is arranged . . . and how you can use it to determine a reliable construction cost estimate. It includes information about how we develop our cost figures and how to completely prepare your estimate.

Commercial/Industrial/Institutional Section

This section lists square foot costs for typical light commercial construction projects. The individual projects are divided into ten common components of construction. An outline of a typical page layout, an explanation of square foot prices, and a Table of Contents are located at the beginning of the section.

Assemblies Section

The cost data in this section has been organized in an "Assemblies" format. These assemblies are the functional elements of a building and are arranged according to the 12 divisions of the UniFormat classification system. For a complete explanation of a typical "Assemblies" page, see "How To Use the Assemblies Cost Tables."

Unit Price Section

All cost data has been divided into the 16 divisions according to the MasterFormat system of classification and numbering as developed by the Construction Specifications Institute (CSI) and Construction Specifications Canada (CSC). For a listing of these divisions and an outline of their subdivisions, see the Unit Price Section Table of Contents.

Estimating tips are included at the beginning of each division.

Reference Section

This section includes information on Reference Numbers, Crew Listings, Historical Cost Indexes, Location Factors, and a listing of Abbreviations. It is visually identified by a vertical gray bar on the edge of pages.

Reference Numbers: Following selected major classifications throughout the book are "reference numbers" shown in bold squares. These numbers refer you to related information in other sections.

In these other sections, you'll find related tables, explanations, and estimating information. Also included are alternate pricing methods and technical data, along with information on design and economy in construction. You'll also find helpful tips on estimating and construction.

It is recommended that you refer to the Reference Section if a "reference number" appears within the section you are estimating.

Crew Listings: This section lists all the crews referenced. For the purposes of this book, a crew is composed of more than one trade classification and/or the addition of equipment to any trade classification. Power equipment is included in the cost of the crew. Costs are shown both with bare labor rates and with the installing contractor's overhead and profit added. For each, the total crew cost per eight-hour day and the composite cost per labor-hour are listed.

Historical Cost Indexes: These indexes provide you with data to adjust construction costs over time. If you know costs for a past project, you can use these indexes to estimate the cost to construct the same project today.

Location Factors: Obviously, costs vary depending on the regional economy. You can adjust the "national average" costs in this book to 930 major cities throughout the U.S. and Canada by using the data in this section.

Abbreviations: A listing of the abbreviations and the terms they represent, is included.

Index

A comprehensive listing of all terms and subjects in this book to help you find what you need quickly.

The Scope of This Book

This book is designed to be comprehensive and easy to use. To that end we have made certain assumptions:

1. We have established material prices based on a "national average."
2. We have computed labor costs based on a 30-city "national average" of open shop wage rates.
3. We have targeted the data for projects of a certain size range.

For a more detailed explanation of how the cost data is developed, see "How To Use the Book: The Details."

Project Size

This book is intended for use by those involved primarily in light commercial construction costing less than $1,000,000. This includes the construction of retail stores, offices, warehouses, apartments and motels.

With reasonable exercise of judgment the figures can be used for any building work. *For other types of projects, such as repair and remodeling or residential buildings, consult the appropriate MEANS publication for more information.*

Quick Start

If you feel you are ready to use this book and don't think you need the detailed instructions that begin on the following page, this Quick Start section is for you.

These steps will allow you to get started estimating in a matter of minutes.

1 First, decide whether you require a Unit Price or Assemblies type estimate. Unit price estimating requires a breakdown of the work to individual items. Assemblies estimates combine individual items or components into building systems.

If you need to estimate each line item separately, follow the instructions for **Unit Prices.**

If you can use an estimate for the entire assembly or system, follow the instructions for **Assemblies.**

2 Find each cost data section you need in the Table of Contents (either for Unit Prices or Assemblies).

Unit Prices: The cost data for Unit Prices has been divided into 16 divisions according to the CSI MasterFormat.

Assemblies: The cost data for Assemblies has been divided into 12 divisions according to the UniFormat.

3 Turn to the indicated section and locate the line item or assemblies table you need for your estimate. Portions of a sample page layout from both the Unit Price Listings and the Assemblies Cost Tables appear below.

Unit Prices: If there is a reference number listed at the beginning of the section, it refers to additional information you may find useful. See the referenced section for additional information.

- Note the crew code designation. You'll find full descriptions of crews in the Crew Listings including labor-hour and equipment costs.

Assemblies: The Assemblies (*not* shown in full here) are generally separated into three parts: 1) an illustration of the system to be estimated; 2) the components and related costs of a typical system; and 3) the costs for similar systems with dimensional and/or size variations. The Assemblies Section also contains reference numbers for additional useful information.

4 Determine the total number of units your job will require.

Unit Prices: Note the unit of measure for the material you're using is listed under "Unit."

- **Bare Costs:** These figures show unit costs for materials and installation. Labor and equipment costs are calculated according to crew costs and average daily output. Bare costs do not contain allowances for overhead, profit or taxes.
- "Labor-hours" allows you to calculate the total labor-hours to complete that task. Just multiply the quantity of work by this figure for an estimate of activity duration.

Assemblies: Note the unit of measure for the assembly or system you're estimating is listed in the Assemblies Table.

5 Then multiply the total units by . . .

Unit Prices: "Total Incl. O&P" which stands for the total cost including the installing contractor's overhead and profit. (See the "How To Use the Unit Price Pages" for a complete explanation.)

Assemblies: The "Total" in the right-hand column, which is the total cost **including the installing contractor's overhead and profit.** (See the "How To Use the Assemblies Cost Tables" section for a complete explanation.)

Material and equipment cost figures include a 10% markup. For labor markups, see the inside back cover of this book. If the work is to be subcontracted, add the general contractor's markup, approximately 10%.

6 The price you calculate will be an estimate for either an individual item of work or a completed assembly or *system.*

7 Compile a list of all items or assemblies included in the total project. Summarize cost information, and add project overhead.

For a more complete explanation of the way costs are derived, please see the following sections.

Editors' Note: We urge you to spend time reading and understanding all of the supporting material and to take into consideration the reference material such as Location Factors and the "reference numbers."

Unit Price Pages

Assemblies Pages

How to Use the Book: The Details

What's Behind the Numbers? The Development of Cost Data

The staff at R.S. Means continuously monitors developments in the construction industry in order to ensure reliable, thorough and up-to-date cost information.

While *overall* construction costs may vary relative to general economic conditions, price fluctuations within the industry are dependent upon many factors. Individual price variations may, in fact, be opposite to overall economic trends. Therefore, costs are continually monitored and complete updates are published yearly. Also, new items are frequently added in response to changes in materials and methods.

Costs—$ (U.S.)

All costs represent U.S. national averages and are given in U.S. dollars. The Means Location Factors can be used to adjust costs to a particular location. The Location Factors for Canada can be used to adjust U.S. national averages to local costs in Canadian dollars.

Material Costs

The R.S. Means staff contacts manufacturers, dealers, distributors, and contractors all across the U.S. and Canada to determine national average material costs. If you have access to current material costs for your specific location, you may wish to make adjustments to reflect differences from the national average. Included within material costs are fasteners for a normal installation. R.S. Means engineers use manufacturers' recommendations, written specifications and/or standard construction practice for size and spacing of fasteners. Adjustments to material costs may be required for your specific application or location. Material costs do not include sales tax.

Labor Costs

Labor costs are based on the average of open shop wage rates from across the U.S. for construction trades for the current year. Rates along with overhead and profit markups are listed on the inside back cover of this book.

- If wage rates in your area vary from those used in this book, or if rate increases are expected within a given year, labor costs should be adjusted accordingly.

Labor costs reflect productivity based on actual working conditions. These figures include time spent during a normal workday on tasks other than actual installation, such as material receiving and handling, mobilization at site, site movement, breaks, and cleanup.

Productivity data is developed over an extended period so as not to be influenced by abnormal variations and reflects a typical average.

Equipment Costs

Equipment costs include not only rental, but also operating costs for equipment under normal use. The operating costs include parts and labor for routine servicing such as repair and replacement of pumps, filters and worn lines. Normal operating expendables such as fuel, lubricants, tires and electricity (where applicable) are also included. Extraordinary operating expendables with highly variable wear patterns such as diamond bits and blades are excluded. These costs are included under materials. Equipment rental rates are obtained from industry sources throughout North America—contractors, suppliers, dealers, manufacturers, and distributors.

Crew Equipment Cost/Day—The power equipment required for each crew is included in the crew cost. The daily cost for crew equipment is based on dividing the weekly bare rental rate by 5 (number of working days per week), and then adding the hourly operating cost times 8 (hours per day). This "Crew Equipment Cost/Day" is listed in Subdivision 01590.

Factors Affecting Costs

Costs can vary depending upon a number of variables. Here's how we have handled the main factors affecting costs.

Quality—The prices for materials and the workmanship upon which productivity is based represent sound construction work. They are also in line with U.S. government specifications.

Overtime—We have made no allowance for overtime. If you anticipate premium time or work beyond normal working hours, be sure to make an appropriate adjustment to your labor costs.

Productivity—The productivity, daily output, and labor-hour figures for each line item are based on working an eight-hour day in daylight hours in moderate temperatures. For work that extends beyond normal work hours or is performed under adverse conditions, productivity may decrease. (See the section in "How To Use the Unit Price Pages" for more on productivity.)

Size of Project—The size, scope of work, and type of construction project will have a significant impact on cost. Economies of scale can reduce costs for large projects. Unit costs can often run higher for small projects. Costs in this book are intended for the size and type of project as previously described in "How the Book Is Built: An Overview." Costs for projects of a significantly different size or type should be adjusted accordingly.

Location—Material prices in this book are for metropolitan areas. However, in dense urban areas, traffic and site storage limitations may increase costs. Beyond a 20-mile radius of large cities, extra trucking or transportation charges may also increase the material costs slightly. On the other hand, lower wage rates may be in effect. Be sure to consider both these factors when preparing an estimate, particularly if the job site is located in a central city or remote rural location.

In addition, highly specialized subcontract items may require travel and per diem expenses for mechanics.

Other factors—
- season of year
- contractor management
- weather conditions
- local union restrictions
- building code requirements
- availability of:
  - adequate energy
  - skilled labor
  - building materials
- owner's special requirements/restrictions
- safety requirements
- environmental considerations

General Conditions—The "Assemblies" and "Commercial/Industrial/Institutional" sections of this book use costs that include the installing contractor's overhead and profit (O&P). The Unit Price section of this book presents cost data in two ways: Bare Costs and Total Cost including O&P. General Conditions, when applicable, should also be added to the Total Cost including O&P. The costs for General Conditions are listed in Division 1 of the Unit Price Section and the Reference Section of this book. General Conditions for the *Installing Contractor* may range from 0% to 10% of the Total Cost including O&P. For the *General* or *Prime Contractor*, costs for General Conditions may range from 5% to 15% of the Total Cost including O&P, with a figure of 10% as the most typical allowance.

Overhead & Profit—Total Cost including O&P for the *Installing Contractor* is shown in the last column on both the Unit Price and the Assemblies pages of this book. This figure is the sum of the bare material cost plus 10% for profit, the base labor cost plus total overhead and profit, and the bare equipment cost plus 10% for profit. Details for the calculation of Overhead and Profit on labor are shown on the inside back cover and in the Reference Section of this book. (See the "How to Use the Unit Price Pages" for an example of this calculation.).

Unpredictable Factors—General business conditions influence "in-place" costs of all items. Substitute materials and construction methods may have to be employed. These may affect the installed cost and/or life cycle costs. Such factors may be difficult to evaluate and cannot necessarily be predicted on the basis of the job's location in a particular section of the country. Thus, where these factors apply, you may find significant, but unavoidable cost variations for which you will have to apply a measure of judgment to your estimate.

Rounding of Costs

In general, all unit prices in excess of $5.00 have been rounded to make them easier to use and still maintain adequate precision of the results. The rounding rules we have chosen are in the following table.

Prices from . . .	Rounded to the nearest . . .
$.01 to $5.00	$.01
$5.01 to $20.00	$.05
$20.01 to $100.00	$.50
$100.01 to $300.00	$1.00
$300.01 to $1,000.00	$5.00
$1,000.01 to $10,000.00	$25.00
$10,000.01 to $50,000.00	$100.00
$50,000.01 and above	$500.00

Final Checklist

Estimating can be a straightforward process provided you remember the basics. Here's a checklist of some of the items you should remember to do before completing your estimate.

Did you remember to . . .

- utilize the appropriate Location Factor
- take into consideration which items have been marked up and by how much
- mark up the entire estimate sufficiently for your purposes
- read the background information on techniques and technical matters that could impact your project time span and cost
- include all components of your project in the final estimate
- double check your figures to be sure of your accuracy
- call R.S. Means if you have any questions about your estimate or the data you've found in our publications

Remember, R.S. Means stands behind its publications. If you have any questions about your estimate . . . about the costs you've used from our books . . . or even about the technical aspects of the job that may affect your estimate, feel free to call the R.S. Means editors at 1-800-334-3509.

Commercial/Industrial/Institutional Section

Table of Contents

Building Types	Page
How To Use	2

Building Types

		Page
M.010	Apartment, 1-3 Story	6
M.050	Bank	8
M.060	Bowling Alley	10
M.070	Bus Terminal	12
M.080	Car Wash	14
M.090	Church	16
M.100	Club, Country	18
M.170	Community Center	20
M.180	Courthouse, 1 Story	22
M.200	Factory, 1 Story	24
M.220	Fire Station, 1 Story	26
M.240	Fraternity/Sorority House	28
M.250	Funeral Home	30

Building Types

		Page
M.260	Garage, Auto Sales	32
M.290	Garage, Repair	34
M.300	Garage, Service Station	36
M.310	Gymnasium	38
M.380	Laundromat	40
M.390	Library	42
M.400	Medical Office, 1 Story	44
M.410	Medical Office, 2 Story	46
M.420	Motel, 1 Story	48
M.430	Motel, 2-3 Story	50
M.440	Movie Theatre	52
M.450	Nursing Home	54
M.460	Office, 2-4 Story	56
M.490	Police Station	58
M.500	Post Office	60

Building Types

		Page
M.510	Racquetball Court	62
M.530	Restaurant	64
M.540	Restaurant, Fast Food	66
M.550	Rink, Hockey/Soccer	68
M.560	School, Elementary	70
M.580	School, Jr. High, 2-3 Story	72
M.600	Store, Convenience	74
M.610	Store, Department, 1 Story	76
M.630	Store, Retail	78
M.640	Supermarket	80
M.660	Telephone Exchange	82
M.670	Town Hall, 1 Story	84
M.680	Town Hall, 2-3 Story	86
M.690	Warehouse	88
M.700	Warehouse, Mini	90

How to Use the Commercial/Industrial/Institutional Section

The following is a detailed explanation of a sample entry in the Commercial/Industrial/Institutional Square Foot Cost Section. Each bold number below corresponds to the item being described on the following page with the appropriate component or cost of the sample entry following in parenthesis.

Prices listed are costs that include overhead and profit of the installing contractor and additional mark-ups for General Conditions and Architects' Fees.

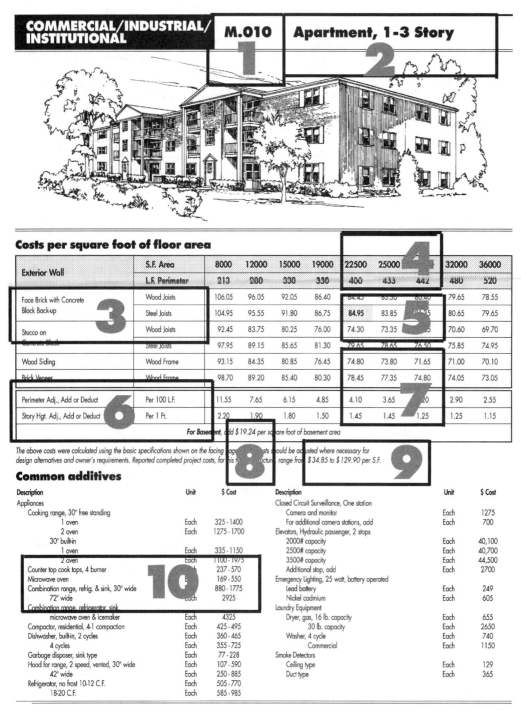

Costs per square foot of floor area

Exterior Wall	S.F. Area	8000	12000	15000	19000	22500	25000		32000	36000
	L.F. Perimeter	213	200	300	330	400	433	442	480	520
Face Brick with Concrete Block Back-up	Wood Joists	106.05	96.05	92.05	86.40	84.45	83.30	80.40	79.65	78.55
	Steel Joists	104.95	95.55	91.80	86.75	84.95	83.85		80.65	79.65
Stucco on Concrete Block	Wood Joists	92.45	83.75	80.25	76.00	74.30	73.35		70.60	69.70
	Steel Joists	97.95	89.15	85.65	81.30	79.65	78.65	76.50	75.85	74.95
Wood Siding	Wood Frame	93.15	84.35	80.85	76.45	74.80	73.80	71.65	71.00	70.10
Brick Veneer	Wood Frame	98.70	89.20	85.40	80.30	78.45	77.35	74.80	74.05	73.05
Perimeter Adj., Add or Deduct	Per 100 L.F.	11.55	7.65	6.15	4.85	4.10	3.65	3.20	2.90	2.55
Story Hgt. Adj., Add or Deduct	Per 1 Ft.	2.20	1.90	1.80	1.50	1.45	1.45	1.25	1.25	1.15

For Basement, add $19.24 per square foot of basement area

The above costs were calculated using the basic specifications shown on the facing page. Costs should be adjusted where necessary for design alternatives and owner's requirements. Reported completed project costs, for this type of structure, range from $34.85 to $129.90 per S.F.

Common additives

Description	Unit	$ Cost	Description	Unit	$ Cost
Appliances			Closed Circuit Surveillance, One station		
Cooking range, 30" free standing			Camera and monitor	Each	1275
1 oven	Each	325 - 1400	For additional camera stations, add	Each	700
2 oven	Each	1275 - 1700	Elevators, Hydraulic passenger, 2 stops		
30" built-in			2000# capacity	Each	40,100
1 oven	Each	335 - 1150	2500# capacity	Each	40,700
2 oven	Each	1100 - 1975	3500# capacity	Each	44,500
Counter top cook tops, 4 burner		237 - 570	Additional stop, add	Each	2700
Microwave oven		169 - 550	Emergency Lighting, 25 watt, battery operated		
Combination range, refrig. & sink, 30" wide		880 - 1775	Lead battery	Each	249
72" wide	Each	2925	Nickel cadmium	Each	605
Combination range, refrigerator, sink, microwave oven & icemaker	Each	4325	Laundry Equipment		
			Dryer, gas, 16 lb. capacity	Each	655
Compactor, residential, 4-1 compaction	Each	425 - 495	30 lb. capacity	Each	2650
Dishwasher, built-in, 2 cycles	Each	360 - 465	Washer, 4 cycle	Each	740
4 cycles	Each	355 - 725	Commercial	Each	1150
Garbage disposer, sink type	Each	77 - 228	Smoke Detectors		
Hood for range, 2 speed, vented, 30" wide	Each	107 - 590	Ceiling type	Each	129
42" wide	Each	250 - 885	Duct type	Each	365
Refrigerator, no frost 10-12 C.F.	Each	505 - 770			
18-20 C.F.	Each	585 - 985			

Important: See the Reference Section for Location Factors

1. Model Number (M.010)

"M" distinguishes this section of the book and stands for model. The number designation is a sequential number.

2. Type of Building (Apartment, 1-3 Story)

There are 43 different types of commercial/industrial/institutional buildings highlighted in this section.

3. Exterior Wall Construction and Building Framing Options (Face Brick with Concrete Block Back-up and Open Web Steel Bar Joists)

Three or more commonly used exterior walls, and in most cases, two typical building framing systems are presented for each type of building. The model selected should be based on the actual characteristics of the building being estimated.

4. Total Square Foot of Floor Area and Base Perimeter Used to Compute Base Costs (22,500 Square Feet and 400 Linear Feet)

Square foot of floor area is the total gross area of all floors at grade, and above, and does not include a basement. The perimeter in linear feet used for the base cost is for a generally rectangular economical building shape.

5. Cost per Square Foot of Floor Area ($84.95)

The highlighted cost is for a building of the selected exterior wall and framing system and floor area. Costs for buildings with floor areas other than those calculated may be interpolated between the costs shown.

6. Building Perimeter and Story Height Adjustments

Square foot costs for a building with a perimeter or floor to floor story height significantly different from the model used to calculate the base cost may be adjusted, add or deduct, to reflect the actual building geometry.

7. Cost per Square Foot of Floor Area for the Perimeter and/or Height Adjustment ($4.10 for Perimeter Difference and $1.45 for Story Height Difference)

Add (or deduct) $4.10 to the base square foot cost for each 100 feet of perimeter difference between the model and the actual building. Add (or deduct) $1.45 to the base square foot cost for each 1 foot of story height difference between the model and the actual building.

8. Optional Cost per Square Foot of Basement Floor Area ($19.24)

The cost of an unfinished basement for the building being estimated is $19.24 times the gross floor area of the basement.

9. Range of Cost per Square Foot of Floor Area for Similar Buildings ($34.85 to $129.90)

Many different buildings of the same type have been built using similar materials and systems. Means historical cost data of actual construction projects indicates a range of $34.85 to $129.90 for this type of building.

10. Common Additives

Common components and/or systems used in this type of building are listed. These costs should be added to the total building cost. Additional selections may be found in the Assemblies Section.

How to Use the Commercial/Industrial/Institutional Section (Continued)

The following is a detailed explanation of a specification and costs for a model building in the Commercial/Industrial/Institutional Square Foot Cost Section. Each bold number below corresponds to the item being described on the following page with the appropriate component of the sample entry following in parenthesis.

Prices listed are costs that include overhead and profit of the installing contractor.

1 Model costs calculated for a 3 story building with 10' story height and 22,500 square feet of floor area

2 Apartment, 1-3 Story

				Unit	Unit Cost	Cost Per S.F.	% Of Sub-Total
1.0 Foundations							
.1	Footings & Foundations	Poured concrete; strip and spread footings and 4' foundation wall		S.F. Ground	5.67	1.89	
.4	Piles & Caissons	N/A		—	—	—	3.5%
.9	Excavation & Backfill	Site preparation for slab and trench for foundation wall and footing		S.F. Ground	.89	.30	
2.0 Substructure							
.1	Slab on Grade	4" reinforced concrete with vapor barrier and granular base		S.F. Slab	2.80	.93	1.5%
.2	Special Substructures	N/A		—	—	—	
3.0 Superstructure							
.1	Columns & Beams	Gypsum board fireproofing on columns; steel columns in 3.5 and 3.7		S.F. Floor	.70	.70	
.4	Structural Walls	N/A		—	—	—	
.5	Elevated Floors	Open web steel joists, slab form, concrete, interior steel columns		S.F. Floor	9.00	6.00	14.4%
.7	Roof	Open web steel joists with rib metal deck, interior steel columns		S.F. Roof	4.12	1.37	
.9	Stairs	Concrete filled metal pan		Flight	4540	1.01	
4.0 Exterior Closure							
.1	Walls	Face brick with concrete block backup	88% of wall	S.F. Wall	13.02	6.11	
.5	Exterior Wall Finishes	N/A		—	—	—	11.8%
.6	Doors	Aluminum and glass		Each	1061	.19	
.7	Windows & Glazed Walls	Aluminum horizontal sliding	12% of wall	Each	262	1.13	
5.0 Roofing							
.1	Roof Coverings	Built-up tar and gravel with flashing		S.F. Roof	2.31		
.7	Insulation	Perlite/EPS composite		S.F. Roof	1.18		1.8%
.9	Openings & Specialties	N/A		—	—		
6.0 Interior Construction							
.1	Partitions	Gypsum board and sound deadening board on metal studs	10 S.F. of Floor/L.F. Partition	S.F. Partition	2.80	2.49	
.4	Interior Doors	15% solid core wood, 85% hollow core wood	80 S.F. Floor/Door	Each	350	4.38	
.5	Wall Finishes	70% paint, 25% vinyl wall covering, 5% ceramic tile		S.F. Surface	.87	1.54	25.0%
.6	Floor Finishes	60% carpet, 30% vinyl composition tile, 10% ceramic		S.F. Floor	4.01	4.01	
.7	Ceiling Finishes	Painted gypsum board on resilient channels		S.F. Ceiling	2.28	2.28	
.9	Interior Surface/Exterior Wall	Painted gypsum board on furring	80% of wall	S.F. Wall	2.47	1.05	
7.0 Conveying							
.1	Elevators	One hydraulic passenger elevator		Each		2.63	4.2%
.2	Special Conveyors	N/A		—	—	—	
8.0 Mechanical							
.1	Plumbing	Kitchen, bathroom and service fixtures, supply and drainage	1 Fixture/200 S.F. Floor	Each	1444	7.22	
.2	Fire Protection	Wet pipe sprinkler system		S.F. Floor	1.31	1.31	
.3	Heating	Oil fired hot water, baseboard radiation		S.F. Floor	3.58	3.58	27.4%
.4	Cooling	Chilled water, air cooled condenser system		S.F. Floor	5.10	5.10	
.5	Special Systems	N/A		—	—	—	
9.0 Electrical							
.1	Service & Distribution	400 ampere service, panel board and feeders		S.F. Floor	.71	.71	
.2	Lighting & Power	Incandescent fixtures, receptacles, switches, A.C. and misc. power		S.F. Floor	3.57	3.57	7.7%
.4	Special Electrical	Alarm systems and emergency lighting		S.F. Floor	.55	.55	
11.0 Special Construction							
.1	Specialties	Kitchen cabinets		S.F. Floor	1.68	1.68	2.7%
12.0 Site Work							
.1	Earthwork	N/A		—	—	—	
.3	Utilities	N/A		—	—	—	0.0%
.5	Roads & Parking	N/A		—	—	—	
.7	Site Improvements	N/A		—	—	—	
					Sub-Total	**62.91**	**100%**
	CONTRACTOR FEES (General Requirements: 10%, Overhead: 5%, Profit: 10%)				25%	15.73	
	ARCHITECT FEES				8%	6.31	
					Total Building Cost	**84.95**	

1 Building Description
(Model costs are calculated for a three-story apartment building with 10' story height and 22,500 square feet of floor area)
The model highlighted is described in terms of building type, number of stories, typical story height and square footage.

2 Type of Building
(Apartment, 1-3 Story)

3 Division 6.0 Interior Construction
(6.4 Interior Doors)
System costs are presented in divisions according to the 12-division UniFormat classifications. Each of the component systems are listed.

4 Specification Highlights
(15% solid core wood; 85% hollow core wood)
All systems in each subdivision are described with the material and proportions used.

5 Quality Criteria
(80 S.F. Floor/Door)
The criteria used in determining quantities for the calculations are shown.

6 Unit (Each)
The unit of measure shown in this column is the unit of measure of the particular system shown that corresponds to the unit cost.

7 Unit Cost ($350)
The cost per unit of measure of each system subdivision.

8 Cost per Square Foot ($4.38)
The cost per square foot for each system is the unit cost of the system times the total number of units divided by the total square feet of building area.

9 % of Sub-Total (25%)
The percent of sub-total is the total cost per square foot of all systems in the division divided by the sub-total cost per square foot of the building.

10 Sub-Total ($62.91)
The sub-total is the total of all the system costs per square foot.

11 Project Fees
(Contractor Fees) (25%)
(Architects' Fees) (8%)
Contractor Fees to cover the general requirements, overhead and profit of the General Contractor are added as a percentage of the sub-total. An Architect's Fee, also as a percentage of the sub-total, is also added. These values vary with the building type.

12 Total Building Cost ($84.95)
The total building cost per square foot of building area is the sum of the square foot costs of all the systems plus the General Contractor's general requirements overhead and profit and the Architect's fee. The total building cost is the amount which appears shaded in the Cost per Square Foot of Floor Area table shown previously.

COMMERCIAL/INDUSTRIAL/INSTITUTIONAL — M.010 — Apartment, 1-3 Story

Costs per square foot of floor area

Exterior Wall	S.F. Area	8000	12000	15000	19000	22500	25000	29000	32000	36000
	L.F. Perimeter	213	280	330	350	400	433	442	480	520
Face Brick with Concrete Block Back-up	Wood Joists	106.05	96.05	92.05	86.40	84.45	83.30	80.40	79.65	78.55
	Steel Joists	104.95	95.55	91.80	86.75	84.95	83.85	81.35	80.65	79.65
Stucco on Concrete Block	Wood Joists	92.45	83.75	80.25	76.00	74.30	73.35	71.25	70.60	69.70
	Steel Joists	97.95	89.15	85.65	81.30	79.65	78.65	76.50	75.85	74.95
Wood Siding	Wood Frame	93.15	84.35	80.85	76.45	74.80	73.80	71.65	71.00	70.10
Brick Veneer	Wood Frame	98.70	89.20	85.40	80.30	78.45	77.35	74.80	74.05	73.05
Perimeter Adj., Add or Deduct	Per 100 L.F.	11.55	7.65	6.15	4.85	4.10	3.65	3.20	2.90	2.55
Story Hgt. Adj., Add or Deduct	Per 1 Ft.	2.20	1.90	1.80	1.50	1.45	1.45	1.25	1.25	1.15

For Basement, add $19.24 per square foot of basement area

The above costs were calculated using the basic specifications shown on the facing page. These costs should be adjusted where necessary for design alternatives and owner's requirements. Reported completed project costs, for this type of structure, range from $34.85 to $129.90 per S.F.

Common additives

Description	Unit	$ Cost
Appliances		
Cooking range, 30" free standing		
1 oven	Each	325 - 1400
2 oven	Each	1275 - 1700
30" built-in		
1 oven	Each	335 - 1150
2 oven	Each	1100 - 1975
Counter top cook tops, 4 burner	Each	237 - 570
Microwave oven	Each	169 - 550
Combination range, refrig. & sink, 30" wide	Each	880 - 1775
72" wide	Each	2925
Combination range, refrigerator, sink,		
microwave oven & icemaker	Each	4325
Compactor, residential, 4-1 compaction	Each	425 - 495
Dishwasher, built-in, 2 cycles	Each	360 - 465
4 cycles	Each	355 - 725
Garbage disposer, sink type	Each	77 - 228
Hood for range, 2 speed, vented, 30" wide	Each	107 - 590
42" wide	Each	250 - 885
Refrigerator, no frost 10-12 C.F.	Each	505 - 770
18-20 C.F.	Each	585 - 985

Description	Unit	$ Cost
Closed Circuit Surveillance, One station		
Camera and monitor	Each	1275
For additional camera stations, add	Each	700
Elevators, Hydraulic passenger, 2 stops		
2000# capacity	Each	40,100
2500# capacity	Each	40,700
3500# capacity	Each	44,500
Additional stop, add	Each	2700
Emergency Lighting, 25 watt, battery operated		
Lead battery	Each	249
Nickel cadmium	Each	605
Laundry Equipment		
Dryer, gas, 16 lb. capacity	Each	655
30 lb. capacity	Each	2650
Washer, 4 cycle	Each	740
Commercial	Each	1150
Smoke Detectors		
Ceiling type	Each	129
Duct type	Each	365

Important: See the Reference Section for Location Factors

Model costs calculated for a 3 story building with 10' story height and 22,500 square feet of floor area

Apartment, 1-3 Story

				Unit	Unit Cost	Cost Per S.F.	% Of Sub-Total
1.0 Foundations							
.1	Footings & Foundations	Poured concrete; strip and spread footings and 4' foundation wall		S.F. Ground	5.67	1.89	
.4	Piles & Caissons	N/A		—	—	—	3.5%
.9	Excavation & Backfill	Site preparation for slab and trench for foundation wall and footing		S.F. Ground	.89	.30	
2.0 Substructure							
.1	Slab on Grade	4" reinforced concrete with vapor barrier and granular base		S.F. Slab	2.80	.93	1.5%
.2	Special Substructures	N/A		—	—	—	
3.0 Superstructure							
.1	Columns & Beams	Gypsum board fireproofing on columns; steel columns in 3.5 and 3.7		S.F. Floor	.70	.70	
.4	Structural Walls	N/A		—	—	—	
.5	Elevated Floors	Open web steel joists, slab form, concrete, interior steel columns		S.F. Floor	9.00	6.00	14.4%
.7	Roof	Open web steel joists with rib metal deck, interior steel columns		S.F. Roof	4.12	1.37	
.9	Stairs	Concrete filled metal pan		Flight	4540	1.01	
4.0 Exterior Closure							
.1	Walls	Face brick with concrete block backup	88% of wall	S.F. Wall	13.02	6.11	
.5	Exterior Wall Finishes	N/A		—	—	—	11.8%
.6	Doors	Aluminum and glass		Each	1061	.19	
.7	Windows & Glazed Walls	Aluminum horizontal sliding	12% of wall	Each	269	1.15	
5.0 Roofing							
.1	Roof Coverings	Built-up tar and gravel with flashing		S.F. Roof	2.31	.77	
.7	Insulation	Perlite/EPS composite		S.F. Roof	1.18	.39	1.8%
.8	Openings & Specialties	N/A		—	—	—	
6.0 Interior Construction							
.1	Partitions	Gypsum board and sound deadening board on metal studs	10 S.F. of Floor/L.F. Partition	S.F. Partition	2.80	2.49	
.4	Interior Doors	15% solid core wood, 85% hollow core wood	80 S.F. Floor/Door	Each	350	4.38	
.5	Wall Finishes	70% paint, 25% vinyl wall covering, 5% ceramic tile		S.F. Surface	.87	1.54	25.0%
.6	Floor Finishes	60% carpet, 30% vinyl composition tile, 10% ceramic tile		S.F. Floor	4.01	4.01	
.7	Ceiling Finishes	Painted gypsum board on resilient channels		S.F. Ceiling	2.28	2.28	
.9	Interior Surface/Exterior Wall	Painted gypsum board on furring	80% of wall	S.F. Wall	2.47	1.05	
7.0 Conveying							
.1	Elevators	One hydraulic passenger elevator		Each	59,175	2.63	4.2%
.2	Special Conveyors	N/A		—	—	—	
8.0 Mechanical							
.1	Plumbing	Kitchen, bathroom and service fixtures, supply and drainage	1 Fixture/200 S.F. Floor	Each	1444	7.22	
.2	Fire Protection	Wet pipe sprinkler system		S.F. Floor	1.31	1.31	
.3	Heating	Oil fired hot water, baseboard radiation		S.F. Floor	3.58	3.58	27.4%
.4	Cooling	Chilled water, air cooled condenser system		S.F. Floor	5.10	5.10	
.5	Special Systems	N/A		—	—	—	
9.0 Electrical							
.1	Service & Distribution	400 ampere service, panel board and feeders		S.F. Floor	.71	.71	
.2	Lighting & Power	Incandescent fixtures, receptacles, switches, A.C. and misc. power		S.F. Floor	3.57	3.57	7.7%
.4	Special Electrical	Alarm systems and emergency lighting		S.F. Floor	.55	.55	
11.0 Special Construction							
.1	Specialties	Kitchen cabinets		S.F. Floor	1.68	1.68	2.7%
12.0 Site Work							
.1	Earthwork	N/A		—	—	—	
.3	Utilities	N/A		—	—	—	0.0%
.5	Roads & Parking	N/A		—	—	—	
.7	Site Improvements	N/A		—	—	—	
				Sub-Total		62.91	100%
	CONTRACTOR FEES (General Requirements: 10%, Overhead: 5%, Profit: 10%)				25%	15.73	
	ARCHITECT FEES				8%	6.31	
				Total Building Cost		84.95	

COMMERCIAL/INDUSTRIAL/INSTITUTIONAL

M.050 Bank

Costs per square foot of floor area

Exterior Wall	S.F. Area	2000	2700	3400	4100	4800	5500	6200	6900	7600
	L.F. Perimeter	180	208	236	256	280	303	317	337	357
Face Brick with Concrete Block Back-up	Steel Frame	141.70	133.20	128.20	123.90	121.30	119.25	116.90	115.50	114.30
	R/Conc. Frame	151.35	142.85	137.85	133.50	130.90	128.85	126.50	125.10	123.95
Precast Concrete Panel	Steel Frame	135.65	128.05	123.55	119.70	117.35	115.55	113.45	112.20	111.15
	R/Conc. Frame	145.30	137.65	133.20	129.35	126.95	125.15	123.10	121.80	120.75
Limestone with Concrete Block Back-up	Steel Frame	153.10	142.95	137.00	131.80	128.65	126.20	123.40	121.65	120.25
	R/Conc. Frame	162.75	152.60	146.65	141.45	138.30	135.85	133.00	131.30	129.90
Perimeter Adj., Add or Deduct	Per 100 L.F.	26.25	19.45	15.45	12.75	10.95	9.55	8.45	7.60	6.90
Story Hgt. Adj., Add or Deduct	Per 1 Ft.	2.50	2.15	1.95	1.75	1.65	1.55	1.45	1.35	1.30

For Basement, add $17.70 per square foot of basement area

The above costs were calculated using the basic specifications shown on the facing page. These costs should be adjusted where necessary for design alternatives and owner's requirements. Reported completed project costs, for this type of structure, range from $80.40 to $197.90 per S.F.

Common additives

Description	Unit	$ Cost
Bulletproof Teller Window, 44" x 60"	Each	3425
60" x 48"	Each	4800
Closed Circuit Surveillance, One station		
Camera and monitor	Each	1275
For additional camera stations, add	Each	700
Counters, Complete	Station	3950
Door & Frame, 3' x 6'-8", bullet resistant steel with vision panel	Each	3025 - 5225
Drive-up Window, Drawer & micr., not incl. glass	Each	5300 - 8975
Emergency Lighting, 25 watt, battery operated		
Lead battery	Each	249
Nickel cadmium	Each	605
Night Depository	Each	7475 - 12,200
Package Receiver, painted	Each	1300
stainless steel	Each	2075
Partitions, Bullet resistant to 8' high	L.F.	214 - 360
Pneumatic Tube Systems, 2 station	Each	22,200
With TV viewer	Each	42,000

Description	Unit	$ Cost
Service Windows, Pass thru, steel		
24" x 36"	Each	1975
48" x 48"	Each	2700
72" x 40"	Each	3425
Smoke Detectors		
Ceiling type	Each	129
Duct type	Each	365
Twenty-four Hour Teller		
Automatic deposit cash & memo	Each	40,600
Vault Front, Door & frame		
1 hour test, 32"x 78"	Opening	3350
2 hour test, 32" door	Opening	3975
40" door	Opening	4300
4 hour test, 32" door	Opening	4050
40" door	Opening	4800
Time lock, two movement, add	Each	1575

Bank

Model costs calculated for a 1 story building with 14' story height and 4,100 square feet of floor area

				Unit	Unit Cost	Cost Per S.F.	% Of Sub-Total
1.0 Foundations							
.1	Footings & Foundations	Poured concrete; strip and spread footings and 4' foundation wall		S.F. Ground	5.21	5.21	
.4	Piles & Caissons	N/A		—	—	—	6.4%
.9	Excavation & Backfill	Site preparation for slab and trench for foundation wall and footing		S.F. Ground	.96	.96	
2.0 Substructure							
.1	Slab on Grade	4" reinforced concrete with vapor barrier and granular base		S.F. Slab	2.80	2.80	2.9%
.2	Special Substructures	N/A		—	—	—	
3.0 Superstructure							
.1	Columns & Beams	Concrete columns		L.F. Columns	55	2.73	
.4	Structural Walls	N/A		—	—	—	
.5	Elevated Floors	N/A		—	—	—	11.5%
.7	Roof	Cast-in-place concrete slab		S.F. Roof	8.32	8.32	
.9	Stairs	N/A		—	—	—	
4.0 Exterior Closure							
.1	Walls	Face brick with concrete block backup	80% of wall	S.F. Wall	17.05	11.92	
.5	Exterior Wall Finishes	N/A		—	—	—	17.5%
.6	Doors	Double aluminum and glass and hollow metal		Each	1829	.89	
.7	Windows & Glazed Walls	Horizontal aluminum sliding	20% of wall	Each	346	4.04	
5.0 Roofing							
.1	Roof Coverings	Built-up tar and gravel with flashing		S.F. Roof	2.45	2.45	
.7	Insulation	Perlite/EPS composite		S.F. Roof	1.18	1.18	4.1%
.8	Openings & Specialties	Gravel stop		L.F. Perimeter	4.99	.31	
6.0 Interior Construction							
.1	Partitions	Gypsum board on metal studs	20 S.F. of Floor/L.F. Partition	S.F. Partition	2.64	1.32	
.4	Interior Doors	Single leaf hollow metal	200 S.F. Floor/Door	Each	492	2.46	
.5	Wall Finishes	50% vinyl wall covering, 50% paint		S.F. Surface	.84	.84	13.9%
.6	Floor Finishes	50% carpet, 50% vinyl composition tile		S.F. Floor	3.91	3.91	
.7	Ceiling Finishes	Mineral fiber tile on concealed zee bars		S.F. Ceiling	3.10	3.10	
.9	Interior Surface/Exterior Wall	Painted gypsum board on furring	80% of wall	S.F. Wall	2.51	1.76	
7.0 Conveying							
.1	Elevators	N/A		—	—	—	0.0%
.2	Special Conveyors	N/A		—	—	—	
8.0 Mechanical							
.1	Plumbing	Toilet and service fixtures, supply and drainage	1 Fixture/580 S.F. Floor	Each	2726	4.70	
.2	Fire Protection	Wet pipe sprinkler system		S.F. Floor	1.74	1.74	
.3	Heating	Included in 8.4		—	—	—	15.5%
.4	Cooling	Single zone rooftop unit, gas heating, electric cooling		S.F. Floor	8.50	8.50	
.5	Special Systems	N/A		—	—	—	
9.0 Electrical							
.1	Service & Distribution	200 ampere service, panel board and feeders		S.F. Floor	1.36	1.36	
.2	Lighting & Power	Fluorescent fixtures, receptacles, switches, A.C. and misc. power		S.F. Floor	5.49	5.49	8.7%
.4	Special Electrical	Alarm systems, emergency lighting, and security television		S.F. Floor	1.48	1.48	
11.0 Special Construction							
.1	Specialties	Vault door, automatic teller, drive up window, closed circuit T.V., night depository		S.F. Floor	18.76	18.76	19.5%
12.0 Site Work							
.1	Earthwork	N/A		—	—	—	
.3	Utilities	N/A		—	—	—	0.0%
.5	Roads & Parking	N/A		—	—	—	
.7	Site Improvements	N/A		—	—	—	
				Sub-Total		96.23	100%
	CONTRACTOR FEES (General Requirements: 10%, Overhead: 5%, Profit: 10%)				25%	24.06	
	ARCHITECT FEES				11%	13.21	
				Total Building Cost		**133.50**	

BUILDING TYPES

COMMERCIAL/INDUSTRIAL/INSTITUTIONAL — M.060 Bowling Alley

Costs per square foot of floor area

Exterior Wall	S.F. Area	12000	14000	16000	18000	20000	22000	24000	26000	28000
	L.F. Perimeter	460	491	520	540	566	593	620	645	670
Concrete Block	Steel Roof Deck	63.00	61.20	59.85	58.60	57.70	57.00	56.40	55.85	55.45
Decorative Concrete Block	Steel Roof Deck	65.80	63.75	62.20	60.80	59.80	58.95	58.25	57.70	57.20
Face Brick on Concrete Block	Steel Roof Deck	69.55	67.20	65.40	63.70	62.55	61.60	60.80	60.10	59.55
Jumbo Brick on Concrete Block	Steel Roof Deck	67.75	65.55	63.85	62.30	61.20	60.30	59.60	58.95	58.40
Stucco on Concrete Block	Steel Roof Deck	63.75	61.90	60.50	59.20	58.30	57.50	56.90	56.35	55.90
Precast Concrete Panel	Steel Roof Deck	64.30	62.40	60.95	59.65	58.70	57.90	57.30	56.70	56.25
Perimeter Adj., Add or Deduct	Per 100 L.F.	2.90	2.50	2.20	1.95	1.75	1.60	1.45	1.35	1.25
Story Hgt. Adj., Add or Deduct	Per 1 Ft.	.65	.60	.55	.50	.50	.45	.45	.45	.45

For Basement, add $17.81 per square foot of basement area

The above costs were calculated using the basic specifications shown on the facing page. These costs should be adjusted where necessary for design alternatives and owner's requirements. Reported completed project costs, for this type of structure, range from $36.00 to $88.10 per S.F.

Common additives

Description	Unit	$ Cost
Bowling Alleys, Incl. alley, pinsetter, scorer, counter & misc. supplies, average	Lane	43,700
For automatic scorer, add (maximum)	Lane	7475
Emergency Lighting, 25 watt, battery operated		
Lead battery	Each	249
Nickel cadmium	Each	605
Lockers, Steel, Single tier, 60" or 72"	Opening	124 - 191
2 tier, 60" or 72" total	Opening	74 - 95
5 tier, box lockers	Opening	35 - 57
Locker bench, lam., maple top only	L.F.	16.35
Pedestals, steel pipe	Each	45
Seating		
Auditorium chair, all veneer	Each	151
Veneer back, padded seat	Each	184
Upholstered, spring seat	Each	186
Sound System		
Amplifier, 250 watts	Each	1500
Speaker, ceiling or wall	Each	125
Trumpet	Each	235

Model costs calculated for a 1 story building with 14' story height and 20,000 square feet of floor area

Bowling Alley

				Unit	Unit Cost	Cost Per S.F.	% Of Sub-Total
1.0 Foundations							
.1	Footings & Foundations	Poured concrete; strip and spread footings and 4' foundation wall		S.F. Ground	1.78	1.78	6.6%
.4	Piles & Caissons	N/A		—	—	—	
.9	Excavation & Backfill	Site preparation for slab and trench for foundation wall and footing		S.F. Ground	1.06	1.06	
2.0 Substructure							
.1	Slab on Grade	4" reinforced concrete with vapor barrier and granular base		S.F. Slab	2.80	2.80	6.5%
.2	Special Substructures	N/A		—	—	—	
3.0 Superstructure							
.1	Columns & Beams	Columns included in 3.7		—	—	—	
.4	Structural Walls	N/A		—	—	—	
.5	Elevated Floors	N/A		—	—	—	11.9%
.7	Roof	Metal deck on open web steel joists with columns and beams		S.F. Roof	5.15	5.15	
.9	Stairs	N/A		—	—	—	
4.0 Exterior Closure							
.1	Walls	Concrete block	90% of wall	S.F. Wall	6.90	2.46	
.5	Exterior Wall Finishes	N/A		—	—	—	11.5%
.6	Doors	Double aluminum and glass and hollow metal		Each	1327	.33	
.7	Windows & Glazed Walls	Horizontal pivoted	10% of wall	Each	1308	2.16	
5.0 Roofing							
.1	Roof Coverings	Built-up tar and gravel with flashing		S.F. Roof	1.86	1.86	
.7	Insulation	Perlite/EPS composite		S.F. Roof	1.18	1.18	7.2%
.8	Openings & Specialties	Gravel stop and hatches		S.F. Roof	.08	.08	
6.0 Interior Construction							
.1	Partitions	Concrete block	50 S.F. Floor/L.F. Partition	S.F. Partition	5.20	1.04	
.4	Interior Doors	Hollow metal single leaf	1000 S.F. Floor/Door	Each	492	.49	
.5	Wall Finishes	Paint and block filler		S.F. Surface	1.70	.68	9.0%
.6	Floor Finishes	Vinyl tile	25% of floor	S.F. Floor	2.00	.50	
.7	Ceiling Finishes	Suspended fiberglass board	25% of area	S.F. Ceiling	2.33	.58	
.9	Interior Surface/Exterior Wall	Paint and block filler	90% of wall	S.F. Wall	1.69	.60	
7.0 Conveying							
.1	Elevators	N/A		—	—	—	0.0%
.2	Special Conveyors	N/A		—	—	—	
8.0 Mechanical							
.1	Plumbing	Toilet and service fixtures, supply and drainage	1 Fixture/2200 S.F. Floor	Each	2288	1.04	
.2	Fire Protection	Sprinklers, light hazard		S.F. Floor	1.09	1.09	
.3	Heating	Included in 8.4		—	—	—	30.2%
.4	Cooling	Single zone rooftop unit, gas heating, electric cooling		S.F. Floor	10.91	10.91	
.5	Special Systems	N/A		—	—	—	
9.0 Electrical							
.1	Service & Distribution	800 ampere service, panel board and feeders		S.F. Floor	1.51	1.51	
.2	Lighting & Power	Fluorescent fixtures, receptacles, switches, A.C. and misc. power		S.F. Floor	5.61	5.61	17.1%
.4	Special Electrical	Alarm systems and emergency lighting		S.F. Floor	.24	.24	
11.0 Special Construction							
.1	Specialties	N/A		—	—	—	0.0%
12.0 Site Work							
.1	Earthwork	N/A		—	—	—	
.3	Utilities	N/A		—	—	—	0.0%
.5	Roads & Parking	N/A		—	—	—	
.7	Site Improvements	N/A		—	—	—	
				Sub-Total		43.15	**100%**
	CONTRACTOR FEES (General Requirements: 10%, Overhead: 5%, Profit: 10%)				25%	10.79	
	ARCHITECT FEES				7%	3.76	
				Total Building Cost		**57.70**	

COMMERCIAL/INDUSTRIAL/INSTITUTIONAL — M.070 — Bus Terminal

Costs per square foot of floor area

Exterior Wall	S.F. Area	6000	8000	10000	12000	14000	16000	18000	20000	22000
	L.F. Perimeter	320	386	453	520	540	597	610	660	710
Face Brick with Concrete Block Back-up	Bearing Walls	85.65	82.25	80.30	78.95	76.50	75.70	74.00	73.45	73.05
	Steel Frame	86.35	83.00	81.05	79.75	77.35	76.55	74.90	74.35	73.95
Decorative Concrete Block	Bearing Walls	81.60	78.60	76.85	75.65	73.55	72.90	71.45	70.95	70.60
	Steel Frame	82.30	79.35	77.60	76.45	74.40	73.75	72.30	71.85	71.50
Precast Concrete Panels	Bearing Walls	81.20	78.25	76.50	75.35	73.30	72.65	71.20	70.75	70.35
	Steel Frame	81.95	79.00	77.30	76.15	74.15	73.50	72.10	71.65	71.25
Perimeter Adj., Add or Deduct	Per 100 L.F.	7.50	5.60	4.50	3.75	3.25	2.80	2.50	2.25	2.05
Story Hgt. Adj., Add or Deduct	Per 1 Ft.	1.25	1.15	1.05	1.00	.90	.90	.80	.75	.75
Basement—Not Applicable										

The above costs were calculated using the basic specifications shown on the facing page. These costs should be adjusted where necessary for design alternatives and owner's requirements. Reported completed project costs, for this type of structure, range from $43.25 to $104.25 per S.F.

Common additives

Description	Unit	$ Cost
Directory Boards, Plastic, glass covered		
30" x 20"	Each	500
36" x 48"	Each	895
Aluminum, 24" x 18"	Each	440
36" x 24"	Each	530
48" x 32"	Each	630
48" x 60"	Each	1375
Emergency Lighting, 25 watt, battery operated		
Lead battery	Each	249
Nickel cadmium	Each	605
Benches, Hardwood	L.F.	74 - 139
Ticket Printer	Each	6900
Turnstiles, One way		
4 arm, 46" dia., manual	Each	390
Electric	Each	1375
High security, 3 arm		
65" dia., manual	Each	3075
Electric	Each	4025
Gate with horizontal bars		
65" dia., 7' high transit type	Each	3525

Model costs calculated for a 1 story building with 14' story height and 12,000 square feet of floor area

Bus Terminal

				Unit	Unit Cost	Cost Per S.F.	% Of Sub-Total
1.0 Foundations							
.1	Footings & Foundations	Poured concrete; strip and spread footings and 4' foundation wall		S.F. Ground	3.17	3.17	
.4	Piles & Caissons	N/A		–	–	–	7.0%
.9	Excavation & Backfill	Site preparation for slab and trench for foundation wall and footing		S.F. Ground	.96	.96	
2.0 Substructure							
.1	Slab on Grade	6" reinforced concrete with vapor barrier and granular base		S.F. Slab	3.43	3.43	5.8%
.2	Special Substructures	N/A		–	–	–	
3.0 Superstructure							
.1	Columns & Beams	Columns included in 3.7		–	–	–	
.4	Structural Walls	N/A		–	–	–	
.5	Elevated Floors	N/A		–	–	–	5.5%
.7	Roof	Metal deck on open web steel joists with columns and beams		S.F. Roof	3.25	3.25	
.9	Stairs	N/A		–	–	–	
4.0 Exterior Closure							
.1	Walls	Face brick with concrete block backup	70% of wall	S.F. Wall	17.05	7.24	
.5	Exterior Wall Finishes	N/A		–	–	–	18.9%
.6	Doors	Double aluminum and glass		Each	2885	.96	
.7	Windows & Glazed Walls	Store front	30% of wall	S.F. Window	16.20	2.95	
5.0 Roofing							
.1	Roof Coverings	Built-up tar and gravel with flashing		S.F. Roof	2.12	2.12	
.7	Insulation	Perlite/EPS composite		S.F. Roof	1.18	1.18	6.0%
.8	Openings & Specialties	Gravel stop		L.F. Perimeter	4.99	.22	
6.0 Interior Construction							
.1	Partitions	Lightweight concrete block	15 S.F. Floor/L.F. Partition	S.F. Partition	5.19	3.46	
.4	Interior Doors	Hollow metal	150 S.F. Floor/Door	Each	492	3.28	
.5	Wall Finishes	Glazed coating		S.F. Surface	1.03	1.37	28.1%
.6	Floor Finishes	Quarry tile and vinyl composition tile		S.F. Floor	4.96	4.96	
.7	Ceiling Finishes	Mineral fiber tile on concealed zee bars		S.F. Ceiling	3.10	3.10	
.9	Interior Surface/Exterior Wall	Glazed coating	70% of wall	S.F. Wall	1.04	.44	
7.0 Conveying							
.1	Elevators	N/A		–	–	–	0.0%
.2	Special Conveyors	N/A		–	–	–	
8.0 Mechanical							
.1	Plumbing	Toilet and service fixtures, supply and drainage	1 Fixture/850 S.F. Floor	Each	1972	2.32	
.2	Fire Protection	Wet pipe sprinkler system		S.F. Floor	1.33	1.33	
.3	Heating	Included in 8.4		–	–	–	20.0%
.4	Cooling	Single zone rooftop unit, gas heating, electric cooling		S.F. Floor	8.13	8.13	
.5	Special Systems	N/A		–	–	–	
9.0 Electrical							
.1	Service & Distribution	400 ampere service, panel board and feeders		S.F. Floor	.93	.93	
.2	Lighting & Power	Fluorescent fixtures, receptacles, switches, A.C. and misc. power		S.F. Floor	3.78	3.78	8.7%
.4	Special Electrical	Alarm systems and emergency lighting		S.F. Floor	.45	.45	
11.0 Special Construction							
.1	Specialties	N/A		–	–	–	0.0%
12.0 Site Work							
.1	Earthwork	N/A		–	–	–	
.3	Utilities	N/A		–	–	–	0.0%
.5	Roads & Parking	N/A		–	–	–	
.7	Site Improvements	N/A		–	–	–	
				Sub-Total		59.03	100%
	CONTRACTOR FEES (General Requirements: 10%, Overhead: 5%, Profit: 10%)				25%	14.76	
	ARCHITECT FEES				7%	5.16	
				Total Building Cost		**78.95**	

BUILDING TYPES

COMMERCIAL/INDUSTRIAL/INSTITUTIONAL — M.080 Car Wash

Costs per square foot of floor area

Exterior Wall	S.F. Area	600	800	1000	1200	1600	2000	2400	3000	4000
	L.F. Perimeter	100	114	128	139	164	189	214	250	314
Brick with Concrete Block Back-up	Steel Frame	186.40	178.70	174.10	170.20	165.95	163.40	161.70	159.85	158.30
	Bearing Walls	185.30	177.60	173.00	169.10	164.85	162.30	160.60	158.75	157.20
Concrete Block	Steel Frame	163.80	159.35	156.70	154.50	152.05	150.60	149.60	148.55	147.65
	Bearing Walls	162.70	158.25	155.60	153.40	150.95	149.50	148.50	147.45	146.55
Galvanized Steel Siding	Steel Frame	164.05	160.15	157.85	155.90	153.75	152.50	151.65	150.70	149.90
Metal Sandwich Panel	Steel Frame	163.15	158.35	155.45	153.00	150.35	148.80	147.70	146.55	145.60
Perimeter Adj., Add or Deduct	Per 100 L.F.	53.10	39.80	31.80	26.55	19.90	15.95	13.25	10.60	8.00
Story Hgt. Adj., Add or Deduct	Per 1 Ft.	3.30	2.80	2.50	2.30	2.00	1.85	1.75	1.65	1.55
Basement—Not Applicable										

The above costs were calculated using the basic specifications shown on the facing page. These costs should be adjusted where necessary for design alternatives and owner's requirements. Reported completed project costs, for this type of structure, range from $77.50 to $220.15 per S.F.

Common additives

Description	Unit	$ Cost
Air Compressors, Electric		
1-1/2 H.P., standard controls	Each	670
Dual controls	Each	780
5 H.P., 115/230 volt, standard control	Each	2175
Dual controls	Each	2275
Emergency Lighting, 25 watt, battery operated		
Lead battery	Each	249
Nickel cadmium	Each	605
Fence, Chain link, 6' high		
9 ga. wire, galvanized	L.F.	13.05
6 ga. wire	L.F.	17.95
Gate	Each	223
Product Dispenser with vapor recovery		
for 6 nozzles	Each	16,500
Laundry Equipment		
Dryer, gas, 16 lb. capacity	Each	655
30 lb. capacity	Each	2650
Washer, 4 cycle	Each	740
Commercial	Each	1150

Description	Unit	$ Cost
Lockers, Steel, single tier, 60" or 72"	Opening	124 - 191
2 tier, 60" or 72" total	Opening	74 - 95
5 tier, box lockers	Opening	35 - 57
Locker bench, lam. maple top only	L.F.	16.35
Pedestals, steel pipe	Each	45
Paving, Bituminous		
Wearing course plus base course	Sq. Yard	7.60
Safe, Office type, 4 hour rating		
30" x 18" x 18"	Each	3925
62" x 33" x 20"	Each	8525
Sidewalks, Concrete 4" thick	S.F.	2.34
Yard Lighting,		
20' aluminum pole with 400 watt high pressure sodium fixture	Each	1870

Model costs calculated for a 1 story building with 12' story height and 800 square feet of floor area

Car Wash

BUILDING TYPES

				Unit	Unit Cost	Cost Per S.F.	% Of Sub-Total
1.0 Foundations							
.1	Footings & Foundations	Poured concrete; strip and spread footings and 4' foundation wall		S.F. Ground	6.59	6.59	
.4	Piles & Caissons	N/A		—	—	—	5.9%
.9	Excavation & Backfill	Site preparation for slab and trench for foundation wall and footing		S.F. Ground	1.35	1.35	
2.0 Substructure							
.1	Slab on Grade	5" reinforced concrete with vapor barrier and granular base		S.F. Slab	3.04	3.04	2.3%
.2	Special Substructures	N/A		—	—	—	
3.0 Superstructure							
.1	Columns & Beams	Steel columns included in 3.7		—	—	—	
.4	Structural Walls	N/A		—	—	—	
.5	Elevated Floors	N/A		—	—	—	2.3%
.7	Roof	Metal deck, open web steel joists, beams, columns		S.F. Roof	3.05	3.05	
.9	Stairs	N/A		—	—	—	
4.0 Exterior Closure							
.1	Walls	Face brick with concrete block backup	70% of wall	S.F. Wall	17.05	20.41	
.5	Exterior Wall Finishes	N/A		—	—	—	25.4%
.6	Doors	Steel overhead hollow metal		Each	1740	8.70	
.7	Windows & Glazed Walls	Horizontal pivoted steel	5% of wall	Each	514	4.88	
5.0 Roofing							
.1	Roof Coverings	Built-up tar and gravel with flashing		S.F. Roof	3.43	3.43	
.7	Insulation	Perlite/EPS composite		S.F. Roof	1.18	1.18	3.5%
.8	Openings & Specialties	N/A		—	—	—	
6.0 Interior Construction							
.1	Partitions	Concrete block	20 S.F. Floor/S.F. Partition	S.F. Partition	4.98	2.49	
.4	Interior Doors	Hollow metal	600 S.F. Floor/Door	Each	492	.82	
.5	Wall Finishes	Paint		S.F. Surface	.68	.68	
.6	Floor Finishes	N/A		—	—	—	3.0%
.7	Ceiling Finishes	N/A		—	—	—	
.9	Interior Surface/Exterior Wall	N/A		—	—	—	
7.0 Conveying							
.1	Elevators	N/A		—	—	—	0.0%
.2	Special Conveyors	N/A		—	—	—	
8.0 Mechanical							
.1	Plumbing	Toilet and service fixtures, supply and drainage	1 Fixture/160 S.F. Floor	Each	4408	27.55	
.2	Fire Protection	N/A		—	—	—	
.3	Heating	Unit heaters		S.F. Floor	9.79	9.79	27.9%
.4	Cooling	N/A		—	—	—	
.5	Special Systems	N/A		—	—	—	
9.0 Electrical							
.1	Service & Distribution	400 ampere service, panel board and feeders		S.F. Floor	12.65	12.65	
.2	Lighting & Power	Fluorescent fixtures, receptacles, switches and misc. power		S.F. Floor	25	25.71	29.7%
.4	Special Electrical	Emergency power		S.F. Floor	1.28	1.28	
11.0 Special Construction							
.1	Specialties	N/A		—	—	—	0.0%
12.0 Site Work							
.1	Earthwork	N/A		—	—	—	
.3	Utilities	N/A		—	—	—	
.5	Roads & Parking	N/A		—	—	—	0.0%
.7	Site Improvements	N/A		—	—	—	
				Sub-Total		133.60	**100%**
	CONTRACTOR FEES (General Requirements: 10%, Overhead: 5%, Profit: 10%)				25%	33.40	
	ARCHITECT FEES				7%	11.70	
				Total Building Cost		**178.70**	

COMMERCIAL/INDUSTRIAL/INSTITUTIONAL — M.090 — Church

Costs per square foot of floor area

Exterior Wall	S.F. Area	2000	7000	12000	17000	22000	27000	32000	37000	42000
	L.F. Perimeter	180	340	470	540	640	740	762	793	860
Decorative Concrete Brick	Wood Arch	138.75	105.40	98.00	92.40	90.30	89.00	86.35	84.60	83.85
	Steel Truss	134.90	101.50	94.10	88.50	86.40	85.15	82.45	80.70	80.00
Stone with Concrete Block Back-up	Wood Arch	159.55	116.60	107.05	99.70	97.00	95.35	91.85	89.55	88.60
	Steel Truss	151.80	108.85	99.30	91.95	89.25	87.60	84.10	81.80	80.80
Face Brick with Concrete Block Back-up	Wood Arch	152.60	112.85	104.00	97.25	94.75	93.20	90.00	87.85	87.00
	Steel Truss	144.85	105.05	96.25	89.50	87.00	85.45	82.25	80.10	79.25
Perimeter Adj., Add or Deduct	Per 100 L.F.	36.30	10.40	6.05	4.25	3.35	2.70	2.25	1.95	1.70
Story Hgt. Adj., Add or Deduct	Per 1 Ft.	2.20	1.15	.95	.75	.70	.65	.60	.50	.50

For Basement, add $18.13 per square foot of basement area

The above costs were calculated using the basic specifications shown on the facing page. These costs should be adjusted where necessary for design alternatives and owner's requirements. Reported completed project costs, for this type of structure, range from $49.45 to $198.75 per S.F.

Common additives

Description	Unit	$ Cost
Altar, Wood, custom design, plain	Each	1925
Deluxe	Each	9700
Granite or marble, average	Each	7775
Deluxe	Each	24,700
Ark, Prefabricated, plain	Each	7275
Deluxe	Each	85,000
Baptistry, Fiberglass, incl. plumbing	Each	3150 - 5825
Bells & Carillons, 48 bells	Each	550,000
24 bells	Each	192,500
Confessional, Prefabricated wood		
Single, plain	Each	2650
Deluxe	Each	6725
Double, plain	Each	5075
Deluxe	Each	15,200
Emergency Lighting, 25 watt, battery operated		
Lead battery	Each	249
Nickel cadmium	Each	605
Lecterns, Wood, plain	Each	550
Deluxe	Each	2600

Description	Unit	$ Cost
Pews/Benches, Hardwood	L.F.	74 - 139
Pulpits, Prefabricated, hardwood	Each	1225 - 7425
Railing, Hardwood	L.F.	149
Steeples, translucent fiberglass		
30" square, 15' high	Each	4150
25' high	Each	4800
Painted fiberglass, 24" square, 14' high	Each	3600
28' high	Each	4225
Aluminum		
20' high, 3'- 6" base	Each	4525
35' high, 8'- 0" base	Each	17,100
60' high, 14'- 0" base	Each	38,600

Model costs calculated for a 1 story building with 24' story height and 17,000 square feet of floor area

Church

			Unit	Unit Cost	Cost Per S.F.	% Of Sub-Total
1.0 Foundations						
.1	Footings & Foundations	Poured concrete; strip and spread footings and 4' foundation wall	S.F. Ground	4.02	4.02	7.0%
.4	Piles & Caissons	N/A	—	—	—	
.9	Excavation & Backfill	Site preparation for slab and trench for foundation wall and footing	S.F. Ground	.89	.89	
2.0 Substructure						
.1	Slab on Grade	4" reinforced concrete with vapor barrier and granular base	S.F. Slab	2.80	2.80	4.0%
.2	Special Substructures	N/A	—	—	—	
3.0 Superstructure						
.1	Columns & Beams	N/A	—	—	—	
.4	Structural Walls	N/A	—	—	—	
.5	Elevated Floors	N/A	—	—	—	18.8%
.7	Roof	Wood deck on laminated wood arches	S.F. Roof	12.42	13.21	
.9	Stairs	N/A	—	—	—	
4.0 Exterior Closure						
.1	Walls	Face brick with concrete block backup 80% of wall (adjusted for end walls)	S.F. Wall	20	12.35	
.5	Exterior Wall Finishes	N/A	—	—	—	22.0%
.6	Doors	Double hollow metal swinging, single hollow metal	Each	1021	.36	
.7	Windows & Glazed Walls	Aluminum, top hinged, in-swinging and curtain wall panels 20% of wall	S.F. Window	17.97	2.74	
5.0 Roofing						
.1	Roof Coverings	Asphalt shingles with flashing	S.F. Roof	1.30	1.30	
.7	Insulation	Polystyrene	S.F. Ground	1.05	1.26	3.9%
.8	Openings & Specialties	Gutters and downspouts	S.F. Ground	.18	.18	
6.0 Interior Construction						
.1	Partitions	Plaster on metal studs 40 S.F. Floor/L.F. Partitions	S.F. Partition	5.68	3.41	
.4	Interior Doors	Hollow metal 400 S.F. Floor/Door	Each	492	1.23	
.5	Wall Finishes	Paint	S.F. Surface	.65	.78	17.2%
.6	Floor Finishes	Carpet	S.F. Floor	4.94	4.94	
.7	Ceiling Finishes	N/A	—	—	—	
.9	Interior Surface/Exterior Wall	Painted plaster 80% of wall	S.F. Wall	2.80	1.71	
7.0 Conveying						
.1	Elevators	N/A	—	—	—	0.0%
.2	Special Conveyors	N/A	—	—	—	
8.0 Mechanical						
.1	Plumbing	Kitchen, toilet and service fixtures, supply and drainage 1 Fixture/2430 S.F. Floor	Each	2162	.89	
.2	Fire Protection	Wet pipe sprinkler system	S.F. Floor	1.33	1.33	
.3	Heating	Oil fired hot water, wall fin radiation	S.F. Floor	4.85	4.85	17.9%
.4	Cooling	Split systems with air cooled condensing units	S.F. Floor	5.41	5.41	
.5	Special Systems	N/A	—	—	—	
9.0 Electrical						
.1	Service & Distribution	200 ampere service, panel board and feeders	S.F. Floor	.29	.29	
.2	Lighting & Power	Fluorescent fixtures, receptacles, switches, A.C. and misc. power	S.F. Floor	4.51	4.51	9.2%
.4	Special Electrical	Alarm systems, sound system and emergency lighting	S.F. Floor	1.64	1.64	
11.0 Special Construction						
.1	Specialties	N/A	—	—	—	0.0%
12.0 Site Work						
.1	Earthwork	N/A	—	—	—	
.3	Utilities	N/A	—	—	—	0.0%
.5	Roads & Parking	N/A	—	—	—	
.7	Site Improvements	N/A	—	—	—	
			Sub-Total		70.10	**100%**
	CONTRACTOR FEES (General Requirements: 10%, Overhead: 5%, Profit: 10%)			25%	17.53	
	ARCHITECT FEES			11%	9.62	
			Total Building Cost		97.25	

BUILDING TYPES

COMMERCIAL/INDUSTRIAL/INSTITUTIONAL — M.100 Club, Country

Costs per square foot of floor area

Exterior Wall	S.F. Area	2000	4000	6000	8000	12000	15000	18000	20000	22000
	L.F. Perimeter	180	280	340	386	460	535	560	600	640
Stone Ashlar Veneer On Concrete Block	Wood Truss	137.45	120.65	112.05	106.85	101.05	99.35	96.90	96.25	95.65
	Steel Joists	134.90	118.10	109.50	104.30	98.50	96.80	94.35	93.65	93.10
Stucco on Concrete Block	Wood Truss	124.65	110.70	103.95	100.00	95.60	94.25	92.45	91.95	91.55
	Steel Joists	122.20	108.25	101.50	97.55	93.15	91.80	90.00	89.50	89.10
Brick Veneer	Wood Frame	131.80	116.20	108.40	103.70	98.50	96.95	94.80	94.15	93.65
Wood Shingles	Wood Frame	125.35	111.15	104.30	100.25	95.75	94.35	92.55	92.00	91.60
Perimeter Adj., Add or Deduct	Per 100 L.F.	23.00	11.50	7.65	5.75	3.80	3.05	2.55	2.30	2.10
Story Hgt. Adj., Add or Deduct	Per 1 Ft.	2.60	2.00	1.60	1.35	1.10	1.05	.90	.85	.85

For Basement, add $16.72 per square foot of basement area

The above costs were calculated using the basic specifications shown on the facing page. These costs should be adjusted where necessary for design alternatives and owner's requirements. Reported completed project costs, for this type of structure, range from $48.85 to $159.90 per S.F.

Common additives

Description	Unit	$ Cost
Bar, Front bar	L.F.	262
Back bar	L.F.	205
Booth, Upholstered, custom, straight	L.F.	134 - 248
"L" or "U" shaped	L.F.	137 - 235
Fireplaces, Brick not incl. chimney or foundation, 30" x 24" opening	Each	1600
Chimney, standard brick		
Single flue 16" x 20"	V.L.F.	51
20" x 20"	V.L.F.	57
2 flue, 20" x 32"	V.L.F.	84
Lockers, Steel, single tier, 60" or 72"	Opening	124 - 191
2 tier, 60" or 72" total	Opening	74 - 95
5 tier, box lockers	Opening	35 - 57
Locker bench, lam. maple top only	L.F.	16.35
Pedestals, steel pipe	Each	45
Refrigerators, Prefabricated, walk-in		
7'-6" High, 6' x 6'	S.F.	117
10' x 10'	S.F.	92
12' x 14'	S.F.	82
12' x 20'	S.F.	72

Description	Unit	$ Cost
Sauna, Prefabricated, complete, 6' x 4'	Each	3875
6' x 9'	Each	5625
8' x 8'	Each	5950
10' x 12'	Each	8675
Smoke Detectors		
Ceiling type	Each	129
Duct type	Each	365
Sound System		
Amplifier, 250 watts	Each	1500
Speaker, ceiling or wall	Each	125
Trumpet	Each	235
Steam Bath, Complete, to 140 C.F.	Each	1050
To 300 C.F.	Each	1150
To 800 C.F.	Each	3300
To 2500 C.F.	Each	5050
Swimming Pool Complete, gunite	S.F.	41 - 50
Tennis Court, Complete with fence		
Bituminous	Each	20,000 - 24,600
Clay	Each	22,300 - 27,500

Model costs calculated for a 1 story building with 12' story height and 6,000 square feet of floor area

Club, Country

			Unit	Unit Cost	Cost Per S.F.	% Of Sub-Total
1.0 Foundations						
.1	Footings & Foundations	Poured concrete; strip and spread footings and 4' foundation wall	S.F. Ground	4.44	4.44	6.7%
.4	Piles & Caissons	N/A	—	—	—	
.9	Excavation & Backfill	Site preparation for slab and trench for foundation wall and footing	S.F. Ground	.96	.96	
2.0 Substructure						
.1	Slab on Grade	4" reinforced concrete with vapor barrier and granular base	S.F. Slab	2.80	2.80	3.5%
.2	Special Substructures	N/A	—	—	—	
3.0 Superstructure						
.1	Columns & Beams	N/A	—	—	—	
.4	Structural Walls	Load bearing partition walls, see item 6.1	—	—	—	
.5	Elevated Floors	N/A	—	—	—	5.6%
.7	Roof	Wood truss with plywood sheathing	S.F. Ground	4.54	4.54	
.9	Stairs	N/A	—	—	—	
4.0 Exterior Closure						
.1	Walls	Stone ashlar veneer on concrete block — 65% of wall	S.F. Wall	19.50	8.62	
.5	Exterior Wall Finishes	N/A	—	—	—	17.9%
.6	Doors	Double aluminum and glass, hollow metal	Each	1550	1.55	
.7	Windows & Glazed Walls	Aluminum horizontal sliding — 35% of wall	Each	269	4.28	
5.0 Roofing						
.1	Roof Coverings	Asphalt shingles	S.F. Roof	.85	1.05	
.7	Insulation	N/A	—	—	—	1.7%
.8	Openings & Specialties	Gutters and downspouts	S.F. Roof	.33	.33	
6.0 Interior Construction						
.1	Partitions	Gypsum board on metal studs, load bearing — 14 S.F. Floor/L.F. Partition	S.F. Partition	2.63	1.88	
.4	Interior Doors	Single leaf wood — 140 S.F. Floor/Door	Each	386	2.76	
.5	Wall Finishes	40% vinyl wall covering, 40% paint, 20% ceramic tile	S.F. Surface	1.58	2.25	19.5%
.6	Floor Finishes	50% carpet, 30% hardwood tile, 20% ceramic tile	S.F. Floor	5.42	5.42	
.7	Ceiling Finishes	Gypsum plaster on wood furring	S.F. Ceiling	2.33	2.33	
.9	Interior Surface/Exterior Wall	Painted gypsum board on furring — 65% of wall	S.F. Wall	2.51	1.11	
7.0 Conveying						
.1	Elevators	N/A	—	—	—	0.0%
.2	Special Conveyors	N/A	—	—	—	
8.0 Mechanical						
.1	Plumbing	Kitchen, toilet and service fixtures, supply and drainage — 1 Fixture/125 S.F. Floor	Each	1571	12.57	
.2	Fire Protection	Wet pipe sprinkler system	S.F. Floor	1.74	1.74	
.3	Heating	Included in 8.4	—	—	—	39.8%
.4	Cooling	Multizone rooftop unit, gas heating, electric cooling	S.F. Floor	17.80	17.80	
.5	Special Systems	N/A	—	—	—	
9.0 Electrical						
.1	Service & Distribution	200 ampere service, panel board and feeders	S.F. Floor	1.13	1.13	
.2	Lighting & Power	Fluorescent fixtures, receptacles, switches, A.C. and misc. power	S.F. Floor	2.81	2.81	5.3%
.4	Special Electrical	Alarm systems and emergency lighting	S.F. Floor	.37	.37	
11.0 Special Construction						
.1	Specialties	N/A	—	—	—	0.0%
12.0 Site Work						
.1	Earthwork	N/A	—	—	—	
.3	Utilities	N/A	—	—	—	0.0%
.5	Roads & Parking	N/A	—	—	—	
.7	Site Improvements	N/A	—	—	—	
			Sub-Total		80.74	100%
	CONTRACTOR FEES (General Requirements: 10%, Overhead: 5%, Profit: 10%)			25%	20.19	
	ARCHITECT FEES			11%	11.12	
			Total Building Cost		**112.05**	

COMMERCIAL/INDUSTRIAL/INSTITUTIONAL M.170 Community Center

Costs per square foot of floor area

Exterior Wall	S.F. Area	4000	6000	8000	10000	12000	14000	16000	18000	20000
	L.F. Perimeter	260	340	420	453	460	510	560	610	600
Face Brick with Concrete Block Back-up	Bearing Walls	94.30	89.65	87.30	83.65	80.20	79.15	78.45	77.85	75.90
	Steel Frame	91.60	87.50	85.45	82.25	79.25	78.35	77.75	77.20	75.55
Decorative Concrete Block	Bearing Walls	83.70	80.00	78.15	75.25	72.55	71.75	71.15	70.70	69.20
	Steel Frame	84.75	81.50	79.90	77.50	75.20	74.55	74.05	73.65	72.40
Tilt Up Concrete Wall Panels	Bearing Walls	84.75	81.30	79.65	77.00	74.60	73.85	73.30	72.85	71.50
	Steel Frame	82.05	79.20	77.75	75.60	73.65	73.05	72.60	72.20	71.15
Perimeter Adj., Add or Deduct	Per 100 L.F.	11.95	8.00	5.95	4.80	3.95	3.40	3.00	2.65	2.40
Story Hgt. Adj., Add or Deduct	Per 1 Ft.	1.55	1.35	1.25	1.10	.90	.85	.85	.80	.70

For Basement, add $18.21 per square foot of basement area

The above costs were calculated using the basic specifications shown on the facing page. These costs should be adjusted where necessary for design alternatives and owner's requirements. Reported completed project costs, for this type of structure, range from $47.05 to $150.05 per S.F.

Common additives

Description	Unit	$ Cost
Bar, Front bar	L.F.	262
Back bar	L.F.	205
Booth, Upholstered, custom straight	L.F.	134 - 248
"L" or "U" shaped	L.F.	137 - 235
Bowling Alleys, incl. alley, pinsetter		
Scorer, counter & misc. supplies, average	Lane	43,700
For automatic scorer, add	Lane	7475
Emergency Lighting, 25 watt, battery operated		
Lead battery	Each	249
Nickel cadmium	Each	605
Kitchen Equipment		
Broiler	Each	3950
Coffee urn, twin 6 gallon	Each	6750
Cooler, 6 ft. long	Each	3075
Dishwasher, 10-12 racks per hr.	Each	2925
Food warmer	Each	700
Freezer, 44 C.F., reach-in	Each	8050
Ice cube maker, 50 lb. per day	Each	1725
Range with 1 oven	Each	2350

Description	Unit	$ Cost
Movie Equipment		
Projector, 35mm	Each	10,900 - 14,500
Screen, wall or ceiling hung	S.F.	5.95 - 9.20
Partitions, Folding leaf, wood		
Acoustic type	S.F.	48 - 78
Seating		
Auditorium chair, all veneer	Each	151
Veneer back, padded seat	Each	184
Upholstered, spring seat	Each	186
Classroom, movable chair & desk	Set	65 - 120
Lecture hall, pedestal type	Each	137 - 435
Sound System		
Amplifier, 250 watts	Each	1500
Speaker, ceiling or wall	Each	125
Trumpet	Each	235
Stage Curtains, Medium weight	S.F.	7.50 - 240
Curtain Track, Light duty	L.F.	47
Swimming Pools, Complete, gunite	S.F.	41 - 50

Community Center

Model costs calculated for a 1 story building with 12' story height and 10,000 square feet of floor area

				Unit	Unit Cost	Cost Per S.F.	% Of Sub-Total
1.0 Foundations							
.1	Footings & Foundations	Poured concrete; strip and spread footings and 4' foundation wall		S.F. Ground	5.14	5.14	
.4	Piles & Caissons	N/A		—	—	—	9.9%
.9	Excavation & Backfill	Site preparation for slab and trench for foundation wall and footing		S.F. Ground	.96	.96	
2.0 Substructure							
.1	Slab on Grade	4" reinforced concrete with vapor barrier and granular base		S.F. Slab	2.80	2.80	4.6%
.2	Special Substructures	N/A		—	—	—	
3.0 Superstructure							
.1	Columns & Beams	N/A		—	—	—	
.4	Structural Walls	Concrete block		S.F. Wall	4.84	.79	
.5	Elevated Floors	N/A		—	—	—	7.5%
.7	Roof	Metal deck on open web steel joists		S.F. Roof	3.82	3.82	
.9	Stairs	N/A		—	—	—	
4.0 Exterior Closure							
.1	Walls	Face brick with concrete block backup	80% of wall	S.F. Wall	17.04	7.41	
.5	Exterior Wall Finishes	N/A		—	—	—	15.2%
.6	Doors	Double aluminum and glass and hollow metal		Each	1411	.57	
.7	Windows & Glazed Walls	Aluminum sliding	20% of wall	Each	392	1.33	
5.0 Roofing							
.1	Roof Coverings	Built-up tar and gravel with flashing		S.F. Roof	2.15	2.15	
.7	Insulation	Perlite/EPS composite		S.F. Roof	1.18	1.18	6.3%
.8	Openings & Specialties	Gravel stop and hatches		S.F. Roof	.54	.54	
6.0 Interior Construction							
.1	Partitions	Gypsum board on metal studs, toilet partitions	14 S.F. Floor/L.F. Partition	S.F. Partition	5.32	3.80	
.4	Interior Doors	Single leaf hollow metal	140 S.F. Floor/Door	Each	492	3.51	
.5	Wall Finishes	Paint		S.F. Surface	.48	.69	
.6	Floor Finishes	50% carpet, 50% vinyl tile		S.F. Floor	4.12	4.12	26.0%
.7	Ceiling Finishes	Mineral fiber tile on concealed zee bars		S.F. Ceiling	3.10	3.10	
.9	Interior Surface/Exterior Wall	Paint	80% of wall	S.F. Wall	1.69	.73	
7.0 Conveying							
.1	Elevators	N/A		—	—	—	0.0%
.2	Special Conveyors	N/A		—	—	—	
8.0 Mechanical							
.1	Plumbing	Kitchen, toilet and service fixtures, supply and drainage	1 Fixture/910 S.F. Floor	Each	4158	4.57	
.2	Fire Protection	Wet pipe sprinkler system		S.F. Floor	1.33	1.33	
.3	Heating	Included in 8.4		—	—	—	21.8%
.4	Cooling	Single zone rooftop unit, gas heating, electric cooling		S.F. Floor	7.48	7.48	
.5	Special Systems	N/A		—	—	—	
9.0 Electrical							
.1	Service & Distribution	200 ampere service, panel board and feeders		S.F. Floor	.57	.57	
.2	Lighting & Power	Incandescent fixtures, receptacles, switches, A.C. and misc. power		S.F. Floor	2.54	2.54	5.6%
.4	Special Electrical	Alarm systems and emergency lighting		S.F. Floor	.34	.34	
11.0 Special Construction							
.1	Specialties	Built-in coat racks, fume hoods, freezer, kitchen equipment		S.F. Floor	1.92	1.92	3.1%
12.0 Site Work							
.1	Earthwork	N/A		—	—	—	
.3	Utilities	N/A		—	—	—	0.0%
.5	Roads & Parking	N/A		—	—	—	
.7	Site Improvements	N/A		—	—	—	
				Sub-Total		61.39	**100%**
	CONTRACTOR FEES (General Requirements: 10%, Overhead: 5%, Profit: 10%)				25%	15.35	
	ARCHITECT FEES				9%	6.91	
				Total Building Cost		**83.65**	

BUILDING TYPES

COMMERCIAL/INDUSTRIAL/INSTITUTIONAL   M.180   Courthouse, 1 Story

Costs per square foot of floor area

Exterior Wall	S.F. Area	16000	23000	30000	37000	44000	51000	58000	65000	72000
	L.F. Perimeter	597	763	821	968	954	1066	1090	1132	1220
Limestone with Concrete Block Back-up	R/Conc. Frame	121.65	117.45	112.85	111.60	108.30	107.50	105.95	104.90	104.50
	Steel Frame	120.55	116.35	111.80	110.55	107.20	106.45	104.85	103.80	103.40
Face Brick with Concrete Block Back-up	R/Conc. Frame	114.90	111.45	107.90	106.85	104.35	103.75	102.55	101.75	101.45
	Steel Frame	113.80	110.35	106.85	105.80	103.30	102.65	101.45	100.65	100.35
Stone with Concrete Block Back-up	R/Conc. Frame	116.10	112.50	108.80	107.70	105.05	104.40	103.15	102.30	101.95
	Steel Frame	115.10	111.50	107.75	106.70	104.05	103.35	102.10	101.25	100.95
Perimeter Adj., Add or Deduct	Per 100 L.F.	4.15	2.85	2.20	1.80	1.50	1.30	1.15	1.05	.90
Story Hgt. Adj., Add or Deduct	Per 1 Ft.	1.45	1.30	1.05	1.00	.85	.85	.75	.70	.65
For Basement, add $17.02 per square foot of basement area										

The above costs were calculated using the basic specifications shown on the facing page. These costs should be adjusted where necessary for design alternatives and owner's requirements. Reported completed project costs, for this type of structure, range from $82.85 to $154.40 per S.F.

Common additives

Description	Unit	$ Cost
Benches, Hardwood	L.F.	74 - 139
Clock System		
20 room	Each	10,300
50 room	Each	24,800
Closed Circuit Surveillance, One station		
Camera and monitor	Each	1275
For additional camera stations, add	Each	700
Directory Boards, Plastic, glass covered		
30" x 20"	Each	500
36" x 48"	Each	895
Aluminum, 24" x 18"	Each	440
36" x 24"	Each	530
48" x 32"	Each	630
48" x 60"	Each	1375
Emergency Lighting, 25 watt, battery operated		
Lead battery	Each	249
Nickel cadmium	Each	605

Description	Unit	$ Cost
Flagpoles, Complete		
Aluminum, 20' high	Each	1000
40' high	Each	2550
70' high	Each	7350
Fiberglass, 23' high	Each	1275
39'-5" high	Each	2575
59' high	Each	6775
Intercom System, 25 station capacity		
Master station	Each	1825
Intercom outlets	Each	107
Handset	Each	293
Safe, Office type, 4 hour rating		
30" x 18" x 18"	Each	3925
62" x 33" x 20"	Each	8525
Smoke Detectors		
Ceiling type	Each	129
Duct type	Each	365

Model costs calculated for a 1 story building with 14' story height and 30,000 square feet of floor area

Courthouse, 1 Story

			Unit	Unit Cost	Cost Per S.F.	% Of Sub-Total
1.0 Foundations						
.1	Footings & Foundations	Poured concrete; strip and spread footings and 4' foundation wall	S.F. Ground	2.09	2.09	
.4	Piles & Caissons	N/A	—	—	—	3.5%
.9	Excavation & Backfill	Site preparation for slab and trench for foundation wall and footing	S.F. Ground	.89	.89	
2.0 Substructure						
.1	Slab on Grade	4" reinforced concrete with vapor barrier and granular base	S.F. Slab	2.80	2.80	3.3%
.2	Special Substructures	N/A	—	—	—	
3.0 Superstructure						
.1	Columns & Beams	Concrete	L.F. Column	35	.88	
.4	Structural Walls	N/A	—	—	—	
.5	Elevated Floors	N/A	—	—	—	14.0%
.7	Roof	Cast-in-place concrete waffle slab	S.F. Roof	10.96	10.96	
.9	Stairs	N/A	—	—	—	
4.0 Exterior Closure						
.1	Walls	Limestone panels with concrete block backup 75% of wall	S.F. Wall	29	8.61	
.5	Exterior Wall Finishes	N/A	—	—	—	12.8%
.6	Doors	Double wood	Each	1397	.33	
.7	Windows & Glazed Walls	Aluminum with insulated glass 25% of wall	Each	446	1.86	
5.0 Roofing						
.1	Roof Coverings	Built-up tar and gravel with flashing	S.F. Roof	1.93	1.93	
.7	Insulation	Perlite/EPS composite	S.F. Roof	1.18	1.18	3.7%
.8	Openings & Specialties	Gravel stop, hatches	S.F. Roof	.05	.05	
6.0 Interior Construction						
.1	Partitions	Plaster on metal studs, toilet partitions 10 S.F. Floor/L.F. Partition	S.F. Partition	6.84	8.21	
.4	Interior Doors	Single leaf wood 100 S.F. Floor/Door	Each	386	3.86	
.5	Wall Finishes	70% paint, 20% wood paneling, 10% vinyl wall covering	S.F. Surface	1.63	3.92	36.0%
.6	Floor Finishes	60% hardwood, 20% carpet, 20% terrazzo	S.F. Floor	8.02	8.02	
.7	Ceiling Finishes	Gypsum plaster on metal lath, suspended	S.F. Ceiling	5.54	5.54	
.9	Interior Surface/Exterior Wall	Painted plaster 75% of wall	S.F. Wall	2.80	.80	
7.0 Conveying						
.1	Elevators	N/A	—	—	—	0.0%
.2	Special Conveyors	N/A	—	—	—	
8.0 Mechanical						
.1	Plumbing	Toilet and service fixtures, supply and drainage 1 Fixture/1110 S.F. Floor	Each	3008	2.71	
.2	Fire Protection	Wet pipe sprinkler system	S.F. Floor	1.09	1.09	
.3	Heating	Included in 8.4	—	—	—	19.7%
.4	Cooling	Multizone unit, gas heating, electric cooling	S.F. Floor	12.78	12.78	
.5	Special Systems	N/A	—	—	—	
9.0 Electrical						
.1	Service & Distribution	400 ampere service, panel board and feeders	S.F. Floor	.53	.53	
.2	Lighting & Power	Fluorescent fixtures, receptacles, switches, A.C. and misc. power	S.F. Floor	5.02	5.02	7.0%
.4	Special Electrical	Alarm systems and emergency lighting	S.F. Floor	.33	.33	
11.0 Special Construction						
.1	Specialties	N/A	—	—	—	0.0%
12.0 Site Work						
.1	Earthwork	N/A	—	—	—	
.3	Utilities	N/A	—	—	—	0.0%
.5	Roads & Parking	N/A	—	—	—	
.7	Site Improvements	N/A	—	—	—	
			Sub-Total		84.39	100%
	CONTRACTOR FEES (General Requirements: 10%, Overhead: 5%, Profit: 10%)			25%	21.10	
	ARCHITECT FEES			7%	7.36	
			Total Building Cost		112.85	

BUILDING TYPES

COMMERCIAL/INDUSTRIAL/INSTITUTIONAL

M.200 Factory, 1 Story

Costs per square foot of floor area

Exterior Wall	S.F. Area	12000	18000	24000	30000	36000	42000	48000	54000	60000
	L.F. Perimeter	460	580	713	730	826	880	965	1006	1045
Concrete Block	Steel Frame	64.50	60.95	59.30	56.95	56.15	55.20	54.80	54.15	53.60
	Bearing Walls	63.70	60.10	58.50	56.15	55.30	54.40	53.95	53.30	52.80
Precast Concrete Panels	Steel Frame	67.75	63.60	61.80	59.00	58.05	56.95	56.45	55.70	55.10
Insulated Metal Panels	Steel Frame	65.35	61.60	59.95	57.50	56.65	55.65	55.20	54.55	54.00
Face Brick on Common Brick	Steel Frame	75.25	69.90	67.60	63.75	62.55	61.10	60.40	59.35	58.50
Tilt-up Concrete Panel	Steel Frame	65.80	62.00	60.30	57.80	56.90	55.90	55.45	54.75	54.20
Perimeter Adj., Add or Deduct	Per 100 L.F.	3.00	2.00	1.55	1.20	1.00	.90	.75	.65	.60
Story Hgt. Adj., Add or Deduct	Per 1 Ft.	.45	.35	.35	.25	.30	.25	.25	.25	.20

For Basement, add $17.95 per square foot of basement area

The above costs were calculated using the basic specifications shown on the facing page. These costs should be adjusted where necessary for design alternatives and owner's requirements. Reported completed project costs, for this type of structure, range from $26.65 to $103.05 per S.F.

Common additives

Description	Unit	$ Cost
Clock System		
20 room	Each	10,300
50 room	Each	24,800
Dock Bumpers, Rubber blocks		
4-1/2" thick, 10" high, 14" long	Each	52
24" long	Each	79
36" long	Each	93
12" high, 14" long	Each	81
24" long	Each	91
36" long	Each	106
6" thick, 10" high, 14" long	Each	78
24" long	Each	101
36" long	Each	128
20" high, 11" long	Each	130
Dock Boards, Heavy		
60" x 60" Aluminum, 5,000# cap.	Each	1125
9000# cap.	Each	1250
15,000# cap.	Each	1375

Description	Unit	$ Cost
Dock Levelers, Hinged 10 ton cap.		
6' x 8'	Each	4075
7' x 8'	Each	4200
Partitions, Woven wire, 10 ga., 1-1/2" mesh		
4' wide x 7' high	Each	114
8' high	Each	129
10' High	Each	154
Platform Lifter, Portable, 6' x 6'		
3000# cap.	Each	6425
4000# cap.	Each	9025
Fixed, 6' x 8', 5000# cap.	Each	8925

Model costs calculated for a 1 story building with 20' story height and 30,000 square feet of floor area

Factory, 1 Story

				Unit	Unit Cost	Cost Per S.F.	% Of Sub-Total
1.0 Foundations							
.1	Footings & Foundations	Poured concrete; strip and spread footings and 4' foundation wall		S.F. Ground	1.92	1.92	6.6%
.4	Piles & Caissons	N/A		—	—	—	
.9	Excavation & Backfill	Site preparation for slab and trench for foundation wall and footing		S.F. Ground	.89	.89	
2.0 Substructure							
.1	Slab on Grade	4" reinforced concrete with vapor barrier and granular base		S.F. Slab	2.80	2.80	6.6%
.2	Special Substructures	N/A		—	—	—	
3.0 Superstructure							
.1	Columns & Beams	Steel columns included in 3.7		—	—	—	
.4	Structural Walls	N/A		—	—	—	
.5	Elevated Floors	N/A		—	—	—	11.0%
.7	Roof	Metal deck, open web steel joists, beams and columns		S.F. Roof	4.68	4.68	
.9	Stairs	N/A		—	—	—	
4.0 Exterior Closure							
.1	Walls	Concrete block	75% of wall	S.F. Wall	4.14	1.51	
.5	Exterior Wall Finishes	N/A		—	—	—	10.1%
.6	Doors	Double aluminum and glass, hollow metal, steel overhead		Each	1257	.63	
.7	Windows & Glazed Walls	Industrial horizontal pivoted steel	25% of wall	Each	562	2.14	
5.0 Roofing							
.1	Roof Coverings	Built-up tar and gravel with flashing		S.F. Roof	1.79	1.79	
.7	Insulation	Perlite/EPS composite		S.F. Roof	1.18	1.18	7.7%
.8	Openings & Specialties	Gravel stop, hatches, gutters and downspouts		S.F. Roof	.30	.30	
6.0 Interior Construction							
.1	Partitions	Concrete block, toilet partitions	60 S.F. Floor/L.F. Partition	S.F. Partition	8.30	1.66	
.4	Interior Doors	Single leaf hollow metal and fire doors	600 S.F. Floor/Door	Each	492	.82	
.5	Wall Finishes	Paint		S.F. Surface	.97	.39	9.6%
.6	Floor Finishes	Vinyl composition tile	10% of floor	S.F. Floor	2.70	.27	
.7	Ceiling Finishes	Fiberglass board on exposed grid system	10% of area	S.F. Ceiling	3.10	.31	
.9	Interior Surface/Exterior Wall	Paint and block filler	75% of wall	S.F. Wall	1.69	.62	
7.0 Conveying							
.1	Elevators	N/A		—	—	—	0.0%
.2	Special Conveyors	N/A		—	—	—	
8.0 Mechanical							
.1	Plumbing	Toilet and service fixtures, supply and drainage	1 Fixture/1000 S.F. Floor	Each	2500	2.50	
.2	Fire Protection	Sprinklers, ordinary hazard		S.F. Floor	1.64	1.64	
.3	Heating	Oil fired hot water, unit heaters		S.F. Floor	4.49	4.49	34.4%
.4	Cooling	Chilled water, air cooled condenser system		S.F. Floor	6.09	6.09	
.5	Special Systems	N/A		—	—	—	
9.0 Electrical							
.1	Service & Distribution	600 ampere service, panel board and feeders		S.F. Floor	.73	.73	
.2	Lighting & Power	High intensity discharge fixtures, receptacles, switches, A.C. and misc. power		S.F. Floor	4.98	4.98	14.0%
.4	Special Electrical	Alarm systems and emergency lighting		S.F. Floor	.24	.24	
11.0 Special Construction							
.1	Specialties	N/A		—	—	—	0.0%
12.0 Site Work							
.1	Earthwork	N/A		—	—	—	
.3	Utilities	N/A		—	—	—	0.0%
.5	Roads & Parking	N/A		—	—	—	
.7	Site Improvements	N/A		—	—	—	
				Sub-Total		42.58	**100%**
	CONTRACTOR FEES (General Requirements: 10%, Overhead: 5%, Profit: 10%)				25%	10.65	
	ARCHITECT FEES				7%	3.72	
				Total Building Cost		**56.95**	

COMMERCIAL/INDUSTRIAL/INSTITUTIONAL — M.220 — Fire Station, 1 Story

Costs per square foot of floor area

Exterior Wall	S.F. Area	4000	4500	5000	5500	6000	6500	7000	7500	8000
	L.F. Perimeter	260	280	300	320	320	336	353	370	386
Face Brick Concrete Block Back-up	Steel Joists	90.75	88.95	87.55	86.35	83.85	82.90	82.10	81.45	80.80
	Precast Conc.	91.85	90.10	88.70	87.50	85.00	84.00	83.25	82.55	81.95
Decorative Concrete Block	Steel Joists	83.30	81.85	80.70	79.70	77.75	77.00	76.35	75.80	75.30
	Precast Conc.	84.55	83.10	82.00	81.00	79.05	78.25	77.65	77.05	76.55
Limestone with Concrete Block Back-up	Steel Joists	98.25	96.15	94.50	93.05	90.00	88.85	87.90	87.15	86.40
	Precast Conc.	99.50	97.40	95.75	94.35	91.30	90.15	89.20	88.40	87.65
Perimeter Adj., Add or Deduct	Per 100 L.F.	11.50	10.25	9.15	8.35	7.65	7.05	6.55	6.15	5.75
Story Hgt. Adj., Add or Deduct	Per 1 Ft.	1.50	1.45	1.35	1.35	1.25	1.20	1.15	1.15	1.10

For Basement, add $20.77 per square foot of basement area

The above costs were calculated using the basic specifications shown on the facing page. These costs should be adjusted where necessary for design alternatives and owner's requirements. Reported completed project costs, for this type of structure, range from $43.10 to $127.50 per S.F.

Common additives

Description	Unit	$ Cost
Appliances		
Cooking range, 30" free standing		
1 oven	Each	325 - 1400
2 oven	Each	1275 - 1700
30" built-in		
1 oven	Each	335 - 1150
2 oven	Each	1100 - 1975
Counter top cook tops, 4 burner	Each	237 - 570
Microwave oven	Each	169 - 550
Combination range, refrig. & sink, 30" wide	Each	880 - 1775
60" wide	Each	2575
72" wide	Each	2925
Combination range refrigerator, sink microwave oven & icemaker	Each	4325
Compactor, residential, 4-1 compaction	Each	425 - 495
Dishwasher, built-in, 2 cycles	Each	360 - 465
4 cycles	Each	355 - 725
Garbage disposer, sink type	Each	77 - 228
Hood for range, 2 speed, vented, 30" wide	Each	107 - 590
42" wide	Each	250 - 885

Description	Unit	$ Cost
Appliances, cont.		
Refrigerator, no frost 10-12 C.F.	Each	505 - 770
14-16 C.F.	Each	530 - 650
18-20 C.F.	Each	585 - 985
Lockers, Steel, single tier, 60" or 72"	Opening	124 - 191
2 tier, 60" or 72" total	Opening	74 - 95
5 tier, box lockers	Opening	35 - 57
Locker bench, lam. maple top only	L.F.	16.35
Pedestals, steel pipe	Each	45
Sound System		
Amplifier, 250 watts	Each	1500
Speaker, ceiling or wall	Each	125
Trumpet	Each	235

Fire Station, 1 Story

Model costs calculated for a 1 story building with 14' story height and 6,000 square feet of floor area

				Unit	Unit Cost	Cost Per S.F.	% Of Sub-Total
1.0 Foundations							
.1	Footings & Foundations	Poured concrete; strip and spread footings and 4' foundation wall		S.F. Ground	3.86	3.86	7.9%
.4	Piles & Caissons	N/A		—	—	—	
.9	Excavation & Backfill	Site preparation for slab and trench for foundation wall and footing		S.F. Ground	1.06	1.06	
2.0 Substructure							
.1	Slab on Grade	4" reinforced concrete with vapor barrier and granular base		S.F. Slab	3.43	3.43	5.5%
.2	Special Substructures	N/A		—	—	—	
3.0 Superstructure							
.1	Columns & Beams	N/A		—	—	—	
.4	Structural Walls	N/A		—	—	—	
.5	Elevated Floors	N/A		—	—	—	3.9%
.7	Roof	Metal deck, open web steel joists, beams		S.F. Roof	2.40	2.40	
.9	Stairs	N/A		—	—	—	
4.0 Exterior Closure							
.1	Walls	Face brick with concrete block backup	75% of wall	S.F. Wall	17.05	9.55	
.5	Exterior Wall Finishes	N/A		—	—	—	21.2%
.6	Doors	Single aluminum and glass, overhead, hollow metal	15% of wall	S.F. Door	20	2.32	
.7	Windows & Glazed Walls	Aluminum insulated glass	10% of wall	Each	562	1.31	
5.0 Roofing							
.1	Roof Coverings	Built-up tar and gravel with flashing		S.F. Roof	2.29	2.29	
.7	Insulation	Perlite/EPS composite		S.F. Roof	1.18	1.18	6.6%
.8	Openings & Specialties	Gravel stop		S.F. Roof	.66	.66	
6.0 Interior Construction							
.1	Partitions	Concrete block, toilet partitions	17 S.F. Floor/L.F. Partition	S.F. Partition	5.30	3.12	
.4	Interior Doors	Single leaf hollow metal	500 S.F. Floor/Door	Each	492	.98	
.5	Wall Finishes	Paint		S.F. Surface	.97	1.14	14.8%
.6	Floor Finishes	50% vinyl tile, 50% paint		S.F. Floor	1.85	1.85	
.7	Ceiling Finishes	Fiberglass board on exposed grid, suspended	50% of area	S.F. Ceiling	3.10	1.55	
.9	Interior Surface/Exterior Wall	Acrylic glazed coating	75% of wall	S.F. Wall	1.04	.58	
7.0 Conveying							
.1	Elevators	N/A		—	—	—	0.0%
.2	Special Conveyors	N/A		—	—	—	
8.0 Mechanical							
.1	Plumbing	Kitchen, toilet and service fixtures, supply and drainage	1 Fixture/375 S.F. Floor	Each	2186	5.83	
.2	Fire Protection	Wet pipe sprinkler system		S.F. Floor	1.74	1.74	
.3	Heating	Included in 8.4		—	—	—	32.9%
.4	Cooling	Rooftop multizone unit system		S.F. Floor	12.79	12.79	
.5	Special Systems	N/A		—	—	—	
9.0 Electrical							
.1	Service & Distribution	200 ampere service, panel board and feeders		S.F. Floor	.82	.82	
.2	Lighting & Power	Fluorescent fixtures, receptacles, switches, A.C. and misc. power		S.F. Floor	3.42	3.42	7.2%
.4	Special Electrical	Alarm systems		S.F. Floor	.24	.24	
11.0 Special Construction							
.1	Specialties	N/A		—	—	—	0.0%
12.0 Site Work							
.1	Earthwork	N/A		—	—	—	
.3	Utilities	N/A		—	—	—	0.0%
.5	Roads & Parking	N/A		—	—	—	
.7	Site Improvements	N/A		—	—	—	
				Sub-Total		62.12	100%
	CONTRACTOR FEES (General Requirements: 10%, Overhead: 5%, Profit: 10%)				25%	15.53	
	ARCHITECT FEES				8%	6.20	
				Total Building Cost		**83.85**	

COMMERCIAL/INDUSTRIAL/INSTITUTIONAL

M.240 Fraternity/Sorority House

Costs per square foot of floor area

Exterior Wall		S.F. Area	4000	5000	6000	8000	10000	12000	14000	16000	18000
		L.F. Perimeter	180	205	230	260	300	340	353	386	420
Cedar Beveled Siding	Wood Frame		94.40	89.35	86.00	80.90	78.20	76.45	74.45	73.35	72.65
Aluminum Siding	Wood Frame		92.25	87.35	84.10	79.20	76.55	74.85	72.95	71.95	71.20
Board and Batten	Wood Frame		92.20	87.95	85.10	80.65	78.25	76.80	75.00	74.05	73.40
Face Brick on Block	Wood Joists		104.70	98.80	94.80	88.40	85.05	82.95	80.20	78.90	78.00
Stucco on Block	Wood Joists		93.65	88.65	85.30	80.25	77.55	75.80	73.85	72.80	72.05
Decorative Block	Wood Joists		97.60	92.30	88.75	83.25	80.30	78.45	76.25	75.10	74.30
Perimeter Adj., Add or Deduct	Per 100 L.F.		8.75	6.95	5.80	4.35	3.50	2.85	2.45	2.20	1.95
Story Hgt. Adj., Add or Deduct	Per 1 Ft.		1.15	1.05	.95	.85	.75	.70	.60	.65	.60
For Basement, add $13.02 per square foot of basement area											

The above costs were calculated using the basic specifications shown on the facing page. These costs should be adjusted where necessary for design alternatives and owner's requirements. Reported completed project costs, for this type of structure, range from $62.25 to $120.85 per S.F.

Common additives

Description	Unit	$ Cost
Appliances		
Cooking range, 30" free standing		
1 oven	Each	325 - 1400
2 oven	Each	1275 - 1700
30" built-in		
1 oven	Each	335 - 1150
2 oven	Each	1100 - 1975
Counter top cook tops, 4 burner	Each	237 - 570
Microwave oven	Each	169 - 550
Combination range, refrig. & sink, 30" wide	Each	880 - 1775
60" wide	Each	2575
72" wide	Each	2925
Combination range, refrigerator, sink,		
microwave oven & icemaker	Each	4325
Compactor, residential, 4-1 compaction	Each	425 - 495
Dishwasher, built-in, 2 cycles	Each	360 - 465
4 cycles	Each	355 - 725
Garbage disposer, sink type	Each	77 - 228
Hood for range, 2 speed, vented, 30" wide	Each	107 - 590
42" wide	Each	250 - 885

Description	Unit	$ Cost
Appliances, cont.		
Refrigerator, no frost 10-12 C.F.	Each	505 - 770
14-16 C.F.	Each	530 - 650
18-20 C.F.	Each	585 - 985
Elevators, Hydraulic passenger, 2 stops		
1500# capacity	Each	39,600
2500# capacity	Each	40,700
3500# capacity	Each	44,500
Laundry Equipment		
Dryer, gas, 16 lb. capacity	Each	655
30 lb. capacity	Each	2650
Washer, 4 cycle	Each	740
Commercial	Each	1150
Sound System		
Amplifier, 250 watts	Each	1500
Speaker, ceiling or wall	Each	125
Trumpet	Each	235

Important: See the Reference Section for Location Factors

Model costs calculated for a 2 story building with 10' story height and 10,000 square feet of floor area

Fraternity/Sorority House

				Unit	Unit Cost	Cost Per S.F.	% Of Sub-Total
1.0 Foundations							
.1	Footings & Foundations	Poured concrete; strip and spread footings and 4' foundation wall		S.F. Ground	3.66	1.83	
.4	Piles & Caissons	N/A		—	—	—	4.1%
.9	Excavation & Backfill	Site preparation for slab and trench for foundation wall and footing		S.F. Ground	1.06	.53	
2.0 Substructure							
.1	Slab on Grade	4" reinforced concrete with vapor barrier and granular base		S.F. Slab	2.80	1.40	2.4%
.2	Special Substructures	N/A		—	—	—	
3.0 Superstructure							
.1	Columns & Beams	N/A		—	—	—	
.4	Structural Walls	Included in 6.1		—	—	—	
.5	Elevated Floors	Plywood on wood joists		S.F. Floor	2.84	1.42	5.4%
.7	Roof	Plywood on wood rafters (pitched)		S.F. Roof	2.20	1.23	
.9	Stairs	Wood		Flight	1113	.44	
4.0 Exterior Closure							
.1	Walls	Cedar bevel siding on wood studs, insulated	80% of wall	S.F. Wall	6.73	3.23	
.5	Exterior Wall Finishes	N/A		—	—	—	10.3%
.6	Doors	Solid core wood		Each	1364	.95	
.7	Windows & Glazed Walls	Double hung wood	20% of wall	Each	356	1.71	
5.0 Roofing							
.1	Roof Coverings	Asphalt shingles with flashing (pitched)		S.F. Ground	1.12	.56	
.7	Insulation	Fiberglass sheets		S.F. Ground	.98	.49	2.1%
.8	Openings & Specialties	Gutters and downspouts		S.F. Ground	.36	.18	
6.0 Interior Construction							
.1	Partitions	Gypsum board on wood studs	25 S.F. Floor/L.F. Partition	S.F. Partition	2.81	.90	
.4	Interior Doors	Single leaf wood	200 S.F. Floor/Door	Each	386	1.93	
.5	Wall Finishes	Paint		S.F. Surface	.48	.31	21.2%
.6	Floor Finishes	50% hardwood, 50% carpet		S.F. Floor	6.04	6.04	
.7	Ceiling Finishes	Gypsum board on wood furring		S.F. Ceiling	2.33	2.33	
.9	Interior Surface/Exterior Wall	Painted gypsum board	80% of wall	S.F. Wall	1.10	.66	
7.0 Conveying							
.1	Elevators	One hydraulic passenger elevator		Each	47,800	4.78	8.3%
.2	Special Conveyors	N/A		—	—	—	
8.0 Mechanical							
.1	Plumbing	Kitchen toilet and service fixtures, supply and drainage	1 Fixture/150 S.F. Floor	Each	676	4.51	
.2	Fire Protection	Wet pipe sprinkler system		S.F. Floor	1.43	1.43	
.3	Heating	Oil fired hot water, baseboard radiation		S.F. Floor	3.17	3.17	24.1%
.4	Cooling	Split system with air cooled condensing unit		S.F. Floor	4.69	4.69	
.5	Special Systems	N/A		—	—	—	
9.0 Electrical							
.1	Service & Distribution	600 ampere service, panel board and feeders		S.F. Floor	2.60	2.60	
.2	Lighting & Power	Fluorescent fixtures, receptacles, switches, A.C. and misc. power		S.F. Floor	5.06	5.06	22.1%
.4	Special Electrical	Alarm, communication system and generator set		S.F. Floor	5.00	5.00	
11.0 Special Construction							
.1	Specialties	N/A		—	—	—	0.0%
12.0 Site Work							
.1	Earthwork	N/A		—	—	—	
.3	Utilities	N/A		—	—	—	0.0%
.5	Roads & Parking	N/A		—	—	—	
.7	Site Improvements	N/A		—	—	—	
				Sub-Total		57.38	100%
	CONTRACTOR FEES (General Requirements: 10%, Overhead: 5%, Profit: 10%)				25%	14.35	
	ARCHITECT FEES				9%	6.47	
				Total Building Cost		**78.20**	

COMMERCIAL/INDUSTRIAL/INSTITUTIONAL   M.250   Funeral Home

Costs per square foot of floor area

Exterior Wall	S.F. Area	4000	5000	6000	7000	8000	9000	10000	11000	12000
	L.F. Perimeter	260	300	340	353	386	420	425	435	460
Vertical Redwood Siding	Wood Frame	88.40	84.25	81.50	78.60	77.05	75.85	74.20	73.00	72.25
Brick Veneer	Wood Frame	93.60	89.10	86.10	82.75	81.05	79.75	77.80	76.35	75.55
Aluminum Siding	Wood Frame	86.80	82.85	80.20	77.50	76.05	74.85	73.35	72.20	71.55
Brick on Block	Wood Truss	98.10	93.35	90.25	86.65	84.85	83.45	81.35	79.75	78.90
Limestone on Block	Wood Truss	107.40	101.95	98.30	93.85	91.70	90.05	87.35	85.35	84.30
Stucco on Block	Wood Truss	89.40	85.35	82.65	79.90	78.40	77.20	75.65	74.50	73.75
Perimeter Adj., Add or Deduct	Per 100 L.F.	5.85	4.65	3.90	3.30	2.90	2.60	2.35	2.15	1.95
Story Hgt. Adj., Add or Deduct	Per 1 Ft.	.90	.80	.80	.70	.65	.65	.60	.55	.50

For Basement, add $17.03 per square foot of basement area

The above costs were calculated using the basic specifications shown on the facing page. These costs should be adjusted where necessary for design alternatives and owner's requirements. Reported completed project costs, for this type of structure, range from $61.15 to $172.65 per S.F.

Common additives

Description	Unit	$ Cost
Autopsy Table, Standard	Each	7225
Deluxe	Each	9150
Directory Boards, Plastic, glass covered		
30" x 20"	Each	500
36" x 48"	Each	895
Aluminum, 24" x 18"	Each	440
36" x 24"	Each	530
48" x 32"	Each	630
48" x 60"	Each	1375
Emergency Lighting, 25 watt, battery operated		
Lead battery	Each	249
Nickel cadmium	Each	605
Mortuary Refrigerator, End operated		
Two capacity	Each	11,200
Six capacity	Each	20,800

Description	Unit	$ Cost
Planters, Precast concrete		
48" diam., 24" high	Each	495
7" diam., 36" high	Each	1000
Fiberglass, 36" diam., 24" high	Each	395
60" diam., 24" high	Each	755
Smoke Detectors		
Ceiling type	Each	129
Duct type	Each	365

Funeral Home

Model costs calculated for a 1 story building with 10' story height and 10,000 square feet of floor area

				Unit	Unit Cost	Cost Per S.F.	% Of Sub-Total
1.0 Foundations							
.1	Footings & Foundations	Poured concrete; strip and spread footings and 4' foundation wall		S.F. Ground	2.58	2.58	6.5%
.4	Piles & Caissons	N/A		—	—	—	
.9	Excavation & Backfill	Site preparation for slab and trench for foundation wall and footing		S.F. Ground	.96	.96	
2.0 Substructure							
.1	Slab on Grade	4" reinforced concrete with vapor barrier and granular base		S.F. Slab	2.80	2.80	5.1%
.2	Special Substructures	N/A		—	—	—	
3.0 Superstructure							
.1	Columns & Beams	N/A		—	—	—	
.4	Structural Walls	N/A		—	—	—	
.5	Elevated Floors	N/A		—	—	—	5.7%
.7	Roof	Plywood on wood rafters (pitched)		S.F. Ground	3.11	3.11	
.9	Stairs	N/A		—	—	—	
4.0 Exterior Closure							
.1	Walls	1" x 4" vertical T & G redwood siding on wood studs	90% of wall	S.F. Wall	7.58	2.90	
.5	Exterior Wall Finishes	N/A		—	—	—	8.8%
.6	Doors	Wood swinging double doors, single leaf hollow metal		Each	1460	.88	
.7	Windows & Glazed Walls	Double hung wood	10% of wall	Each	286	1.01	
5.0 Roofing							
.1	Roof Coverings	Asphalt shingles with flashing		S.F. Roof	1.32	1.32	
.7	Insulation	Fiberglass sheets		S.F. Ground	.88	.88	4.5%
.8	Openings & Specialties	Gutters and downspouts		S.F. Ground	.25	.25	
6.0 Interior Construction							
.1	Partitions	Gypsum board on wood studs with sound deadening board	15 S.F. Floor/L.F. Partition	S.F. Partition	4.59	2.45	
.4	Interior Doors	Single leaf wood	150 S.F. Floor/Door	Each	386	2.57	
.5	Wall Finishes	50% wallpaper, 25% wood paneling, 25% paint		S.F. Surface	1.81	1.93	32.8%
.6	Floor Finishes	70% carpet, 30% terrazzo		S.F. Floor	8.46	8.46	
.7	Ceiling Finishes	Mineral fiberboard on wood furring		S.F. Ceiling	2.03	2.03	
.9	Interior Surface/Exterior Wall	Painted gypsum board on wood furring	90% of wall	S.F. Wall	1.38	.41	
7.0 Conveying							
.1	Elevators	N/A		—	—	—	0.0%
.2	Special Conveyors	N/A		—	—	—	
8.0 Mechanical							
.1	Plumbing	Toilet and service fixtures, supply and drainage	1 Fixture/770 S.F. Floor	Each	4042	5.25	
.2	Fire Protection	Wet pipe sprinkler system		S.F. Floor	1.33	1.33	
.3	Heating	Included in 8.4		—	—	—	30.2%
.4	Cooling	Multizone rooftop unit, gas heating, electric cooling		S.F. Floor	9.86	9.86	
.5	Special Systems	N/A		—	—	—	
9.0 Electrical							
.1	Service & Distribution	200 ampere service, panel board and feeders		S.F. Floor	.50	.50	
.2	Lighting & Power	Fluorescent fixtures, receptacles, switches, A.C. and misc. power		S.F. Floor	2.80	2.80	6.4%
.4	Special Electrical	Alarm systems and emergency lighting		S.F. Floor	.19	.19	
11.0 Special Construction							
.1	Specialties	N/A		—	—	—	0.0%
12.0 Site Work							
.1	Earthwork	N/A		—	—	—	
.3	Utilities	N/A		—	—	—	0.0%
.5	Roads & Parking	N/A		—	—	—	
.7	Site Improvements	N/A		—	—	—	
				Sub-Total		54.47	100%
	CONTRACTOR FEES (General Requirements: 10%, Overhead: 5%, Profit: 10%)				25%	13.62	
	ARCHITECT FEES				9%	6.11	
				Total Building Cost		**74.20**	

BUILDING TYPES

COMMERCIAL/INDUSTRIAL/INSTITUTIONAL

M.260 Garage, Auto Sales

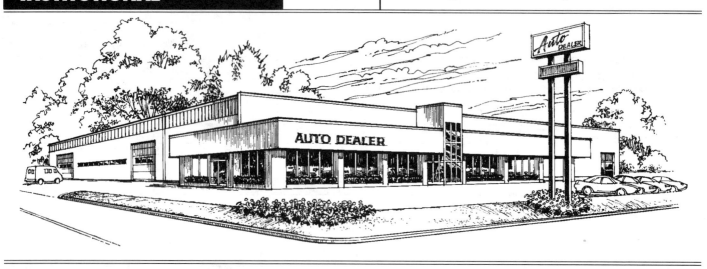

Costs per square foot of floor area

Exterior Wall	S.F. Area	12000	14000	16000	19000	21000	23000	26000	28000	30000
	L.F. Perimeter	440	474	510	556	583	607	648	670	695
Metal Panel Curtain Walls	Steel Frame	63.05	61.35	60.10	58.60	57.75	56.95	56.10	55.55	55.10
Tilt-up Concrete Wall	Steel Frame	61.90	60.25	59.10	57.65	56.85	56.10	55.30	54.80	54.35
Face Brick with Concrete Block Back-up	Bearing Walls	64.60	62.55	61.10	59.25	58.20	57.25	56.25	55.60	55.05
	Steel Frame	67.30	65.25	63.75	61.95	60.90	59.95	58.90	58.25	57.75
Stucco on Concrete Block	Bearing Walls	59.25	57.60	56.45	55.00	54.15	53.40	52.60	52.05	51.65
	Steel Frame	62.20	60.55	59.40	57.95	57.10	56.35	55.55	55.05	54.60
Perimeter Adj., Add or Deduct	Per 100 L.F.	3.70	3.20	2.80	2.35	2.10	1.95	1.70	1.60	1.50
Story Hgt. Adj., Add or Deduct	Per 1 Ft.	.90	.80	.75	.70	.65	.65	.60	.60	.55

For Basement, add $19.45 per square foot of basement area

The above costs were calculated using the basic specifications shown on the facing page. These costs should be adjusted where necessary for design alternatives and owner's requirements. Reported completed project costs, for this type of structure, range from $32.90 to $94.70 per S.F.

Common additives

Description	Unit	$ Cost
Emergency Lighting, 25 watt, battery operated		
Lead battery	Each	249
Nickel cadmium	Each	605
Smoke Detectors		
Ceiling type	Each	129
Duct type	Each	365
Sound System		
Amplifier, 250 watts	Each	1500
Speaker, ceiling or wall	Each	125
Trumpet	Each	235

Important: See the Reference Section for Location Factors

Model costs calculated for a 1 story building with 14' story height and 21,000 square feet of floor area

Garage, Auto Sales

				Unit	Unit Cost	Cost Per S.F.	% Of Sub-Total
1.0 Foundations							
	.1	Footings & Foundations	Poured concrete; strip and spread footings and 4' foundation wall	S.F. Ground	1.86	1.86	6.5%
	.4	Piles & Caissons	N/A	—	—	—	
	.9	Excavation & Backfill	Site preparation for slab and trench for foundation wall and footing	S.F. Ground	.96	.96	
2.0 Substructure							
	.1	Slab on Grade	4" reinforced concrete with vapor barrier and granular base	S.F. Slab	3.43	3.43	7.9%
	.2	Special Substructures	N/A	—	—	—	
3.0 Superstructure							
	.1	Columns & Beams	Steel columns included in 3.7	—	—	—	
	.4	Structural Walls	Metal siding support	S.F. Wall	4.10	1.12	
	.5	Elevated Floors	N/A	—	—	—	14.4%
	.7	Roof	Metal deck, open web steel joists, beams, columns	S.F. Floor	5.10	5.10	
	.9	Stairs	N/A	—	—	—	
4.0 Exterior Closure							
	.1	Walls	Metal panel 70% of wall	S.F. Wall	8.09	2.20	
	.5	Exterior Wall Finishes	N/A	—	—	—	18.0%
	.6	Doors	Double aluminum and glass, hollow metal, steel overhead	Each	2195	1.88	
	.7	Windows & Glazed Walls	Window wall 30% of wall	S.F. Window	31	3.69	
5.0 Roofing							
	.1	Roof Coverings	Built-up tar and gravel with flashing	S.F. Roof	1.77	1.77	
	.7	Insulation	Perlite/EPS composite	S.F. Roof	1.18	1.18	7.2%
	.8	Openings & Specialties	Gravel stop and skylight	L.F. Perimeter	4.99	.14	
6.0 Interior Construction							
	.1	Partitions	Gypsum board on metal studs 28 S.F. Floor/L.F. Partition	S.F. Partition	2.64	1.13	
	.4	Interior Doors	Hollow metal 280 S.F. Floor/Door	Each	492	1.76	
	.5	Wall Finishes	Paint	S.F. Surface	.48	.41	14.6%
	.6	Floor Finishes	50% vinyl tile, 50% paint	S.F. Floor	1.85	1.85	
	.7	Ceiling Finishes	Fiberglass board on exposed grid, suspended 50% of area	S.F. Ceiling	2.33	1.17	
	.9	Interior Surface/Exterior Wall	N/A	—	—	—	
7.0 Conveying							
	.1	Elevators	N/A	—	—	—	0.0%
	.2	Special Conveyors	N/A	—	—	—	
8.0 Mechanical							
	.1	Plumbing	Toilet and service fixtures, supply and drainage 1 Fixture/1500 S.F. Floor	Each	2685	1.79	
	.2	Fire Protection	Wet pipe sprinkler system	S.F. Floor	1.74	1.74	
	.3	Heating	Gas fired hot water, unit heaters (service area)	S.F. Floor	2.04	2.04	20.8%
	.4	Cooling	Single zone rooftop unit, gas heating, electric (office and showroom)	S.F. Floor	3.14	3.14	
	.5	Special Systems	Underfloor garage exhaust system	S.F. Floor	.23	.22	
9.0 Electrical							
	.1	Service & Distribution	200 ampere service, panel board and feeders	S.F. Floor	.24	.24	
	.2	Lighting & Power	Fluorescent fixtures, receptacles, switches, A.C. and misc. power	S.F. Floor	3.44	3.44	8.9%
	.4	Special Electrical	Alarm systems and emergency lighting	S.F. Floor	.18	.18	
11.0 Special Construction							
	.1	Specialties	Hoists, compressor, fuel pump	S.F. Floor	.72	.72	1.7%
12.0 Site Work							
	.1	Earthwork	N/A	—	—	—	
	.3	Utilities	N/A	—	—	—	0.0%
	.5	Roads & Parking	N/A	—	—	—	
	.7	Site Improvements	N/A	—	—	—	
				Sub-Total		43.16	100%
		CONTRACTOR FEES (General Requirements: 10%, Overhead: 5%, Profit: 10%)			25%	10.79	
		ARCHITECT FEES			7%	3.80	
				Total Building Cost		57.75	

COMMERCIAL/INDUSTRIAL/INSTITUTIONAL

M.290 Garage, Repair

Costs per square foot of floor area

Exterior Wall	S.F. Area	2000	4000	6000	8000	10000	12000	14000	16000	18000
	L.F. Perimeter	180	260	340	420	500	580	586	600	610
Concrete Block	Wood Joists	86.20	74.00	69.90	67.90	66.70	65.85	63.85	62.40	61.30
	Steel Joists	86.30	74.50	70.55	68.60	67.45	66.65	64.65	63.25	62.10
Poured Concrete	Wood Joists	92.20	78.50	73.90	71.70	70.30	69.35	66.90	65.15	63.75
	Steel Joists	93.00	79.30	74.75	72.50	71.10	70.20	67.70	66.00	64.60
Insulated Metal Panels	Wood Frame	81.70	70.95	67.30	65.55	64.50	63.75	62.00	60.80	59.80
	Steel Frame	84.25	73.50	69.85	68.15	67.05	66.30	64.55	63.35	62.35
Perimeter Adj., Add or Deduct	Per 100 L.F.	13.70	6.85	4.60	3.40	2.75	2.25	1.95	1.75	1.55
Story Hgt. Adj., Add or Deduct	Per 1 Ft.	1.10	.80	.65	.65	.60	.60	.50	.50	.40

For Basement, add $17.49 per square foot of basement area

The above costs were calculated using the basic specifications shown on the facing page. These costs should be adjusted where necessary for design alternatives and owner's requirements. Reported completed project costs, for this type of structure, range from $41.65 to $125.15 per S.F.

Common additives

Description	Unit	$ Cost
Air Compressors		
Electric 1-1/2 H.P., standard controls	Each	670
Dual controls	Each	780
5 H.P. 115/230 Volt, standard controls	Each	2175
Dual controls	Each	2275
Product Dispenser		
with vapor recovery for 6 nozzles	Each	16,500
Hoists, Single post		
8000# cap., swivel arm	Each	5025
Two post, adjustable frames, 11,000# cap.	Each	6850
24,000# cap.	Each	9750
7500# Frame support	Each	6450
Four post, roll on ramp	Each	5900
Lockers, Steel, single tier, 60" or 72"	Opening	124 - 191
2 tier, 60" or 72" total	Opening	74 - 95
5 tier, box lockers	Opening	35 - 57
Locker bench, lam. maple top only	L.F.	16.35
Pedestals, steel pipe	Each	45
Lube Equipment		
3 reel type, with pumps, no piping	Each	7550
Spray Painting Booth, 26' long, complete	Each	14,400

Important: See the Reference Section for Location Factors

Garage, Repair

Model costs calculated for a 1 story building with 14' story height and 4,000 square feet of floor area

				Unit	Unit Cost	Cost Per S.F.	% Of Sub-Total
1.0 Foundations							
	.1	Footings & Foundations	Poured concrete; strip and spread footings and 4' foundation wall	S.F. Ground	3.78	3.78	8.6%
	.4	Piles & Caissons	N/A	—	—	—	
	.9	Excavation & Backfill	Site preparation for slab and trench for foundation wall and footing	S.F. Ground	.96	.96	
2.0 Substructure							
	.1	Slab on Grade	4" reinforced concrete with vapor barrier and granular base	S.F. Slab	2.80	2.80	5.1%
	.2	Special Substructures	N/A	—	—	—	
3.0 Superstructure							
	.1	Columns & Beams	N/A	—	—	—	
	.4	Structural Walls	N/A	—	—	—	
	.5	Elevated Floors	N/A	—	—	—	5.2%
	.7	Roof	Metal deck on open web steel joists	S.F. Roof	2.87	2.87	
	.9	Stairs	N/A	—	—	—	
4.0 Exterior Closure							
	.1	Walls	Concrete block 80% of wall	L.F. Wall	6.91	5.03	
	.5	Exterior Wall Finishes	N/A	—	—	—	13.9%
	.6	Doors	Steel overhead and hollow metal 15% of wall	S.F. Door	13.19	1.80	
	.7	Windows & Glazed Walls	Hopper type commercial steel 5% of wall	Each	269	.82	
5.0 Roofing							
	.1	Roof Coverings	Built-up tar and gravel	S.F. Roof	2.31	2.31	
	.7	Insulation	Perlite/EPS composite	S.F. Roof	1.18	1.18	6.9%
	.8	Openings & Specialties	Gravel stop and skylight	S.F. Roof	.34	.34	
6.0 Interior Construction							
	.1	Partitions	Concrete block, toilet partitions 50 S.F. Floor/L.F. Partition	S.F. Partition	5.30	1.06	
	.4	Interior Doors	Single leaf hollow metal 3000 S.F. Floor/Door	Each	492	.16	
	.5	Wall Finishes	Paint	S.F. Surface	.80	.32	6.7%
	.6	Floor Finishes	90% metallic floor hardener, 10% vinyl composition tile	S.F. Floor	.67	.67	
	.7	Ceiling Finishes	Gypsum board on wood joists in office and washrooms 10% of area	S.F. Ceiling	2.33	.23	
	.9	Interior Surface/Exterior Wall	Paint 80% of wall	S.F. Wall	1.69	1.23	
7.0 Conveying							
	.1	Elevators	N/A	—	—	—	0.0%
	.2	Special Conveyors	N/A	—	—	—	
8.0 Mechanical							
	.1	Plumbing	Toilet and service fixtures, supply and drainage 1 Fixture/500 S.F. Floor	Each	2400	4.80	
	.2	Fire Protection	Sprinklers, ordinary hazard	S.F. Floor	1.91	1.91	
	.3	Heating	Oil fired hot water, unit heaters	S.F. Floor	4.08	4.08	32.1%
	.4	Cooling	Split systems with air cooled condensing units	S.F. Floor	6.09	6.09	
	.5	Special Systems	Garage exhaust system	S.F. Floor	.89	.89	
9.0 Electrical							
	.1	Service & Distribution	100 ampere service, panel board and feeders	S.F. Floor	.44	.44	
	.2	Lighting & Power	Fluorescent fixtures, receptacles, switches, A.C. and misc. power	S.F. Floor	3.49	3.49	7.8%
	.4	Special Electrical	Alarm systems and emergency lighting	S.F. Floor	.38	.38	
11.0 Special Construction							
	.1	Specialties	Hoists	S.F. Floor	7.54	7.54	13.7%
12.0 Site Work							
	.1	Earthwork	N/A	—	—	—	
	.3	Utilities	N/A	—	—	—	0.0%
	.5	Roads & Parking	N/A	—	—	—	
	.7	Site Improvements	N/A	—	—	—	
				Sub-Total		55.18	100%
		CONTRACTOR FEES (General Requirements: 10%, Overhead: 5%, Profit: 10%)		25%		13.80	
		ARCHITECT FEES		8%		5.52	
				Total Building Cost		**74.50**	

COMMERCIAL/INDUSTRIAL/INSTITUTIONAL   M.300   Garage, Service Station

Costs per square foot of floor area

Exterior Wall	S.F. Area	600	800	1000	1200	1400	1600	1800	2000	2200
	L.F. Perimeter	100	120	126	140	153	160	170	180	190
Face Brick with Concrete Block Back-up	Wood Truss	119.05	109.75	99.60	95.00	91.50	87.65	85.15	83.20	81.65
	Steel Joists	113.70	104.40	94.25	89.65	86.15	82.30	79.80	77.90	76.30
Enameled Sandwich Panel Tile on Concrete Block	Steel Frame	104.75	96.40	87.65	83.55	80.45	77.15	75.00	73.30	71.90
	Steel Joists	121.85	111.85	100.70	95.70	91.90	87.65	84.95	82.85	81.05
Aluminum Siding	Wood Frame	103.15	95.40	87.60	83.85	81.05	78.05	76.15	74.65	73.40
Wood Siding	Wood Frame	103.40	95.65	87.80	84.05	81.25	78.25	76.30	74.80	73.55
Perimeter Adj., Add or Deduct	Per 100 L.F.	54.20	40.65	32.50	27.10	23.20	20.30	18.05	16.30	14.75
Story Hgt. Adj., Add or Deduct	Per 1 Ft.	4.15	3.75	3.15	2.90	2.75	2.50	2.35	2.25	2.15
Basement—Not Applicable										

The above costs were calculated using the basic specifications shown on the facing page. These costs should be adjusted where necessary for design alternatives and owner's requirements. Reported completed project costs, for this type of structure, range from $29.95 to $151.35 per S.F.

Common additives

Description	Unit	$ Cost
Air Compressors		
Electric 1-1/2 H.P., standard controls	Each	670
Dual controls	Each	780
5 H.P. 115/230 volt, standard controls	Each	2175
Dual controls	Each	2275
Product Dispenser		
with vapor recovery for 6 nozzles	Each	16,500
Hoists, Single post		
8000# cap. swivel arm	Each	5025
Two post, adjustable frames, 11,000# cap.	Each	6850
24,000# cap.	Each	9750
7500# cap.	Each	6450
Four post, roll on ramp	Each	5900
Lockers, Steel, single tier, 60" or 72"	Opening	124 - 191
2 tier, 60" or 72" total	Opening	74 - 95
5 tier, box lockers	Each	35 - 57
Locker bench, lam. maple top only	L.F.	16.35
Pedestals, steel pipe	Each	45
Lube Equipment		
3 reel type, with pumps, no piping	Each	7550

Model costs calculated for a 1 story building with 10' story height and 1,400 square feet of floor area

Garage, Service Station

				Unit	Unit Cost	Cost Per S.F.	% Of Sub-Total
1.0 Foundations							
.1	Footings & Foundations	Poured concrete; strip and spread footings and 4' foundation wall		S.F. Ground	6.03	6.03	
.4	Piles & Caissons	N/A		—	—	—	10.9%
.9	Excavation & Backfill	Site preparation for slab and trench for foundation wall and footing		S.F. Ground	1.35	1.35	
2.0 Substructure							
.1	Slab on Grade	4" reinforced concrete with vapor barrier and granular base		S.F. Slab	2.80	2.80	4.1%
.2	Special Substructures	N/A		—	—	—	
3.0 Superstructure							
.1	Columns & Beams	N/A		—	—	—	
.4	Structural Walls	N/A		—	—	—	
.5	Elevated Floors	N/A		—	—	—	6.7%
.7	Roof	Plywood on wood trusses		S.F. Ground	4.54	4.54	
.9	Stairs	N/A		—	—	—	
4.0 Exterior Closure							
.1	Walls	Face brick with concrete block backup	60% of wall	S.F. Wall	17.05	11.18	
.5	Exterior Wall Finishes	N/A		—	—	—	32.4%
.6	Doors	Steel overhead, aluminum & glass and hollow metal	20% of wall	S.F. Door	25	5.47	
.7	Windows & Glazed Walls	Store front and metal top hinged outswinging	20% of wall	S.F. Window	24	5.33	
5.0 Roofing							
.1	Roof Coverings	Asphalt shingles with flashing		S.F. Ground	1.00	1.00	
.7	Insulation	Perlite/EPS composite		S.F. Ground	.88	.88	2.8%
.8	Openings & Specialties	N/A		—	—	—	
6.0 Interior Construction							
.1	Partitions	Concrete block, toilet partitions	25 S.F. Floor/L.F. Partition	S.F. Partition	9.22	2.95	
.4	Interior Doors	Single leaf hollow metal	700 S.F. Floor/Door	Each	492	.70	
.5	Wall Finishes	Paint		S.F. Surface	.97	.62	10.5%
.6	Floor Finishes	Vinyl composition tile	35% of floor area	S.F. Floor	2.71	.95	
.7	Ceiling Finishes	Painted gypsum board on wood joists in sales area & washrooms	35% of floor area	S.F. Ceiling	2.33	.82	
.9	Interior Surface/Exterior Wall	Paint	60% of wall	S.F. Wall	1.69	1.11	
7.0 Conveying							
.1	Elevators	N/A		—	—	—	0.0%
.2	Special Conveyors	N/A		—	—	—	
8.0 Mechanical							
.1	Plumbing	Toilet and service fixtures, supply and drainage	1 Fixture/235 S.F. Floor	Each	1468	6.25	
.2	Fire Protection	N/A		—	—	—	
.3	Heating	Oil fired hot water, wall fin radiation		S.F. Floor	4.08	4.08	23.0%
.4	Cooling	Split systems with air cooled condensing units		S.F. Floor	5.23	5.23	
.5	Special Systems	N/A		—	—	—	
9.0 Electrical							
.1	Service & Distribution	100 ampere service, panel board and feeders		S.F. Floor	2.01	2.01	
.2	Lighting & Power	Fluorescent fixtures, receptacles, switches, A.C. and misc. power		S.F. Floor	3.81	3.81	9.6%
.4	Special Electrical	Alarm systems and emergency lighting		S.F. Floor	.67	.67	
11.0 Special Construction							
.1	Specialties	N/A		—	—	—	0.0%
12.0 Site Work							
.1	Earthwork	N/A		—	—	—	
.3	Utilities	N/A		—	—	—	
.5	Roads & Parking	N/A		—	—	—	0.0%
.7	Site Improvements	N/A		—	—	—	
				Sub-Total		67.78	100%
	CONTRACTOR FEES (General Requirements: 10%, Overhead: 5%, Profit: 10%)				25%	16.95	
	ARCHITECT FEES				8%	6.77	
				Total Building Cost		91.50	

COMMERCIAL/INDUSTRIAL/INSTITUTIONAL M.310 | Gymnasium

Costs per square foot of floor area

Exterior Wall	S.F. Area	12000	16000	20000	25000	30000	35000	40000	45000	50000
	L.F. Perimeter	440	520	600	700	708	780	841	910	979
Reinforced Concrete Block	Lam. Wood Arches	92.00	88.60	86.50	84.85	82.40	81.45	80.60	80.05	79.55
	Rigid Steel Frame	88.30	84.85	82.80	81.10	78.65	77.70	76.85	76.30	75.85
Face Brick with Concrete Block Back-up	Lam. Wood Arches	106.10	101.05	98.00	95.60	91.45	90.00	88.65	87.80	87.10
	Rigid Steel Frame	102.40	97.35	94.30	91.85	87.75	86.30	84.95	84.10	83.40
Metal Sandwich Panels	Lam. Wood Arches	89.40	86.25	84.35	82.85	80.70	79.85	79.10	78.60	78.15
	Rigid Steel Frame	85.70	82.55	80.65	79.10	77.00	76.15	75.35	74.90	74.45
Perimeter Adj., Add or Deduct	Per 100 L.F.	3.60	2.70	2.20	1.75	1.45	1.25	1.10	.95	.90
Story Hgt. Adj., Add or Deduct	Per 1 Ft.	.50	.45	.40	.40	.35	.30	.30	.25	.30
Basement—Not Applicable										

The above costs were calculated using the basic specifications shown on the facing page. These costs should be adjusted where necessary for design alternatives and owner's requirements. Reported completed project costs, for this type of structure, range from $45.50 to $136.55 per S.F.

Common additives

Description	Unit	$ Cost
Bleachers, Telescoping, manual		
To 15 tier	Seat	70 - 100
16-20 tier	Seat	151 - 185
21-30 tier	Seat	159 - 192
For power operation, add	Seat	30 - 47
Gym Divider Curtain, Mesh top		
Manual roll-up	S.F.	7.70
Gym Mats		
2" naugahyde covered	S.F.	3.41
2" nylon	S.F.	5.65
1-1/2" wall pads	S.F.	7.25
1" wrestling mats	S.F.	5.85
Scoreboard		
Basketball, one side	Each	3125 - 21,700
Basketball Backstop		
Wall mtd., 6' extended, fixed	Each	1075 - 1250
Swing up, wall mtd.	Each	1650 - 2575

Description	Unit	$ Cost
Lockers, Steel, single tier, 60" or 72"	Opening	124 - 191
2 tier, 60" or 72" total	Opening	74 - 95
5 tier, box lockers	Opening	35 - 57
Locker bench, lam. maple top only	L.F.	16.35
Pedestals, steel pipe	Each	45
Sound System		
Amplifier, 250 watts	Each	1500
Speaker, ceiling or wall	Each	125
Trumpet	Each	235
Emergency Lighting, 25 watt, battery operated		
Lead battery	Each	249
Nickel cadmium	Each	605

Important: See the Reference Section for Location Factors

Model costs calculated for a 1 story building with 25' story height and 20,000 square feet of floor area

Gymnasium

				Unit	Unit Cost	Cost Per S.F.	% Of Sub-Total
1.0 Foundations							
.1	Footings & Foundations	Poured concrete; strip and spread footings and 4' foundation wall		S.F. Ground	2.13	2.13	4.7%
.4	Piles & Caissons	N/A		—	—	—	
.9	Excavation & Backfill	Site preparation for slab and trench for foundation wall and footing		S.F. Ground	.89	.89	
2.0 Substructure							
.1	Slab on Grade	4" reinforced concrete with vapor barrier and granular base		S.F. Slab	2.80	2.80	4.3%
.2	Special Substructures	N/A		—	—	—	
3.0 Superstructure							
.1	Columns & Beams	N/A		—	—	—	
.4	Structural Walls	N/A		—	—	—	
.5	Elevated Floors	N/A		—	—	—	19.2%
.7	Roof	Wood deck on laminated wood arches		S.F. Ground	12.42	12.42	
.9	Stairs	N/A		—	—	—	
4.0 Exterior Closure							
.1	Walls	Reinforced concrete block (end walls included)	90% of wall	S.F. Wall	6.93	4.68	
.5	Exterior Wall Finishes	N/A		—	—	—	10.8%
.6	Doors	Aluminum and glass, hollow metal, steel overhead		Each	975	.29	
.7	Windows & Glazed Walls	Metal horizontal pivoted	10% of wall	Each	270	2.03	
5.0 Roofing							
.1	Roof Coverings	EPDM, 60 mils, fully adhered		S.F. Ground	1.55	1.55	
.7	Insulation	Polyisocyanurate		—	1.02	1.05	4.0%
.8	Openings & Specialties	N/A		—	—	—	
6.0 Interior Construction							
.1	Partitions	Concrete block, toilet partitions	50 S.F. Floor/L.F. Partition	S.F. Partition	5.80	1.16	
.4	Interior Doors	Single leaf hollow metal	500 S.F. Floor/Door	Each	492	.98	
.5	Wall Finishes	50% paint, 50% ceramic tile		S.F. Surface	2.75	1.10	21.4%
.6	Floor Finishes	90% hardwood, 10% ceramic tile		S.F. Floor	9.01	9.01	
.7	Ceiling Finishes	Mineral fiber tile on concealed zee bars	15% of area	S.F. Ceiling	3.10	.47	
.9	Interior Surface/Exterior Wall	Paint	90% of wall	S.F. Wall	1.69	1.14	
7.0 Conveying							
.1	Elevators	N/A		—	—	—	0.0%
.2	Special Conveyors	N/A		—	—	—	
8.0 Mechanical							
.1	Plumbing	Toilet and service fixtures, supply and drainage	1 Fixture/515 S.F. Floor	Each	3167	6.15	
.2	Fire Protection	Wet pipe sprinkler system		S.F. Floor	1.33	1.33	
.3	Heating	Included in 8.4		—	—	—	24.2%
.4	Cooling	Single zone rooftop unit, gas heating, electric cooling		S.F. Floor	8.13	8.13	
.5	Special Systems	N/A		—	—	—	
9.0 Electrical							
.1	Service & Distribution	400 ampere service, panel board and feeders		S.F. Floor	.57	.57	
.2	Lighting & Power	Fluorescent fixtures, receptacles, switches, A.C. and misc. power		S.F. Floor	4.56	4.56	9.8%
.4	Special Electrical	Alarm systems, sound system and emergency lighting		S.F. Floor	1.19	1.19	
11.0 Special Construction							
.1	Specialties	Bleachers, sauna, weight room		S.F. Floor	1.04	1.04	1.6%
12.0 Site Work							
.1	Earthwork	N/A		—	—	—	
.3	Utilities	N/A		—	—	—	0.0%
.5	Roads & Parking	N/A		—	—	—	
.7	Site Improvements	N/A		—	—	—	
				Sub-Total		64.67	**100%**
	CONTRACTOR FEES (General Requirements: 10%, Overhead: 5%, Profit: 10%)				25%	16.17	
	ARCHITECT FEES				7%	5.66	
				Total Building Cost		**86.50**	

BUILDING TYPES

COMMERCIAL/INDUSTRIAL/INSTITUTIONAL M.380 Laundromat

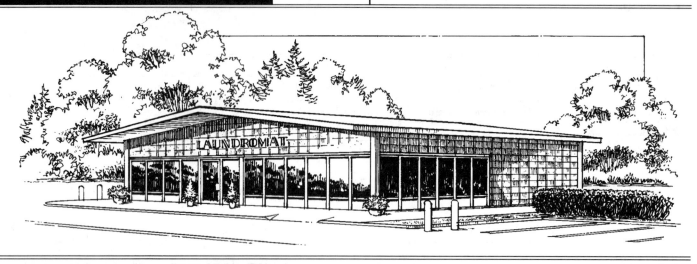

Costs per square foot of floor area

Exterior Wall	S.F. Area	1000	2000	3000	4000	5000	10000	15000	20000	25000
	L.F. Perimeter	126	179	219	253	283	400	490	568	632
Decorative Concrete Block	Steel Frame	111.75	101.05	96.20	93.30	91.35	86.50	84.30	83.05	82.15
	Bearing Walls	117.25	105.90	100.75	97.70	95.60	90.45	88.10	86.80	85.80
Face Brick with Concrete Block Back-up	Steel Frame	131.15	115.80	108.80	104.70	101.85	94.90	91.80	90.00	88.70
	Bearing Walls	131.90	116.30	109.20	105.05	102.15	95.10	91.90	90.10	88.75
Metal Sandwich Panel Precast Concrete Panel	Steel Frame	115.15	104.50	99.70	96.85	94.85	90.05	87.90	86.65	85.75
	Bearing Walls	114.75	104.10	99.30	96.45	94.45	89.65	87.45	86.25	85.30
Perimeter Adj., Add or Deduct	Per 100 L.F.	29.45	14.70	9.80	7.35	5.90	2.90	1.95	1.45	1.15
Story Hgt. Adj., Add or Deduct	Per 1 Ft.	2.00	1.45	1.15	1.05	.90	.60	.50	.45	.40
For Basement, add $17.99 per square foot of basement area										

The above costs were calculated using the basic specifications shown on the facing page. These costs should be adjusted where necessary for design alternatives and owner's requirements. Reported completed project costs, for this type of structure, range from $60.90 to $148.40 per S.F.

Common additives

Description	Unit	$ Cost
Closed Circuit Surveillance, One station		
Camera and monitor	Each	1275
For additional camera stations, add	Each	700
Emergency Lighting, 25 watt, battery operated		
Lead battery	Each	249
Nickel cadmium	Each	605
Laundry Equipment		
Dryers, coin operated 30 lb.	Each	2650
Double stacked	Each	5625
50 lb.	Each	2875
Dry cleaner 20 lb.	Each	32,500
30 lb.	Each	44,500
Washers, coin operated	Each	1150
Washer/extractor 20 lb.	Each	4000
30 lb.	Each	8250
50 lb.	Each	9425
75 lb.	Each	18,700
Smoke Detectors		
Ceiling type	Each	129
Duct type	Each	365

Important: See the Reference Section for Location Factors

Model costs calculated for a 1 story building with 12' story height and 3,000 square feet of floor area

Laundromat

				Unit	Unit Cost	Cost Per S.F.	% Of Sub-Total
1.0 Foundations							
.1	Footings & Foundations	Poured concrete; strip and spread footings and 4' foundation wall		S.F. Ground	4.28	4.28	
.4	Piles & Caissons	N/A		—	—	—	7.4%
.9	Excavation & Backfill	Site preparation for slab and trench for foundation wall and footing		S.F. Ground	1.06	1.06	
2.0 Substructure							
.1	Slab on Grade	5" reinforced concrete with vapor barrier and granular base		S.F. Slab	3.04	3.04	4.2%
.2	Special Substructures	N/A		—	—	—	
3.0 Superstructure							
.1	Columns & Beams	Fireproofing; steel columns included in 3.7		—	—	—	
.4	Structural Walls	N/A		—	—	—	
.5	Elevated Floors	N/A		—	—	—	4.0%
.7	Roof	Metal deck, open web steel joists, beams, columns		S.F. Roof	2.87	2.87	
.9	Stairs	N/A		—	—	—	
4.0 Exterior Closure							
.1	Walls	Decorative concrete block	90% of wall	S.F. Wall	7.98	6.29	
.5	Exterior Wall Finishes	N/A		—	—	—	13.1%
.6	Doors	Double aluminum and glass		Each	2665	.89	
.7	Windows & Glazed Walls	Store front	10% of wall	S.F. Window	25	2.25	
5.0 Roofing							
.1	Roof Coverings	Built-up tar and gravel with flashing		S.F. Roof	2.85	2.85	
.7	Insulation	Perlite/EPS composite		S.F. Roof	1.18	1.18	5.7%
.8	Openings & Specialties	Gravel stop, gutters and downspouts		S.F. Roof	.10	.10	
6.0 Interior Construction							
.1	Partitions	Gypsum board on metal studs	60 S.F. Floor/L.F. Partition	S.F. Partition	2.64	.44	
.4	Interior Doors	Single leaf wood	750 S.F. Floor/Door	Each	386	.51	
.5	Wall Finishes	Paint		S.F. Surface	.48	.16	8.4%
.6	Floor Finishes	Vinyl composition tile		S.F. Floor	1.26	1.26	
.7	Ceiling Finishes	Fiberglass board on exposed grid system		S.F. Ceiling	1.70	1.70	
.9	Interior Surface/Exterior Wall	Painted gypsum board on furring	90% of wall	S.F. Wall	2.47	1.95	
7.0 Conveying							
.1	Elevators	N/A		—	—	—	0.0%
.2	Special Conveyors	N/A		—	—	—	
8.0 Mechanical							
.1	Plumbing	Toilet and service fixtures, supply and drainage	1 Fixture/600 S.F. Floor	Each	10,008	16.68	
.2	Fire Protection	Sprinkler, ordinary hazard		S.F. Floor	1.91	1.91	
.3	Heating	Included in 8.4		—	—	—	35.2%
.4	Cooling	Rooftop single zone unit systems		S.F. Floor	6.65	6.65	
.5	Special Systems	N/A		—	—	—	
9.0 Electrical							
.1	Service & Distribution	200 ampere service, panel board and feeders		S.F. Floor	2.05	2.05	
.2	Lighting & Power	Fluorescent fixtures, receptacles, switches, A.C. and misc. power		S.F. Floor	13.46	13.46	22.0%
.4	Special Electrical	Alarm systems and emergency lighting		S.F. Floor	.34	.34	
11.0 Special Construction							
.1	Specialties	N/A		—	—	—	0.0%
12.0 Site Work							
.1	Earthwork	N/A		—	—	—	
.3	Utilities	N/A		—	—	—	0.0%
.5	Roads & Parking	N/A		—	—	—	
.7	Site Improvements	N/A		—	—	—	
				Sub-Total		71.92	100%
	CONTRACTOR FEES (General Requirements: 10%, Overhead: 5%, Profit: 10%)				25%	17.98	
	ARCHITECT FEES				7%	6.30	
				Total Building Cost		**96.20**	

COMMERCIAL/INDUSTRIAL/INSTITUTIONAL M.390 Library

Costs per square foot of floor area

Exterior Wall	S.F. Area	7000	10000	13000	16000	19000	22000	25000	28000	31000
	L.F. Perimeter	240	300	336	386	411	435	472	510	524
Face Brick with Concrete Block Back-up	R/Conc. Frame	105.25	99.75	95.20	93.05	90.45	88.50	87.50	86.70	85.40
	Steel Frame	103.15	97.70	93.10	91.00	88.35	86.45	85.40	84.60	83.35
Limestone with Concrete Block	R/Conc. Frame	119.15	111.90	105.65	102.85	99.20	96.50	95.15	94.05	92.25
	Steel Frame	117.05	109.85	103.55	100.75	97.10	94.45	93.05	92.00	90.20
Precast Concrete Panels	R/Conc. Frame	101.30	96.35	92.20	90.30	87.95	86.25	85.30	84.60	83.45
	Steel Frame	99.25	94.25	90.10	88.20	85.85	84.15	83.20	82.50	81.40
Perimeter Adj., Add or Deduct	Per 100 L.F.	12.80	8.95	6.90	5.55	4.75	4.05	3.60	3.20	2.85
Story Hgt. Adj., Add or Deduct	Per 1 Ft.	1.95	1.65	1.45	1.35	1.25	1.10	1.05	1.00	.95

For Basement, add $25.85 per square foot of basement area

The above costs were calculated using the basic specifications shown on the facing page. These costs should be adjusted where necessary for design alternatives and owner's requirements. Reported completed project costs, for this type of structure, range from $58.35 to $149.70 per S.F.

Common additives

Description	Unit	$ Cost
Carrels Hardwood	Each	675 - 880
Closed Circuit Surveillance, One station		
Camera and monitor	Each	1275
For additional camera stations, add	Each	700
Elevators, Hydraulic passenger, 2 stops		
1500# capacity	Each	39,600
2500# capacity	Each	40,700
3500# capacity	Each	44,500
Emergency Lighting, 25 watt, battery operated		
Lead battery	Each	249
Nickel cadmium	Each	605
Flagpoles, Complete		
Aluminum, 20' high	Each	1000
40' high	Each	2550
70' high	Each	7350
Fiberglass, 23' high	Each	1275
39'-5" high	Each	2575
59' high	Each	6775

Description	Unit	$ Cost
Library Furnishings		
Bookshelf, 90" high, 10" shelf double face	L.F.	122
single face	L.F.	94
Charging desk, built-in with counter		
Plastic laminated top	L.F.	475
Reading table, laminated		
top 60" x 36"	Each	660

Important: See the Reference Section for Location Factors

Library

Model costs calculated for a 2 story building with 14' story height and 22,000 square feet of floor area

				Unit	Unit Cost	Cost Per S.F.	% Of Sub-Total
1.0 Foundations							
	.1	Footings & Foundations	Poured concrete; strip and spread footings and 4' foundation wall	S.F. Ground	3.64	1.82	
	.4	Piles & Caissons	N/A	—	—	—	3.5%
	.9	Excavation & Backfill	Site preparation for slab and trench for foundation wall and footing	S.F. Ground	.96	.48	
2.0 Substructure							
	.1	Slab on Grade	4" reinforced concrete with vapor barrier and granular base	S.F. Slab	2.80	1.40	2.1%
	.2	Special Substructures	N/A	—	—	—	
3.0 Superstructure							
	.1	Columns & Beams	Concrete columns	L.F. Column	46	1.41	
	.4	Structural Walls	N/A	—	—	—	
	.5	Elevated Floors	Concrete waffle slab	S.F. Floor	11.45	5.72	19.7%
	.7	Roof	Concrete waffle slab	S.F. Roof	10.71	5.36	
	.9	Stairs	Concrete filled metal pan	Flight	4900	.45	
4.0 Exterior Closure							
	.1	Walls	Face brick with concrete block backup 90% of wall	S.F. Wall	17.26	8.60	
	.5	Exterior Wall Finishes	N/A	—	—	—	16.1%
	.6	Doors	Double aluminum and glass, single leaf hollow metal	Each	2885	.26	
	.7	Windows & Glazed Walls	Window wall 10% of wall	S.F. Wall	31	1.72	
5.0 Roofing							
	.1	Roof Coverings	Built-up tar and gravel with flashing	S.F. Roof	1.96	.98	
	.7	Insulation	Perlite/EPS composite	S.F. Roof	1.18	.59	2.6%
	.8	Openings & Specialties	Gravel stop and hatches	L.F. Roof	.24	.12	
6.0 Interior Construction							
	.1	Partitions	Gypsum board on metal studs 30 S.F. Floor/L.F. Partition	S.F. Partition	3.63	1.45	
	.4	Interior Doors	Single leaf wood 300 S.F. Floor/Door	Each	386	1.29	
	.5	Wall Finishes	Paint	S.F. Surface	.47	.38	
	.6	Floor Finishes	50% carpet, 50% vinyl tile	S.F. Floor	3.77	3.77	17.1%
	.7	Ceiling Finishes	Mineral fiber on concealed zee bars	S.F. Ceiling	3.10	3.10	
	.9	Interior Surface/Exterior Wall	Painted gypsum board on furring 90% of wall	S.F. Wall	2.51	1.25	
7.0 Conveying							
	.1	Elevators	One hydraulic passenger elevator	Each	50,160	2.28	3.5%
	.2	Special Conveyors	N/A	—	—	—	
8.0 Mechanical							
	.1	Plumbing	Toilet and service fixtures, supply and drainage 1 Fixture/1835 S.F. Floor	Each	2330	1.27	
	.2	Fire Protection	Wet pipe sprinkler system	S.F. Floor	1.20	1.20	
	.3	Heating	Included in 8.4	—	—	—	25.0%
	.4	Cooling	Multizone unit, gas heating, electric cooling	S.F. Floor	13.85	13.85	
	.5	Special Systems	N/A	—	—	—	
9.0 Electrical							
	.1	Service & Distribution	400 ampere service, panel board and feeders	S.F. Floor	.60	.60	
	.2	Lighting & Power	Fluorescent fixtures, receptacles, switches, A.C. and misc. power	S.F. Floor	5.75	5.75	10.4%
	.4	Special Electrical	Alarm systems and emergency lighting	S.F. Floor	.47	.47	
11.0 Special Construction							
	.1	Specialties	N/A	—	—	—	0.0%
12.0 Site Work							
	.1	Earthwork	N/A	—	—	—	
	.3	Utilities	N/A	—	—	—	0.0%
	.5	Roads & Parking	N/A	—	—	—	
	.7	Site Improvements	N/A	—	—	—	
				Sub-Total		65.57	**100%**
		CONTRACTOR FEES (General Requirements: 10%, Overhead: 5%, Profit: 10%)			25%	16.39	
		ARCHITECT FEES			8%	6.54	
				Total Building Cost		**88.50**	

COMMERCIAL/INDUSTRIAL/INSTITUTIONAL — M.400 Medical Office, 1 Story

Costs per square foot of floor area

Exterior Wall		S.F. Area	4000	5500	7000	8500	10000	11500	13000	14500	16000
		L.F. Perimeter	280	320	380	440	453	503	510	522	560
Face Brick with Concrete Block Back-up		Steel Joists	99.55	94.25	92.25	91.00	88.35	87.55	85.75	84.50	84.05
		Wood Joists	104.20	98.90	96.90	95.65	93.00	92.20	90.45	89.15	88.70
Stucco on Concrete Block		Steel Joists	92.25	88.20	86.60	85.60	83.60	83.00	81.70	80.75	80.40
		Wood Joists	96.95	92.85	91.30	90.25	88.30	87.70	86.35	85.40	85.05
Brick Veneer		Wood Frame	101.85	96.95	95.10	93.90	91.50	90.80	89.10	87.95	87.55
Wood Siding		Wood Frame	96.35	92.35	90.80	89.85	87.90	87.30	86.00	85.10	84.75
Perimeter Adj., Add or Deduct		Per 100 L.F.	9.10	6.65	5.20	4.25	3.70	3.15	2.80	2.50	2.30
Story Hgt. Adj., Add or Deduct		Per 1 Ft.	1.95	1.60	1.50	1.45	1.25	1.20	1.10	1.00	.95
		For Basement, add $16.40 per square foot of basement area									

The above costs were calculated using the basic specifications shown on the facing page. These costs should be adjusted where necessary for design alternatives and owner's requirements. Reported completed project costs, for this type of structure, range from $49.30 to $126.65 per S.F.

Common additives

Description	Unit	$ Cost	Description	Unit	$ Cost
Cabinets, Hospital, base			Directory Boards, Plastic, glass covered		
Laminated plastic	L.F.	252	30" x 20"	Each	500
Stainless steel	L.F.	300	36" x 48"	Each	895
Counter top, laminated plastic	L.F.	47	Aluminum, 24" x 18"	Each	440
Stainless steel	L.F.	109	36" x 24"	Each	530
For drop-in sink, add	Each	485	48" x 32"	Each	630
Nurses station, door type			48" x 60"	Each	1375
Laminated plastic	L.F.	284	Heat Therapy Unit		
Enameled steel	L.F.	253	Humidified, 26" x 78" x 28"	Each	2650
Stainless steel	L.F.	320	Smoke Detectors		
Wall cabinets, laminated plastic	L.F.	184	Ceiling type	Each	129
Enameled steel	L.F.	202	Duct type	Each	365
Stainless steel	L.F.	293	Tables, Examining, vinyl top		
			with base cabinets	Each	1200 - 3325
			Utensil Washer, Sanitizer	Each	8825
			X-Ray, Mobile	Each	10,600 - 60,000

Medical Office, 1 Story

Model costs calculated for a 1 story building with 10' story height and 7,000 square feet of floor area

				Unit	Unit Cost	Cost Per S.F.	% Of Sub-Total
1.0 Foundations							
	.1	Footings & Foundations	Poured concrete; strip and spread footings and 4' foundation wall	S.F. Ground	3.00	3.00	
	.4	Piles & Caissons	N/A	—	—	—	5.6%
	.9	Excavation & Backfill	Site preparation for slab and trench for foundation wall and footing	S.F. Ground	.96	.96	
2.0 Substructure							
	.1	Slab on Grade	4" reinforced concrete with vapor barrier and granular base	S.F. Slab	2.80	2.80	3.9%
	.2	Special Substructures	N/A	—	—	—	
3.0 Superstructure							
	.1	Columns & Beams	N/A	—	—	—	
	.4	Structural Walls	N/A	—	—	—	
	.5	Elevated Floors	N/A	—	—	—	6.4%
	.7	Roof	Plywood on wood trusses	S.F. Ground	4.54	4.54	
	.9	Stairs	N/A	—	—	—	
4.0 Exterior Closure							
	.1	Walls	Face brick with concrete block backup 70% of wall	S.F. Wall	17.24	6.55	
	.5	Exterior Wall Finishes	N/A	—	—	—	16.7%
	.6	Doors	Aluminum and glass doors and entrance with transoms	Each	1021	1.75	
	.7	Windows & Glazed Walls	Wood double hung 30% of wall	Each	373	3.57	
5.0 Roofing							
	.1	Roof Coverings	Asphalt shingles with flashing (Pitched)	S.F. Ground	1.05	1.05	
	.7	Insulation	Fiber glass sheeets	S.F. Ground	.88	.88	3.2%
	.8	Openings & Specialties	Gutters and downspouts	S.F. Ground	.33	.33	
6.0 Interior Construction							
	.1	Partitions	Gypsum bd. & sound deadening bd. on wood studs w/insul. 6 S.F. Floor/L.F. Partition	S.F. Partition	4.60	6.13	
	.4	Interior Doors	Single leaf wood 60 S.F. Floor/Door	Each	386	6.43	
	.5	Wall Finishes	50% paint, 50% vinyl wall covering	S.F. Surface	.84	2.23	31.0%
	.6	Floor Finishes	50% carpet, 50% vinyl composition tile	S.F. Floor	4.26	4.26	
	.7	Ceiling Finishes	Mineral fiber tile on concealed zee bars	S.F. Ceiling	2.03	2.03	
	.9	Interior Surface/Exterior Wall	Painted gypsum board on furring 70% of wall	S.F. Wall	2.47	.94	
7.0 Conveying							
	.1	Elevators	N/A	—	—	—	0.0%
	.2	Special Conveyors	N/A	—	—	—	
8.0 Mechanical							
	.1	Plumbing	Toilet and service fixtures, supply and drainage 1 Fixture/195 S.F. Floor	Each	1524	7.82	
	.2	Fire Protection	Wet pipe sprinkler system	S.F. Floor	1.74	1.74	
	.3	Heating	Included in 8.4	—	—	—	25.5%
	.4	Cooling	Multizone unit, gas heating, electric cooling	S.F. Floor	8.62	8.62	
	.5	Special Systems	N/A	—	—	—	
9.0 Electrical							
	.1	Service & Distribution	200 ampere service, panel board and feeders	S.F. Floor	.82	.82	
	.2	Lighting & Power	Fluorescent fixtures, receptacles, switches, A.C. and misc. power	S.F. Floor	3.82	3.82	7.7%
	.4	Special Electrical	Alarm systems and emergency lighting	S.F. Floor	.86	.86	
11.0 Special Construction							
	.1	Specialties	N/A	—	—	—	0.0%
12.0 Site Work							
	.1	Earthwork	N/A	—	—	—	
	.3	Utilities	N/A	—	—	—	0.0%
	.5	Roads & Parking	N/A	—	—	—	
	.7	Site Improvements	N/A	—	—	—	
				Sub-Total		71.13	**100%**
		CONTRACTOR FEES (General Requirements: 10%, Overhead: 5%, Profit: 10%)			25%	17.78	
		ARCHITECT FEES			9%	7.99	
				Total Building Cost		**96.90**	

BUILDING TYPES

COMMERCIAL/INDUSTRIAL/INSTITUTIONAL — M.410 — Medical Office, 2 Story

Costs per square foot of floor area

Exterior Wall	S.F. Area	4000	5500	7000	8500	10000	11500	13000	14500	16000
	L.F. Perimeter	180	210	240	270	286	311	336	361	386
Face Brick with Concrete Block Back-up	Steel Joists	116.35	110.80	107.60	105.55	103.20	102.00	101.05	100.25	99.65
	Wood Joists	119.35	113.75	110.60	108.50	106.20	104.95	104.05	103.25	102.65
Stucco on Concrete Block	Steel Joists	112.65	107.60	104.75	102.85	100.75	99.70	98.85	98.15	97.60
	Wood Joists	111.75	107.35	104.85	103.20	101.35	100.40	99.70	99.05	98.60
Brick Veneer	Wood Frame	116.50	111.35	108.45	106.55	104.40	103.25	102.40	101.70	101.10
Wood Siding	Wood Frame	111.95	107.50	105.00	103.30	101.50	100.55	99.80	99.15	98.70
Perimeter Adj., Add or Deduct	Per 100 L.F.	15.90	11.55	9.10	7.50	6.35	5.55	4.90	4.40	4.00
Story Hgt. Adj., Add or Deduct	Per 1 Ft.	2.05	2.00	1.75	1.65	1.50	1.40	1.35	1.30	1.25

For Basement, add $18.20 per square foot of basement area

The above costs were calculated using the basic specifications shown on the facing page. These costs should be adjusted where necessary for design alternatives and owner's requirements. Reported completed project costs, for this type of structure, range from $50.60 to $172.80 per S.F.

Common additives

Description	Unit	$ Cost
Cabinets, Hospital, base		
Laminated plastic	L.F.	252
Stainless steel	L.F.	300
Counter top, laminated plastic	L.F.	47
Stainless steel	L.F.	109
For drop-in sink, add	Each	485
Nurses station, door type		
Laminated plastic	L.F.	284
Enameled steel	L.F.	253
Stainless steel	L.F.	320
Wall cabinets, laminated plastic	L.F.	184
Enameled steel	L.F.	202
Stainless steel	L.F.	293
Elevators, Hydraulic passenger, 2 stops		
1500# capacity	Each	39,600
2500# capacity	Each	40,700
3500# capacity	Each	44,500

Description	Unit	$ Cost
Directory Boards, Plastic, glass covered		
30" x 20"	Each	500
36" x 48"	Each	895
Aluminum, 24" x 18"	Each	440
36" x 24"	Each	530
48" x 32"	Each	630
48" x 60"	Each	1375
Emergency Lighting, 25 watt, battery operated		
Lead battery	Each	249
Nickel cadmium	Each	605
Heat Therapy Unit		
Humidified, 26" x 78" x 28"	Each	2650
Smoke Detectors		
Ceiling type	Each	129
Duct type	Each	365
Tables, Examining, vinyl top		
with base cabinets	Each	1200 - 3325
Utensil Washer, Sanitizer	Each	8825
X-Ray, Mobile	Each	10,600 - 60,000

Important: See the Reference Section for Location Factors

Model costs calculated for a 2 story building with 10' story height and 7,000 square feet of floor area

Medical Office, 2 Story

			Unit	Unit Cost	Cost Per S.F.	% Of Sub-Total
1.0 Foundations						
.1	Footings & Foundations	Poured concrete; strip and spread footings and 4' foundation wall	S.F. Ground	4.68	2.34	
.4	Piles & Caissons	N/A	—	—	—	3.6%
.9	Excavation & Backfill	Site preparation for slab and trench for foundation wall and footing	S.F. Ground	.89	.44	
2.0 Substructure						
.1	Slab on Grade	4" reinforced concrete with vapor barrier and granular base	S.F. Slab	2.80	1.40	1.8%
.2	Special Substructures	N/A	—	—	—	
3.0 Superstructure						
.1	Columns & Beams	Included in 3.5 and 3.7	—	—	—	
.4	Structural Walls	N/A	—	—	—	
.5	Elevated Floors	Open web steel joists, slab form, concrete, columns	S.F. Floor	7.36	3.68	8.3%
.7	Roof	Metal deck, open web steel joists, beams, columns	S.F. Roof	2.73	1.37	
.9	Stairs	Concrete filled metal pan	Flight	4540	1.30	
4.0 Exterior Closure						
.1	Walls	Concrete block, insulated 70% of wall	S.F. Floor	3.98	3.98	
.5	Exterior Wall Finishes	Stucco on concrete block 70% of wall	S.F. Wall	4.55	2.18	13.9%
.6	Doors	Aluminum and glass doors with transoms	Each	2885	.82	
.7	Windows & Glazed Walls	Outward projecting metal 30% of wall	Each	269	3.70	
5.0 Roofing						
.1	Roof Coverings	Built-up tar and gravel with flashing	S.F. Roof	2.38	1.19	
.7	Insulation	Perlite/EPS composite	S.F. Roof	1.18	.59	2.6%
.8	Openings & Specialties	Gravel stop and hatches	S.F. Roof	.48	.24	
6.0 Interior Construction						
.1	Partitions	Gypsum bd. & sound deadening bd. on wood studs w/insul. 6 S.F. Floor/L.F. Partition	S.F. Partition	4.60	6.13	
.4	Interior Doors	Single leaf wood 60 S.F. Floor/Door	Each	386	6.43	
.5	Wall Finishes	50% paint, 50% vinyl wall coating, 5% ceramic tile	S.F. Surface	.84	2.23	29.0%
.6	Floor Finishes	50% carpet, 50% vinyl asbestos tile	S.F. Floor	4.26	4.26	
.7	Ceiling Finishes	Mineral fiber tile on concealed zee bars	S.F. Ceiling	2.03	2.03	
.9	Interior Surface/Exterior Wall	Painted gypsum board on furring 70% of wall	S.F. Wall	2.47	1.19	
7.0 Conveying						
.1	Elevators	One hydraulic hospital elevator	Each	59,780	8.54	11.1%
.2	Special Conveyors	N/A	—	—	—	
8.0 Mechanical						
.1	Plumbing	Toilet and service fixtures, supply and drainage 1 Fixture/160 S.F. Floor	Each	1164	7.28	
.2	Fire Protection	Wet pipe sprinkler system	S.F. Floor	1.43	1.43	
.3	Heating	Included in 8.4	—	—	—	22.5%
.4	Cooling	Multizone unit, gas heating, electric cooling	S.F. Floor	8.62	8.62	
.5	Special Systems	N/A	—	—	—	
9.0 Electrical						
.1	Service & Distribution	200 ampere service, panel board and feeders	S.F. Floor	.82	.82	
.2	Lighting & Power	Fluorescent fixtures, receptacles, switches, A.C. and misc. power	S.F. Floor	3.82	3.82	7.2%
.4	Special Electrical	Alarm systems and emergency lighting	S.F. Floor	.86	.86	
11.0 Special Construction						
.1	Specialties	N/A	—	—	—	0.0%
12.0 Site Work						
.1	Earthwork	N/A	—	—	—	
.3	Utilities	N/A	—	—	—	0.0%
.5	Roads & Parking	N/A	—	—	—	
.7	Site Improvements	N/A	—	—	—	
			Sub-Total		76.87	100%
CONTRACTOR FEES (General Requirements: 10%, Overhead: 5%, Profit: 10%)				25%	19.22	
ARCHITECT FEES				9%	8.66	
			Total Building Cost		104.75	

BUILDING TYPES

COMMERCIAL/INDUSTRIAL/INSTITUTIONAL — M.420 — Motel, 1 Story

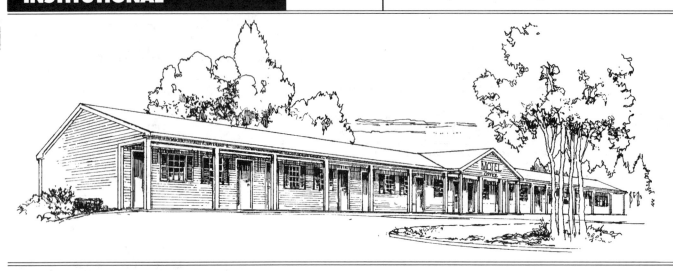

Costs per square foot of floor area

Exterior Wall		S.F. Area	2000	3000	4000	6000	8000	10000	12000	14000	16000
		L.F. Perimeter	240	260	280	380	480	560	580	660	740
Brick Veneer	Wood Frame		104.70	91.75	85.30	82.10	80.40	78.80	76.10	75.55	75.20
Aluminum Siding	Wood Frame		94.85	84.65	79.55	76.90	75.50	74.25	72.15	71.70	71.40
Wood Siding	Wood Frame		95.10	84.85	79.70	77.05	75.65	74.35	72.25	71.80	71.50
Wood Shingles	Wood Frame		97.00	86.20	80.80	78.00	76.55	75.20	73.00	72.55	72.25
Precast Concrete Block	Wood Truss		93.60	83.75	78.80	76.25	74.90	73.65	71.65	71.20	70.95
Brick on Concrete Block	Wood Truss		108.50	94.50	87.50	84.10	82.35	80.60	77.65	77.05	76.70
Perimeter Adj., Add or Deduct	Per 100 L.F.		16.00	10.70	8.00	5.30	4.00	3.20	2.65	2.30	2.00
Story Hgt. Adj., Add or Deduct	Per 1 Ft.		2.00	2.05	1.65	1.50	1.40	1.30	1.15	1.10	1.10

For Basement, add $13.71 per square foot of basement area

The above costs were calculated using the basic specifications shown on the facing page. These costs should be adjusted where necessary for design alternatives and owner's requirements. Reported completed project costs, for this type of structure, range from $40.85 to $193.35 per S.F.

Common additives

Description	Unit	$ Cost
Closed Circuit Surveillance, One station		
Camera and monitor	Each	1275
For additional camera stations, add	Each	700
Emergency Lighting, 25 watt, battery operated		
Lead battery	Each	249
Nickel cadmium	Each	605
Laundry Equipment		
Dryer, gas, 16 lb. capacity	Each	655
30 lb. capacity	Each	2650
Washer, 4 cycle	Each	740
Commercial	Each	1150
Sauna, Prefabricated, complete		
6' x 4'	Each	3875
6' x 6'	Each	4550
6' x 9'	Each	5625
8' x 8'	Each	5950
8' x 10'	Each	6575
10' x 12'	Each	8675
Smoke Detectors		
Ceiling type	Each	129
Duct type	Each	365

Description	Unit	$ Cost
Swimming Pools, Complete, gunite	S.F.	41 - 50
TV Antenna, Master system, 12 outlet	Outlet	215
30 outlet	Outlet	139
100 outlet	Outlet	135

Important: See the Reference Section for Location Factors

Model costs calculated for a 1 story building with 9' story height and 8,000 square feet of floor area

Motel, 1 Story

				Unit	Unit Cost	Cost Per S.F.	% Of Sub-Total
1.0 Foundations							
.1	Footings & Foundations	Poured concrete; strip and spread footings and 4' foundation wall		S.F. Ground	4.53	4.53	
.4	Piles & Caissons	N/A		—	—	—	9.1%
.9	Excavation & Backfill	Site preparation for slab and trench for foundation wall and footing		S.F. Ground	.96	.96	
2.0 Substructure							
.1	Slab on Grade	4" reinforced concrete with vapor barrier and granular base		S.F. Slab	2.80	2.80	4.7%
.2	Special Substructures	N/A		—	—	—	
3.0 Superstructure							
.1	Columns & Beams	N/A		—	—	—	
.4	Structural Walls	N/A		—	—	—	
.5	Elevated Floors	N/A		—	—	—	7.6%
.7	Roof	Plywood on wood trusses		S.F. Ground	4.54	4.54	
.9	Stairs	N/A		—	—	—	
4.0 Exterior Closure							
.1	Walls	Face brick on wood studs with sheathing, insulation and paper	80% of wall	S.F. Wall	13.75	5.94	
.5	Exterior Wall Finishes	N/A		—	—	—	18.9%
.6	Doors	Wood solid core		Each	1021	2.81	
.7	Windows & Glazed Walls	Wood double hung	20% of wall	Each	339	2.62	
5.0 Roofing							
.1	Roof Coverings	Asphalt shingles with flashing (pitched)		S.F. Ground	.90	.90	
.7	Insulation	Fiberglass sheets		S.F. Ground	.88	.88	3.6%
.8	Openings & Specialties	Gutters and downspouts		S.F. Ground	.36	.36	
6.0 Interior Construction							
.1	Partitions	Gypsum bd. and sound deadening bd. on wood studs	9 S.F. Floor/L.F. Partition	S.F. Partition	4.60	4.09	
.4	Interior Doors	Single leaf hollow metal	300 S.F. Floor/Door	Each	344	1.15	
.5	Wall Finishes	90% paint, 10% ceramic tile		S.F. Surface	.89	1.58	26.2%
.6	Floor Finishes	85% carpet, 15% vinyl composition tile		S.F. Floor	5.68	5.68	
.7	Ceiling Finishes	Painted gypsum board on furring		S.F. Ceiling	2.33	2.33	
.9	Interior Surface/Exterior Wall	Painted gypsum board on furring	80% of wall	S.F. Wall	2.47	.93	
7.0 Conveying							
.1	Elevators	N/A		—	—	—	0.0%
.2	Special Conveyors	N/A		—	—	—	
8.0 Mechanical							
.1	Plumbing	Toilet and service fixtures, supply and drainage	1 Fixture/90 S.F. Floor	Each	802	8.92	
.2	Fire Protection	Wet pipe sprinkler system		S.F. Floor	1.74	1.74	
.3	Heating	Included in 8.4		—	—	—	22.8%
.4	Cooling	Through the wall electric heating and cooling units		S.F. Floor	3.11	3.11	
.5	Special Systems	N/A		—	—	—	
9.0 Electrical							
.1	Service & Distribution	100 ampere service, panel board and feeders		S.F. Floor	.48	.48	
.2	Lighting & Power	Fluorescent fixtures, receptacles, switches and misc. power		S.F. Floor	3.54	3.54	7.1%
.4	Special Electrical	Alarm systems		S.F. Floor	.24	.24	
11.0 Special Construction							
.1	Specialties	N/A		—	—	—	0.0%
12.0 Site Work							
.1	Earthwork	N/A		—	—	—	
.3	Utilities	N/A		—	—	—	0.0%
.5	Roads & Parking	N/A		—	—	—	
.7	Site Improvements	N/A		—	—	—	
				Sub-Total		60.13	**100%**
	CONTRACTOR FEES (General Requirements: 10%, Overhead: 5%, Profit: 10%)				25%	15.03	
	ARCHITECT FEES				7%	5.24	
				Total Building Cost		**80.40**	

BUILDING TYPES

COMMERCIAL/INDUSTRIAL/INSTITUTIONAL — M.430 — Motel, 2-3 Story

Costs per square foot of floor area

Exterior Wall	S.F. Area	25000	37000	49000	61000	73000	81000	88000	96000	104000
	L.F. Perimeter	433	593	606	720	835	911	978	1054	1074
Decorative Concrete Block	Wood Joists	76.15	74.35	71.40	70.70	70.25	70.00	69.85	69.70	69.25
	Precast Conc.	82.00	80.15	77.20	76.50	76.05	75.80	75.65	75.50	75.05
Stucco on Concrete Block	Wood Joists	74.70	72.95	70.20	69.55	69.10	68.90	68.75	68.60	68.15
	Precast Conc.	81.05	79.30	76.55	75.90	75.45	75.20	75.10	74.95	74.50
Wood Siding	Wood Frame	74.75	73.00	70.25	69.60	69.15	68.90	68.75	68.60	68.20
Brick Veneer	Wood Frame	77.60	75.60	72.25	71.50	71.00	70.75	70.60	70.40	69.90
Perimeter Adj., Add or Deduct	Per 100 L.F.	2.65	1.75	1.35	1.10	.90	.80	.75	.65	.60
Story Hgt. Adj., Add or Deduct	Per 1 Ft.	.95	.90	.70	.65	.65	.65	.65	.60	.55

For Basement, add $16.73 per square foot of basement area

The above costs were calculated using the basic specifications shown on the facing page. These costs should be adjusted where necessary for design alternatives and owner's requirements. Reported completed project costs, for this type of structure, range from $40.90 to $98.30 per S.F.

Common additives

Description	Unit	$ Cost	Description	Unit	$ Cost
Closed Circuit Surveillance, One station			Sauna, Prefabricated, complete		
Camera and monitor	Each	1275	6' x 4'	Each	3875
For additional camera station, add	Each	700	6' x 6'	Each	4550
Elevators, Hydraulic passenger, 2 stops			6' x 9'	Each	5625
1500# capacity	Each	39,600	8' x 8'	Each	5950
2500# capacity	Each	40,700	8' x 10'	Each	6575
3500# capacity	Each	44,500	10' x 12'	Each	8675
Additional stop, add	Each	2700	Smoke Detectors		
Emergency Lighting, 25 watt, battery operated			Ceiling type	Each	129
Lead battery	Each	249	Duct type	Each	365
Nickel cadmium	Each	605	Swimming Pools, Complete, gunite	S.F.	41 - 50
Laundry Equipment			TV Antenna, Master system, 12 outlet	Outlet	215
Dryer, gas, 16 lb. capacity	Each	655	30 outlet	Outlet	139
30 lb. capacity	Each	2650	100 outlet	Outlet	135
Washer, 4 cycle	Each	740			
Commercial	Each	1150			

Model costs calculated for a 3 story building with 9' story height and 73,000 square feet of floor area

Motel, 2-3 Story

				Unit	Unit Cost	Cost Per S.F.	% Of Sub-Total
1.0 Foundations							
.1	Footings & Foundations	Poured concrete; strip and spread footings and 4' foundation wall		S.F. Ground	3.12	1.04	2.3%
.4	Piles & Caissons	N/A		—	—	—	
.9	Excavation & Backfill	Site preparation for slab and trench for foundation wall and footing		S.F. Ground	.89	.30	
2.0 Substructure							
.1	Slab on Grade	4" reinforced concrete with vapor barrier and granular base		S.F. Slab	2.80	.93	1.6%
.2	Special Substructures	N/A		—	—	—	
3.0 Superstructure							
.1	Columns & Beams	N/A		—	—	—	
.4	Structural Walls	Reinforced concrete block		S.F. Wall	8.53	1.40	14.2%
.5	Elevated Floors	Precast concrete plank		S.F. Floor	6.15	4.10	
.7	Roof	Precast concrete plank		S.F. Roof	5.74	1.91	
.9	Stairs	Concrete filled metal pan		Flight	4540	.75	
4.0 Exterior Closure							
.1	Walls	Face brick with concrete block backup	85% of wall	S.F. Wall	10.02	2.63	10.5%
.5	Exterior Wall Finishes	N/A		—	—	—	
.6	Doors	Aluminum and glass doors and entrance with transom		Each	1046	2.33	
.7	Windows & Glazed Walls	Aluminum sliding	15% of wall	Each	346	1.07	
5.0 Roofing							
.1	Roof Coverings	Built-up tar and gravel with flashing		S.F. Roof	1.89	.63	2.0%
.7	Insulation	Perlite/EPS composite		S.F. Roof	1.18	.39	
.8	Openings & Specialties	Gravel stop, hatches, gutters and downspouts		S.F. Roof	.42	.14	
6.0 Interior Construction							
.1	Partitions	Gypsum board and sound deadening board on wood studs	7 S.F. Floor/L.F. Partition	S.F. Partition	4.60	5.26	36.4%
.4	Interior Doors	Wood hollow core	70 S.F. Floor/Door	Each	344	4.92	
.5	Wall Finishes	90% paint, 10% ceramic tile		S.F. Surface	.89	2.03	
.6	Floor Finishes	85% carpet, 15% vinyl composition tile		S.F. Floor	5.68	5.68	
.7	Ceiling Finishes	Painted gypsum board on furring		S.F. Ceiling	2.33	2.33	
.9	Interior Surface/Exterior Wall	Painted gypsum board on furring	85% of wall	S.F. Wall	2.47	.65	
7.0 Conveying							
.1	Elevators	Two hydraulic passenger elevators		Each	59,130	1.62	2.8%
.2	Special Conveyors	N/A		—	—	—	
8.0 Mechanical							
.1	Plumbing	Toilet and service fixtures, supply and drainage	1 Fixture/180 S.F. Floor	Each	1560	8.67	22.4%
.2	Fire Protection	Sprinklers, light hazard		S.F. Floor	.91	.91	
.3	Heating	Included in 8.4		—	—	—	
.4	Cooling	Through the wall electric heating and cooling units		S.F. Floor	3.25	3.25	
.5	Special Systems	N/A		—	—	—	
9.0 Electrical							
.1	Service & Distribution	400 ampere service, panel board and feeders		S.F. Floor	.28	.28	7.8%
.2	Lighting & Power	Fluorescent fixtures, receptacles, switches and misc. power		S.F. Floor	3.89	3.89	
.4	Special Electrical	Alarm systems and emergency lighting		S.F. Floor	.30	.30	
11.0 Special Construction							
.1	Specialties	N/A		—	—	—	0.0%
12.0 Site Work							
.1	Earthwork	N/A		—	—	—	0.0%
.3	Utilities	N/A		—	—	—	
.5	Roads & Parking	N/A		—	—	—	
.7	Site Improvements	N/A		—	—	—	
				Sub-Total		57.41	100%
	CONTRACTOR FEES (General Requirements: 10%, Overhead: 5%, Profit: 10%)				25%	14.35	
	ARCHITECT FEES				6%	4.29	
				Total Building Cost		76.05	

BUILDING TYPES

COMMERCIAL/INDUSTRIAL/INSTITUTIONAL — M.440 — Movie Theater

Costs per square foot of floor area

Exterior Wall	S.F. Area	9000	10000	12000	13000	14000	15000	16000	18000	20000
	L.F. Perimeter	385	410	460	460	480	500	510	547	583
Decorative Concrete Block	Steel Joists	91.50	89.00	85.30	83.00	81.70	80.55	79.20	77.50	76.10
Painted Concrete Block	Steel Joists	86.55	84.25	80.85	78.90	77.70	76.65	75.55	74.00	72.75
Face Brick on Conc. Block	Steel Joists	98.15	95.35	91.25	88.50	87.00	85.70	84.15	82.25	80.65
Precast Concrete Panels	Steel Joists	95.75	93.05	89.10	86.50	85.10	83.85	82.40	80.55	79.00
Tilt-up Panels	Steel Joists	91.00	88.50	84.85	82.55	81.30	80.15	78.85	77.15	75.75
Metal Sandwich Panels	Steel Joists	83.15	81.00	77.80	76.10	75.00	74.00	73.00	71.60	70.40
Perimeter Adj., Add or Deduct	Per 100 L.F.	5.00	4.55	3.75	3.50	3.25	3.00	2.85	2.50	2.25
Story Hgt. Adj., Add or Deduct	Per 1 Ft.	.75	.70	.65	.60	.60	.60	.55	.50	.50
Basement—Not Applicable										

The above costs were calculated using the basic specifications shown on the facing page. These costs should be adjusted where necessary for design alternatives and owner's requirements. Reported completed project costs, for this type of structure, range from $48.25 to $125.95 per S.F.

Common additives

Description	Unit	$ Cost
Emergency Lighting, 25 watt, battery operated		
Lead battery	Each	249
Nickel cadmium	Each	605
Seating		
Auditorium chair, all veneer	Each	151
Veneer back, padded seat	Each	184
Upholstered, spring seat	Each	186
Classroom, movable chair & desk	Set	65 - 120
Lecture hall, pedestal type	Each	137 - 435
Smoke Detectors		
Ceiling type	Each	129
Duct type	Each	365
Sound System		
Amplifier, 250 watts	Each	1500
Speaker, ceiling or wall	Each	125
Trumpet	Each	235

Movie Theater

Model costs calculated for a 1 story building with 20' story height and 12,000 square feet of floor area

				Unit	Unit Cost	Cost Per S.F.	% Of Sub-Total
1.0 Foundations							
.1	Footings & Foundations	Poured concrete; strip and spread footings and 4' foundation wall		S.F. Ground	2.12	2.12	
.4	Piles & Caissons	N/A		—	—	—	6.5%
.9	Excavation & Backfill	Site preparation for slab and trench for foundation wall and footing		S.F. Ground	2.03	2.03	
2.0 Substructure							
.1	Slab on Grade	4" reinforced concrete with vapor barrier and granular base		S.F. Slab	2.80	2.80	4.4%
.2	Special Substructures	N/A		—	—	—	
3.0 Superstructure							
.1	Columns & Beams	N/A		—	—	—	
.4	Structural Walls	N/A		—	—	—	
.5	Elevated Floors	Open web steel joists, slab form, concrete	mezzanine 2250 S.F.	S.F. Mezz	5.87	1.10	13.1%
.7	Roof	Metal deck on open web steel joists		S.F. Roof	6.35	6.35	
.9	Stairs	Concrete filled metal pan		Flight	5425	.90	
4.0 Exterior Closure							
.1	Walls	Decorative concrete block	100% of wall	S.F. Wall	11.26	8.63	
.5	Exterior Wall Finishes	N/A		—	—	—	14.6%
.6	Doors	Sliding mallfront aluminum and glass and hollow metal		Each	1308	.65	
.7	Windows & Glazed Walls	N/A		—	—	—	
5.0 Roofing							
.1	Roof Coverings	Built-up tar and gravel with flashing		S.F. Roof	1.92	1.92	
.7	Insulation	Perlite/EPS composite		S.F. Roof	1.18	1.18	5.5%
.8	Openings & Specialties	Gravel stop, hatches, gutters and downspouts		S.F. Roof	.42	.42	
6.0 Interior Construction							
.1	Partitions	Concrete block, toilet partitions	40 S.F. Floor/L.F. Partition	S.F. Partition	5.63	2.25	
.4	Interior Doors	Single leaf hollow metal	705 S.F. Floor/Door	Each	492	.70	
.5	Wall Finishes	Paint		S.F. Surface	.98	.78	17.1%
.6	Floor Finishes	Carpet	50% of area	S.F. Floor	5.52	2.76	
.7	Ceiling Finishes	Mineral fiber tile on concealed zee runners suspended		S.F. Ceiling	3.10	3.10	
.9	Interior Surface/Exterior Wall	Paint	100% of wall	S.F. Wall	1.69	1.30	
7.0 Conveying							
.1	Elevators	N/A		—	—	—	0.0%
.2	Special Conveyors	N/A		—	—	—	
8.0 Mechanical							
.1	Plumbing	Toilet and service fixtures, supply and drainage	1 Fixture/500 S.F. Floor	Each	1190	2.38	
.2	Fire Protection	Wet pipe sprinkler system		S.F. Floor	1.33	1.33	
.3	Heating	Included in 8.4		—	—	—	17.5%
.4	Cooling	Single zone rooftop unit, gas heating, electric cooling		S.F. Floor	7.48	7.48	
.5	Special Systems	N/A		—	—	—	
9.0 Electrical							
.1	Service & Distribution	400 ampere service, panel board and feeders		S.F. Floor	.56	.56	
.2	Lighting & Power	Fluorescent fixtures, receptacles, switches, A.C. and misc. power		S.F. Floor	2.45	2.45	5.4%
.4	Special Electrical	Alarm systems, sound system and emergency lighting		S.F. Floor	.42	.42	
11.0 Special Construction							
.1	Specialties	Projection equipment, screen, seating		S.F. Floor	10.17	10.17	15.9%
12.0 Site Work							
.1	Earthwork	N/A		—	—	—	
.3	Utilities	N/A		—	—	—	0.0%
.5	Roads & Parking	N/A		—	—	—	
.7	Site Improvements	N/A		—	—	—	
				Sub-Total		63.78	**100%**
	CONTRACTOR FEES (General Requirements: 10%, Overhead: 5%, Profit: 10%)				25%	15.95	
	ARCHITECT FEES				7%	5.57	
				Total Building Cost		**85.30**	

BUILDING TYPES

COMMERCIAL/INDUSTRIAL/INSTITUTIONAL — M.450 Nursing Home

Costs per square foot of floor area

Exterior Wall	S.F. Area	10000	15000	20000	25000	30000	35000	40000	45000	50000
	L.F. Perimeter	286	370	453	457	513	568	624	680	735
Redwood Siding	Wood Frame	84.55	81.35	79.70	77.40	76.65	76.00	75.60	75.25	74.95
Brick Veneer	Wood Frame	90.00	86.05	84.00	80.90	79.90	79.10	78.60	78.10	77.75
Face Brick with Concrete Block Back-up	Wood Joists	92.35	88.10	85.85	82.40	81.30	80.40	79.85	79.35	78.95
	Steel Joists	99.70	95.40	93.20	89.75	88.60	87.75	87.20	86.70	86.30
Stucco on Concrete Block	Wood Joists	86.30	82.85	81.05	78.55	77.70	76.95	76.55	76.15	75.85
	Steel Joists	93.60	90.20	88.40	85.85	85.00	84.30	83.85	83.50	83.15
Perimeter Adj., Add or Deduct	Per 100 L.F.	4.15	2.75	2.10	1.65	1.40	1.20	1.00	.95	.85
Story Hgt. Adj., Add or Deduct	Per 1 Ft.	.85	.70	.65	.50	.45	.50	.45	.45	.45

For Basement, add $17.19 per square foot of basement area

The above costs were calculated using the basic specifications shown on the facing page. These costs should be adjusted where necessary for design alternatives and owner's requirements. Reported completed project costs, for this type of structure, range from $52.10 to $129.40 per S.F.

Common additives

Description	Unit	$ Cost
Beds, Manual	Each	890 - 1550
Doctors In-Out Register, 200 names	Each	12,200
Elevators, Hydraulic passenger, 2 stops		
1500# capacity	Each	39,600
2500# capacity	Each	40,700
3500# capacity	Each	44,500
Emergency Lighting, 25 watt, battery operated		
Lead battery	Each	249
Nickel cadmium	Each	605
Intercom System, 25 station capacity		
Master station	Each	1825
Intercom outlets	Each	107
Handset	Each	293
Kitchen Equipment		
Broiler	Each	3950
Coffee urn, twin 6 gallon	Each	6750
Cooler, 6 ft. long	Each	3075
Dishwasher, 10-12 racks per hr.	Each	2925
Food warmer	Each	700
Freezer, 44 C.F., reach-in	Each	8050

Description	Unit	$ Cost
Kitchen Equipment, cont.		
Ice cube maker, 50 lb. per day	Each	1725
Range with 1 oven	Each	2350
Laundry Equipment		
Dryer, gas, 16 lb. capacity	Each	655
30 lb. capacity	Each	2650
Washer, 4 cycle	Each	740
Commercial	Each	1150
Nurses Call System		
Single bedside call station	Each	222
Pillow speaker	Each	204
Refrigerator, Prefabricated, walk-in		
7'-6" high, 6' x 6'	S.F.	117
10' x 10'	S.F.	92
12' x 14'	S.F.	82
12' x 20'	S.F.	72
TV Antenna, Master system, 12 outlet	Outlet	215
30 outlet	Outlet	139
100 outlet	Outlet	135
Whirlpool Bath, Mobile, 18" x 24" x 60"	Each	3750
X-Ray, Mobile	Each	10,600 - 60,000

Nursing Home

Model costs calculated for a 2 story building with 10' story height and 25,000 square feet of floor area

				Unit	Unit Cost	Cost Per S.F.	% Of Sub-Total
1.0 Foundations							
.1	Footings & Foundations	Poured concrete; strip and spread footings and 4' foundation wall		S.F. Ground	2.42	1.21	
.4	Piles & Caissons	N/A		—	—	—	3.0%
.9	Excavation & Backfill	Site preparation for slab and trench for foundation wall and footing		S.F. Ground	.96	.48	
2.0 Substructure							
.1	Slab on Grade	4" reinforced concrete with vapor barrier and granular base		S.F. Slab	2.80	1.40	2.5%
.2	Special Substructures	N/A		—	—	—	
3.0 Superstructure							
.1	Columns & Beams	N/A		—	—	—	
.4	Structural Walls	N/A		—	—	—	
.5	Elevated Floors	Plywood on wood joists		S.F. Floor	3.45	1.73	6.0%
.7	Roof	Plywood on wood joists		S.F. Roof	2.63	1.32	
.9	Stairs	Wood		Flight	1180	.28	
4.0 Exterior Closure							
.1	Walls	Vertical T & G redwood siding	85% of wall	S.F. Wall	5.66	1.76	
.5	Exterior Wall Finishes	N/A		—	—	—	5.9%
.6	Doors	Double aluminum & glass doors, single leaf hollow metal		Each	1393	.28	
.7	Windows & Glazed Walls	Wood double hung	15% of wall	Each	346	1.27	
5.0 Roofing							
.1	Roof Coverings	Built-up tar and gravel with flashing		S.F. Roof	1.90	.95	
.7	Insulation	Perlite/EPS composite		S.F. Roof	1.18	.59	3.2%
.8	Openings & Specialties	Gravel stop, hatches, gutters and downspouts		S.F. Roof	.44	.22	
6.0 Interior Construction							
.1	Partitions	Gypsum board on wood studs	8 S.F. Floor/L.F. Partition	S.F. Partition	2.80	2.80	
.4	Interior Doors	Single leaf wood	80 S.F. Floor/Door	Each	386	4.83	
.5	Wall Finishes	50% vinyl wall coverings, 45% paint, 5% ceramic tile		S.F. Surface	1.04	2.08	30.9%
.6	Floor Finishes	50% carpet, 45% vinyl tile, 5% ceramic tile		S.F. Floor	4.45	4.45	
.7	Ceiling Finishes	Painted gypsum board on wood furring		S.F. Ceiling	2.33	2.33	
.9	Interior Surface/Exterior Wall	Painted gypsum board on wood furring	85% of wall	S.F. Wall	2.47	.77	
7.0 Conveying							
.1	Elevators	One hydraulic hospital elevator		Each	59,750	2.39	4.3%
.2	Special Conveyors	N/A		—	—	—	
8.0 Mechanical							
.1	Plumbing	Kitchen, toilet and service fixtures, supply and drainage	1 Fixture/230 S.F. Floor	Each	1865	8.11	
.2	Fire Protection	Sprinkler, light hazard		S.F. Floor	2.38	2.38	
.3	Heating	Oil fired hot water, wall fin radiation		S.F. Floor	3.41	3.41	32.6%
.4	Cooling	Split systems with air cooled condensing units		S.F. Floor	4.27	4.27	
.5	Special Systems	N/A		—	—	—	
9.0 Electrical							
.1	Service & Distribution	600 ampere service, panel board and feeders		S.F. Floor	.86	.86	
.2	Lighting & Power	Incandescent fixtures, receptacles, switches, A.C. and misc. power		S.F. Floor	4.86	4.86	11.6%
.4	Special Electrical	Alarm systems and emergency lighting		S.F. Floor	.77	.77	
11.0 Special Construction							
.1	Specialties	N/A		—	—	—	0.0%
12.0 Site Work							
.1	Earthwork	N/A		—	—	—	
.3	Utilities	N/A		—	—	—	0.0%
.5	Roads & Parking	N/A		—	—	—	
.7	Site Improvements	N/A		—	—	—	
				Sub-Total		55.80	**100%**
	CONTRACTOR FEES (General Requirements: 10%, Overhead: 5%, Profit: 10%)				25%	13.95	
	ARCHITECT FEES				11%	7.65	
				Total Building Cost		**77.40**	

BUILDING TYPES

COMMERCIAL/INDUSTRIAL/INSTITUTIONAL — M.460 — Office, 2-4 Story

Costs per square foot of floor area

Exterior Wall	S.F. Area	10000	22000	34000	46000	58000	63000	68000	73000	78000
	L.F. Perimeter	246	393	443	543	562	590	603	624	645
Face Brick with Concrete Block Back-up	Wood Joists	90.95	78.05	71.35	69.20	66.50	66.05	65.40	64.90	64.55
	Steel Joists	92.30	79.45	72.70	70.60	67.90	67.45	66.80	66.30	65.95
Glass and Metal Curtain Wall	Steel Frame	91.50	79.40	73.20	71.20	68.75	68.35	67.75	67.30	67.00
	R/Conc. Frame	92.95	80.85	74.65	72.70	70.20	69.80	69.20	68.75	68.45
Wood Siding	Wood Frame	79.45	68.80	63.70	62.00	59.95	59.65	59.15	58.75	58.50
Brick Veneer	Wood Frame	84.45	72.45	66.35	64.35	61.95	61.55	60.95	60.50	60.20
Perimeter Adj., Add or Deduct	Per 100 L.F.	10.40	4.70	3.05	2.25	1.80	1.65	1.55	1.45	1.35
Story Hgt. Adj., Add or Deduct	Per 1 Ft.	1.95	1.40	1.00	.90	.75	.70	.70	.65	.65

For Basement, add $19.61 per square foot of basement area

The above costs were calculated using the basic specifications shown on the facing page. These costs should be adjusted where necessary for design alternatives and owner's requirements. Reported completed project costs, for this type of structure, range from $40.85 to $115.65 per S.F.

Common additives

Description	Unit	$ Cost
Clock System		
20 room	Each	10,300
50 room	Each	24,800
Closed Circuit Surveillance, One station		
Camera and monitor	Each	1275
For additional camera stations, add	Each	700
Directory Boards, Plastic, glass covered		
30" x 20"	Each	500
36" x 48"	Each	895
Aluminum, 24" x 18"	Each	440
36" x 24"	Each	530
48" x 32"	Each	630
48" x 60"	Each	1375
Elevators, Hydraulic passenger, 2 stops		
1500# capacity	Each	39,600
2500# capacity	Each	40,700
3500# capacity	Each	44,500
Additional stop, add	Each	2700
Emergency Lighting, 25 watt, battery operated		
Lead battery	Each	249
Nickel cadmium	Each	605

Description	Unit	$ Cost
Smoke Detectors		
Ceiling type	Each	129
Duct type	Each	365
Sound System		
Amplifier, 250 watts	Each	1500
Speaker, ceiling or wall	Each	125
Trumpet	Each	235
TV Antenna, Master system, 12 outlet	Outlet	215
30 outlet	Outlet	139
100 outlet	Outlet	135

Office, 2-4 Story

Model costs calculated for a 3 story building with 12' story height and 58,000 square feet of floor area

				Unit	Unit Cost	Cost Per S.F.	% Of Sub-Total
1.0 Foundations							
.1	Footings & Foundations	Poured concrete; strip and spread footings and 4' foundation wall		S.F. Ground	2.85	.95	2.5%
.4	Piles & Caissons	N/A		—	—	—	
.9	Excavation & Backfill	Site preparation for slab and trench for foundation wall and footing		S.F. Ground	.89	.30	
2.0 Substructure							
.1	Slab on Grade	4" reinforced concrete with vapor barrier and granular base		S.F. Slab	2.80	.93	1.8%
.2	Special Substructures	N/A		—	—	—	
3.0 Superstructure							
.1	Columns & Beams	Fireproofing; interior columns included in 3.5 and 3.7		L.F. Columns	18.57	.50	13.7%
.4	Structural Walls	N/A		—	—	—	
.5	Elevated Floors	Open web steel joists, slab form, concrete, columns		S.F. Floor	7.19	4.79	
.7	Roof	Metal deck, open web steel joists, columns		S.F. Roof	3.26	1.09	
.9	Stairs	Concrete filled metal pan		Flight	4540	.55	
4.0 Exterior Closure							
.1	Walls	Face brick with concrete block backup	80% of wall	S.F. Wall	17.06	4.76	12.4%
.5	Exterior Wall Finishes	N/A		—	—	—	
.6	Doors	Aluminum and glass, hollow metal		Each	1826	.19	
.7	Windows & Glazed Walls	Steel outward projecting	20% of wall	Each	446	1.35	
5.0 Roofing							
.1	Roof Coverings	Built-up tar and gravel with flashing		S.F. Roof	1.89	.63	2.0%
.7	Insulation	Perlite/EPS composite		S.F. Roof	1.18	.39	
.8	Openings & Specialties	N/A		—	—	—	
6.0 Interior Construction							
.1	Partitions	Gypsum board on metal studs, toilet partitions	20 S.F. Floor/L.F. Partition	S.F. Partition	2.63	1.33	25.8%
.4	Interior Doors	Single leaf hollow metal	200 S.F. Floor/Door	Each	492	2.46	
.5	Wall Finishes	60% vinyl wall covering, 40% paint		S.F. Surface	.90	.72	
.6	Floor Finishes	60% carpet, 30% vinyl composition tile, 10% ceramic tile		S.F. Floor	4.78	4.78	
.7	Ceiling Finishes	Mineral fiber tile on concealed zee bars		S.F. Ceiling	3.10	3.10	
.9	Interior Surface/Exterior Wall	Painted gypsum board on furring	80% of wall	S.F. Wall	2.47	.69	
7.0 Conveying							
.1	Elevators	Two hydraulic passenger elevators		Each	61,480	2.12	4.2%
.2	Special Conveyors	N/A		—	—	—	
8.0 Mechanical							
.1	Plumbing	Toilet and service fixtures, supply and drainage	1 Fixture/1320 S.F. Floor	Each	1795	1.36	23.7%
.2	Fire Protection	Standpipes and hose systems		S.F. Floor	.16	.16	
.3	Heating	Included in 8.4		—	—	—	
.4	Cooling	Multizone unit gas heating, electric cooling		S.F. Floor	10.57	10.57	
.5	Special Systems	N/A		—	—	—	
9.0 Electrical							
.1	Service & Distribution	1000 ampere service, panel board and feeders		S.F. Floor	.85	.85	13.9%
.2	Lighting & Power	Fluorescent fixtures, receptacles, switches, A.C. and misc. power		S.F. Floor	6.01	6.01	
.4	Special Electrical	Alarm systems and emergency lighting		S.F. Floor	.18	.18	
11.0 Special Construction							
.1	Specialties	N/A		—	—	—	0.0%
12.0 Site Work							
.1	Earthwork	N/A		—	—	—	0.0%
.3	Utilities	N/A		—	—	—	
.5	Roads & Parking	N/A		—	—	—	
.7	Site Improvements	N/A		—	—	—	
				Sub-Total		50.76	**100%**
	CONTRACTOR FEES (General Requirements: 10%, Overhead: 5%, Profit: 10%)				25%	12.69	
	ARCHITECT FEES				7%	4.45	
				Total Building Cost		**67.90**	

BUILDING TYPES

COMMERCIAL/INDUSTRIAL/INSTITUTIONAL — M.490 — Police Station

Costs per square foot of floor area

Exterior Wall	S.F. Area	7000	9000	11000	13000	15000	17000	19000	21000	23000
	L.F. Perimeter	240	280	303	325	354	372	397	422	447
Limestone with Concrete Block Back-up	Bearing Walls	136.80	127.50	119.55	114.05	110.55	107.10	104.80	102.95	101.40
	R/Conc. Frame	141.55	132.65	125.35	120.25	117.00	113.80	111.65	109.95	108.55
Face Brick with Concrete Block Back-up	Bearing Walls	119.10	111.30	105.20	100.95	98.15	95.55	93.75	92.30	91.15
	R/Conc. Frame	129.95	122.15	116.05	111.80	109.05	106.45	104.60	103.20	102.00
Decorative Concrete Block	Bearing Walls	113.90	106.60	101.05	97.15	94.60	92.25	90.55	89.25	88.20
	R/Conc. Frame	124.75	117.45	111.90	108.00	105.45	103.10	101.40	100.15	99.05
Perimeter Adj., Add or Deduct	Per 100 L.F.	18.10	14.05	11.55	9.75	8.45	7.50	6.65	6.00	5.50
Story Hgt. Adj., Add or Deduct	Per 1 Ft.	3.20	2.90	2.60	2.35	2.20	2.05	1.95	1.90	1.85

For Basement, add $15.49 per square foot of basement area

The above costs were calculated using the basic specifications shown on the facing page. These costs should be adjusted where necessary for design alternatives and owner's requirements. Reported completed project costs, for this type of structure, range from $68.20 to $175.25 per S.F.

Common additives

Description	Unit	$ Cost
Cells Prefabricated, 5'-6' wide, 7'-8' high, 7'-8' deep	Each	9125
Elevators, Hydraulic passenger, 2 stops		
1500# capacity	Each	39,600
2500# capacity	Each	40,700
3500# capacity	Each	44,500
Emergency Lighting, 25 watt, battery operated		
Lead battery	Each	249
Nickel cadmium	Each	605
Flagpoles, Complete		
Aluminum, 20' high	Each	1000
40' high	Each	2550
70' high	Each	7350
Fiberglass, 23' high	Each	1275
39'-5" high	Each	2575
59' high	Each	6775

Description	Unit	$ Cost
Lockers, Steel, Single tier, 60" to 72"	Opening	124 - 191
2 tier, 60" or 72" total	Opening	74 - 95
5 tier, box lockers	Opening	35 - 57
Locker bench, lam. maple top only	L.F.	16.35
Pedestals, steel pipe	Each	45
Safe, Office type, 4 hour rating		
30" x 18" x 18"	Each	3925
62" x 33" x 20"	Each	8525
Shooting Range, Incl. bullet traps, target provisions, and contols, not incl. structural shell	Each	21,000
Smoke Detectors		
Ceiling type	Each	129
Duct type	Each	365
Sound System		
Amplifier, 250 watts	Each	1500
Speaker, ceiling or wall	Each	125
Trumpet	Each	235

Model costs calculated for a 2 story building with 12' story height and 11,000 square feet of floor area

Police Station

BUILDING TYPES

				Unit	Unit Cost	Cost Per S.F.	% Of Sub-Total
1.0 Foundations							
.1	Footings & Foundations	Poured concrete; strip and spread footings and 4' foundation wall		S.F. Ground	4.40	2.20	
.4	Piles & Caissons	N/A		—	—	—	3.1%
.9	Excavation & Backfill	Site preparation for slab and trench for foundation wall and footing		S.F. Ground	.96	.48	
2.0 Substructure							
.1	Slab on Grade	4" reinforced concrete with vapor barrier and granular base		S.F. Slab	2.80	1.40	1.6%
.2	Special Substructures	N/A		—	—	—	
3.0 Superstructure							
.1	Columns & Beams	N/A		—	—	—	
.4	Structural Walls	N/A		—	—	—	
.5	Elevated Floors	Open web steel joists, slab form, concrete		S.F. Floor	6.24	3.12	6.1%
.7	Roof	Metal deck on open web steel joists		S.F. Roof	2.46	1.23	
.9	Stairs	Concrete filled metal pan		Flight	5425	.99	
4.0 Exterior Closure							
.1	Walls	Limestone with concrete block backup	80% of wall	S.F. Wall	29	15.84	
.5	Exterior Wall Finishes	N/A		—	—	—	26.1%
.6	Doors	Hollow metal		Each	1355	.98	
.7	Windows & Glazed Walls	Metal horizontal sliding	20% of wall	Each	692	6.10	
5.0 Roofing							
.1	Roof Coverings	Built-up tar and gravel with flashing		S.F. Roof	2.34	1.17	
.7	Insulation	Perlite/EPS composite		S.F. Roof	1.18	.59	2.2%
.8	Openings & Specialties	Gravel stop		L.F. Perimeter	4.99	.14	
6.0 Interior Construction							
.1	Partitions	Concrete block, toilet partitions	20 S.F. Floor/L.F. Partition	S.F. Partition	4.62	2.75	
.4	Interior Doors	Single leaf kalamein fire door	200 S.F. Floor/Door	Each	492	2.46	
.5	Wall Finishes	90% paint, 10% ceramic tile		S.F. Surface	1.19	1.19	
.6	Floor Finishes	70% vinyl asbestos tile, 20% carpet, 10% ceramic tile		S.F. Floor	3.56	3.56	15.9%
.7	Ceiling Finishes	Mineral fiber tile on concealed zee bars		S.F. Ceiling	3.10	3.10	
.9	Interior Surface/Exterior Wall	Paint	80% of wall	S.F. Wall	1.69	.89	
7.0 Conveying							
.1	Elevators	One hydraulic passenger elevator		Each	47,740	4.34	4.9%
.2	Special Conveyors	N/A		—	—	—	
8.0 Mechanical							
.1	Plumbing	Toilet and service fixtures, supply and drainage	1 Fixture/580 S.F. Floor	Each	2064	3.56	
.2	Fire Protection	Wet pipe sprinkler system		S.F. Floor	1.43	1.43	
.3	Heating	Oil fired hot water, wall fin radiation		S.F. Floor	5.34	5.34	18.9%
.4	Cooling	Split systems with air cooled condensing units		S.F. Floor	6.35	6.35	
.5	Special Systems	N/A		—	—	—	
9.0 Electrical							
.1	Service & Distribution	400 ampere service, panel board and feeders		S.F. Floor	1.02	1.02	
.2	Lighting & Power	Fluorescent fixtures, receptacles, switches, A.C. and misc. power		S.F. Floor	5.60	5.60	8.2%
.4	Special Electrical	Alarm systems and emergency lighting		S.F. Floor	.56	.56	
11.0 Special Construction							
.1	Specialties	Lockers, detention rooms, cells		S.F. Floor	11.37	11.37	13.0%
12.0 Site Work							
.1	Earthwork	N/A		—	—	—	
.3	Utilities	N/A		—	—	—	0.0%
.5	Roads & Parking	N/A		—	—	—	
.7	Site Improvements	N/A		—	—	—	
				Sub-Total		87.76	**100%**
	CONTRACTOR FEES (General Requirements: 10%, Overhead: 5%, Profit: 10%)				25%	21.94	
	ARCHITECT FEES				9%	9.85	
				Total Building Cost		**119.55**	

59

COMMERCIAL/INDUSTRIAL/INSTITUTIONAL — M.500 Post Office

Costs per square foot of floor area

Exterior Wall	S.F. Area	5000	7000	9000	11000	13000	15000	17000	19000	21000
	L.F. Perimeter	300	380	420	486	468	513	540	580	620
Face Brick with Concrete Block Back-up	Steel Frame	84.85	80.60	76.10	74.40	70.05	68.95	67.55	66.80	66.20
	Bearing Walls	83.80	79.60	75.05	73.35	69.00	67.90	66.55	65.75	65.15
Limestone with Concrete Block Back-up	Steel Frame	96.65	91.30	85.30	83.10	77.15	75.70	73.85	72.80	72.00
	Bearing Walls	95.15	89.80	83.80	81.60	75.65	74.20	72.30	71.30	70.50
Decorative Concrete Block	Steel Frame	79.50	75.80	71.95	70.50	66.85	65.95	64.75	64.10	63.60
	Bearing Walls	78.50	74.75	70.95	69.45	65.85	64.90	63.70	63.05	62.55
Perimeter Adj., Add or Deduct	Per 100 L.F.	9.75	6.95	5.45	4.45	3.80	3.25	2.85	2.55	2.35
Story Hgt. Adj., Add or Deduct	Per 1 Ft.	1.55	1.40	1.20	1.15	.95	.90	.80	.80	.75

For Basement, add $16.11 per square foot of basement area

The above costs were calculated using the basic specifications shown on the facing page. These costs should be adjusted where necessary for design alternatives and owner's requirements. Reported completed project costs, for this type of structure, range from $55.80 to $144.10 per S.F.

Common additives

Description	Unit	$ Cost
Closed Circuit Surveillance, One station		
Camera and monitor	Each	1275
For additional camera stations, add	Each	700
Emergency Lighting, 25 watt, battery operated		
Lead battery	Each	249
Nickel cadmium	Each	605
Flagpoles, Complete		
Aluminum, 20' high	Each	1000
40' high	Each	2550
70' high	Each	7350
Fiberglass, 23' high	Each	1275
39'-5" high	Each	2575
59' high	Each	6775

Description	Unit	$ Cost
Mail Boxes, Horizontal, key lock, 15" x 6" x 5"	Each	46
Double 15" x 12" x 5"	Each	77
Quadruple 15" x 12" x 10"	Each	135
Vertical, 6" x 5" x 15", aluminum	Each	40
Bronze	Each	59
Steel, enameled	Each	36
Scales, Dial type, 5 ton cap.		
8' x 6' platform	Each	6450
9' x 7' platform	Each	10,100
Smoke Detectors		
Ceiling type	Each	129
Duct type	Each	365

Model costs calculated for a 1 story building with 14' story height and 13,000 square feet of floor area

Post Office

				Unit	Unit Cost	Cost Per S.F.	% Of Sub-Total
1.0 Foundations							
.1	Footings & Foundations	Poured concrete; strip and spread footings and 4' foundation wall		S.F. Ground	2.90	2.90	
.4	Piles & Caissons	N/A		—	—	—	7.5%
.9	Excavation & Backfill	Site preparation for slab and trench for foundation wall and footing		S.F. Ground	.96	.96	
2.0 Substructure							
.1	Slab on Grade	4" reinforced concrete with vapor barrier and granular base		S.F. Slab	2.80	2.80	5.4%
.2	Special Substructures	N/A		—	—	—	
3.0 Superstructure							
.1	Columns & Beams	Fireproofing, steel columns included in 3.7		L.F. Column	17.96	.06	
.4	Structural Walls	N/A		—	—	—	
.5	Elevated Floors	N/A		—	—	—	9.1%
.7	Roof	Metal deck, open web steel joists, columns		S.F. Roof	4.61	4.61	
.9	Stairs	N/A		—	—	—	
4.0 Exterior Closure							
.1	Walls	Face brick with concrete block backup	80% of wall	S.F. Wall	17.04	6.87	
.5	Exterior Wall Finishes	N/A		—	—	—	18.4%
.6	Doors	Double aluminum & glass, single aluminum, hollow metal, steel overhead		Each	1409	.65	
.7	Windows & Glazed Walls	Double strength window glass	20% of wall	Each	446	1.95	
5.0 Roofing							
.1	Roof Coverings	Built-up tar and gravel with flashing		S.F. Roof	1.99	1.99	
.7	Insulation	Perlite/EPS composite		S.F. Roof	1.18	1.18	6.5%
.8	Openings & Specialties	Gravel stop		L.F. Perimeter	4.99	.18	
6.0 Interior Construction							
.1	Partitions	Concrete block, toilet partitions	15 S.F. Floor/L.F. Partition	S.F. Partition	4.62	4.07	
.4	Interior Doors	Single leaf hollow metal	150 S.F. Floor/Door	Each	492	3.28	
.5	Wall Finishes	Paint		S.F. Surface	.81	1.30	23.3%
.6	Floor Finishes	50% vinyl tile, 50% paint		S.F. Floor	1.85	1.85	
.7	Ceiling Finishes	Mineral fiber tile on concealed zee bars	25% of area	S.F. Ceiling	3.10	.78	
.9	Interior Surface/Exterior Wall	Paint	80% of wall	S.F. Wall	1.69	.68	
7.0 Conveying							
.1	Elevators	N/A		—	—	—	0.0%
.2	Special Conveyors	N/A		—	—	—	
8.0 Mechanical							
.1	Plumbing	Toilet and service fixtures, supply and drainage	1 Fixture/1180 S.F. Floor	Each	1616	1.37	
.2	Fire Protection	Wet pipe sprinkler system		S.F. Floor	1.33	1.33	
.3	Heating	Included in 8.4		—	—	—	17.5%
.4	Cooling	Single zone, gas heating, electric cooling		S.F. Floor	6.28	6.28	
.5	Special Systems	N/A		—	—	—	
9.0 Electrical							
.1	Service & Distribution	400 ampere service, panel board and feeders		S.F. Floor	.87	.87	
.2	Lighting & Power	Fluorescent fixtures, receptacles, switches, A.C. and misc. power		S.F. Floor	4.58	4.58	11.3%
.4	Special Electrical	Alarm systems and emergency lighting		S.F. Floor	.34	.34	
11.0 Special Construction							
.1	Specialties	Cabinets, lockers, shelving		S.F. Floor	.53	.53	1.0%
12.0 Site Work							
.1	Earthwork	N/A		—	—	—	
.3	Utilities	N/A		—	—	—	0.0%
.5	Roads & Parking	N/A		—	—	—	
.7	Site Improvements	N/A		—	—	—	

			Sub-Total	51.41	100%
CONTRACTOR FEES (General Requirements: 10%, Overhead: 5%, Profit: 10%)				25%	12.85
ARCHITECT FEES				9%	5.79
			Total Building Cost	**70.05**	

BUILDING TYPES

COMMERCIAL/INDUSTRIAL/INSTITUTIONAL — M.510 Racquetball Court

Costs per square foot of floor area

Exterior Wall	S.F. Area	5000	10000	15000	21000	25000	30000	40000	50000	60000
	L.F. Perimeter	287	400	500	600	700	700	834	900	1000
Face Brick with Concrete Block Back-up	Steel Frame	135.55	114.45	106.70	101.45	100.45	96.15	93.55	90.90	89.60
	Bearing Walls	131.20	111.25	103.90	98.90	98.00	93.85	91.30	88.70	87.40
Concrete Block	Steel Frame	118.55	102.35	96.50	92.65	91.80	88.95	87.10	85.30	84.40
Brick Veneer	Steel Frame	128.30	108.25	100.90	96.00	95.05	91.20	88.80	86.30	85.15
Galvanized Steel Siding	Steel Frame	111.30	97.15	92.10	88.80	88.05	85.75	84.20	82.70	81.95
Metal Sandwich Panel	Steel Frame	111.05	96.40	91.15	87.75	86.95	84.55	82.90	81.35	80.60
Perimeter Adj., Add or Deduct	Per 100 L.F.	16.40	8.20	5.50	3.95	3.30	2.75	2.05	1.65	1.35
Story Hgt. Adj., Add or Deduct	Per 1 Ft.	3.40	2.35	1.95	1.70	1.65	1.40	1.25	1.05	1.00

For Basement, add $15.50 per square foot of basement area

The above costs were calculated using the basic specifications shown on the facing page. These costs should be adjusted where necessary for design alternatives and owner's requirements. Reported completed project costs, for this type of structure, range from $52.65 to $164.15 per S.F.

Common additives

Description	Unit	$ Cost
Bar, Front Bar	L.F.	262
Back Bar	L.F.	205
Booth, Upholstered, custom straight	L.F.	134 - 248
"L" or "U" shaped	L.F.	137 - 235
Bleachers, Telescoping, manual		
To 15 tier	Seat	70 - 100
21-30 tier	Seat	159 - 192
Courts		
Ceiling	Court	4650
Floor	Court	8525
Walls	Court	16,900
Emergency Lighting, 25 watt, battery operated		
Lead battery	Each	249
Nickel cadmium	Each	605
Kitchen Equipment		
Broiler	Each	3950
Cooler, 6 ft. long, reach-in	Each	3075
Dishwasher, 10-12 racks per hr.	Each	2925
Food warmer, counter 1.2 KW	Each	700
Freezer, reach-in, 44 C.F.	Each	8050
Ice cube maker, 50 lb. per day	Each	1725

Description	Unit	$ Cost
Lockers, Steel, single tier, 60" or 72"	Opening	124 - 191
2 tier, 60" or 72" total	Opening	74 - 95
5 tier, box lockers	Opening	35 - 57
Locker bench, lam. maple top only	L.F.	16.35
Pedestals, steel pipe	Each	45
Sauna, Prefabricated, complete		
6' x 4'	Each	3875
6' x 9'	Each	5625
8' x 8'	Each	5950
8' x 10'	Each	6575
10' x 12'	Each	8675
Sound System		
Amplifier, 250 watts	Each	1500
Speaker, ceiling or wall	Each	125
Trumpet	Each	235
Steam Bath, Complete, to 140 C.F.	Each	1050
To 300 C.F.	Each	1150
To 800 C.F.	Each	3300
To 2500 C.F.	Each	5050

Racquetball Court

Model costs calculated for a 2 story building with 12' story height and 30,000 square feet of floor area

				Unit	Unit Cost	Cost Per S.F.	% Of Sub-Total
1.0 Foundations							
.1	Footings & Foundations	Poured concrete; strip and spread footings and 4' foundation wall		S.F. Ground	3.44	1.72	
.4	Piles & Caissons	N/A		—	—	—	3.3%
.9	Excavation & Backfill	Site preparation for slab and trench for foundation wall and footing		S.F. Ground	.96	.62	
2.0 Substructure							
.1	Slab on Grade	6" reinforced concrete with vapor barrier and granular base		S.F. Slab	5.67	3.69	5.2%
.2	Special Substructures	N/A		—	—	—	
3.0 Superstructure							
.1	Columns & Beams	Steel columns included in 3.5 and 3.7		—	—	—	
.4	Structural Walls	N/A		—	—	—	
.5	Elevated Floors	Open web steel joists, slab form, concrete, columns	50% of area	S.F. Floor	12.84	4.49	9.8%
.7	Roof	Metal deck on open web steel joists, columns		S.F. Roof	4.12	2.06	
.9	Stairs	Concrete filled metal pan		Flight	4540	.45	
4.0 Exterior Closure							
.1	Walls	Face brick with concrete block backup	95% of wall	S.F. Wall	17.26	9.18	
.5	Exterior Wall Finishes	N/A		—	—	—	15.6%
.6	Doors	Aluminum and glass and hollow metal		Each	1824	.24	
.7	Windows & Glazed Walls	Storefront	5% of wall	S.F. Window	61	1.71	
5.0 Roofing							
.1	Roof Coverings	Built-up tar and gravel with flashing		S.F. Roof	2.46	1.23	
.7	Insulation	Perlite/EPS composite		S.F. Roof	1.18	.77	3.0%
.8	Openings & Specialties	Gravel stop and hatches		S.F. Roof	.30	.15	
6.0 Interior Construction							
.1	Partitions	Concrete block, gypsum board on metal studs	25 S.F. Floor/L.F. Partition	S.F. Partition	4.62	2.01	
.4	Interior Doors	Single leaf hollow metal	810 S.F. Floor/Door	Each	492	.61	
.5	Wall Finishes	Paint		S.F. Surface	.77	.62	12.7%
.6	Floor Finishes	80% carpet, 20% ceramic tile	50% of floor area	S.F. Floor	5.20	2.60	
.7	Ceiling Finishes	Mineral fiber tile on concealed zee bars	60% of area	S.F. Ceiling	3.10	1.86	
.9	Interior Surface/Exterior Wall	Painted gypsum board on furring	95% of wall	S.F. Wall	2.47	1.38	
7.0 Conveying							
.1	Elevators	N/A		—	—	—	0.0%
.2	Special Conveyors	N/A		—	—	—	
8.0 Mechanical							
.1	Plumbing	Kitchen, bathroom and service fixtures, supply and drainage	1 Fixture/1000 S.F. Floor	Each	2870	2.87	
.2	Fire Protection	Sprinklers, light hazard		S.F. Floor	.11	.11	
.3	Heating	Included in 8.4		—	—	—	30.6%
.4	Cooling	Multizone unit, gas heating, electric cooling		S.F. Floor	18.80	18.80	
.5	Special Systems	N/A		—	—	—	
9.0 Electrical							
.1	Service & Distribution	400 ampere service, panel board and feeders		S.F. Floor	.46	.46	
.2	Lighting & Power	Fluorescent and high intensity discharge fixtures, receptacles, switches, A.C. and misc. power		S.F. Floor	3.46	3.46	5.8%
.4	Special Electrical	Alarm systems and emergency lighting		S.F. Floor	.21	.21	
11.0 Special Construction							
.1	Specialties	Courts, sauna baths		S.F. Floor	9.94	9.94	14.0%
12.0 Site Work							
.1	Earthwork	N/A		—	—	—	
.3	Utilities	N/A		—	—	—	0.0%
.5	Roads & Parking	N/A		—	—	—	
.7	Site Improvements	N/A		—	—	—	
				Sub-Total		71.24	100%
	CONTRACTOR FEES (General Requirements: 10%, Overhead: 5%, Profit: 10%)				25%	17.81	
	ARCHITECT FEES				8%	7.10	
				Total Building Cost		**96.15**	

BUILDING TYPES

COMMERCIAL/INDUSTRIAL/INSTITUTIONAL M.530 | Restaurant

Costs per square foot of floor area

Exterior Wall	S.F. Area	2000	2800	3500	4200	5000	5800	6500	7200	8000
	L.F. Perimeter	180	212	240	268	300	314	336	344	368
Wood Siding	Wood Frame	127.00	117.75	113.15	110.05	107.55	104.80	103.45	101.70	100.70
Brick Veneer	Wood Frame	133.80	123.40	118.25	114.75	112.00	108.80	107.25	105.20	104.05
Face Brick with Concrete Block Back-up	Wood Joists	137.30	126.40	120.95	117.30	114.40	110.95	109.35	107.15	105.95
	Steel Joists	139.55	127.60	121.65	117.65	114.50	110.60	108.80	106.30	105.00
Stucco on Concrete Block	Wood Joists	128.40	118.90	114.20	111.00	108.50	105.60	104.20	102.40	101.40
	Steel Joists	124.10	114.60	109.85	106.70	104.15	101.30	99.90	98.10	97.10
Perimeter Adj., Add or Deduct	Per 100 L.F.	15.45	11.05	8.80	7.35	6.15	5.35	4.75	4.30	3.85
Story Hgt. Adj., Add or Deduct	Per 1 Ft.	1.75	1.45	1.30	1.20	1.15	1.05	.95	.90	.85

For Basement, add $18.41 per square foot of basement area

The above costs were calculated using the basic specifications shown on the facing page. These costs should be adjusted where necessary for design alternatives and owner's requirements. Reported completed project costs, for this type of structure, range from $71.10 to $166.00 per S.F.

Common additives

Description	Unit	$ Cost
Bar, Front Bar	L.F.	262
Back bar	L.F.	205
Booth, Upholstered, custom straight	L.F.	134 - 248
"L" or "U" shaped	L.F.	137 - 235
Cupola, Stock unit, redwood		
30" square, 37" high, aluminum roof	Each	465
Copper roof	Each	226
Fiberglass, 5'-0" base, 63" high	Each	2825 - 3400
6'-0" base, 63" high	Each	4325 - 4675
Decorative Wood Beams, Non load bearing		
Rough sawn, 4" x 6"	L.F.	5.95
4" x 8"	L.F.	6.90
4" x 10"	L.F.	8.50
4" x 12"	L.F.	9.80
8" x 8"	L.F.	12.45
Emergency Lighting, 25 watt, battery operated		
Lead battery	Each	249
Nickel cadmium	Each	605

Description	Unit	$ Cost
Fireplace, Brick, not incl. chimney or foundation		
30" x 29" opening	Each	1600
Chimney, standard brick		
Single flue, 16" x 20"	V.L.F.	51
20" x 20"	V.L.F.	57
2 Flue, 20" x 24"	V.L.F.	75
20" x 32"	V.L.F.	84
Kitchen Equipment		
Broiler	Each	3950
Coffee urn, twin 6 gallon	Each	6750
Cooler, 6 ft. long	Each	3075
Dishwasher, 10-12 racks per hr.	Each	2925
Food warmer, counter, 1.2 KW	Each	700
Freezer, 44 C.F., reach-in	Each	8050
Ice cube maker, 50 lb. per day	Each	1725
Range with 1 oven	Each	2350
Refrigerators, Prefabricated, walk-in		
7'-6" high, 6' x 6'	S.F.	117
10' x 10'	S.F.	92
12' x 14'	S.F.	82
12' x 20'	S.F.	72

Important: See the Reference Section for Location Factors

Restaurant

Model costs calculated for a 1 story building with 12' story height and 5,000 square feet of floor area

				Unit	Unit Cost	Cost Per S.F.	% Of Sub-Total
1.0 Foundations							
.1	Footings & Foundations	Poured concrete; strip and spread footings and 4' foundation wall		S.F. Ground	3.46	3.46	
.4	Piles & Caissons	N/A		—	—	—	5.6%
.9	Excavation & Backfill	Site preparation for slab and trench for foundation wall and footing		S.F. Ground	1.06	1.06	
2.0 Substructure							
.1	Slab on Grade	4" reinforced concrete with vapor barrier and granular base		S.F. Slab	2.80	2.80	3.5%
.2	Special Substructures	N/A		—	—	—	
3.0 Superstructure							
.1	Columns & Beams	Wood columns		S.F. Ground	.31	.31	
.4	Structural Walls	N/A		—	—	—	
.5	Elevated Floors	N/A		—	—	—	3.1%
.7	Roof	Plywood on wood rafters (pitched)		S.F. Roof	1.94	2.17	
.9	Stairs	N/A		—	—	—	
4.0 Exterior Closure							
.1	Walls	Cedar siding on wood studs with insulation	70% of wall	S.F. Wall	6.88	3.47	
.5	Exterior Wall Finishes	N/A		—	—	—	14.6%
.6	Doors	Aluminum and glass doors and entrance with transom		Each	2803	2.81	
.7	Windows & Glazed Walls	Storefront windows	30% of wall	S.F. Window	25	5.47	
5.0 Roofing							
.1	Roof Coverings	Cedar shingles with flashing (pitched)		S.F. Roof	2.88	3.22	
.7	Insulation	Fiberglass sheets		S.F. Roof	.88	.98	5.7%
.8	Openings & Specialties	Gutters and downspouts and skylight		S.F. Roof	.39	.39	
6.0 Interior Construction							
.1	Partitions	Gypsum board on wood studs, toilet partition	25 S.F. Floor/L.F. Partition	S.F. Partition	682	1.66	
.4	Interior Doors	Hollow core wood	250 S.F. Floor/Door	Each	344	1.38	
.5	Wall Finishes	75% paint, 25% ceramic tile		S.F. Surface	1.50	1.20	18.6%
.6	Floor Finishes	65% carpet, 35% quarry tile		S.F. Floor	6.35	6.35	
.7	Ceiling Finishes	Mineral fiber tile on concealed zee bars		S.F. Ceiling	3.10	3.10	
.9	Interior Surface/Exterior Wall	Painted gypsum board on furring	70% of wall	S.F. Wall	2.51	1.27	
7.0 Conveying							
.1	Elevators	N/A		—	—	—	0.0%
.2	Special Conveyors	N/A		—	—	—	
8.0 Mechanical							
.1	Plumbing	Kitchen, bathroom and service fixtures, supply and drainage	1 Fixture/355 S.F. Floor	Each	2580	7.27	
.2	Fire Protection	Sprinklers, light hazard		S.F. Floor	1.33	1.33	
.3	Heating	Included in 8.4		—	—	—	38.8%
.4	Cooling	Multizone unit, gas heating, electric cooling		S.F. Floor	22	22.60	
.5	Special Systems	N/A		—	—	—	
9.0 Electrical							
.1	Service & Distribution	400 ampere service, panel board and feeders		S.F. Floor	2.35	2.35	
.2	Lighting & Power	Fluorescent fixtures, receptacles, switches, A.C. and misc. power		S.F. Floor	5.20	5.20	10.1%
.4	Special Electrical	Alarm systems and emergency lighting		S.F. Floor	.57	.57	
11.0 Special Construction							
.1	Specialties	N/A		—	—	—	0.0%
12.0 Site Work							
.1	Earthwork	N/A		—	—	—	
.3	Utilities	N/A		—	—	—	0.0%
.5	Roads & Parking	N/A		—	—	—	
.7	Site Improvements	N/A		—	—	—	
				Sub-Total		80.42	100%
	CONTRACTOR FEES (General Requirements: 10%, Overhead: 5%, Profit: 10%)				25%	20.11	
	ARCHITECT FEES				7%	7.02	
				Total Building Cost		**107.55**	

COMMERCIAL/INDUSTRIAL/INSTITUTIONAL — M.540 — Restaurant, Fast Food

Costs per square foot of floor area

Exterior Wall	S.F. Area	2000	2800	3500	4000	5000	5800	6500	7200	8000
	L.F. Perimeter	180	212	240	260	300	314	336	344	368
Face Brick with Concrete Block Back-up	Bearing Walls	116.30	108.25	104.20	102.20	99.35	96.50	95.25	93.45	92.50
	Steel Frame	119.55	111.50	107.40	105.40	102.60	99.75	98.50	96.70	95.70
Concrete Block With Stucco	Bearing Walls	108.80	101.95	98.45	96.75	94.35	92.00	90.95	89.45	88.65
	Steel Frame	111.40	104.65	101.25	99.55	97.20	94.85	93.85	92.40	91.55
Wood Siding	Wood Frame	110.25	103.50	100.10	98.40	96.05	93.80	92.75	91.30	90.50
Brick Veneer	Steel Frame	116.40	108.80	105.05	103.15	100.50	97.85	96.65	95.00	94.10
Perimeter Adj., Add or Deduct	Per 100 L.F.	19.90	14.20	11.35	9.90	7.95	6.85	6.10	5.50	5.00
Story Hgt. Adj., Add or Deduct	Per 1 Ft.	2.65	2.20	2.00	1.90	1.75	1.60	1.50	1.40	1.35
Basement—Not Applicable										

The above costs were calculated using the basic specifications shown on the facing page. These costs should be adjusted where necessary for design alternatives and owner's requirements. Reported completed project costs, for this type of structure, range from $71.00 to $138.15 per S.F.

Common additives

Description	Unit	$ Cost	Description	Unit	$ Cost
Bar, Front Bar	L.F.	262	Refrigerators, Prefabricated, walk-in		
Back bar	L.F.	205	7'-6" High, 6' x 6'	S.F.	117
Booth, Upholstered, custom straight	L.F.	134 - 248	10' x 10'	S.F.	92
"L" or "U" shaped	L.F.	137 - 235	12' x 14'	S.F.	82
Drive-up Window	Each	5300 - 8975	12' x 20'	S.F.	72
Emergency Lighting, 25 watt, battery operated			Serving		
Lead battery	Each	249	Counter top (Stainless steel)	L.F.	109
Nickel cadmium	Each	605	Base cabinets	L.F.	252 - 300
Kitchen Equipment			Sound System		
Broiler	Each	3950	Amplifier, 250 watts	Each	1500
Coffee urn, twin 6 gallon	Each	6750	Speaker, ceiling or wall	Each	125
Cooler, 6 ft. long	Each	3075	Trumpet	Each	235
Dishwasher, 10-12 racks per hr.	Each	2925	Storage		
Food warmer, counter, 1.2 KW	Each	700	Shelving	S.F.	8.10
Freezer, 44 C.F., reach-in	Each	8050	Washing		
Ice cube maker, 50 lb. per day	Each	1725	Stainless steel counter	L.F.	109
Range with 1 oven	Each	2350			

Model costs calculated for a 1 story building with 10' story height and 4,000 square feet of floor area

Restaurant, Fast Food

				Unit	Unit Cost	Cost Per S.F.	% Of Sub-Total
1.0 Foundations							
.1	Footings & Foundations	Poured concrete; strip and spread footings and 4' foundation wall		S.F. Ground	3.92	3.92	
.4	Piles & Caissons	N/A		—	—	—	6.6%
.9	Excavation & Backfill	Site preparation for slab and trench for foundation wall and footing		S.F. Ground	1.06	1.06	
2.0 Substructure							
.1	Slab on Grade	4" reinforced concrete with vapor barrier and granular base		S.F. Slab	2.80	2.80	3.7%
.2	Special Substructures	N/A		—	—	—	
3.0 Superstructure							
.1	Columns & Beams	N/A		—	—	—	
.4	Structural Walls	N/A		—	—	—	
.5	Elevated Floors	N/A		—	—	—	3.8%
.7	Roof	Metal deck on open web steel joists		S.F. Roof	2.87	2.87	
.9	Stairs	N/A		—	—	—	
4.0 Exterior Closure							
.1	Walls	Face brick with concrete block backup	70% of wall	S.F. Wall	14.40	7.76	
.5	Exterior Wall Finishes	N/A		—	—	—	23.7%
.6	Doors	Aluminum and glass		Each	2306	4.61	
.7	Windows & Glazed Walls	Window wall	30% of wall	S.F Window	28	5.58	
5.0 Roofing							
.1	Roof Coverings	Built-up tar and gravel with flashing		S.F. Roof	2.49	2.49	
.7	Insulation	Perlite/EPS composite		S.F. Roof	1.18	1.18	5.2%
.8	Openings & Specialties	Gravel stop and hatches		S.F. Roof	.26	.26	
6.0 Interior Construction							
.1	Partitions	Gypsum board on metal studs	25 S.F. Floor/L.F. Partition	S.F. Partition	3.61	1.30	
.4	Interior Doors	Hollow core wood	1000 S.F. Floor/Door	Each	344	.34	
.5	Wall Finishes	Paint		S.F. Surface	.49	.35	18.2%
.6	Floor Finishes	Quarry tile		S.F. Floor	7.91	7.91	
.7	Ceiling Finishes	Mineral fiber tile on concealed zee bars		S.F. Floor	3.10	3.10	
.9	Interior Surface/Exterior Wall	Paint	70% of wall	S.F. Wall	1.69	.77	
7.0 Conveying							
.1	Elevators	N/A		—	—	—	0.0%
.2	Special Conveyors	N/A		—	—	—	
8.0 Mechanical							
.1	Plumbing	Kitchen, bathroom and service fixtures, supply and drainage	1 Fixture/400 S.F. Floor	Each	2444	6.11	
.2	Fire Protection	Sprinklers, light hazard		S.F. Floor	1.74	1.74	
.3	Heating	Included in 8.4		—	—	—	21.6%
.4	Cooling	Multizone unit, gas heating, electric cooling		S.F. Floor	8.50	8.50	
.5	Special Systems	N/A		—	—	—	
9.0 Electrical							
.1	Service & Distribution	400 ampere service, panel board and feeders		S.F. Floor	2.47	2.47	
.2	Lighting & Power	Fluorescent fixtures, receptacles, switches, A.C. and misc. power		S.F. Floor	5.70	5.70	11.5%
.4	Special Electrical	Alarm systems and emergency lighting		S.F. Floor	.55	.55	
11.0 Special Construction							
.1	Specialties	Walk-in refrigerator		Each	17,263	4.32	5.7%
12.0 Site Work							
.1	Earthwork	N/A		—	—	—	
.3	Utilities	N/A		—	—	—	0.0%
.5	Roads & Parking	N/A		—	—	—	
.7	Site Improvements	N/A		—	—	—	
				Sub-Total		75.69	100%
	CONTRACTOR FEES (General Requirements: 10%, Overhead: 5%, Profit: 10%)				25%	18.92	
	ARCHITECT FEES				8%	7.59	
				Total Building Cost		**102.20**	

BUILDING TYPES

COMMERCIAL/INDUSTRIAL/INSTITUTIONAL   M.550   Rink, Hockey/Indoor Soccer

Costs per square foot of floor area

Exterior Wall	S.F. Area	10000	15000	20000	25000	30000	35000	40000	45000	50000
	L.F. Perimeter	450	500	600	700	740	822	890	920	966
Face Brick with Concrete Block Back-up	Steel Frame	123.15	113.15	109.95	108.05	105.30	104.20	103.15	101.75	100.80
	Lam. Wood Truss	112.40	102.80	99.70	97.90	95.25	94.20	93.20	91.85	90.95
Concrete Block	Steel Frame	97.55	91.55	89.55	88.35	86.70	86.05	85.40	84.65	84.05
	Lam. Wood Truss	99.80	93.80	91.80	90.60	88.95	88.30	87.65	86.85	86.30
Galvanized Steel Siding	Steel Frame	89.10	84.55	82.95	81.95	80.75	80.25	79.75	79.20	78.75
Metal Sandwich Panel	Steel Joists	90.05	85.25	83.55	82.55	81.25	80.70	80.20	79.60	79.15
Perimeter Adj., Add or Deduct	Per 100 L.F.	7.30	4.90	3.65	2.90	2.45	2.10	1.80	1.60	1.45
Story Hgt. Adj., Add or Deduct	Per 1 Ft.	1.15	.05	.75	.70	.65	.60	.55	.50	.50
Basement—Not Applicable										

The above costs were calculated using the basic specifications shown on the facing page. These costs should be adjusted where necessary for design alternatives and owner's requirements. Reported completed project costs, for this type of structure, range from $41.95 to $130.75 per S.F.

Common additives

Description	Unit	$ Cost
Bar, Front Bar	L.F.	262
Back bar	L.F.	205
Booth, Upholstered, custom straight	L.F.	134 - 248
"L" or "U" shaped	L.F.	137 - 235
Bleachers, Telescoping, manual		
To 15 tier	Seat	70 - 100
16-20 tier	Seat	151 - 185
21-30 tier	Seat	159 - 192
For power operation, add	Seat	30 - 47
Emergency Lighting, 25 watt, battery operated		
Lead battery	Each	249
Nickel cadmium	Each	605
Lockers, Steel, single tier, 60" or 72"	Opening	124 - 191
2 tier, 60" or 72" total	Opening	74 - 95
5 tier, box lockers	Opening	35 - 57
Locker bench, lam. maple top only	L.F.	16.35
Pedestals, steel pipe	Each	45

Description	Unit	$ Cost
Rink		
Dasher boards & top guard	Each	142,500
Mats, rubber	S.F.	15.80
Score Board	Each	13,100 - 35,300

Model costs calculated for a 1 story building with 24' story height and 30,000 square feet of floor area

Rink, Hockey/Indoor Soccer

				Unit	Unit Cost	Cost Per S.F.	% Of Sub-Total
1.0 Foundations							
.1	Footings & Foundations	Poured concrete; strip and spread footings and 4' foundation wall		S.F. Ground	2.20	2.20	
.4	Piles & Caissons	N/A		—	—	—	4.8%
.9	Excavation & Backfill	Site preparation for slab and trench for foundation wall and footing		S.F. Ground	.89	.89	
2.0 Substructure							
.1	Slab on Grade	6" reinforced concrete with vapor barrier and granular base		S.F. Slab	3.43	3.43	5.3%
.2	Special Substructures	N/A		—	—	—	
3.0 Superstructure							
.1	Columns & Beams	Wide flange beams and columns		S.F. Ground	7.41	7.41	
.4	Structural Walls	N/A		—	—	—	
.5	Elevated Floors	N/A		—	—	—	18.3%
.7	Roof	Metal deck on steel joist		S.F. Roof	4.46	4.46	
.9	Stairs	N/A		—	—	—	
4.0 Exterior Closure							
.1	Walls	Concrete block	95% of wall	S.F. Wall	6.85	3.85	
.5	Exterior Wall Finishes	N/A		—	—	—	7.9%
.6	Doors	Aluminum and glass, hollow metal, overhead		Each	1779	.47	
.7	Windows & Glazed Walls	Store front	5% of wall	S.F. Window	27	.82	
5.0 Roofing							
.1	Roof Coverings	Elastomeric neoprene membrane with flashing		S.F. Roof	1.43	1.43	
.7	Insulation	Perlite/EPS composite		S.F. Roof	1.18	1.18	4.4%
.8	Openings & Specialties	Hatches		S.F. Roof	.25	.25	
6.0 Interior Construction							
.1	Partitions	Concrete block	140 S.F. Floor/L.F. Partition	S.F. Partition	4.67	.40	
.4	Interior Doors	Hollow metal	2500 S.F. Floor/Door	Each	492	.20	
.5	Wall Finishes	Paint		S.F. Surface	.99	.17	6.5%
.6	Floor Finishes	80% rubber mat, 20% paint	50% of floor area	S.F. Floor	4.34	2.17	
.7	Ceiling Finishes	Mineral fiber tile on concealed zee bar	10% of area	S.F. Ceiling	3.10	.31	
.9	Interior Surface/Exterior Wall	Paint	95% of wall	S.F. Wall	1.69	.95	
7.0 Conveying							
.1	Elevators	N/A		—	—	—	0.0%
.2	Special Conveyors	N/A		—	—	—	
8.0 Mechanical							
.1	Plumbing	Toilet and service fixtures, supply and drainage	1 Fixture/1070 S.F. Floor	Each	3124	2.92	
.2	Fire Protection	Standpipes and hose systems		—	—	—	
.3	Heating	Oil fired hot water, unit heaters	10% of area	S.F. Floor	.41	.41	20.3%
.4	Cooling	Single zone, electric cooling	90% of area	S.F. Floor	9.82	9.82	
.5	Special Systems	N/A		—	—	—	
9.0 Electrical							
.1	Service & Distribution	400 ampere service, panel board and feeders		S.F. Floor	.53	.53	
.2	Lighting & Power	High intensity discharge and fluorescent fixtures, receptacles, switches, A.C. and misc. power		S.F. Floor	4.25	4.25	8.5%
.4	Special Electrical	Alarm systems, emergency lighting and public address		S.F. Floor	.74	.74	
11.0 Special Construction							
.1	Specialties	Dasher boards and rink (including ice making system)		S.F. Floor	15.57	15.57	24.0%
12.0 Site Work							
.1	Earthwork	N/A		—	—	—	
.3	Utilities	N/A		—	—	—	0.0%
.5	Roads & Parking	N/A		—	—	—	
.7	Site Improvements	N/A		—	—	—	
				Sub-Total		64.83	100%
	CONTRACTOR FEES (General Requirements: 10%, Overhead: 5%, Profit: 10%)				25%	16.21	
	ARCHITECT FEES				7%	5.66	
				Total Building Cost		86.70	

COMMERCIAL/INDUSTRIAL/INSTITUTIONAL — M.560 — School, Elementary

Costs per square foot of floor area

Exterior Wall	S.F. Area	25000	30000	35000	40000	45000	50000	55000	60000	65000
	L.F. Perimeter	700	740	823	906	922	994	1033	1100	1166
Face Brick with Concrete Block Back-up	Steel Frame	78.90	76.70	75.85	75.10	73.85	73.35	72.65	72.30	72.00
	Bearing Walls	76.35	74.15	73.25	72.55	71.25	70.80	70.10	69.75	69.45
Stucco on Concrete Block	Steel Frame	76.15	74.30	73.50	72.90	71.85	71.40	70.80	70.50	70.25
	Bearing Walls	73.55	71.70	70.95	70.30	69.25	68.85	68.25	67.90	67.70
Decorative Concrete Block	Steel Frame	77.10	75.10	74.30	73.65	72.50	72.10	71.45	71.10	70.90
	Bearing Walls	74.50	72.55	71.75	71.10	69.95	69.50	68.85	68.55	68.30
Perimeter Adj., Add or Deduct	Per 100 L.F.	2.15	1.80	1.50	1.35	1.20	1.05	1.00	.90	.85
Story Hgt. Adj., Add or Deduct	Per 1 Ft.	.75	.65	.60	.60	.55	.50	.50	.45	.45

For Basement, add $13.26 per square foot of basement area

The above costs were calculated using the basic specifications shown on the facing page. These costs should be adjusted where necessary for design alternatives and owner's requirements. Reported completed project costs, for this type of structure, range from $47.45 to $126.60 per S.F.

Common additives

Description	Unit	$ Cost
Bleachers, Telescoping, manual		
To 15 tier	Seat	70 - 100
16-20 tier	Seat	151 - 185
21-30 tier	Seat	159 - 192
For power operation, add	Seat	30 - 47
Carrels Hardwood	Each	675 - 880
Clock System		
20 room	Each	10,300
50 room	Each	24,800
Emergency Lighting, 25 watt, battery operated		
Lead battery	Each	249
Nickel cadmium	Each	605
Flagpoles, Complete		
Aluminum, 20' high	Each	1000
40' high	Each	2550
Fiberglass, 23' high	Each	1275
39'-5" high	Each	2575
Kitchen Equipment		
Broiler	Each	3950
Cooler, 6 ft. long, reach-in	Each	3075

Description	Unit	$ Cost
Kitchen Equipment, cont.		
Dishwasher, 10-12 racks per hr.	Each	2925
Food warmer, counter, 1.2 KW	Each	700
Freezer, 44 C.F., reach-in	Each	8050
Ice cube maker, 50 lb. per day	Each	1725
Range with 1 oven	Each	2350
Lockers, Steel, single tier, 60" to 72"	Opening	124 - 191
2 tier, 60" to 72" total	Opening	74 - 95
5 tier, box lockers	Opening	35 - 57
Locker bench, lam. maple top only	L.F.	16.35
Pedestals, steel pipe	Each	45
Seating		
Auditorium chair, all veneer	Each	151
Veneer back, padded seat	Each	184
Upholstered, spring seat	Each	186
Classroom, movable chair & desk	Set	65 - 120
Lecture hall, pedestal type	Each	137 - 435
Sound System		
Amplifier, 250 watts	Each	1500
Speaker, ceiling or wall	Each	125
Trumpet	Each	235

Model costs calculated for a 1 story building with 12' story height and 45,000 square feet of floor area

School, Elementary

BUILDING TYPES

				Unit	Unit Cost	Cost Per S.F.	% Of Sub-Total
1.0 Foundations							
.1	Footings & Foundations	Poured concrete; strip and spread footings and 4' foundation wall		S.F. Ground	2.96	2.96	
.4	Piles & Caissons	N/A		—	—	—	7.2%
.9	Excavation & Backfill	Site preparation for slab and trench for foundation wall and footing		S.F. Ground	.89	.89	
2.0 Substructure							
.1	Slab on Grade	4" reinforced concrete with vapor barrier and granular base		S.F. Slab	2.80	2.80	5.3%
.2	Special Substructures	N/A		—	—	—	
3.0 Superstructure							
.1	Columns & Beams	N/A		—	—	—	
.4	Structural Walls	N/A		—	—	—	
.5	Elevated Floors	N/A		—	—	—	4.2%
.7	Roof	Metal deck on open web steel joists		S.F. Roof	2.23	2.23	
.9	Stairs	N/A		—	—	—	
4.0 Exterior Closure							
.1	Walls	Face brick with concrete block backup	70% of wall	S.F. Wall	17.02	2.93	
.5	Exterior Wall Finishes	N/A		—	—	—	8.8%
.6	Doors	Metal and glass	5% of wall	Each	1939	.35	
.7	Windows & Glazed Walls	Steel outward projecting	25% of wall	Each	446	1.43	
5.0 Roofing							
.1	Roof Coverings	Built-up tar and gravel with flashing		S.F. Roof	1.73	1.73	
.7	Insulation	Perlite/EPS composite		S.F. Roof	1.18	1.18	5.6%
.8	Openings & Specialties	Gravel stop		L.F. Perimeter	4.99	.10	
6.0 Interior Construction							
.1	Partitions	Concrete block, toilet partitions	20 S.F. Floor/L.F. Partition	S.F. Partition	4.62	2.96	
.4	Interior Doors	Single leaf kalamein fire doors	700 S.F. Floor/Door	Each	492	.70	
.5	Wall Finishes	75% paint, 15% glazed coating, 10% ceramic tile		S.F. Surface	1.27	1.27	24.5%
.6	Floor Finishes	65% vinyl composition tile, 25% carpet, 10% terrazzo		S.F. Floor	4.61	4.61	
.7	Ceiling Finishes	Mineral fiber tile on concealed zee bars		S.F. Ceiling	3.10	3.10	
.9	Interior Surface/Exterior Wall	Painted gypsum board on furring	70% of wall	S.F. Wall	2.47	.43	
7.0 Conveying							
.1	Elevators	N/A		—	—	—	0.0%
.2	Special Conveyors	N/A		—	—	—	
8.0 Mechanical							
.1	Plumbing	Kitchen, bathroom and service fixtures, supply and drainage	1 Fixture/625 S.F. Floor	Each	2025	3.24	
.2	Fire Protection	Sprinklers, light hazard		S.F. Floor	1.09	1.09	
.3	Heating	Oil fired hot water, wall fin radiation		S.F. Floor	4.85	4.85	29.5%
.4	Cooling	Split systems with air cooled condensing units		S.F. Floor	6.54	6.54	
.5	Special Systems	N/A		—	—	—	
9.0 Electrical							
.1	Service & Distribution	600 ampere service, panel board and feeders		S.F. Floor	.59	.59	
.2	Lighting & Power	Fluorescent fixtures, receptacles, switches, A.C. and misc. power		S.F. Floor	5.41	5.41	13.2%
.4	Special Electrical	Alarm systems, communications systems and emergency lighting		S.F. Floor	1.01	1.01	
11.0 Special Construction							
.1	Specialties	Chalkboards		S.F. Floor	.88	.88	1.7%
12.0 Site Work							
.1	Earthwork	N/A		—	—	—	
.3	Utilities	N/A		—	—	—	0.0%
.5	Roads & Parking	N/A		—	—	—	
.7	Site Improvements	N/A		—	—	—	
				Sub-Total		53.28	**100%**
	CONTRACTOR FEES (General Requirements: 10%, Overhead: 5%, Profit: 10%)				25%	13.32	
	ARCHITECT FEES				7%	4.65	
				Total Building Cost		**71.25**	

71

COMMERCIAL/INDUSTRIAL/INSTITUTIONAL M.580 School, Jr High, 2-3 Story

Costs per square foot of floor area

Exterior Wall	S.F. Area	50000	65000	80000	95000	110000	125000	140000	155000	170000
	L.F. Perimeter	816	1016	1000	1150	1890	2116	2287	2438	2552
Face Brick with Concrete Block Back-up	Steel Frame	82.35	81.50	78.80	78.35	82.40	82.10	81.60	81.05	80.45
	Bearing Walls	79.45	78.55	75.90	75.45	79.45	79.20	78.70	78.10	77.50
Concrete Block Stucco Face	Steel Frame	78.85	78.10	76.10	75.75	78.70	78.45	78.10	77.65	77.20
	Bearing Walls	75.90	75.20	73.20	72.80	75.75	75.55	75.15	74.75	74.25
Decorative Concrete Block	Steel Frame	80.00	79.20	77.00	76.60	79.90	79.65	79.25	78.80	78.25
	Bearing Walls	76.60	75.80	73.60	73.20	76.50	76.25	75.85	75.40	74.85
Perimeter Adj., Add or Deduct	Per 100 L.F.	1.60	1.25	1.00	.85	.70	.65	.55	.55	.50
Story Hgt. Adj., Add or Deduct	Per 1 Ft.	.95	.90	.70	.70	1.00	1.00	.95	.95	.85

For Basement, add $19.39 per square foot of basement area

The above costs were calculated using the basic specifications shown on the facing page. These costs should be adjusted where necessary for design alternatives and owner's requirements. Reported completed project costs, for this type of structure, range from $54.60 to $124.80 per S.F.

Common additives

Description	Unit	$ Cost
Bleachers, Telescoping, manual		
To 15 tier	Seat	70 - 100
16-20 tier	Seat	151 - 185
21-30 tier	Seat	159 - 192
For power operation, add	Seat	30 - 47
Carrels Hardwood	Each	675 - 880
Clock System		
20 room	Each	10,300
50 room	Each	24,800
Elevators, Hydraulic passenger, 2 stops		
1500# capacity	Each	39,600
2500# capacity	Each	40,700
Emergency Lighting, 25 watt, battery operated		
Lead battery	Each	249
Nickel cadmium	Each	605
Flagpoles, Complete		
Aluminum, 20' high	Each	1000
40' high	Each	2550
Fiberglass, 23' high	Each	1275
39'-5" high	Each	2575

Description	Unit	$ Cost
Kitchen Equipment		
Broiler	Each	3950
Cooler, 6 ft. long, reach-in	Each	3075
Dishwasher, 10-12 racks per hr.	Each	2925
Food warmer, counter, 1.2 KW	Each	700
Freezer, 44 C.F., reach-in	Each	8050
Lockers, Steel, single tier, 60" to 72"	Opening	124 - 191
2 tier, 60" to 72" total	Opening	74 - 95
5 tier, box lockers	Opening	35 - 57
Locker bench, lam. maple top only	L.F.	16.35
Pedestals, steel pipe	Each	45
Seating		
Auditorium chair, all veneer	Each	151
Veneer back, padded seat	Each	184
Upholstered, spring seat	Each	186
Classroom, movable chair & desk	Set	65 - 120
Lecture hall, pedestal type	Each	137 - 435
Sound System		
Amplifier, 250 watts	Each	1500
Speaker, ceiling or wall	Each	125
Trumpet	Each	235

Important: See the Reference Section for Location Factors

Model costs calculated for a 2 story building with 12' story height and 110,000 square feet of floor area

School, Jr High, 2-3 Story

				Unit	Unit Cost	Cost Per S.F.	% Of Sub-Total
1.0 Foundations							
.1	Footings & Foundations	Poured concrete; strip and spread footings and 4' foundation wall		S.F. Ground	2.66	1.33	2.9%
.4	Piles & Caissons	N/A		—	—	—	
.9	Excavation & Backfill	Site preparation for slab and trench for foundation wall and footing		S.F. Ground	.89	.44	
2.0 Substructure							
.1	Slab on Grade	4" reinforced concrete with vapor barrier and granular base		S.F. Slab	2.80	1.40	2.3%
.2	Special Substructures	N/A		—	—	—	
3.0 Superstructure							
.1	Columns & Beams	Fireproofing, steel columns included in 3.5 and 3.7		L.F. Column	17.96	.25	
.4	Structural Walls	N/A		—	—	—	
.5	Elevated Floors	Open web steel joists, slab form, concrete, columns		S.F. Floor	13.31	6.66	15.8%
.7	Roof	Metal deck, open web steel joists, columns		S.F. Roof	5.10	2.55	
.9	Stairs	Concrete filled metal pan		Flight	4540	.25	
4.0 Exterior Closure							
.1	Walls	Face brick with concrete block backup	75% of wall	S.F. Wall	17.04	5.27	
.5	Exterior Wall Finishes	N/A		—	—	—	13.9%
.6	Doors	Double aluminum & glass		Each	1288	.34	
.7	Windows & Glazed Walls	Window wall	25% of wall	S.F. Window	28	2.95	
5.0 Roofing							
.1	Roof Coverings	Built-up tar and gravel with flashing		S.F. Roof	1.98	.99	
.7	Insulation	Perlite/EPS composite		S.F. Roof	1.18	.59	2.7%
.8	Openings & Specialties	Gravel stop and hatches		S.F. Roof	.22	.11	
6.0 Interior Construction							
.1	Partitions	Concrete block, toilet partitions	20 S.F. Floor/L.F. Partition	S.F. Partition	4.62	2.75	
.4	Interior Doors	Single leaf kalamein fire doors	750 S.F. Floor/Door	Each	492	.66	
.5	Wall Finishes	50% paint, 40% glazed coatings, 10% ceramic tile		S.F. Surface	1.41	1.41	23.7%
.6	Floor Finishes	50% vinyl composition tile, 30% carpet, 20% terrrazzo		S.F. Floor	5.94	5.94	
.7	Ceiling Finishes	Mineral fiberboard on concealed zee bars		S.F. Ceiling	3.10	3.10	
.9	Interior Surface/Exterior Wall	Painted gypsum board on furring	75% of wall	S.F. Wall	2.47	.76	
7.0 Conveying							
.1	Elevators	One hydraulic passenger elevator		Each	47,300	.43	0.7%
.2	Special Conveyors	N/A		—	—	—	
8.0 Mechanical							
.1	Plumbing	Kitchen, toilet and service fixtures, supply and drainage	1 Fixture/1170 S.F. Floor	Each	2445	2.09	
.2	Fire Protection	Sprinklers, light hazard	10% of area	S.F. Floor	.19	.19	
.3	Heating	Included in 8.4		—	—	—	24.4%
.4	Cooling	Multizone unit, gas heating, electric cooling		S.F. Floor	12.78	12.78	
.5	Special Systems	N/A		—	—	—	
9.0 Electrical							
.1	Service & Distribution	1000 ampere service, panel board and feeders		S.F. Floor	.46	.46	
.2	Lighting & Power	Fluorescent fixtures, receptacles, switches, A.C. and misc. power		S.F. Floor	5.42	5.42	12.4%
.4	Special Electrical	Alarm systems, communications systems and emergency lighting		S.F. Floor	1.75	1.75	
11.0 Special Construction							
.1	Specialties	Chalkboards, laboratory counters		S.F. Floor	.73	.73	1.2%
12.0 Site Work							
.1	Earthwork	N/A		—	—	—	
.3	Utilities	N/A		—	—	—	0.0%
.5	Roads & Parking	N/A		—	—	—	
.7	Site Improvements	N/A		—	—	—	
				Sub-Total		61.60	100%
	CONTRACTOR FEES (General Requirements: 10%, Overhead: 5%, Profit: 10%)				25%	15.40	
	ARCHITECT FEES				7%	5.40	
				Total Building Cost		82.40	

COMMERCIAL/INDUSTRIAL/INSTITUTIONAL — M.600 — Store, Convenience

Costs per square foot of floor area

Exterior Wall	S.F. Area	1000	2000	3000	4000	6000	8000	10000	12000	15000
	L.F. Perimeter	126	179	219	253	310	358	400	438	490
Wood Siding	Wood Frame	73.90	66.20	62.70	60.65	58.15	56.70	55.70	55.00	54.15
Face Brick Veneer	Wood Frame	89.30	76.80	71.15	67.80	63.80	61.45	59.80	58.65	57.30
Stucco on Concrete Block	Steel Frame	82.25	72.35	67.85	65.25	62.10	60.25	58.90	58.00	56.90
Stucco on Concrete Block	Bearing Walls	80.85	70.95	66.45	63.85	60.70	58.85	57.50	56.60	55.50
Metal Sandwich Panel	Steel Frame	85.35	74.65	69.80	67.00	63.55	61.55	60.15	59.15	58.00
Precast Concrete	Steel Frame	97.10	82.75	76.25	72.45	67.85	65.15	63.25	61.90	60.40
Perimeter Adj., Add or Deduct	Per 100 L.F.	21.15	10.55	7.05	5.30	3.55	2.65	2.10	1.75	1.45
Story Hgt. Adj., Add or Deduct	Per 1 Ft.	1.60	1.10	.90	.80	.65	.60	.50	.45	.45

For Basement, add $14.11 per square foot of basement area

The above costs were calculated using the basic specifications shown on the facing page. These costs should be adjusted where necessary for design alternatives and owner's requirements. Reported completed project costs, for this type of structure, range from $40.05 to $138.80 per S.F.

Common additives

Description	Unit	$ Cost
Check Out Counter		
Single belt	Each	2175
Double belt	Each	3675
Emergency Lighting, 25 watt, battery operated		
Lead battery	Each	249
Nickel cadmium	Each	605
Refrigerators, Prefabricated, walk-in		
7'-6" high, 6' x 6'	S.F.	117
10' x 10'	S.F.	92
12' x 14'	S.F.	82
12' x 20'	S.F.	72
Refrigerated Food Cases		
Dairy, multi deck, 12' long	Each	7650
Delicatessen case, single deck, 12' long	Each	5150
Multi deck, 18 S.F. shelf display	Each	6325
Freezer, self-contained chest type, 30 C.F.	Each	3800
Glass door upright, 78 C.F.	Each	7150

Description	Unit	$ Cost
Refrigerated Food Cases, cont.		
Frozen food, chest type, 12' long	Each	5200
Glass door reach-in, 5 door	Each	9850
Island case 12' long, single deck	Each	5875
Multi deck	Each	12,300
Meat cases, 12' long, single deck	Each	4300
Multi deck	Each	7325
Produce, 12' long single deck	Each	5650
Multi deck	Each	6300
Safe, Office type, 4 hour rating		
30" x 18" x 18"	Each	3925
62" x 33" x 20"	Each	8525
Smoke Detectors		
Ceiling type	Each	129
Duct type	Each	365
Sound System		
Amplifier, 250 watts	Each	1500
Speaker, ceiling or wall	Each	125
Trumpet	Each	235

Model costs calculated for a 1 story building with 12' story height and 4,000 square feet of floor area

Store, Convenience

			Unit	Unit Cost	Cost Per S.F.	% Of Sub-Total
1.0 Foundations						
.1	Footings & Foundations	Poured concrete; strip and spread footings and 4' foundation wall	S.F. Ground	2.90	2.90	8.7%
.4	Piles & Caissons	N/A	—	—	—	
.9	Excavation & Backfill	Site preparation for slab and trench for foundation wall and footing	S.F. Ground	1.06	1.06	
2.0 Substructure						
.1	Slab on Grade	4" reinforced concrete with vapor barrier and granular base	S.F. Slab	2.80	2.80	6.2%
.2	Special Substructures	N/A	—	—	—	
3.0 Superstructure						
.1	Columns & Beams	N/A	—	—	—	
.4	Structural Walls	N/A	—	—	—	
.5	Elevated Floors	N/A	—	—	—	10.0%
.7	Roof	Wood truss with plywood sheathing	S.F. Ground	4.54	4.54	
.9	Stairs	N/A	—	—	—	
4.0 Exterior Closure						
.1	Walls	Wood siding on wood studs, insulated — 80% of wall	S.F. Wall	5.88	3.57	
.5	Exterior Wall Finishes	N/A	—	—	—	17.3%
.6	Doors	Double aluminum and glass, solid core wood	Each	1887	1.42	
.7	Windows & Glazed Walls	Storefront — 20% of wall	S.F. Window	28	2.83	
5.0 Roofing						
.1	Roof Coverings	Asphalt shingles	S.F. Roof	1.38	1.38	
.7	Insulation	Fiberglass sheets	S.F. Roof	.88	.88	5.2%
.8	Openings & Specialties	Gutters and downspouts	S.F. Roof	.08	.08	
6.0 Interior Construction						
.1	Partitions	Gypsum board on wood studs — 60 S.F. Floor/L.F. Partition	S.F. Partition	2.82	.47	
.4	Interior Doors	Single leaf wood, hollow metal — 1300 S.F. Floor/Door	Each	604	.47	
.5	Wall Finishes	Paint	S.F. Surface	.66	.22	12.8%
.6	Floor Finishes	Asphalt tile	S.F. Floor	1.79	1.79	
.7	Ceiling Finishes	Mineral fiber tile on wood furring	S.F. Ceiling	2.03	2.03	
.9	Interior Surface/Exterior Wall	Painted gypsum board on furring — 80% of wall	S.F. Wall	1.38	.84	
7.0 Conveying						
.1	Elevators	N/A	—	—	—	0.0%
.2	Special Conveyors	N/A	—	—	—	
8.0 Mechanical						
.1	Plumbing	Toilet and service fixtures, supply and drainage — 1 Fixture/1000 S.F. Floor	Each	1570	1.57	
.2	Fire Protection	Sprinkler, ordinary hazard	S.F. Floor	1.74	1.74	
.3	Heating	Included in 8.4	—	—	—	18.8%
.4	Cooling	Single zone rooftop unit, gas heating, electric cooling	S.F. Floor	5.23	5.23	
.5	Special Systems	N/A	—	—	—	
9.0 Electrical						
.1	Service & Distribution	200 ampere service, panel board and feeders	S.F. Floor	1.53	1.53	
.2	Lighting & Power	Fluorescent fixtures, receptacles, switches, A.C. and misc. power	S.F. Floor	5.46	5.46	21.0%
.4	Special Electrical	Alarm systems and emergency lighting	S.F. Floor	2.52	2.52	
11.0 Special Construction						
.1	Specialties	N/A	—	—	—	0.0%
12.0 Site Work						
.1	Earthwork	N/A	—	—	—	
.3	Utilities	N/A	—	—	—	0.0%
.5	Roads & Parking	N/A	—	—	—	
.7	Site Improvements	N/A	—	—	—	
			Sub-Total		45.33	**100%**
	CONTRACTOR FEES (General Requirements: 10%, Overhead: 5%, Profit: 10%)			25%	11.33	
	ARCHITECT FEES			7%	3.99	
			Total Building Cost		**60.65**	

BUILDING TYPES

COMMERCIAL/INDUSTRIAL/INSTITUTIONAL — M.610 — Store, Department, 1 Story

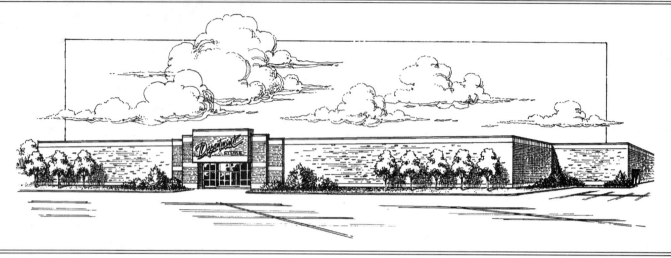

Costs per square foot of floor area

Exterior Wall	S.F. Area	50000	65000	80000	95000	110000	125000	140000	155000	170000
	L.F. Perimeter	920	1065	1167	1303	1333	1433	1533	1633	1733
Face Brick with Concrete Block Back-up	R/Conc. Frame	64.30	62.80	61.60	60.95	60.05	59.60	59.25	58.95	58.75
	Steel Frame	51.30	49.80	48.60	47.95	47.00	46.60	46.20	45.95	45.70
Decorative Concrete Block	R/Conc. Frame	61.55	60.35	59.40	58.90	58.25	57.90	57.60	57.40	57.20
	Steel Joists	49.10	47.85	46.85	46.30	45.55	45.20	44.90	44.65	44.50
Precast Concrete Panels	R/Conc. Frame	62.00	60.80	59.85	59.30	58.60	58.25	57.95	57.75	57.55
	Steel Joists	48.65	47.40	46.45	45.95	45.25	44.90	44.60	44.40	44.20
Perimeter Adj., Add or Deduct	Per 100 L.F.	.95	.75	.65	.50	.45	.40	.35	.30	.25
Story Hgt. Adj., Add or Deduct	Per 1 Ft.	.50	.45	.40	.40	.35	.35	.30	.30	.25

For Basement, add $13.49 per square foot of basement area

The above costs were calculated using the basic specifications shown on the facing page. These costs should be adjusted where necessary for design alternatives and owner's requirements. Reported completed project costs, for this type of structure, range from $34.50 to $76.30 per S.F.

Common additives

Description	Unit	$ Cost
Closed Circuit Surveillance, One station		
Camera and monitor	Each	1275
For additional camera stations, add	Each	700
Directory Boards, Plastic, glass covered		
30" x 20"	Each	500
36" x 48"	Each	895
Aluminum, 24" x 18"	Each	440
36" x 24"	Each	530
48" x 32"	Each	630
48" x 60"	Each	1375
Emergency Lighting, 25 watt, battery operated		
Lead battery	Each	249
Nickel cadmium	Each	605
Safe, Office type, 4 hour rating		
30" x 18" x 18"	Each	3925
62" x 33" x 20"	Each	8525
Sound System		
Amplifier, 250 watts	Each	1500
Speaker, ceiling or wall	Each	125
Trumpet	Each	235

Model costs calculated for a 1 story building with 14' story height and 110,000 square feet of floor area

Store, Department, 1 Story

				Unit	Unit Cost	Cost Per S.F.	% Of Sub-Total
1.0 Foundations							
.1	Footings & Foundations	Poured concrete; strip and spread footings and 4' foundation wall		S.F. Ground	.91	.91	
.4	Piles & Caissons	N/A		—	—	—	3.9%
.9	Excavation & Backfill	Site preparation for slab and trench for foundation wall and footing		S.F. Ground	.85	.85	
2.0 Substructure							
.1	Slab on Grade	4" reinforced concrete with vapor barrier and granular base		S.F. Slab	2.80	2.80	6.2%
.2	Special Substructures	N/A		—	—	—	
3.0 Superstructure							
.1	Columns & Beams	Concrete columns		L.F. Column	35	.45	
.4	Structural Walls	N/A		—	—	—	
.5	Elevated Floors	N/A		—	—	—	31.7%
.7	Roof	Precast concrete beam and plank		S.F. Roof	13.90	13.90	
.9	Stairs	N/A		—	—	—	
4.0 Exterior Closure							
.1	Walls	Face brick with concrete block backup	90% of wall	S.F. Wall	17.22	2.63	
.5	Exterior Wall Finishes	N/A		—	—	—	7.8%
.6	Doors	Sliding electric operated entrance, hollow metal		Each	4823	.27	
.7	Windows & Glazed Walls	Storefront	10% of wall	S.F. Window	37	.62	
5.0 Roofing							
.1	Roof Coverings	Built-up tar and gravel with flashing		S.F. Roof	1.55	1.55	
.7	Insulation	Perlite/EPS composite		S.F. Roof	1.18	1.18	6.2%
.8	Openings & Specialties	Gravel stop and hatches		S.F. Roof	.09	.09	
6.0 Interior Construction							
.1	Partitions	Gypsum board on metal studs	60 S.F. Floor/L.F. Partition	S.F. Partition	3.60	.60	
.4	Interior Doors	Single leaf hollow metal	600 S.F. Floor/Door	Each	492	.82	
.5	Wall Finishes	Paint		S.F. Surface	.48	.16	16.5%
.6	Floor Finishes	50% vinyl tile, 50% carpet		S.F. Floor	4.12	4.12	
.7	Ceiling Finishes	Fiberglass board on exposed grid system, suspended		S.F. Ceiling	1.41	1.41	
.9	Interior Surface/Exterior Wall	Painted gypsum board on furring	90% of wall	S.F. Wall	2.47	.38	
7.0 Conveying							
.1	Elevators	N/A		—	—	—	0.0%
.2	Special Conveyors	N/A		—	—	—	
8.0 Mechanical							
.1	Plumbing	Toilet and service fixtures, supply and drainage	1 Fixture/4075 S.F. Floor	Each	2322	.57	
.2	Fire Protection	Sprinklers, light hazard		S.F. Floor	1.09	1.09	
.3	Heating	Included in 8.4		—	—	—	16.4%
.4	Cooling	Single zone unit gas heating, electric cooling		S.F. Floor	5.77	5.77	
.5	Special Systems	N/A		—	—	—	
9.0 Electrical							
.1	Service & Distribution	1200 ampere service, panel board and feeders		S.F. Floor	.48	.48	
.2	Lighting & Power	Fluorescent fixtures, receptacles, switches, A.C. and misc. power		S.F. Floor	4.39	4.39	11.3%
.4	Special Electrical	Alarm systems and emergency lighting		S.F. Floor	.27	.27	
11.0 Special Construction							
.1	Specialties	N/A		—	—	—	0.0%
12.0 Site Work							
.1	Earthwork	N/A		—	—	—	
.3	Utilities	N/A		—	—	—	0.0%
.5	Roads & Parking	N/A		—	—	—	
.7	Site Improvements	N/A		—	—	—	
				Sub-Total		45.31	**100%**
	CONTRACTOR FEES (General Requirements: 10%, Overhead: 5%, Profit: 10%)				25%	11.33	
	ARCHITECT FEES				6%	3.41	
				Total Building Cost		**60.05**	

BUILDING TYPES

COMMERCIAL/INDUSTRIAL/INSTITUTIONAL
M.630 | Store, Retail

Costs per square foot of floor area

Exterior Wall	S.F. Area	4000	6000	8000	10000	12000	15000	18000	20000	22000
	L.F. Perimeter	260	340	360	410	440	490	540	565	594
Split Face Concrete Block	Steel Joists	77.00	70.45	64.50	62.00	59.75	57.65	56.20	55.35	54.70
Stucco on Concrete Block	Steel Joists	73.10	67.05	61.80	59.50	57.55	55.65	54.40	53.65	53.05
Painted Concrete Block	Steel Joists	71.85	65.70	60.35	58.05	56.00	54.10	52.80	52.05	51.45
Face Brick on Concrete Block	Steel Joists	85.00	77.45	70.05	67.00	64.25	61.65	59.90	58.80	58.00
Painted Reinforced Concrete	Steel Joists	80.70	73.70	67.05	64.30	61.80	59.50	57.90	56.95	56.20
Tilt-up Concrete Panels	Steel Joists	74.00	67.85	62.40	60.10	58.05	56.10	54.80	54.00	53.40
Perimeter Adj., Add or Deduct	Per 100 L.F.	8.80	5.85	4.35	3.55	2.95	2.35	1.95	1.80	1.60
Story Hgt. Adj., Add or Deduct	Per 1 Ft.	1.15	1.00	.80	.75	.65	.55	.50	.50	.50

For Basement, add $14.03 per square foot of basement area

The above costs were calculated using the basic specifications shown on the facing page. These costs should be adjusted where necessary for design alternatives and owner's requirements. Reported completed project costs, for this type of structure, range from $34.85 to $121.30 per S.F.

Common additives

Description	Unit	$ Cost
Emergency Lighting, 25 watt, battery operated		
Lead battery	Each	249
Nickel cadmium	Each	605
Safe, Office type, 4 hour rating		
30" x 18" x 18"	Each	3925
62" x 33" x 20"	Each	8525
Smoke Detectors		
Ceiling type	Each	129
Duct type	Each	365
Sound System		
Amplifier, 250 watts	Each	1500
Speaker, ceiling or wall	Each	125
Trumpet	Each	235

Model costs calculated for a 1 story building with 14' story height and 8,000 square feet of floor area

Store, Retail

				Unit	Unit Cost	Cost Per S.F.	% Of Sub-Total
1.0 Foundations							
.1	Footings & Foundations	Poured concrete; strip and spread footings and 4' foundation wall		S.F. Ground	2.55	2.55	
.4	Piles & Caissons	N/A		—	—	—	7.3%
.9	Excavation & Backfill	Site preparation for slab and trench for foundation wall and footing		S.F. Ground	.96	.96	
2.0 Substructure							
.1	Slab on Grade	4" reinforced concrete with vapor barrier and granular base		S.F. Slab	2.80	2.80	5.9%
.2	Special Substructures	N/A		—	—	—	
3.0 Superstructure							
.1	Columns & Beams	Interior columns included in 3.7		—	—	—	
.4	Structural Walls	N/A		—	—	—	
.5	Elevated Floors	N/A		—	—	—	7.0%
.7	Roof	Metal deck, open web steel joists, beams, interior columns		S.F. Roof	3.36	3.36	
.9	Stairs	N/A		—	—	—	
4.0 Exterior Closure							
.1	Walls	Decorative concrete block	90% of wall	S.F. Wall	10.02	5.68	
.5	Exterior Wall Finishes	N/A		—	—	—	16.2%
.6	Doors	Sliding entrance door and hollow metal service doors		Each	1239	.31	
.7	Windows & Glazed Walls	Storefront windows	10% of wall	S.F. Window	27	1.73	
5.0 Roofing							
.1	Roof Coverings	Built-up tar and gravel with flashing		S.F. Roof	2.02	2.02	
.7	Insulation	Perlite/EPS composite		S.F. Roof	1.18	1.18	7.3%
.8	Openings & Specialties	Gravel stop and hatches		S.F. Roof	.28	.28	
6.0 Interior Construction							
.1	Partitions	Gypsum board on metal studs	60 S.F. Floor/L.F. Partition	S.F. Partition	3.60	.60	
.4	Interior Doors	Single leaf hollow metal	600 S.F. Floor/Door	Each	492	.82	
.5	Wall Finishes	Paint		S.F. Surface	.48	.16	17.5%
.6	Floor Finishes	Vinyl tile		S.F. Floor	2.71	2.71	
.7	Ceiling Finishes	Mineral fiber tile on concealed zee bars		S.F. Ceiling	3.10	3.10	
.9	Interior Surface/Exterior Wall	Paint	90% of wall	S.F. Wall	1.69	.96	
7.0 Conveying							
.1	Elevators	N/A		—	—	—	0.0%
.2	Special Conveyors	N/A		—	—	—	
8.0 Mechanical							
.1	Plumbing	Toilet and service fixtures, supply and drainage	1 Fixture/890 S.F. Floor	Each	3444	3.87	
.2	Fire Protection	Wet pipe sprinkler system		S.F. Floor	1.74	1.74	
.3	Heating	Included in 8.4		—	—	—	23.8%
.4	Cooling	Single zone unit, gas heating, electric cooling		S.F. Floor	5.77	5.77	
.5	Special Systems	N/A		—	—	—	
9.0 Electrical							
.1	Service & Distribution	400 ampere service, panel board and feeders		S.F. Floor	1.41	1.41	
.2	Lighting & Power	Fluorescent fixtures, receptacles, switches, A.C. and misc. power		S.F. Floor	5.39	5.39	15.0%
.4	Special Electrical	Alarm systems and emergency lighting		S.F. Floor	.39	.39	
11.0 Special Construction							
.1	Specialties	N/A		—	—	—	0.0%
12.0 Site Work							
.1	Earthwork	N/A		—	—	—	
.3	Utilities	N/A		—	—	—	0.0%
.5	Roads & Parking	N/A		—	—	—	
.7	Site Improvements	N/A		—	—	—	
				Sub-Total		47.79	100%
	CONTRACTOR FEES (General Requirements: 10%, Overhead: 5%, Profit: 10%)				25%	11.95	
	ARCHITECT FEES				8%	4.76	
				Total Building Cost		64.50	

COMMERCIAL/INDUSTRIAL/INSTITUTIONAL — M.640 Supermarket

Costs per square foot of floor area

Exterior Wall	S.F. Area	6000	8000	12000	16000	20000	25000	32000	38000	45000
	L.F. Perimeter	310	360	460	520	600	656	773	806	900
Face Brick with Concrete Block Back-up	Steel Frame	88.25	82.05	75.85	71.20	69.00	66.20	64.40	62.30	61.40
	Bearing Walls	85.40	79.20	73.00	68.35	66.15	63.35	61.60	59.45	58.55
Stucco on Concrete Block	Steel Frame	78.90	73.95	68.95	65.35	63.60	61.45	60.10	58.50	57.75
	Bearing Walls	79.40	74.40	69.45	65.80	64.10	61.95	60.55	58.95	58.25
Precast Concrete Panels	Steel Frame	86.00	80.10	74.20	69.75	67.70	65.05	63.40	61.40	60.50
Metal Sandwich Panels	Steel Frame	74.40	70.00	65.60	62.50	60.95	59.15	57.95	56.65	56.00
Perimeter Adj., Add or Deduct	Per 100 L.F.	10.60	7.95	5.35	4.00	3.15	2.50	2.05	1.70	1.40
Story Hgt. Adj., Add or Deduct	Per 1 Ft.	1.55	1.30	1.15	.95	.85	.75	.75	.60	.60

For Basement, add $14.17 per square foot of basement area

The above costs were calculated using the basic specifications shown on the facing page. These costs should be adjusted where necessary for design alternatives and owner's requirements. Reported completed project costs, for this type of structure, range from $40.70 to $116.55 per S.F.

Common additives

Description	Unit	$ Cost
Check Out Counter		
Single belt	Each	2175
Double belt	Each	3675
Scanner, registers, guns & memory 2 lanes	Each	13,200
10 lanes	Each	125,500
Power take away	Each	4550
Emergency Lighting, 25 watt, battery operated		
Lead battery	Each	249
Nickel cadmium	Each	605
Refrigerators, Prefabricated, walk-in		
7'-6" High, 6' x 6'	S.F.	117
10' x 10'	S.F.	92
12' x 14'	S.F.	82
12' x 20'	S.F.	72
Refrigerated Food Cases		
Dairy, multi deck, 12' long	Each	7650
Delicatessen case, single deck, 12' long	Each	5150
Multi deck, 18 S.F. shelf display	Each	6325
Freezer, self-contained chest type, 30 C.F.	Each	3800
Glass door upright, 78 C.F.	Each	7150

Description	Unit	$ Cost
Refrigerated Food Cases, cont.		
Frozen food, chest type, 12' long	Each	5200
Glass door reach in, 5 door	Each	9850
Island case 12' long, single deck	Each	5875
Multi deck	Each	12,300
Meat case 12' long, single deck	Each	4300
Multi deck	Each	7325
Produce, 12' long single deck	Each	5650
Multi deck	Each	6300
Safe, Office type, 4 hour rating		
30" x 18" x 18"	Each	3925
62" x 33" x 20"	Each	8525
Smoke Detectors		
Ceiling type	Each	129
Duct type	Each	365
Sound System		
Amplifier, 250 watts	Each	1500
Speaker, ceiling or wall	Each	125
Trumpet	Each	235

Model costs calculated for a 1 story building with 18' story height and 20,000 square feet of floor area

Supermarket

				Unit	Unit Cost	Cost Per S.F.	% Of Sub-Total
1.0 Foundations							
.1	Footings & Foundations	Poured concrete; strip and spread footings and 4' foundation wall		S.F. Ground	2.00	2.00	
.4	Piles & Caissons	N/A		—	—	—	5.8%
.9	Excavation & Backfill	Site preparation for slab and trench for foundation wall and footing		S.F. Ground	.89	.89	
2.0 Substructure							
.1	Slab on Grade	4" reinforced concrete with vapor barrier and granular base		S.F. Slab	2.80	2.80	5.7%
.2	Special Substructures	N/A		—	—	—	
3.0 Superstructure							
.1	Columns & Beams	Interior columns included in 3.7		—	—	—	
.4	Structural Walls	N/A		—	—	—	
.5	Elevated Floors	N/A		—	—	—	7.5%
.7	Roof	Metal deck, open web steel joists, beams, interior columns		S.F. Roof	3.71	3.71	
.9	Stairs	N/A		—	—	—	
4.0 Exterior Closure							
.1	Walls	Face brick with concrete block backup	85% of wall	S.F. Wall	17.06	7.83	
.5	Exterior Wall Finishes	N/A		—	—	—	26.1%
.6	Doors	Sliding entrance doors with electrical operator, hollow metal		Each	3236	1.78	
.7	Windows & Glazed Walls	Storefront windows	15% of wall	S.F. Window	40	3.31	
5.0 Roofing							
.1	Roof Coverings	Built-up tar and gravel with flashing		S.F. Roof	1.80	1.80	
.7	Insulation	Perlite/EPS composite		S.F. Roof	1.18	1.18	6.6%
.8	Openings & Specialties	Gravel stop and hatches		S.F. Roof	.28	.28	
6.0 Interior Construction							
.1	Partitions	50% concrete block, 50% gypsum board on metal studs	50 S.F. Floor/L.F. Partition	S.F. Partition	4.08	.98	
.4	Interior Doors	Single leaf hollow metal	2000 S.F. Floor/Door	Each	492	.25	
.5	Wall Finishes	Paint		S.F. Surface	.65	.31	16.4%
.6	Floor Finishes	Vinyl composition tile		S.F. Floor	2.71	2.71	
.7	Ceiling Finishes	Mineral fiber tile on concealed zee bars		S.F. Ceiling	3.10	3.10	
.9	Interior Surface/Exterior Wall	Paint	85% of wall	S.F. Wall	1.69	.78	
7.0 Conveying							
.1	Elevators	N/A		—	—	—	0.0%
.2	Special Conveyors	N/A		—	—	—	
8.0 Mechanical							
.1	Plumbing	Toilet and service fixtures, supply and drainage	1 Fixture/1820 S.F. Floor	Each	2893	1.59	
.2	Fire Protection	Sprinklers, light hazard		S.F. Floor	1.33	1.33	
.3	Heating	Included in 8.4		—	—	—	17.3%
.4	Cooling	Single zone rooftop unit, gas heating, electric cooling		S.F. Floor	5.60	5.60	
.5	Special Systems	N/A		—	—	—	
9.0 Electrical							
.1	Service & Distribution	400 ampere service, panel board and feeders		S.F. Floor	.77	.77	
.2	Lighting & Power	Fluorescent fixtures, receptacles, switches, A.C. and misc. power		S.F. Floor	6.17	6.17	14.6%
.4	Special Electrical	Alarm systems and emergency lighting		S.F. Floor	.30	.30	
11.0 Special Construction							
.1	Specialties	N/A		—	—	—	0.0%
12.0 Site Work							
.1	Earthwork	N/A		—	—	—	
.3	Utilities	N/A		—	—	—	0.0%
.5	Roads & Parking	N/A		—	—	—	
.7	Site Improvements	N/A		—	—	—	
				Sub-Total		49.47	100%
	CONTRACTOR FEES (General Requirements: 10%, Overhead: 5%, Profit: 10%)				25%	12.37	
	ARCHITECT FEES				7%	4.31	
				Total Building Cost		66.15	

COMMERCIAL/INDUSTRIAL/INSTITUTIONAL — M.660 — Telephone Exchange

Costs per square foot of floor area

Exterior Wall	S.F. Area	2000	3000	4000	5000	6000	7000	8000	9000	10000
	L.F. Perimeter	180	220	260	286	320	353	368	397	425
Face Brick with Concrete Block Back-up	Steel Frame	113.80	101.80	95.80	90.65	88.00	85.95	83.25	82.00	80.90
	Bearing Walls	112.30	100.30	94.30	89.15	86.50	84.45	81.75	80.50	79.40
Limestone with Concrete Block Back-up	Steel Frame	129.25	114.40	106.95	100.50	97.15	94.65	91.15	89.55	88.20
	Bearing Walls	127.75	112.90	105.45	99.00	95.65	93.15	89.65	88.05	86.70
Decorative Concrete Block	Steel Frame	106.85	96.15	90.75	86.25	83.85	82.10	79.70	78.60	77.65
	Bearing Walls	105.35	94.65	89.25	84.75	82.35	80.60	78.20	77.10	76.15
Perimeter Adj., Add or Deduct	Per 100 L.F.	27.15	18.10	13.60	10.85	9.05	7.75	6.80	6.05	5.40
Story Hgt. Adj., Add or Deduct	Per 1 Ft.	3.25	2.65	2.35	2.05	1.90	1.85	1.65	1.60	1.55

For Basement, add $20.35 per square foot of basement area

The above costs were calculated using the basic specifications shown on the facing page. These costs should be adjusted where necessary for design alternatives and owner's requirements. Reported completed project costs, for this type of structure, range from $56.60 to $199.85 per S.F.

Common additives

Description	Unit	$ Cost
Emergency Lighting, 25 watt, battery operated		
Lead battery	Each	249
Nickel cadmium	Each	605
Smoke Detectors		
Ceiling type	Each	129
Duct type	Each	365
Emergency Generators, complete system, gas		
15 kw	Each	12,400
85 kw	Each	27,400
170 kw	Each	70,500
Diesel, 50 kw	Each	23,700
150 kw	Each	41,900
350 kw	Each	63,500

Model costs calculated for a 1 story building with 12' story height and 5,000 square feet of floor area

Telephone Exchange

			Unit	Unit Cost	Cost Per S.F.	% Of Sub-Total
1.0 Foundations						
.1	Footings & Foundations	Poured concrete; strip and spread footings and 4' foundation wall	S.F. Ground	3.72	3.72	7.3%
.4	Piles & Caissons	N/A	—	—	—	
.9	Excavation & Backfill	Site preparation for slab and trench for foundation wall and footing	S.F. Ground	1.06	1.06	
2.0 Substructure						
.1	Slab on Grade	4" reinforced concrete with vapor barrier and granular base	S.F. Slab	2.80	2.80	4.3%
.2	Special Substructures	N/A	—	—	—	
3.0 Superstructure						
.1	Columns & Beams	Steel columns included in 3.7	—	—	—	
.4	Structural Walls	N/A	—	—	—	
.5	Elevated Floors	N/A	—	—	—	6.3%
.7	Roof	Metal deck, open web steel joists, beams, columns	S.F. Roof	4.12	4.12	
.9	Stairs	N/A	—	—	—	
4.0 Exterior Closure						
.1	Walls	Face brick with concrete block backup — 80% of wall	S.F. Wall	17.05	9.36	
.5	Exterior Wall Finishes	N/A	—	—	—	27.2%
.6	Doors	Single aluminum glass with transom	Each	2159	.87	
.7	Windows & Glazed Walls	Outward projecting steel — 20% of wall	Each	1098	7.54	
5.0 Roofing						
.1	Roof Coverings	Built-up tar and gravel with flashing	S.F. Roof	2.20	2.20	
.7	Insulation	Perlite/EPS composite	S.F. Roof	1.18	1.18	5.6%
.8	Openings & Specialties	Gravel stop and hatches	S.F. Roof	.29	.29	
6.0 Interior Construction						
.1	Partitions	Double layer gypsum board on metal studs, toilet partitions — 15 S.F. Floor/L.F. Partition	S.F. Partition	5.07	3.38	
.4	Interior Doors	Single leaf hollow metal — 150 S.F. Floor/Door	Each	492	3.28	
.5	Wall Finishes	Paint	S.F. Surface	.48	.64	24.6%
.6	Floor Finishes	90% carpet, 10% terrazzo	S.F. Floor	5.49	5.49	
.7	Ceiling Finishes	Fiberglass board on exposed grid system, suspended	S.F. Ceiling	2.33	2.33	
.9	Interior Surface/Exterior Wall	Paint — 80% of wall	S.F. Floor	1.69	.93	
7.0 Conveying						
.1	Elevators	N/A	—	—	—	0.0%
.2	Special Conveyors	N/A	—	—	—	
8.0 Mechanical						
.1	Plumbing	Kitchen, toilet and service fixtures, supply and drainage — 1 Fixture/715 S.F. Floor	Each	2452	3.43	
.2	Fire Protection	Wet pipe sprinkler system	S.F. Floor	1.91	1.91	
.3	Heating	Included in 8.4	—	—	—	17.1%
.4	Cooling	Single zone unit, gas heating, electric cooling	S.F. Floor	5.85	5.85	
.5	Special Systems	N/A	—	—	—	
9.0 Electrical						
.1	Service & Distribution	200 ampere service, panel board and feeders	S.F. Floor	.98	.98	
.2	Lighting & Power	Fluorescent fixtures, receptacles, switches, A.C. and misc. power	S.F. Floor	3.10	3.10	7.6%
.4	Special Electrical	Alarm systems and emergency lighting	S.F. Floor	.88	.88	
11.0 Special Construction						
.1	Specialties	N/A	—	—	—	0.0%
12.0 Site Work						
.1	Earthwork	N/A	—	—	—	
.3	Utilities	N/A	—	—	—	0.0%
.5	Roads & Parking	N/A	—	—	—	
.7	Site Improvements	N/A	—	—	—	
			Sub-Total		65.34	**100%**
	CONTRACTOR FEES (General Requirements: 10%, Overhead: 5%, Profit: 10%)			25%	16.34	
	ARCHITECT FEES			11%	8.97	
			Total Building Cost		**90.65**	

BUILDING TYPES

COMMERCIAL/INDUSTRIAL/INSTITUTIONAL   M.670   Town Hall, 1 Story

Costs per square foot of floor area

Exterior Wall	S.F. Area	5000	6500	8000	9500	11000	14000	17500	21000	24000
	L.F. Perimeter	300	360	386	396	435	510	550	620	680
Face Brick with Concrete Block Back-up	Steel Joists	88.40	85.10	81.20	77.80	76.45	74.60	72.05	70.95	70.30
	Wood Joists	87.25	83.90	79.95	76.45	75.10	73.15	70.55	69.45	68.75
Stone with Concrete Block Back-up	Steel Joists	90.05	86.65	82.55	78.95	77.60	75.65	72.95	71.80	71.10
	Wood Joists	88.95	85.45	81.25	77.60	76.20	74.20	71.45	70.25	69.55
Brick Veneer	Wood Frame	85.40	82.35	78.75	75.65	74.40	72.65	70.35	69.30	68.70
Wood Siding	Wood Frame	81.75	79.00	75.80	73.10	72.00	70.45	68.45	67.55	67.00
Perimeter Adj., Add or Deduct	Per 100 L.F.	8.60	6.60	5.40	4.50	3.95	3.05	2.50	2.05	1.75
Story Hgt. Adj., Add or Deduct	Per 1 Ft.	1.60	1.45	1.30	1.10	1.05	.95	.85	.80	.75

For Basement, add $16.65 per square foot of basement area

The above costs were calculated using the basic specifications shown on the facing page. These costs should be adjusted where necessary for design alternatives and owner's requirements. Reported completed project costs, for this type of structure, range from $50.05 to $141.30 per S.F.

Common additives

Description	Unit	$ Cost
Directory Boards, Plastic, glass covered		
30" x 20"	Each	500
36" x 48"	Each	895
Aluminum, 24" x 18"	Each	440
36" x 24"	Each	530
48" x 32"	Each	630
48" x 60"	Each	1375
Emergency Lighting, 25 watt, battery operated		
Lead battery	Each	249
Nickel cadmium	Each	605
Flagpoles, Complete		
Aluminum, 20' high	Each	1000
40' high	Each	2550
70' high	Each	7350
Fiberglass, 23' high	Each	1275
39'-5" high	Each	2575
59' high	Each	6775
Safe, Office type, 4 hour rating		
30" x 18" x 18"	Each	3925
62" x 33" x 20"	Each	8525

Description	Unit	$ Cost
Smoke Detectors		
Ceiling type	Each	129
Duct type	Each	365
Vault Front, Door & frame		
1 Hour test, 32" x 78"	Opening	3350
2 Hour test, 32" door	Opening	3975
40" door	Opening	4300
4 Hour test, 32" door	Opening	4050
40" door	Opening	4800
Time lock movement; two movement	Each	1575

Model costs calculated for a 1 story building with 12' story height and 11,000 square feet of floor area

Town Hall, 1 Story

				Unit	Unit Cost	Cost Per S.F.	% Of Sub-Total
1.0 Foundations							
.1	Footings & Foundations	Poured concrete; strip and spread footings and 4' foundation wall		S.F. Ground	2.46	2.46	
.4	Piles & Caissons	N/A		—	—	—	6.1%
.9	Excavation & Backfill	Site preparation for slab and trench for foundation wall and footing		S.F. Ground	.96	.96	
2.0 Substructure							
.1	Slab on Grade	4" reinforced concrete with vapor barrier and granular base		S.F. Slab	2.80	2.80	5.0%
.2	Special Substructures	N/A		—	—	—	
3.0 Superstructure							
.1	Columns & Beams	Steel columns included in 3.7		—	—	—	
.4	Structural Walls	N/A		—	—	—	
.5	Elevated Floors	N/A		—	—	—	6.3%
.7	Roof	Metal deck, open web steel joists, beams, interior columns		S.F. Roof	3.54	3.54	
.9	Stairs	N/A		—	—	—	
4.0 Exterior Closure							
.1	Walls	Face brick with concrete block backup	70% of wall	S.F. Wall	17.04	5.66	
.5	Exterior Wall Finishes	N/A		—	—	—	15.9%
.6	Doors	Metal and glass with transom		Each	1466	.53	
.7	Windows & Glazed Walls	Metal outward projecting	30% of wall	Each	446	2.76	
5.0 Roofing							
.1	Roof Coverings	Built-up tar and gravel with flashing		S.F. Roof	2.05	2.05	
.7	Insulation	Perlite/EPS composite		S.F. Roof	1.18	1.18	6.1%
.8	Openings & Specialties	Gravel stop and hatches		S.F. Roof	.20	.20	
6.0 Interior Construction							
.1	Partitions	Gypsum board on metal studs, toilet partition	20 S.F. Floor/L.F. Partition	S.F. Partition	3.62	2.18	
.4	Interior Doors	Wood solid core	200 S.F. Floor/Door	Each	386	1.93	
.5	Wall Finishes	90% paint, 10% ceramic tile		S.F. Surface	.89	.89	27.8%
.6	Floor Finishes	70% carpet, 15% terrazzo, 15% vinyl composition tile		S.F. Floor	6.67	6.67	
.7	Ceiling Finishes	Mineral fiber tile on concealed zee bars		S.F. Ceiling	3.10	3.10	
.9	Interior Surface/Exterior Wall	Painted gypsum board on furring	70% of wall	S.F. Wall	2.47	.82	
7.0 Conveying							
.1	Elevators	N/A		—	—	—	0.0%
.2	Special Conveyors	N/A		—	—	—	
8.0 Mechanical							
.1	Plumbing	Kitchen, toilet and service fixtures, supply and drainage	1 Fixture/500 S.F. Floor	Each	2030	4.06	
.2	Fire Protection	Wet pipe sprinkler system		S.F. Floor	1.33	1.33	
.3	Heating	Included in 8.4		—	—	—	20.8%
.4	Cooling	Multizone unit, gas heating, electric cooling		S.F. Floor	6.28	6.28	
.5	Special Systems	N/A		—	—	—	
9.0 Electrical							
.1	Service & Distribution	400 ampere service, panel board and feeders		S.F. Floor	1.02	1.02	
.2	Lighting & Power	Fluorescent fixtures, receptacles, switches, A.C. and misc. power		S.F. Floor	5.23	5.23	12.0%
.4	Special Electrical	Alarm systems and emergency lighting		S.F. Floor	.47	.47	
11.0 Special Construction							
.1	Specialties	N/A		—	—	—	0.0%
12.0 Site Work							
.1	Earthwork	N/A		—	—	—	
.3	Utilities	N/A		—	—	—	0.0%
.5	Roads & Parking	N/A		—	—	—	
.7	Site Improvements	N/A		—	—	—	
				Sub-Total		56.12	100%
	CONTRACTOR FEES (General Requirements: 10%, Overhead: 5%, Profit: 10%)				25%	14.03	
	ARCHITECT FEES				9%	6.30	
				Total Building Cost		76.45	

COMMERCIAL/INDUSTRIAL/INSTITUTIONAL

M.680 Town Hall, 2-3 Story

Costs per square foot of floor area

Exterior Wall	S.F. Area	8000	10000	12000	15000	18000	24000	28000	35000	40000
	L.F. Perimeter	206	233	260	300	320	360	393	451	493
Face Brick with Concrete Block Back-up	Steel Frame	105.75	100.80	97.45	94.10	90.70	86.45	84.85	82.95	82.00
	R/Conc. Frame	106.20	101.25	97.95	94.60	91.20	86.90	85.30	83.40	82.50
Stone with Concrete Block Back-up	Steel Frame	101.55	97.20	94.25	91.30	88.30	84.45	83.00	81.30	80.50
	R/Conc. Frame	108.40	103.25	99.75	96.30	92.70	88.15	86.50	84.50	83.55
Limestone with Concrete Block Back-up	Steel Frame	117.15	111.15	107.05	103.00	98.60	93.10	91.05	88.65	87.45
	R/Conc. Frame	117.60	111.60	107.50	103.45	99.05	93.55	91.50	89.10	87.95
Perimeter Adj., Add or Deduct	Per 100 L.F.	13.35	10.65	8.90	7.10	5.95	4.45	3.85	3.00	2.65
Story Hgt. Adj., Add or Deduct	Per 1 Ft.	2.05	1.85	1.75	1.60	1.40	1.20	1.10	1.00	1.00
For Basement, add $15.54 per square foot of basement area										

The above costs were calculated using the basic specifications shown on the facing page. These costs should be adjusted where necessary for design alternatives and owner's requirements. Reported completed project costs, for this type of structure, range from $50.05 to $141.30 per S.F.

Common additives

Description	Unit	$ Cost
Directory Boards, Plastic, glass covered		
30" x 20"	Each	500
36" x 48"	Each	895
Aluminum, 24" x 18"	Each	440
36" x 24"	Each	530
48" x 32"	Each	630
48" x 60"	Each	1375
Elevators, Hydraulic passenger, 2 stops		
1500# capacity	Each	39,600
2500# capacity	Each	40,700
3500# capacity	Each	44,500
Additional stop, add	Each	2700
Emergency Lighting, 25 watt, battery operated		
Lead battery	Each	249
Nickel cadmium	Each	605

Description	Unit	$ Cost
Flagpoles, Complete		
Aluminum, 20' high	Each	1000
40' high	Each	2550
70' high	Each	7350
Fiberglass, 23' high	Each	1275
39'-5" high	Each	2575
59' high	Each	6775
Safe, Office type, 4 hour rating		
30" x 18" x 18"	Each	3925
62" x 33" x 20"	Each	8525
Smoke Detectors		
Ceiling type	Each	129
Duct type	Each	365
Vault Front, Door & frame		
1 Hour test, 32" x 78"	Opening	3350
2 Hour test, 32" door	Opening	3975
40" door	Opening	4300
4 Hour test, 32" door	Opening	4050
40" door	Opening	4800
Time lock movement; two movement	Each	1575

Important: See the Reference Section for Location Factors

Model costs calculated for a 3 story building with 12' story height and 18,000 square feet of floor area

Town Hall, 2-3 Story

BUILDING TYPES

				Unit	Unit Cost	Cost Per S.F.	% Of Sub-Total
1.0 Foundations							
.1	Footings & Foundations	Poured concrete; strip and spread footings and 4' foundation wall		S.F. Ground	4.56	1.52	2.8%
.4	Piles & Caissons	N/A		—	—	—	
.9	Excavation & Backfill	Site preparation for slab and trench for foundation wall and footing		S.F. Ground	.89	.30	
2.0 Substructure							
.1	Slab on Grade	4" reinforced concrete with vapor barrier and granular base		S.F. Slab	2.80	.93	1.4%
.2	Special Substructures	N/A		—	—	—	
3.0 Superstructure							
.1	Columns & Beams	Wide flange steel columns with gypsum board fireproofing		L.F. Column	44	1.34	
.4	Structural Walls	N/A		—	—	—	
.5	Elevated Floors	Open web steel joists, slab form, concrete		S.F. Floor	10.69	7.13	18.0%
.7	Roof	Metal deck, open web steel joists, beams, interior columns		S.F. Roof	4.16	1.39	
.9	Stairs	Concrete filled metal pan		Flight	5425	1.81	
4.0 Exterior Closure							
.1	Walls	Stone with concrete block backup	70% of wall	S.F. Wall	19.51	8.74	
.5	Exterior Wall Finishes	N/A		—	—	—	16.3%
.6	Doors	Metal and glass with transoms		Each	1371	.38	
.7	Windows & Glazed Walls	Metal outward projecting	10% of wall	Each	617	1.43	
5.0 Roofing							
.1	Roof Coverings	Built-up tar and gravel with flashing		S.F. Roof	2.31	.77	
.7	Insulation	Perlite/EPS composite		S.F. Roof	1.18	.39	1.8%
.8	Openings & Specialties	N/A		—	—	—	
6.0 Interior Construction							
.1	Partitions	Gypsum board on metal studs, toilet partitions	20 S.F. Floor/L.F. Partition	S.F. Partition	3.62	2.04	
.4	Interior Doors	Wood solid core	200 S.F. Floor/Door	Each	386	1.93	
.5	Wall Finishes	90% paint, 10% ceramic tile		S.F. Surface	.89	.89	24.3%
.6	Floor Finishes	70% carpet, 15% terrazzo, 15% vinyl composition tile		S.F. Floor	6.67	6.67	
.7	Ceiling Finishes	Mineral fiber tile on concealed zee bars		S.F. Ceiling	3.10	3.10	
.9	Interior Surface/Exterior Wall	Painted gypsum board on furring	70% of wall	S.F. Wall	2.47	1.11	
7.0 Conveying							
.1	Elevators	Two hydraulic elevators		Each	61,560	6.84	10.6%
.2	Special Conveyors	N/A		—	—	—	
8.0 Mechanical							
.1	Plumbing	Toilet and service fixtures, supply and drainage	1 Fixture/1385 S.F. Floor	Each	1786	1.29	
.2	Fire Protection	Sprinklers, light hazard		S.F. Floor	1.26	1.26	
.3	Heating	Included in 8.4		—	—	—	13.6%
.4	Cooling	Multizone unit, gas heating, electric cooling		S.F. Floor	6.28	6.28	
.5	Special Systems	N/A		—	—	—	
9.0 Electrical							
.1	Service & Distribution	400 ampere service, panel board and feeders		S.F. Floor	.87	.87	
.2	Lighting & Power	Fluorescent fixtures, receptacles, switches, A.C. and misc. power		S.F. Floor	5.89	5.89	11.2%
.4	Special Electrical	Alarm systems and emergency lighting		S.F. Floor	.49	.49	
11.0 Special Construction							
.1	Specialties	N/A		—	—	—	0.0%
12.0 Site Work							
.1	Earthwork	N/A		—	—	—	
.3	Utilities	N/A		—	—	—	0.0%
.5	Roads & Parking	N/A		—	—	—	
.7	Site Improvements	N/A		—	—	—	
				Sub-Total		64.79	**100%**
	CONTRACTOR FEES (General Requirements: 10%, Overhead: 5%, Profit: 10%)				25%	16.20	
	ARCHITECT FEES				9%	7.31	
				Total Building Cost		**88.30**	

COMMERCIAL/INDUSTRIAL/INSTITUTIONAL   M.690   Warehouse

Costs per square foot of floor area

Exterior Wall	S.F. Area	10000	15000	20000	25000	30000	35000	40000	50000	60000
	L.F. Perimeter	410	500	600	700	700	766	833	966	1000
Brick with Concrete Block Back-up	Steel Frame	69.45	62.40	59.20	57.30	53.75	52.45	51.55	50.25	48.20
	Bearing Walls	68.85	61.60	58.30	56.35	52.70	51.40	50.50	49.15	47.10
Concrete Block	Steel Frame	57.15	52.30	50.10	48.80	46.65	45.80	45.20	44.35	43.15
	Bearing Walls	56.25	51.35	49.10	47.80	45.60	44.75	44.10	43.25	42.05
Galvanized Steel Siding	Steel Frame	59.40	54.60	52.35	51.05	48.90	48.10	47.45	46.65	45.40
Metal Sandwich Panels	Steel Frame	59.90	54.60	52.15	50.70	48.25	47.30	46.60	45.65	44.25
Perimeter Adj., Add or Deduct	Per 100 L.F.	6.95	4.65	3.50	2.75	2.30	1.95	1.75	1.40	1.20
Story Hgt. Adj., Add or Deduct	Per 1 Ft.	1.00	.00	.70	.65	.55	.50	.50	.45	.40

For Basement, add $15.46 per square foot of basement area

The above costs were calculated using the basic specifications shown on the facing page. These costs should be adjusted where necessary for design alternatives and owner's requirements. Reported completed project costs, for this type of structure, range from $21.30 to $85.15 per S.F.

Common additives

Description	Unit	$ Cost
Dock Leveler, 10 ton cap.		
6' x 8'	Each	4075
7' x 8'	Each	4200
Emergency Lighting, 25 watt, battery operated		
Lead battery	Each	249
Nickel cadmium	Each	605
Fence, Chain link, 6' high		
9 ga. wire	L.F.	13.05
6 ga. wire	L.F.	17.95
Gate	Each	223
Flagpoles, Complete		
Aluminum, 20' high	Each	1000
40' high	Each	2550
70' high	Each	7350
Fiberglass, 23' high	Each	1275
39'-5" high	Each	2575
59' high	Each	6775
Paving, Bituminous		
Wearing course plus base course	S.Y.	7.60
Sidewalks, Concrete 4" thick	S.F.	2.34

Description	Unit	$ Cost
Sound System		
Amplifier, 250 watts	Each	1500
Speaker, ceiling or wall	Each	125
Trumpet	Each	235
Yard Lighting, 20' aluminum pole with 400 watt high pressure sodium fixture.	Each	1870

Important: See the Reference Section for Location Factors

Warehouse

Model costs calculated for a 1 story building with 24' story height and 30,000 square feet of floor area

				Unit	Unit Cost	Cost Per S.F.	% Of Sub-Total
1.0 Foundations							
.1	Footings & Foundations	Poured concrete; strip and spread footings and 4' foundation wall		S.F. Ground	1.97	1.97	8.4%
.4	Piles & Caissons	N/A		—	—	—	
.9	Excavation & Backfill	Site preparation for slab and trench for foundation wall and footing		S.F. Ground	.89	.89	
2.0 Substructure							
.1	Slab on Grade	5" reinforced concrete with vapor barrier and granular base		S.F. Slab	5.67	5.67	16.6%
.2	Special Substructures	N/A		—	—	—	
3.0 Superstructure							
.1	Columns & Beams	Steel columns included in 3.5 and 3.7		—	—	—	
.4	Structural Walls	N/A		—	—	—	
.5	Elevated Floors	Open web steel joists, slab form, concrete beams, columns, (above office)	10% of area	S.F. Floor	8.43	1.10	13.6%
.7	Roof	Metal deck, open web steel joists, beams, columns		S.F. Roof	3.26	3.26	
.9	Stairs	Steel gate with rails		Flight	4300	.29	
4.0 Exterior Closure							
.1	Walls	Concrete block	95% of wall	S.F. Wall	6.92	3.68	12.5%
.5	Exterior Wall Finishes	N/A		—	—	—	
.6	Doors	Steel overhead, hollow metal	5% of wall	Each	1588	.58	
.7	Windows & Glazed Walls	N/A		—	—	—	
5.0 Roofing							
.1	Roof Coverings	Built-up tar and gravel with flashing		S.F. Roof	1.71	1.71	9.3%
.7	Insulation	Perlite/EPS composite		S.F. Roof	1.18	1.18	
.8	Openings & Specialties	Gravel stop, hatches and skylight		S.F. Roof	.29	.29	
6.0 Interior Construction							
.1	Partitions	Concrete block (office and washrooms)	100 S.F. Floor/L.F. Partition	S.F. Partition	4.63	.37	8.4%
.4	Interior Doors	Single leaf hollow metal	5000 S.F. Floor/Door	Each	492	.10	
.5	Wall Finishes	Paint		S.F. Surface	.81	.13	
.6	Floor Finishes	90% hardener, 10% vinyl composition tile		S.F. Floor	1.05	1.05	
.7	Ceiling Finishes	Suspended mineral tile on zee channels in office area	10% of area	S.F. Ceiling	3.10	.31	
.9	Interior Surface/Exterior Wall	Paint	95% of wall	S.F. Wall	1.69	.90	
7.0 Conveying							
.1	Elevators	N/A		—	—	—	0.0%
.2	Special Conveyors	N/A		—	—	—	
8.0 Mechanical							
.1	Plumbing	Toilet and service fixtures, supply and drainage	1 Fixture/2500 S.F. Floor	Each	2950	1.18	17.3%
.2	Fire Protection	Sprinklers, ordinary hazard		S.F. Floor	1.55	1.55	
.3	Heating	Oil fired hot water, unit heaters	90% of area	S.F. Floor	2.53	2.53	
.4	Cooling	Single zone unit gas, heating, electric cooling	10% of area	S.F. Floor	.63	.63	
.5	Special Systems	N/A		—	—	—	
9.0 Electrical							
.1	Service & Distribution	200 ampere service, panel board and feeders		S.F. Floor	.27	.27	8.9%
.2	Lighting & Power	Fluorescent fixtures, receptacles, switches, A.C. and misc. power		S.F. Floor	2.51	2.51	
.4	Special Electrical	Alarm systems		S.F. Floor	.24	.24	
11.0 Special Construction							
.1	Specialties	Dock boards, dock levelers		S.F. Floor	1.69	1.69	5.0%
12.0 Site Work							
.1	Earthwork	N/A		—	—	—	0.0%
.3	Utilities	N/A		—	—	—	
.5	Roads & Parking	N/A		—	—	—	
.7	Site Improvements	N/A		—	—	—	
				Sub-Total		34.08	100%
	CONTRACTOR FEES (General Requirements: 10%, Overhead: 5%, Profit: 10%)				25%	8.52	
	ARCHITECT FEES				7%	3.00	
				Total Building Cost		45.60	

BUILDING TYPES

COMMERCIAL/INDUSTRIAL/INSTITUTIONAL   M.700   Warehouse, Mini

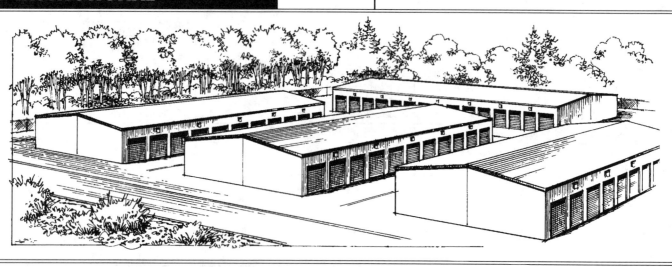

Costs per square foot of floor area

Exterior Wall	S.F. Area	2000	3000	5000	8000	12000	20000	30000	50000	100000
	L.F. Perimeter	179	220	284	350	438	566	693	894	1266
Concrete Block	Steel Frame	100.85	87.95	76.30	68.30	63.85	59.35	56.70	54.20	51.80
	R/Conc. Frame	86.35	79.25	72.60	67.85	65.15	62.45	60.80	59.15	57.60
Metal Sandwich Panel	Steel Frame	79.80	72.40	65.45	60.50	57.75	54.90	53.15	51.50	49.85
Tilt-up Concrete Panel	R/Conc. Frame	90.00	82.65	75.70	70.75	68.05	65.15	63.45	61.75	60.05
Precast Concrete Panel	Steel Frame	80.80	73.15	65.90	60.70	57.80	54.80	53.00	51.20	49.40
	Bearing Wall	94.65	86.35	78.45	72.75	69.60	66.25	64.25	62.25	60.25
Perimeter Adj., Add or Deduct	Per 100 L.F.	19.10	12.75	7.65	4.75	3.15	1.90	1.25	.75	.40
Story Hgt. Adj., Add or Deduct	Per 1 Ft	1.65	1.35	1.05	.80	.65	.50	.45	.35	.25
Basement—Not Applicable											

The above costs were calculated using the basic specifications shown on the facing page. These costs should be adjusted where necessary for design alternatives and owner's requirements. Reported completed project costs, for this type of structure, range from $20.50 to $117.55 per S.F.

Common additives

Description	Unit	$ Cost
Dock Leveler, 10 ton cap.		
6' x 8'	Each	4075
7' x 8'	Each	4200
Emergency Lighting, 25 watt, battery operated		
Lead battery	Each	249
Nickel cadmium	Each	605
Fence, Chain link, 6' high		
9 ga. wire	L.F.	13.05
6 ga. wire	L.F.	17.95
Gate	Each	223
Flagpoles, Complete		
Aluminum, 20' high	Each	1000
40' high	Each	2550
70' high	Each	7350
Fiberglass, 23' high	Each	1275
39'-5" high	Each	2575
59' high	Each	6775
Paving, Bituminous		
Wearing course plus base course	S.Y.	7.60
Sidewalks, Concrete 4" thick	S.F.	2.34

Description	Unit	$ Cost
Sound System		
Amplifier, 250 watts	Each	1500
Speaker, ceiling or wall	Each	125
Trumpet	Each	235
Yard Lighting,	Each	1870
20' aluminum pole with		
400 watt high pressure		
sodium fixture		

Important: See the Reference Section for Location Factors

Model costs calculated for a 1 story building with 12' story height and 20,000 square feet of floor area

Warehouse, Mini

				Unit	Unit Cost	Cost Per S.F.	% Of Sub-Total
1.0 Foundations							
.1	Footings & Foundations	Poured concrete; strip and spread footings and 4' foundation wall		S.F. Ground	.57	.57	3.4%
.4	Piles & Caissons	N/A		—	—	—	
.9	Excavation & Backfill	Site preparation for slab and trench for foundation wall and footing		S.F. Ground	.96	.96	
2.0 Substructure							
.1	Slab on Grade	4" reinforced concrete with vapor barrier and granular base		S.F. Slab	4.72	4.72	10.6%
.2	Special Substructures	N/A		—	—	—	
3.0 Superstructure							
.1	Columns & Beams	Column fireproofing, steel columns included in 3.7		S.F. Floor	2.48	2.48	
.4	Structural Walls	N/A		—	—	—	
.5	Elevated Floors	N/A		—	—	—	17.1%
.7	Roof	Metal deck, open web steel joists, beams, columns		S.F. Roof	5.10	5.10	
.9	Stairs	N/A		—	—	—	
4.0 Exterior Closure							
.1	Walls	Concrete block	70% of wall	S.F. Wall	7.28	1.73	
.5	Exterior Wall Finishes	N/A		—	—	—	14.7%
.6	Doors	Steel overhead, hollow metal	25% of wall	Each	843	1.85	
.7	Windows & Glazed Walls	Aluminum projecting	5% of wall	Each	523	2.96	
5.0 Roofing							
.1	Roof Coverings	Built-up tar and gravel with flashing		S.F. Roof	1.72	1.72	
.7	Insulation	Perlite/EPS composite		S.F. Roof	.69	.69	5.4%
.8	Openings & Specialties	N/A		—	—	—	
6.0 Interior Construction							
.1	Partitions	Concrete block, gypsum board on metal studs	10.65 S.F. Floor/L.F. Partition	S.F. Partition	4.54	7.78	
.4	Interior Doors	Single leaf hollow metal	1835 S.F. Floor/Door	Each	804	2.68	
.5	Wall Finishes	Paint		S.F. Floor	—	—	23.6%
.6	Floor Finishes	Carpet, asphalt tile, ceramic tile		S.F. Floor	—	—	
.7	Ceiling Finishes	N/A		—	—	—	
.9	Interior Surface/Exterior Wall	N/A		—	—	—	
7.0 Conveying							
.1	Elevators	N/A		—	—	—	0.0%
.2	Special Conveyors	N/A		—	—	—	
8.0 Mechanical							
.1	Plumbing	Toilet and service fixtures, supply and drainage	1 Fixture/5000 S.F. Floor	Each	4300	.86	
.2	Fire Protection	Wet pipe sprinkler system		S.F. Floor	1.74	1.74	
.3	Heating	Oil fired hot water, unit heaters		S.F. Floor	4.46	4.46	16.0%
.4	Cooling	N/A		—	—	—	
.5	Special Systems	N/A		—	—	—	
9.0 Electrical							
.1	Service & Distribution	200 ampere service, panel board and feeders		S.F. Floor	.86	.86	
.2	Lighting & Power	Fluorescent fixtures, receptacles, switches and misc. power		S.F. Floor	2.92	2.92	9.2%
.4	Special Electrical	Alarm systems		S.F. Floor	.31	.31	
11.0 Special Construction							
.1	Specialties	N/A		—	—	—	0.0%
12.0 Site Work							
.1	Earthwork	N/A		—	—	—	
.3	Utilities	N/A		—	—	—	0.0%
.5	Roads & Parking	N/A		—	—	—	
.7	Site Improvements	N/A		—	—	—	
				Sub-Total		44.39	100%
	CONTRACTOR FEES (General Requirements: 10%, Overhead: 5%, Profit: 10%)				25%	11.10	
	ARCHITECT FEES				7%	3.86	
				Total Building Cost		59.35	

Assemblies Section

Table of Contents

Table No.	Page
How to Use Assemblies	94

Foundations — 97
Footings & Foundations
- 1.1-120 Spread Footings ... 98
- 1.1-140 Strip Footings ... 98
- 1.1-210 Walls—Cast in Place ... 99
- 1.1-292 Foundation Dampproofing ... 99

Excavation & Backfill
- 1.9-100 Building Excavation & Backfill ... 100

Substructures — 101
Slab on Grade
- 2.1-200 Plain & Reinforced ... 102

Superstructures — 103
Columns, Beams & Joists
- 3.1-114 C.I.P. Columns—Square Tied ... 104
- 3.1-120 Precast Concrete Columns ... 106
- 3.1-130 Steel Columns ... 107
- 3.1-140 Wood Columns ... 110
- 3.1-190 Steel Column Fireproofing ... 111
- 3.1-224 "T" Shaped Precast Beams ... 111
- 3.1-226 "L" Shaped Precast Beams ... 111

Structural Walls
- 3.4-300 Metal Siding Support ... 112

Floor Superstructures
- 3.5-150 C.I.P. Flat Plate ... 113
- 3.5-160 C.I.P. Multispan Joist Slab ... 113
- 3.5-210 Precast Plank ... 114
- 3.5-230 Precast Double "T" Beams ... 115
- 3.5-360 Light Gauge Steel Floor Systems ... 116
- 3.5-420 Deck & Joists on Bearing Walls ... 117
- 3.5-440 Steel Joists on Beam & Wall ... 118
- 3.5-460 Steel Joists, Beams & Slab on Columns ... 119
- 3.5-580 Metal Deck/Concrete Fill ... 121
- 3.5-710 Wood Joist ... 121
- 3.5-720 Wood Beam & Joist ... 121

Roof Superstructures
- 3.7-410 Steel Joists, Beams & Deck on Columns & Walls ... 122
- 3.7-420 Steel Joists, Beams & Deck on Columns ... 123
- 3.7-430 Steel Joists & Deck on Bearing Walls ... 124
- 3.7-440 Steel Joists & Joist Girders on Columns & Walls ... 125
- 3.7-450 Steel Joists & Joist Girders on Columns ... 126
- 3.7-510 Wood/Flat or Pitched ... 127

Stairs
- 3.9-100 Stairs ... 128

Exterior Closure — 129
Walls
- 4.1-110 Cast in Place Concrete ... 130
- 4.1-140 Flat Precast Concrete ... 131
- 4.1-140 Fluted Window or Mullion Precast Concrete ... 132
- 4.1-140 Precast Concrete Specialties ... 132
- 4.1-140 Ribbed Precast Concrete ... 134
- 4.1-160 Tilt Up Concrete Panel ... 135
- 4.1-211 Concrete Block Wall ... 136
- 4.1-212 Split Ribbed Block Wall ... 139
- 4.1-213 Split Face Block Wall ... 140
- 4.1-242 Stone Veneer ... 142
- 4.1-252 Brick Veneer/Wood Stud Backup ... 144
- 4.1-252 Brick Veneer/Metal Stud Backup ... 145
- 4.1-273 Brick Face Cavity Wall ... 147
- 4.1-282 Glass Block ... 152
- 4.1-384 Metal Siding Panel ... 153
- 4.1-412 Wood & Other Siding ... 154

Exterior Wall Finishes
- 4.5-110 Stucco Wall ... 156

Doors
- 4.6-100 Wood, Steel & Aluminum ... 157

Windows & Glazed Walls
- 4.7-110 Wood, Steel & Aluminum ... 160
- 4.7-582 Tubular Alum. Framing ... 162
- 4.7-584 Curtain Wall Panels ... 163

Roofing — 165
Roof Covers
- 5.1-103 Built Up ... 166
- 5.1-220 Single Ply Membrane ... 167
- 5.1-310 Preformed Metal ... 167
- 5.1-330 Formed Metal ... 167
- 5.1-410 Shingle & Tile ... 168
- 5.1-520 Roof Edges ... 168
- 5.1-620 Flashings ... 169

Insulation
- 5.7-101 Roof Deck Rigid Insulation ... 17?

Openings & Specialties
- 5.8-100 Hatches ... 17?
- 5.8-100 Skylights ... 17?
- 5.8-400 Gutters ... 17?
- 5.8-500 Downspouts ... 17?
- 5.8-500 Gravel Stop ... 17?

Interior Construction — 17?
Partitions
- 6.1-210 Concrete Block Partitions ... 17?
- 6.1-510 Drywall Partitions ... 17?
- 6.1-580 Drywall Components ... 17?
- 6.1-680 Plaster Partitions/Components ... 17?
- 6.1-820 Folding Partitions ... 18?
- 6.1-870 Toilet Partitions ... 18?

Doors
- 6.4-100 Special Doors ... 18?

Wall Finishes
- 6.5-100 Paint & Covering ... 18?
- 6.5-100 Paint Trim ... 18?

Floor Finishes
- 6.6-100 Tile & Covering ... 18?

Ceiling Finishes
- 6.7-100 Plaster Ceilings ... 18?
- 6.7-100 Drywall Ceilings ... 18?
- 6.7-100 Acoustical Ceilings ... 18?
- 6.7-810 Acoustical Ceilings ... 18?
- 6.7-820 Plaster Ceilings ... 18?

Conveying Systems — 19?
Elevators
- 7.1-100 Hydraulic ... 19?

Mechanical — 19?
Plumbing
- 8.1-120 Gas Fired Water Heaters–Residential ... 19?
- 8.1-130 Oil Fired Water Heaters–Residential ... 19?
- 8.1-160 Electric Water Heaters–Commercial ... 19?
- 8.1-170 Gas Fired Water Heaters–Commercial ... 19?
- 8.1-180 Oil Fired Water Heaters–Commercial ... 19?
- 8.1-310 Roof Drain Systems ... 19?

Table of Contents

Table No.		Page
Reference	Min. Plumbing Fixture Requirements	199
8.1-410	Bathtub Systems	200
8.1-420	Drinking Fountain Systems	200
8.1-431	Kitchen Sink Systems	201
8.1-432	Laundry Sink Systems	201
8.1-433	Lavatory Systems	202
8.1-434	Laboratory Sink Systems	202
8.1-434	Service Sink Systems	202
8.1-440	Shower Systems	203
8.1-450	Urinal Systems	203
8.1-460	Water Cooler Systems	203
8.1-470	Water Closet Systems	204
8.1-510	Water Closets–Group Systems	204
8.1-560	Group Wash Fountain Systems	204
8.1-620	Two Fixture Bathroom	205
8.1-630	Three Fixture Bathroom	205

Fire Protection

Reference	Sprinkler System Types	206
Reference	Sprinkler System Classification	207
8.2-110	Wet Pipe Sprinkler Systems	208
8.2-120	Dry Pipe Sprinkler Systems	210
8.2-310	Wet Standpipe Risers	212
8.2-320	Dry Standpipe Risers	213
8.2-390	Standpipe Equipment	214
8.2-810	FM200 Fire Suppression	215

Heat Transfer

8.3-011	Factor for Determining Heat Loss	216
8.3-012	Outside Design Temp. Correction Factor	216
8.3-013	Transmission of Heat	217
8.3-014	Transmission of Heat (Low Rate)	217

Heating

8.3-110	Small Hydronic Electric Boilers	218
8.3-120	Large Hydronic Electric Boilers	218
8.3-130	Boilers–Hot Water & Steam	219
8.3-141	Hydronic Heating–Fossil Fuel Unit Heaters	221
8.3-142	Hydronic Heating–Fossil Fuel Fin Tube Radiation	222
8.3-151	Apartment Bldg. Heating–Fin Tube Radiation	223
8.3-161	Commercial Bldg. Heating–Fin Tube Radiation	223
8.3-162	Comm. Bldg. Heating–Terminal Unit Heaters	223

Cooling

8.4-001	Introduction	224
8.4-002	Air Conditioning Requirements	224
8.4-004	Recommended Ventilation Air Changes	225
8.4-005	Ductwork	225
8.4-006	Diffuser Evaluation	225
8.4-009	Sheet Metal Calculator	226
8.4-110	Chilled Water–Air Cooled Condenser	227
8.4-120	Chilled Water–Cooling Tower	229
8.4-210	Rooftop Single Zone Units	231
8.4-230	Self-Contained Water Cooled	233
8.4-240	Self-Contained Air Cooled	235
8.4-250	Split System/Air Cooled Condensing Units	237

Special Systems

8.5-110	Garage Exhaust System	239

Electrical 241

Service & Distribution

9.1-210	Electric Service	242
9.1-310	Feeder Installation	242
9.1-410	Switchgear	242

Lighting & Power

9.2-213	Fluorescent Fixtures (by Wattage)	243
9.2-223	Incandescent Fixtures (by Wattage)	243
9.2-235	H.I.D. Fixture–High Bay (by Wattage)	243
9.2-239	H.I.D. Fixture–High Bay (by Wattage)	244
9.2-242	H.I.D. Fixture–Low Bay (by Wattage)	245
9.2-244	H.I.D. Fixture–Low Bay (by Wattage)	245
9.2-252	Light Poles (Installed)	246
9.2-522	Receptacle (by Wattage)	247
9.2-524	Receptacles	247
9.2-542	Wall Switch by Square Foot	248
9.2-582	Miscellaneous Power	248
9.2-610	Central A.C. Power (by Wattage)	248
9.2-710	Motor Installation	249
9.2-720	Motor Feeder	250

Special Electrical

9.4-100	Communication & Alarm Systems	251
9.4-310	Generators (by KW)	252

Special Construction 253

Specialties

11.1-100	Architectural Specialties	254
11.1-200	Architectural Equipment	255
11.1-500	Furnishings	258
11.1-700	Special Construction	260

Site Work 263

Utilities

12.3-110	Trenching	264
12.3-310	Pipe Bedding	264
12.3-710	Manholes & Catch Basins	264

Roads & Parking

12.5-110	Roadway Pavement	265
12.5-510	Parking Lots	265

How to Use the Assemblies Cost Tables

The following is a detailed explanation of a sample Assemblies Cost Table. Most Assembly Tables are separated into three parts: 1) an illustration of the system to be estimated; 2) the components and related costs of a typical system; and 3) the costs for similar systems with dimensional and/or size variations. For costs of the components that comprise these systems or "assemblies" refer to the Unit Price Section. Next to each bold number below is the item being described with the appropriate component of the sample entry following in parenthesis. In most cases, if the work is to be subcontracted, the general contractor will need to add an additional markup (R.S. Means suggests using 10%) to the "Total" figures.

System/Line Numbers (A6.1-510-2400)

Each Assemblies Cost Line has been assigned a unique identification number based on the UniFormat classification system.

```
         UniFormat Division
              ┌─┴─┐
              6.1 510 2400
               │   │    │
   Means Subdivision   │
       Means Major Classification
           Means Individual Line Number
```

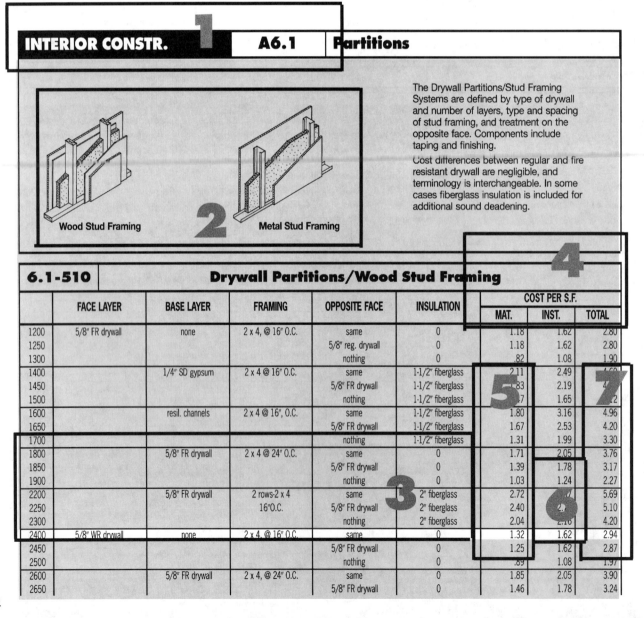

2 Illustration
At the top of most assembly pages is an illustration, a brief description, and the design criteria used to develop the cost.

3 System Description
The components of a typical system are listed in the description to show what has been included in the development of the total system price. The rest of the table contains prices for other similar systems with dimensional and/or size variations.

4 Unit of Measure for Each System (S.F.)
Costs shown in the three right hand columns have been adjusted by the component quantity and unit of measure for the entire system. In this example, "S.F. Cost" is the unit of measure for this system or "assembly."

5 Materials (1.32)
This column contains the Materials Cost of each component. These cost figures are bare costs plus 10% for profit.

6 Installation (1.62)
Installation includes labor and equipment plus the installing contractor's overhead and profit. Equipment costs are the bare rental costs plus 10% for profit. The labor overhead and profit is defined on the inside back cover of this book.

7 Total (2.94)
The figure in this column is the sum of the material and installation costs.

Material Cost	+	Installation Cost	=	Total
$1.32	+	$1.62	=	$2.94

Division 1
Foundations

FOUNDATIONS — A1.1 Footings & Foundations

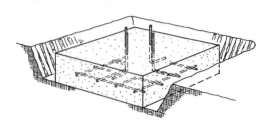

The Spread Footing System includes: excavation; backfill; forms (four uses); all reinforcement; 3,000 p.s.i. concrete (chute placed); and screed finish.

Footing systems are priced per individual unit. The Expanded System Listing at the bottom shows footings that range from 3' square x 12" deep, to 18' square x 52" deep. It is assumed that excavation is done by a truck mounted hydraulic excavator with an operator and oiler.

Backfill is with a dozer, and compaction by air tamp. The excavation and backfill equipment is assumed to operate at 30 C.Y. per hour.

Please see the reference section for further design and cost information.

1.1-120	Spread Footings	COST EACH		
		MAT.	INST.	TOTAL
7090	Spread footings, 3000 psi concrete, chute delivered			
7100	Load 25K, soil capacity 3 KSF, 3'-0" sq. x 12" deep	39.50	55	94.50
7150	Load 50K, soil capacity 3 KSF, 4'-6" sq. x 12" deep	81.50	96	177.50
7200	Load 50K, soil capacity 6 KSF, 3'-0" sq. x 12" deep	39.50	55	94.50
7250	Load 75K, soil capacity 3 KSF, 5'-6" sq. x 13" deep	127	137	264
7300	Load 75K, soil capacity 6 KSF, 4'-0" sq. x 12" deep	66.50	82	148.50
7350	Load 100K, soil capacity 3 KSF, 6'-0" sq. x 14" deep	160	164	324
7410	Load 100K, soil capacity 6 KSF, 4'-6" sq. x 15" deep	99.50	113	212.50
7450	Load 125K, soil capacity 3 KSF, 7'-0" sq. x 17" deep	252	234	486
7500	Load 125K, soil capacity 6 KSF, 5'-0" sq. x 16" deep	127	136	263
7550	Load 150K, soil capacity 3 KSF 7'-6" sq. x 18" deep	305	274	579
7610	Load 150K, soil capacity 6 KSF, 5'-6" sq. x 18" deep	168	172	340
7650	Load 200K, soil capacity 3 KSF, 8'-6" sq. x 20" deep	430	360	790
7700	Load 200K, soil capacity 6 KSF, 6'-0" sq. x 20" deep	219	211	430
7750	Load 300K, soil capacity 3 KSF, 10'-6" sq. x 25" deep	780	595	1,375
7810	Load 300K, soil capacity 6 KSF, 7'-6" sq. x 25" deep	410	355	765
7850	Load 400K, soil capacity 3 KSF, 12'-6" sq. x 28" deep	1,225	885	2,110
7900	Load 400K, soil capacity 6 KSF, 8'-6" sq. x 27" deep	570	465	1,035
8010	Load 500K, soil capacity 6 KSF, 9'-6" sq. x 30" deep	780	605	1,385
8100	Load 600K, soil capacity 6 KSF, 10'-6" sq. x 33" deep	1,050	780	1,830
8200	Load 700K, soil capacity 6 KSF, 11'-6" sq. x 36" deep	1,350	955	2,305
8300	Load 800K, soil capacity 6 KSF, 12'-0" sq. x 37" deep	1,500	1,050	2,550
8400	Load 900K, soil capacity 6 KSF, 13'-0" sq. x 39" deep	1,850	1,250	3,100
8500	Load 1000K, soil capacity 6 KSF, 13'-6" sq. x 41" deep	2,100	1,375	3,475

1.1-140	Strip Footings	COST PER L.F.		
		MAT.	INST.	TOTAL
2100	Strip footing, load 2.6KLF, soil capacity 3KSF, 16"wide x 8"deep plain	4.42	6.65	11.07
2300	Load 3.9 KLF, soil capacity, 3 KSF, 24"wide x 8"deep, plain	5.60	7.45	13.05
2500	Load 5.1KLF, soil capacity 3 KSF, 24"wide x 12"deep, reinf.	8.85	11.15	20
2700	Load 11.1KLF, soil capacity 6 KSF, 24"wide x 12"deep, reinf.	8.85	11.15	20
2900	Load 6.8 KLF, soil capacity 3 KSF, 32"wide x 12"deep, reinf.	10.95	12.25	23.20
3100	Load 14.8 KLF, soil capacity 6 KSF, 32"wide x 12"deep, reinf.	10.95	12.25	23.20
3300	Load 9.3 KLF, soil capacity 3 KSF, 40"wide x 12"deep, reinf.	12.95	13.35	26.30
3500	Load 18.4 KLF, soil capacity 6 KSF, 40"wide x 12"deep, reinf.	13.05	13.45	26.50
4500	Load 10KLF, soil capacity 3 KSF, 48"wide x 16"deep, reinf.	18.50	16.75	35.25
4700	Load 22KLF, soil capacity 6 KSF, 48"wide, 16"deep, reinf.	18.90	17.15	36.05
5700	Load 15KLF, soil capacity 3 KSF, 72"wide x 20"deep, reinf.	32.50	24	56.50
5900	Load 33KLF, soil capacity 6 KSF, 72"wide x 20"deep, reinf.	34.50	25.50	60

FOUNDATIONS — A1.1 Footings & Foundations

1.1-210 Walls, Cast in Place

	WALL HEIGHT (FT.)	PLACING METHOD	CONCRETE (C.Y./L.F.)	REINFORCING (LBS./L.F.)	WALL THICKNESS (IN.)	COST PER L.F. MAT.	COST PER L.F. INST.	COST PER L.F. TOTAL
1500	4'	direct chute	.074	3.3	6	11.50	21	32.50
1520			.099	4.8	8	13.65	21.50	35.15
1540			.123	6.0	10	15.70	22	37.70
1561			.148	7.2	12	17.75	22.50	40.25
1580			.173	8.1	14	19.75	23	42.75
1600			.197	9.44	16	22	23.50	45.50
3000	6'	direct chute	.111	4.95	6	17.25	31.50	48.75
3020			.149	7.20	8	20.50	32.50	53
3040			.184	9.00	10	23.50	33	56.50
3061			.222	10.8	12	26.50	34	60.50
5000	8'	direct chute	.148	6.6	6	23	42	65
5020			.199	9.6	8	27.50	43.50	71
5040			.250	12	10	31.50	44.50	76
5061			.296	14.39	12	35.50	46	81.50
6020	10'	direct chute	.248	12	8	34	54.50	88.50
6040			.307	14.99	10	39	55.50	94.50
6061			.370	17.99	12	44.50	57	101.50
7220	12'	pumped	.298	14.39	8	41	68.50	109.50
7240			.369	17.99	10	47	70.50	117.50
7262			.444	21.59	12	53.50	73	126.50
9220	16'	pumped	.397	19.19	8	55	91.50	146.50
9240			.492	23.99	10	62.50	93.50	156
9260			.593	28.79	12	71	97	168

1.1-292 Foundation Dampproofing

	Description	MAT.	INST.	TOTAL
1000	Foundation dampproofing, bituminous, 1 coat, 4' high	.28	2.11	2.39
1400	8' high	.56	4.23	4.79
1800	12' high	.84	6.55	7.39
2000	2 coats, 4' high	.44	2.63	3.07
2400	8' high	.88	5.25	6.13
2800	12' high	1.32	8.10	9.42
3000	Asphalt with fibers, 1/16" thick, 4' high	.68	2.63	3.31
3400	8' high	1.36	5.25	6.61
3800	12' high	2.04	8.10	10.14
4000	1/8" thick, 4' high	1.24	3.11	4.35
4400	8' high	2.48	6.25	8.73
4800	12' high	3.72	9.55	13.27
5000	Asphalt coated board and mastic, 1/4" thick, 4' high	2.28	2.83	5.11
5400	8' high	4.56	5.65	10.21
5800	12' high	6.85	8.70	15.55
6000	1/2" thick, 4' high	3.42	3.95	7.37
6400	8' high	6.85	7.90	14.75
6800	12' high	10.25	12.05	22.30
7000	Cementitious coating, on walls, 1/8" thick coating, 4' high	6.60	3.76	10.36
7400	8' high	13.20	7.55	20.75
7800	12' high	19.80	11.25	31.05
8000	Cementitious/metallic slurry, 2 coat, 1/4" thick, 2' high	38.50	99.50	138
8400	4' high	77	199	276
8800	6' high	116	299	415

For expanded coverage of these items see *Means Assemblies Cost Data 2000*

FOUNDATIONS — A1.9 — Excavation & Backfill

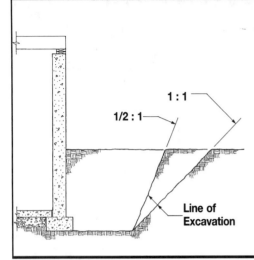

Pricing Assumptions: Two-thirds of excavation is by 2-1/2 C.Y. wheel mounted front end loader and one-third by 1-1/2 C.Y. hydraulic excavator.

Two-mile round trip haul by 12 C.Y. tandem trucks is included for excavation wasted and storage of suitable fill from excavated soil. For excavation in clay, all is wasted and the cost of suitable backfill with two-mile haul is included.

Sand and gravel assumes 15% swell and compaction; common earth assumes 25% swell and 15% compaction; clay assumes 40% swell and 15% compaction (non-clay).

In general, the following items are accounted for in the costs in the table below.
1. Excavation for building or other structure to depth and extent indicated.
2. Backfill compacted in place.
3. Haul of excavated waste.
4. Replacement of unsuitable material with bank run gravel.

Note: Additional excavation and fill beyond this line of general excavation for the building (as required for isolated spread footings, strip footings, etc.) are included in the cost of the appropriate component systems.

1.9-100	Building Excavation & Backfill	MAT.	INST.	TOTAL
2220	Excav & fill, 1000 S.F. 4' sand, gravel, or common earth, on site storage		1.35	1.35
2240	Off site storage		2.95	2.95
2260	Clay excavation, bank run gravel borrow for backfill	.96	2.67	3.63
2280	8' deep, sand, gravel, or common earth, on site storage		3.14	3.14
2300	Off site storage		7.55	7.55
2320	Clay excavation, bank run gravel borrow for backfill	2.31	6.10	8.41
2340	16' deep, sand, gravel, or common earth, on site storage		8.55	8.55
2350	Off site storage		18.30	18.30
2360	Clay excavation, bank run gravel borrow for backfill	6.40	15.65	22.05
3380	4000 S.F., 4' deep, sand, gravel, or common earth, on site storage		1.06	1.06
3400	Off site storage		1.80	1.80
3420	Clay excavation, bank run gravel borrow for backfill	.44	1.67	2.11
3440	8' deep, sand, gravel, or common earth, on site storage		2.30	2.30
3460	Off site storage		4.25	4.25
3480	Clay excavation, bank run gravel borrow for backfill	1.05	3.69	4.74
3500	16' deep, sand, gravel, or common earth, on site storage		5.50	5.50
3520	Off site storage		11.65	11.65
3540	Clay, excavation, bank run gravel borrow for backfill	2.82	8.70	11.52
4560	10,000 S.F., 4' deep, sand gravel, or common earth, on site storage		.96	.96
4580	Off site storage		1.41	1.41
4600	Clay excavation, bank run gravel borrow for backfill	.27	1.34	1.61
4620	8' deep, sand, gravel, or common earth, on site storage		2.03	2.03
4640	Off site storage		3.19	3.19
4660	Clay excavation, bank run gravel borrow for backfill	.64	2.87	3.51
4680	16' deep, sand, gravel, or common earth, on site storage		4.59	4.59
4700	Off site storage		8.15	8.15
4720	Clay excavation, bank run gravel borrow for backfill	1.69	6.55	8.24
5740	30,000 S.F., 4' deep, sand, gravel, or common earth, on site storage		.89	.89
5760	Off site storage		1.15	1.15
5780	Clay excavation, bank run gravel borrow for backfill	.15	1.11	1.26
5860	16' deep, sand, gravel, or common earth, on site storage		4.01	4.01
5880	Off site storage		5.95	5.95
5900	Clay excavation, bank run gravel borrow for backfill	.94	5.10	6.04
6910	100,000 S.F., 4' deep, sand, gravel, or common earth, on site storage		.85	.85
6940	8' deep, sand, gravel, or common earth, on site storage		1.74	1.74
6970	16' deep, sand, gravel, or common earth, on site storage		3.70	3.70
6980	Off site storage		4.72	4.72
6990	Clay excavation, bank run gravel borrow for backfill	.51	4.25	4.76

For information about Means Estimating Seminars, see yellow pages 11 and 12 in back of book

Division 2
Substructures

SUBSTRUCTURES — A2.1 Slab on Grade

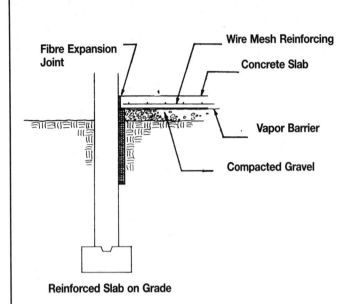

Reinforced Slab on Grade

A Slab on Grade system includes fine grading; 6" of compacted gravel; vapor barrier; 3500 p.s.i. concrete; bituminous fiber expansion joint; all necessary edge forms 4 uses; steel trowel finish; and sprayed on membrane curing compound. Wire mesh reinforcing used in all reinforced slabs.

Non-industrial slabs are for foot traffic only with negligible abrasion. Light industrial slabs are for pneumatic wheels and light abrasion. Industrial slabs are for solid rubber wheels and moderate abrasion. Heavy industrial slabs are for steel wheels and severe abrasion.

2.1-200	Plain & Reinforced	MAT.	INST.	TOTAL
		COST PER S.F.		
2220	Slab on grade, 4" thick, non industrial, non reinforced	1.19	1.35	2.54
2240	Reinforced	1.27	1.53	2.80
2280	Reinforced	1.61	1.82	3.43
2300	Industrial, non reinforced	1.90	2.56	4.46
2320	Reinforced	1.98	2.74	4.72
3340	5" thick, non industrial, non reinforced	1.40	1.38	2.78
3360	Reinforced	1.48	1.56	3.04
3380	Light industrial, non reinforced	1.75	1.67	3.42
3400	Reinforced	1.83	1.85	3.68
3420	Heavy industrial, non reinforced	2.46	2.96	5.42
3440	Reinforced	2.50	3.17	5.67
4460	6" thick, non industrial, non reinforced	1.68	1.36	3.04
4480	Reinforced	1.82	1.61	3.43
4500	Light industrial, non reinforced	2.05	1.65	3.70
4520	Reinforced	2.27	1.98	4.25
4540	Heavy industrial, non reinforced	2.76	3.01	5.77
4560	Reinforced	2.90	3.26	6.16
5580	7" thick, non industrial, non reinforced	1.90	1.39	3.29
5600	Reinforced	2.10	1.65	3.75
5620	Light industrial, non reinforced	2.28	1.68	3.96
5640	Reinforced	2.48	1.94	4.42
5660	Heavy industrial, non reinforced	2.99	2.96	5.95
5680	Reinforced	3.13	3.21	6.34
6700	8" thick, non industrial, non reinforced	2.11	1.41	3.52
6720	Reinforced	2.28	1.63	3.91
6740	Light industrial, non reinforced	2.50	1.70	4.20
6760	Reinforced	2.67	1.92	4.59
6780	Heavy industrial, non reinforced	3.23	2.99	6.22
6800	Reinforced	3.47	3.23	6.70

For information about Means Estimating Seminars, see yellow pages 11 and 12 in back of book

Division 3
Superstructures

SUPERSTRUCTURES — A3.1 — Columns, Beams & Joists

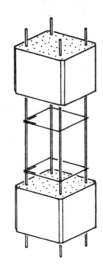

General: It is desirable for purposes of consistency and simplicity to maintain constant column sizes throughout building height. To do this, concrete strength may be varied (higher strength concrete at lower stories and lower strength concrete at upper stories), as well as varying the amount of reinforcing.

The table provides probably minimum column sizes with related costs and weight per lineal foot of story height.

3.1-114 C.I.P. Column, Square Tied

	LOAD (KIPS)	STORY HEIGHT (FT.)	COLUMN SIZE (IN.)	COLUMN WEIGHT (P.L.F.)	CONCRETE STRENGTH (PSI)	COST PER V.L.F. MAT.	COST PER V.L.F. INST.	COST PER V.L.F. TOTAL
0640	100	10	10	96	4000	7.05	21.50	28.55
0680		12	10	97	4000	6.90	21.50	28.40
0720		14	12	142	4000	9	26	35
0840	200	10	12	140	4000	9.10	26	35.10
0860		12	12	142	4000	9	26	35
0900		14	14	196	4000	11.40	30	41.40
0920	300	10	14	192	4000	11.70	30.50	42.20
0960		12	14	194	4000	11.60	30	41.60
0980		14	16	253	4000	13.10	33	46.10
1020	400	10	16	248	4000	14.30	34.50	48.80
1060		12	16	251	4000	14.15	34	48.15
1080		14	16	253	4000	14.05	34	48.05
1200	500	10	18	315	4000	19	41	60
1250		12	20	394	4000	19.40	43	62.40
1300		14	20	397	4000	19.30	43	62.30
1350	600	10	20	388	4000	22.50	47	69.50
1400		12	20	394	4000	22.50	47	69.50
1600		14	20	397	4000	22	46.50	68.50
3400	900	10	24	560	4000	32.50	61.50	94
3800		12	24	567	4000	32	61	93
4000		14	24	571	4000	31.50	60.50	92
7300	300	10	14	192	6000	11.55	30	41.55
7500		12	14	194	6000	11.45	30	41.45
7600		14	14	196	6000	11.40	30	41.40
8000	500	10	16	248	6000	14.30	34.50	48.80
8050		12	16	251	6000	14.15	34	48.15
8100		14	16	253	6000	14.05	34	48.05
8200	600	10	18	315	6000	17.25	39.50	56.75
8300		12	18	319	6000	17.10	39	56.10
8400		14	18	321	6000	16.95	39	55.95
8800	800	10	20	388	6000	19.60	43	62.60
8900		12	20	394	6000	19.40	43	62.40
9000		14	20	397	6000	19.30	43	62.30

SUPERSTRUCTURES   A3.1   Columns, Beams & Joists

3.1-114   C.I.P. Column, Square Tied

	LOAD (KIPS)	STORY HEIGHT (FT.)	COLUMN SIZE (IN.)	COLUMN WEIGHT (P.L.F.)	CONCRETE STRENGTH (PSI)	COST PER V.L.F.		
						MAT.	INST.	TOTAL
9100	900	10	20	388	6000	28	54	82
9300		12	20	394	6000	27.50	53.50	81
9600		14	20	397	6000	27	53	80

3.1-114   C.I.P. Column, Square Tied-Minimum Reinforcing

	LOAD (KIPS)	STORY HEIGHT (FT.)	COLUMN SIZE (IN.)	COLUMN WEIGHT (P.L.F.)	CONCRETE STRENGTH (PSI)	COST PER V.L.F.		
						MAT.	INST.	TOTAL
9912	150	10-14	12	135	4000	8.80	26	34.80
9918	300	10-14	16	240	4000	13	32.50	45.50
9924	500	10-14	20	375	4000	19.10	42.50	61.60
9930	700	10-14	24	540	4000	27.50	55	82.50
9936	1000	10-14	28	740	4000	34.50	65.50	100
9942	1400	10-14	32	965	4000	45	75.50	120.50
9948	1800	10-14	36	1220	4000	54.50	87	141.50

For expanded coverage of these items see *Means Assemblies Cost Data 2000*

SUPERSTRUCTURES A3.1 Columns, Beams & Joists

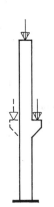

General: Data presented here is for plant produced members transported 50 miles to 100 miles to the site and erected.

Design and pricing assumptions:
Normal wt. concrete, f'c = 5 KSI
Main reinforcement, fy = 60 KSI
Ties, fy = 40 KSI
Minimum design eccentricity, 0.1t.
Concrete encased structural steel haunches are assumed where practical; otherwise galvanized rebar haunches are assumed.
Base plates are integral with columns.
Foundation anchor bolts, nuts and washers are included in price.

3.1-120 Tied, Concentric Loaded Precast Concrete Columns

	LOAD (KIPS)	STORY HEIGHT (FT.)	COLUMN SIZE (IN.)	COLUMN WEIGHT (P.L.F.)	LOAD LEVELS	COST PER V.L.F.		
						MAT.	INST.	TOTAL
0560	100	10	12x12	164	2	35	5.45	40.45
0570		12	12x12	162	2	34.50	4.56	39.06
0580		14	12x12	161	2	33.50	4.56	38.06
0590	150	10	12x12	166	3	34	4.56	38.56
0600		12	12x12	169	3	32.50	4.11	36.61
0610		14	12x12	162	3	33	4.11	37.11
0620	200	10	12x12	168	4	35.50	5	40.50
0630		12	12x12	170	4	35.50	4.56	40.06
0640		14	14x14	220	4	37.50	4.56	42.06

3.1-120 Tied, Eccentric Loaded Precast Concrete Columns

	LOAD (KIPS)	STORY HEIGHT (FT.)	COLUMN SIZE (IN.)	COLUMN WEIGHT (P.L.F.)	LOAD LEVELS	COST PER V.L.F.		
						MAT.	INST.	TOTAL
1130	100	10	12x12	161	2	32	5.45	37.45
1140		12	12x12	159	2	34	4.56	38.56
1150		14	12x12	159	2	33.50	4.56	38.06
1390	600	10	18x18	385	4	63.50	5	68.50
1400		12	18x18	380	4	61	4.56	65.56
1410		14	18x18	375	4	61	4.56	65.56
1480	800	10	20x20	490	4	81	5	86
1490		12	20x20	480	4	79.50	4.56	84.06
1500		14	20x20	475	4	79	4.56	83.56

SUPERSTRUCTURES — A3.1 — Columns, Beams & Joists

(A) Wide Flange
(B) Pipe
(C) Pipe, Concrete Filled
(D) Square Tube
(E) Square Tube Concrete Filled
(F) Rectangular Tube
(G) Rectangular Tube, Concrete Filled

General: The following pages provide data for seven types of steel columns: wide flange, round pipe, round pipe concrete filled, square tube, square tube concrete filled, rectangular tube and rectangular tube concrete filled.

Design Assumptions: Loads are concentric; wide flange and round pipe bearing capacity is for 36 KSI steel. Square and rectangular tubing bearing capacity is for 46 KSI steel.

The effective length factor K=1.1 is used for determining column values in the tables. K=1.1 is within a frequently used range for pinned connections with cross bracing.

How To Use Tables:
a. Steel columns usually extend through two or more stories to minimize splices. Determine floors with splices.
b. Enter Table No. below with load to column at the splice. Use the unsupported height.
c. Determine the column type desired by price or design.

Cost:
a. Multiply number of columns at the desired level by the total height of the column by the cost/VLF.
b. Repeat the above for all tiers.

Please see the reference section for further design and cost information.

3.1-130 Steel Columns

	LOAD (KIPS)	UNSUPPORTED HEIGHT (FT.)	WEIGHT (P.L.F.)	SIZE (IN.)	TYPE	COST PER V.L.F. MAT.	INST.	TOTAL
1000	25	10	13	4	A	10.40	5.15	15.55
1020			7.58	3	B	6.95	5.15	12.10
1040			15	3-1/2	C	8.50	5.15	13.65
1120			20	4x3	G	8.70	5.15	13.85
1200		16	16	5	A	11.85	3.88	15.73
1220			10.79	4	B	9.20	3.88	13.08
1240			36	5-1/2	C	12.85	3.88	16.73
1320			64	8x6	G	17.65	3.88	21.53
1600	50	10	16	5	A	12.80	5.15	17.95
1620			14.62	5	B	13.45	5.15	18.60
1640			24	4-1/2	C	10.20	5.15	15.35
1720			28	6x3	G	11.55	5.15	16.70
1800		16	24	8	A	17.75	3.88	21.63
1840			36	5-1/2	C	12.85	3.88	16.73
1920			64	8x6	G	17.65	3.88	21.53
2000	50	20	28	8	A	19.60	3.88	23.48
2040			49	6-5/8	C	16.05	3.88	19.93
2120			64	8x6	G	16.75	3.88	20.63
2200	75	10	20	6	A	15.95	5.15	21.10
2240			36	4-1/2	C	25.50	5.15	30.65
2320			35	6x4	G	13.05	5.15	18.20
2400		16	31	8	A	23	3.88	26.88
2440			49	6-5/8	C	16.95	3.88	20.83
2520			64	8x6	G	17.65	3.88	21.53
2600		20	31	8	A	21.50	3.88	25.38
2640			81	8-5/8	C	24.50	3.88	28.38
2720			64	8x6	G	16.75	3.88	20.63
2800	100	10	24	8	A	19.15	5.15	24.30
2840			35	4-1/2	C	25.50	5.15	30.65
2920			46	8x4	G	15.95	5.15	21.10
3000		16	31	8	A	23	3.88	26.88
3040			56	6-5/8	C	25	3.88	28.88
3120			64	8x6	G	17.65	3.88	21.53

For expanded coverage of these items see *Means Assemblies Cost Data 2000*

SUPERSTRUCTURES — A3.1 Columns, Beams & Joists

3.1-130 Steel Columns

	LOAD (KIPS)	UNSUPPORTED HEIGHT (FT.)	WEIGHT (P.L.F.)	SIZE (IN.)	TYPE	COST PER V.L.F. MAT.	COST PER V.L.F. INST.	COST PER V.L.F. TOTAL
3200	100	20	40	8	A	28	3.88	31.88
3240			81	8-5/8	C	24.50	3.88	28.38
3320			70	8x6	G	24	3.88	27.88
3400	125	10	31	8	A	25	5.15	30.15
3440			81	8	C	27.50	5.15	32.65
3520			64	8x6	G	19	5.15	24.15
3600	125	16	40	8	A	29.50	3.88	33.38
3640			81	8	C	25.50	3.88	29.38
3720			64	8x6	G	17.65	3.88	21.53
3800		20	48	8	A	33.50	3.88	37.38
3840			81	8	C	24.50	3.88	28.38
3920			60	8x6	G	24	3.88	27.88
4000	150	10	35	8	A	28	5.15	33.15
4040			81	8-5/8	C	27.50	5.15	32.65
4120			64	8x6	G	19	5.15	24.15
4200		16	45	10	A	33.50	3.88	37.38
4240			81	8-5/8	C	25.50	3.88	29.38
4320			70	8x6	G	25	3.88	28.88
4400		20	49	10	A	34.50	3.88	38.38
4440			123	10-3/4	C	34.50	3.88	38.38
4520			86	10x6	G	23.50	3.88	27.38
4600	200	10	45	10	A	36	5.15	41.15
4640			81	8-5/8	C	27.50	5.15	32.65
4720			70	8x6	G	27	5.15	32.15
4800		16	49	10	A	36	3.88	39.88
4840			123	10-3/4	C	36.50	3.88	40.38
4920			85	10x6	G	29.50	3.88	33.38
5200	300	10	61	14	A	48.50	5.15	53.65
5240			169	12-3/4	C	48.50	5.15	53.65
5320			06	10x6	G	40.50	5.15	45.65
5400		16	72	12	A	53	3.88	56.88
5440			169	12-3/4	C	45	3.88	48.88
5600		20	79	12	A	55.50	3.88	59.38
5640			169	12-3/4	C	42.50	3.88	46.38
5800	400	10	79	12	A	63	5.15	68.15
5840			178	12-3/4	C	63	5.15	68.15
6000		16	87	12	A	64.50	3.88	68.38
6040			178	12-3/4	C	58.50	3.88	62.38
6400	500	10	99	14	A	79	5.15	84.15
6600		16	109	14	A	80.50	3.88	84.38
6800		20	120	12	A	84	3.88	87.88
7000	600	10	120	12	A	96	5.15	101.15
7200		16	132	14	A	97.50	3.88	101.38
7400		20	132	14	A	92.50	3.88	96.38
7600	700	10	136	12	A	109	5.15	114.15
7800		16	145	14	A	107	3.88	110.88
8000		20	145	14	A	101	3.88	104.88
8200	800	10	145	14	A	116	5.15	121.15
8300		16	159	14	A	118	3.88	121.88
8400		20	176	14	A	123	3.88	126.88
8800	900	10	159	14	A	127	5.15	132.15
8900		16	176	14	A	130	3.88	133.88
9000		20	193	14	A	135	3.88	138.88

Important: See the Reference Section for critical supporting data - Reference Nos., Crews & Location Factors

SUPERSTRUCTURES — A3.1 Columns, Beams & Joists

3.1-130 Steel Columns

	LOAD (KIPS)	UNSUPPORTED HEIGHT (FT.)	WEIGHT (P.L.F.)	SIZE (IN.)	TYPE	COST PER V.L.F.		
						MAT.	INST.	TOTAL
9100	1000	10	176	14	A	141	5.15	146.15
9200		16	193	14	A	143	3.88	146.88
9300		20	211	14	A	148	3.88	151.88

For expanded coverage of these items see *Means Assemblies Cost Data 2000*

SUPERSTRUCTURES — A3.1 — Columns, Beams & Joists

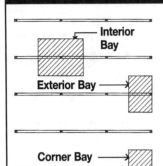

Description: Table below lists costs of columns per S.F. of bay for wood columns of various sizes and unsupported heights and the maximum allowable total load per S.F. by bay size.

Design Assumptions: Columns are concentrically loaded and are not subject to bending.

Fiber stress (f) is 1200 psi maximum.

Modulus of elasticity is 1,760,000. Use table to factor load capacity figures for modulus of elasticity other than 1,760,000.

The cost of columns per S.F. of exterior bay is proportional to the area supported. For exterior bays, multiply the costs below by two. For corner bays, multiply the cost by four.

Modulus of Elasticity	Factor
1,210,000 psi	0.69
1,320,000 psi	0.75
1,430,000 psi	0.81
1,540,000 psi	0.87
1,650,000 psi	0.94
1,760,000 psi	1.00

3.1-140 Wood Columns

	NOMINAL COLUMN SIZE (IN.)	BAY SIZE (FT.)	UNSUPPORTED HEIGHT (FT.)	MATERIAL (BF/M.S.F.)	TOTAL LOAD (P.S.F.)	COST PER S.F. MAT.	COST PER S.F. INST.	COST PER S.F. TOTAL
1000	4 x 4	10 x 8	8	133	100	.19	.12	.31
1050			10	167	60	.24	.15	.39
1200		10 x 10	8	106	80	.15	.10	.25
1250			10	133	50	.19	.12	.31
1400		10 x 15	8	71	50	.10	.06	.16
1450			10	88	30	.13	.08	.21
1600		15 x 15	8	47	30	.07	.04	.11
1650			10	59	15	.08	.05	.13
2000	6 x 6	10 x 15	8	160	230	.32	.13	.45
2050			10	200	210	.40	.17	.57
2200		15 x 15	8	107	150	.21	.09	.30
2250			10	133	140	.26	.11	.37
2400		15 x 20	8	80	110	.16	.07	.23
2450			10	100	100	.20	.08	.28
2600		20 x 20	8	60	80	.12	.05	.17
2650			10	75	70	.15	.06	.21
2800		20 x 25	8	48	60	.09	.04	.13
2850			10	60	50	.12	.05	.17
3400	8 x 8	20 x 20	8	107	160	.23	.08	.31
3450			10	133	160	.29	.10	.39
3600		20 x 25	8	85	130	.12	.08	.20
3650			10	107	130	.15	.10	.25
3800		25 x 25	8	68	100	.15	.05	.20
3850			10	85	100	.19	.06	.25
4200	10 x 10	20 x 25	8	133	210	.30	.10	.40
4250			10	167	210	.37	.12	.49
4400		25 x 25	8	107	160	.24	.08	.32
4450			10	133	160	.30	.10	.40
4700	12 x 12	20 x 25	8	192	310	.43	.13	.56
4750			10	240	310	.53	.16	.69
4900		25 x 25	8	154	240	.34	.10	.44
4950			10	192	240	.43	.13	.56

SUPERSTRUCTURES — A3.1 — Columns, Beams & Joists

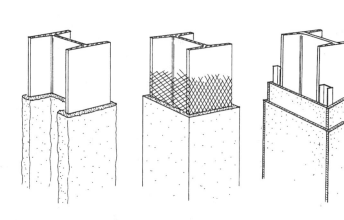

Listed below are costs per V.L.F. for fireproofing by material, column size, thickness and fire rating. Weights listed are for the fireproofing material only.

3.1-190 Steel Column Fireproofing

	ENCASEMENT SYSTEM	COLUMN SIZE (IN.)	THICKNESS (IN.)	FIRE RATING (HRS.)	WEIGHT (P.L.F.)	COST PER V.L.F. MAT.	COST PER V.L.F. INST.	COST PER V.L.F. TOTAL
3000	Concrete	8	1	1	110	4.16	16.10	20.26
3300		14	1	1	258	6.80	23.50	30.30
3400			2	3	325	8.20	27	35.20
3450	Gypsum board	8	1/2	2	8	2.37	10.45	12.82
3550	1 layer	14	1/2	2	18	2.61	11.20	13.81
3600	Gypsum board	8	1	3	14	3.51	13.35	16.86
3650	1/2" fire rated	10	1	3	17	3.81	14.15	17.96
3700	2 layers	14	1	3	22	3.97	14.60	18.57
3750	Gypsum board	8	1-1/2	3	23	4.88	16.85	21.73
3800	1/2" fire rated	10	1-1/2	3	27	5.55	18.65	24.20
3850	3 layers	14	1-1/2	3	35	6.20	20.50	26.70
3900	Sprayed fiber	8	1-1/2	2	6.3	3.12	4.20	7.32
3950	Direct application		2	3	8.3	4.30	5.80	10.10
4050		10	1-1/2	2	7.9	3.77	5.05	8.82
4200		14	1-1/2	2	10.8	4.68	6.30	10.98

3.1-224 "T" Shaped Precast Beams

	SPAN (FT.)	SUPERIMPOSED LOAD (K.L.F.)	SIZE W X D (IN.)	BEAM WEIGHT (P.L.F.)	TOTAL LOAD (K.L.F.)	COST PER L.F. MAT.	COST PER L.F. INST.	COST PER L.F. TOTAL
2300	15	2.8	12x16	260	3.06	102	10	112
2500		8.37	12x28	515	8.89	132	9.55	141.55
8900	45	3.34	12x60	1165	4.51	235	9.10	244.10
9900		6.5	24x60	1915	8.42	305	9.10	314.10

3.1-226 "L" Shaped Precast Beams

	SPAN (FT.)	SUPERIMPOSED LOAD (K.L.F.)	SIZE W X D (IN.)	BEAM WEIGHT (P.L.F.)	TOTAL LOAD (K.L.F.)	COST PER L.F. MAT.	COST PER L.F. INST.	COST PER L.F. TOTAL
2250	15	2.58	12x16	230	2.81	77.50	10	87.50
2400		5.92	12x24	370	6.29	94.50	9.10	103.60
4000	25	2.64	12x28	435	3.08	105	5.45	110.45
4450		6.44	18x36	790	7.23	143	6.85	149.85
5300	30	2.80	12x36	565	3.37	125	5.45	130.45
6400		8.66	24x44	1245	9.90	199	10.05	209.05

For expanded coverage of these items see *Means Assemblies Cost Data 2000*

SUPERSTRUCTURES — A3.4 Structural Walls

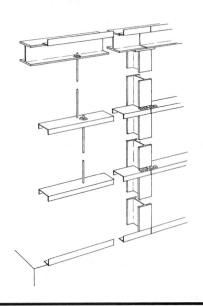

Description: The table below lists costs, $/S.F., for channel girts with sag rods and connector angles top and bottom for various column spacings, building heights and wind loads. Additive costs are shown for wind columns.

How to Use this Table: Add the cost of girts, sag rods, and angles to the framing costs of steel buildings clad in metal or composition siding. If the column spacing is in excess of the column spacing shown, use intermediate wind columns. Additive costs are shown under "Wind Columns".

Column Spacing 20'-0" Girts	$1.98
Wind Columns	.72
Total Costs for Girts	$2.70/S.F.

Design and Pricing Assumptions: Structural steel is A36.

3.4-300 Metal Siding Support

	BLDG. HEIGHT (FT.)	WIND LOAD (P.S.F.)	COL. SPACING (FT.)		INTERMEDIATE COLUMNS	COST PER S.F. MAT.	COST PER S.F. INST.	COST PER S.F. TOTAL
3000	18	20	20			1.10	1.94	3.04
3100					wind cols.	.56	.14	.70
3200		20	25			1.20	1.97	3.17
3300					wind cols.	.45	.11	.56
3400		20	30			1.32	2	3.32
3500					wind cols.	.38	.10	.48
3600		20	35			1.45	2.04	3.49
3700					wind cols.	.32	.08	.40
3800		30	20			1.21	1.97	3.18
3900					wind cols.	.56	.14	.70
4000		30	25			1.32	2	3.32
4100					wind cols.	.45	.11	.56
4200		30	30			1.46	2.04	3.50
4300					wind cols.	.51	.13	.64
4600	30	20	20			1.07	1.36	2.43
4700					wind cols.	.80	.21	1.01
4800		20	25			1.20	1.40	2.60
4900					wind cols.	.64	.17	.81
5000		20	30			1.35	1.44	2.79
5100					wind cols.	.63	.16	.79
5200	30	20	35			1.51	1.48	2.99
5300					wind cols.	.62	.16	.78
5400		30	20			1.21	1.40	2.61
5500					wind cols.	.94	.24	1.18
5600		30	25			1.34	1.43	2.77
5700					wind cols.	.90	.23	1.13
5800		30	30			1.52	1.49	3.01
5900					wind cols.	.85	.21	1.06

SUPERSTRUCTURES | A3.5 | Floors

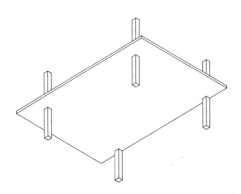

General: Flat Plates: Solid uniform depth concrete two way slab without drops or interior beams. Primary design limit is shear at columns.

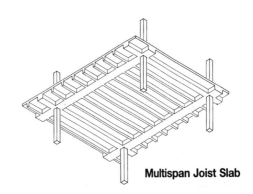

Multispan Joist Slab

General: Combination of thin concrete slab and monolithic ribs at uniform spacing to reduce dead weight and increase rigidity.

3.5-150 Cast in Place Flat Plate

	BAY SIZE (FT.)	SUPERIMPOSED LOAD (P.S.F.)	MINIMUM COL. SIZE (IN.)	SLAB THICKNESS (IN.)	TOTAL LOAD (P.S.F.)	COST PER S.F.		
						MAT.	INST.	TOTAL
3000	15 x 20	40	14	7	127	3.11	3.99	7.10
3400		75	16	7-1/2	169	3.31	4.07	7.38
3600		125	22	8-1/2	231	3.64	4.22	7.86
3800		175	24	8-1/2	281	3.66	4.22	7.88
4200	20 x 20	40	16	7	127	3.10	3.98	7.08
4400		75	20	7-1/2	175	3.34	4.08	7.42
4600		125	24	8-1/2	231	3.64	4.20	7.84
5000		175	24	8-1/2	281	3.68	4.23	7.91
5600	20 x 25	40	18	8-1/2	146	3.62	4.20	7.82
6000		75	20	9	188	3.75	4.24	7.99
6400		125	26	9-1/2	244	4.04	4.39	8.43
6600		175	30	10	300	4.20	4.46	8.66
7000	25 x 25	40	20	9	152	3.74	4.24	7.98
7400		75	24	9-1/2	194	3.97	4.35	8.32
7600		125	30	10	250	4.22	4.45	8.67

3.5-160 Cast in Place Multispan Joist Slab

	BAY SIZE (FT.)	SUPERIMPOSED LOAD (P.S.F.)	MINIMUM COL. SIZE (IN.)	RIB DEPTH (IN.)	TOTAL LOAD (P.S.F.)	COST PER S.F.		
						MAT.	INST.	TOTAL
2000	15 x 15	40	12	8	115	3.23	4.82	8.05
2100		75	12	8	150	3.25	4.84	8.09
2200		125	12	8	200	3.33	4.89	8.22
2300		200	14	8	275	3.43	5.10	8.53
2600	15 x 20	40	12	8	115	3.29	4.82	8.11
2800		75	12	8	150	3.36	4.91	8.27
3000		125	14	8	200	3.50	5.20	8.70
3300		200	16	8	275	3.67	5.30	8.97
3600	20 x 20	40	12	10	120	3.36	4.76	8.12
3900		75	14	10	155	3.54	5.05	8.59
4000		125	16	10	205	3.58	5.10	8.68
4100		200	18	10	280	3.77	5.30	9.07
6200	30 x 30	40	14	14	131	3.80	4.98	8.78
6400		75	18	14	166	3.93	5.15	9.08
6600		125	20	14	216	4.19	5.45	9.64
6700		200	24	16	297	4.52	5.65	10.17

For expanded coverage of these items see *Means Assemblies Cost Data 2000*

SUPERSTRUCTURES — A3.5 Floors

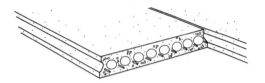

Precast Plank with No Topping

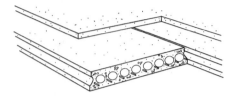

Precast Plank with 2" Concrete Topping

3.5-210 Precast Plank with No Topping

	SPAN (FT.)	SUPERIMPOSED LOAD (P.S.F.)	TOTAL DEPTH (IN.)	DEAD LOAD (P.S.F.)	TOTAL LOAD (P.S.F.)	COST PER S.F. MAT.	INST.	TOTAL
0720	10	40	4	50	90	5.10	1.02	6.12
0750		75	6	50	125	4.93	.81	5.74
0770		100	6	50	150	4.93	.81	5.74
0800	15	40	6	50	90	4.93	.81	5.74
0820		75	6	50	125	4.93	.81	5.74
0850		100	6	50	150	4.93	.81	5.74
0950	25	40	6	50	90	4.93	.81	5.74
0970		75	8	55	130	5.50	.65	6.15
1000		100	8	55	155	5.50	.65	6.15
1200	30	40	8	55	95	5.50	.65	6.15
1300		75	8	55	130	5.50	.65	6.15
1400		100	10	70	170	6.25	.42	6.67
1500	40	40	10	70	110	6.25	.42	6.67
1600		75	12	70	145	5.25	.45	5.70
1700	45	40	12	70	110	5.25	.45	5.70

3.5-210 Precast Plank with 2" Concrete Topping

	SPAN (FT.)	SUPERIMPOSED LOAD (P.S.F.)	TOTAL DEPTH (IN.)	DEAD LOAD (P.S.F.)	TOTAL LOAD (P.S.F.)	COST PER S.F. MAT.	INST.	TOTAL
2000	10	40	6	75	115	5.70	1.99	7.69
2100		75	8	75	150	5.55	1.78	7.33
2200		100	8	75	175	5.55	1.78	7.33
2500	15	40	8	75	115	5.55	1.78	7.33
2600		75	8	75	150	5.55	1.78	7.33
2700		100	8	75	175	5.55	1.78	7.33
3100	25	40	8	75	115	5.55	1.78	7.33
3200		75	8	75	150	5.55	1.78	7.33
3300		100	10	80	180	6.10	1.62	7.72
3400	30	40	10	80	120	6.10	1.62	7.72
3500		75	10	80	155	6.10	1.62	7.72
3600		100	10	80	180	6.10	1.62	7.72
4000	40	40	12	95	135	6.85	1.39	8.24
4500		75	14	95	170	5.85	1.42	7.27
5000	45	40	14	95	135	5.85	1.42	7.27

SUPERSTRUCTURES — A3.5 — Floors

Most widely used for moderate span floors and roofs. At shorter spans, they tend to be competitive with hollow core slabs. They are also used as wall panels.

3.5-230 — Precast Double "T" Beams with No Topping

	SPAN (FT.)	SUPERIMPOSED LOAD (P.S.F.)	DBL. "T" SIZE D (IN.) W (FT.)	CONCRETE "T" TYPE	TOTAL LOAD (P.S.F.)	COST PER S.F. MAT.	COST PER S.F. INST.	COST PER S.F. TOTAL
4300	50	30	20x8	Lt. Wt.	66	5.20	.71	5.91
4400		40	20x8	Lt. Wt.	76	5.55	.65	6.20
4500		50	20x8	Lt. Wt.	86	5.85	.78	6.63
4600		75	20x8	Lt. Wt.	111	6.10	.91	7.01
5600	70	30	32x10	Lt. Wt.	78	6	.49	6.49
5750		40	32x10	Lt. Wt.	88	6.25	.60	6.85
5900		50	32x10	Lt. Wt.	98	6.45	.69	7.14
6000		75	32x10	Lt. Wt.	123	6.80	.84	7.64
6100		100	32x10	Lt. Wt.	148	7.45	1.15	8.60
6200	80	30	32x10	Lt. Wt.	78	6.45	.69	7.14
6300		40	32x10	Lt. Wt.	88	6.80	.84	7.64
6400		50	32x10	Lt. Wt.	98	7.10	1	8.10

3.5-230 — Precast Double "T" Beams With 2" Topping

	SPAN (FT.)	SUPERIMPOSED LOAD (P.S.F.)	DBL. "T" SIZE D (IN.) W (FT.)	CONCRETE "T" TYPE	TOTAL LOAD (P.S.F.)	COST PER S.F. MAT.	COST PER S.F. INST.	COST PER S.F. TOTAL
7100	40	30	18x8	Reg. Wt.	120	4.82	1.51	6.33
7200		40	20x8	Reg. Wt.	130	4.61	1.43	6.04
7300		50	20x8	Reg. Wt.	140	4.90	1.57	6.47
7400		75	20x8	Reg. Wt.	165	5.05	1.62	6.67
7500		100	20x8	Reg. Wt.	190	5.45	1.83	7.28
7550	50	30	24x8	Reg. Wt.	120	5.20	1.51	6.71
7600		40	24x8	Reg. Wt.	130	5.30	1.55	6.85
7750		50	24x8	Reg. Wt.	140	5.30	1.56	6.86
7800		75	24x8	Reg. Wt.	165	5.75	1.77	7.52
7900		100	32x10	Reg. Wt.	189	6.10	1.46	7.56

For expanded coverage of these items see *Means Assemblies Cost Data 2000*

SUPERSTRUCTURES — A3.5 Floors

Description: Table below lists costs for light gauge CEE or PUNCHED DOUBLE joists to suit the span and loading with the minimum thickness subfloor required by the joist spacing.

Design Assumptions:
- Maximum live load deflection is 1/360 of the clear span.
- Maximum total load deflection is 1/240 of the clear span.
- Bending strength is 20,000 psi.

8% allowance has been added to framing quantities for overlaps, double joists at openings under partitions, etc.; 5% added to glued & nailed subfloor for waste.

Maximum span is in feet and is the unsupported clear span.

3.5-360 Light Gauge Steel Floor Systems

	SPAN (FT.)	SUPERIMPOSED LOAD (P.S.F.)	FRAMING DEPTH (IN.)	FRAMING SPAC. (IN.)	TOTAL LOAD (P.S.F.)	COST PER S.F. MAT.	COST PER S.F. INST.	COST PER S.F. TOTAL
1500	15	40	8	16	54	1.97	.89	2.86
1550			8	24	54	2.11	.91	3.02
1600		65	10	16	80	2.57	1.06	3.63
1650			10	24	80	2.60	1.04	3.64
1700		75	10	16	90	2.57	1.06	3.63
1750			10	24	90	2.60	1.04	3.64
1800		100	10	16	116	3.28	1.32	4.60
1850			10	24	116	2.60	1.04	3.64
1900		125	10	16	141	3.28	1.32	4.60
1950			10	24	141	2.60	1.04	3.64
2500	20	40	8	16	55	2.22	.99	3.21
2550			8	24	55	2.11	.91	3.02
2600		65	8	16	80	2.54	1.11	3.65
2650			10	24	80	2.61	1.06	3.67
2700		75	10	16	90	2.44	1.01	3.45
2750			12	24	90	2.87	.92	3.79
2800		100	10	16	115	2.44	1.01	3.45
2850			10	24	116	3.28	1.14	4.42
2900		125	12	16	142	3.69	1.12	4.81
2950			12	24	141	3.54	1.22	4.76
3500	25	40	10	16	55	2.57	1.06	3.63
3550			10	24	55	2.60	1.04	3.64
3600		65	10	16	81	3.28	1.32	4.60
3650			12	24	81	3.15	.98	4.13
3700		75	12	16	92	3.68	1.12	4.80
3750			12	24	91	3.54	1.22	4.76
3800		100	12	16	117	4.12	1.22	5.34
3850		125	12	16	143	4.72	1.59	6.31
4500	30	40	12	16	57	3.68	1.12	4.80
4550			12	24	56	3.14	.98	4.12
4600		65	12	16	82	4.10	1.22	5.32
4650		75	12	16	92	4.10	1.22	5.32

SUPERSTRUCTURES   A3.5   Floors

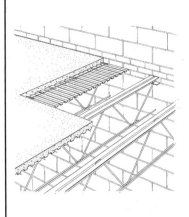

Description: Table below lists cost per S.F. for a floor system on bearing walls using open web steel joists, galvanized steel slab form and 2-1/2" concrete slab reinforced with welded wire fabric.

Design and Pricing Assumptions:
Concrete f'c = 3 KSI placed by pump.
WWF 6 x 6 – W1.4 x W1.4 (10 x 10)
Joists are spaced as shown.
Slab form is 28 gauge galvanized.
Joists costs include appropriate bridging. Deflection is limited to 1/360 of the span. Screeds and steel trowel finish.

Design Loads	Min.	Max.
Joists	3.0 PSF	7.6 PSF
Slab Form	1.0	1.0
2-1/2" Concrete	27.0	27.0
Ceiling	3.0	3.0
Misc.	9.0	9.4
	43.0 PSF	48.0 PSF

3.5-420 Deck & Joists on Bearing Walls

	SPAN (FT.)	SUPERIMPOSED LOAD (P.S.F.)	JOIST SPACING FT./IN.	DEPTH (IN.)	TOTAL LOAD (P.S.F.)	COST PER S.F. MAT.	COST PER S.F. INST.	COST PER S.F. TOTAL
1050	20	40	2-0	14-1/2	83	3	1.87	4.87
1070		65	2-0	16-1/2	109	3.28	1.98	5.26
1100		75	2-0	16-1/2	119	3.28	1.98	5.26
1120		100	2-0	18-1/2	145	3.30	1.98	5.28
1150		125	1-9	18-1/2	170	4.33	2.37	6.70
1170	25	40	2-0	18-1/2	84	3.87	2.22	6.09
1200		65	2-0	20-1/2	109	4.03	2.27	6.30
1220		75	2-0	20-1/2	119	3.68	2.10	5.78
1250		100	2-0	22-1/2	145	3.75	2.12	5.87
1270		125	1-9	22-1/2	170	4.33	2.32	6.65
1300	30	40	2-0	22-1/2	84	3.66	1.97	5.63
1320		65	2-0	24-1/2	110	3.99	2.05	6.04
1350		75	2-0	26-1/2	121	4.15	2.09	6.24
1370		100	2-0	26-1/2	146	4.45	2.17	6.62
1400		125	2-0	24-1/2	172	5.40	2.79	8.19
1420	35	40	2-0	26-1/2	85	5.10	2.67	7.77
1450		65	2-0	28-1/2	111	5.30	2.74	8.04
1470		75	2-0	28-1/2	121	5.30	2.74	8.04
1500		100	1-11	28-1/2	147	5.75	2.92	8.67
1520		125	1-8	28-1/2	172	6.40	3.14	9.54
1550	5/8" gyp. fireproof.							
1560	On metal furring, add					.57	1.47	2.04

For expanded coverage of these items see *Means Assemblies Cost Data 2000*

SUPERSTRUCTURES — A3.5 — Floors

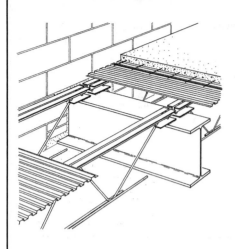

Table below lists costs for a floor system on exterior bearing walls and interior columns and beams using open web steel joists, galvanized steel slab form, 2-1/2" concrete slab reinforced with welded wire fabric.

Design and Pricing Assumptions:
Structural Steel is A36.
Concrete f'c = 3 KSI placed by pump.
WWF 6 x 6 – W1.4 x W1.4 (10 x 10)
Columns are 12' high.
Building is 4 bays long by 4 bays wide.
Joists are 2' O.C. ± and span the long direction of the bay.

Joists at columns have bottom chords extended and are connected to columns.

Slab form is 28 gauge galvanized. Column costs in table are for columns to support 1 floor plus roof loading in a 2-story building; however, column costs are from ground floor to 2nd floor only. Joist costs include appropriate bridging. Deflection is limited to 1/360 of the span. Screeds and steel trowel finish.

Design Loads	Min.	Max.
S.S & Joists	4.4 PSF	11.5 PSF
Slab Form	1.0	1.0
2-1/2" Concrete	27.0	27.0
Ceiling	3.0	3.0
Misc.	7.6	5.5
	43.0 PSF	48.0 PSF

3.5-440 Steel Joists on Beam & Wall

	BAY SIZE (FT.)	SUPERIMPOSED LOAD (P.S.F.)	DEPTH (IN.)	TOTAL LOAD (P.S.F.)	COLUMN ADD	COST PER S.F. MAT.	COST PER S.F. INST.	COST PER S.F. TOTAL
1720	25x25	40	23	84		4.64	2.38	7.02
1730					columns	.19	.06	.25
1750	25x25	65	29	110		4.89	2.47	7.36
1760					columns	.19	.06	.25
1770	25x25	75	26	120		4.81	2.38	7.19
1780					columns	.22	.07	.29
1800	25x25	100	29	145		5.45	2.59	8.04
1810					columns	.22	.07	.29
1820	25x25	125	29	170		5.75	2.68	8.43
1830					columns	.24	.08	.32
2020	30x30	40	29	84		4.65	2.19	6.84
2030					columns	.17	.06	.23
2050	30x30	65	29	110		5.25	2.38	7.63
2060					columns	.17	.06	.23
2070	30x30	75	32	120		5.45	2.42	7.87
2080					columns	.19	.06	.25
2100	30x30	100	35	145		6	2.59	8.59
2110					columns	.23	.07	.30
2120	30x30	125	35	172		7.15	3.28	10.43
2130					columns	.27	.08	.35
2170	30x35	40	29	85		5.30	2.38	7.68
2180					columns	.16	.05	.21
2200	30x35	65	29	111		6.50	3.05	9.55
2210					columns	.19	.06	.25
2220	30x35	75	32	121		6.50	3.05	9.55
2230					columns	.19	.06	.25
2250	30x35	100	35	148		6.45	2.69	9.14
2260					columns	.23	.07	.30
2320	35x35	40	32	85		5.45	2.41	7.86
2330					columns	.17	.06	.23
2350	35x35	65	35	111		6.85	3.16	10.01
2360					columns	.20	.07	.27
2370	35x35	75	35	121		7	3.20	10.20
2380					columns	.20	.07	.27
2400	35x35	100	38	148		7.30	3.32	10.62
2410					columns	.25	.08	.33
2460	5/8 gyp. fireproof.							
2475	On metal furring, add					.57	1.47	2.04

SUPERSTRUCTURES — A3.5 Floors

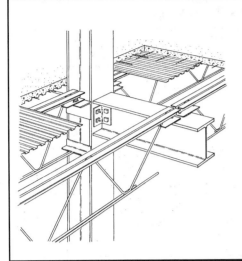

Table below lists costs for a floor system on steel columns and beams using open web steel joists, galvanized steel slab form, and 2-1/2" concrete slab reinforced with welded wire fabric.

Design and Pricing Assumptions:
Structural Steel is A36.
Concrete f'c = 3 KSI placed by pump.
WWF 6 x 6 – W1.4 x W1.4 (10 x 10)
Columns are 12' high.
Building is 4 bays long by 4 bays wide.
Joists are 2' O.C. ± and span the long direction of the bay.

Joists at columns have bottom chords extended and are connected to columns.

Slab form is 28 gauge galvanized. Column costs in table are for columns to support 1 floor plus roof loading in a 2-story building; however, column costs are from ground floor to 2nd floor only. Joist costs include appropriate bridging. Deflection is limited to 1/360 of the span. Screeds and steel trowel finish.

Design Loads	Min.	Max.
S.S. & Joists	6.3 PSF	15.3 PSF
Slab Form	1.0	1.0
2-1/2" Concrete	27.0	27.0
Ceiling	3.0	3.0
Misc.	5.7	1.7
	43.0 PSF	48.0 PSF

3.5-460 Steel Joists, Beams & Slab on Columns

	BAY SIZE (FT.)	SUPERIMPOSED LOAD (P.S.F.)	DEPTH (IN.)	TOTAL LOAD (P.S.F.)	COLUMN ADD	COST PER S.F. MAT.	INST.	TOTAL
2350	15x20	40	17	83		4.20	2.18	6.38
2400					column	.62	.20	.82
2450	15x20	65	19	108		4.62	2.30	6.92
2500					column	.62	.20	.82
2550	15x20	75	19	119		4.84	2.37	7.21
2600					column	.68	.21	.89
2650	15x20	100	19	144		5.15	2.47	7.62
2700					column	.68	.21	.89
2750	15x20	125	19	170		6.05	2.81	8.86
2800					column	.91	.29	1.20
2850	20x20	40	19	83		4.53	2.27	6.80
2900					column	.51	.17	.68
2950	20x20	65	23	109		5	2.43	7.43
3000					column	.68	.21	.89
3100	20x20	75	26	119		5.25	2.51	7.76
3200					column	.68	.21	.89
3400	20x20	100	23	144		5.45	2.56	8.01
3450					column	.68	.21	.89
3500	20x20	125	23	170		6.10	2.77	8.87
3600					column	.82	.26	1.08
3700	20x25	40	44	83		5.45	2.62	8.07
3800					column	.55	.17	.72
3900	20x25	65	26	110		5.95	2.78	8.73
4000					column	.55	.17	.72
4100	20x25	75	26	120		5.60	2.62	8.22
4200					column	.65	.21	.86
4300	20x25	100	26	145		5.90	2.72	8.62
4400					column	.65	.21	.86
4500	20x25	125	29	170		6.65	2.95	9.60
4600					column	.76	.24	1
4700	25x25	40	23	84		5.80	2.72	8.52
4800					column	.52	.17	.69
4900	25x25	65	29	110		6.15	2.85	9
5000					column	.52	.17	.69
5100	25x25	75	26	120		6.15	2.79	8.94
5200					column	.61	.20	.81
5300	25x25	100	29	145		6.90	3.02	9.92
5400					column	.61	.20	.81

For expanded coverage of these items see *Means Assemblies Cost Data 2000*

SUPERSTRUCTURES   A3.5   Floors

3.5-460   Steel Joists, Beams & Slab on Columns

	BAY SIZE (FT.)	SUPERIMPOSED LOAD (P.S.F.)	DEPTH (IN.)	TOTAL LOAD (P.S.F.)	COLUMN ADD	COST PER S.F.		
						MAT.	INST.	TOTAL
5500	25x25	125	32	170		7.25	3.15	10.40
5600					column	.68	.21	.89
5700	25x30	40	29	84		6.10	2.86	8.96
5800					column	.51	.17	.68
5900	25x30	65	29	110		6.40	2.98	9.38
6000					column	.51	.17	.68
6050	25x30	75	29	120		6.40	2.72	9.12
6100					column	.56	.18	.74
6150	25x30	100	29	145		6.95	2.88	9.83
6200					column	.56	.18	.74
6250	25x30	125	32	170		8.05	3.51	11.56
6300					column	.65	.21	.86
6350	30x30	40	29	84		5.95	2.59	8.54
6400					column	.47	.15	.62
6500	30x30	65	29	110		6.80	2.84	9.64
6600					column	.47	.15	.62
6700	30x30	75	32	120		6.95	2.87	9.82
6800					column	.54	.17	.71
6900	30x30	100	35	145		7.70	3.11	10.81
7000					column	.63	.20	.83
7100	30x30	125	35	172		9	3.84	12.84
7200					column	.70	.22	.92
7300	30x35	40	29	85		6.75	2.82	9.57
7400					column	.40	.13	.53
7500	30x35	65	29	111		8.10	3.54	11.64
7600					column	.52	.17	.69
7700	30x35	75	32	121		8.10	3.54	11.64
7800					column	.53	.17	.70
7900	30x35	100	35	148		8.15	3.22	11.37
8000					column	.65	.21	.86
8100	30x35	125	38	173		9.10	3.48	12.58
8200					column	.66	.21	.87
8300	35x35	40	32	85		6.95	2.87	9.82
8400					column	.46	.15	.61
8500	35x35	65	35	111		8.50	3.67	12.17
8600					column	.56	.18	.74
9300	35x35	75	38	121		8.70	3.73	12.43
9400					column	.56	.18	.74
9500	35x35	100	38	148		9.35	3.96	13.31
9600					column	.69	.22	.91
9750	35x35	125	41	173		9.45	3.60	13.05
9800					column	.70	.22	.92
9810	5/8 gyp. fireproof.							
9815	On metal furring, add					.57	1.47	2.04

Important: See the Reference Section for critical supporting data - Reference Nos., Crews & Location Factors

SUPERSTRUCTURES — A3.5 Floors

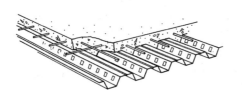

How to Use Tables: Enter any table at superimposed load and support spacing for deck.

Cellular decking tends to be stiffer and has the potential for utility integration with deck structure.

3.5-580 Metal Deck/Concrete Fill

	SUPERIMPOSED LOAD (P.S.F.)	DECK SPAN (FT.)	DECK GAGE DEPTH	SLAB THICKNESS (IN.)	TOTAL LOAD (P.S.F.)	COST PER S.F.		
						MAT.	INST.	TOTAL
0900	125	6	22 1-1/2	4	164	1.50	1.22	2.72
0920		7	20 1-1/2	4	164	1.62	1.30	2.92
0950		8	20 1-1/2	4	165	1.62	1.30	2.92
0970		9	18 1-1/2	4	165	1.86	1.30	3.16
1000		10	18 2	4	165	1.87	1.37	3.24
1020		11	18 3	5	169	2.27	1.47	3.74

3.5-710 Wood Joist

		COST PER S.F.		
		MAT.	INST.	TOTAL
2902	Wood joist, 2" x 8", 12" O.C.	1.61	.92	2.53
2950	16" O.C.	1.36	.79	2.15
3000	24" O.C.	1.44	.72	2.16
3300	2"x10", 12" O.C.	2.09	1.04	3.13
3350	16" O.C.	1.72	.88	2.60
3400	24" O.C.	1.66	.78	2.44
3700	2"x12", 12" O.C.	2.39	1.06	3.45
3750	16" O.C.	1.95	.89	2.84
3800	24" O.C.	1.82	.79	2.61
4100	2"x14", 12" O.C.	3.09	1.15	4.24
4150	16" O.C.	2.48	.96	3.44
4200	24" O.C.	2.18	.84	3.02
7101	Note: Subfloor cost is included in these prices			

3.5-720 Wood Beam & Joist

	BAY SIZE (FT.)	SUPERIMPOSED LOAD (P.S.F.)	GIRDER BEAM (IN.)	JOISTS (IN.)	TOTAL LOAD (P.S.F.)	COST PER S.F.		
						MAT.	INST.	TOTAL
2000	15x15	40	8 x 12 4 x 12	2 x 6 @ 16	53	5.20	2.12	7.32
2050		75	8 x 16 4 x 16	2 x 8 @ 16	90	6.70	2.32	9.02
2100		125	12 x 16 6 x 16	2 x 8 @ 12	144	10.65	2.98	13.63
2150		200	14 x 22 12 x 16	2 x 10 @ 12	227	19.75	4.56	24.31
3000	20x20	40	10 x 14 10 x 12	2 x 8 @ 16	63	7.55	2.13	9.68
3050		75	12 x 16 8 x 16	2 x 10 @ 16	102	9.45	2.39	11.84
3100		125	14 x 22 12 x 16	2 x 10 @ 12	163	16	3.82	19.82

For expanded coverage of these items see *Means Assemblies Cost Data 2000*

SUPERSTRUCTURES — A3.7 Roofs

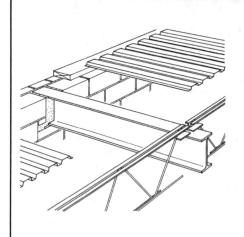

Table below lists the cost per S.F. for a roof system with steel columns, beams, and deck using open web steel joists and 1-1/2" galvanized metal deck. Perimeter of system is supported on bearing walls.

Design and Pricing Assumptions:
Columns are 18' high.
Joists are 5'-0" O.C. and span the long direction of the bay.

Joists at columns have bottom chords extended and are connected to columns. Column costs are not included but are listed separately per S.F. of floor.

Roof deck is 1-1/2", 22 gauge galvanized steel. Joist cost includes appropriate bridging. Deflection is limited to 1/240 of the span. Fireproofing is not included.

Costs/S.F. are based on a building 4 bays long and 4 bays wide.

Design Loads	Min.	Max.
Joists & Beams	3 PSF	5 PSF
Deck	2	2
Insulation	3	3
Roofing	6	6
Misc.	6	6
Total Dead Load	20 PSF	22 PSF

3.7-410 Steel Joists, Beams & Deck on Columns & Walls

	BAY SIZE (FT.)	SUPERIMPOSED LOAD (P.S.F.)	DEPTH (IN.)	TOTAL LOAD (P.S.F.)	COLUMN ADD	COST PER S.F. MAT.	COST PER S.F. INST.	COST PER S.F. TOTAL
3000	25x25	20	18	40		2.05	.68	2.73
3100					columns	.18	.04	.22
3200		30	22	50		2.43	.82	3.25
3300					columns	.23	.06	.29
3400		40	20	60		2.46	.80	3.26
3500					columns	.23	.06	.29
3600	25x30	20	22	40		2.17	.67	2.84
3700					columns	.20	.05	.25
3800		30	20	50		2.53	.87	3.40
3900					columns	.20	.05	.25
4000		40	25	60		2.58	.78	3.36
4100					columns	.23	.06	.29
4200	30x30	20	25	42		2.39	.73	3.12
4300					columns	.16	.04	.20
4400		30	22	52		2.62	.78	3.40
4500					columns	.20	.05	.25
4600		40	28	62		2.72	.82	3.54
4700					columns	.20	.05	.25
4800	30x35	20	22	42		2.50	.76	3.26
4900					columns	.17	.04	.21
5000		30	28	52		2.64	.79	3.43
5100					columns	.17	.04	.21
5200		40	25	62		2.86	.85	3.71
5300					columns	.20	.05	.25
5400	35x35	20	28	42		2.51	.76	3.27
5500					columns	.14	.04	.18

SUPERSTRUCTURES — A3.7 Roofs

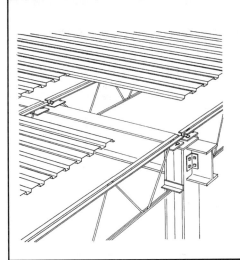

Description: Table below lists the cost per S.F. for a roof system with steel columns, beams, and deck, using open web steel joists and 1-1/2" galvanized metal deck.

Design and Pricing Assumptions:
Columns are 18' high.
Building is 4 bays long by 4 bays wide.
Joists are 5'-0" O.C. and span the long direction of the bay.
Joists at columns have bottom chords extended and are connected to columns. Column costs are not included but are listed separately per S.F. of floor.

Roof deck is 1-1/2", 22 gauge galvanized steel. Joist cost includes appropriate bridging. Deflection is limited to 1/240 of the span. Fireproofing is not included.

Design Loads	Min.	Max.
Joists & Beams	3 PSF	5 PSF
Deck	2	2
Insulation	3	3
Roofing	6	6
Misc.	6	6
Total Dead Load	20 PSF	22 PSF

3.7-420 Steel Joists, Beams, & Deck on Columns

	BAY SIZE (FT.)	SUPERIMPOSED LOAD (P.S.F.)	DEPTH (IN.)	TOTAL LOAD (P.S.F.)	COLUMN ADD	COST PER S.F. MAT.	COST PER S.F. INST.	COST PER S.F. TOTAL
1100	15x20	20	16	40		2.04	.67	2.71
1200					columns	1.02	.27	1.29
1500		40	18	60		2.29	.76	3.05
1600					columns	1.02	.27	1.29
2300	20x25	20	18	40		2.33	.76	3.09
2400					columns	.61	.16	.77
2700		40	20	60		2.64	.85	3.49
2800					columns	.82	.21	1.03
2900	25x25	20	18	40		2.71	.86	3.57
3000					columns	.49	.13	.62
3100		30	22	50		3.04	.98	4.02
3200					columns	.65	.17	.82
3300		40	20	60		3.14	.98	4.12
3400					columns	.65	.17	.82
3500	25x30	20	22	40		2.64	.79	3.43
3600					columns	.55	.14	.69
3900		40	25	60		3.22	.94	4.16
4000					columns	.65	.17	.82
4100	30x30	20	25	42		2.99	.88	3.87
4200					columns	.45	.11	.56
4300		30	22	52		3.30	.96	4.26
4400					columns	.55	.14	.69
4500		40	28	62		3.46	1	4.46
4600					columns	.55	.14	.69
5300	35x35	20	28	42		3.28	.96	4.24
5400					columns	.40	.10	.50
5500		30	25	52		3.66	1.05	4.71
5600					columns	.47	.12	.59
5700		40	28	62		3.96	1.14	5.10
5800					columns	.52	.13	.65

For expanded coverage of these items see *Means Assemblies Cost Data 2000*

SUPERSTRUCTURES — A3.7 Roofs

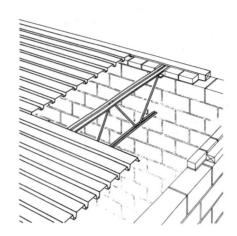

Description: Table below lists cost per S.F. for a roof system using open web steel joists and 1-1/2" galvanized metal deck. The system is assumed supported on bearing walls or other suitable support. Costs for the supports are not included.

Design and Pricing Assumptions:
Joists are 5'-0" O.C.
Roof deck is 1-1/2", 22 gauge galvanized.

3.7-430 Steel Joists & Deck on Bearing Walls

	BAY SIZE (FT.)	SUPERIMPOSED LOAD (P.S.F.)	DEPTH (IN.)	TOTAL LOAD (P.S.F.)		COST PER S.F.		
						MAT.	INST.	TOTAL
1100	20	20	13-1/2	40		1.33	.49	1.82
1200		30	15-1/2	50		1.36	.50	1.86
1300		40	15-1/2	60		1.48	.54	2.02
1400	25	20	17-1/2	40		1.48	.54	2.02
1500		30	17-1/2	50		1.61	.58	2.19
1600		40	19-1/2	60		1.64	.59	2.23
1700	30	20	19-1/2	40		1.64	.58	2.22
1800		30	21-1/2	50		1.68	.60	2.28
1900		40	23-1/2	60		1.82	.64	2.46
2000	35	20	23-1/2	40		1.76	.56	2.32
2100		30	25-1/2	50		1.82	.58	2.40
2200		40	25-1/2	60		1.95	.62	2.57
2300	40	20	25-1/2	41		1.98	.62	2.60
2400		30	25-1/2	51		2.11	.66	2.77
2500		40	25-1/2	61		2.19	.68	2.87
2600	45	20	27-1/2	41		2.48	.92	3.40
2700		30	31-1/2	51		2.63	.97	3.60
2800		40	31-1/2	61		2.79	1.03	3.82
2900	50	20	29-1/2	42		2.80	1.03	3.83
3000		30	31-1/2	52		3.08	1.13	4.21
3100		40	31-1/2	62		3.29	1.21	4.50
3200	60	20	37-1/2	42		3.47	1.01	4.48
3300		30	37-1/2	52		3.88	1.13	5.01
3400		40	37-1/2	62		3.88	1.13	5.01
3500	70	20	41-1/2	42		3.87	1.12	4.99
3600		30	41-1/2	52		4.14	1.19	5.33
3700		40	41-1/2	64		5.10	1.45	6.55
3800	80	20	45-1/2	44		6.40	1.68	8.08
3900		30	45-1/2	54		6.40	1.68	8.08
4000		40	45-1/2	64		7.15	1.86	9.01
4400	100	20	57-1/2	44		5.45	1.33	6.78
4500		30	57-1/2	54		6.60	1.58	8.18
4600		40	57-1/2	65		7.30	1.74	9.04
4700	125	20	69-1/2	44		7.30	1.90	9.20
4800		30	69-1/2	56		8.55	2.21	10.76
4900		40	69-1/2	67		9.85	2.51	12.36

Important: See the Reference Section for critical supporting data - Reference Nos., Crews & Location Factors

SUPERSTRUCTURES A3.7 Roofs

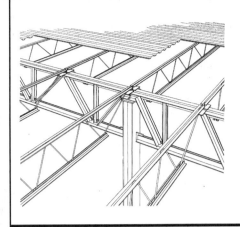

Description: Table below lists costs for a roof system supported on exterior bearing walls and interior columns. Costs include bracing, joist girders, open web steel joists and 1-1/2" galvanized metal deck.

Design and Pricing Assumptions:
  Columns are 18' high.
  Joists are 5'-0" O.C.
Joist girders and joists have bottom chords connected to columns. Roof deck is 1-1/2", 22 gauge galvanized steel. Costs include bridging and bracing. Deflection is limited to 1/240 of the span.

Fireproofing is not included.
Costs/S.F. are based on a building 4 bays long and 4 bays wide.
Costs for bearing walls are not included.

Column costs are not included but are listed separately per S.F. of floor.

3.7-440 Steel Joists, Joist Girders & Deck on Columns & Walls

	BAY SIZE (FT.) GIRD X JOISTS	SUPERIMPOSED LOAD (P.S.F.)	DEPTH (IN.)	TOTAL LOAD (P.S.F.)	COLUMN ADD	COST PER S.F.		
						MAT.	INST.	TOTAL
2350	30x35	20	32-1/2	40		1.93	.71	2.64
2400					columns	.17	.04	.21
2550		40	36-1/2	60		2.18	.78	2.96
2600					columns	.20	.04	.24
3000	35x35	20	36-1/2	40		2.10	.87	2.97
3050					columns	.14	.04	.18
3200		40	36-1/2	60		2.39	.95	3.34
3250					columns	.19	.04	.23
3300	35x40	20	36-1/2	40		2.14	.80	2.94
3350					columns	.15	.04	.19
3500		40	36-1/2	60		2.44	.89	3.33
3550					columns	.16	.04	.20
3900	40x40	20	40-1/2	41		2.65	1.04	3.69
3950					columns	.14	.04	.18
4100		40	40-1/2	61		3.01	1.12	4.13
4150					columns	.14	.04	.18
5100	45x50	20	52-1/2	41		3.03	1.17	4.20
5150					columns	.10	.03	.13
5300		40	52-1/2	61		3.73	1.42	5.15
5350					columns	.15	.04	.19
5400	50x45	20	56-1/2	41		2.91	1.13	4.04
5450					columns	.10	.03	.13
5600		40	56-1/2	61		3.83	1.44	5.27
5650					columns	.14	.03	.17
5700	50x50	20	56-1/2	42		3.30	1.28	4.58
5750					columns	.10	.03	.13
5900		40	59	64		4.77	1.43	6.20
5950					columns	.14	.04	.18
6300	60x50	20	62-1/2	43		3.34	1.29	4.63
6350					columns	.16	.04	.20
6500		40	71	65		4.88	1.47	6.35
6550					columns	.22	.06	.28

For expanded coverage of these items see *Means Assemblies Cost Data 2000*

SUPERSTRUCTURES | A3.7 | Roofs

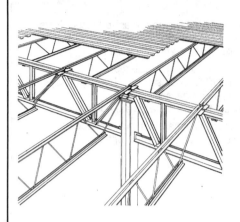

Description: Table below lists the cost per S.F. for a roof system supported on columns. Costs include joist girders, open web steel joists and 1-1/2" galvanized metal deck.

Design and Pricing Assumptions:
Columns are 18' high.
Joists are 5'-0" O.C.
Joist girders and joists have bottom chords connected to columns. Roof deck is 1-1/2", 22 gauge galvanized steel. Costs include bridging and bracing. Deflection is limited to 1/240 of the span. Fireproofing is not included.

Costs/S.F. are based on a building 4 bays long and 4 bays wide.

Costs for columns are not included, but are listed separately per S.F. of area.

3.7-450			Steel Joists & Joist Girders on Columns					
	BAY SIZE (FT.) GIRD X JOISTS	SUPERIMPOSED LOAD (P.S.F.)	DEPTH (IN.)	TOTAL LOAD (P.S.F.)	COLUMN ADD	COST PER S.F.		
						MAT.	INST.	TOTAL
2000	30x30	20	17-1/2	40		2.01	.83	2.84
2050					columns	.55	.14	.69
2100		30	17-1/2	50		2.20	.88	3.08
2150					columns	.55	.14	.69
2200		40	21-1/2	60		2.27	.90	3.17
2250					columns	.55	.14	.69
2300	30x35	20	32-1/2	40		2.08	.79	2.87
2350					columns	.47	.11	.58
2500		40	36-1/2	60		2.37	.88	3.25
2550					columns	.55	.14	.69
3200	35x40	20	36-1/2	40		2.89	1.20	4.09
3250					columns	.41	.10	.51
3400		40	36-1/2	60		3.30	1.34	4.64
3450					columns	.45	.11	.56
3800	40x40	20	40-1/2	41		2.94	1.22	4.16
3850					columns	.40	.10	.50
4000		40	40-1/2	61		3.33	1.35	4.68
4050					columns	.40	.10	.50
4100	40x45	20	40-1/2	41		3.16	1.37	4.53
4150					columns	.35	.09	.44
4200		30	40-1/2	51		3.45	1.47	4.92
4250					columns	.35	.09	.44
4300		40	40-1/2	61		3.88	1.61	5.49
4350					columns	.40	.10	.50
5000	45x50	20	52-1/2	41		3.35	1.40	4.75
5050					columns	.28	.07	.35
5200		40	52-1/2	61		4.17	1.65	5.82
5250					columns	.36	.09	.45
5300	50x45	20	56-1/2	41		3.25	1.36	4.61
5350					columns	.28	.07	.35
5500		40	56-1/2	61		4.29	1.69	5.98
5550					columns	.36	.09	.45
5600	50x50	20	56-1/2	42		3.66	1.50	5.16
5650					columns	.29	.07	.36
5800		40	59	64		4.47	1.37	5.84
5850					columns	.40	.10	.50

SUPERSTRUCTURES — A3.7 Roofs

The table below lists prices per S.F. for roof rafters and sheathing by nominal size and spacing. Sheathing is 5/16" CDX for 12" and 16" spacing and 3/8" CDX for 24" spacing.

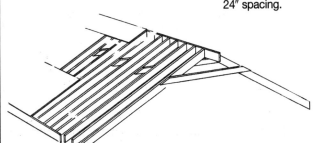

Factors for Converting Inclined to Horizontal

Roof Slope	Approx. Angle	Factor	Roof Slope	Approx. Angle	Factor
Flat	0°	1.000	12 in 12	45.0°	1.414
1 in 12	4.8°	1.003	13 in 12	47.3°	1.474
2 in 12	9.5°	1.014	14 in 12	49.4°	1.537
3 in 12	14.0°	1.031	15 in 12	51.3°	1.601
4 in 12	18.4°	1.054	16 in 12	53.1°	1.667
5 in 12	22.6°	1.083	17 in 12	54.8°	1.734
6 in 12	26.6°	1.118	18 in 12	56.3°	1.803
7 in 12	30.3°	1.158	19 in 12	57.7°	1.873
8 in 12	33.7°	1.202	20 in 12	59.0°	1.943
9 in 12	36.9°	1.250	21 in 12	60.3°	2.015
10 in 12	39.8°	1.302	22 in 12	61.4°	2.088
11 in 12	42.5°	1.357	23 in 12	62.4°	2.162

3.7-510 Wood/Flat or Pitched

		COST PER S.F.		
		MAT.	INST.	TOTAL
2500	Flat rafter, 2"x4", 12" O.C.	.89	.83	1.72
2550	16" O.C.	.78	.72	1.50
2600	24" O.C.	.71	.61	1.32
2900	2"x6", 12" O.C.	1.14	.83	1.97
2950	16" O.C.	.96	.72	1.68
3000	24" O.C.	.83	.61	1.44
3300	2"x8", 12" O.C.	1.42	.90	2.32
3350	16" O.C.	1.17	.77	1.94
3400	24" O.C.	.98	.64	1.62
3700	2"x10", 12" O.C.	1.90	1.02	2.92
3750	16" O.C.	1.53	.86	2.39
3800	24" O.C.	1.20	.70	1.90
4100	2"x12", 12" O.C.	2.20	1.04	3.24
4150	16" O.C.	1.76	.87	2.63
4200	24" O.C.	1.36	.71	2.07
4500	2"x14", 12" O.C.	2.90	1.51	4.41
4550	16" O.C.	2.29	1.23	3.52
4600	24" O.C.	1.72	.95	2.67
4900	3"x6", 12" O.C.	2.52	.89	3.41
4950	16" O.C.	2	.76	2.76
5000	24" O.C.	1.53	.64	2.17
5300	3"x8", 12" O.C.	3.23	.98	4.21
5350	16" O.C.	2.52	.83	3.35
5400	24" O.C.	1.87	.68	2.55
5700	3"x10", 12" O.C.	3.93	1.12	5.05
5750	16" O.C.	3.06	.93	3.99
5800	24" O.C.	2.23	.75	2.98
6100	3"x12", 12" O.C.	4.63	1.35	5.98
6150	16" O.C.	3.58	1.10	4.68
6200	24" O.C.	2.57	.87	3.44
7001	Wood truss, 4 in 12 slope, 24" O.C., 24' to 29' span	3.12	1.25	4.37
7100	30' to 43' span	3.29	1.25	4.54
7200	44' to 60' span	3.48	1.25	4.73

For expanded coverage of these items see *Means Assemblies Cost Data 2000*

SUPERSTRUCTURES — A3.9 Stairs

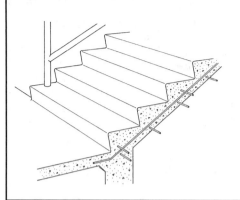

General Design: See reference section for code requirements. Maximum height between landings is 12'; usual stair angle is 20° to 50° with 30° to 35° best. Usual relation of riser to treads is:
- Riser + tread = 17.5.
- 2x (Riser) + tread = 25.
- Riser x tread = 70 or 75.

Maximum riser height is 7" for commercial, 8-1/4" for residential. Usual riser height is 6-1/2" to 7-1/4".

Minimum tread width is 11" for commercial and 9" for residential.

For additional information please see reference section.

Cost Per Flight: Table below lists the cost per flight for 4'-0" wide stairs. Side walls are not included. Railings are included.

3.9-100	Stairs	MAT.	INST.	TOTAL
0470	Stairs, C.I.P. concrete, w/o landing, 12 risers, w/o nosing	695	1,150	1,845
0480	With nosing	1,050	1,300	2,350
0550	W/landing, 12 risers, w/o nosing	775	1,425	2,200
0560	With nosing	1,150	1,575	2,725
0570	16 risers, w/o nosing	965	1,775	2,740
0580	With nosing	1,450	1,975	3,425
0590	20 risers, w/o nosing	1,150	2,150	3,300
0600	With nosing	1,775	2,425	4,200
0610	24 risers, w/o nosing	1,350	2,500	3,850
0620	With nosing	2,075	2,825	4,900
0630	Steel, grate type w/nosing & rails, 12 risers, w/o landing	1,725	410	2,135
0640	With landing	2,250	630	2,880
0660	16 risers, with landing	2,825	770	3,595
0680	20 risers, with landing	3,400	900	4,300
0700	24 risers, with landing	3,950	1,050	5,000
0710	Cement fill metal pan & picket rail, 12 risers, w/o landing	2,250	410	2,660
0720	With landing	2,950	705	3,655
0740	16 risers, with landing	3,700	840	4,540
0760	20 risers, with landing	4,450	975	5,425
0780	24 risers, with landing	5,200	1,125	6,325
0790	Cast iron tread & pipe rail, 12 risers, w/o landing	2,250	410	2,660
0800	With landing	2,950	705	3,655
1120	Wood, prefab box type, oak treads, wood rails 3'-6" wide, 14 risers	955	225	1,180
1150	Prefab basement type, oak treads, wood rails 3'-0" wide, 14 risers	735	180	915

For information about Means Estimating Seminars, see yellow pages 11 and 12 in back of book

Division 4
Exterior Closure

EXTERIOR CLOSURE — A4.1 Walls

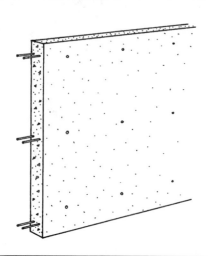

The table below describes a concrete wall system for exterior closure. There are several types of wall finishes priced from plain finish to a finish with 3/4" rustication strip.

Design Assumptions:
Conc. f'c = 3000 to 5000 psi
Reinf. fy = 60,000 psi

4.1-110	Cast In Place Concrete	COST PER S.F.		
		MAT.	INST.	TOTAL
2100	Conc wall reinforced, 8' high, 6" thick, plain finish, 3000 PSI	2.98	8.15	11.13
2700	Aged wood liner, 3000 PSI	4.80	8.50	13.30
3000	Sand blast light 1 side, 3000 PSI	3.20	8.80	12
3700	3/4" bevel rustication strip, 3000 PSI	3.14	9.75	12.89
4000	8" thick, plain finish, 3000 PSI	3.50	8.40	11.90
4100	4000 PSI	3.60	8.40	12
4300	Rub concrete 1 side, 3000 PSI	3.56	9.80	13.36
4550	8" thick, aged wood liner, 3000 PSI	5.30	8.75	14.05
4750	Sand blast light 1 side, 3000 PSI	3.72	9	12.72
5300	3/4" bevel rustication strip, 3000 PSI	3.66	9.95	13.61
5600	10" thick, plain finish, 3000 PSI	4	8.60	12.60
5900	Rub concrete 1 side, 3000 PSI	4.06	10	14.06
6200	Aged wood liner, 3000 PSI	5.80	8.95	14.75
6500	Sand blast light 1 side, 3000 PSI	4.22	9.20	13.42
7100	3/4" bevel rustication strip, 3000 PSI	4.16	10.20	14.36
7400	12" thick, plain finish, 3000 PSI	4.58	8.85	13.43
7700	Rub concrete 1 side, 3000 PSI	4.64	10.25	14.89
8000	Aged wood liner, 3000 PSI	6.40	9.20	15.60
8300	Sand blast light 1 side, 3000 PSI	4.80	9.45	14.25
8900	3/4" bevel rustication strip, 3000 PSI	4.74	10.40	15.14

EXTERIOR CLOSURE — A4.1 — Precast Concrete

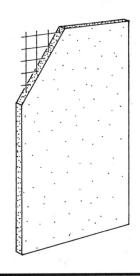

Precast concrete wall panels are either solid or insulated with plain, colored or textured finishes. Transportation is an important cost factor. Prices below are based on delivery within fifty miles of a plant. Engineering data is available from fabricators to assist with construction details. Usual minimum job size for economical use of panels is about 5000 S.F. Small jobs can double the prices below. For large, highly repetitive jobs, deduct up to 15% from the prices below.

4.1-140 Flat Precast Concrete

	THICKNESS (IN.)	PANEL SIZE (FT.)	FINISHES	RIGID INSULATION (IN)	TYPE	COST PER S.F. MAT.	COST PER S.F. INST.	COST PER S.F. TOTAL
3000	4	5x18	smooth gray	none	low rise	5.20	3.48	8.68
3050		6x18				4.35	2.91	7.26
3100		8x20				6.55	1.10	7.65
3150		12x20				6.15	1.04	7.19
3200	6	5x18	smooth gray	2	low rise	6	3.75	9.75
3250		6x18				5.15	3.18	8.33
3300		8x20				7.45	1.40	8.85
3350		12x20				6.80	1.29	8.09
3400	8	5x18	smooth gray	2	low rise	9.50	1.76	11.26
3450		6x18				9.05	1.67	10.72
3500		8x20				8.30	1.55	9.85
3550		12x20				7.55	1.42	8.97
3600	4	4x8	white face	none	low rise	15.95	2.02	17.97
3650		8x8				11.95	1.51	13.46
3700		10x10				10.50	1.33	11.83
3750		20x10				9.50	1.21	10.71
3800	5	4x8	white face	none	low rise	16.30	2.06	18.36
3850		8x8				12.30	1.55	13.85
3900		10x10				10.90	1.38	12.28
3950		20x20				9.95	1.26	11.21
4000	6	4x8	white face	none	low rise	16.85	2.14	18.99
4050		8x8				12.80	1.62	14.42
4100		10x10				11.30	1.43	12.73
4150		20x10				10.35	1.31	11.66
4200	6	4x8	white face	2	low rise	17.70	2.44	20.14
4250		8x8				13.65	1.92	15.57
4300		10x10				12.15	1.73	13.88
4350		20x10				10.35	1.31	11.66
4400	7	4x8	white face	none	low rise	17.25	2.19	19.44
4450		8x8				13.25	1.68	14.93
4500		10x10				11.90	1.51	13.41
4550		20x10				10.90	1.38	12.28
4600	7	4x8	white face	2	low rise	18.10	2.49	20.59
4650		8x8				14.10	1.98	16.08
4700		10x10				12.75	1.81	14.56
4750		20x10				11.75	1.68	13.43

For expanded coverage of these items see *Means Assemblies Cost Data 2000*

EXTERIOR CLOSURE — A4.1 Precast Concrete

4.1-140 Flat Precast Concrete

	THICKNESS (IN.)	PANEL SIZE (FT.)	FINISHES	RIGID INSULATION (IN)	TYPE	COST PER S.F. MAT.	INST.	TOTAL
4800	8	4x8	white face	none	low rise	17.65	2.23	19.88
4850		8x8				13.60	1.72	15.32
4900		10x10				12.25	1.55	13.80
4950		20x10				11.30	1.42	12.72
5000	8	4x8	white face	2	low rise	18.50	2.53	21.03
5050		8x8				14.45	2.02	16.47
5100		10x10				13.10	1.85	14.95
5150		20x10				12.15	1.72	13.87

4.1-140 Fluted Window or Mullion Precast Concrete

	THICKNESS (IN.)	PANEL SIZE (FT.)	FINISHES	RIGID INSULATION (IN)	TYPE	COST PER S.F. MAT.	INST.	TOTAL
5200	4	4x8	smooth gray	none	high rise	11.90	7.95	19.85
5250		8x8				8.50	5.70	14.20
5300		10x10				12.50	2.09	14.59
5350		20x10				10.95	1.84	12.79
5400	5	4x8	smooth gray	none	high rise	12.10	8.10	20.20
5450		8x8				8.80	5.90	14.70
5500		10x10				13	2.18	15.18
5550		20x10				11.50	1.93	13.43
5600	6	4x8	smooth gray	none	high rise	12.45	8.35	20.80
5650		8x8				9.10	6.10	15.20
5700		10x10				13.40	2.24	15.64
5750		20x10				11.90	1.99	13.89
5800	6	4x8	smooth gray	2	high rise	13.30	8.65	21.95
5850		8x8				9.95	6.40	16.35
5900		10x10				14.25	2.54	16.79
5950		20x10				12.75	2.29	15.04
6000	7	4x8	smooth gray	none	high rise	12.70	8.50	21.20
6050		8x8				9.35	6.25	15.60
6100		10x10				13.95	2.34	16.29
6150		20x10				12.25	2.05	14.30
6200	7	4x8	smooth gray	2	high rise	13.55	8.80	22.35
6250		8x8				10.20	6.55	16.75
6300		10x10				14.80	2.64	17.44
6350		20x10				13.10	2.35	15.45
6400	8	4x8	smooth gray	none	high rise	12.85	8.60	21.45
6450		8x8				9.60	6.40	16
6500		10x10				14.40	2.41	16.81
6550		20x10				12.85	2.16	15.01
6600	8	4x8	smooth gray	2	high rise	13.70	8.90	22.60
6650		8x8				10.45	6.70	17.15
6700		10x10				15.25	2.71	17.96
6750		20x10				13.70	2.46	16.16

4.1-140 Precast Concrete Specialties

	TYPE	SIZE				COST PER L.F. MAT.	INST.	TOTAL
6773	Coping, precast	6"wide				11	6.45	17.45
6774	Stock units	10"wide				11.10	6.90	18
6775		12"wide				10.45	7.45	17.90
6776		14" wide				15.30	8.05	23.35

Important: See the Reference Section for critical supporting data - Reference Nos., Crews & Location Factors

EXTERIOR CLOSURE | A4.1 Precast Concrete

4.1-140 Precast Concrete Specialties

	TYPE	SIZE				COST PER L.F. MAT.	INST.	TOTAL
6777	Window sills	6" wide				9.95	6.90	16.85
6778	Precast	10" wide				13.45	8.05	21.50
6779		14" wide				13.35	9.65	23
6780								

For expanded coverage of these items see *Means Assemblies Cost Data 2000*

EXTERIOR CLOSURE — A4.1 Walls

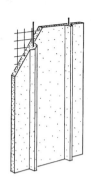

Ribbed Precast Panel

4.1-140 Ribbed Precast Concrete

	THICKNESS (IN.)	PANEL SIZE (FT.)	FINISHES	RIGID INSULATION(IN.)	TYPE	COST PER S.F.		
						MAT.	INST.	TOTAL
6800	4	4x8	aggregate	none	high rise	13.15	7.95	21.10
6850		8x8				9.60	5.80	15.40
6900		10x10				13.75	2.11	15.86
6950		20x10				12.15	1.86	14.01
7000	5	4x8	aggregate	none	high rise	13.40	8.10	21.50
7050		8x8				9.90	6	15.90
7100		10x10				14.25	2.20	16.45
7150		20x10				12.70	1.96	14.66
7200	6	4x8	aggregate	none	high rise	13.65	8.25	21.90
7250		8x8				10.15	6.15	16.30
7300		10x10				14.70	2.26	16.96
7350		20x10				13.20	2.03	15.23
7400	6	4x8	aggregate	2	high rise	14.50	8.55	23.05
7450		8x8				11	6.45	17.45
7500		10x10				15.55	2.56	18.11
7550		20x10				14.05	2.33	16.38
7600	7	4x8	aggregate	none	high rise	13.95	8.45	22.40
7650		8x8				10.40	6.30	16.70
7700		10x10				15.20	2.33	17.53
7750		20x10				13.55	2.09	15.64
7800	7	4x8	aggregate	2	high rise	14.80	8.75	23.55
7850		8x8				11.25	6.60	17.85
7900		10x10				16.05	2.63	18.68
7950		20x10				14.40	2.39	16.79
8000	8	4x8	aggregate	none	high rise	14.20	8.60	22.80
8050		8x8				10.70	6.50	17.20
8100		10x10				15.70	2.41	18.11
8150		20x10				14.20	2.18	16.38
8200	8	4x8	aggregate	2	high rise	15.05	8.90	23.95
8250		8x8				11.55	6.80	18.35
8300		10x10				16.55	2.71	19.26
8350		20x10				15.05	2.48	17.53

EXTERIOR CLOSURE — A4.1 Walls

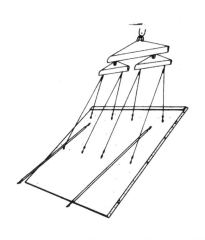

The advantage of tilt up construction is in the low cost of forms and placing of concrete and reinforcing. Tilt up has been used for several types of buildings, including warehouses, stores, offices, and schools. The panels are cast in forms on the ground, or floor slab. Most jobs use 5-1/2" thick solid reinforced concrete panels.

Design Assumptions:
Conc. f'c = 3000 psi
Reinf. fy = 60,000

4.1-160	Tilt-Up Concrete Panel	MAT.	INST.	TOTAL
3200	Tilt up conc panels, broom finish, 5-1/2" thick, 3000 PSI	2.61	2.89	5.50
3250	5000 PSI	2.72	2.85	5.57
3300	6" thick, 3000 PSI	2.88	2.97	5.85
3350	5000 PSI	3	2.91	5.91
3400	7-1/2" thick, 3000 PSI	3.67	3.09	6.76
3450	5000 PSI	3.83	3.04	6.87
3500	8" thick, 3000 PSI	3.95	3.17	7.12
3550	5000 PSI	4.13	3.11	7.24
3700	Steel trowel finish, 5-1/2" thick, 3000 PSI	2.62	2.96	5.58
3750	5000 PSI	2.72	2.91	5.63
3800	6" thick, 3000 PSI	2.88	3.03	5.91
3850	5000 PSI	3	2.97	5.97
3900	7-1/2" thick, 3000 PSI	3.67	3.15	6.82
3950	5000 PSI	3.83	3.10	6.93
4000	8" thick, 3000 PSI	3.95	3.23	7.18
4050	5000 PSI	4.13	3.17	7.30
4200	Exp. aggregate finish, 5-1/2" thick, 3000 PSI	3.09	2.96	6.05
4250	5000 PSI	3.19	2.90	6.09
4300	6" thick, 3000 PSI	3.35	3.02	6.37
4350	5000 PSI	3.47	2.97	6.44
4400	7-1/2" thick, 3000 PSI	4.14	3.15	7.29
4450	5000 PSI	4.30	3.09	7.39
4500	8" thick, 3000 PSI	4.41	3.22	7.63
4550	5000 PSI	4.60	3.17	7.77
4600	Exposed aggregate & vert. rustication 5-1/2" thick, 3000 PSI	4.47	4.76	9.23
4650	5000 PSI	4.57	4.70	9.27
4700	6" thick, 3000 PSI	4.73	4.82	9.55
4750	5000 PSI	4.85	4.77	9.62
4800	7-1/2" thick, 3000 PSI	5.50	4.95	10.45
4850	5000 PSI	5.70	4.89	10.59
4900	8" thick, 3000 PSI	5.80	5	10.80
4950	5000 PSI	6	4.97	10.97
5000	Vertical rib & light sandblast, 5-1/2" thick, 3000 PSI	4.72	3.54	8.26
5100	6" thick, 3000 PSI	4.98	3.60	8.58
5200	7-1/2" thick, 3000 PSI	5.75	3.73	9.48
5300	8" thick, 3000 PSI	6.05	3.80	9.85
6000	Broom finish w/2" urethane insulation, 6" thick, 3000 PSI	2.56	3.59	6.15
6100	Broom finish 2" fiberplank insulation, 6" thick, 3000 PSI	3.08	3.55	6.63
6200	Exposed aggregate w/2" urethane insulation, 6" thick, 3000 PSI	2.97	3.66	6.63
6300	Exposed aggregate 2" fiberplank insulation, 6" thick, 3000 PSI	3.49	3.62	7.11

For expanded coverage of these items see *Means Assemblies Cost Data 2000*

EXTERIOR CLOSURE — A4.1 Walls

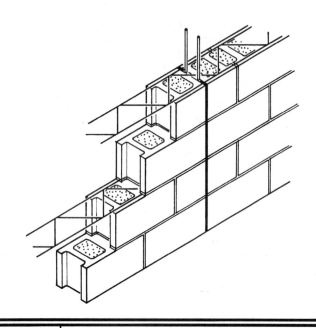

Exterior concrete block walls are defined in the following terms; structural reinforcement, weight, percent solid, size, strength and insulation. Within each of these categories, two to four variations are shown. No costs are included for brick shelf or relieving angles.

4.1-211 Concrete Block Wall - Regular Weight

	TYPE	SIZE (IN.)	STRENGTH (P.S.I.)	CORE FILL		MAT.	INST.	TOTAL
1200	Hollow	4x8x16	2,000	none		1	2.98	3.98
1250			4,500	none		1.01	2.98	3.99
1400		8x8x16	2,000	perlite		1.97	3.62	5.59
1410				styrofoam		2.39	3.42	5.81
1440				none		1.56	3.42	4.98
1450			4,500	perlite		1.99	3.62	5.61
1460				styrofoam		2.41	3.42	5.83
1490				none		1.58	3.42	5
2000	75% solid	4x8x16	2,000	none		.93	3.01	3.94
2100	75% solid	6x8x16	2,000	perlite		1.21	3.31	4.52
2500	Solid	4x8x16	2,000	none		1.43	3.08	4.51
2700		8x8x16	2,000	none		2.19	3.56	5.75

4.1-211 Concrete Block Wall - Lightweight

	TYPE	SIZE (IN.)	WEIGHT (P.C.F.)	CORE FILL		MAT.	INST.	TOTAL
3100	Hollow	8x4x16	105	perlite		1.34	3.42	4.76
3110				styrofoam		1.76	3.22	4.98
3200		4x8x16	105	none		1.16	2.91	4.07
3250			85	none		1.30	2.85	4.15
3300		6x8x16	105	perlite		1.78	3.28	5.06
3310				styrofoam		2.33	3.12	5.45
3340				none		1.50	3.12	4.62
3400		8x8x16	105	perlite		2.26	3.54	5.80
3410				styrofoam		2.68	3.34	6.02
3440				none		1.85	3.34	5.19
3450			85	perlite		2.21	3.45	5.66
4000	75% solid	4x8x16	105	none		1.06	2.95	4.01
4050			85	none		2.20	2.88	5.08
4100		6x8x16	105	perlite		2.15	3.24	5.39
4500	Solid	4x8x16	105	none		1.33	3.05	4.38
4700		8x8x16	105	none		3.50	3.51	7.01

EXTERIOR CLOSURE — A4.1 Walls

4.1-211 Reinforced Concrete Block Wall - Regular Weight

	TYPE	SIZE (IN.)	STRENGTH (P.S.I.)	VERT. REINF & GROUT SPACING		COST PER S.F.		
						MAT.	INST.	TOTAL
5200	Hollow	4x8x16	2,000	#4 @ 48"		1.11	3.31	4.42
5300		6x8x16	2,000	#4 @ 48"		1.53	3.56	5.09
5330				#5 @ 32"		1.66	3.72	5.38
5340				#5 @ 16"		1.98	4.24	6.22
5350			4,500	#4 @ 28"		1.50	3.56	5.06
5390				#5 @ 16"		1.95	4.24	6.19
5400		8x8x16	2,000	#4 @ 48"		1.80	3.80	5.60
5430				#5 @ 32"		1.94	4.06	6
5440				#5 @ 16"		2.32	4.69	7.01
5450		8x8x16	4,500	#4 @ 48"		1.81	3.86	5.67
5490				#5 @ 16"		2.34	4.69	7.03
5500		12x8x16	2,000	#4 @ 48"		2.63	4.87	7.50
5540				#5 @ 16"		3.37	5.75	9.12
6100	75% solid	6x8x16	2,000	#4 @ 48"		1.16	3.51	4.67
6140				#5 @ 16"		1.43	4.01	5.44
6150			4,500	#4 @ 48"		1.99	3.51	5.50
6190				#5 @ 16"		2.26	4.01	6.27
6200		8x8x16	2,000	#4 @ 48"		1.31	3.80	5.11
6230				#5 @ 32"		1.40	3.93	5.33
6240				#5 @ 16"		1.60	4.40	6
6250			4,500	#4 @ 48"		2.89	3.80	6.69
6280				#5 @ 32"		2.98	3.93	6.91
6290				#5 @ 16"		3.18	4.40	7.58
6500	Solid-double Wythe	2-4x8x16	2,000	#4 @ 48" E.W.		3.50	7.05	10.55
6530				#5 @ 16" E.W.		3.92	7.50	11.42
6550			4,500	#4 @ 48" E.W.		3.08	7	10.08
6580				#5 @ 16" E.W.		3.50	7.45	10.95

4.1-211 Reinforced Concrete Block Wall - Lightweight

	TYPE	SIZE (IN.)	WEIGHT (P.C.F.)	VERT REINF. & GROUT SPACING		COST PER S.F.		
						MAT.	INST.	TOTAL
7100	Hollow	8x4x16	105	#4 @ 48"		1.16	3.66	4.82
7140				#5 @ 16"		1.69	4.49	6.18
7150			85	#4 @ 48"		3.14	3.58	6.72
7190				#5 @ 16"		3.67	4.41	8.08
7400		8x8x16	105	#4 @ 48"		2.08	3.78	5.86
7440				#5 @ 16"		2.61	4.61	7.22
7450		8x8x16	85	#4 @ 48"		2.03	3.69	5.72
7490				#5 @ 16"		2.56	4.52	7.08
7800		8x8x24	105	#4 @ 48"		1.60	4.06	5.66
7840				#5 @ 16"		2.13	4.89	7.02
7850			85	#4 @ 48"		3.43	3.47	6.90
7890				#5 @ 16"		3.96	4.30	8.26
8100	75% solid	6x8x16	105	#4 @ 48"		2.10	3.44	5.54
8130				#5 @ 32"		2.19	3.55	5.74
8150			85	#4 @ 48"		2.61	3.36	5.97
8180				#5 @ 32"		2.70	3.47	6.17
8200		8x8x16	105	#4 @ 48"		3	3.71	6.71
8230				#5 @ 32"		3.09	3.84	6.93
8250			85	#4 @ 48"		3.55	3.62	7.17
8280				#5 @ 32"		3.64	3.75	7.39

For expanded coverage of these items see *Means Assemblies Cost Data 2000*

EXTERIOR CLOSURE — A4.1 Walls

4.1-211 Reinforced Concrete Block Wall - Lightweight

	TYPE	SIZE (IN.)	WEIGHT (P.C.F.)	VERT REINF. & GROUT SPACING		COST PER S.F.		
						MAT.	INST.	TOTAL
8500	Solid-double	2-4x8x16	105	#4 @ 48"		3.30	7	10.30
8530	Wythe		105	#5 @ 16"		3.72	7.45	11.17
8600		2-6x8x16	105	#4 @ 48"		5.55	7.45	13
8630			105	#5 @ 16"		5.95	7.90	13.85
8650			85	#4 @ 48"		6.10	7.15	13.25

EXTERIOR CLOSURE — A4.1 Walls

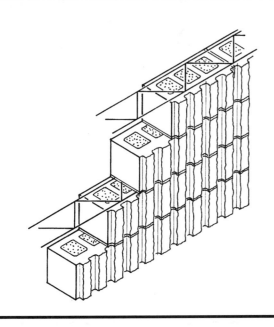

Exterior split ribbed block walls are defined in the following terms; structural reinforcement, weight, percent solid, size, number of ribs and insulation. Within each of these categories two to four variations are shown. No costs are included for brick shelf or relieving angles. Costs include control joints every 20′ and horizontal reinforcing.

4.1-212 Split Ribbed Block Wall - Regular Weight

	TYPE	SIZE (IN.)	RIBS	CORE FILL	MAT.	INST.	TOTAL
1220	Hollow	4x8x16	4	none	2.11	3.69	5.80
1250			8	none	2.46	3.69	6.15
1280			16	none	2	3.74	5.74
1430		8x8x16	8	perlite	4.21	4.38	8.59
1440				styrofoam	4.63	4.18	8.81
1450				none	3.80	4.18	7.98
1530		12x8x16	8	perlite	5.25	5.85	11.10
1540				styrofoam	5.65	5.45	11.10
1550				none	4.60	5.45	10.05
2120	75% solid	4x8x16	4	none	2.64	3.74	6.38
2150			8	none	3.10	3.74	6.84
2180			16	none	2.53	3.80	6.33
2520	Solid	4x8x16	4	none	3.20	3.80	7
2550			8	none	3.75	3.80	7.55
2580			16	none	3.03	3.85	6.88

4.1-212 Reinforced Split Ribbed Block Wall - Regular Weight

	TYPE	SIZE (IN.)	RIBS	VERT. REINF. & GROUT SPACING	MAT.	INST.	TOTAL
5200	Hollow	4x8x16	4	#4 @ 48″	2.22	4.02	6.24
5230			8	#4 @ 48″	2.57	4.02	6.59
5260			16	#4 @ 48″	2.11	4.07	6.18
5430		8x8x16	8	#4 @ 48″	4.03	4.62	8.65
5440				#5 @ 32″	4.18	4.82	9
5450				#5 @ 16″	4.56	5.45	10.01
5530		12x8x16	8	#4 @ 48″	4.93	5.90	10.83
5540				#5 @ 32″	5.15	6.10	11.25
5550				#5 @ 16″	5.65	6.75	12.40
6230	75% solid	6x8x16	8	#4 @ 48″	3.96	4.25	8.21
6240				#5 @ 32″	4.05	4.36	8.41
6250				#5 @ 16″	4.23	4.75	8.98
6330		8x8x16	8	#4 @ 48″	4.96	4.58	9.54
6340				#5 @ 32″	5.05	4.71	9.76
6350				#5 @ 16″	5.25	5.20	10.45

For expanded coverage of these items see *Means Assemblies Cost Data 2000*

EXTERIOR CLOSURE — A4.1 Walls

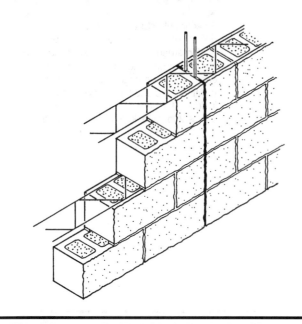

Exterior split face block walls are defined in the following terms; structural reinforcement, weight, percent solid, size, scores and insulation. Within each of these categories two to four variations are shown. No costs are included for brick shelf or relieving angles. Costs include control joints every 20' and horizontal reinforcing.

4.1-213 Split Face Block Wall - Regular Weight

	TYPE	SIZE (IN.)	SCORES	CORE FILL		MAT.	INST.	TOTAL
1200	Hollow	8x4x16	0	perlite		6.15	4.82	10.97
1210				styrofoam		6.55	4.62	11.17
1240				none		5.70	4.62	10.32
1250			1	perlite		6.60	4.82	11.42
1260				styrofoam		7	4.62	11.62
1290				none		6.15	4.62	10.77
1300		12x4x16	0	perlite		8	5.50	13.50
1310				styrofoam		8.40	5.10	13.50
1340				none		7.35	5.10	12.45
1350			1	perlite		8.40	5.50	13.90
1360				styrofoam		8.80	5.10	13.90
1390				none		7.75	5.10	12.85
1400		4x8x16	0	none		2.11	3.64	5.75
1450			1	none		2.37	3.69	6.06
1500		6x8x16	0	perlite		3.01	4.19	7.20
1510				styrofoam		3.56	4.03	7.59
1540				none		2.73	4.03	6.76
1550			1	perlite		3.28	4.25	7.53
1560				styrofoam		3.83	4.09	7.92
1590				none		3	4.09	7.09
1600		8x8x16	0	perlite		3.63	4.52	8.15
1610				styrofoam		4.05	4.32	8.37
1640				none		3.22	4.32	7.54
1650		8x8x16	1	perlite		3.85	4.59	8.44
1660				styrofoam		4.27	4.39	8.66
1690				none		3.44	4.39	7.83
1700		12x8x16	0	perlite		4.65	5.90	10.55
1710				styrofoam		5.05	5.50	10.55
1740				none		3.98	5.50	9.48
1750			1	perlite		4.87	6.05	10.92
1760				styrofoam		5.25	5.65	10.90
1790				none		4.20	5.65	9.85

EXTERIOR CLOSURE — A4.1 Walls

4.1-213 Split Face Block Wall - Regular Weight

	TYPE	SIZE (IN.)	SCORES	CORE FILL		COST PER S.F. MAT.	INST.	TOTAL
1800	75% solid	8x4x16	0	perlite		7.45	4.81	12.26
1840				none		7.20	4.71	11.91
1852		8x4x16	1	perlite		7.90	4.81	12.71
1890				none		7.65	4.71	12.36
2000		4x8x16	0	none		2.64	3.69	6.33
2050			1	none		2.86	3.74	6.60
2400	Solid	8x4x16	0	none		8.80	4.79	13.59
2450			1	none		9.25	4.79	14.04
2900		12x8x16	0	none		6.05	5.75	11.80
2950			1	none		6.25	5.85	12.10

4.1-213 Reinforced Split Face Block Wall - Regular Weight

	TYPE	SIZE (IN.)	SCORES	VERT. REINF. & GROUT SPACING		COST PER S.F. MAT.	INST.	TOTAL
5200	Hollow	8x4x16	0	#4 @ 48"		5.95	5.10	11.05
5210				#5 @ 32"		6.10	5.25	11.35
5240				#5 @ 16"		6.50	5.90	12.40
5250			1	#4 @ 48"		6.40	5.10	11.50
5260				#5 @ 32"		6.55	5.25	11.80
5290				#5 @ 16"		6.95	5.90	12.85
5700		12x8x16	0	#4 @ 48"		4.31	5.95	10.26
5710				#5 @ 32"		4.52	6.15	10.67
5740				#5 @ 16"		5.05	6.80	11.85
5750			1	#4 @ 48"		4.53	6.10	10.63
5760				#5 @ 32"		4.74	6.30	11.04
5790				#5 @ 16"		5.25	6.95	12.20
6000	75% solid	8x4x16	0	#4 @ 48"		7.35	5.05	12.40
6010				#5 @ 32"		7.40	5.15	12.55
6040				#5 @ 16"		7.60	5.65	13.25
6050			1	#4 @ 48"		7.80	5.05	12.85
6060				#5 @ 32"		7.85	5.15	13
6090				#5 @ 16"		8.05	5.65	13.70
6100		12x4x16	0	#4 @ 48"		9.45	5.55	15
6110				#5 @ 32"		9.60	5.75	15.35
6140				#5 @ 16"		9.85	6.15	16
6150			1	#4 @ 48"		9.80	5.55	15.35
6160				#5 @ 32"		9.95	5.75	15.70
6190				#5 @ 16"		10.25	6.20	16.45
6700	Solid-double Wythe	2-4x8x16	0	#4 @ 48" E.W.		7.05	8.40	15.45
6710				#5 @ 32" E.W.		7.20	8.45	15.65
6740				#5 @ 16" E.W.		7.45	8.80	16.25
6750			1	#4 @ 48" E.W.		7.50	8.50	16
6760				#5 @ 32" E.W.		7.65	8.60	16.25
6790				#5 @ 16" E.W.		7.90	8.95	16.85
6800		2-6x8x16	0	#4 @ 48" E.W.		9.10	9.25	18.35
6810				#5 @ 32" E.W.		9.25	9.30	18.55
6840				#5 @ 16" E.W.		9.55	9.65	19.20
6850			1	#4 @ 48" E.W.		9.55	9.40	18.95
6860				#5 @ 32" E.W.		9.70	9.45	19.15
6890				#5 @ 16" E.W.		9.95	9.80	19.75

For expanded coverage of these items see *Means Assemblies Cost Data 2000*

EXTERIOR CLOSURE — A4.1 Walls

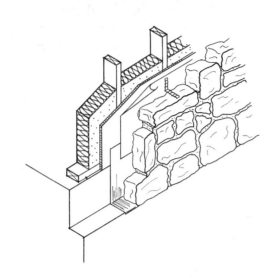

The table below lists costs per S.F. for stone veneer walls on various backup using different stone. Typical components for a system are shown in the component block below.

4.1-242	Stone Veneer	COST PER S.F.		
		MAT.	INST.	TOTAL
2000	Ashlar veneer, 4", 2"x4" stud backup, 16" O.C., 8' high, low priced stone	6.90	9.25	16.15
2100	Metal stud backup, 8' high, 16" O.C.	7.15	9.40	16.55
2150	24" O.C.	6.95	9.15	16.10
2200	Conc. block backup, 4" thick	7.25	11.40	18.65
2300	6" thick	7.60	11.55	19.15
2350	8" thick	7.80	11.70	19.50
2400	10" thick	8.40	11.75	20.15
2500	12" thick	8.55	12.50	21.05
3100	High priced stone, wood stud backup, 10' high, 16" O.C.	12.45	10.45	22.90
3200	Metal stud backup, 10' high, 16" O.C.	12.65	10.60	23.25
3250	24" O.C.	12.45	10.35	22.80
3300	Conc. block backup, 10' high, 4" thick	12.75	12.55	25.30
3350	6" thick	13.10	12.60	25.70
3400	8" thick	13.30	12.85	26.15
3450	10" thick	13.90	12.95	26.85
3500	12" thick	14.05	13.70	27.75
4000	Indiana limestone 2" thick, sawn finish, wood stud backup, 10' high, 16" O.C.	14.05	7.85	21.90
4100	Metal stud backup, 10' high, 16" O.C.	14.30	8	22.30
4150	24" O.C.	14.10	7.75	21.85
4200	Conc. block backup, 4" thick	14.40	10	24.40
4250	6" thick	14.75	10.10	24.85
4300	8" thick	14.95	10.25	25.20
4350	10" thick	15.55	10.35	25.90
4400	12" thick	15.70	11.10	26.80
4450	2" thick, smooth finish, wood stud backup, 8' high, 16" O.C.	14.05	7.85	21.90
4550	Metal stud backup, 8' high, 16" O.C.	14.30	7.90	22.20
4600	24" O.C.	14.10	7.65	21.75
4650	Conc. block backup, 4" thick	14.40	9.90	24.30
4700	6" thick	14.75	10.05	24.80
4750	8" thick	14.95	10.20	25.15
4800	10" thick	15.55	10.35	25.90
4850	12" thick	15.70	11.10	26.80
5350	4" thick, smooth finish, wood stud backup, 8' high, 16" O.C.	18.80	7.85	26.65
5450	Metal stud backup, 8' high, 16" O.C.	19.05	8	27.05
5500	24" O.C.	18.85	7.75	26.60
5550	Conc. block backup, 4" thick	19.15	10	29.15
5600	6" thick	19.50	10.10	29.60
5650	8" thick	19.70	10.25	29.95
5700	10" thick	20.50	10.35	30.85
5750	12" thick	20.50	11.10	31.60

EXTERIOR CLOSURE — A4.1 Walls

4.1-242 Stone Veneer

		COST PER S.F.		
		MAT.	INST.	TOTAL
6000	Granite, grey or pink, 2" thick, wood stud backup, 8' high, 16" O.C.	18.80	13.05	31.85
6100	Metal studs, 8' high, 16" O.C.	19.05	13.20	32.25
6150	24" O.C.	18.85	12.95	31.80
6200	Conc. block backup, 4" thick	19.15	15.15	34.30
6250	6" thick	19.50	15.30	34.80
6300	8" thick	19.70	15.45	35.15
6350	10" thick	20.50	15.55	36.05
6400	12" thick	20.50	16.30	36.80
6900	4" thick, wood stud backup, 8' high, 16" O.C.	30	14.75	44.75
7000	Metal studs, 8' high, 16" O.C.	30.50	14.90	45.40
7050	24" O.C.	30	14.65	44.65
7100	Conc. block backup, 4" thick	30.50	16.85	47.35
7150	6" thick	31	17	48
7200	8" thick	31	17.15	48.15
7250	10" thick	31.50	17.25	48.75
7300	12" thick	32	18	50

For expanded coverage of these items see *Means Assemblies Cost Data 2000*

EXTERIOR CLOSURE   A4.1   Brick Veneer/Stud Backup

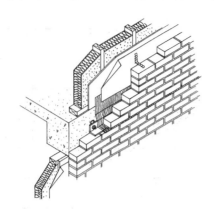

Exterior brick veneer/stud backup walls are defined in the following terms: type of brick and studs, stud spacing and bond. All systems include a back-up wall, a control joint every 20', a brick shelf every 12' of height, ties to the backup and the necessary dampproofing, flashing and insulation.

4.1-252 Brick Veneer/Wood Stud Backup

	FACE BRICK	STUD BACKUP	STUD SPACING (IN.)	BOND		COST PER S.F.		
						MAT.	INST.	TOTAL
1100	Standard	2x4-wood	16	running		5.20	8.30	13.50
1120				common		5.85	9.35	15.20
1140				Flemish		6.30	10.90	17.20
1160				English		6.85	11.50	18.35
1400		2x6-wood	16	running		5.40	8.35	13.75
1420				common		6.05	9.40	15.45
1440				Flemish		6.50	10.95	17.45
1460				English		7.05	11.55	18.60
1700	Glazed	2x4-wood	16	running		9.80	8.55	18.35
1720				common		11.35	9.70	21.05
1740				Flemish		12.40	11.50	23.90
1760				English		13.70	12.15	25.85
2300	Engineer	2x4-wood	16	running		5.30	7.40	12.70
2320				common		5.95	8.30	14.25
2340				Flemish		6.40	9.70	16.10
2360				English		6.95	10.15	17.10
2900	Roman	2x4-wood	16	running		7	7.60	14.60
2920				common		8	8.55	16.55
2940				Flemish		8.65	9.95	18.60
2960				English		9.55	10.65	20.20
4100	Norwegian	2x4-wood	16	running		5.35	5.95	11.30
4120				common		6	6.55	12.55
4140				Flemish		6.45	7.60	14.05
4160				English		7	7.90	14.90

Important: See the Reference Section for critical supporting data - Reference Nos., Crews & Location Factors

EXTERIOR CLOSURE — A4.1 Walls

4.1-252 Brick Veneer/Metal Stud Backup

	FACE BRICK	STUD BACKUP	STUD SPACING (IN.)	BOND		COST PER S.F.		
						MAT.	INST.	TOTAL
5100	Standard	25ga.x6"NLB	24	running		4.76	8.30	13.06
5120				common		5.40	9.35	14.75
5140				Flemish		5.60	10.40	16
5160				English		6.40	11.50	17.90
5200		20ga.x3-5/8"NLB	16	running		4.84	8.35	13.19
5220				common		5.50	9.40	14.90
5240				Flemish		5.90	10.95	16.85
5260				English		6.45	11.55	18
5400		16ga.x3-5/8"LB	16	running		5.25	8.55	13.80
5420				common		5.90	9.60	15.50
5440				Flemish		6.35	11.15	17.50
5460				English		6.90	11.75	18.65
5700	Glazed	25ga.x6"NLB	24	running		9.35	8.55	17.90
5720				common		10.90	9.70	20.60
5740				Flemish		11.95	11.50	23.45
5760				English		13.25	12.15	25.40
5800		20ga.x3-5/8"NLB	24	running		9.35	8.50	17.85
5820				common		10.90	9.65	20.55
5840				Flemish		11.95	11.45	23.40
5860				English		13.25	12.10	25.35
6000		16ga.x3-5/8"LB	16	running		9.85	8.80	18.65
6020				common		11.40	9.95	21.35
6040				Flemish		12.45	11.75	24.20
6060				English		13.75	12.40	26.15
6300	Engineer	25ga.x6"NLB	24	running		4.86	7.40	12.26
6320				common		5.50	8.30	13.80
6340				Flemish		5.95	9.70	15.65
6360				English		6.50	10.15	16.65
6400		20ga.x3-5/8"NLB	16	running		4.94	7.45	12.39
6420				common		5.60	8.35	13.95
6440				Flemish		6.05	9.75	15.80
6460				English		6.60	10.20	16.80
6900	Roman	25ga.x6"NLB	24	running		6.55	7.60	14.15
6920				common		7.55	8.55	16.10
6940				Flemish		8.20	9.95	18.15
6960				English		9.10	10.65	19.75
7000		20ga.x3-5/8"NLB	16	running		6.60	7.65	14.25
7020				common		7.65	8.60	16.25
7040				Flemish		8.30	10	18.30
7060				English		9.20	10.70	19.90
7500	Norman	25ga.x6"NLB	24	running		5.20	6.50	11.70
7520				common		5.95	7.25	13.20
7540				Flemish		9.85	8.40	18.25
7560				English		7.05	8.85	15.90
7600		20ga.x3-5/8"NLB	24	running		5.20	6.50	11.70
7620				common		5.95	7.20	13.15
7640				Flemish		9.85	8.35	18.20
7660				English		7.05	8.80	15.85
8100	Norwegian	25ga.x6"NLB	24	running		4.90	5.95	10.85
8120				common		5.55	6.55	12.10
8140				Flemish		6	7.60	13.60
8160				English		6.60	7.90	14.50

For expanded coverage of these items see *Means Assemblies Cost Data 2000*

EXTERIOR CLOSURE			A4.1	Walls				

4.1-252 Brick Veneer/Metal Stud Backup

	FACE BRICK	STUD BACKUP	STUD SPACING (IN.)	BOND		COST PER S.F.		
						MAT.	INST.	TOTAL
8400	Norwegian	16ga.x3-5/8"LB	16	running		5.40	6.20	11.60
8420				common		6.05	6.85	12.90
8440				Flemish		6.50	7.85	14.35
8460				English		7.10	8.15	15.25

EXTERIOR CLOSURE — A4.1 Walls

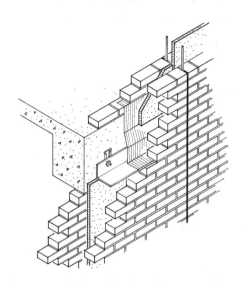

Exterior brick face cavity walls are defined in the following terms: cavity treatment, type of face brick, backup masonry, total thickness and insulation. Seven types of face brick are shown with fourteen types of backup. All systems include a brick shelf, ties to the backups and necessary dampproofing, flashing, and control joints every 20'.

4.1-273 Brick Face Cavity Wall

	FACE BRICK	BACKUP MASONRY	TOTAL THICKNESS (IN.)	CAVITY INSULATION		COST PER S.F. MAT.	INST.	TOTAL
1000	Standard	4"common brick	10	polystyrene		6.15	12.90	19.05
1020				none		6.05	12.55	18.60
1040		6"SCR brick	12	polystyrene		8.50	11.45	19.95
1060				none		8.35	11.15	19.50
1080		4"conc. block	10	polystyrene		4.91	10.35	15.26
1100				none		4.78	10	14.78
1120		6"conc. block	12	polystyrene		5.25	10.60	15.85
1140				none		5.10	10.25	15.35
1160		4"L.W. block	10	polystyrene		5.05	10.30	15.35
1180				none		4.94	9.95	14.89
1200		6"L.W. block	12	polystyrene		5.40	10.50	15.90
1220				none		5.30	10.15	15.45
1240		4"glazed block	10	polystyrene		10.95	11.05	22
1260				none		10.85	10.75	21.60
1280		6"glazed block	12	polystyrene		11	11.05	22.05
1300				none		10.85	10.75	21.60
1320		4"clay tile	10	polystyrene		8.50	9.95	18.45
1340				none		8.35	9.60	17.95
1360		4"glazed tile	10	polystyrene		10.15	13.15	23.30
1380				none		10.05	12.80	22.85
1500	Glazed	4" common brick	10	polystyrene		10.75	13.15	23.90
1520				none		10.60	12.80	23.40
1580		4" conc. block	10	polystyrene		9.50	10.60	20.10
1600				none		9.35	10.25	19.60
1660		4" L.W. block	10	polystyrene		9.65	10.55	20.20
1680				none		9.55	10.20	19.75
1740		4" glazed block	10	polystyrene		15.55	11.30	26.85
1760				none		15.45	11	26.45
1820		4" clay tile	10	polystyrene		13.10	10.20	23.30
1840				none		12.95	9.85	22.80
1860		4" glazed tile	10	polystyrene		12.90	10.50	23.40
1880				none		12.75	10.15	22.90
2000	Engineer	4" common brick	10	polystyrene		6.25	12	18.25
2020				none		6.15	11.65	17.80
2080		4" conc. block	10	polystyrene		5	9.45	14.45
2100				none		4.88	9.15	14.03

For expanded coverage of these items see *Means Assemblies Cost Data 2000*

EXTERIOR CLOSURE — A4.1 Walls

4.1-273 Brick Face Cavity Wall

	FACE BRICK	BACKUP MASONRY	TOTAL THICKNESS (IN.)	CAVITY INSULATION		COST PER S.F. MAT.	INST.	TOTAL
2162	Engineer	4" L.W. block	10	polystyrene		5.15	9.40	14.55
2180				none		5.05	9.05	14.10
2240		4" glazed block	10	polystyrene		11.05	10.20	21.25
2260				none		10.95	9.85	20.80
2320		4" clay tile	10	polystyrene		8.60	9.05	17.65
2340				none		8.45	8.75	17.20
2360		4" glazed tile	10	polystyrene		10.25	12.25	22.50
2380				none		10.15	11.90	22.05
2500	Roman	4" common brick	10	polystyrene		7.95	12.20	20.15
2520				none		7.80	11.85	19.65
2580		4" conc. block	10	polystyrene		6.70	9.65	16.35
2600				none		6.55	9.30	15.85
2660		4" L.W. block	10	polystyrene		6.85	9.60	16.45
2680				none		6.70	9.25	15.95
2740		4" glazed block	10	polystyrene		12.75	10.35	23.10
2760				none		12.60	10.05	22.65
2820		4" clay tile	10	polystyrene		10.30	9.25	19.55
2840				none		10.15	8.90	19.05
2860		4" glazed tile	10	polystyrene		11.95	12.45	24.40
2880				none		11.80	12.10	23.90
3000	Norman	4" common brick	10	polystyrene		6.60	11.10	17.70
3020				none		6.50	10.75	17.25
3080		4" conc. block	10	polystyrene		5.35	8.60	13.95
3100				none		5.25	8.25	13.50
3160		4" L.W. block	10	polystyrene		5.50	8.50	14
3180				none		5.40	8.15	13.55
3240		4" glazed block	10	polystyrene		11.40	9.30	20.70
3260				none		11.30	8.95	20.25
3320		4" clay tile	10	polystyrene		8.95	8.20	17.15
3340				none		8.80	7.85	16.65
3360		4" glazed tile	10	polystyrene		10.60	11.35	21.95
3380				none		10.50	11	21.50
3500	Norwegian	4" common brick	10	polystyrene		6.30	10.55	16.85
3520				none		6.15	10.20	16.35
3580		4" conc. block	10	polystyrene		5.05	8	13.05
3600				none		4.92	7.65	12.57
3660		4" L.W. block	10	polystyrene		5.20	7.95	13.15
3680				none		5.10	7.60	12.70
3740		4" glazed block	10	polystyrene		11.10	8.70	19.80
3760				none		11	8.40	19.40
3820		4" clay tile	10	polystyrene		8.65	7.60	16.25
3840				none		8.50	7.25	15.75
3860		4" glazed tile	10	polystyrene		10.30	10.80	21.10
3880				none		10.20	10.45	20.65
4000	Utility	4" common brick	10	polystyrene		6.05	10	16.05
4020				none		5.90	9.65	15.55
4080		4" conc. block	10	polystyrene		4.79	7.45	12.24
4100				none		4.66	7.10	11.76
4160		4" L.W. block	10	polystyrene		4.95	7.40	12.35
4180				none		4.82	7.05	11.87
4240		4" glazed block	10	polystyrene		10.85	8.15	19
4260				none		10.70	7.85	18.55
4320		4" clay tile	10	polystyrene		8.40	7.05	15.45
4340				none		8.25	6.70	14.95

Important: See the Reference Section for critical supporting data - Reference Nos., Crews & Location Factors

EXTERIOR CLOSURE		A4.1	Walls				
4.1-273		Brick Face Cavity Wall					

	FACE BRICK	BACKUP MASONRY	TOTAL THICKNESS (IN.)	CAVITY INSULATION		COST PER S.F.		
						MAT.	INST.	TOTAL
4360	Utility	4" glazed tile	10	polystyrene		10.05	10.25	20.30
4380				none		9.90	9.90	19.80

For expanded coverage of these items see *Means Assemblies Cost Data 2000*

EXTERIOR CLOSURE — A4.1 Walls

4.1-273 Brick Face Cavity Wall - Insulated Backup

	FACE BRICK	BACKUP MASONRY	TOTAL THICKNESS (IN.)	BACKUP CORE FILL		COST PER S.F. MAT.	COST PER S.F. INST.	COST PER S.F. TOTAL
5100	Standard	6" conc. block	10	perlite		5.40	10.40	15.80
5120				styrofoam		5.95	10.25	16.20
5180		6" L.W. block	10	perlite		5.55	10.30	15.85
5200				styrofoam		6.10	10.15	16.25
5260		6" glazed block	10	perlite		11.50	11.05	22.55
5280				styrofoam		12.05	10.90	22.95
5340		6" clay tile	10	none		8.15	9.90	18.05
5360		8" clay tile	12	none		9.10	10.25	19.35
5600	Glazed	6" conc. block	10	perlite		10	10.65	20.65
5620				styrofoam		10.55	10.50	21.05
5680		6" L.W. block	10	perlite		10.15	10.55	20.70
5700				styrofoam		10.70	10.40	21.10
5760		6" glazed block	10	perlite		16.10	11.30	27.40
5780				styrofoam		16.65	11.15	27.80
5840		6" clay tile	10	none		12.75	10.15	22.90
5860		8" clay tile	8	none		13.70	10.50	24.20
6100	Engineer	6" conc. block	10	perlite		5.50	9.50	15
6120				styrofoam		6.05	9.35	15.40
6180		6" L.W. block	10	perlite		5.65	9.45	15.10
6200				styrofoam		6.20	9.25	15.45
6260		6" glazed block	10	perlite		11.60	10.15	21.75
6280				styrofoam		12.15	10	22.15
6340		6" clay tile	10	none		8.25	9	17.25
6360		8" clay tile	12	none		9.20	9.35	18.55
6600	Roman	6" conc. block	10	perlite		7.15	9.70	16.85
6620				styrofoam		7.70	9.55	17.25
6680		6" L.W. block	10	perlite		7.35	9.60	16.95
6700				styrofoam		7.90	9.45	17.35
6760		6" glazed block	10	perlite		13.30	10.35	23.65
6780				styrofoam		13.85	10.20	24.05
6840		6" clay tile	10	none		9.95	9.20	19.15
6860		8" clay tile	12	none		10.90	9.55	20.45
7100	Norman	6" conc. block	10	perlite		5.85	8.60	14.45
7120				styrofoam		6.40	8.45	14.85
7180		6" L.W. block	10	perlite		6	8.55	14.55
7200				styrofoam		6.55	8.40	14.95
7260		6" glazed block	10	perlite		11.95	9.25	21.20
7280				styrofoam		12.50	9.10	21.60
7340		6" clay tile	10	none		8.60	8.10	16.70
7360		8" clay tile	12	none		9.55	8.45	18
7600	Norwegian	6" conc. block	10	perlite		5.55	8.05	13.60
7620				styrofoam		6.10	7.90	14
7680		6" L.W. block	10	perlite		5.70	7.95	13.65
7700				styrofoam		6.25	7.80	14.05
7760		6" glazed block	10	perlite		11.65	8.70	20.35
7780				styrofoam		12.20	8.55	20.75
7840		6" clay tile	10	none		8.30	7.55	15.85
7860		8" clay tile	12	none		9.25	7.90	17.15
8100	Utility	6" conc. block	10	perlite		5.25	7.50	12.75
8120				styrofoam		5.80	7.35	13.15
8180		6" L.W. block	10	perlite		5.45	7.40	12.85
8200				styrofoam		6	7.25	13.25
8260		6" glazed block	10	perlite		11.40	8.15	19.55
8280				styrofoam		11.95	8	19.95

EXTERIOR CLOSURE — A4.1 Walls

4.1-273 Brick Face Cavity Wall - Insulated Backup

	FACE BRICK	BACKUP MASONRY	TOTAL THICKNESS (IN.)	BACKUP CORE FILL		COST PER S.F.		
						MAT.	INST.	TOTAL
8340	Utility	6" clay tile	10	none		8.05	7	15.05
8360		8" clay tile	12	none		9	7.35	16.35

EXTERIOR CLOSURE — A4.1 Walls

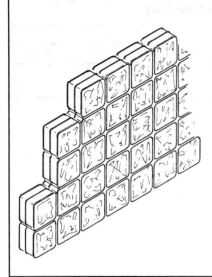

The table below lists costs per S.F. for glass block walls. Included in the costs are the following special accessories required for glass block walls.

Glass block accessories required for proper installation.

Wall ties: Galvanized double steel mesh full length of joint.

Fiberglass expansion joint at sides and top.

Silicone caulking: One gallon does 95 L.F.

Oakum: One lb. does 30 L.F.

Asphalt emulsion: One gallon does 600 L.F.

If block are not set in wall chase, use 2'-0" long wall anchors at 2'-0" O.C.

4.1-282	Glass Block	MAT.	INST.	TOTAL
2300	Glass block 4" thick, 6"x6" plain, under 1,000 S.F.	14.40	12	26.40
2400	1,000 to 5,000 S.F.	13.60	10.40	24
2500	Over 5,000 S.F.	13.70	9.80	23.50
2600	Solar reflective, under 1,000 S.F.	20	16.30	36.30
2700	1,000 to 5,000 S.F.	19	14.05	33.05
2800	Over 5,000 S.F.	19.15	13.20	32.35
3500	8"x8" plain, under 1,000 S.F.	10.30	9	19.30
3600	1,000 to 5,000 S.F.	9.40	7.75	17.15
3700	Over 5,000 S.F.	9.20	7	16.20
3800	Solar reflective, under 1,000 S.F.	14.40	12.10	26.50
3900	1,000 to 5,000 S.F.	13.15	10.35	23.50
4000	Over 5,000 S.F.	12.85	9.30	22.15
5000	12"x12" plain, under 1,000 S.F.	12.30	8.30	20.60
5100	1,000 to 5,000 S.F.	10.50	7	17.50
5200	Over 5,000 S.F.	10.30	6.40	16.70
5300	Solar reflective, under 1,000 S.F.	17.20	11.10	28.30
5400	1,000 to 5,000 S.F.	14.65	9.30	23.95
5600	Over 5,000 S.F.	14.40	8.45	22.85
5800	3" thinline, 6"x6" plain, under 1,000 S.F.	10.40	12	22.40
5900	Over 5,000 S.F.	10.40	12	22.40
6000	Solar reflective, under 1,000 S.F.	14.55	16.30	30.85
6100	Over 5,000 S.F.	13	13.20	26.20
6200	8"x8" plain, under 1,000 S.F.	6.45	9	15.45
6300	Over 5,000 S.F.	5.80	7	12.80
6400	Solar reflective, under 1,000 S.F.	9	12.10	21.10
6500	Over 5,000 S.F.	8.10	9.30	17.40

EXTERIOR CLOSURE   A4.1   Walls

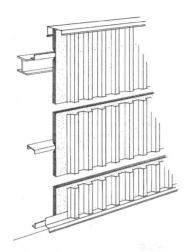

The table below lists costs for metal siding of various descriptions, not including the steel frame, or the structural steel, of a building. Costs are per S.F. including all accessories and insulation.

For steel frame support see System A3.4-300.

4.1-384	Metal Siding Panel	MAT.	INST.	TOTAL
1400	Metal siding aluminum panel, corrugated, .024" thick, natural	1.49	1.85	3.34
1450	Painted	1.71	1.85	3.56
1500	.032" thick, natural	1.82	1.85	3.67
1550	Painted	2.20	1.85	4.05
1600	Ribbed 4" pitch, .032" thick, natural	1.93	1.85	3.78
1650	Painted	2.21	1.85	4.06
1700	.040" thick, natural	2.25	1.85	4.10
1750	Painted	2.58	1.85	4.43
1800	.050" thick, natural	2.56	1.85	4.41
1850	Painted	2.88	1.85	4.73
1900	8" pitch panel, .032" thick, natural	1.84	1.78	3.62
1950	Painted	2.12	1.78	3.90
2000	.040" thick, natural	2.16	1.79	3.95
2050	Painted	2.47	1.80	4.27
2100	.050" thick, natural	2.47	1.79	4.26
2150	Painted	2.81	1.81	4.62
3000	Steel, corrugated or ribbed, 29 Ga. .0135" thick, galvanized	1.24	1.65	2.89
3050	Colored	1.63	1.68	3.31
3100	26 Ga. .0179" thick, galvanized	1.28	1.66	2.94
3150	Colored	1.47	1.69	3.16
3200	24 Ga. .0239" thick, galvanized	1.36	1.67	3.03
3250	Colored	1.66	1.70	3.36
3300	22 Ga. .0299" thick, galvanized	1.54	1.67	3.21
3350	Colored	1.95	1.70	3.65
3400	20 Ga. .0359" thick, galvanized	1.54	1.67	3.21
3450	Colored	2.07	1.80	3.87
4100	Sandwich panels, factory fab., 1" polystyrene, steel core, 26 Ga., galv.	2.65	2.57	5.22
4200	Colored, 1 side	3.85	2.57	6.42
4300	2 sides	4.95	2.57	7.52
4400	2" polystyrene, steel core, 26 Ga., galvanized	3.22	2.57	5.79
4500	Colored, 1 side	4.42	2.57	6.99
4600	2 sides	5.50	2.57	8.07
4700	22 Ga., baked enamel exterior	7.30	2.71	10.01
4800	Polyvinyl chloride exterior	7.70	2.71	10.41
5100	Textured aluminum, 4' x 8' x 5/16" plywood backing, single face	2.40	1.55	3.95
5200	Double face	3.61	1.55	5.16

For expanded coverage of these items see *Means Assemblies Cost Data 2000*

EXTERIOR CLOSURE — A4.1 — Walls

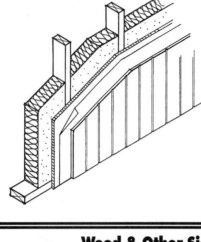

The table below lists costs per S.F. for exterior walls with wood siding. A variety of systems are presented using both wood and metal studs at 16″ and 24″ O.C.

4.1-412	Wood & Other Siding	MAT.	INST.	TOTAL
1400	Wood siding w/2″x4″studs, 16″O.C., insul. wall, 5/8″text 1-11 fir plywood	2.70	2.76	5.46
1450	5/8″ text 1-11 cedar plywood	2.79	2.67	5.46
1500	1″ x 4″ vert T.&G. redwood	4.60	3.76	8.36
1600	1″ x 8″ vert T.&G. redwood	4.19	3.40	7.59
1650	1″ x 5″ rabbetted cedar bev. siding	3.68	3.04	6.72
1700	1″ x 6″ cedar drop siding	3.72	3.06	6.78
1750	1″ x 12″ rough sawn cedar	2.88	2.78	5.66
1800	1″ x 12″ sawn cedar, 1″ x 4″ battens	3.88	3	6.88
1850	1″ x 10″ redwood shiplap siding	4.12	2.91	7.03
1900	18″ no. 1 red cedar shingles, 5-1/2″ exposed	3.45	3.64	7.09
1950	6″ exposed	3.32	3.52	6.84
2000	6-1/2″ exposed	3.19	3.40	6.59
2100	7″ exposed	3.05	3.28	6.33
2150	7-1/2″ exposed	2.92	3.16	6.08
3000	8″ wide aluminum siding	2.00	2.45	5.25
3150	8″ plain vinyl siding	2.13	2.46	4.59
3250	8″ insulated vinyl siding	2.29	2.46	4.75
3300				
3400	2″ x 6″ studs, 16″ O.C., insul. wall, w/ 5/8″ text 1-11 fir plywood	3.02	2.86	5.88
3500	5/8″ text 1-11 cedar plywood	3.11	2.86	5.97
3600	1″ x 4″ vert T.&G. redwood	4.92	3.86	8.78
3700	1″ x 8″ vert T.&G. redwood	4.51	3.50	8.01
3800	1″ x 5″ rabbetted cedar bev siding	4	3.14	7.14
3900	1″ x 6″ cedar drop siding	4.04	3.16	7.20
4000	1″ x 12″ rough sawn cedar	3.20	2.88	6.08
4200	1″ x 12″ sawn cedar, 1″ x 4″ battens	4.20	3.10	7.30
4500	1″ x 10″ redwood shiplap siding	4.44	3.01	7.45
4550	18″ no. 1 red cedar shingles, 5-1/2″ exposed	3.77	3.74	7.51
4600	6″ exposed	3.64	3.62	7.26
4650	6-1/2″ exposed	3.51	3.50	7.01
4700	7″ exposed	3.37	3.38	6.75
4750	7-1/2″ exposed	3.24	3.26	6.50
4800	8″ wide aluminum siding	3.12	2.55	5.67
4850	8″ plain vinyl siding	2.45	2.56	5.01
4900	8″ insulated vinyl siding	2.61	2.56	5.17
4910				
5000	2″ x 6″ studs, 24″ O.C., insul. wall, 5/8″ text 1-11, fir plywood	2.86	2.70	5.56
5050	5/8″ text 1-11 cedar plywood	2.95	2.70	5.65
5100	1″ x 4″ vert T.&G. redwood	4.76	3.70	8.46
5150	1″ x 8″ vert T.&G. redwood	4.35	3.34	7.69
5200	1″ x 5″ rabbetted cedar bev siding	3.84	2.98	6.82
5250	1″ x 6″ cedar drop siding	3.88	3	6.88

EXTERIOR CLOSURE — A4.1 Walls

4.1-412 Wood & Other Siding

		COST PER S.F.		
		MAT.	INST.	TOTAL
5300	1" x 12" rough sawn cedar	3.04	2.72	5.76
5400	1" x 12" sawn cedar, 1" x 4" battens	4.04	2.94	6.98
5450	1" x 10" redwood shiplap siding	4.28	2.85	7.13
5500	18" no. 1 red cedar shingles, 5-1/2" exposed	3.61	3.58	7.19
5550	6" exposed	3.48	3.46	6.94
5650	7" exposed	3.21	3.22	6.43
5700	7-1/2" exposed	3.08	3.10	6.18
5750	8" wide aluminum siding	2.96	2.39	5.35
5800	8" plain vinyl siding	2.29	2.40	4.69
5850	8" insulated vinyl siding	2.45	2.40	4.85
5900	3-5/8" metal studs, 16 Ga., 16" OC insul.wall, 5/8"text 1-11 fir plywood	2.96	2.90	5.86
5950	5/8" text 1-11 cedar plywood	3.05	2.90	5.95
6000	1" x 4" vert T.&G. redwood	4.86	3.90	8.76
6050	1" x 8" vert T.&G. redwood	4.45	3.54	7.99
6100	1" x 5" rabbetted cedar bev siding	3.94	3.18	7.12
6150	1" x 6" cedar drop siding	3.98	3.20	7.18
6200	1" x 12" rough sawn cedar	3.14	2.92	6.06
6250	1" x 12" sawn cedar, 1" x 4" battens	4.14	3.14	7.28
6300	1" x 10" redwood shiplap siding	4.38	3.05	7.43
6350	18" no. 1 red cedar shingles, 5-1/2" exposed	3.71	3.78	7.49
6500	6" exposed	3.58	3.66	7.24
6550	6-1/2" exposed	3.45	3.54	6.99
6600	7" exposed	3.31	3.42	6.73
6650	7-1/2" exposed	3.31	3.33	6.64
6700	8" wide aluminum siding	3.06	2.59	5.65
6750	8" plain vinyl siding	2.39	2.51	4.90
6800	8" insulated vinyl siding	2.55	2.51	5.06
7000	3-5/8" metal studs, 16 Ga. 24" OC insul wall, 5/8" text 1-11 fir plywood	2.75	2.58	5.33
7050	5/8" text 1-11 cedar plywood	2.84	2.58	5.42
7100	1" x 4" vert T.&G. redwood	4.65	3.67	8.32
7150	1" x 8" vert T.&G. redwood	4.24	3.31	7.55
7200	1" x 5" rabbetted cedar bev siding	3.73	2.95	6.68
7250	1" x 6" cedar drop siding	3.77	2.97	6.74
7300	1" x 12" rough sawn cedar	2.93	2.69	5.62
7350	1" x 12" sawn cedar 1" x 4" battens	3.93	2.91	6.84
7400	1" x 10" redwood shiplap siding	4.17	2.82	6.99
7450	18" no. 1 red cedar shingles, 5-1/2" exposed	3.64	3.67	7.31
7500	6" exposed	3.50	3.55	7.05
7550	6-1/2" exposed	3.37	3.43	6.80
7600	7" exposed	3.24	3.31	6.55
7650	7-1/2" exposed	2.97	3.07	6.04
7700	8" wide aluminum siding	2.85	2.36	5.21
7750	8" plain vinyl siding	2.18	2.37	4.55
7800	8" insul. vinyl siding	2.34	2.37	4.71

For expanded coverage of these items see *Means Assemblies Cost Data 2000*

EXTERIOR CLOSURE — A4.5 Exterior Wall Finishes

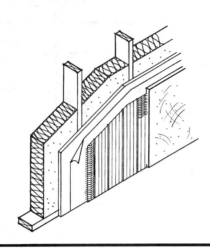

The table below lists costs for some typical stucco walls including all the components as demonstrated in the component block below. Prices are presented for backup walls using wood studs, metal studs and CMU.

4.5-110	Stucco Wall	MAT.	INST.	TOTAL
2100	Cement stucco, 7/8" th., plywood sheathing, stud wall, 2" x 4", 16" O.C.	2.22	3.13	5.35
2200	24" O.C.	2.11	2.99	5.10
2300	2" x 6", 16" O.C.	2.54	3.23	5.77
2400	24" O.C.	2.38	3.07	5.45
2500	No sheathing, metal lath on stud wall, 2" x 4", 16" O.C.	1.46	2.55	4.01
2600	24" O.C.	1.35	2.41	3.76
2700	2" x 6", 16" O.C.	1.78	2.65	4.43
2800	24" O.C.	1.62	2.49	4.11
2900	1/2" gypsum sheathing, 3-5/8" metal studs, 16" O.C.	1.86	2.69	4.55
2950	24" O.C.	1.65	2.46	4.11
3000	Cement stucco, 5/8" th., 2 coats on std. CMU block, 8"x 16", 8" thick	1.81	4.54	6.35
3100	10" Thick	2.49	4.71	7.20
3200	12" Thick	2.55	5.55	8.10
3300	Std. light Wt. block 8" x 16", 8" Thick	2.12	4.32	6.44
3400	10" Thick	2.63	4.45	7.08
3500	12" Thick	2.79	5.30	8.09
3600	3 coat stucco, self furring metal lath 3.4 Lb/SY, on 8" x 16", 8" thick	1.85	4.64	6.49
3700	10" Thick	2.53	4.81	7.34
3800	12" Thick	2.59	5.65	8.24
3900	Lt. Wt. block, 8" Thick	2.16	4.42	6.58
4000	10" Thick	2.67	4.55	7.22
4100	12" Thick	2.83	5.40	8.23

EXTERIOR CLOSURE — A4.6 Doors

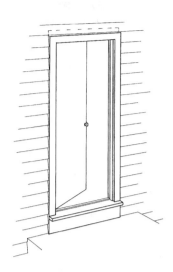

Costs are listed for exterior door systems by material, type and size. Prices between sizes listed can be interpolated with reasonable accuracy. Prices are per opening for a complete door system including frame as illustrated in the component block.

4.6-100 Wood, Steel & Aluminum

	MATERIAL	TYPE	DOORS	SPECIFICATION	OPENING	COST PER OPNG. MAT.	INST.	TOTAL
2350	Birch	solid core	single door	hinged	2'-6" x 6'-8"	855	147	1,002
2400					2'-6" x 7'-0"	855	148	1,003
2450					2'-8" x 7'-0"	860	148	1,008
2500					3'-0" x 7'-0"	870	151	1,021
2550			double door	hinged	2'-6" x 6'-8"	1,575	266	1,841
2600					2'-6" x 7'-0"	1,600	270	1,870
2650					2'-8" x 7'-0"	1,600	270	1,870
2700					3'-0" x 7'-0"	1,625	275	1,900
2750	Wood	combination	storm & screen	hinged	3'-0" x 6'-8"	240	36	276
2800					3'-0" x 7'-0"	264	39.50	303.50
2850		overhead	panels, H.D.	manual oper.	8'-0" x 8'-0"	505	270	775
2900					10'-0" x 10'-0"	760	300	1,060
2950					12'-0" x 12'-0"	1,050	360	1,410
3000					14'-0" x 14'-0"	1,675	415	2,090
3050					20'-0" x 16'-0"	3,475	830	4,305
3100				electric oper.	8'-0" x 8'-0"	1,025	405	1,430
3150					10'-0" x 10'-0"	1,275	435	1,710
3200					12'-0" x 12'-0"	1,575	495	2,070
3250					14'-0" x 14'-0"	2,200	550	2,750
3300					20'-0" x 16'-0"	4,300	1,100	5,400
3350	Steel 18 Ga.	hollow metal	1 door w/frame	no label	2'-6" x 7'-0"	840	153	993
3400					2'-8" x 7'-0"	840	153	993
3450					3'-0" x 7'-0"	840	153	993
3500					3'-6" x 7'-0"	900	161	1,061
3550					4'-0" x 8'-0"	1,100	161	1,261
3600			2 doors w/frame	no label	5'-0" x 7'-0"	1,650	284	1,934
3650					5'-4" x 7'-0"	1,650	284	1,934
3700					6'-0" x 7'-0"	1,650	284	1,934
3750					7'-0" x 7'-0"	1,775	299	2,074
3800					8'-0" x 8'-0"	2,175	300	2,475
3850			1 door w/frame	"A" label	2'-6" x 7'-0"	1,000	184	1,184
3900					2'-8" x 7'-0"	1,000	186	1,186
3950					3'-0" x 7'-0"	1,000	186	1,186
4000					3'-6" x 7'-0"	1,075	190	1,265
4050					4'-0" x 8'-0"	1,125	199	1,324
4100			2 doors w/frame	"A" label	5'-0" x 7'-0"	1,975	340	2,315

For expanded coverage of these items see *Means Assemblies Cost Data 2000*

EXTERIOR CLOSURE — A4.6 Doors

4.6-100 Wood, Steel & Aluminum

	MATERIAL	TYPE	DOORS	SPECIFICATION	OPENING	COST PER OPNG. MAT.	COST PER OPNG. INST.	COST PER OPNG. TOTAL
4150	Steel 18 Ga.	hollow metal	2 doors w/frame	"A" label	5'-4" x 7'-0"	2,000	345	2,345
4200					6'-0" x 7'-0"	2,000	345	2,345
4250					7'-0" x 7'-0"	2,100	355	2,455
4300					8'-0" x 8'-0"	2,100	355	2,455
4350	Steel 24 Ga.	overhead	sectional	manual oper.	8'-0" x 8'-0"	410	270	680
4400					10'-0" x 10'-0"	565	300	865
4450					12'-0" x 12'-0"	760	360	1,120
4500					20'-0" x 14'-0"	2,000	770	2,770
4550				electric oper.	8'-0" x 8'-0"	925	405	1,330
4600					10'-0" x 10'-0"	1,075	435	1,510
4650					12'-0" x 12'-0"	1,275	495	1,770
4700					20'-0" x 14'-0"	2,825	1,050	3,875
4750	Steel	overhead	rolling	manual oper.	8'-0" x 8'-0"	760	420	1,180
4800					10'-0" x 10'-0"	1,000	480	1,480
4850					12'-0" x 12'-0"	1,300	560	1,860
4900					14'-0" x 14'-0"	1,725	840	2,565
4950					20'-0" x 12'-0"	2,125	745	2,870
5000					20'-0" x 16'-0"	2,450	1,125	3,575
5050				electric oper.	8'-0" x 8'-0"	1,500	555	2,055
5100					10'-0" x 10'-0"	1,725	615	2,340
5150					12'-0" x 12'-0"	2,025	695	2,720
5200					14'-0" x 14'-0"	2,450	975	3,425
5250					20'-0" x 12'-0"	2,775	880	3,655
5300					20'-0" x 16'-0"	3,100	1,250	4,350
5350				fire rated	10'-0" x 10'-0"	1,325	610	1,935
5400			rolling grill	manual oper.	10'-0" x 10'-0"	1,650	670	2,320
5450					15'-0" x 8'-0"	1,950	840	2,790
5500		vertical lift	1 door w/frame	motor operator	16'-0" x 16'-0"	16,700	2,950	19,650
5550					32'-0" x 24'-0"	32,700	1,975	34,675
5600	St. Stl. & glass	revolving	stock unit	manual oper.	6'-0" x 7'-0"	29,100	4,475	33,575
5650				auto Cntrls.	6'-10" x 7'-0"	40,800	4,775	45,575
5700	Bronze	revolving	stock unit	manual oper.	6'-10" x 7'-0"	33,900	8,950	42,850
5750				auto Cntrls.	6'-10" x 7'-0"	45,600	9,250	54,850
5800	St. Stl. & glass	balanced	standard	economy	3'-0" x 7'-0"	6,600	750	7,350
5850				premium	3'-0" x 7'-0"	11,500	985	12,485
6000	Aluminum	combination	storm & screen	hinged	3'-0" x 6'-8"	216	38.50	254.50
6050					3'-0" x 7'-0"	238	42.50	280.50
6100		overhead	rolling grill	manual oper.	12'-0" x 12'-0"	2,675	1,175	3,850
6150				motor oper.	12'-0" x 12'-0"	3,525	1,325	4,850
6200	Alum. & Fbrgls.	overhead	heavy duty	manual oper.	12'-0" x 12'-0"	1,400	360	1,760
6250				electric oper.	12'-0" x 12'-0"	1,925	495	2,420
6300	Alum. & glass	w/o transom	narrow stile	w/panic Hrdwre.	3'-0" x 7'-0"	1,025	495	1,520
6350				dbl. door, Hrdwre.	6'-0" x 7'-0"	1,850	815	2,665
6400			wide stile	hdwre.	3'-0" x 7'-0"	1,350	485	1,835
6450				dbl. door, Hdwre.	6'-0" x 7'-0"	2,650	970	3,620
6500			full vision	hdwre.	3'-0" x 7'-0"	1,500	775	2,275
6550				dbl. door, Hdwre.	6'-0" x 7'-0"	2,225	1,100	3,325
6600			non-standard	hdwre.	3'-0" x 7'-0"	950	485	1,435
6650				dbl. door, Hdwre.	6'-0" x 7'-0"	1,900	970	2,870
6700			bronze fin.	hdwre.	3'-0" x 7'-0"	1,000	485	1,485
6750				dbl. door, Hrdwre.	6'-0" x 7'-0"	2,025	970	2,995
6800			black fin.	hdwre.	3'-0" x 7'-0"	1,675	485	2,160
6850				dbl. door, Hdwre.	6'-0" x 7'-0"	3,350	970	4,320
6900		w/transom	narrow stile	hdwre.	3'-0" x 10'-0"	1,350	560	1,910

Important: See the Reference Section for critical supporting data - Reference Nos., Crews & Location Factors

EXTERIOR CLOSURE — A4.6 Doors

4.6-100 Wood, Steel & Aluminum

	MATERIAL	TYPE	DOORS	SPECIFICATION	OPENING	COST PER OPNG. MAT.	INST.	TOTAL
6950	Alum & glass	w/transom		dbl. door, Hdwre.	6'-0" x 10'-0"	1,900	985	2,885
7000			wide stile	hdwre.	3'-0" x 10'-0"	1,550	670	2,220
7050				dbl. door, Hdwre.	6'-0" x 10'-0"	2,050	1,175	3,225
7100			full vision	hdwre.	3'-0" x 10'-0"	1,700	745	2,445
7150				dbl. door, Hdwre.	6'-0" x 10'-0"	2,175	1,300	3,475
7200			non-standard	hdwre.	3'-0" x 10'-0"	1,000	525	1,525
7250				dbl. door, Hdwre.	6'-0" x 10'-0"	2,000	1,050	3,050
7300			bronze fin.	hdwre.	3'-0" x 10'-0"	1,075	525	1,600
7350				dbl. door, Hdwre.	6'-0" x 10'-0"	2,125	1,050	3,175
7400			black fin.	hdwre.	3'-0" x 10'-0"	1,725	525	2,250
7450				dbl. door, Hdwre.	6'-0" x 10'-0"	3,450	1,050	4,500
7500		revolving	stock design	minimum	6'-10" x 7'-0"	16,000	1,800	17,800
7550				average	6'-0" x 7'-0"	19,700	2,225	21,925
7600				maximum	6'-10" x 7'-0"	26,000	2,975	28,975
7650				min., automatic	6'-10" x 7'-0"	27,700	2,100	29,800
7700				avg., automatic	6'-10" x 7'-0"	31,400	2,525	33,925
7750				max., automatic	6'-10" x 7'-0"	37,700	3,275	40,975
7800		balanced	standard	economy	3'-0" x 7'-0"	5,000	745	5,745
7850				premium	3'-0" x 7'-0"	6,225	960	7,185
7900		mall front	sliding panels	alum. fin.	16'-0" x 9'-0"	2,425	395	2,820
7950					24'-0" x 9'-0"	3,525	735	4,260
8000				bronze fin.	16'-0" x 9'-0"	2,825	460	3,285
8050					24'-0" x 9'-0"	4,100	855	4,955
8100			fixed panels	alum. fin.	48'-0" x 9'-0"	6,550	570	7,120
8150				bronze fin.	48'-0" x 9'-0"	7,625	665	8,290
8200		sliding entrance	5' x 7' door	electric oper.	12'-0" x 7'-6"	6,175	735	6,910
8250		sliding patio	temp. glass	economy	6'-0" x 7'-0"	820	135	955
8300			temp. glass	economy	12'-0" x 7'-0"	2,075	180	2,255
8350				premium	6'-0" x 7'-0"	1,225	203	1,428
8400					12'-0" x 7'-0"	3,125	270	3,395

For expanded coverage of these items see *Means Assemblies Cost Data 2000*

EXTERIOR CLOSURE — A4.7 — Windows & Glazed Walls

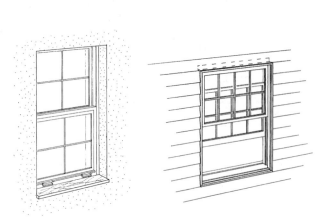

The table below lists window systems by material, type and size. Prices between sizes listed can be interpolated with reasonable accuracy. Prices include frame, hardware, and casing as illustrated in the component block below.

4.7-110 Wood, Steel & Aluminum

	MATERIAL	TYPE	GLAZING	SIZE	DETAIL	COST PER UNIT MAT.	COST PER UNIT INST.	COST PER UNIT TOTAL
3000	Wood	double hung	std. glass	2'-8" x 4'-6"		168	118	286
3050				3'-0" x 5'-6"		221	135	356
3100			insul. glass	2'-8" x 4'-6"		180	118	298
3150				3'-0" x 5'-6"		238	135	373
3200		sliding	std. glass	3'-4" x 2'-7"		203	98	301
3250				4'-4" x 3'-3"		232	107	339
3300				5'-4" x 6'-0"		294	130	424
3350			insul. glass	3'-4" x 2'-7"		247	115	362
3400				4'-4" x 3'-3"		282	125	407
3450				5'-4" x 6'-0"		355	148	503
3500		awning	std. glass	2'-10" x 1'-9"		152	59	211
3600				4'-4" x 2'-8"		248	57.50	305.50
3700			insul. glass	2'-10" x 1'-9"		185	68	253
3800				4'-4" x 2'-8"		305	63	368
3900		casement	std. glass	1'-10" x 3'-2"	1 lite	193	77.50	270.50
3950				4'-2" x 4'-2"	2 lite	395	103	498
4000				5'-11" x 5'-2"	3 lite	610	133	743
4050				7'-11" x 6'-3"	4 lite	860	160	1,020
4100				9'-11" x 6'-3"	5 lite	1,125	186	1,311
4150			insul. glass	1'-10" x 3'-2"	1 lite	203	77.50	280.50
4200				4'-2" x 4'-2"	2 lite	420	103	523
4250				5'-11" x 5'-2"	3 lite	645	133	778
4300				7'-11" x 6'-3"	4 lite	915	160	1,075
4350				9'-11" x 6'-3"	5 lite	1,175	186	1,361
4400		picture	std. glass	4'-6" x 4'-6"		320	133	453
4450				5'-8" x 4'-6"		360	148	508
4500	Wood	picture	insul. glass	4'-6" x 4'-6"		390	155	545
4550				5'-8" x 4'-6"		440	172	612
4600		fixed bay	std. glass	8' x 5'		1,125	244	1,369
4650				9'-9" x 5'-4"		795	345	1,140
4700			insul. glass	8' x 5'		1,750	244	1,994
4750				9'-9" x 5'-4"		860	345	1,205
4800		casement bay	std. glass	8' x 5'		1,125	280	1,405
4850			insul. glass	8' x 5'		1,275	280	1,555
4900		vert. bay	std. glass	8' x 5'		1,275	280	1,555
4950			insul. glass	8' x 5'		1,325	280	1,605
5000	Steel	double hung	1/4" tempered	2'-8" x 4'-6"		575	91.50	666.50
5050				3'-4" x 5'-6"		875	140	1,015

EXTERIOR CLOSURE — A4.7 Windows & Glazed Walls

4.7-110 Wood, Steel & Aluminum

	MATERIAL	TYPE	GLAZING	SIZE	DETAIL	MAT.	INST.	TOTAL
5100	Steel	double hung	insul. glass	2'-8" x 4'-6"		585	105	690
5150				3'-4" x 5'-6"		895	160	1,055
5202		horiz. pivoted	std. glass	2' x 2'		189	30.50	219.50
5250				3' x 3'		425	68.50	493.50
5300				4' x 4'		760	122	882
5350				6' x 4'		1,125	183	1,308
5400			insul. glass	2' x 2'		193	35	228
5450				3' x 3'		435	79	514
5500				4' x 4'		775	140	915
5550				6' x 4'		1,150	210	1,360
5600		picture window	std. glass	3' x 3'		255	68.50	323.50
5650				6' x 4'		680	183	863
5700			insul. glass	3' x 3'		264	79	343
5750				6' x 4'		705	210	915
5800		industrial security	std. glass	2'-9" x 4'-1"		550	85.50	635.50
5850				4'-1" x 5'-5"		1,075	169	1,244
5900			insul. glass	2'-9" x 4'-1"		560	98	658
5950				4'-1" x 5'-5"		1,100	193	1,293
6000		comm. projected	std. glass	3'-9" x 5'-5"		900	155	1,055
6050				6'-9" x 4'-1"		1,225	210	1,435
6100			insul. glass	3'-9" x 5'-5"		920	178	1,098
6150				6'-9" x 4'-1"		1,250	241	1,491
6200		casement	std. glass	4'-2" x 4'-2"	2 lite	655	132	787
6250			insul. glass	4'-2" x 4'-2"		675	152	827
6300			std. glass	5'-11" x 5'-2"	3 lite	1,325	233	1,558
6350			insul. glass	5'-11" x 5'-2"		1,350	267	1,617
6400	Aluminum	projecting	std. glass	3'-1" x 3'-2"		203	67	270
6450				4'-5" x 5'-3"		286	84	370
6500			insul. glass	3'-1" x 3'-2"		244	80.50	324.50
6550				4'-5" x 5'-3"		345	101	446
6600		sliding	std. glass	3' x 2'		153	67	220
6650				5' x 3'		195	74.50	269.50
6700				8' x 4'		280	112	392
6750				9' x 5'		425	168	593
6800			insul. glass	3' x 2'		170	67	237
6850				5' x 3'		272	74.50	346.50
6900				8' x 4'		450	112	562
6950				9' x 5'		680	168	848
7000		single hung	std. glass	2' x 3'		136	67	203
7050				2'-8" x 6'-8"		290	84	374
7100				3'-4" x 5'0"		187	74.50	261.50
7150			insul. glass	2' x 3'		165	67	232
7200				2'-8" x 6'-8"		370	84	454
7250				3'-4" x 5'		262	74.50	336.50
7300		double hung	std. glass	2' x 3'		203	46	249
7350				2'-8" x 6'-8"		605	136	741
7400				3'-4" x 5'		565	127	692
7450			insul. glass	2' x 3'		209	52.50	261.50
7500				2'-8" x 6'-8"		620	156	776
7550				3'-4" x 5'-0"		580	146	726
7600		casements	std. glass	3'-1" x 3'-2"		161	74.50	235.50
7650				4'-5" x 5'-3"		390	180	570
7700			insul. glass	3'-1" x 3'-2"		170	85.50	255.50
7750				4'-5" x 5'-3"		410	207	617

For expanded coverage of these items see *Means Assemblies Cost Data 2000*

EXTERIOR CLOSURE — A4.7 Windows & Glazed Walls

4.7-110 Wood, Steel & Aluminum

	MATERIAL	TYPE	GLAZING	SIZE	DETAIL	COST PER UNIT MAT.	COST PER UNIT INST.	COST PER UNIT TOTAL
7800	Aluminum	hinged swing	std. glass	3' x 4'		365	91.50	456.50
7850				4' x 5'		605	153	758
7900			insul. glass	3' x 4'		375	105	480
7950				4' x 5'		625	175	800
8002		folding type	std. glass	3'-0" x 4'-0"		205	47.50	252.50
8050				4'-0" x 5'-0"		286	47.50	333.50
8100			insul. glass	3'-0" x 4'-0"		257	47.50	304.50
8150				4'-0" x 5'-0"		375	47.50	422.50
8200		picture unit	std. glass	2'-0" x 3'-0"		108	46	154
8250				2'-8" x 6'-8"		320	136	456
8300				3'-4" x 5'-0"		299	127	426
8350			insul. glass	2'-0" x 3'-0"		113	52.50	165.50
8400				2'-8" x 6'-8"		335	156	491
8450				3'-4" x 5'-0"		315	146	461
8500		awning type	std. glass	3'-0" x 3'-0"	2 lite	335	48	383
8550				3'-0" x 4'-0"	3 lite	390	67	457
8600				3'-0" x 5'-4"	4 lite	470	67	537
8650				4'-0" x 5'-4"	4 lite	520	74.50	594.50
8700			insul. glass	3'-0" x 3'-0"	2 lite	355	48	403
8750				3'-0" x 4'-0"	3 lite	450	67	517
8800				3'-0" x 5'-4"	4 lite	555	67	622
8850				4'-0" x 5'-4"	4 lite	620	74.50	694.50
8900		jalousie type	std. glass	1'-7" x 3'-2"		124	67	191
8950				2'-3" x 4'-0"		178	67	245
9000				3'-1" x 2'-0"		122	67	189
9050				3'-1" x 5'-3"		254	67	321

4.7-582 Tubular Aluminum Framing

		COST/S.F. OPNG. MAT.	COST/S.F. OPNG. INST.	COST/S.F. OPNG. TOTAL
1100	Alum flush tube frame, for 1/4"glass, 1-3/4"x4", 5'x6' opng, no inter horizntls	8.90	6.30	15.20
1150	One intermediate horizontal	12.15	7.40	19.55
1200	Two intermediate horizontals	15.35	8.45	23.80
1250	5' x 20' opening, three intermediate horizontals	8	5.25	13.25
1400	1-3/4" x 4-1/2", 5' x 6' opening, no intermediate horizontals	10.10	6.30	16.40
1450	One intermediate horizontal	13.45	7.40	20.85
1500	Two intermediate horizontals	16.85	8.45	25.30
1550	5' x 20' opening, three intermediate horizontals	8.95	5.25	14.20
1700	For insulating glass, 2"x4-1/2", 5'x6' opening, no intermediate horizontals	11.35	6.60	17.95
1750	One intermediate horizontal	14.85	7.75	22.60
1800	Two intermediate horizontals	18.35	8.90	27.25
1850	5' x 20' opening, three intermediate horizontals	9.90	5.50	15.40
2000	Thermal break frame, 2-1/4"x4-1/2", 5'x6' opng, no intermediate horizontals	12.20	6.70	18.90
2050	One intermediate horizontal	16.35	8.05	24.40
2100	Two intermediate horizontals	20.50	9.40	29.90
2150	5' x 20' opening, three intermediate horizontals	11.10	5.75	16.85

Important: See the Reference Section for critical supporting data - Reference Nos., Crews & Location Factors

EXTERIOR CLOSURE — A4.7 — Windows & Glazed Walls

The table below lists costs of curtain wall and spandrel panels per S.F. Costs do not include structural framing used to hang the panels from.

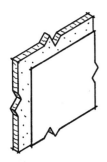
Spandrel Glass Panel

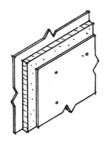
Sandwich Panel

4.7-584	Curtain Wall Panels	MAT.	INST.	TOTAL
1000	Glazing panel, insulating, 1/2" thick, 2 lites 1/8" float, clear	6.80	5.40	12.20
1100	Tinted	10.05	5.40	15.45
1200	5/8" thick units, 2 lites 3/16" float, clear	8.15	5.70	13.85
1300	Tinted	8.40	5.70	14.10
1400	1" thick units, 2 lites, 1/4" float, clear	11.55	6.85	18.40
1500	Tinted	14	6.85	20.85
1600	Heat reflective film inside	16.30	6.05	22.35
1700	Light and heat reflective glass, tinted	18.65	6.05	24.70
2000	Plate glass, 1/4" thick, clear	4.92	4.28	9.20
2050	Tempered	5.85	4.28	10.13
2100	Tinted	4.66	4.28	8.94
2200	3/8" thick, clear	7.75	6.85	14.60
2250	Tempered	11.65	6.85	18.50
2300	Tinted	9.35	6.85	16.20
2400	1/2" thick, clear	15.20	9.35	24.55
2450	Tempered	17.50	9.35	26.85
2500	Tinted	16.30	9.35	25.65
2600	3/4" thick, clear	21	14.65	35.65
2650	Tempered	24.50	14.65	39.15
3000	Spandrel glass, panels, 1/4" plate glass insul w/fiberglass, 1" thick	9.95	4.28	14.23
3100	2" thick	11.60	4.28	15.88
3200	Galvanized steel backing, add	3.44		3.44
3300	3/8" plate glass, 1" thick	16.65	4.28	20.93
3400	2" thick	18.30	4.28	22.58
4000	Polycarbonate, masked, clear or colored, 1/8" thick	5.55	3.02	8.57
4100	3/16" thick	6.70	3.11	9.81
4200	1/4" thick	7.40	3.31	10.71
4300	3/8" thick	13.65	3.42	17.07
5000	Facing panel, textured al, 4' x 8' x 5/16" plywood backing, sgl face	2.40	1.55	3.95
5100	Double face	3.61	1.55	5.16
5200	4' x 10' x 5/16" plywood backing, single face	2.29	1.55	3.84
5300	Double face	3.47	1.55	5.02
5400	4' x 12' x 5/16" plywood backing, single face	3.29	1.55	4.84
5500	Sandwich panel, 22 Ga. galv., both sides 2" insulation, enamel exterior	7.30	2.71	10.01
5600	Polyvinylidene floride exterior finish	7.70	2.71	10.41
5700	26 Ga., galv. both sides, 1" insulation, colored 1 side	3.85	2.57	6.42
5800	Colored 2 sides	4.95	2.57	7.52

Division 5
Roofing

ROOFING — A5.1 Roof Covers

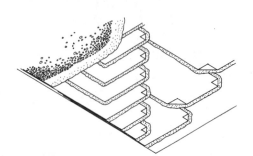

Multiple ply roofing is the most popular covering for minimum pitch roofs. Lines 1200 through 6300 list the costs of the various types, plies and weights per S.F.

5.1-103	Built-Up	MAT.	INST.	TOTAL
1200	Asphalt flood coat w/gravel; not incl. insul, flash., nailers			
1300				
1400	Asphalt base sheets & 3 plies #15 asphalt felt, mopped	.39	.98	1.37
1500	On nailable deck	.43	1.02	1.45
1600	4 plies #15 asphalt felt, mopped	.54	1.07	1.61
1700	On nailable deck	.50	1.13	1.63
1800	Coated glass base sheet, 2 plies glass (type IV), mopped	.42	.98	1.40
1900	For 3 plies	.49	1.07	1.56
2000	On nailable deck	.47	1.13	1.60
2300	4 plies glass fiber felt (type IV), mopped	.59	1.07	1.66
2400	On nailable deck	.54	1.13	1.67
2500	Organic base sheet & 3 plies #15 organic felt, mopped	.46	1.05	1.51
2600	On nailable deck	.44	1.13	1.57
2700	4 plies #15 organic felt, mopped	.53	.98	1.51
2750				
2800	Asphalt flood coat, smooth surface			
2900	Asphalt base sheet & 3 plies #15 asphalt felt, mopped	.39	.89	1.28
3000	On nailable deck	.37	.93	1.30
3100	Coated glass fiber base sheet & 2 plies glass fiber felt, mopped	.36	.85	1.21
3200	On nailable deck	.34	.89	1.23
3300	For 3 plies, mopped	.43	.93	1.36
3400	On nailable deck	.41	.98	1.39
3700	4 plies glass fiber felt (type IV), mopped	.50	.93	1.43
3800	On nailable deck	.48	.98	1.46
3900	Organic base sheet & 3 plies #15 organic felt, mopped	.40	.89	1.29
4000	On nailable decks	.38	.93	1.31
4100	4 plies #15 organic felt, mopped	.47	.98	1.45
4200	Coal tar pitch with gravel surfacing			
4300	4 plies #15 tarred felt, mopped	1.11	1.02	2.13
4400	3 plies glass fiber felt (type IV), mopped	.91	1.13	2.04
4500	Coated glass fiber base sheets 2 plies glass fiber felt, mopped	.91	1.13	2.04
4600	On nailable decks	.81	1.20	2.01
4800	3 plies glass fiber felt (type IV), mopped	1.24	1.02	2.26
4900	On nailable decks	1.14	1.07	2.21

ROOFING — A5.1 Roof Covers

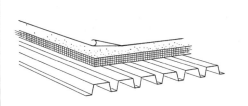

Fully Adhered

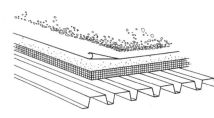

Ballasted

The systems listed below reflect only the cost for the single ply membrane.

For additional components see:

Insulation	A5.7-101
Base Flashing	A5.1-510
Roof Edge	A5.1-520
Roof Openings	A5.8-100

5.1-220	Single Ply Membrane	MAT.	INST.	TOTAL
1000	CSPE (Chlorosulfonated polyethylene), 35 mils, fully adhered	1.34	.52	1.86
2000	EPDM (Ethylene propylene diene monomer), 45 mils, fully adhered	.84	.52	1.36
4000	Modified bit., SBS modified, granule surface cap sheet, mopped, 150 mils	.42	1.07	1.49
4500	APP modified, granule surface cap sheet, torched, 180 mils	.50	.68	1.18
6000	Reinforced PVC, 48 mils, loose laid and ballasted with stone	1.03	.27	1.30
6200	Fully adhered with adhesive	1.27	.52	1.79
6202				

COST PER S.F.

5.1-310	Preformed Metal Roofing	MAT.	INST.	TOTAL
0200	Corrugated roofing, aluminum, mill finish, .0175" thick, .272 P.S.F.	.67	.81	1.48
0250	.0215 thick, .334 P.S.F.	.86	.81	1.67

5.1-330	Formed Metal	MAT.	INST.	TOTAL
1000	Batten seam, formed copper roofing, 3"min slope, 16 oz., 1.2 P.S.F.	4.41	2.68	7.09
1100	18 oz., 1.35 P.S.F.	4.91	2.94	7.85
2000	Zinc copper alloy, 3" min slope, .020" thick, .88 P.S.F.	5.65	2.46	8.11
3000	Flat seam, copper, 1/4" min. slope, 16 oz., 1.2 P.S.F.	3.91	2.46	6.37
3100	18 oz., 1.35 P.S.F.	4.41	2.57	6.98
5000	Standing seam, copper, 2-1/2" min. slope, 16 oz., 1.25 P.S.F.	4.26	2.27	6.53
5100	18 oz., 1.40 P.S.F.	4.76	2.46	7.22
6000	Zinc copper alloy, 2-1/2" min. slope, .020" thick, .87 P.S.F.	5.55	2.46	8.01
6100	.032" thick, 1.39 P.S.F.	7.55	2.68	10.23

For expanded coverage of these items see *Means Assemblies Cost Data 2000*

ROOFING — A5.1 — Roof Covers

Shingles and tiles are practical in applications where the roof slope is more than 3-1/2" per foot of rise. Table below lists the various materials and the weight per S.F.

5.1-410	Shingle & Tile	COST PER S.F.		
		MAT.	INST.	TOTAL
1095	Asphalt roofing			
1100	Strip shingles, 4" slope, inorganic class A 210-235 lb/Sq.	.32	.53	.85
1150	Organic, class C, 235-240 lb./sq.	.42	.58	1
1200	Premium laminated multi-layered, class A, 260-300 lb./Sq.	.51	.80	1.31
1545	Metal roofing			
1550	Alum., shingles, colors, 3" min slope, .019" thick, 0.4 PSF	1.88	.58	2.46
1850	Steel, colors, 3" min slope, 26 gauge, 1.0 PSF	2.15	1.20	3.35
2795	Slate roofing			
2800	4" min. slope, shingles, 3/16" thick, 8.0 PSF	6	1.49	7.49
3495	Wood roofing			
3500	4" min slope, cedar shingles, 16" x 5", 5" exposure 1.6 PSF	1.72	1.16	2.88
4000	Shakes, 18", 8-1/2" exposure, 2.8 PSF	1.13	1.43	2.56
5095	Tile roofing			
5100	Aluminum, mission, 3" min slope, .019" thick, 0.65 PSF	3.96	1.12	5.08
6000	Clay, Americana, 3" minimum slope, 8 PSF	5.50	1.58	7.08
6002				

5.1-520		Roof Edges					
	EDGE TYPE	DESCRIPTION	SPECIFICATION	FACE HEIGHT		COST PER L.F.	
					MAT.	INST.	TOTAL
1000	Aluminum	mill finish	.050" thick	4"	7.45	5.05	12.50
1100				6"	8	5.20	13.20
1300		duranodic	.050" thick	4"	8.55	5.05	13.60
1400				6"	9.25	5.20	14.45
1600		painted	.050" thick	4"	9.15	5.05	14.20
1700				6"	10	5.20	15.20
2000	Copper	plain	16 oz.	4"	6.10	5.05	11.15
2100				6"	7	5.20	12.20
2300			20 oz.	4"	6.65	5.75	12.40
2400				6"	7.40	5.55	12.95
2700	Sheet Metal	galvanized	20 Ga.	4"	6.85	5.95	12.80
2800				6"	8.05	5.95	14
3000			24 Ga.	4"	6.10	5.05	11.15
3100				6"	6.95	5.05	12

Important: See the Reference Section for critical supporting data - Reference Nos., Crews & Location Factors

ROOFING

A5.1 Roof Covers

5.1-620 Flashing

	MATERIAL	BACKING	SIDES	SPECIFICATION	QUANTITY	COST PER S.F.		
						MAT.	INST.	TOTAL
0040	Aluminum	none		.019"		.83	2	2.83
0050				.032"		1.09	2	3.09
0300		fabric	2	.004"		1.02	.88	1.90
0400		mastic		.004"		1.02	.88	1.90
0700	Copper	none		16 oz.	<500 lbs.	3.35	2.53	5.88
0800				24 oz.	<500 lbs.	5.05	2.77	7.82
2000	Copper lead	fabric	1	2 oz.		1.55	.88	2.43
3500	PVC black	none		.010"		.15	.89	1.04
3700				.030"		.32	.89	1.21
4200	Neoprene			1/16"		1.63	.89	2.52
4500	Stainless steel	none		.015"	<500 lbs.	3.41	2.53	5.94
4600	Copper clad				>2000 lbs.	3.29	1.87	5.16
5000	Plain			32 ga.		2.38	1.87	4.25
5009								

For expanded coverage of these items see *Means Assemblies Cost Data 2000*

ROOFING — A5.7 Insulation

5.7-101 Roof Deck Rigid Insulation

		COST PER S.F.		
		MAT.	INST.	TOTAL
0100	Fiberboard low density, 1/2" thick, R1.39			
0150	1" thick R2.78	.35	.32	.67
0300	1 1/2" thick R4.17	.53	.32	.85
0350	2" thick R5.56	.70	.32	1.02
0370	Fiberboard high density, 1/2" thick R1.3	.20	.25	.45
0380	1" thick R2.5	.37	.32	.69
0390	1 1/2" thick R3.8	.61	.32	.93
0410	Fiberglass, 3/4" thick R2.78	.48	.25	.73
0450	15/16" thick R3.70	.63	.25	.88
0550	1-5/16" thick R5.26	1.09	.25	1.34
0650	2 7/16" thick R10	1.32	.32	1.64
1510	Polyisocyanurate 2#/CF density, 1" thick R7.14	.33	.18	.51
1550	1 1/2" thick R10.87	.37	.20	.57
1600	2" thick R14.29	.46	.23	.69
1650	2 1/2" thick R16.67	.53	.24	.77
1700	3" thick R21.74	.65	.25	.90
1750	3 1/2" thick R25	.77	.25	1.02
1800	Tapered for drainage	.40	.18	.58
1810	Expanded polystyrene, 1#/CF density, 3/4" thick R2.89	.19	.17	.36
1820	2" thick R7.69	.35	.20	.55
1830	Extruded polystyrene, 15 PSI compressive strength, 1" thick R5	.23	.17	.40
1835	2" thick R10	.36	.20	.56
1840	3" thick R15	.80	.25	1.05
2550	40 PSI compressive strength, 1" thick R5	.36	.17	.53
2600	2" thick R10	.71	.20	.91
2650	3" thick R15	1.05	.25	1.30
2700	4" thick R20	1.40	.25	1.65
2750	Tapered for drainage	.53	.18	.71
2810	60 PSI compressive strength, 1" thick R5	.44	.18	.62
2850	2" thick R10	.78	.21	.99
2900	Tapered for drainage	.63	.18	.81

ROOFING — A5.8 Openings & Specialties

Roof Hatch

Smoke Hatch

Skylight

5.8-100	Hatches	COST PER OPNG. MAT.	INST.	TOTAL
0200	Roof hatches, with curb, and 1" fiberglass insulation, 2'-6"x3'-0", aluminum	445	97.50	542.50
0300	Galvanized steel 165 lbs.	375	97.50	472.50
0400	Primed steel 164 lbs.	445	97.50	542.50
0500	2'-6"x4'-6" aluminum curb and cover, 150 lbs.	615	109	724
0600	Galvanized steel 220 lbs.	525	109	634
0650	Primed steel 218 lbs.	675	109	784
0800	2'x6"x8'-0" aluminum curb and cover, 260 lbs.	1,100	148	1,248
0900	Galvanized steel, 360 lbs.	1,000	148	1,148
0950	Primed steel 358 lbs.	995	148	1,143
1200	For plexiglass panels, add to the above	380		380
2100	Smoke hatches, unlabeled not incl. hand winch operator, 2'-6"x3', galv	455	118	573
2200	Plain steel, 160 lbs.	540	118	658
2400	2'-6"x8'-0", galvanized steel, 360 lbs.	1,100	162	1,262
2500	Plain steel, 350 lbs.	1,100	162	1,262
3000	4'-0"x8'-0", double leaf low profile, aluminum cover, 359 lb.	1,575	122	1,697
3100	Galvanized steel 475 lbs.	1,375	122	1,497
3200	High profile, aluminum cover, galvanized curb, 361 lbs.	1,425	122	1,547

5.8-100	Skylights	COST PER S.F. MAT.	INST.	TOTAL
5100	Skylights, plastic domes, insul curbs, nom. size to 10 S.F., single glaze	8.85	6.10	14.95
5200	Double glazing	13.20	7.50	20.70
5300	10 S.F. to 20 S.F., single glazing	8.15	2.47	10.62
5400	Double glazing	10.75	3.10	13.85
5500	20 S.F. to 30 S.F., single glazing	9.60	2.10	11.70
5600	Double glazing	10.80	2.47	13.27
5700	30 S.F. to 65 S.F., single glazing	14.45	1.60	16.05
5800	Double glazing	10.10	2.10	12.20
6000	Sandwich panels fiberglass, 9-1/16" thick, 2 S.F. to 10 S.F.	15.35	4.88	20.23
6100	10 S.F. to 18 S.F.	13.75	3.69	17.44
6200	2-3/4" thick, 25 S.F. to 40 S.F.	22	3.31	25.31
6300	40 S.F. to 70 S.F.	18.15	2.96	21.11
6301				

For expanded coverage of these items see *Means Assemblies Cost Data 2000*

ROOFING — A5.8 Openings & Specialties

5.8-400 Gutters

	SECTION	MATERIAL	THICKNESS	SIZE	FINISH	COST PER L.F. MAT.	COST PER L.F. INST.	COST PER L.F. TOTAL
0050	Box	aluminum	.027"	5"	enameled	1.18	2.42	3.60
0100					mill	1.07	2.42	3.49
0200			.032"	5"	enameled	1.32	2.42	3.74
0500		copper	16 Oz.	4"	lead coated	7.65	2.42	10.07
0600					mill	3.95	2.42	6.37
1000		steel galv.	28 Ga.	5"	enameled	1.05	2.42	3.47
1200			26 Ga.	5"	mill	1.05	2.42	3.47
1800		vinyl		4"	colors	.94	2.45	3.39
1900				5"	colors	1.10	2.45	3.55
2300		hemlock or fir		4"x5"	treated	8	2.70	10.70
3000	Half round	copper	16 Oz.	4"	lead coated	5.05	2.42	7.47
3100					mill	3.50	2.42	5.92
3600		steel galv.	28 Ga.	5"	enameled	1.05	2.42	3.47
4102		stainless steel		5"	mill	5.50	2.42	7.92
5000		vinyl		4"	white	.75	2.45	3.20
5002								

5.8-500 Downspouts

	MATERIALS	SECTION	SIZE	FINISH	THICKNESS	COST PER V.L.F. MAT.	COST PER V.L.F. INST.	COST PER V.L.F. TOTAL
0100	Aluminum	rectangular	2"x3"	embossed mill	.020"	.78	1.53	2.31
0150				enameled	.020"	.76	1.53	2.29
0250			3"x4"	enameled	.024"	1.72	2.07	3.79
0300		round corrugated	3"	enameled	.020"	.94	1.53	2.47
0350			4"	enameled	.025"	1.55	2.07	3.62
0500	Copper	rectangular corr.	2"x3"	mill	16 Oz.	3.85	1.53	5.38
0600		smooth		mill	16 Oz.	5.20	1.53	6.73
0700		rectangular corr.	3"x4"	mill	16 Oz.	4.80	2	6.80
1300	Steel	rectangular corr.	2"x3"	galvanized	28 Ga.	.58	1.53	2.11
1350				epoxy coated	24 Ga.	1.11	1.53	2.64
1400		smooth		galvanized	28 Ga.	.88	1.53	2.41
1450		rectangular corr.	3"x4"	galvanized	28 Ga.	1.63	2	3.63
1500				epoxy coated	24 Ga.	2.15	2	4.15
1550		smooth		galvanized	28 Ga.	1.35	2	3.35
1652		round corrugated	3"	galvanized	28 Ga.	.75	1.53	2.28
1700			4"	galvanized	28 Ga.	.99	2	2.99
1750			5"	galvanized	28 Ga.	1.51	2.23	3.74
2000	Steel pipe	round	4"	black	X.H.	5.85	14.50	20.35
2552	S.S. tubing sch.5	rectangular	3"x4"	mill		17.95	2	19.95
2702								

5.8-500 Gravel Stop

	MATERIALS	SECTION	SIZE	FINISH	THICKNESS	COST PER L.F. MAT.	COST PER L.F. INST.	COST PER L.F. TOTAL
5100	Aluminum	extruded	4"	mill	.050"	2.99	2	4.99
5200			4"	duranodic	.050"	4.06	2	6.06
5300			8"	mill	.050"	4.38	2.32	6.70
5400			8"	duranodic	.050"	5.50	2.32	7.82
6000			12"-2 pc.	duranodic	.050"	6.90	2.90	9.80
6100	Stainless	formed	6"	mill	24 Ga.	7.85	2.15	10

For information about Means Estimating Seminars, see yellow pages 11 and 12 in back of book

Division 6
Interior Construction

INTERIOR CONSTR. — A6.1 Block Partitions

The Concrete Block Partition Systems are defined by weight and type of block, thickness, type of finish and number of sides finished. System components include joint reinforcing on alternate courses and vertical control joints.

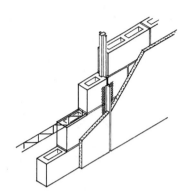

6.1-210 Concrete Block Partitions - Regular Weight

	TYPE	THICKNESS (IN.)	TYPE FINISH	SIDES FINISHED		COST PER S.F. MAT.	INST.	TOTAL
1000	Hollow	4	none	0		1	2.98	3.98
1010			gyp. plaster 2 coat	1		1.34	4.34	5.68
1020				2		1.69	5.70	7.39
1200			portland - 3 coat	1		1.24	4.55	5.79
1400			5/8" drywall	1		1.44	3.97	5.41
1500		6	none	0		1.34	3.20	4.54
1510			gyp. plaster 2 coat	1		1.68	4.56	6.24
1520				2		2.03	5.95	7.98
1700			portland - 3 coat	1		1.58	4.77	6.35
1900			5/8" drywall	1		1.78	4.19	5.97
1910				2		2.22	5.20	7.42
2000		8	none	0		1.56	3.42	4.90
2010			gyp. plaster 2 coat	1		1.90	4.78	6.68
2020			gyp. plaster 2 coat	2		2.25	6.15	8.40
2200			portland - 3 coat	1		1.80	4.99	6.79
2400			5/8" drywall	1		2	4.41	6.41
2410				2		2.44	5.40	7.84
2500		10	none	0		2.15	3.58	5.73
2510			gyp. plaster 2 coat	1		2.49	4.94	7.43
2520				2		2.84	6.30	9.14
2700			portland - 3 coat	1		2.39	5.15	7.54
2900			5/8" drywall	1		2.59	4.57	7.16
2910				2		3.03	5.55	8.58
3000	Solid	2	none	0		.92	2.95	3.87
3010			gyp. plaster	1		1.26	4.31	5.57
3020				2		1.61	5.70	7.31
3200			portland - 3 coat	1		1.16	4.52	5.68
3400			5/8" drywall	1		1.36	3.94	5.30
3410				2		1.80	4.93	6.73
3500		4	none	0		1.43	3.08	4.51
3510			gyp. plaster	1		1.84	4.46	6.30
3520				2		2.12	5.80	7.92
3700			portland - 3 coat	1		1.67	4.65	6.32
3900			5/8" drywall	1		1.87	4.07	5.94
3910				2		2.31	5.05	7.36

INTERIOR CONSTR.	A6.1	Block Partitions

6.1-210 Concrete Block Partitions - Regular Weight

	TYPE	THICKNESS (IN.)	TYPE FINISH	SIDES FINISHED		COST PER S.F.		
						MAT.	INST.	TOTAL
4002	Solid	6	none	0		1.73	3.32	5.05
4010			gyp. plaster	1		2.07	4.68	6.75
4020				2		2.42	6.05	8.47
4200			portland - 3 coat	1		1.97	4.89	6.86
4400			5/8" drywall	1		2.17	4.31	6.48
4410				2		2.61	5.30	7.91

6.1-210 Concrete Block Partitions - Lightweight

	TYPE	THICKNESS (IN.)	TYPE FINISH	SIDES FINISHED		COST PER S.F.		
						MAT.	INST.	TOTAL
5000	Hollow	4	none	0		1.16	2.91	4.07
5010			gyp. plaster	1		1.50	4.27	5.77
5020				2		1.85	5.65	7.50
5200			portland - 3 coat	1		1.40	4.48	5.88
5400			5/8" drywall	1		1.60	3.90	5.50
5410				2		2.04	4.89	6.93
5500		6	none	0		1.50	3.12	4.62
5510			gyp. plaster	1		1.84	4.48	6.32
5520			gyp. plaster	2		2.19	5.85	8.04
5700			portland - 3 coat	1		1.74	4.69	6.43
5900			5/8" drywall	1		1.94	4.11	6.05
5910				2		2.38	5.10	7.48
6000		8	none	0		1.85	3.34	5.19
6010			gyp. plaster	1		2.19	4.70	6.89
6020				2		2.54	6.10	8.64
6200			portland - 3 coat	1		2.09	4.91	7
6400		8	5/8" drywall	1		2.29	4.33	6.62
6410				2		2.73	5.30	8.03
6500		10	none	0		2.35	3.49	5.84
6510			gyp. plaster	1		2.69	4.85	7.54
6520				2		3.04	6.25	9.29
6700			portland - 3 coat	1		2.59	5.05	7.64
6900			5/8" drywall	1		2.79	4.48	7.27
6910				2		3.23	5.45	8.68
7000	Solid	4	none	0		1.33	3.05	4.38
7010			gyp. plaster	1		1.67	4.41	6.08
7020				2		2.02	5.80	7.82
7200			portland - 3 coat	1		1.57	4.62	6.19
7400			5/8" drywall	1		1.77	4.04	5.81
7410				2		2.21	5.05	7.26
7500		6	none	0		2.45	3.27	5.72
7510			gyp. plaster	1		2.84	4.67	7.51
7520				2		3.14	6	9.14
7700			portland - 3 coat	1		2.69	4.84	7.53
7900			5/8" drywall	1		2.89	4.26	7.15
7910				2		3.33	5.25	8.58
8000		8	none	0		3.50	3.51	7.01
8010			gyp. plaster	1		3.84	4.87	8.71
8020				2		4.19	6.25	10.44
8200			portland - 3 coat	1		3.74	5.10	8.84
8400			5/8" drywall	1		3.94	4.50	8.44
8410				2		4.38	5.50	9.88

For expanded coverage of these items see *Means Assemblies Cost Data 2000*

INTERIOR CONSTR. A6.1 Partitions

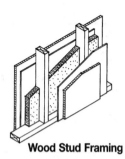
Wood Stud Framing

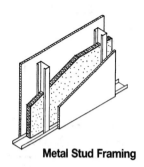
Metal Stud Framing

The Drywall Partitions/Stud Framing Systems are defined by type of drywall and number of layers, type and spacing of stud framing, and treatment on the opposite face. Components include taping and finishing.

Cost differences between regular and fire resistant drywall are negligible, and terminology is interchangeable. In some cases fiberglass insulation is included for additional sound deadening.

6.1-510 Drywall Partitions/Wood Stud Framing

	FACE LAYER	BASE LAYER	FRAMING	OPPOSITE FACE	INSULATION	COST PER S.F.		
						MAT.	INST.	TOTAL
1200	5/8" FR drywall	none	2 x 4, @ 16" O.C.	same	0	1.18	1.62	2.80
1250				5/8" reg. drywall	0	1.18	1.62	2.80
1300				nothing	0	.82	1.08	1.90
1400		1/4" SD gypsum	2 x 4 @ 16" O.C.	same	1-1/2" fiberglass	2.11	2.49	4.60
1450				5/8" FR drywall	1-1/2" fiberglass	1.83	2.19	4.02
1500				nothing	1-1/2" fiberglass	1.47	1.65	3.12
1600		resil. channels	2 x 4 @ 16", O.C.	same	1-1/2" fiberglass	1.80	3.16	4.96
1650				5/8" FR drywall	1-1/2" fiberglass	1.67	2.53	4.20
1700				nothing	1-1/2" fiberglass	1.31	1.99	3.30
1800		5/8" FR drywall	2 x 4 @ 24" O.C.	same	0	1.71	2.05	3.76
1850				5/8" FR drywall	0	1.39	1.78	3.17
1900				nothing	0	1.03	1.24	2.27
2200		5/8" FR drywall	2 rows-2 x 4	same	2" fiberglass	2.72	2.97	5.69
2250			16"O.C.	5/8" FR drywall	2" fiberglass	2.40	2.70	5.10
2300				nothing	2" fiberglass	2.04	2.16	4.20
2400	5/8" WR drywall	none	2 x 4, @ 16" O.C.	same	0	1.32	1.62	2.94
2450				5/8" FR drywall	0	1.25	1.62	2.87
2500				nothing	0	.89	1.08	1.97
2600		5/8" FR drywall	2 x 4, @ 24" O.C.	same	0	1.85	2.05	3.90
2650				5/8" FR drywall	0	1.46	1.78	3.24
2700				nothing	0	1.10	1.24	2.34
2800	5/8 VF drywall	none	2 x 4, @ 16" O.C.	same	0	1.88	1.74	3.62
2850				5/8" FR drywall	0	1.53	1.68	3.21
2900				nothing	0	1.17	1.14	2.31
3000		5/8" FR drywall	2 x 4 , 24" O.C.	same	0	2.41	2.17	4.58
3050				5/8" FR drywall	0	1.74	1.84	3.58
3100				nothing	0	1.38	1.30	2.68
3200	1/2" reg drywall	3/8" reg drywall	2 x 4, @ 16" O.C.	same	0	1.70	2.16	3.86
3252			2 x 4, @ 16"O.C.	5/8"FR drywall	0	1.44	1.89	3.33
3300				nothing	0	1.08	1.35	2.43

6.1-510 Drywall Partitions/Metal Stud Framing

	FACE LAYER	BASE LAYER	FRAMING	OPPOSITE FACE	INSULATION	COST PER S.F.		
						MAT.	INST.	TOTAL
5200	5/8" FR drywall	none	1-5/8" @ 24" O.C.	same	0	.84	1.60	2.44
5250				5/8" reg. drywall	0	.84	1.60	2.44
5300				nothing	0	.48	1.06	1.54
5400			3-5/8" @ 24" O.C.	same	0	.87	1.62	2.49
5450				5/8" reg. drywall	0	.87	1.62	2.49
5500				nothing	0	.51	1.08	1.59

INTERIOR CONSTR. — A6.1 Partitions

6.1-510 Drywall Partitions/Metal Stud Framing

	FACE LAYER	BASE LAYER	FRAMING	OPPOSITE FACE	INSULATION	COST PER S.F. MAT.	INST.	TOTAL
5600	5/8" FR drywall	1/4" SD gypsum	1-5/8" @ 24" O.C.	same	0	1.40	2.20	3.60
5650				5/8" FR drywall	0	1.12	1.90	3.02
5700				nothing	0	.76	1.36	2.12
5800			2-1/2" @ 24" O.C.	same	0	1.41	2.21	3.62
5850				5/8" FR drywall	0	1.13	1.91	3.04
5900				nothing	0	.77	1.37	2.14
6000		5/8" FR drywall	2-1/2" @ 16" O.C.	same	0	1.66	2.23	3.89
6050				5/8" FR drywall	0	1.34	1.96	3.30
6100				nothing	0	.98	1.42	2.40
6200			3-5/8" @ 24" O.C.	same	0	1.51	2.16	3.67
6250				5/8"FR drywall	3-1/2" fiberglass	1.54	2.06	3.60
6300				nothing	0	.83	1.35	2.18
6400	5/8" WR drywall	none	1-5/8" @ 24" O.C.	same	0	.98	1.60	2.58
6450				5/8" FR drywall	0	.91	1.60	2.51
6500				nothing	0	.55	1.06	1.61
6600			3-5/8" @ 24" O.C.	same	0	1.01	1.62	2.63
6650				5/8" FR drywall	0	.94	1.62	2.56
6700				nothing	0	.58	1.08	1.66
6800		5/8" FR drywall	2-1/2" @ 16" O.C.	same	0	1.80	2.23	4.03
6850				5/8" FR drywall	0	1.41	1.96	3.37
6900				nothing	0	1.05	1.42	2.47
7000			3-5/8" @ 24" O.C.	same	0	1.65	2.16	3.81
7050				5/8"FR drywall	3-1/2" fiberglass	1.61	2.06	3.67
7100				nothing	0	.90	1.35	2.25
7200	5/8" VF drywall	none	1-5/8" @ 24" O.C.	same	0	1.54	1.72	3.26
7250				5/8" FR drywall	0	1.19	1.66	2.85
7300				nothing	0	.83	1.12	1.95
7400			3-5/8" @ 24" O.C.	same	0	1.57	1.74	3.31
7450				5/8" FR drywall	0	1.22	1.68	2.90
7500				nothing	0	.86	1.14	2
7600		5/8" FR drywall	2-1/2" @ 16" O.C.	same	0	2.36	2.35	4.71
7650				5/8" FR drywall	0	1.69	2.02	3.71
7700				nothing	0	1.33	1.48	2.81
7800			3-5/8" @ 24" O.C.	same	0	2.21	2.28	4.49
7850				5/8"FR drywall	3-1/2" fiberglass	1.89	2.12	4.01
7900				nothing	0	1.18	1.41	2.59

For expanded coverage of these items see *Means Assemblies Cost Data 2000*

INTERIOR CONSTR.   A6.1   Partitions

6.1-580 Drywall Components

		COST PER S.F.		
		MAT.	INST.	TOTAL
0060	Metal studs, 24" O.C. including track, load bearing, 20 gage, 2-1/2"	.41	.51	.92
0080	3-5/8"	.46	.52	.98
0100	4"	.43	.53	.96
0120	6"	.63	.54	1.17
0140	Metal studs, 24" O.C. including track, load bearing, 18 gage, 2-1/2"	.41	.51	.92
0160	3-5/8"	.46	.52	.98
0180	4"	.43	.53	.96
0200	6"	.63	.54	1.17
0220	16 gage, 2-1/2"	.47	.58	1.05
0240	3-5/8"	.54	.59	1.13
0260	4"	.59	.60	1.19
0280	6"	.73	.62	1.35
0300	Non load bearing, 25 gage, 1-5/8"	.12	.52	.64
0340	3-5/8"	.15	.54	.69
0360	4"	.18	.55	.73
0380	6"	.24	.56	.80
0400	20 gage, 2-1/2"	.22	.53	.75
0420	3-5/8"	.24	.54	.78
0440	4"	.30	.55	.85
0460	6"	.35	.56	.91
0540	Wood studs including blocking, shoe and double top plate, 2"x4", 12" O.C.	.57	.68	1.25
0560	16" O.C.	.46	.54	1
0580	24" O.C.	.35	.43	.78
0600	2"x6", 12" O.C.	.85	.77	1.62
0620	16" O.C.	.68	.60	1.28
0640	24" O.C.	.52	.47	.99
0642	Furring one side only, steel channels, 3/4", 12" O.C.	.20	1.07	1.27
0644	16" O.C.	.18	.95	1.13
0646	24" O.C.	.12	.72	.84
0647	1-1/2", 12" O.C.	.30	1.20	1.50
0648	16" O.C.	.27	1.05	1.32
0649	24" O.C.	.18	.82	1
0650	Wood strips, 1" x 3", on wood, 12" O.C.	.20	.49	.69
0651	16" O.C.	.15	.37	.52
0652	On masonry, 12" O.C.	.20	.55	.75
0653	16" O.C.	.15	.41	.56
0654	On concrete, 12" O.C.	.20	1.04	1.24
0655	16" O.C.	.15	.78	.93
0665	Gypsum board, one face only, exterior sheathing, 1/2"	.31	.48	.79
0680	Fire resistant, 1/2"	.28	.27	.55
0700	5/8"	.32	.27	.59
0720	Sound deadening board 1/4"	.28	.30	.58
0740	Standard drywall 3/8"	.29	.27	.56
0760	1/2"	.29	.27	.56
0780	5/8"	.32	.27	.59
0800	Tongue & groove coreboard 1"	.76	1.13	1.89
0820	Water resistant, 1/2"	.35	.27	.62
0840	5/8"	.39	.27	.66
0860	Add for the following:, foil backing	.11		.11
0880	Fiberglass insulation, 3-1/2"	.35	.17	.52
0900	6"	.43	.20	.63
0920	Rigid insulation 1"	.35	.27	.62
0940	Resilient furring @ 16" O.C.	.16	.85	1.01
0960	Taping and finishing	.04	.27	.31
0980	Texture spray	.06	.31	.37
1000	Thin coat plaster	.08	.34	.42
1040	2"x4" staggered studs 2"x6" plates & blocking	.65	.59	1.24
1050				

INTERIOR CONSTR. — A6.1 Partitions

6.1-680 Plaster Partition Components

		COST PER S.F.		
		MAT.	INST.	TOTAL
0060	Metal studs, 16" O.C., including track, non load bearing, 25 gage, 1-5/8"	.18	.61	.79
0080	2-1/2"	.18	.61	.79
0100	3-1/4"	.21	.63	.84
0120	3-5/8"	.21	.63	.84
0140	4"	.25	.64	.89
0160	6"	.33	.66	.99
0180	Load bearing, 20 gage, 2-1/2"	.56	.70	1.26
0200	3-5/8"	.63	.71	1.34
0220	4"	.68	.73	1.41
0240	6"	.85	.74	1.59
0260	16 gage 2-1/2"	.65	.80	1.45
0280	3-5/8"	.75	.82	1.57
0300	4"	.81	.83	1.64
0320	6"	1	.85	1.85
0340	Wood studs, including blocking, shoe and double plate, 2"x4", 12" O.C.	.57	.68	1.25
0360	16" O.C.	.46	.54	1
0380	24" O.C.	.35	.43	.78
0400	2"x6", 12" O.C.	.85	.77	1.62
0420	16" O.C.	.68	.60	1.28
0440	24" O.C.	.52	.47	.99
0460	Furring one face only, steel channels, 3/4", 12" O.C.	.20	1.07	1.27
0480	16" O.C.	.18	.95	1.13
0500	24" O.C.	.12	.72	.84
0520	1-1/2", 12" O.C.	.30	1.20	1.50
0540	16" O.C.	.27	1.05	1.32
0560	24" O.C.	.18	.82	1
0580	Wood strips 1"x3", on wood., 12" O.C.	.20	.49	.69
0600	16" O.C.	.15	.37	.52
0620	On masonry, 12" O.C.	.20	.55	.75
0640	16" O.C.	.15	.41	.56
0660	On concrete, 12" O.C.	.20	1.04	1.24
0680	16" O.C.	.15	.78	.93
0700	Gypsum lath. plain or perforated, nailed to studs, 3/8" thick	.42	.33	.75
0720	1/2" thick	.43	.35	.78
0740	Clipped to studs, 3/8" thick	.42	.37	.79
0760	1/2" thick	.44	.40	.84
0780	Metal lath, diamond painted, nailed to wood studs, 2.5 lb.	.18	.33	.51
0800	3.4 lb.	.28	.35	.63
0820	Screwed to steel studs, 2.5 lb.	.18	.35	.53
0840	3.4 lb.	.27	.37	.64
0860	Rib painted, wired to steel, 2.75 lb	.28	.37	.65
0880	3.4 lb	.39	.40	.79
0900	4.0 lb	.40	.43	.83
0910				
0920	Gypsum plaster, 2 coats	.42	1.29	1.71
0940	3 coats	.58	1.57	2.15
0960	Perlite or vermiculite plaster, 2 coats	.40	1.48	1.88
0980	3 coats	.65	1.84	2.49
1000	Stucco, 3 coats, 1" thick, on wood framing	.45	1.22	1.67
1020	On masonry	.25	.68	.93
1100	Metal base galvanized and painted 2-1/2" high	.49	1.05	1.54

INTERIOR CONSTR.   A6.1   Partitions

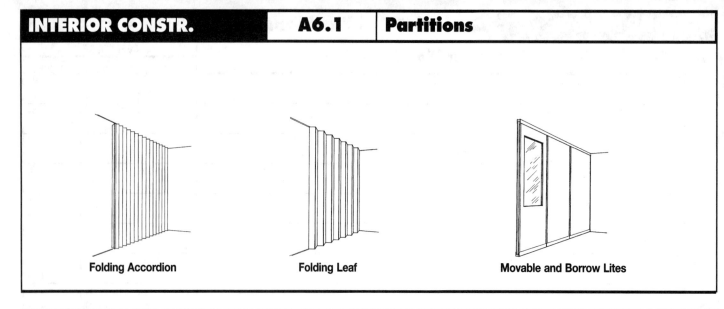

Folding Accordion   Folding Leaf   Movable and Borrow Lites

6.1-820	Partitions	MAT.	INST.	TOTAL
0360	Folding accordion, vinyl covered, acoustical, 3 lb. S.F., 17 ft max. hgt	19.90	5.40	25.30
0380	5 lb. per S.F. 27 ft max height	29.50	5.70	35.20
0400	5.5 lb. per S.F., 17 ft. max height	33	6	39
0420	Commercial, 1.75 lb per S.F., 8 ft. max height	13.20	2.40	15.60
0440	2.0 Lb per S.F., 17 ft. max height	16.05	3.60	19.65
0460	Industrial, 4.0 lb. per S.F. 27 ft. max height	27	7.20	34.20
0480	Vinyl clad wood or steel, electric operation 6 psf	33	3.38	36.38
0500	Wood, non acoustic, birch or mahogany	17.70	1.80	19.50
0560	Folding leaf, aluminum framed acoustical 12 ft.high.,5.5 lb per S.F.,min.	28	9	37
0580	Maximum	34	18	52
0600	6.5 lb. per S.F., minimum	29.50	9	38.50
0620	Maximum	36.50	18	54.50
0640	Steel acoustical, 7.5 per S.F., vinyl faced, minimum	41.50	9	50.50
0660	Maximum	51	18	69
0680	Wood acoustic type, vinyl faced to 18' high 6 psf, minimum	39	9	48
0700	Average	46.50	12	58.50
0720	Maximum	60.50	18	78.50
0740	Formica or hardwood faced, minimum	40.50	9	49.50
0760	Maximum	43	18	61
0780	Wood, low acoustical type to 12 ft. high 4.5 psf	29.50	10.80	40.30
0840	Demountable, trackless wall, cork finish, semi acous, 1-5/8"th, min	21	1.66	22.66
0860	Maximum	25	2.84	27.84
0880	Acoustic, 2" thick, minimum	21.50	1.77	23.27
0900	Maximum	32	2.40	34.40
0920	In-plant modular office system, w/prehung steel door			
0940	3" thick honeycomb core panels			
0960	12' x 12', 2 wall	10.65	.31	10.96
0970	4 wall	10.95	.42	11.37
0980	16' x 16', 2 wall	10.60	.22	10.82
0990	4 wall	7.65	.22	7.87
1000	Gypsum, demountable, 3" to 3-3/4" thick x 9' high, vinyl clad	2.89	1.25	4.14
1020	Fabric clad	8.95	1.36	10.31
1040	1.75 system, vinyl clad hardboard, paper honeycomb core panel			
1060	1-3/4" to 2-1/2" thick x 9' high	6.05	1.25	7.30
1080	Unitized gypsum panel system, 2" to 2-1/2" thick x 9' high			
1100	Vinyl clad gypsum	8.15	1.25	9.40
1120	Fabric clad gypsum	14.65	1.36	16.01
1140	Movable steel walls, modular system			
1160	Unitized panels, 48" wide x 9' high			
1180	Baked enamel, pre-finished	9.55	1	10.55
1200	Fabric clad	14.55	1.07	15.62
1203				

INTERIOR CONSTR.   A6.1   Toilet Partitions

Toilet Units

Entrance Screens

Urinal Screens

6.1-870	Toilet Partitions	MAT.	INST.	TOTAL
0380	Toilet partitions, cubicles, ceiling hung, marble	1,450	269	1,719
0400	Painted metal	420	135	555
0420	Plastic laminate	545	135	680
0460	Stainless steel	1,075	135	1,210
0480	Handicap addition	305		305
0520	Floor and ceiling anchored, marble	1,575	215	1,790
0540	Painted metal	390	108	498
0560	Plastic laminate	515	108	623
0600	Stainless steel	1,225	108	1,333
0620	Handicap addition	278		278
0660	Floor mounted marble	935	179	1,114
0680	Painted metal	355	77	432
0700	Plastic laminate	545	77	622
0740	Stainless steel	1,150	77	1,227
0760	Handicap addition	278		278
0780	Juvenile deduction	38.50		38.50
0820	Floor mounted with handrail marble	1,100	179	1,279
0840	Painted metal	390	90	480
0860	Plastic laminate	575	90	665
0900	Stainless steel	1,200	90	1,290
0920	Handicap addition	278		278
0960	Wall hung, painted metal	495	77	572
1020	Stainless steel	1,200	77	1,277
1040	Handicap addition	278		278
1080	Entrance screens, floor mounted, 54" high, marble	585	59.50	644.50
1100	Painted metal	182	36	218
1140	Stainless steel	645	36	681
1300	Urinal screens, floor mounted, 24" wide, laminated plastic	234	67.50	301.50
1320	Marble	545	80.50	625.50
1340	Painted metal	182	67.50	249.50
1380	Stainless steel	495	67.50	562.50
1428	Wall mounted wedge type, painted metal	214	54	268
1460	Stainless steel	440	54	494

For expanded coverage of these items see *Means Assemblies Cost Data 2000*

INTERIOR CONSTR. A6.4 Doors

Sliding Panel-Mall Front

Rolling Overhead Steel Door

6.4-100	Special Doors	COST PER OPNG.		
		MAT.	INST.	TOTAL
2500	Single leaf, wood, 3'-0"x7'-0"x1 3/8", birch, solid core	284	102	386
2510	Hollow core	247	97.50	344.50
2530	Hollow core, lauan	233	97.50	330.50
2540	Louvered pine	360	97.50	457.50
2550	Paneled pine	360	97.50	457.50
2600	Hollow metal, comm. quality, flush, 3'-0"x7'-0"x1-3/8"	385	107	492
2650	3'-0"x10'-0" openings with panel	630	138	768
2700	Metal fire, comm. quality, 3'-0"x7'-0"x1-3/8"	465	111	576
2800	Kalamein fire, comm. quality, 3'-0"x7'-0"x1-3/4"	385	203	588
3200	Double leaf, wood, hollow core, 2 - 3'-0"x7'-0"x1-3/8"	405	187	592
3300	Hollow metal, comm. quality, B label, 2'-3'-0"x7'-0"x1-3/8"	880	234	1,114
3400	6'-0"x10'-0" opening, with panel	1,200	288	1,488
3500	Double swing door system, 12'-0"x7'-0", mill finish	6,000	1,500	7,500
3700	Black finish	6,525	1,550	8,075
3800	Sliding entrance door and system mill finish	6,975	1,300	8,275
3900	Bronze finish	7,650	1,400	9,050
4000	Black finish	7,975	1,450	9,425
4100	Sliding panel mall front, 16'x9' opening, mill finish	2,425	395	2,820
4200	Bronze finish	3,150	515	3,665
4300	Black finish	3,875	630	4,505
4400	24'x9' opening mill finish	3,525	735	4,260
4500	Bronze finish	4,575	955	5,530
4600	Black finish	5,650	1,175	6,825
4700	48'x9' opening mill finish	6,550	570	7,120
4800	Bronze finish	8,525	740	9,265
4900	Black finish	10,500	910	11,410
5000	Rolling overhead steel door, manual, 8' x 8' high	760	420	1,180
5100	10' x 10' high	1,000	480	1,480
5200	20' x 10' high	2,050	670	2,720
5300	12' x 12' high	1,300	560	1,860
5400	Motor operated, 8' x 8' high	1,500	555	2,055
5500	10' x 10' high	1,725	615	2,340
5600	20' x 10' high	2,775	805	3,580
5700	12' x 12' high	2,025	695	2,720
5800	Roll up grille, aluminum, manual, 10' x 10' high, mill finish	1,850	820	2,670
5900	Bronze anodized	2,950	820	3,770
6000	Motor operated, 10' x 10' high, mill finish	2,700	955	3,655
6100	Bronze anodized	3,800	955	4,755
6200	Steel, manual, 10' x 10' high	1,650	670	2,320
6300	15' x 8' high	1,950	840	2,790
6400	Motor operated, 10' x 10' high	2,500	805	3,305
6500	15' x 8' high	2,800	975	3,775

INTERIOR CONSTR. | A6.5 | Wall Finishes

6.5-100 Paint & Covering

		COST PER S.F.		
		MAT.	INST.	TOTAL
0060	Painting, interior on plaster and drywall, brushwork, primer & 1 coat	.10	.37	.47
0080	Primer & 2 coats	.16	.49	.65
0100	Primer & 3 coats	.21	.60	.81
0120	Walls & ceilings, roller work, primer & 1 coat	.10	.25	.35
0140	Primer & 2 coats	.16	.32	.48
0160	Woodwork incl. puttying, brushwork, primer & 1 coat	.10	.53	.63
0180	Primer & 2 coats	.16	.70	.86
0200	Primer & 3 coats	.21	.96	1.17
0260	Cabinets and casework, enamel, primer & 1 coat	.10	.60	.70
0280	Primer & 2 coats	.15	.74	.89
0300	Masonry or concrete, latex, brushwork, primer & 1 coat	.18	.50	.68
0320	Primer & 2 coats	.25	.72	.97
0340	Addition for block filler	.11	.61	.72
0380	Fireproof paints, intumescent, 1/8" thick 3/4 hour	1.75	.49	2.24
0400	3/16" thick 1 hour	3.90	.74	4.64
0420	7/16" thick 2 hour	5	1.71	6.71
0440	1-1/16" thick 3 hour	8.20	3.42	11.62
0480	Miscellaneous metal brushwork, exposed metal, primer & 1 coat	.07	.60	.67
0500	Gratings, primer & 1 coat	.18	.75	.93
0600	Pipes over 12" diameter	.46	2.40	2.86
0700	Structural steel, brushwork, light framing 300-500 S.F./Ton	.06	.94	1
0720	Heavy framing 50-100 S.F./Ton	.06	.47	.53
0740	Spraywork, light framing 300-500 S.F./Ton	.06	.21	.27
0760	Heavy framing 50-100 S.F./Ton	.06	.23	.29
0800	Varnish, interior wood trim, no sanding sealer & 1 coat	.07	.60	.67
0820	Hardwood floor, no sanding 2 coats	.13	.13	.26
0840	Wall coatings, acrylic glazed coatings, minimum	.26	.46	.72
0860	Maximum	.55	.79	1.34
0880	Epoxy coatings, minimum	.34	.46	.80
0900	Maximum	1.06	1.41	2.47
0940	Exposed epoxy aggregate, troweled on, 1/16" to 1/4" aggregate, minimum	.53	1.02	1.55
0960	Maximum	1.13	1.84	2.97
0980	1/2" to 5/8" aggregate, minimum	1.05	1.84	2.89
1000	Maximum	1.78	3	4.78
1020	1" aggregate, minimum	1.82	2.66	4.48
1040	Maximum	2.77	4.36	7.13
1060	Sprayed on, minimum	.50	.81	1.31
1080	Maximum	.90	1.65	2.55
1100	High build epoxy 50 mil, minimum	.57	.61	1.18
1120	Maximum	.99	2.52	3.51
1140	Laminated epoxy with fiberglass minimum	.63	.81	1.44
1160	Maximum	1.12	1.65	2.77
1180	Sprayed perlite or vermiculite 1/16" thick, minimum	.22	.08	.30
1200	Maximum	.64	.37	1.01
1260	Wall coatings, vinyl plastic, minimum	.29	.33	.62
1280	Maximum	.70	1	1.70
1300	Urethane on smooth surface, 2 coats, minimum	.21	.21	.42
1320	Maximum	.48	.36	.84
1340	3 coats, minimum	.29	.29	.58
1360	Maximum	.64	.51	1.15
1380	Ceramic-like glazed coating, cementitious, minimum	.42	.54	.96
1400	Maximum	.69	.69	1.38
1420	Resin base, minimum	.29	.37	.66
1440	Maximum	.47	.73	1.20
1460	Wall coverings, aluminum foil	.86	.88	1.74
1480	Copper sheets, .025" thick, phenolic backing	5.90	1	6.90
1500	Vinyl backing	4.57	1	5.57
1520	Cork tiles, 12"x12", light or dark, 3/16" thick	2.99	1	3.99
1540	5/16" thick	3.05	1.02	4.07
1560	Basketweave, 1/4" thick	4.68	1	5.68

For expanded coverage of these items see *Means Assemblies Cost Data 2000*

INTERIOR CONSTR. | A6.5 | Wall Finishes

6.5-100 Paint & Covering

		COST PER S.F.		
		MAT.	INST.	TOTAL
1580	Natural, non-directional, 1/2" thick	6.95	1	7.95
1600	12"x36", granular, 3/16" thick	1.01	.63	1.64
1620	1" thick	1.31	.65	1.96
1640	12"x12", polyurethane coated, 3/16" thick	3.16	1	4.16
1660	5/16" thick	4.49	1.02	5.51
1661	Paneling, prefinished plywood, birch	1.21	1.29	2.50
1662	Mahogany, African	2.29	1.35	3.64
1663	Philippine (lauan)	.97	1.08	2.05
1664	Oak or cherry	2.95	1.35	4.30
1665	Rosewood	4.19	1.69	5.88
1666	Teak	2.95	1.35	4.30
1667	Chestnut	4.37	1.44	5.81
1668	Pecan	1.87	1.35	3.22
1669	Walnut	4.76	1.35	6.11
1670	Wood board, knotty pine, finished	1.64	2.26	3.90
1671	Rough sawn cedar	2.02	2.26	4.28
1672	Redwood	4.39	2.26	6.65
1673	Aromatic cedar	3.44	2.42	5.86
1680	Cork wallpaper, paper backed, natural	1.79	.50	2.29
1700	Color	2.22	.50	2.72
1720	Gypsum based, fabric backed, minimum	.71	.30	1.01
1740	Average	.99	.33	1.32
1760	Maximum	1.10	.38	1.48
1780	Vinyl wall covering, fabric back, light weight	.58	.38	.96
1800	Medium weight	.69	.50	1.19
1820	Heavy weight	1.22	.55	1.77
1840	Wall paper, double roll, solid pattern, avg. workmanship	.30	.38	.68
1860	Basic pattern, avg. workmanship	.54	.45	.99
1880	Basic pattern, quality workmanship	1.01	.55	1.56
1900	Grass cloths with lining paper, minimum	.65	.60	1.25
1920	Maximum	2.07	.69	2.76
1940	Ceramic tile, thin set, 4-1/4" x 4-1/4"	2.20	2.36	4.56

6.5-100 Paint Trim

		COST PER L.F.		
		MAT.	INST.	TOTAL
2040	Painting, wood trim, to 6" wide, enamel, primer & 1 coat	.10	.30	.40
2060	Primer & 2 coats	.15	.38	.53
2080	Misc. metal brushwork, ladders	.37	3	3.37
2100	Pipes, to 4" dia.	.06	.63	.69
2120	6" to 8" dia.	.12	1.26	1.38
2140	10" to 12" dia.	.36	1.89	2.25
2160	Railings, 2" pipe	.13	1.50	1.63
2180	Handrail, single	.09	.60	.69

INTERIOR CONSTR. — A6.6 Floor Finishes

6.6-100 Tile & Covering

		COST PER S.F.		
		MAT.	INST.	TOTAL
0060	Carpet tile, nylon, fusion bonded, 18″ x 18″ or 24″ x 24″, 24 oz.	3.11	.35	3.46
0080	35 oz.	3.89	.35	4.24
0100	42 oz.	4.50	.35	4.85
0140	Carpet, tufted, nylon, roll goods, 12′ wide, 26 oz.	2.89	.39	3.28
0160	36 oz.	4.55	.39	4.94
0180	Woven, wool, 36 oz.	7.20	.39	7.59
0200	42 oz.	7.40	.39	7.79
0220	Padding, add to above, minimum	.39	.18	.57
0240	Maximum	.68	.18	.86
0260	Composition flooring, acrylic, 1/4″ thick	1.33	2.59	3.92
0280	3/8″ thick	1.73	3	4.73
0300	Epoxy, minimum	2.32	1.99	4.31
0320	Maximum	2.82	2.75	5.57
0340	Epoxy terrazzo, minimum	4.79	2.93	7.72
0360	Maximum	7.85	3.91	11.76
0380	Mastic, hot laid, 1-1/2″ thick, minimum	3.40	1.95	5.35
0400	Maximum	4.36	2.59	6.95
0420	Neoprene 1/4″ thick, minimum	3.34	2.47	5.81
0440	Maximum	4.60	3.13	7.73
0460	Polyacrylate with ground granite 1/4″, minimum	2.66	1.83	4.49
0480	Maximum	5.15	2.80	7.95
0500	Polyester with colored quart 2 chips 1/16″, minimum	2.35	1.27	3.62
0520	Maximum	4.10	1.99	6.09
0540	Polyurethane with vinyl chips, minimum	6.50	1.27	7.77
0560	Maximum	9.40	1.57	10.97
0600	Concrete topping, granolithic concrete, 1/2″ thick	.18	1.17	1.35
0620	1″ thick	.36	1.20	1.56
0640	2″ thick	.72	1.38	2.10
0660	Heavy duty 3/4″ thick, minimum	.24	1.82	2.06
0680	Maximum	.24	2.16	2.40
0700	For colors, add to above, minimum	.47	.42	.89
0720	Maximum	2.11	.46	2.57
0740	Exposed aggregate finish, minimum	.41	.39	.80
0760	Maximum	1.30	.53	1.83
0780	Abrasives, .25 P.S.F. add to above, minimum	.17	.29	.46
0800	Maximum	.20	.29	.49
0820	Dust on coloring, add, minimum	.64	.19	.83
0840	Maximum	2.23	.39	2.62
0860	1/2″ integral, minimum	.21	1.17	1.38
0880	Maximum	.21	1.17	1.38
0900	Dustproofing, add, minimum	.08	.13	.21
0920	Maximum	.14	.19	.33
0930	Paint	.25	.72	.97
0940	Hardeners, metallic add, minimum	.31	.29	.60
0960	Maximum	.70	.43	1.13
0980	Non-metallic, minimum	.19	.29	.48
1000	Maximum	.52	.43	.95
1020	Integral topping, 3/16″ thick	.06	.69	.75
1040	1/2″ thick	.17	.73	.90
1060	3/4″ thick	.26	.81	1.07
1080	1″ thick	.34	.92	1.26
1100	Terrazzo, minimum	2.61	4.51	7.12
1120	Maximum	4.86	9.80	14.66
1280	Resilient, asphalt tile, 1/8″ thick on concrete, minimum	1.06	.62	1.68
1300	Maximum	1.17	.62	1.79
1320	On wood, add for felt underlay	.20		.20
1340	Cork tile, minimum	3.75	.79	4.54
1360	Maximum	8.90	.79	9.69
1380	Polyethylene, in rolls, minimum	2.46	.90	3.36
1400	Maximum	4.80	.90	5.70

For expanded coverage of these items see *Means Assemblies Cost Data 2000*

INTERIOR CONSTR. — A6.6 Floor Finishes

6.6-100 Tile & Covering

		COST PER S.F.		
		MAT.	INST.	TOTAL
1420	Polyurethane, thermoset, minimum	3.80	2.49	6.29
1440	Maximum	4.51	4.98	9.49
1460	Rubber, sheet goods, minimum	3.40	2.07	5.47
1480	Maximum	5.60	2.76	8.36
1500	Tile, minimum	3.85	.62	4.47
1520	Maximum	6.45	.90	7.35
1580	Vinyl, composition tile, minimum	.76	.50	1.26
1600	Maximum	2.21	.50	2.71
1620	Tile, minimum	1.69	.50	2.19
1640	Maximum	4.79	.50	5.29
1660	Sheet goods, minimum	1.38	1	2.38
1680	Maximum	3.93	1.24	5.17
1720	Tile, ceramic natural clay	4.20	2.46	6.66
1730	Marble, synthetic 12"x12"x5/8"	7.35	7.50	14.85
1740	Porcelain type, minimum	4.52	2.46	6.98
1760	Maximum	4.37	2.36	6.73
1800	Quarry tile, mud set, minimum	2.97	3.21	6.18
1820	Maximum	5.55	4.08	9.63
1840	Thin set, deduct		.64	.64
1860	Terrazzo precast, minimum	3.85	3.32	7.17
1880	Maximum	8.45	3.32	11.77
1900	Non-slip, minimum	16.10	16.70	32.80
1920	Maximum	18.30	23	41.30
1960	Wood, block, end grain factory type, creosoted, 2" thick	2.97	.92	3.89
1980	2-1/2" thick	3.08	2.16	5.24
2000	3" thick	3.03	2.16	5.19
2020	Natural finish, 2" thick	2.99	2.16	5.15
2040	Fir, vertical grain, 1"x4", no finish, minimum	2.38	1.06	3.44
2060	Maximum	2.53	1.06	3.59
2080	Prefinished white oak, prime grade, 2-1/4" wide	6.45	1.59	8.04
2100	3-1/4" wide	8.25	1.46	9.71
2120	Maple strip, sanded and finished, minimum	3.63	2.26	5.89
2140	Maximum	4.29	2.26	6.55
2160	Oak strip, sanded and finished, minimum	3.68	2.26	5.94
2180	Maximum	4.29	2.26	6.55
2200	Parquetry, sanded and finished, minimum	4.05	2.36	6.41
2220	Maximum	6.35	3.37	9.72
2260	Add for sleepers on concrete, treated, 24" O.C., 1"x2"	1.33	1.38	2.71
2280	1"x3"	.89	1.08	1.97
2300	2"x4"	1.08	.55	1.63
2340	Underlayment, plywood, 3/8" thick	.68	.36	1.04
2350	1/2" thick	.88	.37	1.25
2360	5/8" thick	.84	.39	1.23
2370	3/4" thick	1.07	.42	1.49
2380	Particle board, 3/8" thick	.42	.36	.78
2390	1/2" thick	.44	.37	.81
2400	5/8" thick	.51	.39	.90
2410	3/4" thick	.63	.42	1.05
2420	Hardboard, 4' x 4', .215" thick	.45	.36	.81

Important: See the Reference Section for critical supporting data – Reference Nos., Crews & Location Factors

INTERIOR CONSTR. | A6.7 | Ceiling Finishes

Plaster Partitions are defined as follows: type of plaster, type and spacing of framing, type of lath and treatment of opposite face.

Included in the system components are expansion joints. Metal studs are assumed to be nonload bearing.

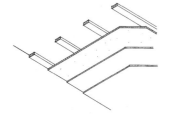

2 Coats of Plaster on Gypsum Lath on Wood Furring

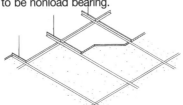

Fiberglass Board on Exposed Suspended Grid System

Plaster and Metal Lath on Metal Furring

6.7-100 Plaster Ceilings

	TYPE	LATH	FURRING	SUPPORT	MAT.	INST.	TOTAL
2400	2 coat gypsum	3/8" gypsum	1"x3" wood, 16" O.C.	wood	1.09	2.77	3.86
2500	Painted			masonry	1.09	2.82	3.91
2600				concrete	1.09	3.16	4.25
2700	3 coat gypsum	3.4# metal	1"x3" wood, 16" O.C.	wood	1.11	2.98	4.09
2800	Painted			masonry	1.11	3.03	4.14
2900				concrete	1.11	3.37	4.48
3000	2 coat perlite	3/8" gypsum	1"x3" wood, 16" O.C.	wood	1.07	2.88	3.95
3100	Painted			masonry	1.07	2.93	4
3200				concrete	1.07	3.27	4.34
3300	3 coat perlite	3.4# metal	1"x3" wood, 16" O.C.	wood	.93	2.95	3.88
3400	Painted			masonry	.93	3	3.93
3500				concrete	.93	3.34	4.27
3600	2 coat gypsum	3/8" gypsum	3/4" CRC, 12" O.C.	1-1/2" CRC, 48"O.C.	1.14	3.39	4.53
3700	Painted		3/4" CRC, 16" O.C.	1-1/2" CRC, 48"O.C.	1.12	3.06	4.18
3800			3/4" CRC, 24" O.C.	1-1/2" CRC, 48"O.C.	1.06	2.79	3.85
3900	2 coat perlite	3/8" gypsum	3/4" CRC, 12" O.C.	1-1/2" CRC, 48"O.C	1.12	3.63	4.75
4000	Painted		3/4" CRC, 16" O.C.	1-1/2" CRC, 48"O.C.	1.10	3.30	4.40
4100			3/4" CRC, 24" O.C.	1-1/2" CRC, 48"O.C.	1.04	3.03	4.07
4200	3 coat gypsum	3.4# metal	3/4" CRC, 12" O.C.	1-1/2" CRC, 36" O.C.	1.49	4.40	5.89
4300	Painted		3/4" CRC, 16" O.C.	1-1/2" CRC, 36" O.C.	1.47	4.07	5.54
4400			3/4" CRC, 24" O.C.	1-1/2" CRC, 36" O.C.	1.41	3.80	5.21
4500	3 coat perlite	3.4# metal	3/4" CRC, 12" O.C.	1-1/2" CRC,36" O.C.	1.56	4.82	6.38
4600	Painted		3/4" CRC, 16" O.C.	1-1/2" CRC, 36" O.C.	1.54	4.49	6.03
4700			3/4" CRC, 24" O.C.	1-1/2" CRC, 36" O.C.	1.48	4.22	5.70

6.7-100 Drywall Ceilings

	TYPE	FINISH	FURRING	SUPPORT	MAT.	INST.	TOTAL
4800	1/2" F.R. drywall	painted and textured	1"x3" wood, 16" O.C.	wood	.63	1.70	2.33
4900				masonry	.63	1.75	2.38
5000				concrete	.63	2.09	2.72
5100	5/8" F.R. drywall	painted and textured	1"x3" wood, 16" O.C.	wood	.67	1.70	2.37
5200				masonry	.67	1.75	2.42
5300				concrete	.67	2.09	2.76
5400	1/2" F.R. drywall	painted and textured	7/8"resil. channels	24" O.C.	.58	1.65	2.23
5500			1"x2" wood	stud clips	.71	1.60	2.31
5602	1/2" F.R. drywall	painted	1-5/8" metal studs	24" O.C.	.60	1.64	2.24

For expanded coverage of these items see *Means Assemblies Cost Data 2000*

INTERIOR CONSTR. — A6.7 Ceiling Finishes

6.7-100 Drywall Ceilings

	TYPE	FINISH	FURRING	SUPPORT		COST PER S.F.		
						MAT.	INST.	TOTAL
5700	5/8" F.R. drywall	painted and textured	1-5/8" metal studs	24" O.C.		.64	1.64	2.28
5702								

6.7-100 Acoustical Ceilings

	TYPE	TILE	GRID	SUPPORT		COST PER S.F.		
						MAT.	INST.	TOTAL
5800	5/8" fiberglass board	24" x 48"	tee	suspended		.92	.78	1.70
5900		24" x 24"	tee	suspended		1.01	.86	1.87
6000	3/4" fiberglass board	24" x 48"	tee	suspended		1.53	.80	2.33
6100		24" x 24"	tee	suspended		1.62	.88	2.50
6500	5/8" mineral fiber	12" x 12"	1"x3" wood, 12" O.C.	wood		.99	1.04	2.03
6600				masonry		.99	1.11	2.10
6700				concrete		.99	1.56	2.55
6800	3/4" mineral fiber	12" x 12"	1"x3" wood, 12" O.C.	wood		1.49	1.04	2.53
6900				masonry		1.49	1.04	2.53
7000				concrete		1.49	1.04	2.53
7100	3/4" mineral fiber on	12" x 12"	25 ga. channels	runners		1.91	1.42	3.33
7102	5/8" F.R. drywall							
7200	5/8" plastic coated	12" x 12"		adhesive backed		1.14	.27	1.41
7201	Mineral fiber							
7202								
7300	3/4" plastic coated	12" x 12"		adhesive backed		1.64	.27	1.91
7301	Mineral fiber							
7302								
7400	3/4" mineral fiber	12" x 12"	conceal 2" bar & channels	suspended		1.73	1.37	3.10
7401								
7402								

6.7-810 Acoustical Ceilings

		COST PER S.F.		
		MAT.	INST.	TOTAL
2480	Ceiling boards, eggcrate, acrylic, 1/2" x 1/2" x 1/2" cubes	1.50	.54	2.04
2500	Polystyrene, 3/8" x 3/8" x 1/2" cubes	1.24	.53	1.77
2520	1/2" x 1/2" x 1/2" cubes	1.67	.54	2.21
2540	Fiberglass boards, plain, 5/8" thick	.56	.43	.99
2560	3/4" thick	1.17	.45	1.62
2580	Grass cloth faced, 3/4" thick	1.82	.54	2.36
2600	1" thick	2.05	.56	2.61
2620	Luminous panels, prismatic, acrylic	1.80	.68	2.48
2640	Polystyrene	.92	.68	1.60
2660	Flat or ribbed, acrylic	3.16	.68	3.84
2680	Polystyrene	2.16	.68	2.84
2700	Drop pan, white, acrylic	4.62	.68	5.30
2720	Polystyrene	3.86	.68	4.54
2740	Mineral fiber boards, 5/8" thick, standard	.65	.40	1.05
2760	Plastic faced	1	.68	1.68
2780	2 hour rating	.88	.40	1.28
2800	Perforated aluminum sheets, .024 thick, corrugated painted	1.84	.55	2.39
2820	Plain	1.65	.54	2.19
3080	Mineral fiber, plastic coated, 12" x 12" or 12" x 24", 5/8" thick	.79	.27	1.06
3100	3/4" thick	1.29	.27	1.56
3120	Fire rated, 3/4" thick, plain faced	1.10	.27	1.37
3140	Mylar faced	1.14	.27	1.41
3160	Add for ceiling primer	.13		.13
3180	Add for ceiling cement	.35		.35
3240	Suspension system, furring, 1" x 3" wood 12" O.C.	.20	.77	.97
3260	T bar suspension system, 2' x 4' grid	.35	.34	.69

INTERIOR CONSTR. — A6.7 Ceiling Finishes

6.7-810 Acoustical Ceilings

		COST PER S.F.		
		MAT.	INST.	TOTAL
3280	2' x 2' grid	.44	.42	.86
3300	Concealed Z bar suspension system 12" module	.39	.52	.91
3320	Add to above for 1-1/2" carrier channels 4' O.C.	.23	.57	.80
3340	Add to above for carrier channels for recessed lighting	.37	.59	.96

6.7-820 Plaster Ceilings

		COST PER S.F.		
		MAT.	INST.	TOTAL
0060	Plaster, gypsum incl. finish, 2 coats			
0080	3 coats	.58	1.75	2.33
0100	Perlite, incl. finish, 2 coats	.40	1.72	2.12
0120	3 coats	.65	2.17	2.82
0140	Thin coat on drywall	.08	.34	.42
0200	Lath, gypsum, 3/8" thick	.42	.46	.88
0220	1/2" thick	.43	.48	.91
0240	5/8" thick	.48	.56	1.04
0260	Metal, diamond, 2.5 lb.	.18	.37	.55
0280	3.4 lb.	.28	.40	.68
0300	Flat rib, 2.75 lb.	.28	.37	.65
0320	3.4 lb.	.39	.40	.79
0440	Furring, steel channels, 3/4" galvanized, 12" O.C.	.20	1.20	1.40
0460	16" O.C.	.18	.87	1.05
0480	24" O.C.	.12	.60	.72
0500	1-1/2" galvanized, 12" O.C.	.30	1.32	1.62
0520	16" O.C.	.27	.97	1.24
0540	24" O.C.	.18	.65	.83
0560	Wood strips, 1"x3", on wood, 12" O.C.	.20	.77	.97
0580	16" O.C.	.15	.58	.73
0600	24" O.C.	.10	.39	.49
0620	On masonry, 12" O.C.	.20	.84	1.04
0640	16" O.C.	.15	.63	.78
0660	24" O.C.	.10	.42	.52
0680	On concrete, 12" O.C.	.20	1.29	1.49
0700	16" O.C.	.15	.97	1.12
0720	24" O.C.	.10	.65	.75
0722				
0740				
0940	Paint on plaster or drywall, roller work, primer + 1 coat	.10	.25	.35
0960	Primer + 2 coats	.16	.32	.48

For information about Means Estimating Seminars, see yellow pages 11 and 12 in back of book

For expanded coverage of these items see *Means Assemblies Cost Data 2000*

Division 7
Conveying Systems

CONVEYING — A7.1 Elevators

The hydraulic elevator obtains its motion from the movement of liquid under pressure in the piston connected to the car bottom. These pistons can provide travel to a maximum rise of 70' and are sized for the intended load. As the rise reaches the upper limits the cost tends to exceed that of a geared electric unit.

7.1-100	Hydraulic	COST EACH		
		MAT.	INST.	TOTAL
1300	Pass. elev., 1500 lb., 2 Floors, 100 FPM	33,400	6,200	39,600
1400	5 Floors, 100 FPM	44,900	16,500	61,400
1600	2000 lb., 2 Floors, 100 FPM	33,900	6,200	40,100
1700	5 floors, 100 FPM	45,400	16,500	61,900
1900	2500 lb., 2 Floors, 100 FPM	34,500	6,200	40,700
2000	5 floors, 100 FPM	46,000	16,500	62,500
2200	3000 lb., 2 Floors, 100 FPM	36,300	6,200	42,500
2300	5 floors, 100 FPM	47,800	16,500	64,300
2500	3500 lb., 2 Floors, 100 FPM	38,300	6,200	44,500
2600	5 floors, 100 FPM	49,800	16,500	66,300
2800	4000 lb., 2 Floors, 100 FPM	39,000	6,200	45,200
2900	5 floors, 100 FPM	50,500	16,500	67,000
3100	4500 lb., 2 Floors, 100 FPM	40,100	6,200	46,300
3200	5 floors, 100 FPM	51,500	16,500	68,000
4000	Hospital elevators, 3500 lb., 2 Floors, 100 FPM	49,200	6,200	55,400
4100	5 floors, 100 FPM	73,000	16,500	89,500
4300	4000 lb., 2 Floors, 100 FPM	49,200	6,200	55,400
4400	5 floors, 100 FPM	73,000	16,500	89,500
4600	4500 lb., 2 Floors, 100 FPM	55,000	6,200	61,200
4800	5 floors, 100 FPM	79,000	16,500	95,500
4900	5000 lb., 2 Floors, 100 FPM	57,500	6,200	63,700
5000	5 floors, 100 FPM	81,500	16,500	98,000
6700	Freight elevators (Class "B"), 3000 lb., 2 Floors, 50 FPM	49,100	7,875	56,975
6800	5 floors, 100 FPM	72,500	20,700	93,200
7000	4000 lb., 2 Floors, 50 FPM	51,500	7,875	59,375
7100	5 floors, 100 FPM	75,000	20,700	95,700
7500	10,000 lb., 2 Floors, 50 FPM	64,500	7,875	72,375
7600	5 floors, 100 FPM	87,500	20,700	108,200
8100	20,000 lb., 2 Floors, 50 FPM	78,000	7,875	85,875
8200	5 Floors, 100 FPM	101,500	20,700	122,200

For information about Means Estimating Seminars, see yellow pages 11 and 12 in back of book

Division 8
Mechanical

MECHANICAL — A8.1 Plumbing

Installation includes piping and fittings within 10' of heater. Gas heaters require vent piping (not included with these units).

8.1-120	Gas Fired Water Heaters - Residential Systems	COST EACH		
		MAT.	INST.	TOTAL
2200	Gas fired water heater, residential, 100° F rise			
2220	20 gallon tank, 25 GPH	490	590	1,080
2260	30 gallon tank, 32 GPH	545	600	1,145
2300	40 gallon tank, 32 GPH	610	675	1,285
2340	50 gallon tank, 63 GPH	880	675	1,555
2380	75 gallon tank, 63 GPH	1,200	755	1,955
2420	100 gallon tank, 63 GPH	1,350	795	2,145
2422				

8.1-130	Oil Fired Water Heaters - Residential Systems	COST EACH		
		MAT.	INST.	TOTAL
2200	Oil fired water heater, residential, 100° F rise			
2220	30 gallon tank, 103 GPH	1,000	560	1,560
2260	50 gallon tank, 145 GPH	1,325	625	1,950
2300	70 gallon tank, 164 GPH	1,900	705	2,605
2340	85 gallon tank, 181 GPH	3,950	720	4,670
2344				

MECHANICAL — A8.1 Plumbing

Systems below include piping and fittings within 10' of heater. Electric water heaters do not require venting.

8.1-160	Electric Water Heaters - Commercial Systems	MAT.	INST.	TOTAL
1800	Electric water heater, commercial, 100° F rise			
1820	50 gallon tank, 9 KW 37 GPH	2,275	505	2,780
1860	80 gal, 12 KW 49 GPH	2,850	625	3,475
1900	36 KW 147 GPH	3,875	675	4,550
1940	120 gal, 36 KW 147 GPH	4,700	730	5,430
1980	150 gal, 120 KW 490 GPH	13,100	780	13,880
2020	200 gal, 120 KW 490 GPH	13,800	800	14,600
2060	250 gal, 150 KW 615 GPH	15,200	935	16,135
2100	300 gal, 180 KW 738 GPH	16,600	980	17,580
2140	350 gal, 30 KW 123 GPH	12,200	1,050	13,250
2180	180 KW 738 GPH	17,100	1,050	18,150
2220	500 gal, 30 KW 123 GPH	16,000	1,250	17,250
2260	240 KW 984 GPH	23,400	1,250	24,650
2300	700 gal, 30 KW 123 GPH	19,400	1,375	20,775
2340	300 KW 1230 GPH	28,900	1,375	30,275
2380	1000 gal, 60 KW 245 GPH	23,400	1,900	25,300
2420	480 KW 1970 GPH	23,300	1,900	25,200
2460	1500 gal, 60 KW 245 GPH	34,300	2,325	36,625
2500	480 KW 1970 GPH	47,200	2,325	49,525

For expanded coverage of these items see *Means Mechanical or Plumbing Cost Data 2000*

MECHANICAL — A8.1 Plumbing

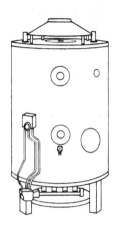

Units may be installed in multiples for increased capacity.

Included below is the heater with self-energizing gas controls, safety pilots, insulated jacket, hi-limit aquastat and pressure relief valve.

Installation includes piping and fittings within 10' of heater. Gas heaters require vent piping (not included in these prices).

8.1-170	Gas Fired Water Heaters - Commercial Systems	MAT.	INST.	TOTAL
1760	Gas fired water heater, commercial, 100° F rise			
1780	75.5 MBH input, 63 GPH	1,375	775	2,150
1820	95 MBH input, 86 GPH	2,100	775	2,875
1860	100 MBH input, 91 GPH	2,275	810	3,085
1900	115 MBH input, 110 GPH	3,000	840	3,840
1980	155 MBH input, 150 GPH	4,400	940	5,340
2020	175 MBH input, 168 GPH	3,825	1,000	4,825
2060	200 MBH input, 192 GPH	3,900	1,150	5,050
2100	240 MBH input, 230 GPH	4,100	1,225	5,325
2140	300 MBH input, 278 GPH	4,475	1,425	5,900
2180	390 MBH input, 374 GPH	5,300	1,425	6,725
2220	500 MBH input, 480 GPH	7,100	1,525	8,625
2260	600 MBH input, 576 GPH	8,375	1,650	10,025

MECHANICAL — A8.1 Plumbing

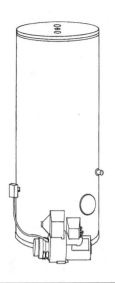

Units may be installed in multiples for increased capacity.

Included below is the heater, wired-in flame retention burners, cadmium cell primary controls, hi-limit controls, ASME pressure relief valves, draft controls, and insulated jacket.

Oil fired water heater systems include piping and fittings within 10' of heater. Oil fired heaters require vent piping (not included in these systems).

8.1-180	Oil Fired Water Heaters - Commercial Systems	COST EACH		
		MAT.	INST.	TOTAL
1800	Oil fired water heater, commercial, 100° F rise			
1820	103 MBH output, 116 GPH	1,175	680	1,855
1900	134 MBH output, 161 GPH	2,425	870	3,295
1940	161 MBH output, 192 GPH	2,675	995	3,670
1980	187 MBH output, 224 GPH	3,000	1,150	4,150
2060	262 MBH output, 315 GPH	3,300	1,325	4,625
2100	341 MBH output, 409 GPH	4,475	1,400	5,875
2140	420 MBH output, 504 GPH	5,025	1,525	6,550
2180	525 MBH output, 630 GPH	6,825	1,725	8,550
2220	630 MBH output, 756 GPH	7,175	1,725	8,900
2260	735 MBH output, 880 GPH	7,950	2,000	9,950
2300	840 MBH output, 1000 GPH	9,050	2,000	11,050
2340	1050 MBH output, 1260 GPH	10,100	1,950	12,050
2380	1365 MBH output, 1640 GPH	12,400	2,200	14,600
2420	1680 MBH output, 2000 GPH	13,800	2,775	16,575
2460	2310 MBH output, 2780 GPH	18,000	3,425	21,425
2500	2835 MBH output, 3400 GPH	21,000	3,425	24,425
2540	3150 MBH output, 3780 GPH	23,000	4,675	27,675

For expanded coverage of these items see *Means Mechanical or Plumbing Cost Data 2000*

MECHANICAL A8.1 Plumbing

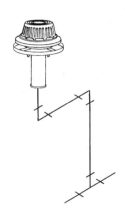

Design Assumptions: Vertical conductor size is based on a maximum rate of rainfall of 4" per hour. To convert roof area to other rates multiply "Max. S.F. Roof Area" shown by four and divide the result by desired local rate. The answer is the local roof area that may be handled by the indicated pipe diameter.

Basic cost is for roof drain, 10' of vertical leader and 10' of horizontal, plus connection to the main.

Pipe Dia.	Max. S.F. Roof Area	Gallons per Min.
2"	544	23
3"	1610	67
4"	3460	144
5"	6280	261
6"	10,200	424
8"	22,000	913

8.1-310	Roof Drain Systems	MAT.	INST.	TOTAL
1880	Roof drain, DWV PVC, 2" diam., piping, 10' high	116	305	421
1920	For each additional foot add	3.20	8.85	12.05
1960	3" diam., 10' high	145	365	510
2000	For each additional foot add	4.16	9.85	14.01
2040	4" diam., 10' high	185	410	595
2080	For each additional foot add	5.65	10.90	16.55
2120	5" diam., 10' high	355	435	790
2160	For each additional foot add	8.25	12.15	20.40
2200	6" diam., 10' high	375	520	895
2240	For each additional foot add	8.15	13.40	21.55
2280	8" diam., 10' high	1,175	790	1,965
2320	For each additional foot add	14.45	15.70	30.15
3940	C.I., soil, single hub, service wt., 2" diam. piping, 10' high	227	320	547
3980	For each additional foot add	5.40	8.30	13.70
4120	3" diam., 10' high	310	345	655
4160	For each additional foot add	6.95	8.70	15.65
4200	4" diam., 10' high	330	375	705
4240	For each additional foot add	6.85	9.50	16.35
4280	5" diam., 10' high	435	390	825
4320	For each additional foot add	12.05	9.95	22
4360	6" diam., 10' high	505	415	920
4400	For each additional foot add	14.20	10.35	24.55
4440	8" diam., 10' high	1,350	865	2,215
4480	For each additional foot add	22.50	17.70	40.20
6040	Steel galv. sch 40 threaded, 2" diam. piping, 10' high	235	310	545
6080	For each additional foot add	3.93	8.15	12.08
6120	3" diam., 10' high	460	445	905
6160	For each additional foot add	7.60	12.15	19.75
6200	4" diam., 10' high	665	570	1,235
6240	For each additional foot add	10.90	14.50	25.40
6280	5" diam., 10' high	845	715	1,560
6320	For each additional foot add	15.95	20	35.95
6360	6" diam, 10' high	1,450	825	2,275
6400	For each additional foot add	21	24.50	45.50
6440	8" diam., 10' high	3,175	1,125	4,300
6480	For each additional foot add	34.50	28	62.50

MECHANICAL — A8.1 Plumbing

Table 8.1-401 Minimum Plumbing Fixture Requirements

Type of Building/Use	Water Closets		Urinals		Lavatories		Bathtubs or Showers		Drinking Fountain	Other
	Persons	Fixtures	Persons	Fixtures	Persons	Fixtures	Persons	Fixtures	Fixtures	Fixtures
Assembly Halls Auditoriums, Theater, Public assembly	1-100 101-200 201-400 Over 400 add 1 fixt. for ea. 500 men; 1 fixt. for ea. 300 women	1 2 3	1-200 201-400 401-600 Over 600 add 1 fixture for each 300 men	1 2 3	1-200 201-400 401-750 Over 750 add 1 fixture for each 500 persons	1 2 3			1 for each 1000 persons	1 service sink
Assembly Public Worship	300 men 150 women	1 1	300 men	1	men women	1 1			1	
Dormitories	Men: 1 for each 10 persons Women: 1 for each 8 persons		1 for each 25 men; over 150 add 1 fixture for each 50 men		1 for ea. 12 persons 1 separate dental lav. for each 50 persons recom.		1 for ea. 8 persons For women add 1 additional for each 30. Over 150 persons add 1 for each 20.		1 for each 75 persons	Laundry trays 1 for each 50 serv. sink 1 for ea. 100
Dwellings Apartments and homes	1 fixture for each unit				1 fixture for each unit		1 fixture for each unit			
Hospitals Indiv. Room, Ward, Waiting room	8 persons	1 1 1			10 persons	1 1 1	20 persons	1 1	1 for 100 patients	1 service sink per floor
Industrial Mfg. plants, Warehouses	1-10 11-25 26-50 51-75 76-100 1 fixture for each additional 30 persons	1 2 3 4 5	0-30 31-80 81-160 161-240	1 2 3 4	1-100 over 100	1 for ea. 10 1 for ea. 15	1 Shower for each 15 persons subject to excessive heat or occupational hazard		1 for each 75 persons	
Public Buildings Businesses, Offices	1-15 16-35 36-55 56-80 81-110 111-150 1 fixture for ea. additional 40 persons	1 2 3 4 5 6	Urinals may be provided in place of water closets but may not replace more than 1/3 required number of men's water closets		1-15 16-35 36-60 61-90 91-125 1 fixture for ea. additional 45 persons	1 2 3 4 5			1 for each 75 persons	1 service sink per floor
Schools Elementary	1 for ea. 30 boys 1 for ea. 25 girls		1 for ea. 25 boys		1 for ea. 35 boys 1 for ea. 35 girls		For gym or pool shower room 1/5 of a class		1 for each 40 pupils	
Schools Secondary	1 for ea. 40 boys 1 for ea. 30 girls		1 for ea. 25 boys		1 for ea. 40 boys 1 for ea. 40 girls		For gym or pool shower room 1/5 of a class		1 for each 50 pupils	

For expanded coverage of these items see *Means Mechanical or Plumbing Cost Data 2000*

MECHANICAL — A8.1 Plumbing

Systems are complete with trim and rough-in (supply, waste and vent) to connect to supply branches and waste mains.

Recessed Bathtub

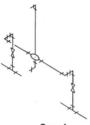
Supply

Waste/Vent

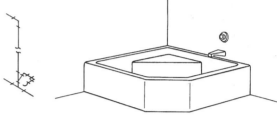
Corner Bathtub

8.1-410	Bathtub Systems	MAT.	INST.	TOTAL
1960	Bathtub, recessed, P.E. on CI., 42" x 37"	865	355	1,220
2000	48" x 42"	1,250	385	1,635
2040	72" x 36"	1,350	425	1,775
2080	Mat bottom, 5' long	460	370	830
2120	5'-6" long	1,050	385	1,435
2160	Corner, 48" x 42"	1,450	370	1,820
2200	Formed steel, enameled, 4'-6" long	385	345	730
2240	5' long	370	350	720

8.1-420	Drinking Fountain Systems	MAT.	INST.	TOTAL
1740	Drinking fountain, one bubbler, wall mounted			
1760	Non recessed			
1800	Bronze, no back	1,075	204	1,279
1840	Cast iron, enameled, low back	495	204	699
1880	Fiberglass, 12" back	595	204	799
1920	Stainless steel, no back	910	204	1,114
1960	Semi-recessed, poly marble	715	204	919
2040	Stainless steel	825	204	1,029
2080	Vitreous china	510	204	714
2120	Full recessed, poly marble	810	204	1,014
2200	Stainless steel	665	204	869
2240	Floor mounted, pedestal type, aluminum	505	276	781
2320	Bronze	1,225	276	1,501
2360	Stainless steel	1,225	276	1,501

MECHANICAL — A8.1 Plumbing

Systems are complete with trim and rough-in (supply, waste and vent) to connect to supply branches and waste mains.

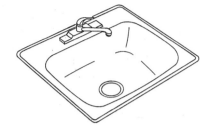
Countertop Single Bowl

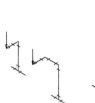
Supply

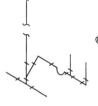
Waste/Vent

Countertop Double Bowl

8.1-431 Kitchen Sink Systems

		\<COST EACH\>		
		MAT.	INST.	TOTAL
1720	Kitchen sink w/trim, countertop, PE on CI, 24"x21", single bowl	271	340	611
1760	30" x 21" single bowl	410	340	750
1800	32" x 21" double bowl	375	365	740
1840	42" x 21" double bowl	545	370	915
1880	Stainless steel, 19" x 18" single bowl	355	340	695
1920	25" x 22" single bowl	385	340	725
1960	33" x 22" double bowl	525	365	890
2000	43" x 22" double bowl	600	370	970
2040	44" x 22" triple bowl	860	385	1,245
2080	44" x 24" corner double bowl	585	370	955
2120	Steel, enameled, 24" x 21" single bowl	176	340	516
2160	32" x 21" double bowl	200	365	565
2240	Raised deck, PE on CI, 32" x 21", dual level, double bowl	460	465	925
2280	42" x 21" dual level, triple bowl	660	505	1,165
2281				
2282				

8.1-432 Laundry Sink Systems

		COST EACH		
		MAT.	INST.	TOTAL
1740	Laundry sink w/trim, PE on CI, black iron frame			
1760	24" x 20", single compartment	435	335	770
1840	48" x 21" double compartment	770	360	1,130
1920	Molded stone, on wall, 22" x 21" single compartment	196	335	531
1960	45"x 21" double compartment	340	360	700
2040	Plastic, on wall or legs, 18" x 23" single compartment	157	325	482
2080	20" x 24" single compartment	186	325	511
2120	36" x 23" double compartment	212	350	562
2160	40" x 24" double compartment	278	350	628
2200	Stainless steel, countertop, 22" x 17" single compartment	345	335	680
2240	19" x 22" single compartment	415	335	750
2280	33" x 22" double compartment	395	360	755

For expanded coverage of these items see *Means Mechanical or Plumbing Cost Data 2000*

MECHANICAL — A8.1 Plumbing

Systems are complete with trim and rough-in (supply, waste and vent) to connect to supply branches and waste mains.

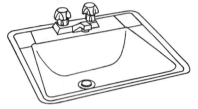
Vanity Top

Supply   Waste/Vent

Wall Hung

8.1-433	Lavatory Systems	MAT.	INST.	TOTAL
1560	Lavatory w/trim, vanity top, PE on CI, 20" x 18", Vanity top by others.	278	310	588
1600	19" x 16" oval	246	310	556
1640	18" round	257	310	567
1680	Cultured marble, 19" x 17"	188	310	498
1720	25" x 19"	218	310	528
1760	Stainless, self-rimming, 25" x 22"	230	310	540
1800	17" x 22"	200	310	510
1840	Steel enameled, 20" x 17"	170	320	490
1880	19" round	168	320	488
1920	Vitreous china, 20" x 16"	261	325	586
1960	19" x 16"	261	325	586
2000	22" x 13"	271	325	596
2040	Wall hung, PE on CI, 18" x 15"	430	340	770
2080	19" x 17"	520	340	860
2120	20" x 18"	395	340	735
2160	Vitreous china, 18" x 15"	395	350	745
2200	19" x 17"	335	350	685
2240	24" x 20"	455	350	805

8.1-434	Laboratory Sink Systems	MAT.	INST.	TOTAL
1580	Laboratory sink w/trim, polyethylene, single bowl,			
1600	Double drainboard, 54" x 24" O.D.	930	435	1,365
1640	Single drainboard, 47" x 24" O.D.	970	435	1,405
1680	70" x 24" O.D.	985	435	1,420
1760	Flanged, 14-1/2" x 14-1/2" O.D.	360	390	750
1800	18-1/2" x 18-1/2" O.D.	325	390	715
1840	23-1/2" x 20-1/2" O.D.	375	390	765
1920	Polypropylene, cup sink, oval, 7" x 4" O.D.	161	345	506
1960	10" x 4-1/2" O.D.	172	345	517
1961				

8.1-434	Service Sink Systems	MAT.	INST.	TOTAL
4260	Service sink w/trim, PE on CI, corner floor, 28" x 28", w/rim guard	650	435	1,085
4300	Wall hung w/rim guard, 22" x 18"	715	505	1,220
4340	24" x 20"	760	505	1,265
4380	Vitreous china, wall hung 22" x 20"	685	505	1,190

MECHANICAL — A8.1 Plumbing

Systems are complete with trim and rough-in (supply, waste and vent) for connection to supply branches and waste mains.

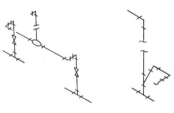
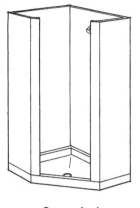

Three Wall — Supply — Waste/Vent — Corner Angle

8.1-440	Shower Systems	MAT.	INST.	TOTAL
1560	Shower, stall, baked enamel, molded stone receptor, 30" square	435	515	950
1600	32" square	405	515	920
1640	Terrazzo receptor, 32" square	1,150	515	1,665
1680	36" square	965	545	1,510
1720	36" corner angle	885	545	1,430
1800	Fiberglass one piece, three walls, 32" square	380	475	855
1840	36" square	425	475	900
1880	Polypropylene, molded stone receptor, 30" square	470	515	985
1920	32" square	455	515	970
1960	Built-in head, arm, bypass, stops and handles	85	134	219

8.1-450	Urinal Systems	MAT.	INST.	TOTAL
2000	Urinal, vitreous china, wall hung	395	375	770
2040	Stall type	605	415	1,020

8.1-460	Water Cooler Systems	MAT.	INST.	TOTAL
1840	Water cooler, electric, wall hung, 8.2 GPH	605	262	867
1880	Dual height, 14.3 GPH	875	268	1,143
1920	Wheelchair type, 7.5 G.P.H.	1,375	262	1,637
1960	Semi recessed, 8.1 G.P.H.	820	262	1,082
2000	Full recessed, 8 G.P.H.	1,275	280	1,555
2040	Floor mounted, 14.3 G.P.H.	640	228	868
2080	Dual height, 14.3 G.P.H.	905	276	1,181
2120	Refrigerated compartment type, 1.5 G.P.H.	1,150	228	1,378

For expanded coverage of these items see *Means Mechanical or Plumbing Cost Data 2000*

MECHANICAL — A8.1 Plumbing

Systems are complete with trim seat and rough-in (supply, waste and vent) for connection to supply branches and waste mains.

One Piece Wall Hung

Supply

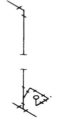
Waste/Vent

Floor Mount

8.1-470 Water Closet Systems

		MAT.	INST.	TOTAL
1800	Water closet, vitreous china, elongated			
1840	Tank type, wall hung, one piece	770	300	1,070
1880	Close coupled two piece	705	300	1,005
1920	Floor mount, one piece	640	330	970
1960	One piece low profile	710	330	1,040
2000	Two piece close coupled	298	330	628
2040	Bowl only with flush valve			
2080	Wall hung	665	345	1,010
2120	Floor mount	495	335	830
2122				

8.1-510 Water Closets, Group

		MAT.	INST.	TOTAL
1760	Water closets, battery mount, wall hung, side by side, first closet	700	355	1,055
1800	Each additional water closet, add	665	335	1,000
3000	Back to back, first pair of closets	1,275	480	1,755
3100	Each additional pair of closets, back to back	1,250	470	1,720

8.1-560 Group Wash Fountain Systems

		MAT.	INST.	TOTAL
1740	Group wash fountain, precast terrazzo			
1760	Circular, 36" diameter	2,525	540	3,065
1800	54" diameter	3,075	585	3,660
1840	Semi-circular, 36" diameter	2,350	540	2,890
1880	54" diameter	2,775	585	3,360
1960	Stainless steel, circular, 36" diameter	2,900	505	3,405
2000	54" diameter	2,775	555	3,330
2040	Semi-circular, 36" diameter	2,025	505	2,530
2080	54" diameter	2,500	555	3,055
2160	Thermoplastic, circular, 36" diameter	2,250	415	2,665
2200	54" diameter	2,575	475	3,050
2240	Semi-circular, 36" diameter	2,075	415	2,490
2280	54" diameter	2,500	475	2,975

MECHANICAL — A8.1 Plumbing

Two Fixture Bathroom Systems consisting of a lavatory, water closet, and rough-in service piping.

- Prices for plumbing and fixtures only.

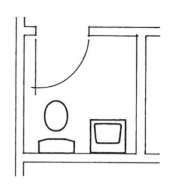

*Common wall is with an adjacent bathroom

8.1-620	Two Fixture Bathroom, Two Wall Plumbing	COST EACH		
		MAT.	INST.	TOTAL
1180	Bathroom, lavatory & water closet, 2 wall plumbing, stand alone	780	850	1,630
1200	Share common plumbing wall*	730	735	1,465

8.1-620	Two Fixture Bathroom, One Wall Plumbing	COST EACH		
		MAT.	INST.	TOTAL
2220	Bathroom, lavatory & water closet, one wall plumbing, stand alone	745	765	1,510
2240	Share common plumbing wall*	670	650	1,320

8.1-630	Three Fixture Bathroom, One Wall Plumbing	COST EACH		
		MAT.	INST.	TOTAL
1150	Bathroom, three fixture, one wall plumbing			
1160	Lavatory, water closet & bathtub			
1170	Stand alone	1,100	995	2,095
1180	Share common plumbing wall *	955	720	1,675

8.1-630	Three Fixture Bathroom, Two Wall Plumbing	COST EACH		
		MAT.	INST.	TOTAL
2130	Bathroom, three fixture, two wall plumbing			
2140	Lavatory, water closet & bathtub			
2160	Stand alone	1,125	1,000	2,125
2180	Long plumbing wall common *	1,000	805	1,805
3610	Lavatory, bathtub & water closet			
3620	Stand alone	1,175	1,150	2,325
3640	Long plumbing wall common *	1,125	1,025	2,150
4660	Water closet, corner bathtub & lavatory			
4680	Stand alone	2,100	1,025	3,125
4700	Long plumbing wall common *	2,000	770	2,770
6100	Water closet, stall shower & lavatory			
6120	Stand alone	1,150	1,300	2,450
6140	Long plumbing wall common *	1,100	1,225	2,325
7060	Lavatory, corner stall shower & water closet			
7080	Stand alone	1,550	1,200	2,750
7100	Short plumbing wall common *	1,450	860	2,310

For expanded coverage of these items see *Means Mechanical or Plumbing Cost Data 2000*

MECHANICAL — A8.2 — Fire Protection

Table 8.2-101 Sprinkler Systems (Automatic)

Sprinkler systems may be classified by type as follows:

1. **Wet Pipe System.** A system employing automatic sprinklers attached to a piping system containing water and connected to a water supply so that water discharges immediately from sprinklers opened by a fire.
2. **Dry Pipe System.** A system employing automatic sprinklers attached to a piping system containing air under pressure, the release of which as from the opening of sprinklers permits the water pressure to open a valve known as a "dry pipe valve". The water then flows into the piping system and out the opened sprinklers.
3. **Pre-Action System.** A system employing automatic sprinklers attached to a piping system containing air that may or may not be under pressure, with a supplemental heat responsive system of generally more sensitive characteristics than the automatic sprinklers themselves, installed in the same areas as the sprinklers; actuation of the heat responsive system, as from a fire, opens a valve which permits water to flow into the sprinkler piping system and to be discharged from any sprinklers which may be open.
4. **Deluge System.** A system employing open sprinklers attached to a piping system connected to a water supply through a valve which is opened by the operation of a heat responsive system installed in the same areas as the sprinklers. When this valve opens, water flows into the piping system and discharges from all sprinklers attached thereto.
5. **Combined Dry Pipe and Pre-Action Sprinkler System.** A system employing automatic sprinklers attached to a piping system containing air under pressure with a supplemental heat responsive system of generally more sensitive characteristics than the automatic sprinklers themselves, installed in the same areas as the sprinklers; operation of the heat responsive system, as from a fire, actuates tripping devices which open dry pipe valves simultaneously and without loss of air pressure in the system. Operation of the heat responsive system also opens approved air exhaust valves at the end of the feed main which facilitates the filling of the system with water which usually precedes the opening of sprinklers. The heat responsive system also serves as an automatic fire alarm system.
6. **Limited Water Supply System.** A system employing automatic sprinklers and conforming to these standards but supplied by a pressure tank of limited capacity.
7. **Chemical Systems.** Systems using halon, carbon dioxide, dry chemical or high expansion foam as selected for special requirements. Agent may extinguish flames by chemically inhibiting flame propagation, suffocate flames by excluding oxygen, interrupting chemical action of oxygen uniting with fuel or sealing and cooling the combustion center.
8. **Firecycle System.** Firecycle is a fixed fire protection sprinkler system utilizing water as its extinguishing agent. It is a time delayed, recycling, preaction type which automatically shuts the water off when heat is reduced below the detector operating temperature and turns the water back on when that temperature is exceeded. The system senses a fire condition through a closed circuit electrical detector system which controls water flow to the fire automatically. Batteries supply up to 90 hour emergency power supply for system operation. The piping system is dry (until water is required) and is monitored with pressurized air. Should any leak in the system piping occur, an alarm will sound, but water will not enter the system until heat is sensed by a firecycle detector.

Area coverage sprinkler systems may be laid out and fed from the supply in any one of several patterns as shown below. It is desirable, if possible, to utilize a central feed and achieve a shorter flow path from the riser to the furthest sprinkler. This permits use of the smallest sizes of pipe possible with resulting savings.

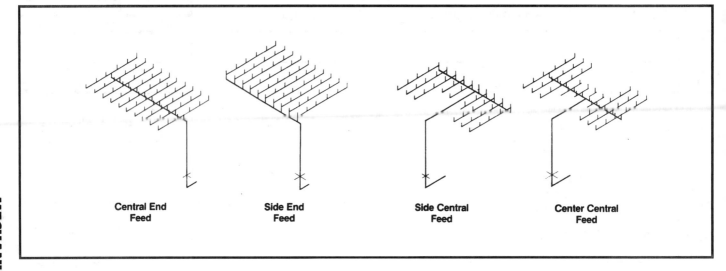

Central End Feed — Side End Feed — Side Central Feed — Center Central Feed

MECHANICAL — A8.2 Fire Protection

Table 8.2-102 System Classification

System Classification
Rules for installation of sprinkler systems vary depending on the classification of occupancy falling into one of three categories as follows:

Light Hazard Occupancy
The protection area allotted per sprinkler should not exceed 200 S.F. with the maximum distance between lines and sprinklers on lines being 15'. The sprinklers do not need to be staggered. Branch lines should not exceed eight sprinklers on either side of a cross main. Each large area requiring more than 100 sprinklers and without a sub-dividing partition should be supplied by feed mains or risers sized for ordinary hazard occupancy.

Included in this group are:
- Auditoriums
- Churches
- Clubs
- Educational
- Hospitals
- Institutional
- Libraries (except large stack rooms)
- Museums
- Nursing Homes
- Offices
- Residential
- Restaurants
- Schools
- Theaters

Ordinary Hazard Occupancy
The protection area allotted per sprinkler shall not exceed 130 S.F. of noncombustible ceiling and 120 S.F. of combustible ceiling. The maximum allowable distance between sprinkler lines and sprinklers on line is 15'. Sprinklers shall be staggered if the distance between heads exceeds 12'. Branch lines should not exceed eight sprinklers on either side of a cross main.

Included in this group are:
- Automotive garages
- Bakeries
- Beverage manufacturing
- Bleacheries
- Boiler houses
- Canneries
- Cement plants
- Clothing factories
- Cold storage warehouses
- Dairy products manufacturing
- Distilleries
- Dry cleaning
- Electric generating stations
- Feed mills
- Grain elevators
- Ice manufacturing
- Laundries
- Machine shops
- Mercantiles
- Paper mills
- Printing and Publishing
- Shoe factories
- Warehouses
- Wood product assembly

Extra Hazard Occupancy
The protection area allotted per sprinkler shall not exceed 90 S.F. of noncombustible ceiling and 80 S.F. of combustible ceiling. The maximum allowable distance between lines and between sprinklers on lines is 12'. Sprinklers on alternate lines shall be staggered if the distance between sprinklers on lines exceeds 8'. Branch lines should not exceed six sprinklers on either side of a cross main.

Included in this group are:
- Aircraft hangars
- Chemical works
- Explosives manufacturing
- Linoleum manufacturing
- Linseed oil mills
- Oil refineries
- Paint shops
- Shade cloth manufacturing
- Solvent extracting
- Varnish works
- Volatile flammable liquid manufacturing & use

For expanded coverage of these items see *Means Mechanical or Plumbing Cost Data 2000*

MECHANICAL — A8.2 Fire Protection

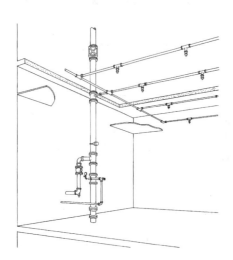

Wet Pipe System. A system employing automatic sprinklers attached to a piping system containing water and connected to a water supply so that water discharges immediately from sprinklers opened by heat from a fire.

All areas are assumed to be open.

8.2-110	Wet Pipe Sprinkler Systems	MAT.	INST.	TOTAL
0520	Wet pipe sprinkler systems, steel, black, sch. 40 pipe			
0530	Light hazard, one floor, 500 S.F.	1.08	1.32	2.40
0560	1000 S.F.	1.56	1.37	2.93
0580	2000 S.F.	1.60	1.39	2.99
0600	5000 S.F.	.77	.97	1.74
0620	10,000 S.F.	.52	.81	1.33
0640	50,000 S.F.	.35	.74	1.09
0660	Each additional floor, 500 S.F.	.53	1.13	1.66
0680	1000 S.F.	.49	1.05	1.54
0700	2000 S.F.	.46	.95	1.41
0720	5000 S.F.	.32	.79	1.11
0740	10,000 S.F.	.30	.75	1.05
0760	50,000 S.F.	.25	.58	.83
1000	Ordinary hazard, one floor, 500 S.F.	1.16	1.42	2.58
1020	1000 S.F.	1.55	1.34	2.89
1040	2000 S.F.	1.64	1.47	3.11
1060	5000 S.F.	.86	1.05	1.91
1080	10,000 S.F.	.64	1.10	1.74
1100	50,000 S.F.	.57	1.07	1.64
1140	Each additional floor, 500 S.F.	.64	1.27	1.91
1160	1000 S.F.	.49	1.03	1.52
1180	2000 S.F.	.52	1.05	1.57
1200	5000 S.F.	.51	.98	1.49
1220	10,000 S.F.	.43	1.04	1.47
1240	50,000 S.F.	.44	.94	1.38
1500	Extra hazard, one floor, 500 S.F.	2.97	2.19	5.16
1520	1000 S.F.	1.90	1.92	3.82
1540	2000 S.F.	1.75	1.97	3.72
1560	5000 S.F.	1.11	1.71	2.82
1580	10,000 S.F.	1.10	1.63	2.73
1600	50,000 S.F.	1.17	1.51	2.68
1660	Each additional floor, 500 S.F.	.83	1.57	2.40
1680	1000 S.F.	.79	1.51	2.30
1700	2000 S.F.	.69	1.50	2.19
1720	5000 S.F.	.56	1.33	1.89
1740	10,000 S.F.	.68	1.24	1.92
1760	50,000 S.F.	.68	1.14	1.82
2020	Grooved steel, black sch. 40 pipe, light hazard, one floor, 2000 S.F.	1.64	1.17	2.81
2060	10,000 S.F.	.65	.73	1.38
2100	Each additional floor, 2000 S.F.	.52	.75	1.27
2150	10,000 S.F.	.33	.64	.97
2200	Ordinary hazard, one floor, 2000 S.F.	1.65	1.26	2.91

MECHANICAL — A8.2 Fire Protection

8.2-110 Wet Pipe Sprinkler Systems

Line	Description	MAT.	INST.	TOTAL
2250	10,000 S.F.	.64	.91	1.55
2300	Each additional floor, 2000 S.F.	.53	.84	1.37
2350	10,000 S.F.	.43	.85	1.28
2400	Extra hazard, one floor, 2000 S.F.	1.82	1.61	3.43
2450	10,000 S.F.	.92	1.16	2.08
2500	Each additional floor, 2000 S.F.	.78	1.23	2.01
2550	10,000 S.F.	.63	1.03	1.66
3050	Grooved steel black sch. 10 pipe, light hazard, one floor, 2000 S.F.	1.62	1.16	2.78
3100	10,000 S.F.	.53	.69	1.22
3150	Each additional floor, 2000 S.F.	.50	.74	1.24
3200	10,000 S.F.	.31	.63	.94
3250	Ordinary hazard, one floor, 2000 S.F.	1.64	1.25	2.89
3300	10,000 S.F.	.62	.90	1.52
3350	Each additional floor, 2000 S.F.	.52	.83	1.35
3400	10,000 S.F.	.41	.84	1.25
3450	Extra hazard, one floor, 2000 S.F.	1.80	1.61	3.41
3500	10,000 S.F.	.88	1.15	2.03
3550	Each additional floor, 2000 S.F.	.76	1.23	1.99
3600	10,000 S.F.	.62	1.02	1.64
4050	Copper tubing, type M, light hazard, one floor, 2000 S.F.	1.51	1.17	2.68
4100	10,000 S.F.	.51	.70	1.21
4150	Each additional floor, 2000 S.F.	.40	.77	1.17
4200	10,000 S.F.	.29	.64	.93
4250	Ordinary hazard, one floor, 2000 S.F.	1.54	1.32	2.86
4300	10,000 S.F.	.59	.81	1.40
4350	Each additional floor, 2000 S.F.	.42	.85	1.27
4400	10,000 S.F.	.37	.74	1.11
4450	Extra hazard, one floor, 2000 S.F.	1.67	1.61	3.28
4500	10,000 S.F.	.98	1.23	2.21
4550	Each additional floor, 2000 S.F.	.63	1.23	1.86
4600	10,000 S.F.	.59	1.09	1.68
5050	Copper tubing, type M, T-drill system, light hazard, one floor			
5060	2000 S.F.	1.51	1.10	2.61
5100	10,000 S.F.	.49	.57	1.06
5150	Each additional floor, 2000 S.F.	.40	.70	1.10
5200	10,000 S.F.	.27	.51	.78
5250	Ordinary hazard, one floor, 2000 S.F.	1.51	1.12	2.63
5300	10,000 S.F.	.58	.71	1.29
5350	Each additional floor, 2000 S.F.	.39	.70	1.09
5400	10,000 S.F.	.37	.65	1.02
5450	Extra hazard, one floor, 2000 S.F.	1.58	1.32	2.90
5500	10,000 S.F.	.85	.90	1.75
5550	Each additional floor, 2000 S.F.	.58	.96	1.54
5600	10,000 S.F.	.46	.76	1.22

For expanded coverage of these items see *Means Mechanical or Plumbing Cost Data 2000*

MECHANICAL — A8.2 — Fire Protection

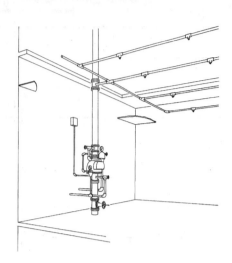

Dry Pipe System: A system employing automatic sprinklers attached to a piping system containing air under pressure, the release of which as from the opening of sprinklers permits the water pressure to open a valve known as a "dry pipe valve". The water then flows into the piping system and out the opened sprinklers.

All areas are assumed to be open.

8.2-120	Dry Pipe Sprinkler Systems	MAT.	INST.	TOTAL
0520	Dry pipe sprinkler systems, steel, black, sch. 40 pipe			
0530	Light hazard, one floor, 500 S.F.	4.18	2.60	6.78
0560	1000 S.F.	2.27	1.51	3.78
0580	2000 S.F.	1.95	1.54	3.49
0600	5000 S.F.	.98	1.03	2.01
0620	10,000 S.F.	.68	.85	1.53
0640	50,000 S.F.	.47	.75	1.22
0660	Each additional floor, 500 S.F.	.76	1.25	2.01
0680	1000 S.F.	.61	1.03	1.64
0700	2000 S.F.	.58	.96	1.54
0720	5000 S.F.	.45	.80	1.25
0740	10,000 S.F.	.42	.76	1.18
0760	50,000 S.F.	.38	.67	1.05
1000	Ordinary hazard, one floor, 500 S.F.	4.23	2.64	6.87
1020	1000 S.F.	2.32	1.53	3.85
1040	2000 S.F.	2	1.63	3.63
1060	5000 S.F.	1.11	1.12	2.23
1080	10,000 S.F.	.85	1.15	2
1100	50,000 S.F.	.77	1.10	1.87
1140	Each additional floor, 500 S.F.	.81	1.29	2.10
1160	1000 S.F.	.70	1.16	1.86
1180	2000 S.F.	.69	1.07	1.76
1200	5000 S.F.	.62	.90	1.52
1220	10,000 S.F.	.54	.92	1.46
1240	50,000 S.F.	.56	.80	1.36
1500	Extra hazard, one floor, 500 S.F.	5.70	3.29	8.99
1520	1000 S.F.	3.27	2.38	5.65
1540	2000 S.F.	2.19	2.08	4.27
1560	5000 S.F.	1.27	1.56	2.83
1580	10,000 S.F.	1.32	1.49	2.81
1600	50,000 S.F.	1.36	1.38	2.74
1660	Each additional floor, 500 S.F.	1.09	1.60	2.69
1680	1000 S.F.	1.05	1.54	2.59
1700	2000 S.F.	.95	1.53	2.48
1720	5000 S.F.	.78	1.33	2.11
1740	10,000 S.F.	.93	1.23	2.16
1760	50,000 S.F.	.95	1.14	2.09
2020	Grooved steel, black, sch. 40 pipe, light hazard, one floor, 2000 S.F.	1.95	1.32	3.27
2060	10,000 S.F.	.71	.74	1.45
2100	Each additional floor, 2000 S.F.	.64	.76	1.40
2150	10,000 S.F.	.45	.65	1.10
2200	Ordinary hazard, one floor, 2000 S.F.	2.01	1.42	3.43

MECHANICAL — A8.2 Fire Protection

8.2-120 Dry Pipe Sprinkler Systems

		COST PER S.F.		
		MAT.	INST.	TOTAL
2250	10,000 S.F.	.85	.96	1.81
2300	Each additional floor, 2000 S.F.	.70	.86	1.56
2350	10,000 S.F.	.60	.87	1.47
2400	Extra hazard, one floor, 2000 S.F.	2.27	1.78	4.05
2450	10,000 S.F.	1.22	1.21	2.43
2500	Each additional floor, 2000 S.F.	1.04	1.26	2.30
2550	10,000 S.F.	.89	1.05	1.94
3050	Grooved steel black sch. 10 pipe, light hazard, one floor, 2000 S.F.	1.93	1.31	3.24
3100	10,000 S.F.	.69	.73	1.42
3150	Each additional floor, 2000 S.F.	.62	.75	1.37
3200	10,000 S.F.	.43	.64	1.07
3250	Ordinary hazard, one floor, 2000 S.F.	2	1.41	3.41
3300	10,000 S.F.	.83	.95	1.78
3350	Each additional floor, 2000 S.F.	.69	.85	1.54
3400	10,000 S.F.	.58	.86	1.44
3450	Extra hazard, one floor, 2000 S.F.	2.25	1.78	4.03
3500	10,000 S.F.	1.18	1.20	2.38
3550	Each additional floor, 2000 S.F.	1.02	1.26	2.28
3600	10,000 S.F.	.88	1.04	1.92
4050	Copper tubing, type M, light hazard, one floor, 2000 S.F.	1.82	1.32	3.14
4100	10,000 S.F.	.67	.74	1.41
4150	Each additional floor, 2000 S.F.	.52	.78	1.30
4200	10,000 S.F.	.41	.65	1.06
4250	Ordinary hazard, one floor, 2000 S.F.	1.90	1.48	3.38
4300	10,000 S.F.	.80	.86	1.66
4350	Each additional floor, 2000 S.F.	.63	.91	1.54
4400	10,000 S.F.	.54	.76	1.30
4450	Extra hazard, one floor, 2000 S.F.	2.12	1.78	3.90
4500	10,000 S.F.	1.29	1.28	2.57
4550	Each additional floor, 2000 S.F.	.89	1.26	2.15
4600	10,000 S.F.	.85	1.11	1.96
5050	Copper tubing, type M, T-drill system, light hazard, one floor			
5060	2000 S.F.	1.82	1.25	3.07
5100	10,000 S.F.	.65	.61	1.26
5150	Each additional floor, 2000 S.F.	.52	.71	1.23
5200	10,000 S.F.	.39	.52	.91
5250	Ordinary hazard, one floor, 2000 S.F.	1.87	1.28	3.15
5300	10,000 S.F.	.79	.76	1.55
5350	Each additional floor, 2000 S.F.	.56	.72	1.28
5400	10,000 S.F.	.52	.64	1.16
5450	Extra hazard, one floor, 2000 S.F.	2.03	1.49	3.52
5500	10,000 S.F.	1.16	.95	2.11
5550	Each additional floor, 2000 S.F.	.80	.97	1.77
5600	10,000 S.F.	.72	.78	1.50

For expanded coverage of these items see *Means Mechanical or Plumbing Cost Data 2000*

MECHANICAL — A8.2 Fire Protection

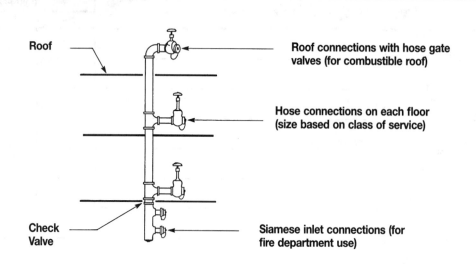

- Roof connections with hose gate valves (for combustible roof)
- Hose connections on each floor (size based on class of service)
- Check Valve
- Siamese inlet connections (for fire department use)

8.2-310	Wet Standpipe Risers, Class I	COST PER FLOOR		
		MAT.	INST.	TOTAL
0550	Wet standpipe risers, Class I, steel black sch. 40, 10' height			
0560	4" diameter pipe, one floor	1,750	1,450	3,200
0580	Additional floors	450	450	900
0600	6" diameter pipe, one floor	3,325	2,500	5,825
0620	Additional floors	835	710	1,545
0640	8" diameter pipe, one floor	5,050	3,025	8,075
0660	Additional floors	1,200	865	2,065

8.2-310	Wet Standpipe Risers, Class II	COST PER FLOOR		
		MAT.	INST.	TOTAL
1030	Wet standpipe risers, Class II, steel black sch. 40, 10' height			
1040	2" diameter pipe, one floor	805	525	1,330
1060	Additional floors	266	203	469
1080	2-1/2" diameter pipe, one floor	1,050	765	1,815
1100	Additional floors	295	236	531

8.2-310	Wet Standpipe Risers, Class III	COST PER FLOOR		
		MAT.	INST.	TOTAL
1530	Wet standpipe risers, Class III, steel black sch. 40, 10' height			
1540	4" diameter pipe, one floor	1,800	1,450	3,250
1560	Additional floors	380	375	755
1580	6" diameter pipe, one floor	3,375	2,500	5,875
1600	Additional floors	860	710	1,570
1620	8" diameter pipe, one floor	5,125	3,025	8,150
1640	Additional floors	1,250	865	2,115

MECHANICAL — A8.2 Fire Protection

Roof connections with hose gate valves (for combustible roof)

Hose connections on each floor (size based on class of service)

Check Valve

Siamese inlet connections (for fire department use)

8.2-320	Dry Standpipe Risers, Class I	COST PER FLOOR		
		MAT.	INST.	TOTAL
0530	Dry standpipe riser, Class I, steel black sch. 40, 10' height			
0540	4" diameter pipe, one floor	1,125	1,175	2,300
0560	Additional floors	405	425	830
0580	6" diameter pipe, one floor	2,425	2,000	4,425
0600	Additional floors	790	685	1,475
0620	8" diameter pipe, one floor	3,700	2,425	6,125
0640	Additional floors	1,175	845	2,020

8.2-320	Dry Standpipe Risers, Class II	COST PER FLOOR		
		MAT.	INST.	TOTAL
1030	Dry standpipe risers, Class II, steel black sch. 40, 10' height			
1040	2" diameter pipe, one floor	645	545	1,190
1060	Additional floor	221	178	399
1080	2-1/2" diameter pipe, one floor	785	640	1,425
1100	Additional floors	250	212	462

8.2-320	Dry Standpipe Risers, Class III	COST PER FLOOR		
		MAT.	INST.	TOTAL
1530	Dry standpipe risers, Class III, steel black sch. 40, 10' height			
1540	4" diameter pipe, one floor	1,150	1,175	2,325
1560	Additional floors	340	385	725
1580	6" diameter pipe, one floor	2,450	2,000	4,450
1600	Additional floors	815	685	1,500
1620	8" diameter pipe, one floor	3,725	2,425	6,150
1640	Additional floor	1,200	845	2,045

For expanded coverage of these items see Means Mechanical or Plumbing Cost Data 2000

MECHANICAL — A8.2 Fire Protection

8.2-390 Standpipe Equipment

		COST EACH		
		MAT.	INST.	TOTAL
0100	Adapters, reducing, 1 piece, FxM, hexagon, cast brass, 2-1/2" x 1-1/2"	26.50		26.50
0200	Pin lug, 1-1/2" x 1"	8.45		8.45
0250	3" x 2-1/2"	33		33
0300	For polished chrome, add 75% mat.			
0400	Cabinets, D.S. glass in door, recessed, steel box, not equipped			
0500	Single extinguisher, steel door & frame	51	65.50	116.50
0550	Stainless steel door & frame	118	65.50	183.50
0600	Valve, 2-1/2" angle, steel door & frame	59.50	44	103.50
0650	Aluminum door & frame	78	44	122
0700	Stainless steel door & frame	109	44	153
0750	Hose rack assy, 2-1/2" x 1-1/2" valve & 100' hose, steel door & frame	119	87.50	206.50
0800	Aluminum door & frame	176	87.50	263.50
0850	Stainless steel door & frame	243	87.50	330.50
0900	Hose rack assy,& extinguisher,2-1/2"x1-1/2" valve & hose,steel door & frame	142	105	247
0950	Aluminum	225	105	330
1000	Stainless steel	315	105	420
1550	Compressor, air, dry pipe system, automatic, 200 gal., 1/3 H.P.	770	225	995
1600	520 gal., 1 H.P.	805	225	1,030
1650	Alarm, electric pressure switch (circuit closer)	127	11.25	138.25
2500	Couplings, hose, rocker lug, cast brass, 1-1/2"	19.65		19.65
2550	2-1/2"	32.50		32.50
3000	Escutcheon plate, for angle valves, polished brass, 1-1/2"	10		10
3050	2-1/2"	25		25
3500	Fire pump, electric, w/controller, fittings, relief valve			
3550	4" pump, 30 H.P., 500 G.P.M.	14,900	1,550	16,450
3600	5" pump, 40 H.P., 1000 G.P.M.	21,800	1,750	23,550
3650	5" pump, 100 H.P., 1000 G.P.M.	24,400	1,950	26,350
3700	For jockey pump system, add	2,675	263	2,938
5000	Hose, per linear foot, synthetic jacket, lined,			
5100	300 lb. test, 1-1/2" diameter	1.46	.20	1.66
5150	2-1/2" diameter	2.39	.24	2.63
5200	500 lb. test, 1-1/2" diameter	1.75	.20	1.95
5250	2-1/2" diameter	3.03	.24	3.27
5500	Nozzle, plain stream, polished brass, 1-1/2" x 10"	22.50		22.50
5550	2-1/2" x 15" x 13/16" or 1-1/2"	40.50		40.50
5600	Heavy duty combination adjustable fog and straight stream w/handle 1-1/2"	273		273
5650	2-1/2" direct connection	390		390
6000	Rack, for 1-1/2" diameter hose 100 ft. long, steel	35.50	26.50	62
6050	Brass	55	26.50	81.50
6500	Reel, steel, for 50 ft. long 1-1/2" diameter hose	64	37.50	101.50
6550	For 75 ft. long 2-1/2" diameter hose	104	37.50	141.50
7050	Siamese, w/plugs & chains, polished brass, sidewalk, 4" x 2-1/2" x 2-1/2"	340	210	550
7100	6" x 2-1/2" x 2-1/2"	565	263	828
7200	Wall type, flush, 4" x 2-1/2" x 2-1/2"	241	105	346
7250	6" x 2-1/2" x 2-1/2"	345	114	459
7300	Projecting, 4" x 2-1/2" x 2-1/2"	220	105	325
7350	6" x 2-1/2" x 2-1/2"	375	114	489
7400	For chrome plate, add 15% mat.			
8000	Valves, angle, wheel handle, 300 Lb., rough brass, 1-1/2"	24.50	24.50	49
8050	2-1/2"	60.50	41.50	102
8100	Combination pressure restricting, 1-1/2"	40	24.50	64.50
8150	2-1/2"	85	41.50	126.50
8200	Pressure restricting, adjustable, satin brass, 1-1/2"	76	24.50	100.50
8250	2-1/2"	107	41.50	148.50
8300	Hydrolator, vent and drain, rough brass, 1-1/2"	45	24.50	69.50
8350	2-1/2"	45	24.50	69.50
8400	Cabinet assy, incls. adapter, rack, hose, and nozzle	425	159	584

Important: See the Reference Section for critical supporting data - Reference Nos., Crews & Location Factors

MECHANICAL — A8.2 Fire Protection

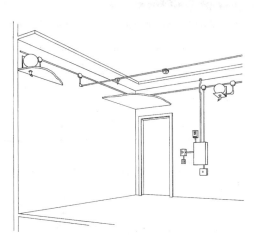

General: Automatic fire protection (suppression) systems other than water sprinklers may be desired for special environments, high risk areas, isolated locations or unusual hazards. Some typical applications would include:

- Paint dip tanks
- Securities vaults
- Electronic data processing
- Tape and data storage
- Transformer rooms
- Spray booths
- Petroleum storage
- High rack storage

Piping and wiring costs are dependent on the individual application and must be added to the component costs shown below.

All areas are assumed to be open.

8.2-810	FM200 Fire Suppression	MAT.	INST.	TOTAL
0020	Detectors with brackets			
0040	Fixed temperature heat detector	31	36	67
0060	Rate of temperature rise detector	37	36	73
0080	Ion detector (smoke) detector	82.50	46.50	129
0200	Extinguisher agent			
0240	200 lb FM200, container	6,375	131	6,506
0280	75 lb carbon dioxide cylinder	1,075	87	1,162
0320	Dispersion nozzle			
0340	FM200 1-1/2" dispersion nozzle	88	20.50	108.50
0380	Carbon dioxide 3" x 5" dispersion nozzle	55	16.15	71.15
0420	Control station			
0440	Single zone control station with batteries	1,525	289	1,814
0470	Multizone (4) control station with batteries	2,700	580	3,280
0500	Electric mechanical release	127	90.50	217.50
0550	Manual pull station	49.50	48.50	98
0640	Battery standby power 10" x 10" x 17"	760	72.50	832.50
0740	Bell signalling device	54.50	36	90.50

COST EACH shown above.

8.2-810	FM200 Systems	MAT.	INST.	TOTAL
0820	Average FM200 system, minimum			1.38
0840	Maximum			2.75

COST PER C.F.

For expanded coverage of these items see *Means Mechanical or Plumbing Cost Data 2000*

MECHANICAL — A8.3 — Heat Transfer

Table 8.3-011 Factor for Determining Heat Loss for Various Types of Buildings

General: While the most accurate estimates of heating requirements would naturally be based on detailed information about the building being considered, it is possible to arrive at a reasonable approximation using the following procedure:

1. Calculate the cubic volume of the room or building.
2. Select the appropriate factor from Table 8.3-021 below. Note that the factors apply only to inside temperatures listed in the first column and to 0°F outside temperature.
3. If the building has bad north and west exposures, multiply the heat loss factor by 1.1.
4. If the outside design temperature is other than 0°F, multiply the factor from Table 8.3-021 by the factor from Table 8.3-022.
5. Multiply the cubic volume by the factor selected from Table 8.3-021. This will give the estimated BTUH heat loss which must be made up to maintain inside temperature.

Building Type	Conditions	Qualifications	Loss Factor*
Factories & Industrial Plants General Office Areas at 70°F	One Story	Skylight in Roof	6.2
		No Skylight in Roof	5.7
	Multiple Story	Two Story	4.6
		Three Story	4.3
		Four Story	4.1
		Five Story	3.9
		Six Story	3.6
	All Walls Exposed	Flat Roof	6.9
		Heated Space Above	5.2
	One Long Warm Common Wall	Flat Roof	6.3
		Heated Space Above	4.7
	Warm Common Walls on Both Long Sides	Flat Roof	5.8
		Heated Space Above	4.1
Warehouses at 60°F	All Walls Exposed	Skylights in Roof	5.5
		No Skylight in Roof	5.1
		Heated Space Above	4.0
	One Long Warm Common Wall	Skylight in Roof	5.0
		No Skylight in Roof	4.9
		Heated Space Above	3.4
	Warm Common Walls on Both Long Sides	Skylight in Roof	4.7
		No Skylight in Roof	4.4
		Heated Space Above	3.0

*Note: This table tends to be conservative particularly for new buildings designed for minimum energy consumption.

Table 8.3-012 Outside Design Temperature Correction Factor (for Degrees Fahrenheit)

Outside Design Temperature	50	40	30	20	10	0	−10	−20	−30
Correction Factor	0.29	0.43	0.57	0.72	0.86	1.00	1.14	1.28	1.43

Figure 8.3-013 and 8.3-014 provide a way to calculate heat transmission of various construction materials from their U values and the TD (Temperature Difference).

1. From the exterior Enclosure Division or elsewhere, determine U values for the construction desired.
2. Determine the coldest design temperature. The difference between this temperature and the desired interior temperature is the TD (temperature difference).
3. Enter Figure 8.3-013 or 8.3-014 at correct U value. Cross horizontally to the intersection with appropriate TD. Read transmission per square foot from bottom of figure.
4. Multiply this value of BTU per hour transmission per square foot of area by the total surface area of that type of construction.

MECHANICAL — A8.3 Heat Transfer

Table 8.3-013 Transmission of Heat

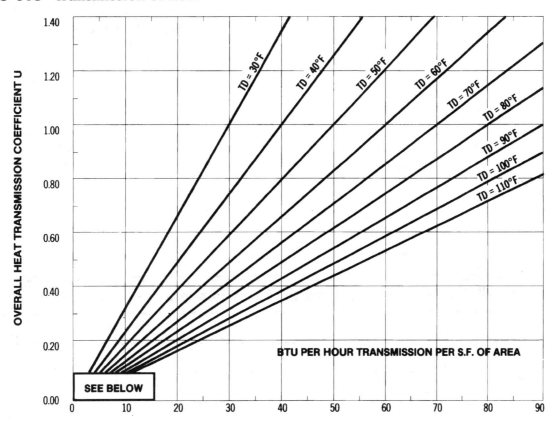

Table 8.3-014 Transmission of Heat (Low Rate)

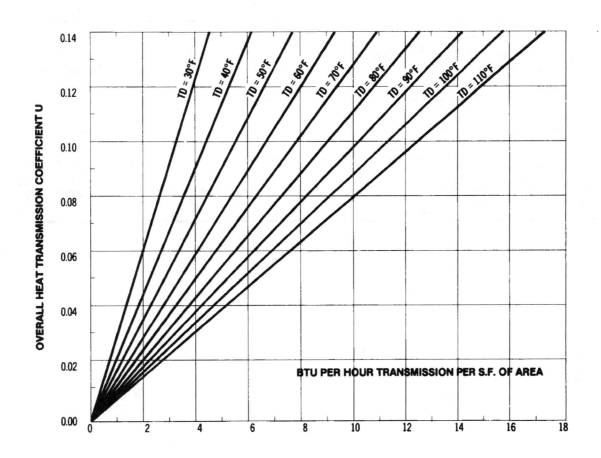

For expanded coverage of these items see *Means Mechanical or Plumbing Cost Data 2000*

MECHANICAL — A8.3 Heating

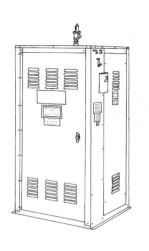

Boiler

Baseboard Radiation

Small Electric Boiler
System Considerations:
1. Terminal units are fin tube baseboard radiation rated at 720 BTU/hr with 200° water temperature or 820 BTU/hr steam.
2. Primary use being for residential or smaller supplementary areas, the floor levels are based on 7-1/2' ceiling heights.
3. All distribution piping is copper for boilers through 205 MBH. All piping for larger systems is steel pipe.

8.3-110	Small Heating Systems, Hydronic, Electric Boilers	MAT.	INST.	TOTAL
1100	Small heating systems, hydronic, electric boilers			
1120	Steam, 1 floor, 1480 S.F., 61 M.B.H.	8.35	3.63	11.98
1160	3,000 S.F., 123 M.B.H.	4.83	3.17	8
1200	5,000 S.F., 205 M.B.H.	3.63	2.92	6.55
1240	2 floors, 12,400 S.F., 512 M.B.H.	2.61	2.92	5.53
1280	3 floors, 24,800 S.F., 1023 M.B.H.	2.27	2.86	5.13
1320	34,750 S.F., 1,433 M.B.H.	1.99	2.78	4.77
1360	Hot water, 1 floor, 1,000 S.F., 41 M.B.H.	7.80	2.02	9.82
1400	2,500 S.F., 103 M.B.H.	4.61	3.63	8.24
1440	2 floors, 4,850 S.F., 205 M.B.H.	4.01	4.36	8.37
1480	3 floors, 9,700 S.F., 410 M.B.H.	3.71	4.52	8.23

8.3-120	Large Heating Systems, Hydronic, Electric Boilers	MAT.	INST.	TOTAL
1230	Large heating systems, hydronic, electric boilers			
1240	9,280 S.F., 150 K.W., 510 M.B.H., 1 floor	2.88	1.38	4.26
1280	14,900 S.F., 240 K.W., 820 M.B.H., 2 floors	2.90	2.26	5.16
1320	18,600 S.F., 300 K.W., 1,024 M.B.H., 3 floors	3.05	2.48	5.53
1360	26,100 S.F., 420 K.W., 1,432 M.B.H., 4 floors	2.91	2.39	5.30
1400	39,100 S.F., 630 K.W., 2,148 M.B.H., 4 floors	2.55	2	4.55
1440	57,700 S.F., 900 K.W., 3,071 M.B.H., 5 floors	2.42	1.98	4.40
1480	111,700 S.F., 1,800 K.W., 6,148 M.B.H., 6 floors	2.12	1.68	3.80
1520	149,000 S.F., 2,400 K.W., 8,191 M.B.H., 8 floors	2.06	1.68	3.74
1560	223,300 S.F., 3,600 K.W., 12,283 M.B.H., 14 floors	2.07	1.93	4

MECHANICAL — A8.3 Heating

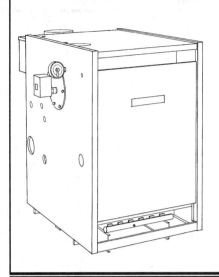

Boiler Selection: The maximum allowable working pressures are limited by ASME "Code for Heating Boilers" to 15 PSI for steam and 160 PSI for hot water heating boilers, with a maximum temperature limitation of 250°F. Hot water boilers are generally rated for a working pressure of 30 PSI. High pressure boilers are governed by the ASME "Code for Power Boilers" which is used almost universally for boilers operating over 15 PSIG. High pressure boilers used for a combination of heating/process loads are usually designed for 150 PSIG.

Boiler ratings are usually indicated as either Gross or Net Output. The Gross Load is equal to the Net Load plus a piping and pickup allowance. When this allowance cannot be determined, divide the gross output rating by 1.25 for a value equal to or greater than the next heat loss requirement of the building.

Table below lists installed cost per boiler and includes insulating jacket, standard controls, burner and safety controls. Costs do not include piping or boiler base pad. Outputs are Gross.

8.3-130	Boilers, Hot Water & Steam	MAT.	INST.	TOTAL
0600	Boiler, electric, steel, hot water, 12 K.W., 41 M.B.H.	3,725	625	4,350
0620	30 K.W., 103 M.B.H.	4,100	680	4,780
0640	60 K.W., 205 M.B.H.	5,200	740	5,940
0660	120 K.W., 410 M.B.H.	6,625	905	7,530
0680	210 K.W., 716 M.B.H.	11,000	1,350	12,350
0700	510 K.W., 1,739 M.B.H.	20,800	2,525	23,325
0720	720 K.W., 2,452 M.B.H.	26,200	2,850	29,050
0740	1,200 K.W., 4,095 M.B.H.	37,000	3,250	40,250
0760	2,100 K.W., 7,167 M.B.H.	58,000	4,100	62,100
0780	3,600 K.W., 12,283 M.B.H.	79,500	6,925	86,425
0820	Steam, 6 K.W., 20.5 M.B.H.	8,700	680	9,380
0840	24 K.W., 81.8 M.B.H.	9,250	740	9,990
0860	60 K.W., 205 M.B.H.	10,700	815	11,515
0880	150 K.W., 512 M.B.H.	13,900	1,250	15,150
0900	510 K.W., 1,740 M.B.H.	22,600	3,075	25,675
0920	1,080 K.W., 3,685 M.B.H.	34,700	4,425	39,125
0940	2,340 K.W., 7,984 M.B.H.	65,500	6,925	72,425
0980	Gas, cast iron, hot water, 80 M.B.H.	1,125	720	1,845
1000	100 M.B.H.	1,300	780	2,080
1020	163 M.B.H.	1,750	1,050	2,800
1040	280 M.B.H.	2,450	1,175	3,625
1060	544 M.B.H.	4,125	2,075	6,200
1080	1,088 M.B.H.	6,950	2,625	9,575
1100	2,000 M.B.H.	11,900	4,125	16,025
1120	2,856 M.B.H.	16,200	5,275	21,475
1140	4,720 M.B.H.	48,300	7,250	55,550
1160	6,970 M.B.H.	61,000	11,800	72,800
1180	For steam systems under 2,856 M.B.H., add 8%			
1240	Steel, hot water, 72 M.B.H.	1,925	380	2,305
1260	101 M.B.H.	2,200	420	2,620
1280	132 M.B.H.	2,500	445	2,945
1300	150 M.B.H.	2,900	505	3,405
1320	240 M.B.H.	4,450	585	5,035
1340	400 M.B.H.	6,350	950	7,300
1360	640 M.B.H.	8,625	1,275	9,900
1380	800 M.B.H.	10,100	1,525	11,625
1400	960 M.B.H.	12,500	1,700	14,200
1420	1,440 M.B.H.	17,400	2,175	19,575
1440	2,400 M.B.H.	28,100	3,800	31,900
1460	3,000 M.B.H.	34,700	5,075	39,775
1520	Oil, cast iron, hot water, 109 M.B.H.	1,475	875	2,350
1540	173 M.B.H.	1,850	1,050	2,900

For expanded coverage of these items see *Means Mechanical or Plumbing Cost Data 2000*

MECHANICAL | A8.3 | Heating

8.3-130 Boilers, Hot Water & Steam

		COST EACH		
		MAT.	INST.	TOTAL
1560	236 M.B.H.	2,875	1,250	4,125
1580	1,084 M.B.H.	7,150	2,800	9,950
1600	1,600 M.B.H.	9,500	4,025	13,525
1620	2,480 M.B.H.	13,400	5,125	18,525
1640	3,550 M.B.H.	17,800	6,150	23,950
1660	Steam systems same price as hot water			
1700	Steel, hot water, 103 M.B.H.	1,425	475	1,900
1720	137 M.B.H.	3,575	555	4,130
1740	225 M.B.H.	4,525	625	5,150
1760	315 M.B.H.	5,250	795	6,045
1780	420 M.B.H.	5,675	1,100	6,775
1800	630 M.B.H.	7,325	1,450	8,775
1820	735 M.B.H.	7,600	1,575	9,175
1840	1,050 M.B.H.	14,900	2,075	16,975
1860	1,365 M.B.H.	17,300	2,350	19,650
1880	1,680 M.B.H.	20,300	2,650	22,950
1900	2,310 M.B.H.	22,400	3,650	26,050
1920	2,835 M.B.H.	25,900	4,475	30,375
1940	3,150 M.B.H.	26,800	5,850	32,650

Important: See the Reference Section for critical supporting data - Reference Nos., Crews & Location Factors

MECHANICAL — A8.3 Heating

Unit Heater

Fossil Fuel Boiler System Considerations:

1. Terminal units are horizontal unit heaters. Quantities are varied to accommodate total heat loss per building.
2. Unit heater selection was determined by their capacity to circulate the building volume a minimum of three times per hour in addition to the BTU output.
3. Systems shown are forced hot water. Steam boilers cost slightly more than hot water boilers. However, this is compensated for by the smaller size or fewer terminal units required with steam.
4. Floor levels are based on 10' story heights.
5. MBH requirements are gross boiler output.

8.3-141	Heating Systems, Unit Heaters	MAT.	INST.	TOTAL
1260	Heating systems, hydronic, fossil fuel, terminal unit heaters,			
1280	Cast iron boiler, gas, 80 M.B.H., 1,070 S.F. bldg.	5.50	4.29	9.79
1320	163 M.B.H., 2,140 S.F. bldg.	3.68	2.91	6.59
1360	544 M.B.H., 7,250 S.F. bldg.	2.46	2.09	4.55
1400	1,088 M.B.H., 14,500 S.F. bldg.	2.14	2.01	4.15
1440	3,264 M.B.H., 43,500 S.F. bldg.	1.81	1.50	3.31
1480	5,032 M.B.H., 67,100 S.F. bldg.	2.16	1.55	3.71
1520	Oil, 109 M.B.H., 1,420 S.F. bldg.	5.35	3.80	9.15
1560	235 M.B.H., 3,150 S.F. bldg.	3.82	2.78	6.60
1600	940 M.B.H., 12,500 S.F. bldg.	2.65	1.81	4.46
1640	1,600 M.B.H., 21,300 S.F. bldg.	2.56	1.74	4.30
1680	2,480 M.B.H., 33,100 S.F. bldg.	2.59	1.57	4.16
1720	3,350 M.B.H., 44,500 S.F. bldg.	2.26	1.61	3.87
1760	Coal, 148 M.B.H., 1,975 S.F. bldg.	4.58	2.50	7.08
1800	300 M.B.H., 4,000 S.F. bldg.	3.53	1.96	5.49
1840	2,360 M.B.H., 31,500 S.F. bldg.	2.43	1.60	4.03
1880	Steel boiler, gas, 72 M.B.H., 1,020 S.F. bldg.	5.50	3.17	8.67
1920	240 M.B.H., 3,200 S.F. bldg.	3.73	2.54	6.27
1960	480 M.B.H., 6,400 S.F. bldg.	3.06	1.89	4.95
2000	800 M.B.H., 10,700 S.F. bldg.	2.64	1.68	4.32
2040	1,960 M.B.H., 26,100 S.F. bldg.	2.33	1.52	3.85
2080	3,000 M.B.H., 40,000 S.F. bldg.	2.29	1.55	3.84
2120	Oil, 97 M.B.H., 1,300 S.F. bldg.	5.25	3.61	8.86
2160	315 M.B.H., 4,550 S.F. bldg.	3.65	1.85	5.50
2200	525 M.B.H., 7,000 S.F. bldg.	3.67	1.87	5.54
2240	1,050 M.B.H., 14,000 S.F. bldg.	3.27	1.89	5.16
2280	2,310 M.B.H. 30,800 S.F. bldg.	2.88	1.62	4.50
2320	3,150 M.B.H., 42,000 S.F. bldg.	2.57	1.67	4.24

For expanded coverage of these items see *Means Mechanical or Plumbing Cost Data 2000*

MECHANICAL   A8.3 Heating

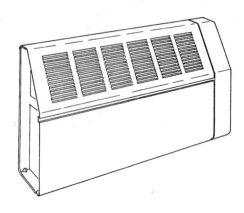

Fin Tube Radiator

Fossil Fuel Boiler System Considerations:

1. Terminal units are commercial steel fin tube radiation. Quantities are varied to accommodate total heat loss per building.
2. Systems shown are forced hot water. Steam boilers cost slightly more than hot water boilers. However, this is compensated for by the smaller size or fewer terminal units required with steam.
3. Floor levels are based on 10' story heights.
4. MBH requirements are gross boiler output.

8.3-142	Heating System, Fin Tube Radiation	COST PER S.F.		
		MAT.	INST.	TOTAL
3230	Heating systems, hydronic, fossil fuel, fin tube radiation			
3240	Cast iron boiler, gas, 80 MBH, 1,070 S.F. bldg.	7.40	6.45	13.85
3280	169 M.B.H., 2,140 S.F. bldg.	4.64	4.09	8.73
3320	544 M.B.H., 7,250 S.F. bldg.	3.78	3.52	7.30
3360	1,088 M.B.H., 14,500 S.F. bldg.	3.52	3.48	7
3400	3,264 M.B.H., 43,500 S.F. bldg.	3.27	3.04	6.31
3440	5,032 M.B.H., 67,100 S.F. bldg.	3.64	3.09	6.73
3480	Oil, 109 M.B.H., 1,420 S.F. bldg.	8.35	7	15.35
3520	235 M.B.H., 3,150 S.F. bldg.	5.10	4.18	9.28
3560	940 M.B.H., 12,500 S.F. bldg.	4.05	3.31	7.36
3600	1,600 M.B.H., 21,300 S.F. bldg.	4.04	3.29	7.33
3640	2,480 M.B.H., 33,100 S.F. bldg.	4.06	3.11	7.17
3680	3,350 M.B.H., 44,500 S.F. bldg.	3.72	3.16	6.88
3720	Coal, 148 M.B.H., 1,975 S.F. bldg.	5.85	3.90	9.75
3760	300 M.B.H., 4,000 S.F. bldg.	4.80	3.34	8.14
3800	2,360 M.B.H., 31,500 S.F. bldg.	3.87	3.12	6.99
3840	Steel boiler, gas, 72 M.B.H., 1,020 S.F. bldg.	8.40	6.15	14.55
3880	240 M.B.H., 3,200 S.F. bldg.	5.10	4.04	9.14
3920	480 M.B.H., 6,400 S.F. bldg.	4.40	3.33	7.73
3960	800 M.B.H., 10,700 S.F. bldg.	4.41	3.53	7.94
4000	1,960 M.B.H., 26,100 S.F. bldg.	3.76	3.05	6.81
4040	3,000 M.B.H., 40,000 S.F. bldg.	3.71	3.06	6.77
4080	Oil, 97 M.B.H., 1,300 S.F. bldg.	5.85	5.75	11.60
4120	315 M.B.H., 4,550 S.F. bldg.	4.88	3.17	8.05
4160	525 M.B.H., 7,000 S.F. bldg.	5.10	3.38	8.48
4200	1,050 M.B.H., 14,000 S.F. bldg.	4.68	3.40	8.08
4240	2,310 M.B.H., 30,800 S.F. bldg.	4.31	3.13	7.44
4280	3,150 M.B.H., 42,000 S.F. bldg.	4	3.19	7.19

MECHANICAL — A8.3 Heating

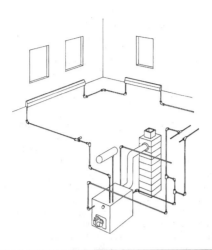

Basis for Heat Loss Estimate, Apartment Type Structures:

1. Masonry walls and flat roof are insulated. U factor is assumed at .08.
2. Window glass area taken as BOCA minimum, 1/10th of floor area. Double insulating glass with 1/4" air space, $U = .65$.
3. Infiltration = 0.3 C.F. per hour per S.F. of net wall.
4. Concrete floor loss is 2 BTUH per S.F.
5. Temperature difference taken as 70° F.
6. Ventilating or makeup air has not been included and must be added if desired. Air shafts are not used.

8.3-151	Apartment Building Heating - Fin Tube Radiation	MAT.	INST.	TOTAL
1740	Heating systems, fin tube radiation, forced hot water			
1760	1,000 S.F. area, 10,000 C.F. volume	4.04	2.72	6.76
1800	10,000 S.F. area, 100,000 C.F. volume	1.20	1.68	2.88
1840	20,000 S.F. area, 200,000 C.F. volume	1.35	1.90	3.25
1880	30,000 S.F. area, 300,000 C.F. volume	1.25	1.84	3.09

8.3-161	Commercial Building Heating - Fin Tube Radiation	MAT.	INST.	TOTAL
1940	Heating systems, fin tube radiation, forced hot water			
1960	1,000 S.F. bldg, one floor	8.50	6.25	14.75
2000	10,000 S.F., 100,000 C.F., total two floors	2.43	2.42	4.85
2040	100,000 S.F., 1,000,000 C.F., total three floors	1.05	1.09	2.14
2080	1,000,000 S.F., 10,000,000 C.F., total five floors	.52	.60	1.12

8.3-162	Commercial Bldg. Heating - Terminal Unit Heaters	MAT.	INST.	TOTAL
1860	Heating systems, terminal unit heaters, forced hot water			
1880	1,000 S.F. bldg., one floor	7.95	5.75	13.70
1920	10,000 S.F. bldg., 100,000 C.F. total two floors	2.07	2.01	4.08
1960	100,000 S.F. bldg., 1,000,000 C.F. total three floors	1.02	1	2.02
2000	1,000,000 S.F. bldg., 10,000,000 C.F. total five floors	.66	.66	1.32

For expanded coverage of these items see *Means Mechanical or Plumbing Cost Data 2000*

MECHANICAL — A8.4 Cooling

Table 8.4-001 Air Conditioning

General: The purpose of air conditioning is to control the environment of a space so that comfort is provided for the occupants and/or conditions are suitable for the processes or equipment contained therein. The several items which should be evaluated to define system objectives are:

- Temperature Control
- Humidity Control
- Cleanliness
- Odor, smoke and fumes
- Ventilation

Efforts to control the above parameters must also include consideration of the degree or tolerance of variation, the noise level introduced, the velocity of air motion and the energy requirements to accomplish the desired results.

The variation in **temperature** and **humidity** is a function of the sensor and the controller. The controller reacts to a signal from the sensor and produces the appropriate suitable response in either the terminal unit, the conductor of the transporting medium (air, steam, chilled water, etc.), or the source (boiler, evaporating coils, etc.).

The **noise level** is a by-product of the energy supplied to moving components of the system. Those items which usually contribute the most noise are pumps, blowers, fans, compressors and diffusers. The level of noise can be partially controlled through use of vibration pads, isolators, proper sizing, shields, baffles and sound absorbing liners.

Some **air motion** is necessary to prevent stagnation and stratification. The maximum acceptable velocity varies with the degree of heating or cooling which is taking place. Most people feel air moving past them at velocities in excess of 25 FPM as an annoying draft. However, velocities up to 45 FPM may be acceptable in certain cases. Ventilation, expressed as air changes per hour and percentage of fresh air, is usually an item regulated by local codes.

Selection of the system to be used for a particular application is usually a trade-off. In some cases the building size, style, or room available for mechanical use limits the range of possibilities. Prime factors influencing the decision are first cost and total life (operating, maintenance and replacement costs). The accuracy with which each parameter is determined will be an important measure of the reliability of the decision and subsequent satisfactory operation of the installed system.

Heat delivery may be desired from an air conditioning system. Heating capability usually is added as follows: A gas fired burner or hot water/steam/electric coils may be added to the air handling unit directly and heat all air equally. For limited or localized heat requirements the water/steam/electric coils may be inserted into the duct branch supplying the cold areas. Gas fired duct furnaces are also available.

Note: When water or steam coils are used the cost of the piping and boiler must also be added. For a rough estimate use the cost per square foot of the appropriate sized hydronic system with unit heaters. This will provide a cost for the boiler and piping, and the unit heaters of the system would equate to the approximate cost of the heating coils. The installed cost of electric and gas heaters, boilers and other heat related items on a unit basis may be located in Section 155 of *Means Mechanical Cost Data*.

Table 8.4-002 Air Conditioning Requirements

BTU's per hour per S.F. of floor area and S.F. per ton of air conditioning.

Type of Building	BTU per S.F.	S.F. per Ton	Type of Building	BTU per S.F.	S.F. per Ton	Type of Building	BTU per S.F.	S.F. per Ton
Apartments, Individual	26	450	Dormitory, Rooms	40	300	Libraries	50	240
Corridors	22	550	Corridors	30	400	Low Rise Office, Exterior	38	320
Auditoriums & Theaters	40	300/18*	Dress Shops	43	280	Interior	33	360
Banks	50	240	Drug Stores	80	150	Medical Centers	28	425
Barber Shops	48	250	Factories	40	300	Motels	28	425
Bars & Taverns	133	90	High Rise Office Ext. Rms.	46	263	Office (small suite)	43	280
Beauty Parlors	66	180	Interior Rooms	37	325	Post Office, Individual Office	42	285
Bowling Alleys	68	175	Hospitals, Core	43	280	Central Area	46	260
Churches	36	330/20*	Perimeter	46	260	Residences	20	600
Cocktail Lounges	68	175	Hotel, Guest Rooms	44	275	Restaurants	60	200
Computer Rooms	141	85	Corridors	30	400	Schools & Colleges	46	260
Dental Offices	52	230	Public Spaces	55	220	Shoe Stores	55	220
Dept. Stores, Basement	34	350	Industrial Plants, Offices	38	320	Shop'g. Ctrs., Supermarkets	34	350
Main Floor	40	300	General Offices	34	350	Retail Stores	48	250
Upper Floor	30	400	Plant Areas	40	300	Specialty	60	200

*Persons per ton
12,000 BTU = 1 ton of air conditioning

MECHANICAL — A8.4 — Air Distribution

Table 8.4-004 Recommended Ventilation Air Changes

Table below lists range of time in minutes per change for various types of facilities.

Assembly Halls	2-10	Dance Halls	2-10	Laundries	1-3
Auditoriums	2-10	Dining Rooms	3-10	Markets	2-10
Bakeries	2-3	Dry Cleaners	1-5	Offices	2-10
Banks	3-10	Factories	2-5	Pool Rooms	2-5
Bars	2-5	Garages	2-10	Recreation Rooms	2-10
Beauty Parlors	2-5	Generator Rooms	2-5	Sales Rooms	2-10
Boiler Rooms	1-5	Gymnasiums	2-10	Theaters	2-8
Bowling Alleys	2-10	Kitchens-Hospitals	2-5	Toilets	2-5
Churches	5-10	Kitchens-Restaurant	1-3	Transformer Rooms	1-5

CFM air required for changes = Volume of room in cubic feet ÷ Minutes per change.

Table 8.4-005 Ductwork

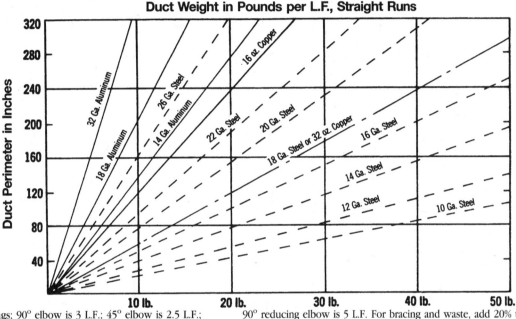

Add to the above for fittings; 90° elbow is 3 L.F.; 45° elbow is 2.5 L.F.; offset is 4 L.F.; transition offset is 6 L.F.; square-to-round transition is 4 L.F.; 90° reducing elbow is 5 L.F. For bracing and waste, add 20% to aluminum and copper, 15% to steel.

Table 8.4-006 Diffuser Evaluation

CFM = V × An × K where V = Outlet velocity in feet per minute. An = Neck area in square feet and K = Diffuser delivery factor. An undersized diffuser for a desired CFM will produce a high velocity and noise level. When air moves past people at a velocity in excess of 25 FPM, an annoying draft is felt. An oversized diffuser will result in low velocity with poor mixing. Consideration must be given to avoid vertical stratification or horizontal areas of stagnation.

For expanded coverage of these items see *Means Mechanical or Plumbing Cost Data 2000*

MECHANICAL — **A8.4** — **Air Distribution**

Table 8.4-009 Sheet Metal Calculator (Weight in Lb./Ft. of Length)

Gauge	26	24	22	20	18	16	Gauge	26	24	22	20	18	16
Wt.-Lb./S.F.	.906	1.156	1.406	1.656	2.156	2.656	Wt.-Lb./S.F.	.906	1.156	1.406	1.656	2.156	2.656
SMACNA Max. Dimension – Long Side		30"	54"	84"	85" Up		SMACNA Max. Dimension – Long Side		30"	54"	84"	85" Up	
Sum-2 sides							Sum-2 Sides						
2	.3	.40	.50	.60	.80	.90	56	9.3	12.0	14.0	16.2	21.3	25.2
3	.5	.65	.80	.90	1.1	1.4	57	9.5	12.3	14.3	16.5	21.7	25.7
4	.7	.85	1.0	1.2	1.5	1.8	58	9.7	12.5	14.5	16.8	22.0	26.1
5	.8	1.1	1.3	1.5	1.9	2.3	59	9.8	12.7	14.8	17.1	22.4	26.6
6	1.0	1.3	1.5	1.7	2.3	2.7	60	10.0	12.9	15.0	17.4	22.8	27.0
7	1.2	1.5	1.8	2.0	2.7	3.2	61	10.2	13.1	15.3	17.7	23.2	27.5
8	1.3	1.7	2.0	2.3	3.0	3.6	62	10.3	13.3	15.5	18.0	23.6	27.9
9	1.5	1.9	2.3	2.6	3.4	4.1	63	10.5	13.5	15.8	18.3	24.0	28.4
10	1.7	2.2	2.5	2.9	3.8	4.5	64	10.7	13.7	16.0	18.6	24.3	28.8
11	1.8	2.4	2.8	3.2	4.2	5.0	65	10.8	13.9	16.3	18.9	24.7	29.3
12	2.0	2.6	3.0	3.5	4.6	5.4	66	11.0	14.1	16.5	19.1	25.1	29.7
13	2.2	2.8	3.3	3.8	4.9	5.9	67	11.2	14.3	16.8	19.4	25.5	30.2
14	2.3	3.0	3.5	4.1	5.3	6.3	68	11.3	14.6	17.0	19.7	25.8	30.6
15	2.5	3.2	3.8	4.4	5.7	6.8	69	11.5	14.8	17.3	20.0	26.2	31.1
16	2.7	3.4	4.0	4.6	6.1	7.2	70	11.7	15.0	17.5	20.3	26.6	31.5
17	2.8	3.7	4.3	4.9	6.5	7.7	71	11.8	15.2	17.8	20.6	27.0	32.0
18	3.0	3.9	4.5	5.2	6.8	8.1	72	12.0	15.4	18.0	20.9	27.4	32.4
19	3.2	4.1	4.8	5.5	7.2	8.6	73	12.2	15.6	18.3	21.2	27.7	32.9
20	3.3	4.3	5.0	5.8	7.6	9.0	74	12.3	15.8	18.5	21.5	28.1	33.3
21	3.5	4.5	5.3	6.1	8.0	9.5	75	12.5	16.1	18.8	21.8	28.5	33.8
22	3.7	4.7	5.5	6.4	8.4	9.9	76	12.7	16.3	19.0	22.0	28.9	34.2
23	3.8	5.0	5.8	6.7	8.7	10.4	77	12.8	16.5	19.3	22.3	29.3	34.7
24	4.0	5.2	6.0	7.0	9.1	10.8	78	13.0	16.7	19.5	22.6	29.6	35.1
25	4.2	5.4	6.3	7.3	9.5	11.3	79	13.2	16.9	19.8	22.9	30.0	35.6
26	4.3	5.6	6.5	7.5	9.9	11.7	80	13.3	17.1	20.0	23.2	30.4	36.0
27	4.5	5.8	6.8	7.8	10.3	12.2	81	13.5	17.3	20.3	23.5	30.8	36.5
28	4.7	6.0	7.0	8.1	10.6	12.6	82	13.7	17.5	20.5	23.8	31.2	36.9
29	4.8	6.2	7.3	8.4	11.0	13.1	83	13.8	17.8	20.8	24.1	31.5	37.4
30	5.0	6.5	7.5	8.7	11.4	13.5	84	14.0	18.0	21.0	24.4	31.9	37.8
31	5.2	6.7	7.8	9.0	11.8	14.0	85	14.2	18.2	21.3	24.7	32.3	38.3
32	5.3	6.9	8.0	9.3	12.2	14.4	86	14.3	18.4	21.5	24.9	32.7	38.7
33	5.5	7.1	8.3	9.6	12.5	14.9	87	14.5	18.6	21.8	25.2	33.1	39.2
34	5.7	7.3	8.5	9.9	12.9	15.3	88	14.7	18.8	22.0	25.5	33.4	39.6
35	5.8	7.5	8.8	10.2	13.3	15.8	89	14.8	19.0	22.3	25.8	33.8	40.1
36	6.0	7.8	9.0	10.4	13.7	16.2	90	15.0	19.3	22.5	26.1	34.2	40.5
37	6.2	8.0	9.3	10.7	14.1	16.7	91	15.2	19.5	22.8	26.4	34.6	41.0
38	6.3	8.2	9.5	11.0	14.4	17.1	92	15.3	19.7	23.0	26.7	35.0	41.4
39	6.5	8.4	9.8	11.3	14.8	17.6	93	15.5	19.9	23.3	27.0	35.3	41.9
40	6.7	8.6	10.0	11.6	15.2	18.0	94	15.7	20.1	23.5	27.3	35.7	42.3
41	6.8	8.8	10.3	11.9	15.6	18.5	95	15.8	20.3	23.8	27.6	36.1	42.8
42	7.0	9.0	10.5	12.2	16.0	18.9	96	16.0	20.5	24.0	27.8	36.5	43.2
43	7.2	9.2	10.8	12.5	16.3	19.4	97	16.2	20.8	24.3	28.1	36.9	43.7
44	7.3	9.5	11.0	12.8	16.7	19.8	98	16.3	21.0	24.5	28.4	37.2	44.1
45	7.5	9.7	11.3	13.1	17.1	20.3	99	16.5	21.2	24.8	28.7	37.6	44.6
46	7.7	9.9	11.5	13.3	17.5	20.7	100	16.7	21.4	25.0	29.0	38.0	45.0
47	7.8	10.1	11.8	13.6	17.9	21.2	101	16.8	21.6	25.3	29.3	38.4	45.5
48	8.0	10.3	12.0	13.9	18.2	21.6	102	17.0	21.8	25.5	29.6	38.8	45.9
49	8.2	10.5	12.3	14.2	18.6	22.1	103	17.2	22.0	25.8	29.9	39.1	46.4
50	8.3	10.7	12.5	14.5	19.0	22.5	104	17.3	22.3	26.0	30.2	39.5	46.8
51	8.5	11.0	12.8	14.8	19.4	23.0	105	17.5	22.5	26.3	30.5	39.9	47.3
52	8.7	11.2	13.0	15.1	19.8	23.4	106	17.7	22.7	26.5	30.7	40.3	47.7
53	8.8	11.4	13.3	15.4	20.1	23.9	107	17.8	22.9	26.8	31.0	40.7	48.2
54	9.0	11.6	13.5	15.7	20.5	24.3	108	18.0	23.1	27.0	31.3	41.0	48.6
55	9.2	11.8	13.8	16.0	20.9	24.8	109	18.2	23.3	27.3	31.6	41.4	49.1
							110	18.3	23.5	27.5	31.9	41.8	49.5

Example: If duct is 34" x 20" x 15' long, 34" is greater than 30" maximum, for 24 ga. so must be 22 ga. 34" + 20" = 54" going across from 54" find 13.5 lb. per foot. 13.5 x 15' = 202.5 lbs.

For S.F. of surface area 202.5 ÷ 1.406 = 144 S.F.
Note: Figures include an allowance for scrap.

MECHANICAL A8.4 Cooling

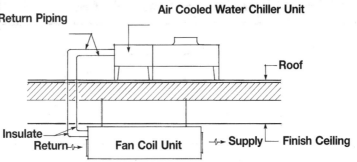

Design Assumptions: The chilled water, air cooled systems priced, utilize reciprocating hermetic compressors and propeller-type condenser fans. Piping with pumps and expansion tanks is included based on a two pipe system. No ducting is included and the fan-coil units are cooling only. Water treatment and balancing are not included. Chilled water piping is insulated. Area distribution is through the use of multiple fan coil units. Fewer but larger fan coil units with duct distribution would be approximately the same S.F. cost.

8.4-110	Chilled Water, Air Cooled Condenser Systems	COST PER S.F.		
		MAT.	INST.	TOTAL
1180	Packaged chiller, air cooled, with fan coil unit			
1200	Apartment corridors, 3,000 S.F., 5.50 ton	4.29	3.59	7.88
1240	6,000 S.F., 11.00 ton	3.38	2.87	6.25
1280	10,000 S.F., 18.33 ton	3	2.17	5.17
1320	20,000 S.F., 36.66 ton	2.45	1.57	4.02
1360	40,000 S.F., 73.33 ton	2.75	1.49	4.24
1440	Banks and libraries, 3,000 S.F., 12.50 ton	6.50	4.20	10.70
1480	6,000 S.F., 25.00 ton	6.10	3.40	9.50
1520	10,000 S.F., 41.66 ton	4.99	2.38	7.37
1560	20,000 S.F., 83.33 ton	5.20	2.02	7.22
1600	40,000 S.F., 167 ton*			
1680	Bars and taverns, 3,000 S.F., 33.25 ton	12.55	4.93	17.48
1720	6,000 S.F., 66.50 ton	11.85	3.81	15.66
1760	10,000 S.F., 110.83 ton	9.80	1.18	10.98
1800	20,000 S.F., 220 ton*			
1840	40,000 S.F., 440 ton*			
1920	Bowling alleys, 3,000 S.F., 17.00 ton	8.35	4.71	13.06
1960	6,000 S.F., 34.00 ton	6.80	3.30	10.10
2000	10,000 S.F., 56.66 ton	6.25	2.40	8.65
2040	20,000 S.F., 113.33 ton	5.80	2.04	7.84
2080	40,000 S.F., 227 ton*			
2160	Department stores, 3,000 S.F., 8.75 ton	6.20	4	10.20
2200	6,000 S.F., 17.50 ton	4.56	3.08	7.64
2240	10,000 S.F., 29.17 ton	3.83	2.23	6.06
2280	20,000 S.F., 58.33 ton	3.44	1.64	5.08
2320	40,000 S.F., 116.66 ton	3.59	1.54	5.13
2400	Drug stores, 3,000 S.F., 20.00 ton	9.60	4.84	14.44
2440	6,000 S.F., 40.00 ton	8.05	3.59	11.64
2480	10,000 S.F., 66.66 ton	8.10	2.91	11.01
2520	20,000 S.F., 133.33 ton	7.15	2.15	9.30
2560	40,000 S.F., 267 ton*			
2640	Factories, 2,000 S.F., 10.00 ton	5.60	4.04	9.64
2680	6,000 S.F., 20.00 ton	5.20	3.24	8.44
2720	10,000 S.F., 33.33 ton	4.22	2.28	6.50
2760	20,000 S.F., 66.66 ton	4.39	1.92	6.31
2800	40,000 S.F., 133.33 ton	3.97	1.57	5.54
2880	Food supermarkets, 3,000 S.F., 8.50 ton	6.05	3.97	10.02
2920	6,000 S.F., 17.00 ton	4.47	3.06	7.53
2960	10,000 S.F., 28.33 ton	3.65	2.17	5.82
3000	20,000 S.F., 56.66 ton	3.32	1.61	4.93
3040	40,000 S.F., 113.33 ton	3.46	1.54	5
3120	Medical centers, 3,000 S.F., 7.00 ton	5.35	3.87	9.22

For expanded coverage of these items see *Means Mechanical or Plumbing Cost Data 2000*

MECHANICAL — A8.4 Cooling

8.4-110 Chilled Water, Air Cooled Condenser Systems

		COST PER S.F.		
		MAT.	INST.	TOTAL
3160	6,000 S.F., 14.00 ton	3.93	2.96	6.89
3200	10,000 S.F., 23.33 ton	3.58	2.26	5.84
3240	20,000 S.F., 46.66 ton	2.98	1.66	4.64
3280	40,000 S.F., 93.33 ton	3.05	1.51	4.56
3360	Offices, 3,000 S.F., 9.50 ton	5.20	3.94	9.14
3400	6,000 S.F., 19.00 ton	5	3.21	8.21
3440	10,000 S.F., 31.66 ton	4.07	2.26	6.33
3480	20,000 S.F., 63.33 ton	4.34	1.96	6.30
3520	40,000 S.F., 126.66 ton	3.88	1.60	5.48
3600	Restaurants, 3,000 S.F., 15.00 ton	7.45	4.36	11.81
3640	6,000 S.F., 30.00 ton	6.35	3.30	9.65
3680	10,000 S.F., 50.00 ton	5.80	2.44	8.24
3720	20,000 S.F., 100.00 ton	5.80	2.14	7.94
3760	40,000 S.F., 200 ton*			
3840	Schools and colleges, 3,000 S.F., 11.50 ton	6.15	4.14	10.29
3880	6,000 S.F., 23.00 ton	5.75	3.34	9.09
3920	10,000 S.F., 38.33 ton	4.68	2.34	7.02
3960	20,000 S.F., 76.66 ton	4.99	2.04	7.03
4000	40,000 S.F., 153 ton*			

Important: See the Reference Section for critical supporting data - Reference Nos., Crews & Location Factors

MECHANICAL — A8.4 Air Conditioning

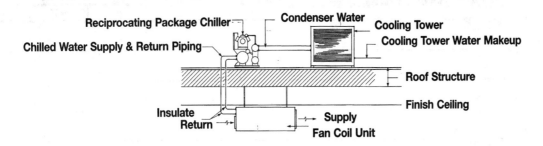

General: Water cooled chillers are available in the same sizes as air cooled units. They are also available in larger capacities.

Design Assumptions: The chilled water systems with water cooled condenser include reciprocating hermetic compressors, water cooling tower, pumps, piping and expansion tanks and are based on a two pipe system. Chilled water piping is insulated. No ducts are included and fan-coil units are cooling only. Area distribution is through use of multiple fan coil units. Fewer but larger fan coil units with duct distribution would be approximately the same S.F. cost. Water treatment and balancing are not included.

*Suggest using multiple chillers.

8.4-120	Chilled Water, Cooling Tower Systems	MAT.	INST.	TOTAL
1300	Packaged chiller, water cooled, with fan coil unit			
1320	Apartment corridors, 4,000 S.F., 7.33 ton	5.35	3.52	8.87
1360	6,000 S.F., 11.00 ton	4.16	3.01	7.17
1400	10,000 S.F., 18.33 ton	3.51	2.24	5.75
1440	20,000 S.F., 26.66 ton	2.79	1.66	4.45
1480	40,000 S.F., 73.33 ton	2.95	1.61	4.56
1520	60,000 S.F., 110.00 ton	2.85	1.64	4.49
1600	Banks and libraries, 4,000 S.F., 16.66 ton	8.05	3.81	11.86
1640	6,000 S.F., 25.00 ton	6.50	3.36	9.86
1680	10,000 S.F., 41.66 ton	5.15	2.50	7.65
1720	20,000 S.F., 83.33 ton	5.60	2.25	7.85
1760	40,000 S.F., 166.66 ton	5.70	2.92	8.62
1800	60,000 S.F., 250.00 ton	5.60	3.08	8.68
1880	Bars and taverns, 4,000 S.F., 44.33 ton	12.65	4.76	17.41
1920	6,000 S.F., 66.50 ton	12.50	4.69	17.19
1960	10,000 S.F., 110.83 ton	12.40	3.71	16.11
2000	20,000 S.F., 221.66 ton	11.95	4.15	16.10
2040	40,000 S.F., 440 ton*			
2080	60,000 S.F., 660 ton*			
2160	Bowling alleys, 4,000 S.F., 22.66 ton	9.45	4.18	13.63
2200	6,000 S.F., 34.00 ton	7.95	3.67	11.62
2240	10,000 S.F., 56.66 ton	6.45	2.69	9.14
2280	20,000 S.F., 113.33 ton	6.65	2.37	9.02
2320	40,000 S.F., 226.66 ton	6.70	3	9.70
2360	60,000 S.F., 340 ton			
2440	Department stores, 4,000 S.F., 11.66 ton	6	3.77	9.77
2480	6,000 S.F., 17.50 ton	5.85	3.11	8.96
2520	10,000 S.F., 29.17 ton	4.22	2.33	6.55
2560	20,000 S.F., 58.33 ton	3.27	1.72	4.99
2600	40,000 S.F., 116.66 ton	3.76	1.69	5.45
2640	60,000 S.F., 175.00 ton	4.54	2.88	7.42
2720	Drug stores, 4,000 S.F., 26.66 ton	9.85	4.30	14.15
2760	6,000 S.F., 40.00 ton	8.20	3.67	11.87
2800	10,000 S.F., 66.66 ton	7.85	3.16	11.01
2840	20,000 S.F., 133.33 ton	7.70	2.47	10.17
2880	40,000 S.F., 266.67 ton	7.80	3.29	11.09
2920	60,000 S.F., 400 ton*			
3000	Factories, 4,000 S.F., 13.33 ton	6.75	3.65	10.40
3040	6,000 S.F., 20.00 ton	5.85	3.14	8.99
3080	10,000 S.F., 33.33 ton	4.69	2.40	7.09
3120	20,000 S.F., 66.66 ton	4.21	2.02	6.23
3160	40,000 S.F., 133.33 ton	4.23	1.74	5.97

For expanded coverage of these items see *Means Mechanical or Plumbing Cost Data 2000*

MECHANICAL — A8.4 Air Conditioning

8.4-120 Chilled Water, Cooling Tower Systems

		COST PER S.F.		
		MAT.	INST.	TOTAL
3200	60,000 S.F., 200.00 ton	4.92	2.97	7.89
3280	Food supermarkets, 4,000 S.F., 11.33 ton	5.85	3.75	9.60
3320	6,000 S.F., 17.00 ton	5.15	3.06	8.21
3360	10,000 S.F., 28.33 ton	4.14	2.32	6.46
3400	20,000 S.F., 56.66 ton	3.38	1.73	5.11
3440	40,000 S.F., 113.33 ton	3.73	1.69	5.42
3480	60,000 S.F., 170.00 ton	4.50	2.89	7.39
3560	Medical centers, 4,000 S.F., 9.33 ton	4.98	3.45	8.43
3600	6,000 S.F., 14.00 ton	4.94	3	7.94
3640	10,000 S.F., 23.33 ton	3.80	2.25	6.05
3680	20,000 S.F., 46.66 ton	2.96	1.69	4.65
3720	40,000 S.F., 93.33 ton	3.38	1.66	5.04
3760	60,000 S.F., 140.00 ton	4.13	2.86	6.99
3840	Offices, 4,000 S.F., 12.66 ton	6.50	3.61	10.11
3880	6,000 S.F., 19.00 ton	5.75	3.22	8.97
3920	10,000 S.F., 31.66 ton	4.60	2.42	7.02
3960	20,000 S.F., 63.33 ton	4.10	2.03	6.13
4000	40,000 S.F., 126.66 ton	4.76	2.80	7.56
4040	60,000 S.F., 190.00 ton	4.73	2.94	7.67
4120	Restaurants, 4,000 S.F., 20.00 ton	8.40	3.88	12.28
4160	6,000 S.F., 30.00 ton	7.10	3.45	10.55
4200	10,000 S.F., 50.00 ton	5.90	2.59	8.49
4240	20,000 S.F., 100.00 ton	6.40	2.37	8.77
4280	40,000 S.F., 200.00 ton	5.95	2.83	8.78
4320	60,000 S.F., 300.00 ton	6.30	3.19	9.49
4400	Schools and colleges, 4,000 S.F., 15.33 ton	7.55	3.75	11.30
4440	6,000 S.F., 23.00 ton	6.10	3.30	9.40
4480	10,000 S.F., 38.33 ton	4.83	2.45	7.28
4520	20,000 S.F., 76.66 ton	5.30	2.22	7.52
4560	40,000 S.F., 153.33 ton	5.35	2.86	8.21
4600	60,000 S.F., 230.00 ton	5.15	2.98	8.13
4603				

Important: See the Reference Section for critical supporting data - Reference Nos., Crews & Location Factors

MECHANICAL — A8.4 — Air Conditioning

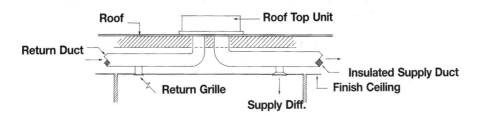

System Description: Rooftop single zone units are electric cooling and gas heat. Duct systems are low velocity, galvanized steel supply and return. Price variations between sizes are due to several factors. Jumps in the cost of the rooftop unit occur when the manufacturer shifts from the largest capacity unit on a small frame to the smallest capacity on the next larger frame, or changes from one compressor to two. As the unit capacity increases for larger areas the duct distribution grows in proportion. For most applications there is a tradeoff point where it is less expensive and more efficient to utilize smaller units with short simple distribution systems. Larger units also require larger initial supply and return ducts which can create a space problem. Supplemental heat may be desired in colder locations. The table below is based on one unit supplying the area listed. The 10,000 S.F. unit for bars and taverns is not listed because a nominal 110 ton unit would be required and this is above the normal single zone rooftop capacity.

*Use multiple units.

8.4-210	Rooftop Single Zone Unit Systems	COST PER S.F.		
		MAT.	INST.	TOTAL
1260	Rooftop, single zone, air conditioner			
1280	Apartment corridors, 500 S.F., .92 ton	4.79	1.69	6.48
1320	1,000 S.F., 1.83 ton	4.77	1.69	6.46
1360	1500 S.F., 2.75 ton	2.81	1.40	4.21
1400	3,000 S.F., 5.50 ton	2.43	1.34	3.77
1440	5,000 S.F., 9.17 ton	2.45	1.21	3.66
1480	10,000 S.F., 18.33 ton	2.61	1.14	3.75
1560	Banks or libraries, 500 S.F., 2.08 ton	10.85	3.83	14.68
1600	1,000 S.F., 4.17 ton	6.40	3.18	9.58
1640	1,500 S.F., 6.25 ton	5.50	3.05	8.55
1680	3,000 S.F., 12.50 ton	5.55	2.75	8.30
1720	5,000 S.F., 20.80 ton	5.90	2.60	8.50
1760	10,000 S.F., 41.67 ton	5.55	2.58	8.13
1840	Bars and taverns, 500 S.F. 5.54 ton	13.05	5.10	18.15
1880	1,000 S.F., 11.08 ton	13.15	4.29	17.44
1920	1,500 S.F., 16.62 ton	11.80	4	15.80
1960	3,000 S.F., 33.25 ton	13.45	3.85	17.30
2000	5,000 S.F., 55.42 ton	12.80	3.85	16.65
2040	10,000 S.F., 110.83 ton*			
2080	Bowling alleys, 500 S.F., 2.83 ton	8.65	4.31	12.96
2120	1,000 S.F., 5.67 ton	7.50	4.16	11.66
2160	1,500 S.F., 8.50 ton	7.55	3.74	11.29
2200	3,000 S.F., 17.00 ton	6.85	3.59	10.44
2240	5,000 S.F., 28.33 ton	7.70	3.51	11.21
2280	10,000 S.F., 56.67 ton	7.40	3.51	10.91
2360	Department stores, 500 S.F., 1.46 ton	7.60	2.68	10.28
2400	1,000 S.F., 2.92 ton	4.47	2.22	6.69
2480	3,000 S.F., 8.75 ton	3.88	1.93	5.81
2520	5,000 S.F., 14.58 ton	3.53	1.85	5.38
2560	10,000 S.F., 29.17 ton	3.96	1.81	5.77
2640	Drug stores, 500 S.F., 3.33 ton	10.20	5.10	15.30
2680	1,000 S.F., 6.67 ton	8.80	4.89	13.69
2720	1,500 S.F., 10.00 ton	8.90	4.40	13.30
2760	3,000 S.F., 20.00 ton	9.50	4.16	13.66
2800	5,000 S.F., 33.33 ton	9.05	4.13	13.18
2840	10,000 S.F., 66.67 ton	8.70	4.13	12.83
2920	Factories, 500 S.F., 1.67 ton	8.65	3.07	11.72
3000	1,500 S.F., 5.00 ton	4.41	2.45	6.86
3040	3,000 S.F., 10.00 ton	4.45	2.20	6.65
3080	5,000 S.F., 16.67 ton	4.05	2.11	6.16

For expanded coverage of these items see *Means Mechanical or Plumbing Cost Data 2000*

MECHANICAL — A8.4 Air Conditioning

8.4-210 Rooftop Single Zone Unit Systems

		COST PER S.F.		
		MAT.	INST.	TOTAL
3120	10,000 S.F., 33.33 ton	4.54	2.07	6.61
3200	Food supermarkets, 500 S.F., 1.42 ton	7.40	2.61	10.01
3240	1,000 S.F., 2.83 ton	4.30	2.15	6.45
3280	1,500 S.F., 4.25 ton	3.75	2.07	5.82
3320	3,000 S.F., 8.50 ton	3.78	1.87	5.65
3360	5,000 S.F., 14.17 ton	3.44	1.79	5.23
3400	10,000 S.F., 28.33 ton	3.85	1.75	5.60
3480	Medical centers, 500 S.F., 1.17 ton	6.10	2.15	8.25
3520	1,000 S.F., 2.33 ton	6.05	2.14	8.19
3560	1,500 S.F., 3.50 ton	3.57	1.78	5.35
3640	5,000 S.F., 11.67 ton	3.11	1.54	4.65
3680	10,000 S.F., 23.33 ton	3.32	1.45	4.77
3760	Offices, 500 S.F., 1.58 ton	8.25	2.91	11.16
3800	1,000 S.F., 3.17 ton	4.85	2.42	7.27
3840	1,500 S.F., 4.75 ton	4.19	2.33	6.52
3880	3,000 S.F., 9.50 ton	4.22	2.09	6.31
3920	5,000 S.F., 15.83 ton	3.84	2.01	5.85
3960	10,000 S.F., 31.67 ton	4.31	1.97	6.28
4000	Restaurants, 500 S.F., 2.50 ton	13.05	4.61	17.66
4040	1,000 S.F., 5.00 ton	6.60	3.67	10.27
4080	1,500 S.F., 7.50 ton	6.65	3.31	9.96
4120	3,000 S.F., 15.00 ton	6.05	3.17	9.22
4160	5,000 S.F., 25.00 ton	7.10	3.12	10.22
4200	10,000 S.F., 50.00 ton	6.50	3.11	9.61
4240	Schools and colleges, 500 S.F., 1.92 ton	10	3.53	13.53
4280	1,000 S.F., 3.83 ton	5.85	2.92	8.77
4360	3,000 S.F., 11.50 ton	5.10	2.53	7.63
4400	5,000 S.F., 19.17 ton	5.45	2.39	7.84
4441	10,000 S.F., 38.33 ton	5.10	2.38	7.48

Important: See the Reference Section for critical supporting data - Reference Nos., Crews & Location Factors

MECHANICAL — A8.4 Air Conditioning

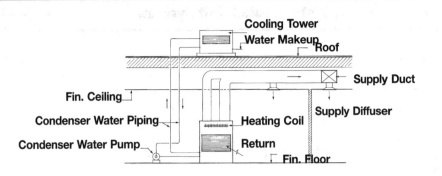

System Description: Self-contained, single package water cooled units include cooling tower, pump, piping allowance. Systems for 1000 S.F. and up include duct and diffusers to provide for even distribution of air. Smaller units distribute air through a supply air plenum, which is integral with the unit.

Returns are not ducted and supplies are not insulated.

Hot water or steam heating coils are included but piping to boiler and the boiler itself is not included.

Where local codes or conditions permit single pass cooling for the smaller units, deduct 10%.

8.4-230	Self-contained, Water Cooled Unit Systems	MAT.	INST.	TOTAL
1280	Self-contained, water cooled unit	2.69	1.33	4.02
1300	Apartment corridors, 500 S.F., .92 ton	2.68	1.33	4.01
1320	1,000 S.F., 1.83 ton	2.67	1.33	4
1360	3,000 S.F., 5.50 ton	2.04	1.13	3.17
1400	5,000 S.F., 9.17 ton	2.07	1.06	3.13
1440	10,000 S.F., 18.33 ton	1.89	.95	2.84
1520	Banks or libraries, 500 S.F., 2.08 ton	5.65	1.28	6.93
1560	1,000 S.F., 4.17 ton	4.63	2.55	7.18
1600	3,000 S.F., 12.50 ton	4.70	2.39	7.09
1640	5,000 S.F., 20.80 ton	4.28	2.15	6.43
1680	10,000 S.F., 41.66 ton	3.74	2.15	5.89
1760	Bars and taverns, 500 S.F., 5.54 ton	11.15	2.15	13.30
1800	1,000 S.F., 11.08 ton	11.85	4.03	15.88
1840	3,000 S.F., 33.25 ton	9.80	3.22	13.02
1880	5,000 S.F., 55.42 ton	9.20	3.40	12.60
1920	10,000 S.F., 110.00 ton	9	3.37	12.37
2000	Bowling alleys, 500 S.F., 2.83 ton	7.70	1.74	9.44
2040	1,000 S.F., 5.66 ton	6.30	3.46	9.76
2080	3,000 S.F., 17.00 ton	5.85	2.92	8.77
2120	5,000 S.F., 28.33 ton	5.30	2.82	8.12
2160	10,000 S.F., 56.66 ton	4.98	2.91	7.89
2200	Department stores, 500 S.F., 1.46 ton	3.97	.89	4.86
2240	1,000 S.F., 2.92 ton	3.23	1.78	5.01
2280	3,000 S.F., 8.75 ton	3.27	1.67	4.94
2320	5,000 S.F., 14.58 ton	2.99	1.50	4.49
2360	10,000 S.F., 29.17 ton	2.72	1.45	4.17
2440	Drug stores, 500 S.F., 3.33 ton	9.10	2.05	11.15
2480	1,000 S.F., 6.66 ton	7.40	4.07	11.47
2520	3,000 S.F., 20.00 ton	6.85	3.45	10.30
2560	5,000 S.F., 33.33 ton	7.20	3.40	10.60
2600	10,000 S.F., 66.66 ton	5.85	3.43	9.28
2680	Factories, 500 S.F., 1.66 ton	4.54	1.03	5.57
2720	1,000 S.F. 3.37 ton	3.74	2.06	5.80
2760	3,000 S.F., 10.00 ton	3.76	1.91	5.67
2800	5,000 S.F., 16.66 ton	3.44	1.72	5.16
2840	10,000 S.F., 33.33 ton	3.13	1.66	4.79
2920	Food supermarkets, 500 S.F., 1.42 ton	3.86	.87	4.73
2960	1,000 S.F., 2.83 ton	4.15	2.05	6.20
3000	3,000 S.F., 8.50 ton	3.16	1.73	4.89
3040	5,000 S.F., 14.17 ton	3.20	1.63	4.83

For expanded coverage of these items see *Means Mechanical or Plumbing Cost Data 2000*

MECHANICAL — A8.4 Air Conditioning

8.4-230 Self-contained, Water Cooled Unit Systems

		COST PER S.F.		
		MAT.	INST.	TOTAL
3080	10,000 S.F., 28.33 ton	2.67	1.42	4.09
3160	Medical centers, 500 S.F., 1.17 ton	3.17	.72	3.89
3200	1,000 S.F., 2.33 ton	3.42	1.69	5.11
3240	3,000 S.F., 7.00 ton	2.60	1.43	4.03
3280	5,000 S.F., 11.66 ton	2.63	1.34	3.97
3320	10,000 S.F., 23.33 ton	2.40	1.21	3.61
3400	Offices, 500 S.F., 1.58 ton	4.31	.97	5.28
3440	1,000 S.F., 3.17 ton	4.63	2.28	6.91
3480	3,000 S.F., 9.50 ton	3.56	1.81	5.37
3520	5,000 S.F., 15.83 ton	3.25	1.63	4.88
3560	10,000 S.F., 31.67 ton	2.96	1.57	4.53
3640	Restaurants, 500 S.F., 2.50 ton	6.80	1.53	8.33
3680	1,000 S.F., 5.00 ton	5.55	3.06	8.61
3720	3,000 S.F., 15.00 ton	5.65	2.87	8.52
3760	5,000 S.F., 25.00 ton	5.15	2.58	7.73
3800	10,000 S.F., 50.00 ton	3.27	2.39	5.66
3880	Schools and colleges, 500 S.F., 1.92 ton	5.20	1.18	6.38
3920	1,000 S.F., 3.83 ton	4.26	2.34	6.60
3960	3,000 S.F., 11.50 ton	4.32	2.20	6.52
4000	5,000 S.F., 19.17 ton	3.95	1.98	5.93
4040	10,000 S.F., 38.33 ton	3.45	1.98	5.43

MECHANICAL — A8.4 Air Conditioning

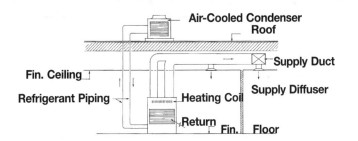

System Description: Self-contained air cooled units with remote air cooled condenser and interconnecting tubing. Systems for 1000 S.F. and up include duct and diffusers. Smaller units distribute air directly.

Returns are not ducted and supplies are not insulated.

Potential savings may be realized by using a single zone rooftop system or through-the-wall unit, especially in the smaller capacities, if the application permits.

Hot water or steam heating coils are included but piping to boiler and the boiler itself is not included.

Condenserless models are available for 15% less where remote refrigerant source is available.

8.4-240	Self-contained, Air Cooled Unit Systems	COST PER S.F. MAT.	INST.	TOTAL
1300	Self-contained, air cooled unit			
1320	Apartment corridors, 500 S.F., .92 ton	4.17	1.74	5.91
1360	1,000 S.F., 1.83 ton	4.10	1.72	5.82
1400	3,000 S.F., 5.50 ton	3.39	1.60	4.99
1440	5,000 S.F., 9.17 ton	2.88	1.52	4.40
1480	10,000 S.F., 18.33 ton	2.40	1.41	3.81
1560	Banks or libraries, 500 S.F., 2.08 ton	8.90	2.19	11.09
1600	1,000 S.F., 4.17 ton	7.65	3.63	11.28
1640	3,000 S.F., 12.50 ton	6.55	3.47	10.02
1680	5,000 S.F., 20.80 ton	5.45	3.20	8.65
1720	10,000 S.F., 41.66 ton	4.99	3.13	8.12
1800	Bars and taverns, 500 S.F., 5.54 ton	16.25	4.77	21.02
1840	1,000 S.F., 11.08 ton	16.85	6.90	23.75
1880	3,000 S.F., 33.25 ton	12.90	6.05	18.95
1920	5,000 S.F., 55.42 ton	12.40	6.10	18.50
1960	10,000 S.F., 110.00 ton	12.50	6.05	18.55
2040	Bowling alleys, 500 S.F., 2.83 ton	12.15	2.99	15.14
2080	1,000 S.F., 5.66 ton	10.45	4.93	15.38
2120	3,000 S.F., 17.00 ton	7.40	4.35	11.75
2160	5,000 S.F., 28.33 ton	6.85	4.26	11.11
2200	10,000 S.F., 56.66 ton	6.60	4.27	10.87
2240	Department stores, 500 S.F., 1.46 ton	6.25	1.54	7.79
2280	1,000 S.F., 2.92 ton	5.40	2.54	7.94
2320	3,000 S.F., 8.75 ton	4.57	2.43	7
2360	5,000 S.F., 14.58 ton	4.57	2.43	7
2400	10,000 S.F., 29.17 ton	3.53	2.19	5.72
2480	Drug stores, 500 S.F., 3.33 ton	14.30	3.52	17.82
2520	1,000 S.F., 6.66 ton	12.25	5.80	18.05
2560	3,000 S.F., 20.00 ton	8.75	5.15	13.90
2600	5,000 S.F., 33.33 ton	8.05	5	13.05
2640	10,000 S.F., 66.66 ton	7.85	5	12.85
2720	Factories, 500 S.F., 1.66 ton	7.25	1.78	9.03
2760	1,000 S.F., 3.33 ton	6.20	2.91	9.11
2800	3,000 S.F., 10.00 ton	5.25	2.78	8.03
2840	5,000 S.F., 16.66 ton	4.33	2.55	6.88
2880	10,000 S.F., 33.33 ton	4.04	2.50	6.54
2960	Food supermarkets, 500 S.F., 1.42 ton	6.10	1.50	7.60
3000	1,000 S.F., 2.83 ton	6.35	2.68	9.03
3040	3,000 S.F., 8.50 ton	5.25	2.48	7.73
3080	5,000 S.F., 14.17 ton	4.45	2.37	6.82

For expanded coverage of these items see *Means Mechanical or Plumbing Cost Data 2000*

MECHANICAL — A8.4 Air Conditioning

8.4-240 Self-contained, Air Cooled Unit Systems

		COST PER S.F.		
		MAT.	INST.	TOTAL
3120	10,000 S.F., 28.33 ton	3.43	2.14	5.57
3200	Medical centers, 500 S.F., 1.17 ton	5.05	1.23	6.28
3240	1,000 S.F., 2.33 ton	5.30	2.20	7.50
3280	3,000 S.F., 7.00 ton	4.32	2.03	6.35
3320	5,000 S.F., 16.66 ton	3.69	1.94	5.63
3360	10,000 S.F., 23.33 ton	3.07	1.79	4.86
3440	Offices, 500 S.F., 1.58 ton	6.80	1.67	8.47
3480	1,000 S.F., 3.16 ton	7.15	2.98	10.13
3520	3,000 S.F., 9.50 ton	4.99	2.63	7.62
3560	5,000 S.F., 15.83 ton	4.15	2.42	6.57
3600	10,000 S.F., 31.66 ton	3.84	2.37	6.21
3680	Restaurants, 500 S.F., 2.50 ton	10.65	2.64	13.29
3720	1,000 S.F., 5.00 ton	9.25	4.36	13.61
3760	3,000 S.F., 15.00 ton	7.85	4.17	12.02
3800	5,000 S.F., 25.00 ton	6.05	3.76	9.81
3840	10,000 S.F., 50.00 ton	6.05	3.74	9.79
3920	Schools and colleges, 500 S.F., 1.92 ton	8.20	2.03	10.23
3960	1,000 S.F., 3.83 ton	7.10	3.34	10.44
4000	3,000 S.F., 11.50 ton	6.05	3.18	9.23
4040	5,000 S.F., 19.17 ton	5	2.95	7.95
4080	10,000 S.F., 38.33 ton	4.60	2.88	7.48

MECHANICAL — A8.4 Cooling

Air Cooled Condensing Unit (diagram showing Refrigerant Piping, Roof, Supply Duct, Supply Diffuser, DX Air Handling Unit, Return Grille, Fin. Ceiling)

General: Split systems offer several important advantages which should be evaluated when a selection is to be made. They provide a greater degree of flexibility in component selection which permits an accurate match-up of the proper equipment size and type with the particular needs of the building. This allows for maximum use of modern energy saving concepts in heating and cooling. Outdoor installation of the air cooled condensing unit allows space savings in the building and also isolates the equipment operating sounds from building occupants.

Design Assumptions: The systems below are comprised of a direct expansion air handling unit and air cooled condensing unit with interconnecting copper tubing. Ducts and diffusers are also included for distribution of air. Systems are priced for cooling only. Heat can be added as desired either by putting hot water/steam coils into the air unit or into the duct supplying the particular area of need. Gas fired duct furnaces are also available. Refrigerant liquid line is insulated.

*Use multiple systems.

8.4-250	Split Systems With Air Cooled Condensing Units	COST PER S.F.		
		MAT.	INST.	TOTAL
1260	Split system, air cooled condensing unit			
1280	Apartment corridors, 1,000 S.F., 1.83 ton	1.62	1.08	2.70
1320	2,000 S.F., 3.66 ton	1.32	1.09	2.41
1360	5,000 S.F., 9.17 ton	1.61	1.36	2.97
1400	10,000 S.F., 18.33 ton	1.86	1.47	3.33
1440	20,000 S.F., 36.66 ton	1.68	1.50	3.18
1520	Banks and libraries, 1,000 S.F., 4.17 ton	3.02	2.49	5.51
1560	2,000 S.F., 8.33 ton	3.68	3.12	6.80
1600	5,000 S.F., 20.80 ton	4.24	3.35	7.59
1640	10,000 S.F., 41.66 ton	3.83	3.42	7.25
1680	20,000 S.F., 83.32 ton	3.71	3.46	7.17
1760	Bars and taverns, 1,000 S.F., 11.08 ton	8.95	4.88	13.83
1800	2,000 S.F., 22.16 ton	12.65	5.70	18.35
1840	5,000 S.F., 55.42 ton	9.15	5.40	14.55
1880	10,000 S.F., 110.84 ton	9.25	5.50	14.75
1920	20,000 S.F., 220 ton*			
2000	Bowling alleys, 1,000 S.F., 5.66 ton	4.22	4.67	8.89
2040	2,000 S.F., 11.33 ton	5	4.24	9.24
2080	5,000 S.F., 28.33 ton	5.75	4.57	10.32
2120	10,000 S.F., 56.66 ton	5.20	4.65	9.85
2160	20,000 S.F., 113.32 ton	5.40	4.90	10.30
2320	Department stores, 1,000 S.F., 2.92 ton	2.13	1.71	3.84
2360	2,000 S.F., 5.83 ton	2.17	2.40	4.57
2400	5,000 S.F., 14.58 ton	2.57	2.19	4.76
2440	10,000 S.F., 29.17 ton	2.96	2.36	5.32
2480	20,000 S.F., 58.33 ton	2.68	2.40	5.08
2560	Drug stores, 1,000 S.F., 6.66 ton	4.97	5.50	10.47
2600	2,000 S.F., 13.32 ton	5.90	4.98	10.88
2640	5,000 S.F., 33.33 ton	6.10	5.45	11.55
2680	10,000 S.F., 66.66 ton	5.80	5.55	11.35
2720	20,000 S.F., 133.32 ton*			
2800	Factories, 1,000 S.F., 3.33 ton	2.43	1.96	4.39
2840	2,000 S.F., 6.66 ton	2.48	2.75	5.23
2880	5,000 S.F., 16.66 ton	3.39	2.70	6.09
2920	10,000 S.F., 33.33 ton	3.07	2.74	5.81
2960	20,000 S.F., 66.66 ton	2.92	2.78	5.70
3040	Food supermarkets, 1,000 S.F., 2.83 ton	2.07	1.66	3.73
3080	2,000 S.F., 5.66 ton	2.11	2.34	4.45
3120	5,000 S.F., 14.66 ton	2.50	2.12	4.62
3160	10,000 S.F., 28.33 ton	2.88	2.29	5.17

For expanded coverage of these items see *Means Mechanical or Plumbing Cost Data 2000*

MECHANICAL | A8.4 Cooling

8.4-250 Split Systems With Air Cooled Condensing Units

		COST PER S.F.		
		MAT.	INST.	TOTAL
3200	20,000 S.F., 56.66 ton	2.60	2.32	4.92
3280	Medical centers, 1,000 S.F., 2.33 ton	1.73	1.34	3.07
3320	2,000 S.F., 4.66 ton	1.73	1.92	3.65
3360	5,000 S.F., 11.66 ton	2.06	1.75	3.81
3400	10,000 S.F., 23.33 ton	2.38	1.88	4.26
3440	20,000 S.F., 46.66 ton	2.15	1.92	4.07
3520	Offices, 1,000 S.F., 3.17 ton	2.32	1.86	4.18
3560	2,000 S.F., 6.33 ton	2.37	2.60	4.97
3600	5,000 S.F., 15.83 ton	2.80	2.36	5.16
3640	10,000 S.F., 31.66 ton	3.22	2.55	5.77
3680	20,000 S.F., 63.32 ton	2.77	2.64	5.41
3760	Restaurants, 1,000 S.F., 5.00 ton	3.74	4.14	7.88
3800	2,000 S.F., 10.00 ton	4.43	3.76	8.19
3840	5,000 S.F., 25.00 ton	5.10	4.05	9.15
3880	10,000 S.F., 50.00 ton	4.59	4.12	8.71
3920	20,000 S.F., 100.00 ton	4.80	4.33	9.13
4000	Schools and colleges, 1,000 S.F., 3.83 ton	2.76	2.29	5.05
4040	2,000 S.F., 7.66 ton	3.38	2.86	6.24
4080	5,000 S.F., 19.17 ton	3.89	3.09	6.98
4120	10,000 S.F., 38.33 ton	3.51	3.14	6.65
4161	20,000 S.F., 76.66 ton	3.36	3.18	6.54

Important: See the Reference Section for critical supporting data - Reference Nos., Crews & Location Factors

MECHANICAL — A8.5 Special

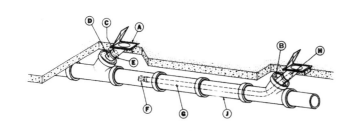

Vitrified Clay Garage Exhaust System

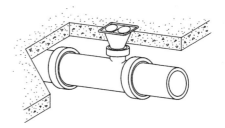
Dual Exhaust System

8.5-110	Garage Exhaust Systems	COST PER BAY		
		MAT.	INST.	TOTAL
1040	Garage, single exhaust, 3" outlet, cars & light trucks, one bay	2,175	585	2,760
1060	Additional bays up to seven bays	310	86	396
1500	4" outlet, trucks, one bay	2,175	585	2,760
1520	Additional bays up to six bays	320	86	406
1600	5" outlet, diesel trucks, one bay	2,225	585	2,810
1650	Additional single bays up to six	380	100	480
1700	Two adjoining bays	2,225	585	2,810
2000	Dual exhaust, 3" outlets, pair of adjoining bays	2,475	655	3,130
2100	Additional pairs of adjoining bays	575	100	675

For information about Means Estimating Seminars, see yellow pages 11 and 12 in back of book

For expanded coverage of these items see *Means Mechanical or Plumbing Cost Data 2000*

Division 9
Electrical

ELECTRICAL — A9.1 Service Distribution

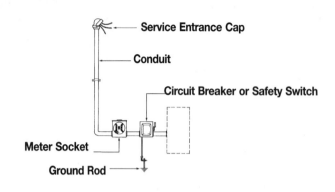

9.1-210	Electric Service, 3 Phase - 4 Wire	COST EACH		
		MAT.	INST.	TOTAL
0200	Service installation, includes breakers, metering, 20' conduit & wire			
0220	3 phase, 4 wire, 120/208 volts, 60 amp	570	445	1,015
0240	100 amps	715	535	1,250
0280	200 amps	1,025	830	1,855
0320	400 amps	2,225	1,525	3,750
0360	600 amps	4,400	2,050	6,450
0400	800 amps	5,700	2,475	8,175
0440	1000 amps	7,300	2,850	10,150
0480	1200 amps	8,725	2,900	11,625
0520	1600 amps	16,800	4,175	20,975
0560	2000 amps	18,600	4,750	23,350
0570	Add 25% for 277/480 volt			

9.1-310	Feeder Installation	COST PER L.F.		
		MAT.	INST.	TOTAL
0200	Feeder installation 600 volt, including conduit and wire, 60 amperes	3.21	5.40	8.61
0240	100 amperes	5.80	7.15	12.95
0280	200 amperes	12.25	11.10	23.35
0320	400 amperes	24.50	22	46.50
0360	600 amperes	51	36	87
0400	800 amperes	64	43	107
0440	1000 amperes	83	55	138
0480	1200 amperes	93.50	56.50	150
0520	1600 amperes	128	86	214
0560	2000 amperes	166	110	276

9.1-410	Switchgear	COST EACH		
		MAT.	INST.	TOTAL
0200	Switchgear inst., incl. swbd., panels & circ bkr, 400 amps, 120/208volt	3,375	1,825	5,200
0240	600 amperes	8,325	2,450	10,775
0280	800 amperes	10,400	3,500	13,900
0320	1200 amperes	13,500	5,375	18,875
0360	1600 amperes	18,200	7,525	25,725
0400	2000 amperes	23,100	9,600	32,700
0410	Add 20% for 277/480 volt			

Important: See the Reference Section for critical supporting data - Reference Nos., Crews & Location Factors

ELECTRICAL — A9.2 — Lighting & Power

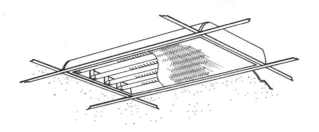

Type C. Recessed, mounted on grid ceiling suspension system, 2' x 4', four 40 watt lamps, acrylic prismatic diffusers.

5.3 watts per S.F. for 100 footcandles.
3 watts per S.F. for 57 footcandles.

9.2-213	Fluorescent Fixtures (by Wattage)	COST PER S.F. MAT.	INST.	TOTAL
0190	Fluorescent fixtures recess mounted in ceiling			
0200	1 watt per S.F., 20 FC, 5 fixtures per 1000 S.F.	.55	.85	1.40
0240	2 watts per S.F., 40 FC, 10 fixtures per 1000 S.F.	1.10	1.67	2.77
0280	3 watts per S.F., 60 FC, 15 fixtures per 1000 S.F	1.63	2.51	4.14
0320	4 watts per S.F., 80 FC, 20 fixtures per 1000 S.F.	2.17	3.34	5.51
0400	5 watts per S.F., 100 FC, 25 fixtures per 1000 S.F.	2.72	4.18	6.90
0402				

9.2-223	Incandescent Fixture (by Wattage)	COST PER S.F. MAT.	INST.	TOTAL
0190	Incandescent fixture recess mounted, type A			
0200	1 watt per S.F., 8 FC, 6 fixtures per 1000 S.F.	.58	.68	1.26
0240	2 watt per S.F., 16 FC, 12 fixtures per 1000 S.F.	1.15	1.37	2.52
0280	3 watt per S.F., 24 FC, 18 fixtures, per 1000 S.F.	1.73	2.02	3.75
0320	4 watt per S.F., 32 FC, 24 fixtures per 1000 S.F.	2.30	2.71	5.01
0400	5 watt per S.F., 40 FC, 30 fixtures per 1000 S.F.	2.89	3.38	6.27

9.2-235	H.I.D. Fixture, High Bay (by Wattage)	COST PER S.F. MAT.	INST.	TOTAL
0190	High intensity discharge fixture, 16' above work plane			
0200	1 watt/S.F., type D, 23 FC, 1 fixture/1000 S.F.	.67	.68	1.35
0240	Type E, 42 FC, 1 fixture/1000 S.F.	.79	.71	1.50
0280	Type G, 52 FC, 1 fixture/1000 S.F.	.79	.71	1.50
0320	Type C, 54 FC, 2 fixture/1000 S.F.	.93	.80	1.73
0400	2 watt/S.F., type D, 45 FC, 2 fixture/1000 S.F.	1.33	1.37	2.70
0440	Type E, 84 FC, 2 fixture/1000 S.F.	1.59	1.45	3.04
0480	Type G, 105 FC, 2 fixture/1000 S.F.	1.59	1.45	3.04
0520	Type C, 108 FC, 4 fixture/1000 S.F.	1.85	1.58	3.43
0600	3 watt/S.F., type D, 68 FC, 3 fixture/1000 S.F.	1.98	2.02	4
0640	Type E, 126 FC, 3 fixture/1000 S.F.	2.37	2.17	4.54
0680	Type G, 157 FC, 3 fixture/1000 S.F.	2.37	2.17	4.54
0720	Type C, 162 FC, 6 fixture/1000 S.F.	2.77	2.40	5.17
0800	4 watt/ S.F., type D, 91 FC, 4 fixture/1000 S.F.	2.87	3.20	6.07
0840	Type E, 168 FC, 4 fixture/1000 S.F.	3.16	2.91	6.07
0880	Type G, 210 FC, 4 fixture/1000 S.F.	3.16	2.91	6.07
0920	Type C, 243 FC, 9 fixture/1000 S.F.	4.04	3.33	7.37
1000	5 watt/S.F., type D, 113 FC, 5 fixture/1000 S.F.	3.31	3.39	6.70
1040	Type E, 210 FC, 5 fixture/1000 S.F.	3.95	3.62	7.57
1080	Type G, 262 FC, 5 fixture/1000 S.F.	3.95	3.62	7.57
1120	Type C, 297 FC, 11 fixture/1000 S.F.	4.96	4.14	9.10

For expanded coverage of these items see *Means Electrical Cost Data 2000*

ELECTRICAL | A9.2 | Lighting & Power

9.2-239 H.I.D. Fixture, High Bay (by Wattage)

		COST PER S.F.		
		MAT.	INST.	TOTAL
0190	High intensity discharge fixture, 30' above work plane			
0200	1 watt/S.F., type D, 23 FC, 1 fixture/1000 S.F.	.75	.87	1.62
0240	Type E, 37 FC, 1 fixture/1000 S.F.	.87	.90	1.77
0280	Type G, 45 FC., 1 fixture/1000 S.F.	.87	.90	1.77
0320	Type F, 50 FC, 1 fixture/1000 S.F.	.74	.71	1.45
0400	2 watt/S.F., type D, 40 FC, 2 fixtures/1000 S.F.	1.48	1.73	3.21
0440	Type E, 74 FC, 2 fixtures/1000 S.F.	1.73	1.81	3.54
0480	Type G, 92 FC, 2 fixtures/1000 S.F.	1.73	1.81	3.54
0520	Type F, 100 FC, 2 fixtures/1000 S.F.	1.48	1.45	2.93
0600	3 watt/S.F., type D, 60 FC, 3 fixtures/1000 S.F.	2.22	2.59	4.81
0640	Type E, 110 FC, 3 fixtures/1000 S.F.	2.60	2.74	5.34
0680	Type G, 138FC, 3 fixtures/1000 S.F.	2.60	2.74	5.34
0720	Type F, 150 FC, 3 fixtures/1000 S.F.	2.21	2.15	4.36
0800	4 watt/ S.F., type D, 80 FC, 4 fixtures/1000 S.F.	2.95	3.44	6.39
0840	Type E, 148 FC, 4 fixtures/1000 S.F.	3.45	3.65	7.10
0880	Type G, 185 FC, 4 fixtures/1000 S.F.	3.45	3.65	7.10
0920	Type F, 200 FC, 4 fixtures/1000 S.F.	2.96	2.88	5.84
1000	5 watt/ S.F., type D, 100 FC 5 fixtures/1000 S.F.	3.70	4.32	8.02
1040	Type E, 185 FC, 5 fixtures/1000 S.F.	4.32	4.55	8.87
1080	Type G, 230 FC, 5 fixtures/1000 S.F.	4.32	4.55	8.87
1120	Type F, 250 FC, 5 fixtrues/1000 S.F.	3.68	3.60	7.28
1121				

Important: See the Reference Section for critical supporting data - Reference Nos., Crews & Location Factors

ELECTRICAL — A9.2 Lighting & Power

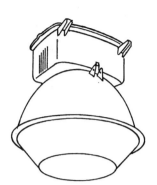

LOW BAY FIXTURES
H. Mercury vapor 250 watt
J. Metal halide 250 watt
K. High pressure sodium 150 watt

9.2-242	H.I.D. Fixture, Low Bay (by Wattage)	MAT.	INST.	TOTAL
0190	High intensity discharge fixture, 8'-10' above work plane			
0200	1 watt/S.F., type H, 19 FC, 4 fixtures/1000 S.F.	1.59	1.45	3.04
0240	Type J, 30 FC, 4 fixtures/1000 S.F.	1.92	1.47	3.39
0280	Type K, 29 FC, 5 fixtures/1000 S.F.	1.85	1.32	3.17
0360	2 watt/S.F. type H, 33 FC, 7 fixtures/1000 S.F.	2.92	2.80	5.72
0400	Type J, 52 FC, 7 fixtures/1000 S.F.	3.42	2.70	6.12
0440	Type K, 63 FC, 11 fixtures/1000 S.F.	4	2.79	6.79
0520	3 watt/S.F., type H, 51 FC, 11 fixtures/1000 S.F.	4.52	4.31	8.83
0560	Type J, 81 FC, 11 fixtures/1000 S.F.	5.30	4.09	9.39
0600	Type K, 92 FC, 16 fixtures/1000 S.F.	5.85	4.11	9.96
0680	4 watt/S.F., type H, 65 FC, 14 fixtures/1000 S.F.	5.85	5.65	11.50
0720	Type J, 103 FC, 14 fixtures/1000 S.F.	6.85	5.40	12.25
0760	Type K, 127 FC, 22 fixtures/1000 S.F.	8	5.55	13.55
0840	5 watt/S.F., type H, 84 FC, 18 fixtures/1000 S.F.	7.45	7.10	14.55
0880	Type J, 133 FC, 18 fixtures/1000 S.F.	8.75	6.80	15.55
0920	Type K, 155 FC, 27 fixtures/1000 S.F.	9.85	6.90	16.75

9.2-244	H.I.D. Fixture, Low Bay (by Wattage)	MAT.	INST.	TOTAL
0190	High intensity discharge fixture, mounted 16' above work plane			
0200	1 watt/S.F., type H, 19 FC, 4 fixtures/1000 S.F.	1.68	1.66	3.34
0240	Type J, 28 FC, 4 fixt./1000 S.F.	2.01	1.67	3.68
0280	Type K, 27 FC, 5 fixt./1000 S.F.	2.08	1.87	3.95
0360	2 watts/S.F., type H, 30 FC, 7 fixt/1000 S.F.	3.09	3.21	6.30
0400	Type J, 48 FC, 7 fixt/1000 S.F.	3.65	3.24	6.89
0440	Type K, 58 FC, 11 fixt/1000 S.F.	4.44	3.83	8.27
0520	3 watts/S.F., type H, 47 FC, 11 fixt/1000 S.F.	4.77	4.88	9.65
0560	Type J, 75 FC, 11 fixt/1000 S.F.	5.65	4.91	10.56
0600	Type K, 85 FC, 16 fixt/1000 S.F.	6.55	5.70	12.25
0680	4 watts/S.F., type H, 60 FC, 14 fixt/1000 S.F.	6.15	6.45	12.60
0720	Type J, 95 FC, 14 fixt/1000 S.F.	7.30	6.45	13.75
0760	Type K, 117 FC, 22 fixt/1000 S.F.	8.90	7.65	16.55
0840	5 watts/S.F., type H, 77 FC, 18 fixt/1000 S.F.	7.95	8.30	16.25
0880	Type J, 122 FC, 18 fixt/1000 S.F.	9.30	8.15	17.45
0920	Type K, 143 FC, 27 fixt/1000 S.F.	10.95	9.55	20.50

For expanded coverage of these items see *Means Electrical Cost Data 2000*

ELECTRICAL — A9.2 Lighting & Power

Light Pole

9.2-252	Light Pole (Installed)	COST EACH		
		MAT.	INST.	TOTAL
0200	Light pole, aluminum, 20' high, 1 arm bracket	795	535	1,330
0240	2 arm brackets	880	535	1,415
0280	3 arm brackets	965	555	1,520
0320	4 arm brackets	1,050	555	1,605
0360	30' high, 1 arm bracket	1,400	670	2,070
0400	2 arm brackets	1,475	670	2,145
0440	3 arm brackets	1,550	690	2,240
0480	4 arm brackets	1,650	690	2,340
0680	40' high, 1 arm bracket	1,675	895	2,570
0720	2 arm brackets	1,775	895	2,670
0760	3 arm brackets	1,850	915	2,765
0800	4 arm brackets	1,925	915	2,840
0840	Steel, 20' high, 1 arm bracket	1,000	570	1,570
0880	2 arm brackets	1,075	570	1,645
0920	3 arm brackets	1,100	590	1,690
0960	4 arm brackets	1,175	590	1,765
1000	30' high, 1 arm bracket	1,150	720	1,870
1040	2 arm brackets	1,225	720	1,945
1080	3 arm brackets	1,250	735	1,985
1120	4 arm brackets	1,350	735	2,085
1320	40' high, 1 arm bracket	1,500	975	2,475
1360	2 arm brackets	1,575	975	2,550
1400	3 arm brackets	1,600	995	2,595
1440	4 arm brackets	1,675	995	2,670

ELECTRICAL — A9.2 Lighting & Power

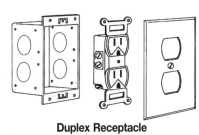

Duplex Receptacle

9.2-522	Receptacle (by Wattage)	COST PER S.F. MAT.	INST.	TOTAL
0190	Receptacles include plate, box, conduit, wire & transformer when required			
0200	2.5 per 1000 S.F., .3 watts per S.F.	.25	.65	.90
0240	With transformer	.29	.68	.97
0280	4 per 1000 S.F., .5 watts per S.F.	.29	.75	1.04
0320	With transformer	.34	.80	1.14
0360	5 per 1000 S.F., .6 watts per S.F.	.34	.89	1.23
0400	With transformer	.41	.95	1.36
0440	8 per 1000 S.F., .9 watts per S.F.	.36	1	1.36
0480	With transformer	.46	1.09	1.55
0520	10 per 1000 S.F., 1.2 watts per S.F.	.37	1.07	1.44
0560	With transformer	.53	1.22	1.75
0600	16.5 per 1000 S.F., 2.0 watts per S.F.	.45	1.33	1.78
0640	With transformer	.73	1.58	2.31
0680	20 per 1000 S.F., 2.4 watts per S.F.	.46	1.45	1.91
0720	With transformer	.78	1.74	2.52

9.2-524	Receptacles	COST PER S.F. MAT.	INST.	TOTAL
0200	Receptacle systems, underfloor duct, 5' on center, low density	3.35	1.42	4.77
0240	High density	3.59	1.83	5.42
0280	7' on center, low density	2.66	1.23	3.89
0320	High density	2.90	1.64	4.54
0400	Poke thru fittings, low density	.87	.65	1.52
0440	High density	1.73	1.29	3.02
0520	Telepoles, using Romex, low density	.63	.43	1.06
0560	High density	1.26	.85	2.11
0600	Using EMT, low density	.66	.56	1.22
0640	High density	1.33	1.15	2.48
0720	Conduit system with floor boxes, low density	.70	.48	1.18
0760	High density	1.40	.99	2.39
0840	Undercarpet power system, 3 conductor with 5 conductor feeder, low density	1.26	.19	1.45
0880	High density	2.46	.36	2.82

For expanded coverage of these items see *Means Electrical Cost Data 2000*

ELECTRICAL — A9.2 — Lighting & Power

Description: Table 9.2-542 includes the cost for switch, plate, box, conduit in slab or EMT exposed and copper wire. Add 20% for exposed conduit.

No power required for switches.

Federal energy guidelines recommend the maximum lighting area controlled per switch shall not exceed 1000 S.F. and that areas over 500 S.F. shall be so controlled that total illumination can be reduced by at least 50%.

9.2-542	Wall Switch by Sq. Ft.	COST PER S.F.		
		MAT.	INST.	TOTAL
0200	Wall switches, 1.0 per 1000 S.F.	.04	.10	.14
0240	1.2 per 1000 S.F.	.04	.11	.15
0280	2.0 per 1000 S.F.	.05	.17	.22
0320	2.5 per 1000 S.F.	.07	.21	.28
0360	5.0 per 1000 S.F.	.18	.45	.63
0400	10.0 per 1000 S.F.	.34	.91	1.25

9.2-582	Miscellaneous Power	COST PER S.F.		
		MAT.	INST.	TOTAL
0200	Miscellaneous power, to .5 watts	.02	.05	.07
0240	.8 watts	.03	.07	.10
0280	1 watt	.04	.09	.13
0320	1.2 watts	.04	.12	.16
0360	1.5 watts	.05	.14	.19
0400	1.8 watts	.06	.15	.21
0440	2 watts	.08	.19	.27
0480	2.5 watts	.10	.23	.33
0520	3 watts	.12	.27	.39

9.2-610	Central A. C. Power (by Wattage)	COST PER S.F.		
		MAT.	INST.	TOTAL
0200	Central air conditioning power, 1 watt	.05	.12	.17
0220	2 watts	.06	.13	.19
0240	3 watts	.07	.15	.22
0280	4 watts	.10	.19	.29
0320	6 watts	.18	.27	.45
0360	8 watts	.22	.28	.50
0400	10 watts	.28	.33	.61

ELECTRICAL — A9.2 Lighting & Power

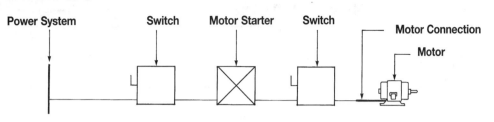

Motor Installation

9.2-710	Motor Installation	MAT.	INST.	TOTAL
0200	Motor installation, single phase, 115V, to and including 1/3 HP motor size	515	440	955
0240	To and incl. 1 HP motor size	535	440	975
0280	To and incl. 2 HP motor size	570	470	1,040
0320	To and incl. 3 HP motor size	630	475	1,105
0360	230V, to and including 1 HP motor size	520	445	965
0400	To and incl. 2 HP motor size	540	445	985
0440	To and incl. 3 HP motor size	595	480	1,075
0520	Three phase, 200V, to and including 1-1/2 HP motor size	590	490	1,080
0560	To and incl. 3 HP motor size	625	535	1,160
0600	To and incl. 5 HP motor size	665	595	1,260
0640	To and incl. 7-1/2 HP motor size	680	605	1,285
0680	To and incl. 10 HP motor size	975	760	1,735
0720	To and incl. 15 HP motor size	1,300	845	2,145
0760	To and incl. 20 HP motor size	1,600	970	2,570
0800	To and incl. 25 HP motor size	1,600	975	2,575
0840	To and incl. 30 HP motor size	2,550	1,150	3,700
0880	To and incl. 40 HP motor size	3,100	1,350	4,450
0920	To and incl. 50 HP motor size	5,475	1,575	7,050
0960	To and incl. 60 HP motor size	5,600	1,675	7,275
1000	To and incl. 75 HP motor size	7,075	1,925	9,000
1040	To and incl. 100 HP motor size	14,900	2,250	17,150
1080	To and incl. 125 HP motor size	15,100	2,475	17,575
1120	To and incl. 150 HP motor size	18,100	2,900	21,000
1160	To and incl. 200 HP motor size	21,800	3,550	25,350
1240	230V, to and including 1-1/2 HP motor size	565	485	1,050
1280	To and incl. 3 HP motor size	605	530	1,135
1320	To and incl. 5 HP motor size	645	590	1,235
1360	To and incl. 7-1/2 HP motor size	645	590	1,235
1400	To and incl. 10 HP motor size	965	720	1,685
1440	To and incl. 15 HP motor size	1,000	785	1,785
1480	To and incl. 20 HP motor size	1,500	945	2,445
1520	To and incl. 25 HP motor size	1,600	970	2,570
1560	To and incl. 30 HP motor size	1,600	975	2,575
1600	To and incl. 40 HP motor size	3,050	1,325	4,375
1640	To and incl. 50 HP motor size	3,150	1,400	4,550
1680	To and incl. 60 HP motor size	5,475	1,600	7,075
1720	To and incl. 75 HP motor size	6,450	1,800	8,250
1760	To and incl. 100 HP motor size	7,375	2,000	9,375
1800	To and incl. 125 HP motor size	15,100	2,325	17,425
1840	To and incl. 150 HP motor size	16,100	2,650	18,750
1880	To and incl. 200 HP motor size	17,400	2,925	20,325
1960	460V, to and including 2 HP motor size	660	490	1,150
2000	To and incl. 5 HP motor size	695	535	1,230
2040	To and incl. 10 HP motor size	725	590	1,315
2080	To and incl. 15 HP motor size	950	675	1,625
2120	To and incl. 20 HP motor size	985	720	1,705
2160	To and incl. 25 HP motor size	985	760	1,745
2200	To and incl. 30 HP motor size	1,275	815	2,090

For expanded coverage of these items see *Means Electrical Cost Data 2000*

ELECTRICAL — A9.2 Lighting & Power

9.2-710 Motor Installation

		COST EACH		
		MAT.	INST.	TOTAL
2240	To and incl. 40 HP motor size	1,575	875	2,450
2280	To and incl. 50 HP motor size	1,750	975	2,725
2320	To and incl. 60 HP motor size	2,675	1,150	3,825
2360	To and incl. 75 HP motor size	3,050	1,250	4,300
2400	To and incl. 100 HP motor size	3,300	1,400	4,700
2440	To and incl. 125 HP motor size	5,600	1,600	7,200
2480	To and incl. 150 HP motor size	6,850	1,775	8,625
2520	To and incl. 200 HP motor size	7,875	2,025	9,900
2600	575V, to and including 2 HP motor size	660	490	1,150
2640	To and incl. 5 HP motor size	695	535	1,230
2680	To and incl. 10 HP motor size	725	590	1,315
2720	To and incl. 20 HP motor size	950	675	1,625
2760	To and incl. 25 HP motor size	985	720	1,705
2800	To and incl. 30 HP motor size	1,275	815	2,090
2840	To and incl. 50 HP motor size	1,350	845	2,195
2880	To and incl. 60 HP motor size	2,675	1,125	3,800
2920	To and incl. 75 HP motor size	2,675	1,150	3,825
2960	To and incl. 100 HP motor size	3,050	1,250	4,300
3000	To and incl. 125 HP motor size	5,550	1,575	7,125
3040	To and incl. 150 HP motor size	5,600	1,600	7,200
3080	To and incl. 200 HP motor size	6,950	1,800	8,750

9.2-720 Motor Feeder

		COST PER L.F.		
		MAT.	INST.	TOTAL
0200	Motor feeder systems, single phase, feed up to 115V 1HP or 230V 2 HP	1.45	3.42	4.87
0240	115V 2HP, 230V 3HP	1.52	3.47	4.99
0280	115V 3HP	1.67	3.61	5.28
0360	Three phase, feed to 200V 3HP, 230V 5HP, 460V 10HP, 575V 10HP	1.50	3.69	5.19
0440	200V 5HP, 230V 7.5HP, 460V 15HP, 575V 20HP	1.61	3.76	5.37
0520	200V 10HP, 230V 10HP, 460V 30HP, 575V 30HP	1.84	3.97	5.81
0600	200V 15HP, 230V 15HP, 460V 40HP, 575V 50HP	2.42	4.55	6.97
0680	200V 20HP, 230V 25HP, 460V 50HP, 575V 60HP	3.55	5.75	9.30
0760	200V 25HP, 230V 30HP, 460V 60HP, 575V 75HP	3.76	5.85	9.61
0840	200V 30HP	4.19	6.05	10.24
0920	230V 40HP, 460V 75HP, 575V 100HP	5.35	6.65	12
1000	200V 40HP	5.95	7.10	13.05
1080	230V 50HP, 460V 100HP, 575V 125HP	7.05	7.80	14.85
1160	200V 50HP, 230V 60HP, 460V 125HP, 575V 150HP	8.15	8.30	16.45
1240	200V 60HP, 460V 150HP	10.20	9.75	19.95
1320	230V 75HP, 575V 200HP	11.75	10.15	21.90
1400	200V 75HP	12.90	10.35	23.25
1480	230V 100HP, 460V 200HP	18.20	12.10	30.30
1560	200V 100HP	24.50	15.10	39.60
1640	230V 125HP	28	16.30	44.30
1720	200V 125HP, 230V 150HP	28	18.75	46.75
1800	200V 150HP	33	20.50	53.50
1880	200V 200HP	49.50	30	79.50
1960	230V 200HP	43	22.50	65.50

Important: See the Reference Section for critical supporting data - Reference Nos., Crews & Location Factors

ELECTRICAL | A9.4 Special Electrical

9.4-100 Communication & Alarm Systems

		COST EACH		
		MAT.	INST.	TOTAL
0200	Communication & alarm systems, includes outlets, boxes, conduit & wire			
0210	Sound system, 6 outlets	4,200	3,650	7,850
0220	12 outlets	5,925	5,850	11,775
0240	30 outlets	10,300	11,000	21,300
0280	100 outlets	31,000	36,900	67,900
0320	Fire detection systems, 12 detectors	2,225	3,025	5,250
0360	25 detectors	3,900	5,125	9,025
0400	50 detectors	7,475	10,100	17,575
0440	100 detectors	13,600	18,100	31,700
0480	Intercom systems, 6 stations	2,400	2,500	4,900
0560	25 stations	8,225	9,575	17,800
0640	100 stations	31,400	35,100	66,500
0680	Master clock systems, 6 rooms	3,775	4,075	7,850
0720	12 rooms	5,600	6,950	12,550
0760	20 rooms	7,500	9,850	17,350
0800	30 rooms	12,000	18,200	30,200
0840	50 rooms	19,100	30,600	49,700
0920	Master TV antenna systems, 6 outlets	2,125	2,575	4,700
0960	12 outlets	3,975	4,800	8,775
1000	30 outlets	7,875	11,100	18,975
1040	100 outlets	26,100	36,300	62,400

For expanded coverage of these items see *Means Electrical Cost Data 2000*

ELECTRICAL — A9.4 Special

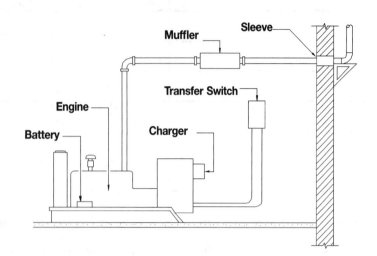

Description: System below tabulates the installed cost for generators by KW. Included in costs are battery, charger, muffler, and transfer switch.

No conduit, wire, or terminations included.

9.4-310	Generators (by kW)	COST PER KW		
		MAT.	INST.	TOTAL
0190	Generator sets, include battery, charger, muffler & transfer switch			
0200	Gas/gasoline operated, 3 phase, 4 wire, 277/480 volt, 7.5 kW	880	141	1,021
0240	11.5 kW	815	108	923
0280	20 kW	550	69.50	619.50
0320	35 kW	375	45	420
0360	80 kW	283	27.50	310.50
0400	100 kW	248	26.50	274.50
0440	125 kW	405	25	430
0480	185 kW	360	18.85	378.85
0560	Diesel engine with fuel tank, 30 kW	585	52.50	637.50
0600	50 kW	430	41.50	471.50
0640	75 kW	375	33.50	408.50
0680	100 kW	315	28.50	343.50
0760	150 kW	257	22.50	279.50
0840	200 kW	209	18.20	227.20
0880	250 kW	183	15.25	198.25
0960	350 kW	169	12.45	181.45
1040	500 kW	167	9.70	176.70

For information about Means Estimating Seminars, see yellow pages 11 and 12 in back of book

Division 11
Special Construction

SPECIAL CONSTR. | A11.1 Specialties

11.1-100 Architectural Specialties/Each

		COST EACH		
		MAT.	INST.	TOTAL
1000	Specialties, bathroom accessories, st. steel, curtain rod, 5' long, 1" diam	22.50	21	43.50
1100	Dispenser, towel, surface mounted	39.50	16.90	56.40
1200	Grab bar, 1-1/4" diam., 12" long	39	11.25	50.25
1300	Mirror, framed with shelf, 18" x 24"	100	13.50	113.50
1340	72" x 24"	325	45	370
1400	Toilet tissue dispenser, surface mounted, single roll	11.65	9	20.65
1500	Towel bar, 18" long	33	11.75	44.75
1600	Canopies, wall hung, prefinished aluminum, 8' x 10'	1,450	860	2,310
1700	Chutes, linen or refuse incl. sprinklers, galv. steel, 18" diam., per floor	810	166	976
1740	Aluminized steel, 36" diam., per floor	1,075	207	1,282
1800	Mail, 8-3/4" x 3-1/2", aluminum & glass, per floor	590	116	706
1840	Bronze or stainless, per floor	920	129	1,049
1900	Control boards, magnetic, porcelain finish, framed, 24" x 18"	160	67.50	227.50
1940	96" x 48"	715	108	823
2000	Directory boards, outdoor, black plastic, 36" x 24"	350	270	620
2040	36" x 36"	730	360	1,090
2100	Indoor, economy, open faced, 18" x 24"	102	77	179
2300	Disappearing stairways, folding, pine, 8'-6" ceiling	91.50	67.50	159
2400	Fire escape, galvanized steel, 8'-0" to 10'-6" ceiling	1,225	540	1,765
2500	Automatic electric, wood, 8' to 9' ceiling	6,075	540	6,615
2600	Display cases, freestanding, glass and aluminum, 3'-6" x 3' x 1'-0" deep	510	67.50	577.50
2700	Wall mounted, glass and aluminum, 3' x 4' x 1'-4" deep	1,050	108	1,158
2900	Fireplace prefabricated, freestanding or wall hung, economy	1,025	208	1,233
2940	Deluxe	2,950	300	3,250
3000	Woodburning stoves, cast iron, economy	840	415	1,255
3040	Deluxe	2,275	675	2,950
3100	Flagpoles, on grade, aluminum, tapered, 20' high	660	340	1,000
3140	70' high	6,500	855	7,355
3240	59' high	6,025	755	6,780
3300	Concrete, internal halyard, 20' high	1,375	273	1,648
3340	100' high	12,800	680	13,480
3400	Lockers, steel, single tier, 5' to 6' high, per opening, min.	103	20.50	123.50
3440	Maximum	167	24	191
3700	Mail boxes, horizontal, rear loaded, aluminum, 5" x 6" x 15" deep	38	7.95	45.95
3740	Front loaded, aluminum, 10" x 12" x 15" deep	134	13.50	147.50
3800	Vertical, front loaded, aluminum, 15" x 5" x 6" deep	32	7.95	39.95
3840	Bronze, duranodic finish	51	7.95	58.95
3900	Letter slot, post office	210	34	244
4000	Mail counter, window, post office, with grille	565	135	700
4100	Medicine cabinets, sliding mirror doors, 20" x 16" x 4-3/4", unlighted	93.50	38.50	132
4140	24" x 19" x 8-1/2", lighted	157	54	211
4400	Partitions, shower stall, single wall, painted steel, 2'-8" x 2'-8"	660	116	776
4440	Fiberglass, 2'-8" x 2'-8"	540	129	669
4500	Double wall, enameled steel, 2'-8" x 2'-8"	675	116	791
4540	Stainless steel, 2'-8" x 2'-8"	1,175	116	1,291
4700	Tub enclosure, sliding panels, tempered glass, aluminum frame	291	145	436
4740	Chrome/brass frame-deluxe	830	194	1,024
5400	Projection screens, wall hung, manual operation, 50 S.F., economy	243	54	297
5440	Electric operation, 100 S.F., deluxe	1,650	270	1,920
5500	Scales, dial type, built in floor, 5 ton capacity, 8' x 6' platform	4,825	1,625	6,450
5540	10 ton capacity, 9' x 7' platform	9,175	2,325	11,500
5600	Truck (including weigh bridge), 20 ton capacity, 24' x 10'	9,225	2,700	11,925
5900	Security gates-scissors type, painted steel, single, 6' high, 5-1/2' wide	119	168	287
5940	Double gate, 7-1/2' high, 14' wide	485	335	820
6000	Shelving, metal industrial, braced, 3' wide, 1' deep	29	5.75	34.75
6040	2' deep	45.50	6.10	51.60
6200	Signs, interior electric exit sign, wall mounted, 6"	52.50	36	88.50
6240	Street, reflective alum., dbl. face, 4 way, w/bracket	78	18	96

SPECIAL CONSTR. — A11.1 Specialties

11.1-100 Architectural Specialties/Each

Line	Description	MAT.	INST.	TOTAL
6300	Letters, cast aluminum, 1/2" deep, 4" high	17.20	15	32.20
6340	1" deep, 10" high	42	15	57
6400	Plaques, cast aluminum, 20" x 30"	830	135	965
6440	Cast bronze, 36" x 48"	3,250	270	3,520
6700	Turnstiles, one way, 4' arm, 46" diam., manual	280	108	388
6740	Electric	935	450	1,385
6800	3 arm, 5'-5' diam. & 7' high, manual	2,525	540	3,065
6840	Electric	3,125	900	4,025

11.1-100 Architectural Specialties/S.F.

Line	Description	MAT.	INST.	TOTAL
7100	Bulletin board, cork sheets, no frame, 1/4" thick	6.20	1.86	8.06
7200	Aluminum frame, 1/4" thick, 3' x 5'	7.15	2.28	9.43
7500	Chalkboards, wall hung, alum, frame & chalktrough	9.75	1.19	10.94
7540	Wood frame & chalktrough	9	1.29	10.29
7600	Sliding board, one board with back panel	31	1.15	32.15
7640	Two boards with back panel	48	1.15	49.15
7700	Liquid chalk type, alum. frame & chalktrough	10.85	1.19	12.04
7740	Wood frame & chalktrough	18.45	1.19	19.64

11.1-100 Architectural Specialties/L.F.

Line	Description	MAT.	INST.	TOTAL
8600	Partitions, hospital curtain, ceiling hung, polyester oxford cloth	12.70	2.64	15.34
8640	Designer oxford cloth	24.50	3.34	27.84

11.1-200 Architectural Equipment/Each

Line	Description	MAT.	INST.	TOTAL
0200	Arch. equip., appliances, range, cook top, 4 burner, economy	189	48	237
0240	Built in, single oven 30" wide, economy	285	48	333
0300	Standing, single oven-21" wide, economy	310	39.50	349.50
0340	Double oven-30" wide, deluxe	1,650	39.50	1,689.50
0400	Compactor, residential, economy	370	54	424
0440	Deluxe	405	90	495
0500	Dish washer, built-in, 2 cycles, economy	271	90.50	361.50
0540	4 or more cycles, deluxe	545	181	726
0600	Garbage disposer, sink type, economy	41	36.50	77.50
0640	Deluxe	191	36.50	227.50
0700	Refrigerator, no frost, 10 to 12 C.F., economy	465	39.50	504.50
0740	21 to 29 C.F., deluxe	2,850	132	2,982
0800	Automotive equipment, compressors, electric, 1-1/2 H.P., std. controls	355	315	670
0840	5 H.P., dual controls	1,800	475	2,275
0900	Hoists, single post, 4 ton capacity, swivel arms	3,850	1,175	5,025
0940	Dual post, 12 ton capacity, adjustable frame	6,600	3,150	9,750
1000	Lube equipment, 3 reel type, with pumps	6,600	950	7,550
1040	Product dispenser, 6 nozzles, w/vapor recovery, not incl. piping, installed	16,500		16,500
1100	Bank equipment, drive up window, drawer & mike, no glazing, economy	4,625	670	5,295
1140	Deluxe	7,625	1,350	8,975
1200	Night depository, economy	6,800	670	7,470
1240	Deluxe	10,800	1,350	12,150
1300	Pneumatic tube systems, 2 station, standard	20,800	1,375	22,175
1340	Teller, automated, 24 hour, single unit	39,200	1,375	40,575
1400	Teller window, bullet proof glazing, 44" x 60"	3,000	430	3,430
1440	Pass through, painted steel, 72" x 40"	2,575	840	3,415
1500	Barber equipment, chair, hydraulic, economy	450	11.25	461.25
1540	Deluxe	2,975	16.90	2,991.90
1700	Church equipment, altar, wood, custom, plain	1,725	193	1,918
1740	Granite, custom, deluxe	22,000	2,700	24,700
1800	Baptistry, fiberglass, economy	2,525	620	3,145
1840	Bells & carillons, keyboard operation	10,200	4,150	14,350

For expanded coverage of these items see *Means Assemblies Cost Data 2000*

SPECIAL CONSTR. | A11.1 | Specialties

11.1-200 Architectural Equipment/Each

		COST EACH		
		MAT.	INST.	TOTAL
1900	Confessional, wood, single, economy	2,200	450	2,650
1940	Double, deluxe	13,800	1,350	15,150
2000	Steeples, translucent fiberglas, 30" square, 15' high	3,300	860	4,160
2100	Checkout counter, single belt	2,125	39.50	2,164.50
2140	Double belt, power take-away	3,625	44	3,669
2600	Dental equipment, central suction system, economy	2,975	242	3,217
2640	Compressor-air, deluxe	22,700	540	23,240
2700	Chair, hydraulic, economy	3,750	540	4,290
2740	Deluxe	7,925	1,075	9,000
2800	Drill console with accessories, economy	3,375	170	3,545
2840	Deluxe	9,625	170	9,795
2900	X-ray unit, portable	4,500	68	4,568
2940	Panoramic unit	20,400	450	20,850
3000	Detention equipment, cell front rolling door, 7/8" bars, 5' x 7' high	4,600	730	5,330
3040	Cells, prefab., including front, 5' x 7' x 7' deep	8,150	975	9,125
3100	Doors and frames, 3' x 7', single plate	3,425	370	3,795
3140	Double plate	4,225	370	4,595
3200	Toilet apparatus, incl wash basin	2,275	410	2,685
3240	Visitor cubicle, vision panel, no intercom	2,600	730	3,330
3300	Dock bumpers, rubber blocks, 4-1/2" thick, 10" high, 14" long	42	10.40	52.40
3340	6" thick, 20" high, 11" long	109	21	130
3400	Dock boards, H.D., 5' x 5', aluminum, 5000 lb. capacity	1,125		1,125
3440	16,000 lb. capacity	1,375		1,375
3500	Dock levelers, hydraulic, 7' x 8', 10 ton capacity	6,550	735	7,285
3540	Dock lifters, platform, 6' x 6', portable, 3000 lb. capacity	6,425		6,425
3600	Dock shelters, truck, scissor arms, economy	1,075	270	1,345
3640	Deluxe	1,625	540	2,165
3700	Kitchen equipment, bake oven, single deck	4,025	65.50	4,090.50
3740	Broiler, without oven	3,875	65.50	3,940.50
3800	Commercial dish washer, semiautomatic, 50 racks/hr.	5,775	400	6,175
3840	Automatic, 275 racks/hr.	23,800	1,500	25,300
3900	Cooler, beverage, reach-in, 6 ft. long	3,000	87	3,087
3940	Food warmer, counter, 1.65 kw	700		700
4000	Fryers, with submerger, single	3,125	74.50	3,199.50
4040	Double	4,250	104	4,354
4100	Kettles, steam jacketed, 20 gallons	5,875	107	5,982
4140	Range, restaurant type, burners, 2 ovens and 24" griddle	4,800	87	4,887
4200	Range hood, incl. carbon dioxide system, economy	3,100	174	3,274
4240	Deluxe	22,100	520	22,620
4300	Laboratory equipment, glassware washer, distilled water, economy	5,275	202	5,477
4340	Deluxe	7,425	365	7,790
4440	Radio isotope	12,100		12,100
4600	Laundry equipment, dryers, gas fired, residential, 16 lb. capacity	560	97	657
4640	Commercial, 30 lb. capacity, single	2,550	97	2,647
4700	Dry cleaners, electric, 20 lb. capacity	30,700	1,825	32,525
4740	30 lb. capacity	42,100	2,425	44,525
4800	Ironers, commercial, 120" with canopy, 8 roll	147,500	5,175	152,675
4840	Institutional, 110", single roll	26,200	1,450	27,650
4900	Washers, residential, 4 cycle	645	97	742
4940	Commercial, coin operated, deluxe	2,725	97	2,822
5000	Library equipment, carrels, metal, economy	204	54	258
5040	Hardwood, deluxe	810	67.50	877.50
5100	Medical equipment, autopsy table, standard	6,925	290	7,215
5140	Deluxe	8,675	485	9,160
5200	Incubators, economy	2,875		2,875
5240	Deluxe	15,700		15,700
5300	Station, scrub-surgical, single, economy	1,075	97	1,172
5340	Dietary, medium, with ice	12,700		12,700
5400	Sterilizers, general purpose, single door, 20" x 20" x 28"	12,100		12,100
5440	Floor loading, double door, 28" x 67" x 52"	161,500		161,500

Important: See the Reference Section for critical supporting data - Reference Nos., Crews & Location Factors

SPECIAL CONSTR. | A11.1 | Specialties

11.1-200 Architectural Equipment/Each

		COST EACH		
		MAT.	INST.	TOTAL
5500	Surgery tables, standard	11,000	480	11,480
5540	Deluxe	24,900	670	25,570
5600	Tables, standard, with base cabinets, economy	1,025	180	1,205
5640	Deluxe	3,050	270	3,320
5700	X-ray, mobile, economy	10,600		10,600
5740	Stationary, deluxe	172,500		172,500
5800	Movie equipment, changeover, economy	385		385
5840	Film transport, incl. platters and autowind, economy	5,500		5,500
5900	Lamphouses, incl. rectifiers, xenon, 1000W	6,475	145	6,620
5940	4000W	10,600	193	10,793
6000	Projector mechanisms, 35 mm, economy	10,900		10,900
6040	Deluxe	14,500		14,500
6100	Sound systems, incl. amplifier, single, economy	2,750	320	3,070
6140	Dual, Dolby/super sound	13,800	725	14,525
6200	Seating, painted steel, upholstered, economy	110	15.45	125.45
6240	Deluxe	260	19.30	279.30
6300	Parking equipment, automatic gates, 8 ft. arm, one way	2,850	525	3,375
6340	Traffic detectors, single treadle	2,100	241	2,341
6400	Booth for attendant, economy	4,500		4,500
6440	Deluxe	22,600		22,600
6500	Ticket printer/dispenser, rate computing	6,625	415	7,040
6540	Key station on pedestal	365	141	506
6700	Frozen food, chest type, 12 ft. long	5,050	160	5,210
7000	Safe, office type, 1 hr. rating, 34" x 20" x 20"	2,200		2,200
7040	4 hr. rating, 62" x 33" x 20"	8,525		8,525
7100	Data storage, 4 hr. rating, 63" x 44" x 16"	12,600		12,600
7140	Jewelers, 63" x 44" x 16"	21,600		21,600
7200	Money, "B" label, 9" x 14" x 14"	490		490
7240	Tool and torch resistive, 24" x 24" x 20"	8,450	99	8,549
7300	Sauna, prefabricated, incl. heater and controls, 7' high, 6' x 4'	3,525	340	3,865
7340	10' x 12'	7,925	750	8,675
7400	Heaters, wall mounted, to 200 C.F.	485		485
7440	Floor standing, to 1000 C.F., 12500 W	1,725	96.50	1,821.50
7500	School equipment, basketball backstops, wall mounted, wood, fixed	590	475	1,065
7540	Suspended type, electrically operated	3,200	895	4,095
7600	Bleachers-telescoping, manual operation, 15 tier, economy (per seat)	56	14.55	70.55
7640	Power operation, 30 tier, deluxe (per seat)	213	26	239
7700	Weight lifting gym, universal, economy	3,100	395	3,495
7740	Deluxe	13,500	790	14,290
7800	Scoreboards, basketball, 1 side, economy	2,700	425	3,125
7840	4 sides, deluxe	29,500	5,825	35,325
8300	Vocational shop equipment, benches, metal	248	108	356
8340	Wood	440	108	548
8400	Dust collector, not incl. ductwork, 6' diam.	2,625	264	2,889
8440	Planer, 13" x 6"	1,150	135	1,285
8500	Waste handling, compactors, single bag, 250 lbs./hr., hand fed	6,250	198	6,448
8540	Heavy duty industrial, 5 C.Y. capacity	15,400	950	16,350
8600	Incinerator, electric, 100 lbs./hr., economy	10,000	1,500	11,500
8640	Gas, 2000 lbs./hr., deluxe	171,000	10,400	181,400
8700	Shredder, no baling, 35 tons/hr.	211,000		211,000
8740	Incl. baling, 50 tons/day	422,000		422,000

For expanded coverage of these items see Means Assemblies Cost Data 2000

SPECIAL CONSTR. — A11.1 Specialties

11.1-200 Architectural Equipment/S.F.

		COST PER S.F.		
		MAT.	INST.	TOTAL
8910	Arch. equip., lab equip., counter tops, acid proof, economy	16.55	6.60	23.15
8940	Stainless steel	66	6.60	72.60
9000	Movie equipment, projection screens, rigid in wall, acrylic, 1/4" thick	35	2.63	37.63
9040	1/2" thick	40.50	3.95	44.45
9100	School equipment, gym mats, naugahyde cover, 2" thick	3.41		3.41
9140	Wrestling, 1" thick, heavy duty	5.85		5.85
9200	Stage equipment, curtains, velour, medium weight	6.60	.90	7.50
9240	Silica based yarn, fireproof	13.20	10.80	24
9300	Stages, portable with steps, folding legs, 8" high	8.25		8.25
9340	Telescoping platforms, aluminum, deluxe	47.50	14.05	61.55

11.1-200 Architectural Equipment/L.F.

		COST PER L.F.		
		MAT.	INST.	TOTAL
9410	Arch. equip., church equip. pews, bench type, hardwood, economy	60.50	13.50	74
9440	Deluxe	121	18	139
9500	Laboratory equipment, cabinets, wall, open	77	27	104
9540	Base, drawer units	269	30	299
9600	Fume hoods, not incl. HVAC, economy	555	100	655
9640	Deluxe incl. fixtures	1,300	225	1,525
9700	Library equipment, book shelf, metal, single face, 90" high x 10" shelf	71.50	22.50	94
9740	Double face, 90" high x 10" shelf	204	49	253
9800	Charging desk, built-in, with counter, plastic laminate	435	38.50	473.50
9900	Stage equipment, curtain track, heavy duty	41.50	30	71.50
9940	Lights, border, quartz, colored	152	14.45	166.45

11.1-500 Furnishings/Each

		COST EACH		
		MAT.	INST.	TOTAL
1000	Furnishings, blinds, exterior, aluminum, louvered, 1'-4" wide x 3'-0" long	32.50	27	59.50
1040	1'-4" wide x 6'-8" long	52.50	30	82.50
1100	Hemlock, solid raised, 1'-4" wide x 3'-0" long	57	27	84
1140	1'-4" wide x 6'-9" long	107	30	137
1200	Polystyrene, louvered, 1'-3" wide x 3'-3" long	44	27	71
1240	1'-3" wide x 6'-8" long	95.50	30	125.50
1300	Interior, wood folding panels, louvered, 7" x 20" (per pair)	39.50	15.90	55.40
1340	18" x 40" (per pair)	123	15.90	138.90
1800	Hospital furniture, beds, manual, economy	890		890
1840	Deluxe	1,550		1,550
1900	All electric, economy	1,475		1,475
1940	Deluxe	3,700		3,700
2000	Patient wall systems, no utilities, economy, per room	895		895
2040	Deluxe, per room	1,650		1,650
2200	Hotel furnishings, standard room set, economy, per room	1,625		1,625
2240	Deluxe, per room	7,875		7,875
2400	Office furniture, standard employee set, economy, per person	360		360
2440	Deluxe, per person	2,125		2,125
2800	Posts, portable, pedestrian traffic control, economy	97.50		97.50
2840	Deluxe	315		315
3000	Restaurant furniture, booth, molded plastic, stub wall and 2 seats, economy	340	135	475
3040	Deluxe	1,500	180	1,680
3100	Upholstered seats, foursome, single-economy	1,150	54	1,204
3140	Foursome, double-deluxe	2,400	90	2,490
3300	Seating, lecture hall, pedestal type, economy	112	24.50	136.50
3340	Deluxe	400	37	437
3400	Auditorium chair, veneer construction	126	24.50	150.50
3440	Fully upholstered, spring seat	161	24.50	185.50

SPECIAL CONSTR.		A11.1	Furnishings			
11.1-500		**Furnishings/S.F.**		COST PER S.F.		
				MAT.	INST.	TOTAL
4010	Furnishings, blinds-interior, venetian-aluminum, stock, 2" slats, economy			1.65	.46	2.11
4040	Custom, 1" slats, deluxe			7.15	.61	7.76
4100	Vertical, PVC or cloth, T&B track, economy			7.25	.59	7.84
4140	Deluxe			9.60	.68	10.28
4300	Draperies, unlined, economy			2.20		2.20
4440	Lightproof, deluxe			6.60		6.60
4700	Floor mats, recessed, inlaid black rubber, 3/8" thick, solid			21	1.28	22.28
4740	Colors, 1/2" thick, perforated			35	1.28	36.28
4800	Link-including nosings, steel-galvanized, 3/8" thick			7.05	1.28	8.33
4840	Vinyl, in colors			18.90	1.28	20.18
5000	Shades, mylar, wood roller, single layer, non-reflective			5.50	.39	5.89
5040	Metal roller, triple layer, heat reflective			7.80	.39	8.19
5100	Vinyl, light weight, 4 ga.			.41	.39	.80
5140	Heavyweight, 6 ga.			1.27	.39	1.66
5200	Vinyl coated cotton, lightproof decorator shades			1.13	.39	1.52
5300	Woven aluminum, 3/8" thick, light and fireproof			4.17	.77	4.94

11.1-500	**Furnishings/L.F.**	COST PER L.F.		
		MAT.	INST.	TOTAL
5510	Furnishings, cabinets, hospital, base, laminated plastic	198	54	252
5540	Stainless steel	248	54	302
5600	Counter top, laminated plastic, no backsplash	34	13.50	47.50
5640	Stainless steel	95	13.50	108.50
5700	Nurses station, door type, laminated plastic	230	54	284
5740	Stainless steel	265	54	319
5900	Household, base, hardwood, one top drawer & one door below x 12" wide	138	22	160
5940	Four drawer x 24" wide	390	24	414
6000	Wall, hardwood, 30" high with one door x 12" wide	118	24.50	142.50
6040	Two doors x 48" wide	286	29.50	315.50
6100	Counter top-laminated plastic, stock, economy	5.40	9	14.40
6140	Custom-square edge, 7/8" thick	11.35	20	31.35
6300	School, counter, wood, 32" high	157	27	184
6340	Metal, 84" high	243	36	279
6500	Dormitory furniture, desk top (built-in), laminated plastc, 24"deep, economy	25.50	10.80	36.30
6540	30" deep, deluxe	141	13.50	154.50
6600	Dressing unit, built-in, economy	189	45	234
6640	Deluxe	565	67.50	632.50
7000	Restaurant furniture, bars, built-in, back bar	151	54	205
7040	Front bar	208	54	262
7200	Wardrobes & coatracks, standing, steel, single pedestal, 30" x 18" x 63"	79.50		79.50
7300	Double face rack, 39" x 26" x 70"	134		134
7340	Wall mounted rack, steel frame & shelves, 12" x 15" x 26"	56.50	3.84	60.34
7400	12" x 15" x 50"	44.50	2.01	46.51

For expanded coverage of these items see *Means Assemblies Cost Data 2000*

SPECIAL CONSTR. | A11.1 | Specialties

11.1-700 Special Construction/Each

		COST EACH		
		MAT.	INST.	TOTAL
1000	Special construction, bowling alley incl. pinsetter, scorer etc., economy	35,500	5,400	40,900
1040	Deluxe	40,400	6,000	46,400
1100	For automatic scorer, economy, add	6,600		6,600
1140	Deluxe, add	7,475		7,475
1200	Chimney, metal, high temp. steel jacket, factory lining, 24" diameter	9,550	2,400	11,950
1240	60" diameter	69,500	10,400	79,900
1300	Poured concrete, brick lining, 10' diam., 200' high			1,175,000
1340	20' diam., 500' high			5,250,000
1400	Radial brick, 3'-6" I.D., 75' high			191,500
1440	7' I.D., 175' high			577,500
1500	Control tower, modular, 12' x 10', incl. instrumentation, economy			450,000
1540	Deluxe			600,000
1600	Garage costs, residential, prefab, wood, single car economy	2,800	540	3,340
1640	Two car deluxe	8,400	1,075	9,475
1700	Grandstands, permanent, closed deck, steel, economy (per seat)			25
1740	Deluxe (per seat)			80
1800	Composite design, economy (per seat)			30
1840	Deluxe (per seat)			90
1900	Hangars, prefab, galv. steel, bottom rolling doors, economy (per plane)	9,250	2,475	11,725
1940	Electrical bi-folding doors, deluxe (per plane)	12,900	3,400	16,300
2000	Ice skating rink, 85' x 200', 55° system, 5 mos., 100 ton			325,000
2040	90° system, 12 mos., 135 ton			253,000
2100	Dash boards, acrylic screens, polyethylene coated plywood	126,500	15,800	142,300
2140	Fiberglass and aluminum construction	165,000	15,800	180,800
2200	Integrated ceilings, radiant electric, 2' x 4' panel, manila finish	88	11.55	99.55
2240	ABS plastic finish	103	22.50	125.50
2300	Kiosks, round, 5' diam., 1/4" fiberglass wall, 8' high	6,050		6,050
2340	Rectangular, 5' x 9', 1" insulated dbl. wall fiberglass, 7'-6" high	10,500		10,500
2400	Portable booth, acoustical, 27 db 1000 hz., 15 S.F. floor	3,175		3,175
2440	55 S.F. flr.	6,500		6,500
2500	Radio towers, guyed, 40 lb. section, 50' high, 70 MPH basic wind speed	1,725	670	2,395
2540	90 lb. section, 400' high, wind load 70 MPH basic wind speed	22,400	7,950	30,350
2600	Self supporting, 60' high, 70 MPH basic wind speed	3,600	1,400	5,000
2640	190' high, wind load 90 MPH basic wind speed	21,400	5,575	26,975
2700	Shelters, aluminum frame, acrylic glazing, 8' high, 3' x 9'	2,800	590	3,390
2740	9' x 12'	4,675	920	5,595
2800	Shielding, lead x-ray protection, radiography room, 1/16" lead, economy	5,050	2,025	7,075
2840	Deluxe	6,600	3,350	9,950
2900	Deep therapy x-ray, 1/4" lead, economy	17,100	6,300	23,400
2940	Deluxe	23,100	8,375	31,475
3000	Shooting range incl. bullet traps, controls, separators, ceilings, economy	12,000	3,500	15,500
3040	Deluxe	22,300	5,900	28,200
3100	Silos, concrete stave, industrial, 12' diam., 35' high	17,300	11,300	28,600
3140	25' diam., 75' high	56,500	24,800	81,300
3200	Steel prefab, 30,000 gal., painted, economy	12,100	2,950	15,050
3240	Epoxy-lined, deluxe	25,100	5,850	30,950
3300	Sport court, squash, regulation, in existing building, economy			14,500
3340	Deluxe			27,100
3400	Racketball, regulation, in existing building, economy	1,925	4,700	6,625
3440	Deluxe	25,700	9,400	35,100
3500	Swimming pool equipment, diving stand, stainless steel, 1 meter	2,925	200	3,125
3540	3 meter	5,175	1,350	6,525
3600	Diving boards, 16 ft. long, aluminum	2,325	200	2,525
3640	Fiberglass	1,725	200	1,925
3700	Filter system, sand, incl. pump, 6000 gal./hr.	990	325	1,315
3800	Lights, underwater, 12 volt with transformer, 300W	165	725	890
3900	Slides, fiberglass with aluminum handrails & ladder, 6' high, straight	1,125	340	1,465
3940	12' high, straight with platform	1,050	450	1,500
4000	Tanks, steel, ground level, 100,000 gal.			98,000
4040	10,000,000 gal.			1,945,000

SPECIAL CONSTR.   A11.1 Specialties

11.1-700 Special Construction/Each

		COST EACH		
		MAT.	INST.	TOTAL
4100	Elevated water, 50,000 gal.		81,500	163,000
4140	1,000,000 gal.		398,500	797,000
4200	Cypress wood, ground level, 3,000 gal.	5,775	4,950	10,725
4240	Redwood, ground level, 45,000 gal.	44,600	13,400	58,000

11.1-700 Special Construction/S.F.

		COST PER S.F.		
		MAT.	INST.	TOTAL
4500	Special construction, acoustical, enclosure, 4" thick, 8 psf panels	27	11.25	38.25
4540	Reverb chamber, 4" thick, parallel walls	33	13.50	46.50
4600	Sound absorbing panels, 2'-6" x 8', painted metal	9.95	3.77	13.72
4640	Vinyl faced	8	3.38	11.38
4700	Flexible transparent curtain, clear	6.15	4.05	10.20
4740	With absorbing foam, 75% coverage	8.60	4.05	12.65
4800	Strip entrance, 2/3 overlap	4.10	6.45	10.55
4840	Full overlap	5.05	7.60	12.65
4900	Audio masking system, plenum mounted, over 10,000 S.F.	.59	.13	.72
4940	Ceiling mounted, under 5,000 S.F.	1.06	.24	1.30
5000	Air supported struc., polyester vinyl fabric, 24oz., warehouse, 5000 S.F.	16.95	.16	17.11
5040	50,000 S.F.	6.80	.13	6.93
5100	Tennis, 7,200 S.F.	14.30	.13	14.43
5140	24,000 S.F.	9.35	.13	9.48
5200	Woven polyethylene, 6 oz., shelter, 3,000 S.F.	6.50	.26	6.76
5240	24,000 S.F.	2.96	.13	3.09
5300	Teflon coated fiberglass, stadium cover, economy	38.50	.07	38.57
5340	Deluxe	44	.09	44.09
5400	Air supported storage tank covers, reinf. vinyl fabric, 12 oz., 400 S.F.	13.20	.23	13.43
5440	18,000 S.F.	3.91	.20	4.11
5500	Anechoic chambers, 7' high, 100 cps cutoff, 25 S.F.			2,000
5540	200 cps cutoff, 100 S.F.			1,025
5600	Audiometric rooms, under 500 S.F.	44	11	55
5640	Over 500 S.F.	42	9	51
5700	Comfort stations, prefab, mobile on steel frame, economy	41		41
5740	Permanent on concrete slab, deluxe	220	22	242
5800	Darkrooms, shell, not including door, 240 S.F., 8' high	37.50	4.50	42
5840	64 S.F., 12' high	86.50	8.45	94.95
5900	Domes, bulk storage, wood framing, wood decking, 50' diam.	27.50	.86	28.36
5940	116' diam.	16.50	.99	17.49
6000	Steel framing, metal decking, 150' diam.	29.50	5.65	35.15
6040	400' diam.	24	4.31	28.31
6100	Geodesic, wood framing, wood panels, 30' diam.	15.55	.84	16.39
6140	60' diam.	12.90	.61	13.51
6200	Aluminum framing, acrylic panels, 40' diam.			80
6240	Aluminum panels, 400' diam.			25
6300	Garden house, prefab, wood, shell only, 48 S.F.	14.80	2.70	17.50
6340	200 S.F.	27	11.25	38.25
6400	Greenhouse, shell-stock, residential, lean-to, 8'-6" long x 3'-10" wide	34	15.90	49.90
6440	Freestanding, 8'-6" long x 13'-6" wide	26.50	5	31.50
6500	Commercial-truss frame, under 2000 S.F., deluxe			38
6540	Over 5,000 S.F., economy			23
6600	Institutional-rigid frame, under 500 S.F., deluxe			99
6640	Over 2,000 S.F., economy			39
6700	Hangar, prefab, galv. roof and walls, bottom rolling doors, economy	8.85	2.24	11.09
6740	Electric bifolding doors, deluxe	10.45	2.92	13.37
6800	Integrated ceilings, Luminaire, suspended, 5' x 5' modules, 50% lighted	3.88	3.80	7.68
6840	100% lighted	5.80	6.85	12.65
6900	Dimensionaire, 2' x 4' module tile system, no air bar	2.35	1.08	3.43
6940	With air bar, deluxe	3.40	1.08	4.48
7000	Music practice room, modular, perforated steel, under 500 S.F.	23.50	7.70	31.20
7040	Over 500 S.F.	17.95	6.75	24.70

For expanded coverage of these items see *Means Assemblies Cost Data 2000*

SPECIAL CONSTR. | A11.1 | Specialties

11.1-700	Special Construction/S.F.	COST PER S.F.		
		MAT.	INST.	TOTAL
7100	Pedestal access floors, stl pnls, no stringers, vinyl covering, >6000 S.F.	11.25	1.20	12.45
7140	With stringers, high pressure laminate covering, under 6000 S.F.	9.70	1.35	11.05
7200	Carpet covering, under 6000 S.F.	11.30	1.35	12.65
7240	Aluminum panels, no stringers, no covering	23	1.08	24.08
7300	Refrigerators, prefab, walk-in, 7'-6" high, 6' x 6'	107	9.85	116.85
7340	12' x 20'	67	4.93	71.93
7400	Shielding, lead, gypsum board, 5/8" thick, 1/16" lead	5.25	3.21	8.46
7440	1/8" lead	10.75	3.67	14.42
7500	Lath, 1/16" thick	5.05	3.73	8.78
7540	1/8" thick	10.50	4.19	14.69
7600	Radio frequency, copper, prefab screen type, economy	26.50	3	29.50
7640	Deluxe	34	3.72	37.72
7700	Swimming pool enclosure, transluscent, freestanding, economy	11	2.70	13.70
7740	Deluxe	275	7.70	282.70
7800	Swimming pools, residential, vinyl liner, metal sides	9.55	3.64	13.19
7840	Concrete sides	11.45	6.70	18.15
7900	Gunite shell, plaster finish, 350 S.F.	18.95	13.80	32.75
7940	800 S.F.	15.25	8	23.25
8000	Motel, gunite shell, plaster finish	23.50	17.45	40.95
8040	Municipal, gunite shell, tile finish, formed gutters	97.50	30	127.50
8100	Tension structures, steel frame, polyester vinyl fabric, 12,000 S.F.	9.65	1.22	10.87
8140	20,800 S.F.	10.35	1.10	11.45

11.1-700	Special Construction/L.F.	COST PER L.F.		
		MAT.	INST.	TOTAL
8500	Spec. const., air curtains, shipping & receiving, 8'high x 5'wide, economy	203	116	319
8540	20' high x 8' wide, heated, deluxe	625	223	848
8600	Customer entrance, 10' high x 5' wide, economy	206	116	322
8640	12' high x 4' wide, heated, deluxe	315	207	522

For information about Means Estimating Seminars, see yellow pages 11 and 12 in back of book

Division 12
Site Work

SITE WORK — A12.3 Utilities

12.3-110 Trenching

		COST PER L.F.		
		MAT.	INST.	TOTAL
1310	Trenching, backhoe, 0 to 1 slope, 2' wide, 2' deep, 3/8 C.Y. bucket		2.04	2.04
1330	4' deep, 3/8 C.Y. bucket		3.43	3.43
1360	10' deep, 1 C.Y. bucket		6.20	6.20
1400	4' wide, 2' deep, 3/8 C.Y. bucket		4.19	4.19
1420	4' deep, 1/2 C.Y. bucket		5.90	5.90
1450	10' deep, 1 C.Y. bucket		12.70	12.70
1480	18' deep, 2-1/2 C.Y. bucket		19.70	19.70
1520	6' wide, 6' deep, 5/8 C.Y. bucket		12.85	12.85
1540	10' deep, 1 C.Y. bucket		17.05	17.05
1570	20' deep, 3-1/2 C.Y. bucket		27	27
1640	8' wide, 12' deep, 1-1/4 C.Y. bucket		22.50	22.50
1680	24' deep, 3-1/2 C.Y. bucket		60	60
1730	10' wide, 20' deep, 3-1/2 C.Y. bucket		48	48
1740	24' deep, 3-1/2 C.Y. bucket		76	76
3500	1 to 1 slope, 2' wide, 2' deep, 3/8 C.Y. bucket		3.14	3.14
3540	4' deep, 3/8 C.Y. bucket		7.10	7.10
3600	10' deep, 1 C.Y. bucket		26.50	26.50
3800	4' wide, 2' deep, 3/8 C.Y. bucket		5.30	5.30
3840	4' deep, 1/2 C.Y. bucket		8.60	8.60
3900	10' deep, 1 C.Y. bucket		28.50	28.50
4030	6' wide, 6' deep, 5/8 C.Y. bucket		16.85	16.85
4050	10' deep, 1 C.Y. bucket		34.50	34.50
4080	20' deep, 3-1/2 C.Y. bucket		88	88
4500	8' wide, 12' deep, 1-1/4 C.Y. bucket		42.50	42.50
4650	24' deep, 3-1/2 C.Y. bucket		182	182
4800	10' wide, 20' deep, 3-1/2 C.Y. bucket		108	108
4850	24' deep, 3-1/2 C.Y. bucket		196	196

12.3-310 Pipe Bedding

		COST PER L.F.		
		MAT.	INST.	TOTAL
1440	Pipe bedding, side slope 0 to 1, 1' wide, pipe size 6" diameter	.26	.40	.66
1460	2' wide, pipe size 8" diameter	.57	.87	1.44
1500	Pipe size 12" diameter	.60	.91	1.51
1600	4' wide, pipe size 20" diameter	1.49	2.26	3.75
1660	Pipe size 30" diameter	1.58	2.40	3.98
1680	6' wide, pipe size 32" diameter	2.77	4.19	6.96
1740	8' wide, pipe size 60" diameter	4.62	7	11.62
1780	12' wide, pipe size 84" diameter	9.05	13.70	22.75

12.3-710 Manholes & Catch Basins

		COST PER EACH		
		MAT.	INST.	TOTAL
1920	Manhole/catch basin, brick, 4' I.D. riser, 4' deep	680	840	1,520
1980	10' deep	1,475	1,950	3,425
3200	Block, 4' I.D. riser, 4' deep	720	675	1,395
3260	10' deep	1,475	1,625	3,100
4620	Concrete, cast-in-place, 4' I.D. riser, 4' deep	765	1,175	1,940
4680	10' deep	1,625	2,800	4,425
5820	Concrete, precast, 4' I.D. riser, 4' deep	735	610	1,345
5880	10' deep	1,425	1,425	2,850
6200	6' I.D. riser, 4' deep	1,300	900	2,200
6260	10' deep	2,625	2,125	4,750

SITE WORK — A12.5 Roads & Parking

12.5-110 Roadway Pavement

Line	Description	MAT.	INST.	TOTAL
1500	Roadways, bituminous conc. paving, 2-1/2" thick, 20' wide	43	32	75
1580	30' wide	49	36.50	85.50
1800	3" thick, 20' wide	44.50	31	75.50
1880	30' wide	51.50	36.50	88
2100	4" thick, 20' wide	47.50	31	78.50
2180	30' wide	55.50	36.50	92
2400	5" thick, 20' wide	53.50	31	84.50
2480	30' wide	65	37.50	102.50
3000	6" thick, 20' wide	72.50	32	104.50
3080	30' wide	93.50	38.50	132
3300	8" thick 20' wide	88.50	32	120.50
3380	30' wide	117	39	156
3600	12" thick 20' wide	106	34.50	140.50
3700	32' wide	151	43.50	194.50

Cost per L.F.

12.5-510 Parking Lots

Line	Description	MAT.	INST.	TOTAL
1500	Parking lot, 90° angle parking, 3" bituminous paving, 6" gravel base	244	176	420
1540	10" gravel base	265	206	471
1560	4" bituminous paving, 6" gravel base	284	175	459
1600	10" gravel base	305	205	510
1620	6" bituminous paving, 6" gravel base	380	182	562
1660	10" gravel base	400	213	613
1800	60° angle parking, 3" bituminous paving, 6" gravel base	279	192	471
1840	10" gravel base	305	227	532
1860	4" bituminous paving, 6" gravel base	325	191	516
1900	10" gravel base	350	226	576
1920	6" bituminous paving, 6" gravel base	435	199	634
1960	10" gravel base	460	235	695
2200	45° angle parking, 3" bituminous paving, 6" gravel base	315	208	523
2240	10" gravel base	345	249	594
2260	4" bituminous paving, 6" gravel base	370	208	578
2300	10" gravel base	395	247	642
2320	6" bituminous paving, 6" gravel base	495	217	712
2360	10" gravel base	525	257	782

Cost per car

For information about Means Estimating Seminars, see yellow pages 11 and 12 in back of book

For expanded coverage of these items see *Means Site Work & Landscape Cost Data 2000*

Unit Price Section

Table of Contents

Div No.		Page
	General Requirements	**271**
01100	Summary	272
01200	Price & Payment Procedures	272
01300	Administrative Requirements	272
01400	Quality Requirements	274
01500	Temporary Facilities & Controls	274
01700	Execution Requirements	284
	Site Construction	**285**
02050	Basic Site Materials & Methods	286
02200	Site Preparation	286
02300	Earthwork	293
02400	Tunneling, Boring & Jacking	295
02500	Utility Services	295
02600	Drainage & Containment	297
02700	Bases, Ballasts, Pavements, & Appurtenances	298
02800	Site Improvements & Amenities	300
02900	Planting	302
	Concrete	**307**
03050	Basic Conc. Mat. & Methods	308
03100	Concrete Forms & Accessories	308
03200	Concrete Reinforcement	309
03300	Cast-In-Place Concrete	310
03400	Precast Concrete	312
03900	Concrete Restor. & Cleaning	312
	Masonry	**313**
04050	Basic Masonry Mat. & Methods	314
04200	Masonry Units	315
04400	Stone	318
04500	Refractories	320
04800	Masonry Assemblies	321
04900	Masonry Restoration & Cleaning	322
	Metals	**323**
05050	Basic Materials & Methods	324
05100	Structural Metal Framing	326
05200	Metal Joists	328
05300	Metal Decking	328
05400	Cold Formed Metal Framing	328
05500	Metal Fabrications	335
05700	Ornamental Metals	336
	Wood & Plastics	**337**
06050	Basic Wd. & Plastic Mat. & Meth.	338
06100	Rough Carpentry	341
06200	Finish Carpentry	352
06400	Architectural Woodwork	356
06600	Plastic Fabrications	364

Div. No.		Page
	Thermal & Moisture Protection	**367**
07100	Dampproofing & Waterproofing	368
07200	Thermal Protection	368
07300	Shingles, Roof Tiles & Roof Cov.	372
07400	Roofing & Siding Panels	374
07500	Membrane Roofing	377
07600	Flashing & Sheet Metal	381
07700	Roof Specialties & Accessories	383
07800	Fire & Smoke Protection	385
07900	Joint Sealers	387
	Doors & Windows	**389**
08100	Metal Doors & Frames	390
08200	Wood & Plastic Doors	392
08300	Specialty Doors	396
08400	Entrances & Storefronts	398
08500	Windows	400
08600	Skylights	406
08700	Hardware	407
08800	Glazing	410
08900	Glazed Curtain Wall	412
	Finishes	**413**
09100	Metal Support Assemblies	414
09200	Plaster & Gypsum Board	414
09300	Tile	418
09400	Terrazzo	420
09500	Ceilings	420
09600	Flooring	422
09700	Wall Finishes	425
09800	Acoustical Treatment	427
09900	Paints & Coatings	428
	Specialties	**441**
10100	Visual Display Boards	442
10150	Compartments & Cubicles	442
10200	Louvers & Vents	444
10260	Wall & Corner Guards	445
10300	Fireplaces & Stoves	445
10340	Manufactured Exterior Specialties	446
10350	Flagpoles	446
10400	Identification Devices	447
10500	Lockers	447
10520	Fire Protection Specialties	448
10530	Protective Covers	448
10550	Postal Specialties	449
10600	Partitions	449
10670	Storage Shelving	450

Div. No.		Page
10750	Telephone Specialties	451
10800	Toilet, Bath, & Laundry Access.	451
	Equipment	**453**
11010	Maintenance Equipment	454
11130	Audio-Visual Equip.	454
11160	Loading Dock Equip.	454
11400	Food Service Equipment	454
11450	Residential Equipment	457
	Furnishings	**459**
12400	Furnishings & Access.	460
	Special Construction	**463**
13030	Special Purpose Rooms	464
13100	Lightning Protection	464
13120	Pre-Engineered Structures	465
13150	Swimming Pools	465
13200	Storage Tanks	466
13280	Hazardous Mat. Remediation	467
13600	Solar & Wind Energy Equipment	471
13800	Building Automation & Control	472
13850	Detection & Alarm	473
13900	Fire Suppression	473
	Conveying Systems	**477**
14100	Dumbwaiters	478
14200	Elevators	478
	Mechanical	**479**
15050	Basic Materials & Methods	480
15100	Building Services Piping	481
15400	Plumbing Fixtures & Equipment	493
15500	Heat Generation Equipment	499
15600	Refrigeration Equipment	502
15700	Heating, Ventilation, & A/C Equip.	503
15800	Air Distribution	506
	Electrical	**511**
16050	Basic Elect. Mat. & Methods	512
16100	Wiring Methods	513
16200	Electrical Power	526
16400	Low-Voltage Distribution	530
16500	Lighting	533
16800	Sound & Video	535

How to Use the Unit Price Pages

The following is a detailed explanation of a sample entry in the Unit Price Section. Next to each bold number below is the item being described with appropriate component of the sample entry following in parenthesis. Some prices are listed as bare costs, others as costs that include overhead and profit of the installing contractor. In most cases, if the work is to be subcontracted, the general contractor will need to add an additional markup (R.S. Means suggests using 10%) to the figures in the column "Total Incl. O&P."

1 Division Number/Title (06100/Rough Carpentry)

Use the Unit Price Section Table of Contents to locate specific items. The sections are classified according to the CSI MasterFormat.

2 Line Numbers (06170 980 5050)

Each unit price line item has been assigned a unique 12-digit code based on the 5-digit CSI MasterFormat classification.

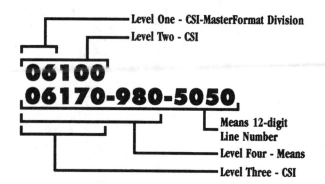

3 Description (Roof Trusses, Etc.)

Each line item is described in detail. Sub-items and additional sizes are indented beneath the appropriate line items. The first line or two after the main item (in boldface) may contain descriptive information that pertains to all line items beneath this boldface listing.

4 Reference Number Information

R06170-100 You'll see reference numbers shown in bold rectangles at the beginning of some sections. These refer to related items in the Reference Section, visually identified by a vertical gray bar on the edge of pages.

The relation may be: (1) an estimating procedure that should be read before estimating, (2) an alternate pricing method, or (3) technical information.

The "R" designates the Reference Section. The numbers refer to the MasterFormat classification system.

It is strongly recommended that you review all reference numbers that appear within the section in which you are working.

Example: The rectangle number above is directing you to refer to the reference number R06170-100. This particular reference number shows how the unit price lines were formulated and costs derived.

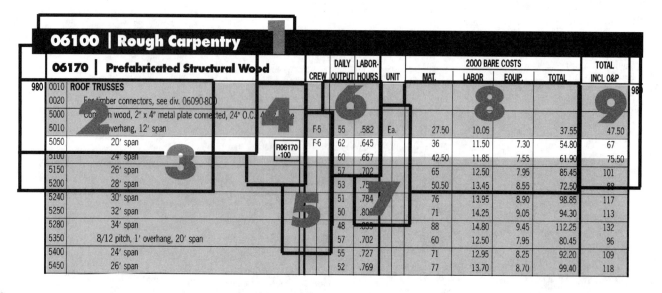

Crew (F-6)

The "Crew" column designates the typical trade or crew used to install the item. If an installation can be accomplished by one trade and requires no power equipment, that trade and the number of workers are listed (for example, "2 Carp"). If an installation requires a composite crew, a crew code designation is listed (for example, "F-6"). You'll find full details on all composite crews in the Crew Listings.

- For a complete list of all trades utilized in this book and their abbreviations, see the inside back cover.

Crews

Crew No.	Bare Costs		Incl. Subs O & P		Cost Per Labor-Hour	
Crew F-6	Hr.	Daily	Hr.	Daily	Bare Costs	Incl. O&P
2 Carpenters	$19.70	$315.20	$33.75	$540.00	$17.79	$30.25
2 Building Laborers	14.45	231.20	24.75	396.00		
1 Equip. Oper. (crane)	20.65	165.20	34.25	274.00		
1 Hyd. Crane, 12 Ton		453.45		498.80	11.34	12.47
40 L.H., Daily Totals		$1165.05		$1708.80	$29.13	$42.72

Productivity: Daily Output (62)/ Labor-Hours (.645)

The "Daily Output" represents the typical number of units the designated crew will install in a normal 8-hour day. To find out the number of days the given crew would require to complete the installation, divide your quantity by the daily output. For example:

Quantity	÷	Daily Output	=	Duration
248 Ea.	÷	62 Ea./ Crew Day	=	4 Crew Days

The "Labor-Hours" figure represents the number of labor-hours required to install one unit of work. To find out the number of labor-hours required for your particular task, multiply the quantity of the item times the number of labor-hours shown. For example:

Quantity	x	Productivity Rate	=	Duration
248 Ea.	x	.645 Labor-Hours/ LF	=	159.96 Labor-Hours

Unit (Ea.)

The abbreviated designation indicates the unit of measure upon which the price, production, and crew are based (Ea. = Each). For a complete listing of abbreviations refer to the Abbreviations Listing in the Reference Section of this book.

Bare Costs:

Mat. (Bare Material Cost) (36.00)

The unit material cost is the "bare" material cost with no overhead and profit included. *Costs shown reflect national average material prices for January of the current year and include delivery to the job site. No sales taxes are included.*

Labor (11.50)

The unit labor cost is derived by multiplying bare labor-hour costs for Crew F-6 by labor-hour units. The bare labor-hour cost is found in the Crew Section under F-6. (If a trade is listed, the hourly labor cost—the wage rate—is found on the inside back cover.)

Labor-Hour Cost Crew F-6	x	Labor-Hour Units	=	Labor
$17.79	x	.645	=	$11.50

Equip. (Equipment) (7.30)

Equipment costs for each crew are listed in the description of each crew. Tools or equipment whose value justifies purchase or ownership by a contractor are considered overhead as shown on the inside back cover. The unit equipment cost is derived by multiplying the bare equipment hourly cost by the labor-hour units.

Equipment Cost Crew F-6	x	Labor-Hour Units	=	Equip.
$11.34	x	.645	=	$7.30

Total (54.80)

The total of the bare costs is the arithmetic total of the three previous columns: mat., labor, and equip.

Material	+	Labor	+	Equip.	=	Total
$36.00	+	$11.50	+	$7.30	=	$54.80

Total Costs Including O&P

This figure is the sum of the bare material cost plus 10% for profit; the bare labor cost plus total overhead and profit (per the inside back cover or, if a crew is listed, from the crew listings); and the bare equipment cost plus 10% for profit.

Material is Bare Material Cost + 10% = $36.00 + $3.60	=	$39.60
Labor for Crew F-6 = Labor-Hour Cost ($30.25) x Labor-Hour Units (.645)	=	$19.51
Equip. is Bare Equip. Cost + 10% = $7.30 + $.73	=	$7.73
Total (Rounded)	=	$67.00

Division 1
General Requirements

Estimating Tips
The General Requirements of any contract are very important to both the bidder and the owner. These lay the ground rules under which the contract will be executed and have a significant influence on the cost of operations. Therefore, it is extremely important to thoroughly read and understand the General Requirements both before preparing an estimate and when the estimate is complete, to ascertain that nothing in the contract is overlooked. Caution should be exercised when applying items listed in Division 1 to an estimate. Many of the items are included in the unit prices listed in the other divisions such as mark-ups on labor and company overhead.

01300 Administrative Requirements
- Before determining a final cost estimate, it is a good practice to review all the items listed in subdivision 01300 to make final adjustments for items that may need customizing to specific job conditions.

01330 Submittal Procedures
- Requirements for initial and periodic submittals can represent a significant cost to the General Requirements of a job. Thoroughly check the submittal specifications when estimating a project to determine any costs that should be included.

01400 Quality Requirements
- All projects will require some degree of Quality Control. This cost is not included in the unit cost of construction listed in each division. Depending upon the terms of the contract, the various costs of inspection and testing can be the responsibility of either the owner or the contractor. Be sure to include the required costs in your estimate.

01500 Temporary Facilities & Controls
- Barricades, access roads, safety nets, scaffolding, security and many more requirements for the execution of a safe project are elements of direct cost. These costs can easily be overlooked when preparing an estimate. When looking through the major classifications of this subdivision, determine which items apply to each division in your estimate.

01559 Equipment Rental
- This subdivision contains transportation, handling, storage, protection and product options and substitutions. Listed in this cost manual are average equipment rental rates for all types of equipment. This is useful information when estimating the time and materials requirement of any particular operation in order to establish a unit or total cost.
- A good rule of thumb is that weekly rental is 3 times daily rental and that monthly rental is 3 times weekly rental.
- The figures in the column for Crew Equipment Cost represent the rental rate used in determining the daily cost of equipment in a crew. It is calculated by dividing the weekly rate by 5 days and adding the hourly operating cost times 8 hours.

01770 Closeout Procedures
- When preparing an estimate, read the specifications to determine the requirements for Contract Closeout thoroughly. Final cleaning, record documentation, operation and maintenance data, warranties and bonds, and spare parts and maintenance materials can all be elements of cost for the completion of a contract. Do not overlook these in your estimate.

01830 Operations & Maintenance
- If maintenance and repair are included in your contract, they require special attention. To estimate the cost to remove and replace any unit usually requires a site visit to determine the accessibility and the specific difficulty at that location. Obstructions, dust control, safety, and often overtime hours must be considered when preparing your estimate.

Reference Numbers
Reference numbers are shown in bold squares at the beginning of some major classifications. These numbers refer to related items in the Reference Section. The reference information may be an estimating procedure, an alternate pricing method or technical information.

Note: Not all subdivisions listed here necessarily appear in this publication.

01100 | Summary

01103 | Models & Renderings

			CREW	DAILY OUTPUT	LABOR-HOURS	UNIT	2000 BARE COSTS MAT.	LABOR	EQUIP.	TOTAL	TOTAL INCL O&P	
500	0010	**RENDERINGS** Color, matted, 20" x 30", eye level,										500
	0020	1 building, minimum				Ea.	1,525			1,525	1,675	
	0050	Average					2,550			2,550	2,800	
	0100	Maximum					3,600			3,600	3,950	

01107 | Professional Consultant

			CREW	DAILY OUTPUT	LABOR-HOURS	UNIT	MAT.	LABOR	EQUIP.	TOTAL	INCL O&P	
100	0011	**ARCHITECTURAL FEES**	R01107-010									100
	0020	For new construction										
	0060	Minimum				Project					4.90%	
200	0011	**CONSTRUCTION MANAGEMENT FEES**										200
	0060	For work to $10,000				Project					10%	
	0070	To $25,000									9%	
	0090	To $100,000									6%	
700	0010	**SURVEYING** Conventional, topographical, minimum	A-7	3.30	7.273	Acre	16	148		164	271	700
	0100	Maximum	A-8	.60	53.333		48	1,075		1,123	1,875	
	0300	Lot location and lines, minimum, for large quantities	A-7	2	12		25	245		270	445	
	0320	Average	"	1.25	19.200		45	390		435	715	
	0400	Maximum, for small quantities	A-8	1	32		72	645		717	1,175	
	0600	Monuments, 3' long	A-7	10	2.400	Ea.	19	49		68	105	
	0800	Property lines, perimeter, cleared land	"	1,000	.024	L.F.	.03	.49		.52	.86	
	0900	Wooded land	A-8	875	.037	"	.05	.74		.79	1.31	
	1100	Crew for building layout, 2 person crew	A-6	1	16	Day		335		335	575	
	1200	3 person crew	A-7	1	24	"		490		490	835	

01200 | Price & Payment Procedures

01250 | Contract Modification Procedures

			CREW	DAILY OUTPUT	LABOR-HOURS	UNIT	MAT.	LABOR	EQUIP.	TOTAL	INCL O&P	
200	0010	**CONTINGENCIES** for estimate at conceptual stage				Project					20%	200
	0150	Final working drawing stage				"					3%	

01290 | Payment Procedures

800	0010	**TAXES** Sales tax, State, average	R01100-090			%	4.71%					800
	0050	Maximum					7.25%					
	0200	Social Security, on first $72,600 of wages	R01100-100					7.65%				
	0300	Unemployment, MA, combined Federal and State, minimum						2.10%				
	0350	Average						7%				
	0400	Maximum						8%				

01300 | Administrative Requirements

01310 | Project Management/Coordination

			CREW	DAILY OUTPUT	LABOR-HOURS	UNIT	MAT.	LABOR	EQUIP.	TOTAL	INCL O&P	
150	0010	**PERMITS** Rule of thumb, most cities, minimum				Job					.50%	150
	0100	Maximum				"					2%	

01300 | Administrative Requirements

01310 | Project Management/Coordination

			CREW	DAILY OUTPUT	LABOR-HOURS	UNIT	2000 BARE COSTS MAT.	LABOR	EQUIP.	TOTAL	TOTAL INCL O&P	
350	0010	**INSURANCE** Builders risk, standard, minimum R01100-040				Job					.22%	350
	0050	Maximum									.59%	
	0200	All-risk type, minimum R01100-060									.25%	
	0250	Maximum									.62%	
	0400	Contractor's equipment floater, minimum R01100-080				Value					.50%	
	0450	Maximum				"					1.50%	
	0600	Public liability, average				Job					1.55%	
	0800	Workers' compensation & employer's liability, average										
	0850	by trade, carpentry, general				Payroll		19.85%				
	0900	Clerical						.53%				
	0950	Concrete						18.81%				
	1000	Electrical						7.03%				
	1050	Excavation						11.41%				
	1100	Glazing						14.43%				
	1150	Insulation						18.52%				
	1200	Lathing						12.26%				
	1250	Masonry						17.75%				
	1300	Painting & decorating						15.27%				
	1350	Pile driving						29.88%				
	1400	Plastering						15.72%				
	1450	Plumbing						8.82%				
	1500	Roofing						34.62%				
	1550	Sheet metal work (HVAC)						12.27%				
	1600	Steel erection, structural						42.56%				
	1650	Tile work, interior ceramic						10.48%				
	1700	Waterproofing, brush or hand caulking						8.32%				
	1800	Wrecking						43.48%				
	2000	Range of 35 trades in 50 states, excl. wrecking, minimum						2%				
	2100	Average						18.10%				
	2200	Maximum						132.92%				
400	0010	**MAIN OFFICE EXPENSE** Average for General Contractors R01100-050										400
	0020	As a percentage of their annual volume										
	0125	Annual volume under 1 million dollars				% Vol.				13.60%		
	0145	Up to 2.5 million dollars								8%		
	0150	Up to 4.0 million dollars								6.80%		
	0200	Up to 7.0 million dollars								5.60%		
620	0010	**OVERHEAD & PROFIT** Allowance to add to items in this R01100-070										620
	0020	book that do not include Subs O&P, average				%				25%		
	0100	Allowance to add to items in this book that										
	0110	do include Subs O&P, minimum				%					5%	
	0150	Average									10%	
	0200	Maximum									15%	
	0290	Typical, by size of project, under $50,000								40%		
	0310	$50,000 to $100,000								35%		
	0320	$100,000 to $500,000								25%		
	0330	$500,000 to $1,000,000								20%		

For expanded coverage of these items see *Means Building Construction Cost Data 2000*

01400 | Quality Requirements

01450 | Quality Control

		Quality Control	CREW	DAILY OUTPUT	LABOR-HOURS	UNIT	2000 BARE COSTS MAT.	LABOR	EQUIP.	TOTAL	TOTAL INCL O&P	
500	0010	**FIELD TESTING**										500
	0015	For concrete building costing $1,000,000, minimum				Project					5,000	
	0020	Maximum									40,000	
	0050	Steel building, minimum									5,000	
	0070	Maximum									15,000	
	0600	Concrete testing, aggregates, abrasion				Ea.					120	
	0650	Absorption									46	
	1800	Compressive strength test, cylinder, delivered to lab/cyl									13	
	1900	Picked up by lab, minimum									15	
	1950	Average									20	
	2000	Maximum									30	
	2200	Compressive strength, cores (not incl. drilling)									40	
	2250	Core drilling, 4" diameter (plus technician)				Inch					25	
	2260	Technician for core drilling				Hr.					50	
	2300	Patching core holes				Ea.					24	

01500 | Temporary Facilities & Controls

01520 | Construction Facilities

		Construction Facilities	CREW	DAILY OUTPUT	LABOR-HOURS	UNIT	2000 BARE COSTS MAT.	LABOR	EQUIP.	TOTAL	TOTAL INCL O&P	
550	0010	**FIELD OFFICE EXPENSE**										550
	0100	Field office expense, office equipment rental average				Month	135			135	149	
	0120	Office supplies, average					80			80	88	
	0140	Telephone bill; avg. bill/month incl. long dist.					225			225	248	
	0160	Field office lights & HVAC					90			90	99	

01540 | Construction Aids

		Construction Aids	CREW	DAILY OUTPUT	LABOR-HOURS	UNIT	2000 BARE COSTS MAT.	LABOR	EQUIP.	TOTAL	TOTAL INCL O&P	
550	0010	**PUMP STAGING**, Aluminum R01540-200										550
	0200	24' long pole section, buy				Ea.	335			335	365	
	0300	18' long pole section, buy					258			258	284	
	0400	12' long pole section, buy					174			174	192	
	0500	6' long pole section, buy					92			92	101	
	0600	6' long splice joint section, buy					68			68	75	
	0700	Pump jack					111			111	123	
	0900	Foldable brace					48.50			48.50	53	
	1000	Workbench/back safety rail support					59			59	64.50	
	1100	Scaffolding planks/workbench, 14" wide x 24' long					545			545	600	
	1200	Plank end safety rail					183			183	202	
	1250	Safety net, 22' long					267			267	294	
	1300	System in place, 50' working height, per use based on 50 uses	2 Carp	84.80	.189	C.S.F.	4.95	3.72		8.67	11.80	
	1400	100 uses		84.80	.189		2.48	3.72		6.20	9.05	
	1500	150 uses		84.80	.189		1.66	3.72		5.38	8.15	
750	0014	**SCAFFOLDING** R01540-100										750
	0015	Steel tubular, rent, 1 use/mo, no plank, erect & dismantle										
	0090	Building exterior, wall face, 1 to 5 stories	3 Carp	24	1	C.S.F.	24.50	19.70		44.20	60.50	
	0200	6 to 12 stories	4 Carp	21.20	1.509		24.50	29.50		54	77.50	
	0310	13 to 20 stories	5 Carp	20	2		24.50	39.50		64	94	
	0460	Building interior, wall face area, up to 16' high	3 Carp	25	.960		24.50	18.90		43.40	59	

01500 | Temporary Facilities & Controls

01540 | Construction Aids

		CREW	DAILY OUTPUT	LABOR-HOURS	UNIT	2000 BARE COSTS MAT.	LABOR	EQUIP.	TOTAL	TOTAL INCL O&P
0560	16' to 40' high	3 Carp	23	1.043	C.S.F.	24.50	20.50		45	61.50
0800	Building interior floor area, up to 30' high		312	.077	C.C.F.	1.87	1.52		3.39	4.66
0900	Over 30' high	4 Carp	275	.116	"	1.87	2.29		4.16	6
0910	Steel tubular, heavy duty shoring, buy									
0920	Frames 5' high 2' wide				Ea.	75			75	82.50
0925	5' high 4' wide					85			85	93.50
0930	6' high 2' wide					86			86	94.50
0935	6' high 4' wide					101			101	111
0940	Accessories									
0945	Cross braces				Ea.	16			16	17.60
0950	U-head, 8" x 8"					16			16	17.60
0955	J-head, 4" x 8"					12			12	13.20
0960	Base plate, 8" x 8"					13			13	14.30
0965	Leveling jack					30.50			30.50	33.50
1000	Steel tubular, regular, buy									
1100	Frames 3' high 5' wide				Ea.	58			58	64
1150	5' high 5' wide					67			67	73.50
1200	6'-4" high 5' wide					84			84	92.50
1350	7'-6" high 6' wide					145			145	160
1500	Accessories cross braces					15			15	16.50
1550	Guardrail post					15			15	16.50
1600	Guardrail 7' section					7.25			7.25	8
1650	Screw jacks & plates					24			24	26.50
1700	Sidearm brackets					28			28	31
1750	8" casters					33			33	36.50
1800	Plank 2" x 10" x 16'-0"					42.50			42.50	47
1900	Stairway section					235			235	259
1910	Stairway starter bar					21			21	23.50
1920	Stairway inside handrail					53			53	58.50
1930	Stairway outside handrail					73			73	80.50
1940	Walk-thru frame guardrail					28			28	31
2000	Steel tubular, regular, rent/mo.									
2100	Frames 3' high 5' wide				Ea.	3.75			3.75	4.13
2150	5' high 5' wide					3.75			3.75	4.13
2200	6'-4" high 5' wide					3.75			3.75	4.13
2250	7'-6" high 6' wide					7			7	7.70
2500	Accessories, cross braces					.60			.60	.66
2550	Guardrail post					1			1	1.10
2600	Guardrail 7' section					.75			.75	.83
2650	Screw jacks & plates					1.50			1.50	1.65
2700	Sidearm brackets					1.50			1.50	1.65
2750	8" casters					6			6	6.60
2800	Outrigger for rolling tower					3			3	3.30
2850	Plank 2" x 10" x 16'-0"					5			5	5.50
2900	Stairway section					10			10	11
2910	Stairway starter bar					.10			.10	.11
2920	Stairway inside handrail					5			5	5.50
2930	Stairway outside handrail					5			5	5.50
2940	Walk-thru frame guardrail					2			2	2.20
3000	Steel tubular, heavy duty shoring, rent/mo.									
3250	5' high 2' & 4' wide				Ea.	5			5	5.50
3300	6' high 2' & 4' wide					5			5	5.50
3500	Accessories, cross braces					1			1	1.10
3600	U - head, 8" x 8"					1			1	1.10
3650	J - head, 4" x 8"					1			1	1.10
3700	Base plate, 8" x 8"					1			1	1.10

For expanded coverage of these items see *Means Building Construction Cost Data 2000*

01500 | Temporary Facilities & Controls

01540 | Construction Aids

			CREW	DAILY OUTPUT	LABOR-HOURS	UNIT	2000 BARE COSTS MAT.	LABOR	EQUIP.	TOTAL	TOTAL INCL O&P	
750	3750	Leveling jack				Ea.	2			2	2.20	750
	5700	Planks, 2"x10"x16'-0" in place, up to 50' high	3 Carp	144	.167			3.28		3.28	5.65	
	5800	In place over 50' high	4 Carp	160	.200			3.94		3.94	6.75	
	6820	Erect and dismantle frames, 1st tier	4 Clab	45	.711			10.30		10.30	17.60	
	6830	Erect and dismantle frames, 2nd tier		93	.344			4.97		4.97	8.50	
	6840	Erect and dismantle frames, 3rd tier		87	.368			5.30		5.30	9.10	
	6850	Erect and dismantle frames, 4th tier		75	.427			6.15		6.15	10.55	
760	0010	**STAGING AIDS** and fall protection equipment										760
	0110	Cost each per day, based on 250 days use				Day	.11			.11	.13	
	0200	Guard post, buy				Ea.	15			15	16.50	
	0210	Cost each per day, based on 250 days use				Day	.06			.06	.07	
	0300	End guard chains, buy per set				Ea.	25			25	27.50	
	0310	Cost per set per day, based on 250 days use				Day	.11			.11	.13	
	1010	Cost each per day, based on 250 days use				"	.02			.02	.03	
	1100	Wood bracket, buy				Ea.	10.60			10.60	11.65	
	1110	Cost each per day, based on 250 days use				Day	.04			.04	.05	
	2010	Cost per pair per day, based on 250 days use				"	.31			.31	.34	
	2100	Steel siderail jack, buy per pair				Pair	69			69	76	
	2110	Cost per pair per day, based on 250 days use				Day	.28			.28	.30	
	3010	Cost each per day, based on 250 days use				"	.17			.17	.19	
	3100	Aluminum scaffolding plank, 20" wide x 24' long, buy				Ea.	660			660	730	
	3110	Cost each per day, based on 250 days use				Day	2.65			2.65	2.91	
	4010	Cost each per day, based on 250 days use				"	.90			.90	.99	
	4100	Rope for safety line, 5/8" x 100' nylon, buy				Ea.	35			35	38.50	
	4110	Cost each per day, based on 250 days use				Day	.14			.14	.15	
	4200	Permanent U-Bolt roof anchor, buy				Ea.	29.50			29.50	32.50	
	4300	Temporary (one use) roof ridge anchor, buy				"	18.60			18.60	20.50	
	5000	Installation (setup and removal) of staging aids										
	5010	Sidewall staging bracket	2 Carp	64	.250	Ea.		4.93		4.93	8.45	
	5020	Guard post with 2 wood rails	"	64	.250			4.93		4.93	8.45	
	5030	End guard chains, set	1 Carp	64	.125			2.46		2.46	4.22	
	5100	Roof shingling bracket		96	.083			1.64		1.64	2.81	
	5200	Ladder jack		64	.125			2.46		2.46	4.22	
	5300	Wood plank, 2x10x16'	2 Carp	80	.200			3.94		3.94	6.75	
	5310	Aluminum scaffold plank, 20" x 24'	"	40	.400			7.90		7.90	13.50	
	5410	Safety rope	1 Carp	40	.200			3.94		3.94	6.75	
	5420	Permanent U-Bolt roof anchor (install only)	2 Carp	40	.400			7.90		7.90	13.50	
	5430	Temporary roof ridge anchor (install only)	1 Carp	64	.125			2.46		2.46	4.22	
800	0010	**TARPAULINS** Cotton duck, 10 oz. to 13.13 oz. per S.Y., minimum				S.F.	.46			.46	.51	800
	0050	Maximum					.55			.55	.61	
	0200	Reinforced polyethylene 3 mils thick, white					.10			.10	.11	
	0300	4 mils thick, white, clear or black					.12			.12	.13	
	0730	Polyester reinforced w/ integral fastening system 11 mils thick					1			1	1.10	
820	0010	**SMALL TOOLS** As % of contractor's work, minimum				Total					.50%	820
	0100	Maximum				"					2%	

Reference: R01540-100

01500 | Temporary Facilities & Controls

01590 | Equipment Rental

		UNIT	HOURLY OPER. COST	RENT PER DAY	RENT PER WEEK	RENT PER MONTH	CREW EQUIPMENT COST/DAY
100	**0010 CONCRETE EQUIPMENT RENTAL**						
0100	without operators						
0200	Bucket, concrete lightweight, 1/2 C.Y.	Ea.	.15	28.50	85	255	18.20
0300	1 C.Y.		.20	42.50	128	385	27.20
0400	1-1/2 C.Y.		.25	45	135	405	29
0500	2 C.Y.		.26	58.50	175	525	37.10
0600	Cart, concrete, self propelled, operator walking, 10 C.F.		1.25	70	210	630	52
0700	Operator riding, 18 C.F.		2.40	107	320	960	83.20
0800	Conveyer for concrete, portable, gas, 16" wide, 26' long		3.60	183	550	1,650	138.80
0900	46' long		3.70	237	710	2,125	171.60
1000	56' long		4	250	750	2,250	182
1100	Core drill, electric, 2-1/2 H.P., 1" to 8" bit diameter		.50	73.50	220	660	48
1150	11 H.P., 8" to 18" cores		.70	91.50	275	825	60.60
1200	Finisher, concrete floor, gas, riding trowel, 48" diameter		2.70	117	350	1,050	91.60
1300	Gas, manual, 3 blade, 36" trowel		2	53.50	160	480	48
1400	4 blade, 48" trowel		2.25	58.50	175	525	53
1500	Float, hand-operated (Bull float) 48" wide		.10	8.35	25	75	5.80
1570	Curb builder, 14 H.P., gas, single screw		1.75	66.50	200	600	54
1590	Double screw		2.35	66.50	200	600	58.80
1600	Grinder, concrete and terrazzo, electric, floor		1.25	66.50	200	600	50
1700	Wall grinder		.60	33.50	100	300	24.80
1800	Mixer, powered, mortar and concrete, gas, 6 C.F., 18 H.P.		2.44	58.50	175	525	54.50
1900	10 C.F., 25 H.P.		3.18	53.50	160	480	57.45
2000	16 C.F.		3.65	75	225	675	74.20
2100	Concrete, stationary, tilt drum, 2 C.Y.		7.75	267	800	2,400	222
2120	Pump, concrete, truck mounted, 4" line, 80' boom		12	935	2,800	8,400	656
2140	5" line, 110' boom		13.50	1,100	3,300	9,900	768
2160	Mud jack, 50 C.F. per hr.		3.15	70	210	630	67.20
2180	225 C.F. per hr.		6.25	335	1,000	3,000	250
2600	Saw, concrete, manual, gas, 18 H.P.		2.75	80	240	720	70
2650	Self-propelled, gas, 30 H.P.		5.50	110	330	990	110
2700	Vibrators, concrete, electric, 60 cycle, 2 H.P.		.25	28.50	85	255	19
2800	3 H.P.		.26	40	120	360	26.10
2900	Gas engine, 5 H.P.		.65	40	120	360	29.20
3000	8 H.P.		1	50	150	450	38
200	**0010 EARTHWORK EQUIPMENT RENTAL** Without operators						
0040	Aggregate spreader, push type 8' to 12' wide	Ea.	1.10	108	325	975	73.80
0050	Augers for truck or trailer mounting, vertical drilling						
0055	Fence post auger, truck mounted	Ea.	10	485	1,450	4,350	370
0060	4" to 36" diam., 54 H.P., gas, 10' spindle travel		16	590	1,775	5,325	483
0070	14' spindle travel		16.40	665	2,000	6,000	531.20
0080	Auger, horizontal boring machine, 12" to 36" diameter, 45 H.P.		7.35	335	1,000	3,000	258.80
0090	12" to 48" diameter, 65 H.P.		12.25	665	2,000	6,000	498
0100	Excavator, diesel hydraulic, crawler mounted, 1/2 C.Y. cap.		11.65	415	1,248	3,750	342.80
0120	5/8 C.Y. capacity		15.50	480	1,438	4,325	411.60
0140	3/4 C.Y. capacity		17.40	530	1,583	4,750	455.80
0150	1 C.Y. capacity		22.25	600	1,800	5,400	538
0200	1-1/2 C.Y. capacity		26.65	835	2,500	7,500	713.20
0300	2 C.Y. capacity		42.15	1,125	3,400	10,200	1,017
0320	2-1/2 C.Y. capacity		59.15	2,000	6,000	18,000	1,673
0340	3-1/2 C.Y. capacity		78.25	2,500	7,500	22,500	2,126
0341	Attachments						
0342	Bucket thumbs		.45	292	875	2,625	178.60
0345	Grapples		.35	300	900	2,700	182.80
0350	Gradall type, truck mounted, 3 ton @ 15' radius, 5/8 C.Y.		25.40	690	2,075	6,225	618.20
0370	1 C.Y. capacity		26.33	1,000	3,036	9,100	817.85
0400	Backhoe-loader, 40 to 45 H.P., 5/8 C.Y. capacity		7.15	217	650	1,950	187.20

GENERAL REQUIREMENTS 1

01500 | Temporary Facilities & Controls

01590 | Equipment Rental

		UNIT	HOURLY OPER. COST	RENT PER DAY	RENT PER WEEK	RENT PER MONTH	CREW EQUIPMENT COST/DAY
0450	45 H.P. to 60 H.P., 3/4 C.Y. capacity	Ea.	8.50	225	675	2,025	203
0460	80 H.P., 1-1/4 C.Y. capacity		12.22	292	877	2,625	273.15
0480	Attachments						
0482	Compactor, 20,000 lb		1.35	192	575	1,725	125.80
0485	Hydraulic hammer, 750 ft-lbs		1.15	300	900	2,700	189.20
0486	Hydraulic hammer, 1200 ft-lbs		1.70	335	1,000	3,000	213.60
0500	Brush chipper, gas engine, 6" cutter head, 35 H.P.		3.85	142	425	1,275	115.80
0550	12" cutter head, 130 H.P.		11.88	193	580	1,750	211.05
0600	15" cutter head, 165 H.P.		15.48	267	800	2,400	283.85
0750	Bucket, clamshell, general purpose, 3/8 C.Y.		.75	61.50	185	555	43
0800	1/2 C.Y.		.85	70	210	630	48.80
0850	3/4 C.Y.		1.10	80	240	720	56.80
0900	1 C.Y.		1.25	100	300	900	70
0950	1-1/2 C.Y.		1.50	133	400	1,200	92
1000	2 C.Y.		1.75	150	450	1,350	104
1010	Bucket, dragline, medium duty, 1/2 C.Y.		.36	40	120	360	26.90
1020	3/4 C.Y.		.65	46.50	140	420	33.20
1030	1 C.Y.		.88	50	150	450	37.05
1040	1-1/2 C.Y.		1.12	66.50	200	600	48.95
1050	2 C.Y.		1.40	83.50	250	750	61.20
1070	3 C.Y.		1.45	117	350	1,050	81.60
1200	Compactor, roller, 2 drum, 2000 lb., operator walking		1.78	133	400	1,200	94.25
1250	Rammer compactor, gas, 1000 lb. blow		.75	50	150	450	36
1300	Vibratory plate, gas, 13" plate, 1000 lb. blow		.75	43.50	130	390	32
1350	24" plate, 5000 lb. blow		1.95	75	225	675	60.60
1750	Extractor, piling, see lines 2500 to 2750						
1860	Grader, self-propelled, 25,000 lb.	Ea.	12.85	555	1,658	4,975	434.40
1910	30,000 lb.		14.50	665	2,000	6,000	516
1920	40,000 lb.		20.90	835	2,500	7,500	667.20
1930	55,000 lb.		29.90	1,400	4,183	12,500	1,076
1950	Hammer, pavement demo., hyd., gas, self-prop., 1000 to 1250 lb.		16.55	410	1,230	3,700	378.40
2000	Diesel 1300 to 1500 lb.		16.60	460	1,375	4,125	407.80
3000	Roller, tandem, gas, 3 to 5 ton		4.15	167	500	1,500	133.20
3050	Diesel, 8 to 12 ton		6	293	880	2,650	224
3100	Towed type, vibratory, gas 12.5 H.P., 2 ton		3.75	133	400	1,200	110
3150	Sheepsfoot, double 60" x 60"		4.80	147	440	1,325	126.40
3200	Pneumatic tire diesel roller, 12 ton		7.60	300	900	2,700	240.80
3250	21 to 25 ton		12.56	350	1,050	3,150	310.50
3300	Sheepsfoot roller, self-propelled, 4 wheel, 130 H.P.		20.95	665	2,000	6,000	567.60
3320	300 H.P.		26.70	785	2,350	7,050	683.60
3350	Vibratory steel drum & pneumatic tire, diesel, 18,000 lb.		11.50	465	1,400	4,200	372
3400	29,000 lb.		12.80	550	1,650	4,950	432.40
3450	Scrapers, towed type, 9 to 12 C.Y. capacity		3.25	91.50	275	825	81
3500	12 to 17 C.Y. capacity		6.10	260	780	2,350	204.80
3550	Scrapers, self-propelled, 4 x 4 drive, 2 engine, 14 C.Y. capacity		60.47	1,900	5,710	17,100	1,626
3600	2 engine, 24 C.Y. capacity		76.64	2,175	6,500	19,500	1,913
3650	Self-loading, 11 C.Y. capacity		27.32	800	2,400	7,200	698.55
3700	22 C.Y. capacity		33.10	1,200	3,600	10,800	984.80
3710	Screening plant 110 hp. w/ 5' x 10' screen		16.28	450	1,350	4,050	400.25
3720	5' x 16' screen		17.65	520	1,560	4,675	453.20
3850	Shovels, see Cranes division 01590-600						
3860	Shovel/backhoe bucket, 1/2 C.Y.	Ea.	1.15	78.50	235	705	56.20
3870	3/4 C.Y.		3.15	127	380	1,150	101.20
3880	1 C.Y.		3.65	173	520	1,550	133.20
3890	1-1/2 C.Y.		4.25	212	635	1,900	161
3910	3 C.Y.		7.55	400	1,200	3,600	300.40
4110	Tractor, crawler, with bulldozer, torque converter, diesel 75 H.P.		9.95	335	1,000	3,000	279.60
4150	105 H.P.		10.75	510	1,535	4,600	393

Important: See the Reference Section for critical supporting data - Reference Nos., Crews, & Location Factors

01500 | Temporary Facilities & Controls

01590 | Equipment Rental

			UNIT	HOURLY OPER. COST	RENT PER DAY	RENT PER WEEK	RENT PER MONTH	CREW EQUIPMENT COST/DAY	
200	4200	140 H.P.	Ea.	17.55	650	1,950	5,850	530.40	200
	4260	200 H.P.		25.85	1,000	3,000	9,000	806.80	
	4310	300 H.P.		34.70	1,375	4,150	12,500	1,108	
	4360	410 H.P.		47.10	1,700	5,100	15,300	1,397	
	4380	700 H.P.		94	3,325	10,000	30,000	2,752	
	4400	Loader, crawler, torque conv., diesel, 1-1/2 C.Y., 80 H.P.		9.87	420	1,255	3,775	329.95	
	4450	1-1/2 to 1-3/4 C.Y., 95 H.P.		10.23	475	1,425	4,275	366.85	
	4510	1-3/4 to 2-1/4 C.Y., 130 H.P.		12.50	585	1,750	5,250	450	
	4530	2-1/2 to 3-1/4 C.Y., 190 H.P.		19.25	1,050	3,150	9,450	784	
	4560	3-1/2 to 5 C.Y., 275 H.P.		26.90	1,375	4,100	12,300	1,035	
	4610	Tractor loader, wheel, torque conv., 4 x 4, 1 to 1-1/4 C.Y., 65 H.P.		8.45	277	830	2,500	233.60	
	4620	1-1/2 to 1-3/4 C.Y., 80 H.P.		9.75	375	1,125	3,375	303	
	4650	1-3/4 to 2 C.Y., 100 H.P.		11.35	405	1,215	3,650	333.80	
	4710	2-1/2 to 3-1/2 C.Y., 130 H.P.		12.60	500	1,500	4,500	400.80	
	4730	3 to 4-1/2 C.Y., 170 H.P.		15.30	700	2,100	6,300	542.40	
	4760	5-1/4 to 5-3/4 C.Y., 270 H.P.		30.15	1,025	3,100	9,300	861.20	
	4810	7 to 8 C.Y., 375 H.P.		45.65	1,425	4,300	12,900	1,225	
	4870	12-1/2 C.Y., 690 H.P.		91.25	2,500	7,500	22,500	2,230	
	4880	Wheeled, skid steer, 10 C.F., 30 H.P. gas		4.55	133	400	1,200	116.40	
	4890	1 C.Y., 78 H.P., diesel		6.35	292	875	2,625	225.80	
	4891	Attachments for all skid steer loaders							
	4892	Auger	Ea.	.12	83.50	250	750	50.95	
	4893	Backhoe		.15	110	330	990	67.20	
	4894	Broom		.16	107	320	960	65.30	
	4895	Forks		.08	36.50	110	330	22.65	
	4896	Grapple		.12	83.50	250	750	50.95	
	4897	Concrete hammer		.25	180	540	1,625	110	
	4898	Tree spade		.36	153	460	1,375	94.90	
	4899	Trencher		.41	122	365	1,100	76.30	
	4900	Trencher, chain, boom type, gas, operator walking, 12 H.P.		2.16	142	426	1,275	102.50	
	4910	Operator riding, 40 H.P.		6.36	267	800	2,400	210.90	
	5000	Wheel type, diesel, 4' deep, 12" wide		12.42	515	1,550	4,650	409.35	
	5100	Diesel, 6' deep, 20" wide		15.65	735	2,200	6,600	565.20	
	5150	Ladder type, diesel, 5' deep, 8" wide		9.40	385	1,150	3,450	305.20	
	5200	Diesel, 8' deep, 16" wide		17.61	665	2,000	6,000	540.90	
	5250	Truck, dump, tandem, 12 ton payload		17.41	375	1,130	3,400	365.30	
	5300	Three axle dump, 16 ton payload		20.28	475	1,425	4,275	447.25	
	5350	Dump trailer only, rear dump, 16-1/2 C.Y.		3.45	177	530	1,600	133.60	
	5400	20 C.Y.		3.51	180	540	1,625	136.10	
	5450	Flatbed, single axle, 1-1/2 ton rating		10.87	143	430	1,300	172.95	
	5500	3 ton rating		11.43	148	445	1,325	180.45	
	5550	Off highway rear dump, 25 ton capacity		20.50	895	2,680	8,050	700	
	5600	35 ton capacity		32.27	1,325	4,000	12,000	1,058	
400	0010	**GENERAL EQUIPMENT RENTAL** Without operators							400
	0150	Aerial lift, scissor type, to 15' high, 1000 lb. cap., electric	Ea.	1.18	83.50	250	750	59.45	
	0160	To 25' high, 2000 lb. capacity		1.80	125	375	1,125	89.40	
	0170	Telescoping boom to 40' high, 750 lb. capacity, gas		6.42	335	1,000	3,000	251.35	
	0180	1000 lb. capacity		7.98	435	1,300	3,900	323.85	
	0195	Air compressor, portable, 6.5 CFM, electric		.12	33.50	100	300	20.95	
	0196	gasoline		.13	43.50	130	390	27.05	
	0200	Air compressor, portable, gas engine, 60 C.F.M.		5.41	58.50	175	525	78.30	
	0300	160 C.F.M.		6.80	73.50	220	660	98.40	
	0400	Diesel engine, rotary screw, 250 C.F.M.		6.50	125	375	1,125	127	
	0500	365 C.F.M.		9.30	167	500	1,500	174.40	
	0600	600 C.F.M.		16	250	750	2,250	278	
	0700	750 C.F.M.		17.60	267	800	2,400	300.80	
	0800	For silenced models, small sizes, add		3%	5%	5%	5%		

GENERAL REQUIREMENTS 1

01500 | Temporary Facilities & Controls
01590 | Equipment Rental

			UNIT	HOURLY OPER. COST	RENT PER DAY	RENT PER WEEK	RENT PER MONTH	CREW EQUIPMENT COST/DAY	
400	0900	Large sizes, add		5%	7%	7%	7%		400
	0920	Air tools and accessories							
	0930	Breaker, pavement, 60 lb.	Ea.	.19	30	90	270	19.50	
	0940	80 lb.		.21	40	120	360	25.70	
	0950	Drills, hand (jackhammer) 65 lb.		.23	28.50	85	255	18.85	
	0960	Track or wagon, swing boom, 4" drifter		10.57	380	1,140	3,425	312.55	
	0970	5" drifter		11.38	660	1,975	5,925	486.05	
	0980	Dust control per drill		2.11	12.65	38	114	24.50	
	0990	Hammer, chipping, 12 lb.		.12	26.50	80	240	16.95	
	1000	Hose, air with couplings, 50' long, 3/4" diameter		.15	5	15	45	4.20	
	1100	1" diameter		.15	6.65	20	60	5.20	
	1200	1-1/2" diameter		.15	10	30	90	7.20	
	1300	2" diameter		.21	16.65	50	150	11.70	
	1400	2-1/2" diameter		.23	20	60	180	13.85	
	1410	3" diameter		.24	25	75	225	16.90	
	1450	Drill, steel, 7/8" x 2'			4	12	36	2.40	
	1460	7/8" x 6'			5	15	45	3	
	1520	Moil points		.82	3.33	10	30	8.55	
	1525	Pneumatic nailer w/accessories		.12	30	90	270	18.95	
	1530	Sheeting driver for 60 lb. breaker		.15	11.65	35	105	8.20	
	1540	For 90 lb. breaker		.15	18.35	55	165	12.20	
	1550	Spade, 25 lb.		.08	8.35	25	75	5.65	
	1560	Tamper, single, 35 lb.		.10	25	75	225	15.80	
	1570	Triple, 140 lb.		1.80	41.50	125	375	39.40	
	1580	Wrenches, impact, air powered, up to 3/4" bolt		.25	22	66	198	15.20	
	1590	Up to 1-1/4" bolt		.35	41.50	125	375	27.80	
	1600	Barricades, barrels, reflectorized, 1 to 50 barrels			2.33	7	21	1.40	
	1610	100 to 200 barrels			1.67	5	15	1	
	1620	Barrels with flashers, 1 to 50 barrels			3.33	10	30	2	
	1630	100 to 200 barrels			2.67	8	24	1.60	
	1640	Barrels with steady burn type C lights			4	12	36	2.40	
	1650	Illuminated board, trailer mounted, with generator		.78	91.50	275	825	61.25	
	1670	Portable, stock, with flashers, 1 to 6 units			.83	2.50	7.50	.50	
	1680	25 to 50 units			.77	2.30	6.90	.45	
	1690	Butt fusion machine, electric		1.50	293	880	2,650	188	
	1695	Electro fusion machine		1.25	127	380	1,150	86	
	1700	Carts, brick, hand powered, 1000 lb. capacity		1.11	23.50	70	210	22.90	
	1800	Gas engine, 1500 lb., 7-1/2' lift		1.65	91.50	275	825	68.20	
	1830	Distributor, asphalt, trailer mtd, 2000 gal., 38 H.P. diesel		7.54	485	1,450	4,350	350.30	
	1840	3000 gal., 38 H.P. diesel		8.12	515	1,550	4,650	374.95	
	1850	Drill, rotary hammer, electric, 1-1/2" diameter		.15	25	75	225	16.20	
	1860	Carbide bit for above			6	18	54	3.60	
	1870	Emulsion sprayer, 65 gal., 5 H.P. gas engine		.63	56.50	170	510	39.05	
	1880	200 gal., 5 H.P. engine		.67	60	180	540	41.35	
	1920	Floodlight, mercury, vapor or quartz, on tripod							
	1930	1000 watt	Ea.	.12	25	75	225	15.95	
	1940	2000 watt		.13	41.50	125	375	26.05	
	1950	Floodlights, trailer mounted with generator, 1-300 watt light		4.96	86.50	260	780	91.70	
	2000	4-300 watt lights		6.10	100	300	900	108.80	
	2020	Forklift, wheeled, for brick, 18', 3000 lb., 2 wheel drive, gas		9.28	167	500	1,500	174.25	
	2040	28', 4000 lb., 4 wheel drive, diesel		6.40	200	600	1,800	171.20	
	2100	Generator, electric, gas engine, 1.5 KW to 3 KW		1.13	40	120	360	33.05	
	2200	5 KW		1.49	58.50	175	525	46.90	
	2300	10 KW		2.40	133	400	1,200	99.20	
	2400	25 KW		6.80	153	460	1,375	146.40	
	2500	Diesel engine, 20 KW		4.35	117	350	1,050	104.80	
	2600	50 KW		6.90	133	400	1,200	135.20	
	2700	100 KW		11.38	193	580	1,750	207.05	

01500 | Temporary Facilities & Controls

01590 | Equipment Rental

			UNIT	HOURLY OPER. COST	RENT PER DAY	RENT PER WEEK	RENT PER MONTH	CREW EQUIPMENT COST/DAY	
400	2800	250 KW	Ea.	29.05	335	1,000	3,000	432.40	400
	2850	Hammer, hydraulic, for mounting on boom, to 500 ft.-lb.		.99	133	400	1,200	87.90	
	2860	500 to 1200 ft.-lb.		2.50	258	775	2,325	175	
	2900	Heaters, space, oil or electric, 50 MBH		.11	26.50	80	240	16.90	
	3000	100 MBH		.12	33.50	100	300	20.95	
	3100	300 MBH		.13	50	150	450	31.05	
	3150	500 MBH		.15	66.50	200	600	41.20	
	3200	Hose, water, suction with coupling, 20' long, 2" diameter		.06	10	30	90	6.50	
	3210	3" diameter		.06	15	45	135	9.50	
	3220	4" diameter		.06	20	60	180	12.50	
	3230	6" diameter		.06	36.50	110	330	22.50	
	3240	8" diameter		.07	45	135	405	27.55	
	3250	Discharge hose with coupling, 50' long, 2" diameter		.05	8.35	25	75	5.40	
	3260	3" diameter		.05	10	30	90	6.40	
	3270	4" diameter		.06	13.35	40	120	8.50	
	3280	6" diameter		.07	33.50	100	300	20.55	
	3290	8" diameter		.09	36.50	110	330	22.70	
	3300	Ladders, extension type, 16' to 36' long			16.65	50	150	10	
	3400	40' to 60' long			30	90	270	18	
	3410	Level, laser type, for pipe laying, self leveling			93.50	281	845	56.20	
	3430	Manual leveling			66.50	200	600	40	
	3440	Rotary beacon with rod and sensor			91.50	275	825	55	
	3460	Builders level with tripod and rod			26.50	80	240	16	
	3500	Light towers, towable, with diesel generator, 2000 watt		1.60	125	375	1,125	87.80	
	3600	4000 watt		1.95	142	425	1,275	100.60	
	3700	Mixer, powered, plaster and mortar, 6 C.F., 7 H.P.		.85	56.50	170	510	40.80	
	3800	10 C.F., 9 H.P.		1.25	78.50	235	705	57	
	3850	Nailer, pneumatic		.12	28.50	85	255	17.95	
	3900	Paint sprayers complete, 8 CFM		.08	45	135	405	27.65	
	4000	17 CFM		.08	61.50	185	555	37.65	
	4020	Pavers, bituminous, rubber tires, 8' wide, 52 H.P., gas		14.95	575	1,725	5,175	464.60	
	4030	8' wide, 64 H.P., diesel		15.50	1,000	3,025	9,075	729	
	4050	Crawler, 10' wide, 78 H.P., gas		22.25	1,400	4,200	12,600	1,018	
	4060	10' wide, 87 H.P., diesel		22.75	1,075	3,200	9,600	822	
	4070	Concrete paver, 12' to 24' wide, 250 H.P.		24.25	1,525	4,550	13,700	1,104	
	4080	Placer-spreader-trimmer, 24' wide, 300 H.P.		32.35	1,775	5,325	16,000	1,324	
	4100	Pump, centrifugal gas pump, 1-1/2", 4 MGPH		.45	35	105	315	24.60	
	4200	2", 8 MGPH		.55	36	108	325	26	
	4300	3", 15 MGPH		1.25	46.50	140	420	38	
	4400	6", 90 MGPH		9.65	133	400	1,200	157.20	
	4500	Submersible electric pump, 1-1/4", 55 GPM		.35	36.50	110	330	24.80	
	4600	1-1/2", 83 GPM		.41	40	120	360	27.30	
	4700	2", 120 GPM		.42	46.50	140	420	31.35	
	4800	3", 300 GPM		.75	58.50	175	525	41	
	4900	4", 560 GPM		1.21	70	210	630	51.70	
	5000	6", 1590 GPM		5.38	217	650	1,950	173.05	
	5100	Diaphragm pump, gas, single, 1-1/2" diameter		.50	26.50	80	240	20	
	5200	2" diameter		.55	41.50	125	375	29.40	
	5300	3" diameter		.75	47.50	143	430	34.60	
	5400	Double, 4" diameter		1.58	88.50	265	795	65.65	
	5500	Trash pump, self-priming, gas, 2" diameter		1.26	40	120	360	34.10	
	5600	Diesel, 4" diameter		1.99	91.50	275	825	70.90	
	5650	Diesel, 6" diameter		5.05	128	385	1,150	117.40	
	5660	Rollers, see division 01590-200							
	5700	Salamanders, L.P. gas fired, 100,000 B.T.U.	Ea.	.75	21.50	65	195	19	
	5720	Sandblaster, portable, open top, 3 C.F. capacity		.20	50	150	450	31.60	
	5730	6 C.F. capacity		.35	63.50	190	570	40.80	
	5740	Accessories for above		.07	16.65	50	150	10.55	

01500 | Temporary Facilities & Controls

01590 | Equipment Rental

			UNIT	HOURLY OPER. COST	RENT PER DAY	RENT PER WEEK	RENT PER MONTH	CREW EQUIPMENT COST/DAY	
400	5750	Sander, floor	Ea.	.15	40.50	121	365	25.40	400
	5760	Edger		.10	25	75	225	15.80	
	5800	Saw, chain, gas engine, 18" long		.55	40	120	360	28.40	
	5900	36" long		1.15	66.50	200	600	49.20	
	6000	Masonry, table mounted, 14" diameter, 5 H.P.		1.80	53.50	160	480	46.40	
	6100	Circular, hand held, electric, 7-1/4" diameter		.18	21.50	65	195	14.45	
	6200	12" diameter		.28	33.50	100	300	22.25	
	6275	Shot blaster, walk behind, 20" wide		1.50	217	650	1,950	142	
	6300	Steam cleaner, 100 gallons per hour		.45	56.50	170	510	37.60	
	6310	200 gallons per hour		.75	66.50	200	600	46	
	6340	Tar Kettle/Pot, 400 gallon		.50	75	225	675	49	
	6350	Torch, cutting, acetylene-oxygen, 150' hose		7.50	33.50	100	300	80	
	6360	Hourly operating cost includes tips and gas		7.65				61.20	
	6420	Recycle flush type			20	60	180	12	
	6450	Toilet, trailers, minimum			33.50	100	300	20	
	6460	Maximum			100	300	900	60	
	6500	Trailers, platform, flush deck, 2 axle, 25 ton capacity		1.40	125	375	1,125	86.20	
	6600	40 ton capacity		1.75	230	690	2,075	152	
	6700	3 axle, 50 ton capacity		2.90	247	740	2,225	171.20	
	6800	75 ton capacity		3.75	325	980	2,950	226	
	6900	Water tank, engine driven discharge, 5000 gallons		8.24	233	700	2,100	205.90	
	7000	10,000 gallons		11	335	1,000	3,000	288	
	7020	Transit with tripod			33.50	100	300	20	
	7030	Trench box, 3000 lbs. 6'x8'		.80	100	300	900	66.40	
	7040	7200 lbs. 6'x20'		1.48	179	537	1,600	119.25	
	7050	8000 lbs., 8' x 16'		1.65	177	530	1,600	119.20	
	7060	9500 lbs., 8'x20'		1.80	190	570	1,700	128.40	
	7065	11,000 lbs., 8'x24'		1.85	241	724	2,175	159.60	
	7070	12,000 lbs., 10' x 20'		2.10	267	800	2,400	176.80	
	7100	Truck, pickup, 3/4 ton, 2 wheel drive		10.65	75	225	675	130.20	
	7200	4 wheel drive		12.08	80	240	720	144.65	
	7300	Tractor, 4 x 2, 30 ton capacity, 195 H.P.		11.20	395	1,180	3,550	325.60	
	7410	250 H.P.		15.25	435	1,300	3,900	382	
	7500	6 x 2, 40 ton capacity, 240 H.P.		17.95	475	1,420	4,250	427.60	
	7600	6 x 4, 45 ton capacity, 240 H.P.		23.15	575	1,730	5,200	531.20	
	7620	Vacuum truck, hazardous material, 2500 gallon		13.95	305	910	2,725	293.60	
	7625	5,000 gallon		15	335	1,000	3,000	320	
	7650	Vacuum, H.E.P.A., 16 gal., wet/dry		.90	55	165	495	40.20	
	7655	55 gal, wet/dry		.85	50	150	450	36.80	
	7690	Large production vacuum loader, 3150 CFM		11.85	725	2,175	6,525	529.80	
	7700	Welder, electric, 200 amp		.81	39	117	350	29.90	
	7800	300 amp		1.10	66.50	200	600	48.80	
	7900	Gas engine, 200 amp		4.40	55	165	495	68.20	
	8000	300 amp		5.25	70	210	630	84	
	8100	Wheelbarrow, any size			8.35	25	75	5	
	8200	Wrecking ball, 4000 lb.		.41	66.50	200	600	43.30	
600	0010	**LIFTING AND HOISTING EQUIPMENT RENTAL**							600
	0100	without operators							
	0600	Crawler, cable, 1/2 C.Y., 15 tons at 12' radius	Ea.	19	540	1,625	4,875	477	
	0700	3/4 C.Y., 20 tons at 12' radius		20.20	550	1,655	4,975	492.60	
	0800	1 C.Y., 25 tons at 12' radius		20.85	585	1,750	5,250	516.80	
	0900	Crawler, cable, 1-1/2 C.Y., 40 tons at 12' radius		30.54	785	2,350	7,050	714.30	
	1000	2 C.Y., 50 tons at 12' radius		35.60	945	2,830	8,500	850.80	
	1100	3 C.Y., 75 tons at 12' radius		43.65	985	2,950	8,850	939.20	
	1200	100 ton capacity, standard boom		42.43	1,325	4,000	12,000	1,139	
	1300	165 ton capacity, standard boom		64.75	2,200	6,600	19,800	1,838	
	1400	200 ton capacity, 150' boom		120.75	2,375	7,100	21,300	2,386	
	1500	450' boom		135	3,025	9,100	27,300	2,900	

01500 | Temporary Facilities & Controls

01590 | Equipment Rental

		UNIT	HOURLY OPER. COST	RENT PER DAY	RENT PER WEEK	RENT PER MONTH	CREW EQUIPMENT COST/DAY
1600	Truck mounted, cable operated, 6 x 4, 20 tons at 10' radius	Ea.	14.07	700	2,100	6,300	532.55
1700	25 tons at 10' radius		20.80	1,075	3,250	9,750	816.40
1800	8 x 4, 30 tons at 10' radius		28.64	615	1,850	5,550	599.10
1900	40 tons at 12' radius		29.35	785	2,350	7,050	704.80
2000	8 x 4, 60 tons at 15' radius		44.87	915	2,750	8,250	908.95
2100	90 tons at 15' radius		49	1,050	3,150	9,450	1,022
2200	115 tons at 15' radius		51.20	1,875	5,660	17,000	1,542
2300	150 tons at 18' radius		76.61	1,575	4,750	14,300	1,563
2400	Truck mounted, hydraulic, 12 ton capacity		22.93	450	1,350	4,050	453.45
2500	25 ton capacity		23.65	615	1,850	5,550	559.20
2550	33 ton capacity		24.37	835	2,500	7,500	694.95
2600	55 ton capacity		34.15	890	2,675	8,025	808.20
2700	80 ton capacity		37.30	1,325	4,000	12,000	1,098
2800	Self-propelled, 4 x 4, with telescoping boom, 5 ton		10.18	325	975	2,925	276.45
2900	12-1/2 ton capacity		16.42	450	1,350	4,050	401.35
3000	15 ton capacity		18.25	490	1,465	4,400	439
3100	25 ton capacity		20.99	660	1,980	5,950	563.90
3200	Derricks, guy, 20 ton capacity, 60' boom, 75' mast		8.75	300	900	2,700	250
3300	100' boom, 115' mast		16.39	520	1,560	4,675	443.10
3400	Stiffleg, 20 ton capacity, 70' boom, 37' mast		11.56	385	1,160	3,475	324.50
3500	100' boom, 47' mast		18.06	630	1,885	5,650	521.50
3600	Hoists, chain type, overhead, manual, 3/4 ton		.06	6.65	20	60	4.50
3900	10 ton		.25	28.50	85	255	19
4000	Hoist and tower, 5000 lb. cap., portable electric, 40' high		4	173	520	1,550	136
4100	For each added 10' section, add			13.35	40	120	8
5200	Jacks, hydraulic, 20 ton		.15	14.65	44	132	10
5500	100 ton		.20	40.50	122	365	26
6000	Jacks, hydraulic, climbing with 50' jackrods						
6010	and control consoles, minimum 3 mo. rental						
6100	30 ton capacity	Ea.	.06	100	300	900	60.50
6150	For each added 10' jackrod section, add			3.33	10	30	2
6300	50 ton capacity			160	480	1,450	96
6350	For each added 10' jackrod section, add			4	12	36	2.40
6500	125 ton capacity			415	1,250	3,750	250
6550	For each added 10' jackrod section, add			28.50	85	255	17
6600	Cable jack, 10 ton capacity with 200' cable			83.50	250	750	50
6650	For each added 50' of cable, add			8.35	25	75	5

01700 | Execution Requirements

01740 | Cleaning

			CREW	DAILY OUTPUT	LABOR-HOURS	UNIT	2000 BARE COSTS				TOTAL INCL O&P	
							MAT.	LABOR	EQUIP.	TOTAL		
500	0010	**CLEANING UP** After job completion, allow, minimum				Job					.30%	500
	0040	Maximum				"					1%	

For information about Means Estimating Seminars, see yellow pages 11 and 12 in back of book

Division 2
Site Construction

Estimating Tips

02200 Site Preparation
- If possible visit the site and take an inventory of the type, quantity and size of the trees. Certain trees may have a landscape resale value or firewood value. Stump disposal can be very expensive, particularly if they cannot be buried at the site. Consider using a bulldozer in lieu of hand cutting trees.
- Estimators should visit the site to determine the need for haul road, access, storage of materials, and security considerations. When estimating for access roads on unstable soil, consider using a geotextile stabilization fabric. It can greatly reduce the quantity of crushed stone or gravel. Sites of limited size and access can cause cost overruns due to lost productivity. Theft and damage is another consideration if the location is isolated. A temporary fence or security guards may be required. Investigate the site thoroughly.

02210 Subsurface Investigation
In preparing estimates on structures involving earthwork or foundations, all information concerning soil characteristics should be obtained. Look particularly for hazardous waste, evidence of prior dumping of debris, and previous stream beds.
- The costs shown for selective demolition do not include rubbish handling or disposal. These items should be estimated separately using Means data or other sources.

02300 Earthwork
- Estimating the actual cost of performing earthwork requires careful consideration of the variables involved. This includes items such as type of soil, whether or not water will be encountered, dewatering, whether or not banks need bracing, disposal of excavated earth, length of haul to fill or spoil sites, etc. If the project has large quantities of cut or fill, consider raising or lowering the site to reduce costs while paying close attention to the effect on site drainage and utilities if doing this.
- If the project has large quantities of fill, creating a borrow pit on the site can significantly lower the costs. It is very important to consider what time of year the project is scheduled for completion. Bad weather can create large cost overruns from dewatering, site repair and lost productivity from cold weather.

02700 Bases, Ballasts, Pavements/Appurtenances
- When estimating paving, keep in mind the project schedule. If an asphaltic paving project is in a colder climate and runs through to the spring, consider placing the base course in the autumn, then topping it in the spring just prior to completion. This could save considerable costs in spring repair. Keep in mind that prices for asphalt and concrete are generally higher in the cold seasons.

02500 Utility Services
02600 Drainage & Containment
- Never assume that the water, sewer and drainage lines will go in at the early stages of the project. Consider the site access needs before dividing the site in half with open trenches, loose pipe, and machinery obstructions. Always inspect the site to establish that the site drawings are complete. Check off all existing utilities on your drawing as you locate them. If you find any discrepancies, mark up the site plan for further research. Differing site conditions can be very costly if discovered later in the project.

02900 Planting
- The timing of planting and guarantee specifications often dictate the costs for establishing tree and shrub growth and a stand of grass or ground cover. Establish the work performance schedule to coincide with the local planting season. Maintenance and growth guarantees can add from 20% to 100% to the total landscaping cost. The cost to replace trees and shrubs can be as high as 5% of the total cost depending on the planting zone, soil conditions and time of year.

Reference Numbers
Reference numbers are shown in bold squares at the beginning of some major classifications. These numbers refer to related items in the Reference Section. The reference information may be an estimating procedure, an alternate pricing method or technical information.

Note: Not all subdivisions listed here necessarily appear in this publication.

02050 | Basic Site Materials & Methods

02060 | Aggregate

		CREW	DAILY OUTPUT	LABOR-HOURS	UNIT	MAT.	LABOR	EQUIP.	TOTAL	TOTAL INCL O&P	
150	0010 **BORROW**										150
	0020 and spread, with 200 H.P. dozer, no compaction										
	0100 Bank run gravel	B-15	600	.047	C.Y.	5.20	.79	2.84	8.83	10.15	
	0200 Common borrow		600	.047		4.77	.79	2.84	8.40	9.70	
	0300 Crushed stone, (1.40 tons per CY), 1-1/2"		600	.047		17.05	.79	2.84	20.68	23.50	
	0320 3/4"		600	.047		18.80	.79	2.84	22.43	25	
	0340 1/2"		600	.047		18.25	.79	2.84	21.88	24.50	
	0360 3/8"		600	.047		15.75	.79	2.84	19.38	22	
	0400 Sand, washed, concrete		600	.047		10.65	.79	2.84	14.28	16.15	
	0500 Dead or bank sand		600	.047		3.50	.79	2.84	7.13	8.30	

02080 | Utility Materials

		CREW	DAILY OUTPUT	LABOR-HOURS	UNIT	MAT.	LABOR	EQUIP.	TOTAL	TOTAL INCL O&P	
800	0010 **UTILITY VAULTS** Precast concrete, 6" thick										800
	0050 5' x 10' x 6' high, I.D.	B-13	2	24	Ea.	1,400	380	280	2,060	2,475	
	0350 Hand hole, precast concrete, 1-1/2" thick										
	0400 1'-0" x 2'-0" x 1'-9", I.D., light duty	B-1	4	6	Ea.	355	90.50		445.50	545	
	0450 4'-6" x 3'-2"-x 2'-0", O.D., heavy duty	B-6	3	8	"	880	128	67.50	1,075.50	1,250	

02200 | Site Preparation

02210 | Subsurface Investigation

		CREW	DAILY OUTPUT	LABOR-HOURS	UNIT	MAT.	LABOR	EQUIP.	TOTAL	TOTAL INCL O&P	
310	0010 **BORINGS** Initial field stake out and determination of elevations	A-6	1	16	Day		335		335	575	310
	0100 Drawings showing boring details				Total		170		170	245	
	0200 Report and recommendations from P.E.						375		375	540	
	0300 Mobilization and demobilization, minimum	B-55	4	4			60.50	166	226.50	285	
	0350 For over 100 miles, per added mile		450	.036	Mile		.54	1.47	2.01	2.53	
	0600 Auger holes in earth, no samples, 2-1/2" diameter		78.60	.204	L.F.		3.08	8.45	11.53	14.50	
	0800 Cased borings in earth, with samples, 2-1/2" diameter		55.50	.288	"	12.35	4.37	11.95	28.67	34	
	1400 Drill rig and crew with truck mounted auger		1	16	Day		242	665	907	1,150	
	1500 For inner city borings add, minimum									10%	
	1510 Maximum									20%	

02220 | Site Demolition

		CREW	DAILY OUTPUT	LABOR-HOURS	UNIT	MAT.	LABOR	EQUIP.	TOTAL	TOTAL INCL O&P	
100	0010 **BUILDING DEMOLITION** Large urban projects, incl. 20 Mi. haul										100
	0500 Small bldgs, or single bldgs, no salvage included, steel	B-3	14,800	.003	C.F.		.05	.11	.16	.21	
	0600 Concrete	"	11,300	.004	"		.07	.15	.22	.28	
	0605 Concrete, plain	B-5	33	1.212	C.Y.		19.30	29	48.30	65	
	0610 Reinforced		25	1.600			25.50	38.50	64	86	
	0615 Concrete walls		34	1.176			18.75	28.50	47.25	63	
	0620 Elevated slabs		26	1.538			24.50	37	61.50	82.50	
	0650 Masonry	B-3	14,800	.003	C.F.		.05	.11	.16	.21	
	0700 Wood	"	14,800	.003	"		.05	.11	.16	.21	
	1000 Single family, one story house, wood, minimum				Ea.				2,300	2,700	
	1020 Maximum								4,000	4,800	
	1200 Two family, two story house, wood, minimum								3,000	3,600	
	1220 Maximum								5,800	7,000	
	1300 Three family, three story house, wood, minimum								4,000	4,800	
	1320 Maximum								7,000	8,400	
	1400 Gutting building, see division 02225-400										

02200 | Site Preparation

02220 | Site Demolition

			CREW	DAILY OUTPUT	LABOR-HOURS	UNIT	2000 BARE COSTS				TOTAL INCL O&P	
							MAT.	LABOR	EQUIP.	TOTAL		
100	5000	For buildings with no interior walls, deduct				Ea.				50%		100
550	0010	**FOOTINGS AND FOUNDATIONS DEMOLITION**										550
	0200	Floors, concrete slab on grade,										
	0240	4" thick, plain concrete	B-9C	500	.080	S.F.		1.19	.36	1.55	2.44	
	0280	Reinforced, wire mesh		470	.085			1.26	.38	1.64	2.59	
	0300	Rods		400	.100			1.49	.45	1.94	3.04	
	0400	6" thick, plain concrete		375	.107			1.58	.48	2.06	3.24	
	0420	Reinforced, wire mesh		340	.118			1.75	.53	2.28	3.57	
	0440	Rods		300	.133			1.98	.60	2.58	4.05	
	1000	Footings, concrete, 1' thick, 2' wide	B-5	300	.133	L.F.		2.13	3.21	5.34	7.15	
	1080	1'-6" thick, 2' wide		250	.160			2.55	3.86	6.41	8.55	
	1120	3' wide		200	.200			3.19	4.82	8.01	10.70	
	1200	Average reinforcing, add								10%	10%	
	2000	Walls, block, 4" thick	1 Clab	180	.044	S.F.		.64		.64	1.10	
	2040	6" thick		170	.047			.68		.68	1.16	
	2080	8" thick		150	.053			.77		.77	1.32	
	2100	12" thick		150	.053			.77		.77	1.32	
	2400	Concrete, plain concrete, 6" thick	B-9	160	.250			3.71	1.13	4.84	7.60	
	2420	8" thick		140	.286			4.24	1.29	5.53	8.65	
	2440	10" thick		120	.333			4.95	1.50	6.45	10.15	
	2500	12" thick		100	.400			5.95	1.80	7.75	12.20	
	2600	For average reinforcing, add								10%	10%	
	4000	For congested sites or small quantities, add up to								200%	200%	
	4200	Add for disposal, on site	B-11A	232	.069	C.Y.		1.18	3.48	4.66	5.80	
	4250	To five miles	B-30	220	.109	"		1.90	7.30	9.20	11.20	
875	0010	**SITE DEMOLITION** No hauling, abandon catch basin or manhole	B-6	7	3.429	Ea.		55	29	84	125	875
	0020	Remove existing catch basin or manhole, masonry	"	4	6			96	51	147	218	
	0030	Catch basin or manhole frames and covers, stored	B-6	13	1.846			29.50	15.60	45.10	67	
	0040	Remove and reset	"	7	3.429			55	29	84	125	
	0600	Fencing, barbed wire, 3 strand	2 Clab	430	.037	L.F.		.54		.54	.92	
	0650	5 strand		280	.057			.83		.83	1.41	
	0700	Chain link, posts & fabric, remove only, 8' to 10' high	B-6	445	.054			.86	.46	1.32	1.96	
	0750	Remove and reset	"	70	.343			5.50	2.90	8.40	12.50	
	1000	Masonry walls, block or tile, solid, remove	B-5	1,800	.022	C.F.		.35	.54	.89	1.19	
	1100	Cavity wall		2,200	.018			.29	.44	.73	.97	
	1200	Brick, solid		900	.044			.71	1.07	1.78	2.38	
	1300	With block back-up		1,130	.035			.56	.85	1.41	1.90	
	1400	Stone, with mortar		900	.044			.71	1.07	1.78	2.38	
	1500	Dry set		1,500	.027			.43	.64	1.07	1.43	
	1710	Pavement removal, bituminous roads, 3" thick	B-38	690	.035	S.Y.		.56	.60	1.16	1.60	
	1750	4" to 6" thick		420	.057			.91	.99	1.90	2.64	
	1800	Bituminous driveways		640	.038			.60	.65	1.25	1.74	
	1900	Concrete to 6" thick, hydraulic hammer, mesh reinforced		255	.094			1.51	1.63	3.14	4.35	
	2000	Rod reinforced		200	.120			1.92	2.08	4	5.55	
	2300	With hand held air equipment, bituminous, to 6" thick	B-39	1,900	.025	S.F.		.37	.09	.46	.74	
	2320	Concrete to 6" thick, no reinforcing		1,200	.040			.59	.15	.74	1.18	
	2340	Mesh reinforced		1,400	.034			.51	.13	.64	1.01	
	2360	Rod reinforced		765	.063			.93	.24	1.17	1.85	
	2400	Curbs, concrete, plain	B-6	360	.067	L.F.		1.07	.56	1.63	2.42	
	2500	Reinforced		275	.087			1.40	.74	2.14	3.17	
	2600	Granite		360	.067			1.07	.56	1.63	2.42	
	2700	Bituminous		528	.045			.73	.38	1.11	1.65	
	2900	Pipe removal, sewer/water, no excavation, 12" diameter		175	.137			2.19	1.16	3.35	4.99	
	2960	24" diameter		120	.200			3.20	1.69	4.89	7.25	
	4000	Sidewalk removal, bituminous, 2-1/2" thick		325	.074	S.Y.		1.18	.62	1.80	2.69	

02200 | Site Preparation

02220 | Site Demolition

		CREW	DAILY OUTPUT	LABOR-HOURS	UNIT	2000 BARE COSTS MAT.	LABOR	EQUIP.	TOTAL	TOTAL INCL O&P
4050	Brick, set in mortar	B-6	185	.130	S.Y.		2.08	1.10	3.18	4.72
4100	Concrete, plain, 4"		160	.150			2.40	1.27	3.67	5.45
4200	Mesh reinforced	▼	150	.160	▼		2.56	1.35	3.91	5.80
5000	Slab on grade removal, plain	B-5	45	.889	C.Y.		14.15	21.50	35.65	47.50
5100	Mesh reinforced		33	1.212			19.30	29	48.30	65
5200	Rod reinforced	▼	25	1.600			25.50	38.50	64	86
5500	For congested sites or small quantities, add up to								200%	200%
5550	For disposal on site, add	B-11A	232	.069			1.18	3.48	4.66	5.80
5600	To 5 miles, add	B-34D	76	.105	▼		1.71	7.40	9.11	11

02225 | Selective Demolition

		CREW	DAILY OUTPUT	LABOR-HOURS	UNIT	MAT.	LABOR	EQUIP.	TOTAL	TOTAL INCL O&P
0010	**CEILING DEMOLITION**									
0200	Drywall, furred and nailed	2 Clab	800	.020	S.F.		.29		.29	.50
1000	Plaster, lime and horse hair, on wood lath, incl. lath		700	.023			.33		.33	.57
1200	Suspended ceiling, mineral fiber, 2'x2' or 2'x4'		1,500	.011			.15		.15	.26
1250	On suspension system, incl. system		1,200	.013			.19		.19	.33
1500	Tile, wood fiber, 12" x 12", glued		900	.018			.26		.26	.44
1540	Stapled		1,500	.011			.15		.15	.26
2000	Wood, tongue and groove, 1" x 4"		1,000	.016			.23		.23	.40
2040	1" x 8"		1,100	.015			.21		.21	.36
2400	Plywood or wood fiberboard, 4' x 8' sheets	▼	1,200	.013	▼		.19		.19	.33
0010	**CUTOUT DEMOLITION** Conc., elev. slab, light reinf., under 6 C.F.	B-9C	65	.615	C.F.		9.15	2.78	11.93	18.70
0050	Light reinforcing, over 6 C.F.	"	75	.533	"		7.90	2.41	10.31	16.20
0200	Slab on grade to 6" thick, not reinforced, under 8 S.F.	B-9	85	.471	S.F.		7	2.12	9.12	14.30
0250	Not reinforced, over 8 S.F.		175	.229	"		3.39	1.03	4.42	6.95
0600	Walls, not reinforced, under 6 C.F.		60	.667	C.F.		9.90	3.01	12.91	20.50
0650	Not reinforced, over 6 C.F.	▼	65	.615			9.15	2.78	11.93	18.70
1000	Concrete, elevated slab, bar reinforced, under 6 C.F.	B-9C	45	.889			13.20	4.01	17.21	27
1050	Bar reinforced, over 6 C.F.	"	50	.800	▼		11.90	3.61	15.51	24.50
1200	Slab on grade to 6" thick, bar reinforced, under 8 S.F.	B-9	75	.533	S.F.		7.90	2.41	10.31	16.20
1250	Bar reinforced, over 8 S.F.	"	105	.381	"		5.65	1.72	7.37	11.60
1400	Walls, bar reinforced, under 6 C.F.	B-9C	50	.800	C.F.		11.90	3.61	15.51	24.50
1450	Bar reinforced, over 6 C.F.	"	55	.727	"		10.80	3.28	14.08	22
2000	Brick, to 4 S.F. opening, not including toothing									
2040	4" thick	B-9C	30	1.333	Ea.		19.80	6	25.80	40.50
2060	8" thick		18	2.222			33	10	43	67.50
2080	12" thick		10	4			59.50	18.05	77.55	122
2400	Concrete block, to 4 S.F. opening, 2" thick		35	1.143			16.95	5.15	22.10	34.50
2420	4" thick		30	1.333			19.80	6	25.80	40.50
2440	8" thick		27	1.481			22	6.70	28.70	45
2460	12" thick	▼	24	1.667			25	7.50	32.50	51
2600	Gypsum block, to 4 S.F. opening, 2" thick	B-9	80	.500			7.45	2.26	9.71	15.20
2620	4" thick		70	.571			8.50	2.58	11.08	17.40
2640	8" thick		55	.727			10.80	3.28	14.08	22
2800	Terra cotta, to 4 S.F. opening, 4" thick		70	.571			8.50	2.58	11.08	17.40
2840	8" thick		65	.615			9.15	2.78	11.93	18.70
2880	12" thick	▼	50	.800	▼		11.90	3.61	15.51	24.50
3000	Toothing masonry cutouts, brick, soft old mortar	1 Brhe	40	.200	V.L.F.		3.14		3.14	5.30
3100	Hard mortar		30	.267			4.19		4.19	7.10
3200	Block, soft old mortar		70	.114			1.79		1.79	3.04
3400	Hard mortar	▼	50	.160	▼		2.51		2.51	4.26
4000	For toothing masonry, see Division 04900-800									
6000	Walls, interior, not including re-framing,									
6010	openings to 5 S.F.									
6100	Drywall to 5/8" thick	A-1	24	.333	Ea.		4.82	2.92	7.74	11.45

Important: See the Reference Section for critical supporting data - Reference Nos., Crews, & Location Factors

02200 | Site Preparation

02225 | Selective Demolition

			CREW	DAILY OUTPUT	LABOR-HOURS	UNIT	MAT.	LABOR	EQUIP.	TOTAL	TOTAL INCL O&P	
320	6200	Paneling to 3/4" thick	A-1	20	.400	Ea.		5.80	3.50	9.30	13.75	320
	6300	Plaster, on gypsum lath		20	.400			5.80	3.50	9.30	13.75	
	6340	On wire lath		14	.571			8.25	5	13.25	19.65	
	7000	Wood frame, not including re-framing, openings to 5 S.F.										
	7200	Floors, sheathing and flooring to 2" thick	A-1	5	1.600	Ea.		23	14	37	55	
	7310	Roofs, sheathing to 1" thick, not including roofing		6	1.333			19.25	11.65	30.90	46	
	7410	Walls, sheathing to 1" thick, not including siding		7	1.143			16.50	10	26.50	39.50	
340	0010	**DOOR DEMOLITION**										340
	0200	Doors, exterior, 1-3/4" thick, single, 3' x 7' high	1 Clab	16	.500	Ea.		7.20		7.20	12.40	
	0220	Double, 6' x 7' high		12	.667			9.65		9.65	16.50	
	0500	Interior, 1-3/8" thick, single, 3' x 7' high		20	.400			5.80		5.80	9.90	
	0520	Double, 6' x 7' high		16	.500			7.20		7.20	12.40	
	0700	Bi-folding, 3' x 6'-8" high		20	.400			5.80		5.80	9.90	
	0720	6' x 6'-8" high		18	.444			6.40		6.40	11	
	0900	Bi-passing, 3' x 6'-8" high		16	.500			7.20		7.20	12.40	
	0940	6' x 6'-8" high		14	.571			8.25		8.25	14.15	
	1500	Remove and reset, minimum	1 Carp	8	1			19.70		19.70	34	
	1520	Maximum	"	6	1.333			26.50		26.50	45	
	2000	Frames, including trim, metal	A-1	8	1			14.45	8.75	23.20	34.50	
	2200	Wood	2 Carp	32	.500			9.85		9.85	16.90	
	2201	Alternate pricing method	A-1	200	.040	L.F.		.58	.35	.93	1.38	
	3000	Special doors, counter doors	2 Carp	6	2.667	Ea.		52.50		52.50	90	
	3300	Glass, sliding, including frames		12	1.333			26.50		26.50	45	
	3400	Overhead, commercial, 12' x 12' high		4	4			79		79	135	
	3500	Residential, 9' x 7' high		8	2			39.50		39.50	67.50	
	3540	16' x 7' high		7	2.286			45		45	77	
	3600	Remove and reset, minimum		4	4			79		79	135	
	3620	Maximum		2.50	6.400			126		126	216	
	3660	Remove and reset elec. garage door opener	1 Carp	8	1			19.70		19.70	34	
	4000	Residential lockset, exterior		30	.267			5.25		5.25	9	
	4200	Deadbolt lock		32	.250			4.93		4.93	8.45	
380	0010	**FLOORING DEMOLITION**										380
	0200	Brick with mortar	2 Clab	475	.034	S.F.		.49		.49	.83	
	0400	Carpet, bonded, including surface scraping		2,000	.008			.12		.12	.20	
	0480	Tackless		9,000	.002			.03		.03	.04	
	0800	Resilient, sheet goods		1,400	.011			.17		.17	.28	
	0900	Vinyl composition tile, 12" x 12"		1,000	.016			.23		.23	.40	
	2000	Tile, ceramic, thin set		675	.024			.34		.34	.59	
	2020	Mud set		625	.026			.37		.37	.63	
	3000	Wood, block, on end	1 Carp	400	.020			.39		.39	.68	
	3200	Parquet		450	.018			.35		.35	.60	
	3400	Strip flooring, interior, 2-1/4" x 25/32" thick		325	.025			.48		.48	.83	
	3500	Exterior, porch flooring, 1" x 4"		220	.036			.72		.72	1.23	
	3800	Subfloor, tongue and groove, 1" x 6"		325	.025			.48		.48	.83	
	3820	1" x 8"		430	.019			.37		.37	.63	
	3840	1" x 10"		520	.015			.30		.30	.52	
	4000	Plywood, nailed		600	.013			.26		.26	.45	
	4100	Glued and nailed		400	.020			.39		.39	.68	
390	0010	**FRAMING DEMOLITION**										390
	3000	Wood framing, beams, 6" x 8"	B-2	275	.145	L.F.		2.16		2.16	3.70	
	3040	6" x 10"		220	.182			2.70		2.70	4.63	
	3080	6" x 12"		185	.216			3.21		3.21	5.50	
	3120	8" x 12"		140	.286			4.24		4.24	7.25	
	3160	10" x 12"		110	.364			5.40		5.40	9.25	

02200 | Site Preparation

02225 | Selective Demolition

			CREW	DAILY OUTPUT	LABOR-HOURS	UNIT	MAT.	2000 BARE COSTS LABOR	EQUIP.	TOTAL	TOTAL INCL O&P	
390	3400	Fascia boards, 1" x 6"	1 Clab	500	.016	L.F.		.23		.23	.40	390
	3440	1" x 8"		450	.018			.26		.26	.44	
	3480	1" x 10"		400	.020			.29		.29	.50	
	3800	Headers over openings, 2 @ 2" x 6"		110	.073			1.05		1.05	1.80	
	3840	2 @ 2" x 8"		100	.080			1.16		1.16	1.98	
	3880	2 @ 2" x 10"		90	.089			1.28		1.28	2.20	
	4230	Joists, 2" x 6"	2 Clab	970	.016			.24		.24	.41	
	4240	2" x 8"		940	.017			.25		.25	.42	
	4250	2" x 10"		910	.018			.25		.25	.44	
	4280	2" x 12"		880	.018			.26		.26	.45	
	5400	Posts, 4" x 4"		800	.020			.29		.29	.50	
	5440	6" x 6"		400	.040			.58		.58	.99	
	5480	8" x 8"		300	.053			.77		.77	1.32	
	5500	10" x 10"		240	.067			.96		.96	1.65	
	5800	Rafters, ordinary, 2" x 6"		850	.019			.27		.27	.47	
	5840	2" x 8"		837	.019			.28		.28	.47	
	6200	Stairs and stringers, minimum		40	.400	Riser		5.80		5.80	9.90	
	6240	Maximum		26	.615	"		8.90		8.90	15.25	
	6600	Studs, 2" x 4"		2,000	.008	L.F.		.12		.12	.20	
	6640	2" x 6"		1,600	.010	"		.14		.14	.25	
	7000	Trusses, 2" x 4" flat wood construction										
	7050	12' span	2 Clab	74	.216	Ea.		3.12		3.12	5.35	
	7150	24' span		66	.242			3.50		3.50	6	
	7200	26' span		64	.250			3.61		3.61	6.20	
	7250	28' span		62	.258			3.73		3.73	6.40	
	7300	30' span		58	.276			3.99		3.99	6.85	
	7350	32' span		56	.286			4.13		4.13	7.05	
	7400	34' span		54	.296			4.28		4.28	7.35	
	7450	36' span		52	.308			4.45		4.45	7.60	
	9500	See Div. 02225-730 for rubbish handling										
400	0010	**GUTTING** Building interior, including disposal, dumpster fees not included										400
	0500	Residential building										
	0560	Minimum	B-16	400	.080	SF Flr.		1.23	1.12	2.35	3.33	
	0580	Maximum	"	360	.089	"		1.37	1.24	2.61	3.70	
	0900	Commercial building										
	1000	Minimum	B-16	350	.091	SF Flr.		1.41	1.28	2.69	3.80	
	1020	Maximum	"	250	.128	"		1.97	1.79	3.76	5.30	
610	0010	**MASONRY DEMOLITION**										610
	1000	Chimney, 16" x 16", soft old mortar	A-1	24	.333	V.L.F.		4.82	2.92	7.74	11.45	
	1020	Hard mortar		18	.444			6.40	3.89	10.29	15.30	
	1080	20" x 20", soft old mortar		12	.667			9.65	5.85	15.50	23	
	1100	Hard mortar		10	.800			11.55	7	18.55	27.50	
	1140	20" x 32", soft old mortar		10	.800			11.55	7	18.55	27.50	
	1160	Hard mortar		8	1			14.45	8.75	23.20	34.50	
	1200	48" x 48", soft old mortar		5	1.600			23	14	37	55	
	1220	Hard mortar		4	2			29	17.50	46.50	69	
	2000	Columns, 8" x 8", soft old mortar		48	.167			2.41	1.46	3.87	5.75	
	2020	Hard mortar		40	.200			2.89	1.75	4.64	6.90	
	2060	16" x 16", soft old mortar		16	.500			7.20	4.38	11.58	17.20	
	2100	Hard mortar		14	.571			8.25	5	13.25	19.65	
	2140	24" x 24", soft old mortar		8	1			14.45	8.75	23.20	34.50	
	2160	Hard mortar		6	1.333			19.25	11.65	30.90	46	
	2200	36" x 36", soft old mortar		4	2			29	17.50	46.50	69	
	2220	Hard mortar		3	2.667			38.50	23.50	62	91.50	
	3000	Copings, precast or masonry, to 8" wide										

02200 | Site Preparation

02225 | Selective Demolition

			CREW	DAILY OUTPUT	LABOR-HOURS	UNIT	MAT.	2000 BARE COSTS LABOR	EQUIP.	TOTAL	TOTAL INCL O&P	
610	3020	Soft old mortar	A-1	180	.044	L.F.		.64	.39	1.03	1.53	610
	3040	Hard mortar	"	160	.050	"		.72	.44	1.16	1.72	
	3100	To 12" wide										
	3120	Soft old mortar	A-1	160	.050	L.F.		.72	.44	1.16	1.72	
	3140	Hard mortar	"	140	.057	"		.83	.50	1.33	1.96	
	4000	Fireplace, brick, 30" x 24" opening										
	4020	Soft old mortar	A-1	2	4	Ea.		58	35	93	138	
	4040	Hard mortar		1.25	6.400			92.50	56	148.50	220	
	4100	Stone, soft old mortar		1.50	5.333			77	46.50	123.50	184	
	4120	Hard mortar		1	8	▼		116	70	186	275	
	5000	Veneers, brick, soft old mortar		140	.057	S.F.		.83	.50	1.33	1.96	
	5020	Hard mortar		125	.064			.92	.56	1.48	2.20	
	5100	Granite and marble, 2" thick		180	.044			.64	.39	1.03	1.53	
	5120	4" thick		170	.047			.68	.41	1.09	1.61	
	5140	Stone, 4" thick		180	.044			.64	.39	1.03	1.53	
	5160	8" thick		175	.046	▼		.66	.40	1.06	1.57	
	5400	Alternate pricing method, stone, 4" thick		60	.133	C.F.		1.93	1.17	3.10	4.58	
	5420	8" thick	▼	85	.094	"		1.36	.82	2.18	3.24	
620	0010	**MILLWORK AND TRIM DEMOLITION**										620
	1000	Cabinets, wood, base cabinets	2 Clab	80	.200	L.F.		2.89		2.89	4.95	
	1020	Wall cabinets		80	.200			2.89		2.89	4.95	
	1100	Steel, painted, base cabinets		60	.267			3.85		3.85	6.60	
	1500	Counter top, minimum		200	.080			1.16		1.16	1.98	
	1510	Maximum		120	.133	▼		1.93		1.93	3.30	
	2000	Paneling, 4' x 8' sheets, 1/4" thick		2,000	.008	S.F.		.12		.12	.20	
	2100	Boards, 1" x 4"		700	.023			.33		.33	.57	
	2120	1" x 6"		750	.021			.31		.31	.53	
	2140	1" x 8"		800	.020	▼		.29		.29	.50	
	3000	Trim, baseboard, to 6" wide		1,200	.013	L.F.		.19		.19	.33	
	3040	12" wide		1,000	.016			.23		.23	.40	
	3100	Ceiling trim		1,000	.016			.23		.23	.40	
	3120	Chair rail		1,200	.013			.19		.19	.33	
	3140	Railings with balusters		240	.067	▼		.96		.96	1.65	
	3160	Wainscoting		700	.023	S.F.		.33		.33	.57	
	4000	Curtain rod	1 Clab	80	.100	L.F.		1.44		1.44	2.48	
690	0010	**ROOFING AND SIDING DEMOLITION**										690
	1200	Wood, boards, tongue and groove, 2" x 6"	2 Clab	960	.017	S.F.		.24		.24	.41	
	1220	2" x 10"		1,040	.015			.22		.22	.38	
	1280	Standard planks, 1" x 6"		1,080	.015			.21		.21	.37	
	1320	1" x 8"		1,160	.014			.20		.20	.34	
	1340	1" x 12"		1,200	.013			.19		.19	.33	
	1350	Plywood, to 1" thick	▼	2,000	.008			.12		.12	.20	
	1360	Flashing, aluminum	1 Clab	290	.028	▼		.40		.40	.68	
	2000	Gutters, aluminum or wood, edge hung	"	240	.033	L.F.		.48		.48	.83	
	2010	Remove and reset, aluminum	1 Shee	125	.064			1.38		1.38	2.32	
	2020	Remove and reset, vinyl	1 Carp	125	.064			1.26		1.26	2.16	
	2100	Built-in	1 Clab	100	.080	▼		1.16		1.16	1.98	
	2500	Roof accessories, plumbing vent flashing		14	.571	Ea.		8.25		8.25	14.15	
	2600	Adjustable metal chimney flashing		9	.889	"		12.85		12.85	22	
	2650	Coping, sheet metal, up to 12" wide	▼	240	.033	L.F.		.48		.48	.83	
	2660	Concrete, up to 12" wide	2 Clab	160	.100	"		1.44		1.44	2.48	
	3000	Roofing, built-up, 5 ply roof, no gravel	B-2	1,600	.025	S.F.		.37		.37	.64	
	3100	Gravel removal, minimum		5,000	.008			.12		.12	.20	
	3120	Maximum		2,000	.020			.30		.30	.51	
	3400	Roof insulation board, up to 2" thick	▼	3,900	.010	▼		.15		.15	.26	

02200 | Site Preparation

02225 | Selective Demolition

			CREW	DAILY OUTPUT	LABOR-HOURS	UNIT	MAT.	LABOR	EQUIP.	TOTAL	TOTAL INCL O&P	
690	3450	Roll roofing, cold adhesive	1 Clab	12	.667	Sq.		9.65		9.65	16.50	690
	4000	Shingles, asphalt strip, 1 layer	B-2	3,500	.011	S.F.		.17		.17	.29	
	4100	Slate		2,500	.016			.24		.24	.41	
	4300	Wood		2,200	.018			.27		.27	.46	
	4500	Skylight to 10 S.F.	1 Clab	8	1	Ea.		14.45		14.45	25	
	5000	Siding, metal, horizontal		444	.018	S.F.		.26		.26	.45	
	5020	Vertical		400	.020			.29		.29	.50	
	5200	Wood, boards, vertical		400	.020			.29		.29	.50	
	5220	Clapboards, horizontal		380	.021			.30		.30	.52	
	5240	Shingles		350	.023			.33		.33	.57	
	5260	Textured plywood		725	.011			.16		.16	.27	
730	0010	**RUBBISH HANDLING** The following are to be added to the										730
	0020	demolition prices										
	0400	Chute, circular, prefabricated steel, 18" diameter	B-1	40	.600	L.F.	19.35	9.05		28.40	37	
	0440	30" diameter	"	30	.800	"	25	12.10		37.10	48	
	0600	Dumpster, weekly rental, 1 dump/week, 6 C.Y. capacity (2 Tons)				Ea.					300	
	0700	10 C.Y. capacity (4 Tons)									375	
	0800	30 C.Y. capacity (10 Tons)									640	
	0840	40 C.Y. capacity (13 Tons)									775	
	1000	Dust partition, 6 mil polyethylene, 4' x 8' panels, 1" x 3" frame	2 Carp	2,000	.008	S.F.	.40	.16		.56	.71	
	1080	2" x 4" frame	"	2,000	.008	"	.50	.16		.66	.82	
	2000	Load, haul to chute & dumping into chute, 50' haul	2 Clab	24	.667	C.Y.		9.65		9.65	16.50	
	2040	100' haul		16.50	.970			14		14	24	
	2080	Over 100' haul, add per 100 L.F.		35.50	.451			6.50		6.50	11.15	
	2120	In elevators, per 10 floors, add		140	.114			1.65		1.65	2.83	
	3000	Loading & trucking, including 2 mile haul, chute loaded	B-16	45	.711			10.95	9.95	20.90	29.50	
	3040	Hand loading truck, 50' haul	"	48	.667			10.25	9.30	19.55	27.50	
	3080	Machine loading truck	B-17	120	.267			4.28	4.74	9.02	12.40	
	5000	Haul, per mile, up to 8 C.Y. truck	B-34B	1,165	.007			.11	.38	.49	.61	
	5100	Over 8 C.Y. truck	"	1,550	.005			.08	.29	.37	.46	
740	0010	**DUMP CHARGES** Typical urban city, tipping fees only										740
	0100	Building construction materials				Ton					55	
	0200	Trees, brush, lumber									45	
	0300	Rubbish only									50	
	0500	Reclamation station, usual charge									80	
840	0010	**WALLS AND PARTITIONS DEMOLITION**										840
	1000	Drywall, nailed	1 Clab	1,000	.008	S.F.		.12		.12	.20	
	1500	Fiberboard, nailed	"	900	.009			.13		.13	.22	
	2200	Metal or wood studs, finish 2 sides, fiberboard	B-1	520	.046			.70		.70	1.20	
	2250	Lath and plaster		260	.092			1.40		1.40	2.39	
	2300	Plasterboard (drywall)		520	.046			.70		.70	1.20	
	2350	Plywood		450	.053			.81		.81	1.38	
	3000	Plaster, lime and horsehair, on wood lath	1 Clab	400	.020			.29		.29	.50	
	3020	On metal lath	"	335	.024			.35		.35	.59	
850	0010	**WINDOW DEMOLITION**										850
	0200	Aluminum, including trim, to 12 S.F.	1 Clab	16	.500	Ea.		7.20		7.20	12.40	
	0240	To 25 S.F.		11	.727			10.50		10.50	18	
	0280	To 50 S.F.		5	1.600			23		23	39.50	
	0320	Storm windows, to 12 S.F.		27	.296			4.28		4.28	7.35	
	0360	To 25 S.F.		21	.381			5.50		5.50	9.45	
	0400	To 50 S.F.		16	.500			7.20		7.20	12.40	
	0600	Glass, minimum		200	.040	S.F.		.58		.58	.99	

02200 | Site Preparation

02225 | Selective Demolition

			CREW	DAILY OUTPUT	LABOR-HOURS	UNIT	MAT.	LABOR	EQUIP.	TOTAL	TOTAL INCL O&P	
850	0620	Maximum	1 Clab	150	.053	S.F.		.77		.77	1.32	850
	2000	Wood, including trim, to 12 S.F.		22	.364	Ea.		5.25		5.25	9	
	2020	To 25 S.F.		18	.444			6.40		6.40	11	
	2060	To 50 S.F.	↓	13	.615			8.90		8.90	15.25	
	5020	Remove and reset window, minimum	1 Carp	6	1.333			26.50		26.50	45	
	5040	Average		4	2			39.50		39.50	67.50	
	5080	Maximum	↓	2	4	↓		79		79	135	

02230 | Site Clearing

			CREW	DAILY OUTPUT	LABOR-HOURS	UNIT	MAT.	LABOR	EQUIP.	TOTAL	TOTAL INCL O&P	
200	0010	**CLEAR AND GRUB** Cut & chip light, trees to 6" diam.	B-7	1	48	Acre		755	1,100	1,855	2,475	200
	0150	Grub stumps and remove	B-30	2	12			209	805	1,014	1,225	
	0200	Cut & chip medium, trees to 12" diam.	B-7	.70	68.571			1,075	1,550	2,625	3,550	
	0250	Grub stumps and remove	B-30	1	24			420	1,600	2,020	2,475	
	0300	Cut & chip heavy, trees to 24" diam.	B-7	.30	160			2,500	3,650	6,150	8,275	
	0350	Grub stumps and remove	B-30	.50	48			835	3,225	4,060	4,925	
	0400	If burning is allowed, reduce cut & chip				↓					40%	
220	0010	**CLEARING** Brush with brush saw	A-1	.25	32	Acre		460	280	740	1,100	220
	0100	By hand	"	.12	66.667			965	585	1,550	2,300	
	0300	With dozer, ball and chain, light clearing	B-11A	2	8			137	405	542	675	
	0400	Medium clearing	"	1.50	10.667	↓		183	540	723	900	

02300 | Earthwork

02305 | Equipment

			CREW	DAILY OUTPUT	LABOR-HOURS	UNIT	MAT.	LABOR	EQUIP.	TOTAL	TOTAL INCL O&P	
250	0010	**MOBILIZATION OR DEMOBILIZATION** Up to 50 miles										250
	0020	Dozer, loader, backhoe or excavator, 70 H.P.- 250 H.P.	B-34K	6	1.333	Ea.		21.50	153	174.50	204	
	0900	Shovel or dragline, 3/4 C.Y.	"	3.60	2.222			36	254	290	340	
	1100	Delivery charge for small equipment on flatbed trailer, minimum									40	
	1150	Maximum				↓					100	

02310 | Grading

			CREW	DAILY OUTPUT	LABOR-HOURS	UNIT	MAT.	LABOR	EQUIP.	TOTAL	TOTAL INCL O&P	
460	0010	**LOAM OR TOPSOIL** Remove and stockpile on site										460
	0700	Furnish and place, truck dumped, screened, 4" deep	B-10S	1,300	.006	S.Y.	2.17	.12	.23	2.52	2.85	
	0800	6" deep	"	820	.010	"	2.78	.19	.37	3.34	3.78	
	0900	Fine grading and seeding, incl. lime, fertilizer & seed,										
	1000	With equipment	B-14	1,000	.048	S.Y.	.29	.75	.20	1.24	1.81	

02315 | Excavation and Fill

			CREW	DAILY OUTPUT	LABOR-HOURS	UNIT	MAT.	LABOR	EQUIP.	TOTAL	TOTAL INCL O&P	
100	0010	**BACKFILL** By hand, no compaction, light soil	1 Clab	14	.571	C.Y.		8.25		8.25	14.15	100
	0100	Heavy soil		11	.727			10.50		10.50	18	
	0300	Compaction in 6" layers, hand tamp, add to above	↓	20.60	.388			5.60		5.60	9.60	
	0500	Air tamp, add	B-9C	190	.211			3.13	.95	4.08	6.40	
	0600	Vibrating plate, add	A-1	60	.133			1.93	1.17	3.10	4.58	
	0800	Compaction in 12" layers, hand tamp, add to above	1 Clab	34	.235			3.40		3.40	5.80	
	1300	Dozer backfilling, bulk, up to 300' haul, no compaction	B-10B	1,200	.007			.13	.67	.80	.96	
	1400	Air tamped	B-11B	240	.067	↓		1.14	4.21	5.35	6.55	

02300 | Earthwork

02315 | Excavation and Fill

			CREW	DAILY OUTPUT	LABOR-HOURS	UNIT	MAT.	LABOR	EQUIP.	TOTAL	TOTAL INCL O&P	
130	0010	**BEDDING** For pipe and conduit, not incl. compaction										130
	0050	Crushed or screened bank run gravel	B-6	150	.160	C.Y.	6.50	2.56	1.35	10.41	12.90	
	0100	Crushed stone 3/4" to 1/2"		150	.160		17.05	2.56	1.35	20.96	24.50	
	0200	Sand, dead or bank		150	.160		3.50	2.56	1.35	7.41	9.65	
	0500	Compacting bedding in trench	A-1	90	.089			1.28	.78	2.06	3.06	
320	0010	**COMPACTION, STRUCTURAL** Steel wheel tandem roller, 5 tons (R02315-300)	B-10E	8	1	Hr.		19.90	16.65	36.55	51.50	320
	0050	Air tamp, 6" to 8" lifts, common fill	B-9	250	.160	C.Y.		2.38	.72	3.10	4.86	
	0060	Select fill	"	300	.133			1.98	.60	2.58	4.05	
	0600	Vibratory plate, 8" lifts, common fill	A-1	200	.040			.58	.35	.93	1.38	
	0700	Select fill	"	216	.037			.54	.32	.86	1.28	
400	0010	**EXCAVATING, BULK BANK MEASURE** Common earth piled (R02315-400)										400
	0020	For loading onto trucks, add								15%	15%	
	0200	Backhoe, hydraulic, crawler mtd., 1 C.Y. cap. = 75 C.Y./hr.	B-12A	600	.013	C.Y.		.28	.90	1.18	1.45	
	0310	Wheel mounted, 1/2 C.Y. cap. = 30 C.Y./hr.	B-12E	240	.033			.69	1.43	2.12	2.71	
	1200	Front end loader, track mtd., 1-1/2 C.Y. cap. = 70 C.Y./hr.	B-10N	560	.014			.28	.59	.87	1.12	
	1500	Wheel mounted, 3/4 C.Y. cap. = 45 C.Y./hr.	B-10R	360	.022			.44	.65	1.09	1.44	
	8000	For hauling excavated material, see div. 02320-200										
440	0010	**EXCAVATING, STRUCTURAL** Hand, pits to 6' deep, sandy soil	1 Clab	8	1	C.Y.		14.45		14.45	25	440
	0100	Heavy soil or clay		4	2			29		29	49.50	
	1100	Hand loading trucks from stock pile, sandy soil		12	.667			9.65		9.65	16.50	
	1300	Heavy soil or clay		8	1			14.45		14.45	25	
	1500	For wet or muck hand excavation, add to above				%				50%	50%	
505	0010	**FILL** Spread dumped material, by dozer, no compaction	B-10B	1,000	.008	C.Y.		.16	.81	.97	1.15	505
	0100	By hand	1 Clab	12	.667	"		9.65		9.65	16.50	
	0500	Gravel fill, compacted, under floor slabs, 4" deep	B-37	10,000	.005	S.F.	.15	.07	.01	.23	.31	
	0600	6" deep		8,600	.006		.23	.09	.02	.34	.42	
	0700	9" deep		7,200	.007		.38	.10	.02	.50	.61	
	0800	12" deep		6,000	.008		.52	.12	.02	.66	.81	
	1000	Alternate pricing method, 4" deep		120	.400	C.Y.	11.25	6.20	1.11	18.56	24	
	1100	6" deep		160	.300		11.25	4.67	.83	16.75	21.50	
	1200	9" deep		200	.240		11.25	3.73	.67	15.65	19.50	
	1300	12" deep		220	.218		11.25	3.39	.61	15.25	18.85	
900	0010	**EXCAVATING, TRENCH** or continuous footing, common earth										900
	0050	1' to 4' deep, 3/8 C.Y. tractor loader/backhoe	B-11C	150	.107	C.Y.		1.83	1.35	3.18	4.57	
	0060	1/2 C.Y. tractor loader/backhoe	B-11M	200	.080			1.37	1.37	2.74	3.81	
	0090	4' to 6' deep, 1/2 C.Y. tractor loader/backhoe	"	200	.080			1.37	1.37	2.74	3.81	
	0100	5/8 C.Y. hydraulic backhoe	B-12Q	250	.032			.66	1.65	2.31	2.91	
	0300	1/2 C.Y. hydraulic excavator, truck mounted	B-12J	200	.040			.83	3.09	3.92	4.77	
	1400	By hand with pick and shovel 2' to 6' deep, light soil	1 Clab	8	1			14.45		14.45	25	
	1500	Heavy soil	"	4	2			29		29	49.50	
940	0010	**EXCAVATING, UTILITY TRENCH** Common earth										940
	0050	Trenching with chain trencher, 12 H.P., operator walking										
	0100	4" wide trench, 12" deep	B-53	800	.010	L.F.		.14	.13	.27	.39	
	1000	Backfill by hand including compaction, add										
	1050	4" wide trench, 12" deep	A-1	800	.010	L.F.		.14	.09	.23	.35	

02320 | Hauling

			CREW	DAILY OUTPUT	LABOR-HOURS	UNIT	MAT.	LABOR	EQUIP.	TOTAL	TOTAL INCL O&P	
200	0011	**HAULING** Excavated or borrow material, loose cubic yards										200
	0015	no loading included, highway haulers										
	0020	6 C.Y. dump truck, 1/4 mile round trip, 5.0 loads/hr.	B-34A	195	.041	C.Y.		.66	1.87	2.53	3.17	
	0200	4 mile round trip, 1.8 loads/hr.	"	70	.114			1.85	5.20	7.05	8.85	

02300 | Earthwork

02320 | Hauling

		CREW	DAILY OUTPUT	LABOR-HOURS	UNIT	MAT.	LABOR	EQUIP.	TOTAL	TOTAL INCL O&P	
200	0310	12 C.Y. dump truck, 1/4 mile round trip 3.7 loads/hr.	B-34B	288	.028	C.Y.		.45	1.55	2	2.46	200
	0500	4 mile round trip, 1.6 loads/hr.	"	125	.064	↓		1.04	3.58	4.62	5.65	

02360 | Soil Treatment

		CREW	DAILY OUTPUT	LABOR-HOURS	UNIT	MAT.	LABOR	EQUIP.	TOTAL	TOTAL INCL O&P	
800	0010	**TERMITE PRETREATMENT**										800
	0020	Slab and walls, residential	1 Skwk	1,200	.007	SF Flr.	.22	.13		.35	.47	
	0400	Insecticides for termite control, minimum		14.20	.563	Gal.	10	11.15		21.15	30	
	0500	Maximum	↓	11	.727	"	17.10	14.35		31.45	43.50	

02370 | Erosion & Sedimentation Control

		CREW	DAILY OUTPUT	LABOR-HOURS	UNIT	MAT.	LABOR	EQUIP.	TOTAL	TOTAL INCL O&P	
550	0010	**EROSION CONTROL** Jute mesh, 100 S.Y. per roll, 4' wide, stapled	B-80A	2,400	.010	S.Y.	.62	.14	.08	.84	1.01	550
	0100	Plastic netting, stapled, 2" x 1" mesh, 20 mil	B-1	2,500	.010		.40	.15		.55	.69	
	0200	Polypropylene mesh, stapled, 6.5 oz./S.Y.		2,500	.010		1	.15		1.15	1.35	
	0300	Tobacco netting, or jute mesh #2, stapled	↓	2,500	.010	↓	.07	.15		.22	.33	
	1000	Silt fence, polypropylene, 3' high, ideal conditions	2 Clab	1,600	.010	L.F.	.30	.14		.44	.58	
	1100	Adverse conditions	"	950	.017	"	.30	.24		.54	.75	

02400 | Tunneling, Boring & Jacking

02441 | Microtunneling

		CREW	DAILY OUTPUT	LABOR-HOURS	UNIT	MAT.	LABOR	EQUIP.	TOTAL	TOTAL INCL O&P	
400	0010	**MICROTUNNELING** Not including excavation, backfill, shoring,										400
	0020	or dewatering, average 50'/day, slurry method										
	0100	24" to 48" outside diameter, minimum				L.F.					600	
	0110	Adverse conditions, add				%					50%	
	1000	Rent microtunneling machine, average monthly lease				Month					80,000	
	1010	Operating technician				Day					600	
	1100	Mobilization and demobilization, minimum				Job					40,000	
	1110	Maximum				"					400,000	

02500 | Utility Services

02510 | Water Distribution

		CREW	DAILY OUTPUT	LABOR-HOURS	UNIT	MAT.	LABOR	EQUIP.	TOTAL	TOTAL INCL O&P	
800	0010	**PIPING, WATER DISTRIBUTION SYSTEMS** Pipe laid in trench,										800
	0020	excavation and backfill not included										
	1400	Ductile Iron, cement lined, class 50 water pipe, 18' lengths										
	1410	Mechanical joint, 4" diameter	B-20	144	.167	L.F.	7.15	2.52		9.67	12.20	
	2650	Polyvinyl chloride pipe, class 160, S.D.R.-26, 1-1/2" diameter		300	.080		.27	1.21		1.48	2.37	
	2700	2" diameter		250	.096		.38	1.45		1.83	2.91	
	2750	2-1/2" diameter		250	.096		.50	1.45		1.95	3.04	
	2800	3" diameter		200	.120		.75	1.81		2.56	3.94	
	2850	4" diameter	↓	200	.120	↓	1.18	1.81		2.99	4.41	

02500 | Utility Services

02520 | Wells

			DAILY	LABOR-		2000 BARE COSTS				TOTAL	
		CREW	OUTPUT	HOURS	UNIT	MAT.	LABOR	EQUIP.	TOTAL	INCL O&P	
900	0010 **WELLS** Domestic water										900
	0100 Drilled, 4" to 6" diameter	B-23	120	.333	L.F.		4.95	16.45	21.40	26.50	
	1500 Pumps, installed in wells to 100' deep, 4" submersible										
	1520 3/4 H.P.	Q-1	2.66	6.015	Ea.	475	119		594	720	
	1600 1 H.P.	"	2.29	6.987		525	138		663	810	
	1800 2 H.P.	Q-22	1.33	12.030		620	238	340	1,198	1,450	

02530 | Sanitary Sewerage

730	0010 **PIPING, DRAINAGE & SEWAGE, CONCRETE** R02510-810										730
	0020 Not including excavation or backfill										
	1020 8" diameter	B-14	224	.214	L.F.	3.82	3.33	.91	8.06	10.85	
	1030 10" diameter	"	216	.222	"	4.23	3.46	.94	8.63	11.60	
	3780 Concrete slotted pipe, class 4 mortar joint										
	3800 12" diameter	B-21	168	.167	L.F.	10.80	2.65	.82	14.27	17.35	
	3840 18" diameter	"	152	.184	"	16.75	2.93	.91	20.59	24.50	
	3900 Class 4 O-ring										
	3940 12" diameter	B-21	168	.167	L.F.	12.35	2.65	.82	15.82	19.05	
	3960 18" diameter	"	152	.184	"	18.55	2.93	.91	22.39	26.50	
780	0010 **PIPING, DRAINAGE & SEWAGE, POLYVINYL CHLORIDE**										780
	0020 Not including excavation or backfill										
	2000 10' lengths, S.D.R. 35, B&S, 4" diameter	B-20	375	.064	L.F.	2.21	.97		3.18	4.09	
	2040 6" diameter		350	.069		2.87	1.04		3.91	4.94	
	2080 8" diameter		335	.072		4.40	1.08		5.48	6.70	
	2120 10" diameter	B-21	330	.085		4.84	1.35	.42	6.61	8.05	
790	0010 **PIPING, DRAINAGE & SEWAGE, VITRIFIED CLAY** C700										790
	0020 Not including excavation or backfill,										
	4030 Extra strength, compression joints, C425										
	5000 4" diameter x 4' long	B-20	265	.091	L.F.	1.79	1.37		3.16	4.32	
	5020 6" diameter x 5' long	"	200	.120		2.93	1.81		4.74	6.35	
	5040 8" diameter x 5' long	B-21	200	.140		4.14	2.23	.69	7.06	9.10	
	5060 10" diameter x 5' long	"	190	.147		6.80	2.34	.73	9.87	12.25	

02540 | Septic Tank Systems

700	0010 **SEPTIC TANKS** Not incl. excav. or piping, precast, 1,000 gallon	B-21	8	3.500	Ea.	490	55.50	17.30	562.80	650	700
	0100 2,000 gallon		5	5.600		1,000	89	27.50	1,116.50	1,275	
	0600 High density polyethylene, 1,000 gallon		6	4.667		800	74.50	23	897.50	1,025	
	0700 1,500 gallon		4	7		1,000	111	34.50	1,145.50	1,325	
	1000 Distribution boxes, concrete, 7 outlets	2 Clab	16	1		86.50	14.45		100.95	120	
	1100 9 outlets	"	8	2		225	29		254	298	
	1150 Leaching field chambers, 13' x 3'-7" x 1'-4", standard	B-13	16	3		665	47.50	35	747.50	850	
	1420 Leaching pit, 6', dia, 3' deep complete					500			500	550	
	2200 Excavation for septic tank, 3/4 C.Y. backhoe	B-12F	145	.055	C.Y.		1.14	3.14	4.28	5.35	
	2400 4' trench for disposal field, 3/4 C.Y. backhoe	"	335	.024	L.F.		.49	1.36	1.85	2.32	
	2600 Gravel fill, run of bank	B-6	150	.160	C.Y.	5.20	2.56	1.35	9.11	11.50	
	2800 Crushed stone, 3/4"	"	150	.160	"	21	2.56	1.35	24.91	29.50	

02550 | Piped Energy Distribution

464	0010 **PIPING, GAS SERVICE & DISTRIBUTION, POLYETHYLENE**										464
	0020 not including excavation or backfill										
	1000 60 psi coils, comp cplg @ 100', 1/2" diameter, SDR 9.3	B-20A	608	.053	L.F.	.36	.93		1.29	1.96	
	1040 1-1/4" diameter, SDR 11		544	.059		.59	1.04		1.63	2.38	
	1100 2" diameter, SDR 11		488	.066		.74	1.15		1.89	2.75	
	1160 3" diameter, SDR 11		408	.078		1.55	1.38		2.93	4.02	
	1500 60 PSI 40' joints with coupling, 3" diameter, SDR 11	B-21A	408	.098		1.55	1.79	.98	4.32	5.75	
	1540 4" diameter, SDR 11		352	.114		3.44	2.07	1.14	6.65	8.50	

02500 | Utility Services

02550 | Piped Energy Distribution

			CREW	DAILY OUTPUT	LABOR-HOURS	UNIT	2000 BARE COSTS MAT.	LABOR	EQUIP.	TOTAL	TOTAL INCL O&P	
464	1600	6" diameter, SDR 11	B-21A	328	.122	L.F.	11.25	2.22	1.22	14.69	17.45	464
	1640	8" diameter, SDR 11	↓	272	.147	↓	15	2.68	1.48	19.16	22.50	

02580 | Elec/Communication Structures

			CREW	DAILY OUTPUT	LABOR-HOURS	UNIT	MAT.	LABOR	EQUIP.	TOTAL	TOTAL INCL O&P	
300	0010	**ELECTRIC & TELEPHONE SITE WORK** Not including excavation										300
	0200	backfill and cast in place concrete										
	4200	Underground duct, banks ready for concrete fill, min. of 7.5"										
	4400	between conduits, ctr. to ctr.(for wire & cable see div. 16120)										
	4601	PVC, type EB, 2 @ 2" diameter	2 Elec	240	.067	L.F.	1.14	1.47		2.61	3.67	
	4800	4 @ 2" diameter		120	.133		2.29	2.95		5.24	7.35	
	5600	4 @ 4" diameter		80	.200		4.86	4.42		9.28	12.60	
	6200	Rigid galvanized steel, 2 @ 2" diameter		180	.089		9.30	1.96		11.26	13.40	
	6400	4 @ 2" diameter		90	.178		18.60	3.93		22.53	27	
	7400	4 @ 4" diameter	↓	34	.471	↓	59	10.40		69.40	81.50	

02600 | Drainage & Containment

02620 | Subdrainage

			CREW	DAILY OUTPUT	LABOR-HOURS	UNIT	MAT.	LABOR	EQUIP.	TOTAL	TOTAL INCL O&P	
210	0010	**PIPING, SUBDRAINAGE, CONCRETE**	R02510-810									210
	0021	Not including excavation and backfill										
	3000	Porous wall concrete underdrain, std. strength, 4" diameter	B-20	335	.072	L.F.	1.78	1.08		2.86	3.82	
	3020	6" diameter	"	315	.076		2.31	1.15		3.46	4.52	
	3040	8" diameter	B-21	310	.090	↓	2.86	1.44	.45	4.75	6.10	
240	0010	**PIPING, SUBDRAINAGE, CORRUGATED METAL**										240
	0021	Not including excavation and backfill										
	2010	Aluminum, perforated										
	2020	6" diameter, 18 ga.	B-14	380	.126	L.F.	2.57	1.97	.53	5.07	6.75	
	2200	8" diameter, 16 ga.		370	.130		3.73	2.02	.55	6.30	8.15	
	2220	10" diameter, 16 ga.	↓	360	.133	↓	4.66	2.07	.56	7.29	9.30	
	3000	Uncoated galvanized, perforated										
	3020	6" diameter, 18 ga.	B-20	380	.063	L.F.	4	.95		4.95	6.05	
	3200	8" diameter, 16 ga.	"	370	.065		5.50	.98		6.48	7.75	
	3220	10" diameter, 16 ga.	B-21	360	.078	↓	8.25	1.24	.38	9.87	11.60	
	4000	Steel, perforated, asphalt coated										
	4020	6" diameter 18 ga.	B-20	380	.063	L.F.	3.20	.95		4.15	5.15	
	4030	8" diameter 18 ga	"	370	.065		5	.98		5.98	7.20	
	4040	10" diameter 16 ga	B-21	360	.078		5.75	1.24	.38	7.37	8.90	
	4050	12" diameter 16 ga		285	.098		6.60	1.56	.49	8.65	10.45	
	4060	18" diameter 16 ga	↓	205	.137	↓	9	2.17	.67	11.84	14.35	
280	0010	**PIPING, SUBDRAINAGE, VITRIFIED CLAY**										280
	4000	Channel pipe, 4" diameter	B-20	430	.056	L.F.	2	.84		2.84	3.65	
	4060	8" diameter	"	295	.081	"	4.50	1.23		5.73	7.05	

02630 | Storm Drainage

			CREW	DAILY OUTPUT	LABOR-HOURS	UNIT	MAT.	LABOR	EQUIP.	TOTAL	TOTAL INCL O&P	
100	0010	**PIPING, STORM DRAINAGE, CORRUGATED METAL**										100
	0020	Not including excavation or backfill										
	2040	8" diameter, 16 ga.	B-14	330	.145	L.F.	7.20	2.26	.62	10.08	12.50	

02600 | Drainage & Containment

02630 | Storm Drainage

		CREW	DAILY OUTPUT	LABOR-HOURS	UNIT	2000 BARE COSTS MAT.	LABOR	EQUIP.	TOTAL	TOTAL INCL O&P	
200	0010	**CATCH BASINS OR MANHOLES** not including footing, excavation,										200
	0020	backfill, frame and cover										
	0050	Brick, 4' inside diameter, 4' deep	D-1	1	16	Ea.	264	286		550	775	
	1110	Precast, 4' I.D., 4' deep	B-22	4.10	7.317		315	119	50.50	484.50	605	
	1600	Frames & covers, C.I., 24" square, 500 lb.	B-6	7.80	3.077	↓	213	49	26	288	345	

02700 | Bases, Ballasts, Pavements & Appurtenances

02720 | Unbound Base Courses & Ballasts

		CREW	DAILY OUTPUT	LABOR-HOURS	UNIT	2000 BARE COSTS MAT.	LABOR	EQUIP.	TOTAL	TOTAL INCL O&P	
200	0010	**BASE COURSE** For roadways and large paved areas										200
	0050	Crushed 3/4" stone base, compacted, 3" deep	B-36B	5,200	.012	S.Y.	2.95	.22	.61	3.78	4.28	
	0100	6" deep		5,000	.013		5.95	.23	.63	6.81	7.60	
	0200	9" deep		4,600	.014		8.90	.25	.69	9.84	10.95	
	0300	12" deep		4,200	.015		12.10	.27	.75	13.12	14.60	
	0301	Crushed 1-1/2" stone base, compacted to 4" deep		6,000	.011		3.14	.19	.53	3.86	4.35	
	0302	6" deep		5,400	.012		4.81	.21	.59	5.61	6.30	
	0303	8" deep		4,500	.014		6.30	.25	.70	7.25	8.10	
	0304	12" deep	↓	3,800	.017	↓	9.60	.30	.83	10.73	12	
	0350	Bank run gravel, spread and compacted										
	0370	6" deep	B-32	6,000	.005	S.Y.	2.20	.10	.26	2.56	2.87	
	0390	9" deep		4,900	.007		3.24	.12	.32	3.68	4.11	
	0400	12" deep	↓	4,200	.008	↓	4.40	.14	.37	4.91	5.50	
	8900	For small and irregular areas, add						50%	50%			
215	0010	**BASE** Prepare and roll sub-base, small areas to 2500 S.Y.	B-32A	1,500	.016	S.Y.		.29	.63	.92	1.18	215

02740 | Flexible Pavement

		CREW	DAILY OUTPUT	LABOR-HOURS	UNIT	2000 BARE COSTS MAT.	LABOR	EQUIP.	TOTAL	TOTAL INCL O&P	
315	0010	**PAVING** Asphaltic concrete R02065-300										315
	0020	6" stone base, 2" binder course, 1" topping	B-25C	9,000	.005	S.F.	1.07	.09	.17	1.33	1.52	
	0300	Binder course, 1-1/2" thick		35,000	.001		.27	.02	.04	.33	.39	
	0400	2" thick		25,000	.002		.35	.03	.06	.44	.51	
	0500	3" thick		15,000	.003		.54	.05	.10	.69	.80	
	0600	4" thick		10,800	.004		.71	.07	.14	.92	1.06	
	0800	Sand finish course, 3/4" thick		41,000	.001		.15	.02	.04	.21	.24	
	0900	1" thick	↓	34,000	.001		.19	.02	.05	.26	.30	
	1000	Fill pot holes, hot mix, 2" thick	B-16	4,200	.008		.39	.12	.11	.62	.74	
	1100	4" thick		3,500	.009		.56	.14	.13	.83	1	
	1120	6" thick	↓	3,100	.010		.76	.16	.14	1.06	1.26	
	1140	Cold patch, 2" thick	B-51	3,000	.016		.45	.24	.06	.75	.97	
	1160	4" thick		2,700	.018		.86	.27	.06	1.19	1.47	
	1180	6" thick	↓	1,900	.025	↓	1.33	.38	.09	1.80	2.22	

02750 | Rigid Pavement

		CREW	DAILY OUTPUT	LABOR-HOURS	UNIT	2000 BARE COSTS MAT.	LABOR	EQUIP.	TOTAL	TOTAL INCL O&P	
100	0010	**CONCRETE PAVEMENT** Including joints, finishing, and curing										100
	0020	Fixed form, 12' pass, unreinforced, 6" thick	B-26	3,000	.029	S.Y.	15.85	.49	.61	16.95	18.95	
	0100	8" thick	"	2,750	.032		22	.53	.67	23.20	26	
	0700	Finishing, broom finish small areas	2 Cefi	120	.133	↓		2.52		2.52	4.11	

02700 | Bases, Ballasts, Pavements & Appurtenances

02770 | Curbs and Gutters

			CREW	DAILY OUTPUT	LABOR-HOURS	UNIT	MAT.	LABOR	EQUIP.	TOTAL	TOTAL INCL O&P	
225	0010	**CURBS** Asphaltic, machine formed, 8" wide, 6" high, 40 L.F./ton	B-27	1,000	.032	L.F.	.54	.48	.07	1.09	1.50	225
	0150	Asphaltic berm, 12" W, 3"-6" H, 35 L.F./ton, before pavement	"	700	.046		.80	.68	.10	1.58	2.17	
	0200	12" W, 1-1/2" to 4" H, 60 L.F. per ton, laid with pavement	B-2	1,050	.038		.49	.57		1.06	1.51	
	0300	Concrete, wood forms, 6" x 18", straight	C-2A	500	.096		2.15	1.83		3.98	5.45	
	0400	6" x 18", radius	"	200	.240		2.26	4.56		6.82	10.25	
	0550	Precast, 6" x 18", straight	B-29	700	.069		6.25	1.08	.88	8.21	9.70	
	0600	6" x 18", radius	"	325	.148		7.75	2.34	1.90	11.99	14.55	
	1000	Granite, split face, straight, 5" x 16"	D-13	500	.096		14.10	1.80	.80	16.70	19.40	
	1100	6" x 18"	"	450	.107		18.55	2	.89	21.44	25	
	1300	Radius curbing, 6" x 18", over 10' radius	B-29	260	.185		22.50	2.92	2.38	27.80	32.50	
	1400	Corners, 2' radius		80	.600	Ea.	76	9.50	7.75	93.25	109	
	1600	Edging, 4-1/2" x 12", straight		300	.160	L.F.	7.05	2.53	2.06	11.64	14.35	
	1800	Curb inlets, (guttermouth) straight		41	1.171	Ea.	169	18.50	15.10	202.60	234	
	2000	Indian granite (belgian block)										
	2100	Jumbo, 10-1/2" x 7-1/2" x 4", grey	D-1	150	.107	L.F.	1.75	1.90		3.65	5.15	
	2150	Pink		150	.107		2.15	1.90		4.05	5.60	
	2200	Regular, 9" x 4-1/2" x 4-1/2", grey		160	.100		1.70	1.79		3.49	4.89	
	2250	Pink		160	.100		2	1.79		3.79	5.20	
	2300	Cubes, 4" x 4" x 4", grey		175	.091		1.65	1.63		3.28	4.58	
	2350	Pink		175	.091		1.75	1.63		3.38	4.69	
	2400	6" x 6" x 6", pink		155	.103		3.60	1.84		5.44	7.10	
	2500	Alternate pricing method for indian granite										
	2550	Jumbo, 10-1/2" x 7-1/2" x 4" (30lb), grey				Ton	100			100	110	
	2600	Pink					125			125	138	
	2650	Regular, 9" x 4-1/2" x 4-1/2" (20lb), grey					120			120	132	
	2700	Pink					140			140	154	
	2750	Cubes, 4" x 4" x 4" (5lb), grey					200			200	220	
	2800	Pink					225			225	248	
	2850	6" x 6" x 6" (25lb), pink					140			140	154	
	2900	For pallets, add					15			15	16.50	

02775 | Sidewalks

			CREW	DAILY OUTPUT	LABOR-HOURS	UNIT	MAT.	LABOR	EQUIP.	TOTAL	TOTAL INCL O&P	
275	0010	**SIDEWALKS, DRIVEWAYS, & PATIOS** No base										275
	0020	Asphaltic concrete, 2" thick	B-37	720	.067	S.Y.	3.16	1.04	.19	4.39	5.45	
	0100	2-1/2" thick	"	660	.073	"	4	1.13	.20	5.33	6.55	
	0300	Concrete, 3000 psi, CIP, 6 x 6 - W1.4 x W1.4 mesh,										
	0310	broomed finish, no base, 4" thick	B-24	600	.040	S.F.	1.05	.71		1.76	2.34	
	0350	5" thick		545	.044		1.40	.78		2.18	2.85	
	0400	6" thick		510	.047		1.63	.83		2.46	3.19	
	0450	For bank run gravel base, 4" thick, add	B-18	2,500	.010		.13	.15	.02	.30	.43	
	0520	8" thick, add	"	1,600	.015		.27	.23	.04	.54	.72	
	1000	Crushed stone, 1" thick, white marble	2 Clab	1,700	.009		.23	.14		.37	.48	
	1050	Bluestone	"	1,700	.009		.18	.14		.32	.43	
	1700	Redwood, prefabricated, 4' x 4' sections	2 Carp	316	.051		7.15	1		8.15	9.55	
	1750	Redwood planks, 1" thick, on sleepers	"	240	.067		4.99	1.31		6.30	7.75	

02778 | Steps

			CREW	DAILY OUTPUT	LABOR-HOURS	UNIT	MAT.	LABOR	EQUIP.	TOTAL	TOTAL INCL O&P	
280	0010	**STEPS** Incl. excav., borrow & concrete base, where applicable										280
	0100	Brick steps	B-24	35	.686	LF Riser	7.60	12.10		19.70	29	
	0200	Railroad ties	2 Clab	25	.640		2.68	9.25		11.93	18.80	
	0300	Bluestone treads, 12" x 2" or 12" x 1-1/2"	B-24	30	.800		19.30	14.15		33.45	45	

02780 | Unit Pavers

			CREW	DAILY OUTPUT	LABOR-HOURS	UNIT	MAT.	LABOR	EQUIP.	TOTAL	TOTAL INCL O&P	
100	0010	**ASPHALT BLOCKS**, 6"x12"x1-1/4", w/bed & neopr. adhesive	D-1	135	.119	S.F.	2.84	2.12		4.96	6.70	100
	0100	3" thick		130	.123		3.98	2.20		6.18	8.10	

02700 | Bases, Ballasts, Pavements & Appurtenances

02780 | Unit Pavers

			Crew	Daily Output	Labor-Hours	Unit	Mat.	Labor	Equip.	Total	Total Incl O&P	
100	0300	Hexagonal tile, 8" wide, 1-1/4" thick	D-1	135	.119	S.F.	2.91	2.12		5.03	6.80	100
	0400	2" thick		130	.123		4.07	2.20		6.27	8.20	
	0500	Square, 8" x 8", 1-1/4" thick		135	.119		2.89	2.12		5.01	6.75	
	0600	2" thick	↓	130	.123	↓	4.05	2.20		6.25	8.15	
200	0010	**BRICK PAVING** 4" x 8" x 1-1/2", without joints (4.5 brick/S.F.)	D-1	110	.145	S.F.	2.04	2.60		4.64	6.65	200
	0100	Grouted, 3/8" joint (3.9 brick/S.F.)		90	.178		2.39	3.17		5.56	8	
	0200	4" x 8" x 2-1/4", without joints (4.5 bricks/S.F.)		110	.145		2.64	2.60		5.24	7.30	
	0300	Grouted, 3/8" joint (3.9 brick/S.F.)	↓	90	.178		2.44	3.17		5.61	8.05	
	0500	Bedding, asphalt, 3/4" thick	B-25	5,130	.017		.30	.28	.35	.93	1.18	
	0540	Course washed sand bed, 1" thick	B-18	5,000	.005		.15	.07	.01	.23	.29	
	0580	Mortar, 1" thick	D-1	300	.053		.36	.95		1.31	2	
	0620	2" thick		200	.080		.36	1.43		1.79	2.81	
	1500	Brick on 1" thick sand bed laid flat, 4.5 per S.F.		100	.160		3.02	2.86		5.88	8.15	
	2000	Brick pavers, laid on edge, 7.2 per S.F.	↓	70	.229	↓	1.99	4.08		6.07	9.10	
800	0010	**STONE PAVERS**										800
	1100	Flagging, bluestone, irregular, 1" thick,	D-1	81	.198	S.F.	4	3.53		7.53	10.35	
	1150	Snapped random rectangular, 1" thick		92	.174		6.05	3.10		9.15	11.90	
	1200	1-1/2" thick		85	.188		7.30	3.36		10.66	13.70	
	1250	2" thick		83	.193		8.50	3.44		11.94	15.20	
	1300	Slate, natural cleft, irregular, 3/4" thick		92	.174		1.80	3.10		4.90	7.25	
	1350	Random rectangular, gauged, 1/2" thick		105	.152		3.90	2.72		6.62	8.90	
	1400	Random rectangular, butt joint, gauged, 1/4" thick	↓	150	.107		4.19	1.90		6.09	7.85	
	1450	For sand rubbed finish, add					2.50			2.50	2.75	
	1550	Granite blocks, 3-1/2" x 3-1/2" x 3-1/2"	D-1	92	.174	↓	5.25	3.10		8.35	11.05	

02785 | Flexible Pavement Coating

			Crew	Daily Output	Labor-Hours	Unit	Mat.	Labor	Equip.	Total	Total Incl O&P	
800	0010	**SEALCOATING** 2 coat coal tar pitch emulsion over 10,000 S.Y.	B-45	5,000	.003	S.Y.	.43	.05	.07	.55	.63	800
	0030	1000 to 10,000 S.Y.	"	3,000	.005		.43	.08	.12	.63	.75	
	0100	Under 1000 S.Y.	B-1	1,050	.023		.43	.35		.78	1.06	
	0300	Petroleum resistant, over 10,000 S.Y.	B-45	5,000	.003		.50	.05	.07	.62	.71	
	0320	1000 to 10,000 S.Y.	"	3,000	.005		.50	.08	.12	.70	.83	
	0400	Under 1000 S.Y.	B-1	1,050	.023	↓	.50	.35		.85	1.14	

02800 | Site Improvements and Amenities

02810 | Irrigation System

			Crew	Daily Output	Labor-Hours	Unit	Mat.	Labor	Equip.	Total	Total Incl O&P	
800	0010	**SPRINKLER IRRIGATION SYSTEM** For lawns										800
	0800	Residential system, custom, 1" supply	B-20	2,000	.012	S.F.	.25	.18		.43	.59	
	0900	1-1/2" supply	"	1,800	.013	"	.28	.20		.48	.66	

02820 | Fences & Gates

			Crew	Daily Output	Labor-Hours	Unit	Mat.	Labor	Equip.	Total	Total Incl O&P	
500	0010	**FENCE, MISC. METAL** Chicken wire, posts @ 4', 1" mesh, 4' high	B-80	410	.059	L.F.	1.10	.89	1.34	3.33	4.21	500
	0100	2" mesh, 6' high		350	.069		1	1.04	1.57	3.61	4.61	
	0200	Galv. steel, 12 ga., 2" x 4" mesh, posts 5' O.C., 3' high		300	.080		1.50	1.21	1.84	4.55	5.75	
	0300	5' high		300	.080		2	1.21	1.84	5.05	6.30	
	0400	14 ga., 1" x 2" mesh, 3' high		300	.080		1.60	1.21	1.84	4.65	5.85	
	0500	5' high	↓	300	.080	↓	2.20	1.21	1.84	5.25	6.50	
	1000	Kennel fencing, 1-1/2" mesh, 6' long, 3'-6" wide, 6'-2" high	2 Clab	4	4	Ea.	250	58		308	375	
	1050	12' long	↓	4	4	↓	300	58		358	430	

02800 | Site Improvements and Amenities

02820 | Fences & Gates

			CREW	DAILY OUTPUT	LABOR-HOURS	UNIT	2000 BARE COSTS MAT.	LABOR	EQUIP.	TOTAL	TOTAL INCL O&P	
500	1200	Top covers, 1-1/2" mesh, 6' long	2 Clab	15	1.067	Ea.	51	15.40		66.40	82.50	500
	1250	12' long	↓	12	1.333	↓	81	19.25		100.25	123	
	1300	For kennel doors, see division 08344-350										
525	0011	**CHAIN LINK FENCE**										525
	0350	Aluminized steel, 9 ga. wire, 3' high	B-1	185	.130	L.F.	4.64	1.96		6.60	8.45	
	0400	6' high		115	.209	"	6.80	3.16		9.96	12.90	
	0450	Add for gate 3' wide, 1-3/8" frame 3' high		12	2	Ea.	45.50	30		75.50	102	
	0490	6' high		10	2.400		92.50	36.50		129	164	
	0500	Add for gate 4' wide, 1-3/8" frame 3' high		10	2.400		62	36.50		98.50	130	
	0540	6' high	↓	8	3	↓	129	45.50		174.50	220	
	0860	Tennis courts, 11 ga. wire, 2 1/2" post 10' O.C., 1-5/8" top rail										
	0900	2-1/2" corner post, 10' high	B-1	95	.253	L.F.	10.30	3.82		14.12	17.90	
	0920	12' high		80	.300	"	12.35	4.54		16.89	21.50	
	1000	Add for gate 3' wide, 1-5/8" frame 10' high		10	2.400	Ea.	129	36.50		165.50	204	
	1040	Aluminized, 11 ga. wire 10' high		95	.253	L.F.	14.40	3.82		18.22	22.50	
	1100	12' high		80	.300	"	14.95	4.54		19.49	24	
	1140	Add for gate 3' wide, 1-5/8" frame, 10' high		10	2.400	Ea.	165	36.50		201.50	243	
	1250	Vinyl covered 11 ga. wire, 10' high		95	.253	L.F.	12.35	3.82		16.17	20	
	1300	12' high		80	.300	"	14.40	4.54		18.94	23.50	
	1400	Add for gate 3' wide, 1-3/8" frame, 10' high	↓	10	2.400	Ea.	185	36.50		221.50	266	
900	0010	**FENCE, WOOD** Basket weave, 3/8" x 4" boards, 2" x 4"										900
	0020	stringers on spreaders, 4" x 4" posts										
	0050	No. 1 cedar, 6' high	B-1	160	.150	L.F.	7.15	2.27		9.42	11.75	
	0860	Open rail fence, split rails, 2 rail 3' high, no. 1 cedar		160	.150		4.33	2.27		6.60	8.65	
	0870	No. 2 cedar		160	.150		3.54	2.27		5.81	7.80	
	0880	3 rail, 4' high, no. 1 cedar		150	.160		6	2.42		8.42	10.75	
	0890	No. 2 cedar		150	.160		4.04	2.42		6.46	8.60	
	1240	Stockade fence, no. 1 cedar, 3-1/4" rails, 6' high		160	.150		8.50	2.27		10.77	13.25	
	1260	8' high		155	.155	↓	11	2.34		13.34	16.10	
	1320	Gate, 3'-6" wide	↓	8	3	Ea.	50	45.50		95.50	133	
925	0010	**FENCE, RAIL** Picket, No. 2 cedar, Gothic, 2 rail, 3' high	B-1	160	.150	L.F.	4.47	2.27		6.74	8.80	925
	0050	Gate, 3'-6" wide		9	2.667	Ea.	38.50	40.50		79	112	
	0400	3 rail, 4' high		150	.160	L.F.	5.15	2.42		7.57	9.80	
	0500	Gate, 3'-6" wide		9	2.667	Ea.	46	40.50		86.50	120	
	1200	Stockade, No. 2 cedar, treated wood rails, 6' high		160	.150	L.F.	5.55	2.27		7.82	10.05	
	1250	Gate, 3' wide		9	2.667	Ea.	45.50	40.50		86	120	
	1300	No. 1 cedar, 3-1/4" cedar rails, 6' high		160	.150	L.F.	13.55	2.27		15.82	18.85	
	1500	Gate, 3' wide		9	2.667	Ea.	108	40.50		148.50	188	
	2700	Prefabricated redwood or cedar, 4' high		160	.150	L.F.	10.60	2.27		12.87	15.55	
	2800	6' high		150	.160	"	14.05	2.42		16.47	19.60	
	3300	Board, shadow box, 1" x 6", treated pine, 6' high		160	.150	L.F.	8.25	2.27		10.52	12.95	
	3400	No. 1 cedar, 6' high		150	.160		16.25	2.42		18.67	22	
	3900	Basket weave, No. 1 cedar, 6' high		160	.150	↓	16.15	2.27		18.42	21.50	
	4200	Gate, 3'-6" wide	↓	9	2.667	Ea.	48	40.50		88.50	122	

02830 | Retaining Walls

			CREW	DAILY OUTPUT	LABOR-HOURS	UNIT	MAT.	LABOR	EQUIP.	TOTAL	TOTAL INCL O&P	
100	0010	**RETAINING WALLS** Aluminized steel bin, excavation										100
	0020	and backfill not included, 10' wide										
	0100	4' high, 5.5' deep	B-13	650	.074	S.F.	14.70	1.17	.86	16.73	19.10	
	0200	8' high, 5.5' deep		615	.078		16.90	1.23	.91	19.04	21.50	
	0300	10' high, 7.7' deep		580	.083		17.75	1.31	.96	20.02	23	
	0400	12' high, 7.7' deep	↓	530	.091	↓	19.15	1.43	1.06	21.64	24.50	

For expanded coverage of these items see *Means Site Work and Landscape Cost Data 2000*

02800 | Site Improvements and Amenities

02830 | Retaining Walls

			CREW	DAILY OUTPUT	LABOR-HOURS	UNIT	MAT.	LABOR	EQUIP.	TOTAL	TOTAL INCL O&P	
100	0500	16' high, 7.7' deep	B-13	515	.093	S.F.	20	1.47	1.09	22.56	26	100
	1800	Concrete gravity wall with vertical face including excavation & backfill										
	1850	No reinforcing										
	1900	6' high, level embankment	C-17C	36	2.306	L.F.	45.50	46.50	11.45	103.45	143	
	2000	33° slope embankment	"	32	2.594	"	42	52.50	12.85	107.35	150	
	2800	Reinforced concrete cantilever, incl. excavation, backfill & reinf.										
	2900	6' high, 33° slope embankment	C-17C	35	2.371	L.F.	42	48	11.75	101.75	141	
400	0010	STONE WALL Including excavation, concrete footing and										400
	0020	stone 3' below grade. Price is exposed face area.										
	0200	Decorative random stone, to 6' high, 1'-6" thick, dry set	D-1	35	.457	S.F.	7.60	8.15		15.75	22	
	0300	Mortar set		40	.400		9.25	7.15		16.40	22.50	
	0500	Cut stone, to 6' high, 1'-6" thick, dry set		35	.457		11.50	8.15		19.65	26.50	
	0600	Mortar set		40	.400		13.50	7.15		20.65	27	
	0800	Retaining wall, random stone, 6' to 10' high, 2' thick, dry set		45	.356		9.50	6.35		15.85	21	
	0900	Mortar set		50	.320		11.50	5.70		17.20	22.50	
	1100	Cut stone, 6' to 10' high, 2' thick, dry set		45	.356		14.75	6.35		21.10	27	
	1200	Mortar set		50	.320		15.75	5.70		21.45	27	

02870 | Site Furnishings

			CREW	DAILY OUTPUT	LABOR-HOURS	UNIT	MAT.	LABOR	EQUIP.	TOTAL	TOTAL INCL O&P	
610	0010	BENCHES Park, precast concrete, w/backs, wood rails, 4' long	2 Clab	5	3.200	Ea.	370	46		416	485	610
	0100	8' long		4	4		765	58		823	940	
	0500	Steel barstock pedestals w/backs, 2" x 3" wood rails, 4' long		10	1.600		755	23		778	870	
	0510	8' long		7	2.286		895	33		928	1,050	
	0800	Cast iron pedestals, back & arms, wood slats, 4' long		8	2		460	29		489	555	
	0820	8' long		5	3.200		765	46		811	920	
	1700	Steel frame, fir seat, 10' long		10	1.600		150	23		173	205	

02900 | Planting

02905 | Transplanting

			CREW	DAILY OUTPUT	LABOR-HOURS	UNIT	MAT.	LABOR	EQUIP.	TOTAL	TOTAL INCL O&P	
725	0010	PLANTING Moving shrubs on site, 12" ball	B-62	28	.857	Ea.		13.70	4.16	17.86	27.50	725
	0100	24" ball	"	22	1.091			17.45	5.30	22.75	35.50	
	0300	Moving trees on site, 36" ball	B-6	3.75	6.400			102	54	156	233	
	0400	60" ball	"	1	24			385	203	588	875	

02910 | Plant Preparation

			CREW	DAILY OUTPUT	LABOR-HOURS	UNIT	MAT.	LABOR	EQUIP.	TOTAL	TOTAL INCL O&P	
500	0010	MULCH										500
	0100	Aged barks, 3" deep, hand spread	1 Clab	100	.080	S.Y.	2	1.16		3.16	4.18	
	0150	Skid steer loader	B-63	13.50	2.963	M.S.F.	220	43	8.60	271.60	325	
	0200	Hay, 1" deep, hand spread	1 Clab	475	.017	S.Y.	.25	.24		.49	.70	
	0250	Power mulcher, small	B-64	180	.089	M.S.F.	18.50	1.35	1.60	21.45	24.50	
	0350	Large	B-65	530	.030	"	18.50	.46	.86	19.82	22	
	0400	Humus peat, 1" deep, hand spread	1 Clab	700	.011	S.Y.	1.11	.17		1.28	1.50	
	0450	Push spreader	A-1	2,500	.003	"	1.39	.05	.03	1.47	1.64	
	0550	Tractor spreader	B-66	700	.011	M.S.F.	104	.22	.27	104.49	115	
	0600	Oat straw, 1" deep, hand spread	1 Clab	475	.017	S.Y.	.28	.24		.52	.73	
	0650	Power mulcher, small	B-64	180	.089	M.S.F.	25	1.35	1.60	27.95	31.50	
	0700	Large	B-65	530	.030	"	25	.46	.86	26.32	29	

Important: See the Reference Section for critical supporting data - Reference Nos., Crews, & Location Factors

02900 | Planting

02910 | Plant Preparation

			CREW	DAILY OUTPUT	LABOR-HOURS	UNIT	2000 BARE COSTS MAT.	LABOR	EQUIP.	TOTAL	TOTAL INCL O&P	
500	0750	Add for asphaltic emulsion	B-45	1,770	.009	Gal.	1.60	.14	.21	1.95	2.22	500
	0800	Peat moss, 1" deep, hand spread	1 Clab	900	.009	S.Y.	1.55	.13		1.68	1.92	
	0850	Push spreader	A-1	2,500	.003	"	1.60	.05	.03	1.68	1.87	
	0950	Tractor spreader	B-66	700	.011	M.S.F.	175	.22	.27	175.49	194	
	1000	Polyethylene film, 6 mil.	2 Clab	2,000	.008	S.Y.	.15	.12		.27	.37	
	1100	Redwood nuggets, 3" deep, hand spread	1 Clab	150	.053	"	6	.77		6.77	7.90	
	1150	Skid steer loader	B-63	13.50	2.963	M.S.F.	600	43	8.60	651.60	745	
	1200	Stone mulch, hand spread, ceramic chips, economy	1 Clab	125	.064	S.Y.	5.75	.92		6.67	7.95	
	1250	Deluxe	"	95	.084	"	8.60	1.22		9.82	11.55	
	1300	Granite chips	B-1	10	2.400	C.Y.	28	36.50		64.50	93	
	1400	Marble chips		10	2.400		105	36.50		141.50	178	
	1500	Onyx gemstone		10	2.400		310	36.50		346.50	400	
	1600	Pea gravel		28	.857		61.50	12.95		74.45	89.50	
	1700	Quartz		10	2.400		135	36.50		171.50	211	
	1800	Tar paper, 15 Lb. felt	1 Clab	800	.010	S.Y.	.40	.14		.54	.69	
	1900	Wood chips, 2" deep, hand spread	"	220	.036	"	1.65	.53		2.18	2.72	
	1950	Skid steer loader	B-63	20.30	1.970	M.S.F.	108	28.50	5.75	142.25	174	

02912 | General Planting

275	0010	GROUND COVER Plants, pachysandra, in prepared beds	B-1	15	1.600	C	15	24		39	58	275
	0200	Vinca minor, 1 yr, bare root		12	2	"	19	30		49	73	
	0600	Stone chips, in 50 lb. bags, Georgia marble		520	.046	Bag	3.36	.70		4.06	4.90	
	0700	Onyx gemstone		260	.092		13.10	1.40		14.50	16.85	
	0800	Quartz		260	.092		5.50	1.40		6.90	8.45	
	0900	Pea gravel, truckload lots		28	.857	Ton	20.50	12.95		33.45	44.50	

02920 | Lawns & Grasses

500	0010	SEEDING Mechanical seeding, 215 lb./acre (R02920-500)	B-66	1.50	5.333	Acre	485	102	125	712	840	500
	0100	44 lb./M.S.Y.	"	2,500	.003	S.Y.	.15	.06	.07	.28	.35	
	0300	Fine grading and seeding incl. lime, fertilizer & seed,										
	0310	with equipment	B-14	1,000	.048	S.Y.	.16	.75	.20	1.11	1.67	
	0600	Limestone hand push spreader, 50 lbs. per M.S.F.	1 Clab	180	.044	M.S.F.	3.25	.64		3.89	4.68	
	0800	Grass seed hand push spreader, 4.5 lbs. per M.S.F.	"	180	.044	"	10	.64		10.64	12.10	
600	0010	SODDING 1" deep, bluegrass sod, on level ground, over 8 M.S.F.	B-63	22	1.818	M.S.F.	165	26.50	5.30	196.80	233	600
	0200	4 M.S.F.		17	2.353		230	34	6.85	270.85	320	
	0300	1000 S.F.		3.50	11.429		230	165	33.50	428.50	575	
	0500	Sloped ground, over 8 M.S.F.		6	6.667		165	96.50	19.40	280.90	370	
	0600	4 M.S.F.		5	8		230	116	23.50	369.50	475	
	0700	1000 S.F.		4	10		230	145	29	404	535	
	1000	Bent grass sod, on level ground, over 6 M.S.F.		20	2		315	29	5.80	349.80	400	
	1100	3 M.S.F.		18	2.222		230	32	6.45	268.45	315	
	1200	Sodding 1000 S.F. or less		14	2.857		345	41.50	8.30	394.80	460	
	1500	Sloped ground, over 6 M.S.F.		15	2.667		315	38.50	7.75	361.25	420	
	1600	3 M.S.F.		13.50	2.963		230	43	8.60	281.60	335	
	1700	1000 S.F.		12	3.333		188	48	9.70	245.70	300	

02930 | Exterior Plants

050	0010	SHRUBS AND TREES Evergreen, in prepared beds, B & B										050
	0100	Arborvitae pyramidal, 4'-5'	B-17	30	1.067	Ea.	35	17.10	18.95	71.05	88.50	
	0150	Globe, 12"-15"	B-1	96	.250		10.05	3.78		13.83	17.50	
	0300	Cedar, blue, 8'-10'	B-17	18	1.778		146	28.50	31.50	206	244	
	0500	Hemlock, canadian, 2-1/2'-3'	B-1	36	.667		14.25	10.10		24.35	33	
	0550	Holly, Savannah, 8'-10' H		9.68	2.479		500	37.50		537.50	615	
	0600	Juniper, andorra, 18"-24"		80	.300		14	4.54		18.54	23	
	0620	Wiltoni, 15"-18"		80	.300		14.50	4.54		19.04	23.50	

For expanded coverage of these items see *Means Site Work and Landscape Cost Data 2000*

02900 | Planting

02930 | Exterior Plants

			CREW	DAILY OUTPUT	LABOR-HOURS	UNIT	2000 BARE COSTS MAT.	LABOR	EQUIP.	TOTAL	TOTAL INCL O&P	
050	0640	Skyrocket, 4-1/2'-5'	B-17	55	.582	Ea.	38.50	9.35	10.35	58.20	69.50	050
	0660	Blue pfitzer, 2'-2-1/2'	B-1	44	.545		16	8.25		24.25	32	
	0680	Ketleerie, 2-1/2'-3'		50	.480		28	7.25		35.25	43.50	
	0700	Pine, black, 2-1/2'-3'		50	.480		29.50	7.25		36.75	45	
	0720	Mugo, 18"-24"		60	.400		30	6.05		36.05	43.50	
	0740	White, 4'-5'	B-17	75	.427		45.50	6.85	7.60	59.95	70	
	0800	Spruce, blue, 18"-24"	B-1	60	.400		28	6.05		34.05	41.50	
	0840	Norway, 4'-5'	B-17	75	.427		56	6.85	7.60	70.45	81.50	
	0900	Yew, denisforma, 12"-15"	B-1	60	.400		22.50	6.05		28.55	35.50	
	1000	Capitata, 18"-24"		30	.800		21	12.10		33.10	43.50	
	1100	Hicksi, 2'-2-1/2'		30	.800		26	12.10		38.10	49	
410	0010	**SHRUBS** Broadleaf evergreen, planted in prepared beds										410
	0100	Andromeda, 15"-18", container	B-1	96	.250	Ea.	14	3.78		17.78	22	
	0200	Azalea, 15" - 18", container		96	.250		18.50	3.78		22.28	27	
	0300	Barberry, 9"-12", container		130	.185		11	2.79		13.79	16.90	
	0400	Boxwood, 15"-18", B & B		96	.250		17.50	3.78		21.28	25.50	
	0500	Euonymus, emerald gaiety, 12" to 15", container		115	.209		12.25	3.16		15.41	18.90	
	0600	Holly, 15"-18", B & B		96	.250		16.75	3.78		20.53	25	
	0900	Mount laurel, 18" - 24", B & B		80	.300		75	4.54		79.54	90.50	
	1000	Paxistema, 9 - 12" high		130	.185		15	2.79		17.79	21.50	
	1100	Rhododendron, 18"-24", container		48	.500		37.50	7.55		45.05	54.50	
	1200	Rosemary, 1 gal container		600	.040		35	.60		35.60	39.50	
	2000	Deciduous, amelanchier, 2'-3', B & B		57	.421		55	6.35		61.35	71.50	
	2100	Azalea, 15"-18", B & B		96	.250		23	3.78		26.78	32	
	2300	Bayberry, 2'-3', B & B		57	.421		22.50	6.35		28.85	35.50	
	2600	Cotoneaster, 15"-18", B & B		80	.300		21	4.54		25.54	31	
	2800	Dogwood, 3'-4', B & B	B-17	40	.800		30	12.85	14.20	57.05	70	
	2900	Euonymus, alatus compacta, 15" to 18", container	B-1	80	.300		33	4.54		37.54	44.50	
	3200	Forsythia, 2'-3', container	"	60	.400		24	6.05		30.05	37	
	3300	Hibiscus, 3'-4', B & B	B-17	75	.427		36	6.85	7.60	50.45	59.50	
	3400	Honeysuckle, 3'-4', B & B	B-1	60	.400		25.50	6.05		31.55	38.50	
	3500	Hydrangea, 2'-3', B & B	"	57	.421		36	6.35		42.35	50.50	
	3600	Lilac, 3'-4', B & B	B-17	40	.800		39	12.85	14.20	66.05	80	
	3900	Privet, bare root, 18"-24"	B-1	80	.300		7.50	4.54		12.04	16	
	4100	Quince, 2'-3', B & B	"	57	.421		19.50	6.35		25.85	32.50	
	4200	Russian olive, 3'-4', B & B	B-17	75	.427		17	6.85	7.60	31.45	38.50	
	4400	Spirea, 3'-4', B & B	B-1	70	.343		27	5.20		32.20	38.50	
	4500	Viburnum, 3'-4', B & B	B-17	40	.800		24	12.85	14.20	51.05	63.50	
680	0010	**PLANT BED PREPARATION**										680
	0100	Backfill planting pit, by hand, on site topsoil	2 Clab	18	.889	C.Y.		12.85		12.85	22	
	0200	Prepared planting mix	"	24	.667			9.65		9.65	16.50	
	0300	Skid steer loader, on site topsoil	B-62	340	.071			1.13	.34	1.47	2.29	
	0400	Prepared planting mix	"	410	.059			.94	.28	1.22	1.89	
	1000	Excavate planting pit, by hand, sandy soil	2 Clab	16	1			14.45		14.45	25	
	1100	Heavy soil or clay	"	8	2			29		29	49.50	
	1200	1/2 C.Y. backhoe, sandy soil	B-11C	150	.107			1.83	1.35	3.18	4.57	
	1300	Heavy soil or clay	"	115	.139			2.39	1.77	4.16	5.95	
	2000	Mix planting soil, incl. loam, manure, peat, by hand	2 Clab	60	.267		23	3.85		26.85	31.50	
	2100	Skid steer loader	B-62	150	.160		23	2.56	.78	26.34	30	
	3000	Pile sod, skid steer loader	"	2,800	.009	S.Y.		.14	.04	.18	.28	
	4000	Remove sod, F.E. loader	B-10S	2,000	.004			.08	.15	.23	.30	
	4100	Sod cutter	B-12K	3,200	.002			.05	.26	.31	.37	
900	0010	**TREES** Deciduous, in prep. beds, balled & burlapped (B&B) R02930-900										900
	0100	Ash, 2" caliper	B-17	8	4	Ea.	100	64	71	235	296	

02900 | Planting

02930 | Exterior Plants

		CREW	DAILY OUTPUT	LABOR-HOURS	UNIT	2000 BARE COSTS MAT.	LABOR	EQUIP.	TOTAL	TOTAL INCL O&P	
900	0200 Beech, 5'-6'	B-17	50	.640	Ea.	200	10.25	11.35	221.60	250	900
	0300 Birch, 6'-8', 3 stems		20	1.600		113	25.50	28.50	167	199	
	0500 Crabapple, 6'-8'		20	1.600		150	25.50	28.50	204	240	
	0600 Dogwood, 4'-5'		40	.800		58	12.85	14.20	85.05	101	
	0700 Eastern redbud 4'-5'		40	.800		115	12.85	14.20	142.05	164	
	0800 Elm, 8'-10'		20	1.600		85	25.50	28.50	139	169	
	0900 Ginkgo, 6'-7'		24	1.333		100	21.50	23.50	145	172	
	1000 Hawthorn, 8'-10', 1" caliper		20	1.600		120	25.50	28.50	174	207	
	1100 Honeylocust, 10'-12', 1-1/2" caliper		10	3.200		120	51.50	57	228.50	281	
	1300 Larch, 8'		32	1		75	16.05	17.75	108.80	129	
	1400 Linden, 8'-10', 1" caliper		20	1.600		120	25.50	28.50	174	207	
	1500 Magnolia, 4'-5'		20	1.600		55	25.50	28.50	109	135	
	1600 Maple, red, 8'-10', 1-1/2" caliper		10	3.200		122	51.50	57	230.50	283	
	1700 Mountain ash, 8'-10', 1" caliper		16	2		150	32	35.50	217.50	258	
	1800 Oak, 2-1/2"-3" caliper		3	10.667		163	171	189	523	675	
	2100 Planetree, 9'-11', 1-1/4" caliper		10	3.200		90	51.50	57	198.50	248	
	2200 Plum, 6'-8', 1" caliper		20	1.600		75	25.50	28.50	129	158	
	2300 Poplar, 9'-11', 1-1/4" caliper		10	3.200		51.50	51.50	57	160	206	
	2500 Sumac, 2'-3'		75	.427		24	6.85	7.60	38.45	46.50	
	2700 Tulip, 5'-6'		40	.800		29.50	12.85	14.20	56.55	69.50	
	2800 Willow, 6'-8', 1" caliper		20	1.600		63	25.50	28.50	117	144	

02945 | Planting Accessories

		CREW	DAILY OUTPUT	LABOR-HOURS	UNIT	MAT.	LABOR	EQUIP.	TOTAL	TOTAL INCL O&P	
310	0010 **EDGING**										310
	0050 Aluminum alloy, including stakes, 1/8" x 4", mill finish	B-1	390	.062	L.F.	1.75	.93		2.68	3.52	
	0051 Black paint		390	.062		2.03	.93		2.96	3.82	
	0052 Black anodized		390	.062		2.34	.93		3.27	4.17	
	0100 Brick, set horizontally, 1-1/2 bricks per L.F.	D-1	370	.043		.63	.77		1.40	2	
	0150 Set vertically, 3 bricks per L.F.	"	135	.119		1.96	2.12		4.08	5.75	
	0200 Corrugated aluminum, roll, 4" wide	1 Carp	650	.012		.28	.24		.52	.73	
	0250 6" wide	"	550	.015		.35	.29		.64	.88	
	0600 Railroad ties, 6" x 8"	2 Carp	170	.094		2.30	1.85		4.15	5.70	
	0650 7" x 9"	"	136	.118		2.55	2.32		4.87	6.80	
	0700 Redwood										
	0750 2" x 4"	2 Carp	330	.048	L.F.	2.67	.96		3.63	4.57	
	0800 Steel edge strips, incl. stakes, 1/4" x 5"	B-1	390	.062		2.75	.93		3.68	4.62	
	0850 3/16" x 4"	"	390	.062		2.17	.93		3.10	3.98	
500	0010 **PLANTERS** Concrete, sandblasted, precast, 48" diameter, 24" high	2 Clab	15	1.067	Ea.	425	15.40		440.40	495	500
	0300 Fiberglass, circular, 36" diameter, 24" high		15	1.067		335	15.40		350.40	395	
	1200 Wood, square, 48" side, 24" high		15	1.067		755	15.40		770.40	855	
	1300 Circular, 48" diameter, 30" high		10	1.600		675	23		698	785	
	1600 Planter/bench, 72"		5	3.200		2,100	46		2,146	2,400	
775	0010 **TREE GUYING** Including stakes, guy wire and wrap										775
	0100 Less than 3" caliper, 2 stakes	2 Clab	35	.457	Ea.	15	6.60		21.60	28	
	0200 3" to 4" caliper, 3 stakes	"	21	.762	"	17.60	11		28.60	38	
	1000 Including arrowhead anchor, cable, turnbuckles and wrap										
	1100 Less than 3" caliper, 3" anchors	2 Clab	20	.800	Ea.	45	11.55		56.55	69.50	
	1200 3" to 6" caliper, 4" anchors		15	1.067		65	15.40		80.40	98	
	1300 6" caliper, 6" anchors		12	1.333		80	19.25		99.25	121	
	1400 8" caliper, 8" anchors		9	1.778		14.90	25.50		40.40	60.50	

For information about Means Estimating Seminars, see yellow pages 11 and 12 in back of book

For expanded coverage of these items see Means Site Work and Landscape Cost Data 2000

Division Notes

		CREW	DAILY OUTPUT	LABOR-HOURS	UNIT	MAT.	LABOR	EQUIP.	TOTAL	TOTAL INCL O&P

Division 3 Concrete

Estimating Tips
General
- Carefully check all the plans and specifications. Concrete often appears on drawings other than structural drawings, including mechanical and electrical drawings for equipment pads. The cost of cutting and patching is often difficult to estimate. See Subdivision 02225 for demolition costs.
- Always obtain concrete prices from suppliers near the job site. A volume discount can often be negotiated depending upon competition in the area. Remember to add for waste, particularly for slabs and footings on grade.

03100 Concrete Forms & Accessories
- A primary cost for concrete construction is forming. Most jobs today are constructed with prefabricated forms. The selection of the forms best suited for the job and the total square feet of forms required for efficient concrete forming and placing are key elements in estimating concrete construction. Enough forms must be available for erection to make efficient use of the concrete placing equipment and crew.
- Concrete accessories for forming and placing depend upon the systems used. Study the plans and specifications to assure that all special accessory requirements have been included in the cost estimate such as anchor bolts, inserts and hangers.

03200 Concrete Reinforcement
- Ascertain that the reinforcing steel supplier has included all accessories, cutting, bending and an allowance for lapping, splicing and waste. A good rule of thumb is 10% for lapping, splicing and waste. Also, 10% waste should be allowed for welded wire fabric.

03300 Cast-in-Place Concrete
- When estimating structural concrete, pay particular attention to requirements for concrete additives, curing methods and surface treatments. Special consideration for climate, hot or cold, must be included in your estimate. Be sure to include requirements for concrete placing equipment and concrete finishing.

03400 Precast Concrete
03500 Cementitious Decks & Toppings
- The cost of hauling precast concrete structural members is often an important factor. For this reason, it is important to get a quote from the nearest supplier. It may become economically feasible to set up precasting beds on the site if the hauling costs are prohibitive.

Reference Numbers
Reference numbers are shown in bold squares at the beginning of some major classifications. These numbers refer to related items in the Reference Section. The reference information may be an estimating procedure, an alternate pricing method or technical information.

Note: Not all subdivisions listed here necessarily appear in this publication.

03050 | Basic Concrete Materials & Methods

03060 | Basic Concrete Materials

			CREW	DAILY OUTPUT	LABOR-HOURS	UNIT	2000 BARE COSTS MAT.	LABOR	EQUIP.	TOTAL	TOTAL INCL O&P	
870	0010	**WINTER PROTECTION** For heated ready mix, add, minimum				C.Y.	3.90			3.90	4.29	870
	0050	Maximum				"	5			5	5.50	
	0100	Protecting concrete and temporary heat, add, minimum	2 Clab	6,000	.003	S.F.	.16	.04		.20	.25	
	0200	Temporary shelter for slab on grade, wood frame and polyethylene										
	0201	sheeting, minimum	2 Carp	10	1.600	M.S.F.	299	31.50		330.50	385	
	0210	Maximum	"	3	5.333	"	360	105		465	575	

03100 | Concrete Forms & Accessories

03110 | Structural C.I.P. Forms

			CREW	DAILY OUTPUT	LABOR-HOURS	UNIT	2000 BARE COSTS MAT.	LABOR	EQUIP.	TOTAL	TOTAL INCL O&P	
430	0010	**FORMS IN PLACE, FOOTINGS** Continuous wall, plywood, 1 use	C-1	375	.085	SFCA	2.16	1.46		3.62	4.88	430
	0150	4 use	"	485	.066	"	.70	1.13		1.83	2.72	
	1500	Keyway, 4 use, tapered wood, 2" x 4"	1 Carp	530	.015	L.F.	.21	.30		.51	.74	
	1550	2" x 6"	"	500	.016	"	.29	.32		.61	.86	
	5000	Spread footings, plywood, 1 use	C-1	305	.105	SFCA	1.69	1.80		3.49	4.94	
	5150	4 use	"	414	.077	"	.55	1.33		1.88	2.88	
445	0010	**FORMS IN PLACE, SLAB ON GRADE**										445
	1000	Bulkhead forms with keyway, wood, 1 use, 2 piece	C-1	510	.063	L.F.	.70	1.08		1.78	2.61	
	1400	Bulkhead forms w/keyway, 1 piece expanded metal, left in place										
	1410	In lieu of 2 piece form	C-1	1,375	.023	L.F.	1.15	.40		1.55	1.95	
	1420	In lieu of 3 piece form		1,200	.027	"	1.15	.46		1.61	2.05	
	2000	Curb forms, wood, 6" to 12" high, on grade, 1 use		215	.149	SFCA	1.39	2.55		3.94	5.90	
	2150	4 use		275	.116	"	.45	2		2.45	3.92	
	3000	Edge forms, wood, 4 use, on grade, to 6" high		600	.053	L.F.	.32	.92		1.24	1.93	
	3050	7" to 12" high		435	.074	SFCA	.75	1.26		2.01	2.99	
	4000	For slab blockouts, to 12" high, 1 use		200	.160	L.F.	.70	2.75		3.45	5.45	
	4100	Plastic (extruded), to 6" high, multiple use, on grade		800	.040	"	.23	.69		.92	1.43	
450	0010	**FORMS IN PLACE, STAIRS** (Slant length x width), 1 use	C-2	165	.291	S.F.	3.68	5.10		8.78	12.80	450
	0150	4 use		190	.253		1.03	4.43		5.46	8.75	
	2000	Stairs, cast on sloping ground (length x width), 1 use		220	.218		2.45	3.82		6.27	9.25	
	2100	4 use		240	.200		.81	3.50		4.31	6.90	
455	0010	**FORMS IN PLACE, WALLS**										455
	0100	Box out for wall openings, to 16" thick, to 10 S.F.	C-2	24	2	Ea.	24	35		59	86.50	
	0150	Over 10 S.F. (use perimeter)	"	280	.171	L.F.	2.09	3		5.09	7.45	
	0250	Brick shelf, 4" w, add to wall forms, use wall area abv shelf										
	0260	1 use	C-2	240	.200	SFCA	2.29	3.50		5.79	8.50	
	0350	4 use		300	.160	"	.92	2.80		3.72	5.80	
	0500	Bulkhead forms, with keyway, 1 use, 2 piece		265	.181	L.F.	2.50	3.17		5.67	8.20	
	0550	3 piece		175	.274	"	3.10	4.81		7.91	11.65	
	0600	Bulkhead forms w/keyway, 1 piece expanded metal, left in place										
	0610	In lieu of 2 piece form	C-1	800	.040	L.F.	1.15	.69		1.84	2.45	
	0620	In lieu of 3 piece form	"	525	.061	"	1.15	1.05		2.20	3.06	
	2000	Job built plywood wall forms, to 8' high, 1 use, below grade	C-2	300	.160	SFCA	1.90	2.80		4.70	6.90	
	2150	4 use, below grade		435	.110		.63	1.93		2.56	4	
	2400	Over 8' to 16' high, 1 use		280	.171		3.68	3		6.68	9.20	
	2550	4 use		395	.122		.69	2.13		2.82	4.40	
	3000	For architectural finish, add		1,820	.026		.60	.46		1.06	1.45	
	3500	Polystyrene (expanded) wall forms										
	3510	To 8' high, 1 use, left in place	1 Carp	295	.027	SFCA	1.60	.53		2.13	2.68	

03100 | Concrete Forms & Accessories

03110 | Structural C.I.P. Forms

			CREW	DAILY OUTPUT	LABOR-HOURS	UNIT	2000 BARE COSTS MAT.	LABOR	EQUIP.	TOTAL	TOTAL INCL O&P	
455	7800	Modular prefabricated plywood, to 8' high, 1 use per month	C-2	1,180	.041	SFCA	1.13	.71		1.84	2.46	455
	7860	4 use per month		1,260	.038		.37	.67		1.04	1.55	
	8000	To 16' high, 1 use per month		715	.067		1.38	1.18		2.56	3.54	
	8060	4 use per month		790	.061		.46	1.06		1.52	2.32	
	8100	Over 16' high, 1 use per month		715	.067		1.66	1.18		2.84	3.84	
	8160	4 use per month		790	.061		.55	1.06		1.61	2.43	
800	0010	SCAFFOLDING See division 01540-750										800

03150 | Concrete Accessories

			CREW	DAILY OUTPUT	LABOR-HOURS	UNIT	MAT.	LABOR	EQUIP.	TOTAL	INCL O&P	
860	0010	WATERSTOP PVC, ribbed 3/16" thick, 4" wide	1 Carp	155	.052	L.F.	.56	1.02		1.58	2.36	860
	0050	6" wide		145	.055		1.21	1.09		2.30	3.19	
	0500	Ribbed, PVC, with center bulb, 9" wide, 3/16" thick		135	.059		1.70	1.17		2.87	3.87	
	0550	3/8" thick		130	.062		1.94	1.21		3.15	4.21	

03200 | Concrete Reinforcement

03210 | Reinforcing Steel

			CREW	DAILY OUTPUT	LABOR-HOURS	UNIT	2000 BARE COSTS MAT.	LABOR	EQUIP.	TOTAL	TOTAL INCL O&P	
600	0010	REINFORCING IN PLACE A615 Grade 60										600
	0500	Footings, #4 to #7	4 Rodm	2.10	15.238	Ton	530	320		850	1,175	
	0550	#8 to #18		3.60	8.889		500	188		688	900	
	0700	Walls, #3 to #7		3	10.667		530	225		755	1,000	
	0750	#8 to #18		4	8		530	169		699	900	
	2400	Dowels, 2 feet long, deformed, #3	2 Rodm	520	.031	Ea.	.23	.65		.88	1.46	
	2410	#4		480	.033		.41	.70		1.11	1.76	
	2420	#5		435	.037		.65	.78		1.43	2.15	
	2430	#6		360	.044		.92	.94		1.86	2.76	

03220 | Welded Wire Fabric

			CREW	DAILY OUTPUT	LABOR-HOURS	UNIT	MAT.	LABOR	EQUIP.	TOTAL	INCL O&P	
200	0010	WELDED WIRE FABRIC ASTM A185										200
	0050	Sheets										
	0100	6 x 6 - W1.4 x W1.4 (10 x 10) 21 lb. per C.S.F.	2 Rodm	35	.457	C.S.F.	6.90	9.65		16.55	25.50	
	0300	6 x 6 - W2.9 x W2.9 (6 x 6) 42 lb. per C.S.F.		29	.552		15.45	11.65		27.10	38.50	
	0500	4 x 4 - W1.4 x W1.4 (10 x 10) 31 lb. per C.S.F.		31	.516		11.30	10.90		22.20	33	
	0750	Rolls										
	0900	2 x 2 - #12 galv. for gunite reinforcing	2 Rodm	6.50	2.462	C.S.F.	20.50	52		72.50	119	
	0950	Material prices for above include 10% lap										

03240 | Fibrous Reinforcing

			CREW	DAILY OUTPUT	LABOR-HOURS	UNIT	MAT.	LABOR	EQUIP.	TOTAL	INCL O&P	
300	0010	FIBROUS REINFORCING										300
	0100	Synthetic fibers				Lb.	3.79			3.79	4.17	
	0110	1-1/2 lb. per C.Y., add to concrete				C.Y.	5.85			5.85	6.45	
	0150	Steel fibers				Lb.	.48			.48	.53	
	0155	25 lb. per C.Y., add to concrete				C.Y.	12			12	13.20	
	0160	50 lb. per C.Y., add to concrete					24			24	26.50	
	0170	75 lb. per C.Y., add to concrete					37			37	40.50	
	0180	100 lb. per C.Y., add to concrete					48			48	53	

For expanded coverage of these items see *Means Concrete & Masonry Cost Data 2000*

03300 | Cast-In-Place Concrete

03310 | Structural Concrete

			CREW	DAILY OUTPUT	LABOR-HOURS	UNIT	2000 BARE COSTS MAT.	LABOR	EQUIP.	TOTAL	TOTAL INCL O&P	
220	0010	**CONCRETE, READY MIX** Regular weight R03310-070										220
	0020	2000 psi				C.Y.	60.50			60.50	66.50	
	0100	2500 psi					61			61	67	
	0150	3000 psi					63			63	69	
	0200	3500 psi					64.50			64.50	70.50	
	0300	4000 psi					67			67	73.50	
	0350	4500 psi					69			69	76	
	0400	5000 psi					71.50			71.50	78.50	
	0411	6000 psi					81.50			81.50	90	
	0412	8000 psi					133			133	146	
	0413	10,000 psi					189			189	208	
	0414	12,000 psi					228			228	251	
	1000	For high early strength cement, add					10%					
	2000	For all lightweight aggregate, add					45%					
240	0010	**CONCRETE IN PLACE** Including forms (4 uses), reinforcing										240
	0050	steel, including finishing unless otherwise indicated										
	0500	Chimney foundations, industrial, minimum R03350-130	C-14C	32.22	3.476	C.Y.	130	64.50	1.18	195.68	254	
	0510	Maximum		23.71	4.724		151	87.50	1.61	240.11	320	
	3800	Footings, spread under 1 C.Y.		38.07	2.942		92.50	54.50	1	148	197	
	3850	Over 5 C.Y.		81.04	1.382		85	25.50	.47	110.97	138	
	3900	Footings, strip, 18" x 9", plain		41.04	2.729		84	50.50	.93	135.43	181	
	3950	36" x 12", reinforced		61.55	1.820		85.50	33.50	.62	119.62	153	
	4000	Foundation mat, under 10 C.Y.		38.67	2.896		116	53.50	.98	170.48	222	
	4050	Over 20 C.Y.		56.40	1.986		103	36.50	.68	140.18	177	
	4520	Handicap access ramp, railing both sides, 3' wide	C-14H	14.58	3.292	L.F.	97	63.50	2.60	163.10	219	
	4525	5' wide		12.22	3.928		111	75.50	3.10	189.60	256	
	4530	With cheek walls and rails both sides, 3' wide		8.55	5.614		99.50	108	4.44	211.94	300	
	4535	5' wide		7.31	6.566		100	126	5.20	231.20	335	
	4760	or reinforcing, over 10,000 S.F., 4" thick slab	C-14F	3,425	.021	S.F.	.84	.37	.01	1.22	1.54	
	4820	6" thick slab	"	3,350	.021	"	1.22	.38	.01	1.61	1.98	
	5000	Slab on grade, incl. textured finish, not incl. forms										
	5001	or reinforcing, 4" thick slab	C-14G	2,873	.019	S.F.	.82	.34	.01	1.17	1.47	
	5010	6" thick		2,590	.022		1.28	.37	.01	1.66	2.04	
	5020	8" thick		2,320	.024		1.66	.42	.02	2.10	2.54	
	6203	Retaining walls, gravity, 4' high				C.Y.	79			79	87	
	6800	Stairs, not including safety treads, free standing, 3'-6" wide	C-14H	83	.578	LF Nose	5.90	11.15	.46	17.51	26.50	
	6850	Cast on ground		125	.384	"	4.10	7.40	.30	11.80	17.60	
	7000	Stair landings, free standing		200	.240	S.F.	2.30	4.62	.19	7.11	10.75	
	7050	Cast on ground		475	.101	"	1.35	1.95	.08	3.38	4.94	
700	0010	**PLACING CONCRETE** and vibrating, including labor & equipment										700
	1400	Elevated slabs, less than 6" thick, pumped	C-20	140	.457	C.Y.		7.30	5.25	12.55	18.10	
	1450	With crane and bucket R03310-090	C-7	95	.758			12.15	9.60	21.75	31	
	1500	6" to 10" thick, pumped	C-20	160	.400			6.40	4.58	10.98	15.85	
	1550	With crane and bucket	C-7	110	.655			10.50	8.30	18.80	27	
	1600	Slabs over 10" thick, pumped	C-20	180	.356			5.65	4.07	9.72	14.05	
	1650	With crane and bucket	C-7	130	.554			8.90	7	15.90	22.50	
	1900	Footings, continuous, shallow, direct chute	C-6	120	.400			6.20	.63	6.83	11.25	
	1950	Pumped	C-20	150	.427			6.80	4.88	11.68	16.85	
	2000	With crane and bucket	C-7	90	.800			12.85	10.15	23	32.50	
	2400	Footings, spread, under 1 C.Y., direct chute	C-6	55	.873			13.55	1.38	14.93	24.50	
	2600	Footings, spread, over 5 C.Y., direct chute		120	.400			6.20	.63	6.83	11.25	
	2900	Foundation mats, over 20 C.Y., direct chute		350	.137			2.13	.22	2.35	3.85	
	4300	Slab on grade, 4" thick, direct chute		110	.436			6.80	.69	7.49	12.25	
	4350	Pumped	C-20	130	.492			7.85	5.65	13.50	19.50	
	4400	With crane and bucket	C-7	110	.655			10.50	8.30	18.80	27	

Important: See the Reference Section for critical supporting data - Reference Nos., Crews, & Location Factors

03300 | Cast-In-Place Concrete

03310 | Structural Concrete

			CREW	DAILY OUTPUT	LABOR-HOURS	UNIT	MAT.	LABOR	EQUIP.	TOTAL	TOTAL INCL O&P	
700	4900	Walls, 8" thick, direct chute	C-6	90	.533	C.Y.		8.30	.84	9.14	15	700
	4950	Pumped	C-20	100	.640			10.20	7.30	17.50	25.50	
	5000	With crane and bucket	C-7	80	.900			14.45	11.40	25.85	37	
	5050	12" thick, direct chute	C-6	100	.480			7.45	.76	8.21	13.50	
	5100	Pumped	C-20	110	.582			9.25	6.65	15.90	23	
	5200	With crane and bucket	C-7	90	.800			12.85	10.15	23	32.50	
	5600	Wheeled concrete dumping, add to placing costs above										
	5610	Walking cart, 50' haul, add	C-18	32	.281	C.Y.		4.13	1.63	5.76	8.85	
	5620	150' haul, add		24	.375			5.50	2.17	7.67	11.80	
	5700	250' haul, add		18	.500			7.35	2.89	10.24	15.75	
	5800	Riding cart, 50' haul, add	C-19	80	.112			1.65	1.04	2.69	3.97	
	5810	150' haul, add		60	.150			2.20	1.39	3.59	5.30	
	5900	250' haul, add		45	.200			2.93	1.85	4.78	7.10	

03350 | Concrete Finishing

			CREW	DAILY OUTPUT	LABOR-HOURS	UNIT	MAT.	LABOR	EQUIP.	TOTAL	TOTAL INCL O&P	
300	0010	**FINISHING FLOORS** Monolithic, screed finish	1 Cefi	900	.009	S.F.		.17		.17	.27	300
	0050	Darby finish		750	.011			.20		.20	.33	
	0100	Screed and float finish		725	.011			.21		.21	.34	
	0150	Screed, float, and broom finish		630	.013			.24		.24	.39	
	0200	Screed, float, and hand trowel		600	.013			.25		.25	.41	
	0250	Machine trowel		550	.015			.27		.27	.45	
	1600	Exposed local aggregate finish, minimum		625	.013		.37	.24		.61	.80	
	1650	Maximum		465	.017		1.18	.33		1.51	1.83	
350	0010	**FINISHING WALLS** Break ties and patch voids	1 Cefi	540	.015	S.F.	.05	.28		.33	.52	350
	0050	Burlap rub with grout	"	450	.018		.05	.34		.39	.61	
	0300	Bush hammer, green concrete	B-39	1,000	.048		.05	.71	.18	.94	1.48	
	0350	Cured concrete	"	650	.074		.05	1.09	.28	1.42	2.23	
600	0010	**SLAB TEXTURE STAMPING**, buy										600
	0020	Approx. 3 S.F.- 5 S.F. each, minimum				Ea.	40			40	44	
	0030	Average				"	44			44	48.50	
	0120	Per S.F. of tool, average				S.F.	48			48	53	
	0200	Commonly used chemicals for texture systems										
	0210	Hardeners w/colors average				S.F.	.40			.40	.44	
	0220	Release agents w/colors, average					.15			.15	.17	
	0225	Clear, average					.10			.10	.11	
	0230	Sealers, clear, average					.10			.10	.11	
	0240	Colors, average					.12			.12	.13	

03390 | Concrete Curing

			CREW	DAILY OUTPUT	LABOR-HOURS	UNIT	MAT.	LABOR	EQUIP.	TOTAL	TOTAL INCL O&P	
200	0010	**CURING** Burlap, 4 uses assumed, 7.5 oz.	2 Clab	55	.291	C.S.F.	3.47	4.20		7.67	11	200
	0100	12 oz.		55	.291		5.65	4.20		9.85	13.45	
	0200	Waterproof curing paper, 2 ply, reinforced		70	.229		4.89	3.30		8.19	11.05	
	0300	Sprayed membrane curing compound		95	.168		3.52	2.43		5.95	8.05	
	0710	Electrically, heated pads, 15 watts/S.F., 20 uses, minimum				S.F.	.16			.16	.18	
	0800	Maximum				"	.27			.27	.29	

For expanded coverage of these items see *Means Concrete & Masonry Cost Data 2000*

03400 | Precast Concrete

03410 | Plant Precast

			CREW	DAILY OUTPUT	LABOR-HOURS	UNIT	2000 BARE COSTS				TOTAL INCL O&P	
							MAT.	LABOR	EQUIP.	TOTAL		
620	0010	**SLABS** Prestressed roof/floor members, solid, grouted, 4" thick	C-11	3,600	.016	S.F.	4.65	.31	.43	5.39	6.10	620
	0050	6" thick		4,500	.012		4.48	.25	.35	5.08	5.75	
	0100	8" thick, hollow		5,600	.010		5	.20	.28	5.48	6.15	
	0150	10" thick		8,800	.006		5.70	.13	.18	6.01	6.65	
	0200	12" thick	▼	8,000	.007	▼	6.15	.14	.20	6.49	7.20	

03480 | Precast Specialties

400	0010	**LINTELS**										400
	0800	Precast concrete, 4" wide, 8" high, to 5' long	D-1	175	.091	L.F.	4.72	1.63		6.35	7.95	
	0850	5'-12' long	D-4	190	.211		4.93	3.43	.57	8.93	11.90	
	1000	6" wide, 8" high, to 5' long		185	.216		6.65	3.53	.59	10.77	14	
	1050	5'-12' long	▼	190	.211	▼	6.45	3.43	.57	10.45	13.60	
800	0010	**STAIRS**, Precast concrete treads on steel stringers, 3' wide	C-12	75	.640	Riser	53.50	12.35	6.05	71.90	86.50	800
	0300	Front entrance, 5' wide with 48" platform, 2 risers		16	3	Flight	276	58	28.50	362.50	435	
	0350	5 risers		12	4		315	77.50	38	430.50	520	
	0500	6' wide, 2 risers	▼	15	3.200		315	62	30	407	485	
	1200	Basement entrance stairs, steel bulkhead doors, minimum	B-51	22	2.182		460	33	7.85	500.85	570	
	1250	Maximum	"	11	4.364	▼	680	65.50	15.70	761.20	880	

03900 | Concrete Restoration & Cleaning

03920 | Concrete Resurfacing

			CREW	DAILY OUTPUT	LABOR-HOURS	UNIT	2000 BARE COSTS				TOTAL INCL O&P	
							MAT.	LABOR	EQUIP.	TOTAL		
400	0012	**FLOOR PATCHING** 1/4" thick, small areas, regular	1 Cefi	170	.047	S.F.	.71	.89		1.60	2.23	400
	0100	Epoxy	"	100	.080	"	1.44	1.51		2.95	4.04	

For information about Means Estimating Seminars, see yellow pages 11 and 12 in back of book

Division 4 Masonry

Estimating Tips

04050 Basic Masonry Materials & Methods

- The terms *mortar* and *grout* are often used interchangeably, and incorrectly. Mortar is used to bed masonry units, seal the entry of air and moisture, provide architectural appearance, and allow for size variations in the units. Grout is used primarily in reinforced masonry construction and is used to bond the masonry to the reinforcing steel. Common mortar types are M(2500 psi), S(1800 psi), N(750 psi), and O(350 psi), and conform to ASTM C270. Grout is either fine or coarse, conforms to ASTM C476, and in-place strengths generally exceed 2500 psi. Mortar and grout are different components of masonry construction and are placed by entirely different methods. An estimator should be aware of their unique uses and costs.

04200 Masonry Units

- The most common types of unit masonry are brick and concrete masonry. The major classifications of brick are building brick (ASTM C62), facing brick (ASTM C216) and glazed brick, fire brick and pavers. Many varieties of texture and appearance can exist within these classifications, and the estimator would be wise to check local custom and availability within the project area. On repair and remodeling jobs, matching the existing brick may be the most important criteria.
- Brick and concrete block are priced by the piece and then converted into a price per square foot of wall. Openings less than two square feet are generally ignored by the estimator because any savings in units used is offset by the cutting and trimming required.
- All masonry walls, whether interior or exterior, require bracing. The cost of bracing walls during construction should be included by the estimator and this bracing must remain in place until permanent bracing is complete. Permanent bracing of masonry walls is accomplished by masonry itself, in the form of pilasters or abutting wall corners, or by anchoring the walls to the structural frame. Accessories in the form of anchors, anchor slots and ties are used, but their supply and installation can be by different trades. For instance, anchor slots on spandrel beams and columns are supplied and welded in place by the steel fabricator, but the ties from the slots into the masonry are installed by the bricklayer. Regardless of the installation method the estimator must be certain that these accessories are accounted for in pricing.

Reference Numbers

Reference numbers are shown in bold squares at the beginning of some major classifications. These numbers refer to related items in the Reference Section. The reference information may be an estimating procedure, an alternate pricing method or technical information.

Note: Not all subdivisions listed here necessarily appear in this publication.

04050 | Basic Masonry Materials & Methods

04060 | Masonry Mortar

			CREW	DAILY OUTPUT	LABOR-HOURS	UNIT	MAT.	LABOR	EQUIP.	TOTAL	TOTAL INCL O&P	
200	0010	**CEMENT** Gypsum 80 lb. bag, T.L. lots R04060-100				Bag	11.40			11.40	12.55	200
	0050	L.T.L. lots					11.90			11.90	13.10	
	0100	Masonry, 70 lb. bag, T.L. lots R04060-200					5.90			5.90	6.50	
	0150	L.T.L. lots					5.70			5.70	6.25	
	0200	White, 70 lb. bag, T.L. lots					16.05			16.05	17.65	
	0250	L.T.L. lots					16.60			16.60	18.25	

04070 | Masonry Grout

			CREW	DAILY OUTPUT	LABOR-HOURS	UNIT	MAT.	LABOR	EQUIP.	TOTAL	TOTAL INCL O&P	
420	0010	**GROUTING** Bond bms. & lintels, 8" dp., pumped, not incl. block										420
	0200	Concrete block cores, solid, 4" thk., by hand, 0.067 C.F./S.F.	D-8	1,100	.036	S.F.	.25	.66		.91	1.40	
	0210	6" thick, pumped, 0.175 C.F. per S.F.	D-4	720	.056		.60	.91	.15	1.66	2.37	
	0250	8" thick, pumped, 0.258 C.F. per S.F.		680	.059		.91	.96	.16	2.03	2.81	
	0300	10" thick, pumped, 0.340 C.F. per S.F.		660	.061		1.19	.99	.17	2.35	3.17	
	0350	12" thick, pumped, 0.422 C.F. per S.F.		640	.063		1.47	1.02	.17	2.66	3.54	

04080 | Masonry Anchor & Reinforcement

			CREW	DAILY OUTPUT	LABOR-HOURS	UNIT	MAT.	LABOR	EQUIP.	TOTAL	TOTAL INCL O&P	
070	0010	**ANCHOR BOLTS** Hooked type with nut and washer, 1/2" diam., 8" long	1 Bric	200	.040	Ea.	.47	.80		1.27	1.87	070
	0030	12" long		190	.042		1.22	.84		2.06	2.77	
	0060	3/4" diameter, 8" long		160	.050		1.71	1		2.71	3.57	
	0070	12" long		150	.053		2.11	1.07		3.18	4.13	
200	0010	**REINFORCING** Steel bars A615, placed horiz., #3 & #4 bars R04080-500	1 Bric	450	.018	Lb.	.28	.36		.64	.91	200
	0050	Placed vertical, #3 & #4 bars		350	.023		.28	.46		.74	1.08	
	0060	#5 & #6 bars		650	.012		.28	.25		.53	.73	
	0200	Joint reinforcing, regular truss, to 6" wide, mill std galvanized		30	.267	C.L.F.	10.20	5.35		15.55	20.50	
	0250	12" wide		20	.400		11.75	8		19.75	26.50	
	0400	Cavity truss with drip section, to 6" wide		30	.267		9.70	5.35		15.05	19.70	
	0450	12" wide		20	.400		11.05	8		19.05	25.50	
650	0010	**WALL TIES** To brick veneer, galv., corrugated, 7/8" x 7", 22 Ga.	1 Bric	10.50	.762	C	4.47	15.25		19.72	31	650
	0100	24 Ga.		10.50	.762		3.97	15.25		19.22	30.50	
	0150	16 Ga.		10.50	.762		13.45	15.25		28.70	41	
	0200	Buck anchors, galv., corrugated, 16 gauge, 2" bend, 8" x 2"		10.50	.762		114	15.25		129.25	151	
	0250	8" x 3"		10.50	.762		99	15.25		114.25	135	
	0600	Cavity wall, Z type, galvanized, 6" long, 1/4" diameter		10.50	.762		20	15.25		35.25	48	
	0650	3/16" diameter		10.50	.762		9.50	15.25		24.75	36.50	
	0800	8" long, 1/4" diameter		10.50	.762		24	15.25		39.25	52.50	
	0850	3/16" diameter		10.50	.762		10.45	15.25		25.70	37.50	
	1000	Rectangular type, galvanized, 1/4" diameter, 2" x 6"		10.50	.762		25.50	15.25		40.75	54.50	
	1050	2" x 8" or 4" x 6"		10.50	.762		29	15.25		44.25	57.50	
	1100	3/16" diameter, 2" x 6"		10.50	.762		16.95	15.25		32.20	44.50	
	1150	2" x 8" or 4" x 6"		10.50	.762		19.40	15.25		34.65	47.50	
	1500	Rigid partition anchors, plain, 8" long, 1" x 1/8"		10.50	.762		48	15.25		63.25	79	
	1550	1" x 1/4"		10.50	.762		94	15.25		109.25	129	
	1580	1-1/2" x 1/8"		10.50	.762		66.50	15.25		81.75	99	
	1600	1-1/2" x 1/4"		10.50	.762		158	15.25		173.25	199	
	1650	2" x 1/8"		10.50	.762		83	15.25		98.25	118	
	1700	2" x 1/4"		10.50	.762		225	15.25		240.25	274	

04090 | Masonry Accessories

			CREW	DAILY OUTPUT	LABOR-HOURS	UNIT	MAT.	LABOR	EQUIP.	TOTAL	TOTAL INCL O&P	
420	0010	**INSULATION** See also division 07210-550										420
	0100	Inserts, styrofoam, plant installed, add to block prices										
	0200	8" x 16" units, 6" thick				S.F.	.75			.75	.83	
	0250	8" thick					.75			.75	.83	
	0300	10" thick					.90			.90	.99	
	0350	12" thick					.95			.95	1.05	

04050 | Basic Masonry Materials & Methods

04090 | Masonry Accessories

		CREW	DAILY OUTPUT	LABOR-HOURS	UNIT	2000 BARE COSTS				TOTAL INCL O&P
						MAT.	LABOR	EQUIP.	TOTAL	
700	0010 SCAFFOLDING & SWING STAGING See division 01540									700

04200 | Masonry Units

04210 | Clay Masonry Units

			CREW	DAILY OUTPUT	LABOR-HOURS	UNIT	2000 BARE COSTS				TOTAL INCL O&P
							MAT.	LABOR	EQUIP.	TOTAL	
100	0010	COMMON BUILDING BRICK C62, TL lots, material only R04210-100										100
	0020	Standard, minimum				M	250			250	275
	0050	Average (select)				"	305			305	335
120	0010	BRICK VENEER Scaffolding not included, truck load lots R04210-550										120
	0015	Material costs incl. 3% brick and 25% mortar waste										
	2000	Standard, sel. common, 4" x 2-2/3" x 8", (6.75/S.F.)	D-8	230	.174	S.F.	2.68	3.18		5.86	8.35
	2020	Standard, red, 4" x 2-2/3" x 8", running bond (6.75/SF)		220	.182		2.68	3.32		6	8.60
	2050	Full header every 6th course (7.88/S.F.)		185	.216		3.13	3.95		7.08	10.15
	2100	English, full header every 2nd course (10.13/S.F.)		140	.286		4.01	5.20		9.21	13.25
	2150	Flemish, alternate header every course (9.00/S.F.)		150	.267		3.57	4.87		8.44	12.15
	2200	Flemish, alt. header every 6th course (7.13/S.F.)		205	.195		2.83	3.57		6.40	9.15
	2250	Full headers throughout (13.50/S.F.)		105	.381		5.35	6.95		12.30	17.65
	2300	Rowlock course (13.50/S.F.)		100	.400		5.35	7.30		12.65	18.25
	2350	Rowlock stretcher (4.50/S.F.)		310	.129		1.80	2.36		4.16	5.95
	2400	Soldier course (6.75/S.F.)		200	.200		2.68	3.66		6.34	9.15
	2450	Sailor course (4.50/S.F.)		290	.138		1.80	2.52		4.32	6.25
	2600	Buff or gray face, running bond, (6.75/S.F.)		220	.182		2.85	3.32		6.17	8.80
	2700	Glazed face brick, running bond		210	.190		10	3.48		13.48	16.90
	2750	Full header every 6th course (7.88/S.F.)		170	.235		11.65	4.30		15.95	20
	3000	Jumbo, 6" x 4" x 12" running bond (3.00/S.F.)		435	.092		3.53	1.68		5.21	6.75
	3050	Norman, 4" x 2-2/3" x 12" running bond, (4.5/S.F.)		320	.125		3.19	2.28		5.47	7.40
	3100	Norwegian, 4" x 3-1/5" x 12" (3.75/S.F.)		375	.107		2.08	1.95		4.03	5.60
	3150	Economy, 4" x 4" x 8" (4.50/S.F.)		310	.129		2.12	2.36		4.48	6.35
	3200	Engineer, 4" x 3-1/5" x 8" (5.63/S.F.)		260	.154		1.88	2.81		4.69	6.85
	3250	Roman, 4" x 2" x 12" (6.00/S.F.)		250	.160		4.52	2.92		7.44	9.90
	3300	SCR, 6" x 2-2/3" x 12" (4.50/S.F.)		310	.129		4.14	2.36		6.50	8.55
	3350	Utility, 4" x 4" x 12" (3.00/S.F.)		450	.089		2.96	1.62		4.58	6
	3400	For cavity wall construction, add						15%			
	3450	For stacked bond, add						10%			
	3500	For interior veneer construction, add						15%			
	3550	For curved walls, add						30%			
300	0010	FACE BRICK C216, TL lots, material only R04210-120										300
	0300	Standard modular, 4" x 2-2/3" x 8", minimum				M	350			350	385
	0350	Maximum				"	315			315	345
	2160	Fire Brick, 2" x 2-2/3" x 9", minimum									
	2165	Maximum				M	900			900	990
	2170	For less than truck load lots, add					10				
	2180	For buff or gray brick, add					15				

04220 | Concrete Masonry Units

220	0010	CONCRETE BLOCK, BACK-UP Scaffolding not included										220
	0020	Sand aggregate, 8" x 16" units, tooled joint 1 side										

For expanded coverage of these items see *Means Concrete & Masonry Cost Data 2000*

04200 | Masonry Units

04220 | Concrete Masonry Units

			CREW	DAILY OUTPUT	LABOR-HOURS	UNIT	MAT.	LABOR	EQUIP.	TOTAL	TOTAL INCL O&P	
220	1000	Reinforced, alternate courses, 4" thick	D-8	435	.092	S.F.	.88	1.68		2.56	3.81	220
	1100	6" thick		415	.096		1.18	1.76		2.94	4.27	
	1150	8" thick		395	.101		1.38	1.85		3.23	4.65	
	1200	10" thick		385	.104		1.92	1.90		3.82	5.35	
	1250	12" thick	D-9	365	.132		2.05	2.35		4.40	6.25	
230	0010	**CONCRETE BLOCK BOND BEAM** Scaffolding not included										230
	0020	Not including grout or reinforcing										
	0100	Regular block, 8" high, 8" thick	D-8	565	.071	L.F.	1.62	1.29		2.91	3.98	
	0150	12" thick	D-9	510	.094	"	2.54	1.68		4.22	5.65	
240	0010	**CONCRETE BLOCK, DECORATIVE** Scaffolding not included										240
	0020	Embossed, simulated brick face										
	0100	8" x 16" units, 4" thick	D-8	400	.100	S.F.	2.15	1.83		3.98	5.45	
	0200	8" thick		340	.118		2.93	2.15		5.08	6.85	
	0250	12" thick		300	.133		3.78	2.44		6.22	8.30	
	0400	Embossed both sides										
	0500	8" thick	D-8	300	.133	S.F.	3.27	2.44		5.71	7.70	
	0550	12" thick	"	275	.145	"	4.28	2.66		6.94	9.20	
	1000	Fluted high strength										
	1100	Flutes 1 side, 8" x 16" x 4" thick	D-8	345	.116	S.F.	2.60	2.12		4.72	6.45	
	1150	Flutes 2 sides, 8" x 16" x 4" thick		335	.119		3.16	2.18		5.34	7.15	
	1200	8" thick		300	.133		4.06	2.44		6.50	8.60	
	1250	For special colors, add					.24			.24	.26	
	1400	Deep grooved, smooth face										
	1450	8" x 16" x 4" thick	D-8	345	.116	S.F.	1.63	2.12		3.75	5.40	
	1500	8" thick	"	300	.133	"	2.80	2.44		5.24	7.20	
	2000	Formblock, incl. inserts & reinforcing										
	2100	8" x 16" x 8" thick	D-8	345	.116	S.F.	2.93	2.12		5.05	6.80	
	2150	12" thick	"	310	.129	"	3.60	2.36		5.96	7.95	
	2500	Ground face										
	2600	8" x 16" x 4" thick	D-8	345	.116	S.F.	3.68	2.12		5.80	7.65	
	2650	6" thick		310	.129		4.03	2.36		6.39	8.40	
	2700	8" thick		290	.138		4.65	2.52		7.17	9.35	
	2750	12" thick	D-9	265	.181		5.60	3.23		8.83	11.65	
	2900	For special colors, add, minimum					15%					
	2950	For special colors, add, maximum					45%					
	4000	Slump block										
	4100	4" face height x 16" x 4" thick	D-1	165	.097	S.F.	3.76	1.73		5.49	7.05	
	4150	6" thick		160	.100		4.61	1.79		6.40	8.05	
	4200	8" thick		155	.103		6.25	1.84		8.09	9.95	
	4250	10" thick		140	.114		9.35	2.04		11.39	13.75	
	4300	12" thick		130	.123		9.95	2.20		12.15	14.65	
	4400	6" face height x 16" x 6" thick		155	.103		3.77	1.84		5.61	7.25	
	4450	8" thick		150	.107		4.94	1.90		6.84	8.65	
	4500	10" thick		130	.123		7.50	2.20		9.70	11.95	
	4550	12" thick		120	.133		7.75	2.38		10.13	12.60	
	5000	Split rib profile units, 1" deep ribs, 8 ribs										
	5100	8" x 16" x 4" thick	D-8	345	.116	S.F.	1.89	2.12		4.01	5.65	
	5150	6" thick		325	.123		2.29	2.25		4.54	6.30	
	5200	8" thick		305	.131		2.74	2.40		5.14	7.10	
	5250	12" thick	D-9	275	.175		3.35	3.12		6.47	9	
	5400	For special deeper colors, 4" thick, add					.17			.17	.19	
	5450	12" thick, add					.34			.34	.37	
	5600	For white, 4" thick, add					.70			.70	.77	
	5650	6" thick, add					.93			.93	1.02	
	5700	8" thick, add					1.13			1.13	1.24	

04200 | Masonry Units

04220 | Concrete Masonry Units

		CREW	DAILY OUTPUT	LABOR-HOURS	UNIT	2000 BARE COSTS MAT.	LABOR	EQUIP.	TOTAL	TOTAL INCL O&P	
240	5750	12" thick, add				S.F.	1.58			1.58	1.74	240
	6000	Split face or scored split face										
	6100	8" x 16" x 4" thick	D-8	350	.114	S.F.	1.95	2.09		4.04	5.70	
	6150	6" thick		315	.127		2.21	2.32		4.53	6.35	
	6200	8" thick		295	.136		4.05	2.48		6.53	8.65	
	6250	12" thick	D-9	270	.178		5.30	3.17		8.47	11.20	
	6400	For special deeper colors, 4" thick, add					.23			.23	.25	
	6450	6" thick, add					.23			.23	.25	
	6500	8" thick, add					.24			.24	.26	
	6550	12" thick, add					.41			.41	.45	
	6650	For white, 4" thick, add					.73			.73	.81	
	6700	6" thick, add					.93			.93	1.02	
	6750	8" thick, add					1.19			1.19	1.31	
	6800	12" thick, add					1.64			1.64	1.80	
	7000	Scored ground face, 2 to 5 scores										
	7100	8" x 16" x 4" thick	D-8	345	.116	S.F.	3.54	2.12		5.66	7.50	
	7150	6" thick		310	.129		4.25	2.36		6.61	8.65	
	7200	8" thick		290	.138		4.76	2.52		7.28	9.50	
	7250	12" thick	D-9	265	.181		6	3.23		9.23	12.10	
	8000	Hexagonal face profile units, 8" x 16" units										
	8100	4" thick, hollow	D-8	345	.116	S.F.	1.99	2.12		4.11	5.80	
	8200	Solid		345	.116		2.83	2.12		4.95	6.70	
	8300	6" thick, hollow		310	.129		2.33	2.36		4.69	6.55	
	8350	8" thick, hollow		290	.138		3.18	2.52		5.70	7.75	
	8500	For stacked bond, add						26%				
	8550	For high rise construction, add per story	D-8	67.80	.590	M.S.F.		10.80		10.80	18.25	
	8600	For scored block, add					10%					
	8650	For honed or ground face, per face, add				Ea.	.25			.25	.27	
	8700	For honed or ground end, per end, add				"	2.49			2.49	2.73	
	8750	For bullnose block, add					10%					
	8800	For special color, add					13%					
250	0010	**CONCRETE BLOCK, EXTERIOR** Not including scaffolding										250
	0020	Reinforced alt courses, tooled joints 2 sides, foam inserts										
	0100	Regular, 8" x 16" x 6" thick	D-8	390	.103	S.F.	1.17	1.87		3.04	4.45	
	0200	8" thick		365	.110		1.36	2		3.36	4.88	
	0250	10" thick		355	.113		1.60	2.06		3.66	5.25	
	0300	12" thick	D-9	330	.145		2.47	2.60		5.07	7.10	
260	0010	**CONCRETE BLOCK FOUNDATION WALL** Scaffolding not included										260
	0050	Normal-weight, trowel cut joints, parged 1/2" thick, no reinforcing										
	0200	Hollow, 8" x 16" x 6" thick	D-8	450	.089	S.F.	1.30	1.62		2.92	4.18	
	0250	8" thick		430	.093		1.50	1.70		3.20	4.53	
	0300	10" thick		420	.095		2.04	1.74		3.78	5.20	
	0350	12" thick	D-9	395	.122		2.18	2.17		4.35	6.05	
	0500	Solid, 8" x 16" block, 6" thick	D-8	440	.091		1.65	1.66		3.31	4.62	
	0550	8" thick	"	415	.096		2.08	1.76		3.84	5.25	
	0600	12" thick	D-9	380	.126		3.24	2.25		5.49	7.40	
280	0010	**CONCRETE BLOCK, LINTELS** Scaffolding not included										280
	0100	Including grout and horizontal reinforcing										
	0200	8" x 8" x 8", 1 #4 bar	D-4	300	.133	L.F.	3.38	2.17	.36	5.91	7.80	
	0250	2 #4 bars		295	.136		3.50	2.21	.37	6.08	8	
	1000	12" x 8" x 8", 1 #4 bar		275	.145		4.79	2.37	.40	7.56	9.70	
	1150	2 #5 bars		270	.148		5.05	2.42	.40	7.87	10.10	
340	0010	**COPING** Stock units										340
	0050	Precast concrete, 10" wide, 4" tapers to 3-1/2", 8" wall	D-1	75	.213	L.F.	10	3.81		13.81	17.45	

For expanded coverage of these items see *Means Concrete & Masonry Cost Data 2000*

04200 | Masonry Units

04220 | Concrete Masonry Units

			CREW	DAILY OUTPUT	LABOR-HOURS	UNIT	2000 BARE COSTS MAT.	LABOR	EQUIP.	TOTAL	TOTAL INCL O&P	
340	0100	12" wide, 3-1/2" tapers to 3", 10" wall	D-1	70	.229	L.F.	10.10	4.08		14.18	18	340
	0150	16" wide, 4" tapers to 3-1/2", 14" wall		60	.267		13.90	4.76		18.66	23.50	
	0300	Limestone for 12" wall, 4" thick		90	.178		13	3.17		16.17	19.65	
	0350	6" thick		80	.200		15.25	3.57		18.82	23	
	0500	Marble, to 4" thick, no wash, 9" wide		90	.178		18.90	3.17		22.07	26.50	
	0550	12" wide		80	.200		28	3.57		31.57	37	
	0700	Terra cotta, 9" wide		90	.178		4.55	3.17		7.72	10.35	
	0750	12" wide		80	.200		7.60	3.57		11.17	14.40	
	0800	Aluminum, for 12" wall		80	.200		10.75	3.57		14.32	17.85	
360	0010	**CORNICES** Brick cornice on existing building										360
	0020	Not including scaffolding										
	0110	Face bricks, 12 brick/S.F., minimum	D-1	30	.533	SF Face	4.70	9.50		14.20	21.50	
	0150	15 brick/S.F., maximum	"	23	.696	"	5.60	12.40		18	27	

04270 | Glass Masonry Units

			CREW	DAILY OUTPUT	LABOR-HOURS	UNIT	MAT.	LABOR	EQUIP.	TOTAL	INCL O&P	
200	0010	**GLASS BLOCK** Scaffolding not included R04270-700										200
	0150	8" x 8" block	D-8	160	.250	S.F.	9.25	4.57		13.82	17.95	
	0200	12" x 12" block	"	175	.229	"	11.10	4.18		15.28	19.25	
	0700	For solar reflective blocks, add					100%					
	0800	Under 1,000 S.F., 4" x 8" blocks	D-8	145	.276	S.F.	12.60	5.05		17.65	22.50	
	1000	Thinline, plain, 3-1/8" thick, under 1,000 S.F., 6" x 6" block		115	.348		9.35	6.35		15.70	21	
	1050	8" x 8" block		160	.250		5.80	4.57		10.37	14.10	
	1400	For cleaning block after installation (both sides), add		1,000	.040		.10	.73		.83	1.35	

04290 | Adobe Masonry Units

			CREW	DAILY OUTPUT	LABOR-HOURS	UNIT	MAT.	LABOR	EQUIP.	TOTAL	INCL O&P	
100	0010	**ADOBE BRICK** Unstabilized, with adobe mortar (Southwestern States)										100
	0260	Brick, 4" x 3" x 8" (6.0 per S.F.)	D-8	266	.150	S.F.	1.80	2.75		4.55	6.65	
	0280	4" x 4" x 8" (4.5 per S.F.)		333	.120		1.83	2.20		4.03	5.75	
	0300	4" x 4" x 14" (2.5 per S.F.)		540	.074		1.13	1.35		2.48	3.53	
	0320	8" x 3" x 16" (3.0 per S.F.)		466	.086		1.52	1.57		3.09	4.33	
	0340	4" x 3" x 12" (4.0 per S.F.)		362	.110		1.86	2.02		3.88	5.45	
	0360	6" x 3" x 12" (4.0 per S.F.)		362	.110		1.67	2.02		3.69	5.25	
	0380	4" x 5" x 16" (1.80 per S.F.)		600	.067		.89	1.22		2.11	3.04	
	0400	8" x 4" x 16" (2.25 per S.F.)		577	.069		1.46	1.27		2.73	3.76	
	0420	10" x 4" x 14" (2.57 per S.F.)		555	.072		1.54	1.32		2.86	3.93	
	0440	Adobe, partially stabilized, add					10%					
	0480	Fully stabilized, add					25%					

04400 | Stone

04412 | Bluestone

			CREW	DAILY OUTPUT	LABOR-HOURS	UNIT	MAT.	LABOR	EQUIP.	TOTAL	INCL O&P	
800	0010	**WINDOW SILL** Bluestone, thermal top, 10" wide, 1-1/2" thick	D-1	85	.188	S.F.	12.75	3.36		16.11	19.75	800
	0050	2" thick		75	.213	"	14.75	3.81		18.56	22.50	
	0100	Cut stone, 5" x 8" plain		48	.333	L.F.	10	5.95		15.95	21	
	0200	Face brick on edge, brick, 8" wide		80	.200		1.90	3.57		5.47	8.15	
	0400	Marble, 9" wide, 1" thick		85	.188		7.10	3.36		10.46	13.50	
	0600	Precast concrete, 4" tapers to 3", 9" wide		70	.229		9.05	4.08		13.13	16.85	
	0650	11" wide		60	.267		12.25	4.76		17.01	21.50	
	0700	13" wide, 3 1/2" tapers to 2 1/2", 12" wall		50	.320		12.15	5.70		17.85	23	

04400 | Stone

04412 | Bluestone

			CREW	DAILY OUTPUT	LABOR-HOURS	UNIT	MAT.	LABOR	EQUIP.	TOTAL	TOTAL INCL O&P	
800	0900	Slate, colored, unfading, honed, 12" wide, 1" thick	D-1	85	.188	L.F.	15	3.36		18.36	22	800
	0950	2" thick	↓	70	.229	↓	21	4.08		25.08	30	

04413 | Granite

			CREW	DAILY OUTPUT	LABOR-HOURS	UNIT	MAT.	LABOR	EQUIP.	TOTAL	TOTAL INCL O&P	
300	0010	**GRANITE** Cut to size										300
	0050	Veneer, polished face, 3/4" to 1-1/2" thick										
	0150	Low price, gray, light gray, etc.	D-10	130	.308	S.F.	18.75	5.75	3.09	27.59	33.50	
	0220	High price, red, black, etc.	"	130	.308	"	32	5.75	3.09	40.84	48	
	0300	1-1/2" to 2-1/2" thick, veneer										
	0350	Low price, gray, light gray, etc.	D-10	130	.308	S.F.	21	5.75	3.09	29.84	36	
	0550	High price, red, black, etc.	"	130	.308	"	36.50	5.75	3.09	45.34	53	
	0700	2-1/2" to 4" thick, veneer										
	0750	Low price, gray, light gray, etc.	D-10	110	.364	S.F.	26	6.75	3.65	36.40	44	
	0950	High price, red, black, etc.	"	110	.364		41.50	6.75	3.65	51.90	61.50	
	1000	For bush hammered finish, deduct					5%					
	1050	Coarse rubbed finish, deduct					10%					
	1100	Honed finish, deduct					5%					
	1150	Thermal finish, deduct				↓	18%					
	2450	For radius under 5', add				L.F.	100%					
	2500	Steps, copings, etc., finished on more than one surface										
	2550	Minimum	D-10	50	.800	C.F.	73	14.90	8	95.90	114	
	2600	Maximum	"	50	.800	"	109	14.90	8	131.90	154	
	2800	Pavers, 4" x 4" x 4" blocks, split face and joints										
	2850	Minimum	D-11	80	.300	S.F.	10.40	5.45		15.85	20.50	
	2900	Maximum	"	80	.300	"	20.50	5.45		25.95	32	
	3500	Curbing, city street type, See Division 02770-225										
	4000	Soffits, 2" thick, minimum	D-13	35	1.371	S.F.	31	25.50	11.45	67.95	90.50	
	4100	Maximum		35	1.371		62.50	25.50	11.45	99.45	125	
	4200	4" thick, minimum		35	1.371		42.50	25.50	11.45	79.45	103	
	4300	Maximum	↓	35	1.371	↓	80	25.50	11.45	116.95	144	

04414 | Limestone

			CREW	DAILY OUTPUT	LABOR-HOURS	UNIT	MAT.	LABOR	EQUIP.	TOTAL	TOTAL INCL O&P	
400	0012	**LIMESTONE,** Cut to size										400
	0020	Veneer facing panels										
	0500	Texture finish, light stick, 4-1/2" thick, 5' x 12'	D-4	300	.133	S.F.	18.95	2.17	.36	21.48	25	
	0750	5" thick, 5' x 14' panels	D-10	275	.145		22.50	2.71	1.46	26.67	31	
	1000	Sugarcube finish, 2" Thick, 3' x 5' panels		275	.145		11.25	2.71	1.46	15.42	18.55	
	1050	3" Thick, 4' x 9' panels		275	.145		11.70	2.71	1.46	15.87	19.05	
	1200	4" Thick, 5' x 11' panels		275	.145		15.60	2.71	1.46	19.77	23.50	
	1400	Sugarcube, textured finish, 4-1/2" thick, 5' x 12'		275	.145		20	2.71	1.46	24.17	28	
	1450	5" thick, 5' x 14' panels		275	.145	↓	23.50	2.71	1.46	27.67	31.50	
	2000	Coping, sugarcube finish, top & 2 sides		30	1.333	C.F.	52	25	13.35	90.35	114	
	2100	Sills, lintels, jambs, trim, stops, sugarcube finish, average		20	2		47	37	20	104	136	
	2150	Detailed		20	2	↓	61.50	37	20	118.50	153	
	2300	Steps, extra hard, 14" wide, 6" rise	↓	50	.800	L.F.	42.50	14.90	8	65.40	80.50	
	3000	Quoins, plain finish, 6"x12"x12"	D-12	25	1.280	Ea.	102	22.50		124.50	150	
	3050	6"x16"x24"	"	25	1.280	"	136	22.50		158.50	188	

04415 | Marble

			CREW	DAILY OUTPUT	LABOR-HOURS	UNIT	MAT.	LABOR	EQUIP.	TOTAL	TOTAL INCL O&P	
500	0010	**MARBLE** Base, 3/4" thick, polished, group A, 4" thick	D-1	60	.267	L.F.	9.30	4.76		14.06	18.30	500
	1000	Facing, polished finish, cut to size, 3/4" to 7/8" thick										
	1050	Average	D-10	130	.308	S.F.	17.95	5.75	3.09	26.79	33	
	1100	Maximum	"	130	.308	"	41.50	5.75	3.09	50.34	59	
	2200	Window sills, 6" x 3/4" thick	D-1	85	.188	L.F.	6.75	3.36		10.11	13.15	
	2500	Flooring, polished tiles, 12" x 12" x 3/8" thick										

For expanded coverage of these items see *Means Concrete & Masonry Cost Data 2000*

04400 | Stone

04415 | Marble

			CREW	DAILY OUTPUT	LABOR-HOURS	UNIT	MAT.	LABOR	EQUIP.	TOTAL	TOTAL INCL O&P	
500	2510	Thin set, average	D-11	90	.267	S.F.	8.30	4.86		13.16	17.40	500
	2600	Maximum		90	.267		30	4.86		34.86	41.50	
	2700	Mortar bed, average		65	.369		8.45	6.75		15.20	20.50	
	2740	Maximum		65	.369		27.50	6.75		34.25	42	
	2780	Travertine, 3/8" thick, average	D-10	130	.308		11.50	5.75	3.09	20.34	25.50	
	2790	Maximum	"	130	.308		28	5.75	3.09	36.84	44	
	3500	Thresholds, 3' long, 7/8" thick, 4" to 5" wide, plain	D-12	24	1.333	Ea.	12.95	23.50		36.45	54	
	3550	Beveled		24	1.333	"	15.10	23.50		38.60	56	
	3700	Window stools, polished, 7/8" thick, 5" wide		85	.376	L.F.	12.10	6.65		18.75	24.50	

04417 | Sandstone

			CREW	DAILY OUTPUT	LABOR-HOURS	UNIT	MAT.	LABOR	EQUIP.	TOTAL	TOTAL INCL O&P	
700	0011	**SANDSTONE OR BROWNSTONE**										700
	0100	Sawed face veneer, 2-1/2" thick, to 2' x 4' panels	D-10	130	.308	S.F.	15.35	5.75	3.09	24.19	30	
	0150	4" thick, to 3'-6" x 8' panels		100	.400		15.35	7.45	4.01	26.81	34	
	0300	Split face, random sizes		100	.400		9.05	7.45	4.01	20.51	27	
	0350	Cut stone trim (limestone)										
	0360	Ribbon stone, 4" thick, 5' pieces	D-8	120	.333	Ea.	111	6.10		117.10	132	
	0370	Cove stone, 4" thick, 5' pieces		105	.381		111	6.95		117.95	134	
	0380	Cornice stone, 10" to 12" wide		90	.444		137	8.10		145.10	165	
	0390	Band stone, 4" thick, 5' pieces		145	.276		71	5.05		76.05	86.50	
	0410	Window and door trim, 3" to 4" wide		160	.250		60	4.57		64.57	74	
	0420	Key stone, 18" long		60	.667		63.50	12.20		75.70	90.50	

04418 | Slate

			CREW	DAILY OUTPUT	LABOR-HOURS	UNIT	MAT.	LABOR	EQUIP.	TOTAL	TOTAL INCL O&P	
800	0010	**SLATE** Pennsylvania, blue gray to gray black; Vermont,										800
	3500	Stair treads, sand finish, 1" thick x 12" wide										
	3600	3 L.F. to 6 L.F.	D-10	120	.333	L.F.	14.60	6.20	3.34	24.14	30	
	3700	Ribbon, sand finish, 1" thick x 12" wide										
	3750	To 6 L.F.	D-10	120	.333	L.F.	9.75	6.20	3.34	19.29	25	

04420 | Collected Stone

			CREW	DAILY OUTPUT	LABOR-HOURS	UNIT	MAT.	LABOR	EQUIP.	TOTAL	TOTAL INCL O&P	
750	0011	**ROUGH STONE WALL**, Dry										750
	0100	Random fieldstone, under 18" thick	D-12	60	.533	C.F.	9.05	9.40		18.45	26	
	0150	Over 18" thick	"	63	.508	"	12.50	8.95		21.45	29	

04500 | Refractories

04550 | Flue Liners

			CREW	DAILY OUTPUT	LABOR-HOURS	UNIT	MAT.	LABOR	EQUIP.	TOTAL	TOTAL INCL O&P	
250	0010	**FLUE LINING** Including mortar joints, 8" x 8"	D-1	125	.128	V.L.F.	2.92	2.28		5.20	7.10	250
	0100	8" x 12"		103	.155		3.80	2.77		6.57	8.90	
	0200	12" x 12"		93	.172		5.70	3.07		8.77	11.45	
	0300	12" x 18"		84	.190		8.75	3.40		12.15	15.35	
	0400	18" x 18"		75	.213		12.05	3.81		15.86	19.70	
	0500	20" x 20"		66	.242		12.65	4.33		16.98	21.50	
	0600	24" x 24"		56	.286		14.85	5.10		19.95	25	
	1000	Round, 18" diameter		66	.242		14.25	4.33		18.58	23	

04500 | Refractories

04550 | Flue Liners

			CREW	DAILY OUTPUT	LABOR-HOURS	UNIT	2000 BARE COSTS MAT.	LABOR	EQUIP.	TOTAL	TOTAL INCL O&P	
250	1100	24" diameter	D-1	47	.340	V.L.F.	46	6.10		52.10	61	250

04580 | Refractory Brick

			CREW	DAILY OUTPUT	LABOR-HOURS	UNIT	MAT.	LABOR	EQUIP.	TOTAL	INCL O&P	
270	0010	**FIREPLACE** For prefabricated fireplace, see div. 10305-100										270
	0100	Brick fireplace, not incl. foundations or chimneys										
	0110	30" x 29" opening, incl. chamber, plain brickwork	D-1	.40	40	Ea.	355	715		1,070	1,600	
	0200	Fireplace box only (110 brick)	"	2	8	"	115	143		258	370	
	0300	For elaborate brickwork and details, add					35%	35%				
	0400	For hearth, brick & stone, add	D-1	2	8	Ea.	132	143		275	385	
	0410	For steel angle, damper, cleanouts, add		4	4		95	71.50		166.50	226	
	0600	Plain brickwork, incl. metal circulator		.50	32		700	570		1,270	1,725	
	0800	Face brick only, standard size, 8" x 2-2/3" x 4"	↓	.30	53.333	M	375	950		1,325	2,025	
	0900	Stone fireplace, fieldstone, add				SF Face	9.50			9.50	10.45	
	1000	Cut stone, add				"	10			10	11	

04800 | Masonry Assemblies

04810 | Unit Masonry Assemblies

			CREW	DAILY OUTPUT	LABOR-HOURS	UNIT	2000 BARE COSTS MAT.	LABOR	EQUIP.	TOTAL	TOTAL INCL O&P	
160	0010	**CHIMNEY** See Div. 03310 for foundation, add to prices below R04210-050										160
	0100	Brick, 16" x 16", 8" flue, scaff. not incl.	D-1	18.20	.879	V.L.F.	16.25	15.70		31.95	44.50	
	0150	16" x 20" with one 8" x 12" flue		16	1		19.70	17.85		37.55	51.50	
	0200	16" x 24" with two 8" x 8" flues		14	1.143		28	20.50		48.50	65.50	
	0250	20" x 20" with one 12" x 12" flue		13.70	1.168		19.75	21		40.75	57	
	0300	20" x 24" with two 8" x 12" flues		12	1.333		31.50	24		55.50	75.50	
	0350	20" x 32" with two 12" x 12" flues	↓	10	1.600	↓	33	28.50		61.50	84.50	
170	0010	**COLUMNS** Brick, scaffolding not included										170
	0050	8" x 8", 9 brick	D-1	56	.286	V.L.F.	3.50	5.10		8.60	12.50	
	0100	12" x 8", 13.5 brick		37	.432		5.25	7.70		12.95	18.80	
	0200	12" x 12", 20 brick		25	.640		7.80	11.40		19.20	28	
	0300	16" x 12", 27 brick		19	.842		10.50	15.05		25.55	37	
	0400	16" x 16", 36 brick		14	1.143		14	20.50		34.50	50	
	0500	20" x 16", 45 brick		11	1.455		17.50	26		43.50	63.50	
	0600	20" x 20", 56 brick	↓	9	1.778	↓	22	31.50		53.50	77.50	
210	0010	**CONCRETE BLOCK, PARTITIONS** Scaffolding not included										210
	1000	Lightweight block, tooled joints, 2 sides, hollow										
	1100	Not reinforced, 8" x 16" x 4" thick	D-8	440	.091	S.F.	.91	1.66		2.57	3.81	
	1150	6" thick		410	.098		1.21	1.78		2.99	4.35	
	1200	8" thick		385	.104		1.52	1.90		3.42	4.90	
	1250	10" thick	↓	370	.108		1.98	1.98		3.96	5.55	
	1300	12" thick	D-9	350	.137	↓	2.12	2.45		4.57	6.50	
	4000	Regular block, tooled joints, 2 sides, hollow										
	4100	Not reinforced, 8" x 16" x 4" thick	D-8	430	.093	S.F.	.77	1.70		2.47	3.72	
	4150	6" thick		400	.100		1.06	1.83		2.89	4.27	
	4200	8" thick		375	.107		1.26	1.95		3.21	4.69	
	4250	10" thick	↓	360	.111		1.80	2.03		3.83	5.40	
	4300	12" thick	D-9	340	.141	↓	1.93	2.52		4.45	6.40	

For expanded coverage of these items see *Means Concrete & Masonry Cost Data 2000*

04800 | Masonry Assemblies

04810 | Unit Masonry Assemblies

			CREW	DAILY OUTPUT	LABOR-HOURS	UNIT	2000 BARE COSTS MAT.	LABOR	EQUIP.	TOTAL	TOTAL INCL O&P	
400	0010	**LINTELS** See division 05120-480										400
650	0016	**WALLS**										650
	0800	4" wall, face, 4" x 2-2/3" x 8"	D-8	215	.186	S.F.	2.65	3.40		6.05	8.65	
	0850	4" thick, as back up, 6.75 bricks per S.F.		240	.167		1.95	3.05		5	7.30	
	0900	8" thick wall, 13.50 brick per S.F.		135	.296		4	5.40		9.40	13.55	
	1000	12" thick wall, 20.25 bricks per S.F.		95	.421		6	7.70		13.70	19.65	
	1050	16" thick wall, 27.00 bricks per S.F.		75	.533		8.10	9.75		17.85	25.50	
	1200	Reinforced, 4" x 2-2/3" x 8", 4" wall		205	.195		1.97	3.57		5.54	8.20	
	1250	8" thick wall, 13.50 brick per S.F.		130	.308		4.01	5.60		9.61	13.90	
	1300	12" thick wall, 20.25 bricks per S.F.		90	.444		6.05	8.10		14.15	20.50	
	1350	16" thick wall, 27.00 bricks per S.F.		70	.571		8.15	10.45		18.60	26.50	

Reference notes: R04210-180, R04210-185, R04210-550

04900 | Masonry Restoration & Cleaning

04930 | Unit Masonry Cleaning

			CREW	DAILY OUTPUT	LABOR-HOURS	UNIT	2000 BARE COSTS MAT.	LABOR	EQUIP.	TOTAL	TOTAL INCL O&P	
200	0010	**CLEAN AND POINT** Smooth brick	1 Bric	300	.027	S.F.	.20	.53		.73	1.12	200
	0100	Rough brick	"	265	.030	"	.21	.60		.81	1.25	
900	0010	**WASHING BRICK** Acid wash, smooth brick	1 Bric	560	.014	S.F.	.23	.29		.52	.73	900
	0050	Rough brick		400	.020		.23	.40		.63	.93	
	0060	Stone, acid wash		600	.013		.23	.27		.50	.70	
	1000	Muriatic acid, price per gallon in 5 gallon lots				Gal.	4			4	4.40	

Reference: R04930-100

For information about Means Estimating Seminars, see yellow pages 11 and 12 in back of book

Important: See the Reference Section for critical supporting data - Reference Nos., Crews, & Location Factors

Division 5 Metals

Estimating Tips

05050 Basic Metal Materials & Methods

- Nuts, bolts, washers, connection angles and plates can add a significant amount to both the tonnage of a structural steel job as well as the estimated cost. As a rule of thumb add 10% to the total weight to account for these accessories.
- Type 2 steel construction, commonly referred to as "simple construction," consists generally of field bolted connections with lateral bracing supplied by other elements of the building, such as masonry walls or x-bracing. The estimator should be aware, however, that shop connections may be accomplished by welding or bolting. The method may be particular to the fabrication shop and may have an impact on the estimated cost.

05200 Metal Joists

- In any given project the total weight of open web steel joists is determined by the loads to be supported and the design. However, economies can be realized in minimizing the amount of labor used to place the joists. This is done by maximizing the joist spacing and therefore minimizing the number of joists required to be installed on the job. Certain spacings and locations may be required by the design, but in other cases maximizing the spacing and keeping it as uniform as possible will keep the costs down.

05300 Metal Deck

- The takeoff and estimating of metal deck involves more than simply the area of the floor or roof and the type of deck specified or shown on the drawings. Many different sizes and types of openings may exist. Small openings for individual pipes or conduits may be drilled after the floor/roof is installed, but larger openings may require special deck lengths as well as reinforcing or structural support. The estimator should determine who will be supplying this reinforcing. Additionally, some deck terminations are part of the deck package, such as screed angles and pour stops, and others will be part of the steel contract, such as angles attached to structural members and cast-in-place angles and plates. The estimator must ensure that all pieces are accounted for in the complete estimate.

05500 Metal Fabrications

- The most economical steel stairs are those that use common materials, standard details and most importantly, a uniform and relatively simple method of field assembly. Commonly available A36 channels and plates are very good choices for the main stringers of the stairs, as are angles and tees for the carrier members. Risers and treads are usually made by specialty shops, and it is most economical to use a typical detail in as many places as possible. The stairs should be pre-assembled and shipped directly to the site. The field connections should be simple and straightforward to be accomplished efficiently and with a minimum of equipment and labor.

Reference Numbers

Reference numbers are shown in bold squares at the beginning of some major classifications. These numbers refer to related items in the Reference Section. The reference information may be an estimating procedure, an alternate pricing method or technical information.

Note: Not all subdivisions listed here necessarily appear in this publication.

05050 | Basic Materials & Methods
05090 | Metal Fastenings

			CREW	DAILY OUTPUT	LABOR-HOURS	UNIT	MAT.	LABOR	EQUIP.	TOTAL	TOTAL INCL O&P	
150	0010	**BOLTS & HEX NUTS** Steel, A307										150
	0100	1/4" diameter, 1/2" long				Ea.	.05			.05	.06	
	0200	1" long					.06			.06	.07	
	0300	2" long					.07			.07	.08	
	0400	3" long					.10			.10	.11	
	0500	4" long					.15			.15	.17	
	0600	3/8" diameter, 1" long					.10			.10	.11	
	0700	2" long					.14			.14	.15	
	0800	3" long					.17			.17	.19	
	0900	4" long					.22			.22	.24	
	1000	5" long					.26			.26	.29	
	1100	1/2" diameter, 1-1/2" long					.22			.22	.24	
	1200	2" long					.25			.25	.28	
	1300	4" long					.37			.37	.41	
	1400	6" long					.49			.49	.54	
	1500	8" long					.66			.66	.73	
	1600	5/8" diameter, 1-1/2" long					.38			.38	.42	
	1700	2" long					.42			.42	.46	
	1800	4" long					.59			.59	.65	
	1900	6" long					.80			.80	.88	
	2000	8" long					1.06			1.06	1.17	
	2100	10" long					1.26			1.26	1.39	
	2200	3/4" diameter, 2" long					.66			.66	.73	
	2300	4" long					.97			.97	1.07	
	2400	6" long					1.23			1.23	1.35	
	2500	8" long					1.54			1.54	1.69	
	2600	10" long					2.07			2.07	2.28	
	2700	12" long					2.36			2.36	2.60	
	2800	1" diameter, 3" long					1.61			1.61	1.77	
	2900	6" long					2.36			2.36	2.60	
	3000	12" long					4.74			4.74	5.20	
	3100	For galvanized, add					75%					
	3200	For stainless, add					350%					
340	0010	**DRILLING** For anchors, up to 4" deep, incl. bit and layout										340
	0050	in concrete or brick walls and floors, no anchor										
	0100	Holes, 1/4" diameter	1 Carp	75	.107	Ea.	.09	2.10		2.19	3.69	
	0150	For each additional inch of depth, add		430	.019		.02	.37		.39	.65	
	0200	3/8" diameter		63	.127		.08	2.50		2.58	4.37	
	0250	For each additional inch of depth, add		340	.024		.02	.46		.48	.81	
	0300	1/2" diameter		50	.160		.07	3.15		3.22	5.50	
	0350	For each additional inch of depth, add		250	.032		.02	.63		.65	1.10	
	0400	5/8" diameter		48	.167		.14	3.28		3.42	5.80	
	0450	For each additional inch of depth, add		240	.033		.03	.66		.69	1.17	
	0500	3/4" diameter		45	.178		.14	3.50		3.64	6.15	
	0550	For each additional inch of depth, add		220	.036		.04	.72		.76	1.27	
	0600	7/8" diameter		43	.186		.17	3.67		3.84	6.50	
	0650	For each additional inch of depth, add		210	.038		.04	.75		.79	1.34	
	0700	1" diameter		40	.200		.20	3.94		4.14	6.95	
	0750	For each additional inch of depth, add		190	.042		.05	.83		.88	1.48	
	0800	1-1/4" diameter		38	.211		.30	4.15		4.45	7.45	
	0850	For each additional inch of depth, add		180	.044		.07	.88		.95	1.58	
	0900	1-1/2" diameter		35	.229		.48	4.50		4.98	8.20	
	0950	For each additional inch of depth, add		165	.048		.12	.96		1.08	1.77	
	1000	For ceiling installations, add					40%					
	1100	Drilling & layout for drywall or plaster walls, no anchor										

05050 | Basic Materials & Methods

05090 | Metal Fastenings

			CREW	DAILY OUTPUT	LABOR-HOURS	UNIT	MAT.	LABOR	EQUIP.	TOTAL	TOTAL INCL O&P	
340	1200	Holes, 1/4" diameter	1 Carp	150	.053	Ea.	.01	1.05		1.06	1.81	340
	1300	3/8" diameter		140	.057		.01	1.13		1.14	1.94	
	1400	1/2" diameter		130	.062		.01	1.21		1.22	2.09	
	1500	3/4" diameter		120	.067		.02	1.31		1.33	2.27	
	1600	1" diameter		110	.073		.03	1.43		1.46	2.48	
	1700	1-1/4" diameter		100	.080		.04	1.58		1.62	2.74	
	1800	1-1/2" diameter	▼	90	.089		.06	1.75		1.81	3.07	
	1900	For ceiling installations, add				▼		40%				
380	0010	**EXPANSION ANCHORS** & shields										380
	0100	Bolt anchors for concrete, brick or stone, no layout and drilling										
	0200	Expansion shields, zinc, 1/4" diameter, 1" long, single	1 Carp	90	.089	Ea.	.90	1.75		2.65	3.99	
	0300	1-3/8" long, double		85	.094		.99	1.85		2.84	4.27	
	0500	2" long, double		80	.100		1.84	1.97		3.81	5.40	
	0700	2-1/2" long, double		75	.107		2.38	2.10		4.48	6.20	
	0900	3" long, double		70	.114		3.52	2.25		5.77	7.75	
	1100	4" long, double	▼	65	.123		7	2.42		9.42	11.85	
	1410	Concrete anchor, w/rod & epoxy cartridge, 1-3/4" diameter x 15" long	E-22	20	1.200		75.50	24.50		100	125	
	1415	18" long		17	1.412		91	29		120	150	
	1420	2" diameter x 18" long		16	1.500		116	30.50		146.50	180	
	1425	24" long		15	1.600		151	32.50		183.50	222	
	1430	Chemical anchor, w/rod & epoxy cartridge, 3/4" diam. x 9-1/2" long		27	.889		11.10	18.15		29.25	43	
	1435	1" diameter x 11-3/4" long		24	1		21	20.50		41.50	58	
	1440	1-1/4" diameter x 14" long	▼	21	1.143	▼	40	23.50		63.50	84	
	2100	Hollow wall anchors for gypsum wall board, plaster or tile										
	2500	3/16" diameter, short				Ea.	.59			.59	.65	
	3000	Toggle bolts, bright steel, 1/8" diameter, 2" long	1 Carp	85	.094		.26	1.85		2.11	3.47	
	3100	4" long		80	.100		.33	1.97		2.30	3.74	
	3200	3/16" diameter, 3" long		80	.100		.37	1.97		2.34	3.79	
	3300	6" long		75	.107		.56	2.10		2.66	4.22	
	3400	1/4" diameter, 3" long		75	.107		.41	2.10		2.51	4.05	
	3500	6" long		70	.114		.61	2.25		2.86	4.53	
	3600	3/8" diameter, 3" long		70	.114		.90	2.25		3.15	4.85	
	3700	6" long		60	.133		1.32	2.63		3.95	5.95	
	3800	1/2" diameter, 4" long		60	.133		2.86	2.63		5.49	7.65	
	3900	6" long	▼	50	.160	▼	3.81	3.15		6.96	9.60	
	4000	Nailing anchors										
	4100	Nylon nailing anchor, 1/4" diameter, 1" long				C	15.90			15.90	17.50	
	4200	1-1/2" long					20.50			20.50	22.50	
	4300	2" long					34			34	37.50	
	4400	Metal nailing anchor, 1/4" diameter, 1" long					22			22	24	
	4500	1-1/2" long					30			30	33	
	4600	2" long				▼	38.50			38.50	42	
	5000	Screw anchors for concrete, masonry,										
	5100	stone & tile, no layout or drilling included										
	5200	Jute fiber, #6, #8, & #10, 1" long				Ea.	.19			.19	.21	
	5300	#12, 1-1/2" long					.28			.28	.31	
	5400	#14, 2" long					.44			.44	.48	
	5500	#16, 2" long					.46			.46	.51	
	5600	#20, 2" long					.74			.74	.81	
	5700	Lag screw shields, 1/4" diameter, short					.43			.43	.47	
	5800	Long					.50			.50	.55	
	5900	3/8" diameter, short					.74			.74	.81	
	6000	Long					.84			.84	.92	
	6100	1/2" diameter, short					1.11			1.11	1.22	
	6200	Long					1.31			1.31	1.44	
	6300	3/4" diameter, short				▼	2.42			2.42	2.66	

For expanded coverage of these items see *Means Building Construction Cost Data 2000*

05050 | Basic Materials & Methods

05090 | Metal Fastenings

		CREW	DAILY OUTPUT	LABOR-HOURS	UNIT	MAT.	LABOR	EQUIP.	TOTAL	TOTAL INCL O&P	
380	6400 Long				Ea.	3.09			3.09	3.40	380
	6600 Lead, #6 & #8, 3/4" long					.18			.18	.20	
	6700 #10 - #14, 1-1/2" long					.27			.27	.30	
	6800 #16 & #18, 1-1/2" long					.37			.37	.41	
	6900 Plastic, #6 & #8, 3/4" long					.04			.04	.04	
	7000 #8 & #10, 7/8" long					.04			.04	.04	
	7100 #10 & #12, 1" long					.05			.05	.06	
	7200 #14 & #16, 1-1/2" long					.06			.06	.07	
460	0005 **LAG SCREWS**										460
	0010 Steel, 1/4" diameter, 2" long	1 Carp	200	.040	Ea.	.07	.79		.86	1.43	
	0100 3/8" diameter, 3" long		150	.053		.18	1.05		1.23	2	
	0200 1/2" diameter, 3" long		130	.062		.30	1.21		1.51	2.41	
	0300 5/8" diameter, 3" long		120	.067		.59	1.31		1.90	2.90	
580	0005 **POWDER ACTUATED** Tools & fasteners										580
	0010 Stud driver, .22 caliber, buy, minimum				Ea.	289			289	320	
	0100 Maximum				"	505			505	555	
	0300 Powder charges for above, low velocity				C	15.15			15.15	16.65	
	0400 Standard velocity					25.50			25.50	28	
	0600 Drive pins & studs, 1/4" & 3/8" diam., to 3" long, minimum					22			22	24	
	0700 Maximum					57.50			57.50	63	
600	0010 **RIVETS**										600
	0100 Aluminum rivet & mandrel, 1/2" grip length x 1/8" diameter				C	4.53			4.53	4.98	
	0200 3/16" diameter					7			7	7.70	
	0300 Aluminum rivet, steel mandrel, 1/8" diameter					4.23			4.23	4.65	
	0400 3/16" diameter					6.35			6.35	7	
	0500 Copper rivet, steel mandrel, 1/8" diameter					5.50			5.50	6.05	
	0600 Monel rivet, steel mandrel, 1/8" diameter					19.30			19.30	21	
	0700 3/16" diameter					40.50			40.50	44.50	
	0800 Stainless rivet & mandrel, 1/8" diameter					12			12	13.20	
	0900 3/16" diameter					18.20			18.20	20	
	1000 Stainless rivet, steel mandrel, 1/8" diameter					7.60			7.60	8.35	
	1100 3/16" diameter					13.75			13.75	15.10	
	1200 Steel rivet and mandrel, 1/8" diameter					4.74			4.74	5.20	
	1300 3/16" diameter					7.15			7.15	7.85	
	1400 Hand riveting tool, minimum				Ea.	76.50			76.50	84	
	1500 Maximum					460			460	505	
	1600 Power riveting tool, minimum					660			660	730	
	1700 Maximum					1,750			1,750	1,925	

05100 | Structural Metal Framing

05120 | Structural Steel

		CREW	DAILY OUTPUT	LABOR-HOURS	UNIT	MAT.	LABOR	EQUIP.	TOTAL	TOTAL INCL O&P	
220	0010 **CEILING SUPPORTS**										220
	1000 Entrance door/folding partition supports	E-4	60	.533	L.F.	12	11.60	1.40	25	37.50	
	1100 Linear accelerator door supports		14	2.286		54.50	49.50	6	110	165	
	1200 Lintels or shelf angles, hung, exterior hot dipped galv.		267	.120		8.20	2.61	.32	11.13	14.50	
	1250 Two coats primer paint instead of galv.		267	.120		7.10	2.61	.32	10.03	13.30	
	1400 Monitor support, ceiling hung, expansion bolted		4	8	Ea.	190	174	21	385	575	

05100 | Structural Metal Framing

05120 | Structural Steel

			CREW	DAILY OUTPUT	LABOR-HOURS	UNIT	2000 BARE COSTS MAT.	LABOR	EQUIP.	TOTAL	TOTAL INCL O&P	
220	1450	Hung from pre-set inserts	E-4	6	5.333	Ea.	205	116	14.05	335.05	470	220
	1600	Motor supports for overhead doors		4	8		96.50	174	21	291.50	475	
	1700	Partition support for heavy folding partitions, without pocket		24	1.333	L.F.	27.50	29	3.51	60.01	91	
	1750	Supports at pocket only		12	2.667		54.50	58	7	119.50	182	
	2000	Rolling grilles & fire door supports		34	.941		23.50	20.50	2.48	46.48	68.50	
	2100	Spider-leg light supports, expansion bolted to ceiling slab		8	4	Ea.	78	87	10.50	175.50	270	
	2150	Hung from pre-set inserts		12	2.667	"	84	58	7	149	214	
	2400	Toilet partition support		36	.889	L.F.	27.50	19.35	2.34	49.19	70.50	
	2500	X-ray travel gantry support		12	2.667	"	93.50	58	7	158.50	225	
260	0005	**COLUMNS**										260
	0800	Steel (lally), concrete filled, extra strong pipe, 3-1/2" diameter	E-2	660	.073	L.F.	18.35	1.56	1.55	21.46	24.50	
	0830	4" diameter		780	.062		20	1.32	1.31	22.63	26	
	0890	5" diameter		1,020	.047		25.50	1.01	1	27.51	31	
	0930	6" diameter		1,200	.040		34.50	.86	.85	36.21	40.50	
	1000	Lightweight units, 3-1/2" diameter		780	.062		2.33	1.32	1.31	4.96	6.55	
	1050	4" diameter		900	.053		3.26	1.15	1.14	5.55	7.05	
	1100	For galvanizing, add				Lb.	.38			.38	.42	
	1300	For web ties, angles, etc., add per added lb.	1 Sswk	945	.008		.60	.18		.78	1.01	
	1500	Steel pipe, extra strong, no concrete, 3" to 5" diameter	E-2	16,000	.003		.69	.06	.06	.81	.95	
	1600	6" to 12" diameter		14,000	.003		.70	.07	.07	.84	.99	
	2400	Structural tubing, rect, 5" to 6" wide, light section		11,200	.004		.63	.09	.09	.81	.97	
	2700	12" x 8" x 1/2" thk wall		24,000	.002		.61	.04	.04	.69	.80	
	2800	Heavy section		32,000	.002		.61	.03	.03	.67	.77	
	8000	Lally columns, to 8', 3-1/2" diameter	2 Carp	24	.667	Ea.	18.65	13.15		31.80	43	
	8080	4" diameter	"	20	.800	"	26	15.75		41.75	55.50	
480	0005	**LINTELS**										480
	0010	Plain steel angles, under 500 lb.	1 Bric	550	.015	Lb.	.46	.29		.75	1	
	0100	500 to 1000 lb.		640	.013	"	.45	.25		.70	.92	
	2000	Steel angles, 3-1/2" x 3", 1/4" thick, 2'-6" long		47	.170	Ea.	6.50	3.40		9.90	12.90	
	2100	4'-6" long		26	.308		11.65	6.15		17.80	23.50	
	2600	4" x 3-1/2", 1/4" thick, 5'-0" long		21	.381		14.90	7.60		22.50	29.50	
	2700	9'-0" long		12	.667		27	13.35		40.35	52	
	3500	For precast concrete lintels, see div. 03480-400										
680	0010	**STRUCTURAL STEEL PROJECTS** Bolted, unless noted otherwise										680
	0200	Apartments, nursing homes, etc., 1 to 2 stories	E-5	10.30	6.990	Ton	1,200	150	107	1,457	1,725	
	0700	Offices, hospitals, etc., steel bearing, 1 to 2 stories		10.30	6.990		1,200	150	107	1,457	1,725	
	3100	Roof trusses, minimum		13	5.538		1,675	119	85	1,879	2,175	
	3200	Maximum		8.30	8.675		2,050	186	133	2,369	2,750	
720	0010	**STRUCTURAL STEEL** Bolted, incl. fabrication										720
	0050	Beams, W 6 x 9	E-2	720	.067	L.F.	6.50	1.43	1.42	9.35	11.45	
	0100	W 8 x 10		720	.067		7.20	1.43	1.42	10.05	12.20	
	0200	Columns, W 6 x 15		540	.089		11.70	1.91	1.89	15.50	18.60	
	0250	W 8 x 31		540	.089		24	1.91	1.89	27.80	32.50	

05200 | Metal Joists

05210 | Steel Joists

		CREW	DAILY OUTPUT	LABOR-HOURS	UNIT	2000 BARE COSTS				TOTAL INCL O&P
						MAT.	LABOR	EQUIP.	TOTAL	
0010	**OPEN WEB JOISTS**									
1000	Bar joists installed, no material included, 15' span	E-2	1,050	.046	L.F.		.98	.97	1.95	2.96
1100	20' span		1,300	.037			.79	.79	1.58	2.38
1200	25' span		1,500	.032			.69	.68	1.37	2.07
1300	30' span		1,650	.029			.62	.62	1.24	1.88
1400	35' span		1,750	.027			.59	.58	1.17	1.77
1500	40' span		1,600	.030			.64	.64	1.28	1.94
1600	45' span		1,665	.029			.62	.61	1.23	1.87
1700	50' span		1,750	.027			.59	.58	1.17	1.77
3000	Add per pound for material, incl. bridging				Lb.	.48			.48	.53

05300 | Metal Decking

05310 | Steel Deck

		CREW	DAILY OUTPUT	LABOR-HOURS	UNIT	2000 BARE COSTS				TOTAL INCL O&P
						MAT.	LABOR	EQUIP.	TOTAL	
0010	**METAL DECKING** Steel decking									
2100	Open type, galv., 1-1/2" deep wide rib, 22 gauge, under 50 squares	E-4	4,500	.007	S.F.	.80	.15	.02	.97	1.20
2600	20 gauge, under 50 squares		3,865	.008		.94	.18	.02	1.14	1.41
2900	18 gauge, under 50 squares		3,800	.008		1.22	.18	.02	1.42	1.72
3700	4-1/2" deep, long span roof, over 50 squares, 20 gauge		2,700	.012		2.18	.26	.03	2.47	2.94
6100	Slab form, steel, 28 gauge, 9/16" deep, uncoated		4,000	.008		.46	.17	.02	.65	.87
6200	Galvanized		4,000	.008		.52	.17	.02	.71	.93
6220	24 gauge, 1" deep, uncoated		3,900	.008		.45	.18	.02	.65	.87
6240	Galvanized		3,900	.008		.56	.18	.02	.76	.99
6300	24 gauge, 1-5/16" deep, uncoated		3,800	.008		.65	.18	.02	.85	1.09
6400	Galvanized		3,800	.008		.71	.18	.02	.91	1.16
6500	22 gauge, 1-5/16" deep, uncoated		3,700	.009		.57	.19	.02	.78	1.02
6600	Galvanized		3,700	.009		.80	.19	.02	1.01	1.27
6700	22 gauge, 3" deep uncoated		3,600	.009		.66	.19	.02	.87	1.14
6800	Galvanized		3,600	.009		.77	.19	.02	.98	1.26

05400 | Cold Formed Metal Framing

05410 | Load-Bearing Metal Studs

		CREW	DAILY OUTPUT	LABOR-HOURS	UNIT	2000 BARE COSTS				TOTAL INCL O&P
						MAT.	LABOR	EQUIP.	TOTAL	
0010	**BRACING**, shear wall X-bracing, per 10' x 10' bay, one face									
0120	Metal strap, 20 ga x 4" wide	2 Carp	18	.889	Ea.	13.95	17.50		31.45	45.50
0130	6" wide		18	.889		21.50	17.50		39	54
0160	18 ga x 4" wide		16	1		23.50	19.70		43.20	59.50
0170	6" wide		16	1		34	19.70		53.70	71.50
0410	Continuous strap bracing, per horizontal row on both faces									
0420	Metal strap, 20 ga x 2" wide, studs 12" O.C.	1 Carp	7	1.143	C.L.F.	35	22.50		57.50	77
0430	16" O.C.		8	1		35	19.70		54.70	72.50
0440	24" O.C.		10	.800		35	15.75		50.75	65.50
0450	18 ga x 2" wide, studs 12" O.C.		6	1.333		57	26.50		83.50	108

05400 | Cold Formed Metal Framing

05410 | Load-Bearing Metal Studs

			CREW	DAILY OUTPUT	LABOR-HOURS	UNIT	2000 BARE COSTS MAT.	LABOR	EQUIP.	TOTAL	TOTAL INCL O&P	
100	0460	16" O.C.	1 Carp	7	1.143	C.L.F.	57	22.50		79.50	102	100
	0470	24" O.C.	↓	8	1	↓	57	19.70		76.70	97	
120	0010	**BRIDGING**, solid between studs w/ 1-1/4" leg track, per stud bay										120
	0200	Studs 12" O.C., 18 ga x 2-1/2" wide	1 Carp	125	.064	Ea.	.55	1.26		1.81	2.77	
	0210	3-5/8" wide		120	.067		.63	1.31		1.94	2.94	
	0220	4" wide		120	.067		.69	1.31		2	3.01	
	0230	6" wide		115	.070		.89	1.37		2.26	3.33	
	0240	8" wide		110	.073		1.18	1.43		2.61	3.75	
	0300	16 ga x 2-1/2" wide		115	.070		.68	1.37		2.05	3.10	
	0310	3-5/8" wide		110	.073		.80	1.43		2.23	3.33	
	0320	4" wide		110	.073		.86	1.43		2.29	3.40	
	0330	6" wide		105	.076		1.09	1.50		2.59	3.77	
	0340	8" wide		100	.080		1.48	1.58		3.06	4.33	
	1200	Studs 16" O.C., 18 ga x 2-1/2" wide		125	.064		.71	1.26		1.97	2.94	
	1210	3-5/8" wide		120	.067		.81	1.31		2.12	3.14	
	1220	4" wide		120	.067		.89	1.31		2.20	3.23	
	1230	6" wide		115	.070		1.14	1.37		2.51	3.60	
	1240	8" wide		110	.073		1.52	1.43		2.95	4.12	
	1300	16 ga x 2-1/2" wide		115	.070		.87	1.37		2.24	3.31	
	1310	3-5/8" wide		110	.073		1.02	1.43		2.45	3.58	
	1320	4" wide		110	.073		1.11	1.43		2.54	3.67	
	1330	6" wide		105	.076		1.40	1.50		2.90	4.11	
	1340	8" wide		100	.080		1.90	1.58		3.48	4.79	
	2200	Studs 24" O.C., 18 ga x 2-1/2" wide		125	.064		1.03	1.26		2.29	3.29	
	2210	3-5/8" wide		120	.067		1.17	1.31		2.48	3.54	
	2220	4" wide		120	.067		1.29	1.31		2.60	3.67	
	2230	6" wide		115	.070		1.65	1.37		3.02	4.16	
	2240	8" wide		110	.073		2.20	1.43		3.63	4.87	
	2300	16 ga x 2-1/2" wide		115	.070		1.27	1.37		2.64	3.74	
	2310	3-5/8" wide		110	.073		1.48	1.43		2.91	4.08	
	2320	4" wide		110	.073		1.60	1.43		3.03	4.21	
	2330	6" wide		105	.076		2.03	1.50		3.53	4.80	
	2340	8" wide	↓	100	.080	↓	2.75	1.58		4.33	5.70	
	3000	Continuous bridging, per row										
	3100	16 ga x 1-1/2" channel thru studs 12" O.C.	1 Carp	6	1.333	C.L.F.	29	26.50		55.50	76.50	
	3110	16" O.C.		7	1.143		29	22.50		51.50	70	
	3120	24" O.C.		8.80	.909		29	17.90		46.90	62	
	4100	2" x 2" angle x 18 ga, studs 12" O.C.		7	1.143		59.50	22.50		82	104	
	4110	16" O.C.		9	.889		59.50	17.50		77	95.50	
	4120	24" O.C.		12	.667		59.50	13.15		72.65	88	
	4200	16 ga, studs 12" O.C.		5	1.600		75	31.50		106.50	137	
	4210	16" O.C.		7	1.143		75	22.50		97.50	121	
	4220	24" O.C.	↓	10	.800	↓	75	15.75		90.75	110	
300	0010	**FRAMING**, boxed headers/beams										300
	0200	Double, 18 ga x 6" deep	2 Carp	220	.073	L.F.	3.12	1.43		4.55	5.90	
	0210	8" deep		210	.076		3.58	1.50		5.08	6.50	
	0220	10" deep		200	.080		4.29	1.58		5.87	7.40	
	0230	12" deep		190	.084		4.93	1.66		6.59	8.25	
	0300	16 ga x 8" deep		180	.089		4.12	1.75		5.87	7.55	
	0310	10" deep		170	.094		4.93	1.85		6.78	8.60	
	0320	12" deep		160	.100		5.40	1.97		7.37	9.35	
	0400	14 ga x 10" deep		140	.114		5.85	2.25		8.10	10.30	
	0410	12" deep		130	.123		6.50	2.42		8.92	11.30	
	1210	Triple, 18 ga x 8" deep		170	.094		5.15	1.85		7	8.85	
	1220	10" deep	↓	165	.097	↓	6.15	1.91		8.06	10	

For expanded coverage of these items see *Means Building Construction Cost Data 2000*

05400 | Cold Formed Metal Framing

05410 | Load-Bearing Metal Studs

			CREW	DAILY OUTPUT	LABOR-HOURS	UNIT	MAT.	LABOR	EQUIP.	TOTAL	TOTAL INCL O&P	
300	1230	12" deep	2 Carp	160	.100	L.F.	7.10	1.97		9.07	11.20	300
	1300	16 ga x 8" deep		145	.110		5.95	2.17		8.12	10.25	
	1310	10" deep		140	.114		7.10	2.25		9.35	11.65	
	1320	12" deep		135	.119		7.80	2.33		10.13	12.55	
	1400	14 ga x 10" deep		115	.139		8.05	2.74		10.79	13.55	
	1410	12" deep		110	.145		9	2.87		11.87	14.80	
400	0010	**FRAMING, STUD WALLS** w/ top & bottom track, no openings,										400
	0020	headers, beams, bridging or bracing										
	4100	8' high walls, 18 ga x 2-1/2" wide, studs 12" O.C.	2 Carp	54	.296	L.F.	5.35	5.85		11.20	15.90	
	4110	16" O.C.		77	.208		4.29	4.09		8.38	11.70	
	4120	24" O.C.		107	.150		3.21	2.95		6.16	8.60	
	4130	3-5/8" wide, studs 12" O.C.		53	.302		6	5.95		11.95	16.80	
	4140	16" O.C.		76	.211		4.80	4.15		8.95	12.40	
	4150	24" O.C.		105	.152		3.60	3		6.60	9.10	
	4160	4" wide, studs 12" O.C.		52	.308		6.50	6.05		12.55	17.60	
	4170	16" O.C.		74	.216		5.20	4.26		9.46	13.05	
	4180	24" O.C.		103	.155		3.92	3.06		6.98	9.55	
	4190	6" wide, studs 12" O.C.		51	.314		8.10	6.20		14.30	19.50	
	4200	16" O.C.		73	.219		6.50	4.32		10.82	14.55	
	4210	24" O.C.		101	.158		4.89	3.12		8.01	10.75	
	4220	8" wide, studs 12" O.C.		50	.320		10.40	6.30		16.70	22.50	
	4230	16" O.C.		72	.222		8.35	4.38		12.73	16.70	
	4240	24" O.C.		100	.160		6.35	3.15		9.50	12.35	
	4300	16 ga x 2-1/2" wide, studs 12" O.C.		47	.340		6.15	6.70		12.85	18.30	
	4310	16" O.C.		68	.235		4.89	4.64		9.53	13.35	
	4320	24" O.C.		94	.170		3.61	3.35		6.96	9.70	
	4330	3-5/8" wide, studs 12" O.C.		46	.348		7.10	6.85		13.95	19.60	
	4340	16" O.C.		66	.242		5.65	4.78		10.43	14.40	
	4350	24" O.C.		92	.174		4.16	3.43		7.59	10.45	
	4360	4" wide, studs 12" O.C.		45	.356		7.70	7		14.70	20.50	
	4370	16" O.C.		65	.246		6.10	4.85		10.95	15.05	
	4380	24" O.C.		90	.178		4.52	3.50		8.02	11	
	4390	6" wide, studs 12" O.C.		44	.364		9.55	7.15		16.70	23	
	4400	16" O.C.		64	.250		7.55	4.93		12.48	16.80	
	4410	24" O.C.		88	.182		5.60	3.58		9.18	12.30	
	4420	8" wide, studs 12" O.C.		43	.372		12.40	7.35		19.75	26	
	4430	16" O.C.		63	.254		9.85	5		14.85	19.40	
	4440	24" O.C.		86	.186		7.35	3.67		11.02	14.35	
	5100	10' high walls, 18 ga x 2-1/2" wide, studs 12" O.C.		54	.296		6.45	5.85		12.30	17.10	
	5110	16" O.C.		77	.208		5.10	4.09		9.19	12.60	
	5120	24" O.C.		107	.150		3.75	2.95		6.70	9.20	
	5130	3-5/8" wide, studs 12" O.C.		53	.302		7.20	5.95		13.15	18.10	
	5140	16" O.C.		76	.211		5.70	4.15		9.85	13.35	
	5150	24" O.C.		105	.152		4.20	3		7.20	9.75	
	5160	4" wide, studs 12" O.C.		52	.308		7.80	6.05		13.85	19	
	5170	16" O.C.		74	.216		6.20	4.26		10.46	14.10	
	5180	24" O.C.		103	.155		4.57	3.06		7.63	10.30	
	5190	6" wide, studs 12" O.C.		51	.314		9.70	6.20		15.90	21.50	
	5200	16" O.C.		73	.219		7.70	4.32		12.02	15.85	
	5210	24" O.C.		101	.158		5.70	3.12		8.82	11.60	
	5220	8" wide, studs 12" O.C.		50	.320		12.45	6.30		18.75	24.50	
	5230	16" O.C.		72	.222		9.90	4.38		14.28	18.40	
	5240	24" O.C.		100	.160		7.35	3.15		10.50	13.50	
	5300	16 ga x 2-1/2" wide, studs 12" O.C.		47	.340		7.45	6.70		14.15	19.70	
	5310	16" O.C.		68	.235		5.85	4.64		10.49	14.40	
	5320	24" O.C.		94	.170		4.25	3.35		7.60	10.45	

Important: See the Reference Section for critical supporting data - Reference Nos., Crews, & Location Factors

05400 | Cold Formed Metal Framing

05410 | Load-Bearing Metal Studs

			CREW	DAILY OUTPUT	LABOR-HOURS	UNIT	MAT.	LABOR	EQUIP.	TOTAL	TOTAL INCL O&P
400	5330	3-5/8" wide, studs 12" O.C.	2 Carp	46	.348	L.F.	8.60	6.85		15.45	21
	5340	16" O.C.		66	.242		6.75	4.78		11.53	15.65
	5350	24" O.C.		92	.174		4.90	3.43		8.33	11.25
	5360	4" wide, studs 12" O.C.		45	.356		9.30	7		16.30	22.50
	5370	16" O.C.		65	.246		7.30	4.85		12.15	16.35
	5380	24" O.C.		90	.178		5.30	3.50		8.80	11.85
	5390	6" wide, studs 12" O.C.		44	.364		11.50	7.15		18.65	25
	5400	16" O.C.		64	.250		9.05	4.93		13.98	18.40
	5410	24" O.C.		88	.182		6.60	3.58		10.18	13.40
	5420	8" wide, studs 12" O.C.		43	.372		14.95	7.35		22.30	29
	5430	16" O.C.		63	.254		11.80	5		16.80	21.50
	5440	24" O.C.		86	.186		8.60	3.67		12.27	15.75
	6190	12' high walls, 18 ga x 6" wide, studs 12" O.C.		41	.390		11.30	7.70		19	25.50
	6200	16" O.C.		58	.276		8.90	5.45		14.35	19.10
	6210	24" O.C.		81	.198		6.50	3.89		10.39	13.80
	6220	8" wide, studs 12" O.C.		40	.400		14.50	7.90		22.40	29.50
	6230	16" O.C.		57	.281		11.45	5.55		17	22
	6240	24" O.C.		80	.200		8.35	3.94		12.29	15.95
	6390	16 ga x 6" wide, studs 12" O.C.		35	.457		13.45	9		22.45	30.50
	6400	16" O.C.		51	.314		10.50	6.20		16.70	22
	6410	24" O.C.		70	.229		7.55	4.50		12.05	16.05
	6420	8" wide, studs 12" O.C.		34	.471		17.50	9.25		26.75	35
	6430	16" O.C.		50	.320		13.70	6.30		20	26
	6440	24" O.C.		69	.232		9.85	4.57		14.42	18.70
	6530	14 ga x 3-5/8" wide, studs 12" O.C.		34	.471		13.30	9.25		22.55	30.50
	6540	16" O.C.		48	.333		10.35	6.55		16.90	22.50
	6550	24" O.C.		65	.246		7.40	4.85		12.25	16.45
	6560	4" wide, studs 12" O.C.		33	.485		14.10	9.55		23.65	32
	6570	16" O.C.		47	.340		11	6.70		17.70	23.50
	6580	24" O.C.		64	.250		7.90	4.93		12.83	17.10
	6730	12 ga x 3-5/8" wide, studs 12" O.C.		31	.516		19.75	10.15		29.90	39
	6740	16" O.C.		43	.372		15.20	7.35		22.55	29.50
	6750	24" O.C.		59	.271		10.65	5.35		16	21
	6760	4" wide, studs 12" O.C.		30	.533		21.50	10.50		32	41.50
	6770	16" O.C.		42	.381		16.60	7.50		24.10	31
	6780	24" O.C.		58	.276		11.60	5.45		17.05	22
	7390	16' high walls, 16 ga x 6" wide, studs 12" O.C.		33	.485		17.35	9.55		26.90	35.50
	7400	16" O.C.		48	.333		13.45	6.55		20	26
	7410	24" O.C.		67	.239		9.55	4.70		14.25	18.55
	7420	8" wide, studs 12" O.C.		32	.500		22.50	9.85		32.35	42
	7430	16" O.C.		47	.340		17.50	6.70		24.20	31
	7440	24" O.C.		66	.242		12.40	4.78		17.18	22
	7560	14 ga x 4" wide, studs 12" O.C.		31	.516		18.30	10.15		28.45	37.50
	7570	16" O.C.		45	.356		14.10	7		21.10	27.50
	7580	24" O.C.		61	.262		9.95	5.15		15.10	19.80
	7590	6" wide, studs 12" O.C.		30	.533		23	10.50		33.50	43.50
	7600	16" O.C.		44	.364		17.90	7.15		25.05	32
	7610	24" O.C.		60	.267		12.65	5.25		17.90	23
	7760	12 ga x 4" wide, studs 12" O.C.		29	.552		28	10.85		38.85	49.50
	7770	16" O.C.		40	.400		21.50	7.90		29.40	37
	7780	24" O.C.		55	.291		14.90	5.75		20.65	26
	7790	6" wide, studs 12" O.C.		28	.571		35	11.25		46.25	58
	7800	16" O.C.		39	.410		27	8.10		35.10	43.50
	7810	24" O.C.		54	.296		18.55	5.85		24.40	30.50
	8590	20' high walls, 14 ga x 6" wide, studs 12" O.C.		29	.552		28.50	10.85		39.35	50
	8600	16" O.C.		42	.381		22	7.50		29.50	37

For expanded coverage of these items see *Means Building Construction Cost Data 2000*

05400 | Cold Formed Metal Framing

05410 | Load-Bearing Metal Studs

			CREW	DAILY OUTPUT	LABOR-HOURS	UNIT	2000 BARE COSTS MAT.	LABOR	EQUIP.	TOTAL	TOTAL INCL O&P	
400	8610	24" O.C.	2 Carp	57	.281	L.F.	15.30	5.55		20.85	26.50	400
	8620	8" wide, studs 12" O.C.		28	.571		35	11.25		46.25	58	
	8630	16" O.C.		41	.390		27	7.70		34.70	43	
	8640	24" O.C.		56	.286		19	5.65		24.65	30.50	
	8790	12 ga x 6" wide, studs 12" O.C.		27	.593		43.50	11.65		55.15	67.50	
	8800	16" O.C.		37	.432		33	8.50		41.50	51	
	8810	24" O.C.		51	.314		22.50	6.20		28.70	35.50	
	8820	8" wide, studs 12" O.C.		26	.615		53	12.10		65.10	79	
	8830	16" O.C.		36	.444		40.50	8.75		49.25	59.50	
	8840	24" O.C.		50	.320		28	6.30		34.30	41.50	

05420 | Cold-Formed Metal Joists

			CREW	DAILY OUTPUT	LABOR-HOURS	UNIT	MAT.	LABOR	EQUIP.	TOTAL	TOTAL INCL O&P	
100	0010	**BRACING**, continuous, per row, top & bottom										100
	0120	Flat strap, 20 ga x 2" wide, joists at 12" O.C.	1 Carp	4.67	1.713	C.L.F.	37	34		71	98.50	
	0130	16" O.C.		5.33	1.501		35.50	29.50		65	89.50	
	0140	24" O.C.		6.66	1.201		34	23.50		57.50	78	
	0150	18 ga x 2" wide, joists at 12" O.C.		4	2		56.50	39.50		96	130	
	0160	16" O.C.		4.67	1.713		55.50	34		89.50	119	
	0170	24" O.C.		5.33	1.501		54.50	29.50		84	111	
120	0010	**BRIDGING**, solid between joists w/ 1-1/4" leg track, per joist bay										120
	0230	Joists 12" O.C., 18 ga track x 6" wide	1 Carp	80	.100	Ea.	.89	1.97		2.86	4.36	
	0240	8" wide		75	.107		1.18	2.10		3.28	4.90	
	0250	10" wide		70	.114		1.52	2.25		3.77	5.55	
	0260	12" wide		65	.123		1.75	2.42		4.17	6.10	
	0330	16 ga track x 6" wide		70	.114		1.09	2.25		3.34	5.05	
	0340	8" wide		65	.123		1.48	2.42		3.90	5.80	
	0350	10" wide		60	.133		1.81	2.63		4.44	6.50	
	0360	12" wide		55	.145		2.15	2.87		5.02	7.25	
	0440	14 ga track x 8" wide		60	.133		1.90	2.63		4.53	6.60	
	0450	10" wide		55	.145		2.34	2.87		5.21	7.50	
	0460	12" wide		50	.160		2.78	3.15		5.93	8.45	
	0550	12 ga track x 10" wide		45	.178		3.55	3.50		7.05	9.90	
	0560	12" wide		40	.200		4.18	3.94		8.12	11.35	
	1230	16" O.C., 18 ga track x 6" wide		80	.100		1.14	1.97		3.11	4.63	
	1240	8" wide		75	.107		1.52	2.10		3.62	5.25	
	1250	10" wide		70	.114		1.95	2.25		4.20	6	
	1260	12" wide		65	.123		2.24	2.42		4.66	6.60	
	1330	16 ga track x 6" wide		70	.114		1.40	2.25		3.65	5.40	
	1340	8" wide		65	.123		1.90	2.42		4.32	6.25	
	1350	10" wide		60	.133		2.33	2.63		4.96	7.05	
	1360	12" wide		55	.145		2.76	2.87		5.63	7.95	
	1440	14 ga track x 8" wide		60	.133		2.44	2.63		5.07	7.20	
	1450	10" wide		55	.145		3	2.87		5.87	8.20	
	1460	12" wide		50	.160		3.56	3.15		6.71	9.30	
	1550	12 ga track x 10" wide		45	.178		4.55	3.50		8.05	11	
	1560	12" wide		40	.200		5.35	3.94		9.29	12.65	
	2230	24" O.C., 18 ga track x 6" wide		80	.100		1.65	1.97		3.62	5.20	
	2240	8" wide		75	.107		2.20	2.10		4.30	6	
	2250	10" wide		70	.114		2.82	2.25		5.07	6.95	
	2260	12" wide		65	.123		3.25	2.42		5.67	7.70	
	2330	16 ga track x 6" wide		70	.114		2.03	2.25		4.28	6.10	
	2340	8" wide		65	.123		2.75	2.42		5.17	7.15	
	2350	10" wide		60	.133		3.37	2.63		6	8.20	
	2360	12" wide		55	.145		3.99	2.87		6.86	9.30	
	2440	14 ga track x 8" wide		60	.133		3.53	2.63		6.16	8.40	

Important: See the Reference Section for critical supporting data - Reference Nos., Crews, & Location Factors

05400 | Cold Formed Metal Framing

05420 | Cold-Formed Metal Joists

			CREW	DAILY OUTPUT	LABOR-HOURS	UNIT	MAT.	LABOR	EQUIP.	TOTAL	TOTAL INCL O&P	
120	2450	10" wide	1 Carp	55	.145	Ea.	4.34	2.87		7.21	9.70	120
	2460	12" wide		50	.160		5.15	3.15		8.30	11.05	
	2550	12 ga track x 10" wide		45	.178		6.60	3.50		10.10	13.25	
	2560	12" wide		40	.200		7.75	3.94		11.69	15.30	
200	0010	FRAMING, BAND JOIST (track) fastened to bearing wall										200
	0220	18 ga track x 6" deep	2 Carp	1,000	.016	L.F.	.72	.32		1.04	1.34	
	0230	8" deep		920	.017		.97	.34		1.31	1.65	
	0240	10" deep		860	.019		1.24	.37		1.61	1.99	
	0320	16 ga track x 6" deep		900	.018		.89	.35		1.24	1.58	
	0330	8" deep		840	.019		1.21	.38		1.59	1.97	
	0340	10" deep		780	.021		1.48	.40		1.88	2.32	
	0350	12" deep		740	.022		1.75	.43		2.18	2.66	
	0430	14 ga track x 8" deep		750	.021		1.55	.42		1.97	2.43	
	0440	10" deep		720	.022		1.91	.44		2.35	2.85	
	0450	12" deep		700	.023		2.27	.45		2.72	3.26	
	0540	12 ga track x 10" deep		670	.024		2.90	.47		3.37	4	
	0550	12" deep		650	.025		3.41	.48		3.89	4.58	
300	0010	FRAMING, BOXED HEADERS/BEAMS										300
	0200	Double, 18 ga x 6" deep	2 Carp	220	.073	L.F.	3.12	1.43		4.55	5.90	
	0210	8" deep		210	.076		3.58	1.50		5.08	6.50	
	0220	10" deep		200	.080		4.29	1.58		5.87	7.40	
	0230	12" deep		190	.084		4.93	1.66		6.59	8.25	
	0300	16 ga x 8" deep		180	.089		4.12	1.75		5.87	7.55	
	0310	10" deep		170	.094		4.93	1.85		6.78	8.60	
	0320	12" deep		160	.100		5.40	1.97		7.37	9.35	
	0400	14 ga x 10" deep		140	.114		5.85	2.25		8.10	10.30	
	0410	12" deep		130	.123		6.50	2.42		8.92	11.30	
	0500	12 ga x 10" deep		110	.145		7.85	2.87		10.72	13.55	
	0510	12" deep		100	.160		8.75	3.15		11.90	15	
	1210	Triple, 18 ga x 8" deep		170	.094		5.15	1.85		7	8.85	
	1220	10" deep		165	.097		6.15	1.91		8.06	10	
	1230	12" deep		160	.100		7.10	1.97		9.07	11.20	
	1300	16 ga x 8" deep		145	.110		5.95	2.17		8.12	10.25	
	1310	10" deep		140	.114		7.10	2.25		9.35	11.65	
	1320	12" deep		135	.119		7.80	2.33		10.13	12.55	
	1400	14 ga x 10" deep		115	.139		8.50	2.74		11.24	14.05	
	1410	12" deep		110	.145		9.45	2.87		12.32	15.30	
	1500	12 ga x 10" deep		90	.178		11.50	3.50		15	18.65	
	1510	12" deep		85	.188		12.85	3.71		16.56	20.50	
410	0010	FRAMING, JOISTS, no band joists (track), web stiffeners, headers,										410
	0020	beams, bridging or bracing										
	0030	Joists (2" flange) and fasteners, materials only										
	0220	18 ga x 6" deep				L.F.	.96			.96	1.05	
	0230	8" deep					1.20			1.20	1.32	
	0240	10" deep					1.41			1.41	1.55	
	0320	16 ga x 6" deep					1.23			1.23	1.35	
	0330	8" deep					1.48			1.48	1.63	
	0340	10" deep					1.74			1.74	1.92	
	0350	12" deep					1.98			1.98	2.18	
	0430	14 ga x 8" deep					1.88			1.88	2.07	
	0440	10" deep					2.24			2.24	2.46	
	0450	12" deep					2.56			2.56	2.82	
	0540	12 ga x 10" deep					3.29			3.29	3.62	
	0550	12" deep					3.75			3.75	4.12	
	1010	Installation of joists to band joists, beams & headers, labor only										

For expanded coverage of these items see *Means Building Construction Cost Data 2000*

05400 | Cold Formed Metal Framing

05420 | Cold-Formed Metal Joists

			CREW	DAILY OUTPUT	LABOR-HOURS	UNIT	2000 BARE COSTS MAT.	LABOR	EQUIP.	TOTAL	TOTAL INCL O&P	
410	1220	18 ga x 6" deep	2 Carp	110	.145	Ea.		2.87		2.87	4.91	410
	1230	8" deep		90	.178			3.50		3.50	6	
	1240	10" deep		80	.200			3.94		3.94	6.75	
	1320	16 ga x 6" deep		95	.168			3.32		3.32	5.70	
	1330	8" deep		70	.229			4.50		4.50	7.70	
	1340	10" deep		60	.267			5.25		5.25	9	
	1350	12" deep		55	.291			5.75		5.75	9.80	
	1430	14 ga x 8" deep		65	.246			4.85		4.85	8.30	
	1440	10" deep		45	.356			7		7	12	
	1450	12" deep		35	.457			9		9	15.45	
	1540	12 ga x 10" deep		40	.400			7.90		7.90	13.50	
	1550	12" deep		30	.533			10.50		10.50	18	
500	0010	**FRAMING, WEB STIFFENERS** at joist bearing, fabricated from										500
	0020	stud piece (1-5/8" flange) to stiffen joist (2" flange)										
	2120	For 6" deep joist, with 18 ga x 2-1/2" stud	1 Carp	120	.067	Ea.	1.19	1.31		2.50	3.56	
	2130	3-5/8" stud		110	.073		1.23	1.43		2.66	3.80	
	2140	4" stud		105	.076		1.23	1.50		2.73	3.93	
	2150	6" stud		100	.080		1.32	1.58		2.90	4.15	
	2160	8" stud		95	.084		1.43	1.66		3.09	4.41	
	2220	8" deep joist, with 2-1/2" stud		120	.067		1.30	1.31		2.61	3.68	
	2230	3-5/8" stud		110	.073		1.33	1.43		2.76	3.91	
	2240	4" stud		105	.076		1.35	1.50		2.85	4.06	
	2250	6" stud		100	.080		1.45	1.58		3.03	4.29	
	2260	8" stud		95	.084		1.64	1.66		3.30	4.64	
	2320	10" deep joist, with 2-1/2" stud		110	.073		1.84	1.43		3.27	4.47	
	2330	3-5/8" stud		100	.080		1.89	1.58		3.47	4.78	
	2340	4" stud		95	.084		1.94	1.66		3.60	4.98	
	2350	6" stud		90	.089		2.06	1.75		3.81	5.25	
	2360	8" stud		85	.094		2.20	1.85		4.05	5.60	
	2420	12" deep joist, with 2-1/2" stud		110	.073		1.94	1.43		3.37	4.59	
	2430	3-5/8" stud		100	.080		1.98	1.58		3.56	4.88	
	2440	4" stud		95	.084		2.02	1.66		3.68	5.05	
	2450	6" stud		90	.089		2.16	1.75		3.91	5.40	
	2460	8" stud		85	.094		2.45	1.85		4.30	5.85	
	3130	For 6" deep joist, with 16 ga x 3-5/8" stud		100	.080		1.30	1.58		2.88	4.12	
	3140	4" stud		95	.084		1.32	1.66		2.98	4.29	
	3150	6" stud		90	.089		1.42	1.75		3.17	4.56	
	3160	8" stud		85	.094		1.59	1.85		3.44	4.93	
	3230	8" deep joist, with 3-5/8" stud		100	.080		1.44	1.58		3.02	4.28	
	3240	4" stud		95	.084		1.45	1.66		3.11	4.43	
	3250	6" stud		90	.089		1.58	1.75		3.33	4.73	
	3260	8" stud		85	.094		1.79	1.85		3.64	5.15	
	3330	10" deep joist, with 3-5/8" stud		85	.094		1.97	1.85		3.82	5.35	
	3340	4" stud		80	.100		2.06	1.97		4.03	5.65	
	3350	6" stud		75	.107		2.20	2.10		4.30	6	
	3360	8" stud		70	.114		2.42	2.25		4.67	6.55	
	3430	12" deep joist, with 3-5/8" stud		85	.094		2.15	1.85		4	5.55	
	3440	4" stud		80	.100		2.16	1.97		4.13	5.75	
	3450	6" stud		75	.107		2.35	2.10		4.45	6.20	
	3460	8" stud		70	.114		2.67	2.25		4.92	6.80	
	4230	For 8" deep joist, with 14 ga x 3-5/8" stud		90	.089		1.97	1.75		3.72	5.15	
	4240	4" stud		85	.094		2.02	1.85		3.87	5.40	
	4250	6" stud		80	.100		2.21	1.97		4.18	5.80	
	4260	8" stud		75	.107		2.39	2.10		4.49	6.25	
	4330	10" deep joist, with 3-5/8" stud		75	.107		2.77	2.10		4.87	6.65	
	4340	4" stud		70	.114		2.76	2.25		5.01	6.90	

Important: See the Reference Section for critical supporting data - Reference Nos., Crews, & Location Factors

05400 | Cold Formed Metal Framing

05420 | Cold-Formed Metal Joists

		CREW	DAILY OUTPUT	LABOR-HOURS	UNIT	MAT.	LABOR	EQUIP.	TOTAL	TOTAL INCL O&P	
500	4350	6" stud	1 Carp	65	.123	Ea.	3.07	2.42		5.49	7.50	500
	4360	8" stud		60	.133		3.23	2.63		5.86	8.05	
	4430	12" deep joist, with 3-5/8" stud		75	.107		2.94	2.10		5.04	6.85	
	4440	4" stud		70	.114		3.02	2.25		5.27	7.20	
	4450	6" stud		65	.123		3.30	2.42		5.72	7.80	
	4460	8" stud		60	.133		3.56	2.63		6.19	8.40	
	5330	For 10" deep joist, with 12 ga x 3-5/8" stud		65	.123		3.15	2.42		5.57	7.60	
	5340	4" stud		60	.133		3.31	2.63		5.94	8.15	
	5350	6" stud		55	.145		3.59	2.87		6.46	8.85	
	5360	8" stud		50	.160		3.94	3.15		7.09	9.75	
	5430	12" deep joist, with 3-5/8" stud		65	.123		3.50	2.42		5.92	8	
	5440	4" stud		60	.133		3.49	2.63		6.12	8.35	
	5450	6" stud		55	.145		3.91	2.87		6.78	9.20	
	5460	8" stud		50	.160		4.50	3.15		7.65	10.35	

05500 | Metal Fabrications

05514 | Ladders

		CREW	DAILY OUTPUT	LABOR-HOURS	UNIT	MAT.	LABOR	EQUIP.	TOTAL	TOTAL INCL O&P	
500	0005	**LADDER**										500
	0010	Steel, 20" wide, bolted to concrete, with cage	E-4	50	.640	V.L.F.	53	13.90	1.68	68.58	88	
	0100	Without cage		85	.376		24.50	8.20	.99	33.69	44	
	0300	Aluminum, bolted to concrete, with cage		50	.640		100	13.90	1.68	115.58	139	
	0400	Without cage		85	.376		55.50	8.20	.99	64.69	78	
	1350	Alternating tread stair, 56/68°, steel, standard paint color	2 Sswk	50	.320		132	6.80		138.80	158	
	1360	Non-standard paint color		50	.320		149	6.80		155.80	177	
	1370	Galvanized steel		50	.320		149	6.80		155.80	177	
	1380	Stainless steel		50	.320		221	6.80		227.80	256	
	1390	68°, aluminum		50	.320		165	6.80		171.80	194	

05517 | Metal Stairs

		CREW	DAILY OUTPUT	LABOR-HOURS	UNIT	MAT.	LABOR	EQUIP.	TOTAL	TOTAL INCL O&P	
350	0005	**FIRE ESCAPE STAIRS**										350
	0010	One story, disappearing, stainless steel	2 Sswk	20	.800	V.L.F.	142	17		159	190	
	0100	Portable ladder				Ea.	50			50	55	
700	0010	**STAIR** Steel, safety nosing, steel stringers										700
	1700	Pre-erected, steel pan tread, 3'-6" wide, 2 line pipe rail	E-2	87	.552	Riser	165	11.85	11.75	188.60	218	
	1800	With flat bar picket rail	"	87	.552		185	11.85	11.75	208.60	240	
	1810	Spiral aluminum, 5'-0" diameter, stock units	E-4	45	.711		185	15.45	1.87	202.32	237	
	1820	Custom units		45	.711		350	15.45	1.87	367.32	420	
	1900	Spiral, cast iron, 4'-0" diameter, ornamental, minimum		45	.711		165	15.45	1.87	182.32	215	
	1920	Maximum		25	1.280		225	28	3.37	256.37	305	

05520 | Handrails & Railings

		CREW	DAILY OUTPUT	LABOR-HOURS	UNIT	MAT.	LABOR	EQUIP.	TOTAL	TOTAL INCL O&P	
700	0005	**RAILING, PIPE**										700
	0010	Aluminum, 2 rail, satin finish, 1-1/4" diameter	E-4	160	.200	L.F.	13	4.35	.53	17.88	23.50	
	0030	Clear anodized		160	.200		16.10	4.35	.53	20.98	27	
	0040	Dark anodized		160	.200		18.15	4.35	.53	23.03	29	
	0080	1-1/2" diameter, satin finish		160	.200		15.55	4.35	.53	20.43	26.50	
	0090	Clear anodized		160	.200		17.35	4.35	.53	22.23	28.50	

For expanded coverage of these items see *Means Building Construction Cost Data 2000*

05500 | Metal Fabrications

05520 | Handrails & Railings

		CREW	DAILY OUTPUT	LABOR-HOURS	UNIT	MAT.	LABOR	EQUIP.	TOTAL	TOTAL INCL O&P
700	0100 Dark anodized	E-4	160	.200	L.F.	19.25	4.35	.53	24.13	30
	0140 Aluminum, 3 rail, 1-1/4" diam., satin finish		137	.234		19.90	5.10	.61	25.61	32.50
	0150 Clear anodized		137	.234		25	5.10	.61	30.71	38
	0160 Dark anodized		137	.234		27.50	5.10	.61	33.21	41
	0200 1-1/2" diameter, satin finish		137	.234		24	5.10	.61	29.71	37
	0210 Clear anodized		137	.234		27	5.10	.61	32.71	40.50
	0220 Dark anodized		137	.234		29.50	5.10	.61	35.21	43
	0500 Steel, 2 rail, on stairs, primed, 1-1/4" diameter		160	.200		9.35	4.35	.53	14.23	19.50
	0520 1-1/2" diameter		160	.200		10.30	4.35	.53	15.18	20.50
	0540 Galvanized, 1-1/4" diameter		160	.200		13	4.35	.53	17.88	23.50
	0560 1-1/2" diameter		160	.200		14.55	4.35	.53	19.43	25
	0580 Steel, 3 rail, primed, 1-1/4" diameter		137	.234		13.95	5.10	.61	19.66	26
	0600 1-1/2" diameter		137	.234		14.80	5.10	.61	20.51	27
	0620 Galvanized, 1-1/4" diameter		137	.234		19.60	5.10	.61	25.31	32
	0640 1-1/2" diameter		137	.234		23	5.10	.61	28.71	36
	0700 Stainless steel, 2 rail, 1-1/4" diam. #4 finish		137	.234		32	5.10	.61	37.71	45.50
	0720 High polish		137	.234		51.50	5.10	.61	57.21	67
	0740 Mirror polish		137	.234		64.50	5.10	.61	70.21	81.50
	0760 Stainless steel, 3 rail, 1-1/2" diam., #4 finish		120	.267		48	5.80	.70	54.50	65
	0770 High polish		120	.267		79.50	5.80	.70	86	99.50
	0780 Mirror finish		120	.267		97	5.80	.70	103.50	119
	0900 Wall rail, alum. pipe, 1-1/4" diam., satin finish		213	.150		7.45	3.27	.40	11.12	15.05
	0905 Clear anodized		213	.150		9.05	3.27	.40	12.72	16.90
	0910 Dark anodized		213	.150		11	3.27	.40	14.67	19
	0915 1-1/2" diameter, satin finish		213	.150		8.25	3.27	.40	11.92	15.95
	0920 Clear anodized		213	.150		10.35	3.27	.40	14.02	18.30
	0925 Dark anodized		213	.150		12.80	3.27	.40	16.47	21
	0930 Steel pipe, 1-1/4" diameter, primed		213	.150		5.70	3.27	.40	9.37	13.15
	0935 Galvanized		213	.150		8.25	3.27	.40	11.92	15.95
	0940 1-1/2" diameter		176	.182		5.85	3.95	.48	10.28	14.75
	0945 Galvanized		213	.150		8.30	3.27	.40	11.97	16
	0955 Stainless steel pipe, 1-1/2" diam., #4 finish		107	.299		25.50	6.50	.79	32.79	41.50
	0960 High polish		107	.299		52	6.50	.79	59.29	70.50
	0965 Mirror polish		107	.299		61	6.50	.79	68.29	81

05580 | Formed Metal Fabrications

		CREW	DAILY OUTPUT	LABOR-HOURS	UNIT	MAT.	LABOR	EQUIP.	TOTAL	TOTAL INCL O&P
600	0005 **LAMP POSTS**									
	0010 Aluminum, 7' high, stock units, post only	1 Carp	16	.500	Ea.	65	9.85		74.85	88.50
	0100 Mild steel, plain	"	16	.500	"	39	9.85		48.85	60

05700 | Ornamental Metal

05720 | Ornamental Railings

		CREW	DAILY OUTPUT	LABOR-HOURS	UNIT	MAT.	LABOR	EQUIP.	TOTAL	TOTAL INCL O&P
700	0005 **RAILINGS, ORNAMENTAL**									
	0010 Aluminum, bronze or stainless, minimum	1 Sswk	24	.333	L.F.	44	7.10		51.10	62.50
	0400 Hand-forged wrought iron, minimum	"	12	.667	"	95	14.15		109.15	132

For information about Means Estimating Seminars, see yellow pages 11 and 12 in back of book

Division 6
Wood & Plastics

Estimating Tips

06050 Basic Wood & Plastic Materials & Methods

- Common to any wood framed structure are the accessory connector items such as screws, nails, adhesives, hangers, connector plates, straps, angles and holdowns. For typical wood framed buildings, such as residential projects, the aggregate total for these items can be significant, especially in areas where seismic loading is a concern. For floor and wall framing, nail quantities can be figured on a "pounds per thousand board feet basis", with 10 to 25 lbs. per MBF the range. Holdowns, hangers and other connectors should be taken off by the piece.

06100 Rough Carpentry

- Lumber is a traded commodity and therefore sensitive to supply and demand in the marketplace. Even in "budgetary" estimating of wood framed projects, it is advisable to call local suppliers for the latest market pricing.
- Common quantity units for wood framed projects are "thousand board feet" (MBF). A board foot is a volume of wood, 1" x 1' x 1', or 144 cubic inches. Board foot quantities are generally calculated using nominal material dimensions—dressed sizes are ignored. Board foot per lineal foot of any stick of lumber can be calculated by dividing the nominal cross sectional area by 12. As an example, 2,000 lineal feet of 2 x 12 equates to 4 MBF by dividing the nominal area, 2 x 12, by 12, which equals 2, and multiplying by 2,000 to give 4,000 board feet. This simple rule applies to all nominal dimensioned lumber.
- Waste is an issue of concern at the quantity takeoff for any area of construction. Framing lumber is sold in even foot lengths, i.e., 10', 12', 14', 16', and depending on spans, wall heights and the grade of lumber, waste is inevitable. A rule of thumb for lumber waste is 5% to 10% depending on material quality and the complexity of the framing.
- Wood in various forms and shapes is used in many projects, even where the main structural framing is steel, concrete or masonry. Plywood as a back-up partition material and 2x boards used as blocking and cant strips around roof edges are two common examples. The estimator should ensure that the costs of all wood materials are included in the final estimate.

06200 Finish Carpentry

- It is necessary to consider the grade of workmanship when estimating labor costs for erecting millwork and interior finish. In practice, there are three grades: premium, custom and economy. The Means daily output for base and case moldings is in the range of 200 to 250 L.F. per carpenter per day. This is appropriate for most average custom grade projects. For premium projects an adjustment to productivity of 25% to 50% should be made depending on the complexity of the job.

Reference Numbers

Reference numbers are shown in bold squares at the beginning of some major classifications. These numbers refer to related items in the Reference Section. The reference information may be an estimating procedure, an alternate pricing method or technical information.

Note: Not all subdivisions listed here necessarily appear in this publication.

06050 | Basic Wood / Plastic Materials / Methods

06055 | Wood & Plastic Laminate

			CREW	DAILY OUTPUT	LABOR-HOURS	UNIT	2000 BARE COSTS				TOTAL INCL O&P	
							MAT.	LABOR	EQUIP.	TOTAL		
740	0010	**COUNTER TOP** Stock, plastic lam., 24" wide w/backsplash, min.	1 Carp	30	.267	L.F.	4.89	5.25		10.14	14.40	740
	0100	Maximum		25	.320		14.25	6.30		20.55	26.50	
	0300	Custom plastic, 7/8" thick, aluminum molding, no splash		30	.267		15.70	5.25		20.95	26.50	
	0400	Cove splash		30	.267		20.50	5.25		25.75	31.50	
	0600	1-1/4" thick, no splash		28	.286		18.50	5.65		24.15	30	
	0700	Square splash		28	.286		23	5.65		28.65	35	
	0900	Square edge, plastic face, 7/8" thick, no splash		30	.267		19.80	5.25		25.05	31	
	1000	With splash		30	.267		26	5.25		31.25	37.50	
	1200	For stainless channel edge, 7/8" thick, add					2.21			2.21	2.43	
	1300	1-1/4" thick, add					2.58			2.58	2.84	
	1500	For solid color suede finish, add					1.96			1.96	2.16	
	1700	For end splash, add				Ea.	12.35			12.35	13.60	
	1900	For cut outs, standard, add, minimum	1 Carp	32	.250		2.63	4.93		7.56	11.35	
	2000	Maximum		8	1		3.14	19.70		22.84	37.50	
	2010	Cut out in blacksplash for elec. wall outlet		38	.211			4.15		4.15	7.10	
	2020	Cut out for sink		20	.400			7.90		7.90	13.50	
	2030	Cut out for stove top		18	.444			8.75		8.75	15	
	2100	Postformed, including backsplash and front edge		30	.267	L.F.	8.55	5.25		13.80	18.40	
	2110	Mitred, add		12	.667	Ea.		13.15		13.15	22.50	
	2200	Built-in place, 25" wide, plastic laminate		25	.320	L.F.	10.80	6.30		17.10	22.50	
	2300	Ceramic tile mosaic		25	.320		24.50	6.30		30.80	38	
	2500	Marble, stock, with splash, 1/2" thick, minimum	1 Bric	17	.471		30	9.40		39.40	49	
	2700	3/4" thick, maximum	"	13	.615		76	12.30		88.30	105	
	2900	Maple, solid, laminated, 1-1/2" thick, no splash	1 Carp	28	.286		31	5.65		36.65	43.50	
	3000	With square splash		28	.286		35	5.65		40.65	48	
	3200	Stainless steel		24	.333	S.F.	73	6.55		79.55	92	
	3400	Recessed cutting block with trim, 16" x 20" x 1"		8	1	Ea.	41	19.70		60.70	79.50	
	3410	Replace cutting block only		8	1	"	29	19.70		48.70	65.50	
	3600	Table tops, plastic laminate, square edge, 7/8" thick		45	.178	S.F.	7	3.50		10.50	13.70	
	3700	1-1/8" thick		40	.200	"	7.20	3.94		11.14	14.70	

06073 | Fire Retardant Treatment

400	0011	**LUMBER TREATMENT**										400
	0400	Fire retardant, wet				M.B.F.	282			282	310	
	0500	KDAT					255			255	281	
	0700	Salt treated, water borne, .40 lb. retention					122			122	134	
	0800	Oil borne, 8 lb. retention					143			143	157	
	1000	Kiln dried lumber, 1" & 2" thick, softwoods					81.50			81.50	90	
	1100	Hardwoods					87			87	95.50	
	1500	For small size 1" stock, add					11			11	12.10	
	1700	For full size rough lumber, add					20%					
600	0010	**PLYWOOD TREATMENT** Fire retardant, 1/4" thick				M.S.F.	204			204	224	600
	0030	3/8" thick					224			224	246	
	0050	1/2" thick					240			240	264	
	0070	5/8" thick					255			255	281	
	0100	3/4" thick					280			280	310	
	0200	For KDAT, add					61			61	67	
	0500	Salt treated water borne, .25 lb., wet, 1/4" thick					112			112	123	
	0530	3/8" thick					117			117	129	
	0550	1/2" thick					122			122	134	
	0570	5/8" thick					133			133	146	
	0600	3/4" thick					138			138	152	
	0800	For KDAT add					61			61	67	
	0900	For .40 lb., per C.F. retention, add					51			51	56	
	1000	For certification stamp, add					30			30	33	

06050 | Basic Wood / Plastic Materials / Methods

06090 | Wood & Plastic Fastenings

			CREW	DAILY OUTPUT	LABOR-HOURS	UNIT	2000 BARE COSTS MAT.	LABOR	EQUIP.	TOTAL	TOTAL INCL O&P	
600	0010	**NAILS** Prices of material only, based on 50# box purchase, copper, plain				Lb.	4.10			4.10	4.51	600
	0400	Stainless steel, plain					5.40			5.40	5.95	
	0500	Box, 3d to 20d, bright					1.13			1.13	1.24	
	0520	Galvanized					1.31			1.31	1.44	
	0600	Common, 3d to 60d, plain					.77			.77	.85	
	0700	Galvanized					.99			.99	1.09	
	0800	Aluminum					3.31			3.31	3.64	
	1000	Annular or spiral thread, 4d to 60d, plain					.66			.66	.73	
	1200	Galvanized					.83			.83	.91	
	1400	Drywall nails, plain					.77			.77	.85	
	1600	Galvanized					1.10			1.10	1.21	
	1800	Finish nails, 4d to 10d, plain					.90			.90	.99	
	2000	Galvanized					1.04			1.04	1.14	
	2100	Aluminum					4.85			4.85	5.35	
	2300	Flooring nails, hardened steel, 2d to 10d, plain					1.22			1.22	1.34	
	2400	Galvanized					1.34			1.34	1.47	
	2500	Gypsum lath nails, 1-1/8", 13 ga. flathead, blued					1.33			1.33	1.46	
	2600	Masonry nails, hardened steel, 3/4" to 3" long, plain					1.61			1.61	1.77	
	2700	Galvanized					1.44			1.44	1.58	
	2900	Roofing nails, threaded, galvanized					1.30			1.30	1.43	
	3100	Aluminum					4.70			4.70	5.15	
	3300	Compressed lead head, threaded, galvanized					1.44			1.44	1.58	
	3600	Siding nails, plain shank, galvanized					1.33			1.33	1.46	
	3800	Aluminum					4			4	4.40	
	5000	Add to prices above for cement coating					.07			.07	.08	
	5200	Zinc or tin plating					.12			.12	.13	
	5500	Vinyl coated sinkers, 8d to 16d					.55			.55	.61	
650	0010	**NAILS** mat. only, for pneumatic tools, framing, per carton of 5000, 2"				Ea.	36.50			36.50	40	650
	0100	2-3/8"					41.50			41.50	45.50	
	0200	Per carton of 4000, 3"					37			37	40.50	
	0300	3-1/4"					39			39	43	
	0400	Per carton of 5000, 2-3/8", galv.					56.50			56.50	62	
	0500	Per carton of 4000, 3", galv.					63.50			63.50	70	
	0600	3-1/4", galv.					79			79	87	
	0700	Roofing, per carton of 7200, 1"					34.50			34.50	38	
	0800	1-1/4"					32			32	35.50	
	0900	1-1/2"					37			37	40.50	
	1000	1-3/4"					44.50			44.50	49	
700	0010	**SHEET METAL SCREWS** Steel, standard, #8 x 3/4", plain				C	2.69			2.69	2.96	700
	0100	Galvanized					3.37			3.37	3.71	
	0300	#10 x 1", plain					3.69			3.69	4.06	
	0400	Galvanized					4.26			4.26	4.69	
	1500	Self-drilling, with washers, (pinch point) #8 x 3/4", plain					4.68			4.68	5.15	
	1600	Galvanized					7			7	7.70	
	1800	#10 x 3/4", plain					6.80			6.80	7.50	
	1900	Galvanized					7.75			7.75	8.55	
	3000	Stainless steel w/aluminum or neoprene washers, #14 x 1", plain					18			18	19.80	
	3100	#14 x 2", plain					24.50			24.50	27	
750	0010	**WOOD SCREWS** #8, 1" long, steel				C	3.29			3.29	3.62	750
	0100	Brass					11			11	12.10	
	0200	#8, 2" long, steel					3.69			3.69	4.06	
	0300	Brass					11.50			11.50	12.65	
	0400	#10, 1" long, steel					4.22			4.22	4.64	
	0500	Brass					22.50			22.50	24.50	

For expanded coverage of these items see *Means Building Construction Cost Data 2000*

06050 | Basic Wood / Plastic Materials / Methods

06090 | Wood & Plastic Fastenings

			DAILY	LABOR-		2000 BARE COSTS				TOTAL	
		CREW	OUTPUT	HOURS	UNIT	MAT.	LABOR	EQUIP.	TOTAL	INCL O&P	
750	0600	#10, 2" long, steel				C	7.50			7.50	8.25	**750**
	0700	Brass					39.50			39.50	43.50	
	0800	#10, 3" long, steel					13.75			13.75	15.10	
	1000	#12, 2" long, steel					4.79			4.79	5.25	
	1100	Brass					16			16	17.60	
	1500	#12, 3" long, steel					15.90			15.90	17.50	
	2000	#12, 4" long, steel					28.50			28.50	31	
800	0010	**TIMBER CONNECTORS** Add up cost of each part for total										**800**
	0020	cost of connection										
	0100	Connector plates, steel, with bolts, straight	2 Carp	75	.213	Ea.	16.35	4.20		20.55	25	
	0110	Tee	"	50	.320		24	6.30		30.30	37.50	
	0200	Bolts, machine, sq. hd. with nut & washer, 1/2" diameter, 4" long	1 Carp	140	.057		.74	1.13		1.87	2.74	
	0300	7-1/2" long		130	.062		.94	1.21		2.15	3.11	
	0500	3/4" diameter, 7-1/2" long		130	.062		1.69	1.21		2.90	3.94	
	0600	15" long		95	.084		2.16	1.66		3.82	5.20	
	0720	Machine bolts, sq. hd. w/nut & wash		150	.053	Lb.	1.88	1.05		2.93	3.87	
	0800	Drilling bolt holes in timber, 1/2" diameter		450	.018	Inch		.35		.35	.60	
	0900	1" diameter		350	.023	"		.45		.45	.77	
	1100	Framing anchors, 2 or 3 dimensional, 10 gauge, no nails incl.		175	.046	Ea.	.39	.90		1.29	1.97	
	1250	Holdowns, 3 gauge base, 10 gauge body		8	1		13.60	19.70		33.30	49	
	1300	Joist and beam hangers, 18 ga. galv., for 2" x 4" joist		175	.046		.49	.90		1.39	2.08	
	1400	2" x 6" to 2" x 10" joist		165	.048		.42	.96		1.38	2.10	
	1600	16 ga. galv., 3" x 6" to 3" x 10" joist		160	.050		2.29	.99		3.28	4.21	
	1700	3" x 10" to 3" x 14" joist		160	.050		2.65	.99		3.64	4.61	
	1800	4" x 6" to 4" x 10" joist		155	.052		2	1.02		3.02	3.94	
	1900	4" x 10" to 4" x 14" joist		155	.052		2.72	1.02		3.74	4.73	
	2000	Two-2" x 6" to two-2" x 10" joists		150	.053		2.19	1.05		3.24	4.21	
	2100	Two-2" x 10" to two-2" x 14" joists		150	.053		2.19	1.05		3.24	4.21	
	2300	3/16" thick, 6" x 8" joist		145	.055		4.80	1.09		5.89	7.15	
	2400	6" x 10" joist		140	.057		5.65	1.13		6.78	8.20	
	2500	6" x 12" joist		135	.059		6.80	1.17		7.97	9.50	
	2700	1/4" thick, 6" x 14" joist		130	.062		8.45	1.21		9.66	11.40	
	2800	Joist anchors, 1/4" x 1-1/4" x 18"		140	.057		3.29	1.13		4.42	5.55	
	2900	Plywood clips, extruded aluminum H clip, for 3/4" panels					.12			.12	.13	
	3000	Galvanized 18 ga. back-up clip					.11			.11	.12	
	3200	Post framing, 16 ga. galv. for 4" x 4" base, 2 piece	1 Carp	130	.062		4.74	1.21		5.95	7.30	
	3300	Cap		130	.062		2.33	1.21		3.54	4.64	
	3500	Rafter anchors, 18 ga. galv., 1-1/2" wide, 5-1/4" long		145	.055		.38	1.09		1.47	2.28	
	3600	10-3/4" long		145	.055		.78	1.09		1.87	2.72	
	3800	Shear plates, 2-5/8" diameter		120	.067		1.41	1.31		2.72	3.80	
	3900	4" diameter		115	.070		3.26	1.37		4.63	5.95	
	4000	Sill anchors, embedded in concrete or block, 18-5/8" long		115	.070		.95	1.37		2.32	3.40	
	4100	Spike grids, 4" x 4", flat or curved		120	.067		.34	1.31		1.65	2.62	
	4400	Split rings, 2-1/2" diameter		120	.067		1.15	1.31		2.46	3.52	
	4500	4" diameter		110	.073		1.77	1.43		3.20	4.40	
	4700	Strap ties, 16 ga., 1-3/8" wide, 12" long		180	.044		.92	.88		1.80	2.51	
	4800	24" long		160	.050		1.42	.99		2.41	3.25	
	5000	Toothed rings, 2-5/8" or 4" diameter		90	.089		.97	1.75		2.72	4.07	
	5200	Truss plates, nailed, 20 gauge, up to 32' span		17	.471	Truss	7	9.25		16.25	23.50	
	5400	Washers, 2" x 2" x 1/8"				Ea.	.22			.22	.24	
	5500	3" x 3" x 3/16"				"	.57			.57	.63	
825	0010	**ROUGH HARDWARE** Average % of carpentry material, minimum					.50%					**825**
	0200	Maximum					1.50%					
850	0010	**BRACING**	R05120-220									**850**
	0300	Let-in, "T" shaped, 22 ga. galv. steel, studs at 16" O.C.	1 Carp	5.80	1.379	C.L.F.	37.50	27		64.50	88	

06050 | Basic Wood / Plastic Materials / Methods

06090 | Wood & Plastic Fastenings

			CREW	DAILY OUTPUT	LABOR-HOURS	UNIT	MAT.	LABOR	EQUIP.	TOTAL	TOTAL INCL O&P	
850	0400	Studs at 24" O.C.	1 Carp	6	1.333	C.L.F.	37.50	26.50		64	86.50	850
	0500	16 ga. galv. steel straps, studs at 16" O.C.		6	1.333		49	26.50		75.50	99	
	0600	Studs at 24" O.C.		6.20	1.290		49	25.50		74.50	97.50	

R05120-220

06100 | Rough Carpentry

06110 | Wood Framing

			CREW	DAILY OUTPUT	LABOR-HOURS	UNIT	MAT.	LABOR	EQUIP.	TOTAL	TOTAL INCL O&P	
100	0010	**BLOCKING**										100
	1950	Miscellaneous, to wood construction										
	2000	2" x 4"	1 Carp	250	.032	L.F.	.39	.63		1.02	1.51	
	2005	Pneumatic nailed		305	.026		.39	.52		.91	1.32	
	2050	2" x 6"		222	.036		.59	.71		1.30	1.87	
	2055	Pneumatic nailed		271	.030		.59	.58		1.17	1.65	
	2100	2" x 8"		200	.040		.83	.79		1.62	2.26	
	2105	Pneumatic nailed		244	.033		.83	.65		1.48	2.02	
	2150	2" x 10"		178	.045		1.22	.89		2.11	2.86	
	2155	Pneumatic nailed		217	.037		1.22	.73		1.95	2.58	
	2200	2" x 12"		151	.053		1.48	1.04		2.52	3.41	
	2205	Pneumatic nailed		185	.043		1.48	.85		2.33	3.08	
	2300	To steel construction										
	2320	2" x 4"	1 Carp	208	.038	L.F.	.39	.76		1.15	1.73	
	2340	2" x 6"		180	.044		.59	.88		1.47	2.15	
	2360	2" x 8"		158	.051		.83	1		1.83	2.62	
	2380	2" x 10"		136	.059		1.22	1.16		2.38	3.33	
	2400	2" x 12"		109	.073		1.48	1.45		2.93	4.10	
150	0010	**BRACING** Let-in, with 1" x 6" boards, studs @ 16" O.C.	1 Carp	1.50	5.333	C.L.F.	49	105		154	234	150
	0200	Studs @ 24" O.C.	"	2.30	3.478	"	49	68.50		117.50	171	
200	0010	**BRIDGING** Wood, for joists 16" O.C., 1" x 3"	1 Carp	1.30	6.154	C.Pr.	26.50	121		147.50	238	200
	0015	Pneumatic nailed		1.70	4.706		26.50	92.50		119	189	
	0100	2" x 3" bridging		1.30	6.154		46	121		167	259	
	0105	Pneumatic nailed		1.70	4.706		46	92.50		138.50	210	
	0300	Steel, galvanized, 18 ga., for 2" x 10" joists at 12" O.C.		1.30	6.154		58.50	121		179.50	273	
	0400	24" O.C.		1.40	5.714		220	113		333	435	
	0600	For 2" x 14" joists at 16" O.C.		1.30	6.154		54.50	121		175.50	268	
	0900	Compression type, 16" O.C., 2" x 8" joists		2	4		126	79		205	274	
	1000	2" x 12" joists		2	4		141	79		220	290	
505	0010	**FRAMING, BEAMS & GIRDERS**										505
	1000	Single, 2" x 6"	2 Carp	700	.023	L.F.	.59	.45		1.04	1.42	
	1005	Pneumatic nailed		812	.020		.59	.39		.98	1.32	
	1020	2" x 8"		650	.025		.83	.48		1.31	1.74	
	1025	Pneumatic nailed		754	.021		.83	.42		1.25	1.63	
	1040	2" x 10"		600	.027		1.22	.53		1.75	2.24	
	1045	Pneumatic nailed		696	.023		1.22	.45		1.67	2.12	
	1060	2" x 12"		550	.029		1.48	.57		2.05	2.60	
	1065	Pneumatic nailed		638	.025		1.48	.49		1.97	2.47	
	1080	2" x 14"		500	.032		2.06	.63		2.69	3.35	
	1085	Pneumatic nailed		580	.028		2.06	.54		2.60	3.20	
	1100	3" x 8"		550	.029		2.33	.57		2.90	3.54	

For expanded coverage of these items see *Means Building Construction Cost Data 2000*

06100 | Rough Carpentry

06110 | Wood Framing

			CREW	DAILY OUTPUT	LABOR-HOURS	UNIT	2000 BARE COSTS MAT.	LABOR	EQUIP.	TOTAL	TOTAL INCL O&P	
505	1120	3" x 10"	2 Carp	500	.032	L.F.	2.91	.63		3.54	4.28	505
	1140	3" x 12"		450	.036		3.49	.70		4.19	5.05	
	1160	3" x 14"	▼	400	.040		4.07	.79		4.86	5.85	
	1180	4" x 8"	F-3	1,000	.040		3.54	.72	.45	4.71	5.60	
	1200	4" x 10"		950	.042		4.42	.75	.48	5.65	6.65	
	1220	4" x 12"		900	.044		5.30	.80	.50	6.60	7.75	
	1240	4" x 14"	▼	850	.047		6.20	.84	.53	7.57	8.80	
	2000	Double, 2" x 6"	2 Carp	625	.026		1.18	.50		1.68	2.16	
	2005	Pneumatic nailed		725	.022		1.18	.43		1.61	2.04	
	2020	2" x 8"		575	.028		1.66	.55		2.21	2.76	
	2025	Pneumatic nailed		667	.024		1.66	.47		2.13	2.63	
	2040	2" x 10"		550	.029		2.44	.57		3.01	3.66	
	2045	Pneumatic nailed		638	.025		2.44	.49		2.93	3.53	
	2060	2" x 12"		525	.030		2.95	.60		3.55	4.28	
	2065	Pneumatic nailed		610	.026		2.95	.52		3.47	4.14	
	2080	2" x 14"		475	.034		4.12	.66		4.78	5.65	
	2085	Pneumatic nailed		551	.029		4.12	.57		4.69	5.50	
	3000	Triple, 2" x 6"		550	.029		1.77	.57		2.34	2.93	
	3005	Pneumatic nailed		638	.025		1.77	.49		2.26	2.80	
	3020	2" x 8"		525	.030		2.49	.60		3.09	3.77	
	3025	Pneumatic nailed		609	.026		2.49	.52		3.01	3.63	
	3040	2" x 10"		500	.032		3.65	.63		4.28	5.10	
	3045	Pneumatic nailed		580	.028		3.65	.54		4.19	4.95	
	3060	2" x 12"		475	.034		4.43	.66		5.09	6	
	3065	Pneumatic nailed		551	.029		4.43	.57		5	5.85	
	3080	2" x 14"		450	.036		6.20	.70		6.90	8	
	3085	Pneumatic nailed	▼	522	.031	▼	6.20	.60		6.80	7.85	
510	0010	**FRAMING, CEILINGS**										510
	6000	Suspended, 2" x 3"	2 Carp	1,000	.016	L.F.	.31	.32		.63	.88	
	6050	2" x 4"		900	.018		.39	.35		.74	1.03	
	6100	2" x 6"		800	.020		.59	.39		.98	1.33	
	6150	2" x 8"	▼	650	.025	▼	.83	.48		1.31	1.74	
515	0010	**FRAMING, COLUMNS**										515
	0100	4" x 4"	2 Carp	390	.041	L.F.	1.11	.81		1.92	2.60	
	0150	4" x 6"		275	.058		2.65	1.15		3.80	4.88	
	0200	4" x 8"		220	.073		3.54	1.43		4.97	6.35	
	0250	6" x 6"		215	.074		5.50	1.47		6.97	8.50	
	0300	6" x 8"		175	.091		5.75	1.80		7.55	9.40	
	0350	6" x 10"	▼	150	.107	▼	7.15	2.10		9.25	11.45	
520	0010	**FRAMING, HEAVY** Mill timber, beams, single 6" x 10"	2 Carp	1.10	14.545	M.B.F.	1,425	287		1,712	2,075	520
	0100	Single 8" x 16"		1.20	13.333		1,625	263		1,888	2,250	
	0200	Built from 2" lumber, multiple 2" x 14"		.90	17.778		885	350		1,235	1,575	
	0210	Built from 3" lumber, multiple 3" x 6"		.70	22.857		1,175	450		1,625	2,050	
	0220	Multiple 3" x 8"		.80	20		1,175	395		1,570	1,950	
	0230	Multiple 3" x 10"		.90	17.778		1,175	350		1,525	1,875	
	0240	Multiple 3" x 12"		1	16		1,175	315		1,490	1,825	
	0250	Built from 4" lumber, multiple 4" x 6"		.80	20		1,325	395		1,720	2,125	
	0260	Multiple 4" x 8"		.90	17.778		1,325	350		1,675	2,050	
	0270	Multiple 4" x 10"		1	16		1,325	315		1,640	2,000	
	0280	Multiple 4" x 12"		1.10	14.545		1,325	287		1,612	1,950	
	0290	Columns, structural grade, 1500f, 4" x 4"		.60	26.667		1,300	525		1,825	2,325	
	0300	6" x 6"		.65	24.615		1,800	485		2,285	2,800	
	0400	8" x 8"	▼	.70	22.857	▼	1,950	450		2,400	2,925	

Important: See the Reference Section for critical supporting data - Reference Nos., Crews, & Location Factors

06100 | Rough Carpentry

06110 | Wood Framing

			Crew	Daily Output	Labor-Hours	Unit	Mat.	Labor	Equip.	Total	Total Incl O&P	
520	0500	10" x 10"	2 Carp	.75	21.333	M.B.F.	2,025	420		2,445	2,950	520
	0600	12" x 12"		.80	20		2,025	395		2,420	2,900	
	0800	Floor planks, 2" thick, T & G, 2" x 6"		1.05	15.238		1,600	300		1,900	2,275	
	0900	2" x 10"		1.10	14.545		760	287		1,047	1,325	
	1100	3" thick, 3" x 6"		1.05	15.238		1,150	300		1,450	1,800	
	1200	3" x 10"		1.10	14.545		1,150	287		1,437	1,775	
	1400	Girders, structural grade, 12" x 12"		.80	20		1,650	395		2,045	2,500	
	1500	10" x 16"		1	16		1,600	315		1,915	2,325	
	2050	Roof planks, see division 06150-600										
	2300	Roof purlins, 4" thick, structural grade	2 Carp	1.05	15.238	M.B.F.	1,125	300		1,425	1,775	
	2500	Roof trusses, add timber connectors, division 06090-800	"	.45	35.556	"	1,100	700		1,800	2,400	
530	0010	**FRAMING, JOISTS**										530
	2000	Joists, 2" x 4"	2 Carp	1,250	.013	L.F.	.39	.25		.64	.86	
	2005	Pneumatic nailed		1,438	.011		.39	.22		.61	.81	
	2100	2" x 6"		1,250	.013		.59	.25		.84	1.08	
	2105	Pneumatic nailed		1,438	.011		.59	.22		.81	1.03	
	2150	2" x 8"		1,100	.015		.83	.29		1.12	1.40	
	2155	Pneumatic nailed		1,265	.013		.83	.25		1.08	1.34	
	2200	2" x 10"		900	.018		1.22	.35		1.57	1.94	
	2205	Pneumatic nailed		1,035	.015		1.22	.30		1.52	1.86	
	2250	2" x 12"		875	.018		1.48	.36		1.84	2.24	
	2255	Pneumatic nailed		1,006	.016		1.48	.31		1.79	2.16	
	2300	2" x 14"		770	.021		2.06	.41		2.47	2.97	
	2305	Pneumatic nailed		886	.018		2.06	.36		2.42	2.88	
	2350	3" x 6"		925	.017		1.75	.34		2.09	2.50	
	2400	3" x 10"		780	.021		2.91	.40		3.31	3.89	
	2450	3" x 12"		600	.027		3.49	.53		4.02	4.74	
	2500	4" x 6"		800	.020		2.65	.39		3.04	3.60	
	2550	4" x 10"		600	.027		4.42	.53		4.95	5.75	
	2600	4" x 12"		450	.036		5.30	.70		6	7.05	
	2605	Sister joist, 2" x 6"		800	.020		.59	.39		.98	1.33	
	2606	Pneumatic nailed		960	.017		.59	.33		.92	1.21	
	3000	Composite wood joist 9-1/2" deep		.90	17.778	M.L.F.	1,400	350		1,750	2,150	
	3010	11-1/2" deep		.88	18.182		1,525	360		1,885	2,300	
	3020	14" deep		.82	19.512		1,700	385		2,085	2,525	
	3030	16" deep		.78	20.513		2,225	405		2,630	3,150	
	4000	Open web joist 12" deep		.88	18.182		1,575	360		1,935	2,350	
	4010	14" deep		.82	19.512		1,825	385		2,210	2,650	
	4020	16" deep		.78	20.513		1,900	405		2,305	2,800	
	4030	18" deep		.74	21.622		1,975	425		2,400	2,900	
545	0010	**FRAMING, MISCELLANEOUS** R06100-010										545
	2000	Firestops, 2" x 4"	2 Carp	780	.021	L.F.	.39	.40		.79	1.12	
	2005	Pneumatic nailed R06110-030		952	.017		.39	.33		.72	1	
	2100	2" x 6"		600	.027		.59	.53		1.12	1.55	
	2105	Pneumatic nailed		732	.022		.59	.43		1.02	1.39	
	5000	Nailers, treated, wood construction, 2" x 4"		800	.020		.66	.39		1.05	1.40	
	5005	Pneumatic nailed		960	.017		.66	.33		.99	1.28	
	5100	2" x 6"		750	.021		1.05	.42		1.47	1.88	
	5105	Pneumatic nailed		900	.018		1.05	.35		1.40	1.76	
	5120	2" x 8"		700	.023		1.47	.45		1.92	2.39	
	5125	Pneumatic nailed		840	.019		1.47	.38		1.85	2.26	
	5200	Steel construction, 2" x 4"		750	.021		.66	.42		1.08	1.44	
	5220	2" x 6"		700	.023		1.05	.45		1.50	1.93	
	5240	2" x 8"		650	.025		1.47	.48		1.95	2.45	

For expanded coverage of these items see Means Building Construction Cost Data 2000

06100 | Rough Carpentry

06110 | Wood Framing

			CREW	DAILY OUTPUT	LABOR-HOURS	UNIT	2000 BARE COSTS MAT.	LABOR	EQUIP.	TOTAL	TOTAL INCL O&P	
545	7000	Rough bucks, treated, for doors or windows, 2" x 6"	2 Carp	400	.040	L.F.	1.05	.79		1.84	2.51	**545**
	7005	Pneumatic nailed		480	.033		1.05	.66		1.71	2.29	
	7100	2" x 8"		380	.042		1.47	.83		2.30	3.04	
	7105	Pneumatic nailed		456	.035		1.47	.69		2.16	2.80	
	8000	Stair stringers, 2" x 10"		130	.123		1.22	2.42		3.64	5.50	
	8100	2" x 12"		130	.123		1.48	2.42		3.90	5.75	
	8150	3" x 10"		125	.128		2.91	2.52		5.43	7.50	
	8200	3" x 12"		125	.128		3.49	2.52		6.01	8.15	
550	0010	**PARTITIONS** Wood stud with single bottom plate and										**550**
	0020	double top plate, no waste, std. & better lumber										
	0180	2" x 4" studs, 8' high, studs 12" O.C.	2 Carp	80	.200	L.F.	4.34	3.94		8.28	11.55	
	0185	12" O.C., pneumatic nailed		96	.167		4.34	3.28		7.62	10.45	
	0200	16" O.C.		100	.160		3.55	3.15		6.70	9.30	
	0205	16" O.C., pneumatic nailed		120	.133		3.55	2.63		6.18	8.40	
	0300	24" O.C.		125	.128		2.76	2.52		5.28	7.35	
	0305	24" O.C., pneumatic nailed		150	.107		2.76	2.10		4.86	6.65	
	0380	10' high, studs 12" O.C.		80	.200		5.15	3.94		9.09	12.40	
	0385	12" O.C., pneumatic nailed		96	.167		5.15	3.28		8.43	11.30	
	0400	16" O.C.		100	.160		4.14	3.15		7.29	9.95	
	0405	16" O.C., pneumatic nailed		120	.133		4.14	2.63		6.77	9.05	
	0500	24" O.C.		125	.128		3.16	2.52		5.68	7.80	
	0505	24" O.C., pneumatic nailed		150	.107		3.16	2.10		5.26	7.05	
	0580	12' high, studs 12" O.C.		65	.246		5.90	4.85		10.75	14.80	
	0585	12" O.C., pneumatic nailed		78	.205		5.90	4.04		9.94	13.40	
	0600	16" O.C.		80	.200		4.74	3.94		8.68	11.95	
	0605	16" O.C., pneumatic nailed		96	.167		4.74	3.28		8.02	10.85	
	0700	24" O.C.		100	.160		3.55	3.15		6.70	9.30	
	0705	24" O.C., pneumatic nailed		120	.133		3.55	2.63		6.18	8.40	
	0780	2" x 6" studs, 8' high, studs 12" O.C.		70	.229		6.50	4.50		11	14.85	
	0785	12" O.C., pneumatic nailed		84	.190		6.50	3.75		10.25	13.60	
	0800	16" O.C.		90	.178		5.30	3.50		8.80	11.85	
	0805	16" O.C., pneumatic nailed		108	.148		5.30	2.92		8.22	10.85	
	0900	24" O.C.		115	.139		4.13	2.74		6.87	9.25	
	0905	24" O.C., pneumatic nailed		138	.116		4.13	2.28		6.41	8.45	
	0980	10' high, studs 12" O.C.		70	.229		7.65	4.50		12.15	16.15	
	0985	12" O.C., pneumatic nailed		84	.190		7.65	3.75		11.40	14.90	
	1000	16" O.C.		90	.178		6.20	3.50		9.70	12.80	
	1005	16" O.C., pneumatic nailed		108	.148		6.20	2.92		9.12	11.80	
	1100	24" O.C.		115	.139		4.72	2.74		7.46	9.90	
	1105	24" O.C., pneumatic nailed		138	.116		4.72	2.28		7	9.10	
	1180	12' high, studs 12" O.C.		55	.291		8.85	5.75		14.60	19.55	
	1185	12" O.C., pneumatic nailed		66	.242		8.85	4.78		13.63	17.95	
	1200	16" O.C.		70	.229		7.10	4.50		11.60	15.50	
	1205	16" O.C., pneumatic nailed		84	.190		7.10	3.75		10.85	14.25	
	1300	24" O.C.		90	.178		5.30	3.50		8.80	11.85	
	1305	24" O.C., pneumatic nailed		108	.148		5.30	2.92		8.22	10.85	
	1400	For horizontal blocking, 2" x 4", add		600	.027		.39	.53		.92	1.33	
	1500	2" x 6", add		600	.027		.59	.53		1.12	1.55	
	1600	For openings, add		250	.064			1.26		1.26	2.16	
	1700	Headers for above openings, material only, add				M.B.F.	620			620	685	
555	0010	**FRAMING, ROOFS**										**555**
	2000	Fascia boards, 2" x 8"	2 Carp	225	.071	L.F.	.83	1.40		2.23	3.31	
	2100	2" x 10"		180	.089		1.22	1.75		2.97	4.34	
	5000	Rafters, to 4 in 12 pitch, 2" x 6", ordinary		1,000	.016		.59	.32		.91	1.19	
	5020	On steep roofs		800	.020		.59	.39		.98	1.33	
	5040	On dormers or complex roofs		590	.027		.59	.53		1.12	1.57	

Reference Nos: R06100-010, R06110-030

06100 | Rough Carpentry

06110	Wood Framing	CREW	DAILY OUTPUT	LABOR-HOURS	UNIT	2000 BARE COSTS MAT.	LABOR	EQUIP.	TOTAL	TOTAL INCL O&P
5060	2" x 8", ordinary	2 Carp	950	.017	L.F.	.83	.33		1.16	1.48
5080	On steep roofs		750	.021		.83	.42		1.25	1.63
5100	On dormers or complex roofs		540	.030		.83	.58		1.41	1.91
5120	2" x 10", ordinary		630	.025		1.22	.50		1.72	2.20
5140	On steep roofs		495	.032		1.22	.64		1.86	2.43
5160	On dormers or complex roofs		425	.038		1.22	.74		1.96	2.61
5180	2" x 12", ordinary		575	.028		1.48	.55		2.03	2.56
5200	On steep roofs		455	.035		1.48	.69		2.17	2.81
5220	On dormers or complex roofs		395	.041		1.48	.80		2.28	2.99
5300	Hip and valley rafters, 2" x 6", ordinary		760	.021		.59	.41		1	1.36
5320	On steep roofs		585	.027		.59	.54		1.13	1.57
5340	On dormers or complex roofs		510	.031		.59	.62		1.21	1.71
5360	2" x 8", ordinary		720	.022		.83	.44		1.27	1.66
5380	On steep roofs		545	.029		.83	.58		1.41	1.90
5400	On dormers or complex roofs		470	.034		.83	.67		1.50	2.06
5420	2" x 10", ordinary		570	.028		1.22	.55		1.77	2.29
5440	On steep roofs		440	.036		1.22	.72		1.94	2.57
5460	On dormers or complex roofs		380	.042		1.22	.83		2.05	2.76
5480	Hip and valley rafters, 2" x 12", ordinary		525	.030		1.48	.60		2.08	2.65
5500	On steep roofs		410	.039		1.48	.77		2.25	2.94
5520	On dormers or complex roofs		355	.045		1.48	.89		2.37	3.14
5540	Hip and valley jacks, 2" x 6", ordinary		600	.027		.59	.53		1.12	1.55
5560	On steep roofs		475	.034		.59	.66		1.25	1.79
5580	On dormers or complex roofs		410	.039		.59	.77		1.36	1.97
5600	2" x 8", ordinary		490	.033		.83	.64		1.47	2.01
5620	On steep roofs		385	.042		.83	.82		1.65	2.31
5640	On dormers or complex roofs		335	.048		.83	.94		1.77	2.52
5660	2" x 10", ordinary		450	.036		1.22	.70		1.92	2.54
5680	On steep roofs		350	.046		1.22	.90		2.12	2.88
5700	On dormers or complex roofs		305	.052		1.22	1.03		2.25	3.11
5720	2" x 12", ordinary		375	.043		1.48	.84		2.32	3.06
5740	On steep roofs		295	.054		1.48	1.07		2.55	3.45
5760	On dormers or complex roofs		255	.063		1.48	1.24		2.72	3.74
5780	Rafter tie, 1" x 4", #3		800	.020		.34	.39		.73	1.06
5790	2" x 4", #3		800	.020		.39	.39		.78	1.11
5800	Ridge board, #2 or better, 1" x 6"		600	.027		.90	.53		1.43	1.89
5820	1" x 8"		550	.029		1.20	.57		1.77	2.30
5840	1" x 10"		500	.032		1.50	.63		2.13	2.73
5860	2" x 6"		500	.032		.59	.63		1.22	1.73
5880	2" x 8"		450	.036		.83	.70		1.53	2.11
5900	2" x 10"		400	.040		1.22	.79		2.01	2.69
5920	Roof cants, split, 4" x 4"		650	.025		1.11	.48		1.59	2.05
5940	6" x 6"		600	.027		5.50	.53		6.03	6.90
5960	Roof curbs, untreated, 2" x 6"		520	.031		.59	.61		1.20	1.69
5980	2" x 12"		400	.040		1.48	.79		2.27	2.97
6000	Sister rafters, 2" x 6"		800	.020		.59	.39		.98	1.33
6020	2" x 8"		640	.025		.83	.49		1.32	1.75
6040	2" x 10"		535	.030		1.22	.59		1.81	2.35
6060	2" x 12"	▼	455	.035	▼	1.48	.69		2.17	2.81
0010	**FRAMING, SILLS**									
2000	Ledgers, nailed, 2" x 4"	2 Carp	755	.021	L.F.	.39	.42		.81	1.15
2050	2" x 6"		600	.027		.59	.53		1.12	1.55
2100	Bolted, not including bolts, 3" x 6"		325	.049		1.75	.97		2.72	3.58
2150	3" x 12"		233	.069		3.49	1.35		4.84	6.15
2600	Mud sills, redwood, construction grade, 2" x 4"	▼	895	.018	▼	2.67	.35		3.02	3.53

For expanded coverage of these items see *Means Building Construction Cost Data 2000*

06100 | Rough Carpentry

06110 | Wood Framing

			CREW	DAILY OUTPUT	LABOR-HOURS	UNIT	MAT.	LABOR	EQUIP.	TOTAL	TOTAL INCL O&P	
560	2620	2" x 6"	2 Carp	780	.021	L.F.	4	.40		4.40	5.10	560
	4000	Sills, 2" x 4"		600	.027		.39	.53		.92	1.33	
	4050	2" x 6"		550	.029		.59	.57		1.16	1.63	
	4080	2" x 8"		500	.032		.83	.63		1.46	1.99	
	4100	2" x 10"		450	.036		1.22	.70		1.92	2.54	
	4120	2" x 12"		400	.040		1.48	.79		2.27	2.97	
	4200	Treated, 2" x 4"		550	.029		.66	.57		1.23	1.70	
	4220	2" x 6"		500	.032		1.05	.63		1.68	2.24	
	4240	2" x 8"		450	.036		1.47	.70		2.17	2.82	
	4260	2" x 10"		400	.040		1.80	.79		2.59	3.33	
	4280	2" x 12"		350	.046		2.33	.90		3.23	4.11	
	4400	4" x 4"		450	.036		1.78	.70		2.48	3.16	
	4420	4" x 6"		350	.046		2.53	.90		3.43	4.32	
	4460	4" x 8"		300	.053		3.37	1.05		4.42	5.50	
	4480	4" x 10"		260	.062		4.59	1.21		5.80	7.15	
565	0010	**FRAMING, SLEEPERS**										565
	0100	On concrete, treated, 1" x 2"	2 Carp	2,350	.007	L.F.	.20	.13		.33	.45	
	0150	1" x 3"		2,000	.008		.20	.16		.36	.49	
	0200	2" x 4"		1,500	.011		.66	.21		.87	1.08	
	0250	2" x 6"		1,300	.012		1.05	.24		1.29	1.58	
570	0010	**FRAMING, SOFFITS & CANOPIES**										570
	1000	Canopy or soffit framing, 1" x 4"	2 Carp	900	.018	L.F.	.60	.35		.95	1.26	
	1020	1" x 6"		850	.019		.90	.37		1.27	1.63	
	1040	1" x 8"		750	.021		1.20	.42		1.62	2.04	
	1100	2" x 4"		620	.026		.39	.51		.90	1.30	
	1120	2" x 6"		560	.029		.59	.56		1.15	1.61	
	1140	2" x 8"		500	.032		.83	.63		1.46	1.99	
	1200	3" x 4"		500	.032		1.07	.63		1.70	2.26	
	1220	3" x 6"		400	.040		1.75	.79		2.54	3.27	
	1240	3" x 10"		300	.053		2.91	1.05		3.96	5	
575	0010	**FRAMING, TREATED LUMBER**										575
	0020	Water-borne salt, C.C.A., A.C.A., wet, .40 P.C.F. retention										
	0100	2" x 4"				M.B.F.	985			985	1,075	
	0110	2" x 6"					1,050			1,050	1,150	
	0120	2" x 8"					1,100			1,100	1,225	
	0130	2" x 10"					1,075			1,075	1,175	
	0140	2" x 12"					1,175			1,175	1,275	
	0200	4" x 4"					1,325			1,325	1,475	
	0210	4" x 6"					1,275			1,275	1,400	
	0220	4" x 8"					1,275			1,275	1,400	
	0250	Add for .60 P.C.F. retention					40%					
	0260	Add for 2.5 P.C.F. retention					200%					
	0270	Add for K.D.A.T.					20%					
590	0010	**FRAMING, WALLS**	R06100-010									590
	2000	Headers over openings, 2" x 6"		2 Carp	360	.044	L.F.	.59	.88		1.47	2.15	
	2005	2" x 6", pneumatic nailed	R06110-030		432	.037		.59	.73		1.32	1.90	
	2050	2" x 8"			340	.047		.83	.93		1.76	2.50	
	2055	2" x 8", pneumatic nailed			408	.039		.83	.77		1.60	2.23	
	2100	2" x 10"			320	.050		1.22	.99		2.21	3.03	
	2105	2" x 10", pneumatic nailed			384	.042		1.22	.82		2.04	2.75	
	2150	2" x 12"			300	.053		1.48	1.05		2.53	3.42	
	2155	2" x 12", pneumatic nailed			360	.044		1.48	.88		2.36	3.12	
	2190	4" x 10"			240	.067		4.42	1.31		5.73	7.10	

Important: See the Reference Section for critical supporting data - Reference Nos., Crews, & Location Factors

06100 | Rough Carpentry

06110 | Wood Framing

			CREW	DAILY OUTPUT	LABOR-HOURS	UNIT	2000 BARE COSTS MAT.	LABOR	EQUIP.	TOTAL	TOTAL INCL O&P	
590	2195	4" x 10", pneumatic nailed	2 Carp	288	.056	L.F.	4.42	1.09		5.51	6.75	590
	2200	4" x 12"		190	.084		5.30	1.66		6.96	8.70	
	2205	4" x 12", pneumatic nailed		228	.070		5.30	1.38		6.68	8.20	
	2240	6" x 10"		165	.097		7.15	1.91		9.06	11.10	
	2245	6" x 10", pneumatic nailed		198	.081		7.15	1.59		8.74	10.60	
	2250	6" x 12"		140	.114		11.20	2.25		13.45	16.20	
	2255	6" x 12", pneumatic nailed		168	.095		11.20	1.88		13.08	15.55	
	5000	Plates, untreated, 2" x 3"		850	.019		.31	.37		.68	.98	
	5005	2" x 3", pneumatic nailed		1,020	.016		.31	.31		.62	.87	
	5020	2" x 4"		800	.020		.39	.39		.78	1.11	
	5025	2" x 4", pneumatic nailed		960	.017		.39	.33		.72	.99	
	5040	2" x 6"		750	.021		.59	.42		1.01	1.37	
	5045	2" x 6", pneumatic nailed		900	.018		.59	.35		.94	1.25	
	5060	Treated, 2" x 3"		850	.019		.55	.37		.92	1.24	
	5065	2" x 3", treated, pneumatic nailed		1,020	.016		.55	.31		.86	1.13	
	5080	2" x 4"		800	.020		.66	.39		1.05	1.40	
	5085	2" x 4", treated, pneumatic nailed		960	.017		.66	.33		.99	1.28	
	5100	2" x 6"		750	.021		1.05	.42		1.47	1.88	
	5103	2" x 6", treated, pneumatic nailed		900	.018		1.05	.35		1.40	1.76	
	5120	Studs, 8' high wall, 2" x 3"		1,200	.013		.31	.26		.57	.79	
	5125	2" x 3", pneumatic nailed		1,440	.011		.31	.22		.53	.72	
	5140	2" x 4"		1,100	.015		.39	.29		.68	.92	
	5146	2" x 4", pneumatic nailed		1,320	.012		.39	.24		.63	.84	
	5160	2" x 6"		1,000	.016		.59	.32		.91	1.19	
	5166	2" x 6", pneumatic nailed		1,200	.013		.59	.26		.85	1.10	
	5180	3" x 4"		800	.020		1.07	.39		1.46	1.86	
	5185	3" x 4", pneumatic nailed		960	.017		1.07	.33		1.40	1.74	
	5200	Installed on second story, 2" x 3"		1,170	.014		.31	.27		.58	.80	
	5205	2" x 3", pneumatic nailed		1,404	.011		.31	.22		.53	.72	
	5220	2" x 4"		1,015	.016		.39	.31		.70	.96	
	5225	2" x 4", pneumatic nailed		1,218	.013		.39	.26		.65	.87	
	5240	2" x 6"		890	.018		.59	.35		.94	1.26	
	5245	2" x 6", pneumatic nailed		1,080	.015		.59	.29		.88	1.15	
	5260	3" x 4"		800	.020		1.07	.39		1.46	1.86	
	5265	3" x 4", pneumatic nailed		960	.017		1.07	.33		1.40	1.74	
	5280	Installed on dormer or gable, 2" x 3"		1,045	.015		.31	.30		.61	.86	
	5285	2" x 3", pneumatic nailed		1,254	.013		.31	.25		.56	.77	
	5300	2" x 4"		905	.018		.39	.35		.74	1.03	
	5305	2" x 4", pneumatic nailed		1,086	.015		.39	.29		.68	.93	
	5320	2" x 6"		800	.020		.59	.39		.98	1.33	
	5325	2" x 6", pneumatic nailed		960	.017		.59	.33		.92	1.21	
	5340	3" x 4"		700	.023		1.07	.45		1.52	1.95	
	5345	3" x 4", pneumatic nailed		840	.019		1.07	.38		1.45	1.82	
	5360	6' high wall, 2" x 3"		970	.016		.31	.32		.63	.90	
	5365	2" x 3", pneumatic nailed		1,164	.014		.31	.27		.58	.80	
	5380	2" x 4"		850	.019		.39	.37		.76	1.07	
	5385	2" x 4", pneumatic nailed		1,020	.016		.39	.31		.70	.96	
	5400	2" x 6"		740	.022		.59	.43		1.02	1.38	
	5405	2" x 6", pneumatic nailed		888	.018		.59	.35		.94	1.26	
	5420	3" x 4"		600	.027		1.07	.53		1.60	2.08	
	5425	3" x 4", pneumatic nailed		720	.022		1.07	.44		1.51	1.93	
	5440	Installed on second story, 2" x 3"		950	.017		.31	.33		.64	.91	
	5445	2" x 3", pneumatic nailed		1,140	.014		.31	.28		.59	.81	
	5460	2" x 4"		810	.020		.39	.39		.78	1.10	
	5465	2" x 4", pneumatic nailed		972	.016		.39	.32		.71	.99	
	5480	2" x 6"		700	.023		.59	.45		1.04	1.42	

For expanded coverage of these items see Means Building Construction Cost Data 2000

06100 | Rough Carpentry

06110 | Wood Framing

		CREW	DAILY OUTPUT	LABOR-HOURS	UNIT	MAT.	LABOR	EQUIP.	TOTAL	TOTAL INCL O&P	
590	5485	2" x 6", pneumatic nailed	2 Carp	840	.019	L.F.	.59	.38		.97	1.29	590
	5500	3" x 4"		550	.029		1.07	.57		1.64	2.16	
	5505	3" x 4", pneumatic nailed		660	.024		1.07	.48		1.55	2	
	5520	Installed on dormer or gable, 2" x 3"		850	.019		.31	.37		.68	.98	
	5525	2" x 3", pneumatic nailed		1,020	.016		.31	.31		.62	.87	
	5540	2" x 4"		720	.022		.39	.44		.83	1.18	
	5545	2" x 4", pneumatic nailed		864	.019		.39	.36		.75	1.06	
	5560	2" x 6"		620	.026		.59	.51		1.10	1.52	
	5565	2" x 6", pneumatic nailed		744	.022		.59	.42		1.01	1.38	
	5580	3" x 4"		480	.033		1.07	.66		1.73	2.31	
	5585	3" x 4", pneumatic nailed		576	.028		1.07	.55		1.62	2.12	
	5600	3' high wall, 2" x 3"		740	.022		.31	.43		.74	1.07	
	5605	2" x 3", pneumatic nailed		888	.018		.31	.35		.66	.95	
	5620	2" x 4"		640	.025		.39	.49		.88	1.27	
	5625	2" x 4", pneumatic nailed		768	.021		.39	.41		.80	1.13	
	5640	2" x 6"		550	.029		.59	.57		1.16	1.63	
	5645	2" x 6", pneumatic nailed		660	.024		.59	.48		1.07	1.47	
	5660	3" x 4"		440	.036		1.07	.72		1.79	2.41	
	5665	3" x 4", pneumatic nailed		528	.030		1.07	.60		1.67	2.20	
	5680	Installed on second story, 2" x 3"		700	.023		.31	.45		.76	1.11	
	5685	2" x 3", pneumatic nailed		840	.019		.31	.38		.69	.98	
	5700	2" x 4"		610	.026		.39	.52		.91	1.32	
	5705	2" x 4", pneumatic nailed		732	.022		.39	.43		.82	1.17	
	5720	2" x 6"		520	.031		.59	.61		1.20	1.69	
	5725	2" x 6", pneumatic nailed		624	.026		.59	.51		1.10	1.52	
	5740	3" x 4"		430	.037		1.07	.73		1.80	2.44	
	5745	3" x 4", pneumatic nailed		516	.031		1.07	.61		1.68	2.23	
	5760	Installed on dormer or gable, 2" x 3"		625	.026		.31	.50		.81	1.20	
	5765	2" x 3", pneumatic nailed		750	.021		.31	.42		.73	1.06	
	5780	2" x 4"		545	.029		.39	.58		.97	1.42	
	5785	2" x 4", pneumatic nailed		654	.024		.39	.48		.87	1.26	
	5800	2" x 6"		465	.034		.59	.68		1.27	1.81	
	5805	2" x 6", pneumatic nailed		558	.029		.59	.56		1.15	1.62	
	5820	3" x 4"		380	.042		1.07	.83		1.90	2.60	
	5825	3" x 4", pneumatic nailed		456	.035		1.07	.69		1.76	2.36	
	8250	For second story & above, add						5%				
	8300	For dormer & gable, add						15%				
600	0010	**FURRING** Wood strips, 1" x 2", on walls, on wood	1 Carp	550	.015	L.F.	.18	.29		.47	.69	600
	0015	On wood, pneumatic nailed		710	.011		.18	.22		.40	.58	
	0300	On masonry		495	.016		.18	.32		.50	.75	
	0400	On concrete		260	.031		.18	.61		.79	1.24	
	0600	1" x 3", on walls, on wood		550	.015		.18	.29		.47	.69	
	0605	On wood, pneumatic nailed		710	.011		.18	.22		.40	.58	
	0700	On masonry		495	.016		.18	.32		.50	.75	
	0800	On concrete		260	.031		.18	.61		.79	1.24	
	0850	On ceilings, on wood		350	.023		.18	.45		.63	.97	
	0855	On wood, pneumatic nailed		450	.018		.18	.35		.53	.80	
	0900	On masonry		320	.025		.18	.49		.67	1.04	
	0950	On concrete		210	.038		.18	.75		.93	1.49	
700	0010	**GROUNDS** For casework, 1" x 2" wood strips, on wood	1 Carp	330	.024	L.F.	.18	.48		.66	1.02	700
	0100	On masonry		285	.028		.18	.55		.73	1.15	
	0200	On concrete		250	.032		.18	.63		.81	1.28	
	0400	For plaster, 3/4" deep, on wood		450	.018		.18	.35		.53	.80	
	0500	On masonry		225	.036		.18	.70		.88	1.40	
	0600	On concrete		175	.046		.18	.90		1.08	1.74	

06100 | Rough Carpentry

06110 | Wood Framing

		CREW	DAILY OUTPUT	LABOR-HOURS	UNIT	2000 BARE COSTS				TOTAL INCL O&P	
						MAT.	LABOR	EQUIP.	TOTAL		
0700	On metal lath	1 Carp	200	.040	L.F.	.18	.79		.97	1.55	700

06150 | Wood Decking

		CREW	DAILY OUTPUT	LABOR-HOURS	UNIT	MAT.	LABOR	EQUIP.	TOTAL	INCL O&P	
0010	**ROOF DECKS**										600
0400	Cedar planks, 3" thick	2 Carp	320	.050	S.F.	5.65	.99		6.64	7.95	
0500	4" thick		250	.064		7.65	1.26		8.91	10.55	
0700	Douglas fir, 3" thick		320	.050		2.12	.99		3.11	4.02	
0800	4" thick		250	.064		2.84	1.26		4.10	5.30	
1000	Hemlock, 3" thick		320	.050		2.12	.99		3.11	4.02	
1100	4" thick		250	.064		2.83	1.26		4.09	5.25	
1300	Western white spruce, 3" thick		320	.050		2.05	.99		3.04	3.95	
1400	4" thick		250	.064		2.73	1.26		3.99	5.15	

06160 | Sheathing

		CREW	DAILY OUTPUT	LABOR-HOURS	UNIT	MAT.	LABOR	EQUIP.	TOTAL	INCL O&P	
0010	**SHEATHING** Plywood on roof, CDX R06160-020										800
0030	5/16" thick	2 Carp	1,600	.010	S.F.	.36	.20		.56	.74	
0035	Pneumatic nailed		1,952	.008		.36	.16		.52	.68	
0050	3/8" thick		1,525	.010		.40	.21		.61	.80	
0055	Pneumatic nailed		1,860	.009		.40	.17		.57	.74	
0100	1/2" thick		1,400	.011		.53	.23		.76	.97	
0105	Pneumatic nailed		1,708	.009		.53	.18		.71	.90	
0200	5/8" thick		1,300	.012		.67	.24		.91	1.15	
0205	Pneumatic nailed		1,586	.010		.67	.20		.87	1.07	
0300	3/4" thick		1,200	.013		.81	.26		1.07	1.34	
0305	Pneumatic nailed		1,464	.011		.81	.22		1.03	1.26	
0500	Plywood on walls with exterior CDX, 3/8" thick		1,200	.013		.40	.26		.66	.90	
0505	Pneumatic nailed		1,488	.011		.40	.21		.61	.81	
0600	1/2" thick		1,125	.014		.53	.28		.81	1.06	
0605	Pneumatic nailed		1,395	.011		.53	.23		.76	.97	
0700	5/8" thick		1,050	.015		.67	.30		.97	1.24	
0705	Pneumatic nailed		1,302	.012		.67	.24		.91	1.14	
0800	3/4" thick		975	.016		.81	.32		1.13	1.44	
0805	Pneumatic nailed		1,209	.013		.81	.26		1.07	1.34	
1000	For shear wall construction, add						20%				
1200	For structural 1 exterior plywood, add				S.F.	10%					
1400	With boards, on roof 1" x 6" boards, laid horizontal	2 Carp	725	.022		1.30	.43		1.73	2.17	
1500	Laid diagonal		650	.025		1.30	.48		1.78	2.26	
1700	1" x 8" boards, laid horizontal		875	.018		1.30	.36		1.66	2.05	
1800	Laid diagonal		725	.022		1.30	.43		1.73	2.17	
2000	For steep roofs, add						40%				
2200	For dormers, hips and valleys, add					5%	50%				
2400	Boards on walls, 1" x 6" boards, laid regular	2 Carp	650	.025		1.30	.48		1.78	2.26	
2500	Laid diagonal		585	.027		1.30	.54		1.84	2.35	
2700	1" x 8" boards, laid regular		765	.021		1.30	.41		1.71	2.14	
2800	Laid diagonal		650	.025		1.30	.48		1.78	2.26	
2850	Gypsum, weatherproof, 1/2" thick		1,125	.014		.28	.28		.56	.79	
2900	Sealed, 4/10" thick		1,100	.015		.44	.29		.73	.97	
3000	Wood fiber, regular, no vapor barrier, 1/2" thick		1,200	.013		.53	.26		.79	1.03	
3100	5/8" thick		1,200	.013		.71	.26		.97	1.23	
3300	No vapor barrier, in colors, 1/2" thick		1,200	.013		.77	.26		1.03	1.30	
3400	5/8" thick		1,200	.013		.95	.26		1.21	1.50	
3600	With vapor barrier one side, white, 1/2" thick		1,200	.013		.54	.26		.80	1.04	
3700	Vapor barrier 2 sides, 1/2" thick		1,200	.013		.82	.26		1.08	1.35	
3800	Asphalt impregnated, 25/32" thick		1,200	.013		.35	.26		.61	.84	

For expanded coverage of these items see *Means Building Construction Cost Data 2000*

06100 | Rough Carpentry

06160 | Sheathing

				Crew	Daily Output	Labor-Hours	Unit	Mat.	2000 Bare Costs Labor	Equip.	Total	Total Incl O&P	
800	3850	Intermediate, 1/2" thick	R06160-020	2 Carp	1,200	.013	S.F.	.29	.26		.55	.77	800
	4000	Wafer board on roof, 1/2" thick			1,455	.011		.40	.22		.62	.81	
	4100	5/8" thick			1,330	.012		.46	.24		.70	.92	
850	0010	**SUBFLOOR** Plywood, CDX, 1/2" thick	R06160-020	2 Carp	1,500	.011	SF Flr.	.53	.21		.74	.94	850
	0015	Pneumatic nailed			1,860	.009		.53	.17		.70	.87	
	0100	5/8" thick			1,350	.012		.67	.23		.90	1.13	
	0105	Pneumatic nailed			1,674	.010		.67	.19		.86	1.05	
	0200	3/4" thick			1,250	.013		.81	.25		1.06	1.32	
	0205	Pneumatic nailed			1,550	.010		.81	.20		1.01	1.24	
	0300	1-1/8" thick, 2-4-1 including underlayment			1,050	.015		2	.30		2.30	2.71	
	0500	With boards, 1" x 10" S4S, laid regular			1,100	.015		1.01	.29		1.30	1.60	
	0600	Laid diagonal			900	.018		1.02	.35		1.37	1.72	
	0800	1" x 8" S4S, laid regular			1,000	.016		.91	.32		1.23	1.54	
	0900	Laid diagonal			850	.019		.91	.37		1.28	1.64	
	1100	Wood fiber, T&G, 2' x 8' planks, 1" thick			1,000	.016		1.26	.32		1.58	1.93	
	1200	1-3/8" thick			900	.018		1.55	.35		1.90	2.30	
	1500	Wafer board, 5/8" thick			1,330	.012	S.F.	.41	.24		.65	.86	
	1600	3/4" thick			1,230	.013	"	.45	.26		.71	.94	
900	0010	**UNDERLAYMENT** Plywood, underlayment grade, 3/8" thick		2 Carp	1,500	.011	SF Flr.	.62	.21		.83	1.04	900
	0015	Pneumatic nailed			1,860	.009		.62	.17		.79	.97	
	0100	1/2" thick			1,450	.011		.80	.22		1.02	1.25	
	0105	Pneumatic nailed			1,798	.009		.80	.18		.98	1.18	
	0200	5/8" thick			1,400	.011		.76	.23		.99	1.23	
	0205	Pneumatic nailed			1,736	.009		.76	.18		.94	1.15	
	0300	3/4" thick			1,300	.012		.97	.24		1.21	1.49	
	0305	Pneumatic nailed			1,612	.010		.97	.20		1.17	1.40	
	0500	Particle board, 3/8" thick			1,500	.011		.38	.21		.59	.78	
	0505	Pneumatic nailed			1,860	.009		.38	.17		.55	.71	
	0600	1/2" thick			1,450	.011		.40	.22		.62	.81	
	0605	Pneumatic nailed			1,798	.009		.40	.18		.58	.74	
	0800	5/8" thick			1,400	.011		.46	.23		.69	.90	
	0805	Pneumatic nailed			1,736	.009		.46	.18		.64	.82	
	0900	3/4" thick			1,300	.012		.57	.24		.81	1.05	
	0905	Pneumatic nailed			1,612	.010		.57	.20		.77	.96	
	1100	Hardboard, underlayment grade, 4' x 4', .215" thick			1,500	.011		.41	.21		.62	.81	

06170 | Prefabricated Structural Wood

			Crew	Daily Output	Labor-Hours	Unit	Mat.	Labor	Equip.	Total	Total Incl O&P	
200	0010	**LAMINATED BEAMS** Fb 2400 psi										200
	0050	3" x 15"	F-3	480	.083	L.F.	8.25	1.49	.95	10.69	12.70	
	0100	3" x 18"		450	.089		9.90	1.59	1.01	12.50	14.70	
	0150	5" x 15"		360	.111		12.40	1.99	1.26	15.65	18.45	
	0200	5" x 18"		290	.138		15.45	2.47	1.56	19.48	23	
	0250	5" x 22-1/2"		220	.182		19.85	3.26	2.06	25.17	30	
	0300	6-3/4" x 18"		320	.125		15.75	2.24	1.42	19.41	22.50	
	0350	6-3/4" x 25-1/2"		260	.154		20.50	2.76	1.74	25	29	
	0400	6-3/4" x 33"		210	.190		26.50	3.42	2.16	32.08	37.50	
	0450	8-3/4" x 34-1/2"		160	.250		37	4.48	2.84	44.32	51	
	0500	For premium appearance, add to S.F. prices					5%					
	0550	For industrial type, deduct					15%					
	0600	For stain and varnish, add					5%					
	0650	For 3/4" laminations, add					25%					
550	0010	**LAMINATED ROOF DECK** Pine or hemlock, 3" thick	2 Carp	425	.038	S.F.	2.84	.74		3.58	4.39	550
	0100	4" thick		325	.049		3.78	.97		4.75	5.80	

06100 | Rough Carpentry

06170 | Prefabricated Structural Wood

			CREW	DAILY OUTPUT	LABOR-HOURS	UNIT	MAT.	LABOR	EQUIP.	TOTAL	TOTAL INCL O&P	
550	0300	Cedar, 3" thick	2 Carp	425	.038	S.F.	3.47	.74		4.21	5.10	550
	0400	4" thick		325	.049		4.42	.97		5.39	6.50	
	0600	Fir, 3" thick		425	.038		2.89	.74		3.63	4.45	
	0700	4" thick		325	.049		3.63	.97		4.60	5.65	
600	0010	**STRUCTURAL JOISTS** Fabricated "I" joists with wood flanges,										600
	0100	Plywood webs, incl. bridging & blocking, panels 24" O.C.										
	1200	15' to 24' span, 50 psf live load	F-5	2,400	.013	SF Flr.	1.44	.23		1.67	1.97	
	1300	55 psf live load		2,250	.014		1.55	.25		1.80	2.12	
	1400	24' to 30' span, 45 psf live load		2,600	.012		1.80	.21		2.01	2.34	
	1500	55 psf live load		2,400	.013		1.80	.23		2.03	2.37	
	1600	Tubular steel open webs, 45 psf, 24" O.C., 40' span	F-3	6,250	.006		1.80	.11	.07	1.98	2.26	
	1700	55' span		7,750	.005		1.75	.09	.06	1.90	2.15	
	1800	70' span		9,250	.004		2.27	.08	.05	2.40	2.68	
	1900	85 psf live load, 26' span		2,300	.017		2.11	.31	.20	2.62	3.07	
980	0010	**ROOF TRUSSES** R06170-100										980
	0020	For timber connectors, see div. 06090-800										
	5000	Common wood, 2" x 4" metal plate connected, 24" O.C., 4/12 slope										
	5010	1' overhang, 12' span	F-5	55	.582	Ea.	27.50	10.05		37.55	47.50	
	5050	20' span	F-6	62	.645		36	11.50	7.30	54.80	67	
	5100	24' span		60	.667		42.50	11.85	7.55	61.90	75.50	
	5150	26' span		57	.702		65	12.50	7.95	85.45	101	
	5200	28' span		53	.755		50.50	13.45	8.55	72.50	88	
	5240	30' span		51	.784		76	13.95	8.90	98.85	117	
	5250	32' span		50	.800		71	14.25	9.05	94.30	113	
	5280	34' span		48	.833		88	14.80	9.45	112.25	132	
	5350	8/12 pitch, 1' overhang, 20' span		57	.702		60	12.50	7.95	80.45	96	
	5400	24' span		55	.727		71	12.95	8.25	92.20	109	
	5450	26' span		52	.769		77	13.70	8.70	99.40	118	
	5500	28' span		49	.816		83	14.50	9.25	106.75	126	
	5550	32' span		45	.889		93	15.80	10.10	118.90	140	
	5600	36' span		41	.976		105	17.35	11.05	133.40	158	
	5650	38' span		40	1		114	17.80	11.35	143.15	168	
	5700	40' span		40	1		130	17.80	11.35	159.15	186	

06180 | Glued-Laminated Construction

			CREW	DAILY OUTPUT	LABOR-HOURS	UNIT	MAT.	LABOR	EQUIP.	TOTAL	TOTAL INCL O&P	
400	0010	**LAMINATED FRAMING** Not including decking										400
	0020	30 lb., short term live load, 15 lb. dead load										
	0200	Straight roof beams, 20' clear span, beams 8' O.C.	F-3	2,560	.016	SF Flr.	1.45	.28	.18	1.91	2.27	
	0300	Beams 16' O.C.		3,200	.013		1.06	.22	.14	1.42	1.71	
	0500	40' clear span, beams 8' O.C.		3,200	.013		2.79	.22	.14	3.15	3.61	
	0600	Beams 16' O.C.		3,840	.010		2.28	.19	.12	2.59	2.96	
	0800	60' clear span, beams 8' O.C.	F-4	2,880	.014		4.79	.25	.28	5.32	6	
	0900	Beams 16' O.C.	"	3,840	.010		3.57	.19	.21	3.97	4.48	
	1100	Tudor arches, 30' to 40' clear span, frames 8' O.C.	F-3	1,680	.024		6.25	.43	.27	6.95	7.90	
	1200	Frames 16' O.C.	"	2,240	.018		4.89	.32	.20	5.41	6.15	
	1400	50' to 60' clear span, frames 8' O.C.	F-4	2,200	.018		6.75	.33	.37	7.45	8.35	
	1500	Frames 16' O.C.		2,640	.015		5.75	.27	.31	6.33	7.10	
	1700	Radial arches, 60' clear span, frames 8' O.C.		1,920	.021		6.30	.37	.42	7.09	8.05	
	1800	Frames 16' O.C.		2,880	.014		4.84	.25	.28	5.37	6.05	
	2000	100' clear span, frames 8' O.C.		1,600	.025		6.50	.45	.51	7.46	8.45	
	2100	Frames 16' O.C.		2,400	.017		5.75	.30	.34	6.39	7.20	
	2300	120' clear span, frames 8' O.C.		1,440	.028		8.70	.50	.56	9.76	11	
	2400	Frames 16' O.C.		1,920	.021		7.90	.37	.42	8.69	9.80	
	2600	Bowstring trusses, 20' O.C., 40' clear span	F-3	2,400	.017		3.91	.30	.19	4.40	5	
	2700	60' clear span	F-4	3,600	.011		3.51	.20	.22	3.93	4.45	

For expanded coverage of these items see *Means Building Construction Cost Data 2000*

06100 | Rough Carpentry

06180 | Glued-Laminated Construction

			CREW	DAILY OUTPUT	LABOR-HOURS	UNIT	2000 BARE COSTS				TOTAL INCL O&P	
							MAT.	LABOR	EQUIP.	TOTAL		
400	2800	100' clear span	F-4	4,000	.010	SF Flr.	4.97	.18	.20	5.35	5.95	400
	2900	120' clear span	↓	3,600	.011		5.35	.20	.22	5.77	6.45	
	3100	For premium appearance, add to S.F. prices					5%					
	3300	For industrial type, deduct					15%					
	3500	For stain and varnish, add					5%					
	3900	For 3/4" laminations, add to straight					25%					
	4100	Add to curved				↓	15%					
	4300	Alternate pricing method: (use nominal footage of										
	4310	components). Straight beams, camber less than 6"	F-3	3.50	11.429	M.B.F.	2,175	205	130	2,510	2,875	
	4400	Columns, including hardware		2	20		2,325	360	227	2,912	3,400	
	4600	Curved members, radius over 32'		2.50	16		2,375	287	181	2,843	3,325	
	4700	Radius 10' to 32'	↓	3	13.333		2,350	239	151	2,740	3,175	
	4900	For complicated shapes, add maximum					100%					
	5100	For pressure treating, add to straight					35%					
	5200	Add to curved				↓	45%					
	6000	Laminated veneer members, southern pine or western species										
	6050	1-3/4" wide x 5-1/2" deep	2 Carp	480	.033	L.F.	3.69	.66		4.35	5.20	
	6100	9-1/2" deep		480	.033		3.86	.66		4.52	5.40	
	6150	14" deep		450	.036		5.50	.70		6.20	7.25	
	6200	18" deep	↓	450	.036	↓	7.15	.70		7.85	9.10	
	6300	Parallel strand members, southern pine or western species										
	6350	1-3/4" wide x 9-1/4" deep	2 Carp	480	.033	L.F.	4.42	.66		5.08	6	
	6400	11-1/4" deep		450	.036		5.45	.70		6.15	7.15	
	6450	14" deep		400	.040		6.45	.79		7.24	8.45	
	6500	3-1/2" wide x 9-1/4" deep		480	.033		8.60	.66		9.26	10.60	
	6550	11-1/4" deep		450	.036		10.60	.70		11.30	12.85	
	6600	14" deep		400	.040		12.65	.79		13.44	15.25	
	6650	7" wide x 9-1/4" deep		450	.036		16.95	.70		17.65	19.85	
	6700	11-1/4" deep		420	.038		21	.75		21.75	24.50	
	6750	14" deep	↓	400	.040	↓	25	.79		25.79	29	

06200 | Finish Carpentry

06220 | Millwork

			CREW	DAILY OUTPUT	LABOR-HOURS	UNIT	2000 BARE COSTS				TOTAL INCL O&P	
							MAT.	LABOR	EQUIP.	TOTAL		
100	0010	**MILLWORK** Rule of thumb: Milled material cost	R06110-030									100
	0020	equals three times cost of lumber										
	1000	Typical finish hardwood milled material										
	1020	1" x 12", custom birch				L.F.	6.10			6.10	6.70	
	1040	Cedar					3.60			3.60	3.96	
	1060	Oak					5.70			5.70	6.30	
	1080	Redwood					5.90			5.90	6.50	
	1100	Southern yellow pine					1.39			1.39	1.53	
	1120	Sugar pine					2.51			2.51	2.76	
	1140	Teak					13.15			13.15	14.45	
	1160	Walnut					9.35			9.35	10.30	
	1180	White pine	↓			↓	3.09			3.09	3.40	
200	0010	**MOLDINGS, BASE**										200
	0500	Base, stock pine, 9/16" x 3-1/2"	1 Carp	240	.033	L.F.	1	.66		1.66	2.23	
	0501	Oak or birch, 9/16" x 3-1/2"		240	.033		2.30	.66		2.96	3.66	
	0550	9/16" x 4-1/2"	↓	200	.040	↓	1.28	.79		2.07	2.76	

06200 | Finish Carpentry

06220 | Millwork

			CREW	DAILY OUTPUT	LABOR-HOURS	UNIT	MAT.	LABOR	EQUIP.	TOTAL	TOTAL INCL O&P	
200	0561	Base shoe, oak, 3/4" x 1"	1 Carp	240	.033	L.F.	.76	.66		1.42	1.97	200
	0570	Base, prefinished, 2-1/2" x 9/16"		242	.033		1	.65		1.65	2.22	
	0580	Shoe, prefinished, 3/8" x 5/8"		266	.030		.42	.59		1.01	1.48	
	0585	Flooring cant strip, 3/4" x 1/2"		500	.016		.22	.32		.54	.78	
400	0010	**MOLDINGS, CASINGS**										400
	0090	Apron, stock pine, 5/8" x 2"	1 Carp	250	.032	L.F.	.96	.63		1.59	2.14	
	0110	5/8" x 3-1/2"		220	.036		1.40	.72		2.12	2.77	
	0300	Band, stock pine, 11/16" x 1-1/8"		270	.030		.38	.58		.96	1.42	
	0350	11/16" x 1-3/4"		250	.032		.60	.63		1.23	1.74	
	0700	Casing, stock pine, 11/16" x 2-1/2"		240	.033		.77	.66		1.43	1.98	
	0701	Oak or birch		240	.033		1.99	.66		2.65	3.32	
	0750	11/16" x 3-1/2"		215	.037		1.48	.73		2.21	2.89	
	0760	Door & window casing, exterior, 1-1/4" x 2"		200	.040		1.40	.79		2.19	2.89	
	0770	Finger jointed, 1-1/4" x 2"		200	.040		1.26	.79		2.05	2.74	
	4600	Mullion casing, stock pine, 5/16" x 2"		200	.040		.62	.79		1.41	2.03	
	4601	Oak or birch		200	.040		.72	.79		1.51	2.14	
	4700	Teak, custom, nominal 1" x 1"		215	.037		1.11	.73		1.84	2.48	
	4800	Nominal 1" x 3"		200	.040		3.26	.79		4.05	4.94	
450	0010	**MOLDINGS, CEILINGS**										450
	0600	Bed, stock pine, 9/16" x 1-3/4"	1 Carp	270	.030	L.F.	.60	.58		1.18	1.66	
	0650	9/16" x 2"		240	.033		.58	.66		1.24	1.77	
	1200	Cornice molding, stock pine, 9/16" x 1-3/4"		330	.024		.62	.48		1.10	1.50	
	1300	9/16" x 2-1/4"		300	.027		.85	.53		1.38	1.84	
	2400	Cove scotia, stock pine, 9/16" x 1-3/4"		270	.030		.54	.58		1.12	1.59	
	2401	Oak or birch, 9/16" x 1-3/4"		270	.030		.81	.58		1.39	1.89	
	2500	11/16" x 2-3/4"		255	.031		1.15	.62		1.77	2.33	
	2600	Crown, stock pine, 9/16" x 3-5/8"		250	.032		1.51	.63		2.14	2.74	
	2700	11/16" x 4-5/8"		220	.036		1.91	.72		2.63	3.33	
500	0010	**MOLDINGS, EXTERIOR**										500
	1500	Cornice, boards, pine, 1" x 2"	1 Carp	330	.024	L.F.	.24	.48		.72	1.08	
	1600	1" x 4"		250	.032		.60	.63		1.23	1.74	
	1700	1" x 6"		250	.032		.86	.63		1.49	2.03	
	1800	1" x 8"		200	.040		.85	.79		1.64	2.29	
	1900	1" x 10"		180	.044		1.12	.88		2	2.73	
	2000	1" x 12"		180	.044		1.14	.88		2.02	2.75	
	2200	Three piece, built-up, pine, minimum		80	.100		1.47	1.97		3.44	5	
	2300	Maximum		65	.123		5.05	2.42		7.47	9.70	
	3000	Corner board, sterling pine, 1" x 4"		200	.040		.66	.79		1.45	2.08	
	3100	1" x 6"		200	.040		.96	.79		1.75	2.41	
	3200	2" x 6"		165	.048		2	.96		2.96	3.84	
	3300	2" x 8"		165	.048		2.66	.96		3.62	4.57	
	3350	Fascia, sterling pine, 1" x 6"		250	.032		.96	.63		1.59	2.14	
	3370	1" x 8"		225	.036		1.50	.70		2.20	2.85	
	3372	2" x 6"		225	.036		2	.70		2.70	3.40	
	3374	2" x 8"		200	.040		2.66	.79		3.45	4.28	
	3376	2" x 10"		180	.044		3.33	.88		4.21	5.15	
	3395	Grounds, 1" x 1" redwood		300	.027		.18	.53		.71	1.10	
	3400	Trim, exterior, sterling pine, back band		250	.032		.57	.63		1.20	1.71	
	3500	Casing		250	.032		.60	.63		1.23	1.74	
	3600	Crown		250	.032		1.20	.63		1.83	2.40	
	3700	Porch rail with balusters		22	.364		9	7.15		16.15	22	
	3800	Screen		395	.020		.32	.40		.72	1.03	
	4100	Verge board, sterling pine, 1" x 4"		200	.040		.47	.79		1.26	1.87	
	4200	1" x 6"		200	.040		.72	.79		1.51	2.14	
	4300	2" x 6"		165	.048		1.40	.96		2.36	3.18	
	4400	2" x 8"		165	.048		1.86	.96		2.82	3.69	

For expanded coverage of these items see *Means Building Construction Cost Data 2000*

06200 | Finish Carpentry

06220 | Millwork

			CREW	DAILY OUTPUT	LABOR-HOURS	UNIT	2000 BARE COSTS				TOTAL INCL O&P	
							MAT.	LABOR	EQUIP.	TOTAL		
500	4700	For redwood trim, add					200%					500
	5000	Casing/fascia, rough-sawn cedar										
	5100	1" x 2"	1 Carp	275	.029	L.F.	.23	.57		.80	1.23	
	5200	1" x 6"		250	.032		.70	.63		1.33	1.85	
	5300	1" x 8"		230	.035		.90	.69		1.59	2.16	
	5400	2" x 4"		220	.036		.90	.72		1.62	2.22	
	5500	2" x 6"		220	.036		1.37	.72		2.09	2.74	
	5600	2" x 8"		200	.040		1.82	.79		2.61	3.35	
	5700	2" x 10"		180	.044		2.27	.88		3.15	4	
	5800	2" x 12"		170	.047		2.72	.93		3.65	4.58	
700	0010	**MOLDINGS, TRIM**										700
	0200	Astragal, stock pine, 11/16" x 1-3/4"	1 Carp	255	.031	L.F.	.78	.62		1.40	1.92	
	0250	1-5/16" x 2-3/16"		240	.033		2.73	.66		3.39	4.13	
	0800	Chair rail, stock pine, 5/8" x 2-1/2"		270	.030		.84	.58		1.42	1.92	
	0900	5/8" x 3-1/2"		240	.033		1.26	.66		1.92	2.52	
	1000	Closet pole, stock pine, 1-1/8" diameter		200	.040		.80	.79		1.59	2.23	
	1100	Fir, 1-5/8" diameter		200	.040		1.19	.79		1.98	2.66	
	1150	Corner, inside, 5/16" x 1"		225	.036		.38	.70		1.08	1.62	
	1160	Outside, 1-1/16" x 1-1/16"		240	.033		.96	.66		1.62	2.19	
	1161	1-5/16" x 1-5/16"		240	.033		1.07	.66		1.73	2.31	
	3300	Half round, stock pine, 1/4" x 1/2"		270	.030		.19	.58		.77	1.21	
	3350	1/2" x 1"		255	.031		.34	.62		.96	1.43	
	3400	Handrail, fir, single piece, stock, hardware not included										
	3450	1-1/2" x 1-3/4"	1 Carp	80	.100	L.F.	1.21	1.97		3.18	4.71	
	3470	Pine, 1-1/2" x 1-3/4"		80	.100		1.06	1.97		3.03	4.55	
	3500	1-1/2" x 2-1/2"		76	.105		1.43	2.07		3.50	5.10	
	3600	Lattice, stock pine, 1/4" x 1-1/8"		270	.030		.32	.58		.90	1.35	
	3700	1/4" x 1-3/4"		250	.032		.35	.63		.98	1.47	
	3800	Miscellaneous, custom, pine, 1" x 1"		270	.030		.98	.58		1.56	2.08	
	3900	1" x 3"		240	.033		.78	.66		1.44	1.99	
	4100	Birch or oak, nominal 1" x 1"		240	.033		.46	.66		1.12	1.64	
	4200	Nominal 1" x 3"		215	.037		1.59	.73		2.32	3.01	
	4400	Walnut, nominal 1" x 1"		215	.037		.76	.73		1.49	2.10	
	4500	Nominal 1" x 3"		200	.040		2.29	.79		3.08	3.87	
	4700	Teak, nominal 1" x 1"		215	.037		1.05	.73		1.78	2.41	
	4800	Nominal 1" x 3"		200	.040		3	.79		3.79	4.65	
	4900	Quarter round, stock pine, 1/4" x 1/4"		275	.029		.20	.57		.77	1.20	
	4950	3/4" x 3/4"		255	.031		.41	.62		1.03	1.51	
	5600	Wainscot moldings, 1-1/8" x 9/16", 2' high, minimum		76	.105	S.F.	5.90	2.07		7.97	10.05	
	5700	Maximum		65	.123	"	13.45	2.42		15.87	18.95	
800	0010	**MOLDINGS, WINDOW AND DOOR**										800
	2800	Door moldings, stock, decorative, 1-1/8" wide, plain	1 Carp	17	.471	Set	28.50	9.25		37.75	47	
	2900	Detailed		17	.471	"	68	9.25		77.25	91	
	3150	Door trim set, 1 head and 2 sides, pine, 2-1/2 wide		5.90	1.356	Opng.	12.70	26.50		39.20	60	
	3170	4-1/2" wide		5.30	1.509	"	24	29.50		53.50	77	
	3200	Glass beads, stock pine, 1/4" x 11/16"		285	.028	L.F.	.30	.55		.85	1.28	
	3250	3/8" x 1/2"		275	.029		.36	.57		.93	1.38	
	3270	3/8" x 7/8"		270	.030		.40	.58		.98	1.44	
	4850	Parting bead, stock pine, 3/8" x 3/4"		275	.029		.29	.57		.86	1.30	
	4870	1/2" x 3/4"		255	.031		.36	.62		.98	1.46	
	5000	Stool caps, stock pine, 11/16" x 3-1/2"		200	.040		1.31	.79		2.10	2.79	
	5100	1-1/16" x 3-1/4"		150	.053		1.97	1.05		3.02	3.97	
	5300	Threshold, oak, 3' long, inside, 5/8" x 3-5/8"		32	.250	Ea.	6.20	4.93		11.13	15.30	
	5400	Outside, 1-1/2" x 7-5/8"		16	.500	"	25	9.85		34.85	44.50	
	5900	Window trim sets, including casings, header, stops,										
	5910	stool and apron, 2-1/2" wide, minimum	1 Carp	13	.615	Opng.	15.30	12.10		27.40	38	

06200 | Finish Carpentry

06220 | Millwork

			DAILY	LABOR-		2000 BARE COSTS				TOTAL	
		CREW	OUTPUT	HOURS	UNIT	MAT.	LABOR	EQUIP.	TOTAL	INCL O&P	
800	5950 Average	1 Carp	10	.800	Opng.	26.50	15.75		42.25	56	800
	6000 Maximum	▼	6	1.333	▼	38	26.50		64.50	86.50	
900	0010 **SOFFITS** Wood fiber, no vapor barrier, 15/32" thick	2 Carp	525	.030	S.F.	.75	.60		1.35	1.86	900
	0100 5/8" thick		525	.030		.81	.60		1.41	1.92	
	0300 As above, 5/8" thick, with factory finish		525	.030		.83	.60		1.43	1.94	
	0500 Hardboard, 3/8" thick, slotted		525	.030		1	.60		1.60	2.13	
	1000 Exterior AC plywood, 1/4" thick		420	.038		.70	.75		1.45	2.06	
	1100 1/2" thick	▼	420	.038	▼	.97	.75		1.72	2.36	

06250 | Prefinished Paneling

			DAILY	LABOR-		2000 BARE COSTS				TOTAL	
		CREW	OUTPUT	HOURS	UNIT	MAT.	LABOR	EQUIP.	TOTAL	INCL O&P	
200	0010 **PANELING, HARDBOARD**										200
	0050 Not incl. furring or trim, hardboard, tempered, 1/8" thick	2 Carp	500	.032	S.F.	.30	.63		.93	1.41	
	0100 1/4" thick		500	.032		.38	.63		1.01	1.50	
	0300 Tempered pegboard, 1/8" thick		500	.032		.38	.63		1.01	1.50	
	0400 1/4" thick		500	.032		.41	.63		1.04	1.53	
	0600 Untempered hardboard, natural finish, 1/8" thick		500	.032		.32	.63		.95	1.43	
	0700 1/4" thick		500	.032		.31	.63		.94	1.42	
	0900 Untempered pegboard, 1/8" thick		500	.032		.32	.63		.95	1.43	
	1000 1/4" thick		500	.032		.36	.63		.99	1.48	
	1200 Plastic faced hardboard, 1/8" thick		500	.032		.53	.63		1.16	1.66	
	1300 1/4" thick		500	.032		.71	.63		1.34	1.86	
	1500 Plastic faced pegboard, 1/8" thick		500	.032		.50	.63		1.13	1.63	
	1600 1/4" thick		500	.032		.62	.63		1.25	1.76	
	1800 Wood grained, plain or grooved, 1/4" thick, minimum		500	.032		.47	.63		1.10	1.60	
	1900 Maximum		425	.038	▼	.89	.74		1.63	2.25	
	2100 Moldings for hardboard, wood or aluminum, minimum		500	.032	L.F.	.32	.63		.95	1.43	
	2200 Maximum	▼	425	.038	"	.90	.74		1.64	2.26	
500	0010 **PANELING, PLYWOOD**										500
	2400 Plywood, prefinished, 1/4" thick, 4' x 8' sheets										
	2410 with vertical grooves. Birch faced, minimum	2 Carp	500	.032	S.F.	.72	.63		1.35	1.87	
	2420 Average		420	.038		1.10	.75		1.85	2.50	
	2430 Maximum		350	.046		1.62	.90		2.52	3.32	
	2600 Mahogany, African		400	.040		2.08	.79		2.87	3.64	
	2700 Philippine (Lauan)		500	.032		.88	.63		1.51	2.05	
	2900 Oak or Cherry, minimum		500	.032		1.74	.63		2.37	2.99	
	3000 Maximum		400	.040		2.68	.79		3.47	4.30	
	3200 Rosewood		320	.050		3.81	.99		4.80	5.90	
	3400 Teak		400	.040		2.68	.79		3.47	4.30	
	3600 Chestnut		375	.043		3.97	.84		4.81	5.80	
	3800 Pecan		400	.040		1.70	.79		2.49	3.22	
	3900 Walnut, minimum		500	.032		2.27	.63		2.90	3.58	
	3950 Maximum		400	.040		4.33	.79		5.12	6.10	
	4000 Plywood, prefinished, 3/4" thick, stock grades, minimum		320	.050		1.03	.99		2.02	2.82	
	4100 Maximum		224	.071		4.48	1.41		5.89	7.35	
	4300 Architectural grade, minimum		224	.071		3.30	1.41		4.71	6.05	
	4400 Maximum		160	.100		5.05	1.97		7.02	8.95	
	4600 Plywood, "A" face, birch, V.C., 1/2" thick, natural		450	.036		1.55	.70		2.25	2.90	
	4700 Select		450	.036		1.70	.70		2.40	3.07	
	4900 Veneer core, 3/4" thick, natural		320	.050		1.65	.99		2.64	3.51	
	5000 Select		320	.050		1.85	.99		2.84	3.73	
	5200 Lumber core, 3/4" thick, natural		320	.050		2.47	.99		3.46	4.41	
	5500 Plywood, knotty pine, 1/4" thick, A2 grade		450	.036		1.34	.70		2.04	2.67	
	5600 A3 grade		450	.036		1.70	.70		2.40	3.07	
	5800 3/4" thick, veneer core, A2 grade		320	.050		1.75	.99		2.74	3.62	
	5900 A3 grade	▼	320	.050	▼	1.96	.99		2.95	3.85	

WOOD & PLASTICS   6

For expanded coverage of these items see *Means Building Construction Cost Data 2000*

06200 | Finish Carpentry

06250 | Prefinished Paneling

		CREW	DAILY OUTPUT	LABOR-HOURS	UNIT	2000 BARE COSTS MAT.	LABOR	EQUIP.	TOTAL	TOTAL INCL O&P	
500	6100 Aromatic cedar, 1/4" thick, plywood	2 Carp	400	.040	S.F.	1.72	.79		2.51	3.24	500
	6200 1/4" thick, particle board	↓	400	.040	↓	.83	.79		1.62	2.26	

06260 | Board Paneling

		CREW	DAILY OUTPUT	LABOR-HOURS	UNIT	MAT.	LABOR	EQUIP.	TOTAL	INCL O&P	
400	0010 **PANELING, BOARDS**										400
	6400 Wood board paneling, 3/4" thick, knotty pine	2 Carp	300	.053	S.F.	1.25	1.05		2.30	3.18	
	6500 Rough sawn cedar		300	.053		1.60	1.05		2.65	3.56	
	6700 Redwood, clear, 1" x 4" boards		300	.053		3.75	1.05		4.80	5.95	
	6900 Aromatic cedar, closet lining, boards	↓	275	.058	↓	2.89	1.15		4.04	5.15	

06270 | Closet/Utility Wood Shelving

		CREW	DAILY OUTPUT	LABOR-HOURS	UNIT	MAT.	LABOR	EQUIP.	TOTAL	INCL O&P	
200	0010 **SHELVING** Pine, clear grade, no edge band, 1" x 8"	1 Carp	115	.070	L.F.	1.71	1.37		3.08	4.23	200
	0100 1" x 10"		110	.073		2.31	1.43		3.74	4.99	
	0200 1" x 12"	↓	105	.076		4.18	1.50		5.68	7.15	
	0400 For lumber edge band, by hand, add					1.46			1.46	1.61	
	0420 By machine, add					.94			.94	1.03	
	0600 Plywood, 3/4" thick with lumber edge, 12" wide	1 Carp	75	.107		1.24	2.10		3.34	4.96	
	0700 24" wide		70	.114	↓	2.35	2.25		4.60	6.45	
	0900 Bookcase, clear grade pine, shelves 12" O.C., 8" deep		70	.114	S.F.	3.35	2.25		5.60	7.55	
	1000 12" deep shelves		65	.123	"	4.19	2.42		6.61	8.75	
	1200 Adjustable closet rod and shelf, 12" wide, 3' long		20	.400	Ea.	40.50	7.90		48.40	58	
	1300 8' long		15	.533	"	57	10.50		67.50	80.50	
	1500 Prefinished shelves with supports, stock, 8" wide		75	.107	L.F.	3.65	2.10		5.75	7.60	
	1600 10" wide	↓	70	.114	"	4.07	2.25		6.32	8.35	
	1800 Custom, high quality dadoed pine shelving units, minimum				S.F.					28	
	1900 Maximum				"					40	

06400 | Architectural Woodwork

06410 | Custom Cabinets

		CREW	DAILY OUTPUT	LABOR-HOURS	UNIT	MAT.	LABOR	EQUIP.	TOTAL	INCL O&P	
100	0010 **CABINETS** Corner china cabinets, stock pine,										100
	0020 80" high, unfinished, minimum	2 Carp	6.60	2.424	Ea.	430	48		478	550	
	0100 Maximum	"	4.40	3.636	"	950	71.50		1,021.50	1,175	
	0700 Kitchen base cabinets, hardwood, not incl. counter tops,										
	0710 24" deep, 35" high, prefinished										
	0800 One top drawer, one door below, 12" wide	2 Carp	24.80	.645	Ea.	125	12.70		137.70	160	
	0820 15" wide		24	.667		177	13.15		190.15	217	
	0840 18" wide		23.30	.687		188	13.55		201.55	230	
	0860 21" wide		22.70	.705		198	13.90		211.90	242	
	0880 24" wide		22.30	.717		223	14.15		237.15	269	
	1000 Four drawers, 12" wide		24.80	.645		305	12.70		317.70	355	
	1020 15" wide		24	.667		310	13.15		323.15	365	
	1040 18" wide		23.30	.687		330	13.55		343.55	385	
	1060 24" wide		22.30	.717		355	14.15		369.15	415	
	1200 Two top drawers, two doors below, 27" wide		22	.727		266	14.35		280.35	315	
	1220 30" wide		21.40	.748		283	14.75		297.75	335	
	1240 33" wide		20.90	.766		296	15.10		311.10	350	
	1260 36" wide	↓	20.30	.788	↓	305	15.55		320.55	360	

06400 | Architectural Woodwork

06410 | Custom Cabinets

			CREW	DAILY OUTPUT	LABOR-HOURS	UNIT	MAT.	LABOR	EQUIP.	TOTAL	TOTAL INCL O&P	
100	1280	42" wide	2 Carp	19.80	.808	Ea.	325	15.90		340.90	385	100
	1300	48" wide		18.90	.847		340	16.70		356.70	405	
	1500	Range or sink base, two doors below, 30" wide		21.40	.748		222	14.75		236.75	270	
	1520	33" wide		20.90	.766		239	15.10		254.10	289	
	1540	36" wide		20.30	.788		250	15.55		265.55	300	
	1560	42" wide		19.80	.808		267	15.90		282.90	320	
	1580	48" wide		18.90	.847		281	16.70		297.70	340	
	1800	For sink front units, deduct					53			53	58.50	
	2000	Corner base cabinets, 36" wide, standard	2 Carp	18	.889		195	17.50		212.50	245	
	2100	Lazy Susan with revolving door	"	16.50	.970		320	19.10		339.10	385	
	4000	Kitchen wall cabinets, hardwood, 12" deep with two doors										
	4050	12" high, 30" wide	2 Carp	24.80	.645	Ea.	122	12.70		134.70	157	
	4100	36" wide		24	.667		144	13.15		157.15	182	
	4400	15" high, 30" wide		24	.667		144	13.15		157.15	181	
	4420	33" wide		23.30	.687		154	13.55		167.55	192	
	4440	36" wide		22.70	.705		157	13.90		170.90	197	
	4450	42" wide		22.70	.705		174	13.90		187.90	215	
	4700	24" high, 30" wide		23.30	.687		166	13.55		179.55	206	
	4720	36" wide		22.70	.705		183	13.90		196.90	225	
	4740	42" wide		22.30	.717		168	14.15		182.15	209	
	5000	30" high, one door, 12" wide		22	.727		107	14.35		121.35	143	
	5020	15" wide		21.40	.748		126	14.75		140.75	163	
	5040	18" wide		20.90	.766		139	15.10		154.10	179	
	5060	24" wide		20.30	.788		152	15.55		167.55	194	
	5300	Two doors, 27" wide		19.80	.808		209	15.90		224.90	258	
	5320	30" wide		19.30	.829		185	16.35		201.35	231	
	5340	36" wide		18.80	.851		211	16.75		227.75	262	
	5360	42" wide		18.50	.865		231	17.05		248.05	283	
	5380	48" wide		18.40	.870		260	17.15		277.15	315	
	6000	Corner wall, 30" high, 24" wide		18	.889		132	17.50		149.50	175	
	6050	30" wide		17.20	.930		165	18.35		183.35	214	
	6100	36" wide		16.50	.970		176	19.10		195.10	227	
	6500	Revolving Lazy Susan		15.20	1.053		249	20.50		269.50	310	
	7000	Broom cabinet, 84" high, 24" deep, 18" wide		10	1.600		350	31.50		381.50	440	
	7500	Oven cabinets, 84" high, 24" deep, 27" wide		8	2		530	39.50		569.50	655	
	7750	Valance board trim		396	.040	L.F.	7.60	.80		8.40	9.70	
	9000	For deluxe models of all cabinets, add					40%					
	9500	For custom built in place, add					25%	10%				
	9550	Rule of thumb, kitchen cabinets not including										
	9560	appliances & counter top, minimum	2 Carp	30	.533	L.F.	80.50	10.50		91	107	
	9600	Maximum	"	25	.640	"	225	12.60		237.60	270	
210	0010	**CASEWORK, FRAMES**										210
	0050	Base cabinets, counter storage, 36" high, one bay										
	0100	18" wide	1 Carp	2.70	2.963	Ea.	91	58.50		149.50	200	
	0400	Two bay, 36" wide		2.20	3.636		139	71.50		210.50	276	
	1100	Three bay, 54" wide		1.50	5.333		165	105		270	360	
	2800	Book cases, one bay, 7' high, 18" wide		2.40	3.333		107	65.50		172.50	231	
	3500	Two bay, 36" wide		1.60	5		155	98.50		253.50	340	
	4100	Three bay, 54" wide		1.20	6.667		257	131		388	510	
	6100	Wall mounted cabinet, one bay, 24" high, 18" wide		3.60	2.222		59	44		103	140	
	6800	Two bay, 36" wide		2.20	3.636		86	71.50		157.50	218	
	7400	Three bay, 54" wide		1.70	4.706		107	92.50		199.50	277	
	8400	30" high, one bay, 18" wide		3.60	2.222		64	44		108	146	
	9000	Two bay, 36" wide		2.15	3.721		85	73.50		158.50	220	
	9400	Three bay, 54" wide		1.60	5		106	98.50		204.50	286	

For expanded coverage of these items see Means Building Construction Cost Data 2000

06400 | Architectural Woodwork

06410 | Custom Cabinets

		CREW	DAILY OUTPUT	LABOR-HOURS	UNIT	2000 BARE COSTS MAT.	LABOR	EQUIP.	TOTAL	TOTAL INCL O&P	
210	9800	Wardrobe, 7' high, single, 24" wide	1 Carp	2.70	2.963	Ea.	118	58.50		176.50	230	210
	9950	Partition, adjustable shelves & drawers, 48" wide	↓	1.40	5.714	↓	225	113		338	440	
220	0010	**CABINET DOORS**										220
	2000	Glass panel, hardwood frame										
	2200	12" wide, 18" high	1 Carp	34	.235	Ea.	14.85	4.64		19.49	24.50	
	2400	24" high		33	.242		19.40	4.78		24.18	29.50	
	2600	30" high		32	.250		24.50	4.93		29.43	35.50	
	2800	36" high		30	.267		29.50	5.25		34.75	41.50	
	3000	48" high		23	.348		39	6.85		45.85	54.50	
	3200	60" high		17	.471		49	9.25		58.25	70	
	3400	72" high		15	.533		59	10.50		69.50	83	
	3600	15" wide x 18" high		33	.242		15.30	4.78		20.08	25	
	3800	24" high		32	.250		19.40	4.93		24.33	30	
	4000	30" high		30	.267		24.50	5.25		29.75	36	
	4250	36" high		28	.286		29.50	5.65		35.15	42	
	4300	48" high		22	.364		39	7.15		46.15	55	
	4350	60" high		16	.500		49	9.85		58.85	71	
	4400	72" high		14	.571		59	11.25		70.25	84.50	
	4450	18" wide, 18" high		32	.250		15.30	4.93		20.23	25.50	
	4500	24" high		30	.267		19.40	5.25		24.65	30.50	
	4550	30" high		29	.276		24.50	5.45		29.95	36.50	
	4600	36" high		27	.296		29.50	5.85		35.35	42.50	
	4650	48" high		21	.381		39	7.50		46.50	55.50	
	4700	60" high		15	.533		49	10.50		59.50	72	
	4750	72" high	↓	13	.615	↓	59	12.10		71.10	86	
	5000	Hardwood, raised panel										
	5100	12" wide, 18" high	1 Carp	16	.500	Ea.	20.50	9.85		30.35	39.50	
	5150	24" high		15.50	.516		26.50	10.15		36.65	46.50	
	5200	30" high		15	.533		32.50	10.50		43	54	
	5250	36" high		14	.571		40	11.25		51.25	63.50	
	5300	48" high		11	.727		53	14.35		67.35	83	
	5320	60" high		8	1		66	19.70		85.70	107	
	5340	72" high		7	1.143		79.50	22.50		102	126	
	5360	15" wide x 18" high		15.50	.516		26.50	10.15		36.65	46.50	
	5380	24" high		15	.533		33	10.50		43.50	54.50	
	5400	30" high		14.50	.552		43	10.85		53.85	66	
	5420	36" high		13.50	.593		50	11.65		61.65	75	
	5440	48" high		10.50	.762		65	15		80	97	
	5460	60" high		7.50	1.067		83	21		104	128	
	5480	72" high		6.50	1.231		99	24.50		123.50	151	
	5500	18" wide, 18" high		15	.533		29.50	10.50		40	50.50	
	5550	24" high		14.50	.552		39	10.85		49.85	61	
	5600	30" high		14	.571		50	11.25		61.25	74.50	
	5650	36" high		13	.615		59	12.10		71.10	86	
	5700	48" high		10	.800		79.50	15.75		95.25	115	
	5750	60" high		7	1.143		99	22.50		121.50	148	
	5800	72" high	↓	6	1.333	↓	118	26.50		144.50	175	
	6000	Plastic laminate on particle board										
	6100	12" wide, 18" high	1 Carp	25	.320	Ea.	13	6.30		19.30	25	
	6120	24" high		24	.333		17	6.55		23.55	30	
	6140	30" high		23	.348		21	6.85		27.85	35	
	6160	36" high		21	.381		26	7.50		33.50	41.50	
	6200	48" high		16	.500		34	9.85		43.85	54.50	
	6250	60" high		13	.615		43	12.10		55.10	68.50	
	6300	72" high		12	.667		52	13.15		65.15	79.50	
	6320	15" wide x 18" high	↓	24.50	.327	↓	16	6.45		22.45	28.50	

06400 | Architectural Woodwork

06410 | Custom Cabinets

			CREW	DAILY OUTPUT	LABOR-HOURS	UNIT	MAT.	LABOR	EQUIP.	TOTAL	TOTAL INCL O&P	
220	6340	24" high	1 Carp	23.50	.340	Ea.	22	6.70		28.70	35.50	220
	6360	30" high		22.50	.356		27	7		34	41.50	
	6380	36" high		20.50	.390		32	7.70		39.70	48	
	6400	48" high		15.50	.516		43	10.15		53.15	65	
	6450	60" high		12.50	.640		54	12.60		66.60	81	
	6480	72" high		11.50	.696		65	13.70		78.70	95	
	6500	18" wide, 18" high		24	.333		19	6.55		25.55	32.50	
	6550	24" high		23	.348		26	6.85		32.85	40.50	
	6600	30" high		22	.364		32	7.15		39.15	47.50	
	6650	36" high		20	.400		39	7.90		46.90	56.50	
	6700	48" high		15	.533		52	10.50		62.50	75	
	6750	60" high		12	.667		65	13.15		78.15	94	
	6800	72" high		11	.727		77	14.35		91.35	109	
	7000	Plywood, with edge band										
	7010	12" wide, 18" high	1 Carp	27	.296	Ea.	17.50	5.85		23.35	29.50	
	7100	24" high		26	.308		23.50	6.05		29.55	36.50	
	7120	30" high		25	.320		30	6.30		36.30	44	
	7140	36" high		23	.348		35.50	6.85		42.35	51.50	
	7180	48" high		18	.444		47	8.75		55.75	66.50	
	7200	60" high		15	.533		58.50	10.50		69	82	
	7250	72" high		14	.571		70.50	11.25		81.75	97	
	7300	15" wide x 18" high		26.50	.302		22.50	5.95		28.45	34.50	
	7350	24" high		25.50	.314		29	6.20		35.20	42.50	
	7400	30" high		24.50	.327		36	6.45		42.45	50.50	
	7450	36" high		22.50	.356		43.50	7		50.50	60	
	7500	48" high		17.50	.457		58	9		67	79.50	
	7550	60" high		14.50	.552		73	10.85		83.85	99	
	7600	72" high		13.50	.593		88	11.65		99.65	117	
	7650	18" wide, 18" high		26	.308		25.50	6.05		31.55	38.50	
	7700	24" high		25	.320		34.50	6.30		40.80	49	
	7750	30" high		24	.333		44	6.55		50.55	60	
230	0010	**CABINET HARDWARE**										230
	1000	Catches, minimum	1 Carp	235	.034	Ea.	.72	.67		1.39	1.94	
	1020	Average		119.40	.067		2.19	1.32		3.51	4.67	
	1040	Maximum		80	.100		4.19	1.97		6.16	8	
	2000	Door/drawer pulls, handles										
	2200	Handles and pulls, projecting, metal, minimum	1 Carp	160	.050	Ea.	1.54	.99		2.53	3.38	
	2220	Average		95.24	.084		3.16	1.65		4.81	6.30	
	2240	Maximum		68	.118		7.35	2.32		9.67	12.05	
	2300	Wood, minimum		160	.050		1.54	.99		2.53	3.38	
	2320	Average		95.24	.084		2.08	1.65		3.73	5.10	
	2340	Maximum		68	.118		4.20	2.32		6.52	8.60	
	2600	Flush, metal, minimum		160	.050		1.47	.99		2.46	3.31	
	2620	Average		95.24	.084		3.15	1.65		4.80	6.30	
	2640	Maximum		68	.118		10.50	2.32		12.82	15.50	
	3000	Drawer tracks/glides, minimum		48	.167	Pr.	5.75	3.28		9.03	12	
	3020	Average		32	.250		9.75	4.93		14.68	19.20	
	3040	Maximum		24	.333		16.80	6.55		23.35	30	
	4000	Cabinet hinges, minimum		160	.050		1.43	.99		2.42	3.26	
	4020	Average		95.24	.084		2.57	1.65		4.22	5.65	
	4040	Maximum		68	.118		6.30	2.32		8.62	10.90	
240	0010	**DRAWERS**										240
	0100	Solid hardwood front										
	1000	4" high, 12" wide	1 Carp	17	.471	Ea.	19.10	9.25		28.35	37	
	1200	18" wide		16	.500		25	9.85		34.85	44.50	

For expanded coverage of these items see *Means Building Construction Cost Data 2000*

06400 | Architectural Woodwork

06410 | Custom Cabinets

			CREW	DAILY OUTPUT	LABOR-HOURS	UNIT	MAT.	LABOR	EQUIP.	TOTAL	TOTAL INCL O&P	
240	1400	24" wide	1 Carp	15	.533	Ea.	32.50	10.50		43	54	240
	1600	6" high, 12" wide		16	.500		25	9.85		34.85	44.50	
	1800	18" wide		15	.533		32.50	10.50		43	54	
	2000	24" wide		14	.571		40	11.25		51.25	63.50	
	2200	9" high, 12" wide		15	.533		32.50	10.50		43	54	
	2400	18" wide		14	.571		40	11.25		51.25	63.50	
	2600	24" wide		13	.615		49.50	12.10		61.60	75.50	
	2800	Plastic laminate on particle board front										
	3000	4" high, 12" wide	1 Carp	17	.471	Ea.	19.95	9.25		29.20	38	
	3200	18" wide		16	.500		23	9.85		32.85	42.50	
	3600	24" wide		15	.533		27.50	10.50		38	48	
	3800	6" high, 12" wide		16	.500		23	9.85		32.85	42.50	
	4000	18" wide		15	.533		28.50	10.50		39	49	
	4500	24" wide		14	.571		34.50	11.25		45.75	57.50	
	4800	9" high, 12" wide		15	.533		27	10.50		37.50	47.50	
	5000	18" wide		14	.571		27	11.25		38.25	49	
	5200	24" wide		13	.615		43.50	12.10		55.60	68.50	
	5400	Plywood, flush panel front										
	6000	4" high, 12" wide	1 Carp	17	.471	Ea.	20.50	9.25		29.75	38.50	
	6200	18" wide	"	16	.500	"	25	9.85		34.85	44.50	
400	0010	**VANITIES**										400
	8000	Vanity bases, 2 doors, 30" high, 21" deep, 24" wide	2 Carp	20	.800	Ea.	124	15.75		139.75	163	
	8050	30" wide		16	1		130	19.70		149.70	177	
	8100	36" wide		13.33	1.200		146	23.50		169.50	201	
	8150	48" wide		11.43	1.400		184	27.50		211.50	249	
	9000	For deluxe models of all vanities, add to above					40%					
	9500	For custom built in place, add to above					25%	10%				

06430 | Stairs & Railings

			CREW	DAILY OUTPUT	LABOR-HOURS	UNIT	MAT.	LABOR	EQUIP.	TOTAL	TOTAL INCL O&P	
500	0010	**RAILING** Custom design, architectural grade, hardwood, minimum	1 Carp	38	.211	L.F.	12	4.15		16.15	20.50	500
	0100	Maximum		30	.267		46	5.25		51.25	60	
	0300	Stock interior railing with spindles 6" O.C., 4' long		40	.200		29	3.94		32.94	38.50	
	0400	8' long		48	.167		27	3.28		30.28	35	
620	0011	**STAIRS, PREFABRICATED**										620
	0100	Box stairs, prefabricated, 3'-0" wide										
	0110	Oak treads, no handrails, 2' high	2 Carp	5	3.200	Flight	216	63		279	345	
	0200	4' high		4	4		430	79		509	610	
	0300	6' high		3.50	4.571		620	90		710	835	
	0400	8' high		3	5.333		775	105		880	1,025	
	0600	With pine treads for carpet, 2' high		5	3.200		88.50	63		151.50	205	
	0700	4' high		4	4		164	79		243	315	
	0800	6' high		3.50	4.571		240	90		330	420	
	0900	8' high		3	5.333		274	105		379	480	
	1100	For 4' wide stairs, add					25%					
	1500	Prefabricated stair rail with balusters, 5 risers	2 Carp	15	1.067	Ea.	229	21		250	288	
	1700	Basement stairs, prefabricated, soft wood,										
	1710	open risers, 3' wide, 8' high	2 Carp	4	4	Flight	575	79		654	765	
	1900	Open stairs, prefabricated prefinished poplar, metal stringers,										
	1910	treads 3'-6" wide, no railings										
	2000	3' high	2 Carp	5	3.200	Flight	229	63		292	360	
	2100	4' high		4	4		485	79		564	670	
	2200	6' high		3.50	4.571		555	90		645	770	
	2300	8' high		3	5.333		730	105		835	980	
	2500	For prefab. 3 piece wood railings & balusters, add for										
	2600	3' high stairs	2 Carp	15	1.067	Ea.	31.50	21		52.50	70.50	

06400 | Architectural Woodwork

06430 | Stairs & Railings

			CREW	DAILY OUTPUT	LABOR-HOURS	UNIT	MAT.	LABOR	EQUIP.	TOTAL	TOTAL INCL O&P	
620	2700	4' high stairs	2 Carp	14	1.143	Ea.	51	22.50		73.50	95	620
	2800	6' high stairs		13	1.231		63	24.50		87.50	111	
	2900	8' high stairs		12	1.333		96.50	26.50		123	151	
	3100	For 3'-6" x 3'-6" platform, add		4	4		72.50	79		151.50	215	
	3300	Curved stairways, 3'-3" wide, prefabricated, oak, unfinished,										
	3310	incl. curved balustrade system, open one side										
	3400	9' high	2 Carp	.70	22.857	Flight	6,375	450		6,825	7,800	
	3500	10' high		.70	22.857		7,200	450		7,650	8,700	
	3700	Open two sides, 9' high		.50	32		10,000	630		10,630	12,100	
	3800	10' high		.50	32		10,800	630		11,430	13,000	
	4000	Residential, wood, oak treads, prefabricated		1.50	10.667		930	210		1,140	1,375	
	4200	Built in place		.44	36.364		1,325	715		2,040	2,675	
	4400	Spiral, oak, 4'-6" diameter, unfinished, prefabricated,										
	4500	incl. railing, 9' high	2 Carp	1.50	10.667	Flight	4,000	210		4,210	4,750	
630	0010	**STAIR PARTS** Balusters, turned, 30" high, pine, minimum R06430-100	1 Carp	28	.286	Ea.	4.10	5.65		9.75	14.15	630
	0100	Maximum		26	.308		9	6.05		15.05	20.50	
	0300	30" high birch balusters, minimum		28	.286		6.30	5.65		11.95	16.60	
	0400	Maximum		26	.308		10.20	6.05		16.25	21.50	
	0600	42" high, pine balusters, minimum		27	.296		6	5.85		11.85	16.60	
	0700	Maximum		25	.320		13	6.30		19.30	25	
	0900	42" high birch balusters, minimum		27	.296		7.65	5.85		13.50	18.40	
	1000	Maximum		25	.320		27	6.30		33.30	40.50	
	1050	Baluster, stock pine, 1-1/16" x 1-1/16"		240	.033	L.F.	1.97	.66		2.63	3.30	
	1100	1-5/8" x 1-5/8"		220	.036	"	2.18	.72		2.90	3.63	
	1200	Newels, 3-1/4" wide, starting, minimum		7	1.143	Ea.	33.50	22.50		56	75.50	
	1300	Maximum		6	1.333		126	26.50		152.50	184	
	1500	Landing, minimum		5	1.600		73.50	31.50		105	135	
	1600	Maximum		4	2		179	39.50		218.50	264	
	1800	Railings, oak, built-up, minimum		60	.133	L.F.	5.55	2.63		8.18	10.60	
	1900	Maximum		55	.145		15.75	2.87		18.62	22.50	
	2100	Add for sub rail		110	.073		4.20	1.43		5.63	7.05	
	2300	Risers, beech, 3/4" x 7-1/2" high		64	.125		5.75	2.46		8.21	10.55	
	2400	Fir, 3/4" x 7-1/2" high		64	.125		1.58	2.46		4.04	5.95	
	2600	Oak, 3/4" x 7-1/2" high		64	.125		4.99	2.46		7.45	9.70	
	2800	Pine, 3/4" x 7-1/2" high		66	.121		1.57	2.39		3.96	5.80	
	2850	Skirt board, pine, 1" x 10"		55	.145		1.73	2.87		4.60	6.80	
	2900	1" x 12"		52	.154		2.10	3.03		5.13	7.50	
	3000	Treads, 1-1/16" x 9-1/2" wide, 3' long, oak		18	.444	Ea.	23	8.75		31.75	40.50	
	3100	4' long, oak		17	.471		29	9.25		38.25	48	
	3300	1-1/16" x 11-1/2" wide, 3' long, oak		18	.444		23.50	8.75		32.25	40.50	
	3400	6' long, oak		14	.571		56	11.25		67.25	81	
	3600	Beech treads, add					40%					
	3800	For mitered return nosings, add				L.F.	8.40			8.40	9.25	

06440 | Wood Ornaments

			CREW	DAILY OUTPUT	LABOR-HOURS	UNIT	MAT.	LABOR	EQUIP.	TOTAL	TOTAL INCL O&P	
150	0010	**BEAMS, DECORATIVE** Rough sawn cedar, non-load bearing, 4" x 4"	2 Carp	180	.089	L.F.	1.30	1.75		3.05	4.43	150
	0100	4" x 6"		170	.094		2.50	1.85		4.35	5.95	
	0200	4" x 8"		160	.100		3.21	1.97		5.18	6.90	
	0300	4" x 10"		150	.107		4.46	2.10		6.56	8.50	
	0400	4" x 12"		140	.114		5.40	2.25		7.65	9.80	
	0500	8" x 8"		130	.123		7.55	2.42		9.97	12.45	
	0600	Plastic beam, "hewn finish", 6" x 2"		240	.067		2.78	1.31		4.09	5.30	
	0601	6" x 4"		220	.073		3.24	1.43		4.67	6	
	1100	Beam connector plates see div. 06090-800										

For expanded coverage of these items see *Means Building Construction Cost Data 2000*

06400 | Architectural Woodwork

06440 | Wood Ornaments

			CREW	DAILY OUTPUT	LABOR-HOURS	UNIT	2000 BARE COSTS MAT.	LABOR	EQUIP.	TOTAL	TOTAL INCL O&P	
350	0010	**GRILLES** and panels, hardwood, sanded										350
	0020	2' x 4' to 4' x 8', custom designs, unfinished, minimum	1 Carp	38	.211	S.F.	12	4.15		16.15	20.50	
	0050	Average		30	.267		26	5.25		31.25	37.50	
	0100	Maximum		19	.421		40	8.30		48.30	58	
	0300	As above, but prefinished, minimum		38	.211		12	4.15		16.15	20.50	
	0400	Maximum		19	.421		45	8.30		53.30	63.50	
400	0010	**LOUVERS** Redwood, 2'-0" diameter, full circle	1 Carp	16	.500	Ea.	105	9.85		114.85	133	400
	0100	Half circle		16	.500		104	9.85		113.85	131	
	0200	Octagonal		16	.500		86	9.85		95.85	111	
	0300	Triangular, 5/12 pitch, 5'-0" at base		16	.500		180	9.85		189.85	215	
500	0010	**FIREPLACE MANTELS** 6" molding, 6' x 3'-6" opening, minimum	1 Carp	5	1.600	Opng.	130	31.50		161.50	196	500
	0100	Maximum		5	1.600		157	31.50		188.50	227	
	0300	Prefabricated pine, colonial type, stock, deluxe		2	4		760	79		839	970	
	0400	Economy		3	2.667		256	52.50		308.50	370	
550	0010	**FIREPLACE MANTEL BEAMS** Rough texture wood, 4" x 8"	1 Carp	36	.222	L.F.	4.24	4.38		8.62	12.15	550
	0100	4" x 10"		35	.229	"	5.30	4.50		9.80	13.55	
	0300	Laminated hardwood, 2-1/4" x 10-1/2" wide, 6' long		5	1.600	Ea.	95.50	31.50		127	159	
	0400	8' long		5	1.600	"	133	31.50		164.50	200	
	0600	Brackets for above, rough sawn		12	.667	Pr.	8.75	13.15		21.90	32	
	0700	Laminated		12	.667	"	13.25	13.15		26.40	37	
700	0011	**COLUMNS**										700
	0050	Aluminum, round colonial, 6" diameter	2 Carp	80	.200	V.L.F.	16	3.94		19.94	24.50	
	0100	8" diameter		62.25	.257		19	5.05		24.05	29.50	
	0200	10" diameter		55	.291		23.50	5.75		29.25	36	
	0250	Fir, stock units, hollow round, 6" diameter		80	.200		13	3.94		16.94	21	
	0300	8" diameter		80	.200		15	3.94		18.94	23.50	
	0350	10" diameter		70	.229		19	4.50		23.50	28.50	
	0400	Solid turned, to 8' high, 3-1/2" diameter		80	.200		7	3.94		10.94	14.45	
	0500	4-1/2" diameter		75	.213		10	4.20		14.20	18.20	
	0600	5-1/2" diameter		70	.229		14	4.50		18.50	23	
	0800	Square columns, built-up, 5" x 5"		65	.246		13	4.85		17.85	22.50	
	0900	Solid, 3-1/2" x 3-1/2"		130	.123		6	2.42		8.42	10.75	
	1600	Hemlock, tapered, T & G, 12" diam, 10' high		100	.160		30	3.15		33.15	38.50	
	1700	16' high		65	.246		53	4.85		57.85	67	
	1900	10' high, 14" diameter		100	.160		77	3.15		80.15	90	
	2000	18' high		65	.246		73	4.85		77.85	89	
	2200	18" diameter, 12' high		65	.246		103	4.85		107.85	121	
	2300	20' high		50	.320		99	6.30		105.30	120	
	2500	20" diameter, 14' high		40	.400		122	7.90		129.90	148	
	2600	20' high		35	.457		126	9		135	154	
	2800	For flat pilasters, deduct					33%					
	3000	For splitting into halves, add				Ea.	60			60	66	
	4000	Rough sawn cedar posts, 4" x 4"	2 Carp	250	.064	V.L.F.	2.43	1.26		3.69	4.83	
	4100	4" x 6"		235	.068		3.62	1.34		4.96	6.30	
	4200	6" x 6"		220	.073		5.45	1.43		6.88	8.45	
	4300	8" x 8"		200	.080		5.60	1.58		7.18	8.85	

06445 | Simulated Wood Ornaments

			CREW	DAILY OUTPUT	LABOR-HOURS	UNIT	MAT.	LABOR	EQUIP.	TOTAL	TOTAL INCL O&P	
100	0010	**MILLWORK, HIGH DENSITY POLYMER**										100
	0100	Base, 9/16" x 3-3/16"	1 Carp	230	.035	L.F.	1.22	.69		1.91	2.51	
	0200	Casing, fluted, 5/8" x 3-1/4"		215	.037		1.22	.73		1.95	2.60	
	0300	Chair rail, 9/16" x 2-1/4"		260	.031		.62	.61		1.23	1.72	
	0400	5/8" x 3-1/8"		230	.035		1.17	.69		1.86	2.46	
	0500	Corner, inside, 1/2" x 1-1/8"		220	.036		.61	.72		1.33	1.90	
	0600	Cove, 13/16" x 3-3/4"		260	.031		1.22	.61		1.83	2.38	
	0700	Crown, 3/4" x 3-13/16"		260	.031		1.22	.61		1.83	2.38	

06400 | Architectural Woodwork

06445 | Simulated Wood Ornaments

		CREW	DAILY OUTPUT	LABOR-HOURS	UNIT	2000 BARE COSTS				TOTAL INCL O&P	
						MAT.	LABOR	EQUIP.	TOTAL		
100	0800 Half round, 15/16" x 2"	1 Carp	240	.033	L.F.	.67	.66		1.33	1.87	100

06470 | Screen, Blinds & Shutters

		CREW	DAILY OUTPUT	LABOR-HOURS	UNIT	MAT.	LABOR	EQUIP.	TOTAL	TOTAL INCL O&P	
100	0010 SHUTTERS, EXTERIOR Aluminum, louvered, 1'-4" wide, 3'-0" long	1 Carp	10	.800	Pr.	29.50	15.75		45.25	59.50	100
	0200 4'-0" long		10	.800		32.50	15.75		48.25	63	
	0300 5'-4" long		10	.800		37.50	15.75		53.25	68.50	
	0400 6'-8" long		9	.889		48	17.50		65.50	82.50	
	1000 Pine, louvered, primed, each 1'-2" wide, 3'-3" long		10	.800	↓	41	15.75		56.75	72	
	1001 Pine, louvered, primed, each 1'-2" wide, 3'-3" long		20	.400	Ea.	24	7.90		31.90	40	
	1100 4'-7" long		10	.800	Pr.	49	15.75		64.75	81	
	1101 4'-7" long		20	.400	Ea.	34.50	7.90		42.40	51.50	
	1250 Each 1'-4" wide, 3'-0" long		10	.800	Pr.	37.50	15.75		53.25	68.50	
	1251 Each 1'-4" wide, 3'-0" long		20	.400	Ea.	25.50	7.90		33.40	41.50	
	1350 5'-3" long		10	.800	Pr.	53	15.75		68.75	85.50	
	1351 5'-3" long		20	.400	Ea.	31	7.90		38.90	47.50	
	1500 Each 1'-6" wide, 3'-3" long		10	.800	Pr.	45	15.75		60.75	76.50	
	1600 4'-7" long		10	.800	"	61	15.75		76.75	94	
	1601 4'-7" long		20	.400	Ea.	38	7.90		45.90	55.50	
	1620 Hemlock, louvered, 1'-2" wide, 5'-7" long		10	.800	Pr.	60.50	15.75		76.25	93.50	
	1630 Each 1'-4" wide, 2'-2" long		10	.800		38	15.75		53.75	68.50	
	1670 4'-3" long		10	.800		50	15.75		65.75	81.50	
	1680 5'-3" long		10	.800		56.50	15.75		72.25	89	
	1690 5'-11" long		10	.800		61.50	15.75		77.25	95	
	1700 Door blinds, 6'-9" long, each 1'-3" wide		9	.889		71.50	17.50		89	109	
	1710 1'-6" wide		9	.889		90	17.50		107.50	129	
	2500 Polystyrene, solid raised panel, each 1'-4" wide, 3'-3" long		10	.800		52	15.75		67.75	84	
	2600 3'-11" long		10	.800		59.50	15.75		75.25	92.50	
	2700 4'-7" long		10	.800		62.50	15.75		78.25	96	
	2800 5'-3" long		10	.800		70.50	15.75		86.25	105	
	2900 6'-8" long		9	.889		97.50	17.50		115	137	
	3500 Polystyrene, solid raised panel, each 3'-3" wide, 3'-0" long		10	.800		126	15.75		141.75	165	
	3600 3'-11" long		10	.800		156	15.75		171.75	199	
	3700 4'-7" long		10	.800		178	15.75		193.75	223	
	3800 5'-3" long		10	.800		198	15.75		213.75	244	
	3900 6'-8" long		9	.889		222	17.50		239.50	275	
	4500 Polystyrene, louvered, each 1'-2" wide, 3'-3" long		10	.800		40	15.75		55.75	71	
	4600 4'-7" long		10	.800		50	15.75		65.75	81.50	
	4750 5'-3" long		10	.800		53	15.75		68.75	85.50	
	4850 6'-8" long		9	.889		87	17.50		104.50	126	
	6000 Vinyl, louvered, each 1'-2" x 4'-7" long		10	.800		52	15.75		67.75	84	
	6200 Each 1'-4" x 6'-8" long	↓	9	.889	↓	79	17.50		96.50	117	
	8000 PVC exterior rolling shutters										
	8100 including crank control	1 Carp	8	1	Ea.	360	19.70		379.70	430	
	8500 Insulative - 6' x 6'8" stock unit	"	8	1	"	515	19.70		534.70	600	

For expanded coverage of these items see *Means Building Construction Cost Data 2000*

06600 | Plastic Fabrications

06620 | Non-Structural Plastics

		CREW	DAILY OUTPUT	LABOR-HOURS	UNIT	2000 BARE COSTS MAT.	LABOR	EQUIP.	TOTAL	TOTAL INCL O&P
810 0010	**SOLID SURFACE COUNTERTOPS**, Acrylic polymer									810
0020	Pricing for orders of 100 L.F. or greater									
0100	25" wide, solid colors	2 Carp	28	.571	L.F.	42	11.25		53.25	65.50
0200	Patterned colors		28	.571		53	11.25		64.25	78
0300	Premium patterned colors		28	.571		66.50	11.25		77.75	92.50
0400	With silicone attached 4" backsplash, solid colors		27	.593		46	11.65		57.65	70.50
0500	Patterned colors		27	.593		58	11.65		69.65	84
0600	Premium patterned colors		27	.593		72.50	11.65		84.15	99.50
0700	With hard seam attached 4" backsplash, solid colors		23	.696		46	13.70		59.70	74
0800	Patterned colors		23	.696		58	13.70		71.70	87.50
0900	Premium patterned colors	▼	23	.696	▼	72.50	13.70		86.20	103
1000	Pricing for order of 51 - 99 L.F.									
1100	25" wide, solid colors	2 Carp	24	.667	L.F.	48	13.15		61.15	75.50
1200	Patterned colors		24	.667		61	13.15		74.15	89.50
1300	Premium patterned colors		24	.667		76.50	13.15		89.65	107
1400	With silicone attached 4" backsplash, solid colors		23	.696		53	13.70		66.70	81.50
1500	Patterned colors		23	.696		67	13.70		80.70	97
1600	Premium patterned colors		23	.696		83.50	13.70		97.20	115
1700	With hard seam attached 4" backsplash, solid colors		20	.800		53	15.75		68.75	85
1800	Patterned colors		20	.800		67	15.75		82.75	101
1900	Premium patterned colors	▼	20	.800	▼	83.50	15.75		99.25	119
2000	Pricing for order of 1 - 50 L.F.									
2100	25" wide, solid colors	2 Carp	20	.800	L.F.	56.50	15.75		72.25	89
2200	Patterned colors		20	.800		71.50	15.75		87.25	106
2300	Premium patterned colors		20	.800		89.50	15.75		105.25	126
2400	With silicone attached 4" backsplash, solid colors		19	.842		62	16.60		78.60	96.50
2500	Patterned colors		19	.842		78.50	16.60		95.10	115
2600	Premium patterned colors		19	.842		98	16.60		114.60	137
2700	With hard seam attached 4" backsplash, solid colors		4	4		62	79		141	203
2800	Patterned colors		15	1.067		78.50	21		99.50	123
2900	Premium patterned colors	▼	15	1.067	▼	98	21		119	144
3000	Sinks, pricing for order of 100 or greater units									
3100	Single bowl, hard seamed, solid colors, 13" x 17"	1 Carp	3	2.667	Ea.	283	52.50		335.50	400
3200	10" x 15"		7	1.143		131	22.50		153.50	183
3300	Cutouts for sinks	▼	8	1	▼		19.70		19.70	34
3400	Sinks, pricing for order of 51 - 99 units									
3500	Single bowl, hard seamed, solid colors, 13" x 17"	1 Carp	2.55	3.137	Ea.	325	62		387	460
3600	10" x 15"		6	1.333		150	26.50		176.50	210
3700	Cutouts for sinks	▼	7	1.143	▼		22.50		22.50	38.50
3800	Sinks, pricing for order of 1 - 50 units									
3900	Single bowl, hard seamed, solid colors, 13" x 17"	1 Carp	2	4	Ea.	380	79		459	555
4000	10" x 15"		4.55	1.758		176	34.50		210.50	254
4100	Cutouts for sinks		5.25	1.524			30		30	51.50
4200	Cooktop cutouts, pricing for 100 or greater units		4	2		20.50	39.50		60	90
4300	51 - 99 units		3.40	2.353		23.50	46.50		70	106
4400	1 - 50 units	▼	3	2.667	▼	27.50	52.50		80	121
850 0010	**VANITY TOPS**									850
0015	Solid surface, center bowl, 17" x 19"	1 Carp	12	.667	Ea.	168	13.15		181.15	208
0020	19" x 25"		12	.667		203	13.15		216.15	246
0030	19" x 31"		12	.667		247	13.15		260.15	295
0040	19" x 37"		12	.667		287	13.15		300.15	340
0050	22" x 25"		10	.800		229	15.75		244.75	278
0060	22" x 31"		10	.800		267	15.75		282.75	320
0070	22" x 37"		10	.800		310	15.75		325.75	365
0080	22" x 43"		10	.800		355	15.75		370.75	415
0090	22" x 49"	▼	10	.800		390	15.75		405.75	455

06600 | Plastic Fabrications

06620 | Non-Structural Plastics

			CREW	DAILY OUTPUT	LABOR-HOURS	UNIT	2000 BARE COSTS				TOTAL INCL O&P
							MAT.	LABOR	EQUIP.	TOTAL	
850	0110	22" x 55"	1 Carp	8	1	Ea.	445	19.70		464.70	525	850
	0120	22" x 61"		8	1		510	19.70		529.70	595
	0130	22" x 68"		8	1		650	19.70		669.70	750
	0140	22" x 73"		8	1		735	19.70		754.70	845
	0150	22" x 85"		8	1		850	19.70		869.70	970
	0160	Offset bowl, left or right, 22" x 49"		10	.800		470	15.75		485.75	540
	0170	22" x 55"		8	1		535	19.70		554.70	620
	0180	22" x 61"		8	1		605	19.70		624.70	700
	0190	22" x 68"		8	1		710	19.70		729.70	815
	0200	22" x 73"		8	1		875	19.70		894.70	995
	0210	22" x 85"		8	1		1,025	19.70		1,044.70	1,150
	0220	Double bowl, 22" x 61"		8	1		555	19.70		574.70	650
	0230	Corner top/bowl, 22" x 49"	▼	8	1	▼	435	19.70		454.70	510
	0240	For aggregate colors, add					35%				
	0250	For faucets and fittings see 15410-300									

For information about Means Estimating Seminars, see yellow pages 11 and 12 in back of book

For expanded coverage of these items see *Means Building Construction Cost Data 2000*

Division Notes

	CREW	DAILY OUTPUT	LABOR-HOURS	UNIT	MAT.	LABOR	EQUIP.	TOTAL	TOTAL INCL O&P

Division 7
Thermal & Moisture Protection

Estimating Tips

07100 Dampproofing & Waterproofing
- Be sure of the job specifications before pricing this subdivision. The difference in cost between waterproofing and dampproofing can be great. Waterproofing will hold back standing water. Dampproofing prevents the transmission of water vapor. Also included in this section are vapor retarding membranes.

07200 Thermal Protection
- Insulation and fireproofing products are measured by area, thickness, volume or R value. Specifications may only give what the specific R value should be in a certain situation. The estimator may need to choose the type of insulation to meet that R value.

07300 Shingles, Roof Tiles & Roof Coverings
07400 Roofing & Siding Panels
- Many roofing and siding products are bought and sold by the square. One square is equal to an area that measures 100 square feet. This simple change in unit of measure could create a large error if the estimator is not observant. Accessories and fasteners necessary for a complete installation must be figured into any calculations for both material and labor.

07500 Membrane Roofing
07600 Flashing & Sheet Metal
07700 Roof Specialties & Accessories
- The items in these subdivisions compose a roofing system. No one component completes the installation and all must be estimated. Built-up or single ply membrane roofing systems are made up of many products and installation trades. Wood blocking at roof perimeters or penetrations, parapet coverings, reglets, roof drains, gutters, downspouts, sheet metal flashing, skylights, smoke vents or roof hatches all need to be considered along with the roofing material. Several different installation trades will need to work together on the roofing system. Inherent difficulties in the scheduling and coordination of various trades must be accounted for when estimating labor costs.

07900 Joint Sealers
- To complete the weather-tight shell the sealants and caulkings must be estimated. Where different materials meet—at expansion joints, at flashing penetrations, and at hundreds of other locations throughout a construction project—they provide another line of defense against water penetration. Often, an entire system is based on the proper location and placement of caulking or sealants. The detail drawings that are included as part of a set of architectural plans, show typical locations for these materials. When caulking or sealants are shown at typical locations, this means the estimator must include them for all the locations where this detail is applicable. Be careful to keep different types of sealants separate, and remember to consider backer rods and primers if necessary.

Reference Numbers
Reference numbers are shown in bold squares at the beginning of some major classifications. These numbers refer to related items in the Reference Section. The reference information may be an estimating procedure, an alternate pricing method or technical information.

Note: Not all subdivisions listed here necessarily appear in this publication.

07100 | Dampproofing & Waterproofing

07110 | Dampproofing

				CREW	DAILY OUTPUT	LABOR-HOURS	UNIT	MAT.	LABOR	EQUIP.	TOTAL	TOTAL INCL O&P	
100	0010	**BITUMINOUS ASPHALT COATING** For foundation											100
	0030	Brushed on, below grade, 1 coat		1 Rofc	665	.012	S.F.	.06	.21		.27	.45	
	0100	2 coat			500	.016		.10	.27		.37	.62	
	0300	Sprayed on, below grade, 1 coat, 25.6 S.F./gal.			830	.010		.07	.16		.23	.39	
	0400	2 coat, 20.5 S.F./gal.			500	.016		.14	.27		.41	.67	
	0600	Troweled on, asphalt with fibers, 1/16″ thick			500	.016		.15	.27		.42	.68	
	0700	1/8″ thick			400	.020		.28	.34		.62	.94	
	1000	1/2″ thick			350	.023		.92	.39		1.31	1.74	
200	0010	**CEMENT PARGING** 2 coats, 1/2″ thick, regular P.C.	R07110-010	D-1	250	.064	S.F.	.15	1.14		1.29	2.10	200
	0100	Waterproofed Portland cement		″	250	.064	″	.17	1.14		1.31	2.12	

07190 | Water Repellents

			CREW	DAILY OUTPUT	LABOR-HOURS	UNIT	MAT.	LABOR	EQUIP.	TOTAL	TOTAL INCL O&P	
700	0010	**RUBBER COATING** Water base liquid, roller applied	2 Rofc	7,000	.002	S.F.	.55	.04		.59	.68	700
	0200	Silicone or stearate, sprayed on CMU, 1 coat	1 Rofc	4,000	.002		.27	.03		.30	.36	
	0300	2 coats	″	3,000	.003		.54	.05		.59	.68	

07200 | Thermal Protection

07210 | Building Insulation

			CREW	DAILY OUTPUT	LABOR-HOURS	UNIT	MAT.	LABOR	EQUIP.	TOTAL	TOTAL INCL O&P	
150	0010	**BLOWN-IN INSULATION** Ceilings, with open access										150
	0020	Cellulose, 3-1/2″ thick, R13	G-4	5,000	.005	S.F.	.13	.07	.05	.25	.32	
	0030	5-3/16″ thick, R19		3,800	.006		.19	.10	.07	.36	.45	
	0050	6-1/2″ thick, R22		3,000	.008		.24	.12	.09	.45	.57	
	1000	Fiberglass, 5″ thick, R11		3,800	.006		.16	.10	.07	.33	.42	
	1050	6″ thick, R13		3,000	.008		.17	.12	.09	.38	.50	
	1100	8-1/2″ thick, R19		2,200	.011		.23	.16	.12	.51	.67	
	1300	12″ thick, R26		1,500	.016		.33	.24	.18	.75	.97	
	2000	Mineral wool, 4″ thick, R12		3,500	.007		.16	.10	.08	.34	.45	
	2050	6″ thick, R17		2,500	.010		.18	.15	.11	.44	.57	
	2100	9″ thick, R23		1,750	.014		.27	.21	.16	.64	.83	
	2500	Wall installation, incl. drilling & patching from outside, two 1″										
	2510	diam. holes @ 16″ O.C., top & mid-point of wall, add to above										
	2700	For masonry	G-4	415	.058	S.F.	.06	.87	.65	1.58	2.29	
	2800	For wood siding		840	.029		.06	.43	.32	.81	1.17	
	2900	For stucco/plaster		665	.036		.06	.55	.41	1.02	1.45	
350	0010	**FLOOR INSULATION, NONRIGID** Including										350
	0020	spring type wire fasteners										
	2000	Fiberglass, blankets or batts, paper or foil backing										
	2100	1 side, 3-1/2″ thick, R11	1 Carp	700	.011	S.F.	.27	.23		.50	.69	
	2150	6″ thick, R19		600	.013		.36	.26		.62	.85	
	2200	8-1/2″ thick, R30		550	.015		.61	.29		.90	1.16	
500	0010	**POURED INSULATION** Cellulose fiber, R3.8 per inch	1 Carp	200	.040	C.F.	.45	.79		1.24	1.85	500
	0080	Fiberglass wool, R4 per inch		200	.040		.33	.79		1.12	1.71	
	0100	Mineral wool, R3 per inch		200	.040		.30	.79		1.09	1.68	
	0300	Polystyrene, R4 per inch		200	.040		1.93	.79		2.72	3.47	
	0400	Vermiculite or perlite, R2.7 per inch		200	.040		1.45	.79		2.24	2.95	
550	0010	**MASONRY INSULATION** Vermiculite or perlite, poured										550
	0100	In cores of concrete block, 4″ thick wall, .115 CF/SF	D-1	4,800	.003	S.F.	.17	.06		.23	.28	

07200 | Thermal Protection

07210 | Building Insulation

			CREW	DAILY OUTPUT	LABOR-HOURS	UNIT	MAT.	LABOR	EQUIP.	TOTAL	TOTAL INCL O&P	
550	0700	Foamed in place, urethane in 2-5/8" cavity	G-2	1,035	.023	S.F.	.38	.38	.24	1	1.32	550
	0800	For each 1" added thickness, add	"	2,372	.010		.12	.16	.11	.39	.53	
600	0600	**PERIMETER INSULATION**, polystyrene, expanded, 1" thick, R4	1 Carp	680	.012	S.F.	.17	.23		.40	.59	600
	0700	2" thick, R8		675	.012	"	.32	.23		.55	.75	
700	0010	**REFLECTIVE INSULATION**, aluminum foil on reinforced scrim		19	.421	C.S.F.	13.90	8.30		22.20	29.50	700
	0100	Reinforced with woven polyolefin		19	.421		16.90	8.30		25.20	33	
	0500	With single bubble air space, R8.8		15	.533		26	10.50		36.50	46.50	
	0600	With double bubble air space, R9.8		15	.533		28	10.50		38.50	49	
900	0010	**WALL INSULATION, RIGID**										900
	0040	Fiberglass, 1.5#/CF, unfaced, 1" thick, R4.1	1 Carp	1,000	.008	S.F.	.25	.16		.41	.55	
	0060	1-1/2" thick, R6.2		1,000	.008		.34	.16		.50	.64	
	0080	2" thick, R8.3		1,000	.008		.41	.16		.57	.72	
	0120	3" thick, R12.4		800	.010		.49	.20		.69	.88	
	0370	3#/CF, unfaced, 1" thick, R4.3		1,000	.008		.32	.16		.48	.62	
	0390	1-1/2" thick, R6.5		1,000	.008		.63	.16		.79	.96	
	0400	2" thick, R8.7		890	.009		.77	.18		.95	1.15	
	0420	2-1/2" thick, R10.9		800	.010		.96	.20		1.16	1.40	
	0440	3" thick, R13		800	.010		1.14	.20		1.34	1.59	
	0520	Foil faced, 1" thick, R4.3		1,000	.008		.76	.16		.92	1.11	
	0540	1-1/2" thick, R6.5		1,000	.008		1.02	.16		1.18	1.39	
	0560	2" thick, R8.7		890	.009		1.27	.18		1.45	1.70	
	0580	2-1/2" thick, R10.9		800	.010		1.50	.20		1.70	1.99	
	0600	3" thick, R13		800	.010		1.64	.20		1.84	2.14	
	0670	6#/CF, unfaced, 1" thick, R4.3		1,000	.008		.73	.16		.89	1.07	
	0690	1-1/2" thick, R6.5		890	.009		1.13	.18		1.31	1.54	
	0700	2" thick, R8.7		800	.010		1.58	.20		1.78	2.08	
	0721	2-1/2" thick, R10.9		800	.010		1.74	.20		1.94	2.25	
	0741	3" thick, R13		730	.011		2.08	.22		2.30	2.66	
	0821	Foil faced, 1" thick, R4.3		1,000	.008		1.03	.16		1.19	1.40	
	0840	1-1/2" thick, R6.5		890	.009		1.48	.18		1.66	1.93	
	0850	2" thick, R8.7		800	.010		1.93	.20		2.13	2.46	
	0880	2-1/2" thick, R10.9		800	.010		2.32	.20		2.52	2.89	
	0900	3" thick, R13		730	.011		2.77	.22		2.99	3.42	
	1500	Foamglass, 1-1/2" thick, R4.5		800	.010		1.45	.20		1.65	1.94	
	1550	3" thick, R9		730	.011		2.71	.22		2.93	3.35	
	1600	Isocyanurate, 4' x 8' sheet, foil faced, both sides										
	1610	1/2" thick, R3.9	1 Carp	800	.010	S.F.	.27	.20		.47	.64	
	1620	5/8" thick, R4.5		800	.010		.28	.20		.48	.65	
	1630	3/4" thick, R5.4		800	.010		.29	.20		.49	.66	
	1640	1" thick, R7.2		800	.010		.32	.20		.52	.69	
	1650	1-1/2" thick, R10.8		730	.011		.35	.22		.57	.76	
	1660	2" thick, R14.4		730	.011		.44	.22		.66	.85	
	1670	3" thick, R21.6		730	.011		1.05	.22		1.27	1.52	
	1680	4" thick, R28.8		730	.011		1.29	.22		1.51	1.79	
	1700	Perlite, 1" thick, R2.77		800	.010		.24	.20		.44	.60	
	1750	2" thick, R5.55		730	.011		.48	.22		.70	.90	
	1900	Extruded polystyrene, 25 PSI compressive strength, 1" thick, R5		800	.010		.30	.20		.50	.67	
	1940	2" thick R10		730	.011		.59	.22		.81	1.02	
	1960	3" thick, R15		730	.011		.86	.22		1.08	1.32	
	2100	Expanded polystyrene, 1" thick, R3.85		800	.010		.12	.20		.32	.47	
	2120	2" thick, R7.69		730	.011		.32	.22		.54	.72	
	2140	3" thick, R11.49		730	.011		.49	.22		.71	.91	
950	0010	**WALL OR CEILING INSUL., NON-RIGID**										950
	0040	Fiberglass, kraft faced, batts or blankets										
	0060	3-1/2" thick, R11, 11" wide	1 Carp	1,150	.007	S.F.	.22	.14		.36	.47	
	0140	6" thick, R19, 11" wide		1,000	.008		.31	.16		.47	.61	

For expanded coverage of these items see *Means Building Construction Cost Data 2000*

07200 | Thermal Protection

07210 | Building Insulation

		CREW	DAILY OUTPUT	LABOR-HOURS	UNIT	2000 BARE COSTS MAT.	LABOR	EQUIP.	TOTAL	TOTAL INCL O&P	
950	0200	9" thick, R30, 15" wide	1 Carp	1,150	.007	S.F.	.56	.14		.70	.85	950
	0240	12" thick, R38, 15" wide	↓	1,000	.008	↓	.72	.16		.88	1.06	
	0400	Fiberglass, foil faced, batts or blankets										
	0420	3-1/2" thick, R11, 15" wide	1 Carp	1,600	.005	S.F.	.32	.10		.42	.52	
	0460	6" thick, R19, 15" wide		1,350	.006		.39	.12		.51	.63	
	0500	9" thick, R30, 15" wide	↓	1,150	.007		.67	.14		.81	.97	
	0800	Fiberglass, unfaced, batts or blankets										
	0820	3-1/2" thick, R11, 15" wide	1 Carp	1,350	.006	S.F.	.20	.12		.32	.42	
	0860	6" thick, R19, 15" wide		1,150	.007		.33	.14		.47	.59	
	0900	9" thick, R30, 15" wide		1,000	.008		.56	.16		.72	.89	
	0940	12" thick, R38, 15" wide	↓	1,000	.008	↓	.72	.16		.88	1.06	
	1300	Mineral fiber batts, kraft faced										
	1320	3-1/2" thick, R12	1 Carp	1,600	.005	S.F.	.24	.10		.34	.43	
	1340	6" thick, R19		1,600	.005		.37	.10		.47	.58	
	1380	10" thick, R30		1,350	.006	↓	.58	.12		.70	.84	
	1850	Friction fit wire insulation supports, 16" O.C.	↓	960	.008	Ea.	.05	.16		.21	.34	
	1900	For foil backing, add				S.F.	.04			.04	.04	

07220 | Roof & Deck Insulation

		CREW	DAILY OUTPUT	LABOR-HOURS	UNIT	MAT.	LABOR	EQUIP.	TOTAL	TOTAL INCL O&P	
700	0010	**ROOF DECK INSULATION**										700
	0020	Fiberboard low density, 1/2" thick R1.39	1 Rofc	1,000	.008	S.F.	.17	.14		.31	.44	
	0030	1" thick R2.78		800	.010		.32	.17		.49	.67	
	0080	1 1/2" thick R4.17		800	.010		.48	.17		.65	.85	
	0100	2" thick R5.56		800	.010		.64	.17		.81	1.02	
	0110	Fiberboard high density, 1/2" thick R1.3		1,000	.008		.18	.14		.32	.45	
	0120	1" thick R2.5		800	.010		.34	.17		.51	.69	
	0130	1-1/2" thick R3.8		800	.010		.55	.17		.72	.93	
	0200	Fiberglass, 3/4" thick R2.78		1,000	.008		.44	.14		.58	.73	
	0400	15/16" thick R3.70		1,000	.008		.57	.14		.71	.88	
	0460	1-1/16" thick R4.17		1,000	.008		.72	.14		.86	1.04	
	0600	1-5/16" thick R5.26		1,000	.008		.99	.14		1.13	1.34	
	0650	2-1/16" thick R8.33		800	.010		1.06	.17		1.23	1.49	
	0700	2-7/16" thick R10		800	.010		1.20	.17		1.37	1.64	
	1500	Foamglass, 1-1/2" thick R4.5		800	.010		1.41	.17		1.58	1.87	
	1530	3" thick R9		700	.011	↓	2.81	.19		3	3.45	
	1600	Tapered for drainage		600	.013	B.F.	.91	.23		1.14	1.42	
	1650	Perlite, 1/2" thick R1.32		1,050	.008	S.F.	.25	.13		.38	.52	
	1655	3/4" thick R2.08		800	.010		.30	.17		.47	.65	
	1660	1" thick R2.78		800	.010		.24	.17		.41	.58	
	1670	1-1/2" thick R4.17		800	.010		.36	.17		.53	.72	
	1680	2" thick R5.56		700	.011		.48	.19		.67	.89	
	1685	2-1/2" thick R6.67		700	.011	↓	.71	.19		.90	1.14	
	1690	Tapered for drainage		800	.010	B.F.	.54	.17		.71	.91	
	1700	Polyisocyanurate, 2#/CF density, 3/4" thick, R5.1		1,500	.005	S.F.	.29	.09		.38	.49	
	1705	1" thick R7.14		1,400	.006		.30	.10		.40	.51	
	1715	1-1/2" thick R10.87		1,250	.006		.34	.11		.45	.57	
	1725	2" thick R14.29		1,100	.007		.42	.12		.54	.69	
	1735	2-1/2" thick R16.67		1,050	.008		.48	.13		.61	.77	
	1745	3" thick R21.74		1,000	.008		.59	.14		.73	.90	
	1755	3-1/2" thick R25		1,000	.008	↓	.70	.14		.84	1.02	
	1765	Tapered for drainage	↓	1,400	.006	B.F.	.36	.10		.46	.58	
	1900	Extruded Polystyrene										
	1910	15 PSI compressive strength, 1" thick, R5	1 Rofc	1,500	.005	S.F.	.21	.09		.30	.40	
	1920	2" thick, R10		1,250	.006		.33	.11		.44	.56	
	1930	3" thick R15	↓	1,000	.008	↓	.73	.14		.87	1.05	

07200 | Thermal Protection

07220 | Roof & Deck Insulation

		CREW	DAILY OUTPUT	LABOR-HOURS	UNIT	MAT.	LABOR	EQUIP.	TOTAL	TOTAL INCL O&P
1932	4" thick R20	1 Rofc	1,000	.008	S.F.	1.01	.14		1.15	1.36
1934	Tapered for drainage		1,500	.005	B.F.	.33	.09		.42	.53
1940	25 PSI compressive strength, 1" thick R5		1,500	.005	S.F.	.31	.09		.40	.51
1942	2" thick R10		1,250	.006		.63	.11		.74	.89
1944	3" thick R15		1,000	.008		.94	.14		1.08	1.28
1946	4" thick R20		1,000	.008		1.05	.14		1.19	1.40
1948	Tapered for drainage		1,500	.005	B.F.	.38	.09		.47	.59
1950	40 psi compressive strength, 1" thick R5		1,500	.005	S.F.	.33	.09		.42	.53
1952	2" thick R10		1,250	.006		.65	.11		.76	.91
1954	3" thick R15		1,000	.008		.95	.14		1.09	1.30
1956	4" thick R20		1,000	.008		1.27	.14		1.41	1.65
1958	Tapered for drainage		1,400	.006	B.F.	.48	.10		.58	.71
1960	60 PSI compressive strength, 1" thick R5		1,450	.006	S.F.	.40	.09		.49	.62
1962	2" thick R10		1,200	.007		.71	.11		.82	.99
1964	3" thick R15		975	.008		1.06	.14		1.20	1.43
1966	4" thick R20		950	.008		1.47	.14		1.61	1.89
1968	Tapered for drainage		1,400	.006	B.F.	.57	.10		.67	.81
2010	Expanded polystyrene, 1#/CF density, 3/4" thick R2.89		1,500	.005	S.F.	.17	.09		.26	.36
2020	1" thick R3.85		1,500	.005		.17	.09		.26	.36
2100	2" thick R7.69		1,250	.006		.32	.11		.43	.55
2110	3" thick R11.49		1,250	.006		.48	.11		.59	.73
2120	4" thick R15.38		1,200	.007		.53	.11		.64	.79
2130	5" thick R19.23		1,150	.007		.67	.12		.79	.96
2140	6" thick R23.26		1,150	.007		.79	.12		.91	1.09
2150	Tapered for drainage		1,500	.005	B.F.	.32	.09		.41	.52
2400	Composites with 2" EPS									
2410	1" fiberboard	1 Rofc	950	.008	S.F.	.74	.14		.88	1.08
2420	7/16" oriented strand board		800	.010		.87	.17		1.04	1.28
2430	1/2" plywood		800	.010		.94	.17		1.11	1.35
2440	1" perlite		800	.010		.78	.17		.95	1.18
2450	Composites with 1 1/2" polyisocyanurate									
2460	1" fiberboard	1 Rofc	800	.010	S.F.	.80	.17		.97	1.20
2470	1" perlite		850	.009		.83	.16		.99	1.21
2480	7/16" oriented strand board		800	.010		.95	.17		1.12	1.37

07260 | Vapor Retarders

		CREW	DAILY OUTPUT	LABOR-HOURS	UNIT	MAT.	LABOR	EQUIP.	TOTAL	TOTAL INCL O&P
0010	**BUILDING PAPER** Aluminum and kraft laminated, foil 1 side	1 Carp	37	.216	Sq.	3.55	4.26		7.81	11.20
0100	Foil 2 sides		37	.216		5.70	4.26		9.96	13.55
0300	Asphalt, two ply, 30#, for subfloors		19	.421		10.90	8.30		19.20	26
0400	Asphalt felt sheathing paper, 15#		37	.216		2.59	4.26		6.85	10.15
0450	Housewrap, exterior, spun bonded polypropylene									
0470	Small roll	1 Carp	3,800	.002	S.F.	.10	.04		.14	.18
0480	Large roll	"	4,000	.002	"	.09	.04		.13	.17
0500	Material only, 3' x 111.1' roll				Ea.	33			33	36.50
0520	9' x 111.1' roll				"	94			94	103
0600	Polyethylene vapor barrier, standard, .002" thick	1 Carp	37	.216	Sq.	1.09	4.26		5.35	8.50
0700	.004" thick		37	.216		2.36	4.26		6.62	9.90
0900	.006" thick		37	.216		2.86	4.26		7.12	10.45
1200	.010" thick		37	.216		6.15	4.26		10.41	14.05
1500	Red rosin paper, 5 sq rolls, 4 lb per square		37	.216		1.55	4.26		5.81	9
1600	5 lbs. per square		37	.216		2	4.26		6.26	9.50
1800	Reinf. waterproof, .002" polyethylene backing, 1 side		37	.216		4.92	4.26		9.18	12.70
1900	2 sides		37	.216		6.50	4.26		10.76	14.45
3000	Building wrap, spunbonded polyethylene	2 Carp	8,000	.002	S.F.	.10	.04		.14	.18

For expanded coverage of these items see Means Building Construction Cost Data 2000

07300 | Shingles, Roof Tiles & Roof Coverings

07310 | Shingles

			CREW	DAILY OUTPUT	LABOR-HOURS	UNIT	MAT.	LABOR	EQUIP.	TOTAL	TOTAL INCL O&P	
050	0010	**ALUMINUM** Shingles, mill finish, .019" thick	1 Carp	5	1.600	Sq.	150	31.50		181.50	219	050
	0100	.020" thick	"	5	1.600		145	31.50		176.50	213	
	0300	For colors, add					15.15			15.15	16.65	
	0600	Ridge cap, .024" thick	1 Carp	170	.047	L.F.	1.85	.93		2.78	3.63	
	0700	End wall flashing, .024" thick		170	.047		1.26	.93		2.19	2.98	
	0900	Valley section, .024" thick		170	.047		2.30	.93		3.23	4.12	
	1000	Starter strip, .024" thick		400	.020		1.20	.39		1.59	2	
	1200	Side wall flashing, .024" thick		170	.047		1.25	.93		2.18	2.97	
100	0010	**ASPHALT SHINGLES**										100
	0100	Standard strip shingles										
	0150	Inorganic, class A, 210-235 lb/sq	1 Rofc	5.50	1.455	Sq.	26	25		51	74.50	
	0155	Pneumatic nailed		7	1.143		26	19.50		45.50	65	
	0200	Organic, class C, 235-240 lb/sq		5	1.600		35.50	27.50		63	90	
	0205	Pneumatic nailed		6.25	1.280		35.50	22		57.50	79.50	
	0250	Standard, laminated multi-layered shingles										
	0300	Class A, 240-260 lb/sq	1 Rofc	4.50	1.778	Sq.	34	30.50		64.50	94	
	0305	Pneumatic nailed		5.63	1.422		34	24.50		58.50	82.50	
	0350	Class C, 260-300 lb/square, 4 bundles/square		4	2		49	34		83	118	
	0355	Pneumatic nailed		5	1.600		49	27.50		76.50	105	
	0400	Premium, laminated multi-layered shingles										
	0450	Class A, 260-300 lb, 4 bundles/sq	1 Rofc	3.50	2.286	Sq.	43	39		82	120	
	0455	Pneumatic nailed		4.37	1.831		43	31		74	106	
	0500	Class C, 300-385 lb/square, 5 bundles/square		3	2.667		65.50	45.50		111	157	
	0505	Pneumatic nailed		3.75	2.133		65.50	36.50		102	140	
	0800	#15 felt underlayment		64	.125		2.59	2.13		4.72	6.80	
	0825	#30 felt underlayment		58	.138		5.45	2.35		7.80	10.40	
	0850	Self adhering polyethylene and rubberized asphalt underlayment		22	.364		37	6.20		43.20	52.50	
	0900	Ridge shingles		330	.024	L.F.	.72	.41		1.13	1.56	
	0905	Pneumatic nailed		412.50	.019	"	.72	.33		1.05	1.41	
	1000	For steep roofs (7 to 12 pitch or greater), add						50%				
500	0010	**FIBER CEMENT** shingles, 16" x 9.35", 500 lb per square	1 Carp	4	2	Sq.	244	39.50		283.50	335	500
	0200	Shakes, 16" x 9.35", 550 lb per square		2.20	3.636	"	221	71.50		292.50	365	
	0300	Hip & ridge, 4.75 x 14"		1	8	C.L.F.	600	158		758	930	
	0400	Hexagonal, 16" x 16"		3	2.667	Sq.	165	52.50		217.50	272	
	0500	Square, 16" x 16"		3	2.667		148	52.50		200.50	253	
	2000	For steep roofs (7/12 pitch or greater), add						50%				
800	0010	**SLATE**, Buckingham, Virginia, black										800
	0100	3/16" - 1/4" thick	1 Rots	1.75	4.571	Sq.	540	78		618	740	
	0200	1/4" thick		1.75	4.571		720	78		798	935	
	0900	Pennsylvania black, Bangor, #1 clear		1.75	4.571		435	78		513	625	
	1200	Vermont, unfading, green, mottled green		1.75	4.571		395	78		473	580	
	1300	Semi-weathering green & gray		1.75	4.571		296	78		374	470	
	1400	Purple		1.75	4.571		390	78		468	575	
	1500	Black or gray		1.75	4.571		355	78		433	535	
	2700	Ridge shingles, slate		200	.040	L.F.	8.50	.68		9.18	10.60	
980	0010	**WOOD** 16" No. 1 red cedar shingles, 5" exposure, on roof	1 Carp	2.50	3.200	Sq.	151	63		214	274	980
	0015	Pneumatic nailed		3.25	2.462	"	151	48.50		199.50	249	
	0200	7-1/2" exposure, on walls		2.05	3.902	Sq.	101	77		178	243	
	0205	Pneumatic nailed		2.67	2.996		101	59		160	212	
	0300	18" No. 1 red cedar perfections, 5-1/2" exposure, on roof		2.75	2.909		164	57.50		221.50	279	
	0305	Pneumatic nailed		3.57	2.241		164	44		208	257	
	0500	7-1/2" exposure, on walls		2.25	3.556		121	70		191	253	
	0505	Pneumatic nailed		2.92	2.740		121	54		175	226	
	0600	Resquared, and rebutted, 5-1/2" exposure, on roof		3	2.667		199	52.50		251.50	310	
	0605	Pneumatic nailed		3.90	2.051		199	40.50		239.50	288	

07300 | Shingles, Roof Tiles & Roof Coverings

07310 | Shingles

			CREW	DAILY OUTPUT	LABOR-HOURS	UNIT	MAT.	LABOR	EQUIP.	TOTAL	TOTAL INCL O&P	
980	0900	7-1/2" exposure, on walls	1 Carp	2.45	3.265	Sq.	146	64.50		210.50	271	980
	0905	Pneumatic nailed		3.18	2.516		146	49.50		195.50	246	
	1000	Add to above for fire retardant shingles, 16" long					30			30	33	
	1050	18" long					28.50			28.50	31.50	
	1060	Preformed ridge shingles	1 Carp	400	.020	L.F.	1.65	.39		2.04	2.50	
	1100	Hand-split red cedar shakes, 1/2" thick x 24" long, 10" exp. on roof		2.50	3.200	Sq.	138	63		201	260	
	1105	Pneumatic nailed		3.25	2.462		138	48.50		186.50	235	
	1110	3/4" thick x 24" long, 10" exp. on roof		2.25	3.556		138	70		208	272	
	1115	Pneumatic nailed		2.92	2.740		138	54		192	245	
	1200	1/2" thick, 18" long, 8-1/2" exp. on roof		2	4		97.50	79		176.50	242	
	1205	Pneumatic nailed		2.60	3.077		97.50	60.50		158	211	
	1210	3/4" thick x 18" long, 8 1/2" exp. on roof		1.80	4.444		97.50	87.50		185	257	
	1215	Pneumatic nailed		2.34	3.419		97.50	67.50		165	222	
	1255	10" exp. on walls		2	4		110	79		189	256	
	1260	10" exposure on walls, pneumatic nailed		2.60	3.077		110	60.50		170.50	225	
	1700	Add to above for fire retardant shakes, 24" long					30			30	33	
	1800	18" long					30			30	33	
	1810	Ridge shakes	1 Carp	350	.023	L.F.	2.35	.45		2.80	3.35	
	2000	White cedar shingles, 16" long, extras, 5" exposure, on roof		2.40	3.333	Sq.	112	65.50		177.50	237	
	2005	Pneumatic nailed		3.12	2.564		112	50.50		162.50	211	
	2050	5" exposure on walls		2	4		112	79		191	259	
	2055	Pneumatic nailed		2.60	3.077		112	60.50		172.50	228	
	2100	7-1/2" exposure, on walls		2	4		80	79		159	224	
	2105	Pneumatic nailed		2.60	3.077		80	60.50		140.50	193	
	2150	"B" grade, 5" exposure on walls		2	4		117	79		196	264	
	2155	Pneumatic nailed		2.60	3.077		117	60.50		177.50	233	
	2300	For 15# organic felt underlayment on roof, 1 layer, add		64	.125		2.59	2.46		5.05	7.05	
	2400	2 layers, add		32	.250		5.20	4.93		10.13	14.15	
	2600	For steep roofs (7/12 pitch or greater), add to above						50%				
	3000	Ridge shakes or shingle wood	1 Carp	280	.029	L.F.	2.40	.56		2.96	3.60	

07320 | Roof Tiles

			CREW	DAILY OUTPUT	LABOR-HOURS	UNIT	MAT.	LABOR	EQUIP.	TOTAL	TOTAL INCL O&P	
100	0010	**ALUMINUM** Tiles with accessories, .032" thick, mission tile	1 Carp	2.50	3.200	Sq.	355	63		418	500	100
	0200	Spanish tiles	"	3	2.667	"	355	52.50		407.50	480	
200	0010	**CLAY TILE** ASTM C1167, GR 1, severe weathering, acces. incl.										200
	0200	Lanai tile or Classic tile, 158 pc per sq	1 Rots	1.65	4.848	Sq.	490	83		573	695	
	0300	Americana, 158 pc per sq, most colors		1.65	4.848		495	83		578	700	
	0350	Green, gray or brown		1.65	4.848		490	83		573	695	
	0400	Blue		1.65	4.848		490	83		573	695	
	0600	Spanish tile, 171 pc per sq, red		1.80	4.444		305	76		381	480	
	0800	Blend		1.80	4.444		420	76		496	600	
	0900	Glazed white		1.80	4.444		500	76		576	690	
	1100	Mission tile, 192 pc per sq, machine scored finish, red		1.15	6.957		635	119		754	915	
	1700	French tile, 133 pc per sq, smooth finish, red		1.35	5.926		575	101		676	820	
	1750	Blue or green		1.35	5.926		685	101		786	945	
	1800	Norman black 317 pc per sq		1	8		805	137		942	1,150	
	2200	Williamsburg tile, 158 pc per sq, aged cedar		1.35	5.926		490	101		591	730	
	2250	Gray or green		1.35	5.926		490	101		591	730	
	2350	Ridge shingles, clay tile		200	.040	L.F.	8.65	.68		9.33	10.75	
	3000	For steep roofs (7/12 pitch or greater), add to above				Sq.		50%				
300	0010	**CONCRETE TILE** Including installation of accessories										300
	0020	Corrugated, 13" x 16-1/2", 90 per sq, 950 lb per sq										
	0050	Earthtone colors, nailed to wood deck	1 Rots	1.35	5.926	Sq.	103	101		204	300	
	0150	Blues		1.35	5.926		114	101		215	315	
	0200	Greens		1.35	5.926		114	101		215	315	
	0250	Premium colors		1.35	5.926		153	101		254	355	

For expanded coverage of these items see *Means Building Construction Cost Data 2000*

07300 | Shingles, Roof Tiles & Roof Coverings

07320 | Roof Tiles

		CREW	DAILY OUTPUT	LABOR-HOURS	UNIT	2000 BARE COSTS				TOTAL INCL O&P
						MAT.	LABOR	EQUIP.	TOTAL	
0500	Shakes, 13" x 16-1/2", 90 per sq, 950 lb per sq									
0600	All colors, nailed to wood deck	1 Rots	1.50	5.333	Sq.	185	91		276	375
1500	Accessory pieces, ridge & hip, 10" x 16-1/2", 8 lbs. each				Ea.	2.25			2.25	2.48
1700	Rake, 6-1/2" x 16-3/4", 9 lbs. each					2.25			2.25	2.48
1800	Mansard hip, 10" x 16-1/2", 9.2 lbs. each					2.25			2.25	2.48
1900	Hip starter, 10" x 16-1/2", 10.5 lbs. each					9.50			9.50	10.45
2000	3 or 4 way apex, 10" each side, 11.5 lbs. each					10.25			10.25	11.30

07400 | Roofing & Siding Panels

07410 | Metal Roof & Wall Panels

		CREW	DAILY OUTPUT	LABOR-HOURS	UNIT	2000 BARE COSTS				TOTAL INCL O&P
						MAT.	LABOR	EQUIP.	TOTAL	
0010	**ALUMINUM ROOFING** Corrugated or ribbed, .0155" thick, natural	G-3	1,200	.027	S.F.	.61	.48		1.09	1.48
0300	Painted	"	1,200	.027	"	.87	.48		1.35	1.77
0010	**STEEL ROOFING** on steel frame, corrugated or ribbed, 30 ga galv	G-3	1,100	.029	S.F.	.75	.52		1.27	1.72
0100	28 ga		1,050	.030		.79	.55		1.34	1.80
0300	26 ga		1,000	.032		.86	.58		1.44	1.93
0400	24 ga		950	.034		1.02	.61		1.63	2.15
0600	Colored, 28 ga		1,050	.030		1.05	.55		1.60	2.08
0700	26 ga		1,000	.032		1.12	.58		1.70	2.21
0710	Flat profile, 1-3/4" standing seams, 10" wide, standard finish, 26 ga		1,000	.032		2.30	.58		2.88	3.51
0715	24 ga		950	.034		2.67	.61		3.28	3.97
0720	22 ga		900	.036		3.29	.64		3.93	4.71
0725	Zinc aluminum alloy finish, 26 ga		1,000	.032		1.80	.58		2.38	2.96
0730	24 ga		950	.034		2.15	.61		2.76	3.40
0735	22 ga		900	.036		2.47	.64		3.11	3.81
0740	12" wide, standard finish, 26 ga		1,000	.032		2.29	.58		2.87	3.50
0745	24 ga		950	.034		2.66	.61		3.27	3.96
0750	Zinc aluminum alloy finish, 26 ga		1,000	.032		1.70	.58		2.28	2.85
0755	24 ga		950	.034		2.03	.61		2.64	3.26
0840	Flat profile, 1" x 3/8" batten, 12" wide, standard finish, 26 ga		1,000	.032		1.79	.58		2.37	2.95
0845	24 ga		950	.034		2.10	.61		2.71	3.34
0850	22 ga		900	.036		2.52	.64		3.16	3.86
0855	Zinc aluminum alloy finish, 26 ga		1,000	.032		1.51	.58		2.09	2.64
0860	24 ga		950	.034		1.68	.61		2.29	2.88
0865	22 ga		900	.036		1.95	.64		2.59	3.24
0870	16-1/2" wide, standard finish, 24 ga		950	.034		2.13	.61		2.74	3.37
0875	22 ga		900	.036		2.39	.64		3.03	3.72
0880	Zinc aluminum alloy finish, 24 ga		950	.034		1.60	.61		2.21	2.79
0885	22 ga		900	.036		1.78	.64		2.42	3.05
0890	Flat profile, 2" x 2" batten, 12" wide, standard finish, 26 ga		1,000	.032		2.05	.58		2.63	3.24
0895	24 ga		950	.034		2.45	.61		3.06	3.73
0900	22 ga		900	.036		3	.64		3.64	4.39
0905	Zinc aluminum alloy finish, 26 ga		1,000	.032		1.69	.58		2.27	2.84
0910	24 ga		950	.034		1.83	.61		2.44	3.04
0915	22 ga		900	.036		2.25	.64		2.89	3.57
0920	16-1/2" wide, standard finish, 24 ga		950	.034		2.27	.61		2.88	3.53
0925	22 ga		900	.036		2.64	.64		3.28	3.99
0930	Zinc aluminum alloy finish, 24 ga		950	.034		1.80	.61		2.41	3.01
0935	22 ga		900	.036		2.05	.64		2.69	3.35

Important: See the Reference Section for critical supporting data - Reference Nos., Crews, & Location Factors

07400 | Roofing & Siding Panels

07420 | Plastic Roof & Wall Panels

			CREW	DAILY OUTPUT	LABOR-HOURS	UNIT	2000 BARE COSTS MAT.	LABOR	EQUIP.	TOTAL	TOTAL INCL O&P	
770	0010	**FIBERGLASS** Corrugated panels, roofing, 8 oz per SF	G-3	1,000	.032	S.F.	2.17	.58		2.75	3.37	770
	0300	Corrugated siding, 6 oz per SF		880	.036		1.88	.65		2.53	3.18	
	0400	8 oz per SF		880	.036		2.17	.65		2.82	3.50	
	0600	12 oz. siding, textured		880	.036		3.68	.65		4.33	5.15	
	0900	Flat panels, 6 oz per SF, clear or colors		880	.036		1.68	.65		2.33	2.96	
	1300	8 oz per SF, clear or colors		880	.036		2.17	.65		2.82	3.50	

07460 | Siding

			CREW	DAILY OUTPUT	LABOR-HOURS	UNIT	MAT.	LABOR	EQUIP.	TOTAL	TOTAL INCL O&P	
100	0011	**ALUMINUM SIDING**										100
	6040	.024 thick smooth white single 8" wide	2 Carp	515	.031	S.F.	1.23	.61		1.84	2.40	
	6060	Double 4" pattern		515	.031		1.17	.61		1.78	2.34	
	6080	Double 5" pattern		550	.029		1.17	.57		1.74	2.27	
	6120	Embossed white, 8" wide		515	.031		1.45	.61		2.06	2.65	
	6140	Double 4" pattern		515	.031		1.32	.61		1.93	2.50	
	6160	Double 5" pattern		550	.029		1.32	.57		1.89	2.43	
	6170	Vertical, embossed white, 12" wide		590	.027		1.32	.53		1.85	2.37	
	6320	.019 thick, insulated, smooth white, 8" wide		515	.031		1.18	.61		1.79	2.35	
	6340	Double 4" pattern		515	.031		1.16	.61		1.77	2.33	
	6360	Double 5" pattern		550	.029		1.16	.57		1.73	2.26	
	6400	Embossed white, 8" wide		515	.031		1.35	.61		1.96	2.54	
	6420	Double 4" pattern		515	.031		1.37	.61		1.98	2.56	
	6440	Double 5" pattern		550	.029		1.37	.57		1.94	2.49	
	6500	Shake finish 10" wide white		550	.029		1.45	.57		2.02	2.58	
	6600	Vertical pattern, 12" wide, white		590	.027		1.37	.53		1.90	2.43	
	6640	For colors add					.08			.08	.09	
	6700	Accessories, white										
	6720	Starter strip 2-1/8"	2 Carp	610	.026	L.F.	.19	.52		.71	1.10	
	6740	Sill trim		450	.036		.30	.70		1	1.53	
	6760	Inside corner		610	.026		.92	.52		1.44	1.90	
	6780	Outside corner post		610	.026		1.57	.52		2.09	2.62	
	6800	Door & window trim		440	.036		.29	.72		1.01	1.55	
	6820	For colors add					.08			.08	.09	
	6900	Soffit & fascia 1' overhang solid	2 Carp	110	.145		2.12	2.87		4.99	7.25	
	6920	Vented		110	.145		2.12	2.87		4.99	7.25	
	6940	2' overhang solid		100	.160		3.11	3.15		6.26	8.80	
	6960	Vented		100	.160		3.11	3.15		6.26	8.80	
300	0010	**FASCIA** Aluminum, reverse board and batten,										300
	0100	.032" thick, colored, no furring included	1 Shee	145	.055	S.F.	2.10	1.19		3.29	4.31	
	0200	Residential type, aluminum	1 Carp	200	.040	L.F.	1.15	.79		1.94	2.62	
	0300	Steel, galv and enameled, stock, no furring, long panels	1 Shee	145	.055	S.F.	2.19	1.19		3.38	4.41	
	0600	Short panels	"	115	.070	"	3.31	1.50		4.81	6.15	
500	0010	**FIBER CEMENT SIDING**										500
	0020	Lap siding, 5/16" thick, 6" wide, smooth texture	2 Carp	415	.039	S.F.	.88	.76		1.64	2.27	
	0025	Woodgrain texture		415	.039		.88	.76		1.64	2.27	
	0030	7-1/2" wide, smooth texture		425	.038		.88	.74		1.62	2.24	
	0035	Woodgrain texture		425	.038		.88	.74		1.62	2.24	
	0040	8" wide, smooth texture		425	.038		.87	.74		1.61	2.23	
	0045	Roughsawn texture		425	.038		.87	.74		1.61	2.23	
	0050	9-1/2" wide, smooth texture		440	.036		.84	.72		1.56	2.16	
	0055	Woodgrain texture		440	.036		.84	.72		1.56	2.16	
	0060	12" wide, smooth texture		455	.035		.81	.69		1.50	2.09	
	0065	Woodgrain texture		455	.035		.81	.69		1.50	2.09	
	0070	Panel siding, 5/16" thick, smooth texture		750	.021		.73	.42		1.15	1.52	
	0075	Stucco texture		750	.021		.73	.42		1.15	1.52	
	0080	Grooved woodgrain texture		750	.021		.73	.42		1.15	1.52	

For expanded coverage of these items see *Means Building Construction Cost Data 2000*

07400 | Roofing & Siding Panels

07460 | Siding

			CREW	DAILY OUTPUT	LABOR-HOURS	UNIT	2000 BARE COSTS MAT.	LABOR	EQUIP.	TOTAL	TOTAL INCL O&P	
500	0085	V - grooved woodgrain texture	2 Carp	750	.021	S.F.	.73	.42		1.15	1.52	500
	0090	Wood starter strip	↓	400	.040	L.F.	.20	.79		.99	1.57	
600	0010	**VINYL SIDING** Solid PVC panels, 8" to 10" wide, plain	1 Carp	255	.031	S.F.	.62	.62		1.24	1.74	600
	2000	Smooth, white, single, 8" wide	2 Carp	495	.032		.62	.64		1.26	1.77	
	2020	Dutch lap, 10" wide		550	.029		.64	.57		1.21	1.68	
	2100	Double 4" pattern, 8" wide		495	.032		.55	.64		1.19	1.70	
	2120	Double 5" pattern, 10" wide		550	.029		.52	.57		1.09	1.55	
	2200	Embossed, white, single, 8" wide		495	.032		.66	.64		1.30	1.82	
	2220	10" wide		550	.029		.67	.57		1.24	1.72	
	2300	Double 4" pattern, 8" wide		495	.032		.57	.64		1.21	1.72	
	2320	5" pattern, 10" wide		550	.029		.58	.57		1.15	1.62	
	2400	Shake finish, 10" wide, white		550	.029		1.90	.57		2.47	3.07	
	2600	Vertical pattern, double 5", 10" wide, white	↓	550	.029	↓	1.30	.57		1.87	2.41	
	2620											
	2700	For colors, add				S.F.	.08			.08	.09	
	2720	1/4" extruded polystyrene fan folded insulation	2 Carp	2,000	.008	"	.14	.16		.30	.42	
	3000	Accessories, starter strip		700	.023	L.F.	.23	.45		.68	1.02	
	3100	"J" channel, 1/2"		700	.023		.23	.45		.68	1.02	
	3120	5/8"		700	.023		.24	.45		.69	1.03	
	3140	3/4"		695	.023		.26	.45		.71	1.07	
	3160	1"		690	.023		.28	.46		.74	1.09	
	3180	1-1/8"		685	.023		.27	.46		.73	1.09	
	3190	1-1/4"		680	.024		.29	.46		.75	1.11	
	3200	Under sill trim		500	.032		.26	.63		.89	1.37	
	3300	Outside corner post, 3" face, pocket 5/8"		700	.023		1.05	.45		1.50	1.92	
	3320	7/8"		690	.023		1.05	.46		1.51	1.93	
	3340	1-1/4"		680	.024		1.07	.46		1.53	1.97	
	3400	Inside corner post, pocket 5/8"		700	.023		.56	.45		1.01	1.39	
	3420	7/8"		690	.023		.62	.46		1.08	1.46	
	3440	1-1/4"		680	.024		.66	.46		1.12	1.52	
	3500	Door & window trim, 2-1/2" face, pocket 5/8"		510	.031		.54	.62		1.16	1.65	
	3520	7/8"		500	.032		.53	.63		1.16	1.66	
	3540	1-1/4"		490	.033		.61	.64		1.25	1.77	
	3600	Soffit & fascia, 1' overhang, solid		120	.133		1.33	2.63		3.96	5.95	
	3620	Vented		120	.133		1.35	2.63		3.98	6	
	3700	2' overhang, solid		110	.145		1.97	2.87		4.84	7.10	
	3720	Vented	↓	110	.145	↓	1.97	2.87		4.84	7.10	
800	0010	**STEEL SIDING**, Beveled, vinyl coated, 8" wide, including fasteners	1 Carp	265	.030	S.F.	1.12	.59		1.71	2.25	800
	0050	10" wide	"	275	.029		1.19	.57		1.76	2.29	
	0080	Galv, corrugated or ribbed, on steel frame, 30 gauge	G-3	800	.040		.75	.72		1.47	2.05	
	0100	28 gauge		795	.040		.79	.72		1.51	2.10	
	0300	26 gauge		790	.041		.86	.73		1.59	2.19	
	0400	24 gauge		785	.041		1.03	.73		1.76	2.37	
	0600	22 gauge		770	.042		1.17	.75		1.92	2.56	
	0700	Colored, corrugated/ribbed, on steel frame, 10 yr fnsh, 28 ga.		800	.040		1.08	.72		1.80	2.41	
	0900	26 gauge		795	.040		.94	.72		1.66	2.26	
	1000	24 gauge		790	.041		1.11	.73		1.84	2.46	
	1020	20 gauge	↓	785	.041	↓	1.37	.73		2.10	2.75	
900	0010	**WOOD SIDING, BOARDS**										900
	2000	Board & batten, cedar, "B" grade, 1" x 10"	1 Carp	400	.020	S.F.	1.98	.39		2.37	2.86	
	2200	Redwood, clear, vertical grain, 1" x 10"		400	.020		3.59	.39		3.98	4.63	
	2400	White pine, #2 & better, 1" x 10"		400	.020		.70	.39		1.09	1.45	
	2410	Board & batten siding, white pine #2, 1" x 12"		450	.018		.80	.35		1.15	1.48	
	3200	Wood, cedar bevel, A grade, 1/2" x 6"	↓	250	.032	↓	1.90	.63		2.53	3.17	

07400 | Roofing & Siding Panels

07460 | Siding

			CREW	DAILY OUTPUT	LABOR-HOURS	UNIT	MAT.	LABOR	EQUIP.	TOTAL	TOTAL INCL O&P	
900	3300	1/2" x 8"	1 Carp	275	.029	S.F.	1.56	.57		2.13	2.70	900
	3500	3/4" x 10", clear grade		300	.027		2.92	.53		3.45	4.11	
	3600	"B" grade		300	.027		2.89	.53		3.42	4.08	
	3800	Cedar, rough sawn, 1" x 4", A grade, natural		240	.033		2.75	.66		3.41	4.16	
	3900	Stained		240	.033		3.10	.66		3.76	4.54	
	4100	1" x 12", board & batten, #3 & Btr., natural		260	.031		2.08	.61		2.69	3.33	
	4200	Stained		260	.031		2.42	.61		3.03	3.70	
	4400	1" x 8" channel siding, #3 & Btr., natural		250	.032		2.02	.63		2.65	3.30	
	4500	Stained		250	.032		2.30	.63		2.93	3.61	
	4700	Redwood, clear, beveled, vertical grain, 1/2" x 4"		200	.040		3.22	.79		4.01	4.89	
	4750	1/2" x 6"		225	.036		2.70	.70		3.40	4.17	
	4800	1/2" x 8"		250	.032		2.19	.63		2.82	3.49	
	5000	3/4" x 10"		300	.027		3.57	.53		4.10	4.83	
	5200	Channel siding, 1" x 10", B grade		285	.028		2.30	.55		2.85	3.48	
	5250	Redwood, T&G boards, B grade, 1" x 4"	2 Carp	300	.053		2.74	1.05		3.79	4.81	
	5270	1" x 8"	"	375	.043		2.36	.84		3.20	4.04	
	5400	White pine, rough sawn, 1" x 8", natural	1 Carp	275	.029		.69	.57		1.26	1.74	
	5500	Stained	"	275	.029		1.02	.57		1.59	2.10	
	5600	Tongue and groove, 1" x 8", horizontal	2 Carp	375	.043		.63	.84		1.47	2.13	
950	0010	**WOOD PRODUCT SIDING**										950
	0030	Lap siding, hardboard, 7/16" x 8", primed										
	0050	Wood grain texture finish	2 Carp	650	.025	S.F.	1.04	.48		1.52	1.97	
	0100	Panels, 7/16" thick, smooth, textured or grooved, primed		700	.023		.74	.45		1.19	1.58	
	0200	Stained		700	.023		.88	.45		1.33	1.74	
	0700	Particle board, overlaid, 3/8" thick		750	.021		.63	.42		1.05	1.41	
	0900	Plywood, medium density overlaid, 3/8" thick		750	.021		1.02	.42		1.44	1.84	
	1000	1/2" thick		700	.023		1.19	.45		1.64	2.08	
	1100	3/4" thick		650	.025		1.56	.48		2.04	2.55	
	1600	Texture 1-11, cedar, 5/8" thick, natural		675	.024		1.09	.47		1.56	2	
	1700	Factory stained		675	.024		1.74	.47		2.21	2.71	
	1900	Texture 1-11, fir, 5/8" thick, natural		675	.024		1.01	.47		1.48	1.91	
	2000	Factory stained		675	.024		1.14	.47		1.61	2.05	
	2050	Texture 1-11, S.Y.P., 5/8" thick, natural		675	.024		.85	.47		1.32	1.74	
	2100	Factory stained		675	.024		.93	.47		1.40	1.82	
	2200	Rough sawn cedar, 3/8" thick, natural		675	.024		1.14	.47		1.61	2.05	
	2300	Factory stained		675	.024		1.26	.47		1.73	2.19	
	2500	Rough sawn fir, 3/8" thick, natural		675	.024		.61	.47		1.08	1.47	
	2600	Factory stained		675	.024		.68	.47		1.15	1.55	
	2800	Redwood, textured siding, 5/8" thick		675	.024		1.89	.47		2.36	2.88	

07500 | Membrane Roofing

07510 | Built-Up Bituminous Roofing

			CREW	DAILY OUTPUT	LABOR-HOURS	UNIT	MAT.	LABOR	EQUIP.	TOTAL	TOTAL INCL O&P	
050	0010	**ASPHALT** Coated felt, #30, 2 sq per roll, not mopped	1 Rofc	58	.138	Sq.	5.45	2.35		7.80	10.40	050
	0200	#15, 4 sq per roll, plain or perforated, not mopped		58	.138		2.59	2.35		4.94	7.25	
	0250	Perforated		58	.138		2.59	2.35		4.94	7.25	
	0300	Roll roofing, smooth, #65		15	.533		12	9.10		21.10	30	
	0500	#90		15	.533		13.65	9.10		22.75	32	
	0520	Mineralized		15	.533		14.75	9.10		23.85	33	

For expanded coverage of these items see *Means Building Construction Cost Data 2000*

07500 | Membrane Roofing

07510 | Built-Up Bituminous Roofing

			CREW	DAILY OUTPUT	LABOR-HOURS	UNIT	2000 BARE COSTS MAT.	LABOR	EQUIP.	TOTAL	TOTAL INCL O&P	
050	0540	D.C. (Double coverage), 19" selvage edge	1 Rofc	10	.800	Sq.	27.50	13.65		41.15	55.50	050
	0580	Adhesive (lap cement)				Gal.	3.68			3.68	4.05	
300	0010	**BUILT-UP ROOFING**										300
	0120	Asphalt flood coat with gravel/slag surfacing, not including										
	0140	Insulation, flashing or wood nailers										
	0200	Asphalt base sheet, 3 plies #15 asphalt felt, mopped	G-1	22	2.545	Sq.	35.50	41	19	95.50	137	
	0350	On nailable decks		21	2.667		38.50	43	19.90	101.40	145	
	0500	4 plies #15 asphalt felt, mopped		20	2.800		49.50	45	21	115.50	161	
	0550	On nailable decks		19	2.947		45	47.50	22	114.50	162	
	0700	Coated glass base sheet, 2 plies glass (type IV), mopped		22	2.545		37.50	41	19	97.50	139	
	0850	3 plies glass, mopped		20	2.800		44	45	21	110	156	
	0950	On nailable decks		19	2.947		42	47.50	22	111.50	159	
	1100	4 plies glass fiber felt (type IV), mopped		20	2.800		53	45	21	119	166	
	1150	On nailable decks		19	2.947		48.50	47.50	22	118	166	
	1200	Coated & saturated base sheet, 3 plies #15 asph. felt, mopped		20	2.800		42	45	21	108	153	
	1250	On nailable decks		19	2.947		39.50	47.50	22	109	156	
	1300	4 plies #15 asphalt felt, mopped		22	2.545		48	41	19	108	151	
	2000	Asphalt flood coat, smooth surface										
	2200	Asphalt base sheet & 3 plies #15 asphalt felt, mopped	G-1	24	2.333	Sq.	35.50	37.50	17.45	90.45	128	
	2400	On nailable decks		23	2.435		33.50	39	18.20	90.70	130	
	2600	4 plies #15 asphalt felt, mopped		24	2.333		41.50	37.50	17.45	96.45	135	
	2700	On nailable decks		23	2.435		39.50	39	18.20	96.70	137	
	2900	Coated glass fiber base sheet, mopped, and 2 plies of										
	2910	glass fiber felt (type IV)	G-1	25	2.240	Sq.	32	36	16.75	84.75	121	
	3100	On nailable decks		24	2.333		30.50	37.50	17.45	85.45	123	
	3200	3 plies, mopped		23	2.435		39	39	18.20	96.20	136	
	3300	On nailable decks		22	2.545		36.50	41	19	96.50	138	
	3800	4 plies glass fiber felt (type IV), mopped		23	2.435		45.50	39	18.20	102.70	143	
	3900	On nailable decks		22	2.545		43.50	41	19	103.50	145	
	4000	Coated & saturated base sheet, 3 plies #15 asph. felt, mopped		24	2.333		36.50	37.50	17.45	91.45	129	
	4200	On nailable decks		23	2.435		34.50	39	18.20	91.70	131	
	4300	4 plies #15 organic felt, mopped		22	2.545		42.50	41	19	102.50	145	
	4500	Coal tar pitch with gravel/slag surfacing										
	4600	4 plies #15 tarred felt, mopped	G-1	21	2.667	Sq.	101	43	19.90	163.90	213	
	4800	3 plies glass fiber felt (type IV), mopped	"	19	2.947	"	83	47.50	22	152.50	204	
	5000	Coated glass fiber base sheet, and 2 plies of										
	5010	glass fiber felt, (type IV), mopped	G-1	19	2.947	Sq.	82.50	47.50	22	152	204	
	5300	On nailable decks		18	3.111		73	50	23	146	200	
	5600	4 plies glass fiber felt (type IV), mopped		21	2.667		113	43	19.90	175.90	226	
	5800	On nailable decks		20	2.800		103	45	21	169	221	
400	0010	**CANTS** 4" x 4", treated timber, cut diagonally	1 Rofc	325	.025	L.F.	.80	.42		1.22	1.66	400
	0100	Foamglass		325	.025		1.92	.42		2.34	2.89	
	0300	Mineral or fiber, trapezoidal, 1"x 4" x 48"		325	.025		.17	.42		.59	.97	
	0400	1-1/2" x 5-5/8" x 48"		325	.025		.29	.42		.71	1.10	
700	0010	**FELT** Glass fibered, #15, no mopping	1 Rofc	58	.138	Sq.	3	2.35		5.35	7.70	700
	0300	Base sheet, #45, channel vented		58	.138		16.50	2.35		18.85	22.50	
	0400	#50, coated		58	.138		7.65	2.35		10	12.80	
	0500	Cap, mineral surfaced		58	.138		16.50	2.35		18.85	22.50	
	0600	Flashing membrane, #65		16	.500		21	8.50		29.50	39.50	
	0800	Coal tar fibered, #15, no mopping		58	.138		7.90	2.35		10.25	13.10	
	0900	Asphalt felt, #15, 4 sq per roll, no mopping		58	.138		2.59	2.35		4.94	7.25	
	1100	#30, 2 sq per roll		58	.138		5.45	2.35		7.80	10.40	
	1200	Double coated, #33		58	.138		5.90	2.35		8.25	10.90	
	1400	#40, base sheet		58	.138		5.50	2.35		7.85	10.45	
	1450	Coated and saturated		58	.138		6.45	2.35		8.80	11.50	
	1500	Tarred felt, organic, #15, 4 sq rolls		58	.138		7.80	2.35		10.15	13	

Important: See the Reference Section for critical supporting data - Reference Nos., Crews, & Location Factors

07500 | Membrane Roofing

07510 | Built-Up Bituminous Roofing

			CREW	DAILY OUTPUT	LABOR-HOURS	UNIT	MAT.	LABOR	EQUIP.	TOTAL	TOTAL INCL O&P	
700	1550	#30, 2 sq roll	1 Rofc	58	.138	Sq.	15.60	2.35		17.95	21.50	700
	1700	Add for mopping above felts, per ply, asphalt, 24 lb per sq	G-1	192	.292		3.12	4.70	2.18	10	14.55	
	1800	Coal tar mopping, 30 lb per sq		186	.301		8.45	4.85	2.25	15.55	21	
	1900	Flood coat, with asphalt, 60 lb per sq		60	.933		7.80	15.05	6.95	29.80	44.50	
	2000	With coal tar, 75 lb per sq		56	1		21	16.10	7.45	44.55	61.50	

07520 | Cold Applied Bituminous Roofing

			CREW	DAILY OUTPUT	LABOR-HOURS	UNIT	MAT.	LABOR	EQUIP.	TOTAL	TOTAL INCL O&P	
200	0010	**COLD APPLIED** 3-ply system (components listed below)	G-5	50	.800	Sq.		12.60	3.43	16.03	27.50	200
	0100	Spunbond poly. fabric, 1.35 oz/SY, 36"W, 10.8 Sq/roll				Ea.	123			123	135	
	0200	49" wide, 14.6 Sq./roll					169			169	186	
	0300	2.10 oz./S.Y., 36" wide, 10.8 Sq./roll					185			185	203	
	0400	49" wide, 14.6 Sq./roll					250			250	275	
	0500	Base & finish coat, 3 gal./Sq., 5 gal./can				Gal.	3.25			3.25	3.58	
	0600	Coating, ceramic granules, 1/2 Sq./bag				Ea.	11.50			11.50	12.65	
	0700	Aluminum, 2 gal./Sq.				Gal.	9.25			9.25	10.20	
	0800	Emulsion, fibered or non-fibered, 4 gal./Sq.				"	4.25			4.25	4.68	

07530 | Elastomeric Membrane Roofing

			CREW	DAILY OUTPUT	LABOR-HOURS	UNIT	MAT.	LABOR	EQUIP.	TOTAL	TOTAL INCL O&P	
800	0010	**SINGLE-PLY MEMBRANE**										800
	0800	Chlorosulfonated polyethylene-hypalon (CSPE), 45 mils,										
	0900	0.29 P.S.F., fully adhered	G-5	26	1.538	Sq.	122	24	6.60	152.60	186	
	1100	Loose-laid & ballasted with stone (10 P.S.F.)		51	.784		129	12.35	3.36	144.71	169	
	1200	Partially adhered with fastening strips		35	1.143		124	18	4.90	146.90	175	
	1300	Plates with adhesive attachment		35	1.143		122	18	4.90	144.90	173	
	3500	Ethylene propylene diene monomer (EPDM), 45 mils, 0.28 P.S.F.										
	3600	Loose-laid & ballasted with stone (10 P.S.F.)	G-5	51	.784	Sq.	62.50	12.35	3.36	78.21	95	
	3700	Partially adhered		35	1.143		53.50	18	4.90	76.40	98	
	3800	Fully adhered with adhesive		26	1.538		76	24	6.60	106.60	136	
	4500	60 mils, 0.40 P.S.F.										
	4600	Loose-laid & ballasted with stone (10 P.S.F.)	G-5	51	.784	Sq.	74.50	12.35	3.36	90.21	109	
	4700	Partially adhered		35	1.143		65	18	4.90	87.90	110	
	4800	Fully adhered with adhesive		26	1.538		87	24	6.60	117.60	148	
	4810	45 mil, .28 PSF, membrane only					34			34	37	
	4820	60 mil, .40 PSF, membrane only					44			44	48.50	
	4850	Seam tape for membrane, 4" x 100' roll				Ea.	67			67	73.50	
	4900	Batten strips, 10' sections					2.45			2.45	2.70	
	4910	Cover tape for batten strips, 6" x 100' roll					126			126	139	
	4930	Plate anchors				M	123			123	135	
	4970	Adhesive for fully adhered systems, 60 S.F./gal.				Gal.	13.60			13.60	14.95	
	7500	Polyisobutylene (PIB), 100 mils, 0.57 P.S.F.										
	7600	Loose-laid & ballasted with stone/gravel (10 P.S.F.)	G-5	51	.784	Sq.	123	12.35	3.36	138.71	162	
	7700	Partially adhered with adhesive		35	1.143		154	18	4.90	176.90	208	
	7800	Hot asphalt attachment		35	1.143		147	18	4.90	169.90	201	
	7900	Fully adhered with contact cement		26	1.538		159	24	6.60	189.60	227	
	8160											
	8200	Polyvinyl chloride (PVC), heat welded seams										
	8700	Reinforced, 48 mils, 0.33 P.S.F.										
	8750	Loose-laid & ballasted with stone/gravel (12 P.S.F.)	G-5	51	.784	Sq.	93.50	12.35	3.36	109.21	130	
	8800	Partially adhered with mechanical fasteners		35	1.143		83.50	18	4.90	106.40	131	
	8850	Fully adhered with adhesive		26	1.538		116	24	6.60	146.60	179	
	8860	Reinforced, 60 mils, .40 P.S.F.										
	8870	Loose-laid & ballasted with stone/gravel (12 P.S.F.)	G-5	51	.784	Sq.	89.50	12.35	3.36	105.21	125	
	8880	Partially adhered with mechanical fasteners		35	1.143		80	18	4.90	102.90	127	
	8890	Fully adhered with adhesive		26	1.538		112	24	6.60	142.60	175	

For expanded coverage of these items see *Means Building Construction Cost Data 2000*

07500 | Membrane Roofing

07550 | Modified Bit. Membrane Roofing

			CREW	DAILY OUTPUT	LABOR-HOURS	UNIT	MAT.	LABOR	EQUIP.	TOTAL	TOTAL INCL O&P
500	0010	**MODIFIED BITUMEN ROOFING**									
	0020	Base sheet, #15 glass fiber felt, nailed to deck (R07550-030)	1 Rofc	58	.138	Sq.	3.73	2.35		6.08	8.50
	0030	Spot mopped to deck	G-1	295	.190		4.56	3.06	1.42	9.04	12.25
	0040	Fully mopped to deck	"	192	.292		6.10	4.70	2.18	12.98	17.90
	0050	#15 organic felt, nailed to deck	1 Rofc	58	.138		3.32	2.35		5.67	8.05
	0060	Spot mopped to deck	G-1	295	.190		4.15	3.06	1.42	8.63	11.85
	0070	Fully mopped to deck	"	192	.292		5.70	4.70	2.18	12.58	17.45
	0080	SBS modified, granule surf cap sheet, polyester rein., mopped									
	0600	150 mils	G-1	2,000	.028	S.F.	.38	.45	.21	1.04	1.49
	1100	160 mils		2,000	.028		.57	.45	.21	1.23	1.70
	1500	Glass fiber reinforced, mopped, 160 mils		2,000	.028		.36	.45	.21	1.02	1.47
	1600	Smooth surface cap sheet, mopped, 145 mils		2,100	.027		.36	.43	.20	.99	1.42
	1700	Smooth surface flashing, 145 mils		1,260	.044		.36	.72	.33	1.41	2.09
	1800	150 mils		1,260	.044		.35	.72	.33	1.40	2.08
	1900	Granular surface flashing, 150 mils		1,260	.044		.38	.72	.33	1.43	2.11
	2000	160 mils		1,260	.044		.57	.72	.33	1.62	2.32
	2100	APP mod., smooth surf. cap sheet, poly. reinf., torched, 160 mils	G-5	2,100	.019		.35	.30	.08	.73	1.04
	2150	170 mils		2,100	.019		.39	.30	.08	.77	1.08
	2200	Granule surface cap sheet, poly. reinf., torched, 180 mils		2,000	.020		.45	.31	.09	.85	1.18
	2250	Smooth surface flashing, torched, 160 mils		1,260	.032		.35	.50	.14	.99	1.47
	2300	170 mils		1,260	.032		.39	.50	.14	1.03	1.51
	2350	Granule surface flashing, torched, 180 mils		1,260	.032		.45	.50	.14	1.09	1.58
	2400	Fibrated aluminum coating	1 Rofc	3,800	.002		.09	.04		.13	.17

07580 | Roll Roofing

			CREW	DAILY OUTPUT	LABOR-HOURS	UNIT	MAT.	LABOR	EQUIP.	TOTAL	TOTAL INCL O&P
200	0010	**ROLL ROOFING**									
	0100	Asphalt, mineral surface									
	0200	1 ply #15 organic felt, 1 ply mineral surfaced									
	0300	Selvage roofing, lap 19", nailed & mopped	G-1	27	2.074	Sq.	33.50	33.50	15.50	82.50	116
	0400	3 plies glass fiber felt (type IV), 1 ply mineral surfaced									
	0500	Selvage roofing, lapped 19", mopped	G-1	25	2.240	Sq.	48.50	36	16.75	101.25	139
	0600	Coated glass fiber base sheet, 2 plies of glass fiber									
	0700	Felt (type IV), 1 ply mineral surfaced selvage									
	0800	Roofing, lapped 19", mopped	G-1	25	2.240	Sq.	53.50	36	16.75	106.25	144
	0900	On nailable decks	"	24	2.333	"	50.50	37.50	17.45	105.45	145
	1000	3 plies glass fiber felt (type III), 1 ply mineral surfaced									
	1100	Selvage roofing, lapped 19", mopped	G-1	25	2.240	Sq.	48.50	36	16.75	101.25	139

07590 | Roof Maintenance & Repairs

			CREW	DAILY OUTPUT	LABOR-HOURS	UNIT	MAT.	LABOR	EQUIP.	TOTAL	TOTAL INCL O&P
300	0010	**ROOF COATINGS** Asphalt				Gal.	2.95			2.95	3.25
	0200	Asphalt base, fibered aluminum coating					8.90			8.90	9.75
	0300	Asphalt primer, 5 gallon					3.27			3.27	3.60
	0600	Coal tar pitch, 200 lb. barrels				Ton	565			565	620
	0700	Tar roof cement, 5 gal. lots				Gal.	5.75			5.75	6.35
	0800	Glass fibered roof & patching cement, 5 gallon				"	3.50			3.50	3.85
	0900	Reinforcing glass membrane, 450 S.F./roll				Ea.	45.50			45.50	50
	1000	Neoprene roof coating, 5 gal, 2 gal/sq				Gal.	19.95			19.95	22
	1100	Roof patch & flashing cement, 5 gallon					17.45			17.45	19.20
	1200	Roof resaturant, glass fibered, 3 gal/sq					7.30			7.30	8.05
	1300	Mineral rubber, 3 gal/sq					4.71			4.71	5.20

07600 | Flashing & Sheet Metal

07610 | Sheet Metal Roofing

			CREW	DAILY OUTPUT	LABOR-HOURS	UNIT	2000 BARE COSTS MAT.	LABOR	EQUIP.	TOTAL	TOTAL INCL O&P	
300	0010	COPPER ROOFING Batten seam, over 10 sq, 16 oz, 130 lb/sq	1 Shee	1.10	7.273	Sq.	395	156		551	700	300
	0200	18 oz, 145 lb per sq		1	8		440	172		612	775	
	0400	Standing seam, over 10 squares, 16 oz, 125 lb per sq		1.30	6.154		380	132		512	645	
	0600	18 oz, 140 lb per sq		1.20	6.667		425	143		568	710	
	0900	Flat seam, over 10 squares, 16 oz, 115 lb per sq	↓	1.20	6.667		350	143		493	625	
	1200	For abnormal conditions or small areas, add					25%	100%				
	1300	For lead-coated copper, add				↓	25%					
500	0010	LEAD ROOFING 5 lb. per SF, batten seam	1 Shee	1.20	6.667	Sq.	375	143		518	655	500
	0100	Flat seam	"	1.30	6.154	"	375	132		507	640	
900	0010	ZINC Copper alloy roofing, batten seam, .020" thick	1 Shee	1.20	6.667	Sq.	510	143		653	800	900
	0100	.027" thick		1.15	6.957		615	150		765	935	
	0300	.032" thick		1.10	7.273		695	156		851	1,025	
	0400	.040" thick	↓	1.05	7.619		820	164		984	1,175	
	0600	For standing seam construction, deduct					2%					
	0700	For flat seam construction, deduct				↓	3%					

07650 | Flexible Flashing

			CREW	DAILY OUTPUT	LABOR-HOURS	UNIT	MAT.	LABOR	EQUIP.	TOTAL	TOTAL INCL O&P	
600	0010	FLASHING Aluminum, mill finish, .013" thick	1 Shee	145	.055	S.F.	.34	1.19		1.53	2.37	600
	0030	.016" thick		145	.055		.50	1.19		1.69	2.55	
	0060	.019" thick		145	.055		.75	1.19		1.94	2.83	
	0100	.032" thick		145	.055		.99	1.19		2.18	3.09	
	0200	.040" thick		145	.055		1.36	1.19		2.55	3.50	
	0300	.050" thick		145	.055	↓	1.80	1.19		2.99	3.98	
	0325	Mill finish 5" x 7" step flashing, .016" thick		1,920	.004	Ea.	.10	.09		.19	.26	
	0350	Mill finish 12" x 12" step flashing, .016" thick	↓	1,600	.005	"	.40	.11		.51	.62	
	0400	Painted finish, add				S.F.	.24			.24	.26	
	0500	Fabric-backed 2 sides, .004" thick	1 Shee	330	.024		.93	.52		1.45	1.90	
	0700	.005" thick		330	.024		1.09	.52		1.61	2.08	
	0750	Mastic-backed, self adhesive		460	.017		2.22	.37		2.59	3.07	
	0800	Mastic-coated 2 sides, .004" thick		330	.024		.93	.52		1.45	1.90	
	1000	.005" thick		330	.024		1.09	.52		1.61	2.08	
	1100	.016" thick	↓	330	.024		1.20	.52		1.72	2.20	
	1300	Asphalt flashing cement, 5 gallon				Gal.	3.34			3.34	3.67	
	1600	Copper, 16 oz, sheets, under 1000 lbs.	1 Shee	115	.070	S.F.	3.05	1.50		4.55	5.90	
	1700	Over 4000 lbs.		155	.052		2.85	1.11		3.96	5	
	1900	20 oz sheets, under 1000 lbs.		110	.073		3.82	1.56		5.38	6.85	
	2000	Over 4000 lbs.		145	.055		3.55	1.19		4.74	5.90	
	2200	24 oz sheets, under 1000 lbs.		105	.076		4.60	1.64		6.24	7.80	
	2500	32 oz sheets, under 1000 lbs.		100	.080	↓	6.10	1.72		7.82	9.60	
	2700	W shape for valleys, 16 oz, 24" wide		100	.080	L.F.	5.90	1.72		7.62	9.40	
	2800	Copper, paperbacked 1 side, 2 oz		330	.024	S.F.	.86	.52		1.38	1.83	
	2900	3 oz		330	.024		1.12	.52		1.64	2.11	
	3100	Paperbacked 2 sides, 2 oz		330	.024		.88	.52		1.40	1.85	
	3150	3 oz		330	.024		1.11	.52		1.63	2.10	
	3200	5 oz		330	.024		1.70	.52		2.22	2.75	
	3400	Mastic-backed 2 sides, copper, 2 oz		330	.024		1.02	.52		1.54	2	
	3500	3 oz		330	.024		1.25	.52		1.77	2.26	
	3700	5 oz		330	.024		1.85	.52		2.37	2.92	
	3800	Fabric-backed 2 sides, copper, 2 oz		330	.024		1.09	.52		1.61	2.08	
	4000	3 oz		330	.024		1.35	.52		1.87	2.37	
	4100	5 oz		330	.024		1.90	.52		2.42	2.97	
	4300	Copper-clad stainless steel, .015" thick, under 500 lbs.		115	.070		3.10	1.50		4.60	5.95	
	4600	.018" thick, under 500 lbs.		100	.080		4.10	1.72		5.82	7.40	
	4750	EPDM Cured		285	.028		.95	.60		1.55	2.07	
	4800	Uncured	↓	285	.028	↓	1.06	.60		1.66	2.19	

For expanded coverage of these items see *Means Building Construction Cost Data 2000*

07600 | Flashing & Sheet Metal

07650 | Flexible Flashing

			CREW	DAILY OUTPUT	LABOR-HOURS	UNIT	2000 BARE COSTS MAT.	LABOR	EQUIP.	TOTAL	TOTAL INCL O&P	
600	4900	Fabric, asphalt-saturated cotton, specification grade	1 Rofc	35	.229	S.Y.	1.93	3.90		5.83	9.35	600
	5000	Utility grade		35	.229		1.22	3.90		5.12	8.60	
	5200	Open-mesh fabric, saturated, 40 oz per S.Y.		35	.229		1.35	3.90		5.25	8.75	
	5300	Close-mesh fabric, saturated, 17 oz per S.Y.		35	.229		1.42	3.90		5.32	8.80	
	5500	Fiberglass, resin-coated		35	.229		1.16	3.90		5.06	8.55	
	5600	Asphalt-coated, 40 oz per S.Y.		35	.229		7.90	3.90		11.80	15.95	
	5800	Lead, 2.5 lb. per SF, up to 12" wide		135	.059	S.F.	3.25	1.01		4.26	5.45	
	5900	Over 12" wide		135	.059		3.25	1.01		4.26	5.45	
	6100	Lead-coated copper, fabric-backed, 2 oz	1 Shee	330	.024		1.41	.52		1.93	2.43	
	6200	5 oz		330	.024		1.66	.52		2.18	2.71	
	6400	Mastic-backed 2 sides, 2 oz		330	.024		1.10	.52		1.62	2.09	
	6500	5 oz		330	.024		1.39	.52		1.91	2.41	
	6700	Paperbacked 1 side, 2 oz		330	.024		.95	.52		1.47	1.93	
	6800	3 oz		330	.024		1.12	.52		1.64	2.11	
	7000	Paperbacked 2 sides, 2 oz		330	.024		.98	.52		1.50	1.96	
	7100	5 oz		330	.024		1.58	.52		2.10	2.62	
	7150	Neoprene, 60 mil		285	.028		1.07	.60		1.67	2.20	
	7160	Self-curing		285	.028		1.40	.60		2	2.56	
	7170	Uncured		285	.028		1.35	.60		1.95	2.51	
	7300	Polyvinyl chloride, black, .010" thick	1 Rofc	285	.028		.14	.48		.62	1.04	
	7400	.020" thick		285	.028		.19	.48		.67	1.10	
	7600	.030" thick		285	.028		.29	.48		.77	1.21	
	7700	.056" thick		285	.028		.70	.48		1.18	1.66	
	7900	Black or white for exposed roofs, .060" thick		285	.028		1.55	.48		2.03	2.59	
	8060	PVC tape, 5" x 45 mils, for joint covers, 100 L.F./roll				Ea.	79.50			79.50	87	
	8100	Rubber, butyl, 1/32" thick	1 Rofc	285	.028	S.F.	.70	.48		1.18	1.66	
	8200	1/16" thick		285	.028		1.05	.48		1.53	2.04	
	8300	Neoprene, cured, 1/16" thick		285	.028		1.48	.48		1.96	2.52	
	8400	1/8" thick		285	.028		2.99	.48		3.47	4.18	
	8500	Shower pan, bituminous membrane, 7 oz	1 Shee	155	.052		1.08	1.11		2.19	3.06	
	8550	3 ply copper and fabric, 3 oz		155	.052		1.60	1.11		2.71	3.63	
	8600	7 oz		155	.052		3.30	1.11		4.41	5.50	
	8650	Copper, 16 oz		100	.080		3.05	1.72		4.77	6.25	
	8700	Lead on copper and fabric, 5 oz		155	.052		1.66	1.11		2.77	3.70	
	8800	7 oz		155	.052		2.87	1.11		3.98	5.05	
	8900	Stainless steel sheets, 32 ga, .010" thick		155	.052		2.16	1.11		3.27	4.25	
	9000	28 ga, .015" thick		155	.052		2.55	1.11		3.66	4.67	
	9100	26 ga, .018" thick		155	.052		3.16	1.11		4.27	5.35	
	9200	24 ga, .025" thick		155	.052		4.10	1.11		5.21	6.40	
	9290	For mechanically keyed flashing, add					40%					
	9300	Stainless steel, paperbacked 2 sides, .005" thick	1 Shee	330	.024	S.F.	1.97	.52		2.49	3.05	
	9320	Steel sheets, galvanized, 20 gauge		130	.062		.72	1.32		2.04	3.02	
	9340	30 gauge		160	.050		.30	1.08		1.38	2.14	
	9400	Terne coated stainless steel, .015" thick, 28 ga		155	.052		3.97	1.11		5.08	6.25	
	9500	.018" thick, 26 ga		155	.052		4.48	1.11		5.59	6.80	
	9600	Zinc and copper alloy (brass), .020" thick		155	.052		3.20	1.11		4.31	5.40	
	9700	.027" thick		155	.052		4.28	1.11		5.39	6.60	
	9800	.032" thick		155	.052		5	1.11		6.11	7.35	
	9900	.040" thick		155	.052		6.10	1.11		7.21	8.55	

07700 | Roof Specialties & Accessories

07710 | Manufactured Roof Specialties

			CREW	DAILY OUTPUT	LABOR-HOURS	UNIT	2000 BARE COSTS MAT.	LABOR	EQUIP.	TOTAL	TOTAL INCL O&P	
400	0010	**DOWNSPOUTS** Aluminum 2" x 3", .020" thick, embossed	1 Shee	190	.042	L.F.	.71	.91		1.62	2.31	400
	0100	Enameled		190	.042		.69	.91		1.60	2.29	
	0300	Enameled, .024" thick, 2" x 3"		180	.044		1.09	.96		2.05	2.81	
	0400	3" x 4"		140	.057		1.56	1.23		2.79	3.79	
	0600	Round, corrugated aluminum, 3" diameter, .020" thick		190	.042		.85	.91		1.76	2.47	
	0700	4" diameter, .025" thick		140	.057		1.41	1.23		2.64	3.62	
	0900	Wire strainer, round, 2" diameter		155	.052	Ea.	1.75	1.11		2.86	3.80	
	1000	4" diameter		155	.052		1.82	1.11		2.93	3.87	
	1200	Rectangular, perforated, 2" x 3"		145	.055		2.15	1.19		3.34	4.37	
	1300	3" x 4"		145	.055		3.10	1.19		4.29	5.40	
	1500	Copper, round, 16 oz., stock, 2" diameter		190	.042	L.F.	4.94	.91		5.85	7	
	1600	3" diameter		190	.042		3.79	.91		4.70	5.70	
	1800	4" diameter		145	.055		4.60	1.19		5.79	7.05	
	1900	5" diameter		130	.062		6.55	1.32		7.87	9.45	
	2100	Rectangular, corrugated copper, stock, 2" x 3"		190	.042		3.50	.91		4.41	5.40	
	2200	3" x 4"		145	.055		4.36	1.19		5.55	6.80	
	2400	Rectangular, plain copper, stock, 2" x 3"		190	.042		4.71	.91		5.62	6.75	
	2500	3" x 4"		145	.055		6.20	1.19		7.39	8.80	
	2700	Wire strainers, rectangular, 2" x 3"		145	.055	Ea.	2.76	1.19		3.95	5.05	
	2800	3" x 4"		145	.055		4.37	1.19		5.56	6.80	
	3000	Round, 2" diameter		145	.055		2.59	1.19		3.78	4.85	
	3100	3" diameter		145	.055		3.62	1.19		4.81	6	
	3300	4" diameter		145	.055		5.60	1.19		6.79	8.15	
	3400	5" diameter		115	.070		8.10	1.50		9.60	11.45	
	3600	Lead-coated copper, round, stock, 2" diameter		190	.042	L.F.	4.95	.91		5.86	7	
	3700	3" diameter		190	.042		4.51	.91		5.42	6.50	
	3900	4" diameter		145	.055		5.65	1.19		6.84	8.25	
	4300	Rectangular, corrugated, stock, 2" x 3"		190	.042		4.09	.91		5	6.05	
	4500	Plain, stock, 2" x 3"		190	.042		6.95	.91		7.86	9.20	
	4600	3" x 4"		145	.055		7.55	1.19		8.74	10.30	
	4800	Steel, galvanized, round, corrugated, 2" or 3" diam, 28 ga		190	.042		.68	.91		1.59	2.28	
	4900	4" diameter, 28 gauge		145	.055		.90	1.19		2.09	2.99	
	5700	Rectangular, corrugated, 28 gauge, 2" x 3"		190	.042		.53	.91		1.44	2.11	
	5800	3" x 4"		145	.055		1.48	1.19		2.67	3.63	
	6000	Rectangular, plain, 28 gauge, galvanized, 2" x 3"		190	.042		.80	.91		1.71	2.41	
	6100	3" x 4"		145	.055		1.23	1.19		2.42	3.35	
	6300	Epoxy painted, 24 gauge, corrugated, 2" x 3"		190	.042		1.01	.91		1.92	2.64	
	6400	3" x 4"		145	.055		1.95	1.19		3.14	4.15	
	6600	Wire strainers, rectangular, 2" x 3"		145	.055	Ea.	1.62	1.19		2.81	3.78	
	6700	3" x 4"		145	.055		2.61	1.19		3.80	4.87	
	6900	Round strainers, 2" or 3" diameter		145	.055		1.20	1.19		2.39	3.32	
	7000	4" diameter		145	.055		1.42	1.19		2.61	3.56	
	7200	5" diameter		145	.055		2.20	1.19		3.39	4.42	
	7300	6" diameter		115	.070		2.63	1.50		4.13	5.40	
	8200	Vinyl, rectangular, 2" x 3"		210	.038	L.F.	.72	.82		1.54	2.17	
	8300	Round, 2-1/2"		220	.036	"	.72	.78		1.50	2.11	
450	0010	**DRIP EDGE**, aluminum, .016" thick, 5" wide, mill finish	1 Carp	400	.020	L.F.	.20	.39		.59	.90	450
	0100	White finish		400	.020		.22	.39		.61	.92	
	0200	8" wide, mill finish		400	.020		.30	.39		.69	1.01	
	0300	Ice belt, 28" wide, mill finish		100	.080		3.42	1.58		5	6.45	
	0310	Vented, mill finish		400	.020		1.38	.39		1.77	2.20	
	0320	Painted finish		400	.020		1.50	.39		1.89	2.33	
	0400	Galvanized, 5" wide		400	.020		.22	.39		.61	.92	
	0500	8" wide, mill finish		400	.020		.33	.39		.72	1.04	
	0510	Rake edge, aluminum, 1-1/2" x 1-1/2"		400	.020		.13	.39		.52	.82	
	0520	3-1/2" x 1-1/2"		400	.020		.19	.39		.58	.89	

For expanded coverage of these items see Means Building Construction Cost Data 2000

07700 | Roof Specialties & Accessories

07710 | Manufactured Roof Specialties

			CREW	DAILY OUTPUT	LABOR-HOURS	UNIT	MAT.	LABOR	EQUIP.	TOTAL	TOTAL INCL O&P	
500	0010	**ELBOWS** Aluminum, 2" x 3", embossed	1 Shee	100	.080	Ea.	.90	1.72		2.62	3.89	500
	0100	Enameled		100	.080		1.63	1.72		3.35	4.69	
	0200	3" x 4", .025" thick, embossed		100	.080		3.15	1.72		4.87	6.35	
	0300	Enameled		100	.080		3.15	1.72		4.87	6.35	
	0400	Round corrugated, 3", embossed, .020" thick		100	.080		1.95	1.72		3.67	5.05	
	0500	4", .025" thick		100	.080		2.95	1.72		4.67	6.15	
	0600	Copper, 16 oz. round, 2" diameter		100	.080		11	1.72		12.72	15	
	0700	3" diameter		100	.080		5.05	1.72		6.77	8.45	
	0800	4" diameter		100	.080		9.50	1.72		11.22	13.35	
	1000	2" x 3" corrugated		100	.080		5.05	1.72		6.77	8.50	
	1100	3" x 4" corrugated		100	.080		9.75	1.72		11.47	13.65	
	1300	Vinyl, 2-1/2" diameter, 45° or 75°		100	.080		2	1.72		3.72	5.10	
	1400	Tee Y junction	▼	75	.107	▼	8.50	2.29		10.79	13.20	
550	0010	**GRAVEL STOP** Aluminum, .050" thick, 4" face height, mill finish	1 Shee	145	.055	L.F.	2.72	1.19		3.91	4.99	550
	0080	Duranodic finish		145	.055		3.69	1.19		4.88	6.05	
	0100	Painted		145	.055		4.26	1.19		5.45	6.70	
	1350	Galv steel, 24 ga., 4" leg, plain, with continuous cleat, 4" face		145	.055		1.50	1.19		2.69	3.65	
	1500	Polyvinyl chloride, 6" face height		135	.059		3.28	1.27		4.55	5.75	
	1800	Stainless steel, 24 ga., 6" face height	▼	135	.059	▼	7.15	1.27		8.42	10	
650	0010	**GUTTERS** Aluminum, stock units, 5" box, .027" thick, plain	1 Shee	120	.067	L.F.	.97	1.43		2.40	3.49	650
	0020	Inside corner		25	.320	Ea.	5.15	6.90		12.05	17.25	
	0030	Outside corner		25	.320	"	5.15	6.90		12.05	17.25	
	0100	Enameled		120	.067	L.F.	1.07	1.43		2.50	3.60	
	0110	Inside corner		25	.320	Ea.	5.20	6.90		12.10	17.30	
	0120	Outside corner		25	.320	"	5.20	6.90		12.10	17.30	
	0600	5" x 6" combination fascia & gutter, .032" thick, enameled	▼	60	.133	L.F.	3.45	2.87		6.32	8.65	
	3000	Vinyl, O.G., 4" wide	1 Carp	110	.073		.85	1.43		2.28	3.39	
	3100	5" wide		110	.073		1	1.43		2.43	3.55	
	3200	4" half round, stock units	▼	110	.073	▼	.68	1.43		2.11	3.20	
	3250	Joint connectors				Ea.	1.36			1.36	1.50	
	3300	Wood, clear treated cedar, fir or hemlock, 3" x 4"	1 Carp	100	.080	L.F.	6.30	1.58		7.88	9.60	
	3400	4" x 5"	"	100	.080	"	7.30	1.58		8.88	10.70	
700	0010	**GUTTER GUARD** 6" wide strip, aluminum mesh	1 Carp	500	.016	L.F.	.37	.32		.69	.95	700
	0100	Vinyl mesh	"	500	.016	"	.22	.32		.54	.78	
750	0010	**REGLET** Aluminum, .025" thick, in concrete parapet	1 Carp	225	.036	L.F.	.92	.70		1.62	2.21	750
	0100	Copper, 10 oz.		225	.036		1.59	.70		2.29	2.95	
	0300	16 oz.		225	.036		2.12	.70		2.82	3.53	
	0400	Galvanized steel, 24 gauge		225	.036		.78	.70		1.48	2.06	
	0600	Stainless steel, .020" thick		225	.036		1.57	.70		2.27	2.93	
	0700	Zinc and copper alloy, 20 oz.		225	.036		1.75	.70		2.45	3.13	
	0900	Counter flashing for above, 12" wide, .032 aluminum	1 Shee	150	.053		1.19	1.15		2.34	3.25	
	1000	Copper, 10 oz.		150	.053		3.32	1.15		4.47	5.60	
	1200	16 oz.		150	.053		3.70	1.15		4.85	6	
	1300	Galvanized steel, .020" thick		150	.053		.62	1.15		1.77	2.62	
	1500	Stainless steel, .020" thick		150	.053		2.76	1.15		3.91	4.98	
	1600	Zinc and copper alloy, 20 oz.	▼	150	.053	▼	3.11	1.15		4.26	5.35	

07720 | Roof Accessories

			CREW	DAILY OUTPUT	LABOR-HOURS	UNIT	MAT.	LABOR	EQUIP.	TOTAL	TOTAL INCL O&P	
480	0010	**PITCH POCKETS**										480
	0100	Adjustable, 4" to 7", welded corners, 4" deep	1 Rofc	48	.167	Ea.	10.50	2.84		13.34	16.85	
	0200	Side extenders, 6"	"	240	.033	"	1.70	.57		2.27	2.93	
700	0010	**ROOF HATCHES** With curb, 1" fiberglass insulation, 2'-6" x 3'-0"										700
	0500	Aluminum curb and cover	G-3	10	3.200	Ea.	405	57.50		462.50	545	

07700 | Roof Specialties & Accessories

07720 | Roof Accessories

		Crew	Daily Output	Labor-Hours	Unit	2000 Bare Costs Mat.	Labor	Equip.	Total	Total Incl O&P	
700	0520	Galvanized steel curb and aluminum cover	G-3	10	3.200	Ea.	340	57.50		397.50	475	700
	0540	Galvanized steel curb and cover		10	3.200		405	57.50		462.50	545	
	0600	2'-6" x 4'-6", aluminum curb and cover		9	3.556		560	64		624	725	
	0800	Galvanized steel curb and aluminum cover		9	3.556		475	64		539	635	
	0900	Galvanized steel curb and cover		9	3.556		615	64		679	785	
	1200	2'-6" x 8'-0", aluminum curb and cover		6.60	4.848		995	87		1,082	1,250	
	1400	Galvanized steel curb and aluminum cover		6.60	4.848		920	87		1,007	1,150	
	1500	Galvanized steel curb and cover		6.60	4.848		905	87		992	1,150	
	1800	For plexiglass panels, 2'-6" x 3'-0", add to above					345			345	380	
870	0010	**VENTS, ONE-WAY** For insul. decks, 1 per M.S.F., plastic, min.	1 Rofc	40	.200	Ea.	12.50	3.41		15.91	20	870
	0100	Maximum		20	.400		28.50	6.80		35.30	44	
	0300	Aluminum		30	.267		12.50	4.55		17.05	22	
	0800	Polystyrene baffles, 12" wide for 16" O.C. rafter spacing	1 Carp	90	.089		.45	1.75		2.20	3.50	
	0900	For 24" O.C. rafter spacing	"	110	.073		1.05	1.43		2.48	3.60	

07800 | Fire & Smoke Protection

07812 | Cementitious Fireproofing

		Crew	Daily Output	Labor-Hours	Unit	2000 Bare Costs Mat.	Labor	Equip.	Total	Total Incl O&P	
600	0010	**SPRAYED** Mineral fiber or cementitious for fireproofing,										600
	0050	not incl tamping or canvas protection										
	0100	1" thick, on flat plate steel	G-2	3,000	.008	S.F.	.42	.13	.08	.63	.77	
	0200	Flat decking		2,400	.010		.42	.16	.10	.68	.84	
	0400	Beams		1,500	.016		.42	.26	.17	.85	1.08	
	0500	Corrugated or fluted decks		1,250	.019		.63	.31	.20	1.14	1.44	
	0700	Columns, 1-1/8" thick		1,100	.022		.47	.35	.23	1.05	1.37	
	0800	2-3/16" thick		700	.034		.88	.56	.36	1.80	2.30	
	0850	For tamping, add						10%				
	0900	For canvas protection, add	G-2	5,000	.005	S.F.	.06	.08	.05	.19	.26	

07840 | Firestopping

		Crew	Daily Output	Labor-Hours	Unit	2000 Bare Costs Mat.	Labor	Equip.	Total	Total Incl O&P	
100	0010	**FIRESTOPPING** R07800-030										100
	0100	Metallic piping, non insulated										
	0110	Through walls, 2" diameter	1 Carp	16	.500	Ea.	9.60	9.85		19.45	27.50	
	0120	4" diameter		14	.571		14.65	11.25		25.90	35.50	
	0130	6" diameter		12	.667		19.70	13.15		32.85	44	
	0140	12" diameter		10	.800		35	15.75		50.75	65.50	
	0150	Through floors, 2" diameter		32	.250		5.80	4.93		10.73	14.85	
	0160	4" diameter		28	.286		8.35	5.65		14	18.85	
	0170	6" diameter		24	.333		11	6.55		17.55	23.50	
	0180	12" diameter		20	.400		18.50	7.90		26.40	34	
	0190	Metallic piping, insulated										
	0200	Through walls, 2" diameter	1 Carp	16	.500	Ea.	13.60	9.85		23.45	32	
	0210	4" diameter		14	.571		18.65	11.25		29.90	40	
	0220	6" diameter		12	.667		23.50	13.15		36.65	48.50	
	0230	12" diameter		10	.800		39	15.75		54.75	69.50	
	0240	Through floors, 2" diameter		32	.250		9.80	4.93		14.73	19.25	
	0250	4" diameter		28	.286		12.35	5.65		18	23.50	
	0260	6" diameter		24	.333		15	6.55		21.55	28	

For expanded coverage of these items see *Means Building Construction Cost Data 2000*

07800 | Fire & Smoke Protection

07840 | Firestopping

		Crew	Daily Output	Labor-Hours	Unit	Mat.	2000 Bare Costs Labor	Equip.	Total	Total Incl O&P
0270	12" diameter	1 Carp	20	.400	Ea.	18.50	7.90		26.40	34
0280	Non metallic piping, non insulated									
0290	Through walls, 2" diameter	1 Carp	12	.667	Ea.	39.50	13.15		52.65	66
0300	4" diameter		10	.800		49.50	15.75		65.25	81.50
0310	6" diameter		8	1		69	19.70		88.70	110
0330	Through floors, 2" diameter		16	.500		31	9.85		40.85	51
0340	4" diameter		6	1.333		38.50	26.50		65	87
0350	6" diameter		6	1.333		46	26.50		72.50	95.50
0370	Ductwork, insulated & non insulated, round									
0380	Through walls, 6" diameter	1 Carp	12	.667	Ea.	20	13.15		33.15	44.50
0390	12" diameter		10	.800		40	15.75		55.75	71
0400	18" diameter		8	1		65	19.70		84.70	106
0410	Through floors, 6" diameter		16	.500		11	9.85		20.85	29
0420	12" diameter		14	.571		20	11.25		31.25	41.50
0430	18" diameter		12	.667		35	13.15		48.15	61
0440	Ductwork, insulated & non insulated, rectangular									
0450	With stiffener/closure angle, through walls, 6" x 12"	1 Carp	8	1	Ea.	16.65	19.70		36.35	52.50
0460	12" x 24"		6	1.333		22	26.50		48.50	69.50
0470	24" x 48"		4	2		63	39.50		102.50	137
0480	With stiffener/closure angle, through floors, 6" x 12"		10	.800		9	15.75		24.75	37
0490	12" x 24"		8	1		16.20	19.70		35.90	52
0500	24" x 48"		6	1.333		32	26.50		58.50	80
0510	Multi trade openings									
0520	Through walls, 6" x 12"	1 Carp	2	4	Ea.	35	79		114	174
0530	12" x 24"	"	1	8		141	158		299	425
0540	24" x 48"	2 Carp	1	16		565	315		880	1,150
0550	48" x 96"	"	.75	21.333		2,275	420		2,695	3,225
0560	Through floors, 6" x 12"	1 Carp	2	4		35	79		114	174
0570	12" x 24"	"	1	8		141	158		299	425
0580	24" x 48"	2 Carp	.75	21.333		565	420		985	1,350
0590	48" x 96"	"	.50	32		2,275	630		2,905	3,575
0600	Structural penetrations, through walls									
0610	Steel beams, W8 x 10	1 Carp	8	1	Ea.	22	19.70		41.70	58
0620	W12 x 14		6	1.333		35	26.50		61.50	83.50
0630	W21 x 44		5	1.600		70	31.50		101.50	131
0640	W36 x 135		3	2.667		170	52.50		222.50	277
0650	Bar joists, 18" deep		6	1.333		32	26.50		58.50	80
0660	24" deep		6	1.333		40	26.50		66.50	89
0670	36" deep		5	1.600		60	31.50		91.50	120
0680	48" deep		4	2		70	39.50		109.50	145
0690	Construction joints, floor slab at exterior wall									
0700	Precast, brick, block or drywall exterior									
0710	2" wide joint	1 Carp	125	.064	L.F.	5	1.26		6.26	7.65
0720	4" wide joint	"	75	.107	"	10	2.10		12.10	14.60
0730	Metal panel, glass or curtain wall exterior									
0740	2" wide joint	1 Carp	40	.200	L.F.	11.85	3.94		15.79	19.75
0750	4" wide joint	"	25	.320	"	16.15	6.30		22.45	28.50
0760	Floor slab to drywall partition									
0770	Flat joint	1 Carp	100	.080	L.F.	4.90	1.58		6.48	8.10
0780	Fluted joint		50	.160		10	3.15		13.15	16.40
0790	Etched fluted joint		75	.107		6.50	2.10		8.60	10.75
0800	Floor slab to concrete/masonry partition									
0810	Flat joint	1 Carp	75	.107	L.F.	11	2.10		13.10	15.70
0820	Fluted joint	"	50	.160	"	13	3.15		16.15	19.70
0830	Concrete/CMU wall joints									
0840	1" wide	1 Carp	100	.080	L.F.	6	1.58		7.58	9.30

07800 | Fire & Smoke Protection

07840 | Firestopping

		CREW	DAILY OUTPUT	LABOR-HOURS	UNIT	MAT.	LABOR	EQUIP.	TOTAL	TOTAL INCL O&P
0850	2" wide	1 Carp	75	.107	L.F.	11	2.10		13.10	15.70
0860	4" wide	↓	50	.160	↓	21	3.15		24.15	28.50
0870	Concrete/CMU floor joints									
0880	1" wide	1 Carp	200	.040	L.F.	3	.79		3.79	4.65
0890	2" wide		150	.053		5.50	1.05		6.55	7.85
0900	4" wide	↓	100	.080	↓	10.50	1.58		12.08	14.25

(R07800-030)

07900 | Joint Sealers

07920 | Joint Sealants

		CREW	DAILY OUTPUT	LABOR-HOURS	UNIT	MAT.	LABOR	EQUIP.	TOTAL	TOTAL INCL O&P
0010	**CAULKING AND SEALANTS**									
0020	Acoustical sealant, elastomeric, cartridges				Ea.	2.05			2.05	2.26
0100	Acrylic latex caulk, white									
0200	11 fl. oz cartridge				Ea.	1.99			1.99	2.19
0500	1/4" x 1/2"	1 Bric	248	.032	L.F.	.16	.65		.81	1.27
0600	1/2" x 1/2"		250	.032		.32	.64		.96	1.44
0800	3/4" x 3/4"		230	.035		.73	.70		1.43	1.98
0900	3/4" x 1"		200	.040		.98	.80		1.78	2.42
1000	1" x 1"	↓	180	.044	↓	1.22	.89		2.11	2.84
1400	Butyl based, bulk				Gal.	22			22	24
1500	Cartridges				"	26.50			26.50	29.50
1700	Bulk, in place 1/4" x 1/2", 154 L.F./gal.	1 Bric	230	.035	L.F.	.14	.70		.84	1.34
1800	1/2" x 1/2", 77 L.F./gal.	"	180	.044	"	.29	.89		1.18	1.81
2000	Latex acrylic based, bulk				Gal.	23			23	25.50
2100	Cartridges				"	26			26	28.50
2200	Bulk in place, 1/4" x 1/2", 154 L.F./gal.	1 Bric	230	.035	L.F.	.15	.70		.85	1.34
2300	Polysulfide compounds, 1 component, bulk				Gal.	43.50			43.50	47.50
2400	Cartridges				"	46			46	51
2600	1 or 2 component, in place, 1/4" x 1/4", 308 L.F./gal.	1 Bric	145	.055	L.F.	.14	1.10		1.24	2.02
2700	1/2" x 1/4", 154 L.F./gal.		135	.059		.28	1.19		1.47	2.32
2900	3/4" x 3/8", 68 L.F./gal.		130	.062		.64	1.23		1.87	2.78
3000	1" x 1/2", 38 L.F./gal.	↓	130	.062	↓	1.14	1.23		2.37	3.33
3200	Polyurethane, 1 or 2 component				Gal.	49			49	54
3300	Cartridges				"	46.50			46.50	51
3500	Bulk, in place, 1/4" x 1/4"	1 Bric	150	.053	L.F.	.16	1.07		1.23	1.98
3600	1/2" x 1/4"		145	.055		.32	1.10		1.42	2.22
3800	3/4" x 3/8", 68 L.F./gal.		130	.062		.72	1.23		1.95	2.87
3900	1" x 1/2"	↓	110	.073	↓	1.27	1.45		2.72	3.86
4100	Silicone rubber, bulk				Gal.	34			34	37.50
4200	Cartridges				"	40			40	44

For information about Means Estimating Seminars, see yellow pages 11 and 12 in back of book

For expanded coverage of these items see *Means Building Construction Cost Data 2000*

Division Notes

		CREW	DAILY OUTPUT	LABOR-HOURS	UNIT	2000 BARE COSTS MAT.	LABOR	EQUIP.	TOTAL	TOTAL INCL O&P

Division 8
Doors & Windows

Estimating Tips

08100 Metal Doors & Frames
- Most metal doors and frames look alike, but there may be significant differences among them. When estimating these items be sure to choose the line item that most closely compares to the specification or door schedule requirements regarding:
  - type of metal
  - metal gauge
  - door core material
  - fire rating
  - finish

08200 Wood & Plastic Doors
- Wood and plastic doors vary considerably in price. The primary determinant is the veneer material. Lauan, birch and oak are the most common veneers. Other variables include the following:
  - hollow or solid core
  - fire rating
  - flush or raised panel
  - finish
- If the specifications require compliance with AWI (Architectural Woodwork Institute) standards or acoustical standards, the cost of the door may increase substantially. All wood doors are priced pre-mortised for hinges and predrilled for cylindrical locksets.

08300 Specialty Doors
- There are many varieties of special doors, and they are usually priced per each. Add frames, hardware or operators required for a complete installation.

08510 Steel Windows
- Most metal windows are delivered preglazed. However, some metal windows are priced without glass. Refer to 08800 Glazing for glass pricing. The grade C indicates commercial grade windows, usually ASTM C-35.

08550 Wood Windows
- All wood windows are priced preglazed. The two glazing options priced are single pane float glass and insulating glass 1/2" thick. Add the cost of screens and grills if required.

08700 Hardware
- Hardware costs add considerably to the cost of a door. The most efficient method to determine the hardware requirements for a project is to review the door schedule. This schedule, in conjunction with the specifications, is all you should need to take off the door hardware.
- Door hinges are priced by the pair, with most doors requiring 1-1/2 pairs per door. The hinge prices do not include installation labor because it is included in door installation. Hinges are classified according to the frequency of use.

08800 Glazing
- Different openings require different types of glass. The three most common types are:
  - float
  - tempered
  - insulating
- Most exterior windows are glazed with insulating glass. Entrance doors and window walls, where the glass is less than 18" from the floor, are generally glazed with tempered glass. Interior windows and some residential windows are glazed with float glass.

08900 Glazed Curtain Wall
- Glazed curtain walls consist of the metal tube framing and the glazing material. The cost data in this subdivision is presented for the metal tube framing alone or the composite wall. If your estimate requires a detailed takeoff of the framing, be sure to add the glazing cost.

Reference Numbers
Reference numbers are shown in bold squares at the beginning of some major classifications. These numbers refer to related items in the Reference Section. The reference information may be an estimating procedure, an alternate pricing method or technical information.

Note: Not all subdivisions listed here necessarily appear in this publication.

08100 | Metal Doors & Frames

08110 | Steel Doors & Frames

		CREW	DAILY OUTPUT	LABOR-HOURS	UNIT	2000 BARE COSTS MAT.	LABOR	EQUIP.	TOTAL	TOTAL INCL O&P
200	**0010 COMMERCIAL STEEL DOORS**									**200**
0015	Flush, full panel, hollow core R08110-120									
0020	1-3/8" thick, 20 ga., 2'-0"x 6'-8"	2 Carp	20	.800	Ea.	150	15.75		165.75	192
0040	2'-8" x 6'-8"		18	.889		155	17.50		172.50	201
0060	3'-0" x 6'-8"		17	.941		158	18.55		176.55	205
0100	3'-0" x 7'-0"	↓	17	.941		163	18.55		181.55	211
0120	For vision lite, add					51.50			51.50	57
0140	For narrow lite, add					62.50			62.50	69
0160	For bottom louver, add				↓	64			64	70.50
0230	For baked enamel finish, add					30%	15%			
0260	For galvanizing, add					15%				
0320	Half glass, 20 ga., 2'-0" x 6'-8"	2 Carp	20	.800	Ea.	202	15.75		217.75	249
0340	2'-8" x 6'-8"		18	.889		215	17.50		232.50	267
0360	3'-0" x 6'-8"		17	.941		221	18.55		239.55	275
0400	3'-0" x 7'-0"		17	.941		219	18.55		237.55	273
0500	Hollow core, 1-3/4" thick, full panel, 20 ga., 2'-8" x 6'-8"		18	.889		162	17.50		179.50	209
0520	3'-0" x 6'-8"		17	.941		165	18.55		183.55	214
0640	3'-0" x 7'-0"		17	.941		170	18.55		188.55	219
0680	4'-0" x 7'-0"		15	1.067		340	21		361	410
0700	4'-0" x 8'-0"		13	1.231		360	24.50		384.50	435
1000	18 ga., 2'-8" x 6'-8"		17	.941		187	18.55		205.55	238
1020	3'-0" x 6'-8"		16	1		185	19.70		204.70	237
1120	3'-0" x 7'-0"		17	.941		189	18.55		207.55	240
1180	4'-0" x 7'-0"		14	1.143		246	22.50		268.50	310
1200	4'-0" x 8'-0"		17	.941		291	18.55		309.55	350
1230	Half glass, 20 ga., 2'-8" x 6'-8"		20	.800		235	15.75		250.75	286
1240	3'-0" x 6'-8"		18	.889		235	17.50		252.50	289
1260	3'-0" x 7'-0"		18	.889		246	17.50		263.50	300
1280	4'-0" x 7'-0"		16	1		535	19.70		554.70	625
1300	4'-0" x 8'-0"		13	1.231		560	24.50		584.50	655
1320	18 ga., 2'-8" x 6'-8"		18	.889		158	17.50		175.50	204
1340	3'-0" x 6'-8"		17	.941		253	18.55		271.55	310
1360	3'-0" x 7'-0"		17	.941		263	18.55		281.55	320
1380	4'-0" x 7'-0"		15	1.067		370	21		391	445
1400	4'-0" x 8'-0"		14	1.143		430	22.50		452.50	510
1720	Insulated, 1-3/4" thick, full panel, 18 ga., 3'-0" x 6'-8"		15	1.067		245	21		266	305
1740	2'-8" x 7'-0"		16	1		256	19.70		275.70	315
1760	3'-0" x 7'-0"		15	1.067		252	21		273	315
1800	4'-0" x 8'-0"		13	1.231		305	24.50		329.50	375
1820	Half glass, 18 ga., 3'-0" x 6'-8"		16	1		315	19.70		334.70	380
1840	2'-8" x 7'-0"		17	.941		330	18.55		348.55	390
1860	3'-0" x 7'-0"		16	1		260	19.70		279.70	320
1900	4'-0" x 8'-0"	↓	14	1.143	↓	490	22.50		512.50	575
250	**0010 DOOR FRAMES**									**250**
0020	Steel channels with anchors and bar stops									
0100	6" channel @ 8.2#/L.F., 3' x 7' door, weighs 150#	E-4	13	2.462	Ea.	108	53.50	6.45	167.95	232
0200	8" channel @ 11.5#/L.F., 6' x 8' door, weighs 275#		9	3.556		198	77.50	9.35	284.85	380
0300	8' x 12' door, weighs 400#	↓	6.50	4.923		288	107	12.95	407.95	540
0800	For frames without bar stops, light sections, deduct					15%				
0900	Heavy sections, deduct				↓	10%				
300	**0010 FIRE DOOR**									**300**
0015	Steel, flush, "B" label, 90 minute									
0020	Full panel, 20 ga., 2'-0" x 6'-8"	2 Carp	20	.800	Ea.	173	15.75		188.75	217
0040	2'-8" x 6'-8"	↓	18	.889	↓	180	17.50		197.50	228

08100 | Metal Doors & Frames

08110 | Steel Doors & Frames

			CREW	DAILY OUTPUT	LABOR-HOURS	UNIT	MAT.	LABOR	EQUIP.	TOTAL	TOTAL INCL O&P	
300	0060	3'-0" x 6'-8"	2 Carp	17	.941	Ea.	180	18.55		198.55	230	300
	0080	3'-0" x 7'-0"		17	.941		187	18.55		205.55	238	
	0140	18 ga., 3'-0" x 6'-8"		16	1		200	19.70		219.70	253	
	0160	2'-8" x 7'-0"		17	.941		213	18.55		231.55	266	
	0180	3'-0" x 7'-0"		16	1		207	19.70		226.70	261	
	0200	4'-0" x 7'-0"		15	1.067		261	21		282	325	
	0220	For "A" label, 3 hour, 18 ga., use same price as "B" label										
	0240	For vision lite, add				Ea.	35			35	38.50	
	0520	Flush, "B" label 90 min., composite, 20 ga., 2'-0" x 6'-8"	2 Carp	18	.889		233	17.50		250.50	286	
	0540	2'-8" x 6'-8"		17	.941		237	18.55		255.55	293	
	0560	3'-0" x 6'-8"		16	1		240	19.70		259.70	298	
	0580	3'-0" x 7'-0"		16	1		248	19.70		267.70	305	
	0640	Flush, "A" label 3 hour, composite, 18 ga., 3'-0" x 6'-8"		15	1.067		260	21		281	320	
	0660	2'-8" x 7'-0"		16	1		270	19.70		289.70	330	
	0680	3'-0" x 7'-0"		15	1.067		267	21		288	330	
	0700	4'-0" x 7'-0"		14	1.143		320	22.50		342.50	395	
600	0011	**RESIDENTIAL DOOR**										600
	2510	Bi-passing closet, incl. hardware, no frame or trim incl.										
	2511	Mirrored, metal frame, 6'-8" x 4'-0" wide	2 Carp	10	1.600	Opng.	164	31.50		195.50	235	
	2512	5'-0" wide		10	1.600		296	31.50		327.50	380	
	2513	6'-0" wide		10	1.600		325	31.50		356.50	415	
	2514	7'-0" wide		9	1.778		380	35		415	475	
	2515	8'-0" wide		9	1.778		415	35		450	520	
	2611	Mirrored, metal, 8'-0" x 4'-0" wide		10	1.600		284	31.50		315.50	370	
	2612	5'-0" wide		10	1.600		320	31.50		351.50	405	
	2613	6'-0" wide		10	1.600		350	31.50		381.50	440	
	2614	7'-0" wide		9	1.778		380	35		415	475	
	2615	8'-0" wide		9	1.778		415	35		450	520	
820	0010	**STEEL FRAMES, KNOCK DOWN** R08110-100										820
	0020	18 ga., up to 5-3/4" deep										
	0025	6'-8" high, 3'-0" wide, single	2 Carp	16	1	Ea.	61.50	19.70		81.20	102	
	0040	6'-0" wide, double		14	1.143		74.50	22.50		97	121	
	0100	7'-0" high, 3'-0" wide, single		16	1		63.50	19.70		83.20	104	
	0140	6'-0" wide, double		14	1.143		77.50	22.50		100	124	
	1000	18 ga., up to 4-7/8" deep, 7'-0" H, 3'-0" W, single		16	1		65.50	19.70		85.20	106	
	1140	6'-0" wide, double		14	1.143		79	22.50		101.50	126	
	2800	16 ga., up to 3-7/8" deep, 7'-0" high, 3'-0" wide, single		16	1		70	19.70		89.70	111	
	2840	6'-0" wide, double		14	1.143		86	22.50		108.50	133	
	3600	5-3/4" deep, 7'-0" high, 4'-0" wide, single		15	1.067		73	21		94	117	
	3640	8'-0" wide, double		12	1.333		93.50	26.50		120	148	
	3700	8'-0" high, 4'-0" wide, single		15	1.067		91	21		112	136	
	3740	8'-0" wide, double		12	1.333		92.50	26.50		119	147	
	4000	6-3/4" deep, 7'-0" high, 4'-0" wide, single		15	1.067		82	21		103	127	
	4040	8'-0"		12	1.333		101	26.50		127.50	156	
	4100	8'-0" high, 4'-0" wide, single		15	1.067		95	21		116	141	
	4140	8'-0" wide, double		12	1.333		113	26.50		139.50	169	
	4400	8-3/4" deep, 7'-0" high, 4'-0" wide, single		15	1.067		93	21		114	138	
	4440	8'-0" wide, double		12	1.333		120	26.50		146.50	177	
	4500	8'-0" high, 4'-0" wide, single		15	1.067		104	21		125	150	
	4540	8'-0" wide, double		12	1.333		113	26.50		139.50	170	
	4900	For welded frames, add					32			32	35	
	5400	16 ga., "B" label, up to 5-3/4" deep, 7'-0" high, 4'-0" wide, single	2 Carp	15	1.067		80.50	21		101.50	125	
	5440	8'-0" wide, double		12	1.333		122	26.50		148.50	180	
	5800	6-3/4" deep, 7'-0" high, 4'-0" wide, single		15	1.067		86.50	21		107.50	132	
	5840	8'-0" wide, double		12	1.333		116	26.50		142.50	172	
	6200	8-3/4" deep, 7'-0" high, 4'-0" wide, single		15	1.067		96.50	21		117.50	142	

For expanded coverage of these items see *Means Building Construction Cost Data 2000*

08100 | Metal Doors & Frames

08110 | Steel Doors & Frames

		CREW	DAILY OUTPUT	LABOR-HOURS	UNIT	MAT.	LABOR	EQUIP.	TOTAL	TOTAL INCL O&P
6240	8'-0" wide, double	2 Carp	12	1.333	Ea.	138	26.50		164.50	197
6300	For "A" label use same price as "B" label									
6400	For baked enamel finish, add					30%	15%			
6500	For galvanizing, add					15%				
7900	Transom lite frames, fixed, add	2 Carp	155	.103	S.F.	25	2.03		27.03	31
8000	Movable, add	"	130	.123	"	30	2.42		32.42	37

R08110-100

08200 | Wood & Plastic Doors

08210 | Wood Doors

		CREW	DAILY OUTPUT	LABOR-HOURS	UNIT	MAT.	LABOR	EQUIP.	TOTAL	TOTAL INCL O&P
0010	**KALAMEIN**									
0020	Interior, flush type, 3' x 7'	2 Carp	4.30	3.721	Opng.	173	73.50		246.50	315
0010	**PRE-HUNG DOORS**									
0300	Exterior, wood, comb. storm & screen, 6'-9" x 2'-6" wide	2 Carp	15	1.067	Ea.	208	21		229	265
0320	2'-8" wide		15	1.067		208	21		229	265
0340	3'-0" wide		15	1.067		218	21		239	276
0360	For 7'-0" high door, add					28			28	30.50
0370	For aluminum storm doors, see division 08280-800									
1600	Entrance door, flush, birch, solid core									
1620	4-5/8" solid jamb, 1-3/4" x 6'-8" x 2'-8" wide	2 Carp	16	1	Ea.	191	19.70		210.70	244
1640	3'-0" wide	"	16	1		199	19.70		218.70	252
1680	For 7'-0" high door, add					13.25			13.25	14.60
2000	Entrance door, colonial, 6 panel pine									
2020	4-5/8" solid jamb, 1-3/4" x 6'-8" x 2'-8" wide	2 Carp	16	1	Ea.	440	19.70		459.70	520
2040	3'-0" wide	"	16	1		465	19.70		484.70	545
2060	For 7'-0" high door, add					13.25			13.25	14.60
2200	For 5-5/8" solid jamb, add					13.25			13.25	14.60
4000	Interior, passage door, 4-5/8" solid jamb									
4400	Lauan, flush, solid core, 1-3/8" x 6'-8" x 2'-6" wide	2 Carp	20	.800	Ea.	135	15.75		150.75	175
4420	2'-8" wide		20	.800		138	15.75		153.75	179
4440	3'-0" wide		19	.842		161	16.60		177.60	206
4600	Hollow core, 1-3/8" x 6'-8" x 2'-6" wide		20	.800		102	15.75		117.75	139
4620	2'-8" wide		20	.800		102	15.75		117.75	139
4640	3'-0" wide		19	.842		104	16.60		120.60	143
4700	For 7'-0" high door, add					8.95			8.95	9.80
5000	Birch, flush, solid core, 1-3/8" x 6'-8" x 2'-6" wide	2 Carp	20	.800		153	15.75		168.75	195
5020	2'-8" wide		20	.800		156	15.75		171.75	198
5040	3'-0" wide		19	.842		162	16.60		178.60	208
5200	Hollow core, 1-3/8" x 6'-8" x 2'-6" wide		20	.800		122	15.75		137.75	161
5220	2'-8" wide		20	.800		146	15.75		161.75	188
5240	3'-0" wide		19	.842		127	16.60		143.60	169
5280	For 7'-0" high door, add					9.40			9.40	10.35
5500	Hardboard paneled, 1-3/8" x 6'-8" x 2'-6" wide	2 Carp	20	.800		123	15.75		138.75	162
5520	2'-8" wide		20	.800		125	15.75		140.75	164
5540	3'-0" wide		19	.842		127	16.60		143.60	169
6000	Pine paneled, 1-3/8" x 6'-8" x 2'-6" wide		20	.800		210	15.75		225.75	258
6020	2'-8" wide		20	.800		221	15.75		236.75	270
6500	For 5-5/8" solid jamb, add					21.50			21.50	24
6520	For split jamb, deduct					12.05			12.05	13.25

Important: See the Reference Section for critical supporting data - Reference Nos., Crews, & Location Factors

08200 | Wood & Plastic Doors

08210 | Wood Doors

			CREW	DAILY OUTPUT	LABOR-HOURS	UNIT	2000 BARE COSTS MAT.	LABOR	EQUIP.	TOTAL	TOTAL INCL O&P	
850	0010	**TIN CLAD**										850
	0020	3 ply, 6' x 7', double sliding, manual with hardware	2 Carp	1	16	Opng.	1,325	315		1,640	2,000	
	1000	For electric operator, add	1 Elec	2	4	"	2,350	88.50		2,438.50	2,725	
910	0010	**WOOD DOORS, DECORATOR**										910
	3000	Solid wood, 1-3/4" thick stile and rail										
	3020	Mahogany, 3'-0" x 7'-0", minimum	2 Carp	14	1.143	Ea.	440	22.50		462.50	525	
	3030	Maximum		10	1.600		640	31.50		671.50	760	
	3040	3'-6" x 8'-0", minimum		10	1.600		645	31.50		676.50	765	
	3050	Maximum		8	2		750	39.50		789.50	895	
	3100	Pine, 3'-0" x 7'-0", minimum		14	1.143		289	22.50		311.50	355	
	3120	3'-6" x 8'-0", minimum		10	1.600		570	31.50		601.50	685	
	3130	Maximum		8	2		1,550	39.50		1,589.50	1,775	
	3200	Red oak, 3'-0" x 7'-0", minimum		14	1.143		650	22.50		672.50	755	
	3210	Maximum		10	1.600		1,150	31.50		1,181.50	1,325	
	3220	3'-6" x 8'-0", minimum		10	1.600		825	31.50		856.50	960	
	3230	Maximum		8	2		1,350	39.50		1,389.50	1,575	
	4000	Hand carved door, mahogany										
	4020	3'-0" x 7'-0", minimum	2 Carp	14	1.143	Ea.	645	22.50		667.50	750	
	4040	3'-6" x 8'-0", minimum		10	1.600		1,050	31.50		1,081.50	1,200	
	4050	Maximum		8	2		2,200	39.50		2,239.50	2,500	
	4200	Red oak, 3'-0" x 7'-0", minimum		14	1.143		1,275	22.50		1,297.50	1,450	
	4210	Maximum		11	1.455		2,975	28.50		3,003.50	3,325	
	4220	3'-6" x 8'-0", minimum		10	1.600		2,550	31.50		2,581.50	2,850	
	4280	For 6'-8" high door, deduct from 7'-0" door					23			23	25.50	
	4400	For custom finish, add					101			101	112	
	4600	Side light, mahogany, 7'-0" x 1'-6" wide, minimum	2 Carp	18	.889		255	17.50		272.50	310	
	4610	Maximum		14	1.143		695	22.50		717.50	800	
	4620	8'-0" x 1'-6" wide, minimum		14	1.143		315	22.50		337.50	385	
	4630	Maximum		10	1.600		805	31.50		836.50	940	
	4640	Side light, oak, 7'-0" x 1'-6" wide, minimum		18	.889		345	17.50		362.50	410	
	4650	Maximum		14	1.143		800	22.50		822.50	920	
	4660	8'-0" x 1-6" wide, minimum		14	1.143		415	22.50		437.50	500	
	4670	Maximum		10	1.600		935	31.50		966.50	1,075	
	6520	Interior cafe doors, 2'-6" opening, stock, panel pine		16	1		146	19.70		165.70	194	
	6540	3'-0" opening		16	1		151	19.70		170.70	200	
	6550	Louvered pine										
	6560	2'-6" opening	2 Carp	16	1	Ea.	127	19.70		146.70	173	
	8000	3'-0" opening		16	1		135	19.70		154.70	183	
	8010	2'-6" opening, hardwood		16	1		171	19.70		190.70	222	
	8020	3'-0" opening		16	1		201	19.70		220.70	255	
	8800	Pre-hung doors, see division 08210-720										
920	0010	**WOOD DOORS, PANELED**										920
	0020	Interior, six panel, hollow core, 1-3/8" thick										
	0040	Molded hardboard, 2'-0" x 6'-8"	2 Carp	17	.941	Ea.	41	18.55		59.55	77	
	0060	2'-6" x 6'-8"		17	.941		44.50	18.55		63.05	80.50	
	0080	3'-0" x 6'-8"		17	.941		49	18.55		67.55	85.50	
	0140	Embossed print, molded hardboard, 2'-0" x 6'-8"		17	.941		44.50	18.55		63.05	80.50	
	0160	2'-6" x 6'-8"		17	.941		44.50	18.55		63.05	80.50	
	0180	3'-0" x 6'-8"		17	.941		49	18.55		67.55	85.50	
	0540	Six panel, solid, 1-3/8" thick, pine, 2'-0" x 6'-8"		15	1.067		107	21		128	154	
	0560	2'-6" x 6'-8"		14	1.143		120	22.50		142.50	171	
	0580	3'-0" x 6'-8"		13	1.231		138	24.50		162.50	194	
	1020	Two panel, bored rail, solid, 1-3/8" thick, pine, 1'-6" x 6'-8"		16	1		195	19.70		214.70	249	
	1040	2'-0" x 6'-8"		15	1.067		257	21		278	320	
	1060	2'-6" x 6'-8"		14	1.143		293	22.50		315.50	365	

For expanded coverage of these items see *Means Building Construction Cost Data 2000*

08200 | Wood & Plastic Doors

08210 | Wood Doors

			CREW	DAILY OUTPUT	LABOR-HOURS	UNIT	2000 BARE COSTS MAT.	LABOR	EQUIP.	TOTAL	TOTAL INCL O&P	
920	1340	Two panel, solid, 1-3/8" thick, fir, 2'-0" x 6'-8"	2 Carp	15	1.067	Ea.	107	21		128	154	920
	1360	2'-6" x 6'-8"		14	1.143		120	22.50		142.50	171	
	1380	3'-0" x 6'-8"		13	1.231		293	24.50		317.50	365	
	1740	Five panel, solid, 1-3/8" thick, fir, 2'-0" x 6'-8"		15	1.067		192	21		213	247	
	1760	2'-6" x 6'-8"		14	1.143		293	22.50		315.50	365	
	1780	3'-0" x 6'-8"		13	1.231		294	24.50		318.50	365	
960	0010	**WOOD FRAMES**										960
	0400	Exterior frame, incl. ext. trim, pine, 5/4 x 4-9/16" deep	2 Carp	375	.043	L.F.	3.89	.84		4.73	5.70	
	0420	5-3/16" deep		375	.043		6.60	.84		7.44	8.75	
	0440	6-9/16" deep		375	.043		5.70	.84		6.54	7.70	
	0600	Oak, 5/4 x 4-9/16" deep		350	.046		8.35	.90		9.25	10.70	
	0620	5-3/16" deep		350	.046		9.40	.90		10.30	11.90	
	0640	6-9/16" deep		350	.046		10.45	.90		11.35	13.05	
	0800	Walnut, 5/4 x 4-9/16" deep		350	.046		9.80	.90		10.70	12.35	
	0820	5-3/16" deep		350	.046		14.25	.90		15.15	17.20	
	0840	6-9/16" deep		350	.046		16.80	.90		17.70	20	
	1000	Sills, 8/4 x 8" deep, oak, no horns		100	.160		11.85	3.15		15	18.40	
	1020	2" horns		100	.160		13.20	3.15		16.35	19.90	
	1040	3" horns		100	.160		15.20	3.15		18.35	22	
	1100	8/4 x 10" deep, oak, no horns		90	.178		15.95	3.50		19.45	23.50	
	1120	2" horns		90	.178		17.75	3.50		21.25	25.50	
	1140	3" horns		90	.178		19.35	3.50		22.85	27.50	
	2000	Exterior, colonial, frame & trim, 3' opng., in-swing, minimum		22	.727	Ea.	276	14.35		290.35	330	
	2010	Average		21	.762		235	15		250	284	
	2020	Maximum		20	.800		930	15.75		945.75	1,050	
	2100	5'-4" opening, in-swing, minimum		17	.941		310	18.55		328.55	375	
	2120	Maximum		15	1.067		930	21		951	1,050	
	2140	Out-swing, minimum		17	.941		320	18.55		338.55	380	
	2160	Maximum		15	1.067		965	21		986	1,075	
	2400	6'-0" opening, in-swing, minimum		16	1		299	19.70		318.70	365	
	2420	Maximum		10	1.600		965	31.50		996.50	1,100	
	2460	Out-swing, minimum		16	1		320	19.70		339.70	385	
	2480	Maximum		10	1.600		1,200	31.50		1,231.50	1,375	
	2600	For two sidelights, add, minimum		30	.533	Opng.	310	10.50		320.50	360	
	2620	Maximum		20	.800	"	985	15.75		1,000.75	1,100	
	2700	Custom birch frame, 3'-0" opening		16	1	Ea.	177	19.70		196.70	229	
	2750	6'-0" opening		16	1		252	19.70		271.70	310	
	2900	Exterior, modern, plain trim, 3' opng., in-swing, minimum		26	.615		26.50	12.10		38.60	50.50	
	2920	Average		24	.667		34	13.15		47.15	59.50	
	2940	Maximum		22	.727		38.50	14.35		52.85	67	
	3000	Interior frame, pine, 11/16" x 3-5/8" deep		375	.043	L.F.	3.81	.84		4.65	5.65	
	3020	4-9/16" deep		375	.043		5.10	.84		5.94	7.10	
	3200	Oak, 11/16" x 3-5/8" deep		350	.046		4.73	.90		5.63	6.75	
	3220	4-9/16" deep		350	.046		5.10	.90		6	7.15	
	3240	5-3/16" deep		350	.046		5.25	.90		6.15	7.35	
	3400	Walnut, 11/16" x 3-5/8" deep		350	.046		8.40	.90		9.30	10.80	
	3420	4-9/16" deep		350	.046		8.40	.90		9.30	10.80	
	3440	5-3/16" deep		350	.046		8.75	.90		9.65	11.15	
	3600	Pocket door frame		16	1	Ea.	71	19.70		90.70	112	
	3800	Threshold, oak, 5/8" x 3-5/8" deep		200	.080	L.F.	3.07	1.58		4.65	6.10	
	3820	4-5/8" deep		190	.084		3.91	1.66		5.57	7.15	
	3840	5-5/8" deep		180	.089		4.69	1.75		6.44	8.15	
	4000	For casing see division 06220-400										

Reference: R08210-100

08200 | Wood & Plastic Doors

08260 | Sliding Wood & Plastic Doors

		CREW	DAILY OUTPUT	LABOR-HOURS	UNIT	2000 BARE COSTS MAT.	LABOR	EQUIP.	TOTAL	TOTAL INCL O&P
0010	**GLASS, SLIDING**									
0012	Vinyl clad, 1" insul. glass, 6'-0" x 6'-10" high	2 Carp	4	4	Opng.	750	79		829	955
0030	6'-0" x 8'-0" high		4	4	Ea.	1,575	79		1,654	1,850
0100	8'-0" x 6'-10" high		4	4	Opng.	1,675	79		1,754	1,975
0500	3 leaf, 9'-0" x 6'-10" high		3	5.333		1,525	105		1,630	1,850
0600	12'-0" x 6'-10" high		3	5.333		1,875	105		1,980	2,250
0010	**GLASS, SLIDING**									
0020	Wood, 5/8" tempered insul. glass, 6' wide, premium	2 Carp	4	4	Ea.	930	79		1,009	1,150
0100	Economy		4	4		680	79		759	880
0150	8' wide, wood, premium		3	5.333		1,075	105		1,180	1,350
0200	Economy		3	5.333		795	105		900	1,050
0250	12' wide, wood, premium		2.50	6.400		1,725	126		1,851	2,125
0300	Economy		2.50	6.400		1,125	126		1,251	1,475
0350	Aluminum, 5/8" tempered insulated glass, 6' wide									
0400	Premium	2 Carp	4	4	Ea.	1,225	79		1,304	1,475
0450	Economy		4	4		505	79		584	695
0500	8' wide, premium		3	5.333		1,225	105		1,330	1,525
0550	Economy		3	5.333		1,025	105		1,130	1,300
0600	12' wide, premium		2.50	6.400		2,025	126		2,151	2,475
0650	Economy		2.50	6.400		1,275	126		1,401	1,625

08280 | Wood/Plastic Storm/Screen Doors

		CREW	DAILY OUTPUT	LABOR-HOURS	UNIT	MAT.	LABOR	EQUIP.	TOTAL	TOTAL INCL O&P
0010	**STORM DOORS & FRAMES** Aluminum, residential,									
0020	combination storm and screen									
0400	Clear anodic coating, 6'-8" x 2'-6" wide	2 Carp	15	1.067	Ea.	138	21		159	187
0420	2'-8" wide		14	1.143		165	22.50		187.50	220
0440	3'-0" wide		14	1.143		165	22.50		187.50	220
0500	For 7' door height, add					5%				
1000	Mill finish, 6'-8" x 2'-6" wide	2 Carp	15	1.067	Ea.	194	21		215	249
1020	2'-8" wide		14	1.143		194	22.50		216.50	252
1040	3'-0" wide		14	1.143		210	22.50		232.50	270
1100	For 7'-0" door, add					5%				
1500	White painted, 6'-8" x 2'-6" wide	2 Carp	15	1.067		189	21		210	244
1520	2'-8" wide		14	1.143		177	22.50		199.50	233
1540	3'-0" wide		14	1.143		196	22.50		218.50	255
1541	Storm door, painted, alum., insul., 6'-8" x 2'-6" wide		14	1.143		210	22.50		232.50	270
1545	2'-8" wide		14	1.143		210	22.50		232.50	270
1600	For 7'-0" door, add					5%				
1800	Aluminum screen door, minimum, 6'-8" x 2'-8" wide	2 Carp	14	1.143		92	22.50		114.50	140
1810	3'-0" wide		14	1.143		194	22.50		216.50	252
1820	Average, 6'-8" x 2'-8" wide		14	1.143		153	22.50		175.50	207
1830	3'-0" wide		14	1.143		194	22.50		216.50	252
1840	Maximum, 6'-8" x 2'-8" wide		14	1.143		305	22.50		327.50	375
1850	3'-0" wide		14	1.143		210	22.50		232.50	270
2000	Wood door & screen, see division 08210-930									
2020										

For expanded coverage of these items see *Means Building Construction Cost Data 2000*

08300 | Specialty Doors

08310 | Access Doors & Panels

			CREW	DAILY OUTPUT	LABOR-HOURS	UNIT	MAT.	LABOR	EQUIP.	TOTAL	TOTAL INCL O&P
150	0010	**BULKHEAD CELLAR DOORS**									150
	0020	Steel, not incl. sides, 44" x 62"	1 Carp	5.50	1.455	Ea.	187	28.50		215.50	255
	0100	52" x 73"		5.10	1.569		208	31		239	282
	0500	With sides and foundation plates, 57" x 45" x 24"		4.70	1.702		244	33.50		277.50	325
	0600	42" x 49" x 51"	↓	4.30	1.860	↓	294	36.50		330.50	390
300	0010	**FLOOR, COMMERCIAL**									300
	0020	Aluminum tile, steel frame, one leaf, 2' x 2' opng.	2 Sswk	3.50	4.571	Opng.	350	97		447	575
	0050	3'-6" x 3'-6" opening		3.50	4.571		635	97		732	885
	0500	Double leaf, 4' x 4' opening		3	5.333		930	113		1,043	1,250
	0550	5' x 5' opening	↓	3	5.333	↓	1,375	113		1,488	1,725
350	0010	**FLOOR, INDUSTRIAL**									350
	0020	Steel 300 psf L.L., single leaf, 2' x 2', 175#	2 Sswk	6	2.667	Opng.	465	56.50		521.50	620
	0050	3' x 3' opening, 300#		5.50	2.909		640	62		702	825
	0300	Double leaf, 4' x 4' opening, 455#		5	3.200		960	68		1,028	1,175
	0350	5' x 5' opening, 645#	↓	4.50	3.556	↓	1,250	75.50		1,325.50	1,525

08330 | Coiling Doors & Grilles

			CREW	DAILY OUTPUT	LABOR-HOURS	UNIT	MAT.	LABOR	EQUIP.	TOTAL	TOTAL INCL O&P
130	0010	**COUNTER DOORS**									130
	0020	Manual, incl. frm and hdwe, galv. stl., 4' roll-up, 6' long	2 Carp	2	8	Opng.	630	158		788	965
	0300	Galvanized steel, UL label		1.80	8.889		765	175		940	1,150
	0600	Stainless steel, 4' high roll-up, 6' long		2	8		1,050	158		1,208	1,425
	0700	10' long		1.80	8.889		1,550	175		1,725	2,000
	2000	Aluminum, 4' high, 4' long		2.20	7.273		700	143		843	1,025
	2020	6' long		2	8		775	158		933	1,125
	2040	8' long		1.90	8.421		895	166		1,061	1,275
	2060	10' long		1.80	8.889		1,075	175		1,250	1,475
	2080	14' long		1.40	11.429		1,550	225		1,775	2,075
	2100	6' high, 4' long		2	8		775	158		933	1,125
	2120	6' long		1.60	10		910	197		1,107	1,350
	2140	10' long	↓	1.40	11.429	↓	1,250	225		1,475	1,750
640	0010	**COILING GRILLE**									640
	2020	Aluminum, manual operated, mill finish	2 Sswk	82	.195	S.F.	16.90	4.15		21.05	27
	2040	Bronze anodized		82	.195	"	27	4.15		31.15	37.50
	2060	Steel, manual operated, 10' x 10' high		1	16	Opng.	1,500	340		1,840	2,325
	2080	15' x 8' high	↓	.80	20	"	1,775	425		2,200	2,800
	3000	For safety edge bottom bar, electric, add				L.F.	31			31	34.50
	8000	For motor operation, add	2 Sswk	5	3.200	Opng.	760	68		828	975
720	0010	**ROLLING SERVICE DOORS** Steel, manual, 20 ga., incl. hardware									720
	0050	8' x 8' high	2 Sswk	1.60	10	Ea.	690	213		903	1,175
	0100	10' x 10' high		1.40	11.429		910	243		1,153	1,475
	0120	8' x 8' high, class A fire door		1.40	11.429		905	243		1,148	1,475
	0130	12' x 12' high, standard		1.20	13.333		1,150	283		1,433	1,800
	0140	12' x 12' high, class A fire door		1	16		1,600	340		1,940	2,425
	0160	10' x 20' high, standard		.50	32		1,350	680		2,030	2,825
	0180	10' x 20' high, class A fire door	↓	.40	40	↓	2,400	850		3,250	4,325
	3000	For 18 ga. doors, add				S.F.	.69			.69	.76
	3300	For enamel finish, add				"	.81			.81	.89
	3600	For safety edge bottom bar, pneumatic, add				L.F.	10.65			10.65	11.70
	3700	Electric, add					20			20	22
	4000	For weatherstripping, extruded rubber, jambs, add					6.90			6.90	7.60
	4100	Hood, add				↓	4.38			4.38	4.82

08300 | Specialty Doors

08330 | Coiling Doors & Grilles

			CREW	DAILY OUTPUT	LABOR-HOURS	UNIT	MAT.	LABOR	EQUIP.	TOTAL	TOTAL INCL O&P	
720	4200	Sill, add				L.F.	2.50			2.50	2.75	720
	4500	Motor operators, to 14' x 14' opening	2 Sswk	5	3.200	Ea.	665	68		733	865	
	4700	For fire door, additional fusible link, add				"	12.50			12.50	13.75	

08344 | Industrial Doors

			CREW	DAILY OUTPUT	LABOR-HOURS	UNIT	MAT.	LABOR	EQUIP.	TOTAL	TOTAL INCL O&P	
200	0010	**DOUBLE ACTING, SWING**										200
	1000	.063" aluminum, 7'-0" high, 4'-0" wide	2 Carp	4.20	3.810	Pr.	355	75		430	520	
	1050	6'-8" wide	"	4	4	"	490	79		569	670	
	2000	Solid core wood, 3/4" thick, metal frame, stainless steel										
	2010	base plate, 7' high opening, 4' wide	2 Carp	4	4	Pr.	720	79		799	925	
	2050	7' wide	"	3.80	4.211	"	1,025	83		1,108	1,275	
300	0010	**GLASS DOOR, SWING**										300
	0020	Including hardware, 1/2" thick, tempered, 3' x 7' opening	2 Glaz	2	8	Opng.	1,600	155		1,755	2,000	
	0100	6' x 7' opening	"	1.40	11.429	"	3,175	221		3,396	3,875	
600	0010	**SHOCK ABSORBING DOORS**										600
	0020	Rigid, no frame, 1-1/2" thick, 5' x 7'	2 Sswk	1.90	8.421	Opng.	1,125	179		1,304	1,600	
	0100	8' x 8'		1.80	8.889		1,600	189		1,789	2,125	
	0500	Flexible, no frame, insulated, .16" thick, economy, 5' x 7'		2	8		1,375	170		1,545	1,850	
	0600	Deluxe		1.90	8.421		2,100	179		2,279	2,650	
	1000	8' x 8' opening, economy		2	8		2,150	170		2,320	2,700	
	1100	Deluxe		1.90	8.421		2,775	179		2,954	3,400	

08348 | Sound Control Doors

			CREW	DAILY OUTPUT	LABOR-HOURS	UNIT	MAT.	LABOR	EQUIP.	TOTAL	TOTAL INCL O&P	
100	0010	**ACOUSTICAL DOORS**										100
	0020	Including framed seals, 3' x 7', wood, 27 STC rating	2 Carp	1.50	10.667	Ea.	330	210		540	725	
	0100	Steel, 40 STC rating		1.50	10.667		1,275	210		1,485	1,750	
	0200	45 STC rating		1.50	10.667		1,675	210		1,885	2,200	
	0300	48 STC rating		1.50	10.667		2,125	210		2,335	2,700	
	0400	52 STC rating		1.50	10.667		2,550	210		2,760	3,150	

08360 | Overhead Doors

			CREW	DAILY OUTPUT	LABOR-HOURS	UNIT	MAT.	LABOR	EQUIP.	TOTAL	TOTAL INCL O&P	
550	0010	**OVERHEAD, COMMERCIAL** Frames not included										550
	1000	Stock, sectional, heavy duty, wood, 1-3/4" thick, 8' x 8' high	2 Carp	2	8	Ea.	460	158		618	775	
	1100	10' x 10' high		1.80	8.889		690	175		865	1,050	
	1200	12' x 12' high		1.50	10.667		960	210		1,170	1,400	
	1300	Chain hoist, 14' x 14' high		1.30	12.308		1,525	242		1,767	2,100	
	1400	12' x 16' high		1	16		1,350	315		1,665	2,025	
	1500	20' x 8' high		1.30	12.270		1,250	242		1,492	1,800	
	1600	20' x 16' high		.65	24.615		3,150	485		3,635	4,300	
	1800	Center mullion openings, 8' high		4	4		490	79		569	675	
	1900	20' high		2	8		915	158		1,073	1,275	
	2100	For medium duty custom door, deduct					5%	5%				
	2150	For medium duty stock doors, deduct					10%	5%				
	2300	Fiberglass and aluminum, heavy duty, sectional, 12' x 12' high	2 Carp	1.50	10.667	Ea.	1,275	210		1,485	1,750	
	2450	Chain hoist, 20' x 20' high		.50	32		4,400	630		5,030	5,900	
	2600	Steel, 24 ga. sectional, manual, 8' x 8' high		2	8		375	158		533	680	
	2650	10' x 10' high		1.80	8.889		515	175		690	865	
	2700	12' x 12' high		1.50	10.667		690	210		900	1,125	
	2800	Chain hoist, 20' x 14' high		.70	22.857		1,825	450		2,275	2,775	
	2850	For 1-1/4" rigid insulation and 26 ga. galv.										
	2860	back panel, add				S.F.	1.78			1.78	1.96	
	2900	For electric trolley operator, 1/3 H.P., to 12' x 12', add	1 Carp	2	4	Ea.	470	79		549	650	
	2950	Over 12' x 12', 1/2 H.P., add	"	1	8	"	745	158		903	1,100	

For expanded coverage of these items see *Means Building Construction Cost Data 2000*

08400 | Entrances & Storefronts

08411 | Aluminum Framed Storefront

			CREW	DAILY OUTPUT	LABOR-HOURS	UNIT	2000 BARE COSTS MAT.	LABOR	EQUIP.	TOTAL	TOTAL INCL O&P	
100	0010	**ALUMINUM FRAMES**										100
	0020	Entrance, 3' x 7' opening, clear anodized finish	2 Sswk	7	2.286	Opng.	175	48.50		223.50	288	
	0040	Bronze finish		7	2.286		273	48.50		321.50	395	
	0060	Black finish		7	2.286		310	48.50		358.50	435	
	0200	3'-6" x 7'-0", mill finish		7	2.286		242	48.50		290.50	365	
	0220	Bronze finish		7	2.286		293	48.50		341.50	415	
	0240	Black finish		7	2.286		320	48.50		368.50	445	
	0500	6' x 7' opening, clear finish		6	2.667		273	56.50		329.50	410	
	0520	Bronze finish		6	2.667		310	56.50		366.50	450	
	0540	Black finish		6	2.667		365	56.50		421.50	510	
	0600	7'-0" x 7'-0", mill finish		6	2.667	Ea.	254	56.50		310.50	390	
	0620	Bronze finish		6	2.667		288	56.50		344.50	425	
	0640	Black finish		6	2.667		340	56.50		396.50	485	
	2000	Transoms, 3'-0" x 3'-0", mill finish		80	.200		96	4.25		100.25	114	
	2020	Bronze finish		80	.200		106	4.25		110.25	125	
	2040	Black finish		80	.200		121	4.25		125.25	141	
	2200	3'-6" x 3'-0", mill finish		80	.200		111	4.25		115.25	130	
	2220	Bronze finish		80	.200		126	4.25		130.25	147	
	2240	Black finish		80	.200		138	4.25		142.25	160	
	2500	6'-0" x 3'-0", mill finish		65	.246		111	5.25		116.25	132	
	2520	Bronze finish		65	.246		128	5.25		133.25	151	
	2540	Black finish		65	.246		157	5.25		162.25	182	
	2700	7'-0" x 3'-0", mill finish		65	.246		118	5.25		123.25	140	
	2720	Bronze finish		65	.246		137	5.25		142.25	160	
	2740	Black finish		65	.246		156	5.25		161.25	181	
120	0010	**ALUMINUM DOORS** Commercial entrance										120
	0800	Narrow stile, no glazing, standard hardware, pair of 2'-6" x 7'-0"	2 Carp	1.70	9.412	Pr.	635	185		820	1,025	
	1000	3'-0" x 7'-0", single		3	5.333	Ea.	420	105		525	640	
	1050	Bronze finish		3	5.333		320	105		425	535	
	1100	Black finish		3	5.333		485	105		590	715	
	1200	Pair of 3'-0" x 7'-0"		1.70	9.412	Pr.	840	185		1,025	1,250	
	1500	3'-6" x 7'-0", single		3	5.333	Ea.	450	105		555	675	
	1550	Bronze finish		3	5.333		340	105		445	555	
	1600	Black finish		3	5.333		515	105		620	745	
	2000	Medium stile, pair of 2'-6" x 7'-0"		1.70	9.412	Pr.	695	185		880	1,075	
	2100	3'-0" x 7'-0", single		3	5.333	Ea.	550	105		655	785	
	2200	Pair of 3'-0" x 7'-0"		1.70	9.412	Pr.	1,050	185		1,235	1,475	
	2300	3'-6" x 7'-0", single		5.33	3.002	Ea.	645	59		704	810	
	5000	Flush panel doors, pair of 2'-6" x 7'-0"	2 Sswk	2	8	Pr.	845	170		1,015	1,275	
	5050	3'-0" x 7'-0", single		2.50	6.400	Ea.	425	136		561	735	
	5100	Pair of 3'-0" x 7'-0"		2	8	Pr.	870	170		1,040	1,300	
	5150	3'-6" x 7'-0", single		2.50	6.400	Ea.	520	136		656	840	
140	0010	**ALUMINUM DOORS & FRAMES** Entrance, narrow stile, including										140
	0015	hardware & closer, clear finish, not incl. glass, 2'-6" x 7'-0" opng.	2 Sswk	2	8	Ea.	475	170		645	860	
	0020	3'-0" x 7'-0" opening		2	8		415	170		585	795	
	0030	3'-6" x 7'-0" opening		2	8		430	170		600	805	
	0100	3'-0" x 10'-0" opening, 3' high transom		1.80	8.889		680	189		869	1,125	
	0200	3'-6" x 10'-0" opening, 3' high transom		1.80	8.889		670	189		859	1,100	
	0280	5'-0" x 7'-0" opening		2	8		710	170		880	1,125	
	0300	6'-0" x 7'-0" opening		1.30	12.308	Pr.	690	262		952	1,275	
	0400	6'-0" x 10'-0" opening, 3' high transom		1.10	14.545		625	310		935	1,300	
	0420	7'-0" x 7'-0" opening		1	16		795	340		1,135	1,550	
	0500	Wide stile, 2'-6" x 7'-0" opening		2	8	Ea.	700	170		870	1,100	
	0520	3'-0" x 7'-0" opening		2	8		625	170		795	1,025	

08400 | Entrances & Storefronts

08411 | Aluminum Framed Storefront

			CREW	DAILY OUTPUT	LABOR-HOURS	UNIT	2000 BARE COSTS MAT.	LABOR	EQUIP.	TOTAL	TOTAL INCL O&P	
140	0540	3'-6" x 7'-0" opening	2 Sswk	2	8	Ea.	655	170		825	1,050	140
	0560	5'-0" x 7'-0" opening		2	8		1,000	170		1,170	1,425	
	0580	6'-0" x 7'-0" opening		1.30	12.308	Pr.	960	262		1,222	1,575	
	0600	7'-0" x 7'-0" opening		1	16	"	1,100	340		1,440	1,875	
	1100	For full vision doors, with 1/2" glass, add				Leaf	55%					
	1200	For non-standard size, add					67%					
	1300	Light bronze finish, add					36%					
	1400	Dark bronze finish, add					18%					
	1500	For black finish, add					36%					
	1600	Concealed panic device, add					895			895	985	
	1700	Electric striker release, add				Opng.	230			230	253	
	1800	Floor check, add				Leaf	685			685	750	
	1900	Concealed closer, add				"	455			455	500	
	2000	Flush 3' x 7' Insulated, 12"x 12" lite, clear finish	2 Sswk	2	8	Ea.	865	170		1,035	1,275	
600	0010	**STAINLESS STEEL AND GLASS** Entrance unit, narrow stiles										600
	0020	3' x 7' opening, including hardware, minimum	2 Sswk	1.60	10	Opng.	4,425	213		4,638	5,300	
	0050	Average		1.40	11.429		4,800	243		5,043	5,750	
	0100	Maximum		1.20	13.333		5,150	283		5,433	6,200	
	1000	For solid bronze entrance units, statuary finish, add					60%					
	1100	Without statuary finish, add					45%					
	2000	Balanced doors, 3' x 7', economy	2 Sswk	.90	17.778	Ea.	6,000	380		6,380	7,350	
	2100	Premium	"	.70	22.857	"	10,300	485		10,785	12,400	
650	0010	**STOREFRONT SYSTEMS** Aluminum frame, clear 3/8" plate glass,										650
	0020	incl. 3' x 7' door with hardware (400 sq. ft. max. wall)										
	0500	Wall height to 12' high, commercial grade	2 Glaz	150	.107	S.F.	11.45	2.06		13.51	16	
	0600	Institutional grade		130	.123		15.25	2.38		17.63	20.50	
	0700	Monumental grade		115	.139		22	2.69		24.69	28.50	
	1000	6' x 7' door with hardware, commercial grade		135	.119		11.70	2.29		13.99	16.65	
	1100	Institutional grade		115	.139		16	2.69		18.69	22	
	1200	Monumental grade		100	.160		29.50	3.10		32.60	37.50	
	1500	For bronze anodized finish, add					15%					
	1600	For black anodized finish, add					30%					
	1700	For stainless steel framing, add to monumental					75%					

08460 | Automatic Entrance Doors

			CREW	DAILY OUTPUT	LABOR-HOURS	UNIT	MAT.	LABOR	EQUIP.	TOTAL	TOTAL INCL O&P	
600	0010	**SLIDING ENTRANCE** 12' x 7'-6" opng., 5' x 7' door, 2 way traf.,										600
	0020	mat activated, panic pushout, incl. operator & hardware,										
	0030	not including glass or glazing	2 Glaz	.70	22.857	Opng.	5,625	440		6,065	6,900	
650	0010	**SLIDING PANELS**										650
	0020	Mall fronts, aluminum & glass, 15' x 9' high	2 Glaz	1.30	12.308	Opng.	2,200	238		2,438	2,825	
	0100	24' x 9' high		.70	22.857		3,200	440		3,640	4,250	
	0200	48' x 9' high, with fixed panels		.90	17.778		5,950	345		6,295	7,125	
	0500	For bronze finish, add					17%					

08480 | Balanced Entrance Doors

			CREW	DAILY OUTPUT	LABOR-HOURS	UNIT	MAT.	LABOR	EQUIP.	TOTAL	TOTAL INCL O&P	
150	0010	**BALANCED DOORS**										150
	0020	Hardware & frame, alum. & glass, 3' x 7', econ.	2 Sswk	.90	17.778	Ea.	4,550	380		4,930	5,750	
	0150	Premium	"	.70	22.857	"	5,650	485		6,135	7,175	

For expanded coverage of these items see *Means Building Construction Cost Data 2000*

08500 | Windows

08510 | Steel Windows

			CREW	DAILY OUTPUT	LABOR-HOURS	UNIT	2000 BARE COSTS MAT.	LABOR	EQUIP.	TOTAL	TOTAL INCL O&P	
700	0010	**SCREENS**										700
	0020	For metal sash, aluminum or bronze mesh, flat screen	2 Sswk	1,200	.013	S.F.	2.94	.28		3.22	3.79	
	0500	Wicket screen, inside window	"	1,000	.016	"	4.49	.34		4.83	5.60	
	0600	Residential, aluminum mesh and frame, 2' x 3'	2 Carp	32	.500	Ea.	10.85	9.85		20.70	29	
	0610	Rescreen		50	.320		8.35	6.30		14.65	19.95	
	0620	3' x 5'		32	.500		23.50	9.85		33.35	43	
	0630	Rescreen		45	.356		20.50	7		27.50	34.50	
	0640	4' x 8'		25	.640		44.50	12.60		57.10	70.50	
	0650	Rescreen		40	.400		33	7.90		40.90	49.50	
	0660	Patio door		25	.640		126	12.60		138.60	161	
	0680	Rescreening		1,600	.010	S.F.	1.06	.20		1.26	1.51	
	0800	Security screen, aluminum frame with stainless steel cloth	2 Sswk	1,200	.013		15.85	.28		16.13	18	
	0900	Steel grate, painted, on steel frame		1,600	.010		8.50	.21		8.71	9.75	
	1000	For solar louvers, add		160	.100		16.40	2.13		18.53	22	
750	0010	**STEEL SASH** Custom units, glazing and trim not included										750
	0100	Casement, 100% vented	2 Sswk	200	.080	S.F.	34	1.70		35.70	41	
	0200	50% vented		200	.080		29	1.70		30.70	35.50	
	0300	Fixed		200	.080		19.35	1.70		21.05	25	
	1000	Projected, commercial, 40% vented		200	.080		35	1.70		36.70	42	
	1100	Intermediate, 50% vented		200	.080		36.50	1.70		38.20	43.50	
	1500	Industrial, horizontally pivoted		200	.080		37.50	1.70		39.20	45	
	1600	Fixed		200	.080		21	1.70		22.70	26.50	
	2000	Industrial security sash, 50% vented		200	.080		39	1.70		40.70	46.50	
	2100	Fixed		200	.080		32.50	1.70		34.20	39.50	
	2500	Picture window		200	.080		20.50	1.70		22.20	26	
	3000	Double hung		200	.080		38	1.70		39.70	45.50	
	5000	Mullions for above, open interior face		240	.067	L.F.	6.90	1.42		8.32	10.40	
	5100	With interior cover		240	.067	"	11.65	1.42		13.07	15.60	
	6000	Double glazing for above, add	2 Glaz	200	.080	S.F.	9	1.55		10.55	12.45	
	6100	Triple glazing for above, add	"	85	.188	"	8.45	3.64		12.09	15.35	
770	0010	**STEEL WINDOWS** Stock, including frame, trim and insul. glass										770
	1000	Custom units, double hung, 2'-8" x 4'-6" opening	2 Sswk	12	1.333	Ea.	485	28.50		513.50	590	
	1100	2'-4" x 3'-9" opening		12	1.333		400	28.50		428.50	495	
	1500	Commercial projected, 3'-9" x 5'-5" opening		10	1.600		850	34		884	1,000	
	1600	6'-9" x 4'-1" opening		7	2.286		1,125	48.50		1,173.50	1,325	
	2000	Intermediate projected, 2'-9" x 4'-1" opening		12	1.333		475	28.50		503.50	580	
	2100	4'-1" x 5'-5" opening		10	1.600		960	34		994	1,125	

08520 | Aluminum Windows

			CREW	DAILY OUTPUT	LABOR-HOURS	UNIT	MAT.	LABOR	EQUIP.	TOTAL	TOTAL INCL O&P	
100	0010	**ALUMINUM SASH**										100
	0020	Stock, grade C, glaze & trim not incl., casement	2 Sswk	200	.080	S.F.	26.50	1.70		28.20	32.50	
	0050	Double hung		200	.080		25.50	1.70		27.20	31.50	
	0100	Fixed casement		200	.080		9.65	1.70		11.35	14	
	0150	Picture window		200	.080		11	1.70		12.70	15.45	
	0200	Projected window		200	.080		22.50	1.70		24.20	28	
	0250	Single hung		200	.080		11.50	1.70		13.20	16	
	0300	Sliding		200	.080		14.85	1.70		16.55	19.70	
	1000	Mullions for above, tubular		240	.067	L.F.	3.94	1.42		5.36	7.10	
	3000	Double glazing for above, add	2 Glaz	200	.080	S.F.	8.65	1.55		10.20	12.10	
	3100	Triple glazing for above, add	"	85	.188	"	10.15	3.64		13.79	17.25	
120	0010	**ALUMINUM WINDOWS** Incl. frame and glazing, grade C										120
	1000	Stock units, casement, 3'-1" x 3'-2" opening	2 Sswk	10	1.600	Ea.	258	34		292	350	
	1040	Insulating glass	"	10	1.600		258	34		292	350	
	1050	Add for storms					52			52	57	

08500 | Windows

08520 | Aluminum Windows

			CREW	DAILY OUTPUT	LABOR-HOURS	UNIT	2000 BARE COSTS MAT.	LABOR	EQUIP.	TOTAL	TOTAL INCL O&P	
120	1600	Projected, with screen, 3'-1" x 3'-2" opening	2 Sswk	10	1.600	Ea.	184	34		218	270	120
	1650	Insulating glass	"	10	1.600		172	34		206	257	
	1700	Add for storms					49			49	54	
	2000	4'-5" x 5'-3" opening	2 Sswk	8	2		260	42.50		302.50	370	
	2050	Insulating glass	"	8	2		299	42.50		341.50	415	
	2100	Add for storms					68			68	74.50	
	2500	Enamel finish windows, 3'-1" x 3'-2"	2 Sswk	10	1.600		165	34		199	249	
	2550	Insulating glass		10	1.600		180	34		214	265	
	2600	4'-5" x 5'-3"		8	2		248	42.50		290.50	355	
	2700	Insulating glass		8	2		325	42.50		367.50	440	
	3000	Single hung, 2' x 3' opening, enameled, standard glazed		10	1.600		124	34		158	203	
	3100	Insulating glass		10	1.600		150	34		184	232	
	3300	2'-8" x 6'-8" opening, standard glazed		8	2		263	42.50		305.50	375	
	3400	Insulating glass		8	2		340	42.50		382.50	455	
	3700	3'-4" x 5'-0" opening, standard glazed		9	1.778		170	38		208	262	
	3800	Insulating glass		9	1.778		238	38		276	335	
	4000	Sliding aluminum, 3' x 2' opening, standard glazed		10	1.600		139	34		173	220	
	4100	Insulating glass		10	1.600		154	34		188	237	
	4300	5' x 3' opening, standard glazed		9	1.778		177	38		215	270	
	4400	Insulating glass		9	1.778		247	38		285	345	
	4600	8' x 4' opening, standard glazed		6	2.667		255	56.50		311.50	390	
	4700	Insulating glass		6	2.667		410	56.50		466.50	560	
	5000	9' x 5' opening, standard glazed		4	4		385	85		470	595	
	5100	Insulating glass		4	4		615	85		700	850	
	5500	Sliding, with thermal barrier and screen, 6' x 4', 2 track		8	2		525	42.50		567.50	660	
	5700	4 track		8	2		640	42.50		682.50	790	
	6000	For above units with bronze finish, add					12%					
	6200	For installation in concrete openings, add					5%					
500	0010	**JALOUSIES**										500
	0020	Aluminum incl. glazing & screens, stock, 1'-7" x 3'-2"	2 Sswk	10	1.600	Ea.	113	34		147	191	
	0100	2'-3" x 4'-0"		10	1.600		162	34		196	245	
	0200	3'-1" x 2'-0"		10	1.600		111	34		145	189	
	0300	3'-1" x 5'-3"		10	1.600		231	34		265	320	
	1000	Mullions for above, 2'-0" long		80	.200		8.70	4.25		12.95	17.95	
	1100	5'-3" long		80	.200		14.90	4.25		19.15	25	

08550 | Wood Windows

			CREW	DAILY OUTPUT	LABOR-HOURS	UNIT	MAT.	LABOR	EQUIP.	TOTAL	TOTAL INCL O&P	
100	0010	**AWNING WINDOW** Including frame, screen, and exterior trim										100
	0100	Average quality, builders model, 34" x 22", double insulated glass	1 Carp	10	.800	Ea.	220	15.75		235.75	269	
	0200	Low E glass		10	.800		220	15.75		235.75	269	
	0300	40" x 28", double insulated glass		9	.889		370	17.50		387.50	440	
	0400	Low E Glass		9	.889		370	17.50		387.50	440	
	0500	48" x 36", double insulated glass		8	1		685	19.70		704.70	790	
	0600	Low E glass		8	1		685	19.70		704.70	790	
	1000	34" x 22"		10	.800		193	15.75		208.75	239	
	1100	40" x 22"		10	.800		213	15.75		228.75	261	
	1200	36" x 28"		9	.889		227	17.50		244.50	280	
	1300	36" x 36"		9	.889		286	17.50		303.50	345	
	1400	48" x 28"		8	1		270	19.70		289.70	330	
	1500	60" x 36"		8	1		370	19.70		389.70	445	
	2000	Metal clad, deluxe, double insulated glass, 34" x 22"		10	.800		184	15.75		199.75	230	
	2100	40" x 22"		10	.800		201	15.75		216.75	249	
	2200	36" x 25"		9	.889		205	17.50		222.50	256	
	2300	40" x 30"		9	.889		267	17.50		284.50	325	
	2400	48" x 28"		8	1		241	19.70		260.70	299	

For expanded coverage of these items see *Means Building Construction Cost Data 2000*

08500 | Windows
08550 | Wood Windows

			Crew	Daily Output	Labor-Hours	Unit	Mat.	Labor	Equip.	Total	Total Incl O&P	
100	2500	60" x 36"	1 Carp	8	1	Ea.	237	19.70		256.70	295	100
150	0010	**BOW-BAY WINDOW** Including frame, screens and grills,										150
	0020	end panels operable										
	1000	Bow type, casement, wood, bldrs mdl, 8' x 5' dbl insltd glass, 4 panel	2 Carp	10	1.600	Ea.	850	31.50		881.50	990	
	1050	Low E glass		10	1.600		1,100	31.50		1,131.50	1,250	
	1100	10'-0" x 5'-0", double insulated glass, 6 panels		6	2.667		1,125	52.50		1,177.50	1,350	
	1200	Low E glass, 6 panels		6	2.667		1,200	52.50		1,252.50	1,425	
	1300	Vinyl clad, bldrs model, double insulated glass, 6'-0" x 4'-0", 3 panel		10	1.600		925	31.50		956.50	1,075	
	1340	9'-0" x 4'-0", 4 panel		8	2		1,225	39.50		1,264.50	1,425	
	1380	10'-0" x 6'-0", 5 panels		7	2.286		2,075	45		2,120	2,350	
	1420	12'-0" x 6'-0", 6 panels		6	2.667		2,650	52.50		2,702.50	3,000	
	1600	Metal clad, casement, bldrs mdl, 6'-0" x 4'-0", dbl insltd gls, 3 panels		10	1.600		770	31.50		801.50	900	
	1640	9'-0" x 4'-0", 4 panels		8	2		1,075	39.50		1,114.50	1,275	
	1680	10'-0" x 5'-0", 5 panels		7	2.286		1,475	45		1,520	1,700	
	1720	12'-0" x 6'-0", 6 panels		6	2.667		2,075	52.50		2,127.50	2,375	
	2000	Bay window, casement, builders model, 8' x 5' dbl insul glass, 4 panels		10	1.600		1,300	31.50		1,331.50	1,475	
	2050	Low E glass,		10	1.600		1,575	31.50		1,606.50	1,775	
	2100	12'-0" x 6'-0", double insulated glass, 6 panels		6	2.667		1,625	52.50		1,677.50	1,875	
	2200	Low E glass		6	2.667		1,675	52.50		1,727.50	1,950	
	2280	6'-0" x 4'-0"		11	1.455		955	28.50		983.50	1,100	
	2300	Vinyl clad, premium, insulating glass, 8'-0" x 5'-0"		10	1.600		1,125	31.50		1,156.50	1,275	
	2340	10'-0" x 5'-0"		8	2		1,600	39.50		1,639.50	1,825	
	2380	10'-0" x 6'-0"		7	2.286		1,650	45		1,695	1,900	
	2420	12'-0" x 6'-0"		6	2.667		1,975	52.50		2,027.50	2,275	
	2600	Metal clad, deluxe, insul. glass, 8'-0" x 5'-0" high, 4 panels		10	1.600		1,150	31.50		1,181.50	1,300	
	2640	10'-0" x 5'-0" high, 5 panels		8	2		1,225	39.50		1,264.50	1,425	
	2680	10'-0" x 6'-0" high, 5 panels		7	2.286		1,450	45		1,495	1,675	
	2720	12'-0" x 6'-0" high, 6 panels		6	2.667		2,025	52.50		2,077.50	2,325	
	3000	Double hung, bldrs. model, bay, 8' x 4' high, std. glazed		10	1.600		910	31.50		941.50	1,050	
	3050	Insulating glass		10	1.600		975	31.50		1,006.50	1,125	
	3100	9'-0" x 5'-0" high, standard glazed		6	2.667		980	52.50		1,032.50	1,175	
	3200	Insulating glass		6	2.667		1,025	52.50		1,077.50	1,225	
	3300	Vinyl clad, premium, insulating glass, 7'-0" x 4'-6"		10	1.600		945	31.50		976.50	1,075	
	3340	8'-0" x 4'-6"		8	2		960	39.50		999.50	1,125	
	3380	8'-0" x 5'-0"		7	2.286		1,000	45		1,045	1,175	
	3420	9'-0" x 5'-0"		6	2.667		1,025	52.50		1,077.50	1,225	
	3600	Metal clad, deluxe, insul. glass, 7'-0" x 4'-0" high		10	1.600		870	31.50		901.50	1,025	
	3640	8'-0" x 4'-0" high		8	2		900	39.50		939.50	1,075	
	3680	8'-0" x 5'-0" high		7	2.286		935	45		980	1,100	
	3720	9'-0" x 5'-0" high		6	2.667		990	52.50		1,042.50	1,200	
	7000	Drip cap, premolded vinyl, 8' long		30	.533		67.50	10.50		78	92	
	7040	12' long		26	.615		73.50	12.10		85.60	102	
200	0010	**CASEMENT WINDOW** Including frame, screen, and exterior trim										200
	0100	Avg. quality, bldrs. model, 2'-0" x 3'-0" H, standard glazed	1 Carp	10	.800	Ea.	159	15.75		174.75	202	
	0150	Insulating glass		10	.800		196	15.75		211.75	242	
	0200	2'-0" x 4'-6" high, standard glazed		9	.889		207	17.50		224.50	257	
	0250	Insulating glass		9	.889		266	17.50		283.50	325	
	0300	2'-3" x 6'-0" high, standard glazed		8	1		226	19.70		245.70	282	
	0350	Insulating glass		8	1		299	19.70		318.70	365	
	8000	Solid vinyl, premium, insulating glass, 2'-0" x 3'-0" high		10	.800		162	15.75		177.75	206	
	8020	2'-0" x 4'-0" high		9	.889		199	17.50		216.50	249	
	8040	2'-0" x 5'-0" high		8	1		229	19.70		248.70	286	
	8100	Metal clad, deluxe, insulating glass, 2'-0" x 3'-0" high		10	.800		160	15.75		175.75	203	
	8120	2'-0" x 4'-0" high		9	.889		193	17.50		210.50	242	

Important: See the Reference Section for critical supporting data - Reference Nos., Crews, & Location Factors

08500 | Windows

08550 | Wood Windows

			CREW	DAILY OUTPUT	LABOR-HOURS	UNIT	2000 BARE COSTS				TOTAL INCL O&P
							MAT.	LABOR	EQUIP.	TOTAL	
200	8140	2'-0" x 5'-0" high	1 Carp	8	1	Ea.	219	19.70		238.70	275
	8160	2'-0" x 6'-0" high	↓	8	1	↓	252	19.70		271.70	310
	8200	For multiple leaf units, deduct for stationary sash									
	8220	2' high				Ea.	16.65			16.65	18.30
	8240	4'-6" high					19.20			19.20	21
	8260	6' high					25.50			25.50	28
	8300	For installation, add per leaf				↓		15%			
250	0010	**DOUBLE HUNG** Including frame, screen, and exterior trim									
	0100	Avg. quality, bldrs. model, 2'-0" x 3'-0" high, standard glazed	1 Carp	10	.800	Ea.	107	15.75		122.75	144
	0150	Insulating glass		10	.800		151	15.75		166.75	194
	0200	3'-0" x 4'-0" high, standard glazed		9	.889		140	17.50		157.50	184
	0250	Insulating glass		9	.889		185	17.50		202.50	233
	0300	4'-0" x 4'-6" high, standard glazed		8	1		172	19.70		191.70	223
	0350	Insulating glass		8	1		227	19.70		246.70	284
	1000	Vinyl clad, premium, insulating glass, 2'-6" x 3'-0"		10	.800		240	15.75		255.75	291
	1100	3'-0" x 3'-6"		10	.800		291	15.75		306.75	345
	1200	3'-0" x 4'-0"		9	.889		315	17.50		332.50	375
	1300	3'-0" x 4'-6"		9	.889		330	17.50		347.50	390
	1400	3'-0" x 5'-0"		8	1		340	19.70		359.70	410
	1500	3'-6" x 6'-0"		8	1		570	19.70		589.70	660
	1540	Solid vinyl, average quality, insulated glass, 2'-0" x 3'-0"		10	.800		121	15.75		136.75	160
	1542	3'-0" x 4'-0"		9	.889		146	17.50		163.50	191
	1544	4'-0" x 4'-6"		8	1		175	19.70		194.70	227
	1560	Premium, insulating glass, 2'-6" x 3'-0"		10	.800		137	15.75		152.75	178
	1562	3'-0" x 3'-6"		9	.889		159	17.50		176.50	205
	1564	3'-0" x 4'-0"		9	.889		169	17.50		186.50	216
	1566	3'-0" x 4'-6"		9	.889		173	17.50		190.50	221
	1568	3'-0" x 5'-0"		8	1		178	19.70		197.70	230
	1570	3'-6" x 6'-0"		8	1		195	19.70		214.70	249
	2000	Metal clad, deluxe, insulating glass, 2'-6" x 3'-0" high		10	.800		170	15.75		185.75	214
	2100	3'-0" x 3'-6" high		10	.800		201	15.75		216.75	248
	2200	3'-0" x 4'-0" high		9	.889		214	17.50		231.50	265
	2300	3'-0" x 4'-6" high		9	.889		231	17.50		248.50	284
	2400	3'-0" x 5'-0" high		8	1		247	19.70		266.70	305
	2500	3'-6" x 6'-0" high	↓	8	1	↓	300	19.70		319.70	365
650	0010	**PALLADIAN WINDOWS**									
	0020	Aluminum clad, including frame and exterior trim									
	0040	3'-2" x 2'-0" high	2 Carp	11	1.455	Ea.	905	28.50		933.50	1,050
	0060	3'-2" x 4'-10"		11	1.455		1,025	28.50		1,053.50	1,175
	0080	3'-2" x 6'-4"		10	1.600		1,225	31.50		1,256.50	1,400
	0100	4'-0" x 4'-0"	↓	10	1.600		1,000	31.50		1,031.50	1,150
	0120	4'-0" x 5'-4"	3 Carp	10	2.400		1,150	47.50		1,197.50	1,350
	0140	4'-0" x 6'-0"		9	2.667		1,200	52.50		1,252.50	1,425
	0160	4'-0" x 7'-4"		9	2.667		1,325	52.50		1,377.50	1,550
	0180	5'-5" x 4'-10"		9	2.667		1,300	52.50		1,352.50	1,525
	0200	5'-5" x 6'-10"		9	2.667		1,550	52.50		1,602.50	1,800
	0220	5'-5" x 7'-9"		9	2.667		1,675	52.50		1,727.50	1,950
	0240	6'-0" x 7'-11"		8	3		2,225	59		2,284	2,550
	0260	8'-0" x 6'-0"	↓	8	3		1,900	59		1,959	2,200
670	0010	**PICTURE WINDOW** Including frame and exterior trim									
	0100	Average quality, bldrs. model, 3'-6" x 4'-0" high, standard glazed	2 Carp	12	1.333	Ea.	238	26.50		264.50	305
	0150	Insulating glass		12	1.333		200	26.50		226.50	265
	0200	4'-0" x 4'-6" high, standard glazed		11	1.455		198	28.50		226.50	267
	0250	Insulating glass		11	1.455		202	28.50		230.50	271
	0300	5'-0" x 4'-0" high, standard glazed	↓	11	1.455		200	28.50		228.50	269

For expanded coverage of these items see *Means Building Construction Cost Data 2000*

08500 | Windows

08550 | Wood Windows

			CREW	DAILY OUTPUT	LABOR-HOURS	UNIT	2000 BARE COSTS MAT.	LABOR	EQUIP.	TOTAL	TOTAL INCL O&P	
670	0350	Insulating glass	2 Carp	11	1.455	Ea.	250	28.50		278.50	325	670
	0400	6'-0" x 4'-6" high, standard glazed		10	1.600		280	31.50		311.50	365	
	0450	Insulating glass		10	1.600		345	31.50		376.50	435	
	1000	Vinyl clad, premium, insulating glass, 4'-0" x 4'-0"		12	1.333		420	26.50		446.50	510	
	1100	4'-0" x 6'-0"		11	1.455		575	28.50		603.50	685	
	1200	5'-0" x 6'-0"		10	1.600		750	31.50		781.50	880	
	1300	6'-0" x 6'-0"		10	1.600		765	31.50		796.50	895	
	2000	Metal clad, deluxe, insulating glass, 4'-0" x 4'-0" high		12	1.333		252	26.50		278.50	320	
	2100	4'-0" x 6'-0" high		11	1.455		370	28.50		398.50	460	
	2200	5'-0" x 6'-0" high		10	1.600		410	31.50		441.50	505	
	2300	6'-0" x 6'-0" high	▼	10	1.600	▼	470	31.50		501.50	570	
750	0010	**SLIDING WINDOW** Including frame, screen, and exterior trim										750
	0100	Average quality, bldrs. model, 3'-0" x 3'-0" high, standard glazed	1 Carp	10	.800	Ea.	128	15.75		143.75	168	
	0120	Insulating glass		10	.800		161	15.75		176.75	205	
	0200	4'-0" x 3'-6" high, standard glazed		9	.889		152	17.50		169.50	197	
	0220	Insulating glass		9	.889		190	17.50		207.50	239	
	0300	6'-0" x 5'-0" high, standard glazed		8	1		279	19.70		298.70	340	
	0320	Insulating glass		8	1		335	19.70		354.70	400	
	2000	Metal clad, deluxe, insulating glass, 3'-0" x 3'-0" high		10	.800		260	15.75		275.75	315	
	2050	4'-0" x 3'-6" high		9	.889		320	17.50		337.50	380	
	2100	5'-0" x 4'-0" high		9	.889		385	17.50		402.50	450	
	2150	6'-0" x 5'-0" high	▼	8	1	▼	470	19.70		489.70	550	
770	0010	**WEATHERSTRIPPING** See division 08720										770
800	0010	**WINDOW GRILLE OR MUNTIN** Snap-in type										800
	0020	Colonial or diamond pattern										
	2000	Wood, awning window, glass size 28" x 16" high	1 Carp	30	.267	Ea.	22	5.25		27.25	33	
	2060	44" x 24" high		32	.250		28	4.93		32.93	39.50	
	2100	Casement, glass size, 20" x 36" high		30	.267		27.50	5.25		32.75	39	
	2180	20" x 56" high		32	.250	▼	36.50	4.93		41.43	49	
	2200	Double hung, glass size, 16" x 24" high		24	.333	Set	57.50	6.55		64.05	75	
	2280	32" x 32" high		34	.235	"	104	4.64		108.64	122	
	2500	Picture, glass size, 48" x 48" high		30	.267	Ea.	110	5.25		115.25	130	
	2580	60" x 68" high		28	.286	"	122	5.65		127.65	144	
	2600	Sliding, glass size, 14" x 36" high		24	.333	Set	31.50	6.55		38.05	46	
	2680	36" x 36" high	▼	22	.364	"	49.50	7.15		56.65	67	
820	0010	**WOOD SASH** Including glazing but not including trim										820
	0050	Custom, 5'-0" x 4'-0", 1" dbl. glazed, 3/16" thick lites	2 Carp	3.20	5	Ea.	187	98.50		285.50	375	
	0100	1/4" thick lites		5	3.200		192	63		255	320	
	0200	1" thick, triple glazed		5	3.200		440	63		503	595	
	0300	7'-0" x 4'-6" high, 1" double glazed, 3/16" thick lites		4.30	3.721		445	73.50		518.50	615	
	0400	1/4" thick lites		4.30	3.721		505	73.50		578.50	680	
	0500	1" thick, triple glazed		4.30	3.721		575	73.50		648.50	760	
	0600	8'-6" x 5'-0" high, 1" double glazed, 3/16" thick lites		3.50	4.571		605	90		695	820	
	0700	1/4" thick lites		3.50	4.571		660	90		750	880	
	0800	1" thick, triple glazed	▼	3.50	4.571	▼	665	90		755	885	
	0900	Window frames only, based on perimeter length				L.F.	3.69			3.69	4.06	
	7000	Sash, single lite, 2'-0" x 2'-0" high	1 Carp	20	.400	Ea.	35.50	7.90		43.40	52.50	
	7050	2'-6" x 2'-0" high		19	.421		38	8.30		46.30	56	
	7100	2'-6" x 2'-6" high		18	.444		40.50	8.75		49.25	59.50	
	7150	3'-0" x 2'-0" high	▼	17	.471	▼	51	9.25		60.25	72	

08500 | Windows

08580 | Special Function Windows

		Crew	Daily Output	Labor-Hours	Unit	Mat.	Labor	Equip.	Total	Total Incl O&P
0010	**STORM WINDOWS** Aluminum, residential									
0300	Basement, mill finish, incl. fiberglass screen									
0320	1'-10" x 1'-0" high	2 Carp	30	.533	Ea.	25	10.50		35.50	45
0340	2'-9" x 1'-6" high		30	.533		27	10.50		37.50	47.50
0360	3'-4" x 2'-0" high	▼	30	.533	▼	32.50	10.50		43	54
1600	Double-hung, combination, storm & screen									
1700	Custom, clear anodic coating, 2'-0" x 3'-5" high	2 Carp	30	.533	Ea.	64	10.50		74.50	88.50
1720	2'-6" x 5'-0" high		28	.571		86	11.25		97.25	114
1740	4'-0" x 6'-0" high		25	.640		182	12.60		194.60	222
1800	White painted, 2'-0" x 3'-5" high		30	.533		76.50	10.50		87	102
1820	2'-6" x 5'-0" high		28	.571		123	11.25		134.25	154
1840	4'-0" x 6'-0" high		25	.640		220	12.60		232.60	264
2000	Average quality, clear anodic coating, 2'-0" x 3'-5" high		30	.533		65	10.50		75.50	89.50
2020	2'-6" x 5'-0" high		28	.571		82	11.25		93.25	110
2040	4'-0" x 6'-0" high		25	.640		96.50	12.60		109.10	128
2400	White painted, 2'-0" x 3'-5" high		30	.533		64	10.50		74.50	88.50
2420	2'-6" x 5'-0" high		28	.571		71	11.25		82.25	97.50
2440	4'-0" x 6'-0" high		25	.640		77.50	12.60		90.10	107
2600	Mill finish, 2'-0" x 3'-5" high		30	.533		58.50	10.50		69	82.50
2620	2'-6" x 5'-0" high		28	.571		65	11.25		76.25	91.50
2640	4'-0" x 6-8" high	▼	25	.640	▼	73	12.60		85.60	102
4000	Picture window, storm, 1 lite, white or bronze finish									
4020	4'-6" x 4'-6" high	2 Carp	25	.640	Ea.	98	12.60		110.60	130
4040	5'-8" x 4'-6" high		20	.800		111	15.75		126.75	149
4400	Mill finish, 4'-6" x 4'-6" high		25	.640		98	12.60		110.60	130
4420	5'-8" x 4'-6" high	▼	20	.800	▼	111	15.75		126.75	149
4600	3 lite, white or bronze finish									
4620	4'-6" x 4'-6" high	2 Carp	25	.640	Ea.	119	12.60		131.60	153
4640	5'-8" x 4'-6" high		20	.800		133	15.75		148.75	173
4800	Mill finish, 4'-6" x 4'-6" high		25	.640		105	12.60		117.60	137
4820	5'-8" x 4'-6" high	▼	20	.800	▼	111	15.75		126.75	149
5000	Sliding glass door, storm 6' x 6'-8", standard	1 Glaz	2	4		600	77.50		677.50	790
5100	Economy	"	2	4	▼	265	77.50		342.50	420
6000	Sliding window, storm, 2 lite, white or bronze finish									
6020	3'-4" x 2'-7" high	2 Carp	28	.571	Ea.	82.50	11.25		93.75	110
6040	4'-4" x 3'-3" high		25	.640		112	12.60		124.60	145
6060	5'-4" x 6'-0" high	▼	20	.800		180	15.75		195.75	225
6400	3 lite, white or bronze finish									
6420	4'-4" x 3'-3" high	2 Carp	25	.640	Ea.	130	12.60		142.60	165
6440	5'-4" x 6'-0" high		20	.800		235	15.75		250.75	285
6460	6'-0" x 6'-0" high		18	.889		236	17.50		253.50	289
6800	Mill finish, 4'-4" x 3'-3" high		25	.640		112	12.60		124.60	145
6820	5'-4" x 6'-0" high		20	.800		235	15.75		250.75	286
6840	6'-0" x 6-0" high	▼	18	.889	▼	243	17.50		260.50	298
9000	Magnetic interior storm window									
9100	3/16" plate glass	1 Glaz	107	.075	S.F.	3.79	1.45		5.24	6.55

08590 | Window Restoration & Replace

		Crew	Daily Output	Labor-Hours	Unit	Mat.	Labor	Equip.	Total	Total Incl O&P
0010	**SOLID VINYL REPLACEMENT WINDOWS** R08550-200									
0020	White, double hung, up to 83 united inches	2 Carp	8	2	Ea.	145	39.50		184.50	227
0040	84 to 93		8	2		145	39.50		184.50	227
0060	94 to 101		6	2.667		145	52.50		197.50	249
0080	102 to 111		6	2.667		167	52.50		219.50	273
0100	112 to 120		6	2.667	▼	186	52.50		238.50	295
0120	For each united inch over 120, add		800	.020	Inch	2.17	.39		2.56	3.07
0140	Casement windows, one operating sash, 42 to 60 united inches	▼	8	2	Ea.	165	39.50		204.50	249

For expanded coverage of these items see *Means Building Construction Cost Data 2000*

08500 | Windows

08590 | Window Restoration & Replace

		CREW	DAILY OUTPUT	LABOR-HOURS	UNIT	2000 BARE COSTS MAT.	LABOR	EQUIP.	TOTAL	TOTAL INCL O&P
0160	61 to 70	2 Carp	8	2	Ea.	188	39.50		227.50	275
0180	71 to 80		8	2		204	39.50		243.50	292
0200	81 to 96		8	2		217	39.50		256.50	305
0220	Two operating sash, 58 to 78 united inches		8	2		299	39.50		338.50	400
0240	79 to 88		8	2		335	39.50		374.50	440
0260	89 to 98		8	2		365	39.50		404.50	470
0280	99 to 108		6	2.667		385	52.50		437.50	515
0300	109 to 121		6	2.667		420	52.50		472.50	555
0320	Three operating sash, 73 to 108 united inches		8	2		450	39.50		489.50	565
0340	109 to 118		8	2		480	39.50		519.50	600
0360	119 to 128		6	2.667		525	52.50		577.50	665
0380	129 to 138		6	2.667		570	52.50		622.50	715
0400	139 to 156		6	2.667		605	52.50		657.50	755
0420	Four operating sash, 89 to 98 united inches		8	2		600	39.50		639.50	730
0440	99 to 108		8	2		640	39.50		679.50	775
0460	109 to 118		6	2.667		695	52.50		747.50	855
0480	119 to 128		6	2.667		755	52.50		807.50	920
0500	129 to 138		6	2.667		810	52.50		862.50	980
0520	139 to 148		6	2.667		870	52.50		922.50	1,050
0540	149 to 168		4	4		925	79		1,004	1,150
0560	Fixed picture window, up to 63 united inches		8	2		125	39.50		164.50	205
0580	64 to 83		8	2		167	39.50		206.50	252
0600	84 to 101		8	2		213	39.50		252.50	300
0620	For each united inch over 101, add		900	.018	Inch	2.32	.35		2.67	3.15
0640	Picture window opt., low E glazing, up to 101 united inches				Ea.	17.70			17.70	19.45
0660	102 to 124					22.50			22.50	25
0680	124 and over					34.50			34.50	38
0700	Options, low E glazing, up to 101 united inches					8.85			8.85	9.70
0720	102 to 124					22.50			22.50	25
0740	124 and over					34.50			34.50	38
0760	Muntins, between glazing, square, per lite					1.67			1.67	1.84
0780	Diamond shape, per full or partial diamond					2.78			2.78	3.06
0800	Celluose fiber insulation, poured into sash balance cavity	1 Carp	36	.222	C.F.	.45	4.38		4.83	8
0820	Silicone caulking at perimeter	"	800	.010	L.F.	.13	.20		.33	.48

08600 | Skylights

08620 | Unit Skylights

		CREW	DAILY OUTPUT	LABOR-HOURS	UNIT	2000 BARE COSTS MAT.	LABOR	EQUIP.	TOTAL	TOTAL INCL O&P
0010	**SKYLIGHT** Plastic domes, flush or curb mounted, ten or									
0100	more units, curb not included									
0300	Nominal size under 10 S.F., double	G-3	130	.246	S.F.	12	4.43		16.43	20.50
0400	Single		160	.200		8.05	3.60		11.65	14.95
0600	10 S.F. to 20 S.F., double		315	.102		9.80	1.83		11.63	13.85
0700	Single		395	.081		7.40	1.46		8.86	10.60
0900	20 S.F. to 30 S.F., double		395	.081		9.80	1.46		11.26	13.25
1000	Single		465	.069		8.75	1.24		9.99	11.70
1200	30 S.F. to 65 S.F., double		465	.069		9.20	1.24		10.44	12.20
1300	Single		610	.052		13.15	.94		14.09	16.05
1500	For insulated 4" curbs, double, add					25%				
1600	Single, add					30%				

08600 | Skylights

08620 | Unit Skylights

			CREW	DAILY OUTPUT	LABOR-HOURS	UNIT	MAT.	LABOR	EQUIP.	TOTAL	TOTAL INCL O&P	
800	1800	For integral insulated 9" curbs, double, add					30%					800
	1900	Single, add					40%					
	2120	Ventilating insulated plexiglass dome with										
	2130	curb mounting, 36" x 36"	G-3	12	2.667	Ea.	340	48		388	455	
	2150	52" x 52"		12	2.667		510	48		558	645	
	2160	28" x 52"		10	3.200		400	57.50		457.50	540	
	2170	36" x 52"		10	3.200		430	57.50		487.50	575	
	2180	For electric opening system, add					256			256	281	
	2210	Operating skylight, with thermopane glass, 24" x 48"	G-3	10	3.200		500	57.50		557.50	650	
	2220	32" x 48"	"	9	3.556		525	64		589	685	
	2310	Non venting insulated plexiglass dome skylight with										
	2320	Flush mount 22" x 46"	G-3	15.23	2.101	Ea.	279	38		317	370	
	2330	30" x 30"		16	2		256	36		292	340	
	2340	46" x 46"		13.91	2.301		470	41.50		511.50	590	
	2350	Curb mount 22" x 46"		15.23	2.101		244	38		282	335	
	2360	30" x 30"		16	2		233	36		269	315	
	2370	46" x 46"		13.91	2.301		440	41.50		481.50	550	
	2381	Non-insulated flush mount 22" x 46"		15.23	2.101		188	38		226	270	
	2382	30" x 30"		16	2		171	36		207	249	
	2383	46" x 46"		13.91	2.301		320	41.50		361.50	420	
	2384	Curb mount 22" x 46"		15.23	2.101		159	38		197	239	
	2385	30" x 30"		16	2		153	36		189	230	
	2400	Sandwich panels, fiberglass, for walls, 1-9/16" thick, to 250 SF		200	.160	S.F.	13.95	2.88		16.83	20	
	2500	250 SF and up		265	.121		12.50	2.17		14.67	17.45	
	2700	As above, but for roofs, 2-3/4" thick, to 250 SF		295	.108		20	1.95		21.95	25.50	
	2800	250 SF and up		330	.097		16.50	1.74		18.24	21	

08700 | Hardware

08710 | Door Hardware

			CREW	DAILY OUTPUT	LABOR-HOURS	UNIT	MAT.	LABOR	EQUIP.	TOTAL	TOTAL INCL O&P	
100	0010	**AUTOMATIC OPENERS COMMERCIAL** Swing doors, single	2 Skwk	.80	20	Ea.	2,225	395		2,620	3,125	100
	0100	Single operating pair		.50	32	Pr.	4,100	630		4,730	5,600	
	0400	For double simultaneous doors, one way, add		1.20	13.333		345	263		608	830	
	0500	Two way, add		.90	17.778		430	350		780	1,075	
	1000	Sliding doors, 3' wide, including track & hanger, single		.60	26.667	Opng.	3,775	525		4,300	5,075	
	1300	Bi-parting		.50	32		4,575	630		5,205	6,100	
	1450	Activating carpet, single door, one way add		2.20	7.273		700	144		844	1,025	
	1550	Two way, add		1.30	12.308		1,000	243		1,243	1,525	
	1750	Handicap opener, button operating	1 Carp	1.50	5.333	Ea.	1,300	105		1,405	1,600	
150	0010	**AVERAGE** Percentage for hardware, total job cost, minimum									.75%	150
	0050	Maximum									3.50%	
	0500	Total hardware for building, average distribution					85%	15%				
	1000	Door hardware, apartment, interior				Door	107			107	118	
	1500	Hospital bedroom, minimum					243			243	268	
	2000	Maximum					535			535	585	
	2100	Pocket door				Ea.	109			109	120	
	2250	School, single exterior, incl. lever, not incl. panic device				Door	360			360	395	
	2500	Single interior, regular use, no lever included					240			240	264	
	2550	Including handicap lever					325			325	360	

For expanded coverage of these items see *Means Building Construction Cost Data 2000*

08700 | Hardware

08710 | Door Hardware

			CREW	DAILY OUTPUT	LABOR-HOURS	UNIT	2000 BARE COSTS				TOTAL INCL O&P	
							MAT.	LABOR	EQUIP.	TOTAL		
150	2600	Heavy use, incl. lever and closer				Door	420			420	460	150
	2850	Stairway, single interior				↓	600			600	660	
	3100	Double exterior, with panic device				Pr.	845			845	930	
	3600	Toilet, public, single interior				Door	132			132	145	
300	0010	**DOOR CLOSER** Rack and pinion	1 Carp	6.50	1.231	Ea.	108	24.50		132.50	161	300
	0020	Adjustable backcheck, 3 way mount, all sizes, regular arm		6	1.333	"	114	26.50		140.50	171	
	0040	Hold open arm		6	1.333	Ea.	125	26.50		151.50	182	
	0100	Fusible link		6.50	1.231		97	24.50		121.50	149	
	0200	Non sized, regular arm		6	1.333		108	26.50		134.50	164	
	0240	Hold open arm		6	1.333		135	26.50		161.50	193	
	0400	4 way mount, non sized, regular arm		6	1.333		148	26.50		174.50	208	
	0440	Hold open arm	↓	6	1.333	↓	160	26.50		186.50	221	
340	0010	**DOORSTOPS** Holder and bumper, floor or wall	1 Carp	32	.250	Ea.	27	4.93		31.93	38.50	340
	1300	Wall bumper, 4" diameter, with rubber pad, aluminum		32	.250		6.60	4.93		11.53	15.70	
	1600	Door bumper, floor type, aluminum		32	.250		3.77	4.93		8.70	12.60	
	1900	Plunger type, door mounted	↓	32	.250		23	4.93		27.93	34	
400	0010	**ENTRANCE LOCKS** Cylinder, grip handle, deadlocking latch	1 Carp	9	.889	Ea.	103	17.50		120.50	143	400
	0020	Deadbolt		8	1		125	19.70		144.70	171	
	0100	Push and pull plate, dead bolt	↓	8	1		119	19.70		138.70	165	
	0900	For handicapped lever, add				↓	130			130	143	
500	0010	**HASP** Steel, 3" assembly	1 Carp	26	.308	Ea.	2.39	6.05		8.44	13.05	500
	0020	4-1/2"		13	.615		3.04	12.10		15.14	24.50	
	0040	6"	↓	12.50	.640	↓	4.81	12.60		17.41	27	
520	0010	**HINGES** Full mortise, avg. freq., steel base, 4-1/2" x 4-1/2", USP				Pr.	17.90			17.90	19.70	520
	0100	5" x 5", USP					29			29	32	
	0200	6" x 6", USP					60.50			60.50	67	
	0400	Brass base, 4-1/2" x 4-1/2", US10					37			37	40.50	
	0500	5" x 5", US10					52.50			52.50	57.50	
	0600	6" x 6", US10					87			87	96	
	0800	Stainless steel base, 4-1/2" x 4-1/2", US32				↓	61			61	67.50	
	0900	For non removable pin, add				Ea.	2.14			2.14	2.35	
	0910	For floating pin, driven tips, add					3.90			3.90	4.29	
	0930	For hospital type tip on pin, add					11.45			11.45	12.60	
	0940	For steeple type tip on pin, add				↓	7.65			7.65	8.40	
	0950	Full mortise, high frequency, steel base, 3-1/2" x 3-1/2", US26D				Pr.	19.40			19.40	21.50	
	1000	4-1/2" x 4-1/2", USP					37.50			37.50	41.50	
	1100	5" x 5", USP					40			40	44	
	1200	6" x 6", USP					97			97	107	
	1400	Brass base, 3-1/2" x 3-1/2", US4					35			35	38.50	
	1430	4-1/2" x 4-1/2", US10					41.50			41.50	46	
	1500	5" x 5", US10					74.50			74.50	82	
	1600	6" x 6", US10					124			124	136	
	1800	Stainless steel base, 4-1/2" x 4-1/2", US32				↓	94.50			94.50	104	
	1930	For hospital type tip on pin, add				Ea.	15.05			15.05	16.55	
	1950	Full mortise, low frequency, steel base, 3-1/2" x 3-1/2", US26D				Pr.	12.25			12.25	13.45	
	2000	4-1/2" x 4-1/2", USP					13.45			13.45	14.80	
	2100	5" x 5", USP					21.50			21.50	23.50	
	2200	6" x 6", USP					42			42	46.50	
	2300	4-1/2" x 4-1/2", US3					9.85			9.85	10.80	
	2310	5" x 5", US3					30.50			30.50	33.50	
	2400	Brass bass, 4-1/2" x 4-1/2", US10					30.50			30.50	34	
	2500	5" x 5", US10					46			46	50.50	
	2800	Stainless steel base, 4-1/2" x 4-1/2", US32				↓	51			51	56	
550	0010	**KICK PLATE** 6" high, for 3' door, stainless steel	1 Carp	15	.533	Ea.	14.25	10.50		24.75	33.50	550
	0500	Bronze	↓	15	.533	↓	20	10.50		30.50	40	

08700 | Hardware

08710 | Door Hardware

			CREW	DAILY OUTPUT	LABOR-HOURS	UNIT	2000 BARE COSTS MAT.	LABOR	EQUIP.	TOTAL	TOTAL INCL O&P	
550	2000	Aluminum, .050, with 3 beveled edges, 10" x 28"	1 Carp	15	.533	Ea.	14.70	10.50		25.20	34	550
	2010	10" x 30"		15	.533		15.75	10.50		26.25	35.50	
	2020	10" x 34"		15	.533		17.80	10.50		28.30	37.50	
	2040	10" x 38"		15	.533		19.65	10.50		30.15	39.50	
650	0010	**LOCKSET** Standard duty, cylindrical, with sectional trim										650
	0020	Non-keyed, passage	1 Carp	12	.667	Ea.	37	13.15		50.15	63.50	
	0100	Privacy		12	.667		45	13.15		58.15	72.50	
	0400	Keyed, single cylinder function		10	.800		66	15.75		81.75	100	
	0420	Hotel		8	1		90	19.70		109.70	133	
	0500	Lever handled, keyed, single cylinder function		10	.800		91	15.75		106.75	127	
	0600	Bedroom, bathroom and inner office doors		10	.800		114	15.75		129.75	153	
	0900	Apartment, office and corridor doors		10	.800		143	15.75		158.75	184	
	1400	Keyed, single cylinder function		10	.800		156	15.75		171.75	199	
	1700	Residential, interior door, minimum		16	.500		12.60	9.85		22.45	31	
	1720	Maximum		8	1		33.50	19.70		53.20	71	
	1800	Exterior, minimum		14	.571		28	11.25		39.25	50.50	
	1810	Average		8	1		59.50	19.70		79.20	99.50	
	1820	Maximum		8	1		118	19.70		137.70	164	
750	0010	**PANIC DEVICE** For rim locks, single door, exit only	1 Carp	6	1.333	Ea.	330	26.50		356.50	410	750
	1000	Mortise, bar, exit only		4	2	"	435	39.50		474.50	550	
	4000	Double doors, exit only		2	4	Pr.	665	79		744	865	
	4500	Exit & entrance		2	4	"	760	79		839	970	
780	0010	**PUSH-PULL PLATE**										780
	0100	Push plate, .050 thick, 4" x 16", aluminum	1 Carp	12	.667	Ea.	5.05	13.15		18.20	28	
	0500	Bronze		12	.667		11.30	13.15		24.45	35	
	1500	Pull handle and push bar, aluminum		11	.727		108	14.35		122.35	144	
	2000	Bronze		10	.800		140	15.75		155.75	181	
	4000	Door pull, designer style, cast aluminum, minimum		12	.667		61	13.15		74.15	89.50	
	5000	Maximum		8	1		282	19.70		301.70	345	

08720 | Weatherstripping & Seals

			CREW	DAILY OUTPUT	LABOR-HOURS	UNIT	MAT.	LABOR	EQUIP.	TOTAL	TOTAL INCL O&P	
300	0010	**WEATHERSTRIPPING** Window, double hung, 3' x 5', zinc	1 Carp	7.20	1.111	Opng.	10.70	22		32.70	49.50	300
	0100	Bronze		7.20	1.111		19.45	22		41.45	59	
	0200	Vinyl V strip		7	1.143		3.50	22.50		26	42.50	
	0500	As above but heavy duty, zinc		4.60	1.739		13	34.50		47.50	73	
	0600	Bronze		4.60	1.739		22.50	34.50		57	83.50	
	1000	Doors, wood frame, interlocking, for 3' x 7' door, zinc		3	2.667		11.65	52.50		64.15	103	
	1100	Bronze		3	2.667		18.35	52.50		70.85	110	
	1300	6' x 7' opening, zinc		2	4		12.75	79		91.75	149	
	1400	Bronze		2	4		24	79		103	162	
	1700	Wood frame, spring type, bronze										
	1800	3' x 7' door	1 Carp	7.60	1.053	Opng.	15.45	20.50		35.95	52.50	
	1900	6' x 7' door		7	1.143		16.45	22.50		38.95	56.50	
	1920	Felt, 3' x 7' door		14	.571		1.91	11.25		13.16	21.50	
	1930	6' x 7' door		13	.615		2.07	12.10		14.17	23.50	
	1950	Rubber, 3' x 7' door		7.60	1.053		4.14	20.50		24.64	40	
	1951	Rubber, 3' x 7' door		119	.067	L.F.	.24	1.32		1.56	2.54	
	1960	6' x 7' door		7	1.143	Opng.	4.72	22.50		27.22	43.50	
	2200	Metal frame, spring type, bronze										
	2300	3' x 7' door	1 Carp	3	2.667	Opng.	25.50	52.50		78	118	
	2400	6' x 7' door	"	2.50	3.200	"	35	63		98	146	
	2500	For stainless steel, spring type, add					133%					
	2700	Metal frame, extruded sections, 3' x 7' door, aluminum	1 Carp	2	4	Opng.	34	79		113	172	
	2800	Bronze		2	4		85.50	79		164.50	229	
	3100	6' x 7' door, aluminum		1.20	6.667		43	131		174	273	

For expanded coverage of these items see *Means Building Construction Cost Data 2000*

08700 | Hardware

08720 | Weatherstripping & Seals

			CREW	DAILY OUTPUT	LABOR-HOURS	UNIT	2000 BARE COSTS MAT.	LABOR	EQUIP.	TOTAL	TOTAL INCL O&P	
300	3200	Bronze	1 Carp	1.20	6.667	Opng.	101	131		232	335	300
	3500	Threshold weatherstripping										
	3650	Door sweep, flush mounted, aluminum	1 Carp	25	.320	Ea.	10	6.30		16.30	22	
	3700	Vinyl		25	.320	"	11.85	6.30		18.15	24	
	4000	Astragal for double doors, aluminum		4	2	Opng.	16.95	39.50		56.45	86	
	4100	Bronze		4	2	"	27.50	39.50		67	98	
	5000	Garage door bottom weatherstrip, 12' aluminum, clear		14	.571	Ea.	16.05	11.25		27.30	37	
	5010	Bronze		14	.571		61	11.25		72.25	87	
	5050	Bottom protection, 12' aluminum, clear		14	.571		19.45	11.25		30.70	41	
	5100	Bronze		14	.571		76	11.25		87.25	103	
800	0010	**THRESHOLD** 3' long door saddles, aluminum	1 Carp	48	.167	L.F.	3.38	3.28		6.66	9.35	800
	0100	Aluminum, 8" wide, 1/2" thick		12	.667	Ea.	29	13.15		42.15	54.50	
	0500	Bronze		60	.133	L.F.	27.50	2.63		30.13	34.50	
	0600	Bronze, panic threshold, 5" wide, 1/2" thick		12	.667	Ea.	57	13.15		70.15	85	
	0700	Rubber, 1/2" thick, 5-1/2" wide		20	.400		28.50	7.90		36.40	45	
	0800	2-3/4" wide		20	.400		13.25	7.90		21.15	28	

08750 | Window Hardware

			CREW	DAILY OUTPUT	LABOR-HOURS	UNIT	MAT.	LABOR	EQUIP.	TOTAL	TOTAL INCL O&P	
400	0010	**WINDOW HARDWARE**										400
	1000	Handles, surface mounted, aluminum	1 Carp	24	.333	Ea.	1.79	6.55		8.34	13.20	
	1020	Brass		24	.333		2.08	6.55		8.63	13.55	
	1040	Chrome		24	.333		1.90	6.55		8.45	13.35	
	1500	Recessed, aluminum		12	.667		1.04	13.15		14.19	23.50	
	1520	Brass		12	.667		1.15	13.15		14.30	24	
	1540	Chrome		12	.667		1.09	13.15		14.24	23.50	
	2000	Latches, aluminum		20	.400		1.49	7.90		9.39	15.15	
	2020	Brass		20	.400		1.79	7.90		9.69	15.45	
	2040	Chrome		20	.400		1.68	7.90		9.58	15.35	

08770 | Door/Window Accessories

			CREW	DAILY OUTPUT	LABOR-HOURS	UNIT	MAT.	LABOR	EQUIP.	TOTAL	TOTAL INCL O&P	
550	0010	**DETECTION SYSTEMS** See division 13851-065										550
560	0010	**DOOR ACCESSORIES**										560
	1000	Knockers, brass, standard	1 Carp	16	.500	Ea.	35	9.85		44.85	55.50	
	1100	Deluxe		10	.800		108	15.75		123.75	146	
	4000	Security chain, standard		18	.444		6	8.75		14.75	21.50	
	4100	Deluxe		18	.444		36	8.75		44.75	54.50	

08800 | Glazing

08810 | Glass

			CREW	DAILY OUTPUT	LABOR-HOURS	UNIT	2000 BARE COSTS MAT.	LABOR	EQUIP.	TOTAL	TOTAL INCL O&P	
260	0010	**FLOAT GLASS** 3/16" thick, clear, plain	2 Glaz (R08810-010)	130	.123	S.F.	3.55	2.38		5.93	7.85	260
	0200	Tempered, clear		130	.123		4.24	2.38		6.62	8.60	
	0300	Tinted		130	.123		5.30	2.38		7.68	9.80	
	0600	1/4" thick, clear, plain		120	.133		4.47	2.58		7.05	9.20	
	0700	Tinted		120	.133		4.24	2.58		6.82	8.95	
	0800	Tempered, clear		120	.133		5.30	2.58		7.88	10.15	
	0900	Tinted		120	.133		7.35	2.58		9.93	12.40	
	1600	3/8" thick, clear, plain		75	.213		7.05	4.13		11.18	14.60	

08800 | Glazing

08810 | Glass

			CREW	DAILY OUTPUT	LABOR-HOURS	UNIT	2000 BARE COSTS MAT.	LABOR	EQUIP.	TOTAL	TOTAL INCL O&P	
260	1700	Tinted	2 Glaz	75	.213	S.F.	8.50	4.13		12.63	16.20	260
	1800	Tempered, clear		75	.213		10.60	4.13		14.73	18.50	
	1900	Tinted		75	.213		13.20	4.13		17.33	21.50	
	2200	1/2" thick, clear, plain		55	.291		13.80	5.65		19.45	24.50	
	2300	Tinted		55	.291		14.85	5.65		20.50	25.50	
	2400	Tempered, clear		55	.291		15.90	5.65		21.55	27	
	2500	Tinted		55	.291		19.85	5.65		25.50	31.50	
	2800	5/8" thick, clear, plain		45	.356		14.85	6.90		21.75	27.50	
	2900	Tempered, clear		45	.356		16.95	6.90		23.85	30	
	8900	For low emissivity coating for 3/16" and 1/4" only, add to above					3					
270	0010	FULL VISION Window system with 3/4" glass mullions, 10' high	H-2	130	.185	S.F.	41.50	3.27		44.77	51.50	270
	0100	10' to 20' high, minimum		110	.218		44	3.87		47.87	55	
	0150	Average		100	.240		47.50	4.25		51.75	59.50	
	0200	Maximum		80	.300		53.50	5.30		58.80	68	
300	0010	GLAZING VARIABLES										300
	0500	For high rise glazing, exterior, add per S.F. per story				S.F.					.08	
	0600	For glass replacement, add				"		100%				
	0700	For gasket settings, add				L.F.	3.18			3.18	3.50	
	0800	For concrete reglet settings, add				S.F.	20%	25%				
	0900	For sloped glazing, add				"		25%				
	2000	Fabrication, polished edges, 1/4" thick				Inch	.27			.27	.30	
	2100	1/2" thick					.70			.70	.77	
	2500	Mitered edges, 1/4" thick					.70			.70	.77	
	2600	1/2" thick					1.12			1.12	1.23	
460	0010	INSULATING GLASS 2 lites 1/8" float, 1/2" thk, under 15 S.F.										460
	0100	Tinted	2 Glaz	95	.168	S.F.	9.15	3.26		12.41	15.45	
	0280	Double glazed, 5/8" thk unit, 3/16" float, 15-30 S.F., clear		90	.178		7.40	3.44		10.84	13.85	
	0400	1" thk. dbl. glazed, 1/4" float, 30-70 S.F., clear		75	.213		10.50	4.13		14.63	18.40	
	0500	Tinted		75	.213		12.70	4.13		16.83	21	
	2000	Both lites, light & heat reflective		85	.188		16.95	3.64		20.59	24.50	
	2500	Heat reflective, film inside, 1" thick unit, clear		85	.188		14.85	3.64		18.49	22.50	
	2600	Tinted		85	.188		16	3.64		19.64	23.50	
	3000	Film on weatherside, clear, 1/2" thick unit		95	.168		10.60	3.26		13.86	17.05	
	3100	5/8" thick unit		90	.178		13.45	3.44		16.89	20.50	
	3200	1" thick unit		85	.188		14.60	3.64		18.24	22	
850	0010	WINDOW GLASS Clear float, stops, putty bed, 1/8" thick	2 Glaz	480	.033	S.F.	2.92	.65		3.57	4.28	850
	0500	3/16" thick, clear		480	.033		3.60	.65		4.25	5.05	
	0600	Tinted		480	.033		4.08	.65		4.73	5.55	
	0700	Tempered		480	.033		4.94	.65		5.59	6.50	

08830 | Mirrors

			CREW	DAILY OUTPUT	LABOR-HOURS	UNIT	MAT.	LABOR	EQUIP.	TOTAL	TOTAL INCL O&P	
100	0010	MIRRORS No frames, wall type, 1/4" plate glass, polished edge										100
	0100	Up to 5 S.F.	2 Glaz	125	.128	S.F.	5.85	2.48		8.33	10.50	
	0200	Over 5 S.F.		160	.100		5.65	1.94		7.59	9.40	
	0500	Door type, 1/4" plate glass, up to 12 S.F.		160	.100		6.05	1.94		7.99	9.85	
	1000	Float glass, up to 10 S.F., 1/8" thick		160	.100		3.57	1.94		5.51	7.15	
	1100	3/16" thick		150	.107		4.15	2.06		6.21	8	
	1500	12" x 12" wall tiles, square edge, clear		195	.082		1.52	1.59		3.11	4.30	
	1600	Veined		195	.082		3.87	1.59		5.46	6.90	
	2010	Bathroom, unframed, laminated		160	.100		10.10	1.94		12.04	14.30	

For expanded coverage of these items see *Means Building Construction Cost Data 2000*

08900 | Glazed Curtain Wall

08911 | Glazed Aluminum Curtain Wall

		CREW	DAILY OUTPUT	LABOR-HOURS	UNIT	MAT.	LABOR	EQUIP.	TOTAL	TOTAL INCL O&P
0010	**TUBE FRAMING** For window walls and store fronts, aluminum, stock									
0050	Plain tube frame, mill finish, 1-3/4" x 1-3/4"	2 Glaz	103	.155	L.F.	6.05	3.01		9.06	11.65
0150	1-3/4" x 4"		98	.163		7.60	3.16		10.76	13.60
0200	1-3/4" x 4-1/2"		95	.168		8.40	3.26		11.66	14.65
0250	2" x 6"		89	.180		12.75	3.48		16.23	19.80
0350	4" x 4"		87	.184		12.45	3.56		16.01	19.60
0400	4-1/2" x 4-1/2"		85	.188		13.95	3.64		17.59	21.50
0450	Glass bead		240	.067		1.61	1.29		2.90	3.91
1000	Flush tube frame, mill finish, 1/4" glass, 1-3/4" x 4", open header		80	.200		7.45	3.87		11.32	14.60
1050	Open sill		82	.195		6.50	3.78		10.28	13.40
1100	Closed back header		83	.193		10.45	3.73		14.18	17.65
1150	Closed back sill		85	.188		9.90	3.64		13.54	16.95
1200	Vertical mullion, one piece		75	.213		11.05	4.13		15.18	19
1250	Two piece		73	.219		11.85	4.24		16.09	20
1300	90° or 180° vertical corner post		75	.213		18.60	4.13		22.73	27.50
1400	1-3/4" x 4-1/2", open header		80	.200		9.10	3.87		12.97	16.40
1450	Open sill		82	.195		7.50	3.78		11.28	14.50
1500	Closed back header		83	.193		11.50	3.73		15.23	18.85
1550	Closed back sill		85	.188		10.70	3.64		14.34	17.80
1600	Vertical mullion, one piece		75	.213		11.95	4.13		16.08	20
1650	Two piece		73	.219		12.65	4.24		16.89	21
1700	90° or 180° vertical corner post		75	.213		12.95	4.13		17.08	21
2000	Flush tube frame, mill fin. for ins. glass, 2" x 4-1/2", open header		75	.213		10.60	4.13		14.73	18.50
2050	Open sill		77	.208		9.30	4.02		13.32	16.85
2100	Closed back header		78	.205		11.55	3.97		15.52	19.30
2150	Closed back sill		80	.200		11.40	3.87		15.27	18.95
2200	Vertical mullion, one piece		70	.229		12.80	4.42		17.22	21.50
2250	Two piece		68	.235		13.70	4.55		18.25	22.50
2300	90° or 180° vertical corner post		70	.229		12.95	4.42		17.37	21.50
5000	Flush tube frame, mill fin., thermal brk., 2-1/4"x 4-1/2", open header		74	.216		11.70	4.18		15.88	19.80
5050	Open sill		75	.213		10.20	4.13		14.33	18.05
5100	Vertical mullion, one piece		69	.232		14.10	4.49		18.59	23
5150	Two piece		67	.239		15.05	4.62		19.67	24
5200	90° or 180° vertical corner post		69	.232		13.55	4.49		18.04	22.50
6980	Door stop (snap in)		380	.042		2.18	.81		2.99	3.75
7000	For joints, 90°, clip type, add				Ea.	18.65			18.65	20.50
7050	Screw spline joint, add					14.05			14.05	15.45
7100	For joint other than 90°, add					29.50			29.50	32.50
8000	For bronze anodized aluminum, add					15%				
8050	For stainless steel materials, add					350%				
8100	For monumental grade, add					50%				
8150	For steel stiffener, add	2 Glaz	200	.080	L.F.	7.15	1.55		8.70	10.45
8200	For 2 to 5 stories, add per story				Story		5%			

R089-200

For information about Means Estimating Seminars, see yellow pages 11 and 12 in back of book

Division 9 Finishes

Estimating Tips

General
- Room Finish Schedule: A complete set of plans should contain a room finish schedule. If one is not available, it would be well worth the time and effort to put one together. A room finish schedule should contain the room number, room name (for clarity), floor materials, base materials, wainscot materials, wainscot height, wall materials (for each wall), ceiling materials and special instructions.
- Surplus Finishes: Review the specifications to determine if there is any requirement to provide certain amounts of extra materials for the owner's maintenance department. In some cases the owner may require a substantial amount of materials, especially when it is a special order item or long lead time item.

09200 Plaster & Gypsum Board
- Lath is estimated by the square yard for both gypsum and metal lath, plus usually 5% allowance for waste. Furring, channels and accessories are measured by the linear foot. An extra foot should be allowed for each accessory miter or stop.
- Plaster is also estimated by the square yard. Deductions for openings vary by preference, from zero deduction to 50% of all openings over 2 feet in width. Some estimators deduct a percentage of the total yardage for openings. The estimator should allow one extra square foot for each linear foot of horizontal interior or exterior angle located below the ceiling level. Also, double the areas of small radius work.
- Each room should be measured, perimeter times maximum wall height. Ceiling areas are equal to length times width.
- Drywall accessories, studs, track, and acoustical caulking are all measured by the linear foot. Drywall taping is figured by the square foot. Gypsum wallboard is estimated by the square foot. No material deductions should be made for door or window openings under 32 S.F. Coreboard can be obtained in a 1" thickness for solid wall and shaft work. Additions should be made to price out the inside or outside corners.
- Different types of partition construction should be listed separately on the quantity sheets. There may be walls with studs of various widths, double studded, and similar or dissimilar surface materials. Shaft work is usually different construction from surrounding partitions requiring separate quantities and pricing of the work.

09300 Tile
09400 Terrazzo
- Tile and terrazzo areas are taken off on a square foot basis. Trim and base materials are measured by the linear foot. Accent tiles are listed per each. Two basic methods of installation are used. Mud set is approximately 30% more expensive than the thin set. In terrazzo work, be sure to include the linear footage of embedded decorative strips, grounds, machine rubbing and power cleanup.

09600 Flooring
- Wood flooring is available in strip, parquet, or block configuration. The latter two types are set in adhesives with quantities estimated by the square foot. The laying pattern will influence labor costs and material waste. In addition to the material and labor for laying wood floors, the estimator must make allowances for sanding and finishing these areas unless the flooring is prefinished.
- Most of the various types of flooring are all measured on a square foot basis. Base is measured by the linear foot. If adhesive materials are to be quantified, they are estimated at a specified coverage rate by the gallon depending upon the specified type and the manufacturer's recommendations.
- Sheet flooring is measured by the square yard. Roll widths vary, so consideration should be given to use the most economical width, as waste must be figured into the total quantity. Consider also the installation methods available, direct glue down or stretched.

09700 Wall Finishes
- Wall coverings are estimated by the square foot. The area to be covered is measured, length by height of wall above baseboards, to calculate the square footage of each wall. This figure is divided by the number of square feet in the single roll which is being used. Deduct, in full, the areas of openings such as doors and windows. Where a pattern match is required allow 25%-30% waste. One gallon of paste should be sufficient to hang 12 single rolls of light to medium weight paper.

09800 Acoustical Treatment
- Acoustical systems fall into several categories. The takeoff of these materials is by the square foot of area with a 5% allowance for waste. Do not forget about scaffolding, if applicable, when estimating these systems.

09900 Paints & Coatings
- Painting is one area where bids vary to a greater extent than almost any other section of a project. This arises from the many methods of measuring surfaces to be painted. The estimator should check the plans and specifications carefully to be sure of the required number of coats.
- Protection of adjacent surfaces is not included in painting costs. When considering the method of paint application, an important factor is the amount of protection and masking required. These must be estimated separately and may be the determining factor in choosing the method of application.

Reference Numbers
Reference numbers are shown in bold squares at the beginning of some major classifications. These numbers refer to related items in the Reference Section. The reference information may be an estimating procedure, an alternate pricing method or technical information.

Note: Not all subdivisions listed here necessarily appear in this publication.

09100 | Metal Support Assemblies

09110 | Non-Load Bearing Wall Framing

			CREW	DAILY OUTPUT	LABOR-HOURS	UNIT	2000 BARE COSTS MAT.	LABOR	EQUIP.	TOTAL	TOTAL INCL O&P
100	0010	**METAL STUDS, PARTITIONS**, 10' high, with runners R09110-610										100
	2000	Non-load bearing, galvanized, 25 ga. 1-5/8" wide, 16" O.C.	1 Carp	450	.018	S.F.	.15	.35		.50	.76
	2100	24" O.C.		520	.015		.11	.30		.41	.64
	2200	2-1/2" wide, 16" O.C.		440	.018		.16	.36		.52	.79
	2250	24" O.C.		510	.016		.12	.31		.43	.66
	2300	3-5/8" wide, 16" O.C.		430	.019		.19	.37		.56	.84
	2350	24" O.C.		500	.016		.14	.32		.46	.69
	2400	4" wide, 16" O.C.		420	.019		.23	.38		.61	.89
	2450	24" O.C.		490	.016		.17	.32		.49	.73
	2500	6" wide, 16" O.C.		410	.020		.30	.38		.68	.99
	2550	24" O.C.		480	.017		.22	.33		.55	.80
	2600	20 ga. studs, 1-5/8" wide, 16" O.C.		450	.018		.25	.35		.60	.88
	2650	24" O.C.		520	.015		.19	.30		.49	.73
	2700	2-1/2" wide, 16" O.C.		440	.018		.27	.36		.63	.91
	2750	24" O.C.		510	.016		.20	.31		.51	.75
	2800	3-5/8" wide, 16" O.C.		430	.019		.29	.37		.66	.95
	2850	24" O.C.		500	.016		.22	.32		.54	.78
	2900	4" wide, 16" O.C.		420	.019		.37	.38		.75	1.04
	2950	24" O.C.		490	.016		.27	.32		.59	.85
	3000	6" wide, 16" O.C.		410	.020		.43	.38		.81	1.13
	3050	24" O.C.		480	.017		.32	.33		.65	.91
	5000	Load bearing studs, see division 05410-400									

09130 | Acoustical Suspension

			CREW	DAILY OUTPUT	LABOR-HOURS	UNIT	MAT.	LABOR	EQUIP.	TOTAL	TOTAL INCL O&P
100	0010	**CEILING SUSPENSION SYSTEMS** For boards and tile										100
	0050	Class A suspension system, 15/16" T bar, 2' x 4' grid	1 Carp	800	.010	S.F.	.32	.20		.52	.69
	0300	2' x 2' grid	"	650	.012		.40	.24		.64	.86
	0350	For 9/16" grid, add					.13			.13	.14
	0360	For fire rated grid, add					.07			.07	.08
	0370	For colored grid, add					.15			.15	.17
	0400	Concealed Z bar suspension system, 12" module	1 Carp	520	.015		.35	.30		.65	.91
	0600	1-1/2" carrier channels, 4' O.C., add		470	.017		.21	.34		.55	.80
	0650	1-1/2" x 3-1/2" channels		470	.017		.39	.34		.73	1
	0700	Carrier channels for ceilings with									
	0900	recessed lighting fixtures, add	1 Carp	460	.017	S.F.	.34	.34		.68	.96
	5000	Wire hangers, #12 wire	"	300	.027	Ea.	.38	.53		.91	1.32

09200 | Plaster & Gypsum Board

09205 | Furring & Lathing

			CREW	DAILY OUTPUT	LABOR-HOURS	UNIT	MAT.	LABOR	EQUIP.	TOTAL	TOTAL INCL O&P
530	0010	**FURRING** Beams & columns, 7/8" galvanized channels,										530
	0030	12" O.C.	1 Lath	155	.052	S.F.	.20	.99		1.19	1.84
	0050	16" O.C.		170	.047		.16	.90		1.06	1.66
	0070	24" O.C.		185	.043		.11	.83		.94	1.48
	0100	Ceilings, on steel, 7/8" channels, galvanized, 12" O.C.		210	.038		.18	.73		.91	1.40
	0300	16" O.C.		290	.028		.16	.53		.69	1.05
	0400	24" O.C.		420	.019		.11	.37		.48	.72
	0600	1-5/8" channels, galvanized, 12" O.C.		190	.042		.28	.81		1.09	1.62
	0700	16" O.C.		260	.031		.25	.59		.84	1.24
	0900	24" O.C.		390	.021		.17	.39		.56	.83

09200 | Plaster & Gypsum Board

09205 | Furring & Lathing

			CREW	DAILY OUTPUT	LABOR-HOURS	UNIT	MAT.	LABOR	EQUIP.	TOTAL	TOTAL INCL O&P	
530	1000	Walls, 7/8" channels, galvanized, 12" O.C.	1 Lath	235	.034	S.F.	.18	.65		.83	1.27	530
	1200	16" O.C.		265	.030		.16	.58		.74	1.13	
	1300	24" O.C.		350	.023		.11	.44		.55	.84	
	1500	1-5/8" channels, galvanized, 12" O.C.		210	.038		.28	.73		1.01	1.50	
	1600	16" O.C.		240	.033		.25	.64		.89	1.32	
	1800	24" O.C.		305	.026		.17	.50		.67	1	
	8000	Suspended ceilings, including carriers										
	8200	1-1/2" carriers, 24" O.C. with:										
	8300	7/8" channels, 16" O.C.	1 Lath	165	.048	S.F.	.58	.93		1.51	2.16	
	8320	24" O.C.		200	.040		.53	.77		1.30	1.84	
	8400	1-5/8" channels, 16" O.C.		155	.052		.67	.99		1.66	2.35	
	8420	24" O.C.		190	.042		.59	.81		1.40	1.96	
	8600	2" carriers, 24" O.C. with:										
	8700	7/8" channels, 16" O.C.	1 Lath	155	.052	S.F.	.16	.99		1.15	1.80	
	8720	24" O.C.		190	.042		.11	.81		.92	1.44	
	8800	1-5/8" channels, 16" O.C.		145	.055		.63	1.06		1.69	2.43	
	8820	24" O.C.		180	.044		.54	.85		1.39	2	
540	0010	**GYPSUM LATH** Plain or perforated, nailed, 3/8" thick R09205-050	1 Lath	85	.094	S.Y.	3.42	1.81		5.23	6.70	540
	0100	1/2" thick, nailed		80	.100		3.51	1.92		5.43	7	
	0300	Clipped to steel studs, 3/8" thick		75	.107		3.42	2.05		5.47	7.10	
	0400	1/2" thick		70	.114		3.60	2.19		5.79	7.55	
	0600	Firestop gypsum base, to steel studs, 3/8" thick		70	.114		3.51	2.19		5.70	7.45	
	0700	1/2" thick		65	.123		3.96	2.36		6.32	8.25	
	0900	Foil back, to steel studs, 3/8" thick		75	.107		3.60	2.05		5.65	7.30	
	1000	1/2" thick		70	.114		3.96	2.19		6.15	7.95	
	1500	For ceiling installations, add		216	.037			.71		.71	1.16	
	1600	For columns and beams, add		170	.047			.90		.90	1.48	
560	0011	**METAL LATH** R09205-060										560
	3600	2.5 lb. diamond painted, on wood framing, on walls	1 Lath	85	.094	S.Y.	1.47	1.81		3.28	4.58	
	3700	On ceilings		75	.107		1.47	2.05		3.52	4.97	
	4200	3.4 lb. diamond painted, wired to steel framing		75	.107		2.26	2.05		4.31	5.85	
	4300	On ceilings		60	.133		2.26	2.56		4.82	6.70	
	5100	Rib lath, painted, wired to steel, on walls, 2.5 lb.		75	.107		2.28	2.05		4.33	5.85	
	5200	3.4 lb.		70	.114		3.20	2.19		5.39	7.10	
	5700	Suspended ceiling system, incl. 3.4 lb. diamond lath, painted		15	.533		8.65	10.25		18.90	26.50	
	5800	Galvanized		15	.533		8.90	10.25		19.15	26.50	
700	0010	**ACCESSORIES, PLASTER** Casing bead, expanded flange, galvanized	1 Lath	2.70	2.963	C.L.F.	27	57		84	123	700
	0900	Channels, cold rolled, 16 ga., 3/4" deep, galvanized					18			18	19.80	
	1200	1-1/2" deep, 16 ga., galvanized					27.50			27.50	30.50	
	1620	Corner bead, expanded bullnose, 3/4" radius, #10, galvanized	1 Lath	2.60	3.077		20	59		79	119	
	1650	#1, galvanized		2.55	3.137		35	60		95	137	
	1670	Expanded wing, 2-3/4" wide, galv. #1		2.65	3.019		20	58		78	117	
	1700	Inside corner, (corner rite) 3" x 3", painted		2.60	3.077		17.65	59		76.65	116	
	1750	Strip-ex, 4" wide, painted		2.55	3.137		14.75	60		74.75	115	
	1800	Expansion joint, 3/4" grounds, limited expansion, galv., 1 piece		2.70	2.963		65	57		122	165	
	2100	Extreme expansion, galvanized, 2 piece		2.60	3.077		120	59		179	229	

09210 | Gypsum Plaster

			CREW	DAILY OUTPUT	LABOR-HOURS	UNIT	MAT.	LABOR	EQUIP.	TOTAL	TOTAL INCL O&P	
100	0010	**GYPSUM PLASTER** 80# bag, less than 1 ton R09210-105				Bag	14.20			14.20	15.65	100
	0300	2 coats, no lath included, on walls	J-1	105	.381	S.Y.	3.42	6.65	.52	10.59	15.45	
	0400	On ceilings		92	.435		3.42	7.60	.59	11.61	17.10	
	0900	3 coats, no lath included, on walls		87	.460		4.77	8	.63	13.40	19.35	
	1000	On ceilings		78	.513		4.77	8.95	.70	14.42	21	
	1600	For irregular or curved surfaces, add						30%				

For expanded coverage of these items see Means Building Construction Cost Data 2000

09200 | Plaster & Gypsum Board

09210 | Gypsum Plaster

			CREW	DAILY OUTPUT	LABOR-HOURS	UNIT	MAT.	LABOR	EQUIP.	TOTAL	TOTAL INCL O&P	
100	1800	For columns & beams, add R09210-105						50%				100
500	0010	**PERLITE OR VERMICULITE PLASTER** 100 lb. bags Under 200 bags				Bag	13.05			13.05	14.35	500
	0300	2 coats, no lath included, on walls	J-1	92	.435	S.Y.	3.24	7.60	.59	11.43	16.90	
	0400	On ceilings R09210-115		79	.506		3.24	8.85	.69	12.78	19.05	
	0900	3 coats, no lath included, on walls		74	.541		5.30	9.45	.74	15.49	22.50	
	1000	On ceilings	↓	63	.635		5.30	11.10	.86	17.26	25.50	
	1700	For irregular or curved surfaces, add to above						30%				
	1800	For columns and beams, add to above						50%				
	1900	For soffits, add to ceiling prices				↓		40%				
900	0010	**THIN COAT** Plaster, 1 coat veneer, not incl. lath	J-1	3,600	.011	S.F.	.07	.19	.02	.28	.42	900
	1000	In 50 lb. bags				Bag	8.60			8.60	9.45	

09220 | Portland Cement Plaster

			CREW	DAILY OUTPUT	LABOR-HOURS	UNIT	MAT.	LABOR	EQUIP.	TOTAL	TOTAL INCL O&P	
200	0010	**STUCCO**, 3 coats 1" thick, float finish, with mesh, on wood frame	J-2	135	.356	S.Y.	3.65	6.30	.41	10.36	14.95	200
	0100	On masonry construction, no mesh incl.	J-1	200	.200		2.03	3.49	.27	5.79	8.40	
	0150	2 coats, 3/4" thick, float finish, no lath incl. R09220-300	"	110	.364		1.92	6.35	.49	8.76	13.25	
	0300	For trowel finish, add	1 Plas	170	.047			.88		.88	1.47	
	0600	For coloring and special finish, add, minimum	J-1	685	.058		.36	1.02	.08	1.46	2.19	
	0700	Maximum		200	.200		1.26	3.49	.27	5.02	7.55	
	1000	Exterior stucco, with bonding agent, 3 coats, on walls, no mesh incl.		200	.200		3.25	3.49	.27	7.01	9.75	
	1200	Ceilings		180	.222		3.25	3.88	.30	7.43	10.40	
	1300	Beams		80	.500		3.25	8.75	.68	12.68	18.95	
	1500	Columns	↓	100	.400		3.25	7	.54	10.79	15.85	
	1600	Mesh, painted, nailed to wood, 1.8 lb.	1 Lath	60	.133		3.03	2.56		5.59	7.50	
	1800	3.6 lb.		55	.145		1.62	2.79		4.41	6.35	
	1900	Wired to steel, painted, 1.8 lb.		53	.151		3.03	2.90		5.93	8.10	
	2100	3.6 lb.	↓	50	.160	↓	1.62	3.07		4.69	6.85	

09250 | Gypsum Board

			CREW	DAILY OUTPUT	LABOR-HOURS	UNIT	MAT.	LABOR	EQUIP.	TOTAL	TOTAL INCL O&P	
300	0010	**BLUEBOARD** For use with thin coat										300
	0100	plaster application (see division 09210-900)										
	1000	3/8" thick, on walls or ceilings, standard, no finish included	2 Carp	1,900	.008	S.F.	.27	.17		.44	.58	
	1100	With thin coat plaster finish		875	.018		.34	.36		.70	.99	
	1400	On beams, columns, or soffits, standard, no finish included		675	.024		.31	.47		.78	1.14	
	1450	With thin coat plaster finish		475	.034		.39	.66		1.05	1.57	
	3000	1/2" thick, on walls or ceilings, standard, no finish included		1,900	.008		.27	.17		.44	.58	
	3100	With thin coat plaster finish		875	.018		.34	.36		.70	.99	
	3300	Fire resistant, no finish included		1,900	.008		.27	.17		.44	.58	
	3400	With thin coat plaster finish		875	.018		.34	.36		.70	.99	
	3450	On beams, columns, or soffits, standard, no finish included		675	.024		.31	.47		.78	1.14	
	3500	With thin coat plaster finish		475	.034		.39	.66		1.05	1.57	
	3700	Fire resistant, no finish included		675	.024		.31	.47		.78	1.14	
	3800	With thin coat plaster finish		475	.034		.39	.66		1.05	1.57	
	5000	5/8" thick, on walls or ceilings, fire resistant, no finish included		1,900	.008		.30	.17		.47	.61	
	5100	With thin coat plaster finish		875	.018		.37	.36		.73	1.03	
	5500	On beams, columns, or soffits, no finish included		675	.024		.35	.47		.82	1.18	
	5600	With thin coat plaster finish		475	.034		.43	.66		1.09	1.61	
	6000	For high ceilings, over 8' high, add		3,060	.005			.10		.10	.18	
	6500	For over 3 stories high, add per story	↓	6,100	.003	↓		.05		.05	.09	
700	0010	**DRYWALL** Gypsum plasterboard, nailed or screwed to studs										700
	0100	unless otherwise noted										
	0150	3/8" thick, on walls, standard, no finish included	2 Carp	2,000	.008	S.F.	.26	.16		.42	.56	
	0200	On ceilings, standard, no finish included		1,800	.009		.26	.18		.44	.59	
	0250	On beams, columns, or soffits, no finish included		675	.024		.30	.47		.77	1.13	
	0300	1/2" thick, on walls, standard, no finish included	↓	2,000	.008	↓	.26	.16		.42	.56	

09200 | Plaster & Gypsum Board

09250 | Gypsum Board

		CREW	DAILY OUTPUT	LABOR-HOURS	UNIT	MAT.	LABOR	EQUIP.	TOTAL	TOTAL INCL O&P
0350	Taped and finished (level 4 finish)	2 Carp	965	.017	S.F.	.30	.33		.63	.89
0390	With compound skim coat (level 5 finish)		775	.021		.32	.41		.73	1.05
0400	Fire resistant, no finish included		2,000	.008		.25	.16		.41	.55
0450	Taped and finished (level 4 finish)		965	.017		.29	.33		.62	.88
0490	With compound skim coat (level 5 finish)		775	.021		.31	.41		.72	1.04
0500	Water resistant, no finish included		2,000	.008		.32	.16		.48	.62
0550	Taped and finished (level 4 finish)		965	.017		.36	.33		.69	.96
0590	With compound skim coat (level 5 finish)		775	.021		.38	.41		.79	1.12
0600	Prefinished, vinyl, clipped to studs		900	.018		.56	.35		.91	1.22
1000	On ceilings, standard, no finish included		1,800	.009		.26	.18		.44	.59
1050	Taped and finished (level 4 finish)		765	.021		.30	.41		.71	1.04
1090	With compound skim coat (level 5 finish)		610	.026		.32	.52		.84	1.24
1100	Fire resistant, no finish included		1,800	.009		.25	.18		.43	.58
1150	Taped and finished (level 4 finish)		765	.021		.29	.41		.70	1.03
1195	With compound skim coat (level 5 finish)		610	.026		.31	.52		.83	1.23
1200	Water resistant, no finish included		1,800	.009		.32	.18		.50	.65
1250	Taped and finished (level 4 finish)		765	.021		.36	.41		.77	1.11
1290	With compound skim coat (level 5 finish)		610	.026		.38	.52		.90	1.31
1500	On beams, columns, or soffits, standard, no finish included		675	.024		.30	.47		.77	1.13
1550	Taped and finished (level 4 finish)		475	.034		.35	.66		1.01	1.52
1590	With compound skim coat (level 5 finish)		540	.030		.37	.58		.95	1.40
1600	Fire resistant, no finish included		675	.024		.29	.47		.76	1.12
1650	Taped and finished (level 4 finish)		475	.034		.33	.66		.99	1.51
1690	With compound skim coat (level 5 finish)		540	.030		.36	.58		.94	1.39
1700	Water resistant, no finish included		675	.024		.37	.47		.84	1.20
1750	Taped and finished (level 4 finish)		475	.034		.41	.66		1.07	1.60
1790	With compound skim coat (level 5 finish)		540	.030		.44	.58		1.02	1.48
2000	5/8" thick, on walls, standard, no finish included		2,000	.008		.29	.16		.45	.59
2050	Taped and finished (level 4 finish)		965	.017		.33	.33		.66	.92
2090	With compound skim coat (level 5 finish)		775	.021		.35	.41		.76	1.09
2100	Fire resistant, no finish included		2,000	.008		.29	.16		.45	.59
2150	Taped and finished (level 4 finish)		965	.017		.33	.33		.66	.92
2195	With compound skim coat (level 5 finish)		775	.021		.35	.41		.76	1.09
2200	Water resistant, no finish included		2,000	.008		.35	.16		.51	.66
2250	Taped and finished (level 4 finish)		965	.017		.39	.33		.72	.99
2290	With compound skim coat (level 5 finish)		775	.021		.41	.41		.82	1.15
2300	Prefinished, vinyl, clipped to studs		900	.018		.65	.35		1	1.31
3000	On ceilings, standard, no finish included		1,800	.009		.29	.18		.47	.62
3050	Taped and finished (level 4 finish)		765	.021		.33	.41		.74	1.07
3090	With compound skim coat (level 5 finish)		615	.026		.35	.51		.86	1.27
3100	Fire resistant, no finish included		1,800	.009		.29	.18		.47	.62
3150	Taped and finished (level 4 finish)		765	.021		.33	.41		.74	1.07
3190	With compound skim coat (level 5 finish)		615	.026		.35	.51		.86	1.27
3200	Water resistant, no finish included		1,800	.009		.35	.18		.53	.69
3250	Taped and finished (level 4 finish)		765	.021		.39	.41		.80	1.14
3290	With compound skim coat (level 5 finish)		615	.026		.41	.51		.92	1.33
3500	On beams, columns, or soffits, no finish included		675	.024		.33	.47		.80	1.17
3550	Taped and finished (level 4 finish)		475	.034		.38	.66		1.04	1.56
3590	With compound skim coat (level 5 finish)		380	.042		.40	.83		1.23	1.86
3600	Fire resistant, no finish included		675	.024		.33	.47		.80	1.17
3650	Taped and finished (level 4 finish)		475	.034		.38	.66		1.04	1.56
3690	With compound skim coat (level 5 finish)		380	.042		.40	.83		1.23	1.86
3700	Water resistant, no finish included		675	.024		.40	.47		.87	1.24
3750	Taped and finished (level 4 finish)		475	.034		.45	.66		1.11	1.63
3790	With compound skim coat (level 5 finish)		380	.042		.47	.83		1.30	1.94
4000	Fireproofing, beams or columns, 2 layers, 1/2" thick, incl finish		330	.048		.54	.96		1.50	2.23

For expanded coverage of these items see Means Building Construction Cost Data 2000

09200 | Plaster & Gypsum Board

09250 | Gypsum Board

			CREW	DAILY OUTPUT	LABOR-HOURS	UNIT	MAT.	LABOR	EQUIP.	TOTAL	TOTAL INCL O&P	
700	4050	5/8" thick	2 Carp	300	.053	S.F.	.66	1.05		1.71	2.53	700
	4100	3 layers, 1/2" thick		225	.071		.87	1.40		2.27	3.36	
	4150	5/8" thick		210	.076		.99	1.50		2.49	3.66	
	5200	For high ceilings, over 8' high, add	↓	3,060	.005			.10		.10	.18	
	5270	For textured spray, add	2 Lath	1,600	.010		.05	.19		.24	.37	
	5300	For over 3 stories high, add per story	2 Carp	6,100	.003	↓		.05		.05	.09	
	5350	For finishing corners, inside or outside, add	"	1,100	.015	L.F.	.06	.29		.35	.56	
	5500	For acoustical sealant, add per bead	1 Carp	500	.016	"	.03	.32		.35	.57	
	5550	Sealant, 1 quart tube				Ea.	4.89			4.89	5.40	
	5600	Sound deadening board, 1/4" gypsum	2 Carp	1,800	.009	S.F.	.25	.18		.43	.58	
	5650	1/2" wood fiber	"	1,800	.009	"	.34	.18		.52	.67	

09270 | Drywall Accessories

			CREW	DAILY OUTPUT	LABOR-HOURS	UNIT	MAT.	LABOR	EQUIP.	TOTAL	TOTAL INCL O&P	
100	0010	**ACCESSORIES, DRYWALL** Casing bead, galvanized steel	1 Carp	2.90	2.759	C.L.F.	13.80	54.50		68.30	108	100
	0100	Vinyl		3	2.667		16	52.50		68.50	108	
	0400	Corner bead, galvanized steel, 1-1/4" x 1-1/4"		3.50	2.286		9	45		54	87	
	0600	Vinyl corner bead		4	2		21	39.50		60.50	90.50	
	0900	Furring channel, galv. steel, 7/8" deep, standard		2.60	3.077		17.35	60.50		77.85	123	
	1000	Resilient		2.55	3.137		18.65	62		80.65	127	
	1100	J trim, galvanized steel, 1/2" wide		3	2.667		11.85	52.50		64.35	103	
	1120	5/8" wide	↓	2.95	2.712	↓	12.50	53.50		66	105	
	1160	Screws #6 x 1" A				M	6.65			6.65	7.30	
	1170	#6 x 1-5/8" A				"	8.65			8.65	9.55	
	1500	Z stud, galvanized steel, 1-1/2" wide	1 Carp	2.60	3.077	C.L.F.	24	60.50		84.50	130	

09300 | Tile

09310 | Ceramic Tile

			CREW	DAILY OUTPUT	LABOR-HOURS	UNIT	MAT.	LABOR	EQUIP.	TOTAL	TOTAL INCL O&P	
100	0010	**CERAMIC TILE**										100
	0050	Base, using 1' x 4" high pc. with 1" x 1" tiles, mud set	D-7	82	.195	L.F.	3.99	3.38		7.37	9.90	
	0100	Thin set	"	128	.125		3.79	2.17		5.96	7.70	
	0300	For 6" high base, 1" x 1" tile face, add					.63			.63	.69	
	0400	For 2" x 2" tile face, add to above					.33			.33	.36	
	0600	Cove base, 4-1/4" x 4-1/4" high, mud set	D-7	91	.176		3.02	3.05		6.07	8.25	
	0700	Thin set		128	.125		3.02	2.17		5.19	6.85	
	0900	6" x 4-1/4" high, mud set		100	.160		2.54	2.77		5.31	7.30	
	1000	Thin set		137	.117		2.54	2.02		4.56	6.05	
	1200	Sanitary cove base, 6" x 4-1/4" high, mud set		93	.172		3.52	2.98		6.50	8.70	
	1300	Thin set		124	.129		3.52	2.24		5.76	7.50	
	1500	6" x 6" high, mud set		84	.190		3	3.30		6.30	8.65	
	1600	Thin set	↓	117	.137		3	2.37		5.37	7.15	
	1800	Bathroom accessories, average		82	.195	Ea.	9.50	3.38		12.88	15.95	
	1900	Bathtub, 5', rec. 4-1/4" x 4-1/4" tile wainscot, adhesive set 6' high		2.90	5.517		140	95.50		235.50	310	
	2100	7' high wainscot		2.50	6.400		160	111		271	355	
	2200	8' high wainscot		2.20	7.273	↓	170	126		296	390	
	2400	Bullnose trim, 4-1/4" x 4-1/4", mud set		82	.195	L.F.	2.74	3.38		6.12	8.50	
	2500	Thin set		128	.125		2.74	2.17		4.91	6.50	
	2700	6" x 4-1/4" bullnose trim, mud set	↓	84	.190	↓	2.30	3.30		5.60	7.90	

09300 | Tile

09310 | Ceramic Tile

		Crew	Daily Output	Labor Hours	Unit	Mat.	Labor	Equip.	Total	Total Incl O&P
2800	Thin set	D-7	124	.129	L.F.	2.30	2.24		4.54	6.15
3000	Floors, natural clay, random or uniform, thin set, color group 1		183	.087	S.F.	3.55	1.52		5.07	6.35
3100	Color group 2		183	.087		3.82	1.52		5.34	6.65
3300	Porcelain type, 1 color, color group 2, 1" x 1"		183	.087		4.11	1.52		5.63	7
3400	2" x 2" or 2" x 1", thin set		190	.084		3.97	1.46		5.43	6.75
3600	For random blend, 2 colors, add					.75			.75	.83
3700	4 colors, add					1.08			1.08	1.19
4300	Specialty tile, 4-1/4" x 4-1/4" x 1/2", decorator finish	D-7	183	.087		8.95	1.52		10.47	12.30
4500	Add for epoxy grout, 1/16" joint, 1" x 1" tile		800	.020		.53	.35		.88	1.14
4600	2" x 2" tile		820	.020		.50	.34		.84	1.10
4800	Pregrouted sheets, walls, 4-1/4" x 4-1/4", 6" x 4-1/4"									
4810	and 8-1/2" x 4-1/4", 4 S.F. sheets, silicone grout	D-7	240	.067	S.F.	4.07	1.16		5.23	6.35
5100	Floors, unglazed, 2 S.F. sheets,									
5110	urethane adhesive	D-7	180	.089	S.F.	4.05	1.54		5.59	6.95
5400	Walls, interior, thin set, 4-1/4" x 4-1/4" tile		190	.084		2	1.46		3.46	4.56
5500	6" x 4-1/4" tile		190	.084		2.25	1.46		3.71	4.84
5700	8-1/2" x 4-1/4" tile		190	.084		3.19	1.46		4.65	5.85
5800	6" x 6" tile		200	.080		2.63	1.39		4.02	5.15
5810	8" x 8" tile		225	.071		3.08	1.23		4.31	5.40
5820	12" x 12" tile		300	.053		2.92	.92		3.84	4.71
5830	16" x 16" tile		500	.032		3.17	.55		3.72	4.39
6000	Decorated wall tile, 4-1/4" x 4-1/4", minimum		270	.059		4	1.03		5.03	6.05
6100	Maximum		180	.089		42	1.54		43.54	48.50
6600	Crystalline glazed, 4-1/4" x 4-1/4", mud set, plain		100	.160		3.15	2.77		5.92	7.95
6700	4-1/4" x 4-1/4", scored tile		100	.160		3.97	2.77		6.74	8.85
6900	6" x 6" plain		93	.172		4.02	2.98		7	9.25
7000	For epoxy grout, 1/16" joints, 4-1/4" tile, add		800	.020		.33	.35		.68	.92
7200	For tile set in dry mortar, add		1,735	.009			.16		.16	.26
7300	For tile set in portland cement mortar, add		290	.055			.96		.96	1.55

09330 | Quarry Tile

		Crew	Daily Output	Labor Hours	Unit	Mat.	Labor	Equip.	Total	Total Incl O&P
0010	QUARRY TILE Base, cove or sanitary, 2" or 5" high, mud set									
0100	1/2" thick	D-7	110	.145	L.F.	3.70	2.52		6.22	8.15
0300	Bullnose trim, red, mud set, 6" x 6" x 1/2" thick		120	.133		3.81	2.31		6.12	7.95
0400	4" x 4" x 1/2" thick		110	.145		3.83	2.52		6.35	8.30
0600	4" x 8" x 1/2" thick, using 8" as edge		130	.123		3.63	2.13		5.76	7.45
0700	Floors, mud set, 1,000 S.F. lots, red, 4" x 4" x 1/2" thick		120	.133	S.F.	3.74	2.31		6.05	7.85
0900	6" x 6" x 1/2" thick		140	.114		2.70	1.98		4.68	6.20
1000	4" x 8" x 1/2" thick		130	.123		3.74	2.13		5.87	7.55
1300	For waxed coating, add					.60			.60	.66
1500	For colors other than green, add					.35			.35	.39
1600	For abrasive surface, add					.42			.42	.46
1800	Brown tile, imported, 6" x 6" x 3/4"	D-7	120	.133		4.44	2.31		6.75	8.60
1900	8" x 8" x 1"		110	.145		5.05	2.52		7.57	9.65
2100	For thin set mortar application, deduct		700	.023			.40		.40	.64
2700	Stair tread, 6" x 6" x 3/4", plain		50	.320		4.04	5.55		9.59	13.45
2800	Abrasive		47	.340		4.59	5.90		10.49	14.60
3000	Wainscot, 6" x 6" x 1/2", thin set, red		105	.152		3.42	2.64		6.06	8.05
3100	Colors other than green		105	.152		3.81	2.64		6.45	8.45
3300	Window sill, 6" wide, 3/4" thick		90	.178	L.F.	4.41	3.08		7.49	9.85
3400	Corners		80	.200	Ea.	4.81	3.47		8.28	10.90

09370 | Metal Tile

		Crew	Daily Output	Labor Hours	Unit	Mat.	Labor	Equip.	Total	Total Incl O&P
0010	METAL TILE 4' x 4' sheet, 24 ga., tile pattern, nailed									
0200	Stainless steel	2 Carp	512	.031	S.F.	21.50	.62		22.12	25

For expanded coverage of these items see *Means Building Construction Cost Data 2000*

09300 | Tile

09370 | Metal Tile

			CREW	DAILY OUTPUT	LABOR-HOURS	UNIT	2000 BARE COSTS MAT.	LABOR	EQUIP.	TOTAL	TOTAL INCL O&P	
100	0400	Aluminized steel	2 Carp	512	.031	S.F.	11.70	.62		12.32	13.90	100

09400 | Terrazzo

09420 | Precast Terrazzo

			CREW	DAILY OUTPUT	LABOR-HOURS	UNIT	2000 BARE COSTS MAT.	LABOR	EQUIP.	TOTAL	TOTAL INCL O&P	
900	0010	TERRAZZO, PRECAST Base, 6" high, straight	1 Mstz	35	.229	L.F.	8.35	4.37		12.72	16.25	900
	0100	Cove		30	.267		9	5.10		14.10	18.15	
	0300	8" high base, straight		30	.267		7.75	5.10		12.85	16.80	
	0400	Cove		25	.320		12	6.10		18.10	23	
	0600	For white cement, add					.33			.33	.36	
	0700	For 16 ga. zinc toe strip, add					.99			.99	1.09	
	0900	Curbs, 4" x 4" high	1 Mstz	19	.421		23	8.05		31.05	38	
	1000	8" x 8" high	"	15	.533		28.50	10.20		38.70	47.50	
	1200	Floor tiles, non-slip, 1" thick, 12" x 12"	D-1	29	.552	S.F.	14.60	9.85		24.45	33	
	1300	1-1/4" thick, 12" x 12"		29	.552		16.75	9.85		26.60	35	
	1500	16" x 16"		23	.696		18.20	12.40		30.60	41	
	1600	1-1/2" thick, 16" x 16"		21	.762		16.65	13.60		30.25	41.50	
	4800	Wainscot, 12" x 12" x 1" tiles	1 Mstz	12	.667		5	12.75		17.75	26	
	4900	16" x 16" x 1-1/2" tiles	"	8	1		11.70	19.10		30.80	44	

09450 | Cast-in-Place Terrazzo

			CREW	DAILY OUTPUT	LABOR-HOURS	UNIT	2000 BARE COSTS MAT.	LABOR	EQUIP.	TOTAL	TOTAL INCL O&P	
100	0010	TERRAZZO, CAST IN PLACE Cove base, 6" high, 16ga. zinc R09400-100	1 Mstz	20	.400	L.F.	2.64	7.65		10.29	15.30	100
	0100	Curb, 6" high and 6" wide		6	1.333		4.44	25.50		29.94	46.50	
	0300	Divider strip for floors, 14 ga., 1-1/4" deep, zinc		375	.021		.91	.41		1.32	1.66	
	0400	Brass		375	.021		1.70	.41		2.11	2.53	
	0600	Heavy top strip 1/4" thick, 1-1/4" deep, zinc		300	.027		1.54	.51		2.05	2.52	
	1200	For thin set floors, 16 ga., 1/2" x 1/2", zinc		350	.023		.80	.44		1.24	1.59	
	1500	Floor, bonded to concrete, 1-3/4" thick, gray cement	J-3	130	.123	S.F.	2.37	2.13	.96	5.46	7.10	
	1600	White cement, mud set		130	.123		2.70	2.13	.96	5.79	7.50	
	1800	Not bonded, 3" total thickness, gray cement		115	.139		2.96	2.41	1.08	6.45	8.35	
	1900	White cement, mud set		115	.139		3.23	2.41	1.08	6.72	8.65	
200	0010	TILE OR TERRAZZO BASE Scratch coat only	1 Mstz	150	.053	S.F.	.31	1.02		1.33	1.99	200
	0500	Scratch and brown coat only	"	75	.107	"	.56	2.04		2.60	3.92	

09500 | Ceilings

09510 | Acoustical Ceilings

			CREW	DAILY OUTPUT	LABOR-HOURS	UNIT	2000 BARE COSTS MAT.	LABOR	EQUIP.	TOTAL	TOTAL INCL O&P	
700	0010	SUSPENDED ACOUSTIC CEILING TILES, Not including										700
	0100	suspension system										
	0300	Fiberglass boards, film faced, 2' x 2' or 2' x 4', 5/8" thick	1 Carp	625	.013	S.F.	.51	.25		.76	.99	
	0400	3/4" thick		600	.013		1.06	.26		1.32	1.62	
	0500	3" thick, thermal, R11		450	.018		1.25	.35		1.60	1.98	
	0600	Glass cloth faced fiberglass, 3/4" thick		500	.016		1.65	.32		1.97	2.36	

09500 | Ceilings

09510 | Acoustical Ceilings

		CREW	DAILY OUTPUT	LABOR-HOURS	UNIT	MAT.	LABOR	EQUIP.	TOTAL	TOTAL INCL O&P
700	0700	1" thick	1 Carp	485	.016	S.F.	1.86	.32		2.18	2.61
	0820	1-1/2" thick, nubby face		475	.017		2.21	.33		2.54	3
	1110	Mineral fiber tile, lay-in, 2' x 2' or 2' x 4', 5/8" thick, fine texture		625	.013		.41	.25		.66	.88
	1115	Rough textured		625	.013		1	.25		1.25	1.53
	1125	3/4" thick, fine textured		600	.013		1.12	.26		1.38	1.68
	1130	Rough textured		600	.013		1.40	.26		1.66	1.99
	1135	Fissured		600	.013		1.65	.26		1.91	2.27
	1150	Tegular, 5/8" thick, fine textured		470	.017		.99	.34		1.33	1.66
	1155	Rough textured		470	.017		1.29	.34		1.63	1.99
	1165	3/4" thick, fine textured		450	.018		1.40	.35		1.75	2.14
	1170	Rough textured		450	.018		1.58	.35		1.93	2.34
	1175	Fissured		450	.018		2.46	.35		2.81	3.31
	1180	For aluminum face, add					4.49			4.49	4.94
	1185	For plastic film face, add					.69			.69	.76
	1190	For fire rating, add					.34			.34	.37
	1300	Mirror faced panels, 15/16" thick, 2' x 2'	1 Carp	500	.016		9.10	.32		9.42	10.55
	1900	Eggcrate, acrylic, 1/2" x 1/2" x 1/2" cubes		500	.016		1.36	.32		1.68	2.04
	2100	Polystyrene eggcrate, 3/8" x 3/8" x 1/2" cubes		510	.016		1.13	.31		1.44	1.77
	2200	1/2" x 1/2" x 1/2" cubes		500	.016		1.52	.32		1.84	2.21
	2400	Luminous panels, prismatic, acrylic		400	.020		1.64	.39		2.03	2.48
	2500	Polystyrene		400	.020		.84	.39		1.23	1.60
	2700	Flat white acrylic		400	.020		2.87	.39		3.26	3.84
	2800	Polystyrene		400	.020		1.96	.39		2.35	2.84
	3000	Drop pan, white, acrylic		400	.020		4.20	.39		4.59	5.30
	3100	Polystyrene		400	.020		3.51	.39		3.90	4.54
	3600	Perforated aluminum sheets, .024" thick, corrugated, painted		490	.016		1.67	.32		1.99	2.39
	3700	Plain		500	.016		1.50	.32		1.82	2.19
760	0010	**SUSPENDED CEILINGS, COMPLETE** Including standard									
	0100	suspension system but not incl. 1-1/2" carrier channels									
	0600	Fiberglass ceiling board, 2' x 4' x 5/8", plain faced,	1 Carp	500	.016	S.F.	1.02	.32		1.34	1.66
	0700	Offices, 2' x 4' x 3/4"		380	.021		1.13	.41		1.54	1.95
	1800	Tile, Z bar suspension, 5/8" mineral fiber tile		150	.053		1.47	1.05		2.52	3.42
	1900	3/4" mineral fiber tile		150	.053		1.57	1.05		2.62	3.53
900	0010	**CEILING TILE,** Stapled or cemented									
	0100	12" x 12" or 12" x 24", not including furring									
	0600	Mineral fiber, vinyl coated, 5/8" thick	1 Carp	1,000	.008	S.F.	.72	.16		.88	1.06
	0700	3/4" thick		1,000	.008		1.17	.16		1.33	1.56
	0900	Fire rated, 3/4" thick, plain faced		1,000	.008		1	.16		1.16	1.37
	1000	Plastic coated face		1,000	.008		1.04	.16		1.20	1.41
	1200	Aluminum faced, 5/8" thick, plain		1,000	.008		1.09	.16		1.25	1.47
	3300	For flameproofing, add					.09			.09	.10
	3400	For sculptured 3 dimensional, add					.24			.24	.26
	3900	For ceiling primer, add					.12			.12	.13
	4000	For ceiling cement, add					.32			.32	.35

For expanded coverage of these items see *Means Building Construction Cost Data 2000*

09600 | Flooring

09631 | Brick Flooring

		CREW	DAILY OUTPUT	LABOR-HOURS	UNIT	MAT.	LABOR	EQUIP.	TOTAL	TOTAL INCL O&P	
100	0010 **FLOORING**										100
	0020 Acid proof shales, red, 8" x 3-3/4" x 1-1/4" thick	D-7	.43	37.209	M	695	645		1,340	1,825	
	0050 2-1/4" thick	D-1	.40	40		755	715		1,470	2,025	
	0200 Acid proof clay brick, 8" x 3-3/4" x 2-1/4" thick	"	.40	40		755	715		1,470	2,025	
	0260 Cast ceramic, pressed, 4" x 8" x 1/2", unglazed	D-7	100	.160	S.F.	4.94	2.77		7.71	9.95	
	0270 Glazed		100	.160		6.60	2.77		9.37	11.75	
	0280 Hand molded flooring, 4" x 8" x 3/4", unglazed		95	.168		6.55	2.92		9.47	11.95	
	0290 Glazed		95	.168		8.20	2.92		11.12	13.75	
	0300 8" hexagonal, 3/4" thick, unglazed		85	.188		7.15	3.26		10.41	13.20	
	0310 Glazed		85	.188		12.95	3.26		16.21	19.50	
	0450 Acid proof joints, 1/4" wide	D-1	65	.246		1.13	4.39		5.52	8.70	
	0500 Pavers, 8" x 4", 1" to 1-1/4" thick, red	D-7	95	.168		2.88	2.92		5.80	7.90	
	0510 Ironspot	"	95	.168		4.07	2.92		6.99	9.20	
	0540 1-3/8" to 1-3/4" thick, red	D-1	95	.168		2.78	3.01		5.79	8.15	
	0560 Ironspot		95	.168		4.02	3.01		7.03	9.50	
	0580 2-1/4" thick, red		90	.178		2.83	3.17		6	8.45	
	0590 Ironspot		90	.178		4.38	3.17		7.55	10.15	
	0800 For sidewalks and patios with pavers, see division 02780-200										
	0870 For epoxy joints, add	D-1	600	.027	S.F.	2.15	.48		2.63	3.18	
	0880 For Furan underlayment, add	"	600	.027		1.78	.48		2.26	2.77	
	0890 For waxed surface, steam cleaned, add	D-5	1,000	.008		.15	.13		.28	.38	

09635 | Marble Flooring

		CREW	DAILY OUTPUT	LABOR-HOURS	UNIT	MAT.	LABOR	EQUIP.	TOTAL	TOTAL INCL O&P	
100	0010 **MARBLE** Thin gauge tile, 12" x 6", 3/8", White Carara	D-7	60	.267	S.F.	8.75	4.62		13.37	17.15	100
	0100 Travertine		60	.267		9.65	4.62		14.27	18.10	
	0200 12" x 12" x 3/8", thin set, floors		60	.267		6.70	4.62		11.32	14.85	
	0300 On walls		52	.308		8.75	5.35		14.10	18.30	

09637 | Stone Flooring

		CREW	DAILY OUTPUT	LABOR-HOURS	UNIT	MAT.	LABOR	EQUIP.	TOTAL	TOTAL INCL O&P	
100	0010 **SLATE TILE** Vermont, 6" x 6" x 1/4" thick, thin set	D-7	180	.089	S.F.	4.14	1.54		5.68	7.05	100
200	0010 **SLATE & STONE FLOORS** See division 02780-800										200

09643 | Wood Block Flooring

		CREW	DAILY OUTPUT	LABOR-HOURS	UNIT	MAT.	LABOR	EQUIP.	TOTAL	TOTAL INCL O&P	
100	0010 **WOOD BLOCK FLOORING** End grain flooring, coated, 2" thick	1 Carp	295	.027	S.F.	2.70	.53		3.23	3.89	100
	0400 Natural finish, 1" thick, fir		125	.064		2.80	1.26		4.06	5.25	
	0600 1-1/2" thick, pine		125	.064		2.75	1.26		4.01	5.20	
	0700 2" thick, pine		125	.064		2.72	1.26		3.98	5.15	

09648 | Wood Strip Flooring

		CREW	DAILY OUTPUT	LABOR-HOURS	UNIT	MAT.	LABOR	EQUIP.	TOTAL	TOTAL INCL O&P	
100	0010 **WOOD** Fir, vertical grain, 1" x 4", not incl. finish, B & better	1 Carp	255	.031	S.F.	2.30	.62		2.92	3.59	100
	0100 C grade & better		255	.031		2.16	.62		2.78	3.44	
	0300 Flat grain, 1" x 4", not incl. finish, B & better		255	.031		2.63	.62		3.25	3.95	
	0400 C & better		255	.031		2.53	.62		3.15	3.84	
	4000 Maple, strip, 25/32" x 2-1/4", not incl. finish, select		170	.047		3.10	.93		4.03	5	
	4100 #2 & better		170	.047		2.65	.93		3.58	4.51	
	4300 33/32" x 3-1/4", not incl. finish, #1 grade		170	.047		3.25	.93		4.18	5.15	
	4400 #2 & better		170	.047		2.90	.93		3.83	4.78	
	4600 Oak, white or red, 25/32" x 2-1/4", not incl. finish										
	4700 #1 common	1 Carp	170	.047	S.F.	2.70	.93		3.63	4.56	
	4900 Select quartered, 2-1/4" wide		170	.047		3.02	.93		3.95	4.91	
	5000 Clear		170	.047		3.25	.93		4.18	5.15	
	5200 Parquetry, standard, 5/16" thick, not incl. finish, oak, minimum		160	.050		3.04	.99		4.03	5.05	
	5300 Maximum		100	.080		5.10	1.58		6.68	8.35	

09600 | Flooring

09648 | Wood Strip Flooring

		CREW	DAILY OUTPUT	LABOR-HOURS	UNIT	2000 BARE COSTS MAT.	LABOR	EQUIP.	TOTAL	TOTAL INCL O&P
5500	Teak, minimum	1 Carp	160	.050	S.F.	4.15	.99		5.14	6.25
5600	Maximum		100	.080		7.25	1.58		8.83	10.70
5650	13/16" thick, select grade oak, minimum		160	.050		8	.99		8.99	10.50
5700	Maximum		100	.080		12.15	1.58		13.73	16.05
5800	Custom parquetry, including finish, minimum		100	.080		13.40	1.58		14.98	17.45
5900	Maximum		50	.160		17.75	3.15		20.90	25
6100	Prefinished, white oak, prime grade, 2-1/4" wide		170	.047		5.85	.93		6.78	8.05
6200	3-1/4" wide		185	.043		7.50	.85		8.35	9.70
6400	Ranch plank		145	.055		7.25	1.09		8.34	9.85
6500	Hardwood blocks, 9" x 9", 25/32" thick		160	.050		4.85	.99		5.84	7.05
6700	Parquetry, 5/16" thick, oak, minimum		160	.050		3.30	.99		4.29	5.30
6800	Maximum		100	.080		8.10	1.58		9.68	11.60
7000	Walnut or teak, parquetry, minimum		160	.050		4.50	.99		5.49	6.65
7100	Maximum		100	.080		7.85	1.58		9.43	11.35
7200	Acrylic wood parquet blocks, 12" x 12" x 5/16",									
7210	irradiated, set in epoxy	1 Carp	160	.050	S.F.	6.55	.99		7.54	8.90
7400	Yellow pine, 3/4" x 3-1/8", T & G, C & better, not incl. finish	"	200	.040		2.15	.79		2.94	3.72
7500	Refinish wood floor, sand, 2 cts poly, wax, soft wood, min.	1 Clab	400	.020		.65	.29		.94	1.21
7600	Hard wood, max		130	.062		.98	.89		1.87	2.60
7800	Sanding and finishing, 2 coats polyurethane		295	.027		.65	.39		1.04	1.38
7900	Subfloor and underlayment, see division 06160									
8015	Transition molding, 2 1/4" wide, 5' long	1 Carp	19.20	.417	Ea.	10.35	8.20		18.55	25.50
8300	Floating floor, wood composition strip, complete.	1 Clab	133	.060	S.F.	4.16	.87		5.03	6.05
8310	Floating floor components, T & G wood composite strips					3.87			3.87	4.26
8320	Film					.09			.09	.10
8330	Foam					.19			.19	.21
8340	Adhesive					.06			.06	.07
8350	Installation kit					.16			.16	.18
8360	Trim, 2" wide x 3' long				L.F.	2.25			2.25	2.48
8370	Reducer moulding				"	4.15			4.15	4.57

09658 | Resilient Tile Flooring

		CREW	DAILY OUTPUT	LABOR-HOURS	UNIT	MAT.	LABOR	EQUIP.	TOTAL	TOTAL INCL O&P
0010	**RESILIENT FLOORING**									
0800	Base, cove, rubber or vinyl, .080" thick									
1100	Standard colors, 2-1/2" high	1 Tilf	315	.025	L.F.	.40	.49		.89	1.23
1150	4" high		315	.025		.44	.49		.93	1.27
1200	6" high		315	.025		.73	.49		1.22	1.59
1450	1/8" thick, standard colors, 2-1/2" high		315	.025		.45	.49		.94	1.29
1500	4" high		315	.025		.65	.49		1.14	1.50
1550	6" high		315	.025		.81	.49		1.30	1.68
1600	Corners, 2-1/2" high		315	.025	Ea.	1.02	.49		1.51	1.91
1630	4" high		315	.025		1.07	.49		1.56	1.97
1660	6" high		315	.025		1.38	.49		1.87	2.31
1700	Conductive flooring, rubber tile, 1/8" thick		315	.025	S.F.	2.50	.49		2.99	3.54
1800	Homogeneous vinyl tile, 1/8" thick		315	.025		3.42	.49		3.91	4.55
2200	Cork tile, standard finish, 1/8" thick		315	.025		3.41	.49		3.90	4.54
2250	3/16" thick		315	.025		3.64	.49		4.13	4.79
2300	5/16" thick		315	.025		4.56	.49		5.05	5.80
2350	1/2" thick		315	.025		5.45	.49		5.94	6.80
2500	Urethane finish, 1/8" thick		315	.025		4.33	.49		4.82	5.55
2550	3/16" thick		315	.025		5.15	.49		5.64	6.50
2600	5/16" thick		315	.025		6.05	.49		6.54	7.45
2650	1/2" thick		315	.025		8.10	.49		8.59	9.70
3700	Polyethylene, in rolls, no base incl., landscape surfaces		275	.029		2.24	.56		2.80	3.36
3800	Nylon action surface, 1/8" thick		275	.029		2.41	.56		2.97	3.55
3900	1/4" thick		275	.029		3.47	.56		4.03	4.72

For expanded coverage of these items see Means Building Construction Cost Data 2000

09600 | Flooring

09658 | Resilient Tile Flooring

		CREW	DAILY OUTPUT	LABOR-HOURS	UNIT	2000 BARE COSTS MAT.	LABOR	EQUIP.	TOTAL	TOTAL INCL O&P
4000	3/8" thick	1 Tilf	275	.029	S.F.	4.36	.56		4.92	5.70
5900	Rubber, sheet goods, 36" wide, 1/8" thick		120	.067		3.09	1.28		4.37	5.45
5950	3/16" thick		100	.080		4.39	1.54		5.93	7.30
6000	1/4" thick		90	.089		5.05	1.71		6.76	8.35
6050	Tile, marbleized colors, 12" x 12", 1/8" thick		400	.020		3.50	.38		3.88	4.47
6100	3/16" thick		400	.020		4.75	.38		5.13	5.80
6300	Special tile, plain colors, 1/8" thick		400	.020		3.83	.38		4.21	4.83
6350	3/16" thick		400	.020		5.15	.38		5.53	6.25
7000	Vinyl composition tile, 12" x 12", 1/16" thick		500	.016		.69	.31		1	1.26
7050	Embossed		500	.016		.86	.31		1.17	1.45
7100	Marbleized		500	.016		.86	.31		1.17	1.45
7150	Solid		500	.016		.97	.31		1.28	1.57
7200	3/32" thick, embossed		500	.016		.85	.31		1.16	1.44
7250	Marbleized		500	.016		.98	.31		1.29	1.58
7300	Solid		500	.016		1.43	.31		1.74	2.07
7350	1/8" thick, marbleized		500	.016		.94	.31		1.25	1.53
7400	Solid		500	.016		2.01	.31		2.32	2.71
7450	Conductive		500	.016		3.33	.31		3.64	4.16
7500	Vinyl tile, 12" x 12", .050" thick, minimum		500	.016		1.54	.31		1.85	2.19
7550	Maximum		500	.016		3.01	.31		3.32	3.81
7600	1/8" thick, minimum		500	.016		1.94	.31		2.25	2.63
7650	Solid colors		500	.016		4.35	.31		4.66	5.30
7700	Marbleized or Travertine pattern		500	.016		3.11	.31		3.42	3.92
7750	Florentine pattern		500	.016		3.57	.31		3.88	4.43
7800	Maximum		500	.016		7.35	.31		7.66	8.55
8000	Vinyl sheet goods, backed, .065" thick, minimum		250	.032		1.25	.61		1.86	2.38
8050	Maximum		200	.040		2.32	.77		3.09	3.79
8100	.080" thick, minimum		230	.035		1.50	.67		2.17	2.73
8150	Maximum		200	.040		2.65	.77		3.42	4.16
8200	.125" thick, minimum		230	.035		1.63	.67		2.30	2.87
8250	Maximum		200	.040		3.57	.77		4.34	5.15
8700	Adhesive cement, 1 gallon does 200 to 300 S.F.				Gal.	13.85			13.85	15.20
8800	Asphalt primer, 1 gallon per 300 S.F.					8.40			8.40	9.25
8900	Emulsion, 1 gallon per 140 S.F.					9			9	9.90
8950	Latex underlayment, liquid, fortified					27.50			27.50	30

09673 | Composition Flooring

		CREW	DAILY OUTPUT	LABOR-HOURS	UNIT	MAT.	LABOR	EQUIP.	TOTAL	TOTAL INCL O&P
0010	**COMPOSITION FLOORING** Acrylic, 1/4" thick	C-6	520	.092	S.F.	1.21	1.43	.15	2.79	3.92
0600	Epoxy, with colored quartz chips, broadcast, minimum		675	.071		2.11	1.10	.11	3.32	4.31
0700	Maximum		490	.098		2.56	1.52	.15	4.23	5.55
0900	Trowelled, minimum		560	.086		2.72	1.33	.14	4.19	5.40
1000	Maximum		480	.100		3.96	1.55	.16	5.67	7.15
1200	Heavy duty epoxy topping, 1/4" thick,									
1300	500 to 1,000 S.F.	C-6	420	.114	S.F.	4.09	1.77	.18	6.04	7.70
1500	1,000 to 2,000 S.F.		450	.107		2.99	1.66	.17	4.82	6.30
1600	Over 10,000 S.F.		480	.100		2.67	1.55	.16	4.38	5.75
1800	Epoxy terrazzo, 1/4" thick, chemical resistant, minimum	J-3	200	.080		4.35	1.39	.62	6.36	7.70
1900	Maximum	"	150	.107		7.10	1.85	.83	9.78	11.75

09680 | Carpet

		CREW	DAILY OUTPUT	LABOR-HOURS	UNIT	MAT.	LABOR	EQUIP.	TOTAL	TOTAL INCL O&P
0010	**CARPET** Commercial grades, direct cement									
0700	Nylon, level loop, 26 oz., light to medium traffic	1 Tilf	75	.107	S.Y.	12.90	2.05		14.95	17.50
0720	28 oz., light to medium traffic		75	.107		14.25	2.05		16.30	19
0900	32 oz., medium traffic		75	.107		18.30	2.05		20.35	23.50

09600 | Flooring

09680 | Carpet

			CREW	DAILY OUTPUT	LABOR-HOURS	UNIT	MAT.	LABOR	EQUIP.	TOTAL	TOTAL INCL O&P	
800	1100	40 oz., medium to heavy traffic	1 Tilf	75	.107	S.Y.	27	2.05		29.05	33.50	800
	2920	Nylon plush, 30 oz., medium traffic		57	.140		13.50	2.69		16.19	19.20	
	3000	36 oz., medium traffic		75	.107		17	2.05		19.05	22	
	3100	42 oz., medium to heavy traffic		70	.114		16.70	2.19		18.89	22	
	3200	46 oz., medium to heavy traffic		70	.114		23.50	2.19		25.69	29.50	
	3300	54 oz., heavy traffic		70	.114		26.50	2.19		28.69	32.50	
	4500	50 oz., medium to heavy traffic		75	.107		60	2.05		62.05	69.50	
	4700	Patterned, 32 oz., medium to heavy traffic		70	.114		59	2.19		61.19	68.50	
	4900	48 oz., heavy traffic		70	.114		60.50	2.19		62.69	70	
	5000	For less than full roll, add					25%					
	5100	For small rooms, less than 12' wide, add						25%				
	5200	For large open areas (no cuts), deduct						25%				
	5600	For bound carpet baseboard, add	1 Tilf	300	.027	L.F.	1.02	.51		1.53	1.95	
	5610	For stairs, not incl. price of carpet, add	"	30	.267	Riser		5.10		5.10	8.30	
	5620	For borders and patterns, add to labor						18%				
	8950	For tackless, stretched installation, add padding to above										
	9000	Sponge rubber pad, minimum	1 Tilf	150	.053	S.Y.	2.61	1.02		3.63	4.53	
	9100	Maximum		150	.053		6.95	1.02		7.97	9.30	
	9200	Felt pad, minimum		150	.053		2.78	1.02		3.80	4.72	
	9300	Maximum		150	.053		5.20	1.02		6.22	7.35	
	9400	Bonded urethane pad, minimum		150	.053		3.15	1.02		4.17	5.15	
	9500	Maximum		150	.053		5.60	1.02		6.62	7.80	
	9600	Prime urethane pad, minimum		150	.053		1.83	1.02		2.85	3.67	
	9700	Maximum		150	.053		3.35	1.02		4.37	5.35	
	9850	For "branded" fiber, add					25%					
900	0010	**CARPET TILE**										900
	0100	Tufted nylon, 18" x 18", hard back, 20 oz.	1 Tilf	150	.053	S.Y.	16.85	1.02		17.87	20	
	0110	26 oz.		150	.053		29	1.02		30.02	33	
	0200	Cushion back, 20 oz.		150	.053		21	1.02		22.02	25	
	0210	26 oz.		150	.053		33	1.02		34.02	38	

09700 | Wall Finishes

09710 | Acoustical Wall Treatment

			CREW	DAILY OUTPUT	LABOR-HOURS	UNIT	MAT.	LABOR	EQUIP.	TOTAL	TOTAL INCL O&P	
100	0010	**WALL COVERING** Including sizing, add 10%-30% waste at takeoff										100
	0050	Aluminum foil	1 Pape	275	.029	S.F.	.78	.53		1.31	1.74	
	0100	Copper sheets, .025" thick, vinyl backing		240	.033		4.15	.60		4.75	5.55	
	0300	Phenolic backing		240	.033		5.35	.60		5.95	6.90	
	0600	Cork tiles, light or dark, 12" x 12" x 3/16"		240	.033		2.72	.60		3.32	3.99	
	0700	5/16" thick		235	.034		2.77	.61		3.38	4.07	
	0900	1/4" basketweave		240	.033		4.25	.60		4.85	5.70	
	1000	1/2" natural, non-directional pattern		240	.033		6.35	.60		6.95	7.95	
	1100	3/4" natural, non-directional pattern		240	.033		8.05	.60		8.65	9.85	
	1200	Granular surface, 12" x 36", 1/2" thick		385	.021		.92	.38		1.30	1.64	
	1300	1" thick		370	.022		1.19	.39		1.58	1.96	
	1500	Polyurethane coated, 12" x 12" x 3/16" thick		240	.033		2.87	.60		3.47	4.16	
	1600	5/16" thick		235	.034		4.08	.61		4.69	5.50	
	1800	Cork wallpaper, paperbacked, natural		480	.017		1.63	.30		1.93	2.29	

For expanded coverage of these items see *Means Building Construction Cost Data 2000*

09700 | Wall Finishes

09710 | Acoustical Wall Treatment

			CREW	DAILY OUTPUT	LABOR-HOURS	UNIT	MAT.	LABOR	EQUIP.	TOTAL	TOTAL INCL O&P	
100	1900	Colors	1 Pape	480	.017	S.F.	2.02	.30		2.32	2.72	100
	2100	Flexible wood veneer, 1/32" thick, plain woods		100	.080		1.64	1.44		3.08	4.21	
	2200	Exotic woods	↓	95	.084	↓	2.49	1.52		4.01	5.25	
	2400	Gypsum-based, fabric-backed, fire										
	2500	resistant for masonry walls, minimum, 21 oz./S.Y.	1 Pape	800	.010	S.F.	.65	.18		.83	1.01	
	2600	Average		720	.011		.90	.20		1.10	1.32	
	2700	Maximum, (small quantities)	↓	640	.013		1	.23		1.23	1.48	
	2750	Acrylic, modified, semi-rigid PVC, .028" thick	2 Carp	330	.048		.82	.96		1.78	2.54	
	2800	.040" thick	"	320	.050		1.08	.99		2.07	2.88	
	3000	Vinyl wall covering, fabric-backed, lightweight, (12-15 oz./S.Y.)	1 Pape	640	.013		.53	.23		.76	.96	
	3300	Medium weight, type 2, (20-24 oz./S.Y.)		480	.017		.63	.30		.93	1.19	
	3400	Heavy weight, type 3, (28 oz./S.Y.)	↓	435	.018	↓	1.11	.33		1.44	1.77	
	3600	Adhesive, 5 gal. lots, (18SY/Gal.)				Gal.	8.30			8.30	9.15	
	3700	Wallpaper, average workmanship, solid pattern, low cost paper	1 Pape	640	.013	S.F.	.27	.23		.50	.68	
	3900	basic patterns (matching required), avg. cost paper		535	.015		.49	.27		.76	.99	
	4000	Paper at $50 per double roll, quality workmanship	↓	435	.018	↓	.92	.33		1.25	1.56	
	4100	Linen wall covering, paper backed										
	4150	Flame treatment, minimum				S.F.	.65			.65	.71	
	4180	Maximum					1.17			1.17	1.29	
	4200	Grass cloths with lining paper, minimum	1 Pape	400	.020		.59	.36		.95	1.25	
	4300	Maximum	"	350	.023	↓	1.88	.41		2.29	2.76	

09770 | Special Wall Surfaces

			CREW	DAILY OUTPUT	LABOR-HOURS	UNIT	MAT.	LABOR	EQUIP.	TOTAL	TOTAL INCL O&P	
400	0010	**FIBERGLASS REINFORCED PLASTIC** panels, .090" thick, on walls										400
	0020	Adhesive mounted, embossed surface	2 Carp	640	.025	S.F.	1.21	.49		1.70	2.17	
	0030	Smooth surface		640	.025		1.38	.49		1.87	2.36	
	0040	Fire rated, embossed surface		640	.025		1.59	.49		2.08	2.59	
	0050	Nylon rivet mounted, on drywall, embossed surface		480	.033		1.11	.66		1.77	2.35	
	0060	Smooth surface		480	.033		1.28	.66		1.94	2.54	
	0070	Fire rated, embossed surface		480	.033		1.49	.66		2.15	2.77	
	0080	On masonry, embossed surface		320	.050		1.11	.99		2.10	2.91	
	0090	Smooth surface		320	.050		1.28	.99		2.27	3.10	
	0100	Fire rated, embossed surface		320	.050		1.49	.99		2.48	3.33	
	0110	Nylon rivet and adhesive mounted, on drywall, embossed surface		240	.067		1.30	1.31		2.61	3.68	
	0120	Smooth surface		240	.067		1.47	1.31		2.78	3.87	
	0130	Fire rated, embossed surface		240	.067		1.68	1.31		2.99	4.10	
	0140	On masonry, embossed surface		190	.084		1.30	1.66		2.96	4.27	
	0150	Smooth surface		190	.084		1.47	1.66		3.13	4.46	
	0160	Fire rated, embossed surface	↓	190	.084	↓	1.68	1.66		3.34	4.69	
	0170	For moldings add	1 Carp	250	.032	L.F.	.22	.63		.85	1.32	
	0180	On ceilings, for lay in grid system, embossed surface		400	.020	S.F.	1.21	.39		1.60	2.01	
	0190	Smooth surface		400	.020		1.38	.39		1.77	2.20	
	0200	Fire rated, embossed surface	↓	400	.020	↓	1.59	.39		1.98	2.43	
700	0010	**RAISED PANEL SYSTEM**, 3/4" MDO										700
	0100	Standard, paint grade	2 Carp	300	.053	S.F.	6.75	1.05		7.80	9.25	
	0110	Oak veneer		300	.053		10.25	1.05		11.30	13.10	
	0120	Maple veneer		300	.053		16.65	1.05		17.70	20	
	0130	Cherry veneer		300	.053		20	1.05		21.05	24	
	0300	Class I fire rated, paint grade		300	.053		6.75	1.05		7.80	9.25	
	0310	Oak veneer		300	.053		10.25	1.05		11.30	13.10	
	0320	Maple veneer		300	.053		16.65	1.05		17.70	20	
	0330	Cherry veneer	↓	300	.053	↓	20	1.05		21.05	24	
	5000	For prefinished paneling, see division 06250-500 & 06250-200										
750	0010	**SLATWALL PANELS AND ACCESSORIES**										750
	0100	Slatwall panel, 4' x 8' x 3/4" T, MDF, paint grade	1 Carp	500	.016	S.F.	1.15	.32		1.47	1.81	

09700 | Wall Finishes

09770 | Special Wall Surfaces

			CREW	DAILY OUTPUT	LABOR-HOURS	UNIT	2000 BARE COSTS MAT.	LABOR	EQUIP.	TOTAL	TOTAL INCL O&P	
750	0110	Melamine finish	1 Carp	500	.016	S.F.	1.65	.32		1.97	2.36	750
	0120	High pressure plastic laminate finish	↓	500	.016		1.95	.32		2.27	2.69	
	0130	Aluminum channel inserts, add					1.40			1.40	1.54	
	0200	Accessories, corner forms, 8' L				L.F.	3.10			3.10	3.41	
	0210	T-connector, 8' L					4.50			4.50	4.95	
	0220	J-mold, 8' L					1.15			1.15	1.27	
	0230	Edge cap, 8' L					.75			.75	.83	
	0240	Finish end cap, 8' L				↓	2.75			2.75	3.03	
	0300	Display hook, 4" L				Ea.	.85			.85	.94	
	0310	6" L					.90			.90	.99	
	0320	8" L					1			1	1.10	
	0330	10" L					1.10			1.10	1.21	
	0340	12" L					1.20			1.20	1.32	
	0350	Acrylic, 4" L					.70			.70	.77	
	0360	6" L					.80			.80	.88	
	0370	8" L					.85			.85	.94	
	0380	10" L					.95			.95	1.05	
	0400	Waterfall hanger, metal, 12" - 16"					5.40			5.40	5.95	
	0410	Acrylic					7.70			7.70	8.45	
	0500	Shelf bracket, metal, 8"					4.20			4.20	4.62	
	0510	10"					4.50			4.50	4.95	
	0520	12"					4.80			4.80	5.30	
	0530	14"					5.40			5.40	5.95	
	0540	16"					6			6	6.60	
	0550	Acrylic, 8"					2.50			2.50	2.75	
	0560	10"					2.80			2.80	3.08	
	0570	12"					3.10			3.10	3.41	
	0580	14"					3.50			3.50	3.85	
	0600	Shelf, acrylic, 12" x 16" x 1/4"					21			21	23	
	0610	12" x 24" x 1/4"				↓	35			35	38.50	

09800 | Acoustical Treatment

09820 | Acoustical Insul/Sealants

			CREW	DAILY OUTPUT	LABOR-HOURS	UNIT	2000 BARE COSTS MAT.	LABOR	EQUIP.	TOTAL	TOTAL INCL O&P	
500	0010	**SOUND ATTENUATION** Blanket, 1" thick	1 Carp	925	.009	S.F.	.21	.17		.38	.52	500
	0500	1-1/2" thick		920	.009		.24	.17		.41	.55	
	1000	2" thick		915	.009		.31	.17		.48	.64	
	1500	3" thick	↓	910	.009	↓	.40	.17		.57	.74	

09840 | Acoustical Wall Treatment

			CREW	DAILY OUTPUT	LABOR-HOURS	UNIT	2000 BARE COSTS MAT.	LABOR	EQUIP.	TOTAL	TOTAL INCL O&P	
100	0010	**SOUND ABSORBING PANELS** Perforated steel facing, painted with										100
	0100	fiberglass or mineral filler, no backs, 2-1/4" thick, modular										
	0200	space units, ceiling or wall hung, white or colored	1 Carp	100	.080	S.F.	7.45	1.58		9.03	10.85	
	0300	Fiberboard sound deadening panels, 1/2" thick	"	600	.013	"	.28	.26		.54	.76	
	0500	Fiberglass panels, 4' x 8' x 1" thick, with										
	0600	glass cloth face for walls, cemented	1 Carp	155	.052	S.F.	4.69	1.02		5.71	6.90	
	0700	1-1/2" thick, dacron covered, inner aluminum frame,										
	0710	wall mounted	1 Carp	300	.027	S.F.	7.15	.53		7.68	8.80	

For expanded coverage of these items see *Means Building Construction Cost Data 2000*

09900 | Paints & Coatings

09910 | Paints & Coatings

		CREW	DAILY OUTPUT	LABOR-HOURS	UNIT	2000 BARE COSTS				TOTAL INCL O&P
						MAT.	LABOR	EQUIP.	TOTAL	
100 0010	**CABINETS AND CASEWORK**									**100**
1000	Primer coat, oil base, brushwork	1 Pord	650	.012	S.F.	.04	.22		.26	.42
2000	Paint, oil base, brushwork, 1 coat		650	.012		.05	.22		.27	.43
2500	2 coats		400	.020		.10	.36		.46	.71
3000	Stain, brushwork, wipe off		650	.012		.04	.22		.26	.42
4000	Shellac, 1 coat, brushwork		650	.012		.05	.22		.27	.42
4500	Varnish, 3 coats, brushwork, sand after 1st coat	↓	325	.025		.16	.44		.60	.92
5000	For latex paint, deduct					10%				
300 0010	**DOORS AND WINDOWS, EXTERIOR**		R09910-100							**300**
0100	Door frames & trim, only									
0110	Brushwork, primer	1 Pord	512	.016	L.F.	.05	.28		.33	.53
0120	Finish coat, exterior latex	R09910-220	512	.016		.05	.28		.33	.53
0130	Primer & 1 coat, exterior latex		300	.027	↓	.10	.48		.58	.91
0135	2 coats, exterior latex, both sides		15	.533	Ea.	4.70	9.55		14.25	21
0140	Primer & 2 coats, exterior latex	↓	265	.030	L.F.	.16	.54		.70	1.07
0150	Doors, flush, both sides, incl. frame & trim									
0160	Roll & brush, primer	1 Pord	10	.800	Ea.	3.83	14.35		18.18	28
0170	Finish coat, exterior latex		10	.800		4	14.35		18.35	28.50
0180	Primer & 1 coat, exterior latex		7	1.143		7.85	20.50		28.35	42.50
0190	Primer & 2 coats, exterior latex		5	1.600		11.85	28.50		40.35	61
0200	Brushwork, stain, sealer & 2 coats polyurethane	↓	4	2	↓	14.25	36		50.25	75.50
0210	Doors, French, both sides, 10-15 lite, incl. frame & trim									
0220	Brushwork, primer	1 Pord	6	1.333	Ea.	1.91	24		25.91	42
0230	Finish coat, exterior latex		6	1.333		2	24		26	42
0240	Primer & 1 coat, exterior latex		3	2.667		3.92	48		51.92	84.50
0250	Primer & 2 coats, exterior latex		2	4		5.80	72		77.80	126
0260	Brushwork, stain, sealer & 2 coats polyurethane	↓	2.50	3.200	↓	5.10	57.50		62.60	102
0270	Doors, louvered, both sides, incl. frame & trim									
0280	Brushwork, primer	1 Pord	7	1.143	Ea.	3.83	20.50		24.33	38
0290	Finish coat, exterior latex		7	1.143		4	20.50		24.50	38.50
0300	Primer & 1 coat, exterior latex		4	2		7.85	36		43.85	68.50
0310	Primer & 2 coats, exterior latex		3	2.667		11.60	48		59.60	93
0320	Brushwork, stain, sealer & 2 coats polyurethane	↓	4.50	1.778	↓	14.25	32		46.25	68.50
0330	Doors, panel, both sides, incl. frame & trim									
0340	Roll & brush, primer	1 Pord	6	1.333	Ea.	3.83	24		27.83	44
0350	Finish coat, exterior latex		6	1.333		4	24		28	44.50
0360	Primer & 1 coat, exterior latex		3	2.667		7.85	48		55.85	88.50
0370	Primer & 2 coats, exterior latex		2.50	3.200		11.60	57.50		69.10	109
0380	Brushwork, stain, sealer & 2 coats polyurethane	↓	3	2.667	↓	14.25	48		62.25	95.50
0400	Windows, per ext. side, based on 15 SF									
0410	1 to 6 lite									
0420	Brushwork, primer	1 Pord	13	.615	Ea.	.76	11.05		11.81	19.30
0430	Finish coat, exterior latex		13	.615		.79	11.05		11.84	19.30
0440	Primer & 1 coat, exterior latex		8	1		1.55	17.95		19.50	31.50
0450	Primer & 2 coats, exterior latex		6	1.333		2.29	24		26.29	42.50
0460	Stain, sealer & 1 coat varnish	↓	7	1.143		2.01	20.50		22.51	36
0470	7 to 10 lite									
0480	Brushwork, primer	1 Pord	11	.727	Ea.	.76	13.05		13.81	23
0490	Finish coat, exterior latex		11	.727		.79	13.05		13.84	23
0500	Primer & 1 coat, exterior latex		7	1.143		1.55	20.50		22.05	35.50
0510	Primer & 2 coats, exterior latex		5	1.600		2.29	28.50		30.79	50.50
0520	Stain, sealer & 1 coat varnish	↓	6	1.333	↓	2.01	24		26.01	42
0530	12 lite									
0540	Brushwork, primer	1 Pord	10	.800	Ea.	.76	14.35		15.11	25
0550	Finish coat, exterior latex		10	.800		.79	14.35		15.14	25
0560	Primer & 1 coat, exterior latex	↓	6	1.333		1.55	24		25.55	41.50

09900 | Paints & Coatings

09910 | Paints & Coatings

			CREW	DAILY OUTPUT	LABOR-HOURS	UNIT	MAT.	LABOR	EQUIP.	TOTAL	TOTAL INCL O&P	
300	0570	Primer & 2 coats, exterior latex	1 Pord	5	1.600	Ea.	2.29	28.50		30.79	50.50	300
	0580	Stain, sealer & 1 coat varnish	↓	6	1.333		2	24		26	42	
	0590	For oil base paint, add				↓	10%					
310	0010	**DOORS & WINDOWS, INTERIOR LATEX**										310
	0100	Doors flush, both sides, incl. frame & trim										
	0110	Roll & brush, primer	1 Pord	10	.800	Ea.	3.32	14.35		17.67	27.50	
	0120	Finish coat, latex		10	.800		3.73	14.35		18.08	28	
	0130	Primer & 1 coat latex		7	1.143		7.05	20.50		27.55	42	
	0140	Primer & 2 coats latex		5	1.600		10.60	28.50		39.10	59.50	
	0160	Spray, both sides, primer		20	.400		3.50	7.20		10.70	15.85	
	0170	Finish coat, latex		20	.400		3.92	7.20		11.12	16.30	
	0180	Primer & 1 coat latex		11	.727		7.45	13.05		20.50	30	
	0190	Primer & 2 coats latex	↓	8	1	↓	11.20	17.95		29.15	42.50	
	0200	Doors, French, both sides, 10-15 lite, incl. frame & trim										
	0210	Roll & brush, primer	1 Pord	6	1.333	Ea.	1.66	24		25.66	42	
	0220	Finish coat, latex		6	1.333		1.87	24		25.87	42	
	0230	Primer & 1 coat latex		3	2.667		3.53	48		51.53	84	
	0240	Primer & 2 coats latex	↓	2	4	↓	5.30	72		77.30	126	
	0260	Doors, louvered, both sides, incl. frame & trim										
	0270	Roll & brush, primer	1 Pord	7	1.143	Ea.	3.32	20.50		23.82	37.50	
	0280	Finish coat, latex		7	1.143		3.73	20.50		24.23	38	
	0290	Primer & 1 coat, latex		4	2		6.85	36		42.85	67.50	
	0300	Primer & 2 coats, latex		3	2.667		10.80	48		58.80	92	
	0320	Spray, both sides, primer		20	.400		3.50	7.20		10.70	15.85	
	0330	Finish coat, latex		20	.400		3.92	7.20		11.12	16.30	
	0340	Primer & 1 coat, latex		11	.727		7.45	13.05		20.50	30	
	0350	Primer & 2 coats, latex	↓	8	1	↓	11.45	17.95		29.40	42.50	
	0360	Doors, panel, both sides, incl. frame & trim										
	0370	Roll & brush, primer	1 Pord	6	1.333	Ea.	3.50	24		27.50	44	
	0380	Finish coat, latex		6	1.333		3.73	24		27.73	44	
	0390	Primer & 1 coat, latex		3	2.667		7.05	48		55.05	88	
	0400	Primer & 2 coats, latex		2.50	3.200		10.80	57.50		68.30	108	
	0420	Spray, both sides, primer		10	.800		3.50	14.35		17.85	28	
	0430	Finish coat, latex		10	.800		3.92	14.35		18.27	28.50	
	0440	Primer & 1 coat, latex		5	1.600		7.45	28.50		35.95	56	
	0450	Primer & 2 coats, latex	↓	4	2	↓	11.45	36		47.45	72.50	
	0460	Windows, per interior side, base on 15 SF										
	0470	1 to 6 lite										
	0480	Brushwork, primer	1 Pord	13	.615	Ea.	.66	11.05		11.71	19.15	
	0490	Finish coat, enamel		13	.615		.74	11.05		11.79	19.25	
	0500	Primer & 1 coat enamel		8	1		1.39	17.95		19.34	31.50	
	0510	Primer & 2 coats enamel	↓	6	1.333	↓	2.13	24		26.13	42.50	
	0530	7 to 10 lite										
	0540	Brushwork, primer	1 Pord	11	.727	Ea.	.66	13.05		13.71	22.50	
	0550	Finish coat, enamel		11	.727		.74	13.05		13.79	23	
	0560	Primer & 1 coat enamel		7	1.143		1.39	20.50		21.89	35.50	
	0570	Primer & 2 coats enamel		5	1.600	↓	2.13	28.50		30.63	50.50	
	0590	12 lite										
	0600	Brushwork, primer	1 Pord	10	.800	Ea.	.66	14.35		15.01	24.50	
	0610	Finish coat, enamel		10	.800		.74	14.35		15.09	25	
	0620	Primer & 1 coat enamel		6	1.333		1.39	24		25.39	41.50	
	0630	Primer & 2 coats enamel	↓	5	1.600		2.13	28.50		30.63	50.50	
	0650	For oil base paint, add				↓	10%					
320	0010	**DOORS AND WINDOWS, INTERIOR ALKYD (OIL BASE)**										320
	0500	Flush door & frame, 3' x 7', oil, primer, brushwork	1 Pord	10	.800	Ea.	1.94	14.35		16.29	26	

For expanded coverage of these items see Means Building Construction Cost Data 2000

09900 | Paints & Coatings

09910 | Paints & Coatings

			CREW	DAILY OUTPUT	LABOR-HOURS	UNIT	MAT.	LABOR	EQUIP.	TOTAL	TOTAL INCL O&P	
320	1000	Paint, 1 coat	1 Pord	10	.800	Ea.	1.86	14.35		16.21	26	320
	1200	2 coats		7	1.143		3.07	20.50		23.57	37.50	
	1400	Stain, brushwork, wipe off		18	.444		.90	8		8.90	14.30	
	1600	Shellac, 1 coat, brushwork		25	.320		.96	5.75		6.71	10.65	
	1800	Varnish, 3 coats, brushwork, sand after 1st coat		9	.889		3.36	15.95		19.31	30	
	2000	Panel door & frame, 3' x 7', oil, primer, brushwork		6	1.333		1.63	24		25.63	42	
	2200	Paint, 1 coat		6	1.333		1.86	24		25.86	42	
	2400	2 coats		3	2.667		5.35	48		53.35	86	
	2600	Stain, brushwork, panel door, 3' x 7', not incl. frame		16	.500		.90	9		9.90	16	
	2800	Shellac, 1 coat, brushwork		22	.364		.96	6.55		7.51	11.95	
	3000	Varnish, 3 coats, brushwork, sand after 1st coat		7.50	1.067		3.36	19.15		22.51	35.50	
	3020	French door, incl. 3' x 7', 6 lites, frame & trim										
	3022	Paint, 1 coat, over existing paint	1 Pord	5	1.600	Ea.	3.72	28.50		32.22	52	
	3024	2 coats, over existing paint		5	1.600		7.25	28.50		35.75	56	
	3026	Primer & 1 coat		3.50	2.286		7	41		48	76	
	3028	Primer & 2 coats		3	2.667		10.70	48		58.70	92	
	3032	Varnish or polyurethane, 1 coat		5	1.600		4.09	28.50		32.59	52.50	
	3034	2 coats, sanding between		3	2.667		8.20	48		56.20	89	
	4400	Windows, including frame and trim, per side										
	4600	Colonial type, 6/6 lites, 2' x 3', oil, primer, brushwork	1 Pord	14	.571	Ea.	.26	10.25		10.51	17.40	
	5800	Paint, 1 coat		14	.571		.29	10.25		10.54	17.40	
	6000	2 coats		9	.889		.57	15.95		16.52	27	
	6200	3' x 5' opening, 6/6 lites, primer coat, brushwork		12	.667		.64	11.95		12.59	20.50	
	6400	Paint, 1 coat		12	.667		.73	11.95		12.68	21	
	6600	2 coats		7	1.143		1.43	20.50		21.93	35.50	
	6800	4' x 8' opening, 6/6 lites, primer coat, brushwork		8	1		1.37	17.95		19.32	31.50	
	7000	Paint, 1 coat		8	1		1.57	17.95		19.52	31.50	
	7200	2 coats		5	1.600		3.04	28.50		31.54	51.50	
	8000	Single lite type, 2' x 3', oil base, primer coat, brushwork		33	.242		.26	4.35		4.61	7.55	
	8200	Paint, 1 coat		33	.242		.29	4.35		4.64	7.55	
	8400	2 coats		20	.400		.57	7.20		7.77	12.65	
	8600	3' x 5' opening, primer coat, brushwork		20	.400		.64	7.20		7.84	12.70	
	8800	Paint, 1 coat		20	.400		.73	7.20		7.93	12.80	
	9000	2 coats		13	.615		1.43	11.05		12.48	20	
	9200	4' x 8' opening, primer coat, brushwork		14	.571		1.37	10.25		11.62	18.60	
	9400	Paint, 1 coat		14	.571		1.57	10.25		11.82	18.80	
	9600	2 coats		8	1		3.04	17.95		20.99	33.50	
400	0010	**FENCES**										400
	0100	Chain link or wire metal, one side, water base										
	0110	Roll & brush, first coat	1 Pord	960	.008	S.F.	.05	.15		.20	.31	
	0120	Second coat		1,280	.006		.05	.11		.16	.24	
	0130	Spray, first coat		2,275	.004		.05	.06		.11	.17	
	0140	Second coat		2,600	.003		.05	.06		.11	.15	
	0150	Picket, water base										
	0160	Roll & brush, first coat	1 Pord	865	.009	S.F.	.06	.17		.23	.34	
	0170	Second coat		1,050	.008		.06	.14		.20	.29	
	0180	Spray, first coat		2,275	.004		.06	.06		.12	.17	
	0190	Second coat		2,600	.003		.06	.06		.12	.15	
	0200	Stockade, water base										
	0210	Roll & brush, first coat	1 Pord	1,040	.008	S.F.	.06	.14		.20	.29	
	0220	Second coat		1,200	.007		.06	.12		.18	.26	
	0230	Spray, first coat		2,275	.004		.06	.06		.12	.17	
	0240	Second coat		2,600	.003		.06	.06		.12	.15	
500	0010	**FLOORS**, Interior										500
	0100	Concrete										

09900 | Paints & Coatings

09910 | Paints & Coatings

		CREW	DAILY OUTPUT	LABOR-HOURS	UNIT	2000 BARE COSTS MAT.	LABOR	EQUIP.	TOTAL	TOTAL INCL O&P	
500	0120	1st coat	1 Pord	975	.008	S.F.	.11	.15		.26	.37	500
	0130	2nd coat		1,150	.007		.07	.12		.19	.29	
	0140	3rd coat		1,300	.006		.06	.11		.17	.24	
	0150	Roll, latex, block filler										
	0160	1st coat	1 Pord	2,600	.003	S.F.	.14	.06		.20	.25	
	0170	2nd coat		3,250	.002		.08	.04		.12	.16	
	0180	3rd coat		3,900	.002		.06	.04		.10	.13	
	0190	Spray, latex, block filler										
	0200	1st coat	1 Pord	2,600	.003	S.F.	.12	.06		.18	.22	
	0210	2nd coat		3,250	.002		.07	.04		.11	.14	
	0220	3rd coat		3,900	.002		.05	.04		.09	.12	
620	0010	**MISCELLANEOUS, EXTERIOR**										620
	0015	For painting metals, see Div. 05120-680										
	0100	Railing, ext., decorative wood, incl. cap & baluster										
	0110	newels & spindles @ 12" O.C.										
	0120	Brushwork, stain, sand, seal & varnish										
	0130	First coat	1 Pord	90	.089	L.F.	.40	1.60		2	3.10	
	0140	Second coat	"	120	.067	"	.40	1.20		1.60	2.44	
	0150	Rough sawn wood, 42" high, 2"x2" verticals, 6" O.C.										
	0160	Brushwork, stain, each coat	1 Pord	90	.089	L.F.	.13	1.60		1.73	2.80	
	0170	Wrought iron, 1" rail, 1/2" sq. verticals										
	0180	Brushwork, zinc chromate, 60" high, bars 6" O.C.										
	0190	Primer	1 Pord	130	.062	L.F.	.47	1.10		1.57	2.36	
	0200	Finish coat		130	.062		.15	1.10		1.25	2	
	0210	Additional coat		190	.042		.17	.76		.93	1.45	
	0220	Shutters or blinds, single panel, 2'x4', paint all sides										
	0230	Brushwork, primer	1 Pord	20	.400	Ea.	.55	7.20		7.75	12.60	
	0240	Finish coat, exterior latex		20	.400		.42	7.20		7.62	12.45	
	0250	Primer & 1 coat, exterior latex		13	.615		.85	11.05		11.90	19.40	
	0260	Spray, primer		35	.229		.81	4.10		4.91	7.75	
	0270	Finish coat, exterior latex		35	.229		.90	4.10		5	7.85	
	0280	Primer & 1 coat, exterior latex		20	.400		.86	7.20		8.06	12.95	
	0290	For louvered shutters, add				S.F.	10%					
	0300	Stair stringers, exterior, metal										
	0310	Roll & brush, zinc chromate, to 14", each coat	1 Pord	320	.025	L.F.	.05	.45		.50	.80	
	0320	Rough sawn wood, 4" x 12"										
	0330	Roll & brush, exterior latex, each coat	1 Pord	215	.037	L.F.	.06	.67		.73	1.18	
	0340	Trellis/lattice, 2"x2" @ 3" O.C. with 2"x8" supports										
	0350	Spray, latex, per side, each coat	1 Pord	475	.017	S.F.	.06	.30		.36	.57	
	0450	Decking, Ext., sealer, alkyd, brushwork, sealer coat		1,140	.007		.05	.13		.18	.26	
	0460	1st coat		1,140	.007		.05	.13		.18	.26	
	0470	2nd coat		1,300	.006		.04	.11		.15	.22	
	0500	Paint, alkyd, brushwork, primer coat		1,140	.007		.07	.13		.20	.28	
	0510	1st coat		1,140	.007		.06	.13		.19	.28	
	0520	2nd coat		1,300	.006		.05	.11		.16	.23	
	0600	Sand paint, alkyd, brushwork, 1 coat		150	.053		.08	.96		1.04	1.69	
630	0010	**MISCELLANEOUS, INTERIOR**										630
	2400	Floors, conc./wood, oil base, primer/sealer coat, brushwork	2 Pord	1,950	.008	S.F.	.05	.15		.20	.31	
	2450	Roller		5,200	.003		.06	.06		.12	.15	
	2600	Spray		6,000	.003		.06	.05		.11	.14	
	2650	Paint 1 coat, brushwork		1,950	.008		.05	.15		.20	.30	
	2800	Roller		5,200	.003		.05	.06		.11	.15	
	2850	Spray		6,000	.003		.05	.05		.10	.14	
	3000	Stain, wood floor, brushwork, 1 coat		4,550	.004		.04	.06		.10	.16	

For expanded coverage of these items see *Means Building Construction Cost Data 2000*

09900 | Paints & Coatings

09910 | Paints & Coatings

		CREW	DAILY OUTPUT	LABOR-HOURS	UNIT	2000 BARE COSTS MAT.	LABOR	EQUIP.	TOTAL	TOTAL INCL O&P
3200	Roller	2 Pord	5,200	.003	S.F.	.05	.06		.11	.14
3250	Spray		6,000	.003		.04	.05		.09	.13
3400	Varnish, wood floor, brushwork		4,550	.004		.05	.06		.11	.17
3450	Roller		5,200	.003		.06	.06		.12	.15
3600	Spray		6,000	.003		.06	.05		.11	.15
3800	Grilles, per side, oil base, primer coat, brushwork	1 Pord	520	.015		.09	.28		.37	.55
3850	Spray		1,140	.007		.09	.13		.22	.31
3880	Paint 1 coat, brushwork		520	.015		.10	.28		.38	.57
3900	Spray		1,140	.007		.11	.13		.24	.33
3920	Paint 2 coats, brushwork		325	.025		.19	.44		.63	.95
3940	Spray		650	.012		.22	.22		.44	.61
4250	Paint 1 coat, brushwork	2 Pord	1,300	.012	L.F.	.10	.22		.32	.48
4500	Louvers, one side, primer, brushwork	1 Pord	524	.015	S.F.	.05	.27		.32	.52
4520	Paint one coat, brushwork		520	.015		.05	.28		.33	.52
4530	Spray		1,140	.007		.06	.13		.19	.27
4540	Paint two coats, brushwork		325	.025		.10	.44		.54	.85
4550	Spray		650	.012		.11	.22		.33	.49
4560	Paint three coats, brushwork		270	.030		.14	.53		.67	1.05
4570	Spray		500	.016		.16	.29		.45	.66
5000	Pipe, to 4" diameter, primer or sealer coat, oil base, brushwork	2 Pord	1,250	.013	L.F.	.06	.23		.29	.44
5100	Spray		2,165	.007		.05	.13		.18	.28
5200	Paint 1 coat, brushwork		1,250	.013		.06	.23		.29	.44
5300	Spray		2,165	.007		.05	.13		.18	.28
5350	Paint 2 coats, brushwork		775	.021		.10	.37		.47	.73
5400	Spray		1,240	.013		.11	.23		.34	.51
5450	To 8" diameter, primer or sealer coat, brushwork		620	.026		.11	.46		.57	.89
5500	Spray		1,085	.015		.18	.26		.44	.64
5550	Paint 1 coat, brushwork		620	.026		.16	.46		.62	.94
5600	Spray		1,085	.015		.17	.26		.43	.63
5650	Paint 2 coats, brushwork		385	.042		.20	.75		.95	1.46
5700	Spray		620	.026		.23	.46		.69	1.02
5750	To 12" diameter, primer or sealer coat, brushwork		415	.039		.17	.69		.86	1.33
5800	Spray		725	.022		.19	.40		.59	.86
5850	Paint 1 coat, brushwork		415	.039		.16	.69		.85	1.32
6000	Spray		725	.022		.17	.40		.57	.85
6200	Paint 2 coats, brushwork		260	.062		.31	1.10		1.41	2.18
6250	Spray		415	.039		.34	.69		1.03	1.52
6300	To 16" diameter, primer or sealer coat, brushwork		310	.052		.22	.93		1.15	1.80
6350	Spray		540	.030		.25	.53		.78	1.16
6400	Paint 1 coat, brushwork		310	.052		.21	.93		1.14	1.78
6450	Spray		540	.030		.23	.53		.76	1.15
6500	Paint 2 coats, brushwork		195	.082		.41	1.47		1.88	2.91
6550	Spray		310	.052		.45	.93		1.38	2.05
6600	Radiators, per side, primer, brushwork	1 Pord	520	.015	S.F.	.05	.28		.33	.52
6620	Paint one coat, brushwork		520	.015		.05	.28		.33	.52
6640	Paint two coats, brushwork		340	.024		.10	.42		.52	.81
6660	Paint three coats, brushwork		283	.028		.14	.51		.65	1.01
7000	Trim, wood, incl. puttying, under 6" wide									
7200	Primer coat, oil base, brushwork	1 Pord	650	.012	L.F.	.02	.22		.24	.39
7250	Paint, 1 coat, brushwork		650	.012		.02	.22		.24	.40
7400	2 coats		400	.020		.05	.36		.41	.65
7450	3 coats		325	.025		.07	.44		.51	.82
7500	Over 6" wide, primer coat, brushwork		650	.012		.04	.22		.26	.42
7550	Paint, 1 coat, brushwork		650	.012		.05	.22		.27	.42
7600	2 coats		400	.020		.10	.36		.46	.70
7650	3 coats		325	.025		.14	.44		.58	.90

09900 | Paints & Coatings

09910 | Paints & Coatings

		CREW	DAILY OUTPUT	LABOR-HOURS	UNIT	MAT.	LABOR	EQUIP.	TOTAL	TOTAL INCL O&P
630 8000	Cornice, simple design, primer coat, oil base, brushwork	1 Pord	650	.012	S.F.	.04	.22		.26	.42
8250	Paint, 1 coat		650	.012		.05	.22		.27	.42
8300	2 coats		400	.020		.10	.36		.46	.70
8350	Ornate design, primer coat		350	.023		.04	.41		.45	.73
8400	Paint, 1 coat		350	.023		.05	.41		.46	.73
8450	2 coats		400	.020		.10	.36		.46	.70
8600	Balustrades, primer coat, oil base, brushwork		520	.015		.04	.28		.32	.51
8650	Paint, 1 coat		520	.015		.05	.28		.33	.51
8700	2 coats		325	.025		.10	.44		.54	.84
8900	Trusses and wood frames, primer coat, oil base, brushwork		800	.010		.04	.18		.22	.35
8950	Spray		1,200	.007		.04	.12		.16	.25
9000	Paint 1 coat, brushwork		750	.011		.05	.19		.24	.37
9200	Spray		1,200	.007		.05	.12		.17	.26
9220	Paint 2 coats, brushwork		500	.016		.10	.29		.39	.58
9240	Spray		600	.013		.11	.24		.35	.52
9260	Stain, brushwork, wipe off		600	.013		.04	.24		.28	.45
9280	Varnish, 3 coats, brushwork		275	.029		.16	.52		.68	1.05
9350	For latex paint, deduct					10%				
700 0010	**SIDING EXTERIOR**, Alkyd (oil base)	R09910-100								
0500	Spray	2 Pord	4,550	.004	S.F.	.08	.06		.14	.19
0800	Paint 2 coats, brushwork		1,300	.012		.10	.22		.32	.48
1000	Spray		4,550	.004		.14	.06		.20	.26
1200	Stucco, rough, oil base, paint 2 coats, brushwork		1,300	.012		.10	.22		.32	.48
1400	Roller		1,625	.010		.11	.18		.29	.41
1600	Spray		2,925	.005		.11	.10		.21	.28
1800	Texture 1-11 or clapboard, oil base, primer coat, brushwork		1,300	.012		.09	.22		.31	.46
2000	Spray		4,550	.004		.09	.06		.15	.20
2100	Paint 1 coat, brushwork		1,300	.012		.07	.22		.29	.45
2200	Spray		4,550	.004		.07	.06		.13	.19
2400	Paint 2 coats, brushwork		810	.020		.15	.35		.50	.75
2600	Spray		2,600	.006		.16	.11		.27	.36
3000	Stain 1 coat, brushwork		1,520	.011		.04	.19		.23	.37
3200	Spray		5,320	.003		.05	.05		.10	.14
3400	Stain 2 coats, brushwork		950	.017		.09	.30		.39	.59
4000	Spray		3,050	.005		.10	.09		.19	.26
4200	Wood shingles, oil base primer coat, brushwork		1,300	.012		.08	.22		.30	.46
4400	Spray		3,900	.004		.07	.07		.14	.20
4600	Paint 1 coat, brushwork		1,300	.012		.06	.22		.28	.44
4800	Spray		3,900	.004		.08	.07		.15	.20
5000	Paint 2 coats, brushwork		810	.020		.12	.35		.47	.72
5200	Spray		2,275	.007		.12	.13		.25	.34
5800	Stain 1 coat, brushwork		1,500	.011		.04	.19		.23	.37
6000	Spray		3,900	.004		.04	.07		.11	.17
6500	Stain 2 coats, brushwork		950	.017		.09	.30		.39	.59
7000	Spray		2,660	.006		.12	.11		.23	.31
8000	For latex paint, deduct					10%				
710 0010	**SIDING, MISC.**									
0100	Aluminum siding									
0110	Brushwork, primer	2 Pord	2,275	.007	S.F.	.05	.13		.18	.27
0120	Finish coat, exterior latex		2,275	.007		.04	.13		.17	.25
0130	Primer & 1 coat exterior latex		1,300	.012		.10	.22		.32	.48
0140	Primer & 2 coats exterior latex		975	.016		.14	.29		.43	.64
0150	Mineral Fiber shingles									
0160	Brushwork, primer	2 Pord	1,495	.011	S.F.	.09	.19		.28	.41
0170	Finish coat, industrial enamel		1,495	.011		.09	.19		.28	.42
0180	Primer & 1 coat enamel		810	.020		.18	.35		.53	.79

For expanded coverage of these items see *Means Building Construction Cost Data 2000*

09900 | Paints & Coatings

09910 | Paints & Coatings

		CREW	DAILY OUTPUT	LABOR-HOURS	UNIT	MAT.	LABOR	EQUIP.	TOTAL	TOTAL INCL O&P	
710	0190 Primer & 2 coats enamel	2 Pord	540	.030	S.F.	.27	.53		.80	1.19	710
	0200 Roll, primer		1,625	.010		.09	.18		.27	.39	
	0210 Finish coat, industrial enamel		1,625	.010		.10	.18		.28	.40	
	0220 Primer & 1 coat enamel		975	.016		.20	.29		.49	.70	
	0230 Primer & 2 coats enamel		650	.025		.30	.44		.74	1.06	
	0240 Spray, primer		3,900	.004		.07	.07		.14	.20	
	0250 Finish coat, industrial enamel		3,900	.004		.08	.07		.15	.21	
	0260 Primer & 1 coat enamel		2,275	.007		.16	.13		.29	.38	
	0270 Primer & 2 coats enamel		1,625	.010		.24	.18		.42	.55	
	0280 Waterproof sealer, first coat		4,485	.004		.06	.06		.12	.18	
	0290 Second coat		5,235	.003		.06	.05		.11	.16	
	0300 Rough wood incl. shingles, shakes or rough sawn siding										
	0310 Brushwork, primer	2 Pord	1,280	.013	S.F.	.11	.22		.33	.49	
	0320 Finish coat, exterior latex		1,280	.013		.07	.22		.29	.45	
	0330 Primer & 1 coat exterior latex		960	.017		.18	.30		.48	.70	
	0340 Primer & 2 coats exterior latex		700	.023		.25	.41		.66	.95	
	0350 Roll, primer		2,925	.005		.15	.10		.25	.32	
	0360 Finish coat, exterior latex		2,925	.005		.08	.10		.18	.25	
	0370 Primer & 1 coat exterior latex		1,790	.009		.23	.16		.39	.53	
	0380 Primer & 2 coats exterior latex		1,300	.012		.31	.22		.53	.72	
	0390 Spray, primer		3,900	.004		.13	.07		.20	.26	
	0400 Finish coat, exterior latex		3,900	.004		.06	.07		.13	.19	
	0410 Primer & 1 coat exterior latex		2,600	.006		.19	.11		.30	.39	
	0420 Primer & 2 coats exterior latex		2,080	.008		.25	.14		.39	.51	
	0430 Waterproof sealer, first coat		4,485	.004		.11	.06		.17	.23	
	0440 Second coat		4,485	.004		.06	.06		.12	.18	
	0450 Smooth wood incl. butt, T&G, beveled, drop or B&B siding										
	0460 Brushwork, primer	2 Pord	2,325	.007	S.F.	.08	.12		.20	.30	
	0470 Finish coat, exterior latex		1,280	.013		.07	.22		.29	.45	
	0480 Primer & 1 coat exterior latex		800	.020		.15	.36		.51	.76	
	0490 Primer & 2 coats exterior latex		630	.025		.22	.46		.68	1	
	0500 Roll, primer		2,275	.007		.09	.13		.22	.31	
	0510 Finish coat, exterior latex		2,275	.007		.07	.13		.20	.29	
	0520 Primer & 1 coat exterior latex		1,300	.012		.16	.22		.38	.55	
	0530 Primer & 2 coats exterior latex		975	.016		.24	.29		.53	.75	
	0540 Spray, primer		4,550	.004		.07	.06		.13	.19	
	0550 Finish coat, exterior latex		4,550	.004		.06	.06		.12	.18	
	0560 Primer & 1 coat exterior latex		2,600	.006		.13	.11		.24	.33	
	0570 Primer & 2 coats exterior latex		1,950	.008		.20	.15		.35	.47	
	0580 Waterproof sealer, first coat		5,230	.003		.06	.05		.11	.16	
	0590 Second coat		5,980	.003		.06	.05		.11	.15	
	0600 For oil base paint, add					10%					
800	0010 **TRIM, EXTERIOR**										800
	0100 Door frames & trim (see Doors, interior or exterior)										
	0110 Fascia, latex paint, one coat coverage										
	0120 1" x 4", brushwork	1 Pord	640	.013	L.F.	.02	.22		.24	.39	
	0130 Roll		1,280	.006		.02	.11		.13	.21	
	0140 Spray		2,080	.004		.01	.07		.08	.13	
	0150 1" x 6" to 1" x 10", brushwork		640	.013		.06	.22		.28	.43	
	0160 Roll		1,230	.007		.06	.12		.18	.25	
	0170 Spray		2,100	.004		.05	.07		.12	.16	
	0180 1" x 12", brushwork		640	.013		.06	.22		.28	.43	
	0190 Roll		1,050	.008		.06	.14		.20	.29	
	0200 Spray		2,200	.004		.05	.07		.12	.16	
	0210 Gutters & downspouts, metal, zinc chromate paint										
	0220 Brushwork, gutters, 5", first coat	1 Pord	640	.013	L.F.	.05	.22		.27	.43	

09900 | Paints & Coatings

09910 | Paints & Coatings

		Crew	Daily Output	Labor-Hours	Unit	2000 Bare Costs Mat.	Labor	Equip.	Total	Total Incl O&P	
800	0230 Second coat	1 Pord	960	.008	L.F.	.05	.15		.20	.30	**800**
	0240 Third coat		1,280	.006		.04	.11		.15	.23	
	0250 Downspouts, 4", first coat		640	.013		.05	.22		.27	.43	
	0260 Second coat		960	.008		.05	.15		.20	.30	
	0270 Third coat	↓	1,280	.006	↓	.04	.11		.15	.23	
	0280 Gutters & downspouts, wood										
	0290 Brushwork, gutters, 5", primer	1 Pord	640	.013	L.F.	.05	.22		.27	.43	
	0300 Finish coat, exterior latex		640	.013		.05	.22		.27	.42	
	0310 Primer & 1 coat exterior latex		400	.020		.10	.36		.46	.71	
	0320 Primer & 2 coats exterior latex		325	.025		.16	.44		.60	.91	
	0330 Downspouts, 4", primer		640	.013		.05	.22		.27	.43	
	0340 Finish coat, exterior latex		640	.013		.05	.22		.27	.42	
	0350 Primer & 1 coat exterior latex		400	.020		.10	.36		.46	.71	
	0360 Primer & 2 coats exterior latex	↓	325	.025	↓	.08	.44		.52	.83	
	0370 Molding, exterior, up to 14" wide										
	0380 Brushwork, primer	1 Pord	640	.013	L.F.	.06	.22		.28	.44	
	0390 Finish coat, exterior latex		640	.013		.06	.22		.28	.43	
	0400 Primer & 1 coat exterior latex		400	.020		.12	.36		.48	.74	
	0410 Primer & 2 coats exterior latex		315	.025		.12	.46		.58	.90	
	0420 Stain & fill		1,050	.008		.05	.14		.19	.29	
	0430 Shellac		1,850	.004		.05	.08		.13	.19	
	0440 Varnish	↓	1,275	.006	↓	.06	.11		.17	.26	
910	0350 **WALLS, MASONRY (CMU), EXTERIOR**										**910**
	0360 Concrete masonry units (CMU), smooth surface										
	0370 Brushwork, latex, first coat	1 Pord	640	.013	S.F.	.03	.22		.25	.41	
	0380 Second coat		960	.008		.03	.15		.18	.28	
	0390 Waterproof sealer, first coat		736	.011		.06	.20		.26	.40	
	0400 Second coat		1,104	.007		.04	.13		.17	.26	
	0410 Roll, latex, paint, first coat		1,465	.005		.04	.10		.14	.20	
	0420 Second coat		1,790	.004		.03	.08		.11	.16	
	0430 Waterproof sealer, first coat		1,680	.005		.06	.09		.15	.21	
	0440 Second coat		2,060	.004		.04	.07		.11	.16	
	0450 Spray, latex, paint, first coat		1,950	.004		.03	.07		.10	.15	
	0460 Second coat		2,600	.003		.02	.06		.08	.12	
	0470 Waterproof sealer, first coat		2,245	.004		.08	.06		.14	.19	
	0480 Second coat	↓	2,990	.003	↓	.03	.05		.08	.11	
	0490 Concrete masonry unit (CMU), porous										
	0500 Brushwork, latex, first coat	1 Pord	640	.013	S.F.	.07	.22		.29	.44	
	0510 Second coat		960	.008		.03	.15		.18	.29	
	0520 Waterproof sealer, first coat		736	.011		.08	.20		.28	.41	
	0530 Second coat		1,104	.007		.04	.13		.17	.26	
	0540 Roll latex, first coat		1,465	.005		.05	.10		.15	.21	
	0550 Second coat		1,790	.004		.03	.08		.11	.17	
	0560 Waterproof sealer, first coat		1,680	.005		.08	.09		.17	.22	
	0570 Second coat		2,060	.004		.04	.07		.11	.17	
	0580 Spray latex, first coat		1,950	.004		.04	.07		.11	.16	
	0590 Second coat		2,600	.003		.02	.06		.08	.12	
	0600 Waterproof sealer, first coat		2,245	.004		.06	.06		.12	.18	
	0610 Second coat	↓	2,990	.003	↓	.03	.05		.08	.12	
920	0010 **WALLS AND CEILINGS**, Interior										**920**
	0100 Concrete, dry wall or plaster, oil base, primer or sealer coat										
	0200 Smooth finish, brushwork	1 Pord	1,150	.007	S.F.	.04	.12		.16	.26	
	0240 Roller		2,040	.004		.04	.07		.11	.16	
	0300 Sand finish, brushwork		975	.008		.04	.15		.19	.29	
	0340 Roller	↓	1,150	.007	↓	.05	.12		.17	.26	

For expanded coverage of these items see *Means Building Construction Cost Data 2000*

09900 | Paints & Coatings

09910 | Paints & Coatings

			CREW	DAILY OUTPUT	LABOR-HOURS	UNIT	2000 BARE COSTS MAT.	LABOR	EQUIP.	TOTAL	TOTAL INCL O&P	
920	0380	Spray	1 Pord	2,275	.004	S.F.	.03	.06		.09	.15	920
	0400	Paint 1 coat, smooth finish, brushwork		1,200	.007		.04	.12		.16	.25	
	0440	Roller		1,300	.006		.04	.11		.15	.23	
	0480	Spray		2,275	.004		.04	.06		.10	.15	
	0500	Sand finish, brushwork		1,050	.008		.04	.14		.18	.27	
	0540	Roller		1,600	.005		.04	.09		.13	.20	
	0580	Spray		2,100	.004		.04	.07		.11	.15	
	0800	Paint 2 coats, smooth finish, brushwork		680	.012		.08	.21		.29	.44	
	0840	Roller		800	.010		.08	.18		.26	.39	
	0880	Spray		1,625	.005		.07	.09		.16	.23	
	0900	Sand finish, brushwork		605	.013		.08	.24		.32	.49	
	0940	Roller		1,020	.008		.08	.14		.22	.32	
	0980	Spray		1,700	.005		.07	.08		.15	.22	
	1200	Paint 3 coats, smooth finish, brushwork		510	.016		.12	.28		.40	.60	
	1240	Roller		650	.012		.13	.22		.35	.51	
	1280	Spray		1,625	.005		.11	.09		.20	.27	
	1300	Sand finish, brushwork		454	.018		.12	.32		.44	.66	
	1340	Roller		680	.012		.13	.21		.34	.49	
	1380	Spray		1,133	.007		.11	.13		.24	.33	
	1600	Glaze coating, 5 coats, spray, clear		900	.009		.60	.16		.76	.93	
	1640	Multicolor		900	.009		.81	.16		.97	1.16	
	1700	For latex paint, deduct					10%					
	1800	For ceiling installations, add						25%				
	2000	Masonry or concrete block, oil base, primer or sealer coat										
	2100	Smooth finish, brushwork	1 Pord	1,224	.007	S.F.	.04	.12		.16	.25	
	2180	Spray		2,400	.003		.06	.06		.12	.17	
	2200	Sand finish, brushwork		1,089	.007		.07	.13		.20	.29	
	2280	Spray		2,400	.003		.06	.06		.12	.17	
	2400	Paint 1 coat, smooth finish, brushwork		1,100	.007		.07	.13		.20	.29	
	2480	Spray		2,400	.003		.06	.06		.12	.17	
	2500	Sand finish, brushwork		979	.008		.07	.15		.22	.31	
	2580	Spray		2,400	.003		.06	.06		.12	.17	
	2800	Paint 2 coats, smooth finish, brushwork		756	.011		.13	.19		.32	.47	
	2880	Spray		1,360	.006		.12	.11		.23	.32	
	2900	Sand finish, brushwork		672	.012		.13	.21		.34	.51	
	2980	Spray		1,360	.006		.12	.11		.23	.32	
	3200	Paint 3 coats, smooth finish, brushwork		560	.014		.20	.26		.46	.65	
	3280	Spray		1,088	.007		.19	.13		.32	.42	
	3300	Sand finish, brushwork		498	.016		.20	.29		.49	.70	
	3380	Spray		1,088	.007		.19	.13		.32	.42	
	3600	Glaze coating, 5 coats, spray, clear		900	.009		.60	.16		.76	.93	
	3620	Multicolor		900	.009		.81	.16		.97	1.16	
	4000	Block filler, 1 coat, brushwork		425	.019		.12	.34		.46	.69	
	4100	Silicone, water repellent, 2 coats, spray		2,000	.004		.24	.07		.31	.38	
	4120	For latex paint, deduct					10%					

09930 | Stains/Transp. Finishes

			CREW	DAILY OUTPUT	LABOR-HOURS	UNIT	MAT.	LABOR	EQUIP.	TOTAL	TOTAL INCL O&P	
100	0010	VARNISH 1 coat + sealer, on wood trim, no sanding included	1 Pord	400	.020	S.F.	.06	.36		.42	.67	100
	0100	Hardwood floors, 2 coats, no sanding included, roller	"	1,890	.004	"	.12	.08		.20	.26	

09963 | Glazed Coatings

			CREW	DAILY OUTPUT	LABOR-HOURS	UNIT	MAT.	LABOR	EQUIP.	TOTAL	TOTAL INCL O&P	
200	0010	WALL COATINGS Acrylic glazed coatings, minimum	1 Pord	525	.015	S.F.	.24	.27		.51	.72	200
	0100	Maximum		305	.026		.50	.47		.97	1.34	
	0300	Epoxy coatings, minimum		525	.015		.31	.27		.58	.80	
	0400	Maximum		170	.047		.96	.84		1.80	2.47	

09900 | Paints & Coatings

09963 | Glazed Coatings

			CREW	DAILY OUTPUT	LABOR-HOURS	UNIT	2000 BARE COSTS MAT.	LABOR	EQUIP.	TOTAL	TOTAL INCL O&P	
200	0600	Exposed aggregate, troweled on, 1/16" to 1/4", minimum	1 Pord	235	.034	S.F.	.48	.61		1.09	1.55	200
	0700	Maximum (epoxy or polyacrylate)		130	.062		1.03	1.10		2.13	2.97	
	0900	1/2" to 5/8" aggregate, minimum		130	.062		.95	1.10		2.05	2.89	
	1000	Maximum		80	.100		1.62	1.80		3.42	4.78	
	1200	1" aggregate size, minimum		90	.089		1.65	1.60		3.25	4.48	
	1300	Maximum		55	.145		2.52	2.61		5.13	7.15	
	1500	Exposed aggregate, sprayed on, 1/8" aggregate, minimum		295	.027		.45	.49		.94	1.31	
	1600	Maximum		145	.055		.82	.99		1.81	2.55	
	1800	High build epoxy, 50 mil, minimum		390	.021		.52	.37		.89	1.18	
	1900	Maximum		95	.084		.90	1.51		2.41	3.51	
	2100	Laminated epoxy with fiberglass, minimum		295	.027		.57	.49		1.06	1.44	
	2200	Maximum		145	.055		1.02	.99		2.01	2.77	
	2400	Sprayed perlite or vermiculite, 1/16" thick, minimum		2,935	.003		.20	.05		.25	.30	
	2500	Maximum		640	.013		.58	.22		.80	1.01	
	2700	Vinyl plastic wall coating, minimum		735	.011		.26	.20		.46	.62	
	2800	Maximum		240	.033		.64	.60		1.24	1.70	
	3000	Urethane on smooth surface, 2 coats, minimum		1,135	.007		.19	.13		.32	.42	
	3100	Maximum		665	.012		.44	.22		.66	.84	
	3300	3 coat, minimum		840	.010		.26	.17		.43	.58	
	3400	Maximum		470	.017		.58	.31		.89	1.15	
	3600	Ceramic-like glazed coating, cementitious, minimum		440	.018		.38	.33		.71	.96	
	3700	Maximum		345	.023		.63	.42		1.05	1.38	
	3900	Resin base, minimum		640	.013		.26	.22		.48	.66	
	4000	Maximum		330	.024		.43	.44		.87	1.20	

09990 | Paint Restoration

			CREW	DAILY OUTPUT	LABOR-HOURS	UNIT	MAT.	LABOR	EQUIP.	TOTAL	TOTAL INCL O&P	
500	0010	**SCRAPE AFTER FIRE DAMAGE**										500
	0050	Boards, 1" x 4"	1 Pord	336	.024	L.F.		.43		.43	.71	
	0060	1" x 6"		260	.031			.55		.55	.92	
	0070	1" x 8"		207	.039			.69		.69	1.16	
	0080	1" x 10"		174	.046			.83		.83	1.38	
	0500	Framing, 2" x 4"		265	.030			.54		.54	.90	
	0510	2" x 6"		221	.036			.65		.65	1.08	
	0520	2" x 8"		190	.042			.76		.76	1.26	
	0530	2" x 10"		165	.048			.87		.87	1.45	
	0540	2" x 12"		144	.056			1		1	1.66	
	1000	Heavy framing, 3" x 4"		226	.035			.64		.64	1.06	
	1010	4" x 4"		210	.038			.68		.68	1.14	
	1020	4" x 6"		191	.042			.75		.75	1.25	
	1030	4" x 8"		165	.048			.87		.87	1.45	
	1040	4" x 10"		144	.056			1		1	1.66	
	1060	4" x 12"		131	.061			1.10		1.10	1.83	
	2900	For sealing, minimum		825	.010	S.F.	.11	.17		.28	.41	
	2920	Maximum		460	.017	"	.24	.31		.55	.78	
800	0010	**SANDING** and puttying interior trim, compared to										800
	0100	Painting 1 coat, on quality work				L.F.		100%				
	0300	Medium work						50%				
	0400	Industrial grade						25%				
	0500	Surface protection, placement and removal										
	0510	Basic drop cloths	1 Pord	6,400	.001	S.F.		.02		.02	.04	
	0520	Masking with paper		800	.010		.02	.18		.20	.32	
	0530	Volume cover up (using plastic sheathing, or building paper)		16,000	.001			.01		.01	.01	
900	0010	**SURFACE PREPARATION, EXTERIOR**										900
	0015	Doors, per side, not incl. frames or trim										
	0020	Scrape & sand										
	0030	Wood, flush	1 Pord	616	.013	S.F.		.23		.23	.39	

For expanded coverage of these items see *Means Building Construction Cost Data 2000*

09900 | Paints & Coatings

09990 | Paint Restoration

		CREW	DAILY OUTPUT	LABOR-HOURS	UNIT	MAT.	LABOR	EQUIP.	TOTAL	TOTAL INCL O&P
0040	Wood, detail	1 Pord	496	.016	S.F.		.29		.29	.48
0050	Wood, louvered		280	.029			.51		.51	.86
0060	Wood, overhead		616	.013			.23		.23	.39
0070	Wire brush									
0080	Metal, flush	1 Pord	640	.013	S.F.		.22		.22	.37
0090	Metal, detail		520	.015			.28		.28	.46
0100	Metal, louvered		360	.022			.40		.40	.67
0110	Metal or fibr., overhead		640	.013			.22		.22	.37
0120	Metal, roll up		560	.014			.26		.26	.43
0130	Metal, bulkhead		640	.013			.22		.22	.37
0140	Power wash, based on 2500 lb. operating pressure									
0150	Metal, flush	B-9	2,240	.018	S.F.		.27	.08	.35	.54
0160	Metal, detail		2,120	.019			.28	.09	.37	.57
0170	Metal, louvered		2,000	.020			.30	.09	.39	.61
0180	Metal or fibr., overhead		2,400	.017			.25	.08	.33	.50
0190	Metal, roll up		2,400	.017			.25	.08	.33	.50
0200	Metal, bulkhead		2,200	.018			.27	.08	.35	.55
0400	Windows, per side, not incl. trim									
0410	Scrape & sand									
0420	Wood, 1-2 lite	1 Pord	320	.025	S.F.		.45		.45	.75
0430	Wood, 3-6 lite		280	.029			.51		.51	.86
0440	Wood, 7-10 lite		240	.033			.60		.60	1
0450	Wood, 12 lite		200	.040			.72		.72	1.20
0460	Wood, Bay / Bow		320	.025			.45		.45	.75
0470	Wire brush									
0480	Metal, 1-2 lite	1 Pord	480	.017	S.F.		.30		.30	.50
0490	Metal, 3-6 lite		400	.020			.36		.36	.60
0500	Metal, Bay / Bow		480	.017			.30		.30	.50
0510	Power wash, based on 2500 lb. operating pressure									
0520	1-2 lite	B-9	4,400	.009	S.F.		.14	.04	.18	.28
0530	3-6 lite		4,320	.009			.14	.04	.18	.29
0540	7-10 lite		4,240	.009			.14	.04	.18	.29
0550	12 lite		4,160	.010			.14	.04	.18	.29
0560	Bay / Bow		4,400	.009			.14	.04	.18	.28
0600	Siding, scrape and sand, light=10-30%, med.=30-70%									
0610	Heavy=70-100%, % of surface to sand									
0620	For Chemical Washing, see Division 04930-900									
0630	For Steam Cleaning, see Division 04930-750									
0640	For Sand Blasting, see Division 04930-700									
0650	Texture 1-11, light	1 Pord	480	.017	S.F.		.30		.30	.50
0660	Med.		440	.018			.33		.33	.54
0670	Heavy		360	.022			.40		.40	.67
0680	Wood shingles, shakes, light		440	.018			.33		.33	.54
0690	Med.		360	.022			.40		.40	.67
0700	Heavy		280	.029			.51		.51	.86
0710	Clapboard, light		520	.015			.28		.28	.46
0720	Med.		480	.017			.30		.30	.50
0730	Heavy		400	.020			.36		.36	.60
0740	Wire brush									
0750	Aluminum, light	1 Pord	600	.013	S.F.		.24		.24	.40
0760	Med.		520	.015			.28		.28	.46
0770	Heavy		440	.018			.33		.33	.54
0780	Pressure wash, based on 2500 lb.. operating pressure									
0790	Stucco	B-9	3,080	.013	S.F.		.19	.06	.25	.39
0800	Aluminum or vinyl		3,200	.013			.19	.06	.25	.38
0810	Siding, masonry, brick & block		2,400	.017			.25	.08	.33	.50

09900 | Paints & Coatings

09990 | Paint Restoration

			CREW	DAILY OUTPUT	LABOR-HOURS	UNIT	MAT.	2000 BARE COSTS LABOR	EQUIP.	TOTAL	TOTAL INCL O&P	
900	1300	Miscellaneous, wire brush										900
	1310	Metal, pedestrian gate	1 Pord	100	.080	S.F.		1.44		1.44	2.40	
910	0010	**SURFACE PREPARATION, INTERIOR**										910
	0020	Doors										
	0030	Scrape & sand										
	0040	Wood, flush	1 Pord	616	.013	S.F.		.23		.23	.39	
	0050	Wood, detail		496	.016			.29		.29	.48	
	0060	Wood, louvered	↓	280	.029	↓		.51		.51	.86	
	0070	Wire brush										
	0080	Metal, flush	1 Pord	640	.013	S.F.		.22		.22	.37	
	0090	Metal, detail		520	.015			.28		.28	.46	
	0100	Metal, louvered	↓	360	.022	↓		.40		.40	.67	
	0110	Hand wash										
	0120	Wood, flush	1 Pord	2,160	.004	S.F.		.07		.07	.11	
	0130	Wood, detailed		2,000	.004			.07		.07	.12	
	0140	Wood, louvered		1,360	.006			.11		.11	.18	
	0150	Metal, flush		2,160	.004			.07		.07	.11	
	0160	Metal, detail		2,000	.004			.07		.07	.12	
	0170	Metal, louvered	↓	1,360	.006	↓		.11		.11	.18	
	0400	Windows, per side, not incl. trim										
	0410	Scrape & sand										
	0420	Wood, 1-2 lite	1 Pord	360	.022	S.F.		.40		.40	.67	
	0430	Wood, 3-6 lite		320	.025			.45		.45	.75	
	0440	Wood, 7-10 lite		280	.029			.51		.51	.86	
	0450	Wood, 12 lite		240	.033			.60		.60	1	
	0460	Wood, Bay / Bow	↓	360	.022	↓		.40		.40	.67	
	0470	Wire brush										
	0480	Metal, 1-2 lite	1 Pord	520	.015	S.F.		.28		.28	.46	
	0490	Metal, 3-6 lite		440	.018			.33		.33	.54	
	0500	Metal, Bay / Bow	↓	520	.015	↓		.28		.28	.46	
	0600	Walls, sanding, light=10-30%										
	0610	Med.=30-70%, heavy=70-100%, % of surface to sand										
	0620	For Chemical Washing, see Division 04930-900										
	0630	For Steam Cleaning, see Division 04930-750										
	0640	For Sand Blasting, see Division 04930-700										
	0650	Walls, sand										
	0660	Drywall, gypsum, plaster, light	1 Pord	3,077	.003	S.F.		.05		.05	.08	
	0670	Drywall, gypsum, plaster, med.		2,160	.004			.07		.07	.11	
	0680	Drywall, gypsum, plaster, heavy		923	.009			.16		.16	.26	
	0690	Wood, T&G, light		2,400	.003			.06		.06	.10	
	0700	Wood, T&G, med.		1,600	.005			.09		.09	.15	
	0710	Wood, T&G, heavy	↓	800	.010	↓		.18		.18	.30	
	0720	Walls, wash										
	0730	Drywall, gypsum, plaster	1 Pord	3,200	.002	S.F.		.04		.04	.07	
	0740	Wood, T&G		3,200	.002			.04		.04	.07	
	0750	Masonry, brick & block, smooth		2,800	.003			.05		.05	.09	
	0760	Masonry, brick & block, coarse	↓	2,000	.004	↓		.07		.07	.12	

For information about Means Estimating Seminars, see yellow pages 11 and 12 in back of book

For expanded coverage of these items see *Means Building Construction Cost Data 2000*

Division Notes

		CREW	DAILY OUTPUT	LABOR-HOURS	UNIT	2000 BARE COSTS				TOTAL INCL O&P
						MAT.	LABOR	EQUIP.	TOTAL	

Division 10 Specialties

Estimating Tips
General
- The items in this division are usually priced per square foot or each.
- Many items in Division 10 require some type of support system that is not usually furnished with the item. Examples of these systems include blocking for the attachment of grab bars and support angles for ceiling hung toilet partitions. The required blocking or supports must be added to the estimate in the appropriate division.
- Some items in Division 10, such as lockers, may require assembly before installation. Verify the amount of assembly required. Assembly can often exceed installation time.

10100 Visual Display Boards
- Toilet partitions are priced by the stall. A stall consists of a side wall, pilaster and door with hardware. Toilet tissue holders and grab bars are extra.

10600 Partitions
- The required acoustical rating of a folding partition can have a significant impact on costs. Verify the sound transmission coefficient rating of the panel priced to the specification requirements.

Reference Numbers
Reference numbers are shown in bold squares at the beginning of some major classifications. These numbers refer to related items in the Reference Section. The reference information may be an estimating procedure, an alternate pricing method or technical information.

Note: Not all subdivisions listed here necessarily appear in this publication.

10100 | Visual Display Boards

10110 | Chalkboards

			CREW	DAILY OUTPUT	LABOR-HOURS	UNIT	MAT.	LABOR	EQUIP.	TOTAL	TOTAL INCL O&P	
240	0010	**CHALKBOARDS** Porcelain enamel steel										240
	0100	Freestanding, reversible										
	0120	Economy, wood frame, 4' x 6'										
	0140	Chalkboard both sides				Ea.	460			460	510	
	0160	Chalkboard one side, cork other side				"	400			400	440	
	0200	Standard, lightweight satin finished aluminum, 4' x 6'										
	0220	Chalkboard both sides				Ea.	475			475	525	
	0240	Chalkboard one side, cork other side				"	420			420	460	
	0300	Deluxe, heavy duty extruded aluminum, 4' x 6'										
	0320	Chalkboard both sides				Ea.	1,025			1,025	1,150	
	0340	Chalkboard one side, cork other side				"	1,200			1,200	1,325	
	3900	Wall hung										
	4000	Aluminum frame and chalktrough										
	4300	3' x 5'	2 Carp	15	1.067	Ea.	186	21		207	241	
	4600	4' x 12'	"	13	1.231	"	410	24.50		434.50	495	
	4700	Wood frame and chalktrough										
	4800	3' x 4'	2 Carp	16	1	Ea.	130	19.70		149.70	177	
	5300	4' x 8'	"	13	1.231	"	264	24.50		288.50	330	
	5400	Liquid chalk, white porcelain enamel, wall hung										
	5420	Deluxe units, aluminum trim and chalktrough										
	5450	4' x 4'	2 Carp	16	1	Ea.	181	19.70		200.70	234	
	5550	4' x 12'	"	12	1.333	"	435	26.50		461.50	525	
	5700	Wood trim and chalktrough										
	5900	4' x 4'	2 Carp	16	1	Ea.	375	19.70		394.70	450	
	6200	4' x 8'	"	14	1.143	"	540	22.50		562.50	635	

10120 | Tack & Visual Aid Boards

			CREW	DAILY OUTPUT	LABOR-HOURS	UNIT	MAT.	LABOR	EQUIP.	TOTAL	TOTAL INCL O&P	
940	0010	**BULLETIN BOARD** Cork sheets, unbacked, no frame, 1/4" thick	2 Carp	290	.055	S.F.	5.65	1.09		6.74	8.05	940
	2120	Prefabricated, 1/4" cork, 3' x 5' with aluminum frame		16	1	Ea.	97.50	19.70		117.20	141	
	2140	Wood frame	↓	16	1	"	126	19.70		145.70	172	
	2300	Glass enclosed cabinets, alum., cork panel, hinged doors										
	2600	4' x 7', 3 door	2 Carp	10	1.600	Ea.	1,150	31.50		1,181.50	1,325	

10150 | Compartments & Cubicles

10155 | Toilet Compartments

			CREW	DAILY OUTPUT	LABOR-HOURS	UNIT	MAT.	LABOR	EQUIP.	TOTAL	TOTAL INCL O&P	
100	0010	**PARTITIONS, TOILET**										100
	0100	Cubicles, ceiling hung, marble	2 Marb	2	8	Ea.	1,325	159		1,484	1,725	
	0200	Painted metal	2 Carp	4	4		385	79		464	555	
	0250	Phenolic		4	4		730	79		809	935	
	0300	Plastic laminate on particle board		4	4		495	79		574	680	
	0500	Stainless steel	↓	4	4		965	79		1,044	1,200	
	0600	For handicap units, incl. 52" grab bars, add ♿					277			277	305	
	0800	Floor & ceiling anchored, marble	2 Marb	2.50	6.400		1,450	127		1,577	1,800	
	1000	Painted metal	2 Carp	5	3.200		355	63		418	500	
	1050	Phenolic		5	3.200		785	63		848	975	
	1100	Plastic laminate on particle board		5	3.200		470	63		533	625	
	1300	Stainless steel	↓	5	3.200	↓	1,125	63		1,188	1,325	

Important: See the Reference Section for critical supporting data - Reference Nos., Crews, & Location Factors

10150 | Compartments & Cubicles

10155 | Toilet Compartments

		CREW	DAILY OUTPUT	LABOR-HOURS	UNIT	MAT.	LABOR	EQUIP.	TOTAL	TOTAL INCL O&P
1400	For handicap units, incl. 52" grab bars, add				Ea.	253			253	278
1600	Floor mounted, marble	2 Marb	3	5.333		850	106		956	1,125
1700	Painted metal	2 Carp	7	2.286		325	45		370	430
1750	Phenolic		7	2.286		730	45		775	875
1800	Plastic laminate on particle board		7	2.286		495	45		540	620
2000	Stainless steel		7	2.286		1,050	45		1,095	1,225
2100	For handicap units, incl. 52" grab bars, add					253			253	278
2200	For juvenile units, deduct					35			35	38.50
2400	Floor mounted, headrail braced, marble	2 Marb	3	5.333		990	106		1,096	1,275
2500	Painted metal	2 Carp	6	2.667		355	52.50		407.50	480
2550	Phenolic		6	2.667		760	52.50		812.50	925
2600	Plastic laminate on particle board		6	2.667		525	52.50		577.50	665
2800	Stainless steel		6	2.667		1,100	52.50		1,152.50	1,300
2900	For handicap units, incl. 52" grab bars, add					253			253	278
3000	Wall hung partitions, painted metal	2 Carp	7	2.286		450	45		495	570
3300	Stainless steel	"	7	2.286		1,075	45		1,120	1,275
3400	For handicap units, incl. 52" grab bars, add					253			253	278
4000	Screens, entrance, floor mounted, 58" high, 48" wide									
4100	Marble	2 Marb	9	1.778	Ea.	535	35.50		570.50	645
4200	Painted metal	2 Carp	15	1.067		166	21		187	218
4300	Plastic laminate on particle board		15	1.067		294	21		315	360
4500	Stainless steel		15	1.067		585	21		606	680
4600	Urinal screen, 18" wide, ceiling braced, marble	D-1	6	2.667		535	47.50		582.50	665
4700	Painted metal	2 Carp	8	2		126	39.50		165.50	207
4800	Plastic laminate on particle board		8	2		235	39.50		274.50	325
5000	Stainless steel		8	2		420	39.50		459.50	530
5100	Floor mounted, head rail braced									
5200	Marble	D-1	6	2.667	Ea.	495	47.50		542.50	625
5300	Painted metal	2 Carp	8	2		166	39.50		205.50	250
5400	Plastic laminate on particle board		8	2		212	39.50		251.50	300
5600	Stainless steel		8	2		450	39.50		489.50	565
5700	Pilaster, flush, marble	D-1	9	1.778		605	31.50		636.50	720
5800	Painted metal	2 Carp	10	1.600		216	31.50		247.50	292
5900	Plastic laminate on particle board		10	1.600		270	31.50		301.50	350
6100	Stainless steel		10	1.600		370	31.50		401.50	460
6200	Post braced, marble	D-1	9	1.778		595	31.50		626.50	710
6300	Painted metal	2 Carp	10	1.600		222	31.50		253.50	299
6400	Plastic laminate on particle board		10	1.600		276	31.50		307.50	360
6600	Stainless steel		10	1.600		365	31.50		396.50	460
6700	Wall hung, bracket supported									
6800	Painted metal	2 Carp	10	1.600	Ea.	225	31.50		256.50	300
6900	Plastic laminate on particle board		10	1.600		108	31.50		139.50	173
7100	Stainless steel		10	1.600		335	31.50		366.50	425
7400	Flange supported, painted metal		10	1.600		165	31.50		196.50	236
7500	Plastic laminate on particle board		10	1.600		232	31.50		263.50	310
7700	Stainless steel		10	1.600		390	31.50		421.50	485
7800	Wedge type, painted metal		10	1.600		195	31.50		226.50	268
8100	Stainless steel		10	1.600		400	31.50		431.50	495

10185 | Shower/Dressing Compartments

		CREW	DAILY OUTPUT	LABOR-HOURS	UNIT	MAT.	LABOR	EQUIP.	TOTAL	TOTAL INCL O&P
0010	**PARTITIONS, SHOWER** Floor mounted, no plumbing									
0100	Cabinet, incl. base, no door, painted steel, 1" thick walls	2 Shee	5	3.200	Ea.	600	69		669	775
0300	With door, fiberglass		4.50	3.556		490	76.50		566.50	670
0600	Galvanized and painted steel, 1" thick walls		5	3.200		635	69		704	815
0800	Stall, 1" thick wall, no base, enameled steel		5	3.200		610	69		679	790
1500	Circular fiberglass, cabinet 36" diameter,		4	4		505	86		591	700

For expanded coverage of these items see *Means Building Construction Cost Data 2000*

10150 | Compartments & Cubicles

10185 | Shower/Dressing Compartments

		CREW	DAILY OUTPUT	LABOR-HOURS	UNIT	2000 BARE COSTS MAT.	LABOR	EQUIP.	TOTAL	TOTAL INCL O&P
1700	One piece, 36" diameter, less door	2 Shee	4	4	Ea.	430	86		516	620
1800	With door		3.50	4.571		695	98.50		793.50	930
2400	Glass stalls, with doors, no receptors, chrome on brass		3	5.333		1,050	115		1,165	1,350
2700	Anodized aluminum	↓	4	4		720	86		806	940
3200	Receptors, precast terrazzo, 32" x 32"	2 Marb	14	1.143		266	22.50		288.50	330
3300	48" x 34"		9.50	1.684		385	33.50		418.50	475
3500	Plastic, simulated terrazzo receptor, 32" x 32"		14	1.143		85.50	22.50		108	133
3600	32" x 48"		12	1.333		125	26.50		151.50	183
3800	Precast concrete, colors, 32" x 32"		14	1.143		152	22.50		174.50	206
3900	48" x 48"	↓	8	2		187	39.50		226.50	273
4100	Shower doors, economy plastic, 24" wide	1 Shee	9	.889		90	19.10		109.10	132
4200	Tempered glass door, economy		8	1		174	21.50		195.50	228
4400	Folding, tempered glass, aluminum frame		6	1.333		213	28.50		241.50	283
4700	Deluxe, tempered glass, chrome on brass frame, minimum		8	1		220	21.50		241.50	279
4800	Maximum		1	8		640	172		812	990
4850	On anodized aluminum frame, minimum		2	4		103	86		189	258
4900	Maximum	↓	1	8	↓	350	172		522	675
5100	Shower enclosure, tempered glass, anodized alum. frame									
5120	2 panel & door, corner unit, 32" x 32"	1 Shee	2	4	Ea.	350	86		436	530
5140	Neo-angle corner unit, 16" x 24" x 16"	"	2	4		610	86		696	820
5200	Shower surround, 3 wall, polypropylene, 32" x 32"	1 Carp	4	2		258	39.50		297.50	350
5220	PVC, 32" x 32"		4	2		226	39.50		265.50	315
5240	Fiberglass		4	2		271	39.50		310.50	365
5250	2 wall, polypropylene, 32" x 32"		4	2		187	39.50		226.50	274
5270	PVC		4	2		226	39.50		265.50	315
5290	Fiberglass	↓	4	2		251	39.50		290.50	345
5300	Tub doors, tempered glass & frame, minimum	1 Shee	8	1		148	21.50		169.50	200
5400	Maximum		6	1.333		340	28.50		368.50	425
5600	Chrome plated, brass frame, minimum		8	1		193	21.50		214.50	250
5700	Maximum		6	1.333		385	28.50		413.50	475
5900	Tub/shower enclosure, temp. glass, alum. frame, minimum		2	4		264	86		350	435
6200	Maximum		1.50	5.333		515	115		630	760
6500	On chrome-plated brass frame, minimum		2	4		365	86		451	550
6600	Maximum	↓	1.50	5.333		755	115		870	1,025
6800	Tub surround, 3 wall, polypropylene	1 Carp	4	2		148	39.50		187.50	231
6900	PVC		4	2		226	39.50		265.50	315
7000	Fiberglass, minimum		4	2		251	39.50		290.50	345
7100	Maximum	↓	3	2.667	↓	430	52.50		482.50	565

10200 | Louvers & Vents

10210 | Wall Louvers

		CREW	DAILY OUTPUT	LABOR-HOURS	UNIT	2000 BARE COSTS MAT.	LABOR	EQUIP.	TOTAL	TOTAL INCL O&P
0010	**LOUVERS** Aluminum with screen, residential, 8" x 8"	1 Carp	38	.211	Ea.	6.95	4.15		11.10	14.70
0100	12" x 12"		38	.211		8.75	4.15		12.90	16.70
0200	12" x 18"		35	.229		12.70	4.50		17.20	21.50
0250	14" x 24"		30	.267		16.35	5.25		21.60	27
0300	18" x 24"		27	.296		18.85	5.85		24.70	31
0500	24" x 30"		24	.333		27.50	6.55		34.05	42
0700	Triangle, adjustable, small		20	.400		23	7.90		30.90	38.50
0800	Large	↓	15	.533	↓	61.50	10.50		72	85.50

10200 | Louvers & Vents

10210 | Wall Louvers

		CREW	DAILY OUTPUT	LABOR-HOURS	UNIT	2000 BARE COSTS				TOTAL INCL O&P
						MAT.	LABOR	EQUIP.	TOTAL	
2100	Midget, aluminum, 3/4" deep, 1" diameter	1 Carp	85	.094	Ea.	.56	1.85		2.41	3.80
2150	3" diameter		60	.133		1.32	2.63		3.95	5.95
2200	4" diameter		50	.160		1.66	3.15		4.81	7.25
2250	6" diameter		30	.267		2.25	5.25		7.50	11.50
2300	Ridge vent strip, mill finish	1 Shee	155	.052	L.F.	1.95	1.11		3.06	4.02
2400	Under eaves vent, aluminum, mill finish, 16" x 4"	1 Carp	48	.167	Ea.	1.49	3.28		4.77	7.30
2500	16" x 8"		48	.167		1.79	3.28		5.07	7.60
7000	Vinyl gable vent, 8" x 8"		38	.211		9.50	4.15		13.65	17.55
7020	12" x 12"		38	.211		17	4.15		21.15	26
7080	12" x 18"		35	.229		22	4.50		26.50	31.50
7200	18" x 24"		30	.267		26	5.25		31.25	37.50

10260 | Wall & Corner Guards

10265 | Wall & Corner Guards

		CREW	DAILY OUTPUT	LABOR-HOURS	UNIT	2000 BARE COSTS				TOTAL INCL O&P
						MAT.	LABOR	EQUIP.	TOTAL	
0010	**WALLGUARD**									
0400	Rub rail, vinyl, adhesive mounted	1 Carp	185	.043	L.F.	3.67	.85		4.52	5.50
0500	Neoprene, aluminum backing, 1-1/2" x 2"		110	.073		6.15	1.43		7.58	9.20
1000	Trolley rail, PVC, clipped to wall, 5" high		185	.043		5.10	.85		5.95	7.05
1050	8" high		180	.044		5.95	.88		6.83	8.05
1200	Bed bumper, vinyl acrylic, alum. retainer, 21" long		10	.800	Ea.	33.50	15.75		49.25	64
1300	53" long with aligner		9	.889	"	100	17.50		117.50	140
1400	Bumper, vinyl cover, alum. retain., cush. mnt., 1-1/2" x 2-3/4"		80	.100	L.F.	10.10	1.97		12.07	14.50
1500	2" x 4-1/4"		80	.100		13.50	1.97		15.47	18.25
1600	Surface mounted, 1-3/4" x 3-5/8"		80	.100		9.10	1.97		11.07	13.40
2000	Crash rail, vinyl cover, alum. retainer, 1" x 4"		110	.073		10.30	1.43		11.73	13.80
2100	1" x 8"		90	.089		12.85	1.75		14.60	17.15
2150	Vinyl inserts, aluminum plate, 1" x 2-1/2"		110	.073		10.05	1.43		11.48	13.55
2200	1" x 5"		90	.089		14.50	1.75		16.25	18.95
3000	Handrail/bumper, vinyl cover, alum. retainer									
3010	Bracket mounted, flat rail, 5-1/2"	1 Carp	80	.100	L.F.	15.50	1.97		17.47	20.50
3100	6-1/2"		80	.100		19.05	1.97		21.02	24.50
3200	Bronze bracket, 1-3/4" diam. rail		80	.100		15.50	1.97		17.47	20.50

10300 | Fireplaces & Stoves

10305 | Manufactured Fireplaces

		CREW	DAILY OUTPUT	LABOR-HOURS	UNIT	2000 BARE COSTS				TOTAL INCL O&P
						MAT.	LABOR	EQUIP.	TOTAL	
0010	**FIREPLACE, PREFABRICATED** Free standing or wall hung									
0100	with hood & screen, minimum	1 Carp	1.30	6.154	Ea.	930	121		1,051	1,225
0150	Average		1	8		1,125	158		1,283	1,500
0200	Maximum		.90	8.889		2,675	175		2,850	3,250
0500	Chimney dbl. wall, all stainless, over 8'-6", 7" diam., add		33	.242	V.L.F.	31.50	4.78		36.28	42.50
0600	10" diameter, add		32	.250		45	4.93		49.93	58

For expanded coverage of these items see *Means Building Construction Cost Data 2000*

10300 | Fireplaces & Stoves

10305 | Manufactured Fireplaces

		CREW	DAILY OUTPUT	LABOR-HOURS	UNIT	2000 BARE COSTS MAT.	LABOR	EQUIP.	TOTAL	TOTAL INCL O&P	
100	0700	12" diameter, add	1 Carp	31	.258	V.L.F.	58.50	5.10		63.60	73	100
	0800	14" diameter, add		30	.267	↓	74	5.25		79.25	90.50	
	1000	Simulated brick chimney top, 4' high, 16" x 16"		10	.800	Ea.	160	15.75		175.75	203	
	1100	24" x 24"		7	1.143	"	298	22.50		320.50	370	
	1500	Simulated logs, gas fired, 40,000 BTU, 2' long, minimum		7	1.143	Set	415	22.50		437.50	500	
	1600	Maximum		6	1.333		605	26.50		631.50	715	
	1700	Electric, 1,500 BTU, 1'-6" long, minimum		7	1.143		110	22.50		132.50	161	
	1800	11,500 BTU, maximum		6	1.333	↓	239	26.50		265.50	310	
	2000	Fireplace, built-in, 36" hearth, radiant		1.30	6.154	Ea.	480	121		601	735	
	2100	Recirculating, small fan		1	8		685	158		843	1,025	
	2150	Large fan		.90	8.889		1,275	175		1,450	1,700	
	2200	42" hearth, radiant		1.20	6.667		610	131		741	895	
	2300	Recirculating, small fan		.90	8.889		800	175		975	1,175	
	2350	Large fan		.80	10		1,525	197		1,722	2,025	
	2400	48" hearth, radiant		1.10	7.273		1,125	143		1,268	1,500	
	2500	Recirculating, small fan		.80	10		1,400	197		1,597	1,900	
	2550	Large fan		.70	11.429		2,175	225		2,400	2,775	
	3000	See through, including doors		.80	10		1,800	197		1,997	2,325	
	3200	Corner (2 wall)	↓	1	8	↓	890	158		1,048	1,250	

10340 | Manufactured Exterior Specialties

10342 | Cupolas

		CREW	DAILY OUTPUT	LABOR-HOURS	UNIT	2000 BARE COSTS MAT.	LABOR	EQUIP.	TOTAL	TOTAL INCL O&P	
100	0010	**CUPOLA** Stock units, pine, painted, 18" sq., 28" high, alum. roof	1 Carp	4.10	1.951	Ea.	99	38.50		137.50	175	100
	0100	Copper roof		3.80	2.105		141	41.50		182.50	226	
	0300	23" square, 33" high, aluminum roof		3.70	2.162		232	42.50		274.50	330	
	0400	Copper roof		3.30	2.424		233	48		281	340	
	0600	30" square, 37" high, aluminum roof		3.70	2.162		355	42.50		397.50	465	
	0700	Copper roof		3.30	2.424		365	48		413	480	
	0900	Hexagonal, 31" wide, 46" high, copper roof		4	2		470	39.50		509.50	590	
	1000	36" wide, 50" high, copper roof	↓	3.50	2.286		560	45		605	695	
	1200	For deluxe stock units, add to above					25%					
	1400	For custom built units, add to above					50%	50%				
	1600	Fiberglass, 5'-0" base, 63" high minimum	F-3	6	6.667		2,325	120	75.50	2,520.50	2,825	
	1650	Maximum		4	10		2,725	179	113	3,017	3,400	
	1700	6'-0" base, 63" high, minimum		5	8		3,625	143	90.50	3,858.50	4,325	
	1750	Maximum	↓	3	13.333	↓	3,725	239	151	4,115	4,675	

10350 | Flagpoles

10355 | Flagpoles

		CREW	DAILY OUTPUT	LABOR-HOURS	UNIT	2000 BARE COSTS MAT.	LABOR	EQUIP.	TOTAL	TOTAL INCL O&P	
400	0010	**FLAGPOLE**, Ground set										400
	0050	Not including base or foundation										

Important: See the Reference Section for critical supporting data - Reference Nos., Crews, & Location Factors

10350 | Flagpoles

10355 | Flagpoles

			CREW	DAILY OUTPUT	LABOR-HOURS	UNIT	2000 BARE COSTS MAT.	LABOR	EQUIP.	TOTAL	TOTAL INCL O&P	
400	0100	Aluminum, tapered, ground set 20' high	K-1	2	8	Ea.	600	142	90	832	1,000	400
	0200	25' high		1.70	9.412		725	167	106	998	1,200	
	0300	30' high		1.50	10.667		995	190	120	1,305	1,550	
	0500	40' high	↓	1.20	13.333		1,775	237	150	2,162	2,550	

10400 | Identification Devices

10410 | Directories

			CREW	DAILY OUTPUT	LABOR-HOURS	UNIT	2000 BARE COSTS MAT.	LABOR	EQUIP.	TOTAL	TOTAL INCL O&P	
100	0010	**DIRECTORY BOARDS**										100
	0050	Plastic, glass covered, 30" x 20"	2 Carp	3	5.333	Ea.	289	105		394	500	
	0100	36" x 48"		2	8		565	158		723	895	
	0900	Outdoor, weatherproof, black plastic, 36" x 24"		2	8		315	158		473	620	
	1000	36" x 36"	↓	1.50	10.667	↓	665	210		875	1,100	

10430 | Exterior Signage

			CREW	DAILY OUTPUT	LABOR-HOURS	UNIT	2000 BARE COSTS MAT.	LABOR	EQUIP.	TOTAL	TOTAL INCL O&P	
200	0012	**SIGNS** Plaques										200
	3900	Plaques, custom, 20" x 30", for up to 450 letters, cast aluminum	2 Carp	4	4	Ea.	755	79		834	965	
	4000	Cast bronze		4	4		1,000	79		1,079	1,225	
	4200	30" x 36", up to 900 letters cast aluminum		3	5.333		1,525	105		1,630	1,850	
	4300	Cast bronze	↓	3	5.333		1,950	105		2,055	2,325	
	5100	Exit signs, 24 ga. alum., 14" x 12" surface mounted	1 Carp	30	.267		12.95	5.25		18.20	23.50	
	5200	10" x 7"	"	20	.400		7.90	7.90		15.80	22	
	6400	Replacement sign faces, 6" or 8"	1 Clab	50	.160	↓	21	2.31		23.31	27.50	

10500 | Lockers

10505 | Metal Lockers

			CREW	DAILY OUTPUT	LABOR-HOURS	UNIT	2000 BARE COSTS MAT.	LABOR	EQUIP.	TOTAL	TOTAL INCL O&P	
500	0011	**LOCKERS** Steel, baked enamel										500
	0110	Single tier box locker, 12" x 15" x 72"	1 Shee	8	1	Ea.	132	21.50		153.50	182	
	0120	18" x 15" x 72"		8	1		154	21.50		175.50	207	
	0130	12" x 18" x 72"		8	1		136	21.50		157.50	187	
	0140	18" x 18" x 72"		8	1		136	21.50		157.50	187	
	0410	Double tier, 12" x 15" x 36"		21	.381		114	8.20		122.20	139	
	0420	18" x 15" x 36"		21	.381		116	8.20		124.20	141	
	0430	12" x 18" x 36"		21	.381		99	8.20		107.20	123	
	0440	18" x 18" x 36"		21	.381		96.50	8.20		104.70	120	
	0500	Two person, 18" x 15" x 72"		8	1		182	21.50		203.50	237	
	0510	18" x 18" x 72"		8	1		201	21.50		222.50	258	
	0520	Duplex, 15" x 15" x 72"		8	1		216	21.50		237.50	275	
	0530	15" x 21" x 72"		8	1		259	21.50		280.50	320	
	1100	Wire meshed wardrobe, floor. mtd., open front varsity type		7.50	1.067	↓	147	23		170	201	
	3250	Rack w/ 24 wire mesh baskets		1.50	5.333	Set	270	115		385	490	
	3260	30 baskets	↓	1.25	6.400	↓	232	138		370	485	

For expanded coverage of these items see *Means Building Construction Cost Data 2000*

10500 | Lockers

10505 | Metal Lockers

		CREW	DAILY OUTPUT	LABOR-HOURS	UNIT	2000 BARE COSTS MAT.	LABOR	EQUIP.	TOTAL	TOTAL INCL O&P	
500	3270	36 baskets	1 Shee	.95	8.421	Set	325	181		506	665	500
	3280	42 baskets	↓	.80	10	↓	365	215		580	770	
	3600	For hanger rods, add				Ea.	1.65			1.65	1.82	

10520 | Fire Protection Specialties

10525 | Fire Prot. Specialties

			CREW	DAILY OUTPUT	LABOR-HOURS	UNIT	2000 BARE COSTS MAT.	LABOR	EQUIP.	TOTAL	TOTAL INCL O&P	
200	0010	**FIRE EQUIPMENT CABINETS** Not equipped, 20 ga. steel box,										200
	0040	recessed, D.S. glass in door, box size given										
	1000	Portable extinguisher, single, 8" x 12" x 27", alum. door & frame	Q-12	8	2	Ea.	75.50	39.50		115	149	
	1100	Steel door and frame	"	8	2	"	46	39.50		85.50	117	
	3000	Hose rack assy., 1-1/2" valve & 100' hose, 24" x 40" x 5-1/2"										
	3100	Aluminum door and frame	Q-12	6	2.667	Ea.	159	53		212	262	
	3200	Steel door and frame	"	6	2.667	"	101	53		154	199	
300	0010	**FIRE EXTINGUISHERS**										300
	0120	CO_2, portable with swivel horn, 5 lb.				Ea.	101			101	111	
	0140	With hose and "H" horn, 10 lb.				"	150			150	165	
	1000	Dry chemical, pressurized										
	1040	Standard type, portable, painted, 2-1/2 lb.				Ea.	27.50			27.50	30.50	
	1080	10 lb.					67			67	73.50	
	1100	20 lb.					90			90	99	
	1120	30 lb.					145			145	160	
	2000	ABC all purpose type, portable, 2-1/2 lb.					27.50			27.50	30.50	
	2080	9-1/2 lb.				↓	65			65	71.50	

10530 | Protective Covers

10535 | Awning & Canopies

			CREW	DAILY OUTPUT	LABOR-HOURS	UNIT	2000 BARE COSTS MAT.	LABOR	EQUIP.	TOTAL	TOTAL INCL O&P	
200	0010	**CANOPIES** Wall hung, .032", aluminum, prefinished, 8' x 10'	K-2	1.30	18.462	Ea.	1,325	370	139	1,834	2,300	200
	0300	8' x 20'	"	1.10	21.818	"	2,625	440	164	3,229	3,900	
	2300	Aluminum entrance canopies, flat soffit, .032"										
	2500	3'-6" x 4'-0", clear anodized	2 Carp	4	4	Ea.	550	79		629	740	
	4700	Canvas awnings, including canvas, frame & lettering										
	5000	Minimum	2 Carp	100	.160	S.F.	38	3.15		41.15	47	
	5300	Average	↓	90	.178	↓	45.50	3.50		49	56.50	
	5500	Maximum		80	.200		92.50	3.94		96.44	109	

Important: See the Reference Section for critical supporting data - Reference Nos., Crews, & Location Factors

10550 | Postal Specialties

10555 | Mail Delivery Systems

			CREW	DAILY OUTPUT	LABOR-HOURS	UNIT	2000 BARE COSTS MAT.	LABOR	EQUIP.	TOTAL	TOTAL INCL O&P	
600	0010	**MAIL BOXES** Horiz., key lock, 5"H x 6"W x 15"D, alum., rear load	1 Carp	34	.235	Ea.	34.50	4.64		39.14	46	600
	0100	Front loading		34	.235		38.50	4.64		43.14	50.50	
	0200	Double, 5"H x 12"W x 15"D, rear loading		26	.308		61	6.05		67.05	77.50	
	0300	Front loading		26	.308		66	6.05		72.05	83.50	
	0500	Quadruple, 10"H x 12"W x 15"D, rear loading		20	.400		110	7.90		117.90	135	
	0600	Front loading		20	.400		121	7.90		128.90	148	
	1600	Vault type, horizontal, for apartments, 4" x 5"		34	.235		36	4.64		40.64	47.50	
	1700	Alphabetical directories, 120 names		10	.800		140	15.75		155.75	181	
700	0010	**MAIL CHUTES** Aluminum & glass, 14-1/4" wide, 4-5/8" deep	2 Shee	4	4	Floor	635	86		721	845	700
	0100	8-5/8" deep		3.80	4.211	"	705	90.50		795.50	930	
	0600	Lobby collection boxes, aluminum		5	3.200	Ea.	1,750	69		1,819	2,050	
	0700	Bronze or stainless		4.50	3.556	"	2,150	76.50		2,226.50	2,500	

10600 | Partitions

10615 | Demountable Partitions

			CREW	DAILY OUTPUT	LABOR-HOURS	UNIT	2000 BARE COSTS MAT.	LABOR	EQUIP.	TOTAL	TOTAL INCL O&P	
100	0010	**PARTITIONS, MOVABLE OFFICE** Demountable, add for doors										100
	0100	Do not deduct door openings from total L.F.										
	0900	Demountable gypsum system on 2" to 2-1/2"										
	1000	steel studs, 9' high, 3" to 3-3/4" thick										
	1200	Vinyl clad gypsum	2 Carp	48	.333	L.F.	23.50	6.55		30.05	37.50	
	1300	Fabric clad gypsum		44	.364		73.50	7.15		80.65	93	
	1500	Steel clad gypsum		40	.400		78	7.90		85.90	99.50	
	1600	1.75 system, aluminum framing, vinyl clad hardboard,										
	1800	paper honeycomb core panel, 1-3/4" to 2-1/2" thick										
	1900	9' high	2 Carp	48	.333	L.F.	49.50	6.55		56.05	66	
	2100	7' high		60	.267		45.50	5.25		50.75	59	
	2200	5' high		80	.200		39	3.94		42.94	50	
	2250	Unitized gypsum system										
	2300	Unitized panel, 9' high, 2" to 2-1/2" thick										
	2350	Vinyl clad gypsum	2 Carp	48	.333	L.F.	67	6.55		73.55	85	
	2400	Fabric clad gypsum	"	44	.364	"	120	7.15		127.15	144	
	2500	Unitized mineral fiber system										
	2510	Unitized panel, 9' high, 2-1/4" thick, aluminum frame										
	2550	Vinyl clad mineral fiber	2 Carp	48	.333	L.F.	85	6.55		91.55	105	
	2600	Fabric clad mineral fiber	"	44	.364	"	109	7.15		116.15	131	
	2800	Movable steel walls, modular system										
	2900	Unitized panels, 9' high, 48" wide										
	3100	Baked enamel, pre-finished	2 Carp	60	.267	L.F.	78	5.25		83.25	95	
	3200	Fabric clad steel	"	56	.286	"	119	5.65		124.65	141	
	5500	For acoustical partitions, add, minimum				S.F.	1.38			1.38	1.52	
	5550	Maximum				"	5.05			5.05	5.55	
	5700	For doors, see div. 08100 & 08200										
	5800	For door hardware, see div. 08700										
	6100	In-plant modular office system, w/prehung steel door										
	6200	3" thick honeycomb core panels										
	6250	12' x 12', 2 wall	2 Clab	3.80	4.211	Ea.	3,225	61		3,286	3,650	
	6300	4 wall		1.90	8.421		4,975	122		5,097	5,675	
	6350	16' x 16', 2 wall		3.60	4.444		4,825	64		4,889	5,400	
	6400	4 wall		1.80	8.889		6,925	128		7,053	7,850	

For expanded coverage of these items see *Means Building Construction Cost Data 2000*

10600 | Partitions

10630 | Port. Partitions/Screens/Panels

		CREW	DAILY OUTPUT	LABOR-HOURS	UNIT	2000 BARE COSTS MAT.	LABOR	EQUIP.	TOTAL	TOTAL INCL O&P	
100	0010 **PARTITIONS, PORTABLE** Divider panels, free standing, fiber core										100
	0020 Fabric face straight										
	1500 6"-0" high	2 Carp	125	.128	L.F.	82.50	2.52		85.02	95.50	
	1600 6'-0" long, 5'-0" high		162	.099		69	1.95		70.95	79	
	3100 Curved, 3'-0" long, 5'-0" high		90	.178		85	3.50		88.50	99.50	
	3150 6'-0" high		75	.213		98.50	4.20		102.70	115	
	3200 Economical panels, fabric face, 4'-0" long, 5'-0" high		132	.121		135	2.39		137.39	153	
	3250 6'-0" high		112	.143		23	2.81		25.81	30	
	3300 5'-0" long, 5'-0" high		150	.107		27.50	2.10		29.60	34	
	3350 6'-0" high		125	.128		33.50	2.52		36.02	41.50	
	3380 3'-0" curved, 5'-0" high		90	.178		85	3.50		88.50	99.50	
	3390 6'-0" high		75	.213		98.50	4.20		102.70	115	
	3450 Acoustical panels, 60 to 90 NRC, 3'-0" long, 5'-0" high		90	.178		79	3.50		82.50	93	
	3550 6'-0" high		75	.213		77	4.20		81.20	91.50	
	3600 5'-0" long, 5'-0" high		150	.107		55	2.10		57.10	64	
	3650 6'-0" high		125	.128		61	2.52		63.52	72	
	3700 6'-0" long, 5'-0" high		162	.099		53.50	1.95		55.45	62	
	3750 6'-0" high		138	.116		56.50	2.28		58.78	66.50	
	3800 Economy acoustical panels, 40 NRC, 4'-0" long, 5'-0" high		132	.121		34	2.39		36.39	41	
	3850 6'-0" high		112	.143		38	2.81		40.81	47	
	3900 5'-0" long, 6'-0" high		125	.128		33.50	2.52		36.02	41.50	
	3950 6'-0" long, 5'-0" high		162	.099		33	1.95		34.95	40	

10651 | Accordion Folding Partitions

		CREW	DAILY OUTPUT	LABOR-HOURS	UNIT	MAT.	LABOR	EQUIP.	TOTAL	INCL O&P	
100	0010 **PARTITIONS, FOLDING ACCORDION**										100
	0100 Vinyl covered, over 150 S.F., frame not included										
	0300 Residential, 1.25 lb. per S.F., 8' maximum height	2 Carp	300	.053	S.F.	12	1.05		13.05	15	
	0400 Commercial, 1.75 lb. per S.F., 8' maximum height		225	.071		12	1.40		13.40	15.60	
	0900 Acoustical, 3 lb. per S.F., 17' maximum height		100	.160		18.10	3.15		21.25	25.50	
	1200 5 lb. per S.F., 20' maximum height		95	.168		26.50	3.32		29.82	35	
	1500 Vinyl clad wood or steel, electric operation, 5.0 psf		160	.100		30	1.97		31.97	36.50	
	1900 Wood, non-acoustic, birch or mahogany, to 10' high		300	.053		16.10	1.05		17.15	19.50	

10653 | Folding Panel Partitions

		CREW	DAILY OUTPUT	LABOR-HOURS	UNIT	MAT.	LABOR	EQUIP.	TOTAL	INCL O&P	
200	0010 **PARTITIONS, FOLDING LEAF** Acoustic, wood										200
	0100 Vinyl faced, to 18' high, 6 psf, minimum	2 Carp	60	.267	S.F.	35.50	5.25		40.75	48	
	0150 Average		45	.356		42.50	7		49.50	58.50	
	0200 Maximum		30	.533		55	10.50		65.50	78.50	
	0400 Formica or hardwood finish, minimum		60	.267		36.50	5.25		41.75	49.50	
	0500 Maximum		30	.533		39	10.50		49.50	61	
	0600 Wood, low acoustical type, 4.5 psf, to 14' high		50	.320		26.50	6.30		32.80	40.50	

10670 | Storage Shelving

10674 | Storage Shelving

		CREW	DAILY OUTPUT	LABOR-HOURS	UNIT	MAT.	LABOR	EQUIP.	TOTAL	INCL O&P	
500	0010 **SHELVING** Metal, industrial, cross-braced, 3' wide, 12" deep	1 Sswk	175	.046	SF Shlf	8.80	.97		9.77	11.60	500
	0100 24" deep		330	.024		6.85	.52		7.37	8.55	
	2200 Wide span, 1600 lb. capacity per shelf, 6' wide, 24" deep		380	.021		6.55	.45		7	8.10	
	2400 36" deep		440	.018		5.80	.39		6.19	7.10	

10670 | Storage Shelving

	10674	Storage Shelving	CREW	DAILY OUTPUT	LABOR-HOURS	UNIT	2000 BARE COSTS MAT.	LABOR	EQUIP.	TOTAL	TOTAL INCL O&P	
500	3000	Residential, vinyl covered wire, wardrobe, 12" deep	1 Carp	195	.041	L.F.	2.72	.81		3.53	4.37	500
	3100	16" deep		195	.041		2.85	.81		3.66	4.52	
	3200	Standard, 6" deep		195	.041		2.52	.81		3.33	4.15	
	3300	9" deep		195	.041		2.52	.81		3.33	4.15	
	3400	12" deep		195	.041		2.52	.81		3.33	4.15	
	3500	16" deep		195	.041		2.58	.81		3.39	4.22	
	3600	20" deep		195	.041		2.63	.81		3.44	4.27	
	3700	Support bracket		80	.100	Ea.	1.32	1.97		3.29	4.83	

10750 | Telephone Specialties

	10755	Telephone Enclosures	CREW	DAILY OUTPUT	LABOR-HOURS	UNIT	2000 BARE COSTS MAT.	LABOR	EQUIP.	TOTAL	TOTAL INCL O&P	
400	0010	**TELEPHONE ENCLOSURE**										400
	0300	Shelf type, wall hung, minimum	2 Carp	5	3.200	Ea.	705	63		768	890	
	0400	Maximum		5	3.200		2,425	63		2,488	2,775	
	0600	Booth type, painted steel, indoor or outdoor, minimum		1.50	10.667		2,975	210		3,185	3,625	
	0700	Maximum (stainless steel)		1.50	10.667		9,850	210		10,060	11,200	
	1900	Outdoor, drive-up type, wall mounted		4	4		775	79		854	990	
	2000	Post mounted, stainless steel posts		3	5.333		1,200	105		1,305	1,500	

10800 | Toilet/Bath/Laundry Accessories

	10820	Bath Accessories	CREW	DAILY OUTPUT	LABOR-HOURS	UNIT	2000 BARE COSTS MAT.	LABOR	EQUIP.	TOTAL	TOTAL INCL O&P	
100	0010	**BATH ACCESSORIES**										100
	0200	Curtain rod, stainless steel, 5' long, 1" diameter	1 Carp	13	.615	Ea.	20.50	12.10		32.60	43.50	
	0300	1-1/4" diameter	"	13	.615	"	24.50	12.10		36.60	48	
	0500	Dispenser units, combined soap & towel dispensers,										
	0510	mirror and shelf, flush mounted	1 Carp	10	.800	Ea.	271	15.75		286.75	325	
	0600	Towel dispenser and waste receptacle,										
	0610	18 gallon capacity	1 Carp	10	.800	Ea.	271	15.75		286.75	325	
	0800	Grab bar, straight, 1-1/4" diameter, stainless steel, 18" long		24	.333		35.50	6.55		42.05	50.50	
	1100	36" long		20	.400		45	7.90		52.90	63	
	3000	Mirror, with stainless steel 3/4" square frame, 18" x 24"		20	.400		61	7.90		68.90	80.50	
	3100	36" x 24"		15	.533		108	10.50		118.50	137	
	3300	72" x 24"		6	1.333		196	26.50		222.50	261	
	3500	With 5" stainless steel shelf, 18" x 24"		20	.400		90.50	7.90		98.40	114	
	3600	36" x 24"		15	.533		132	10.50		142.50	163	
	3800	72" x 24"		6	1.333		295	26.50		321.50	370	
	4200	Napkin/tampon dispenser, recessed		15	.533		365	10.50		375.50	420	
	4300	Robe hook, single, regular		36	.222		9.60	4.38		13.98	18.05	
	4400	Heavy duty, concealed mounting		36	.222		14.20	4.38		18.58	23	
	4600	Soap dispenser, chrome, surface mounted, liquid		20	.400		41	7.90		48.90	59	
	5000	Recessed stainless steel, liquid		10	.800		66.50	15.75		82.25	101	

For expanded coverage of these items see *Means Building Construction Cost Data 2000*

10800 | Toilet/Bath/Laundry Accessories

10820 | Bath Accessories

			CREW	DAILY OUTPUT	LABOR-HOURS	UNIT	MAT.	LABOR	EQUIP.	TOTAL	TOTAL INCL O&P	
100	5600	Shelf, stainless steel, 5" wide, 18 ga., 24" long	1 Carp	24	.333	Ea.	35.50	6.55		42.05	50.50	100
	6100	Toilet tissue dispenser, surface mounted, SS, single roll		30	.267		10.60	5.25		15.85	20.50	
	6200	Double roll		24	.333		15.65	6.55		22.20	28.50	
	6400	Towel bar, stainless steel, 18" long		23	.348		30	6.85		36.85	45	
	6500	30" long		21	.381		34.50	7.50		42	51	
	6700	Towel dispenser, stainless steel, surface mounted		16	.500		36	9.85		45.85	56.50	
	6800	Flush mounted, recessed		10	.800		138	15.75		153.75	179	
	7400	Tumbler holder, tumbler only		30	.267		20.50	5.25		25.75	31.50	
	7500	Soap, tumbler & toothbrush		30	.267		20.50	5.25		25.75	31.50	
	7700	Wall urn ash receiver, surface mount, 11" long		12	.667		104	13.15		117.15	137	
	8000	Waste receptacles, stainless steel, with top, 13 gallon	▼	10	.800	▼	188	15.75		203.75	234	
400	0010	**MEDICINE CABINETS** With mirror, st. st. frame, 16" x 22", unlighted	1 Carp	14	.571	Ea.	65	11.25		76.25	91	400
	0100	Wood frame		14	.571		95	11.25		106.25	124	
	0300	Sliding mirror doors, 20" x 16" x 4-3/4", unlighted		7	1.143		85	22.50		107.50	132	
	0400	24" x 19" x 8-1/2", lighted		5	1.600		143	31.50		174.50	211	
	0600	Triple door, 30" x 32", unlighted, plywood body		7	1.143		211	22.50		233.50	271	
	0700	Steel body		7	1.143		278	22.50		300.50	345	
	0900	Oak door, wood body, beveled mirror, single door		7	1.143		124	22.50		146.50	175	
	1000	Double door		6	1.333		315	26.50		341.50	390	
	1200	Hotel cabinets, stainless, with lower shelf, unlighted		10	.800		170	15.75		185.75	214	
	1300	Lighted	▼	5	1.600	▼	252	31.50		283.50	330	

For information about Means Estimating Seminars, see yellow pages 11 and 12 in back of book

Division 11 Equipment

Estimating Tips
General
- The items in this division are usually priced per square foot or each. Many of these items are purchased by the owner for installation by the contractor. Check the specifications for responsibilities, and include time for receiving, storage, installation and mechanical and electrical hook-ups in the appropriate divisions.
- Many items in Division 11 require some type of support system that is not usually furnished with the item. Examples of these systems include blocking for the attachment of casework and support angles for ceiling hung projection screens. The required blocking or supports must be added to the estimate in the appropriate division.
- Some items in Division 11 may require assembly or electrical hook-ups. Verify the amount of assembly required or the need for a hard electrical connection and add the appropriate costs.

Reference Numbers
Reference numbers are shown in bold squares at the beginning of some major classifications. These numbers refer to related items in the Reference Section. The reference information may be an estimating procedure, an alternate pricing method or technical information.

Note: Not all subdivisions listed here necessarily appear in this publication.

11010 | Maintenance Equipment

11013 | Floor/Wall Cleaning Equipment

		CREW	DAILY OUTPUT	LABOR-HOURS	UNIT	2000 BARE COSTS				TOTAL INCL O&P
						MAT.	LABOR	EQUIP.	TOTAL	
0010	VACUUM CLEANING									
0020	Central, 3 inlet, residential	1 Skwk	.90	8.889	Total	685	176		861	1,050
0400	5 inlet system, residential		.50	16		840	315		1,155	1,450
0600	7 inlet system, commercial		.40	20		1,050	395		1,445	1,825
0800	9 inlet system, residential	▼	.30	26.667		1,200	525		1,725	2,225
4010	Rule of thumb: First 1200 S.F., installed				▼					1,044
4020	For each additional S.F., add				S.F.					.17

(800)

11130 | Audio-Visual Equipment

11136 | Projection Screens

		CREW	DAILY OUTPUT	LABOR-HOURS	UNIT	2000 BARE COSTS				TOTAL INCL O&P
						MAT.	LABOR	EQUIP.	TOTAL	
0010	PROJECTION SCREENS Wall or ceiling hung, matte white									
0100	Manually operated, economy	2 Carp	500	.032	S.F.	4.41	.63		5.04	5.95
0300	Intermediate		450	.036		5.15	.70		5.85	6.85
0400	Deluxe	▼	400	.040	▼	7.15	.79		7.94	9.20

(500)

11160 | Loading Dock Equipment

11161 | Loading Dock Equipment

		CREW	DAILY OUTPUT	LABOR-HOURS	UNIT	2000 BARE COSTS				TOTAL INCL O&P
						MAT.	LABOR	EQUIP.	TOTAL	
0010	DOCK BUMPERS Bolts not included									
0020	2" x 6" to 4" x 8", average	1 Carp	.30	26.667	M.B.F.	930	525		1,455	1,925

(200)

11400 | Food Service Equipment

11410 | Food Prep Equipment

		CREW	DAILY OUTPUT	LABOR-HOURS	UNIT	2000 BARE COSTS				TOTAL INCL O&P
						MAT.	LABOR	EQUIP.	TOTAL	
0011	KITCHEN EQUIPMENT									
1050	Butter pat dispenser	1 Clab	13	.615	Ea.	410	8.90		418.90	470
1100	Bread dispenser, counter top	"	13	.615		480	8.90		488.90	545
1650	Cabinet, heated, 1 compartment, reach-in	R-18	5.60	4.643		2,175	82		2,257	2,500
1655	Pass-thru roll-in		5.60	4.643		3,025	82		3,107	3,450
1660	2 compartment, reach-in	▼	4.80	5.417		4,575	95.50		4,670.50	5,175
1670	Mobile					2,175			2,175	2,375
1700	Choppers, 5 pounds	R-18	7	3.714		1,550	65.50		1,615.50	1,800
1720	16 pounds		5	5.200		1,925	92		2,017	2,275
1740	35 to 40 pounds	▼	4	6.500		3,800	115		3,915	4,375
1840	Coffee brewer, 5 burners	1 Plum	3	2.667		950	58.50		1,008.50	1,150
1860	Single, 3 gallon	"	3	2.667	▼	4,375	58.50		4,433.50	4,925

(150)

11400 | Food Service Equipment

11410 | Food Prep Equipment

		CREW	DAILY OUTPUT	LABOR-HOURS	UNIT	MAT.	LABOR	EQUIP.	TOTAL	TOTAL INCL O&P	
150	1900	Cup and glass dispenser, drop in	1 Clab	4	2	Ea.	850	29		879	985	150
	1920	Disposable cup, drop in		16	.500		240	7.20		247.20	275	
	2650	Dish dispenser, drop in, 12"		11	.727		1,000	10.50		1,010.50	1,125	
	2660	Mobile		10	.800		1,675	11.55		1,686.55	1,875	
	2950	Dishwasher hood, canopy type	L-3A	10	1.200	L.F.	3,600	26		3,626	4,025	
	2960	Pant leg type	"	2.50	4.800	Ea.	4,800	104		4,904	5,450	
	2970	Exhaust hood, sst, gutter on all sides, 4' x 4' x 2'	1 Carp	1.80	4.444		2,650	87.50		2,737.50	3,050	
	2980	4' x 4' x 7'		1.60	5		4,150	98.50		4,248.50	4,725	
	3600	Well, hot food, built-in, rectangular, 12" x 20"	R-30	10	2.600		256	45.50		301.50	360	
	3610	Circular, 7 qt		10	2.600		214	45.50		259.50	315	
	3620	Refrigerated, 2 compartments		10	2.600		1,550	45.50		1,595.50	1,775	
	3630	3 compartments		9	2.889		1,550	50.50		1,600.50	1,775	
	3640	4 compartments		8	3.250		2,125	57		2,182	2,425	
	3850	40 quarts	L-7	5.40	4.815		7,225	81		7,306	8,100	
	4040	80 quarts		3.90	6.667		11,300	112		11,412	12,600	
	4080	130 quarts		2.20	11.818		16,900	199		17,099	18,900	
	4100	Floor type, 20 quarts		15	1.733		2,975	29		3,004	3,325	
	4120	60 quarts		14	1.857		8,650	31.50		8,681.50	9,575	
	4140	80 quarts		12	2.167		11,200	36.50		11,236.50	12,400	
	4160	140 quarts		8.60	3.023		19,900	51		19,951	22,000	
	4600	Freezer, pre-fab, 8' x 8' w/refrigeration	2 Carp	.45	35.556		6,750	700		7,450	8,625	
	4620	8' x 12'		.35	45.714		8,900	900		9,800	11,400	
	4640	8' x 16'		.25	64		11,400	1,250		12,650	14,800	
	4660	8' x 20'		.17	94.118		13,000	1,850		14,850	17,500	
	4680	Reach-in, 1 compartment	Q-1	4	4		1,950	79		2,029	2,275	
	4700	2 compartment	"	3	5.333		3,175	105		3,280	3,675	
	5700	Hot chocolate dispenser	1 Plum	4	2		715	44		759	860	
	5900	250 pounds per day	Q-1	1.20	13.333		1,925	263		2,188	2,550	
	6060	With bin		1.20	13.333		3,525	263		3,788	4,325	
	6090	1000 pounds per day, with bin		1	16		7,650	315		7,965	8,925	
	6100	Ice flakers, 300 pounds per day		1.60	10		3,750	198		3,948	4,450	
	6120	600 pounds per day		.95	16.842		3,625	335		3,960	4,525	
	6130	1000 pounds per day		.75	21.333		6,300	420		6,720	7,625	
	6140	2000 pounds per day		.65	24.615		11,800	485		12,285	13,800	
	6160	Ice storage bin, 500 pound capacity	Q-5	1	16		1,500	320		1,820	2,175	
	6180	1000 pound	"	.56	28.571		2,175	570		2,745	3,350	
	6200	Iced tea brewer	1 Plum	3.44	2.326		640	51		691	790	
	6300	Juice dispenser, concentrate	R-18	4.50	5.778		815	102		917	1,075	
	6690	Milk dispenser, bulk, 2 flavor	R-30	8	3.250		925	57		982	1,125	
	6695	3 flavor	"	8	3.250		1,375	57		1,432	1,625	
	6700	Peelers, small	R-18	8	3.250		1,725	57.50		1,782.50	1,975	
	6720	Large	"	6	4.333		3,000	76.50		3,076.50	3,425	
	6750	Pot sink, 3 compartment	1 Plum	7.25	1.103	L.F.	435	24		459	515	
	6760	Pot washer, small		1.60	5	Ea.	12,600	110		12,710	14,100	
	6770	Large		1.20	6.667		31,500	146		31,646	34,900	
	6800	Pulper/extractor, close coupled, 5 HP		1.90	4.211		2,775	92.50		2,867.50	3,200	
	6850	Mobile rack w/pan slide					850			850	935	
	8320	Refrigerator, reach-in, 1 compartment	R-18	7.80	3.333		1,750	59		1,809	2,000	
	8330	2 compartment		6.20	4.194		2,300	74		2,374	2,650	
	8340	3 compartment		5.60	4.643		3,375	82		3,457	3,825	
	8350	Pre-fab, with refrigeration, 8' x 8'	2 Carp	.45	35.556		5,475	700		6,175	7,225	
	8360	8' x 12'		.35	45.714		6,725	900		7,625	8,950	
	8370	8' x 16'		.25	64		7,700	1,250		8,950	10,600	
	8380	8' x 20'		.17	94.118		10,100	1,850		11,950	14,300	
	8390	Pass-thru/roll-in, 1 compartment	R-18	7.80	3.333		2,925	59		2,984	3,325	
	8400	2 compartment		6.24	4.167		4,050	73.50		4,123.50	4,600	

For expanded coverage of these items see *Means Building Construction Cost Data 2000*

11400 | Food Service Equipment

11410 | Food Prep Equipment

		CREW	DAILY OUTPUT	LABOR-HOURS	UNIT	2000 BARE COSTS MAT.	LABOR	EQUIP.	TOTAL	TOTAL INCL O&P
8410	3 compartment	R-18	5.60	4.643	Ea.	5,550	82		5,632	6,250
8420	Walk-in, alum, door & floor only, no refrig, 6' x 6' x 7'-6"	2 Carp	1.40	11.429		7,050	225		7,275	8,125
8430	10' x 6' x 7'-6"		.55	29.091		10,100	575		10,675	12,100
8440	12' x 14' x 7'-6"		.25	64		13,400	1,250		14,650	17,000
8450	12' x 20' x 7'-6"	▼	.17	94.118		16,200	1,850		18,050	21,000
8460	Refrigerated cabinets, mobile					2,150			2,150	2,350
8470	Refrigerator/freezer, reach-in, 1 compartment	R-18	5.60	4.643		4,475	82		4,557	5,050
8480	2 compartment		4.80	5.417		6,150	95.50		6,245.50	6,900
8580	Slicer with table	▼	9	2.889		2,800	51		2,851	3,175
8600	Stainless steel shelving, louvered 4-tier, 20" x 3'	1 Clab	6	1.333		1,025	19.25		1,044.25	1,175
8605	20" x 4'		6	1.333		1,225	19.25		1,244.25	1,375
8610	20" x 6'		6	1.333		1,675	19.25		1,694.25	1,875
8615	24" x 3'		6	1.333		1,075	19.25		1,094.25	1,225
8620	24" x 4'		6	1.333		1,300	19.25		1,319.25	1,450
8625	24" x 6'		6	1.333		1,800	19.25		1,819.25	2,000
8630	Flat 4-tier, 20" x 3'		6	1.333		955	19.25		974.25	1,075
8635	20" x 4'		6	1.333		1,150	19.25		1,169.25	1,300
8640	20" x 5'		6	1.333		1,325	19.25		1,344.25	1,475
8645	24" x 3'		6	1.333		1,025	19.25		1,044.25	1,150
8650	24" x 4'		6	1.333		1,275	19.25		1,294.25	1,425
8655	24" x 6'		6	1.333		2,425	19.25		2,444.25	2,700
8700	Galvanized shelving, louvered 4-tier, 20" x 3'		6	1.333		365	19.25		384.25	440
8705	20" x 4'		6	1.333		415	19.25		434.25	490
8710	20" x 6'		6	1.333		615	19.25		634.25	710
8715	24" x 3'		6	1.333		400	19.25		419.25	475
8720	24" x 4'		6	1.333		455	19.25		474.25	535
8725	24" x 6'		6	1.333		625	19.25		644.25	725
8730	Flat 4-tier, 20" x 3'		6	1.333		375	19.25		394.25	445
8735	20" x 4'		6	1.333		425	19.25		444.25	505
8740	20" x 6'		6	1.333		575	19.25		594.25	670
8745	24" x 3'		6	1.333		395	19.25		414.25	470
8750	24" x 4'		6	1.333		455	19.25		474.25	535
8755	24" x 6'		6	1.333		625	19.25		644.25	720
8760	Stainless steel dunnage rack, 24" x 3'		8	1		470	14.45		484.45	540
8765	24" x 4'		8	1		545	14.45		559.45	625
8770	Galvanized dunnage rack, 24" x 3'		8	1		128	14.45		142.45	166
8775	24" x 4'	▼	8	1	▼	145	14.45		159.45	185
8800	Serving counter, straight	1 Carp	40	.200	L.F.	425	3.94		428.94	475
8820	Curved section	"	30	.267	"	605	5.25		610.25	680
8825	Solid surface, see section 06610-810									
8830	Soft serve ice cream machine, medium	R-18	11	2.364	Ea.	6,300	41.50		6,341.50	7,025
8840	Large	"	9	2.889		9,500	51		9,551	10,500
9160	Pop-up, 2 slot					455			455	500
9170	Trash compactor, small, up to 125 lb. compacted weight	L-4	4	4		14,300	69		14,369	15,900
9175	Large, up to 175 lb. compacted weight	"	3	5.333		17,500	92		17,592	19,500
9180	Tray and silver dispenser, mobile	1 Clab	16	.500	▼	630	7.20		637.20	705

11450 | Residential Equipment

11454 | Residential Appliances

		CREW	DAILY OUTPUT	LABOR-HOURS	UNIT	2000 BARE COSTS MAT.	LABOR	EQUIP.	TOTAL	TOTAL INCL O&P
0010	**RESIDENTIAL APPLIANCES**									
0020	Cooking range, 30" free standing, 1 oven, minimum	2 Clab	10	1.600	Ea.	259	23		282	325
0050	Maximum		4	4		1,175	58		1,233	1,400
0150	2 oven, minimum		10	1.600		1,125	23		1,148	1,275
0200	Maximum	▼	10	1.600		1,500	23		1,523	1,700
0350	Built-in, 30" wide, 1 oven, minimum	1 Elec	6	1.333		259	29.50		288.50	335
0400	Maximum	2 Carp	2	8		795	158		953	1,150
0500	2 oven, conventional, minimum		4	4		880	79		959	1,100
0550	1 conventional, 1 microwave, maximum	▼	2	8		1,550	158		1,708	1,975
0700	Free-standing, 1 oven, 21" wide range, minimum	2 Clab	10	1.600		281	23		304	350
0750	21" wide, maximum	"	4	4		465	58		523	615
0900	Counter top cook tops, 4 burner, standard, minimum	1 Elec	6	1.333		172	29.50		201.50	237
0950	Maximum		3	2.667		430	59		489	570
1050	As above, but with grille and griddle attachment, minimum		6	1.333		400	29.50		429.50	490
1100	Maximum		3	2.667		515	59		574	660
1250	Microwave oven, minimum		4	2		87.50	44		131.50	169
1300	Maximum	▼	2	4		365	88.50		453.50	550
1500	Combination range, refrigerator and sink, 30" wide, minimum	L-1	2	5		635	110		745	880
1550	Maximum		1	10		1,275	220		1,495	1,775
1570	60" wide, average		1.40	7.143		2,100	157		2,257	2,575
1590	72" wide, average	▼	1.20	8.333	▼	2,375	183		2,558	2,925
1640	Combination range, refrigerator, sink, microwave									
1660	oven and ice maker	L-1	.80	12.500	Ea.	3,525	275		3,800	4,325
1750	Compactor, residential size, 4 to 1 compaction, minimum	1 Carp	5	1.600		335	31.50		366.50	425
1800	Maximum	"	3	2.667		370	52.50		422.50	495
2000	Deep freeze, 15 to 23 C.F., minimum	2 Clab	10	1.600		370	23		393	445
2050	Maximum		5	3.200		510	46		556	640
2200	30 C.F., minimum		8	2		730	29		759	855
2250	Maximum	▼	3	5.333		850	77		927	1,075
2450	Dehumidifier, portable, automatic, 15 pint					185			185	204
2550	40 pint					220			220	242
2750	Dishwasher, built-in, 2 cycles, minimum	L-1	4	2.500		247	55		302	360
2800	Maximum		2	5		258	110		368	465
2950	4 or more cycles, minimum		4	2.500		239	55		294	355
2960	Average		4	2.500		320	55		375	440
3000	Maximum	▼	2	5		495	110		605	725
3200	Dryer, automatic, minimum	L-2	3	5.333		251	92		343	435
3250	Maximum	"	2	8		545	138		683	835
3300	Garbage disposer, sink type, minimum	L-1	10	1		37.50	22		59.50	77.50
3350	Maximum	"	10	1		174	22		196	228
3550	Heater, electric, built-in, 1250 watt, ceiling type, minimum	1 Elec	4	2		68	44		112	148
3600	Maximum		3	2.667		111	59		170	219
3700	Wall type, minimum		4	2		96.50	44		140.50	179
3750	Maximum		3	2.667		128	59		187	238
3900	1500 watt wall type, with blower		4	2		119	44		163	204
3950	3000 watt	▼	3	2.667		244	59		303	365
4150	Hood for range, 2 speed, vented, 30" wide, minimum	L-3	5	2		35	40.50		75.50	107
4200	Maximum		3	3.333		430	67.50		497.50	590
4300	42" wide, minimum		5	2		164	40.50		204.50	250
4330	Custom		5	2		575	40.50		615.50	700
4350	Maximum	▼	3	3.333		700	67.50		767.50	885
4500	For ventless hood, 2 speed, add					13.40			13.40	14.75
4650	For vented 1 speed, deduct from maximum					26			26	28.50
4850	Humidifier, portable, 8 gallons per day					92.50			92.50	102
5000	15 gallons per day					149			149	164
5200	Icemaker, automatic, 20 lb. per day	1 Plum	7	1.143	▼	505	25		530	595

For expanded coverage of these items see *Means Building Construction Cost Data 2000*

11450 | Residential Equipment

11454 | Residential Appliances

		CREW	DAILY OUTPUT	LABOR-HOURS	UNIT	2000 BARE COSTS MAT.	LABOR	EQUIP.	TOTAL	TOTAL INCL O&P	
500	5350	51 lb. per day	1 Plum	2	4	Ea.	785	88		873	1,000	500
	5380	Oven, built in, standard	1 Elec	4	2		360	44		404	470	
	5390	Deluxe	"	2	4		815	88.50		903.50	1,050	
	5500	Refrigerator, no frost, 10 C.F. to 12 C.F. minimum	2 Clab	10	1.600		420	23		443	505	
	5600	Maximum		6	2.667		640	38.50		678.50	770	
	5750	14 C.F. to 16 C.F., minimum		9	1.778		440	25.50		465.50	530	
	5800	Maximum		5	3.200		520	46		566	650	
	5950	18 C.F. to 20 C.F., minimum		8	2		485	29		514	585	
	6000	Maximum		4	4		805	58		863	985	
	6150	21 C.F. to 29 C.F., minimum		7	2.286		790	33		823	925	
	6200	Maximum		3	5.333		2,600	77		2,677	2,975	
	6400	Sump pump cellar drainer, pedestal, 1/3 H.P., molded PVC base	1 Plum	3	2.667		86.50	58.50		145	192	
	6450	Solid brass	"	2	4		179	88		267	340	
	6460	Sump pump, see also division 15440-940										
	6650	Washing machine, automatic, minimum	1 Plum	3	2.667	Ea.	320	58.50		378.50	445	
	6700	Maximum	"	1	8		620	176		796	970	
	6900	Water heater, electric, glass lined, 30 gallon, minimum	L-1	5	2		215	44		259	310	
	6950	Maximum		3	3.333		299	73.50		372.50	450	
	7100	80 gallon, minimum		2	5		415	110		525	635	
	7150	Maximum		1	10		575	220		795	995	
	7180	Water heater, gas, glass lined, 30 gallon, minimum	2 Plum	5	3.200		284	70		354	425	
	7220	Maximum		3	5.333		395	117		512	630	
	7260	50 gallon, minimum		2.50	6.400		515	140		655	800	
	7300	Maximum		1.50	10.667		720	234		954	1,175	
	7310	Water heater, see also division 15480-200										
	7350	Water softener, automatic, to 30 grains per gallon	2 Plum	5	3.200	Ea.	420	70		490	575	
	7400	To 100 grains per gallon	"	4	4		510	88		598	710	
	7450	Vent kits for dryers	1 Carp	10	.800		9.55	15.75		25.30	37.50	
550	0010	**DISAPPEARING STAIRWAY** No trim included										550
	0020	One piece, yellow pine, 8'-0" ceiling	2 Carp	4	4	Ea.	850	79		929	1,075	
	0030	9'-0" ceiling		4	4		875	79		954	1,100	
	0040	10'-0" ceiling		3	5.333		920	105		1,025	1,175	
	0050	11'-0" ceiling		3	5.333		1,075	105		1,180	1,375	
	0060	12'-0" ceiling		3	5.333		1,125	105		1,230	1,425	
	0100	Custom grade, pine, 8'-6" ceiling, minimum	1 Carp	4	2		100	39.50		139.50	178	
	0150	Average		3.50	2.286		105	45		150	193	
	0200	Maximum		3	2.667		185	52.50		237.50	294	
	0500	Heavy duty, pivoted, from 7'-7" to 12'-10" floor to floor		3	2.667		315	52.50		367.50	440	
	0600	16'-0" ceiling		2	4		1,025	79		1,104	1,275	
	0800	Economy folding, pine, 8'-6" ceiling		4	2		83	39.50		122.50	159	
	0900	9'-6" ceiling		4	2		93.50	39.50		133	171	
	1000	Fire escape, galvanized steel, 8'-0" to 10'-4" ceiling	2 Carp	1	16		1,125	315		1,440	1,775	
	1010	10'-6" to 13'-6" ceiling		1	16		1,400	315		1,715	2,100	
	1100	Automatic electric, aluminum, floor to floor height, 8' to 9'		1	16		5,500	315		5,815	6,625	

For information about Means Estimating Seminars, see yellow pages 11 and 12 in back of book

Division 12 Furnishings

Estimating Tips
General
- The items in this division are usually priced per square foot or each. Most of these items are purchased by the owner and placed by the supplier. Do not assume the items in Division 12 will be purchased and installed by the supplier. Check the specifications for responsibilities and include receiving, storage, installation and mechanical and electrical hook-ups in the appropriate divisions.

- Some items in this division require some type of support system that is not usually furnished with the item. Examples of these systems include blocking for the attachment of casework and heavy drapery rods. The required blocking must be added to the estimate in the appropriate division.

Reference Numbers
Reference numbers are shown in bold squares at the beginning of some major classifications. These numbers refer to related items in the Reference Section. The reference information may be an estimating procedure, an alternate pricing method or technical information.

Note: Not all subdivisions listed here necessarily appear in this publication.

12400 | Furnishings & Accessories

12492 | Blinds and Shades

		CREW	DAILY OUTPUT	LABOR-HOURS	UNIT	2000 BARE COSTS MAT.	LABOR	EQUIP.	TOTAL	TOTAL INCL O&P
100	0010 **BLINDS, INTERIOR**									100
	0020 Horizontal, 1" aluminum slats, solid color, stock	1 Carp	590	.014	S.F.	2.60	.27		2.87	3.32
	0090 Custom, minimum		590	.014		2.06	.27		2.33	2.73
	0100 Maximum		440	.018		6.50	.36		6.86	7.75
	0450 Stock, minimum		590	.014		2.66	.27		2.93	3.39
	0500 Maximum		440	.018		4.94	.36		5.30	6.05
	3000 Wood folding panels with movable louvers, 7" x 20" each		17	.471	Pr.	36	9.25		45.25	55.50
	3300 8" x 28" each		17	.471		52	9.25		61.25	73
	3450 9" x 36" each		17	.471		62	9.25		71.25	84
	3600 10" x 40" each		17	.471		70	9.25		79.25	93
	4000 Fixed louver type, stock units, 8" x 20" each		17	.471		54	9.25		63.25	75.50
	4150 10" x 28" each		17	.471		74	9.25		83.25	97.50
	4300 12" x 36" each		17	.471		94	9.25		103.25	119
	4450 18" x 40" each		17	.471		112	9.25		121.25	139
	5000 Insert panel type, stock, 7" x 20" each		17	.471		13.90	9.25		23.15	31
	5150 8" x 28" each		17	.471		25.50	9.25		34.75	44
	5300 9" x 36" each		17	.471		32.50	9.25		41.75	51.50
	5450 10" x 40" each		17	.471		35	9.25		44.25	54
	5600 Raised panel type, stock, 10" x 24" each		17	.471		93	9.25		102.25	118
	5650 12" x 26" each		17	.471		108	9.25		117.25	135
	5700 14" x 30" each		17	.471		122	9.25		131.25	150
	5750 16" x 36" each		17	.471		137	9.25		146.25	167
	6000 For custom built pine, add					22%				
	6500 For custom built hardwood blinds, add					42%				
600	0011 **SHADES** Basswood roll-up, stain finish, 3/8" slats	1 Carp	300	.027	S.F.	9	.53		9.53	10.80	600
	5011 Insulative shades		125	.064		7.55	1.26		8.81	10.45
	6011 Solar screening, fiberglass		85	.094		3.61	1.85		5.46	7.15
	8011 Interior insulative shutter									
	8111 Stock unit, 15" x 60"	1 Carp	17	.471	Pr.	7.55	9.25		16.80	24

12493 | Curtains and Drapes

		CREW	DAILY OUTPUT	LABOR-HOURS	UNIT	MAT.	LABOR	EQUIP.	TOTAL	TOTAL INCL O&P
200	0010 **DRAPERY HARDWARE**									200
	0030 Standard traverse, per foot, minimum	1 Carp	59	.136	L.F.	1.40	2.67		4.07	6.10
	0100 Maximum		51	.157	"	7.50	3.09		10.59	13.55
	0200 Decorative traverse, 28"-48", minimum		22	.364	Ea.	12	7.15		19.15	25.50
	0220 Maximum		21	.381		28	7.50		35.50	44
	0300 48"-84", minimum		20	.400		16	7.90		23.90	31
	0320 Maximum		19	.421		45.50	8.30		53.80	64
	0400 66"-120", minimum		18	.444		18	8.75		26.75	35
	0420 Maximum		17	.471		60	9.25		69.25	82
	0500 84"-156", minimum		16	.500		20	9.85		29.85	39
	0520 Maximum		15	.533		67.50	10.50		78	92
	0600 130"-240", minimum		14	.571		24	11.25		35.25	46
	0620 Maximum		13	.615		93	12.10		105.10	123
	0700 Slide rings, each, minimum					.65			.65	.71
	0720 Maximum					1.40			1.40	1.54
	3000 Ripplefold, snap-a-pleat system, 3' or less, minimum	1 Carp	15	.533		39.50	10.50		50	61
	3020 Maximum	"	14	.571		56	11.25		67.25	81
	3200 Each additional foot, add, minimum				L.F.	2			2	2.20
	3220 Maximum				"	5.45			5.45	6
	4000 Traverse rods, adjustable, 28" to 48"	1 Carp	22	.364	Ea.	15.20	7.15		22.35	29
	4020 48" to 84"		20	.400		24.50	7.90		32.40	40.50
	4040 66" to 120"		18	.444		26	8.75		34.75	43.50
	4060 84" to 156"		16	.500		30	9.85		39.85	49.50
	4080 100" to 180"		14	.571		29.50	11.25		40.75	51.50

12400 | Furnishings & Accessories

12493 | Curtains and Drapes

		CREW	DAILY OUTPUT	LABOR-HOURS	UNIT	2000 BARE COSTS				TOTAL INCL O&P
						MAT.	LABOR	EQUIP.	TOTAL	
4100	228" to 312"	1 Carp	13	.615	Ea.	45.50	12.10		57.60	71
4500	Curtain rod, 28" to 48", single		22	.364		3.91	7.15		11.06	16.55
4510	Double		22	.364		6.65	7.15		13.80	19.60
4520	48" to 86", single		20	.400		6.70	7.90		14.60	21
4530	Double		20	.400		11.15	7.90		19.05	26
4540	66" to 120", single		18	.444		11.20	8.75		19.95	27.50
4550	Double		18	.444		17.50	8.75		26.25	34.50
4600	Valance, pinch pleated fabric, 12" deep, up to 54" long, minimum					28			28	31
4610	Maximum					70			70	77
4620	Up to 77" long, minimum					43			43	47.50
4630	Maximum					113			113	124
5000	Stationary rods, first 2 feet					6.95			6.95	7.65

For information about Means Estimating Seminars, see yellow pages 11 and 12 in back of book

For expanded coverage of these items see *Means Building Construction Cost Data 2000*

Division Notes

Division 13
Special Construction

Estimating Tips
General
- The items and systems in this division are usually estimated, purchased, supplied and installed as a unit by one or more subcontractors. The estimator must ensure that all parties are operating from the same set of specifications and assumptions and that all necessary items are estimated and will be provided. Many times the complex items and systems are covered but the more common ones such as excavation or a crane are overlooked for the very reason that everyone assumes nobody could miss them. The estimator should be the central focus and be able to ensure that all systems are complete.
- Another area where problems can develop in this division is at the interface between systems. The estimator must ensure, for instance, that anchor bolts, nuts and washers are estimated and included for the air-supported structures and pre-engineered buildings to be bolted to their foundations. Utility supply is a common area where essential items or pieces of equipment can be missed or overlooked due to the fact that each subcontractor may feel it is the others' responsibility. The estimator should also be aware of certain items which may be supplied as part of a package but installed by others, and ensure that the installing contractor's estimate includes the cost of installation. Conversely, the estimator must also ensure that items are not costed by two different subcontractors, resulting in an inflated overall estimate.

13120 Pre-Engineered Structures
- The foundations and floor slab, as well as rough mechanical and electrical, should be estimated, as this work is required for the assembly and erection of the structure. Generally, as noted in the book, the pre-engineered building comes as a shell and additional features must be included by the estimator. Here again, the estimator must have a clear understanding of the scope of each portion of the work and all the necessary interfaces.

13200 Storage Tanks
- The prices in this subdivision for above and below ground storage tanks do not include foundations or hold-down slabs. The estimator should refer to Divisions 2 and 3 for foundation system pricing. In addition to the foundations, required tank accessories such as tank gauges, leak detection devices, and additional manholes and piping must be added to the tank prices.

Reference Numbers
Reference numbers are shown in bold squares at the beginning of some major classifications. These numbers refer to related items in the Reference Section. The reference information may be an estimating procedure, an alternate pricing method or technical information.

Note: Not all subdivisions listed here necessarily appear in this publication.

13030 | Special Purpose Rooms

13035 | Prefab Enclosures

			CREW	DAILY OUTPUT	LABOR-HOURS	UNIT	MAT.	LABOR	EQUIP.	TOTAL	TOTAL INCL O&P	
800	0010	**SAUNA** Prefabricated, incl. heater & controls, 7' high, 6' x 4', C/C	L-7	2.20	11.818	Ea.	3,200	199		3,399	3,875	800
	0050	6' x 4', C/P		2	13		2,525	219		2,744	3,150	
	0400	6' x 5', C/C		2	13		3,200	219		3,419	3,900	
	0450	6' x 5', C/P		2	13		2,875	219		3,094	3,525	
	0600	6' x 6', C/C		1.80	14.444		3,750	244		3,994	4,550	
	0650	6' x 6', C/P		1.80	14.444		3,000	244		3,244	3,725	
	0800	6' x 9', C/C		1.60	16.250		4,675	274		4,949	5,625	
	0850	6' x 9', C/P		1.60	16.250		4,675	274		4,949	5,625	
	1000	8' x 12', C/C		1.10	23.636		6,300	400		6,700	7,600	
	1050	8' x 12', C/P		1.10	23.636		5,275	400		5,675	6,475	
	1400	8' x 10', C/C		1.20	21.667		5,400	365		5,765	6,575	
	1450	8' x 10', C/P		1.20	21.667		4,725	365		5,090	5,825	
	1600	10' x 12', C/C		1	26		7,200	440		7,640	8,675	
	1650	10' x 12', C/P		1	26		6,375	440		6,815	7,775	
	1700	Door only, cedar, 2'x6', with tempered insulated glass window	2 Carp	3.40	4.706		275	92.50		367.50	465	
	1800	Prehung, incl. jambs, pulls & hardware	"	12	1.333		325	26.50		351.50	405	
	2500	Heaters only (incl. above), wall mounted, to 200 C.F.					440			440	485	
	2750	To 300 C.F.					465			465	515	
	3000	Floor standing, to 720 C.F., 10,000 watts, w/controls	1 Elec	3	2.667		890	59		949	1,075	
	3250	To 1,000 C.F., 16,000 watts	"	3	2.667		1,575	59		1,634	1,825	
940	0010	**STEAM BATH** Heater, timer & head, single, to 140 C.F.	1 Plum	1.20	6.667	Ea.	740	146		886	1,050	940
	0500	To 300 C.F.	"	1.10	7.273		800	160		960	1,150	
	2700	Conversion unit for residential tub, including door					2,225			2,225	2,450	

13100 | Lightning Protection

13101 | Lightning Protection

			CREW	DAILY OUTPUT	LABOR-HOURS	UNIT	MAT.	LABOR	EQUIP.	TOTAL	TOTAL INCL O&P	
055	0010	**LIGHTNING PROTECTION**										055
	0200	Air terminals & base, copper										
	0400	3/8" diameter x 10" (to 75' high)	1 Elec	8	1	Ea.	21.50	22		43.50	59.50	
	0500	1/2" diameter x 12" (over 75' high)		8	1		26	22		48	64.50	
	1000	Aluminum, 1/2" diameter x 12" (to 75' high)		8	1		17	22		39	54.50	
	1100	5/8" diameter x 12" (over 75' high)		8	1		22	22		44	60	
	2000	Cable, copper, 220 lb. per thousand ft. (to 75' high)		320	.025	L.F.	1.07	.55		1.62	2.08	
	2100	375 lb. per thousand ft. (over 75' high)		230	.035		1.69	.77		2.46	3.12	
	2500	Aluminum, 101 lb. per thousand ft. (to 75' high)		280	.029		.46	.63		1.09	1.54	
	2600	199 lb. per thousand ft. (over 75' high)		240	.033		.75	.74		1.49	2.04	
	3000	Arrester, 175 volt AC to ground		8	1	Ea.	36	22		58	75.50	
	3100	650 volt AC to ground		6.70	1.194	"	70	26.50		96.50	120	

13120 | Pre-Engineered Structures

13128 | Pre-Eng. Structures

			CREW	DAILY OUTPUT	LABOR-HOURS	UNIT	2000 BARE COSTS				TOTAL INCL O&P
							MAT.	LABOR	EQUIP.	TOTAL	
540	0010	**GREENHOUSE** Shell only, stock units, not incl. 2' stub walls,										540
	0020	foundation, floors, heat or compartments										
	0300	Residential type, free standing, 8'-6" long x 7'-6" wide	2 Carp	59	.271	SF Flr.	35	5.35		40.35	47.50	
	0400	10'-6" wide		85	.188		27	3.71		30.71	36	
	0600	13'-6" wide		108	.148		24	2.92		26.92	31.50	
	0700	17'-0" wide		160	.100		27	1.97		28.97	33	
	0900	Lean-to type, 3'-10" wide		34	.471		31	9.25		40.25	50	
	1000	6'-10" wide		58	.276		24	5.45		29.45	36	
	1100	Wall mounted, to existing window, 3' x 3'	1 Carp	4	2	Ea.	335	39.50		374.50	440	
	1120	4' x 5'	"	3	2.667	"	500	52.50		552.50	640	
	1200	Deluxe quality, free standing, 7'-6" wide	2 Carp	55	.291	SF Flr.	69	5.75		74.75	86	
	1220	10'-6" wide		81	.198		64	3.89		67.89	77	
	1240	13'-6" wide		104	.154		60	3.03		63.03	71	
	1260	17'-0" wide		150	.107		51	2.10		53.10	59.50	
	1400	Lean-to type, 3'-10" wide		31	.516		80	10.15		90.15	105	
	1420	6'-10" wide		55	.291		75	5.75		80.75	92.50	
	1440	8'-0" wide		97	.165		70	3.25		73.25	82.50	
	1500	Commercial, custom, truss frame, incl. equip., plumbing, elec.,										
	1510	benches and controls, under 2,000 S.F., minimum				SF Flr.					30	
	1550	Maximum									38	
	1700	Over 5,000 S.F., minimum									23	
	1750	Maximum									30	
880	0010	**SWIMMING POOL ENCLOSURE** Translucent, free standing,										880
	0020	not including foundations, heat or light										
	0200	Economy, minimum	2 Carp	200	.080	SF Hor.	10	1.58		11.58	13.70	
	0300	Maximum		100	.160		21	3.15		24.15	28.50	
	0400	Deluxe, minimum		100	.160		23	3.15		26.15	31	
	0600	Maximum		70	.229		250	4.50		254.50	283	

13150 | Swimming Pools

13151 | Swimming Pools

			CREW	DAILY OUTPUT	LABOR-HOURS	UNIT	2000 BARE COSTS				TOTAL INCL O&P
							MAT.	LABOR	EQUIP.	TOTAL	
200	0010	**SWIMMING POOLS** Residential in-ground, vinyl lined, concrete sides										200
	0020	Sides including equipment, sand bottom	B-52	300	.187	SF Surf	10.40	3.05	1.31	14.76	18.15	
	0100	Metal or polystyrene sides	B-14	410	.117		8.70	1.82	.50	11.02	13.20	
	0200	Add for vermiculite bottom	R13128-520				.67			.67	.74	
	0500	Gunite bottom and sides, white plaster finish										
	0600	12' x 30' pool	B-52	145	.386	SF Surf	17.25	6.30	2.70	26.25	33	
	0720	16' x 32' pool		155	.361		15.50	5.90	2.53	23.93	30	
	0750	20' x 40' pool		250	.224		13.85	3.66	1.57	19.08	23.50	
	0810	Concrete bottom and sides, tile finish										
	0820	12' x 30' pool	B-52	80	.700	SF Surf	17.40	11.45	4.90	33.75	44	
	0830	16' x 32' pool		95	.589		14.40	9.65	4.13	28.18	37	
	0840	20' x 40' pool		130	.431		11.45	7.05	3.02	21.52	28	
	1600	For water heating system, see division 15510-880										
	1700	Filtration and deck equipment only, as % of total				Total				20%	20%	
	1800	Deck equipment, rule of thumb, 20' x 40' pool				SF Pool					1.30	
	3000	Painting pools, preparation + 3 coats, 20' x 40' pool, epoxy	2 Pord	.33	48.485	Total	595	870		1,465	2,100	
	3100	Rubber base paint, 18 gallons	"	.33	48.485	"	450	870		1,320	1,950	

13150 | Swimming Pools

13151 | Swimming Pools

		CREW	DAILY OUTPUT	LABOR-HOURS	UNIT	2000 BARE COSTS MAT.	LABOR	EQUIP.	TOTAL	TOTAL INCL O&P
0010	**SWIMMING POOL EQUIPMENT** Diving stand, stainless steel, 3 meter	2 Carp	.40	40	Ea.	4,725	790		5,515	6,525
0600	Diving boards, 16' long, aluminum		2.70	5.926		2,100	117		2,217	2,525
0700	Fiberglass		2.70	5.926		1,575	117		1,692	1,925
0900	Filter system, sand or diatomite type, incl. pump, 6,000 gal./hr.	2 Plum	1.80	8.889	Total	900	195		1,095	1,325
1020	Add for chlorination system, 800 S.F. pool	"	3	5.333	Ea.	196	117		313	410
1200	Ladders, heavy duty, stainless steel, 2 tread	2 Carp	7	2.286		790	45		835	940
1500	4 tread	"	6	2.667		1,350	52.50		1,402.50	1,600
2100	Lights, underwater, 12 volt, with transformer, 300 watt	1 Elec	.40	20		150	440		590	890
2200	110 volt, 500 watt, standard	"	.40	20		148	440		588	890
3000	Pool covers, reinforced vinyl	3 Clab	1,800	.013	S.F.	.35	.19		.54	.72
3100	Vinyl water tube		1,800	.013		.25	.19		.44	.61
3200	Maximum		1,600	.015		.45	.22		.67	.87
3300	Slides, tubular, fiberglass, aluminum handrails & ladder, 5'-0", straight	2 Carp	1.60	10	Ea.	1,025	197		1,222	1,475
3320	8'-0", curved	"	3	5.333	"	11,000	105		11,105	12,300

13200 | Storage Tanks

13201 | Storage Tanks

		CREW	DAILY OUTPUT	LABOR-HOURS	UNIT	2000 BARE COSTS MAT.	LABOR	EQUIP.	TOTAL	TOTAL INCL O&P
0010	**UNDERGROUND STORAGE TANKS**									
0210	Fiberglass, underground, single wall, U.L. listed, not including									
0220	manway or hold-down strap									
0230	1,000 gallon capacity	Q-5	2.46	6.504	Ea.	1,825	129		1,954	2,225
0250	4,000 gallon capacity	Q-7	3.55	9.014		3,425	179		3,604	4,075
0270	8,000 gallon capacity		2.29	13.974		4,800	278		5,078	5,725
0280	10,000 gallon capacity		2	16		5,200	320		5,520	6,250
0500	For manway, fittings and hold-downs, add					20%	15%			
1150	For hold-downs 500-4000 gal, add	Q-7	16	2	Set	160	40		200	242
1160	For hold-downs 5000-15000 gal, add		8	4		320	79.50		399.50	480
1170	For hold-downs 20,000 gal, add		5.33	6.004		480	119		599	725
1180	For hold-downs 25,000 gal, add		4	8		640	159		799	965
1190	For hold-downs 30,000 gal, add		2.60	12.308		960	245		1,205	1,450
2210	Fiberglass, underground, single wall, U.L. listed, including									
2220	hold-down straps, no manways									
2230	1,000 gallon capacity	Q-5	1.88	8.511	Ea.	1,975	169		2,144	2,450
2250	4,000 gallon capacity	Q-7	2.90	11.034		3,575	220		3,795	4,325
2270	8,000 gallon capacity		1.78	17.978		5,125	360		5,485	6,225
2280	10,000 gallon capacity		1.60	20		5,525	400		5,925	6,725
5000	Steel underground, sti-P3, set in place, not incl. hold-down bars.									
5500	Excavation, pad, pumps and piping not included									
5510	Single wall, 500 gallon capacity, 7 gauge shell	Q-5	2.70	5.926	Ea.	820	118		938	1,100
5520	1,000 gallon capacity, 7 gauge shell	"	2.50	6.400		1,350	127		1,477	1,700
5530	2,000 gallon capacity, 1/4" thick shell	Q-7	4.60	6.957		2,175	138		2,313	2,625
5535	2,500 gallon capacity, 7 gauge shell	Q-5	3	5.333		2,375	106		2,481	2,800
5540	5,000 gallon capacity, 1/4" thick shell	Q-7	3.20	10		4,450	199		4,649	5,225
5560	10,000 gallon capacity, 1/4" thick shell		2	16		7,175	320		7,495	8,425
5610	25,000 gallon capacity, 3/8" thick shell		1.30	24.615		17,600	490		18,090	20,200
5630	40,000 gallon capacity, 3/8" thick shell		.90	35.556		30,700	710		31,410	34,900
5640	50,000 gallon capacity, 3/8" thick shell		.80	40		38,300	795		39,095	43,500
6200	Steel, underground, 360°, double wall, U.L. listed,									
6210	with sti-P3 corrosion protection,									

13200 | Storage Tanks

13201	Storage Tanks	CREW	DAILY OUTPUT	LABOR-HOURS	UNIT	2000 BARE COSTS MAT.	LABOR	EQUIP.	TOTAL	TOTAL INCL O&P
6220	(dielectric coating, cathodic protection, electrical									
6230	isolation) 30 year warranty,									
6240	not incl. manholes or hold-downs.									
6260	1,000 gallon capactiy	Q-5	2.25	7.111	Ea.	3,175	142		3,317	3,725
6280	3,000 gallon capacity	Q-7	3.90	8.205		5,250	163		5,413	6,050
6300	5,000 gallon capacity		2.91	10.997		7,700	219		7,919	8,825
6310	6,000 gallon capacity		2.42	13.223		9,150	263		9,413	10,500
6330	10,000 gallon capacity		1.82	17.582		10,700	350		11,050	12,400
6400	For hold-downs 500-2000 gal, add		16	2	Set	262	40		302	355
6410	For hold-downs 3000-6000 gal, add		12	2.667		325	53		378	450
6420	For hold-downs 8000-12,000 gal, add		11	2.909		610	58		668	765

13280 | Hazardous Material Remediation

13281	Hazardous Material Remediation	CREW	DAILY OUTPUT	LABOR-HOURS	UNIT	2000 BARE COSTS MAT.	LABOR	EQUIP.	TOTAL	TOTAL INCL O&P
0010	**BULK ASBESTOS REMOVAL**									
0020	Includes disposable tools and 1 suit and respirator/day/worker									
0100	Beams, W 10 x 19	A-9	235	.272	L.F.		5.60		5.60	9.80
0110	W 12 x 22		210	.305			6.25		6.25	10.95
0120	W 14 x 26		180	.356			7.30		7.30	12.80
0130	W 16 x 31		160	.400			8.20		8.20	14.40
0140	W 18 x 40		140	.457			9.40		9.40	16.45
0150	W 24 x 55		110	.582			11.95		11.95	21
0160	W 30 x 108		85	.753			15.50		15.50	27
0170	W 36 x 150		72	.889			18.30		18.30	32
0200	Boiler insulation		480	.133	S.F.	.06	2.74		2.80	4.87
0210	With metal lath add				%				50%	
0300	Boiler breeching or flue insulation	A-9	520	.123	S.F.		2.53		2.53	4.43
0310	For active boiler, add				%				100%	
0400	Duct or AHU insulation	A-10B	440	.073	S.F.		1.50		1.50	2.63
0500	Duct vibration isolation joints, up to 24 Sq. In. duct	A-9	56	1.143	Ea.		23.50		23.50	41
0520	25 Sq. In. to 48 Sq. In. duct		48	1.333			27.50		27.50	48
0530	49 Sq. In. to 76 Sq. In. duct		40	1.600			33		33	57.50
0600	Pipe insulation, air cell type, up to 4" diameter pipe		900	.071	L.F.		1.46		1.46	2.56
0610	4" to 8" diameter pipe		800	.080			1.64		1.64	2.88
0620	10" to 12" diameter pipe		700	.091			1.88		1.88	3.29
0630	14" to 16" diameter pipe		550	.116			2.39		2.39	4.19
0650	Over 16" diameter pipe		650	.098	S.F.		2.02		2.02	3.55
0700	With glove bag up to 3" diameter pipe		100	.640	L.F.	3.15	13.15		16.30	26.50
1000	Pipe fitting insulation up to 4" diameter pipe		320	.200	Ea.		4.11		4.11	7.20
1100	6" to 8" diameter pipe		304	.211			4.33		4.33	7.60
1110	10" to 12" diameter pipe		192	.333			6.85		6.85	12
1120	14" to 16" diameter pipe		128	.500			10.30		10.30	18
1130	Over 16" diameter pipe		176	.364	S.F.		7.50		7.50	13.10
1200	With glove bag, up to 8" diameter pipe		40	1.600	Ea.	6.55	33		39.55	64.50
2000	Scrape foam fireproofing from flat surface		2,400	.027	S.F.		.55		.55	.96
2100	Irregular surfaces		1,200	.053			1.10		1.10	1.92
3000	Remove cementitious material from flat surface		1,800	.036			.73		.73	1.28
3100	Irregular surface		1,400	.046			.94		.94	1.65

13280 | Hazardous Material Remediation

13281 | Hazardous Material Remediation

			CREW	DAILY OUTPUT	LABOR-HOURS	UNIT	2000 BARE COSTS MAT.	LABOR	EQUIP.	TOTAL	TOTAL INCL O&P
120	4000	Scrape acoustical coating/fireproofing, from ceiling	A-9	3,200	.020	S.F.		.41		.41	.72
	5000	Remove VAT from floor by hand	↓	2,400	.027			.55		.55	.96
	5100	By machine	A-11	4,800	.013	↓		.27	.01	.28	.49
	5150	For 2 layers, add				%				50%	
	6000	Remove contaminated soil from crawl space by hand	A-9	400	.160	C.F.		3.29		3.29	5.75
	6100	With large production vacuum loader	A-12	700	.091	"		1.88	.76	2.64	4.12
	7000	Radiator backing, not including radiator removal	A-9	1,200	.053	S.F.		1.10		1.10	1.92
	8000	Cement-asbestos transite board	2 Asbe	2,000	.008		.08	.16		.24	.38
	8100	Transite shingle siding	A-10D	750	.043		.08	.84	.93	1.85	2.54
	8200	Shingle roofing	A-10B	2,000	.016		.07	.33		.40	.66
	8250	Built-up, no gravel, non-friable	B-2	1,400	.029		.07	.42		.49	.81
	8300	Asbestos millboard	2 Asbe	1,000	.016	↓	.08	.33		.41	.66
	9000	For type C (supplied air) respirator equipment, add				%					10%
125	0010	**ASBESTOS ABATEMENT WORK AREA** Containment and preparation.									
	0100	Pre-cleaning, HEPA vacuum and wet wipe, flat surfaces	A-10	12,000	.005	S.F.		.11		.11	.19
	0200	Protect carpeted area, 2 layers 6 mil poly on 3/4" plywood	"	1,000	.064		1.50	1.32		2.82	3.95
	0300	Separation barrier, 2" x 4" @ 16", 1/2" plywood ea. side, 8' high	2 Carp	400	.040		1.25	.79		2.04	2.73
	0310	12' high		320	.050		1.40	.99		2.39	3.23
	0320	16' high		200	.080		1.50	1.58		3.08	4.35
	0400	Personnel decontam. chamber, 2" x 4" @ 16", 3/4" ply ea. side		280	.057		2.50	1.13		3.63	4.68
	0450	Waste decontam. chamber, 2" x 4" studs @ 16", 3/4" ply ea. side	↓	360	.044	↓	3	.88		3.88	4.80
	0500	Cover surfaces with polyethelene sheeting									
	0501	Including glue and tape									
	0550	Floors, each layer, 6 mil	A-10	8,000	.008	S.F.	.10	.16		.26	.40
	0551	4 mil		9,000	.007		.06	.15		.21	.33
	0560	Walls, each layer, 6 mil		6,000	.011		.09	.22		.31	.48
	0561	4 mil	↓	7,000	.009	↓	.07	.19		.26	.41
	0570	For heights above 12', add						20%			
	0575	For heights above 20', add						30%			
	0580	For fire retardant poly, add					100%				
	0590	For large open areas, deduct					10%	20%			
	0600	Seal floor penetrations with foam firestop to 36 Sq. In.	2 Carp	200	.080	Ea.	6.25	1.58		7.83	9.60
	0610	36 Sq. In. to 72 Sq. In.		125	.128		12.50	2.52		15.02	18.05
	0615	72 Sq. In. to 144 Sq. In.		80	.200		25	3.94		28.94	34.50
	0620	Wall penetrations, to 36 square inches		180	.089		6.25	1.75		8	9.90
	0630	36 Sq. In. to 72 Sq. In.		100	.160		12.50	3.15		15.65	19.15
	0640	72 Sq. In. to 144 Sq. In.	↓	60	.267	↓	25	5.25		30.25	36.50
	0800	Caulk seams with latex	1 Carp	230	.035	L.F.	.15	.69		.84	1.34
	0900	Set up neg. air machine, 1-2k C.F.M. /25 M.C.F. volume	1 Asbe	4.30	1.860	Ea.		38		38	67
130	0010	**DEMOLITION IN ASBESTOS CONTAMINATED AREA**									
	0200	Ceiling, including suspension system, plaster and lath	A-9	2,100	.030	S.F.		.63		.63	1.10
	0210	Finished plaster, leaving wire lath		585	.109			2.25		2.25	3.94
	0220	Suspended acoustical tile		3,500	.018			.38		.38	.66
	0230	Concealed tile grid system		3,000	.021			.44		.44	.77
	0240	Metal pan grid system		1,500	.043			.88		.88	1.54
	0250	Gypsum board		2,500	.026	↓		.53		.53	.92
	0260	Lighting fixtures up to 2' x 4'		72	.889	Ea.		18.30		18.30	32
	0400	Partitions, non load bearing									
	0410	Plaster, lath, and studs	A-9	690	.093	S.F.	.55	1.91		2.46	3.95
	0450	Gypsum board and studs	"	1,390	.046	"		.95		.95	1.66
	9000	For type C (supplied air) respirator equipment, add				%					10%
135	0010	**ASBESTOS ABATEMENT EQUIPMENT** and supplies, buy	R13281-120								
	0200	Air filtration device, 2000 C.F.M.				Ea.	2,500			2,500	2,750
	0250	Large volume air sampling pump, minimum					450			450	495
	0260	Maximum	↓				1,000			1,000	1,100

Important: See the Reference Section for critical supporting data - Reference Nos., Crews, & Location Factors

13280 | Hazardous Material Remediation

13281 | Hazardous Material Remediation

		CREW	DAILY OUTPUT	LABOR-HOURS	UNIT	2000 BARE COSTS MAT.	LABOR	EQUIP.	TOTAL	TOTAL INCL O&P	
135	0300	Airless sprayer unit, 2 gun				Ea.	2,000			2,000	2,200	135
	0350	Light stand, 500 watt	R13281-120				250			250	275	
	0400	Personal respirators										
	0410	Negative pressure, 1/2 face, dual operation, min.				Ea.	22.50			22.50	25	
	0420	Maximum					24			24	26.50	
	0450	P.A.P.R., full face, minimum					400			400	440	
	0460	Maximum					700			700	770	
	0470	Supplied air, full face, incl. air line, minimum					450			450	495	
	0480	Maximum					600			600	660	
	0500	Personnel sampling pump, minimum					450			450	495	
	0510	Maximum					750			750	825	
	1500	Power panel, 20 unit, incl. G.F.I.					1,800			1,800	1,975	
	1600	Shower unit, including pump and filters					1,125			1,125	1,250	
	1700	Supplied air system (type C)					10,000			10,000	11,000	
	1750	Vacuum cleaner, HEPA, 16 gal., stainless steel, wet/dry					1,000			1,000	1,100	
	1760	55 gallon					2,200			2,200	2,425	
	1800	Vacuum loader, 9-18 ton/hr					90,000			90,000	99,000	
	1900	Water atomizer unit, including 55 gal. drum					230			230	253	
	2000	Worker protection, whole body, foot, head cover & gloves, plastic					30			30	33	
	2500	Respirator, single use					10.50			10.50	11.55	
	2550	Cartridge for respirator					37.50			37.50	41.50	
	2570	Glove bag, 7 mil, 50" x 64"					8.50			8.50	9.35	
	2580	10 mil, 44" x 60"					8.40			8.40	9.25	
	3000	HEPA vacuum for work area, minimun					1,000			1,000	1,100	
	3050	Maximum					2,775			2,775	3,050	
	6000	Disposable polyethelene bags, 6 mil, 3 C.F.					1.15			1.15	1.27	
	6300	Disposable fiber drums, 3 C.F.					6.50			6.50	7.15	
	6400	Pressure sensitive caution lables, 3" x 5"					8.90			8.90	9.80	
	6450	11" x 17"					6.50			6.50	7.15	
	6500	Negative air machine, 1800 C.F.M.					775			775	855	
140	0010	**DECONTAMINATION CONTAINMENT AREA DEMOLITION** and clean-up										140
	0100	Spray exposed substrate with surfactant (bridging)										
	0200	Flat surfaces	A-9	6,000	.011	S.F.	.35	.22		.57	.77	
	0250	Irregular surfaces		4,000	.016	"	.30	.33		.63	.91	
	0300	Pipes, beams, and columns		2,000	.032	L.F.	.55	.66		1.21	1.76	
	1000	Spray encapsulate polyethelene sheeting		8,000	.008	S.F.	.30	.16		.46	.62	
	1100	Roll down polyethelene sheeting		8,000	.008	"		.16		.16	.29	
	1500	Bag polyethelene sheeting		400	.160	Ea.	.75	3.29		4.04	6.60	
	2000	Fine clean exposed substrate, with nylon brush		2,400	.027	S.F.		.55		.55	.96	
	2500	Wet wipe substrate		4,800	.013			.27		.27	.48	
	2600	Vacuum surfaces, fine brush		6,400	.010			.21		.21	.36	
	3000	Structural demolition										
	3100	Wood stud walls	A-9	2,800	.023	S.F.		.47		.47	.82	
	3500	Window manifolds, not incl. window replacement		4,200	.015			.31		.31	.55	
	3600	Plywood carpet protection		2,000	.032			.66		.66	1.15	
	4000	Remove custom decontamination facility	A-10A	8	3	Ea.	15	62		77	126	
	4100	Remove portable decontamination facility	3 Asbe	12	2	"	12.50	41		53.50	86	
	5000	HEPA vacuum, shampoo carpeting	A-9	4,800	.013	S.F.	.05	.27		.32	.54	
	9000	Final cleaning of protected surfaces	A-10A	8,000	.003	"		.06		.06	.11	
145	0010	**OSHA TESTING**										145
	0100	Certified technician, minimum				Day					300	
	0110	Maximum				"					500	
	0200	Personal sampling, PCM analysis, minimum	1 Asbe	8	1	Ea.	2.75	20.50		23.25	39	
	0210	Maximum	"	4	2	"	3	41		44	75.50	
	0300	Industrial hygenist, minimum				Day					400	

13280 | Hazardous Material Remediation

13281 | Hazardous Material Remediation

			CREW	DAILY OUTPUT	LABOR-HOURS	UNIT	MAT.	LABOR	EQUIP.	TOTAL	TOTAL INCL O&P	
145	0310	Maximum				Day					550	145
	1000	Cleaned area samples	1 Asbe	8	1	Ea.	2.63	20.50		23.13	39	
	1100	PCM air sample analysis, minimum		8	1		30	20.50		50.50	69	
	1110	Maximum		4	2		3.09	41		44.09	75.50	
	1200	TEM air sample analysis, minimum								100	125	
	1210	Maximum								400	500	
150	0010	**ENCAPSULATION WITH SEALANTS**										150
	0100	Ceilings and walls, minimum	A-9	21,000	.003	S.F.	.25	.06		.31	.39	
	0110	Maximum		10,600	.006		.40	.12		.52	.66	
	0200	Columns and beams, minimum		13,300	.005		.25	.10		.35	.45	
	0210	Maximum		5,325	.012		.45	.25		.70	.93	
	0300	Pipes to 12" diameter including minor repairs, minimum		800	.080	L.F.	.35	1.64		1.99	3.27	
	0310	Maximum		400	.160	"	1	3.29		4.29	6.85	
155	0010	**WASTE PACKAGING, HANDLING, & DISPOSAL**										155
	0100	Collect and bag bulk material, 3 C.F. bags, by hand	A-9	400	.160	Ea.	1.15	3.29		4.44	7	
	0200	Large production vacuum loader	A-12	880	.073		.80	1.50	.60	2.90	4.16	
	1000	Double bag and decontaminate	A-9	960	.067		2.30	1.37		3.67	4.93	
	2000	Containerize bagged material in drums, per 3 C.F. drum	"	800	.080		6.50	1.64		8.14	10.05	
	3000	Cart bags 50' to dumpster	2 Asbe	400	.040			.82		.82	1.44	
	5000	Disposal charges, not including haul, minimum				C.Y.					50	
	5020	Maximum				"					175	
	5100	Remove refrigerant from system	1 Plum	40	.200	Lb.		4.39		4.39	7.25	
	9000	For type C (supplied air) respirator equipment, add				%					10%	
440	0010	**REMOVAL** Existing lead paint, by chemicals, per application	R13281-460									440
	0020	See also, Div. 13280, Haz. Mat'l. Abatement										
	0050	Baseboard, to 6" wide	1 Pord	64	.125	L.F.	1.42	2.24		3.66	5.30	
	0070	To 12" wide		32	.250	"	2.78	4.49		7.27	10.55	
	0200	Balustrades, one side		28	.286	S.F.	3.16	5.15		8.31	12.05	
	1400	Cabinets, simple design		32	.250		2.76	4.49		7.25	10.55	
	1420	Ornate design		25	.320		3.55	5.75		9.30	13.50	
	1600	Cornice, simple design		60	.133		1.48	2.39		3.87	5.60	
	1620	Ornate design		20	.400		4.37	7.20		11.57	16.80	
	2800	Doors, one side, flush		84	.095		1.07	1.71		2.78	4.03	
	2820	Two panel		80	.100		1.11	1.80		2.91	4.22	
	2840	Four panel		45	.178		1.96	3.19		5.15	7.45	
	2880	For trim, one side, add		64	.125	L.F.	1.42	2.24		3.66	5.30	
	3000	Fence, picket, one side		30	.267	S.F.	2.97	4.79		7.76	11.25	
	3200	Grilles, one side, simple design		30	.267		2.97	4.79		7.76	11.25	
	3220	Ornate design		25	.320		3.55	5.75		9.30	13.50	
	4400	Pipes, to 4" diameter		90	.089	L.F.	1.01	1.60		2.61	3.77	
	4420	To 8" diameter		50	.160		1.76	2.87		4.63	6.75	
	4440	To 12" diameter		36	.222		2.46	3.99		6.45	9.35	
	4460	To 16" diameter		20	.400		4.40	7.20		11.60	16.85	
	4500	For hangers, add		40	.200	Ea.	2.21	3.59		5.80	8.45	
	4800	Siding		90	.089	S.F.	1.01	1.60		2.61	3.77	
	5000	Trusses, open		55	.145	SF Face	1.61	2.61		4.22	6.15	
	6200	Windows, one side only, double hung, 1/1 light, 24" x 48" high		4	2	Ea.	22	36		58	84.50	
	6220	30" x 60" high		3	2.667		29.50	48		77.50	113	
	6240	36" x 72" high		2.50	3.200		35.50	57.50		93	135	
	6280	40" x 80" high		2	4		44.50	72		116.50	169	
	6400	Colonial window, 6/6 light, 24" x 48" high		2	4		44.50	72		116.50	169	
	6420	30" x 60" high		1.50	5.333		59.50	95.50		155	225	
	6440	36" x 72" high		1	8		89	144		233	340	
	6480	40" x 80" high		1	8		89	144		233	340	
	6600	8/8 light, 24" x 48" high		2	4		44.50	72		116.50	169	

Important: See the Reference Section for critical supporting data - Reference Nos., Crews, & Location Factors

13280 | Hazardous Material Remediation

13281	Hazardous Material Remediation	CREW	DAILY OUTPUT	LABOR-HOURS	UNIT	2000 BARE COSTS				TOTAL INCL O&P
						MAT.	LABOR	EQUIP.	TOTAL	
6620	40" x 80" high	1 Pord	1	8	Ea.	89	144		233	340
6800	12/12 light, 24" x 48" high	R13281-460	1	8		89	144		233	340
6820	40" x 80" high		.75	10.667		119	191		310	450
6840	Window frame & trim items, included in pricing above									
0010	**LEAD PAINT ENCAPSULATION**, water based polymer coating, 14 mil DFT									
0020	Interior, brushwork, trim, under 6"	1 Pord	240	.033	L.F.	2.09	.60		2.69	3.30
0030	6" to 12" wide		180	.044		2.77	.80		3.57	4.38
0040	Balustrades		300	.027		1.67	.48		2.15	2.64
0050	Pipe to 4" diameter		500	.016		1	.29		1.29	1.58
0060	To 8" diameter		375	.021		1.33	.38		1.71	2.10
0070	To 12" diameter		250	.032		2	.57		2.57	3.16
0080	To 16" diameter		170	.047		2.94	.84		3.78	4.64
0090	Cabinets, ornate design		200	.040	S.F.	2.50	.72		3.22	3.95
0100	Simple design		250	.032	"	2	.57		2.57	3.16
0110	Doors, 3' x 7', both sides, incl. frame & trim									
0120	Flush	1 Pord	6	1.333	Ea.	25.50	24		49.50	68
0130	French, 10-15 lite		3	2.667		5.10	48		53.10	85.50
0140	Panel		4	2		30.50	36		66.50	94
0150	Louvered		2.75	2.909		28	52		80	118
0160	Windows, per interior side, per 15 S.F.									
0170	1 to 6 lite	1 Pord	14	.571	Ea.	17.70	10.25		27.95	36.50
0180	7 to 10 lite		7.50	1.067		19.50	19.15		38.65	53.50
0190	12 lite		5.75	1.391		26	25		51	70.50
0200	Radiators		8	1		62.50	17.95		80.45	98.50
0210	Grilles, vents		275	.029	S.F.	1.82	.52		2.34	2.87
0220	Walls, roller, drywall or plaster		1,000	.008		.49	.14		.63	.78
0230	With spunbonded reinforcing fabric		720	.011		.56	.20		.76	.95
0240	Wood		800	.010		.62	.18		.80	.98
0250	Ceilings, roller, drywall or plaster		900	.009		.56	.16		.72	.89
0260	Wood		700	.011		.71	.21		.92	1.12
0270	Exterior, brushwork, gutters and downspouts		300	.027	L.F.	1.67	.48		2.15	2.64
0280	Columns		400	.020	S.F.	1.25	.36		1.61	1.98
0290	Spray, siding		600	.013	"	.83	.24		1.07	1.31
0300	Miscellaneous									
0310	Electrical conduit, brushwork, to 2" diameter	1 Pord	500	.016	L.F.	1	.29		1.29	1.58
0320	Brick, block or concrete, spray		500	.016	S.F.	1	.29		1.29	1.58
0330	Steel, flat surfaces and tanks to 12"		500	.016		1	.29		1.29	1.58
0340	Beams, brushwork		400	.020		1.25	.36		1.61	1.98
0350	Trusses		400	.020		1.25	.36		1.61	1.98

13600 | Solar and Wind Energy Equipment

13630	Solar Collector Components	CREW	DAILY OUTPUT	LABOR-HOURS	UNIT	2000 BARE COSTS				TOTAL INCL O&P
						MAT.	LABOR	EQUIP.	TOTAL	
0010	**SOLAR ENERGY**									
0020	System/Package prices, not including connecting									
0030	pipe, insulation, or special heating/plumbing fixtures									
0500	Hot water, standard package, low temperature									
0540	1 collector, circulator, fittings, 65 gal. tank	Q-1	.50	32	Ea.	1,350	630		1,980	2,550
0580	2 collectors, circulator, fittings, 120 gal. tank	"	.40	40	"	1,925	790		2,715	3,425

13600 | Solar and Wind Energy Equipment

13630 | Solar Collector Components

			CREW	DAILY OUTPUT	LABOR-HOURS	UNIT	MAT.	LABOR	EQUIP.	TOTAL	TOTAL INCL O&P	
200	0701	Solar energy, medium temperature package										200
	0720	1 collector, circulator, fittings, 80 gal. tank	Q-1	.50	32	Ea.	1,475	630		2,105	2,675	
	0740	2 collectors, circulator, fittings, 120 gal. tank	"	.40	40		2,075	790		2,865	3,600	
	0980	For each additional 120 gal. tank, add					720			720	795	

13800 | Building Automation & Control

13834 | Electric/Electronic Control

			CREW	DAILY OUTPUT	LABOR-HOURS	UNIT	MAT.	LABOR	EQUIP.	TOTAL	TOTAL INCL O&P	
200	0010	CONTROL SYSTEMS, ELECTRONIC										200
	0020	For electronic costs, add to division 13836-200				Ea.					15%	
	1000	Electronic hydronic controller	1 Plum	8	1	"	31	22		53	70.50	

13836 | Pneumatic Controls

			CREW	DAILY OUTPUT	LABOR-HOURS	UNIT	MAT.	LABOR	EQUIP.	TOTAL	TOTAL INCL O&P	
200	0010	CONTROL SYSTEMS, PNEUMATIC (Sub's quote incl. mat. & labor)										200
	0011	and nominal 50 Ft. of tubing. Add control panelboard if req'd.										
	0100	Heating and Ventilating, split system										
	0200	Mixed air control, economizer cycle, panel readout, tubing										
	0220	Up to 10 tons	Q-19	.68	35.294	Ea.	2,650	730		3,380	4,125	
	0240	For 10 to 20 tons		.63	37.915		2,875	780		3,655	4,450	
	0260	For over 20 tons		.58	41.096		3,100	850		3,950	4,825	
	0300	Heating coil, hot water, 3 way valve,										
	0320	Freezestat, limit control on discharge, readout	Q-5	.69	23.088	Ea.	2,000	460		2,460	2,950	
	0500	Cooling coil, chilled water, room										
	0520	Thermostat, 3 way valve	Q-5	2	8	Ea.	885	159		1,044	1,250	
	0600	Cooling tower, fan cycle, damper control,										
	0620	Control system including water readout in/out at panel	Q-19	.67	35.821	Ea.	3,525	740		4,265	5,100	
	1000	Unit ventilator, day/night operation,										
	1100	freezestat, ASHRAE, cycle 2	Q-19	.91	26.374	Ea.	1,950	545		2,495	3,050	
	3000	Boiler room combustion air, damper to 5 SF, controls		1.36	17.582		1,750	365		2,115	2,550	
	3500	Fan coil, heating and cooling valves, 4 pipe control system		3	8		795	165		960	1,150	
	3600	Heat exchanger system controls		.86	27.907		1,700	575		2,275	2,825	
	3900	Multizone control (one per zone), includes thermostat, damper										
	3910	motor and reset of discharge temperature	Q-5	.51	31.373	Ea.	1,775	625		2,400	2,975	
	4000	Pneumatic thermostat, including controlling room radiator valve	"	2.43	6.593		530	131		661	800	
	4060	Pump control system	Q-19	3	8		815	165		980	1,175	
	4500	Air supply for pneumatic control system										
	4600	Tank mounted duplex compressor, starter, alternator,										
	4620	piping, dryer, PRV station and filter										
	4640	3/4 HP	Q-19	.64	37.383	Ea.	6,875	770		7,645	8,825	
	4650	1 HP		.61	39.539		7,500	815		8,315	9,600	
	4680	3 HP		.55	43.956		10,900	905		11,805	13,500	
	4690	5 HP		.42	57.143		19,000	1,175		20,175	22,900	
	4800	Main air supply, includes 3/8" copper main and labor	Q-5	1.82	8.791	C.L.F.	220	175		395	530	
	4810	If poly tubing used, deduct									30%	
	8600	VAV boxes, incl. thermostat, damper motor, reheat coil & tubing	Q-5	1.46	10.989	Ea.	820	219		1,039	1,250	
	8610	If no reheat coil, deduct				"					204	

13800 | Building Automation & Control

13838 | Pneumatic/Electric Controls

		CREW	DAILY OUTPUT	LABOR-HOURS	UNIT	2000 BARE COSTS MAT.	LABOR	EQUIP.	TOTAL	TOTAL INCL O&P	
0010	**CONTROL COMPONENTS**										200
5000	Thermostats										
5030	Manual	1 Shee	8	1	Ea.	24.50	21.50		46	63.50	
5040	1 set back, electric, timed	"	8	1	"	79	21.50		100.50	124	

13850 | Detection & Alarm

13851 | Detection & Alarm

		CREW	DAILY OUTPUT	LABOR-HOURS	UNIT	2000 BARE COSTS MAT.	LABOR	EQUIP.	TOTAL	TOTAL INCL O&P	
0010	**DETECTION SYSTEMS**, not including wires & conduits										065
0100	Burglar alarm, battery operated, mechanical trigger	1 Elec	4	2	Ea.	249	44		293	345	
0200	Electrical trigger		4	2		297	44		341	400	
0400	For outside key control, add		8	1		70.50	22		92.50	114	
0600	For remote signaling circuitry, add		8	1		112	22		134	159	
0800	Card reader, flush type, standard		2.70	2.963		835	65.50		900.50	1,025	
1000	Multi-code		2.70	2.963		1,075	65.50		1,140.50	1,275	
1200	Door switches, hinge switch		5.30	1.509		52.50	33.50		86	113	
1400	Magnetic switch		5.30	1.509		62	33.50		95.50	123	
2800	Ultrasonic motion detector, 12 volt		2.30	3.478		206	77		283	355	
3000	Infrared photoelectric detector		2.30	3.478		170	77		247	315	
3200	Passive infrared detector		2.30	3.478		254	77		331	405	
3420	Switchmats, 30" x 5'		5.30	1.509		76	33.50		109.50	138	
3440	30" x 25'		4	2		182	44		226	273	
3460	Police connect panel		4	2		219	44		263	315	
3480	Telephone dialer		5.30	1.509		345	33.50		378.50	435	
3500	Alarm bell		4	2		69.50	44		113.50	149	
3520	Siren		4	2		131	44		175	217	
5200	Smoke detector, ceiling type		6.20	1.290		75	28.50		103.50	129	
5600	Strobe and horn		5.30	1.509		95	33.50		128.50	160	
5800	Fire alarm horn		6.70	1.194		36.50	26.50		63	83	
6600	Drill switch		8	1		86.50	22		108.50	131	
6800	Master box		2.70	2.963		3,100	65.50		3,165.50	3,500	
7800	Remote annunciator, 8 zone lamp		1.80	4.444		175	98		273	355	
8000	12 zone lamp	2 Elec	2.60	6.154		300	136		436	550	
8200	16 zone lamp	"	2.20	7.273		300	161		461	595	
8400	Standpipe or sprinkler alarm, alarm device	1 Elec	8	1		125	22		147	174	
8600	Actuating device	"	8	1		290	22		312	355	

13900 | Fire Suppression

13910 | Basic Fire Protection Matl/Methd

		CREW	DAILY OUTPUT	LABOR-HOURS	UNIT	2000 BARE COSTS MAT.	LABOR	EQUIP.	TOTAL	TOTAL INCL O&P	
0010	**FIRE HOSE AND EQUIPMENT**										400
0200	Adapters, rough brass, straight hose threads										
2200	Hose, less couplings										
2260	Synthetic jacket, lined, 300 lb. test, 1-1/2" diameter	Q-12	2,600	.006	L.F.	1.33	.12		1.45	1.66	

SPECIAL CONSTRUCTION 13

13900 | Fire Suppression

13910 | Basic Fire Protection Matl/Methd

			CREW	DAILY OUTPUT	LABOR-HOURS	UNIT	2000 BARE COSTS MAT.	LABOR	EQUIP.	TOTAL	TOTAL INCL O&P	
400	2280	2-1/2" diameter	Q-12	2,200	.007	L.F.	2.17	.14		2.31	2.63	400
	2360	High strength, 500 lb. test, 1-1/2" diameter		2,600	.006		1.59	.12		1.71	1.95	
	2380	2-1/2" diameter	▼	2,200	.007	▼	2.75	.14		2.89	3.27	
	2600	Hose rack, swinging, for 1-1/2" diameter hose,										
	2620	Enameled steel, 50' & 75' lengths of hose	Q-12	20	.800	Ea.	32.50	15.90		48.40	62	
	2640	100' and 125' lengths of hose		20	.800		32.50	15.90		48.40	62	
	2680	Chrome plated, 50' and 75' lengths of hose		20	.800		50	15.90		65.90	81.50	
	2700	100' and 125' lengths of hose	▼	20	.800	▼	50	15.90		65.90	81.50	
	2990	Hose reel, swinging, for 1-1/2" polyester neoprene lined hose										
	3000	50' long	Q-12	14	1.143	Ea.	58	22.50		80.50	102	
	3020	100' long		14	1.143		79	22.50		101.50	125	
	3060	For 2-1/2" cotton rubber hose, 75' long		14	1.143		94	22.50		116.50	142	
	3100	150' long	▼	14	1.143	▼	107	22.50		129.50	156	
	3750	Hydrants, wall, w/caps, single, flush, polished brass										
	3800	2-1/2" x 2-1/2"	Q-12	5	3.200	Ea.	91.50	63.50		155	205	
	3840	2-1/2" x 3"		5	3.200		122	63.50		185.50	239	
	3860	3" x 3"	▼	4.80	3.333	▼	149	66		215	274	
	5600	Nozzles, brass										
	5620	Adjustable fog, 3/4" booster line				Ea.	69			69	76	
	5630	1" booster line					80.50			80.50	88.50	
	5640	1-1/2" leader line					84.50			84.50	93	
	5660	2-1/2" direct connection					165			165	182	
	5680	2-1/2" playpipe nozzle				▼	136			136	150	
	5780	For chrome plated, add					8%					
	5850	Electrical fire, adjustable fog, no shock										
	5900	1-1/2"				Ea.	263			263	289	
	5920	2-1/2"					360			360	395	
	5980	For polished chrome, add				▼	6%					
	6200	Heavy duty, comb. adj. fog and str. stream, with handle										
	6210	1" booster line				Ea.	251			251	277	
	6240	1-1/2"					248			248	273	
	6260	2-1/2", for playpipe					284			284	315	
	6280	2-1/2" direct connection					355			355	390	
	6300	2-1/2" playpipe combination				▼	510			510	560	
	7140	Standpipe connections, wall, w/plugs & chains										
	7160	Single, flush, brass, 2-1/2" x 2-1/2"	Q-12	5	3.200	Ea.	65	63.50		128.50	177	
	7180	2-1/2" x 3"	"	5	3.200	"	71	63.50		134.50	183	
	7240	For polished chrome, add					15%					
	7280	Double, flush, polished brass										
	7300	2-1/2" x 2-1/2" x 4"	Q-12	5	3.200	Ea.	219	63.50		282.50	345	
	8800	Sidewalk siamese unit, polished brass, two way										
	8820	2-1/2" x 2-1/2" x 4"	Q-12	2.50	6.400	Ea.	310	127		437	550	
800	0010	**FIRE VALVES**										800
	0020	Angle, combination pressure adjust/restricting, rough brass										
	0030	1-1/2"	1 Spri	12	.667	Ea.	36.50	14.70		51.20	64.50	
	0040	2-1/2"	"	7	1.143	"	77.50	25		102.50	127	
	0042	Nonpressure adjustable/restricting, rough brass										
	0044	1-1/2"	1 Spri	12	.667	Ea.	23.50	14.70		38.20	50.50	
	0046	2-1/2"	"	7	1.143	"	44	25		69	89.50	

13930 | Wet-Pipe Fire Supp. Sprinklers

			CREW	DAILY OUTPUT	LABOR-HOURS	UNIT	MAT.	LABOR	EQUIP.	TOTAL	TOTAL INCL O&P	
400	0010	**SPRINKLER SYSTEM COMPONENTS**										400
	0800	Air compressor for dry pipe system, automatic, complete										
	0820	280 gal. system capacity, 3/4 HP	1 Spri	1.30	6.154	Ea.	700	136		836	995	
	1100	Alarm, electric pressure switch (circuit closer)	"	26	.308	"	115	6.80		121.80	138	

13900 | Fire Suppression

13930 | Wet-Pipe Fire Supp. Sprinklers

		CREW	DAILY OUTPUT	LABOR-HOURS	UNIT	2000 BARE COSTS MAT.	LABOR	EQUIP.	TOTAL	TOTAL INCL O&P	
400	2600	Sprinkler heads, not including supply piping										400
	2640	Dry, pendent, 1/2" orifice, 3/4" or 1" NPT										
	2700	11" to 12-3/4" length	1 Spri	14	.571	Ea.	33.50	12.60		46.10	58	
	3700	Standard spray, pendent or upright, brass, 135° to 286°F										
	3730	1/2" NPT, 7/16" orifice	1 Spri	16	.500	Ea.	5.30	11		16.30	24	
	5600	Concealed, complete with cover plate										
	5620	1/2" NPT, 1/2" orifice, 135°F to 212°F	1 Spri	9	.889	Ea.	10.15	19.60		29.75	43.50	
	6200	Valves										
	6210	Alarm, includes										
	6220	retard chamber, trim, gauges, alarm line strainer										
	6260	3" size	Q-12	3	5.333	Ea.	690	106		796	935	
	6280	4" size	"	2	8	"	700	159		859	1,050	

13960 | CO2 Fire Extinguishing

		CREW	DAILY OUTPUT	LABOR-HOURS	UNIT	MAT.	LABOR	EQUIP.	TOTAL	TOTAL INCL O&P	
200	0010	**AUTOMATIC FIRE SUPPRESSION SYSTEMS**										200
	0040	For detectors and control stations, see division 13850										
	0100	Control panel, single zone with batteries (2 zones det., 1 suppr.)	1 Elec	1	8	Ea.	1,375	177		1,552	1,825	
	0150	Multizone (4) with batteries (8 zones det., 4 suppr.)	"	.50	16		2,450	355		2,805	3,275	
	1000	Dispersion nozzle, CO2, 3" x 5"	1 Plum	18	.444		50	9.75		59.75	71	
	1100	FM200, 1-1/2"	"	14	.571		80	12.55		92.55	109	
	2000	Extinguisher, CO2 system, high pressure, 75 lb. cylinder	Q-1	6	2.667		975	52.50		1,027.50	1,150	
	2400	FM200 system, filled, with mounting bracket										
	2460	26 lb. container	Q-1	8	2	Ea.	1,600	39.50		1,639.50	1,825	
	2500	63 lb. container		6	2.667		2,800	52.50		2,852.50	3,150	
	2520	101 lb. container		5	3.200		3,750	63		3,813	4,225	
	2540	196 lb. container		4	4		5,800	79		5,879	6,500	
	3000	Electro/mechanical release	L-1	4	2.500		115	55		170	218	
	3400	Manual pull station	1 Plum	6	1.333		45	29.50		74.50	98	
	4000	Pneumatic damper release	"	8	1		175	22		197	230	
	6000	Average FM200 system, minimum				C.F.	1.25			1.25	1.38	
	6020	Maximum				"	2.50			2.50	2.75	

For information about Means Estimating Seminars, see yellow pages 11 and 12 in back of book

Division Notes

		CREW	DAILY OUTPUT	LABOR-HOURS	UNIT	MAT.	LABOR	EQUIP.	TOTAL	TOTAL INCL O&P

Division 14
Conveying Systems

Estimating Tips
General
- Many products in Division 14 will require some type of support or blocking for installation not included with the item itself. Examples are supports for conveyors or tube systems, attachment points for lifts, and footings for hoists or cranes. Add these supports in the appropriate division.

14100 Dumbwaiters
14200 Elevators
- Dumbwaiters and elevators are estimated and purchased in a method similar to buying a car. The manufacturer has a base unit with standard features. Added to this base unit price will be whatever options the owner or specifications require. Increased load capacity, additional vertical travel, additional stops, higher speed, and cab finish options are items to be considered. When developing an estimate for dumbwaiters and elevators, remember that some items needed by the installers may have to be included as part of the general contract.

Examples are:
- shaftway
- rail support brackets
- machine room
- electrical supply
- sill angles
- electrical connections
- pits
- roof penthouses
- pit ladders

Check the job specifications and drawings before pricing.

14300 Escalators & Moving Walks
- Escalators and moving walks are specialty items installed by specialty contractors. There are numerous options associated with these items. For specific options contact a manufacturer or contractor. In a method similar to estimating dumbwaiters and elevators, you should verify the extent of general contract work and add items as necessary.

14400 Lifts
14500 Material Handling
14600 Hoists & Cranes
- Products such as correspondence lifts, conveyors, chutes, pneumatic tube systems, material handling cranes and hoists as well as other items specified in this subdivision may require trained installers. The general contractor might not have any choice as to who will perform the installation or when it will be performed. Long lead times are often required for these products, making early decisions in scheduling necessary.

Reference Numbers
Reference numbers are shown in bold squares at the beginning of some major classifications. These numbers refer to related items in the Reference Section. The reference information may be an estimating procedure, an alternate pricing method or technical information.

Note: Not all subdivisions listed here necessarily appear in this publication.

14100 | Dumbwaiters

14110	Manual Dumbwaiters	CREW	DAILY OUTPUT	LABOR-HOURS	UNIT	2000 BARE COSTS MAT.	LABOR	EQUIP.	TOTAL	TOTAL INCL O&P	
400	0010	**DUMBWAITERS** 2 stop, hand, minimum	2 Elev	.75	21.333	Ea.	2,225	490		2,715	3,250	400
	0100	Maximum		.50	32	"	5,050	735		5,785	6,750	
	0300	For each additional stop, add	▼	.75	21.333	Stop	800	490		1,290	1,700	

14120	Electric Dumbwaiters										
400	0010	**DUMBWAITERS** 2 stop, electric, minimum	2 Elev	.13	123	Ea.	5,600	2,825		8,425	10,800	400
	0100	Maximum		.11	145	"	16,800	3,325		20,125	24,000	
	0600	For each additional stop, add	▼	.54	29.630	Stop	2,475	680		3,155	3,825	

14200 | Elevators

14210	Electric Traction Elevators	CREW	DAILY OUTPUT	LABOR-HOURS	UNIT	2000 BARE COSTS MAT.	LABOR	EQUIP.	TOTAL	TOTAL INCL O&P	
100	0012	**ELEVATORS OR LIFTS**										100
	7000	Residential, cab type, 1 floor, 2 stop, minimum	2 Elev	.20	80	Ea.	7,850	1,825		9,675	11,700	
	7100	Maximum		.10	160		13,300	3,675		16,975	20,700	
	7200	2 floor, 3 stop, minimum		.12	133		11,700	3,050		14,750	17,900	
	7300	Maximum		.06	266		19,000	6,100		25,100	31,000	
	7700	Stair climber (chair lift), single seat, minimum		1	16		3,800	365		4,165	4,775	
	7800	Maximum	▼	.20	80	▼	5,225	1,825		7,050	8,775	

For information about Means Estimating Seminars, see yellow pages 11 and 12 in back of book

Division 15 Mechanical

Estimating Tips

15100 Building Services Piping
This subdivision is primarily basic pipe and related materials. The pipe may be used by any of the mechanical disciplines, i.e., plumbing, fire protection, heating, and air conditioning.

- The piping section lists the add to labor for elevated pipe installation. These adds apply to all elevated pipe, fittings, valves, insulation, etc., that are placed above 10' high. CAUTION: the correct percentage may vary for the same pipe. For example, the percentage add for the basic pipe installation should be based on the maximum height that the craftsman must install for that particular section. If the pipe is to be located 14' above the floor but it is suspended on threaded rod from beams, the bottom flange of which is 18' high (4' rods), then the height is actually 18' and the add is 20%. The pipe coverer, however, does not have to go above the 14' and so his add should be 10%.

- Most pipe is priced first as straight pipe with a joint (coupling, weld, etc.) every 10' and a hanger usually every 10'. There are exceptions with hanger spacing such as: for cast iron pipe (5') and plastic pipe (3 per 10'). Following each type of pipe there are several lines listing sizes and the amount to be subtracted to delete couplings and hangers. This is for pipe that is to be buried or supported together on trapeze hangers. The reason that the couplings are deleted is that these runs are usually long and frequently longer lengths of pipe are used. By deleting the couplings the estimator is expected to look up and add back the correct reduced number of couplings.

- When preparing an estimate it may be necessary to approximate the fittings. Fittings usually run between 25% and 50% of the cost of the pipe. The lower percentage is for simpler runs, and the higher number is for complex areas like mechanical rooms.

15400 Plumbing Fixtures & Equipment
- Plumbing fixture costs usually require two lines, the fixture itself and its "rough-in, supply and waste".
- Remember that gas and oil fired units need venting.

15500 Heat Generation Equipment
- When estimating the cost of an HVAC system, check to see who is responsible for providing and installing the temperature control system. It is possible to overlook controls, assuming that they would be included in the electrical estimate.
- When looking up a boiler be careful on specified capacity. Some manufacturers rate their products on output while others use input.
- Include HVAC insulation for pipe, boiler and duct (wrap and liner).
- Be careful when looking up mechanical items to get the correct pressure rating and connection type (thread, weld, flange).

15700 Heating/Ventilation/Air Conditioning Equipment
- Combination heating and cooling units are sized by the air conditioning requirements. (See Reference No. R15710-020 for preliminary sizing guide.)
- A ton of air conditioning is nominally 400 CFM.

- Rectangular duct is taken off by the linear foot for each size, but its cost is usually estimated by the pound. Remember that SMACNA standards now base duct on internal pressure.
- Prefabricated duct is estimated and purchased like pipe: straight sections and fittings.
- Note that cranes or other lifting equipment are not included on any lines in Division 15. For example, if a crane is required to lift a heavy piece of pipe into place high above a gym floor, or to put a rooftop unit on the roof of a four-story building, etc., it must be added. Due to the potential for extreme variation—from nothing additional required, to a major crane or helicopter—we feel that including a nominal amount for "lifting contingency" would be useless and detract from the accuracy of the estimate. When using equipment rental from Means do not forget to include the cost of the operator(s).

Reference Numbers
Reference numbers are shown in bold squares at the beginning of some major classifications. These numbers refer to related items in the Reference Section. The reference information may be an estimating procedure, an alternate pricing method or technical information.

Note: Not all subdivisions listed here necessarily appear in this publication.

15050 | Basic Materials & Methods

15051 | Mechanical General

			CREW	DAILY OUTPUT	LABOR-HOURS	UNIT	2000 BARE COSTS MAT.	LABOR	EQUIP.	TOTAL	TOTAL INCL O&P	
100	0010	**BOILERS, GENERAL** Prices do not include flue piping, elec. wiring,										100
	0020	gas or oil piping, boiler base, pad, or tankless unless noted										
700	0010	**PIPING** See also divisions 02500 & 02600 for site work										700
	1000	Add to labor for elevated installation										
	1080	10' to 15' high						10%				
	1100	15' to 20' high						20%				
	1120	20' to 25' high						25%				
	1140	25' to 30' high						35%				
	1160	30' to 35' high						40%				
	1180	35' to 40' high						50%				
	1200	Over 40' high						55%				

15055 | Selective Mech Demolition

			CREW	DAILY OUTPUT	LABOR-HOURS	UNIT	MAT.	LABOR	EQUIP.	TOTAL	TOTAL INCL O&P	
300	0010	**HVAC DEMOLITION**										300
	0100	Air conditioner, split unit, 3 ton	Q-5	2	8	Ea.		159		159	263	
	0150	Package unit, 3 ton	Q-6	3	8	"		153		153	253	
	0260	Baseboard, hydronic fin tube, 1/2"	Q-5	117	.137	L.F.		2.72		2.72	4.50	
	0300	Boiler, electric	Q-19	2	12	Ea.		248		248	410	
	0340	Gas or oil, steel, under 150 MBH	Q-6	3	8	"		153		153	253	
	1000	Ductwork, 4" high, 8" wide	1 Clab	200	.040	L.F.		.58		.58	.99	
	1100	6" high, 8" wide		165	.048			.70		.70	1.20	
	1200	10" high, 12" wide		125	.064			.92		.92	1.58	
	1300	12"-14" high, 16"-18" wide		85	.094			1.36		1.36	2.33	
	1500	30" high, 36" wide		56	.143			2.06		2.06	3.54	
	2800	Heat pump, package unit, 3 ton	Q-5	2.40	6.667	Ea.		133		133	219	
	2840	Split unit, 3 ton		2	8			159		159	263	
	2950	Tank, steel, oil, 275 gal., above ground		10	1.600			32		32	52.50	
	2960	Remove and reset		3	5.333			106		106	175	
600	0010	**PLUMBING DEMOLITION**										600
	1020	Fixtures, including 10' piping										
	1100	Bath tubs, cast iron	1 Plum	4	2	Ea.		44		44	72.50	
	1120	Fiberglass		6	1.333			29.50		29.50	48.50	
	1140	Steel		5	1.600			35		35	58	
	1200	Lavatory, wall hung		10	.800			17.55		17.55	29	
	1220	Counter top		8	1			22		22	36.50	
	1300	Sink, steel or cast iron, single		8	1			22		22	36.50	
	1320	Double		7	1.143			25		25	41.50	
	1400	Water closet, floor mounted		8	1			22		22	36.50	
	1420	Wall mounted		7	1.143			25		25	41.50	
	2000	Piping, metal, to 2" diameter		200	.040	L.F.		.88		.88	1.45	
	2050	2" to 4" diameter		150	.053			1.17		1.17	1.94	
	2100	4" to 8" diameter	2 Plum	100	.160			3.51		3.51	5.80	
	2250	Water heater, 40 gal.	1 Plum	6	1.333	Ea.		29.50		29.50	48.50	
	3000	Submersible sump pump		24	.333			7.30		7.30	12.10	
	6000	Remove and reset fixtures, minimum		6	1.333			29.50		29.50	48.50	
	6100	Maximum		4	2			44		44	72.50	

15082 | Duct Insulation

			CREW	DAILY OUTPUT	LABOR-HOURS	UNIT	MAT.	LABOR	EQUIP.	TOTAL	TOTAL INCL O&P	
200	0010	**INSULATION**										200
	3000	Ductwork										
	3020	Blanket type, fiberglass, flexible										
	3030	Fire resistant liner, black coating one side										
	3050	1/2" thick, 2 lb. density	Q-14	380	.042	S.F.	.29	.78		1.07	1.68	
	3060	1" thick, 1-1/2 lb. density	"	350	.046	"	.39	.84		1.23	1.91	

Important: See the Reference Section for critical supporting data - Reference Nos., Crews, & Location Factors

15050 | Basic Materials & Methods

15082 | Duct Insulation

		Crew	Daily Output	Labor-Hours	Unit	Mat.	Labor	Equip.	Total	Total Incl O&P
3140	FRK vapor barrier wrap, .75 lb. density									
3160	1" thick	Q-14	350	.046	S.F.	.27	.84		1.11	1.78
3170	1-1/2" thick	"	320	.050	"	.24	.92		1.16	1.87
3210	Vinyl jacket, same as FRK									
3490	Board type, fiberglass liner, 3 lb. density									
3500	Fire resistant, black pigmented, 1 side									
3520	1" thick	Q-14	150	.107	S.F.	1.16	1.97		3.13	4.73
3540	1-1/2" thick		130	.123		1.43	2.27		3.70	5.55
3560	2" thick	▼	120	.133	▼	1.72	2.46		4.18	6.20
3600	FRK vapor barrier									
3620	1" thick	Q-14	150	.107	S.F.	1.27	1.97		3.24	4.85
3630	1-1/2" thick	"	130	.123	"	1.64	2.27		3.91	5.80
4000	Pipe covering (price copper tube one size less than IPS)									
6600	Fiberglass, with all service jacket									
6840	1" wall, 1/2" iron pipe size	Q-14	240	.067	L.F.	1.04	1.23		2.27	3.29
6860	3/4" iron pipe size		230	.070		1.19	1.28		2.47	3.56
6870	1" iron pipe size		220	.073		1.21	1.34		2.55	3.68
6900	2" iron pipe size		200	.080		1.62	1.48		3.10	4.36
6910	2-1/2" iron pipe size		190	.084		1.78	1.55		3.33	4.68
6920	3" iron pipe size		180	.089		2.01	1.64		3.65	5.10
6940	4" iron pipe size		150	.107		2.58	1.97		4.55	6.30
6960	6" iron pipe size	▼	120	.133	▼	3.17	2.46		5.63	7.80
7660	For fittings, add 3 L.F. for each fitting									
7680	plus 4 L.F. for each flange of the fitting									
7700	For equipment or congested areas, add						20%			
7879	Rubber tubing, flexible closed cell foam									
8300	3/4" wall, 1/4" iron pipe size	1 Asbe	90	.089	L.F.	.86	1.82		2.68	4.14
8330	1/2" iron pipe size		89	.090		1.12	1.84		2.96	4.46
8340	3/4" iron pipe size		89	.090		1.38	1.84		3.22	4.75
8350	1" iron pipe size		88	.091		1.56	1.86		3.42	4.98
8360	1-1/4" iron pipe size		87	.092		2.11	1.89		4	5.60
8380	2" iron pipe size		86	.093		2.77	1.91		4.68	6.40
8390	2-1/2" iron pipe size		86	.093		3.73	1.91		5.64	7.45
8400	3" iron pipe size		85	.094		4.23	1.93		6.16	8.05
8420	4" iron pipe size		80	.100		5.35	2.05		7.40	9.50
8440	6" iron pipe size		80	.100		7.85	2.05		9.90	12.25
8444	1" wall, 1/2" iron pipe size		86	.093		2.18	1.91		4.09	5.75
8445	3/4" iron pipe size		84	.095		2.64	1.95		4.59	6.30
8446	1" iron pipe size		84	.095		3.07	1.95		5.02	6.80
8447	1-1/4" iron pipe size		82	.098		3.47	2		5.47	7.30
8448	1-1/2" iron pipe size		82	.098		4.03	2		6.03	7.95
8449	2" iron pipe size		80	.100		5.40	2.05		7.45	9.55
8450	2-1/2" iron pipe size	▼	80	.100	▼	7	2.05		9.05	11.30
8456	Rubber insulation tape, 1/8" x 2" x 30'				Ea.	12.95			12.95	14.25

15100 | Building Services Piping

15107 | Metal Pipe & Fittings

		Crew	Daily Output	Labor-Hours	Unit	Mat.	Labor	Equip.	Total	Total Incl O&P
0010	**PIPE, CAST IRON** Soil, on hangers 5' O.C.									
0020	Single hub, service wt., lead & oakum joints 10' O.C.									

For expanded coverage of these items see Means Mechanical or Plumbing Cost Data 2000

15100 | Building Services Piping

15107 | Metal Pipe & Fittings

		CREW	DAILY OUTPUT	LABOR-HOURS	UNIT	MAT.	LABOR	EQUIP.	TOTAL	TOTAL INCL O&P	
320	2120	2" diameter	Q-1	63	.254	L.F.	4.92	5		9.92	13.70	**320**
	2140	3" diameter		60	.267		6.35	5.25		11.60	15.65	
	2160	4" diameter	↓	55	.291		6.25	5.75		12	16.35	
	2180	5" diameter	Q-2	76	.316		10.95	6		16.95	22	
	2200	6" diameter	"	73	.329		12.90	6.25		19.15	24.50	
	2320	For service weight, double hub, add					10%					
	2340	For extra heavy, single hub, add					48%	4%				
	2360	For extra heavy, double hub, add				↓	71%	4%				
	2400	Lead for caulking, (1#/dia in)				Lb.	.98			.98	1.08	
	2420	Oakum for caulking, (1/8#/dia in)				"	4.66			4.66	5.15	
	4000	No hub, couplings 10' O.C.										
	4100	1-1/2" diameter	Q-1	71	.225	L.F.	5.60	4.45		10.05	13.50	
	4120	2" diameter		67	.239		5.70	4.72		10.42	14.10	
	4140	3" diameter		64	.250		7.15	4.94		12.09	16.05	
	4160	4" diameter	↓	58	.276		9.30	5.45		14.75	19.20	
	4180	5" diameter	Q-2	83	.289		12.80	5.50		18.30	23	
	4200	6" diameter	"	79	.304	↓	15.45	5.80		21.25	26.50	
360	0010	**PIPE, CAST IRON, FITTINGS** Soil										**360**
	0040	Hub and spigot, service weight, lead & oakum joints										
	0080	1/4 bend, 2"	Q-1	16	1	Ea.	6.10	19.75		25.85	39	
	0120	3"		14	1.143		10.60	22.50		33.10	49	
	0140	4"	↓	13	1.231		16.55	24.50		41.05	58	
	0160	5"	Q-2	18	1.333		23	25.50		48.50	67.50	
	0180	6"	"	17	1.412		29	27		56	76.50	
	0500	Sanitary tee, 2"	Q-1	10	1.600		9.75	31.50		41.25	62.50	
	0540	3"		9	1.778		18.70	35		53.70	78.50	
	0620	4"	↓	8	2		22	39.50		61.50	89.50	
	0700	5"	Q-2	12	2		43.50	38		81.50	111	
	0800	6"	"	11	2.182		49.50	41.50		91	123	
	5990	No hub										
	6000	Cplg. & labor required at joints not incl. in fitting										
	6010	price. Add 1 coupling per joint for installed price										
	6020	1/4 Bend, 1-1/2"				Ea.	4.47			4.47	4.92	
	6060	2"					4.85			4.85	5.35	
	6080	3"					6.35			6.35	6.95	
	6120	4"					9.70			9.70	10.65	
	6140	5"					22.50			22.50	25	
	6160	6"					24			24	26.50	
	6184	1/4 Bend, long sweep, 1-1/2"					10.95			10.95	12.05	
	6186	2"					10.95			10.95	12.05	
	6188	3"					13.05			13.05	14.35	
	6189	4"					21			21	23	
	6190	5"					38.50			38.50	42	
	6191	6"					46.50			46.50	51.50	
	6192	8"					114			114	125	
	6193	10"					205			205	225	
	6380	Sanitary Tee, tapped, 1-1/2"					8.55			8.55	9.40	
	6382	2" x 1-1/2"					8.55			8.55	9.40	
	6384	2"					9.45			9.45	10.40	
	6386	3" x 2"					12.10			12.10	13.30	
	6388	3"					22.50			22.50	24.50	
	6390	4" x 1-1/2"					10.75			10.75	11.85	
	6392	4" x 2"					12.10			12.10	13.30	
	6394	6" x 1-1/2"					25			25	27.50	
	6396	6" x 2"					25.50			25.50	28	

Important: See the Reference Section for critical supporting data - Reference Nos., Crews, & Location Factors

15100 | Building Services Piping

15107 | Metal Pipe & Fittings

			CREW	DAILY OUTPUT	LABOR-HOURS	UNIT	MAT.	LABOR	EQUIP.	TOTAL	TOTAL INCL O&P	
360	6459	Sanitary Tee, 1-1/2"				Ea.	6.15			6.15	6.80	360
	6460	2"					6.75			6.75	7.40	
	6470	3"					8.20			8.20	9	
	6472	4"					12.70			12.70	13.95	
	6474	5"					36			36	39.50	
	6476	6"					37			37	40.50	
	8000	Coupling, standard (by CISPI Mfrs.)										
	8020	1-1/2"	Q-1	48	.333	Ea.	3.82	6.60		10.42	15.10	
	8040	2"		44	.364		3.82	7.20		11.02	16.05	
	8080	3"		38	.421		4.54	8.30		12.84	18.75	
	8120	4"		33	.485		5.40	9.60		15	22	
	8160	5"	Q-2	44	.545		13.25	10.35		23.60	31.50	
	8180	6"	"	40	.600		13.80	11.40		25.20	34	
420	0010	**PIPE, COPPER** Solder joints										420
	1000	Type K tubing, couplings & clevis hangers 10' O.C.										
	1180	3/4" diameter	1 Plum	74	.108	L.F.	1.85	2.37		4.22	5.95	
	1200	1" diameter		66	.121		2.37	2.66		5.03	7	
	1240	1-1/2" diameter		50	.160		3.35	3.51		6.86	9.50	
	1260	2" diameter		40	.200		4.90	4.39		9.29	12.65	
	1280	2-1/2" diameter	Q-1	60	.267		6.90	5.25		12.15	16.30	
	1300	3" diameter	"	54	.296		9.35	5.85		15.20	19.90	
	2000	Type L tubing, couplings & hangers 10' O.C.										
	2140	1/2" diameter	1 Plum	81	.099	L.F.	1.14	2.17		3.31	4.84	
	2180	3/4" diameter		76	.105		1.50	2.31		3.81	5.45	
	2200	1" diameter		68	.118		2.03	2.58		4.61	6.50	
	2220	1-1/4" diameter		58	.138		2.42	3.03		5.45	7.65	
	2240	1-1/2" diameter		52	.154		2.93	3.38		6.31	8.80	
	2260	2" diameter		42	.190		4.20	4.18		8.38	11.50	
	2280	2-1/2" diameter	Q-1	62	.258		5.90	5.10		11	14.95	
	2300	3" diameter		56	.286		7.85	5.65		13.50	18	
	2320	3-1/2" diameter		43	.372		9.95	7.35		17.30	23	
	2340	4" diameter		39	.410		12.40	8.10		20.50	27	
	3000	Type M tubing, couplings & hangers 10' O.C.										
	3140	1/2" diameter	1 Plum	84	.095	L.F.	.98	2.09		3.07	4.53	
	3180	3/4" diameter		78	.103		1.26	2.25		3.51	5.10	
	3200	1" diameter		70	.114		1.74	2.51		4.25	6.05	
	3220	1-1/4" diameter		60	.133		2.08	2.93		5.01	7.15	
	3240	1-1/2" diameter		54	.148		2.67	3.25		5.92	8.35	
	3260	2" diameter		44	.182		3.93	3.99		7.92	10.90	
	3280	2-1/2" diameter	Q-1	64	.250		5.35	4.94		10.29	14.05	
	3300	3" diameter		58	.276		6.75	5.45		12.20	16.40	
	3320	3-1/2" diameter		45	.356		8.50	7		15.50	21	
	3340	4" diameter		40	.400		11.15	7.90		19.05	25.50	
	4000	Type DWV tubing, couplings & hangers 10' O.C.										
	4100	1-1/4" diameter	1 Plum	60	.133	L.F.	2.06	2.93		4.99	7.10	
	4120	1-1/2" diameter		54	.148		2.45	3.25		5.70	8.10	
	4140	2" diameter		44	.182		3.04	3.99		7.03	9.95	
	4160	3" diameter	Q-1	58	.276		4.74	5.45		10.19	14.20	
	4180	4" diameter	"	40	.400		8.25	7.90		16.15	22	
460	0010	**PIPE, COPPER, FITTINGS** Wrought unless otherwise noted										460
	0040	Solder joints, copper x copper										
	0100	1/2"	1 Plum	20	.400	Ea.	.18	8.80		8.98	14.70	
	0120	3/4"		19	.421		.39	9.25		9.64	15.75	
	0130	1"		16	.500		.91	11		11.91	19.15	
	0140	1-1/4"		15	.533		1.39	11.70		13.09	21	

For expanded coverage of these items see Means Mechanical or Plumbing Cost Data 2000

15100 | Building Services Piping

15107 | Metal Pipe & Fittings

		CREW	DAILY OUTPUT	LABOR-HOURS	UNIT	2000 BARE COSTS				TOTAL INCL O&P
						MAT.	LABOR	EQUIP.	TOTAL	
0150	1-1/2"	1 Plum	13	.615	Ea.	2.16	13.50		15.66	25
0160	2"	↓	11	.727		3.93	15.95		19.88	31
0170	2-1/2"	Q-1	13	1.231		7.50	24.50		32	48.50
0180	3"		11	1.455		10.50	28.50		39	59
0190	3-1/2"		10	1.600		34.50	31.50		66	90
0200	4"	↓	9	1.778		27	35		62	87.50
0450	Tee, 1/4"	1 Plum	14	.571		1.10	12.55		13.65	21.50
0480	1/2"		13	.615		.30	13.50		13.80	23
0500	3/4"		12	.667		.73	14.65		15.38	25
0510	1"		10	.800		2.12	17.55		19.67	31.50
0520	1-1/4"		9	.889		3.21	19.50		22.71	36
0530	1-1/2"		8	1		4.45	22		26.45	41.50
0540	2"	↓	7	1.143		6.95	25		31.95	49
0550	2-1/2"	Q-1	8	2		14	39.50		53.50	81
0560	3"		7	2.286		21.50	45		66.50	98
0580	4"	↓	5	3.200		51.50	63		114.50	161
0612	Tee, reducing on the outlet, 1/4"	1 Plum	15	.533		1.78	11.70		13.48	21.50
0613	3/8"		15	.533		1.67	11.70		13.37	21
0614	1/2"		14	.571		1.46	12.55		14.01	22
0615	5/8"		13	.615		2.47	13.50		15.97	25
0616	3/4"		12	.667		1.97	14.65		16.62	26
0617	1"		11	.727		2.47	15.95		18.42	29
0618	1-1/4"		10	.800		3.63	17.55		21.18	33
0619	1-1/2"		9	.889		3.31	19.50		22.81	36
0620	2"	↓	8	1		5.25	22		27.25	42.50
0621	2-1/2"	Q-1	9	1.778		16.30	35		51.30	76
0622	3"		8	2		19.65	39.50		59.15	87
0623	4"		6	2.667		69	52.50		121.50	163
0624	5"	↓	5	3.200		161	63		224	281
0625	6"	Q-2	7	3.429		220	65		285	350
0626	8"	"	6	4		895	76		971	1,100
0630	Tee, reducing on the run, 1/4"	1 Plum	15	.533		2.21	11.70		13.91	22
0631	3/8"		15	.533		3.13	11.70		14.83	23
0632	1/2"		14	.571		2.64	12.55		15.19	23.50
0633	5/8"		13	.615		2.97	13.50		16.47	26
0634	3/4"		12	.667		1.46	14.65		16.11	25.50
0635	1"		11	.727		2.52	15.95		18.47	29.50
0636	1-1/4"		10	.800		4.09	17.55		21.64	33.50
0637	1-1/2"		9	.889		7.05	19.50		26.55	40.50
0638	2"	↓	8	1		9	22		31	46.50
0639	2-1/2"	Q-1	9	1.778		22	35		57	82
0640	3"		8	2		33	39.50		72.50	102
0641	4"		6	2.667		69	52.50		121.50	163
0642	5"	↓	5	3.200		161	63		224	281
0643	6"	Q-2	7	3.429		220	65		285	350
0644	8"	"	6	4	↓	895	76		971	1,100
2000	DWV, solder joints, copper x copper									
2030	90° Elbow, 1-1/4"	1 Plum	13	.615	Ea.	1.61	13.50		15.11	24.50
2050	1-1/2"		12	.667		4.12	14.65		18.77	28.50
2070	2"	↓	10	.800		5.85	17.55		23.40	35.50
2090	3"	Q-1	10	1.600		11.45	31.50		42.95	64.50
2100	4"	"	9	1.778		36.50	35		71.50	98.50
2250	Tee, Sanitary, 1-1/4"	1 Plum	9	.889		3.17	19.50		22.67	36
2290	2"	"	7	1.143		4.61	25		29.61	46.50
2310	3"	Q-1	7	2.286	↓	16.45	45		61.45	92.50

Important: See the Reference Section for critical supporting data - Reference Nos., Crews, & Location Factors

15100 | Building Services Piping

15107 | Metal Pipe & Fittings

		CREW	DAILY OUTPUT	LABOR-HOURS	UNIT	2000 BARE COSTS				TOTAL INCL O&P
						MAT.	LABOR	EQUIP.	TOTAL	
0010	**PIPE, STEEL**									
0050	Schedule 40, threaded, with couplings, and clevis type									
0060	hangers sized for covering, 10' O.C.									
0540	Black, 1/4" diameter	1 Plum	66	.121	L.F.	1.45	2.66		4.11	6
0570	3/4" diameter		61	.131		1.54	2.88		4.42	6.45
0580	1" diameter	↓	53	.151		1.83	3.31		5.14	7.50
0590	1-1/4" diameter	Q-1	89	.180		2.17	3.55		5.72	8.25
0600	1-1/2" diameter		80	.200		2.42	3.95		6.37	9.20
0610	2" diameter	↓	64	.250		3.28	4.94		8.22	11.75
2000	Welded, sch. 40, on yoke & roll hangers, sized for covering,									
2010	10' O.C. (no hangers incl. for 14" diam. and up)									
2080	2-1/2" diameter	Q-15	47	.340	L.F.	4.20	6.70	1.04	11.94	16.85
2090	3" diameter		43	.372		5.05	7.35	1.13	13.53	18.95
2100	3-1/2" diameter		39	.410		6.05	8.10	1.25	15.40	21.50
2110	4" diameter		37	.432		6.80	8.55	1.32	16.67	23
2120	5" diameter	↓	32	.500		9.80	9.90	1.52	21.22	29
2130	6" diameter	Q-16	36	.667	↓	16.60	13.65	1.35	31.60	42
0010	**PIPE, STEEL, FITTINGS** Threaded									
0020	Cast Iron									
0040	Standard weight, black									
0060	90° Elbow, straight									
0100	3/4"	1 Plum	14	.571	Ea.	2.01	12.55		14.56	22.50
0110	1"	"	13	.615		2.38	13.50		15.88	25
0130	1-1/2"	Q-1	20	.800		4.68	15.80		20.48	31
0140	2"	"	18	.889		7.30	17.55		24.85	37
0500	Tee, straight									
0530	1/2"	1 Plum	9	.889	Ea.	3	19.50		22.50	36
0540	3/4"		9	.889		3.51	19.50		23.01	36.50
0550	1"	↓	8	1		3.12	22		25.12	40
0570	1-1/2"	Q-1	13	1.231		7.40	24.50		31.90	48
0580	2"	"	11	1.455		10.30	28.50		38.80	59
6000	For galvanized elbows, tees, and couplings add				↓	20%				
0010	**PIPE, GROOVED-JOINT STEEL FITTINGS & VALVES**									
0020	Pipe includes coupling & clevis type hanger 10' O.C.									
1000	Schedule 40, black									
1040	3/4" diameter	1 Plum	71	.113	L.F.	2.04	2.47		4.51	6.35
1050	1" diameter		63	.127		2.17	2.79		4.96	7
1060	1-1/4" diameter		58	.138		2.64	3.03		5.67	7.90
1070	1-1/2" diameter		51	.157		2.97	3.44		6.41	8.95
1080	2" diameter	↓	40	.200		3.46	4.39		7.85	11.05
1090	2-1/2" diameter	Q-1	57	.281		4.68	5.55		10.23	14.30
1100	3" diameter		50	.320		5.70	6.30		12	16.70
1110	4" diameter	↓	45	.356	↓	8.05	7		15.05	20.50
4000	Elbow, 90° or 45°, painted									
4040	1" diameter	1 Plum	50	.160	Ea.	12.10	3.51		15.61	19.10
4050	1-1/4" diameter		40	.200		12.10	4.39		16.49	20.50
4060	1-1/2" diameter		33	.242		12.10	5.30		17.40	22
4070	2" diameter	↓	25	.320		12.10	7		19.10	25
4080	2-1/2" diameter	Q-1	40	.400		12.10	7.90		20	26.50
4100	4" diameter	"	25	.640		23.50	12.65		36.15	47
4250	For galvanized elbows, add				↓	26%				
4690	Tee, painted									
4700	3/4" diameter	1 Plum	38	.211	Ea.	18.65	4.62		23.27	28
4740	1" diameter	↓	33	.242	↓	18.65	5.30		23.95	29.50

For expanded coverage of these items see Means Mechanical or Plumbing Cost Data 2000

15100 | Building Services Piping

15107 | Metal Pipe & Fittings

		CREW	DAILY OUTPUT	LABOR-HOURS	UNIT	2000 BARE COSTS MAT.	LABOR	EQUIP.	TOTAL	TOTAL INCL O&P	
690	4750	1-1/4" diameter	1 Plum	27	.296	Ea.	18.65	6.50		25.15	31.50	690
	4760	1-1/2" diameter		22	.364		18.65	8		26.65	33.50	
	4770	2" diameter	▼	17	.471		18.65	10.35		29	37.50	
	4780	2-1/2" diameter	Q-1	27	.593		18.65	11.70		30.35	40	
	4800	4" diameter	"	17	.941	▼	40	18.60		58.60	74.50	
	4900	For galvanized tees, add					24%					

15108 | Plastic Pipe & Fittings

		CREW	DAILY OUTPUT	LABOR-HOURS	UNIT	MAT.	LABOR	EQUIP.	TOTAL	TOTAL INCL O&P	
520	0011	**PIPE, PLASTIC**										520
	1800	PVC, couplings 10' O.C., hangers 3 per 10'										
	1820	Schedule 40										
	1860	1/2" diameter	1 Plum	54	.148	L.F.	1.90	3.25		5.15	7.50	
	1870	3/4" diameter		51	.157		2.06	3.44		5.50	7.95	
	1880	1" diameter		46	.174		2.57	3.82		6.39	9.15	
	1890	1-1/4" diameter		42	.190		2.61	4.18		6.79	9.75	
	1900	1-1/2" diameter	▼	36	.222		2.81	4.88		7.69	11.15	
	1910	2" diameter	Q-1	59	.271		2.83	5.35		8.18	11.95	
	1920	2-1/2" diameter		56	.286		3.52	5.65		9.17	13.20	
	1930	3" diameter		53	.302		4.05	5.95		10	14.30	
	1940	4" diameter	▼	48	.333	▼	5.50	6.60		12.10	16.95	
	4100	DWW type, schedule 40, couplings 10' O.C., hangers 3 per 10'										
	4120	ABS										
	4140	1-1/4" diameter	1 Plum	42	.190	L.F.	2.58	4.18		6.76	9.75	
	4150	1-1/2" diameter	"	36	.222		2.56	4.88		7.44	10.85	
	4160	2" diameter	Q-1	59	.271		2.66	5.35		8.01	11.75	
	4170	3" diameter		53	.302		3.28	5.95		9.23	13.45	
	4180	4" diameter	▼	48	.333	▼	4.44	6.60		11.04	15.80	
	5360	CPVC, couplings 10' O.C., hangers 3 per 10'										
	5380	Schedule 40										
	5460	1/2" diameter	1 Plum	54	.148	L.F.	2.63	3.25		5.88	8.30	
	5470	3/4" diameter		51	.157		3.04	3.44		6.48	9.05	
	5480	1" diameter		46	.174		3.98	3.82		7.80	10.70	
	5490	1-1/4" diameter		42	.190		4.57	4.18		8.75	11.95	
	5500	1-1/2" diameter	▼	36	.222		5.20	4.88		10.08	13.75	
	5510	2" diameter	Q-1	59	.271	▼	6.05	5.35		11.40	15.55	
	9000	Perforated										
	9040	4" diameter				L.F.	.42			.42	.46	
560	0010	**PIPE, PLASTIC, FITTINGS**										560
	2700	PVC (white), schedule 40, socket joints										
	2760	90° elbow, 1/2"	1 Plum	22	.364	Ea.	.23	8		8.23	13.45	
	2770	3/4"		21	.381		.25	8.35		8.60	14.15	
	2780	1"		18	.444		.45	9.75		10.20	16.65	
	2790	1-1/4"		17	.471		.80	10.35		11.15	18	
	2800	1-1/2"	▼	16	.500		.85	11		11.85	19.10	
	2810	2"	Q-1	28	.571		1.34	11.30		12.64	20	
	2820	2-1/2"		22	.727		4.06	14.35		18.41	28.50	
	2830	3"		17	.941		4.86	18.60		23.46	36	
	2840	4"	▼	14	1.143		8.70	22.50		31.20	47	
	3180	Tee, 1/2"	1 Plum	14	.571		.28	12.55		12.83	21	
	3190	3/4"		13	.615		.32	13.50		13.82	23	
	3200	1"		12	.667		.59	14.65		15.24	24.50	
	3210	1-1/4"		11	.727		.94	15.95		16.89	27.50	
	3220	1-1/2"	▼	10	.800		1.13	17.55		18.68	30	
	3230	2"	Q-1	17	.941		1.65	18.60		20.25	32.50	
	3240	2-1/2"	▼	14	1.143		5.45	22.50		27.95	43.50	

15100 | Building Services Piping

15108 | Plastic Pipe & Fittings

		CREW	DAILY OUTPUT	LABOR-HOURS	UNIT	2000 BARE COSTS MAT.	LABOR	EQUIP.	TOTAL	TOTAL INCL O&P	
3250	3"	Q-1	11	1.455	Ea.	7.15	28.50		35.65	55.50	
3260	4"		9	1.778		12.90	35		47.90	72	
3380	Coupling, 1/2"	1 Plum	22	.364		.18	8		8.18	13.40	
3390	3/4"		21	.381		.24	8.35		8.59	14.10	
3400	1"		18	.444		.41	9.75		10.16	16.60	
3410	1-1/4"		17	.471		.57	10.35		10.92	17.75	
3420	1-1/2"		16	.500		.61	11		11.61	18.80	
3430	2"	Q-1	28	.571		.95	11.30		12.25	19.70	
3440	2-1/2"		20	.800		2.08	15.80		17.88	28.50	
3450	3"		19	.842		3.27	16.65		19.92	31	
3460	4"		16	1		4.72	19.75		24.47	37.50	
4500	DWV, ABS, non pressure, socket joints										
4540	1/4 Bend, 1-1/4"	1 Plum	17	.471	Ea.	.99	10.35		11.34	18.20	
4560	1-1/2"	"	16	.500		.53	11		11.53	18.75	
4570	2"	Q-1	28	.571		.87	11.30		12.17	19.60	
4580	3"		17	.941		2.05	18.60		20.65	33	
4590	4"		14	1.143		3.57	22.50		26.07	41.50	
4600	6"		8	2		22.50	39.50		62	90.50	
4650	1/8 Bend, same as 1/4 Bend										
4800	Tee, sanitary										
4820	1-1/4"	1 Plum	11	.727	Ea.	1.25	15.95		17.20	28	
4830	1-1/2"	"	10	.800		.77	17.55		18.32	30	
4840	2"	Q-1	17	.941		1.10	18.60		19.70	31.50	
4850	3"		11	1.455		2.79	28.50		31.29	50.50	
4860	4"		9	1.778		6.90	35		41.90	65.50	
4862	Tee, sanitary, reducing, 2" x 1-1/2"		17	.941		1.63	18.60		20.23	32.50	
4864	3" x 2"		11	1.455		2.36	28.50		30.86	50	
4868	4" x 3"		10	1.600		7.70	31.50		39.20	60.50	
4870	Combination Y and 1/8 bend										
4872	1-1/2"	1 Plum	10	.800	Ea.	2.34	17.55		19.89	31.50	
4874	2"	Q-1	17	.941		2.84	18.60		21.44	33.50	
4876	3"		11	1.455		5	28.50		33.50	53	
4878	4"		9	1.778		9.85	35		44.85	69	
4880	3" x 1-1/2"		11	1.455		5.85	28.50		34.35	54	
4882	4" x 3"		10	1.600		8.80	31.50		40.30	61.50	
4900	Wye, 1-1/4"	1 Plum	11	.727		1.44	15.95		17.39	28	
4902	1-1/2"	"	10	.800		1.47	17.55		19.02	30.50	
4904	2"	Q-1	17	.941		1.51	18.60		20.11	32	
4906	3"		11	1.455		3.47	28.50		31.97	51.50	
4908	4"		9	1.778		8.20	35		43.20	67	
4910	6"		5	3.200		31	63		94	139	
4918	3" x 1-1/2"		11	1.455		3.43	28.50		31.93	51.50	
4920	4" x 3"		10	1.600		5.70	31.50		37.20	58.50	
4922	6" x 4"		6	2.667		25.50	52.50		78	116	
4930	Double Wye, 1-1/2"	1 Plum	8	1		3.12	22		25.12	40	
4932	2"	Q-1	12	1.333		4.01	26.50		30.51	48	
4934	3"		8	2		10.35	39.50		49.85	77	
4936	4"		6	2.667		21	52.50		73.50	110	
4940	2" x 1-1/2"		11	1.455		4.01	28.50		32.51	52	
4942	3" x 2"		8	2		7.70	39.50		47.20	74	
4944	4" x 3"		7	2.286		16.65	45		61.65	93	
4946	6" x 4"		5	3.200		34.50	63		97.50	142	
4950	Reducer bushing, 2" x 1-1/2"		30	.533		.51	10.55		11.06	17.95	
4952	3" x 1-1/2"		24	.667		1.85	13.15		15	24	
4954	4" x 2"		20	.800		4.27	15.80		20.07	30.50	
4956	6" x 4"		17	.941		11.40	18.60		30	43	

For expanded coverage of these items see Means Mechanical or Plumbing Cost Data 2000

15100 | Building Services Piping

15108 | Plastic Pipe & Fittings

		CREW	DAILY OUTPUT	LABOR-HOURS	UNIT	MAT.	LABOR	EQUIP.	TOTAL	TOTAL INCL O&P	
560	4960	Couplings, 1-1/2"	1 Plum	16	.500	Ea.	.56	11		11.56	18.75	560
	4962	2"	Q-1	28	.571		.33	11.30		11.63	19	
	4963	3"		22	.727		1.15	14.35		15.50	25.50	
	4964	4"		17	.941		1.77	18.60		20.37	32.50	
	4966	6"		12	1.333		11.85	26.50		38.35	56.50	
	4970	2" x 1-1/2"		30	.533		.34	10.55		10.89	17.75	
	4972	3" x 1-1/2"		24	.667		1.15	13.15		14.30	23.50	
	4974	4" x 3"		19	.842		4.53	16.65		21.18	32.50	
	4978	Closet flange, 4"	1 Plum	32	.250		4.69	5.50		10.19	14.20	
	4980	4" x 3"	"	34	.235		4.02	5.15		9.17	12.95	
	5500	CPVC, Schedule 80, threaded joints										
	5540	90° Elbow, 1/4"	1 Plum	20	.400	Ea.	5.35	8.80		14.15	20.50	
	5560	1/2"		18	.444		2.09	9.75		11.84	18.45	
	5570	3/4"		17	.471		2.67	10.35		13.02	20	
	5580	1"		15	.533		4.24	11.70		15.94	24	
	5590	1-1/4"		14	.571		9.20	12.55		21.75	30.50	
	5600	1-1/2"		13	.615		10.25	13.50		23.75	34	
	5610	2"	Q-1	22	.727		12.40	14.35		26.75	37.50	
	5620	2-1/2"		18	.889		28.50	17.55		46.05	60.50	
	5630	3"		14	1.143		32	22.50		54.50	73	
	6000	Coupling, 1/4"	1 Plum	20	.400		5.70	8.80		14.50	21	
	6020	1/2"		18	.444		2.21	9.75		11.96	18.60	
	6030	3/4"		17	.471		3.09	10.35		13.44	20.50	
	6040	1"		15	.533		4.16	11.70		15.86	24	
	6050	1-1/4"		14	.571		6.25	12.55		18.80	27.50	
	6060	1-1/2"		13	.615		7.85	13.50		21.35	31	
	6070	2"	Q-1	22	.727		9.15	14.35		23.50	34	
	6080	2-1/2"		20	.800		20.50	15.80		36.30	48.50	
	6090	3"		19	.842		22	16.65		38.65	52	

15110 | Valves

		CREW	DAILY OUTPUT	LABOR-HOURS	UNIT	MAT.	LABOR	EQUIP.	TOTAL	TOTAL INCL O&P	
160	0010	**VALVES, BRONZE**										160
	1380	Ball, 150 psi, threaded										
	1450	1/2" size	1 Plum	22	.364	Ea.	6.40	8		14.40	20.50	
	1460	3/4" size		20	.400		10.60	8.80		19.40	26	
	1470	1" size		19	.421		13.30	9.25		22.55	30	
	1480	1-1/4" size		15	.533		22.50	11.70		34.20	44.50	
	1490	1-1/2" size		13	.615		28.50	13.50		42	54	
	1500	2" size		11	.727		35.50	15.95		51.45	66	
	1600	Butterfly, 175 psi, full port, solder or threaded ends										
	1610	Stainless steel disc and stem										
	1620	1/4" size	1 Plum	24	.333	Ea.	5.75	7.30		13.05	18.45	
	1630	3/8" size		24	.333		5.75	7.30		13.05	18.45	
	1640	1/2" size		22	.364		6.10	8		14.10	19.90	
	1650	3/4" size		20	.400		9.85	8.80		18.65	25.50	
	1660	1" size		19	.421		12.25	9.25		21.50	29	
	1670	1-1/4" size		15	.533		19.45	11.70		31.15	41	
	1680	1-1/2" size		13	.615		25	13.50		38.50	50	
	1690	2" size		11	.727		31.50	15.95		47.45	61	
	1750	Check, swing, class 150, regrinding disc, threaded										
	1850	1/2" size	1 Plum	24	.333	Ea.	21.50	7.30		28.80	35.50	
	1860	3/4" size		20	.400		28.50	8.80		37.30	46	
	1870	1" size		19	.421		43	9.25		52.25	63	
	1880	1-1/4" size		15	.533		59.50	11.70		71.20	85	
	1900	2" size		11	.727		103	15.95		118.95	140	

15100 | Building Services Piping

15110 | Valves

			CREW	DAILY OUTPUT	LABOR-HOURS	UNIT	MAT.	LABOR	EQUIP.	TOTAL	TOTAL INCL O&P
160	2850	Gate, N.R.S., soldered, 125 psi										160
	2920	1/2" size	1 Plum	24	.333	Ea.	14.05	7.30		21.35	27.50
	2940	3/4" size		20	.400		16.20	8.80		25	32.50
	2950	1" size		19	.421		23	9.25		32.25	41
	2960	1-1/4" size		15	.533		40.50	11.70		52.20	64
	2970	1-1/2" size		13	.615		45.50	13.50		59	72.50
	2980	2" size		11	.727		54.50	15.95		70.45	86.50
	4250	Threaded, class 150									
	4340	3/4" size	1 Plum	20	.400	Ea.	23	8.80		31.80	40
	4350	1" size		19	.421		30.50	9.25		39.75	49
	4360	1-1/4" size		15	.533		41	11.70		52.70	64.50
	4370	1-1/2" size		13	.615		51	13.50		64.50	78.50
	5600	Relief, pressure & temperature, self-closing, ASME, threaded									
	5650	1" size	1 Plum	24	.333	Ea.	92	7.30		99.30	113
	5660	1-1/4" size	"	20	.400	"	184	8.80		192.80	217
	6400	Pressure, water, ASME, threaded									
	6440	3/4" size	1 Plum	28	.286	Ea.	45	6.25		51.25	60
	6450	1" size		24	.333		91	7.30		98.30	112
	6460	1-1/4" size		20	.400		131	8.80		139.80	159
	6470	1-1/2" size		18	.444		206	9.75		215.75	242
	6480	2" size		16	.500		298	11		309	345
	6490	2-1/2" size		15	.533		830	11.70		841.70	935
	6900	Reducing, water pressure									
	6940	1/2" size	1 Plum	24	.333	Ea.	103	7.30		110.30	126
	6950	3/4" size		20	.400		103	8.80		111.80	129
	6960	1" size		19	.421		160	9.25		169.25	191
	8350	Tempering, water, sweat connections									
	8400	1/2" size	1 Plum	24	.333	Ea.	39	7.30		46.30	55
	8440	3/4" size	"	20	.400	"	48	8.80		56.80	67.50
200	0010	**VALVES, IRON BODY**										200
	1650	Gate, 125 lb., N.R.S.									
	2150	Flanged									
	2240	2-1/2" size	Q-1	5	3.200	Ea.	270	63		333	400
	2260	3" size		4.50	3.556		300	70		370	450
	2280	4" size		3	5.333		435	105		540	655
	3550	OS&Y, 125 lb., flanged									
	3660	3" size	Q-1	4.50	3.556	Ea.	231	70		301	370
	3680	4" size	"	3	5.333		248	105		353	445
	3700	6" size	Q-2	3	8		545	152		697	850
	5450	Swing check, 125 lb., threaded									
	5500	2" size	1 Plum	11	.727	Ea.	320	15.95		335.95	375
	5950	Flanged									
	6040	2-1/2" size	Q-1	5	3.200	Ea.	168	63		231	289
	6050	3" size		4.50	3.556		181	70		251	315
	6060	4" size		3	5.333		286	105		391	490
	6070	6" size	Q-2	3	8		505	152		657	805

15120 | Piping Specialties

			CREW	DAILY OUTPUT	LABOR-HOURS	UNIT	MAT.	LABOR	EQUIP.	TOTAL	TOTAL INCL O&P
320	0010	**EXPANSION TANKS**										320
	1505	Fiberglass and steel single / double wall storage, see Div 13201									
	1510	Tank leak detection systems, see Div 13851-350									
	2000	Steel, liquid expansion, ASME, painted, 15 gallon capacity	Q-5	17	.941	Ea.	330	18.75		348.75	395
	2040	30 gallon capacity		12	1.333		370	26.50		396.50	450
	2120	100 gallon capacity		6	2.667		670	53		723	825

For expanded coverage of these items see *Means Mechanical or Plumbing Cost Data 2000*

15100 | Building Services Piping

15120 | Piping Specialties

			CREW	DAILY OUTPUT	LABOR-HOURS	UNIT	MAT.	LABOR	EQUIP.	TOTAL	TOTAL INCL O&P	
320	3000	Steel ASME expansion, rubber diaphragm, 19 gal. cap. accept.	Q-5	12	1.333	Ea.	1,250	26.50		1,276.50	1,425	320
	3020	31 gallon capacity		8	2		1,400	40		1,440	1,600	
	3080	119 gallon capacity	▼	4	4	▼	2,200	79.50		2,279.50	2,550	
520	0010	**HYDRONIC HEATING CONTROL VALVES**										520
	0050	Hot water, nonelectric, thermostatic										
	0100	Radiator supply, 1/2" diameter	1 Stpi	24	.333	Ea.	35.50	7.35		42.85	51	
	0120	3/4" diameter		20	.400		36	8.85		44.85	54	
	0140	1" diameter		19	.421		44.50	9.30		53.80	64.50	
	0160	1-1/4" diameter	▼	15	.533		63	11.80		74.80	88.50	
	0500	For low pressure steam, add					25%					
	1000	Manual, radiator supply										
	1010	1/2" pipe size, angle union	1 Stpi	24	.333	Ea.	20.50	7.35		27.85	34.50	
	1020	3/4" pipe size, angle union		20	.400		25.50	8.85		34.35	42.50	
	1030	1" pipe size, angle union		19	.421		33	9.30		42.30	51.50	
	1140	Balance and stop valve 1/2" size		22	.364		20.50	8.05		28.55	36	
	1150	3/4" size		20	.400		13.45	8.85		22.30	29.50	
	1160	1" size		19	.421		16.90	9.30		26.20	34	
	1170	1-1/4" size	▼	15	.533	▼	21	11.80		32.80	42.50	
	8000	System balancing and shut-off										
	8020	Butterfly, quarter turn, calibrated, threaded or solder										
	8040	Bronze, -30° F to +350° F, pressure to 175 psi										
	8060	1/2" size	1 Stpi	22	.364	Ea.	9.35	8.05		17.40	23.50	
	8070	3/4" size		20	.400		8.50	8.85		17.35	24	
	8080	1" size		19	.421		11.15	9.30		20.45	27.50	
	8090	1-1/4" size		15	.533		18.35	11.80		30.15	39.50	
	8110	2" size	▼	11	.727		27.50	16.05		43.55	56.50	
840	0010	**STRAINERS, Y TYPE** Iron body										840
	0100	1/2" pipe size	1 Stpi	20	.400	Ea.	6.65	8.85		15.50	22	
	0120	3/4" pipe size		18	.444		7.95	9.80		17.75	25	
	0140	1" pipe size		16	.500		10.75	11.05		21.80	30	
	0150	1-1/4" pipe size		15	.533		14.75	11.80		26.55	35.50	
	0180	2" pipe size		8	1		27.50	22		49.50	66.50	
	1000	Flanged, 125 lb., 1-1/2" pipe size		11	.727		86	16.05		102.05	121	
	1020	2" pipe size	▼	8	1		64	22		86	107	
	1030	2-1/2" pipe size	Q-5	5	3.200		72.50	63.50		136	185	
	1040	3" pipe size		4.50	3.556		84.50	71		155.50	210	
	1060	4" pipe size	▼	3	5.333	▼	155	106		261	345	
940	0010	**WATER SUPPLY METERS**										940
	2000	Domestic/commercial, bronze										
	2020	Threaded										
	2080	3/4" diameter, to 30 GPM	1 Plum	14	.571	Ea.	115	12.55		127.55	148	
	2100	1" diameter, to 50 GPM	"	12	.667	"	157	14.65		171.65	196	
	2300	Threaded/flanged										
	2340	1-1/2" diameter, to 100 GPM	1 Plum	8	1	Ea.	485	22		507	570	
	2360	2" diameter, to 160 GPM	"	6	1.333	"	645	29.50		674.50	760	
	2600	Flanged, compound										
	2601	Water supply meters, bronze, flanged, compound										
	2640	3" diameter, 320 GPM	Q-1	3	5.333	Ea.	2,900	105		3,005	3,375	
	2660	4" diameter, to 500 GPM	"	1.50	10.667	"	4,525	211		4,736	5,325	

15140 | Domestic Water Piping

100	0010	**BACKFLOW PREVENTER** Includes valves										100
	0020	and four test cocks, corrosion resistant, automatic operation										

Important: See the Reference Section for critical supporting data - Reference Nos., Crews, & Location Factors

15100 | Building Services Piping

15140 | Domestic Water Piping

			CREW	DAILY OUTPUT	LABOR-HOURS	UNIT	2000 BARE COSTS MAT.	LABOR	EQUIP.	TOTAL	TOTAL INCL O&P	
100	4100	Threaded, valves are ball										100
	4120	3/4" pipe size	1 Plum	16	.500	Ea.	685	11		696	775	
	4140	1" pipe size		14	.571		730	12.55		742.55	820	
	4150	1-1/4" pipe size		12	.667		1,025	14.65		1,039.65	1,150	
	4160	1-1/2" pipe size		10	.800		990	17.55		1,007.55	1,100	
	4180	2" pipe size		7	1.143		1,050	25		1,075	1,200	
	5000	Flanged, bronze, valves are OS&Y										
	5060	2-1/2" pipe size	Q-1	5	3.200	Ea.	2,725	63		2,788	3,100	
	5080	3" pipe size		4.50	3.556		3,375	70		3,445	3,850	
	5100	4" pipe size		3	5.333		4,800	105		4,905	5,450	
	5120	6" pipe size	Q-2	3	8		10,200	152		10,352	11,500	
800	0010	**WATER HAMMER ARRESTORS / SHOCK ABSORBERS**										800
	0490	Copper										
	0500	3/4" male I.P.S. For 1 to 11 fixtures	1 Plum	12	.667	Ea.	16.05	14.65		30.70	41.50	
	0600	1" male I.P.S., For 12 to 32 fixtures		8	1		41	22		63	81.50	
	0700	1-1/4" male I.P.S. For 33 to 60 fixtures		8	1		47.50	22		69.50	89	

15155 | Drainage Specialties

			CREW	DAILY OUTPUT	LABOR-HOURS	UNIT	MAT.	LABOR	EQUIP.	TOTAL	TOTAL INCL O&P	
170	0010	**CLEANOUT TEE**										170
	0100	Cast iron, B&S, with countersunk plug										
	0220	3" pipe size	1 Plum	3.60	2.222	Ea.	87	49		136	176	
	0240	4" pipe size	"	3.30	2.424		108	53		161	207	
	0500	For round smooth access cover, same price										
	4000	Plastic, tees and adapters. Add plugs										
	4010	ABS, DWV										
	4020	Cleanout tee, 1-1/2" pipe size	1 Plum	15	.533	Ea.	5.45	11.70		17.15	25.50	
300	0010	**DRAINS**										300
	0140	Cornice, C.I., 45° or 90° outlet										
	0200	3" and 4" pipe size	Q-1	12	1.333	Ea.	98	26.50		124.50	152	
	0260	For galvanized body, add					19.60			19.60	21.50	
	0280	For polished bronze dome, add					16.65			16.65	18.30	
	2000	Floor, medium duty, C.I., deep flange, 7" dia top										
	2040	2" and 3" pipe size	Q-1	12	1.333	Ea.	60	26.50		86.50	110	
	2080	For galvanized body, add					25			25	27.50	
	2120	For polished bronze top, add					33			33	36.50	
	2500	Heavy duty, cleanout & trap w/bucket, C.I., 15" top										
	2540	2", 3", and 4" pipe size	Q-1	6	2.667	Ea.	1,475	52.50		1,527.50	1,700	
	2560	For galvanized body, add					440			440	485	
	2580	For polished bronze top, add					465			465	510	
	3860	Roof, flat metal deck, C.I. body, 12" C.I. dome										
	3890	3" pipe size	Q-1	14	1.143	Ea.	117	22.50		139.50	167	
400	0010	**INTERCEPTORS**										400
	0150	Grease, cast iron, 4 GPM, 8 lb. fat capacity	1 Plum	4	2	Ea.	355	44		399	465	
	0200	7 GPM, 14 lb. fat capacity		4	2		490	44		534	615	
	1040	15 GPM, 30 lb. fat capacity		4	2		855	44		899	1,025	
	1060	20 GPM, 40 lb. fat capacity		3	2.667		1,050	58.50		1,108.50	1,250	
	3000	Hair, cast iron, 1-1/4" and 1-1/2" pipe connection		8	1		132	22		154	183	
	3100	For chrome-plated cast iron, add					81.50			81.50	89.50	
	4000	Oil, fabricated steel, 10 GPM, 2" pipe size	1 Plum	4	2		745	44		789	895	
	4100	15 GPM, 2" or 3" pipe size		4	2		1,025	44		1,069	1,200	
	4120	20 GPM, 2" or 3" pipe size		3	2.667		1,225	58.50		1,283.50	1,450	
	6000	Solids, precious metals recovery, C.I., 1-1/4" to 2" pipe		4	2		200	44		244	292	
	6100	Dental Lab., large, C.I., 1-1/2" to 2" pipe		3	2.667		700	58.50		758.50	865	

For expanded coverage of these items see *Means Mechanical* or *Plumbing Cost Data 2000*

15100 | Building Services Piping

15155 | Drainage Specialties

			CREW	DAILY OUTPUT	LABOR-HOURS	UNIT	2000 BARE COSTS MAT.	LABOR	EQUIP.	TOTAL	TOTAL INCL O&P	
740	0010	**SINK WASTE TREATMENT** System for commercial kitchens										740
	0100	includes clock timer, & fittings										
	0200	System less chemical, wall mounted cabinet	1 Plum	16	.500	Ea.	279	11		290	325	
	2000	Chemical, 1 gallon, add					30			30	33	
	2100	6 gallons, add					134			134	147	
	2200	15 gallons, add					365			365	405	
	2300	30 gallons, add					690			690	755	
	2400	55 gallons, add					1,175			1,175	1,275	
780	0010	**TRAPS**										780
	0030	Cast iron, service weight										
	0050	Running P trap, without vent										
	1100	2"	Q-1	16	1	Ea.	15.95	19.75		35.70	50	
	1150	4"	"	13	1.231		50	24.50		74.50	95.50	
	1160	6"	Q-2	17	1.412		232	27		259	300	
	3000	P trap, B&S, 2" pipe size	Q-1	16	1		13.20	19.75		32.95	47	
	3040	3" pipe size	"	14	1.143		19.75	22.50		42.25	59.50	
	3800	Drum trap, 4" x 5", 1-1/2" tapping	Q-2	17	1.412		13.50	27		40.50	59.50	
	4700	Copper, drainage, drum trap										
	4840	3" x 6" swivel, 1-1/2" pipe size	1 Plum	16	.500	Ea.	22	11		33	42.50	
	5100	P trap, standard pattern										
	5200	1-1/4" pipe size	1 Plum	18	.444	Ea.	10.75	9.75		20.50	28	
	5240	1-1/2" pipe size		17	.471		10.75	10.35		21.10	29	
	5260	2" pipe size		15	.533		16.60	11.70		28.30	37.50	
	5280	3" pipe size		11	.727		40	15.95		55.95	70.50	
	6710	ABS DWV P trap, solvent weld joint										
	6720	1-1/2" pipe size	1 Plum	18	.444	Ea.	5.75	9.75		15.50	22.50	
	6722	2" pipe size		17	.471		9.35	10.35		19.70	27.50	
	6724	3" pipe size		15	.533		42.50	11.70		54.20	66.50	
	6726	4" pipe size		14	.571		87.50	12.55		100.05	117	
	6760	PP DWV, dilution trap, 1-1/2" pipe size		16	.500		92.50	11		103.50	120	
	6770	P trap, 1-1/2" pipe size		17	.471		25	10.35		35.35	44.50	
	6780	2" pipe size		16	.500		34	11		45	55.50	
	6790	3" pipe size		14	.571		78.50	12.55		91.05	107	
	6800	4" pipe size		13	.615		100	13.50		113.50	133	
	6860	PVC DWV hub x hub, basin trap, 1-1/4" pipe size		18	.444		4.22	9.75		13.97	21	
	6870	Sink P trap, 1-1/2" pipe size		18	.444		4.22	9.75		13.97	21	
	6880	Tubular S trap, 1-1/2" pipe size		17	.471		9.70	10.35		20.05	28	
	6890	PVC sch. 40 DWV, drum trap										
	6900	1-1/2" pipe size	1 Plum	16	.500	Ea.	9.90	11		20.90	29	
	6910	P trap, 1-1/2" pipe size		18	.444		1.91	9.75		11.66	18.25	
	6920	2" pipe size		17	.471		2.60	10.35		12.95	19.95	
	6930	3" pipe size		15	.533		12.95	11.70		24.65	33.50	
	6940	4" pipe size		14	.571		31.50	12.55		44.05	55	
	6950	P trap w/clean out, 1-1/2" pipe size		18	.444		4.64	9.75		14.39	21.50	
	6960	2" pipe size		17	.471		7.90	10.35		18.25	26	
940	0010	**VENT FLASHING**										940
	1000	Aluminum with lead ring										
	1040	2" pipe	1 Plum	18	.444	Ea.	6.85	9.75		16.60	23.50	
	1060	4" pipe	"	16	.500	"	9.15	11		20.15	28.50	
	1350	Copper with neoprene ring										
	1430	1-1/2" pipe	1 Plum	20	.400	Ea.	14.05	8.80		22.85	30	
	1440	2" pipe		18	.444		14.85	9.75		24.60	32.50	
	1460	4" pipe		16	.500		19.20	11		30.20	39	

15100 | Building Services Piping

15180 | Heating and Cooling Piping

		CREW	DAILY OUTPUT	LABOR-HOURS	UNIT	2000 BARE COSTS MAT.	LABOR	EQUIP.	TOTAL	TOTAL INCL O&P
0010	**PUMPS, CIRCULATING** Heated or chilled water application									
0600	Bronze, sweat connections, 1/40 HP, in line									
0640	3/4" size	Q-1	16	1	Ea.	110	19.75		129.75	154
1000	Flange connection, 3/4" to 1-1/2" size									
1040	1/12 HP	Q-1	6	2.667	Ea.	300	52.50		352.50	415
1060	1/8 HP		6	2.667		520	52.50		572.50	655
1100	1/3 HP		6	2.667		570	52.50		622.50	710
1140	2" size, 1/6 HP		5	3.200		740	63		803	920
2000	Cast iron, flange connection									
2040	3/4" to 1-1/2" size, in line, 1/12 HP	Q-1	6	2.667	Ea.	197	52.50		249.50	305
2060	1/8 HP		6	2.667		330	52.50		382.50	445
2100	1/3 HP		6	2.667		365	52.50		417.50	490
2101	Pumps, circulating, 3/4" to 1-1/2" size, 1/3 HP		6	2.667		365	52.50		417.50	490
2140	2" size, 1/6 HP		5	3.200		400	63		463	545
2180	2-1/2" size, 1/4 HP		5	3.200		525	63		588	685
2220	3" size, 1/4 HP		4	4		535	79		614	715
2260	1/3 HP		4	4		720	79		799	925
2300	1/2 HP		4	4		750	79		829	955
2340	3/4 HP		4	4		865	79		944	1,075
2380	1 HP		4	4		1,250	79		1,329	1,500
4000	Close coupled, end suction, bronze impeller									
4040	1-1/2" size, 1-1/2 HP, to 40 GPM	Q-1	3	5.333	Ea.	1,025	105		1,130	1,325
4090	2" size, 2 HP, to 50 GPM		3	5.333		1,050	105		1,155	1,350
4100	2" size, 3 HP, to 90 GPM		2.30	6.957		1,150	137		1,287	1,500
4190	2-1/2" size, 3 HP, to 150 GPM		2	8		1,250	158		1,408	1,625
4300	3" size, 5 HP, to 225 GPM		1.80	8.889		1,350	176		1,526	1,800
4410	3" size, 10 HP, to 350 GPM		1.60	10		1,950	198		2,148	2,475
4420	4" size, 7-1/2 HP, to 350 GPM		1.60	10		1,900	198		2,098	2,400
5000	Base mounted, bronze impeller, coupling guard									
5190	2-1/2" size, 3 HP, to 150 GPM	Q-1	1.80	8.889	Ea.	1,750	176		1,926	2,225
5300	3" size, 5 HP, to 225 GPM		1.60	10		1,875	198		2,073	2,375
5410	4" size, 5 HP, to 350 GPM		1.50	10.667		2,025	211		2,236	2,575

15400 | Plumbing Fixtures & Equipment

15410 | Plumbing Fixtures

		CREW	DAILY OUTPUT	LABOR-HOURS	UNIT	2000 BARE COSTS MAT.	LABOR	EQUIP.	TOTAL	TOTAL INCL O&P
0010	**CARRIERS/SUPPORTS** For plumbing fixtures									
0500	Drinking fountain, wall mounted									
0600	Plate type with studs, top back plate	1 Plum	7	1.143	Ea.	26.50	25		51.50	70.50
0700	Top front and back plate	"	7	1.143	"	33	25		58	77.50
3000	Lavatory, concealed arm									
3050	Floor mounted, single									
3100	High back fixture	1 Plum	6	1.333	Ea.	128	29.50		157.50	190
3200	Flat slab fixture		6	1.333		149	29.50		178.50	213
3220	Paraplegic		6	1.333		136	29.50		165.50	199
3250	Floor mounted, back to back									
3300	High back fixtures	1 Plum	5	1.600	Ea.	182	35		217	258
3400	Flat slab fixtures		5	1.600		185	35		220	261
3430	Paraplegic		5	1.600		191	35		226	268
4600	Sink, floor mounted									

For expanded coverage of these items see *Means Mechanical or Plumbing Cost Data 2000*

15400 | Plumbing Fixtures & Equipment

15410 | Plumbing Fixtures

			CREW	DAILY OUTPUT	LABOR-HOURS	UNIT	MAT.	LABOR	EQUIP.	TOTAL	TOTAL INCL O&P	
200	4650	Exposed arm system										200
	4700	Single heavy fixture	1 Plum	5	1.600	Ea.	298	35		333	385	
	5400	Wall mounted, exposed arms, single heavy fixture		5	1.600		126	35		161	197	
	6000	Urinal, floor mounted, 2" or 3" coupling, blowout type		6	1.333		136	29.50		165.50	198	
	6100	With fixture or hanger bolts, blowout or washout		6	1.333		96.50	29.50		126	155	
	6300	Wall mounted, plate type system		6	1.333		97	29.50		126.50	155	
	6980	Water closet, siphon jet										
	7000	Horizontal, adjustable, caulk										
	7040	Single, 4" pipe size	1 Plum	6	1.333	Ea.	205	29.50		234.50	275	
	7050	4" pipe size, paraplegic		6	1.333		205	29.50		234.50	275	
	7100	Double, 4" pipe size		5	1.600		385	35		420	480	
	7110	4" pipe size, paraplegic		5	1.600		205	35		240	284	
	8201	Water closet, residential, vert. centerline, floor mount										
	8300	4" copper sweat, 4" vent	1 Plum	6	1.333	Ea.	185	29.50		214.50	252	
	9000	Water cooler (electric), floor mounted										
	9100	Plate type with bearing plate, single	1 Plum	6	1.333	Ea.	108	29.50		137.50	168	
300	0010	**FAUCETS/FITTINGS**										300
	0150	Bath, faucets, diverter spout combination, sweat	1 Plum	8	1	Ea.	65	22		87	108	
	0200	For integral stops, IPS unions, add					68.50			68.50	75.50	
	0500	Drain, central lift, 1-1/2" IPS male	1 Plum	20	.400		35.50	8.80		44.30	54	
	0600	Trip lever, 1-1/2" IPS male		20	.400		36.50	8.80		45.30	55	
	1000	Kitchen sink faucets, top mount, cast spout		10	.800		48	17.55		65.55	82	
	1100	For spray, add		24	.333		12	7.30		19.30	25.50	
	2000	Laundry faucets, shelf type, IPS or copper unions		12	.667		39.50	14.65		54.15	67.50	
	2020											
	2100	Lavatory faucet, centerset, without drain	1 Plum	10	.800	Ea.	51	17.55		68.55	85	
	2200	For pop-up drain, add		16	.500		14.20	11		25.20	34	
	2800	Self-closing, center set		10	.800		102	17.55		119.55	142	
	3000	Service sink faucet, cast spout, pail hook, hose end		14	.571		67.50	12.55		80.05	94.50	
	4000	Shower by-pass valve with union		18	.444		47.50	9.75		57.25	68	
	4200	Shower thermostatic mixing valve, concealed		8	1		226	22		248	286	
	4300	For inlet strainer, check, and stops, add					43			43	47.50	
	5000	Sillcock, compact, brass, IPS or copper to hose	1 Plum	24	.333		4.94	7.30		12.24	17.55	

15411 | Commercial/Indust Fixtures

			CREW	DAILY OUTPUT	LABOR-HOURS	UNIT	MAT.	LABOR	EQUIP.	TOTAL	TOTAL INCL O&P	
700	0010	**URINALS**										700
	3000	Wall hung, vitreous china, with hanger & self-closing valve										
	3100	Siphon jet type	Q-1	3	5.333	Ea.	278	105		383	480	
	3120	Blowout type		3	5.333		315	105		420	525	
	3300	Rough-in, supply, waste & vent		2.83	5.654		83	112		195	277	
	5000	Stall type, vitreous china, includes valve		2.50	6.400		450	126		576	705	
	5100	3" seam cover, add		12	1.333		118	26.50		144.50	174	
	6980	Rough-in, supply, waste and vent		1.99	8.040		99.50	159		258.50	375	
840	0010	**WASH FOUNTAINS** Rigging not included										840
	1900	Group, foot control										
	2000	Precast terrazzo, circular, 36" diam., 5 or 6 persons	Q-2	3	8	Ea.	2,225	152		2,377	2,700	
	2100	54" diameter for 8 or 10 persons		2.50	9.600		2,725	183		2,908	3,300	
	2400	Semi-circular, 36" diam. for 3 persons		3	8		2,050	152		2,202	2,525	
	2500	54" diam. for 4 or 5 persons		2.50	9.600		2,450	183		2,633	3,000	
	5610	Group, infrared control, barrier free										
	5614	Precast terrazzo										
	5620	Semi-circular 36" diam. for 3 persons	Q-2	3	8	Ea.	2,850	152		3,002	3,400	
	5630	46" diam. for 4 persons		2.80	8.571		3,075	163		3,238	3,675	
	5640	Circular, 54" diam. for 8 persons, button control		2.50	9.600		4,750	183		4,933	5,525	

15400 | Plumbing Fixtures & Equipment

15413 | Electric Water Coolers

			CREW	DAILY OUTPUT	LABOR-HOURS	UNIT	MAT.	LABOR	EQUIP.	TOTAL	TOTAL INCL O&P
900	0010	**WATER COOLER**									900
	0100	Wall mounted, non-recessed									
	0180	8.2 GPH	Q-1	4	4	Ea.	520	79		599	700
	0600	For hot and cold water, add					133			133	146
	1000	Dual height, 8.2 GPH	Q-1	3.80	4.211		715	83		798	920
	2600	Wheelchair type, 8 GPH	"	4	4		1,225	79		1,304	1,475
	4600	Floor mounted, flush-to-wall									
	4680	8.2 GPH	1 Plum	3	2.667	Ea.	530	58.50		588.50	675
	9800	For supply, waste & vent, all coolers	"	2.21	3.620	"	32	79.50		111.50	167

15418 | Resi/Comm/Industrial Fixtures

			CREW	DAILY OUTPUT	LABOR-HOURS	UNIT	MAT.	LABOR	EQUIP.	TOTAL	TOTAL INCL O&P
100	0010	**BATHS** R15100-420									100
	0100	Tubs, recessed porcelain enamel on cast iron, with trim									
	0180	48" x 42"	Q-1	4	4	Ea.	1,025	79		1,104	1,250
	0300	Mat bottom, 4' long		5.50	2.909		875	57.50		932.50	1,050
	0380	5' long		4.40	3.636		305	72		377	455
	0480	Above floor drain, 5' long		4	4		545	79		624	730
	0560	Corner 48" x 44"		4.40	3.636		1,200	72		1,272	1,450
	2000	Enameled formed steel, 4'-6" long		5.80	2.759		240	54.50		294.50	355
	2200	5' long		5.50	2.909		224	57.50		281.50	340
	4600	Module tub & showerwall surround, molded fiberglass									
	4610	5' long x 34" wide x 76" high	Q-1	4	4	Ea.	575	79		654	760
	4750	Handicap with 1-1/2" OD grab bar, antiskid bottom									
	4760	60" x 32-3/4" x 72" high	Q-1	4	4	Ea.	620	79		699	810
	4770	60" x 30" x 71" high with molded seat		3.50	4.571		780	90.50		870.50	1,000
	5300	Whirlpool, porcelain enamel on cast iron, 72" x 36"		1	16		2,600	315		2,915	3,375
	6000	Whirlpool, bath with vented overflow, molded fiberglass									
	6100	66" x 48" x 24"	Q-1	1	16	Ea.	1,975	315		2,290	2,700
	6400	72" x 36" x 24"		1	16		1,925	315		2,240	2,650
	6500	60" x 30" x 21"		1	16		1,700	315		2,015	2,375
	7000	Redwood hot tub system									
	7050	4' diameter x 4' deep	Q-1	1	16	Ea.	1,350	315		1,665	2,000
	7150	6' diameter x 4' deep		.80	20		2,150	395		2,545	3,025
	7200	8' diameter x 4' deep		.80	20		3,175	395		3,570	4,150
	9600	Rough-in, supply, waste and vent, for all above tubs, add		2.07	7.729		102	153		255	365
400	0010	**LAUNDRY SINKS** With trim									400
	0020	Porcelain enamel on cast iron, black iron frame									
	0050	24" x 20", single compartment	Q-1	6	2.667	Ea.	340	52.50		392.50	455
	3000	Plastic, on wall hanger or legs									
	3100	20" x 24", single compartment	Q-1	6.50	2.462	Ea.	108	48.50		156.50	200
	3200	36" x 23", double compartment		5.50	2.909		127	57.50		184.50	235
	9600	Rough-in, supply, waste and vent, for all laundry sinks		2.14	7.477		60	148		208	310
450	0010	**LAVATORIES** With trim, white unless noted otherwise									450
	0500	Vanity top, porcelain enamel on cast iron									
	0600	20" x 18"	Q-1	6.40	2.500	Ea.	199	49.50		248.50	300
	0640	33" x 19" oval	"	6.40	2.500	"	345	49.50		394.50	455
	0860	For color, add					25%				
	1000	Cultured marble, 19" x 17", single bowl	Q-1	6.40	2.500	Ea.	117	49.50		166.50	211
	1120	25" x 22", single bowl		6.40	2.500		153	49.50		202.50	251
	1900	Stainless steel, self-rimming, 25" x 22", single bowl, ledge		6.40	2.500		155	49.50		204.50	253
	1960	17" x 22", single bowl		6.40	2.500		128	49.50		177.50	223
	2600	Steel, enameled, 20" x 17", single bowl		5.80	2.759		101	54.50		155.50	201
	2900	Vitreous china, 20" x 16", single bowl		5.40	2.963		184	58.50		242.50	299
	3200	22" x 13", single bowl		5.40	2.963		193	58.50		251.50	310

For expanded coverage of these items see *Means Mechanical or Plumbing Cost Data 2000*

15400 | Plumbing Fixtures & Equipment

15418 | Resi/Comm/Industrial Fixtures

			CREW	DAILY OUTPUT	LABOR-HOURS	UNIT	2000 BARE COSTS MAT.	LABOR	EQUIP.	TOTAL	TOTAL INCL O&P	
450	3580	Rough-in, supply, waste and vent for all above lavatories	Q-1	2.30	6.957	Ea.	52.50	137		189.50	285	450
	4000	Wall hung										
	4040	Porcelain enamel on cast iron, 16" x 14", single bowl	Q-1	8	2	Ea.	345	39.50		384.50	445	
	4180	20" x 18", single bowl	"	8	2	"	200	39.50		239.50	286	
	4580	For color, add					30%					
	6000	Vitreous china, 18" x 15", single bowl with backsplash	Q-1	7	2.286	Ea.	197	45		242	291	
	6060	19" x 17", single bowl		7	2.286		146	45		191	235	
	6960	Rough-in, supply, waste and vent for above lavatories		1.66	9.639		160	190		350	490	
500	0010	**SHOWERS**										500
	1500	Stall, with drain only. Add for valve and door/curtain										
	1520	32" square	Q-1	2	8	Ea.	315	158		473	605	
	1530	36" square		2	8		410	158		568	710	
	1540	Terrazzo receptor, 32" square		2	8		990	158		1,148	1,350	
	1560	36" square		1.80	8.889		820	176		996	1,200	
	1580	36" corner angle		1.80	8.889		750	176		926	1,125	
	3000	Fiberglass, one piece, with 3 walls, 32" x 32" square		2.40	6.667		290	132		422	540	
	3100	36" x 36" square		2.40	6.667		330	132		462	585	
	3200	Handicap, 1-1/2" O.D. grab bars, nonskid floor										
	3210	48" x 34-1/2" x 72" corner seat	Q-1	2.40	6.667	Ea.	340	132		472	595	
	3250	64" x 65-3/4" x 81-1/2" fold. seat, whlchr.		1.80	8.889		2,675	176		2,851	3,250	
	4000	Polypropylene, with molded-stone floor, 30" x 30"		2	8		370	158		528	670	
	4200	Rough-in, supply, waste and vent for above showers		2.05	7.805		53.50	154		207.50	315	
	5000	Built-in, head, arm, 4 GPM valve	1 Plum	4	2		71	44		115	151	
	5200	Head, arm, by-pass, integral stops, handles		3.60	2.222		179	49		228	278	
	5800	Mixing valve, built-in		6	1.333		93	29.50		122.50	151	
600	0010	**SINKS** With faucets and drain										600
	2000	Kitchen, counter top style, P.E. on C.I., 24" x 21" single bowl	Q-1	5.60	2.857	Ea.	185	56.50		241.50	298	
	2100	31" x 22" single bowl		5.60	2.857		315	56.50		371.50	440	
	2200	32" x 21" double bowl		4.80	3.333		276	66		342	415	
	3000	Stainless steel, self rimming, 19" x 18" single bowl		5.60	2.857		263	56.50		319.50	385	
	3100	25" x 22" single bowl		5.60	2.857		290	56.50		346.50	415	
	3300	43" x 22" double bowl		4.80	3.333		475	66		541	635	
	4000	Steel, enameled, with ledge, 24" x 21" single bowl		5.60	2.857		99	56.50		155.50	203	
	4100	32" x 21" double bowl		4.80	3.333		116	66		182	237	
	4960	For color sinks except stainless steel, add					10%					
	4980	For rough-in, supply, waste and vent, counter top sinks	Q-1	2.14	7.477		60	148		208	310	
	5000	Kitchen, raised deck, P.E. on C.I.										
	5100	32" x 21", dual level, double bowl	Q-1	2.60	6.154	Ea.	350	122		472	585	
	5790	For rough-in, supply, waste & vent, sinks		1.85	8.649		60	171		231	350	
	6650	Service, floor, corner, P.E. on C.I., 28" x 28"		4.40	3.636		475	72		547	640	
	6750	Vinyl coated rim guard, add					61.50			61.50	68	
	6760	Mop sink, molded stone, 24" x 36"	1 Plum	3.33	2.402		187	52.50		239.50	293	
	6770	Mop sink, molded stone, 24" x 36", w/rim 3 sides	"	3.33	2.402		410	52.50		462.50	535	
	6790	For rough-in, supply, waste & vent, floor service sinks	Q-1	1.64	9.756		114	193		307	445	
900	0010	**WATER CLOSETS**										900
	0150	Tank type, vitreous china, incl. seat, supply pipe w/stop										
	0200	Wall hung, one piece	Q-1	5.30	3.019	Ea.	445	59.50		504.50	590	
	0960	For rough-in, supply, waste, vent and carrier		2.73	5.861		255	116		371	470	
	1000	Floor mounted, one piece		5.30	3.019		435	59.50		494.50	580	
	1020	One piece, low profile		5.30	3.019		500	59.50		559.50	650	
	1100	Two piece, close coupled, water saver		5.30	3.019		126	59.50		185.50	238	
	1960	For color, add					30%					
	1980	For rough-in, supply, waste and vent	Q-1	3.05	5.246	Ea.	125	104		229	310	
	3000	Bowl only, with flush valve, seat										

Important: See the Reference Section for critical supporting data - Reference Nos., Crews, & Location Factors

15400 | Plumbing Fixtures & Equipment

15418 | Resi/Comm/Industrial Fixtures

			CREW	DAILY OUTPUT	LABOR-HOURS	UNIT	MAT.	LABOR	EQUIP.	TOTAL	TOTAL INCL O&P	
900	3100	Wall hung	Q-1	5.80	2.759	Ea.	335	54.50		389.50	455	900
	3200	For rough-in, supply, waste and vent, single WC		2.56	6.250		261	123		384	490	
	3300	Floor mounted		5.80	2.759		299	54.50		353.50	420	
	3400	For rough-in, supply, waste and vent, single WC	↓	2.84	5.634	↓	131	111		242	330	

15440 | Plumbing Pumps

			CREW	DAILY OUTPUT	LABOR-HOURS	UNIT	MAT.	LABOR	EQUIP.	TOTAL	TOTAL INCL O&P	
240	0010	**PUMPS, PRESSURE BOOSTER SYSTEM**										240
	0200	Pump system, with diaphragm tank, control, press. switch										
	0300	1 HP pump	Q-1	1.30	12.308	Ea.	2,950	243		3,193	3,650	
	0400	1-1/2 HP pump		1.25	12.800		2,975	253		3,228	3,700	
	0420	2 HP pump		1.20	13.333		3,050	263		3,313	3,775	
	0440	3 HP pump	↓	1.10	14.545		3,100	287		3,387	3,875	
	0460	5 HP pump	Q-2	1.50	16		3,425	305		3,730	4,275	
	0480	7-1/2 HP pump		1.42	16.901		3,825	320		4,145	4,725	
	0500	10 HP pump	↓	1.34	17.910		4,000	340		4,340	4,975	
	1000	Pump/ energy storage system, diaphragm tank, 3 HP pump										
	1100	motor, PRV, switch, gauge, control center, flow switch										
	1200	125 lb. working pressure	Q-2	.70	34.286	Ea.	8,575	650		9,225	10,500	
	1300	250 lb. working pressure	"	.64	37.500	"	9,375	715		10,090	11,500	
800	0010	**PUMPS, SEWAGE EJECTOR** With operating and level controls										800
	0100	Simplex system incl. tank, cover, pump 15' head										
	0500	37 gal PE tank, 12 GPM, 1/2 HP, 2" discharge	Q-1	3.20	5	Ea.	365	99		464	565	
	0510	3" discharge		3.10	5.161		390	102		492	600	
	0600	45 gal. coated stl tank, 12 GPM, 1/2 HP, 2" discharge		3	5.333		645	105		750	885	
	0610	3" discharge		2.90	5.517		675	109		784	920	
	0700	70 gal. PE tank, 12 GPM, 1/2 HP, 2" discharge		2.60	6.154		730	122		852	1,000	
	0710	3" discharge		2.40	6.667		780	132		912	1,075	
	0730	87 GPM, 0.7 HP, 2" discharge		2.50	6.400		980	126		1,106	1,275	
	0740	3" discharge		2.30	6.957		1,000	137		1,137	1,325	
	0760	134 GPM, 1 HP, 2" discharge		2.20	7.273		1,025	144		1,169	1,350	
	0770	3" discharge	↓	2	8		1,100	158		1,258	1,450	
	1040	Duplex system incl. tank, covers, pumps										
	1060	110 gal. fiberglass tank, 24 GPM, 1/2 HP, 2" discharge	Q-1	1.60	10	Ea.	1,425	198		1,623	1,900	
	1080	3" discharge		1.40	11.429		1,475	226		1,701	2,000	
	1100	174 GPM, .7 HP, 2" discharge		1.50	10.667		1,875	211		2,086	2,400	
	1120	3" discharge		1.30	12.308		1,950	243		2,193	2,525	
	1140	268 GPM, 1 HP, 2" discharge		1.20	13.333		2,025	263		2,288	2,650	
	1160	3" discharge	↓	1	16		2,100	315		2,415	2,825	
	1260	135 gal. coated stl. tank, 24 GPM, 1/2 HP, 2" discharge	Q-2	1.70	14.118		1,450	269		1,719	2,050	
	2000	3" discharge		1.60	15		1,525	285		1,810	2,150	
	2640	174 GPM, .7 HP, 2" discharge		1.60	15		1,900	285		2,185	2,550	
	2660	3" discharge		1.50	16		2,000	305		2,305	2,700	
	2700	268 GPM, 1 HP, 2" discharge		1.30	18.462		2,050	350		2,400	2,825	
	3040	3" discharge	↓	1.10	21.818	↓	2,150	415		2,565	3,050	
900	0010	**PUMPS, PEDESTAL SUMP** With float control										900
	0400	Molded PVC base, 21 GPM at 15' head, 1/3 HP	1 Plum	5	1.600	Ea.	86.50	35		121.50	153	
	0800	Iron base, 21 GPM at 15' head, 1/3 HP		5	1.600		110	35		145	179	
	1200	Solid brass, 21 GPM at 15' head, 1/3 HP	↓	5	1.600	↓	179	35		214	255	

15470 | Domstc Water Filt Equip

			CREW	DAILY OUTPUT	LABOR-HOURS	UNIT	MAT.	LABOR	EQUIP.	TOTAL	TOTAL INCL O&P	
400	0010	**WATER FILTERS** Purification and treatment										400
	1000	Cartridge style, dirt and rust type	1 Plum	12	.667	Ea.	26	14.65		40.65	53	
	1200	Replacement cartridge		32	.250		.81	5.50		6.31	9.95	
	1600	Taste and odor type	↓	12	.667	↓	26	14.65		40.65	52.50	

For expanded coverage of these items see *Means Mechanical or Plumbing Cost Data 2000*

15400 | Plumbing Fixtures & Equipment

15470 | Domstc Water Filt Equip

		CREW	DAILY OUTPUT	LABOR-HOURS	UNIT	MAT.	LABOR	EQUIP.	TOTAL	TOTAL INCL O&P
400	1700	Replacement cartridge	1 Plum	32	.250	Ea.	1.90	5.50		7.40	11.15
	8000	Commercial, fully automatic or push button automatic									
	8200	Iron removal, 660 GPH, 1" pipe size	Q-1	1.50	10.667	Ea.	1,525	211		1,736	2,025
	8240	1500 GPH, 1-1/4" pipe size	"	1	16	"	2,575	315		2,890	3,350

15480 | Domestic Water Heaters

			CREW	DAILY OUTPUT	LABOR-HOURS	UNIT	MAT.	LABOR	EQUIP.	TOTAL	TOTAL INCL O&P
200	0010	**WATER HEATERS**										200
	1000	Residential, electric, glass lined tank, 5 yr, 10 gal., single element	1 Plum	2.30	3.478	Ea.	188	76.50		264.50	335
	1060	30 gallon, double element		2.20	3.636		239	80		319	395
	1100	52 gallon, double element		2	4		296	88		384	470
	2000	Gas fired, glass lined tank, 5 yr, vent not incl., 20 gallon		2.10	3.810		265	83.50		348.50	430
	2040	30 gallon		2	4		315	88		403	490
	2100	75 gallon		1.50	5.333		805	117		922	1,075
	3000	Oil fired, glass lined tank, 5 yr, vent not included, 30 gallon		2	4		745	88		833	965
	3040	50 gallon		1.80	4.444		1,000	97.50		1,097.50	1,250
	4000	Commercial, 100° rise. NOTE: for each size tank, a range of									
	4010	heaters between the ones shown are available									
	4020	Electric									
	4100	5 gal., 3 kW, 12 GPH, 208V	1 Plum	2	4	Ea.	1,250	88		1,338	1,525
	4120	10 gal., 6 kW, 25 GPH, 208V		2	4		1,375	88		1,463	1,675
	4140	50 gal., 9 kW, 37 GPH, 208V		1.80	4.444		1,900	97.50		1,997.50	2,250
	4160	50 gal., 36 kW, 148 GPH, 208V		1.80	4.444		2,900	97.50		2,997.50	3,350
	4200	80 gal., 36 kW, 148 GPH, 208V		1.50	5.333		3,250	117		3,367	3,775
	4220	100 gal., 36 kW, 148 GPH, 208V		1.20	6.667		3,875	146		4,021	4,525
	4240	120 gal., 36 kW, 148 GPH, 208V		1.20	6.667		3,975	146		4,121	4,625
	4280	150 gal., 120 kW, 490 GPH, 480V		1	8		11,600	176		11,776	13,100
	4300	200 gal., 15 kW, 61 GPH, 480V	Q-1	1.70	9.412		8,975	186		9,161	10,200
	6000	Gas fired, flush jacket, std. controls, vent not incl.									
	6040	75 MBH input, 73 GPH	1 Plum	1.40	5.714	Ea.	950	125		1,075	1,250
	6060	98 MBH input, 95 GPH		1.40	5.714		1,625	125		1,750	1,975
	6080	120 MBH input, 110 GPH		1.20	6.667		1,775	146		1,921	2,200
	6120	140 MBH input, 130 GPH		1	8		3,075	176		3,251	3,675
	6140	155 MBH input, 150 GPH		.80	10		3,700	220		3,920	4,450
	6180	200 MBH input, 192 GPH		.60	13.333		3,175	293		3,468	3,975
	6220	260 MBH input, 250 GPH	Q-1	.80	20		3,600	395		3,995	4,600
	6240	360 MBH input, 360 GPH		.80	20		4,225	395		4,620	5,300
	6260	500 MBH input, 480 GPH		.70	22.857		5,850	450		6,300	7,200
	6280	725 MBH input, 690 GPH		.60	26.667		7,025	525		7,550	8,600
	8000	Oil fired, flush jacket, std. controls, vent not incl.									
	8060	103 MBH gross output, 116 GPH	1 Plum	1.10	7.273	Ea.	895	160		1,055	1,250
	8080	122 MBH gross output, 141 GPH		1	8		1,875	176		2,051	2,375
	8120	168 MBH gross output, 192 GPH		.60	13.333		2,175	293		2,468	2,850
	8140	195 MBH gross output, 224 GPH		.50	16		2,375	350		2,725	3,200
	8180	262 MBH gross output, 315 GPH	Q-1	.70	22.857		2,650	450		3,100	3,675
	8200	315 MBH gross output, 409 GPH		.70	22.857		3,625	450		4,075	4,750
	8220	420 MBH gross output, 504 GPH		.60	26.667		4,125	525		4,650	5,425
	8240	525 MBH gross output, 630 GPH		.50	32		5,650	630		6,280	7,275
940	0010	**WATER HEATER PACKAGED SYSTEMS**										940
	1000	Car wash package, continuous duty, high recovery, gas fired									
	1040	100° rise, 180 MBH input, 174 GPH	1 Plum	3	2.667	Ea.	2,875	58.50		2,933.50	3,275
	1060	280 MBH input, 270 GPH		2.50	3.200		3,075	70		3,145	3,500
	1100	480 MBH input, 464 GPH		2	4		3,725	88		3,813	4,250
	1120	605 MBH input, 584 GPH		1.50	5.333		4,250	117		4,367	4,875
	1140	700 MBH input, 676 GPH		1	8		4,425	176		4,601	5,175
	3000	Combination dishwasher & general purpose, 2 temp. gas fired									

15400 | Plumbing Fixtures & Equipment

15480 | Domestic Water Heaters

		CREW	DAILY OUTPUT	LABOR-HOURS	UNIT	MAT.	LABOR	EQUIP.	TOTAL	TOTAL INCL O&P	
940	3040	124 GPH @ 140° rise; 434 GPH @ 40° rise	1 Plum	2	4	Ea.	2,925	88		3,013	3,375	940
	3060	193 GPH @ 140° rise; 677 GPH @ 40° rise	"	1.60	5	"	3,350	110		3,460	3,875	
	5000	Coin laundry units, gas fired, 100° rise										
	5020	Single heater,										
	5040	280 MBH input, 270 GPH	1 Plum	1.60	5	Ea.	3,350	110		3,460	3,875	
	5060	400 MBH input, 386 GPH	"	1.30	6.154	"	4,425	135		4,560	5,075	

15500 | Heat Generation Equipment

15510 | Heating Boilers and Accessories

		CREW	DAILY OUTPUT	LABOR-HOURS	UNIT	MAT.	LABOR	EQUIP.	TOTAL	TOTAL INCL O&P	
300	0010	**BOILERS, ELECTRIC, ASME** Standard controls and trim R15500-050	Q-19	1.20	20	Ea.	7,900	415		8,315	9,375	300
	1000	Steam, 6 KW, 20.5 MBH										
	1160	60 KW, 205 MBH		1	24		9,775	495		10,270	11,500	
	2000	Hot water, 12 KW, 41 MBH		1.30	18.462		3,400	380		3,780	4,350	
	2040	24 KW, 82 MBH		1.20	20		3,625	415		4,040	4,675	
	2070	36 KW, 123 MBH		1.20	20		3,950	415		4,365	5,025	
400	0010	**BOILERS, GAS FIRED** Natural or propane, standard controls										400
	1000	Cast iron, with insulated jacket										
	2000	Steam, gross output, 81 MBH	Q-7	1.40	22.857	Ea.	1,125	455		1,580	2,000	
	2080	203 MBH		.90	35.556		1,900	710		2,610	3,275	
	2180	400 MBH		.56	56.838		3,150	1,125		4,275	5,325	
	2280	1275 MBH		.31	102		7,675	2,025		9,700	11,800	
	2380	3060 MBH		.16	200		15,200	3,975		19,175	23,300	
	2400	3570 MBH		.15	210		16,900	4,200		21,100	25,500	
	3000	Hot water, gross output, 80 MBH		1.46	21.918		1,000	435		1,435	1,850	
	3020	100 MBH		1.35	23.704		1,175	470		1,645	2,075	
	3080	203 MBH		1	32		1,775	635		2,410	3,025	
	3180	400 MBH		.64	50		2,975	995		3,970	4,925	
	3280	1,275 MBH		.36	89.888		7,400	1,800		9,200	11,100	
	3340	2,312 MBH		.22	148		12,600	2,950		15,550	18,700	
	3400	3,808 MBH		.16	195		17,800	3,875		21,675	26,000	
	4000	Steel, insulating jacket										
	6000	Hot water, including burner & one zone valve, gross output										
	6010	51.2 MBH	Q-6	2	12	Ea.	1,575	230		1,805	2,100	
	6020	72 MBH		2	12		1,750	230		1,980	2,300	
	6040	89 MBH		1.90	12.632		1,775	242		2,017	2,375	
	6060	105 MBH		1.80	13.333		2,000	256		2,256	2,625	
	6080	132 MBH		1.70	14.118		2,275	271		2,546	2,950	
	6100	155 MBH		1.50	16		2,650	305		2,955	3,400	
	6110	186 MBH		1.40	17.143		3,200	330		3,530	4,050	
	6140	292 MBH		1.20	20		4,575	385		4,960	5,650	
	7000	For tankless water heater on smaller gas units, add					10%					
	7990	Special feature gas fired boilers										
	8000	Pulse combustion, standard controls / trim, 44,000 BTU	Q-5	1.50	10.667	Ea.	2,275	212		2,487	2,850	
	8050	88,000 BTU		1.40	11.429		2,700	227		2,927	3,325	
	8080	134,000 BTU		1.20	13.333		3,500	265		3,765	4,275	
500	0010	**BOILERS, OIL FIRED** Standard controls, flame retention burner										500
	1000	Cast iron, with insulated flush jacket										
	2000	Steam, gross output, 109 MBH	Q-7	1.20	26.667	Ea.	1,325	530		1,855	2,350	
	2060	207 MBH		.90	35.556		1,825	710		2,535	3,175	

For expanded coverage of these items see *Means Mechanical or Plumbing Cost Data 2000*

15500 | Heat Generation Equipment

15510 | Heating Boilers and Accessories

			CREW	DAILY OUTPUT	LABOR-HOURS	UNIT	MAT.	LABOR	EQUIP.	TOTAL	TOTAL INCL O&P	
500	2180	1,084 MBH	Q-7	.38	85.106	Ea.	6,500	1,700		8,200	9,950	500
	3000	Hot water, same price as steam										
	5000	Steel, insulated jacket, burner										
	7000	Hot water, gross output, 103 MBH	Q-6	1.60	15	Ea.	1,300	288		1,588	1,900	
	7020	122 MBH		1.45	16.506		2,100	315		2,415	2,825	
	7060	168 MBH		1.30	18.405		3,625	355		3,980	4,575	
	7080	225 MBH		1.22	19.704		4,125	380		4,505	5,150	
700	0010	**BOILERS, PACKAGED SCOTCH MARINE** Steam or hot water										700
	1000	Packaged fire tube, #2 oil, gross output										
	1020	3348 MBH, 100 HP	Q-7	.21	152	Ea.	28,100	3,025		31,125	35,900	
	1120	To fire #6, add	"	.83	38.554	"	6,350	765		7,115	8,275	
	2000	Packaged water tube, #2 oil, gross output										
	2040	1200 MBH	Q-7	.50	64	Ea.	11,700	1,275		12,975	15,000	
	2060	1600 MBH		.40	80		14,700	1,600		16,300	18,700	
	2080	2400 MBH		.30	106		18,800	2,125		20,925	24,200	
	2100	3200 MBH		.25	128		21,500	2,550		24,050	27,800	
	2120	4800 MBH		.20	160		28,100	3,175		31,275	36,200	
	2140	For gas fired, add		.40	80		1,900	1,600		3,500	4,700	
880	0010	**SWIMMING POOL HEATERS** Not including wiring, external										880
	0020	piping, base or pad,										
	0060	Gas fired, input, 115 MBH	Q-6	3	8	Ea.	1,500	153		1,653	1,900	
	0100	135 MBH		2	12		1,675	230		1,905	2,225	
	0160	155 MBH		1.50	16		1,775	305		2,080	2,450	
	0220	235 MBH		.70	34.286		2,825	655		3,480	4,175	
	0240	285 MBH		.60	40		3,625	765		4,390	5,275	
	0280	500 MBH		.40	60		5,600	1,150		6,750	8,050	
	0370	1,200 MBH		.21	114		11,100	2,200		13,300	15,900	
	0400	1,800 MBH		.14	171		15,600	3,275		18,875	22,500	
	0420	3,000 MBH		.11	218		25,300	4,175		29,475	34,700	
	0440	3,750 MBH		.09	266		31,400	5,100		36,500	43,100	
	2000	Electric, 12 KW, 4,800 gallon pool	Q-19	3	8		1,475	165		1,640	1,900	
	2020	15 KW, 7,200 gallon pool		2.80	8.571		1,475	177		1,652	1,925	
	2040	24 KW, 9,600 gallon pool		2.40	10		1,975	206		2,181	2,525	
	2100	55 KW, 24,000 gallon pool		1.20	20		2,825	415		3,240	3,775	

15530 | Furnaces

			CREW	DAILY OUTPUT	LABOR-HOURS	UNIT	MAT.	LABOR	EQUIP.	TOTAL	TOTAL INCL O&P	
400	0010	**FURNACES** Hot air heating, blowers, standard controls										400
	0020	not including gas, oil or flue piping										
	3000	Gas, AGA certified, direct drive models										
	3040	60 MBH input	Q-9	3.80	4.211	Ea.	580	81.50		661.50	780	
	3100	100 MBH input		3.20	5		655	97		752	885	
	3130	150 MBH input		2.80	5.714		875	111		986	1,150	
	3140	200 MBH input		2.60	6.154		1,850	119		1,969	2,225	
	4000	For starter plenum, add		16	1		61.50	19.35		80.85	100	
	6000	Oil, UL listed, atomizing gun type burner										
	6020	56 MBH output	Q-9	3.60	4.444	Ea.	745	86		831	965	
	6040	95 MBH output		3.40	4.706		790	91		881	1,025	
	6080	151 MBH output		3	5.333		1,225	103		1,328	1,500	
	6100	200 MBH input		2.60	6.154		1,950	119		2,069	2,350	

15540 | Fuel-Fired Heaters

300	0010	**DUCT FURNACES** Includes burner, controls, stainless steel										300
	0020	heat exchanger. Gas fired, electric ignition										

Important: See the Reference Section for critical supporting data - Reference Nos., Crews, & Location Factors

15500 | Heat Generation Equipment

15540 | Fuel-Fired Heaters

			CREW	DAILY OUTPUT	LABOR-HOURS	UNIT	MAT.	LABOR	EQUIP.	TOTAL	TOTAL INCL O&P	
300	0030	Indoor installation										300
	0080	100 MBH output	Q-5	5	3.200	Ea.	1,000	63.50		1,063.50	1,200	
	0100	120 MBH output		4	4		1,150	79.50		1,229.50	1,375	
	0130	200 MBH output		2.70	5.926		1,575	118		1,693	1,925	
	0140	240 MBH output		2.30	6.957		1,650	138		1,788	2,025	
	0160	280 MBH output		2	8		1,825	159		1,984	2,300	
	0180	320 MBH output	↓	1.60	10		2,000	199		2,199	2,525	
	0300	For powered venter and adapter, add					250			250	275	
900	0010	**SPACE HEATERS** Cabinet, grilles, fan, controls, burner,										900
	0020	thermostat, no piping. For flue see 15550										
	1000	Gas fired, floor mounted										
	1100	60 MBH output	Q-5	10	1.600	Ea.	530	32		562	640	
	1140	100 MBH output		8	2		580	40		620	700	
	1180	180 MBH output	↓	6	2.667	↓	790	53		843	960	
	1500	Rooftop mounted, gravity vent, stainless steel exchanger										
	1520	75 MBH output	Q-6	4	6	Ea.	3,275	115		3,390	3,800	
	1560	120 MBH output		3.30	7.273		3,575	139		3,714	4,150	
	1600	190 MBH output		2.60	9.231		4,150	177		4,327	4,875	
	1620	225 MBH output		2.30	10.435		4,225	200		4,425	4,950	
	1640	300 MBH output		1.90	12.632		5,025	242		5,267	5,925	
	1660	375 MBH output		1.40	17.143		5,700	330		6,030	6,825	
	1700	600 MBH output		1	24		7,000	460		7,460	8,450	
	1720	750 MBH output		.80	30		12,000	575		12,575	14,200	
	1760	1200 MBH output	↓	.30	80		16,300	1,525		17,825	20,500	
	1800	For power vent, add					10%					
	2000	Suspension mounted, propeller fan, 20 MBH output	Q-5	8.50	1.882		400	37.50		437.50	500	
	2100	130 MBH output		5	3.200		730	63.50		793.50	910	
	2240	320 MBH output	↓	2	8		1,500	159		1,659	1,925	
	2500	For powered venter and adapter, add					243			243	267	
	5000	Wall furnace, 17.5 MBH output	Q-5	6	2.667		470	53		523	605	
	5020	24 MBH output		5	3.200		485	63.50		548.50	635	
	5040	35 MBH output	↓	4	4	↓	650	79.50		729.50	845	

15550 | Breechings, Chimneys & Stacks

			CREW	DAILY OUTPUT	LABOR-HOURS	UNIT	MAT.	LABOR	EQUIP.	TOTAL	TOTAL INCL O&P	
440	0010	**VENT CHIMNEY** Prefab metal, U.L. listed										440
	0020	Gas, double wall, galvanized steel										
	0080	3" diameter	Q-9	72	.222	V.L.F.	3.38	4.30		7.68	10.95	
	0100	4" diameter		68	.235		4.14	4.55		8.69	12.25	
	0120	5" diameter		64	.250		4.91	4.84		9.75	13.55	
	0140	6" diameter		60	.267		5.70	5.15		10.85	15	
	0160	7" diameter		56	.286		8.40	5.55		13.95	18.60	
	0180	8" diameter		52	.308		9.35	5.95		15.30	20.50	
	0200	10" diameter		48	.333		19.75	6.45		26.20	32.50	
	7800	All fuel, double wall, stainless steel, 6" diameter		60	.267		28	5.15		33.15	39	
	7802	7" diameter		56	.286		36	5.55		41.55	49	
	7806	10" diameter	↓	48	.333	↓	61	6.45		67.45	78.50	

For expanded coverage of these items see *Means Mechanical or Plumbing Cost Data 2000*

15600 | Refrigeration Equipment

15620 | Packaged Water Chillers

		CREW	DAILY OUTPUT	LABOR-HOURS	UNIT	2000 BARE COSTS MAT.	LABOR	EQUIP.	TOTAL	TOTAL INCL O&P
100 0010	**ABSORPTION WATER CHILLERS**									100
0020	Steam or hot water, water cooled R15700-040									
0050	100 ton	Q-7	.13	240	Ea.	110,500	4,775		115,275	129,500
0100	148 ton	"	.12	258	"	112,000	5,125		117,125	132,000
3000	Gas fired, air cooled									
3270	10 ton	Q-5	.40	40	Ea.	11,700	795		12,495	14,100
600 0010	**CENTRIFUGAL/RECIP. WATER CHILLERS**, With standard controls									600
0020	Centrifugal liquid chiller, water cooled									
0030	not including water tower									
0200	Packaged unit, water cooled, not incl. tower									
0240	200 ton	Q-7	.13	251	Ea.	85,500	5,025		90,525	102,500
0490	Reciprocating, packaged w/integral air cooled condenser, 15 ton cool		.37	86.486		12,300	1,725		14,025	16,400
0500	20 ton cooling		.34	94.955		15,500	1,900		17,400	20,200
0510	25 ton cooling		.33	97.859		17,600	1,950		19,550	22,600
0980	Water cooled, multiple compress., semi-hermetic, tower not incl.									
1000	15 ton cooling	Q-6	.36	65.934	Ea.	16,000	1,275		17,275	19,600
1040	25 ton cooling	Q-7	.36	89.888		18,700	1,800		20,500	23,600
1060	30 ton cooling		.31	101		19,900	2,025		21,925	25,300
1100	50 ton cooling		.28	113		25,100	2,275		27,375	31,400
1130	70 ton cooling		.23	139		34,100	2,775		36,875	42,100
1150	90 ton cooling		.19	164		47,600	3,275		50,875	58,000
1160	100 ton cooling		.18	179		48,700	3,575		52,275	59,500
4000	Packaged chiller, with remote air cooled condensers incl.									
4020	15 ton cooling	Q-7	.30	108	Ea.	13,800	2,150		15,950	18,800
4040	25 ton cooling		.25	125		19,000	2,500		21,500	25,200
4050	30 ton cooling		.24	133		21,600	2,675		24,275	28,200
4070	50 ton cooling		.21	153		31,400	3,050		34,450	39,600
4080	60 ton cooling		.19	164		34,200	3,275		37,475	43,000

15640 | Packaged Cooling Towers

		CREW	DAILY OUTPUT	LABOR-HOURS	UNIT	MAT.	LABOR	EQUIP.	TOTAL	TOTAL INCL O&P
400 0010	**COOLING TOWERS** Packaged units									400
0070	Galvanized steel									
0080	Draw thru, single flow									
0100	Belt drive, 60 tons	Q-6	90	.267	TonAC	76	5.10		81.10	92
0150	95 tons		100	.240		67	4.60		71.60	81
0200	110 tons		109	.220		65.50	4.22		69.72	79
1000	For higher capacities, use multiples									
5000	Fiberglass									
5010	Draw thru									
5100	60 tons	Q-6	1.50	16	Ea.	2,775	305		3,080	3,550
5120	125 tons		.99	24.242		5,675	465		6,140	7,025
5140	300 tons		.43	55.814		13,300	1,075		14,375	16,400
5160	600 tons		.22	109		24,200	2,100		26,300	30,100
5180	1000 tons		.15	160		41,500	3,075		44,575	50,500
6000	Stainless steel									
6010	Draw thru									
6100	60 tons	Q-6	1.50	16	Ea.	7,200	305		7,505	8,425
6120	110 tons		.99	24.242		11,500	465		11,965	13,500
6140	300 tons		.43	55.814		31,900	1,075		32,975	36,900
6160	600 tons		.22	109		49,400	2,100		51,500	58,000
6180	1000 tons		.15	160		80,500	3,075		83,575	93,500

15660 | Liquid Coolers/Evap Condensers

		CREW	DAILY OUTPUT	LABOR-HOURS	UNIT	MAT.	LABOR	EQUIP.	TOTAL	TOTAL INCL O&P
100 0010	**CONDENSERS** Ratings are for 30°F TD, R-22									100
0080	Air cooled, belt drive, propeller fan									

15600 | Refrigeration Equipment

15660 | Liquid Coolers/Evap Condensers

			CREW	DAILY OUTPUT	LABOR-HOURS	UNIT	MAT.	LABOR	EQUIP.	TOTAL	TOTAL INCL O&P	
100	0220	45 ton	Q-6	.70	34.286	Ea.	9,900	655		10,555	12,000	100
	0240	50 ton		.69	34.985		10,200	670		10,870	12,300	
	0260	54 ton	↓	.63	37.795	↓	10,600	725		11,325	12,800	

15700 | Heating/Ventilating/Air Conditioning Equipment

15720 | Air Handling Units

			CREW	DAILY OUTPUT	LABOR-HOURS	UNIT	MAT.	LABOR	EQUIP.	TOTAL	TOTAL INCL O&P	
200	0010	**CENTRAL STATION AIR-HANDLING UNIT** Chilled water										200
	1000	Modular, capacity at 700 FPM face velocity										
	1100	1300 CFM	Q-5	1.20	13.333	Ea.	2,325	265		2,590	3,025	
	1200	1900 CFM		1.10	14.545		2,600	289		2,889	3,325	
	1300	3200 CFM	↓	.80	20		3,075	400		3,475	4,050	
	1400	5400 CFM	Q-6	.80	30	↓	4,650	575		5,225	6,050	

15730 | Unitary Air Conditioning Equip

			CREW	DAILY OUTPUT	LABOR-HOURS	UNIT	MAT.	LABOR	EQUIP.	TOTAL	TOTAL INCL O&P	
200	0010	**COMPUTER ROOM UNITS**										200
	1000	Air cooled, includes remote condenser but not										
	1020	interconnecting tubing or refrigerant										
	1080	3 ton	Q-5	.50	32	Ea.	7,025	635		7,660	8,775	
	1120	5 ton		.45	35.556		9,400	710		10,110	11,500	
	1160	6 ton		.30	53.333		17,300	1,050		18,350	20,900	
	1200	8 ton		.27	59.259		19,600	1,175		20,775	23,600	
	1240	10 ton		.25	64		20,400	1,275		21,675	24,500	
	1280	15 ton		.22	72.727		22,800	1,450		24,250	27,400	
	1290	18 ton	↓	.20	80		25,900	1,600		27,500	31,100	
	1320	20 ton	Q-6	.29	82.759	↓	27,200	1,575		28,775	32,500	
500	0010	**PACKAGED TERMINAL AIR CONDITIONER** Cabinet, wall sleeve,										500
	0100	louver, electric heat, thermostat, manual changeover, 208 V										
	0200	6,000 BTUH cooling, 8800 BTU heat	Q-5	6	2.667	Ea.	915	53		968	1,100	
	0220	9,000 BTUH cooling, 13,900 BTU heat		5	3.200		935	63.50		998.50	1,125	
	0240	12,000 BTUH cooling, 13,900 BTU heat		4	4		970	79.50		1,049.50	1,200	
	0260	15,000 BTUH cooling, 13,900 BTU heat		3	5.333		1,050	106		1,156	1,325	
	0320	30,000 BTUH cooling, 10 KW heat		1.40	11.429		1,550	227		1,777	2,075	
	0340	36,000 BTUH cooling, 10 KW heat		1.25	12.800		1,625	255		1,880	2,225	
	0360	42,000 BTUH cooling, 10 KW heat		1	16		2,050	320		2,370	2,800	
	0380	48,000 BTUH cooling, 10 KW heat	↓	.90	17.778	↓	2,225	355		2,580	3,025	
600	0010	**ROOF TOP AIR CONDITIONERS** Standard controls, curb, economizer										600
	1000	Single zone, electric cool, gas heat										
	1140	5 ton cooling, 112 MBH heating	Q-5	.56	28.520	Ea.	4,675	570		5,245	6,100	
	1160	10 ton cooling, 200 MBH heating	Q-6	.67	35.982		9,500	690		10,190	11,600	
	1180	15 ton cooling, 270 MBH heating	"	.57	42.032		12,500	805		13,305	15,100	
	1220	30 ton cooling, 540 MBH heating	Q-7	.47	68.376		29,100	1,350		30,450	34,300	
	1260	50 ton cooling, 810 MBH heating		.28	113		46,000	2,275		48,275	54,500	
	1270	80 ton cooling, 1000 MBH heating	↓	.17	182	↓	86,500	3,650		90,150	101,500	
800	0010	**WINDOW UNIT AIR CONDITIONERS**										800
	4000	Portable/window, 15 amp 125V grounded receptacle required										
	4060	5000 BTUH	1 Carp	8	1	Ea.	219	19.70		238.70	275	
	4340	6000 BTUH	↓	8	1	↓	265	19.70		284.70	325	

For expanded coverage of these items see *Means Mechanical or Plumbing Cost Data 2000*

15700 | Heating/Ventilating/Air Conditioning Equipment

15730 | Unitary Air Conditioning Equip

			CREW	DAILY OUTPUT	LABOR-HOURS	UNIT	MAT.	LABOR	EQUIP.	TOTAL	TOTAL INCL O&P	
800	4480	8000 BTUH	1 Carp	6	1.333	Ea.	279	26.50		305.50	350	800
	4500	10,000 BTUH	↓	6	1.333		380	26.50		406.50	460	
	4520	12,000 BTUH	L-2	8	2	↓	470	34.50		504.50	575	
840	0010	**SELF-CONTAINED SINGLE PACKAGE**										840
	0100	Air cooled, for free blow or duct, including remote condenser										
	0220	5 ton cooling	Q-6	1.20	20	Ea.	5,525	385		5,910	6,700	
	0221	Self-contained, air cooled, 5 ton cooling		1.20	20			385		385	635	
	0230	7.5 ton cooling	↓	.90	26.667		7,500	510		8,010	9,100	
	0240	10 ton cooling	Q-7	1	32		9,400	635		10,035	11,400	
	0250	15 ton cooling		.95	33.684		13,300	670		13,970	15,700	
	0260	20 ton cooling		.90	35.556		15,300	710		16,010	18,000	
	0270	25 ton cooling	↓	.85	37.647	↓	18,400	750		19,150	21,600	
	1000	Water cooled for free blow or duct, not including tower										
	1160	20 ton cooling	Q-7	.80	40	Ea.	13,300	795		14,095	15,900	
	1170	25 ton cooling		.75	42.667		15,700	850		16,550	18,600	
	1180	30 ton cooling		.70	45.714		17,600	910		18,510	20,900	
	1200	40 ton cooling	↓	.40	80	↓	22,300	1,600		23,900	27,100	

15740 | Heat Pumps

			CREW	DAILY OUTPUT	LABOR-HOURS	UNIT	MAT.	LABOR	EQUIP.	TOTAL	TOTAL INCL O&P	
100	0010	**HEAT PUMPS** (Not including interconnecting tubing)										100
	1000	Air to air, split system, not including curbs, pads, or ductwork										
	1060	5 ton cooling, 27 MBH heat @ 0°F	Q-5	.50	32	Ea.	4,275	635		4,910	5,750	
	1080	7.5 ton cooling, 33 MBH heat @ 0°F	"	.30	53.333		6,575	1,050		7,625	8,975	
	1100	10 ton cooling, 50 MBH heat @ 0°F	Q-6	.38	63.158		8,500	1,200		9,700	11,400	
	1120	15 ton cooling, 64 MBH heat @ 0°F		.26	92.308		11,700	1,775		13,475	15,800	
	1130	20 ton cooling, 85 MBH heat @ 0°F	↓	.20	120		17,100	2,300		19,400	22,700	
	1160	30 ton cooling, 163 MBH heat @ 0°F	Q-7	.17	188		23,700	3,750		27,450	32,300	
	1200	50 ton cooling, 220 MBH heat @ 0°F	"	.10	320	↓	47,400	6,375		53,775	62,500	
	2000	Water source to air, single package										
	2220	5 ton cooling, 29 MBH heat @ 75°F	Q-5	.90	17.778	Ea.	2,125	355		2,480	2,925	
	2240	7.5 ton cooling, 35 MBH heat @ 75°F		.60	26.667		2,425	530		2,955	3,525	
	2260	10 ton cooling, 50 MBH heat @ 75°F	↓	.53	30.189		7,600	600		8,200	9,350	
	2280	15 ton cooling, 64 MBH heat @ 75°F	Q-6	.47	51.064		8,000	980		8,980	10,400	
	2300	20 ton cooling, 100 MBH heat @ 75°F		.41	58.537		8,250	1,125		9,375	10,900	
	2320	30 ton cooling, (twin 15 ton units)		.23	102		16,000	1,950		17,950	20,800	
	2340	40 ton cooling, (twin 20 ton units)		.20	117		16,500	2,250		18,750	21,800	
	2360	50 ton cooling, (twin 15 + 20 ton unit)	↓	.15	160	↓	24,300	3,075		27,375	31,800	
	3960	For supplementary heat coil, add					10%					

15750 | Humidity Control Equipment

			CREW	DAILY OUTPUT	LABOR-HOURS	UNIT	MAT.	LABOR	EQUIP.	TOTAL	TOTAL INCL O&P	
500	0010	**HUMIDIFIERS**										500
	0520	Steam, room or duct, filter, regulators, auto. controls, 220 V										
	0580	33 lb. per hour	Q-5	4	4	Ea.	2,250	79.50		2,329.50	2,600	
	0620	100 lb. per hour	"	3	5.333	"	3,300	106		3,406	3,825	

15765 | Fan Coil Unit/Unit Ventilators

			CREW	DAILY OUTPUT	LABOR-HOURS	UNIT	MAT.	LABOR	EQUIP.	TOTAL	TOTAL INCL O&P	
200	0010	**FAN COIL AIR CONDITIONING** Cabinet mounted, filters, controls										200
	0100	Chilled water, 1/2 ton cooling	Q-5	8	2	Ea.	730	40		770	870	
	0120	1 ton cooling		6	2.667		830	53		883	1,000	
	0180	3 ton cooling		4	4		1,675	79.50		1,754.50	1,950	
	0940	Direct expansion, for use w/air cooled condensing, 1.5 ton cooling		5	3.200		440	63.50		503.50	590	
	1000	5 ton cooling	↓	3	5.333	↓	1,000	106		1,106	1,275	
	1510	For condensing unit add see division 15670.										

15700 | Heating/Ventilating/Air Conditioning Equipment

15765 | Fan Coil Unit/Unit Ventilators

			CREW	DAILY OUTPUT	LABOR-HOURS	UNIT	2000 BARE COSTS MAT.	LABOR	EQUIP.	TOTAL	TOTAL INCL O&P	
600	0010	**HEATING & VENTILATING UNITS** Classroom										600
	0020	Includes filter, heating/cooling coils, standard controls										
	0080	750 CFM, 2 tons cooling	Q-6	2	12	Ea.	2,950	230		3,180	3,625	
	0100	1000 CFM, 2-1/2 tons cooling		1.60	15		3,475	288		3,763	4,300	
	0120	1250 CFM, 3 tons cooling		1.40	17.143		3,600	330		3,930	4,500	
	0140	1500 CFM, 4 tons cooling		.80	30		3,850	575		4,425	5,175	
	0500	For electric heat, add					35%					
	1000	For no cooling, deduct					25%	10%				

15766 | Fin Tube Radiation

			CREW	DAILY OUTPUT	LABOR-HOURS	UNIT	MAT.	LABOR	EQUIP.	TOTAL	TOTAL INCL O&P	
190	0010	**HYDRONIC HEATING** Terminal units, not incl. main supply pipe										190
	1000	Radiation										
	1150	Fin tube, wall hung, 14" slope top cover, with damper										
	1200	1-1/4" copper tube, 4-1/4" alum. fin	Q-5	38	.421	L.F.	34.50	8.40		42.90	52	
	1255	2" steel tube, 4-1/4" steel fin		32	.500		33.50	9.95		43.45	53.50	
	1310	Baseboard, pkgd, 1/2" copper tube, alum. fin, 7" high		60	.267		6.75	5.30		12.05	16.20	
	1320	3/4" copper tube, alum. fin, 7" high		58	.276		7.15	5.50		12.65	17	
	1340	1" copper tube, alum. fin, 8-7/8" high		56	.286		14.80	5.70		20.50	25.50	
	1360	1-1/4" copper tube, alum. fin, 8-7/8" high		54	.296		22	5.90		27.90	34	
	3950	Unit heaters, propeller, 115 V 2 psi steam, 60°F entering air										
	4000	Horizontal, 12 MBH	Q-5	12	1.333	Ea.	330	26.50		356.50	410	
	4060	43.9 MBH		8	2		440	40		480	550	
	4240	286.9 MBH		2	8		1,175	159		1,334	1,575	
	4250	326.0 MBH		1.90	8.421		1,375	168		1,543	1,800	
	4260	364 MBH		1.80	8.889		1,525	177		1,702	1,975	
	4270	404 MBH		1.60	10		1,925	199		2,124	2,450	
	4300	For vertical diffuser, add					136			136	149	
	4310	Vertical flow, 40 MBH	Q-5	11	1.455		365	29		394	450	
	4326	131.0 MBH	"	4	4		685	79.50		764.50	880	
	4354	420 MBH, (460 V)	Q-6	1.80	13.333		2,425	256		2,681	3,075	
	4358	500 MBH, (460 V)		1.71	14.035		2,750	269		3,019	3,475	
	4362	570 MBH, (460 V)		1.40	17.143		2,850	330		3,180	3,700	
	4366	620 MBH, (460 V)		1.30	18.462		3,275	355		3,630	4,175	
	4370	960 MBH, (460 V)		1.10	21.818		5,750	420		6,170	7,025	

15768 | Infrared Heaters

			CREW	DAILY OUTPUT	LABOR-HOURS	UNIT	MAT.	LABOR	EQUIP.	TOTAL	TOTAL INCL O&P	
600	0010	**INFRA-RED UNIT**										600
	0020	Gas fired, unvented, electric ignition, 100% shutoff. Piping and										
	0030	wiring not included										
	0060	Input, 15 MBH	Q-5	7	2.286	Ea.	345	45.50		390.50	455	
	0120	45 MBH		5	3.200		390	63.50		453.50	530	
	0160	60 MBH		4	4		455	79.50		534.50	630	
	0180	75 MBH		3	5.333		485	106		591	710	
	0220	105 MBH		2	8		680	159		839	1,025	

15770 | Floor-Heating & Snow-Melting Eq.

			CREW	DAILY OUTPUT	LABOR-HOURS	UNIT	MAT.	LABOR	EQUIP.	TOTAL	TOTAL INCL O&P	
200	0010	**ELECTRIC HEATING**, equipments are not including wires & conduits										200
	0400	Cable heating, radiant heat plaster, no controls, in South	1 Elec	130	.062	S.F.	7.25	1.36		8.61	10.20	
	0600	In North		90	.089		7.25	1.96		9.21	11.20	
	0800	Cable on 1/2" board not incl. controls, tract housing		90	.089		6.10	1.96		8.06	9.90	
	1000	Custom housing		80	.100		7.10	2.21		9.31	11.40	
	1100	Rule of thumb: Baseboard units, including control		4.40	1.818	kW	75	40		115	148	
	1300	Baseboard heaters, 2' long, 375 watt		8	1	Ea.	34	22		56	73.50	
	1600	4' long, 750 watt		6.70	1.194		48	26.50		74.50	96	
	2000	6' long, 1125 watt		5	1.600		65	35.50		100.50	130	
	2800	10' long, 1875 watt		3.30	2.424		134	53.50		187.50	235	

For expanded coverage of these items see *Means Mechanical or Plumbing Cost Data 2000*

15700 | Heating/Ventilating/Air Conditioning Equipment

15770 | Floor-Heating & Snow-Melting Eq.

		CREW	DAILY OUTPUT	LABOR-HOURS	UNIT	MAT.	LABOR	EQUIP.	TOTAL	TOTAL INCL O&P
2950	Wall heaters with fan, 120 to 277 volt									
2970	surface mounted, residential, 750 watt	1 Elec	7	1.143	Ea.	105	25.50		130.50	158
2990	1250 watt		6	1.333		120	29.50		149.50	180
3000	1500 watt		5	1.600		120	35.50		155.50	190
3010	2000 watt		5	1.600		123	35.50		158.50	193
3050	2500 watt		4	2		255	44		299	355
3070	4000 watt		3.50	2.286		260	50.50		310.50	370
5000	Radiant heating ceiling panels, 2' x 4', 500 watt		16	.500		174	11.05		185.05	209
5050	750 watt		16	.500		192	11.05		203.05	229
5300	Infra-red quartz heaters, 120 volts, 1000 watts		6.70	1.194		119	26.50		145.50	174
5350	1500 watt		5	1.600		119	35.50		154.50	189
5400	240 volts, 1500 watt		5	1.600		119	35.50		154.50	189
5450	2000 watt		4	2		119	44		163	204

15800 | Air Distribution

15810 | Ducts

		CREW	DAILY OUTPUT	LABOR-HOURS	UNIT	MAT.	LABOR	EQUIP.	TOTAL	TOTAL INCL O&P
0010	**DUCTWORK**									
0020	Fabricated rectangular, includes fittings, joints, supports,	R15810 -050								
0030	allowance for flexible connections, no insulation									
0031	NOTE: Fabrication and installation are combined									
0040	as LABOR cost. Approx. 25% fittings assumed.									
0050	Add to labor for elevated installation									
0051	of fabricated ductwork									
0052	10' to 15' high						6%			
0053	15' to 20' high						12%			
0054	20' to 25' high						15%			
0100	Aluminum, alloy 3003-H14, under 100 lb.	Q-10	75	.320	Lb.	2.80	6.40		9.20	13.95
0140	1,000 to 2,000 lb.		120	.200		1.20	4.01		5.21	8.10
0150	2,000 to 5,000 lb.		130	.185		1.15	3.71		4.86	7.50
0500	Galvanized steel, under 200 lb.		235	.102		1.05	2.05		3.10	4.61
0520	200 to 500 lb.		245	.098		.85	1.97		2.82	4.26
0540	500 to 1,000 lb.		255	.094		.41	1.89		2.30	3.64
0560	1,000 to 2,000 lb.		265	.091		.41	1.82		2.23	3.52
0570	2,000 to 5,000 lb.		275	.087		.35	1.75		2.10	3.35
1300	Flexible, coated fiberglass fabric on corr. resist. metal helix									
1400	pressure to 12" (WG) UL-181									
1500	Non-insulated, 3" diameter	Q-9	400	.040	L.F.	.95	.77		1.72	2.36
1520	4" diameter		360	.044		.95	.86		1.81	2.50
1540	5" diameter		320	.050		1.11	.97		2.08	2.85
1560	6" diameter		280	.057		1.25	1.11		2.36	3.25
1561	Ductwork, flexible, non-insulated, 6" diameter		280	.057		1.25	1.11		2.36	3.25
1580	7" diameter		240	.067		1.64	1.29		2.93	3.98
1600	8" diameter		200	.080		1.69	1.55		3.24	4.47
1620	9" diameter		180	.089		1.91	1.72		3.63	5
1640	10" diameter		160	.100		2.17	1.94		4.11	5.65
1900	Insulated, 1" thick with 3/4 lb., PE jacket, 3" diameter		380	.042		1.64	.81		2.45	3.18
1920	5" diameter		300	.053		1.92	1.03		2.95	3.85
1940	6" diameter		260	.062		2.15	1.19		3.34	4.38

15800 | Air Distribution

15810 | Ducts

			CREW	DAILY OUTPUT	LABOR-HOURS	UNIT	2000 BARE COSTS MAT.	LABOR	EQUIP.	TOTAL	TOTAL INCL O&P	
600	1960	7" diameter	Q-9	220	.073	L.F.	2.57	1.41		3.98	5.20	600
	1980	8" diameter	R15810-050	180	.089		2.63	1.72		4.35	5.80	
	2000	9" diameter		160	.100		3.16	1.94		5.10	6.75	
	2020	10" diameter		140	.114		3.42	2.21		5.63	7.50	
	3490	Rigid fiberglass duct board, foil reinf. kraft facing										
	3500	Rectangular, 1" thick, alum. faced, (FRK), std. weight	Q-10	350	.069	SF Surf	.81	1.38		2.19	3.21	

15820 | Duct Accessories

			CREW	DAILY OUTPUT	LABOR-HOURS	UNIT	MAT.	LABOR	EQUIP.	TOTAL	TOTAL INCL O&P	
300	0010	**DUCT ACCESSORIES**										300
	0050	Air extractors, 12" x 4"	1 Shee	24	.333	Ea.	14.45	7.15		21.60	28	
	0100	8" x 6"		22	.364		14.45	7.80		22.25	29	
	0200	20" x 8"		16	.500		32.50	10.75		43.25	54	
	3000	Fire damper, curtain type, 1-1/2 hr rated, vertical, 6" x 6"		24	.333		19.70	7.15		26.85	33.50	
	3020	8" x 6"		22	.364		19.70	7.80		27.50	34.50	
	3240	16" x 14"		18	.444		31.50	9.55		41.05	50.50	
	5990	Multi-blade dampers, opposed blade, 8" x 6"		24	.333		18.40	7.15		25.55	32	
	5994	8" x 8"		22	.364		19.20	7.80		27	34	
	5996	10" x 10"		21	.381		21.50	8.20		29.70	38	
	6000	12" x 12"		21	.381		25	8.20		33.20	41.50	
	6020	12" x 18"		18	.444		33.50	9.55		43.05	53	
	6030	14" x 10"		20	.400		21.50	8.60		30.10	38.50	
	6031	14" x 14"		17	.471		29.50	10.10		39.60	49.50	
	6033	16" x 12"		17	.471		29.50	10.10		39.60	49.50	
	6035	16" x 16"		16	.500		37	10.75		47.75	58.50	
	6037	18" x 14"		16	.500		37	10.75		47.75	58.50	
	6038	18" x 18"		15	.533		44	11.45		55.45	68	
	6070	20" x 16"		14	.571		44	12.30		56.30	69	
	6072	20" x 20"		13	.615		53	13.25		66.25	80.50	
	6074	22" x 18"		14	.571		53	12.30		65.30	78.50	
	6076	24" x 16"		11	.727		52	15.65		67.65	83.50	
	6078	24" x 20"		8	1		61.50	21.50		83	105	
	6080	24" x 24"		8	1		72	21.50		93.50	116	
	6110	26" x 26"		6	1.333		84.50	28.50		113	142	
	6133	30" x 30"	Q-9	6.60	2.424		112	47		159	203	
	6135	32" x 32"	"	6.40	2.500		128	48.50		176.50	223	
	7500	Variable volume modulating motorized damper, incl. elect. mtr.										
	7504	8" x 6"	1 Shee	15	.533	Ea.	101	11.45		112.45	130	
	7506	10" x 6"		14	.571		101	12.30		113.30	132	
	7510	10" x 10"		13	.615		104	13.25		117.25	137	
	7520	12" x 12"		12	.667		106	14.35		120.35	141	
	7522	12" x 16"		11	.727		110	15.65		125.65	148	
	7524	16" x 10"		12	.667		107	14.35		121.35	142	
	7526	16" x 14"		10	.800		112	17.20		129.20	152	
	7528	16" x 18"		9	.889		119	19.10		138.10	164	
	7542	18" x 18"		8	1		123	21.50		144.50	173	
	7544	20" x 14"		8	1		126	21.50		147.50	175	
	7546	20" x 18"		7	1.143		134	24.50		158.50	189	
	7560	24" x 12"		8	1		127	21.50		148.50	177	
	7562	24" x 18"		7	1.143		142	24.50		166.50	198	
	7568	28" x 10"		7	1.143		211	24.50		235.50	274	
	7590	30" x 14"		5	1.600		224	34.50		258.50	305	
	7610	30" x 24"		3.80	2.105		258	45.50		303.50	360	
	7700	For thermostat, add		8	1		25.50	21.50		47	64.50	
	8000	Multi-blade dampers, parallel blade										
	8100	8" x 8"	1 Shee	24	.333	Ea.	52.50	7.15		59.65	70	
	8160	18" x 12"	"	18	.444	"	57	9.55		66.55	79	

For expanded coverage of these items see *Means Mechanical or Plumbing Cost Data 2000*

15800 | Air Distribution

15820 | Duct Accessories

		CREW	DAILY OUTPUT	LABOR-HOURS	UNIT	2000 BARE COSTS				TOTAL INCL O&P
						MAT.	LABOR	EQUIP.	TOTAL	
9000	Silencers, noise control for air flow, duct				MCFM	41.50			41.50	45.50

15830 | Fans

		CREW	DAILY OUTPUT	LABOR-HOURS	UNIT	MAT.	LABOR	EQUIP.	TOTAL	TOTAL INCL O&P
0010	**FANS**									
0020	Air conditioning and process air handling									
2500	Ceiling fan, right angle, extra quiet, 0.10" S.P.									
2520	95 CFM	Q-20	20	1	Ea.	93	19.90		112.90	136
2540	210 CFM	"	19	1.053	"	120	21		141	167
5000	Utility set, centrifugal, V belt drive, motor									
5020	1/4" S.P., 1200 CFM, 1/4 HP	Q-20	6	3.333	Ea.	2,925	66.50		2,991.50	3,325
5040	1520 CFM, 1/3 HP		5	4		2,925	79.50		3,004.50	3,350
5060	1850 CFM, 1/2 HP		4	5		2,950	99.50		3,049.50	3,400
5080	2180 CFM, 3/4 HP		3	6.667		2,950	133		3,083	3,450
6000	Propeller exhaust, wall shutter, 1/4" S.P.									
6020	Direct drive, two speed									
6100	375 CFM, 1/10 HP	Q-20	10	2	Ea.	198	40		238	284
6140	1000 CFM, 1/8 HP		8	2.500		305	50		355	420
6160	1890 CFM, 1/4 HP		7	2.857		274	57		331	395
6650	Residential, bath exhaust, grille, back draft damper									
6660	50 CFM	Q-20	24	.833	Ea.	13.25	16.60		29.85	42.50
6670	110 CFM	"	22	.909	"	47	18.10		65.10	82.50
7000	Roof exhauster, centrifugal, aluminum housing, 12" galvanized									
7020	curb, bird screen, back draft damper, 1/4" S.P.									
7100	Direct drive, 320 CFM, 11" sq. damper	Q-20	7	2.857	Ea.	380	57		437	515
7120	600 CFM, 11" sq. damper		6	3.333		390	66.50		456.50	540
7140	815 CFM, 13" sq. damper		5	4		390	79.50		469.50	565
7160	1450 CFM, 13" sq. damper		4.20	4.762		510	95		605	720
7180	2050 CFM, 16" sq. damper		4	5		510	99.50		609.50	725
7200	V-belt drive, 1650 CFM, 12" sq. damper		6	3.333		650	66.50		716.50	825
7220	2750 CFM, 21" sq. damper		5	4		750	79.50		829.50	960
7230	3500 CFM, 21" sq. damper		4.50	4.444		830	88.50		918.50	1,050
7240	4910 CFM, 23" sq. damper		4	5		1,025	99.50		1,124.50	1,300

15850 | Air Outlets & Inlets

		CREW	DAILY OUTPUT	LABOR-HOURS	UNIT	MAT.	LABOR	EQUIP.	TOTAL	TOTAL INCL O&P
0010	**DIFFUSERS** Aluminum, opposed blade damper unless noted									
0100	Ceiling, linear, also for sidewall									
0120	2" wide	1 Shee	32	.250	L.F.	25	5.40		30.40	36.50
0160	4" wide		26	.308	"	33.50	6.60		40.10	48
0500	Perforated, 24" x 24" lay-in panel size, 6" x 6"		16	.500	Ea.	79.50	10.75		90.25	105
0520	8" x 8"		15	.533		82	11.45		93.45	109
0530	9" x 9"		14	.571		84	12.30		96.30	113
0590	16" x 16"		11	.727		107	15.65		122.65	145
1000	Rectangular, 1 to 4 way blow, 6" x 6"		16	.500		46	10.75		56.75	69
1010	8" x 8"		15	.533		54	11.45		65.45	79
1014	9" x 9"		15	.533		56	11.45		67.45	81
1016	10" x 10"		15	.533		67	11.45		78.45	93
1020	12" x 6"		15	.533		57	11.45		68.45	82
1040	12" x 9"		14	.571		69.50	12.30		81.80	97
1060	12" x 12"		12	.667		82	14.35		96.35	114
1070	14" x 6"		13	.615		61	13.25		74.25	90
1074	14" x 14"		12	.667		101	14.35		115.35	135
1150	18" x 18"		9	.889		149	19.10		168.10	197
1170	24" x 12"		10	.800		134	17.20		151.20	177
1180	24" x 24"		7	1.143		236	24.50		260.50	300

15800 | Air Distribution

15850 | Air Outlets & Inlets

			CREW	DAILY OUTPUT	LABOR-HOURS	UNIT	2000 BARE COSTS MAT.	LABOR	EQUIP.	TOTAL	TOTAL INCL O&P	
300	1500	Round, butterfly damper, 6" diameter	1 Shee	18	.444	Ea.	15.45	9.55		25	33	300
	1520	8" diameter		16	.500		16.35	10.75		27.10	36	
	2000	T bar mounting, 24" x 24" lay-in frame, 6" x 6"		16	.500		75.50	10.75		86.25	101	
	2020	9" x 9"		14	.571		84	12.30		96.30	113	
	2040	12" x 12"		12	.667		111	14.35		125.35	146	
	2060	15" x 15"		11	.727		143	15.65		158.65	184	
	2080	18" x 18"		10	.800		149	17.20		166.20	193	
	6000	For steel diffusers instead of aluminum, deduct					10%					
500	0010	**GRILLES**										500
	0020	Aluminum										
	1000	Air return, 6" x 6"	1 Shee	26	.308	Ea.	12.35	6.60		18.95	25	
	1020	10" x 6"		24	.333		14.95	7.15		22.10	28.50	
	1080	16" x 8"		22	.364		21.50	7.80		29.30	36.50	
	1100	12" x 12"		22	.364		21.50	7.80		29.30	36.50	
	1120	24" x 12"		18	.444		38.50	9.55		48.05	58	
	1180	16" x 16"		22	.364		30.50	7.80		38.30	46.50	
700	0010	**REGISTERS**										700
	0980	Air supply										
	3000	Baseboard, hand adj. damper, enameled steel										
	3012	8" x 6"	1 Shee	26	.308	Ea.	8.30	6.60		14.90	20.50	
	3020	10" x 6"		24	.333		9.30	7.15		16.45	22.50	
	3060	12" x 6"		23	.348		9.80	7.50		17.30	23.50	
	4000	Floor, toe operated damper, enameled steel										
	4020	4" x 8"	1 Shee	32	.250	Ea.	15.90	5.40		21.30	26.50	
	4040	4" x 12"	"	26	.308	"	18.10	6.60		24.70	31	

15854 | Ventilators

			CREW	DAILY OUTPUT	LABOR-HOURS	UNIT	MAT.	LABOR	EQUIP.	TOTAL	INCL O&P	
300	0010	**VENTILATORS** Base, damper & bird screen, CFM in 5 MPH wind										300
	4200	Stationary mushroom, aluminum, 16" orifice diameter	Q-9	10	1.600	Ea.	246	31		277	325	
	4240	38" orifice diameter		8	2		760	38.50		798.50	900	
	4250	42" orifice diameter		7.60	2.105		1,000	40.50		1,040.50	1,175	
	4260	50" orifice diameter		7	2.286		1,200	44		1,244	1,400	

15860 | Air Cleaning Devices

			CREW	DAILY OUTPUT	LABOR-HOURS	UNIT	MAT.	LABOR	EQUIP.	TOTAL	INCL O&P	
100	0010	**AIR FILTERS**										100
	0050	Activated charcoal type, full flow				MCFM	600			600	660	
	2000	Electronic air cleaner, duct mounted										
	2150	400 - 1000 CFM	1 Shee	2.30	3.478	Ea.	545	75		620	720	
	2200	1000 - 1400 CFM		2.20	3.636		700	78		778	895	
	2250	1400 - 2000 CFM		2.10	3.810		760	82		842	980	
	2950	Mechanical media filtration units										
	3000	High efficiency type, with frame, non-supported				MCFM	45			45	49.50	
	3100	Supported type					55			55	60.50	
	4000	Medium efficiency, extended surface					5			5	5.50	
	4500	Permanent washable					20			20	22	
	5000	Renewable disposable roll					120			120	132	
	5500	Throwaway glass or paper media type				Ea.	4.60			4.60	5.05	

For information about Means Estimating Seminars, see yellow pages 11 and 12 in back of book

For expanded coverage of these items see *Means Mechanical or Plumbing Cost Data 2000*

Division Notes

		CREW	DAILY OUTPUT	LABOR-HOURS	UNIT	2000 BARE COSTS				TOTAL INCL O&P
						MAT.	LABOR	EQUIP.	TOTAL	

Division 16 Electrical

Estimating Tips

16060 Grounding & Bonding
- When taking off grounding system, identify separately the type and size of wire and list each unique type of ground connection.

16100 Wiring Methods
- Conduit should be taken off in three main categories: power distribution, branch power, and branch lighting, so the estimator can concentrate on systems and components, therefore making it easier to ensure all items have been accounted for.
- For cost modifications for elevated conduit installation, add the percentages to labor according to the height of installation and only the quantities exceeding the different height levels, not to the total conduit quantities.
- Remember that aluminum wiring of equal ampacity is larger in diameter than copper and may require larger conduit.
- If more than three wires at a time are being pulled, deduct percentages from the labor hours of that grouping of wires.
- The estimator should take the weights of materials into consideration when completing a takeoff. Topics to consider include: How will the materials be supported? What methods of support are available? How high will the support structure have to reach? Will the final support structure be able to withstand the total burden? Is the support material included or separate from the fixture, equipment and material specified?

16200 Electrical Power
- Do not overlook the costs for equipment used in the installation. If scaffolding or highlifts are available in the field, contractors may use them in lieu of the proposed ladders and rolling staging.

16400 Low-Voltage Distribution
- Supports and concrete pads may be shown on drawings for the larger equipment, or the support system may be just a piece of plywood for the back of a panelboard. In either case, it must be included in the costs.

16500 Lighting
- Fixtures should be taken off room by room, using the fixture schedule, specifications, and the ceiling plan. For large concentrations of lighting fixtures in the same area deduct the percentages from labor hours.

16700 Communications
16800 Sound & Video
- When estimating material costs for special systems, it is always prudent to obtain manufacturers' quotations for equipment prices and special installation requirements which will affect the total costs.

Reference Numbers
Reference numbers are shown in bold squares at the beginning of some major classifications. These numbers refer to related items in the Reference Section. The reference information may be an estimating procedure, an alternate pricing method or technical information.

Note: Not all subdivisions listed here necessarily appear in this publication.

16050 | Basic Electrical Materials & Methods

16055 | Selective Demolition

		CREW	DAILY OUTPUT	LABOR-HOURS	UNIT	2000 BARE COSTS				TOTAL INCL O&P
						MAT.	LABOR	EQUIP.	TOTAL	
0010	**ELECTRICAL DEMOLITION**									
0020	Conduit to 15' high, including fittings & hangers									
0100	Rigid galvanized steel, 1/2" to 1" diameter	1 Elec	242	.033	L.F.		.73		.73	1.20
0120	1-1/4" to 2"	"	200	.040	"		.88		.88	1.45
0270	Armored cable, (BX) avg. 50' runs									
0280	#14, 2 wire	1 Elec	690	.012	L.F.		.26		.26	.42
0290	#14, 3 wire		571	.014			.31		.31	.51
0300	#12, 2 wire		605	.013			.29		.29	.48
0310	#12, 3 wire		514	.016			.34		.34	.56
0320	#10, 2 wire		514	.016			.34		.34	.56
0330	#10, 3 wire		425	.019			.42		.42	.68
0340	#8, 3 wire		342	.023			.52		.52	.85
0350	Non metallic sheathed cable (Romex)									
0360	#14, 2 wire	1 Elec	720	.011	L.F.		.25		.25	.40
0370	#14, 3 wire		657	.012			.27		.27	.44
0380	#12, 2 wire		629	.013			.28		.28	.46
0390	#10, 3 wire		450	.018			.39		.39	.64
0400	Wiremold raceway, including fittings & hangers									
0440	No. 4000	1 Elec	217	.037	L.F.		.81		.81	1.33
0460	No. 6000	"	166	.048	"		1.07		1.07	1.74
0500	Channels, steel, including fittings & hangers									
0520	3/4" x 1-1/2"	1 Elec	308	.026	L.F.		.57		.57	.94
0540	1-1/2" x 1-1/2"		269	.030			.66		.66	1.08
0560	1-1/2" x 1-7/8"		229	.035			.77		.77	1.26
1180	400 amp	2 Elec	6.80	2.353	Ea.		52		52	85
1210	Panel boards, incl. removal of all breakers,									
1220	pipe terminations & wire connections									
1720	Junction boxes, 4" sq. & oct.	1 Elec	80	.100	Ea.		2.21		2.21	3.62
1760	Switch box		107	.075			1.65		1.65	2.70
1780	Receptacle & switch plates		257	.031			.69		.69	1.13
1800	Wire, THW-THWN-THHN, removed from									
1810	in place conduit, to 15' high									
1830	#14	1 Elec	65	.123	C.L.F.		2.72		2.72	4.45
1840	#12		55	.145			3.21		3.21	5.25
1850	#10		45.50	.176			3.89		3.89	6.35
2000	Interior fluorescent fixtures, incl. supports									
2010	& whips, to 15' high									
2100	Recessed drop-in 2' x 2', 2 lamp	2 Elec	35	.457	Ea.		10.10		10.10	16.55
2140	2' x 4', 4 lamp	"	30	.533	"		11.80		11.80	19.30
2180	Surface mount, acrylic lens & hinged frame									
2220	2' x 2', 2 lamp	2 Elec	44	.364	Ea.		8.05		8.05	13.15
2260	2' x 4', 4 lamp	"	33	.485	"		10.70		10.70	17.55
2300	Strip fixtures, surface mount									
2320	4' long, 1 lamp	2 Elec	53	.302	Ea.		6.65		6.65	10.90
2380	8' long, 2 lamp	"	40	.400	"		8.85		8.85	14.45
2600	Exterior fixtures, incandescent, wall mount									
2620	100 Watt	2 Elec	50	.320	Ea.		7.05		7.05	11.55

16060 | Grounding & Bonding

		CREW	DAILY OUTPUT	LABOR-HOURS	UNIT	MAT.	LABOR	EQUIP.	TOTAL	TOTAL INCL O&P
0010	**GROUNDING**									
0030	Rod, copper clad, 8' long, 1/2" diameter	1 Elec	5.50	1.455	Ea.	15.85	32		47.85	70
0050	3/4" diameter		5.30	1.509		30	33.50		63.50	87.50
0080	10' long, 1/2" diameter		4.80	1.667		17.50	37		54.50	80
0100	3/4" diameter		4.40	1.818		29.50	40		69.50	98
0130	15' long, 3/4" diameter		4	2		79	44		123	160

16050 | Basic Electrical Materials & Methods

16060 | Grounding & Bonding

			CREW	DAILY OUTPUT	LABOR-HOURS	UNIT	2000 BARE COSTS MAT.	LABOR	EQUIP.	TOTAL	TOTAL INCL O&P	
800	0260	Wire ground bare armored, #8-1 conductor	1 Elec	2	4	C.L.F.	63.50	88.50		152	215	800
	0270	#6-1 conductor		1.80	4.444		80	98		178	249	
	0390	Bare copper wire, #8 stranded		11	.727		14	16.05		30.05	42	
	0400	#6		10	.800		21	17.70		38.70	52.50	
	0600	#2	2 Elec	10	1.600		51	35.50		86.50	115	
	0800	3/0		6.60	2.424		135	53.50		188.50	236	
	1000	4/0		5.70	2.807		168	62		230	286	
	1200	250 kcmil	3 Elec	7.20	3.333		199	73.50		272.50	340	
	1800	Water pipe ground clamps, heavy duty										
	2000	Bronze, 1/2" to 1" diameter	1 Elec	8	1	Ea.	11.70	22		33.70	49	
	2100	1-1/4" to 2" diameter		8	1		15.30	22		37.30	53	
	2200	2-1/2" to 3" diameter		6	1.333		41	29.50		70.50	93	
	2800	Brazed connections, #6 wire		12	.667		10.75	14.75		25.50	36	
	3000	#2 wire		10	.800		14.40	17.70		32.10	45	
	3100	3/0 wire		8	1		21.50	22		43.50	60	
	3200	4/0 wire		7	1.143		24.50	25.50		50	68.50	
	3400	250 kcmil wire		5	1.600		29	35.50		64.50	89.50	

16100 | Wiring Methods

16120 | Conductors & Cables

			CREW	DAILY OUTPUT	LABOR-HOURS	UNIT	2000 BARE COSTS MAT.	LABOR	EQUIP.	TOTAL	TOTAL INCL O&P	
120	0010	**ARMORED CABLE**										120
	0050	600 volt, copper (BX) #14, 2 conductor, solid	1 Elec	2.40	3.333	C.L.F.	46.50	73.50		120	173	
	0100	3 conductor, solid		2.20	3.636		61	80.50		141.50	198	
	0120	4 conductor, solid		2	4		82	88.50		170.50	235	
	0150	#12, 2 conductor, solid		2.30	3.478		48.50	77		125.50	179	
	0200	3 conductor, solid		2	4		72.50	88.50		161	225	
	0220	4 conductor, solid		1.80	4.444		101	98		199	272	
	0250	#10, 2 conductor, solid		2	4		85.50	88.50		174	239	
	0300	3 conductor, solid		1.60	5		112	111		223	305	
	0320	4 conductor, solid		1.40	5.714		168	126		294	390	
	0350	#8, 3 conductor, solid		1.30	6.154		183	136		319	425	
	0370	4 conductor, stranded		1.10	7.273		285	161		446	580	
	9010	600 volt, copper (MC) steel clad, #14, 2 wire		2.40	3.333		46.50	73.50		120	172	
	9020	3 wire		2.20	3.636		60	80.50		140.50	197	
	9030	4 wire		2	4		83.50	88.50		172	237	
	9040	#12, 2 wire		2.30	3.478		49.50	77		126.50	181	
	9050	3 wire		2	4		73	88.50		161.50	226	
	9060	4 wire		1.80	4.444		95	98		193	266	
	9070	#10, 2 wire		2	4		85	88.50		173.50	239	
	9080	3 wire		1.60	5		128	111		239	320	
	9090	4 wire		1.40	5.714		197	126		323	425	
	9100	#8, 2 wire, stranded		1.80	4.444		162	98		260	340	
	9110	3 wire, stranded		1.30	6.154		230	136		366	475	
	9120	4 wire, stranded		1.10	7.273		300	161		461	595	
400	0010	**FIBER OPTICS**										400
	0020	Fiber optics cable only. Added costs depend on the type of fiber										
	0030	special connectors, optical modems, and networking parts.										
	0040	Specialized tools & techniques cause installation costs to vary.										

For expanded coverage of these items see *Means Electrical Cost Data 2000*

16100 | Wiring Methods

16120 | Conductors & Cables

			CREW	DAILY OUTPUT	LABOR-HOURS	UNIT	2000 BARE COSTS MAT.	LABOR	EQUIP.	TOTAL	TOTAL INCL O&P	
400	0070	Cable, minimum, bulk simplex	1 Elec	8	1	C.L.F.	24	22		46	62.50	400
	0080	Cable, maximum, bulk plenum quad	"	2.29	3.493	"	117	77		194	255	
	0150	Fiber optic jumper				Ea.	55.50			55.50	61	
	0200	Fiber optic pigtail					30			30	33	
	0300	Fiber optic connector	1 Elec	24	.333		16.90	7.35		24.25	30.50	
	0350	Fiber optic finger splice		32	.250		32	5.55		37.55	44	
	0400	Transceiver (low cost bi-directional)		8	1		375	22		397	450	
	0450	Multi-channel rack enclosure (10 modules)		2	4		550	88.50		638.50	750	
	0500	Fiber optic patch panel (12 ports)		6	1.333		178	29.50		207.50	244	
550	0010	**NON-METALLIC SHEATHED CABLE** 600 volt										550
	0100	Copper with ground wire, (Romex)										
	0150	#14, 2 conductor	1 Elec	2.70	2.963	C.L.F.	13.40	65.50		78.90	122	
	0200	3 conductor		2.40	3.333		22.50	73.50		96	146	
	0220	4 conductor		2.20	3.636		46	80.50		126.50	182	
	0250	#12, 2 conductor		2.50	3.200		18.90	70.50		89.40	137	
	0300	3 conductor		2.20	3.636		32	80.50		112.50	166	
	0320	4 conductor		2	4		68	88.50		156.50	220	
	0350	#10, 2 conductor		2.20	3.636		32	80.50		112.50	166	
	0400	3 conductor		1.80	4.444		48.50	98		146.50	214	
	0420	4 conductor		1.60	5		89.50	111		200.50	280	
	0450	#8, 3 conductor		1.50	5.333		97.50	118		215.50	300	
	0480	4 conductor		1.40	5.714		167	126		293	390	
	0500	#6, 3 conductor		1.40	5.714		150	126		276	370	
	0550	SE type SER aluminum cable, 3 RHW and										
	0600	1 bare neutral, 3 #8 & 1 #8	1 Elec	1.60	5	C.L.F.	96	111		207	287	
	0650	3 #6 & 1 #6	"	1.40	5.714		109	126		235	325	
	0700	3 #4 & 1 #6	2 Elec	2.40	6.667		122	147		269	375	
	0750	3 #2 & 1 #4		2.20	7.273		179	161		340	460	
	0800	3 #1/0 & 1 #2		2	8		271	177		448	585	
	0850	3 #2/0 & 1 #1		1.80	8.889		320	196		516	670	
	0900	3 #4/0 & 1 #2/0		1.60	10		455	221		676	860	
	1450	UF underground feeder cable, copper with ground, #14, 2 conductor	1 Elec	4	2		16.10	44		60.10	90	
	1500	#12, 2 conductor		3.50	2.286		22.50	50.50		73	108	
	1550	#10, 2 conductor		3	2.667		36	59		95	136	
	1600	#14, 3 conductor		3.50	2.286		23.50	50.50		74	108	
	1650	#12, 3 conductor		3	2.667		33.50	59		92.50	134	
	1700	#10, 3 conductor		2.50	3.200		50	70.50		120.50	171	
	2400	SEU service entrance cable, copper 2 conductors, #8 + #8 neut.		1.50	5.333		83	118		201	285	
	2600	#6 + #8 neutral		1.30	6.154		115	136		251	350	
	2800	#6 + #6 neutral		1.30	6.154		125	136		261	360	
	3000	#4 + #6 neutral	2 Elec	2.20	7.273		164	161		325	445	
	3200	#4 + #4 neutral		2.20	7.273		185	161		346	465	
	3400	#3 + #5 neutral		2.10	7.619		203	168		371	500	
	3600	#3 + #3 neutral		2.10	7.619		234	168		402	530	
	3800	#2 + #4 neutral		2	8		269	177		446	585	
	4000	#1 + #1 neutral		1.90	8.421		445	186		631	795	
	4200	1/0 + 1/0 neutral		1.80	8.889		495	196		691	865	
	4400	2/0 + 2/0 neutral		1.70	9.412		590	208		798	985	
	4600	3/0 + 3/0 neutral		1.60	10		725	221		946	1,150	
	4800	Aluminum 2 conductors, #8 + #8 neutral	1 Elec	1.60	5		78.50	111		189.50	268	
	5000	#6 + #6 neutral	"	1.40	5.714		79	126		205	294	
	5100	#4 + #6 neutral	2 Elec	2.50	6.400		102	141		243	345	
	5200	#4 + #4 neutral		2.40	6.667		102	147		249	355	
	5300	#2 + #4 neutral		2.30	6.957		136	154		290	400	
	5400	#2 + #2 neutral		2.20	7.273		136	161		297	410	

16100 | Wiring Methods

16120 | Conductors & Cables

		CREW	DAILY OUTPUT	LABOR-HOURS	UNIT	2000 BARE COSTS				TOTAL INCL O&P	
						MAT.	LABOR	EQUIP.	TOTAL		
550	5450	1/0 + #2 neutral	2 Elec	2.10	7.619	C.L.F.	207	168		375	505	**550**
	5500	1/0 + 1/0 neutral		2	8		207	177		384	515	
	5550	2/0 + #1 neutral		1.90	8.421		238	186		424	565	
	5600	2/0 + 2/0 neutral		1.80	8.889		238	196		434	580	
	5800	3/0 + 1/0 neutral		1.70	9.412		305	208		513	675	
	6000	3/0 + 3/0 neutral		1.70	9.412		305	208		513	675	
	6200	4/0 + 2/0 neutral		1.60	10		335	221		556	725	
	6400	4/0 + 4/0 neutral	▼	1.60	10	▼	335	221		556	725	
	6500	Service entrance cap for copper SEU										
	6600	100 amp	1 Elec	12	.667	Ea.	6.90	14.75		21.65	31.50	
	6700	150 amp		10	.800		12.10	17.70		29.80	42.50	
	6800	200 amp	▼	8	1	▼	18.25	22		40.25	56	
750	0010	**SPECIAL WIRES & FITTINGS**										**750**
	0100	Fixture TFFN 600 volt 90°C stranded, #18	1 Elec	13	.615	C.L.F.	4.25	13.60		17.85	27	
	0150	#16		13	.615		5.40	13.60		19	28.50	
	0500	Thermostat, no jacket, twisted, #18-2 conductor		8	1		10.10	22		32.10	47	
	0550	#18-3 conductor		7	1.143		14	25.50		39.50	57	
	0600	#18-4 conductor		6.50	1.231		19.45	27		46.45	66	
	0650	#18-5 conductor		6	1.333		23.50	29.50		53	73.50	
	0700	#18-6 conductor		5.50	1.455		29	32		61	84	
	0750	#18-7 conductor		5	1.600		32	35.50		67.50	93	
	0800	#18-8 conductor		4.80	1.667		37	37		74	101	
	0900	TV antenna lead-in, 300 ohm, #20-2 conductor		7	1.143		13.60	25.50		39.10	56.50	
	0950	Coaxial, feeder outlet		7	1.143		14.40	25.50		39.90	57.50	
	1000	Coaxial, main riser		6	1.333		21	29.50		50.50	71	
	1100	Sound, shielded with drain, #22-2 conductor		8	1		9.40	22		31.40	46.50	
	1150	#22-3 conductor		7.50	1.067		17.90	23.50		41.40	58	
	1200	#22-4 conductor		6.50	1.231		22	27		49	68.50	
	1250	Nonshielded, #22-2 conductor		10	.800		7.50	17.70		25.20	37.50	
	1300	#22-3 conductor		9	.889		10.30	19.65		29.95	43.50	
	1350	#22-4 conductor		8	1		13	22		35	50.50	
	1400	Microphone cable		8	1		51.50	22		73.50	92.50	
	2200	Telephone twisted, PVC insulation, #22-2 conductor		10	.800		6	17.70		23.70	35.50	
	2250	#22-3 conductor		9	.889		7.90	19.65		27.55	40.50	
	2300	#22-4 conductor		8	1		10	22		32	47	
	2350	#18-2 conductor	▼	9	.889	▼	7.30	19.65		26.95	40	
755	0010	**MODULAR FLEXIBLE WIRING SYSTEM**										**755**
	0050	Standard selector cable, 3 conductor, 15' long	1 Elec	40	.200	Ea.	38.50	4.42		42.92	50	
	0100	Conversion Module		16	.500		14	11.05		25.05	33.50	
	0150	Cable set, 3 conductor, 15' long		24	.333		15	7.35		22.35	28.50	
	0200	Switching assembly, 3 conductor, 10' long	▼	32	.250	▼	29.50	5.55		35.05	41.50	
800	0010	**UNDERCARPET**										**800**
	0020	Power System										
	0100	Cable flat, 3 conductor, #12, w/attached bottom shield	1 Elec	982	.008	L.F.	4.25	.18		4.43	4.97	
	0200	Shield, top, steel		1,768	.005	"	4.60	.10		4.70	5.20	
	0250	Splice, 3 conductor		48	.167	Ea.	14.10	3.68		17.78	21.50	
	0300	Top shield		96	.083		1.25	1.84		3.09	4.39	
	0350	Tap		40	.200		18.10	4.42		22.52	27	
	0400	Insulating patch, splice, tap, & end		48	.167		44	3.68		47.68	54	
	0450	Fold		230	.035			.77		.77	1.26	
	0500	Top shield, tap & fold		96	.083		1.25	1.84		3.09	4.39	
	0700	Transition, block assembly		77	.104		65.50	2.30		67.80	76	
	0750	Receptacle frame & base		32	.250		36	5.55		41.55	48.50	
	0800	Cover receptacle		120	.067		3.05	1.47		4.52	5.75	
	0850	Cover blank	▼	160	.050		3.60	1.11		4.71	5.75	

For expanded coverage of these items see *Means Electrical Cost Data 2000*

16100 | Wiring Methods

16120 | Conductors & Cables

			CREW	DAILY OUTPUT	LABOR-HOURS	UNIT	2000 BARE COSTS MAT.	LABOR	EQUIP.	TOTAL	TOTAL INCL O&P
800	0860	Receptacle, direct connected, single	1 Elec	25	.320	Ea.	77	7.05		84.05	96
	0870	Dual		16	.500		126	11.05		137.05	157
	0880	Combination Hi & Lo, tension		21	.381		93	8.40		101.40	116
	0900	Box, floor with cover		20	.400		78	8.85		86.85	100
	0920	Floor service w/barrier		4	2		220	44		264	315
	1000	Wall, surface, with cover		20	.400		51.50	8.85		60.35	71
	1100	Wall, flush, with cover		20	.400		36	8.85		44.85	54
	2500	Telephone System									
	2510	Transition fitting wall box, surface	1 Elec	24	.333	Ea.	37.50	7.35		44.85	53
	2520	Flush		24	.333		37.50	7.35		44.85	53
	2530	Flush, for PC board		24	.333		37.50	7.35		44.85	53
	2540	Floor service box		4	2		200	44		244	293
	2550	Cover, surface					12.30			12.30	13.55
	2560	Flush					12.30			12.30	13.55
	2570	Flush for PC board					12.30			12.30	13.55
	2700	Floor fitting w/duplex jack & cover	1 Elec	21	.381		38	8.40		46.40	56
	2720	Low profile		53	.151		13.20	3.34		16.54	19.95
	2740	Miniature w/duplex jack		53	.151		20.50	3.34		23.84	28
	2760	25 pair kit		21	.381		40	8.40		48.40	58
	2780	Low profile		53	.151		13.50	3.34		16.84	20.50
	2800	Call director kit for 5 cable		19	.421		63.50	9.30		72.80	85
	2820	4 pair kit		19	.421		75.50	9.30		84.80	98
	2840	3 pair kit		19	.421		80	9.30		89.30	103
	2860	Comb. 25 pair & 3 cond power		21	.381		63.50	8.40		71.90	84
	2880	5 cond power		21	.381		72.50	8.40		80.90	94
	2900	PC board, 8-3 pair		161	.050		55	1.10		56.10	62.50
	2920	6-4 pair		161	.050		55	1.10		56.10	62.50
	2940	3 pair adapter		161	.050		50	1.10		51.10	57
	2950	Plug		77	.104		2.30	2.30		4.60	6.30
	2960	Couplers		321	.025		6.55	.55		7.10	8.10
	3000	Bottom shield for 25 pr. cable		4,420	.002	L.F.	.65	.04		.69	.78
	3020	4 pair		4,420	.002		.31	.04		.35	.41
	3040	Top shield for 25 pr. cable		4,420	.002		.65	.04		.69	.78
	3100	Cable assembly, double-end, 50', 25 pr.		11.80	.678	Ea.	193	15		208	237
	3110	3 pair		23.60	.339		57	7.50		64.50	75
	3120	4 pair		23.60	.339		63.50	7.50		71	82.50
	3140	Bulk 3 pair		1,473	.005	L.F.	.90	.12		1.02	1.19
	3160	4 pair		1,473	.005	"	1	.12		1.12	1.30
	3500	Data System									
	3520	Cable 25 conductor w/conn. 40', 75 ohm	1 Elec	14.50	.552	Ea.	59	12.20		71.20	85
	3530	Single lead		22	.364		161	8.05		169.05	190
	3540	Dual lead		22	.364		200	8.05		208.05	233
	3560	Shields same for 25 cond. as 25 pair tele.									
	3570	Single & dual, none req'd.									
	3590	BNC coax connectors, Plug	1 Elec	40	.200	Ea.	8.55	4.42		12.97	16.65
	3600	TNC coax connectors, Plug	"	40	.200	"	11	4.42		15.42	19.35
	3700	Cable-bulk									
	3710	Single lead	1 Elec	1,473	.005	L.F.	2.15	.12		2.27	2.57
	3720	Dual lead	"	1,473	.005	"	3.10	.12		3.22	3.61
	3730	Hand tool crimp				Ea.	287			287	315
	3740	Hand tool notch				"	15.35			15.35	16.90
	3750	Boxes & floor fitting same as telephone									
900	0010	**WIRE**									
	0020	600 volt type THW, copper, solid, #14	1 Elec	13	.615	C.L.F.	4.08	13.60		17.68	27
	0030	#12		11	.727		5.25	16.05		21.30	32.50
	0040	#10		10	.800		8.40	17.70		26.10	38.50

16100 | Wiring Methods

16120 | Conductors & Cables

			CREW	DAILY OUTPUT	LABOR-HOURS	UNIT	2000 BARE COSTS MAT.	LABOR	EQUIP.	TOTAL	TOTAL INCL O&P	
900	0050	Stranded, #14	1 Elec	13	.615	C.L.F.	4.40	13.60		18	27.50	900
	0051	Wire, 600 volt, stranded										
	0100	#12	1 Elec	11	.727	C.L.F.	6.05	16.05		22.10	33	
	0120	#10		10	.800		9.30	17.70		27	39.50	
	0140	#8	▼	8	1		15.60	22		37.60	53	
	0400	250 kcmil	3 Elec	6	4		216	88.50		304.50	385	
	0420	300 kcmil		5.70	4.211		257	93		350	435	
	0450	350 kcmil		5.40	4.444		297	98		395	485	
	0480	400 kcmil		5.10	4.706		345	104		449	550	
	0490	500 kcmil	▼	4.80	5		420	111		531	640	
	0530	Aluminum, stranded, #8	1 Elec	9	.889		12.35	19.65		32	45.50	
	0540	#6	"	8	1		15.75	22		37.75	53.50	
	0560	#4	2 Elec	13	1.231		19.70	27		46.70	66	
	0580	#2		10.60	1.509		26.50	33.50		60	84	
	0600	#1		9	1.778		39	39.50		78.50	108	
	0620	1/0		8	2		46.50	44		90.50	124	
	0640	2/0		7.20	2.222		55	49		104	141	
	0680	3/0		6.60	2.424		68.50	53.50		122	163	
	0700	4/0	▼	6.20	2.581		76	57		133	177	
	0720	250 kcmil	3 Elec	8.70	2.759		92.50	61		153.50	202	
	0740	300 kcmil		8.10	2.963		128	65.50		193.50	248	
	0760	350 kcmil		7.50	3.200		130	70.50		200.50	259	
	0780	400 kcmil		6.90	3.478		152	77		229	293	
	0800	500 kcmil		6	4		168	88.50		256.50	330	
	0850	600 kcmil		5.70	4.211		213	93		306	385	
	0880	700 kcmil		5.10	4.706		245	104		349	440	
	0900	750 kcmil	▼	4.80	5		248	111		359	455	
	0920	Type THWN-THHN, copper, solid, #14	1 Elec	13	.615		4.08	13.60		17.68	27	
	0940	#12		11	.727		5.25	16.05		21.30	32.50	
	0960	#10		10	.800		8.40	17.70		26.10	38.50	
	1000	Stranded, #14		13	.615	▼	4.40	13.60		18	27.50	
	1010	2 wire, #14, install wire only		655	.012	L.F.	.09	.27		.36	.54	
	1020	3 wire, #14, install wire only		440	.018	"	.13	.40		.53	.81	
	1200	#12		11	.727	C.L.F.	6.05	16.05		22.10	33	
	1250	#10		10	.800		9.30	17.70		27	39.50	
	1300	#8		8	1		15.60	22		37.60	53	
	1350	#6	▼	6.50	1.231		25.50	27		52.50	72.50	
	1400	#4	2 Elec	10.60	1.509	▼	41.50	33.50		75	100	

16131 | Cable Trays

105	0010	**CABLE TRAY LADDER TYPE** w/ ftngs & supports, 4" dp., to 15' elev.										105
	0160	Galvanized steel tray										
	0170	4" rung spacing, 6" wide	2 Elec	98	.163	L.F.	8.15	3.61		11.76	14.85	
	0800	6" rung spacing, 6" wide		100	.160		7.40	3.54		10.94	13.95	
	0850	9" wide		94	.170		8.15	3.76		11.91	15.10	
	0860	12" wide		88	.182		8.55	4.02		12.57	15.95	
	0880	24" wide	▼	80	.200	▼	10.90	4.42		15.32	19.25	
130	0010	**CABLE TRAY, COVERS AND DIVIDERS** To 15' high										130
	0100	Covers, ventilated galv. steel, straight, 6" wide tray size	2 Elec	520	.031	L.F.	3.15	.68		3.83	4.58	
	0200	9" wide tray size		460	.035		3.90	.77		4.67	5.55	
	0300	12" wide tray size		400	.040		4.65	.88		5.53	6.55	
	0400	18" wide tray size		300	.053		6.35	1.18		7.53	8.95	
	0500	24" wide tray size		220	.073		7.80	1.61		9.41	11.25	
	3250	Covers, aluminum, straight, 6" wide tray size		520	.031		3.05	.68		3.73	4.46	
	3270	9" wide tray size	▼	460	.035	▼	3.70	.77		4.47	5.35	

For expanded coverage of these items see *Means Electrical Cost Data 2000*

16100 | Wiring Methods

16131 | Cable Trays

			CREW	DAILY OUTPUT	LABOR-HOURS	UNIT	2000 BARE COSTS MAT.	LABOR	EQUIP.	TOTAL	TOTAL INCL O&P	
130	3290	12" wide tray size	2 Elec	400	.040	L.F.	4.40	.88		5.28	6.30	130
	3330	24" wide tray size	↓	260	.062	↓	7.35	1.36		8.71	10.30	

16132 | Conduit & Tubing

			CREW	DAILY OUTPUT	LABOR-HOURS	UNIT	MAT.	LABOR	EQUIP.	TOTAL	TOTAL INCL O&P	
205	0010	**CONDUIT** To 15' high, includes 2 terminations, 2 elbows and										205
	0020	11 beam clamps per 100 L.F. [R16132-210]										
	1750	Rigid galvanized steel, 1/2" diameter	1 Elec	90	.089	L.F.	1.52	1.96		3.48	4.88	
	1770	3/4" diameter		80	.100		1.81	2.21		4.02	5.60	
	1800	1" diameter		65	.123		2.53	2.72		5.25	7.25	
	1850	1-1/2" diameter		55	.145		4.02	3.21		7.23	9.65	
	1870	2" diameter		45	.178		5.35	3.93		9.28	12.35	
	2500	Steel, intermediate conduit (IMC), 1/2" diameter		100	.080		1.21	1.77		2.98	4.22	
	2530	3/4" diameter		90	.089		1.44	1.96		3.40	4.79	
	2550	1" diameter		70	.114		1.98	2.53		4.51	6.30	
	2600	1-1/2" diameter		60	.133		2.98	2.95		5.93	8.10	
	2630	2" diameter		50	.160		3.87	3.54		7.41	10.05	
	2650	2-1/2" diameter	↓	40	.200		7.65	4.42		12.07	15.70	
	2670	3" diameter	2 Elec	60	.267		9.90	5.90		15.80	20.50	
	2700	3-1/2" diameter		54	.296		12.05	6.55		18.60	24	
	2730	4" diameter	↓	50	.320		14	7.05		21.05	27	
	5000	Electric metallic tubing (EMT), 1/2" diameter	1 Elec	170	.047		.36	1.04		1.40	2.10	
	5020	3/4" diameter		130	.062		.55	1.36		1.91	2.82	
	5040	1" diameter		115	.070		.93	1.54		2.47	3.53	
	5080	1-1/2" diameter		90	.089		1.76	1.96		3.72	5.15	
	5100	2" diameter		80	.100		2.23	2.21		4.44	6.10	
	5120	2-1/2" diameter		60	.133		5.95	2.95		8.90	11.35	
	5140	3" diameter	2 Elec	100	.160		6.45	3.54		9.99	12.90	
	5160	3-1/2" diameter		90	.178		8.35	3.93		12.28	15.65	
	5180	4" diameter	↓	80	.200		9.45	4.42		13.87	17.65	
	9100	PVC, #40, 1/2" diameter	1 Elec	190	.042		.57	.93		1.50	2.15	
	9110	3/4" diameter		145	.055		.69	1.22		1.91	2.75	
	9120	1" diameter		125	.064		.99	1.41		2.40	3.39	
	9140	1-1/2" diameter		100	.080		1.55	1.77		3.32	4.60	
	9150	2" diameter		90	.089		2.01	1.96		3.97	5.45	
	9160	2-1/2" diameter	↓	65	.123		3.14	2.72		5.86	7.90	
	9170	3" diameter	2 Elec	110	.145	↓	3.70	3.21		6.91	9.35	
	9880	Heat bender, to 6" diameter				Ea.	960			960	1,050	
	9910	15' to 20' high, add						10%				
	9920	20' to 25' high, add						20%				
	9980	Allow. for cond. ftngs., 5% min.-20% max.										
230	0010	**CONDUIT IN CONCRETE SLAB** Including terminations,										230
	0020	fittings and supports										
	3230	PVC, schedule 40, 1/2" diameter	1 Elec	270	.030	L.F.	.35	.65		1	1.45	
	3250	3/4" diameter		230	.035		.42	.77		1.19	1.73	
	3270	1" diameter		200	.040		.56	.88		1.44	2.07	
	3330	1-1/2" diameter		140	.057		.96	1.26		2.22	3.13	
	3350	2" diameter		120	.067		1.23	1.47		2.70	3.76	
	4350	Rigid galvanized steel, 1/2" diameter		200	.040		1.27	.88		2.15	2.84	
	4400	3/4" diameter		170	.047		1.56	1.04		2.60	3.41	
	4450	1" diameter		130	.062		2.28	1.36		3.64	4.73	
	4600	1-1/2" diameter		100	.080		3.59	1.77		5.36	6.85	
	4800	2" diameter	↓	90	.089	↓	4.79	1.96		6.75	8.45	
240	0010	**CONDUIT IN TRENCH** Includes terminations and fittings										240
	0200	Rigid galvanized steel, 2" diameter	1 Elec	150	.053	L.F.	4.63	1.18		5.81	7.05	
	0600	3" diameter	2 Elec	160	.100		10.40	2.21		12.61	15.05	
	0800	3-1/2" diameter	"	140	.114		13.20	2.53		15.73	18.65	

16100 | Wiring Methods

16132 | Conduit & Tubing

			CREW	DAILY OUTPUT	LABOR-HOURS	UNIT	2000 BARE COSTS MAT.	LABOR	EQUIP.	TOTAL	TOTAL INCL O&P
250	0010	**CONDUIT FITTINGS FOR RIGID GALVANIZED STEEL**										250
	2280	LB, LR or LL fittings & covers, 1/2" diameter	1 Elec	16	.500	Ea.	7.90	11.05		18.95	27
	2290	3/4" diameter		13	.615		9.55	13.60		23.15	33
	2300	1" diameter		11	.727		14.10	16.05		30.15	42
	2350	1-1/2" diameter		6	1.333		26.50	29.50		56	77
	2370	2" diameter		5	1.600		44	35.50		79.50	106
	2380	2-1/2" diameter		4	2		88	44		132	170
	2390	3" diameter		3.50	2.286		114	50.50		164.50	208
	2400	3-1/2" diameter		3	2.667		183	59		242	298
	2410	4" diameter		2.50	3.200		212	70.50		282.50	350
	5280	Service entrance cap, 1/2" diameter		16	.500		6.10	11.05		17.15	25
	5300	3/4" diameter		13	.615		7.05	13.60		20.65	30.50
	5320	1" diameter		10	.800		6.30	17.70		24	36
	5360	1-1/2" diameter		6.50	1.231		11.25	27		38.25	57
	5380	2" diameter		5.50	1.455		34	32		66	90
	5400	2-1/2" diameter		4	2		105	44		149	189
	5420	3" diameter		3.40	2.353		161	52		213	262
	5440	3-1/2" diameter		3	2.667		196	59		255	315
	5460	4" diameter		2.70	2.963		299	65.50		364.50	435
300	0010	**ELECTRICAL NONMETALLIC TUBING (ENT)**										300
	0050	Flexible, 1/2" diameter	1 Elec	270	.030	L.F.	.29	.65		.94	1.39
	0100	3/4" diameter		230	.035		.40	.77		1.17	1.70
	0200	1" diameter		145	.055		.64	1.22		1.86	2.69
	0210	1-1/4" diameter		125	.064		.85	1.41		2.26	3.25
	0220	1-1/2" diameter		100	.080		1.17	1.77		2.94	4.18
	0230	2" diameter		75	.107		1.35	2.36		3.71	5.35
	0300	Connectors, to outlet box, 1/2" diameter		230	.035	Ea.	.54	.77		1.31	1.85
	0310	3/4" diameter		210	.038		.99	.84		1.83	2.47
	0320	1" diameter		200	.040		1.56	.88		2.44	3.17
	0400	Couplings, to conduit, 1/2" diameter		145	.055		.59	1.22		1.81	2.64
	0410	3/4" diameter		130	.062		.78	1.36		2.14	3.08
	0420	1" diameter		125	.064		1.34	1.41		2.75	3.78

16133 | Multi-outlet Assemblies

			CREW	DAILY OUTPUT	LABOR-HOURS	UNIT	MAT.	LABOR	EQUIP.	TOTAL	TOTAL INCL O&P
800	0010	**SURFACE RACEWAY**										800
	0100	No. 500	1 Elec	100	.080	L.F.	.63	1.77		2.40	3.58
	0110	No. 700		100	.080		.71	1.77		2.48	3.67
	0200	No. 1000		90	.089		1.30	1.96		3.26	4.64
	0400	No. 1500, small pancake		90	.089		1.33	1.96		3.29	4.67
	0600	No. 2000, base & cover, blank		90	.089		1.29	1.96		3.25	4.63
	0800	No. 3000, base & cover, blank		75	.107		2.53	2.36		4.89	6.65
	1000	No. 4000, base & cover, blank		65	.123		4.30	2.72		7.02	9.20
	1200	No. 6000, base & cover, blank		50	.160		7.05	3.54		10.59	13.55
	2400	Fittings, elbows, No. 500		40	.200	Ea.	1.14	4.42		5.56	8.50
	2800	Elbow cover, No. 2000		40	.200		2.30	4.42		6.72	9.80
	2880	Tee, No. 500		42	.190		2.21	4.21		6.42	9.35
	2900	No. 2000		27	.296		7.05	6.55		13.60	18.45
	3000	Switch box, No. 500		16	.500		8.10	11.05		19.15	27
	3400	Telephone outlet, No. 1500		16	.500		8.60	11.05		19.65	27.50
	3600	Junction box, No. 1500		16	.500		6	11.05		17.05	24.50
	3800	Plugmold wired sections, No. 2000									
	4000	1 circuit, 6 outlets, 3 ft. long	1 Elec	8	1	Ea.	21.50	22		43.50	59.50
	4100	2 circuits, 8 outlets, 6 ft. long	"	5.30	1.509	"	36	33.50		69.50	94
	9300	Non-metallic, straight section									

For expanded coverage of these items see *Means Electrical Cost Data 2000*

16100 | Wiring Methods

16133 | Multi-outlet Assemblies

			CREW	DAILY OUTPUT	LABOR-HOURS	UNIT	2000 BARE COSTS MAT.	LABOR	EQUIP.	TOTAL	TOTAL INCL O&P	
800	9310	No. 400, base & cover, blank	1 Elec	160	.050	L.F.	1.07	1.11		2.18	2.99	800
	9320	Base & cover w/ adhesive		160	.050		1.24	1.11		2.35	3.17	
	9340	No. 800, base & cover, blank		145	.055		1.24	1.22		2.46	3.35	
	9350	Base & cover w/ adhesive		145	.055		1.44	1.22		2.66	3.57	
	9370	No. 2300, base & cover, blank		130	.062		1.76	1.36		3.12	4.16	
	9380	Base & cover w/ adhesive		130	.062		2.03	1.36		3.39	4.45	
	9400	Fittings, elbows, No. 400		50	.160	Ea.	1.24	3.54		4.78	7.15	
	9410	No. 800		45	.178		1.29	3.93		5.22	7.85	
	9420	No. 2300		40	.200		1.41	4.42		5.83	8.80	
	9430	Tees, No. 400		35	.229		1.63	5.05		6.68	10.05	
	9440	No. 800		32	.250		1.68	5.55		7.23	10.90	
	9450	No. 2300		30	.267		1.71	5.90		7.61	11.55	
	9460	Cover clip, No. 400		80	.100		.33	2.21		2.54	3.98	
	9470	No. 800		72	.111		.30	2.46		2.76	4.35	
	9480	No. 2300		64	.125		.50	2.76		3.26	5.05	
	9490	Blank end, No. 400		50	.160		.47	3.54		4.01	6.30	
	9510	No. 2300		40	.200		.80	4.42		5.22	8.15	
	9520	Round fixture box 5.5" dia x 1"		25	.320		8	7.05		15.05	20.50	
	9530	Device box, 1 gang		30	.267		3.53	5.90		9.43	13.55	
	9540	2 gang		25	.320		5.25	7.05		12.30	17.30	

16134 | Wireway & Aux Gutters

			CREW	DAILY OUTPUT	LABOR-HOURS	UNIT	MAT.	LABOR	EQUIP.	TOTAL	TOTAL INCL O&P	
150	0010	**WIREWAY** to 15' high										150
	0100	Screw cover, with fittings and supports, 2-1/2" x 2-1/2"	1 Elec	45	.178	L.F.	8.65	3.93		12.58	15.95	
	0200	4" x 4"	"	40	.200		9.55	4.42		13.97	17.75	
	0400	6" x 6"	2 Elec	60	.267		16.10	5.90		22	27.50	
	0600	8" x 8"	"	40	.400		29	8.85		37.85	46.50	

16136 | Boxes

			CREW	DAILY OUTPUT	LABOR-HOURS	UNIT	MAT.	LABOR	EQUIP.	TOTAL	TOTAL INCL O&P	
600	0010	**OUTLET BOXES**										600
	0020	Pressed steel, octagon, 4"	1 Elec	20	.400	Ea.	1.46	8.85		10.31	16.05	
	0060	Covers, blank		64	.125		.60	2.76		3.36	5.20	
	0100	Extension		40	.200		2.33	4.42		6.75	9.80	
	0150	Square, 4"		20	.400		1.98	8.85		10.83	16.65	
	0200	Extension		40	.200		2.44	4.42		6.86	9.95	
	0250	Covers, blank		64	.125		.69	2.76		3.45	5.30	
	0300	Plaster rings		64	.125		1.11	2.76		3.87	5.75	
	0550	Handy box		27	.296		1.71	6.55		8.26	12.60	
	0560	Covers, device		64	.125		.56	2.76		3.32	5.15	
	0650	Switchbox		27	.296		2.26	6.55		8.81	13.20	
	0700	Masonry, 1 gang, 2-1/2" deep		27	.296		5.15	6.55		11.70	16.40	
	0710	3-1/2" deep		27	.296		4.48	6.55		11.03	15.65	
	0750	2 gang, 2-1/2" deep		20	.400		7.95	8.85		16.80	23	
	0760	3-1/2" deep		20	.400		7.55	8.85		16.40	23	
	1100	Concrete, floor, 1 gang		5.30	1.509		57.50	33.50		91	118	
	1400	Cast, 1 gang, FS (2" deep), 1/2" hub		12	.667		11.55	14.75		26.30	36.50	
	1410	3/4" hub		12	.667		12.30	14.75		27.05	37.50	
	1420	FD (2-11/16" deep), 1/2" hub		12	.667		13.90	14.75		28.65	39.50	
	1430	3/4" hub		12	.667		15.10	14.75		29.85	40.50	
	1450	2 gang, FS, 1/2" hub		10	.800		20.50	17.70		38.20	52	
	1460	3/4" hub		10	.800		21.50	17.70		39.20	52.50	
	1470	FD, 1/2" hub		10	.800		25	17.70		42.70	57	
	1480	3/4" hub		10	.800		25	17.70		42.70	56.50	
	1500	3 gang, FS, 3/4" hub		9	.889		32	19.65		51.65	67.50	
	1510	Switch cover, 1 gang, FS		64	.125		3.08	2.76		5.84	7.90	

16100 | Wiring Methods

16136 | Boxes

			CREW	DAILY OUTPUT	LABOR-HOURS	UNIT	MAT.	LABOR	EQUIP.	TOTAL	TOTAL INCL O&P	
600	1520	2 gang	1 Elec	53	.151	Ea.	5.05	3.34		8.39	11.05	600
	1530	Duplex receptacle cover, 1 gang, FS		64	.125		3.10	2.76		5.86	7.95	
	1540	2 gang, FS		53	.151		5.10	3.34		8.44	11.05	
	1550	Weatherproof switch cover		64	.125		4.82	2.76		7.58	9.80	
	1600	Weatherproof receptacle cover		64	.125		4.82	2.76		7.58	9.80	
	1750	FSC, 1 gang, 1/2" hub		11	.727		12.80	16.05		28.85	40.50	
	1760	3/4" hub		11	.727		14.05	16.05		30.10	42	
	1770	2 gang, 1/2" hub		9	.889		23	19.65		42.65	57	
	1780	3/4" hub		9	.889		24.50	19.65		44.15	58.50	
	1790	FDC, 1 gang, 1/2" hub		11	.727		15.30	16.05		31.35	43.50	
	1800	3/4" hub		11	.727		16.50	16.05		32.55	44.50	
	1810	2 gang, 1/2" hub		9	.889		28	19.65		47.65	63	
	1820	3/4" hub		9	.889		26.50	19.65		46.15	61	
620	0010	**OUTLET BOXES, PLASTIC**										620
	0050	4" diameter, round with 2 mounting nails	1 Elec	25	.320	Ea.	1.70	7.05		8.75	13.40	
	0100	Bar hanger mounted		25	.320		2.90	7.05		9.95	14.75	
	0200	4", square with 2 mounting nails		25	.320		2.62	7.05		9.67	14.45	
	0300	Plaster ring		64	.125		.92	2.76		3.68	5.55	
	0400	Switch box with 2 mounting nails, 1 gang		30	.267		1.07	5.90		6.97	10.85	
	0500	2 gang		25	.320		2.10	7.05		9.15	13.85	
	0600	3 gang		20	.400		3.32	8.85		12.17	18.10	
700	0010	**PULL BOXES & CABINETS**										700
	0100	Sheet metal, pull box, NEMA 1, type SC, 6" W x 6" H x 4" D	1 Elec	8	1	Ea.	8.35	22		30.35	45	
	0200	8" W x 8" H x 4" D		8	1		11.45	22		33.45	48.50	
	0300	10" W x 12" H x 6" D		5.30	1.509		20	33.50		53.50	76.50	
	0800	12" W x 16" H x 6" D		4.70	1.702		28	37.50		65.50	92	
	1000	20" W x 20" H x 6" D		3.60	2.222		56	49		105	142	
	7000	Cabinets, current transformer										
	7050	Single door, 24" H x 24" W x 10" D	1 Elec	1.60	5	Ea.	107	111		218	299	
	7100	30" H x 24" W x 10" D		1.30	6.154		120	136		256	355	
	7150	36" H x 24" W x 10" D		1.10	7.273		134	161		295	410	
	7200	30" H x 30" W x 10" D		1	8		140	177		317	445	
	7250	36" H x 30" W x 10" D		.90	8.889		189	196		385	530	
	7300	36" H x 36" W x 10" D		.80	10		206	221		427	585	
	7500	Double door, 48" H x 36" W x 10" D		.60	13.333		360	295		655	875	
	7550	24" H x 24" W x 12" D		1	8		193	177		370	500	

16139 | Residential Wiring

			CREW	DAILY OUTPUT	LABOR-HOURS	UNIT	MAT.	LABOR	EQUIP.	TOTAL	TOTAL INCL O&P	
700	0010	**RESIDENTIAL WIRING**										700
	0020	20' avg. runs and #14/2 wiring incl. unless otherwise noted										
	1000	Service & panel, includes 24' SE-AL cable, service eye, meter,										
	1010	Socket, panel board, main bkr., ground rod, 15 or 20 amp										
	1020	1-pole circuit breakers, and misc. hardware										
	1100	100 amp, with 10 branch breakers	1 Elec	1.19	6.723	Ea.	400	149		549	680	
	1110	With PVC conduit and wire		.92	8.696		425	192		617	785	
	1120	With RGS conduit and wire		.73	10.959		545	242		787	995	
	1200	200 amp, with 18 branch breakers	2 Elec	1.80	8.889		810	196		1,006	1,200	
	1220	With PVC conduit and wire		1.46	10.959		870	242		1,112	1,350	
	1230	With RGS conduit and wire		1.24	12.903		1,150	285		1,435	1,750	
	1800	Lightning surge suppressor for above services, add	1 Elec	32	.250		39.50	5.55		45.05	52.50	
	2000	Switch devices										
	2100	Single pole, 15 amp, Ivory, with a 1-gang box, cover plate,										
	2110	Type NM (Romex) cable	1 Elec	17.10	.468	Ea.	6.65	10.35		17	24.50	
	2120	Type MC (BX) cable		14.30	.559		16.80	12.35		29.15	38.50	

For expanded coverage of these items see *Means Electrical Cost Data 2000*

16100 | Wiring Methods

16139 | Residential Wiring

			CREW	DAILY OUTPUT	LABOR-HOURS	UNIT	2000 BARE COSTS				TOTAL INCL O&P
							MAT.	LABOR	EQUIP.	TOTAL	
700	2130	EMT & wire	1 Elec	5.71	1.401	Ea.	17.45	31		48.45	69.50	700
	2150	3-way, #14/3, type NM cable		14.55	.550		9.85	12.15		22	30.50
	2170	Type MC cable		12.31	.650		21	14.35		35.35	46.50
	2180	EMT & wire		5	1.600		19.65	35.50		55.15	79.50
	2250	S.P., 20 amp, #12/2, type NM cable		13.33	.600		11.65	13.25		24.90	34.50
	2270	Type MC cable		11.43	.700		21	15.45		36.45	48.50
	2280	EMT & wire		4.85	1.649		23	36.50		59.50	84.50
	2290	S.P. rotary dimmer, 600W, no wiring		17	.471		15.55	10.40		25.95	34
	2300	S.P. rotary dimmer, 600W, type NM cable		14.55	.550		18.25	12.15		30.40	40
	2320	Type MC cable		12.31	.650		28.50	14.35		42.85	54.50
	2330	EMT & wire		5	1.600		30	35.50		65.50	91
	2350	3-way rotary dimmer, type NM cable		13.33	.600		15.65	13.25		28.90	38.50
	2370	Type MC cable		11.43	.700		26	15.45		41.45	54
	2380	EMT & wire		4.85	1.649		27.50	36.50		64	89.50
	2400	Interval timer wall switch, 20 amp, 1-30 min., #12/2									
	2410	Type NM cable	1 Elec	14.55	.550	Ea.	28.50	12.15		40.65	51
	2420	Type MC cable		12.31	.650		35.50	14.35		49.85	62.50
	2430	EMT & wire		5	1.600		39.50	35.50		75	102
	2500	Decorator style									
	2510	S.P., 15 amp, type NM cable	1 Elec	17.10	.468	Ea.	10.05	10.35		20.40	28
	2520	Type MC cable		14.30	.559		20	12.35		32.35	42
	2530	EMT & wire		5.71	1.401		21	31		52	73.50
	2550	3-way, #14/3, type NM cable		14.55	.550		13.25	12.15		25.40	34.50
	2570	Type MC cable		12.31	.650		24.50	14.35		38.85	50.50
	2580	EMT & wire		5	1.600		23	35.50		58.50	83.50
	2650	S.P., 20 amp, #12/2, type NM cable		13.33	.600		15.05	13.25		28.30	38
	2670	Type MC cable		11.43	.700		24.50	15.45		39.95	52.50
	2680	EMT & wire		4.85	1.649		26	36.50		62.50	88.50
	2700	S.P., slide dimmer, type NM cable		17.10	.468		25.50	10.35		35.85	45
	2720	Type MC cable		14.30	.559		35.50	12.35		47.85	59
	2730	EMT & wire		5.71	1.401		37	31		68	91.50
	2750	S.P., touch dimmer, type NM cable		17.10	.468		21.50	10.35		31.85	40.50
	2770	Type MC cable		14.30	.559		31.50	12.35		43.85	54.50
	2780	EMT & wire		5.71	1.401		33	31		64	87
	2800	3-way touch dimmer, type NM cable		13.33	.600		39	13.25		52.25	64
	2820	Type MC cable		11.43	.700		49	15.45		64.45	79.50
	2830	EMT & wire		4.85	1.649		50.50	36.50		87	115
	3000	Combination devices									
	3100	S.P. switch/15 amp recpt., Ivory, 1-gang box, plate									
	3110	Type NM cable	1 Elec	11.43	.700	Ea.	14.45	15.45		29.90	41.50
	3120	Type MC cable		10	.800		24.50	17.70		42.20	56
	3130	EMT & wire		4.40	1.818		26	40		66	94
	3150	S.P. switch/pilot light, type NM cable		11.43	.700		15.05	15.45		30.50	42
	3170	Type MC cable		10	.800		25	17.70		42.70	56.50
	3180	EMT & wire		4.43	1.806		26.50	40		66.50	95
	3190	2-S.P. switches, 2-#14/2, no wiring		14	.571		5.45	12.65		18.10	26.50
	3200	2-S.P. switches, 2-#14/2, type NM cables		10	.800		16.45	17.70		34.15	47
	3220	Type MC cable		8.89	.900		33	19.90		52.90	69
	3230	EMT & wire		4.10	1.951		28	43		71	101
	3250	3-way switch/15 amp recpt., #14/3, type NM cable		10	.800		20.50	17.70		38.20	51.50
	3270	Type MC cable		8.89	.900		31.50	19.90		51.40	67.50
	3280	EMT & wire		4.10	1.951		30.50	43		73.50	104
	3300	2-3 way switches, 2-#14/3, type NM cables		8.89	.900		27	19.90		46.90	62
	3320	Type MC cable		8	1		46	22		68	86.50
	3330	EMT & wire		4	2		35	44		79	111
	3350	S.P. switch/20 amp recpt., #12/2, type NM cable		10	.800		24.50	17.70		42.20	55.50

16100 | Wiring Methods

16139 | Residential Wiring

		CREW	DAILY OUTPUT	LABOR-HOURS	UNIT	2000 BARE COSTS MAT.	LABOR	EQUIP.	TOTAL	TOTAL INCL O&P
3370	Type MC cable	1 Elec	8.89	.900	Ea.	31	19.90		50.90	67
3380	EMT & wire	↓	4.10	1.951	↓	35.50	43		78.50	110
4000	Receptacle devices									
4010	Duplex outlet, 15 amp recpt., Ivory, 1-gang box, plate									
4015	Type NM cable	1 Elec	14.55	.550	Ea.	5.20	12.15		17.35	25.50
4020	Type MC cable		12.31	.650		15.30	14.35		29.65	40.50
4030	EMT & wire		5.33	1.501		16	33		49	72
4050	With #12/2, type NM cable		12.31	.650		6.30	14.35		20.65	30.50
4070	Type MC cable		10.67	.750		15.65	16.55		32.20	44
4080	EMT & wire		4.71	1.699		17.50	37.50		55	81
4100	20 amp recpt., #12/2, type NM cable		12.31	.650		9.50	14.35		23.85	34
4120	Type MC cable		10.67	.750		18.85	16.55		35.40	48
4130	EMT & wire		4.71	1.699		20.50	37.50		58	84.50
4500	Weather-proof cover for above receptacles, add	↓	32	.250	↓	4.40	5.55		9.95	13.90
4550	Air conditioner outlet, 20 amp-240 volt recpt.									
4560	30' of #12/2, 2 pole circuit breaker									
4570	Type NM cable	1 Elec	10	.800	Ea.	39.50	17.70		57.20	72
4580	Type MC cable		9	.889		51.50	19.65		71.15	88.50
4590	EMT & wire		4	2		50	44		94	128
4600	Decorator style, type NM cable		10	.800		43	17.70		60.70	76.50
4620	Type MC cable		9	.889		55.50	19.65		75.15	93
4630	EMT & wire	↓	4	2	↓	54	44		98	132
4650	Dryer outlet, 30 amp-240 volt recpt., 20' of #10/3									
4660	2 pole circuit breaker									
4670	Type NM cable	1 Elec	6.41	1.248	Ea.	47.50	27.50		75	97.50
4680	Type MC cable		5.71	1.401		57	31		88	114
4690	EMT & wire	↓	3.48	2.299	↓	53	51		104	142
4700	Range outlet, 50 amp-240 volt recpt., 30' of #8/3									
4710	Type NM cable	1 Elec	4.21	1.900	Ea.	69	42		111	145
4720	Type MC cable		4	2		98.50	44		142.50	181
4730	EMT & wire		2.96	2.703		67.50	59.50		127	172
4750	Central vacuum outlet, Type NM cable		6.40	1.250		41.50	27.50		69	90.50
4770	Type MC cable		5.71	1.401		58.50	31		89.50	115
4780	EMT & wire	↓	3.48	2.299	↓	52.50	51		103.50	141
4800	30 amp-110 volt locking recpt., #10/2 circ. bkr.									
4810	Type NM cable	1 Elec	6.20	1.290	Ea.	48	28.50		76.50	99
4820	Type MC cable		5.40	1.481		70.50	32.50		103	131
4830	EMT & wire	↓	3.20	2.500	↓	61	55.50		116.50	158
4900	Low voltage outlets									
4910	Telephone recpt., 20' of 4/C phone wire	1 Elec	26	.308	Ea.	6.75	6.80		13.55	18.50
4920	TV recpt., 20' of RG59U coax wire, F type connector	"	16	.500	"	11.45	11.05		22.50	30.50
4950	Door bell chime, transformer, 2 buttons, 60' of bellwire									
4970	Economy model	1 Elec	11.50	.696	Ea.	48.50	15.35		63.85	78.50
4980	Custom model	"	11.50	.696	"	83.50	15.35		98.85	117
6000	Lighting outlets									
6050	Wire only (for fixture), type NM cable	1 Elec	32	.250	Ea.	3.51	5.55		9.06	12.90
6070	Type MC cable		24	.333		11.40	7.35		18.75	24.50
6080	EMT & wire		10	.800		11.15	17.70		28.85	41.50
6100	Box (4"), and wire (for fixture), type NM cable		25	.320		7.50	7.05		14.55	19.80
6120	Type MC cable		20	.400		15.40	8.85		24.25	31.50
6130	EMT & wire	↓	11	.727	↓	15.10	16.05		31.15	43
6200	Fixtures (use with lines 6050 or 6100 above)									
6210	Canopy style, economy grade	1 Elec	40	.200	Ea.	25.50	4.42		29.92	35.50
6220	Custom grade		40	.200		46	4.42		50.42	58
6250	Dining room chandelier, economy grade		19	.421		76	9.30		85.30	98.50
6260	Custom grade	↓	19	.421	↓	225	9.30		234.30	263

For expanded coverage of these items see *Means Electrical Cost Data 2000*

16100 | Wiring Methods
16139 | Residential Wiring

			CREW	DAILY OUTPUT	LABOR-HOURS	UNIT	2000 BARE COSTS				TOTAL INCL O&P
							MAT.	LABOR	EQUIP.	TOTAL	
700	6310	Kitchen fixture (fluorescent), economy grade	1 Elec	30	.267	Ea.	51.50	5.90		57.40	66
	6320	Custom grade		25	.320		161	7.05		168.05	189
	6350	Outdoor, wall mounted, economy grade		30	.267		27	5.90		32.90	39
	6360	Custom grade		30	.267		101	5.90		106.90	121
	6410	Outdoor PAR floodlights, 1 lamp, 150 watt		20	.400		20	8.85		28.85	36.50
	6420	2 lamp, 150 watt each		20	.400		33.50	8.85		42.35	51.50
	6430	For infrared security sensor, add		32	.250		87	5.55		92.55	105
	6450	Outdoor, quartz-halogen, 300 watt flood		20	.400		38	8.85		46.85	56.50
	6600	Recessed downlight, round, pre-wired, 50 or 75 watt trim		30	.267		35	5.90		40.90	48
	6610	With shower light trim		30	.267		42	5.90		47.90	55.50
	6620	With wall washer trim		28	.286		52.50	6.30		58.80	68.50
	6640	For direct contact with insulation, add					1.60			1.60	1.76
	6700	Porcelain lamp holder	1 Elec	40	.200		3.50	4.42		7.92	11.10
	6710	With pull switch		40	.200		3.75	4.42		8.17	11.40
	6750	Fluorescent strip, 1-20 watt tube, wrap around diffuser, 24"		24	.333		51.50	7.35		58.85	68.50
	6760	1-40 watt tube, 48"		24	.333		65	7.35		72.35	83.50
	6770	2-40 watt tubes, 48"		20	.400		78	8.85		86.85	100
	6780	With residential ballast		20	.400		89.50	8.85		98.35	113
	6800	Bathroom heat lamp, 1-250 watt		28	.286		30.50	6.30		36.80	44
	6810	2-250 watt lamps		28	.286		51	6.30		57.30	66.50
	6820	For timer switch, see line 2400									
	6900	Outdoor post lamp, incl. post, fixture, 35' of #14/2									
	6910	Type NMC cable	1 Elec	3.50	2.286	Ea.	165	50.50		215.50	265
	6920	Photo-eye, add		27	.296		29	6.55		35.55	42.50
	6950	Clock dial time switch, 24 hr., w/enclosure, type NM cable		11.43	.700		50.50	15.45		65.95	81
	6970	Type MC cable		11	.727		60.50	16.05		76.55	93.50
	6980	EMT & wire		4.85	1.649		61.50	36.50		98	127
	7000	Alarm systems									
	7050	Smoke detectors, box, #14/3, type NM cable	1 Elec	14.55	.550	Ea.	28	12.15		40.15	50.50
	7070	Type MC cable		12.31	.650		36.50	14.35		50.85	64
	7080	EMT & wire		5	1.600		35.50	35.50		71	97
	7090	For relay output to security system, add					11.75			11.75	12.95
	8000	Residential equipment									
	8050	Disposal hook-up, incl. switch, outlet box, 3' of flex									
	8060	20 amp-1 pole circ. bkr., and 25' of #12/2									
	8070	Type NM cable	1 Elec	10	.800	Ea.	19.10	17.70		36.80	50
	8080	Type MC cable		8	1		30	22		52	69
	8090	EMT & wire		5	1.600		31.50	35.50		67	93
	8100	Trash compactor or dishwasher hook-up, incl. outlet box,									
	8110	3' of flex, 15 amp-1 pole circ. bkr., and 25' of #14/2									
	8130	Type MC cable	1 Elec	8	1	Ea.	25.50	22		47.50	64
	8140	EMT & wire	"	5	1.600	"	26	35.50		61.50	86.50
	8150	Hot water sink dispensor hook-up, use line 8100									
	8200	Vent/exhaust fan hook-up, type NM cable	1 Elec	32	.250	Ea.	3.51	5.55		9.06	12.90
	8220	Type MC cable		24	.333		11.40	7.35		18.75	24.50
	8230	EMT & wire		10	.800		11.15	17.70		28.85	41.50
	8250	Bathroom vent fan, 50 CFM (use with above hook-up)									
	8260	Economy model	1 Elec	15	.533	Ea.	21.50	11.80		33.30	43
	8280	Custom model	"	12	.667	"	111	14.75		125.75	146
	8300	Bathroom or kitchen vent fan, 110 CFM									
	8310	Economy model	1 Elec	15	.533	Ea.	53	11.80		64.80	78
	8320	Low noise model	"	15	.533	"	72	11.80		83.80	98.50
	8350	Paddle fan, variable speed (w/o lights)									
	8360	Economy model (AC motor)	1 Elec	10	.800	Ea.	94.50	17.70		112.20	133
	8370	Custom model (AC motor)		10	.800		165	17.70		182.70	211
	8390	Remote speed switch for above, add		12	.667		21	14.75		35.75	47

16100 | Wiring Methods

16139 | Residential Wiring

		CREW	DAILY OUTPUT	LABOR-HOURS	UNIT	MAT.	LABOR	EQUIP.	TOTAL	TOTAL INCL O&P	
700	8500	Whole house exhaust fan, ceiling mount, 36", variable speed										700
	8510	Remote switch, incl. shutters, 20 amp-1 pole circ. bkr.										
	8520	30' of #12/2, type NM cable	1 Elec	4	2	Ea.	675	44		719	820	
	8530	Type MC cable		3.50	2.286		690	50.50		740.50	845	
	8540	EMT & wire	↓	3	2.667	↓	690	59		749	855	
	8600	Whirlpool tub hook-up, incl. timer switch, outlet box										
	8610	3' of flex, 20 amp-1 pole GFI circ. bkr.										
	8620	30' of #12/2, type NM cable	1 Elec	5	1.600	Ea.	74.50	35.50		110	140	
	8630	Type MC cable		4.20	1.905		82	42		124	160	
	8640	EMT & wire	↓	3.40	2.353	↓	84	52		136	178	
	8650	Hot water heater hook-up, incl. 1-2 pole circ. bkr., box;										
	8660	3' of flex, 20' of #10/2, type NM cable	1 Elec	5	1.600	Ea.	19.20	35.50		54.70	79	
	8670	Type MC cable		4.20	1.905		33.50	42		75.50	106	
	8680	EMT & wire	↓	3.40	2.353	↓	28	52		80	116	
	9000	Heating/air conditioning										
	9050	Furnace/boiler hook-up, incl. firestat, local on-off switch										
	9060	Emergency switch, and 40' of type NM cable	1 Elec	4	2	Ea.	42	44		86	119	
	9070	Type MC cable		3.50	2.286		59	50.50		109.50	148	
	9080	EMT & wire	↓	1.50	5.333	↓	60	118		178	259	
	9100	Air conditioner hook-up, incl. local 60 amp disc. switch										
	9110	3' sealtite, 40 amp, 2 pole circuit breaker										
	9130	40' of #8/2, type NM cable	1 Elec	3.50	2.286	Ea.	136	50.50		186.50	232	
	9140	Type MC cable		3	2.667		182	59		241	298	
	9150	EMT & wire	↓	1.30	6.154	↓	146	136		282	380	
	9200	Heat pump hook-up, 1-40 & 1-100 amp 2 pole circ. bkr.										
	9210	Local disconnect switch, 3' sealtite										
	9220	40' of #8/2 & 30' of #3/2										
	9230	Type NM cable	1 Elec	1.30	6.154	Ea.	325	136		461	575	
	9240	Type MC cable		1.08	7.407		445	164		609	760	
	9250	EMT & wire	↓	.94	8.511	↓	335	188		523	680	
	9500	Thermostat hook-up, using low voltage wire										
	9520	Heating only	1 Elec	24	.333	Ea.	5.75	7.35		13.10	18.35	
	9530	Heating/cooling	"	20	.400	"	7.10	8.85		15.95	22.50	

16140 | Wiring Devices

		CREW	DAILY OUTPUT	LABOR-HOURS	UNIT	MAT.	LABOR	EQUIP.	TOTAL	TOTAL INCL O&P	
500	0010	**LOW VOLTAGE SWITCHING**										500
	3600	Relays, 120 V or 277 V standard	1 Elec	12	.667	Ea.	26	14.75		40.75	52.50	
	3800	Flush switch, standard		40	.200		9.05	4.42		13.47	17.20	
	4000	Interchangeable		40	.200		11.80	4.42		16.22	20.50	
	4100	Surface switch, standard		40	.200		6.60	4.42		11.02	14.55	
	4200	Transformer 115 V to 25 V		12	.667		93	14.75		107.75	126	
	4400	Master control, 12 circuit, manual		4	2		94	44		138	176	
	4500	25 circuit, motorized		4	2		102	44		146	185	
	4600	Rectifier, silicon		12	.667		30.50	14.75		45.25	57.50	
	4800	Switchplates, 1 gang, 1, 2 or 3 switch, plastic		80	.100		3	2.21		5.21	6.90	
	5000	Stainless steel		80	.100		8.10	2.21		10.31	12.50	
	5400	2 gang, 3 switch, stainless steel	↓	53	.151	↓	15.65	3.34		18.99	22.50	
910	0010	**WIRING DEVICES**										910
	0200	Toggle switch, quiet type, single pole, 15 amp	1 Elec	40	.200	Ea.	4.63	4.42		9.05	12.35	
	1650	Dimmer switch, 120 volt, incandescent, 600 watt, 1 pole		16	.500		10.80	11.05		21.85	30	
	2460	Receptacle, duplex, 120 volt, grounded, 15 amp		40	.200		1.14	4.42		5.56	8.50	
	2470	20 amp		27	.296		9.50	6.55		16.05	21	
	2490	Dryer, 30 amp		15	.533		11.70	11.80		23.50	32	
	2500	Range, 50 amp		11	.727		10.45	16.05		26.50	38	
	2600	Wall plates, stainless steel, 1 gang		80	.100		2.05	2.21		4.26	5.90	

For expanded coverage of these items see *Means Electrical Cost Data 2000*

16100 | Wiring Methods

16140 | Wiring Devices

		Crew	Daily Output	Labor-Hours	Unit	Mat.	Labor	Equip.	Total	Total Incl O&P	
910	2800 2 gang	1 Elec	53	.151	Ea.	4.10	3.34		7.44	9.95	910
	3200 Lampholder, keyless		26	.308		8.80	6.80		15.60	21	
	3400 Pullchain with receptacle		22	.364		8.90	8.05		16.95	23	

16150 | Wiring Connections

		Crew	Daily Output	Labor-Hours	Unit	Mat.	Labor	Equip.	Total	Total Incl O&P	
275	0010 **MOTOR CONNECTIONS**										275
	0020 Flexible conduit and fittings, 115 volt, 1 phase, up to 1 HP motor	1 Elec	8	1	Ea.	4.30	22		26.30	40.50	
	0050 2 HP motor		6.50	1.231		4.40	27		31.40	49.50	
	0100 3 HP motor		5.50	1.455		7.10	32		39.10	60.50	
	0120 230 volt, 10 HP motor, 3 phase		4.20	1.905		7.70	42		49.70	77.50	
	0150 15 HP motor		3.30	2.424		15.40	53.50		68.90	104	
	0200 25 HP motor		2.70	2.963		17.50	65.50		83	126	
	1500 460 volt, 5 HP motor, 3 phase		8	1		4.77	22		26.77	41.50	
	1520 10 HP motor		8	1		4.77	22		26.77	41.50	
	1530 25 HP motor		6	1.333		8.80	29.50		38.30	57.50	

16200 | Electrical Power

16210 | Elect Utility Services

		Crew	Daily Output	Labor-Hours	Unit	Mat.	Labor	Equip.	Total	Total Incl O&P	
600	0010 **METER CENTERS AND SOCKETS**										600
	0100 Sockets, single position, 4 terminal, 100 amp	1 Elec	3.20	2.500	Ea.	27.50	55.50		83	121	
	0200 150 amp		2.30	3.478		30	77		107	159	
	0300 200 amp		1.90	4.211		40	93		133	196	
	0400 20 amp		3.20	2.500		27.50	55.50		83	121	
	0500 Double position, 4 terminal, 100 amp		2.80	2.857		116	63		179	231	
	0600 150 amp		2.10	3.810		120	84		204	270	
	0700 200 amp		1.70	4.706		288	104		392	485	
	2590 Basic meter device										
	2600 1P 3W 120/240V 4 jaw 125A sockets, 3 meter	2 Elec	1	16	Ea.	410	355		765	1,025	
	2610 4 meter		.90	17.778		490	395		885	1,175	
	2620 5 meter		.80	20		615	440		1,055	1,400	
	2630 6 meter		.60	26.667		705	590		1,295	1,750	
	2640 7 meter		.56	28.571		900	630		1,530	2,025	
	2650 8 meter		.52	30.769		985	680		1,665	2,175	
	2660 10 meter		.48	33.333		1,225	735		1,960	2,550	
	2680 Rainproof 1P 3W 120/240V 4 jaw 125A sockets										
	2690 3 meter	2 Elec	1	16	Ea.	410	355		765	1,025	
	2700 4 meter		.90	17.778		490	395		885	1,175	
	2710 6 meter		.60	26.667		705	590		1,295	1,750	
	2720 7 meter		.56	28.571		900	630		1,530	2,025	
	2730 8 meter		.52	30.769		985	680		1,665	2,175	
	2750 1P 3W 120/240V 4 jaw sockets										
	2760 with 125A circuit breaker, 3 meter	2 Elec	1	16	Ea.	770	355		1,125	1,425	
	2770 4 meter		.90	17.778		970	395		1,365	1,725	
	2780 5 meter		.80	20		1,200	440		1,640	2,050	
	2790 6 meter		.60	26.667		1,425	590		2,015	2,550	
	2800 7 meter		.56	28.571		1,725	630		2,355	2,925	
	2810 8 meter		.52	30.769		1,925	680		2,605	3,225	
	2820 10 meter		.48	33.333		2,425	735		3,160	3,875	

16200 | Electrical Power

16210 | Elect Utility Services

		CREW	DAILY OUTPUT	LABOR-HOURS	UNIT	2000 BARE COSTS MAT.	LABOR	EQUIP.	TOTAL	TOTAL INCL O&P
3250	1P 3W 120/240V 4 jaw sockets									
3260	with 200A circuit breaker, 3 meter	2 Elec	1	16	Ea.	1,150	355		1,505	1,850
3270	4 meter		.90	17.778		1,550	395		1,945	2,350
3290	6 meter		.60	26.667		2,300	590		2,890	3,500
3300	7 meter		.56	28.571		2,700	630		3,330	4,000
3310	8 meter		.56	28.571		3,125	630		3,755	4,450

16230 | Generator Assemblies

		CREW	DAILY OUTPUT	LABOR-HOURS	UNIT	MAT.	LABOR	EQUIP.	TOTAL	TOTAL INCL O&P
0010	**GENERATOR SET**									
0020	Gas or gasoline operated, includes battery,									
0050	charger, muffler & transfer switch									
0200	3 phase 4 wire, 277/480 volt, 7.5 kW	R-3	.83	24.096	Ea.	6,000	530	167	6,697	7,650
0300	11.5 kW		.71	28.169		8,500	620	195	9,315	10,600
0400	20 kW		.63	31.746		10,000	700	219	10,919	12,400
0500	35 kW		.55	36.364		12,000	800	251	13,051	14,800
2000	Diesel engine, including battery, charger,									
2010	muffler, automatic transfer switch & day tank, 30 kW	R-3	.55	36.364	Ea.	16,000	800	251	17,051	19,200
2100	50 kW		.42	47.619		19,700	1,050	330	21,080	23,700
2200	75 kW		.35	57.143		25,700	1,250	395	27,345	30,800
2300	100 kW		.31	64.516		28,500	1,425	445	30,370	34,200

16260 | Static Power Converters

		CREW	DAILY OUTPUT	LABOR-HOURS	UNIT	MAT.	LABOR	EQUIP.	TOTAL	TOTAL INCL O&P
0010	**UNINTERRUPTIBLE POWER SUPPLY/CONDITIONER TRANSFORMERS**									
0100	Volt. regulating, isolating trans., w/invert. & 10 min. battery pack									
0110	Single-phase, 120 V, 0.35 kVA	1 Elec	2.29	3.493	Ea.	900	77		977	1,125
0120	0.5 kVA		2	4		950	88.50		1,038.50	1,200
0130	For additional 55 min. battery, add to .35 kVA		2.29	3.493		525	77		602	705
0140	Add to 0.5 kVA		1.14	7.018		550	155		705	860
0150	Single-phase, 120 V, 0.75 kVA		.80	10		1,225	221		1,446	1,700
0160	1.0 kVA		.80	10		1,800	221		2,021	2,325
0170	1.5 kVA	2 Elec	1.14	14.035		2,850	310		3,160	3,625
0180	2 kVA	"	.89	17.978		3,150	395		3,545	4,125
0190	3 kVA	R-3	.63	31.746		3,750	700	219	4,669	5,525
0200	5 kVA		.42	47.619		5,975	1,050	330	7,355	8,625
0210	7.5 kVA		.33	60.606		7,525	1,325	420	9,270	10,900
0220	10 kVA		.28	71.429		7,675	1,575	495	9,745	11,500
0230	15 kVA		.22	90.909		10,600	2,000	630	13,230	15,700
0500	For options & accessories add to above, minimum									10%
0520	Maximum									35%
0600	For complex & special design systems to meet specific									
0610	requirements, obtain quote from vendor.									

16270 | Transformers

		CREW	DAILY OUTPUT	LABOR-HOURS	UNIT	MAT.	LABOR	EQUIP.	TOTAL	TOTAL INCL O&P
0010	**BUCK-BOOST TRANSFORMER**									
0100	Single phase, 120/240 V primary, 12/24 V secondary									
0200	0.10 kVA	1 Elec	8	1	Ea.	49	22		71	90
0400	0.25 kVA		5.70	1.404		72	31		103	130
0600	0.50 kVA		4	2		98	44		142	181
0800	0.75 kVA		3.10	2.581		126	57		183	233
1200	1.5 kVA		1.80	4.444		193	98		291	375
1600	3.0 kVA		1.40	5.714		315	126		441	550
2000	3 phase, 240 V primary, 208/120 V secondary, 15 kVA	2 Elec	2.40	6.667		1,200	147		1,347	1,575
2200	30 kVA		1.60	10		1,325	221		1,546	1,825
2400	45 kVA		1.40	11.429		1,600	253		1,853	2,200
2600	75 kVA		1.20	13.333		1,950	295		2,245	2,625

For expanded coverage of these items see *Means Electrical Cost Data 2000*

16200 | Electrical Power

16270 | Transformers

		CREW	DAILY OUTPUT	LABOR-HOURS	UNIT	MAT.	LABOR	EQUIP.	TOTAL	TOTAL INCL O&P
100 3000	150 kVA	R-3	1.10	18.182	Ea.	3,200	400	126	3,726	4,325
3200	225 kVA	"	1	20		4,175	440	138	4,753	5,475
200 0010	**DRY TYPE TRANSFORMER**									
0050	Single phase, 240/480 volt primary, 120/240 volt secondary									
0100	1 kVA	1 Elec	2	4	Ea.	147	88.50		235.50	305
0300	2 kVA		1.60	5		221	111		332	425
0500	3 kVA		1.40	5.714		274	126		400	505
0700	5 kVA		1.20	6.667		375	147		522	655
1100	10 kVA	2 Elec	1.60	10		670	221		891	1,100
1300	15 kVA		1.20	13.333		915	295		1,210	1,475
1700	37.5 kVA		.80	20		1,525	440		1,965	2,400
1900	50 kVA		.70	22.857		1,800	505		2,305	2,800
2110	100 kVA	R-3	.90	22.222		3,100	490	154	3,744	4,375
2310	Ventilated, 3 kVA	1 Elec	1	8		425	177		602	760
2900	9 kVA	"	.70	11.429		715	253		968	1,200
3100	15 kVA	2 Elec	1.10	14.545		930	320		1,250	1,550
3300	30 kVA		.90	17.778		1,100	395		1,495	1,875
3700	75 kVA		.70	22.857		2,025	505		2,530	3,050
3900	112.5 kVA	R-3	.90	22.222		2,675	490	154	3,319	3,925
4100	150 kVA		.85	23.529		3,500	520	163	4,183	4,875
4500	300 kVA		.55	36.364		5,975	800	251	7,026	8,150
4700	500 kVA		.45	44.444		9,900	980	305	11,185	12,800
5380	3 phase, 5 kV primary 277/480 volt secondary									
5400	High voltage, 112.5 kVA	R-3	.85	23.529	Ea.	8,025	520	163	8,708	9,850
5440	500 kVA		.35	57.143		17,000	1,250	395	18,645	21,200
5480	2000 kVA		.25	80		37,700	1,750	555	40,005	45,000
5590	15 kV primary 277/480 volt secondary									
5600	High voltage, 112.5 kVA	R-3	.85	23.529	Ea.	12,300	520	163	12,983	14,500
5640	500 kVA		.35	57.143		23,600	1,250	395	25,245	28,400
5680	2000 kVA		.25	80		42,900	1,750	555	45,205	50,500
610 0010	**TRANSFORMER, LIQUID-FILLED** Pad mounted									
0020	5 kV or 15 kV primary, 277/480 volt secondary, 3 phase									
0050	225 kVA	R-3	.55	36.364	Ea.	8,400	800	251	9,451	10,800
0100	300 kVA		.45	44.444		9,975	980	305	11,260	12,900
0200	500 kVA		.40	50		12,600	1,100	345	14,045	16,100
0250	750 kVA		.38	52.632		16,300	1,150	365	17,815	20,200

16280 | Power Filters & Conditioners

		CREW	DAILY OUTPUT	LABOR-HOURS	UNIT	MAT.	LABOR	EQUIP.	TOTAL	TOTAL INCL O&P
100 0010	**AUTOMATIC VOLTAGE REGULATORS**									
0100	Computer grade, solid state, variable trans. volt. regulator									
0110	Single-phase, 120 V, 8.6 kVA	2 Elec	1.33	12.030	Ea.	4,400	266		4,666	5,275
0120	17.3 kVA		1.14	14.035		5,200	310		5,510	6,225
0130	208/240 V, 7.5/8.6 kVA		1.33	12.030		4,400	266		4,666	5,275
0140	13.5/15.6 kVA		1.33	12.030		5,200	266		5,466	6,150
0150	27.0/31.2 kVA		1.14	14.035		6,550	310		6,860	7,725
0210	Two-phase, single control, 208/240 V, 15.0/17.3 kVA		1.14	14.035		5,200	310		5,510	6,225
0220	Individual phase control, 15.0/17.3 kVA		1.14	14.035		5,200	310		5,510	6,225
0230	30.0/34.6 kVA	3 Elec	1.33	18.045		6,550	400		6,950	7,875
0310	Three-phase single control, 208/240 V, 26/30 kVA	2 Elec	1	16		5,200	355		5,555	6,300
0320	380/480 V, 24/30 kVA	"	1	16		5,200	355		5,555	6,300
0330	43/54 kVA	3 Elec	1.33	18.045		9,450	400		9,850	11,100
0340	Individual phase control, 208 V, 26 kVA	"	1.33	18.045		5,200	400		5,600	6,375
0350	52 kVA	R-3	.91	21.978		6,550	485	152	7,187	8,175
0360	340/480 V, 24/30 kVA	2 Elec	1	16		5,200	355		5,555	6,300
0370	43/54 kVA	"	1	16		6,550	355		6,905	7,800
0380	48/60 kVA	3 Elec	1.33	18.045		9,500	400		9,900	11,200

16200 | Electrical Power

16280 | Power Filters & Conditioners

			CREW	DAILY OUTPUT	LABOR-HOURS	UNIT	2000 BARE COSTS MAT.	LABOR	EQUIP.	TOTAL	TOTAL INCL O&P	
100	0390	86/108 kVA	R-3	.91	21.978	Ea.	10,600	485	152	11,237	12,600	100
	0500	Standard grade, solid state, variable transformer volt. regulator										
	0510	Single-phase, 115 V, 2.3 kVA	1 Elec	2	4	Ea.	1,625	88.50		1,713.50	1,950	
	0520	4.2 kVA		2.29	3.493		2,600	77		2,677	2,975	
	0530	6.6 kVA		1.14	7.018		3,200	155		3,355	3,750	
	0540	13.0 kVA		1.14	7.018		4,125	155		4,280	4,800	
	0550	16.6 kVA	2 Elec	1.23	13.008		4,875	287		5,162	5,850	
	0610	230 V, 8.3 kVA		1.33	12.030		4,125	266		4,391	4,975	
	0620	21.4 kVA		1.23	13.008		4,875	287		5,162	5,850	
	0630	29.9 kVA		1.23	13.008		4,875	287		5,162	5,850	
	0710	460 V, 9.2 kVA		1.33	12.030		4,125	266		4,391	4,975	
	0720	20.7 kVA		1.23	13.008		4,875	287		5,162	5,850	
	0810	Three-phase, 230 V, 13.1 kVA	3 Elec	1.41	17.021		4,125	375		4,500	5,175	
	0820	19.1 kVA		1.41	17.021		4,875	375		5,250	6,000	
	0830	25.1 kVA		1.60	15		4,875	330		5,205	5,925	
	0840	57.8 kVA	R-3	.95	21.053		8,850	465	145	9,460	10,600	
	0850	74.9 kVA	"	.91	21.978		8,850	485	152	9,487	10,700	
	0910	460 V, 14.3 kVA	3 Elec	1.41	17.021		4,125	375		4,500	5,175	
	0920	19.1 kVA		1.41	17.021		4,875	375		5,250	6,000	
	0930	27.9 kVA		1.50	16		4,875	355		5,230	5,950	
	0940	59.8 kVA	R-3	1	20		8,850	440	138	9,428	10,600	
	0950	79.7 kVA		.95	21.053		9,900	465	145	10,510	11,800	
	0960	118 kVA		.95	21.053		10,400	465	145	11,010	12,400	
	1000	Laboratory grade, precision, electronic voltage regulator										
	1110	Single-phase, 115 V, 0.5 kVA	1 Elec	2.29	3.493	Ea.	1,350	77		1,427	1,600	
	1120	1.0 kVA		2	4		1,450	88.50		1,538.50	1,725	
	1130	3.0 kVA		.80	10		2,000	221		2,221	2,550	
	1140	6.0 kVA	2 Elec	1.46	10.959		3,725	242		3,967	4,500	
	1150	10.0 kVA	3 Elec	1	24		4,825	530		5,355	6,200	
	1160	15.0 kVA	"	1.50	16		5,525	355		5,880	6,650	
	1210	230 V, 3.0 kVA	1 Elec	.80	10		2,275	221		2,496	2,850	
	1220	6.0 kVA	2 Elec	1.46	10.959		3,800	242		4,042	4,575	
	1230	10.0 kVA	3 Elec	1.71	14.035		5,050	310		5,360	6,050	
	1240	15.0 kVA	"	1.60	15		5,700	330		6,030	6,825	
340	0010	**COMPUTER ISOLATION TRANSFORMER**										340
	0100	Computer grade										
	0110	Single-phase, 120/240 V, 0.5 kVA	1 Elec	4	2	Ea.	400	44		444	515	
	0120	1.0 kVA		2.67	2.996		520	66		586	680	
	0130	2.5 kVA		2	4		780	88.50		868.50	1,000	
	0140	5 kVA		1.14	7.018		830	155		985	1,175	
360	0010	**COMPUTER REGULATOR TRANSFORMER**										360
	0100	Ferro-resonant, constant voltage, variable transformer										
	0110	Single-phase, 240 V, 0.5 kVA	1 Elec	2.67	2.996	Ea.	440	66		506	595	
	0120	1.0 kVA		2	4		605	88.50		693.50	810	
	0130	2.0 kVA		1	8		1,025	177		1,202	1,425	
	0210	Plug-in unit 120 V, 0.14 kVA		8	1		254	22		276	315	
	0220	0.25 kVA		8	1		297	22		319	360	
	0230	0.5 kVA		8	1		440	22		462	520	
	0240	1.0 kVA		5.33	1.501		605	33		638	720	
	0250	2.0 kVA		4	2		1,025	44		1,069	1,200	
600	0010	**POWER CONDITIONER TRANSFORMER**										600
	0100	Electronic solid state, buck-boost, transformer, w/tap switch										
	0110	Single-phase, 115 V, 3.0 kVA, + or - 3% accuracy	2 Elec	1.60	10	Ea.	2,375	221		2,596	2,975	
	0120	208, 220, 230, or 240 V, 5.0 kVA, + or - 1.5% accuracy	3 Elec	1.60	15		3,075	330		3,405	3,925	
	0130	5.0 kVA, + or - 6% accuracy	2 Elec	1.14	14.035		2,750	310		3,060	3,525	
	0140	7.5 kVA, + or - 1.5% accuracy	3 Elec	1.50	16		3,900	355		4,255	4,875	

For expanded coverage of these items see *Means Electrical Cost Data 2000*

16200 | Electrical Power

16280 | Power Filters & Conditioners

			CREW	DAILY OUTPUT	LABOR-HOURS	UNIT	MAT.	LABOR	EQUIP.	TOTAL	TOTAL INCL O&P	
600	0150	7.5 kVA, + or - 6% accuracy	3 Elec	1.60	15	Ea.	3,250	330		3,580	4,125	600
	0160	10.0 kVA, + or - 1.5% accuracy		1.33	18.045		5,200	400		5,600	6,375	
	0170	10.0 kVA, + or - 6% accuracy	↓	1.41	17.021	↓	4,425	375		4,800	5,500	
840	0010	**TRANSIENT VOLTAGE SUPPRESSOR TRANSFORMER**										840
	0110	Single-phase, 120 V, 1.8 kVA	1 Elec	4	2	Ea.	490	44		534	610	
	0120	3.6 kVA		4	2		865	44		909	1,025	
	0130	7.2 kVA		3.20	2.500		1,125	55.50		1,180.50	1,325	
	0150	240 V, 3.6 kVA		4	2		865	44		909	1,025	
	0160	7.2 kVA		4	2		1,125	44		1,169	1,300	
	0170	14.4 kVA		3.20	2.500		1,425	55.50		1,480.50	1,675	
	0210	Plug-in unit, 120 V, 1.8 kVA	↓	8	1	↓	470	22		492	550	

16290 | Power Measure & Control

			CREW	DAILY OUTPUT	LABOR-HOURS	UNIT	MAT.	LABOR	EQUIP.	TOTAL	TOTAL INCL O&P	
860	0010	**VOLTAGE MONITOR SYSTEMS** (test equipment)										860
	0100	AC voltage monitor system, 120/240 V, one-channel				Ea.	3,000			3,000	3,300	
	0110	Modem adapter					375			375	415	
	0120	Add-on detector only					1,575			1,575	1,725	
	0150	AC voltage remote monitor sys., 3 channel, 120, 230, or 480 V					5,450			5,450	6,000	
	0160	With internal modem					5,750			5,750	6,325	
	0170	Combination temperature and humidity probe					845			845	930	
	0180	Add-on detector only					3,950			3,950	4,350	
	0190	With internal modem				↓	4,300			4,300	4,725	

16400 | Low-Voltage Distribution

16410 | Encl Switches & Circuit Breakers

			CREW	DAILY OUTPUT	LABOR-HOURS	UNIT	MAT.	LABOR	EQUIP.	TOTAL	TOTAL INCL O&P	
200	0010	**CIRCUIT BREAKERS** (in enclosure)										200
	0100	Enclosed (NEMA 1), 600 volt, 3 pole, 30 amp	1 Elec	3.20	2.500	Ea.	365	55.50		420.50	490	
	0200	60 amp		2.80	2.857		365	63		428	505	
	0400	100 amp		2.30	3.478		415	77		492	585	
	0600	225 amp	↓	1.50	5.333		965	118		1,083	1,275	
	0700	400 amp	2 Elec	1.60	10	↓	1,650	221		1,871	2,175	
800	0010	**SAFETY SWITCHES**										800
	0100	General duty 240 volt, 3 pole NEMA 1, fusible, 30 amp	1 Elec	3.20	2.500	Ea.	65.50	55.50		121	163	
	0200	60 amp		2.30	3.478		111	77		188	248	
	0300	100 amp		1.90	4.211		191	93		284	360	
	0400	200 amp	↓	1.30	6.154		410	136		546	670	
	0500	400 amp	2 Elec	1.80	8.889	↓	1,025	196		1,221	1,450	

16415 | Transfer Switches

			CREW	DAILY OUTPUT	LABOR-HOURS	UNIT	MAT.	LABOR	EQUIP.	TOTAL	TOTAL INCL O&P	
600	0010	**AUTOMATIC TRANSFER SWITCHES**										600
	0100	Switches, enclosed 480 volt, 3 pole, 30 amp	1 Elec	2.30	3.478	Ea.	2,900	77		2,977	3,325	
	0200	60 amp	"	1.90	4.211	"	2,900	93		2,993	3,350	

16420 | Enclosed Controllers

			CREW	DAILY OUTPUT	LABOR-HOURS	UNIT	MAT.	LABOR	EQUIP.	TOTAL	TOTAL INCL O&P	
200	0010	**CONTACTORS, AC** Enclosed (NEMA 1)										200
	0050	Lighting, 600 volt 3 pole, electrically held										

16400 | Low-Voltage Distribution

16420 | Enclosed Controllers

			CREW	DAILY OUTPUT	LABOR-HOURS	UNIT	MAT.	LABOR	EQUIP.	TOTAL	TOTAL INCL O&P	
200	0100	20 amp	1 Elec	4	2	Ea.	178	44		222	269	200
	0200	30 amp		3.60	2.222		189	49		238	288	
	0300	60 amp		3	2.667		375	59		434	510	
	0400	100 amp		2.50	3.200		625	70.50		695.50	800	
	0500	200 amp	↓	1.40	5.714		1,475	126		1,601	1,800	
	0600	300 amp	2 Elec	1.60	10		3,125	221		3,346	3,775	
	0800	600 volt 3 pole, mechanically held, 30 amp	1 Elec	3.60	2.222		285	49		334	395	
	0900	60 amp		3	2.667		565	59		624	715	
	1000	75 amp		2.80	2.857		760	63		823	940	
	1100	100 amp		2.50	3.200		800	70.50		870.50	995	
	1200	150 amp		2	4		2,200	88.50		2,288.50	2,550	
	1300	200 amp		1.40	5.714		2,200	126		2,326	2,600	
	1500	Magnetic with auxiliary contact, size 00, 9 amp		4	2		153	44		197	241	
	1600	Size 0, 18 amp		4	2		182	44		226	273	
	1700	Size 1, 27 amp		3.60	2.222		207	49		256	310	
	1800	Size 2, 45 amp		3	2.667		385	59		444	515	
	1900	Size 3, 90 amp		2.50	3.200		620	70.50		690.50	795	
	2000	Size 4, 135 amp	↓	2.30	3.478		1,425	77		1,502	1,675	
	2100	Size 5, 270 amp	2 Elec	1.80	8.889		2,975	196		3,171	3,600	
	2310	Size 8, 1215 amp	"	.80	20		18,200	440		18,640	20,800	
	2500	Magnetic, 240 volt, 1-2 pole, .75 HP motor	1 Elec	4	2		123	44		167	208	
	2520	2 HP motor		3.60	2.222		138	49		187	233	
	2540	5 HP motor		2.50	3.200		335	70.50		405.50	485	
	2560	10 HP motor		1.40	5.714		550	126		676	810	
	2600	240 volt or less, 3 pole, .75 HP motor		4	2		123	44		167	208	
	2620	5 HP motor		3.60	2.222		153	49		202	249	
	2640	10 HP motor		3.60	2.222		177	49		226	276	
	2660	15 HP motor		2.50	3.200		355	70.50		425.50	505	
	2700	25 HP motor		2.50	3.200		355	70.50		425.50	505	
	2720	30 HP motor	2 Elec	2.80	5.714		590	126		716	855	
	2740	40 HP motor		2.80	5.714		590	126		716	855	
	2760	50 HP motor		1.60	10		590	221		811	1,000	
	2800	75 HP motor		1.60	10		1,375	221		1,596	1,875	
	2820	100 HP motor		1	16		1,375	355		1,730	2,100	
	2860	150 HP motor		1	16		2,950	355		3,305	3,825	
	2880	200 HP motor	↓	1	16		2,950	355		3,305	3,825	
	3000	600 volt, 3 pole, 5 HP motor	1 Elec	4	2		153	44		197	241	
	3020	10 HP motor		3.60	2.222		177	49		226	276	
	3040	25 HP motor		3	2.667		355	59		414	485	
	3100	50 HP motor	↓	2.50	3.200		590	70.50		660.50	765	
	3160	100 HP motor	2 Elec	2.80	5.714		1,375	126		1,501	1,725	
	3220	200 HP motor	"	1.60	10	↓	2,950	221		3,171	3,600	
220	0010	**CONTROL STATIONS**										220
	0050	NEMA 1, heavy duty, stop/start	1 Elec	8	1	Ea.	113	22		135	160	
	0100	Stop/start, pilot light		6.20	1.290		153	28.50		181.50	215	
	0200	Hand/off/automatic	↓	6.20	1.290	↓	83.50	28.50		112	139	

16440 | Swbds, Panels & Cont Centers

			CREW	DAILY OUTPUT	LABOR-HOURS	UNIT	MAT.	LABOR	EQUIP.	TOTAL	TOTAL INCL O&P	
500	0010	**LOAD CENTERS** (residential type)										500
	0100	3 wire, 120/240V, 1 phase, including 1 pole plug-in breakers										
	0200	100 amp main lugs, indoor, 8 circuits	1 Elec	1.40	5.714	Ea.	113	126		239	330	
	0300	12 circuits		1.20	6.667		157	147		304	415	
	0400	Rainproof, 8 circuits		1.40	5.714		135	126		261	355	
	0500	12 circuits	↓	1.20	6.667		190	147		337	450	
	0600	200 amp main lugs, indoor, 16 circuits	R-1A	1.80	8.889		262	164		426	560	
	0700	20 circuits	↓	1.50	10.667		330	197		527	690	

For expanded coverage of these items see Means Electrical Cost Data 2000

16400 | Low-Voltage Distribution

16440 | Swbds, Panels & Cont Centers

		CREW	DAILY OUTPUT	LABOR-HOURS	UNIT	2000 BARE COSTS MAT.	LABOR	EQUIP.	TOTAL	TOTAL INCL O&P	
500	0800	24 circuits	R-1A	1.30	12.308	Ea.	430	227		657	855	500
	1200	Rainproof, 16 circuits		1.80	8.889		310	164		474	620	
	1300	20 circuits		1.50	10.667		375	197		572	740	
	1400	24 circuits		1.30	12.308		560	227		787	995	
	1500	30 circuits		1.20	13.333		625	246		871	1,100	
	1600	42 circuits		.80	20		820	370		1,190	1,525	
	1800	400 amp main lugs, indoor, 42 circuits		.72	22.222		960	410		1,370	1,725	
	1900	Rainproof, 42 circuit		.72	22.222		1,150	410		1,560	1,925	
	2200	Plug in breakers, 20 amp, 1 pole, 4 wire, 120/208 volts										
	2210	125 amp main lugs, indoor, 12 circuits	1 Elec	1.20	6.667	Ea.	247	147		394	510	
	2300	18 circuits		.80	10		350	221		571	745	
	2400	Rainproof, 12 circuits		1.20	6.667		283	147		430	550	
	2500	18 circuits		.80	10		380	221		601	780	
	2600	200 amp main lugs, indoor, 24 circuits	R-1A	1.30	12.308		470	227		697	900	
	2700	30 circuits		1.20	13.333		535	246		781	1,000	
	2800	36 circuits		1	16		650	295		945	1,200	
	2900	42 circuits		.80	20		710	370		1,080	1,400	
	3000	Rainproof, 24 circuits		1.30	12.308		520	227		747	955	
	3100	30 circuits		1.20	13.333		585	246		831	1,050	
	3200	36 circuits		1	16		815	295		1,110	1,375	
	3300	42 circuits		.80	20		880	370		1,250	1,575	
640	0010	**MOTOR CONTROL CENTER** Consists of starters & structures										640
	0050	Starters, class 1, type B, comb. MCP, FVNR, with										
	0100	control transformer, 10 HP, size 1, 12" high	1 Elec	2.70	2.963	Ea.	1,050	65.50		1,115.50	1,250	
	0200	25 HP, size 2, 18" high	2 Elec	4	4		1,200	88.50		1,288.50	1,475	
	0300	50 HP, size 3, 24" high		2	8		1,850	177		2,027	2,325	
	0350	75 HP, size 4, 24" high		1.60	10		2,450	221		2,671	3,050	
	0400	100 HP, size 4, 30" high		1.40	11.429		3,250	253		3,503	4,000	
	0500	200 HP, size 5, 48" high		1	16		4,850	355		5,205	5,900	
660	0010	**MOTOR STARTERS & CONTROLS**										660
	0050	Magnetic, FVNR, with enclosure and heaters, 480 volt										
	0080	2 HP, size 00	1 Elec	3.50	2.286	Ea.	165	50.50		215.50	264	
	0100	5 HP, size 0		2.30	3.478		199	77		276	345	
	0200	10 HP, size 1		1.60	5		224	111		335	425	
	0300	25 HP, size 2	2 Elec	2.20	7.273		420	161		581	730	
	0400	50 HP, size 3	"	1.80	8.889		685	196		881	1,075	
	1400	Combination, with fused switch, 5 HP, size 0	1 Elec	1.80	4.444		495	98		593	705	
	1600	10 HP, size 1	"	1.30	6.154		530	136		666	800	
	1800	25 HP, size 2	2 Elec	2	8		860	177		1,037	1,225	
700	0010	**PANELBOARD & LOAD CENTER CIRCUIT BREAKERS**										700
	0050	Bolt-on, 10,000 amp IC, 120 volt, 1 pole										
	0100	15 to 50 amp	1 Elec	10	.800	Ea.	10	17.70		27.70	40	
	0200	60 amp		8	1		10	22		32	47	
	0300	70 amp		8	1		19	22		41	57	
	0350	240 volt, 2 pole										
	0400	15 to 50 amp	1 Elec	8	1	Ea.	22	22		44	60	
	0500	60 amp		7.50	1.067		22	23.50		45.50	62.50	
	0600	80 to 100 amp		5	1.600		56.50	35.50		92	120	
	0700	3 pole, 15 to 60 amp		6.20	1.290		69.50	28.50		98	123	
	0800	70 amp		5	1.600		87.50	35.50		123	155	
	0900	80 to 100 amp		3.60	2.222		99.50	49		148.50	190	
	1000	22,000 amp I.C., 240 volt, 2 pole, 70 - 225 amp		2.70	2.963		430	65.50		495.50	575	
	1100	3 pole, 70 - 225 amp		2.30	3.478		475	77		552	650	
	2000	Plug-in panel or load center, 120/240 volt, to 60 amp, 1 pole		12	.667		7.75	14.75		22.50	32.50	
	2010	2 pole		9	.889		17.50	19.65		37.15	51.50	

Important: See the Reference Section for critical supporting data - Reference Nos., Crews, & Location Factors

16400 | Low-Voltage Distribution

16440 | Swbds, Panels & Cont Centers

			CREW	DAILY OUTPUT	LABOR-HOURS	UNIT	2000 BARE COSTS MAT.	LABOR	EQUIP.	TOTAL	TOTAL INCL O&P	
700	2020	3 pole	1 Elec	7.50	1.067	Ea.	60.50	23.50		84	105	700
	2030	100 amp, 2 pole		6	1.333		81	29.50		110.50	137	
	2040	3 pole		4.50	1.778		90	39.50		129.50	164	
	2050	150 amp, 2 pole	↓	3	2.667	↓	113	59		172	221	
720	0010	**PANELBOARDS** (Commercial use)										720
	0050	NQOD, w/20 amp 1 pole bolt-on circuit breakers										
	0600	4 wire, 120/208 volts, 100 amp main lugs, 12 circuits	1 Elec	1	8	Ea.	440	177		617	775	
	0650	16 circuits		.75	10.667		505	236		741	940	
	0700	20 circuits		.65	12.308		590	272		862	1,100	
	0750	24 circuits		.60	13.333		640	295		935	1,175	
	0800	30 circuits	↓	.53	15.094		740	335		1,075	1,350	
	0850	225 amp main lugs, 32 circuits	2 Elec	.90	17.778		830	395		1,225	1,550	
	0900	34 circuits		.84	19.048		850	420		1,270	1,625	
	0950	36 circuits		.80	20		870	440		1,310	1,675	
	1000	42 circuits	↓	.68	23.529	↓	975	520		1,495	1,925	
	1200	NEHB,w/20 amp, 1 pole bolt-on circuit breakers										
	1250	4 wire, 277/480 volts, 100 amp main lugs, 12 circuits	1 Elec	.88	9.091	Ea.	840	201		1,041	1,250	
	1300	20 circuits	"	.60	13.333		1,250	295		1,545	1,850	
	1350	225 amp main lugs, 24 circuits	2 Elec	.90	17.778		1,425	395		1,820	2,225	
	1400	30 circuits		.80	20		1,725	440		2,165	2,625	
	1450	36 circuits		.72	22.222		2,000	490		2,490	3,000	
	1500	42 circuits		.60	26.667		2,300	590		2,890	3,500	
	1510	225 amp main lugs, NEMA 7, 12 circuits		.90	17.778		3,475	395		3,870	4,450	
	1590	24 circuits	↓	.30	53.333	↓	4,200	1,175		5,375	6,525	

16500 | Lighting

16510 | Interior Luminaires

			CREW	DAILY OUTPUT	LABOR-HOURS	UNIT	2000 BARE COSTS MAT.	LABOR	EQUIP.	TOTAL	TOTAL INCL O&P	
440	0010	**INTERIOR LIGHTING FIXTURES** Including lamps, mounting										440
	0030	hardware and connections										
	0100	Fluorescent, C.W. lamps, troffer, recess mounted in grid, RS										
	0130	grid ceiling mount										
	0200	Acrylic lens, 1'W x 4'L, two 40 watt	1 Elec	5.70	1.404	Ea.	46	31		77	101	
	0300	2'W x 2'L, two U40 watt		5.70	1.404		50	31		81	106	
	0600	2'W x 4'L, four 40 watt		4.70	1.702		56	37.50		93.50	123	
	0910	Acrylic lens, 1'W x 4'L, two 32 watt		5.70	1.404		55	31		86	111	
	0930	2'W x 2'L, two U32 watt		5.70	1.404		75	31		106	133	
	0940	2'W x 4'L, two 32 watt		5.30	1.509		67	33.50		100.50	128	
	0950	2'W x 4'L, three 32 watt		5	1.600		72	35.50		107.50	137	
	0960	2'W x 4'L, four 32 watt	↓	4.70	1.702	↓	75	37.50		112.50	144	
	1000	Surface mounted, RS										
	2100	Strip fixture										
	2200	4' long, one 40 watt RS	1 Elec	8.50	.941	Ea.	26	21		47	62.50	
	2300	4' long, two 40 watt RS	"	8	1		28	22		50	67	
	2600	8' long, one 75 watt, SL	2 Elec	13.40	1.194		39	26.50		65.50	86	
	2700	8' long, two 75 watt, SL	"	12.40	1.290		47	28.50		75.50	98	
	3535	Downlight, recess mounted	1 Elec	8	1		80	22		102	124	
	3540	Wall washer, recess mounted		8	1		80	22		102	124	
	3550	Direct/indirect, 4' long, stl., pendent mtd.		5	1.600		110	35.50		145.50	179	
	3560	4' long, alum., pendent mtd.	↓	5	1.600	↓	315	35.50		350.50	405	

For expanded coverage of these items see *Means Electrical Cost Data 2000*

16500 | Lighting

16510 | Interior Luminaires

			Crew	Daily Output	Labor-Hours	Unit	Mat.	Labor	Equip.	Total	Total Incl O&P	
440	3565	Prefabricated cove, 4' long, stl. continuous row	1 Elec	5	1.600	Ea.	173	35.50		208.50	248	440
	3570	4' long, alum. continuous row	↓	5	1.600	↓	325	35.50		360.50	420	
	3580	Mercury vapor, integral ballast, ceiling, recess mounted,										
	3590	prismatic glass lens, floating door										
	3600	2'W x 2'L, 250 watt DX lamp	1 Elec	3.20	2.500	Ea.	250	55.50		305.50	365	
	3700	2'W x 2'L, 400 watt DX lamp		2.90	2.759		255	61		316	380	
	3800	Surface mtd., prismatic lens, 2'W x 2'L, 250 watt DX lamp		2.70	2.963		250	65.50		315.50	380	
	3900	2'W x 2'L, 400 watt DX lamp	↓	2.40	3.333		283	73.50		356.50	430	
	4000	High bay, aluminum reflector										
	4030	Single unit, 400 watt DX lamp	2 Elec	4.60	3.478	Ea.	248	77		325	400	
	4100	Single unit, 1000 watt DX lamp		4	4		390	88.50		478.50	575	
	4200	Twin unit, two 400 watt DX lamps	↓	3.20	5		420	111		531	640	
	4210	Low bay, aluminum reflector, 250W DX lamp	1 Elec	3.20	2.500	↓	255	55.50		310.50	370	
	4220	Metal halide, integral ballast, ceiling, recess mounted										
	4230	prismatic glass lens, floating door										
	4240	2'W x 2'L, 250 watt	1 Elec	3.20	2.500	Ea.	245	55.50		300.50	360	
	4250	2'W x 2'L, 400 watt	2 Elec	5.80	2.759		285	61		346	415	
	4260	Surface mounted, 2'W x 2'L, 250 watt	1 Elec	2.70	2.963		245	65.50		310.50	375	
	4270	400 watt	2 Elec	4.80	3.333	↓	290	73.50		363.50	440	
	4280	High bay, aluminum reflector,										
	4290	Single unit, 400 watt	2 Elec	4.60	3.478	Ea.	340	77		417	500	
	4300	Single unit, 1000 watt		4	4		485	88.50		573.50	680	
	4310	Twin unit, 400 watt	↓	3.20	5		680	111		791	925	
	4320	Low bay, aluminum reflector, 250W DX lamp	1 Elec	3.20	2.500	↓	325	55.50		380.50	450	
	4450	Incandescent, high hat can, round alzak reflector, prewired										
	4470	100 watt	1 Elec	8	1	Ea.	53	22		75	94.50	
	4480	150 watt	"	8	1	"	77	22		99	121	
	5200	Ceiling, surface mounted, opal glass drum										
	5300	8", one 60 watt lamp	1 Elec	10	.800	Ea.	32	17.70		49.70	64	
	5400	10", two 60 watt lamps		8	1		36	22		58	75.50	
	5500	12", four 60 watt lamps		6.70	1.194		50	26.50		76.50	98	
	6900	Mirror light, fluorescent, RS, acrylic enclosure, two 40 watt		8	1		77	22		99	121	
	6910	One 40 watt		8	1		60	22		82	102	
	6920	One 20 watt	↓	12	.667	↓	47.50	14.75		62.25	76.50	

16520 | Exterior Luminaires

			Crew	Daily Output	Labor-Hours	Unit	Mat.	Labor	Equip.	Total	Total Incl O&P	
300	0010	**EXTERIOR FIXTURES** With lamps										300
	0400	Quartz, 500 watt	1 Elec	5.30	1.509	Ea.	51	33.50		84.50	111	
	0800	Wall pack, mercury vapor, 175 watt		4	2		210	44		254	305	
	1000	250 watt		4	2		213	44		257	305	
	1100	Low pressure sodium, 35 watt		4	2		208	44		252	300	
	1150	55 watt		4	2		280	44		324	385	
	6420	Wood pole, 4-1/2" x 5-1/8", 8' high		6	1.333		230	29.50		259.50	300	
	6440	12' high		5.70	1.404		315	31		346	395	
	6460	20' high	↓	4	2	↓	440	44		484	560	
	6500	Bollard light, lamp & ballast, 42" high with polycarbonate lens										
	6700	Mercury vapor, 175 watt	1 Elec	3	2.667	Ea.	505	59		564	650	
	7200	Incandescent, 150 watt	"	3	2.667	"	390	59		449	525	
	7380	Landscape recessed uplight, incl. housing, ballast, transformer										
	7390	& reflector										
	7420	Incandescent, 250 watt	1 Elec	5	1.600	Ea.	400	35.50		435.50	500	
	7440	Quartz, 250 watt	"	5	1.600	"	380	35.50		415.50	480	

16530 | Emergency Lighting

			Crew	Daily Output	Labor-Hours	Unit	Mat.	Labor	Equip.	Total	Total Incl O&P	
320	0010	**EXIT AND EMERGENCY LIGHTING**										320
	0080	Exit light ceiling or wall mount, incandescent, single face	1 Elec	8	1	Ea.	35.50	22		57.50	75.50	

16500 | Lighting

16530 | Emergency Lighting

		Crew	Daily Output	Labor-Hours	Unit	Mat.	Labor	Equip.	Total	Total Incl O&P
320	0100 Double face	1 Elec	6.70	1.194	Ea.	44	26.50		70.50	91.50
	0120 Explosion proof		3.80	2.105		370	46.50		416.50	480
	0150 Fluorescent, single face		8	1		62	22		84	104
	0160 Double face		6.70	1.194		65	26.50		91.50	115
	0300 Emergency light units, battery operated									
	0350 Twin sealed beam light, 25 watt, 6 volt each									
	0500 Lead battery operated	1 Elec	4	2	Ea.	160	44		204	249
	0700 Nickel cadmium battery operated		4	2		480	44		524	605
	0780 Additional remote mount, sealed beam, 25W 6V		26.70	.300		20	6.60		26.60	33
	0790 Twin sealed beam light, 25W 6V each		26.70	.300		40	6.60		46.60	55

16550 | Special Purpose Ltg.

		Crew	Daily Output	Labor-Hours	Unit	Mat.	Labor	Equip.	Total	Total Incl O&P
820	0010 **TRACK LIGHTING**									
	0080 Track, 1 circuit, 4' section	1 Elec	6.70	1.194	Ea.	33	26.50		59.50	79.50
	0100 8' section		5.30	1.509		55	33.50		88.50	115
	0200 12' section		4.40	1.818		87	40		127	161
	0300 3 circuits, 4' section		6.70	1.194		44	26.50		70.50	91.50
	0400 8' section		5.30	1.509		68.50	33.50		102	130
	0500 12' section		4.40	1.818		136	40		176	216
	1000 Feed kit, surface mounting		16	.500		8	11.05		19.05	27
	1100 End cover		24	.333		3	7.35		10.35	15.35
	1200 Feed kit, stem mounting, 1 circuit		16	.500		22.50	11.05		33.55	43
	1300 3 circuit		16	.500		22.50	11.05		33.55	43
	2000 Electrical joiner, for continuous runs, 1 circuit		32	.250		10.50	5.55		16.05	20.50
	2100 3 circuit		32	.250		24	5.55		29.55	35.50
	2200 Fixtures, spotlight, 75W PAR halogen		16	.500		82.50	11.05		93.55	109
	2210 50W MR16 halogen		16	.500		98.50	11.05		109.55	126
	3000 Wall washer, 250 watt tungsten halogen		16	.500		93.50	11.05		104.55	121
	3100 Low voltage, 25/50 watt, 1 circuit		16	.500		90	11.05		101.05	117
	3120 3 circuit		16	.500		94	11.05		105.05	121

16585 | Lamps

		Crew	Daily Output	Labor-Hours	Unit	Mat.	Labor	Equip.	Total	Total Incl O&P
600	0010 **LAMPS**									
	0080 Fluorescent, rapid start, cool white, 2' long, 20 watt	1 Elec	1	8	C	274	177		451	590
	0100 4' long, 40 watt		.90	8.889		253	196		449	600
	1000 Metal halide, mogul base, 175 watt		.30	26.667		4,225	590		4,815	5,625
	1100 250 watt		.30	26.667		4,775	590		5,365	6,225
	1350 High pressure sodium, 70 watt		.30	26.667		4,225	590		4,815	5,600
	1370 150 watt		.30	26.667		4,525	590		5,115	5,950

16800 | Sound & Video

16820 | Sound Reinforcement

		Crew	Daily Output	Labor-Hours	Unit	Mat.	Labor	Equip.	Total	Total Incl O&P
840	0010 **SOUND SYSTEM** not including rough-in wires, cables & conduits									
	2000 Intercom, 25 station capacity, master station	2 Elec	2	8	Ea.	1,375	177		1,552	1,825
	2020 11 station capacity	"	4	4		675	88.50		763.50	890
	2200 Remote station	1 Elec	8	1		110	22		132	157
	2400 Intercom outlets		8	1		64.50	22		86.50	107
	2600 Handset		4	2		200	44		244	293

For expanded coverage of these items see *Means Electrical Cost Data 2000*

16800 | Sound & Video

16820 | Sound Reinforcement

			CREW	DAILY OUTPUT	LABOR-HOURS	UNIT	2000 BARE COSTS MAT.	LABOR	EQUIP.	TOTAL	TOTAL INCL O&P	
840	2800	Emergency call system, 12 zones, annunciator	1 Elec	1.30	6.154	Ea.	645	136		781	930	840
	3000	Bell		5.30	1.509		67	33.50		100.50	128	
	3200	Light or relay		8	1		33.50	22		55.50	73	
	3400	Transformer		4	2		147	44		191	235	
	3600	House telephone, talking station		1.60	5		315	111		426	525	
	3800	Press to talk, release to listen		5.30	1.509		73.50	33.50		107	136	
	4000	System-on button					44			44	48.50	
	4200	Door release	1 Elec	4	2		77.50	44		121.50	158	
	4400	Combination speaker and microphone		8	1		134	22		156	183	
	4600	Termination box		3.20	2.500		42	55.50		97.50	137	
	4800	Amplifier or power supply		5.30	1.509		485	33.50		518.50	585	
	5000	Vestibule door unit		16	.500	Name	89	11.05		100.05	116	
	5200	Strip cabinet		27	.296	Ea.	168	6.55		174.55	196	
	5400	Directory		16	.500		79	11.05		90.05	105	
	6000	Master door, button buzzer type, 100 unit	2 Elec	.54	29.630		805	655		1,460	1,950	
	6020	200 unit	"	.30	53.333		1,500	1,175		2,675	3,600	

16850 | Television Equipment

			CREW	DAILY OUTPUT	LABOR-HOURS	UNIT	MAT.	LABOR	EQUIP.	TOTAL	TOTAL INCL O&P	
600	0010	T.V. SYSTEMS not including rough-in wires, cables & conduits										600
	0100	Master TV antenna system										
	0200	VHF reception & distribution, 12 outlets	1 Elec	6	1.333	Outlet	152	29.50		181.50	215	
	0800	VHF & UHF reception & distribution, 12 outlets		6	1.333	"	151	29.50		180.50	214	
	5000	T.V. Antenna only, minimum		6	1.333	Ea.	33	29.50		62.50	84.50	
	5100	Maximum		4	2	"	140	44		184	227	

For information about Means Estimating Seminars, see yellow pages 11 and 12 in back of book

Reference Section

All the reference information is in one section making it easy to find what you need to know... and easy to use the book on a daily basis. This section is visually identified by a vertical gray bar on the edge of pages.

In the reference number information that follows, you'll see the background that relates to the "reference numbers" that appeared in the Unit Price Sections. You'll find reference tables, explanations and estimating information that support how we arrived at the unit price data. Also included are alternate pricing methods, technical data and estimating procedures along with information on design and economy in construction.

Also in this Reference Section, we've included Change Orders, information on pricing changes to contract documents; Crew Listings, a full listing of all the crews, equipment and their costs; Historical Cost Indexes for cost comparisons over time; City Cost Indexes and Location Factors for adjusting costs to the region you are in; and an explanation of all Abbreviations used in the book.

Table of Contents

Reference Numbers

R011	Overhead & Misc. Data	538
R012	Special Project Procedures	545
R015	Construction Aids	546
R020	Paving & Surfacing	547
R023	Earthwork	548
R025	Sewerage & Drainage	550
R029	Landscaping	550
R033	Cast-in-Place Concrete	551
R034	Precast Concrete	552
R035	Cementitious Decks & Toppings	554
R040	Mortar & Masonry Accessories	555
R042	Unit Masonry	556
R044	Stone	561
R049	Restoration & Cleaning	562
R051	Structural Metal Framing	562
R061	Rough Carpentry	563
R064	Architectural Woodwork	566
R071	Waterproofing & Dampproofing	567
R073	Shingles & Roofing Tiles	567

Reference Numbers (cont.)

R075	Membrane Roofing	567
R081	Metal Doors & Frames	568
R082	Wood & Plastic Doors	569
R085	Metal Windows	569
R088	Glazing	570
R091	Metal Support Assemblies	570
R092	Plaster & Gypsum Board	571
R094	Terrazzo	572
R096	Flooring	573
R097	Wall Finishes	573
R099	Paints & Coatings	574
R131	Pre-Engineered Structures & Aquatic Facilities	575
R132	Hazardous Material Remediation	575
R151	Plumbing	576
R155	Heating	577
R157	Air Conditioning & Ventilation	578
R158	Air Distribution	578
R161	Wiring Methods	579

Crew Listings — 580

Historical Cost Indexes — 603

Location Factors — 611

Abbreviations — 617

General Requirements | R011 | Overhead & Miscellaneous Data

R01100-005 Tips for Accurate Estimating

1. Use pre-printed or columnar forms for orderly sequence of dimensions and locations and for recording telephone quotations.
2. Use only the front side of each paper or form except for certain pre-printed summary forms.
3. Be consistent in listing dimensions: For example, length x width x height. This helps in rechecking to ensure that, the total length of partitions is appropriate for the building area.
4. Use printed (rather than measured) dimensions where given.
5. Add up multiple printed dimensions for a single entry where possible.
6. Measure all other dimensions carefully.
7. Use each set of dimensions to calculate multiple related quantities.
8. Convert foot and inch measurements to decimal feet when listing. Memorize decimal equivalents to .01 parts of a foot (1/8″ equals approximately .01′).
9. Do not "round off" quantities until the final summary.
10. Mark drawings with different colors as items are taken off.
11. Keep similar items together, different items separate.
12. Identify location and drawing numbers to aid in future checking for completeness.
13. Measure or list everything on the drawings or mentioned in the specifications.
14. It may be necessary to list items not called for to make the job complete.
15. Be alert for: Notes on plans such as N.T.S. (not to scale); changes in scale throughout the drawings; reduced size drawings; discrepancies between the specifications and the drawings.
16. Develop a consistent pattern of performing an estimate. For example:
    a. Start the quantity takeoff at the lower floor and move to the next higher floor.
    b. Proceed from the main section of the building to the wings.
    c. Proceed from south to north or vice versa, clockwise or counterclockwise.
    d. Take off floor plan quantities first, elevations next, then detail drawings.
17. List all gross dimensions that can be either used again for different quantities, or used as a rough check of other quantities for verification (exterior perimeter, gross floor area, individual floor areas, etc.).
18. Utilize design symmetry or repetition (repetitive floors, repetitive wings, symmetrical design around a center line, similar room layouts, etc.). Note: Extreme caution is needed here so as not to omit or duplicate an area.
19. Do not convert units until the final total is obtained. For instance, when estimating concrete work, keep all units to the nearest cubic foot, then summarize and convert to cubic yards.
20. When figuring alternatives, it is best to total all items involved in the basic system, then total all items involved in the alternates. Therefore you work with positive numbers in all cases. When adds and deducts are used, it is often confusing whether to add or subtract a portion of an item; especially on a complicated or involved alternate.

R01100-040 Builder's Risk Insurance

Builder's Risk Insurance is insurance on a building during construction. Premiums are paid by the owner or the contractor. Blasting, collapse and underground insurance would raise total insurance costs above those listed. Floater policy for materials delivered to the job runs $.75 to $1.25 per $100 value. Contractor equipment insurance runs $.50 to $1.50 per $100 value. Insurance for miscellaneous tools to $1,500 value runs from $3.00 to $7.50 per $100 value.

Tabulated below are New England Builder's Risk insurance rates in dollars per $100 value for $1,000 deductible. For $25,000 deductible, rates can be reduced 13% to 34%. On contracts over $1,000,000, rates may be lower than those tabulated. Policies are written annually for the total completed value in place. For "all risk" insurance (excluding flood, earthquake and certain other perils) add $.025 to total rates below.

Coverage	Frame Construction (Class 1)		Brick Construction (Class 4)		Fire Resistive (Class 6)	
	Range	Average	Range	Average	Range	Average
Fire Insurance	$.300 to $.420	$.394	$.132 to $.189	$.174	$.052 to $.080	$.070
Extended Coverage	.115 to .150	.144	.080 to .105	.101	.081 to .105	.100
Vandalism	.012 to .016	.015	.008 to .011	.011	.008 to .011	.010
Total Annual Rate	$.427 to $.586	$.553	$.220 to $.305	$.286	$.141 to $.196	$.180

General Requirements

R011 Overhead & Miscellaneous Data

R01100-050 General Contractor's Overhead

The table below shows a contractor's overhead as a percentage of direct cost in two ways. The figures on the right are for the overhead, markup based on both material and labor. The figures on the left are based on the entire overhead applied only to the labor. This figure would be used if the owner supplied the materials or if a contract is for labor only. Note: Some of these markups are included in the labor rates shown on Reference Table R01100-070.

Items of General Contractor's Indirect Costs	% of Direct Costs	
	As a Markup of Labor Only	As a Markup of Both Material and Labor
Field Supervision	6.0%	2.9%
Main Office Expense (see details below)	16.2	7.7
Tools and Minor Equipment	1.0	0.5
Workers' Compensation & Employers' Liability. See R01100-060	18.1	8.6
Field Office, Sheds, Photos, Etc.	1.5	0.7
Performance and Payment Bond, 0.7% to 1.5%. See R01100-080	2.3	1.1
Unemployment Tax See R01100-100 (Combined Federal and State)	7.0	3.3
Social Security and Medicare, See R01100-100	7.7	3.7
Sales Tax — add if applicable 42/80 x % as markup of total direct costs including both material and labor. See R01100-090		
Sub Total	59.8%	28.5%
*Builder's Risk Insurance ranges from .141% to .586%. See R01100-040	0.6	0.3
*Public Liability Insurance	3.2	1.5
Grand Total	63.6%	30.3%

*Paid by Owner or Contractor

Main Office Expense

A General Contractor's main office expense consists of many items not detailed in the front portion of the book. The percentage of main office expense declines with increased annual volume of the contractor. Typical main office expense ranges from 2% to 20% with the median about 7.2% of total volume. This equals about 7.7% of direct costs. The following are approximate percentages of total overhead for different items usually included in a General Contractor's main office overhead. With different accounting procedures, these percentages may vary.

Item	Typical Range	Average
Managers', clerical and estimators' salaries	40 % to 55 %	48%
Profit sharing, pension and bonus plans	2 to 20	12
Insurance	5 to 8	6
Estimating and project management (not including salaries)	5 to 9	7
Legal, accounting and data processing	0.5 to 5	3
Automobile and light truck expense	2 to 8	5
Depreciation of overhead capital expenditures	2 to 6	4
Maintenance of office equipment	0.1 to 1.5	1
Office rental	3 to 5	4
Utilities including phone and light	1 to 3	2
Miscellaneous	5 to 15	8
Total		100%

General Requirements — R011 — Overhead & Miscellaneous Data

R01100-060 Workers' Compensation Insurance Rates by Trade

The table below tabulates the national averages for Workers' Compensation insurance rates by trade and type of building. The average "Insurance Rate" is multiplied by the "% of Building Cost" for each trade. This produces the "Workers' Compensation Cost" by % of total labor cost, to be added for each trade by building type to determine the weighted average Workers' Compensation rate for the building types analyzed.

Trade	Insurance Rate (% Labor Cost) Range	Insurance Rate (% Labor Cost) Average	% of Building Cost Office Bldgs.	% of Building Cost Schools & Apts.	% of Building Cost Mfg.	Workers' Compensation Office Bldgs.	Workers' Compensation Schools & Apts.	Workers' Compensation Mfg.
Excavation, Grading, etc.	4.0% to 26.6%	11.4%	4.8%	4.9%	4.5%	.55	.56%	.51%
Piles & Foundations	8.1 to 80.1	29.9	7.1	5.2	8.7	2.12	1.55	2.60
Concrete	7.5 to 39.4	18.8	5.0	14.8	3.7	.94	2.78	.70
Masonry	5.8 to 48.6	17.8	6.9	7.5	1.9	1.23	1.34	.34
Structural Steel	8.1 to 132.9	42.6	10.7	3.9	17.6	4.56	1.66	7.50
Miscellaneous & Ornamental Metals	5.4 to 34.0	13.8	2.8	4.0	3.6	.39	.55	.50
Carpentry & Millwork	7.0 to 49.9	19.9	3.7	4.0	0.5	.74	.80	.10
Metal or Composition Siding	6.9 to 36.7	18.1	2.3	0.3	4.3	.42	.05	.78
Roofing	8.1 to 88.8	34.6	2.3	2.6	3.1	.80	.90	1.07
Doors & Hardware	3.8 to 28.8	11.7	0.9	1.4	0.4	.11	.16	.05
Sash & Glazing	5.7 to 30.0	14.4	3.5	4.0	1.0	.50	.58	.14
Lath & Plaster	5.3 to 37.4	15.7	3.3	6.9	0.8	.52	1.08	.13
Tile, Marble & Floors	3.5 to 29.5	10.5	2.6	3.0	0.5	.27	.32	.05
Acoustical Ceilings	4.0 to 26.7	12.3	2.4	0.2	0.3	.30	.02	.04
Painting	6.1 to 41.1	15.3	1.5	1.6	1.6	.23	.24	.24
Interior Partitions	7.0 to 49.9	19.9	3.9	4.3	4.4	.78	.86	.88
Miscellaneous Items	2.6 to 112.5	19.0	5.2	3.7	9.7	.99	.70	1.84
Elevators	2.0 to 19.1	8.5	2.1	1.1	2.2	.18	.09	.19
Sprinklers	2.8 to 20.6	9.0	0.5	—	2.0	.05	—	.18
Plumbing	3.0 to 15.9	8.8	4.9	7.2	5.2	.43	.63	.46
Heat., Vent., Air Conditioning	3.9 to 25.8	12.3	13.5	11.0	12.9	1.66	1.35	1.59
Electrical	2.9 to 11.6	7.0	10.1	8.4	11.1	.71	.59	.78
Total	2.0% to 132.9%	—	100.0%	100.0%	100.0%	18.48%	16.81%	20.67%
Overall Weighted Average 18.65%								

Workers' Compensation Insurance Rates by States

The table below lists the weighted average Workers' Compensation base rate for each state with a factor comparing this with the national average of 18.1%.

State	Weighted Average	Factor	State	Weighted Average	Factor	State	Weighted Average	Factor
Alabama	30.4%	168	Kentucky	19.8%	109	North Dakota	16.7%	92
Alaska	11.3	62	Louisiana	27.3	151	Ohio	16.2	90
Arizona	14.5	80	Maine	21.8	120	Oklahoma	21.9	121
Arkansas	14.2	78	Maryland	11.9	66	Oregon	19.3	107
California	18.6	103	Massachusetts	26.6	147	Pennsylvania	22.9	127
Colorado	26.5	146	Michigan	20.6	114	Rhode Island	22.4	124
Connecticut	21.1	117	Minnesota	37.7	208	South Carolina	13.8	76
Delaware	12.3	68	Mississippi	23.8	131	South Dakota	17.9	99
District of Columbia	25.2	139	Missouri	15.8	87	Tennessee	13.0	72
Florida	28.0	155	Montana	37.4	207	Texas	24.7	136
Georgia	24.9	138	Nebraska	15.7	87	Utah	13.0	72
Hawaii	16.7	92	Nevada	19.4	107	Vermont	14.9	82
Idaho	11.9	66	New Hampshire	23.2	128	Virginia	11.0	61
Illinois	23.3	129	New Jersey	11.0	61	Washington	12.0	66
Indiana	7.5	41	New Mexico	23.1	128	West Virginia	14.1	78
Iowa	12.1	67	New York	17.0	94	Wisconsin	13.9	77
Kansas	10.1	56	North Carolina	14.1	78	Wyoming	8.9	49
Weighted Average for U.S. is 18.7% of payroll = 100%								

Rates in the following table are the base or manual costs per $100 of payroll for Workers' Compensation in each state. Rates are usually applied to straight time wages only and not to premium time wages and bonuses.

The weighted average skilled worker rate for 35 trades is 18.1%. For bidding purposes, apply the full value of Workers' Compensation directly to total labor costs, or if labor is 38%, materials 42% and overhead and profit 20% of total cost, carry 38/80 x 18.1% = 8.6% of cost (before overhead and profit) into overhead. Rates vary not only from state to state but also with the experience rating of the contractor.

Rates are the most current available at the time of publication.

General Requirements | R011 | Overhead & Miscellaneous Data

R01100-060 Workers' Compensation Insurance Rates by Trade and State (cont.)

State	Carpentry — 3 stories or less	Carpentry — interior cab. work	Carpentry — general	Concrete Work — NOC	Concrete Work — flat (flr., sdwk.)	Electrical Wiring — inside	Excavation — earth NOC	Excavation — rock	Glaziers	Insulation Work	Lathing	Masonry	Painting & Decorating	Pile Driving	Plastering	Plumbing	Roofing	Sheet Metal Work (HVAC)	Steel Erection — door & sash	Steel Erection — inter., ornam.	Steel Erection — structure	Steel Erection — NOC	Tile Work — (interior ceramic)	Waterproofing	Wrecking
	5651	5437	5403	5213	5221	5190	6217	6217	5462	5479	5443	5022	5474	6003	5480	5183	5551	5538	5102	5102	5040	5057	5348	9014	5701
AL	36.41	14.51	34.52	32.46	16.19	9.57	20.98	20.98	24.55	20.09	15.28	29.78	21.80	64.38	33.79	11.96	57.62	25.78	16.41	16.41	54.68	66.67	16.19	6.92	54.68
AK	9.30	6.63	9.58	9.26	5.69	5.31	8.16	8.16	9.44	8.72	6.37	9.33	8.41	36.08	11.46	4.46	16.70	6.03	10.43	10.43	18.69	18.69	6.28	5.07	25.07
AZ	14.89	9.11	26.40	13.69	8.09	6.28	8.66	8.66	12.90	15.63	7.83	13.69	10.23	19.88	12.27	7.77	19.67	10.13	10.02	10.02	36.62	23.31	7.24	4.94	50.31
AR	16.12	10.60	15.38	16.45	5.97	7.93	9.21	9.21	12.58	15.29	12.66	9.32	11.10	19.14	13.01	6.58	30.74	10.28	8.19	8.19	27.11	25.39	7.80	4.58	27.11
CA	28.28	9.07	28.28	13.18	13.18	9.93	7.88	7.88	16.37	23.34	10.90	14.55	19.45	21.18	18.12	11.59	39.79	15.33	13.55	13.55	23.93	21.91	7.74	19.45	21.91
CO	32.22	13.54	20.95	20.96	15.13	9.20	15.99	15.99	15.98	21.31	14.68	31.06	21.85	41.53	37.40	14.50	59.10	12.54	11.83	11.83	73.46	43.49	15.04	12.26	73.46
CT	19.25	13.14	24.07	20.48	14.17	6.68	8.42	8.42	21.25	34.89	13.64	25.45	18.84	24.54	17.60	12.25	33.98	13.93	13.67	13.67	50.86	35.83	13.94	5.02	50.86
DE	13.83	13.83	11.56	9.28	6.29	5.82	8.54	8.54	11.18	11.56	10.51	9.85	13.41	11.64	10.51	5.41	22.89	10.29	10.21	10.21	26.69	10.21	7.54	9.85	25.69
DC	17.51	10.80	18.88	29.27	16.85	8.76	17.96	17.96	28.00	22.14	12.07	29.67	16.85	42.12	17.06	12.83	22.39	15.78	33.66	33.66	66.43	42.52	16.42	4.66	66.43
FL	36.73	23.28	29.09	29.92	16.15	10.51	15.27	15.27	17.81	25.18	24.27	28.72	28.92	49.35	28.79	13.37	53.20	17.12	18.67	18.67	43.71	50.61	12.56	7.89	43.71
GA	31.04	16.02	32.71	20.67	15.12	9.82	19.55	19.55	21.97	19.15	26.66	23.03	21.20	40.88	19.50	11.14	40.72	18.65	13.74	13.74	36.03	53.09	13.08	10.16	36.03
HI	17.38	11.62	28.20	13.74	12.27	7.10	7.73	7.73	20.55	21.19	10.94	17.87	11.33	20.15	15.61	6.06	33.24	8.02	11.11	11.11	31.71	21.70	9.78	11.54	31.71
ID	12.68	6.50	16.29	9.46	9.04	5.37	6.87	6.87	8.58	11.01	8.97	9.38	8.83	18.27	8.94	5.04	21.34	7.76	7.51	7.51	26.68	24.89	5.35	7.44	26.68
IL	20.76	12.31	21.64	35.48	12.83	8.86	8.69	8.69	23.09	24.90	14.24	20.34	13.15	37.63	13.78	11.68	35.88	14.58	21.53	21.53	58.63	52.47	13.87	6.26	58.63
IN	6.86	3.76	6.99	7.50	3.50	2.89	4.01	4.01	5.67	7.82	3.97	5.79	6.06	12.19	5.30	2.96	12.65	4.33	5.55	5.55	27.19	12.61	3.49	3.67	27.19
IA	10.22	5.19	12.01	9.18	6.42	3.84	5.56	5.56	11.62	16.49	5.95	10.93	7.58	16.30	9.26	5.01	17.38	6.09	9.80	9.80	45.02	27.20	5.60	4.16	45.02
KS	11.44	7.93	9.21	9.68	6.17	4.12	4.14	4.14	6.93	16.00	7.06	11.86	8.63	13.22	9.21	5.11	22.98	5.57	5.38	5.38	16.49	22.99	3.90	3.77	16.49
KY	16.47	15.26	24.34	22.12	9.53	8.95	13.67	13.67	15.96	14.68	10.62	22.00	19.83	30.44	13.94	10.38	32.83	15.92	13.21	13.21	45.21	33.73	10.60	8.85	45.21
LA	29.63	22.95	26.01	20.86	14.80	10.48	24.73	24.73	23.88	20.82	14.00	21.27	27.63	54.39	21.42	13.78	49.11	21.35	15.51	15.51	66.41	36.94	12.82	10.26	66.41
ME	13.43	10.42	43.59	24.29	11.12	10.02	14.75	14.75	13.35	15.39	14.03	17.26	16.07	31.44	17.95	11.12	32.07	17.70	15.08	15.08	37.84	61.81	11.66	9.37	37.84
MD	10.55	5.95	10.55	11.35	5.05	5.15	9.25	9.25	13.20	13.05	6.45	11.35	6.75	27.45	6.65	5.55	22.80	7.00	9.15	9.15	26.80	18.50	6.35	3.50	26.80
MA	15.25	11.38	22.82	39.38	16.79	6.14	10.72	10.72	16.78	27.76	14.02	29.03	14.95	29.58	13.75	8.49	67.48	13.20	16.07	16.07	80.76	79.51	16.46	6.62	62.87
MI	16.08	9.61	17.85	24.50	12.37	5.71	10.99	10.99	9.62	20.95	11.61	19.18	17.86	64.20	18.33	8.19	43.08	10.91	14.52	14.52	36.41	31.20	13.08	10.42	41.58
MN	27.66	28.76	49.94	30.20	22.59	8.28	19.73	19.73	30.01	37.20	23.93	29.89	19.91	80.08	23.93	15.88	88.83	13.91	27.21	27.21	132.92	31.83	29.54	8.71	132.92
MS	21.77	16.71	28.09	17.88	11.02	10.84	13.03	13.03	18.18	21.21	14.41	22.67	17.90	77.98	20.06	10.84	36.25	20.16	14.52	14.52	37.42	42.77	12.76	8.68	37.42
MO	18.72	7.52	11.84	13.28	10.66	5.79	11.05	11.05	8.97	19.99	9.67	13.57	11.77	16.88	14.22	7.10	25.74	9.76	13.82	13.82	47.98	31.96	5.83	5.71	47.98
MT	30.63	15.19	47.53	32.78	24.74	11.01	26.63	26.63	25.04	38.54	19.54	48.55	41.06	53.68	27.00	15.03	85.06	16.93	33.95	33.95	86.98	55.25	13.30	12.10	86.98
NE	16.19	9.23	15.94	17.31	12.73	5.43	10.19	10.19	12.36	16.79	9.34	18.07	11.58	22.01	14.62	9.06	33.91	13.86	10.75	10.75	25.03	26.96	7.88	5.47	25.03
NV	19.86	10.88	14.94	14.89	11.46	9.64	9.26	9.26	13.64	22.78	14.60	14.48	16.01	27.25	17.93	10.76	33.53	18.03	16.53	16.53	45.36	32.56	18.03	7.43	45.36
NH	16.38	9.82	19.68	28.99	9.29	7.10	13.87	13.87	12.87	43.98	10.90	25.97	17.94	28.09	17.33	9.74	58.76	12.48	13.27	13.27	55.90	49.67	12.82	7.37	55.90
NJ	10.40	8.85	10.40	10.83	8.37	4.23	7.95	7.95	7.36	12.78	7.88	10.77	11.49	11.32	7.88	5.48	29.41	6.20	8.03	8.03	22.52	12.96	5.74	5.37	29.92
NM	25.35	13.09	20.45	19.66	13.39	6.34	10.95	10.95	14.58	31.49	16.25	20.00	18.61	26.97	16.87	12.10	43.41	15.64	21.26	21.26	67.82	29.91	9.67	10.65	67.82
NY	15.39	6.82	14.50	24.92	15.16	6.89	9.69	9.69	16.92	15.82	13.00	18.61	12.55	27.49	13.03	9.03	32.78	14.13	14.12	14.12	27.11	24.09	10.56	7.48	32.53
NC	14.43	9.91	15.37	16.82	6.37	7.64	7.73	7.73	9.52	12.92	8.00	12.94	10.40	17.95	14.47	6.93	25.39	10.26	13.32	13.32	39.03	16.68	7.68	3.86	39.03
ND	13.89	13.89	13.89	12.17	12.17	6.16	9.91	9.91	9.85	9.38	13.16	12.20	11.42	29.82	13.16	9.91	31.37	8.05	13.89	13.89	29.82	29.82	9.63	31.37	29.82
OH	13.27	13.27	13.27	13.45	13.45	6.02	13.45	13.45	18.28	16.31	16.31	15.62	18.28	13.45	16.31	7.96	28.62	24.12	13.45	13.45	13.45	13.45	13.08	13.45	13.45
OK	26.16	11.79	19.33	15.51	11.50	7.41	18.45	18.45	13.55	18.00	12.05	16.80	17.10	36.26	18.29	8.72	42.26	12.45	11.17	11.17	69.45	46.76	11.42	8.20	69.45
OR	29.51	9.28	20.03	16.50	10.61	6.63	12.94	12.94	14.95	19.68	12.27	12.43	19.83	22.05	13.85	8.38	38.11	11.85	11.01	11.01	61.69	25.54	9.92	12.10	61.69
PA	15.80	15.80	19.85	28.18	11.84	7.80	12.74	12.74	15.69	19.85	18.94	18.96	22.05	32.12	18.94	11.57	43.30	12.58	24.71	24.71	59.05	24.71	11.83	18.96	81.34
RI	21.32	11.02	18.77	22.11	15.55	5.86	12.16	12.16	17.02	20.95	11.89	20.66	19.86	41.76	14.19	7.25	29.48	10.52	13.18	13.18	78.79	49.67	14.34	9.90	78.79
SC	19.10	13.03	19.24	12.88	6.19	7.02	8.32	8.32	12.22	10.69	7.27	9.55	11.19	16.64	18.89	6.88	28.98	11.42	9.49	9.49	21.80	23.66	6.23	4.65	21.80
SD	14.61	9.94	20.92	21.64	9.04	5.72	11.73	11.73	14.77	17.73	10.89	13.30	13.25	32.14	16.96	9.50	29.29	11.35	11.51	11.51	46.57	33.20	9.85	5.59	46.57
TN	13.22	9.52	11.99	12.39	7.53	5.00	8.68	8.68	8.96	15.04	8.51	12.73	10.88	15.62	12.19	5.91	24.81	8.76	7.36	7.36	45.06	13.95	5.89	4.98	45.06
TX	28.29	18.95	28.29	25.59	19.49	11.60	18.16	18.16	14.17	25.84	13.47	23.12	18.29	43.91	20.99	12.88	47.24	22.92	14.29	14.29	50.47	31.01	9.94	11.38	54.81
UT	11.17	11.17	11.17	17.68	7.99	7.05	6.38	6.38	9.55	10.95	12.32	13.75	15.29	17.24	10.60	6.41	26.27	6.30	8.99	8.99	23.51	23.51	6.06	5.83	26.53
VT	11.93	7.61	14.88	34.92	9.35	4.51	7.91	7.91	11.95	11.32	8.84	13.76	7.82	19.81	10.86	6.71	24.32	10.29	10.29	10.29	32.12	35.79	6.45	8.56	32.12
VA	10.00	6.46	8.65	9.29	5.47	3.45	6.13	6.13	10.17	8.77	7.59	8.07	8.11	24.38	8.20	5.60	17.80	6.74	10.42	10.42	19.28	30.25	6.78	2.57	19.28
WA	8.78	8.78	8.78	8.53	8.53	3.68	8.82	8.82	8.71	9.93	8.78	11.54	10.23	20.99	11.78	5.41	23.39	3.94	16.21	16.21	16.21	16.21	9.81	10.23	16.21
WV	15.43	15.43	15.43	20.87	20.87	5.01	9.14	9.14	7.33	7.33	16.88	16.02	16.88	10.56	16.88	7.40	12.04	7.33	14.98	14.98	13.27	14.98	16.02	4.87	13.27
WI	10.47	10.13	19.98	8.91	8.50	5.73	7.00	7.00	9.67	13.88	13.48	15.43	12.10	17.04	11.81	5.95	29.35	8.64	11.52	11.52	36.51	18.72	8.45	3.92	36.51
WY	8.13	8.13	8.13	8.13	8.13	8.13	8.13	8.13	8.13	8.13	8.13	8.13	8.13	8.13	8.13	8.13	8.13	8.13	8.13	8.13	8.13	8.13	8.13	8.13	8.13
AVG.	18.12	11.65	19.85	18.81	11.46	7.03	11.41	11.41	14.43	18.52	12.26	17.75	15.27	29.88	15.72	8.82	34.62	12.27	13.77	13.77	42.56	31.55	10.48	8.32	43.48

General Requirements — R011 Overhead & Miscellaneous Data

R01100-060 Workers' Compensation (cont.) (Canada in Canadian dollars)

Province		Alberta	British Columbia	Manitoba	Ontario	New Brunswick	Newfndld. & Labrador	Northwest Territories	Nova Scotia	Prince Edward Island	Quebec	Saskatchewan	Yukon
Carpentry—3 stories or less	Rate	2.65	6.89	5.35	8.19	3.76	5.36	2.75	6.34	5.68	13.85	5.38	2.35
	Code	25401	70600	40102	723	422	403	4-41	4013	401	80110	B12-02	4-042
Carpentry—interior cab. work	Rate	2.48	2.68	5.35	8.19	3.85	5.36	2.75	6.34	3.74	13.85	3.86	2.35
	Code	42133	60412	40102	723	427	403	4-41	4013	402	80110	B11-25	4-042
CARPENTRY—general	Rate	2.65	6.89	5.35	8.19	3.76	5.36	2.75	6.34	5.68	13.85	5.38	2.35
	Code	25401	70600	40102	723	422	403	4-41	4013	401	80110	B12-02	4-042
CONCRETE WORK—NOC	Rate	3.26	6.89	6.72	14.29	3.76	5.36	2.75	7.87	5.68	16.29	7.99	3.25
	Code	42104	70604	40110	745	422	403	4-41	4222	401	80100	B14-04	2-032
CONCRETE WORK—flat (flr. sidewalk)	Rate	3.26	6.89	6.72	14.29	3.76	5.36	2.75	7.87	5.68	16.29	7.99	3.25
	Code	42104	70604	40110	745	422	403	4-41	4222	401	80100	B14-04	2-032
ELECTRICAL Wiring—inside	Rate	1.61	4.61	3.50	3.61	1.49	4.34	2.00	3.19	3.74	7.41	3.86	2.35
	Code	42124	71100	40203	704	426	400	4-46	4261	402	80170	B11-05	4-041
EXCAVATION—earth NOC	Rate	2.03	4.13	5.35	4.95	2.72	5.36	2.50	3.92	3.95	7.87	4.69	3.25
	Code	40604	72607	40706	711	421	403	4-43	4214	404	80030	R11-06	2-016
EXCAVATION—rock	Rate	2.03	4.13	5.35	4.95	2.72	5.36	2.50	3.92	3.95	7.87	4.69	3.25
	Code	40604	72607	40706	711	421	403	4-43	4214	404	80030	R11-06	2-016
GLAZIERS	Rate	1.93	2.03	3.50	10.27	3.85	4.09	2.75	7.75	3.74	17.29	7.99	2.35
	Code	42121	60236	40109	751	423	402	4-41	4233	402	80150	B13-04	4-042
INSULATION WORK	Rate	2.28	6.86	5.35	10.27	3.85	4.09	2.75	7.75	5.68	15.05	5.38	3.25
	Code	42184	70504	40102	751	423	402	4-41	4234	401	80120	B12-07	2-035
LATHING	Rate	5.05	6.86	5.35	8.19	3.85	4.09	2.75	6.14	3.74	15.05	7.99	3.25
	Code	42135	70500	40102	723	427	402	4-41	4271	402	80120	B13-02	2-036
MASONRY	Rate	3.26	6.89	5.35	16.52	3.85	5.36	2.75	7.75	5.68	16.29	7.99	3.25
	Code	42102	70602	40102	741	423	403	4-41	4231	401	80100	B15-01	2-032
PAINTING & DECORATING	Rate	2.63	6.86	6.72	9.76	3.85	4.09	2.75	6.14	3.74	15.05	5.38	3.25
	Code	42111	70501	40105	719	427	402	4-41	4275	402	80120	B12-01	2-036
PILE DRIVING	Rate	3.26	17.50	5.35	5.76	3.76	9.60	2.50	7.87	5.68	7.87	5.38	3.25
	Code	42159	72502	40706	732	422	404	4-43	4221	401	80030	B12-10	2-030
PLASTERING	Rate	5.05	6.86	6.72	9.76	3.85	4.09	2.75	6.14	3.74	15.05	7.99	3.25
	Code	42135	70502	40108	719	427	402	4-41	4271	402	80120	B13-02	2-036
PLUMBING	Rate	1.61	4.07	3.50	4.42	1.88	3.57	2.00	3.70	3.74	8.03	3.86	2.35
	Code	42122	70712	40204	707	424	401	4-46	4241	402	80160	B11-01	4-039
ROOFING	Rate	6.56	6.89	6.72	12.05	7.72	5.36	2.75	9.30	5.68	22.52	7.99	3.25
	Code	42118	70600	40403	728	430	403	4-41	4235	401	80130	B15-02	2-031
SHEET METAL WORK (HVAC)	Rate	1.61	4.07	5.35	4.42	1.88	3.57	2.00	7.75	3.74	8.03	3.86	2.35
	Code	42117	70714	40402	707	424	401	4-46	4236	402	80160	B11-07	4-040
STEEL ERECTION—door & sash	Rate	2.28	17.50	5.35	20.96	3.76	9.60	2.75	7.87	5.68	32.75	7.99	3.25
	Code	42106	72509	40502	748	422	404	4-41	4223	401	80080	B15-03	2-012
STEEL ERECTION—inter., ornam.	Rate	2.28	17.50	5.35	20.96	3.76	5.36	2.75	7.87	5.68	32.75	7.99	3.25
	Code	42106	72509	40502	748	422	403	4-41	4223	401	80080	B15-03	2-012
STEEL ERECTION—structure	Rate	2.28	17.50	5.35	20.96	3.76	9.60	2.75	10.79	5.68	32.75	7.99	3.25
	Code	42106	72509	40502	748	422	404	4-41	4227	401	80080	B15-04	2-012
STEEL ERECTION—NOC	Rate	2.28	17.50	5.35	20.96	3.76	9.60	2.75	10.79	5.68	32.75	7.99	3.25
	Code	42106	72509	40502	748	422	404	4-41	4227	401	80080	B15-03	2-012
TILE WORK—inter. (ceramic)	Rate	2.53	6.86	2.31	9.76	3.85	4.09	2.75	6.14	3.74	15.05	7.99	3.25
	Code	42113	70506	40103	719	427	402	4-41	4276	402	80120	B13-01	2-034
WATERPROOFING	Rate	2.63	6.89	5.35	8.19	3.85	5.36	2.75	7.75	3.74	22.52	3.86	3.25
	Code	42139	70620	40102	723	423	403	4-41	4239	402	80130	B11-17	2-030
WRECKING	Rate	12.86	6.89	6.72	20.96	2.72	5.36	2.50	3.92	5.68	26.98	7.99	3.25
	Code	42108	70600	40106	748	421	403	4-43	4211	401	80220	B14-07	2-030

General Requirements

R011 Overhead & Miscellaneous Data

R01100-070 Contractor's Overhead & Profit

Below are the **average** installing contractor's percentage mark-ups applied to base labor rates to arrive at typical billing rates.

Column A: Labor rates are based on average open shop wages for 7 major U.S. regions. Base rates including fringe benefits are listed hourly and daily. These figures are the sum of the wage rate and employer-paid fringe benefits such as vacation pay, and employer-paid health costs.

Column B: Workers' Compensation rates are the national average of state rates established for each trade.

Column C: Column C lists average fixed overhead figures for all trades. Included are Federal and State Unemployment costs set at 7.0%; Social Security Taxes (FICA) set at 7.65%; Builder's Risk Insurance costs set at 0.34%; and Public Liability costs set at 1.55%. All the percentages except those for Social Security Taxes vary from state to state as well as from company to company.

Columns D and E: Percentages in Columns D and E are based on the presumption that the installing contractor has annual billing of $1,000,000 and up. Overhead percentages may increase with smaller annual billing. The overhead percentages for any given contractor may vary greatly and depend on a number of factors, such as the contractor's annual volume, engineering and logistical support costs, and staff requirements. The figures for overhead and profit will also vary depending on the type of job, the job location, and the prevailing economic conditions. All factors should be examined very carefully for each job.

Column F: Column F lists the total of Columns B, C, D, and E.

Column G: Column G is Column A (hourly base labor rate) multiplied by the percentage in Column F (O&P percentage).

Column H: Column H is the total of Column A (hourly base labor rate) plus Column G (Total O&P).

Column I: Column I is Column H multiplied by eight hours.

		A		B	C	D	E	F	G	H	I
		Base Rate Incl. Fringes		Workers' Comp. Ins.	Average Fixed Overhead	Overhead	Profit	Total Overhead & Profit		Rate with O & P	
Abbr.	Trade	Hourly	Daily					%	Amount	Hourly	Daily
Skwk	Skilled Workers Average (35 trades)	$19.75	$158.00	18.1%	16.5%	27.0%	10.0%	71.6%	$14.15	$33.90	$271.20
	Helpers Average (5 trades)	14.80	118.40	19.7		25.0		71.2	10.55	25.35	202.80
	Foreman Average, Inside ($.50 over trade)	20.25	162.00	18.1		27.0		71.6	14.50	34.75	278.00
	Foreman Average, Outside ($2.00 over trade)	21.75	174.00	18.1		27.0		71.6	15.55	37.30	298.40
Clab	Common Building Laborers	14.45	115.60	19.9		25.0		71.4	10.30	24.75	198.00
Asbe	Asbestos Workers	20.50	164.00	18.5		30.0		75.0	15.40	35.90	287.20
Boil	Boilermakers	22.10	176.80	16.2		30.0		72.7	16.05	38.15	305.20
Bric	Bricklayers	20.00	160.00	17.8		25.0		69.3	13.85	33.85	270.80
Brhe	Bricklayer Helpers	15.70	125.60	17.8		25.0		69.3	10.90	26.60	212.80
Carp	Carpenters	19.70	157.60	19.9		25.0		71.4	14.05	33.75	270.00
Cefi	Cement Finishers	18.90	151.20	11.5		25.0		63.0	11.90	30.80	246.40
Elec	Electricians	22.10	176.80	7.0		30.0		63.5	14.05	36.15	289.20
Elev	Elevator Constructors	22.90	183.20	8.5		30.0		65.0	14.90	37.80	302.40
Eqhv	Equipment Operators, Crane or Shovel	20.65	165.20	11.4		28.0		65.9	13.60	34.25	274.00
Eqmd	Equipment Operators, Medium Equipment	19.90	159.20	11.4		28.0		65.9	13.10	33.00	264.00
Eqlt	Equipment Operators, Light Equipment	19.10	152.80	11.4		28.0		65.9	12.60	31.70	253.60
Eqol	Equipment Operators, Oilers	17.00	136.00	11.4		28.0		65.9	11.20	28.20	225.60
Eqmm	Equipment Operators, Master Mechanics	21.05	168.40	11.4		28.0		65.9	13.85	34.90	279.20
Glaz	Glaziers	19.35	154.80	14.4		25.0		65.9	12.75	32.10	256.80
Lath	Lathers	19.20	153.60	12.3		25.0		63.8	12.25	31.45	251.60
Marb	Marble Setters	19.85	158.80	17.8		25.0		69.3	13.75	33.60	268.80
Mill	Millwrights	20.60	164.80	11.7		25.0		63.2	13.00	33.60	268.80
Mstz	Mosaic and Terrazzo Workers	19.10	152.80	10.5		25.0		62.0	11.85	30.95	247.60
Pord	Painters, Ordinary	17.95	143.60	15.3		25.0		66.8	12.00	29.95	239.60
Psst	Painters, Structural Steel	18.65	149.20	51.1		25.0		102.6	19.15	37.80	302.40
Pape	Paper Hangers	18.05	144.40	15.3		25.0		66.8	12.05	30.10	240.80
Pile	Pile Drivers	19.45	155.60	29.9		30.0		86.4	16.80	36.25	290.00
Plas	Plasterers	18.65	149.20	15.7		25.0		67.2	12.55	31.20	249.60
Plah	Plasterer Helpers	15.65	125.20	15.7		25.0		67.2	10.50	26.15	209.20
Plum	Plumbers	21.95	175.60	8.8		30.0		65.3	14.35	36.30	290.40
Rodm	Rodmen (Reinforcing)	21.10	168.80	31.6		28.0		86.1	18.15	39.25	314.00
Rofc	Roofers, Composition	17.05	136.40	34.6		25.0		86.1	14.70	31.75	254.00
Rots	Roofers, Tile and Slate	17.10	136.80	34.6		25.0		86.1	14.70	31.80	254.40
Rohe	Roofer Helpers (Composition)	12.75	102.00	34.6		25.0		86.1	11.00	23.75	190.00
Shee	Sheet Metal Workers	21.50	172.00	12.3		30.0		68.8	14.80	36.30	290.40
Spri	Sprinkler Installers	22.05	176.40	9.0		30.0		65.5	14.45	36.50	292.00
Stpi	Steamfitters or Pipefitters	22.10	176.80	8.8		30.0		65.3	14.45	36.55	292.40
Ston	Stone Masons	19.50	156.00	17.8		25.0		69.3	13.50	33.00	264.00
Sswk	Structural Steel Workers	21.25	170.00	42.6		28.0		97.1	20.65	41.90	335.20
Tilf	Tile Layers (Floor)	19.20	153.60	10.5		25.0		62.0	11.90	31.10	248.80
Tilh	Tile Layer Helpers	15.45	123.60	10.5		25.0		62.0	9.60	25.05	200.40
Trlt	Truck Drivers, Light	15.85	126.80	15.6		25.0		67.1	10.65	26.50	212.00
Trhv	Truck Drivers, Heavy	16.20	129.60	15.6		25.0		67.1	10.85	27.05	216.40
Sswl	Welders, Structural Steel	21.25	170.00	42.6		28.0		97.1	20.65	41.90	335.20
Wrck	*Wrecking	14.90	119.20	43.5	▼	25.0	▼	95.0	14.15	29.05	232.40

*Not included in Averages.

General Requirements — R011 | Overhead & Miscellaneous Data

R01100-080 Performance Bond

This table shows the cost of a Performance Bond for a construction job scheduled to be completed in 12 months. Add 1% of the premium cost per month for jobs requiring more than 12 months to complete. The rates are "standard" rates offered to contractors that the bonding company considers financially sound and capable of doing the work. Preferred rates are offered by some bonding companies based upon financial strength of the contractor. Actual rates vary from contractor to contractor and from bonding company to bonding company. Contractors should prequalify through a bonding agency before submitting a bid on a contract that requires a bond.

Contract Amount	Building Construction Class B Projects			Highways & Bridges Class A New Construction			Class A-1 Highway Resurfacing		
First $ 100,000 bid	$25.00 per M			$15.00 per M			$9.40 per M		
Next 400,000 bid	$ 2,500	plus	$15.00 per M	$ 1,500	plus	$10.00 per M	$ 940	plus	$7.20 per M
Next 2,000,000 bid	8,500	plus	10.00 per M	5,500	plus	7.00 per M	3,820	plus	5.00 per M
Next 2,500,000 bid	28,500	plus	7.50 per M	19,500	plus	5.50 per M	15,820	plus	4.50 per M
Next 2,500,000 bid	47,250	plus	7.00 per M	33,250	plus	5.00 per M	28,320	plus	4.50 per M
Over 7,500,000 bid	64,750	plus	6.00 per M	45,750	plus	4.50 per M	39,570	plus	4.00 per M

R01100-090 Sales Tax by State (United States)

State sales tax on materials is tabulated below (5 states have no sales tax). Many states allow local jurisdictions, such as a county or city, to levy additional sales tax.

Some projects may be sales tax exempt, particularly those constructed with public funds.

State	Tax (%)	State	Tax (%)	State	Tax (%)	State	Tax (%)
Alabama	4	Illinois	6.25	Montana	0	Rhode Island	7
Alaska	0	Indiana	5	Nebraska	4.5	South Carolina	5
Arizona	5	Iowa	5	Nevada	6.875	South Dakota	4
Arkansas	4.625	Kansas	4.9	New Hampshire	0	Tennessee	6
California	7.25	Kentucky	6	New Jersey	6	Texas	6.25
Colorado	3	Louisiana	4	New Mexico	5	Utah	4.75
Connecticut	6	Maine	6	New York	4	Vermont	5
Delaware	0	Maryland	5	North Carolina	4	Virginia	4.5
District of Columbia	5.75	Massachusetts	5	North Dakota	5	Washington	6.5
Florida	6	Michigan	6	Ohio	5	West Virginia	6
Georgia	4	Minnesota	6.5	Oklahoma	4.5	Wisconsin	5
Hawaii	4	Mississippi	7	Oregon	0	Wyoming	4
Idaho	5	Missouri	4.225	Pennsylvania	6	Average	4.71 %

Sales Tax by Province (Canada)

GST - a value-added tax, which the federal government imposes on most goods and services provided in or imported into Canada.
PST - a retail sales tax, which five of the provinces impose on the price of most goods and some services.

QST - a value-added tax, similar to the federal GST, which Quebec imposes.
HST - Three provinces have combined their retail sales tax with the federal GST into the Harmonized Sales Tax.

Province	PST (%)	QST (%)	GST (%)	HST (%)
Alberta	0	0	7	0
British Columbia	7	0	7	0
Manitoba	7	0	7	0
New Brunswick	0	0	0	15
Newfoundland	0	0	0	15
Northwest Territories	0	0	7	0
Nova Scotia	0	0	0	15
Ontario	8	0	7	0
Prince Edward Island	10	0	7	0
Quebec	0	6.5	7	0
Saskatchewan	7	0	7	0
Yukon	0	0	7	0

General Requirements — R011 — Overhead & Miscellaneous Data

R01100-100 Unemployment Taxes and Social Security Taxes

Mass. State Unemployment tax ranges from 1.325% to 7.225% plus an experience rating assessment the following year, on the first $10,800 of wages. Federal Unemployment tax is 6.2% of the first $7,000 of wages. This is reduced by a credit for payment to the state. The minimum Federal Unemployment tax is .8% after all credits.

Combined rates in Mass. thus vary from 2.125% to 8.025% of the first $10,800 of wages. Combined average U.S. rate is about 7.0% of the first $7,000. Contractors with permanent workers will pay less since the average annual wages for skilled workers is $28.75 x 2,000 hours or about $57,500 per year. The average combined rate for U.S. would thus be 7.0% x $7,000 ÷ $57,500 = 0.9% of total wages for permanent employees.

Rates vary not only from state to state but also with the experience rating of the contractor.

Social Security (FICA) for 2000 is estimated at time of publication to be 7.65% of wages up to $72,600.

R01107-010 Architectural Fees

Tabulated below are typical percentage fees by project size, for good professional architectural service. Fees may vary from those listed depending upon degree of design difficulty and economic conditions in any particular area.

Rates can be interpolated horizontally and vertically. Various portions of the same project requiring different rates should be adjusted proportionately. For alterations, add 50% to the fee for the first $500,000 of project cost and add 25% to the fee for project cost over $500,000.

Architectural fees tabulated below include Structural, Mechanical and Electrical Engineering Fees. They do not include the fees for special consultants such as kitchen planning, security, acoustical, interior design, etc.

Building Types	Total Project Size in Thousands of Dollars						
	100	250	500	1,000	5,000	10,000	50,000
Factories, garages, warehouses, repetitive housing	9.0%	8.0%	7.0%	6.2%	5.3%	4.9%	4.5%
Apartments, banks, schools, libraries, offices, municipal buildings	12.2	12.3	9.2	8.0	7.0	6.6	6.2
Churches, hospitals, homes, laboratories, museums, research	15.0	13.6	12.7	11.9	9.5	8.8	8.0
Memorials, monumental work, decorative furnishings	—	16.0	14.5	13.1	10.0	9.0	8.3

General Requirements — R012 — Special Project Procedures

R01250-010 Repair and Remodeling

Cost figures are based on new construction utilizing the most cost-effective combination of labor, equipment and material with the work scheduled in proper sequence to allow the various trades to accomplish their work in an efficient manner.

The costs for repair and remodeling work must be modified due to the following factors that may be present in any given repair and remodeling project.

1. Equipment usage curtailment due to the physical limitations of the project, with only hand-operated equipment being used.
2. Increased requirement for shoring and bracing to hold up the building while structural changes are being made and to allow for temporary storage of construction materials on above-grade floors.
3. Material handling becomes more costly due to having to move within the confines of an enclosed building. For multi-story construction, low capacity elevators and stairwells may be the only access to the upper floors.
4. Large amount of cutting and patching and attempting to match the existing construction is required. It is often more economical to remove entire walls rather than create many new door and window openings. This sort of trade-off has to be carefully analyzed.
5. Cost of protection of completed work is increased since the usual sequence of construction usually cannot be accomplished.
6. Economies of scale usually associated with new construction may not be present. If small quantities of components must be custom fabricated due to job requirements, unit costs will naturally increase. Also, if only small work areas are available at a given time, job scheduling between trades becomes difficult and subcontractor quotations may reflect the excessive start-up and shut-down phases of the job.
7. Work may have to be done on other than normal shifts and may have to be done around an existing production facility which has to stay in production during the course of the repair and remodeling.
8. Dust and noise protection of adjoining non-construction areas can involve substantial special protection and alter usual construction methods.
9. Job may be delayed due to unexpected conditions discovered during demolition or removal. These delays ultimately increase construction costs.
10. Piping and ductwork runs may not be as simple as for new construction. Wiring may have to be snaked through walls and floors.
11. Matching "existing construction" may be impossible because materials may no longer be manufactured. Substitutions may be expensive.
12. Weather protection of existing structure requires additional temporary structures to protect building at openings.
13. On small projects, because of local conditions, it may be necessary to pay a tradesman for a minimum of four hours for a task that is completed in one hour.

All of the above areas can contribute to increased costs for a repair and remodeling project. Each of the above factors should be considered in the planning, bidding and construction stage in order to minimize the increased costs associated with repair and remodeling jobs.

General Requirements | R015 | Construction Aids

R01540-100 Steel Tubular Scaffolding

On new construction, tubular scaffolding is efficient up to 60' high or five stories. Above this it is usually better to use a hung scaffolding if construction permits. Swing scaffolding operations may interfere with tenants. In this case, the tubular is more practical at all heights.

In repairing or cleaning the front of an existing building the cost of tubular scaffolding per S.F. of building front increases as the height increases above the first tier. The first tier cost is relatively high due to leveling and alignment.

The minimum efficient crew for erection is three workers. For heights over 50', a crew of four is more efficient. Use two or more on top and two at the bottom for handing up or hoisting. Four workers can erect and dismantle about nine frames per hour up to five stories. From five to eight stories they will average six frames per hour. With 7' horizontal spacing this will run about 400 S.F. and 265 S.F. of wall surface, respectively. Time for placing planks must be added to the above. On heights above 50', five planks can be placed per labor-hour.

The cost per 1,000 S.F. of building front in the table below was developed by pricing the materials required for a typical tubular scaffolding system eleven frames long and two frames high. Planks were figured five wide for standing plus two wide for materials.

Frames are 5' wide and usually spaced 7' O.C. horizontally. Sidewalk frames are 6' wide. Rental rates will be lower for jobs over three months duration.

For jobs under twenty-five frames, add 50% to rental cost. These figures do not include accessories which are listed separately below. Large quantities for long periods can reduce rental rates by 20%.

Item	Unit	Monthly Rent	Per 1,000 S.F. of Building Front	
			No. of Pieces	Rental per Month
5' Wide Standard Frame, 6'-4" High	Ea.	$ 3.75	24	$ 90.00
Leveling Jack & Plate		1.50	24	36.00
Cross Brace		.60	44	26.40
Side Arm Bracket, 21"		1.50	12	18.00
Guardrail Post		1.00	12	12.00
Guardrail, 7' section		.75	22	16.50
Stairway Section		10.00	2	20.00
Stairway Starter Bar		.10	1	.10
Stairway Inside Handrail		5.00	2	10.00
Stairway Outside Handrail		5.00	2	10.00
Walk-Thru Frame Guardrail		2.00	2	4.00
			Total	$243.00
			Per C.S.F., 1 Use/Mo.	$ 24.30

Scaffolding is often used as falsework over 15' high during construction of cast-in-place concrete beams and slabs. Two foot wide scaffolding is generally used for heavy beam construction. The span between frames depends upon the load to be carried with a maximum span of 5'.

Heavy duty shoring frames with a capacity of 10,000#/leg can be spaced up to 10' O.C. depending upon form support design and loading.

Scaffolding used as horizontal shoring requires less than half the material required with conventional shoring.

On new construction, erection is done by carpenters.

Rolling towers supporting horizontal shores can reduce labor and speed the job. For maintenance work, catwalks with spans up to 70' can be supported by the rolling towers.

Site Construction — R020 Paving & Surfacing

R02065-300 Bituminous Paving

City	Bituminous Asphalt per Ton*	Pavement (3") 6.13 S.Y./ton				Sidewalks (2") 9.2 S.Y./ton			
		Cost per S.Y.			Per Ton	Cost per S.Y.			Per Ton
		Material*	Installation	Total	Total	Material*	Installation	Total	Total
Atlanta	$25.00	$4.08	$ .68	$4.76	$29.15	$2.72	$1.11	$3.83	$35.24
Baltimore	30.50	4.98	.72	5.70	34.94	3.32	1.25	4.57	42.04
Boston	36.25	5.91	.89	6.80	41.69	3.94	1.83	5.77	53.08
Buffalo	31.80	5.19	.86	6.05	37.09	3.46	1.73	5.19	47.75
Chicago	32.50	5.30	.91	6.21	38.06	3.53	1.89	5.42	49.86
Cincinnati	33.25	5.42	.76	6.18	37.88	3.61	1.39	5.00	46.00
Cleveland	30.00	4.89	.82	5.71	34.98	3.26	1.58	4.84	44.53
Columbus	26.50	4.32	.76	5.08	31.14	2.88	1.39	4.27	39.28
Dallas	25.00	4.08	.69	4.77	29.23	2.72	1.15	3.87	35.60
Denver	25.75	4.20	.70	4.90	30.03	2.80	1.18	3.98	36.62
Detroit	29.00	4.73	.82	5.55	34.02	3.15	1.59	4.74	43.61
Houston	30.50	4.98	.68	5.66	34.72	3.32	1.13	4.45	40.94
Indianapolis	27.80	4.54	.77	5.31	32.52	3.02	1.41	4.43	40.76
Kansas City	24.50	4.00	.78	4.78	29.29	2.66	1.45	4.11	37.81
Los Angeles	30.00	4.89	.88	5.77	35.38	3.26	1.80	5.06	46.55
Memphis	29.30	4.78	.65	5.43	33.29	3.18	1.02	4.20	38.64
Milwaukee	26.50	4.32	.86	5.18	31.77	2.88	1.74	4.62	42.50
Minneapolis	27.25	4.45	.82	5.27	32.28	2.96	1.58	4.54	41.77
Nashville	28.00	4.57	.68	5.25	32.15	3.04	1.11	4.15	38.18
New Orleans	37.00	6.04	.64	6.68	40.96	4.02	.99	5.01	46.09
New York City	44.00	7.18	1.04	8.22	50.38	4.78	2.33	7.11	65.41
Philadelphia	27.75	4.53	.82	5.35	32.82	3.02	1.60	4.62	42.50
Phoenix	23.75	3.87	.69	4.56	27.95	2.58	1.15	3.73	34.32
Pittsburgh	32.25	5.26	.80	6.06	37.12	3.51	1.51	5.02	46.18
St. Louis	25.50	4.16	.84	5.00	30.64	2.77	1.65	4.42	40.66
San Antonio	27.25	4.45	.63	5.08	31.14	2.96	.95	3.91	35.97
San Diego	28.50	4.65	.88	5.53	33.91	3.10	1.80	4.90	45.08
San Francisco	31.50	5.14	.91	6.05	37.08	3.42	1.89	5.31	48.85
Seattle	32.75	5.34	.84	6.18	37.87	3.56	1.65	5.21	47.93
Washington, D.C.	35.00	5.71	.71	6.42	39.37	3.80	1.23	5.03	46.28
Average	$29.80	$4.87	$ .78	$5.65	$34.63	$3.24	$1.47	$4.71	$43.34

Assumed density is 145 lb. per C.F.
*Includes delivery within 20 miles

Table below shows quantities and bare costs for 1000 S.Y. of Bituminous Paving.

Item	Roads and Parking Areas, 3" Thick (02740-300-0460)		Sidewalks, 2" Thick (02775-275-0010)	
	Quantities	Cost	Quantities	Cost
Bituminous asphalt	163 tons @ $29.80 per ton	$4,857.40	109 tons @ $29.80 per ton	$3,248.20
Installation using	Crew B-25B @ $3,575.40 /4900SY/ day x 1000	729.67	Crew B-37 @ $880.00 /720 SY/day x 1000	1,222.22
Total per 1000 S.Y.		$5,587.07		$4,470.42
Total per S.Y.		$ 5.65		$ 4.47
Total per Ton		$ 34.28		$ 41.01

Site Construction — R023 Earthwork

R02315-300 Compacting Backfill

Compaction of fill in embankments, around structures, in trenches, and under slabs is important to control settlement. Factors affecting compaction are:

1. Soil gradation
2. Moisture content
3. Equipment used
4. Depth of fill per lift
5. Density required

The costs for testing and soil analyses are listed in Division 01450-500. Also, see Division 02315 for further backfill, borrow, and compaction costs.

Example:

Compact granular fill around a building foundation using a 21" wide x 24" vibratory plate in 8" lifts. Operator moves at 50 FPM working a 50 minute hour to develop 95% Modified Proctor Density with 4 passes.

Production Rate:

$$\frac{1.75'' \text{ plate wide} \times 50 \text{ F.P.M.} \times 50 \text{ min./hr} \times 67'' \text{ lift}}{27 \text{ C.F. per C.Y.}} = 108.5 \text{ C.Y./hr.}$$

Production Rate for 4 Passes:

$$\frac{108.5 \text{ C.Y.}}{4 \text{ passes}} = 27.125 \text{ C.Y./hr.} \times 8 \text{ hrs.} = 217 \text{ C.Y./day}$$

	Compacting 217 C.Y. with 21" Wide Vibratory Plate	L.H./Day	Hourly Cost	Daily Cost	CY Cost
1	Laborer	8	$14.45	$115.60	$.54
1	Vibratory Plate Compactor			60.60	.28
	Total for 217 C.Y./day			$176.20	$.82

Site Construction — R023 Earthwork

R02315-400 Excavating

The selection of equipment used for structural excavation and bulk excavation or for grading is determined by the following factors.
1. Quantity of material.
2. Type of material.
3. Depth or height of cut.
4. Length of haul.
5. Condition of haul road.
6. Accessibility of site.
7. Moisture content and dewatering requirements.
8. Availability of excavating and hauling equipment.

Some additional costs must be allowed for hand trimming the sides and bottom of concrete pours and other excavation below the general excavation.

Number of B.C.Y. per truck = 1.5 C.Y. bucket × 8 passes = 12 loose C.Y.

$$= 12 \times \frac{100}{118} = 10.2 \text{ B.C.Y. per truck}$$

Truck Haul Cycle:
Load truck 8 passes	=	4 minutes
Haul distance 1 mile	=	9 minutes
Dump time	=	2 minutes
Return 1 mile	=	7 minutes
Spot under machine	=	1 minute
		23 minute cycle

When planning excavation and fill, the following should also be considered.
1. Swell factor.
2. Compaction factor.
3. Moisture content.
4. Density requirements.

A typical example for scheduling and estimating the cost of excavation of a 15' deep basement on a dry site when the material must be hauled off the site, is outlined below.

Assumptions:
1. Swell factor, 18%.
2. No mobilization or demobilization.
3. Allowance included for idle time and moving on job.
4. No dewatering, sheeting, or bracing.
5. No truck spotter or hand trimming.

Fleet Haul Production per day in B.C.Y.

$$4 \text{ trucks} \times \frac{50 \text{ min. hour}}{23 \text{ min. haul cycle}} \times 8 \text{ hrs.} \times 10.2 \text{ B.C.Y.}$$

$$= 4 \times 2.2 \times 8 \times 10.2 = 718 \text{ B.C.Y./day}$$

	Excavating Cost with a 1-1/2 C.Y. Hydraulic Excavator 15' Deep, 2 Mile Round Trip Haul	L.H./Day	Hourly Cost	Daily Cost	Subtotal	Unit Price
1	Equipment Operator	8	$20.65	$ 165.20		
1	Oiler	8	17.00	136.00		
4	Truck Drivers	32	16.20	518.40	$ 819.60	$1.14
1	Hydraulic Excavator			713.20		
4	Dump Trucks			1,789.00	2,502.20	3.48
Total for 720 BCY				$3,321.80	$3,321.80	$4.61

Description	1-1/2 C.Y. Hyd. Backhoe 15' Deep		1-1/2 C.Y. Power Shovel 7' Bank		1-1/2 C.Y. Dragline 7' Deep		2-1/2 C.Y. Trackloader Stockpile	
Operator (and Oiler, if required)		$ 301.20		$ 301.20		$ 301.20		$ 301.20
Truck Drivers	3 Ea.	388.80	4 Ea.	518.40	3 Ea.	388.80	4 Ea.	518.40
Equipment Rental		713.20		875.30		899.75		865.10
20 C.Y. Trailer Dump Trucks	3 Ea.	1,341.75	4 Ea.	1,789.00	3 Ea.	1,341.75	4 Ea.	1,789.00
Total Cost per Day		$2,744.95		$3,483.90		$2,931.50		$3,473.70
Daily Production, C.Y. Bank Measure		720.00		960.00		640.00		1000.00
Cost per C.Y.		$ 3.81		$ 3.63		$ 4.58		$ 3.47

Add the mobilization and demobilization costs to the total excavation costs. When equipment is rented for more than three days, there is often no mobilization charge by the equipment dealer. On larger jobs outside of urban areas, scrapers can move earth economically provided a dump site or fill area and adequate haul roads are available. Excavation within sheeting bracing or cofferdam bracing is usually done with a clamshell and production is low, since the clamshell may have to be guided by hand between the bracing. When excavating or filling an area enclosed with a wellpoint system, add 10% to 15% to the cost to allow for restricted access. When estimating earth excavation quantities for structures, allow work space outside the building footprint for construction of the foundation, and a slope of 1:1 unless sheeting is used.

Site Construction — R023 Earthwork

R02315-450 Excavating Equipment

The table below lists THEORETICAL hourly production in C.Y./hr. bank measure for some typical excavation equipment. Figures assume 50 minute hours, 83% job efficiency, 100% operator efficiency, 90° swing and properly sized hauling units, which must be modified for adverse digging and loading conditions. Actual production costs in the front of the book average about 50% of the theoretical values listed here.

Equipment	Soil Type	B.C.Y. Weight	% Swell	1 C.Y.	1-1/2 C.Y.	2 C.Y.	2-1/2 C.Y.	3 C.Y.	3-1/2 C.Y.	4 C.Y.
Hydraulic Excavator "Backhoe" 15' Deep Cut	Moist loam, sandy clay	3400 lb.	40%	85	125	175	220	275	330	380
	Sand and gravel	3100	18	80	120	160	205	260	310	365
	Common earth	2800	30	70	105	150	190	240	280	330
	Clay, hard, dense	3000	33	65	100	130	170	210	255	300
Power Shovel Optimum Cut (Ft.)	Moist loam, sandy clay	3400	40	170 (6.0)	245 (7.0)	295 (7.8)	335 (8.4)	385 (8.8)	435 (9.1)	475 (9.4)
	Sand and gravel	3100	18	165 (6.0)	225 (7.0)	275 (7.8)	325 (8.4)	375 (8.8)	420 (9.1)	460 (9.4)
	Common earth	2800	30	145 (7.8)	200 (9.2)	250 (10.2)	295 (11.2)	335 (12.1)	375 (13.0)	425 (13.8)
	Clay, hard, dense	3000	33	120 (9.0)	175 (10.7)	220 (12.2)	255 (13.3)	300 (14.2)	335 (15.1)	375 (16.0)
Drag Line Optimum Cut (Ft.)	Moist loam, sandy clay	3400	40	130 (6.6)	180 (7.4)	220 (8.0)	250 (8.5)	290 (9.0)	325 (9.5)	385 (10.0)
	Sand and gravel	3100	18	130 (6.6)	175 (7.4)	210 (8.0)	245 (8.5)	280 (9.0)	315 (9.5)	375 (10.0)
	Common earth	2800	30	110 (8.0)	160 (9.0)	190 (9.9)	220 (10.5)	250 (11.0)	280 (11.5)	310 (12.0)
	Clay, hard, dense	3000	33	90 (9.3)	130 (10.7)	160 (11.8)	190 (12.3)	225 (12.8)	250 (13.3)	280 (12.0)

Equipment	Soil Type	B.C.Y. Weight	% Swell	Wheel Loaders				Track Loaders		
				3 C.Y.	4 C.Y.	6 C.Y.	8 C.Y.	2-1/4 C.Y.	3 C.Y.	4 C.Y.
Loading Tractors	Moist loam, sandy clay	3400	40	260	340	510	690	135	180	250
	Sand and gravel	3100	18	245	320	480	650	130	170	235
	Common earth	2800	30	230	300	460	620	120	155	220
	Clay, hard, dense	3000	33	200	270	415	560	110	145	200
	Rock, well-blasted	4000	50	180	245	380	520	100	130	180

Site Construction — R025 Sewerage & Drainage

R02510-810 Concrete Pipe

Prices given are for inside 20 mile delivery zone. Add $1.75 per ton of pipe for each additional 10 miles. Minimum truckload is 10 tons. The non-reinforced pipe listed in the front of the book is designation ASTM C14-59 extra strength. The reinforced pipe listed is ASTM C76-65T class 3, no gaskets. The installation cost given includes shaping bottom of the trench, placing the pipe, and backfilling and tamping to the top of the pipe only.

Site Construction — R029 Landscaping

R02920-500 Seeding

The type of grass is determined by light, shade and moisture content of soil plus intended use. Fertilizer should be disked 4" before seeding. For steep slopes disk five tons of mulch and lay two tons of hay or straw on surface per acre after seeding. Surface mulch can be staked, lightly disked or tar emulsion sprayed. Material for mulch can be wood chips, peat moss, partially rotted hay or straw, wood fibers and sprayed emulsions. Hemp seed blankets with fertilizer are also available. For spring seeding, watering is necessary. Late fall seeding may have to be reseeded in the spring. Hydraulic seeding, power mulching, and aerial seeding can be used on large areas.

Concrete — R033 Cast-In-Place Concrete

R03310-090 Placing Ready Mixed Concrete

For ground pours allow for 5% waste when figuring quantities.

Prices in the front of the book assume normal deliveries. If deliveries are made before 8 A.M. or after 5 P.M. or on Saturday afternoons add $25 per C.Y. Large volume discounts are not included in prices in front of book.

For the lower floors without truck access, concrete may be wheeled in rubber tired buggies, conveyer handled, crane handled or pumped. Pumping is economical if there is top steel. Conveyers are more efficient for thick slabs. Concrete pump with an operator can be rented from $620 per day for up to 25 C.Y. to $1,500 per day for a 400 C.Y. pour. Figures include travel time if done at straight time. Pumping lightweight concrete costs an extra $2.40 per C.Y.

At higher floors the rubber tired buggies may be hoisted by a hoisting tower then wheeled to location. Placement by a conveyer is limited to three floors and is best for high volume pours. Pumped concrete is best when building has no crane access. Concrete may be pumped directly as high as thirty-six stories using special pumping techniques. Normal maximum height is about fifteen stories.

Best pumping aggregate is screened and graded bank gravel rather than crushed stone.

Pumping downward is more difficult than pumping upwards. Horizontal distance from pump to pour may increase preparation time prior to pour. Placing by cranes, either mobile, climbing or tower types continues as the most efficient method for high rise concrete buildings.

	Cost per C.Y. for Wheeled Concrete, Dumped Only (Add to appropriate placing cost)							
	10 C.F. Walking Cart				18 C.F. Riding Cart			
Item	Hourly Cost	Wheeled up to 50 ft.	Wheeled up to 150 ft.	Wheeled up to 250 ft.	Hourly Cost	Wheeled up to 50 ft.	Wheeled up to 150 ft.	Wheeled up to 250 ft.
Laborer	$14.45	$3.61	$4.83	$6.43	$14.45	$1.45	$1.88	$2.46
.125 Labor foreman	2.06	.51	.69	.92	2.06	.21	.27	.35
Concrete cart	6.50	1.63	2.17	2.89	10.40	1.04	1.35	1.77
Total Cost/C.Y.		$5.75	$7.69	$10.24		$2.70	$3.50	$4.58
Hourly production		4 C.Y.				10 C.Y.		

Concrete | R034 | Precast Concrete

R03450-010 Precast Concrete Wall Panels

Panels are either solid or insulated with plain, colored or textured finishes. Transportation is an important cost factor. Prices below are based on delivery within 50 miles of a plant including fabricators' overhead and profit. Engineering data is available from fabricators to assist with construction details. Usual minimum job size for economical use of panels is about 5000 S.F. Small jobs can double the prices below. For large, highly repetitive jobs, deduct up to 15% from the prices below.

Panel Cost and Maximum Size Base price for panels based on 50 S.F. or larger panels is as follows:

Thickness	Cost per S.F.	Maximum Size	Thickness	Cost per S.F.	Maximum Size
3"	$ 7.90	50 S.F.	6"	$12.00	300 S.F.
4"	8.70	150 S.F.	7"	12.95	300 S.F.
5"	10.45	200 S.F.	8"	13.45	300 S.F.

The above prices are for gray smooth form or board finish one side. Add $1.60 per S.F. for white facing; $8.50 per C.F. for solid white for full thickness. For fluted surface (not broken) add $.50 to $1.05 per S.F. For broken rib finish add $.80 to $1.60 to fluted surface. Add $.35 to $1.35 per S.F. for exposed local aggregate and $1.00 to $4.00 per S.F. for special facing aggregates.

Sandblasting runs $.55 to $1.00 per S.F., and bush hammering runs $1.75 to $3.45 per S.F. Loose hardware (not included above) runs $1.00 to $2.00 per S.F.

2" thick panels cost about the same as 3" thick panels and maximum panel size is less. For building panels faced with granite, marble or stone, add the material prices from Division 04400 to the plain panel price above. There is a growing trend toward aggregate facings and broken rib finish rather than plain gray concrete panels.

Composite Panels Add to above panel prices for core insulation, mesh and shear ties per S.F.

Type	1" Thick	1-1/2" Thick	2" Thick
Fiberglass	$1.65	$2.30	$2.95
Polystyrene (E.P.S. Board)	1.10	1.55	2.00

Erection Table shows cost ranges for erection including Subcontractor's O & P using a six man crew, crane, operator & oiler.

Panel Size L x H	Area per Panel	Low Rise Hyd. 55 Ton Crane @ $3626 Daily Cost				High Rise		55 Ton Crane $3,626 Daily Cost		90 Ton Crane $3385 Daily Cost		150 Ton Crane $3983 Daily Cost	
		Daily Production Range		Erection Cost		Daily Production Range							
		Pieces	Area	Per Piece	Per S.F.	Pieces	Area	Piece Cost	S.F. Cost	Piece Cost	S.F. Cost	Piece Cost	S.F. Cost
4' x 4'	16 S.F.	10	160 S.F.	$339	$21.15	9	144 S.F.	$376	$23.51	$403	$25.18	$443	$27.66
		20	320	169	10.60	18	288	188	11.75	201	12.59	221	13.83
4' x 8'	32	10	320	339	10.60	9	288	376	11.75	403	12.59	443	13.83
		19	608	178	5.55	17	544	199	6.22	213	6.67	234	7.32
8' x 8'	64	9	576	376	5.90	8	512	423	6.61	453	7.08	498	7.78
		18	1152	188	2.95	16	1024	212	3.31	227	3.54	249	3.89
10' x 10'	100	8	800	423	4.25	7	700	484	4.84	518	5.18	569	5.69
		15	1500	226	2.25	14	1400	242	2.42	259	2.59	285	2.85
15' x 10'	150	7	1050	484	3.20	6	900	—	—	604	4.03	664	4.43
		12	1800	282	1.90	11	1650	—	—	330	2.20	362	2.41
20 x 10'	200	6	1200	564	2.80	5	1000	—	—	725	3.63	797	3.98
		8	1600	423	2.10	7	1400	—	—	518	2.59	569	2.85
30' x 10'	300	5	1500	677	2.25	4	1200	—	—	907	3.02	996	3.32
		7	2100	484	1.60	7	2100	—	—	518	1.73	569	1.90

Total Cost in Place for Low Rise Construction Including Subcontractor's O & P

Description	Gray		White Face		Exposed Aggregate	
	4' x 8' x 4"	20' x 10' x 6"	4' x 8' x 4"	20' x 10' x 6"	4' x 8' x 4"	20' x 10' x 6"
Panel, steel form, broomed finish	$ 8.70	$12.00	$10.25	$13.55	$10.70	$13.80
Caulking, grout, etc. (runs about $2.50 per L.F.)	.95	.36	.90	.36	.90	.36
Erect, plumb, align (from above)	10.60	2.80	10.60	2.80	10.60	2.80
Total in place per S.F.	$20.25	$15.16	$21.75	$16.71	$22.20	$16.96

No allowance has been made for supporting steel framework. On one story buildings, panels may rest on grade beams and require only wind bracing and fasteners. On multi-story buildings panels can span from column to column and floor to floor. Plastic designed steel framed structures may have large deflections which slow down erection and raise costs.

Large panels are more economical than small panels on a S.F. basis. When figuring areas include all protrusions, returns, etc. Overhangs can triple erection costs. Panels over 45' have been produced. Larger flat units should be prestressed. Vacuum lifting of smooth finish panels eliminates inserts and can speed erection.

Concrete | R034 | Precast Concrete

R03470-020 Tilt Up Concrete Panels

The advantage of tilt up construction is in the low cost of forms and the placing of concrete and reinforcing. Panels up to 75' high and 5-1/2" thick have been tilted using strongbacks. Tilt up has been used for one to five story buildings and is well suited for warehouses, stores, offices, schools and residences.

The panels are cast in forms on the floor slab. Most jobs use 5-1/2" thick solid reinforced concrete panels. Sandwich panels with a layer of insulating materials are also used. Where dampness is a factor, lightweight aggregate is used at an added cost of $.50 per square foot. Optimum panel size is 300 to 500 S.F.

Slabs are usually poured with 3000 psi concrete which permits tilting seven days after pouring. Slabs may be stacked on top of each other and are separated from each other by either two coats of bond breaker or a film of polyethylene. Use of high early strength cement allows tilting two days after a pour. Tilting up is done with a roller outrigger crane with a capacity of at least 1-1/2 times the weight of the panel at the required reach. Exterior precast columns can be set at the same time as the panels; interior precast columns can be set first and the panels clipped directly to them. The use of cast-in-place concrete columns is diminishing due to shrinkage problems. Structural steel columns are sometimes used if crane rails are planned. Panels can be clipped to the columns or lowered between the flanges. Steel channels with anchors may be used as edge forms for the slab. When the panels are lifted the channels form an integral steel column to take structural loads. Roof loads can be carried directly by the panels for wall heights to 14'. For soft ground requiring mats for cranes, add 100% to costs below.

Below are typical costs per S.F., for panels of 300 to 500 S.F. 20' high, not including contractor's overhead and profit.

Item	5-1/2" Thick			7-1/2" Thick		
	Material	Installation	Total	Material	Installation	Total
Prepare pouring surface	$ .02	$ .05	$ .07	$ .02	$ .05	$ .07
Erect, strip side forms	.08	.24	.32	.14	.33	.47
Place concrete, 3000 psi	1.06	.09	1.15	1.44	.33	1.77
Steel trowel finish & curing	$ .02	.30	.32	.02	.30	.32
Reinforcing, inserts & misc. items	1.02	.42	1.44	1.41	.56	1.97
Panel erection and aligning	—	.68	.68	—	.68	.68
Total per S.F. of Wall	$ 2.20	$ 1.78	$ 3.98	$ 3.03	$ 2.25	$ 5.28
Site precast concrete columns, add	1.10	.98	2.08	1.06	.98	2.04
Total per S.F. of Wall	$ 3.30	$ 2.76	$ 6.06	$ 4.09	$ 3.23	$ 7.32
Panels only, per C.Y.	$118.80	$96.12	$214.92	$130.90	$97.20	$228.10

Requirements of local building codes may be a limiting factor and should be checked. Building floor slabs should be poured first and should be a minimum of 5" thick with 100% compaction of soil or 6" thick with less than 100% compaction.

Setting times as fast as nine minutes per panel have been observed, but a safer expectation would be four panels per hour with a crane and a four person setting crew. If crane erects from inside building, some provision must be made to get crane out after walls are erected. Good yarding procedure is important to minimize delays. Equalizing three point lifting beams and self-releasing pick-up hooks speed erection. If panels must be carried to their final location, setting time per panel will be increased and erection costs may fall in the erection cost range of architectural precast wall panels. Placing panels into slots formed in continuous footers will speed erection.

Reinforcing should be with #5 bars with vertical bars on the bottom. If surface is to be sandblasted, stainless steel chairs should be used to prevent rust staining.

Use of a broom finish is popular since the unavoidable surface blemishes are concealed. Many West Coast jobs have used exposed aggregate panels which cost an additional $.50 to $3.60 per S.F. depending on the aggregate and construction method employed. Eastern prices for exposed aggregate tend to be considerably higher. Panel connections run $1.60 to $6.00 per L.F. Precast columns run from three to five times the C.Y. price of the panels only.

Concrete — R035 Cementitious Decks & Toppings

R03520-010 Lightweight Concrete

Vermiculite or Perlite come in bags of 4 C.F. under various trade names. Weight is about 8 lbs. per C.F. For insulating roof fill use 1:6 mix. For structural deck use 1:4 mix over gypsum boards, steeltex, steel centering, etc. supported by closely spaced joists or bulb trees. For structural slabs use 1:3:2 vermiculite sand concrete over steeltex, metal lath, steel centering, etc. on joists spaced 2'-0" O.C. for maximum L.L. of 80#/S.F. Use same mix for slab base fill over steel flooring or regular reinforced concrete slab when tile, terrazzo or other finish is to be laid over.

For slabs on grade use 1:3:2 mix when tile, etc. finish is to be laid over. If radiant heating units are installed use a 1:6 mix for a base. After coils are in place, cover with a regular granolithic finish (mix 1:3:2) to a minimum depth of 1-1/2" over top of units.

Reinforce all slabs with 6 x 6 or 10 x 10 welded wire mesh.

Vermiculite concrete can be purchased ready mixed but the following breakdown is included for field mix. Prices given below are for a one story building and assume 50 C.Y. or more. For less than 50 C.Y. add 10%. For over one story add $3.50 per C.Y. Screed finish cost is included below. Ready mix 1:6 costs $92.05 per C.Y. delivered in truckload lots.

See R3310-070 for prices of ready mix lightweight concrete.

Quantities per C.Y., Field Mix			1:6 Mix Insulating Roof Fill		1:3:2 Mix Lightweight Structural Concrete	
Portland cement @	$ 7.10	per bag	5.0 bags	$ 35.50	6.2 bags	$ 44.02
Vermiculite or Perlite @	9.85	per bag	7.5 bags	73.88		
treated type @	11.50	per bag			4.7 bags	54.05
Sand @	12.75	per C.Y.			12.5 C.F.	5.90
Plant and Water				6.46		6.46
Labor, machine mix, hoist and place, Crew C-8 @	$28.50	per L.H.	1.12 L.H.	31.92	.91 L.H.	25.94
Total in place, per C.Y.				$147.76		$136.37

Masonry | R040 | Mortar & Masonry Accessories

R04060-100 Cement Mortar (material only)

Type N - 1:1:6 mix by volume. Use everywhere above grade except as noted below.
- 1:3 mix using conventional masonry cement which saves handling two separate bagged materials.

Type M - 1:1/4:3 mix by volume, or 1 part cement, 1/4 (10% by wt.) lime, 3 parts sand. Use for heavy loads and where earthquakes or hurricanes may occur. Also for reinforced brick, sewers, manholes and everywhere below grade.

Cost and Mix Proportions of Various Types of Mortar

Components	Type Mortar and Mix Proportions by Volume										
	M		S		N		O		K	PM	PL
	1:1:6	1:1/4:3	1/2:1:4	1:1/2:4	1:3	1:1:6	1:3	1:2:9	1:3:12	1:1:6	1:1/2:4
Portland cement @ $7.10 per bag	$ 7.10	$ 7.10	$ 3.55	$ 7.10	—	$ 7.10	—	$ 7.10	$ 7.10	$ 7.10	$ 7.10
Masonry cement @ $5.90 per bag	5.90	—	5.90	—	5.90	—	$5.90	—	—	$ 5.90	—
Lime @ $5.60 per 50 lb. bag	—	1.40	—	2.80	—	5.60	—	11.20	16.80	—	2.80
Masonry sand @ $17.50 per C.Y.*	3.89	1.94	2.59	2.59	1.94	3.89	1.94	5.83	7.78	3.89	2.59
Mixing machine incl. fuel**	1.72	.86	1.15	1.15	.86	1.72	.86	2.58	3.45	1.72	1.15
Total for Materials	$18.61	$11.30	$13.19	$13.64	$8.70	$18.31	$8.70	$26.71	$35.13	$18.61	$13.64
Total C.F.	6	3	4	4	3	6	3	9	12	6	4
Approximate Cost per C.F.	$ 3.10	$ 3.77	$ 3.30	$ 3.41	$2.90	$ 3.05	$2.90	$ 2.97	$ 2.93	$ 3.10	$ 3.41

*Includes 10 mile haul
**Based on a daily rental, 10 C.F., 25 H.P. mixer, mix 200 C.F./Day

Mix Proportions by Volume, Compressive Strength and Cost of Mortar

Where Used	Mortar Type	Allowable Proportions by Volume				Compressive Strength @ 28 days	Cost per Cubic Foot
		Portland Cement	Masonry Cement	Hydrated Lime	Masonry Sand		
Plain Masonry	M	1	1	—	6		$3.10
		1	—	1/4	3	2500 psi	3.77
	S	1/2	1	—	4		3.30
		1	—	1/4 to 1/2	4	1800 psi	3.41
	N	—	1	—	3		2.90
		1	—	1/2 to 1-1/4	6	750 psi	3.05
	O	—	1	—	3		2.90
		1	—	1-1/4 to 2-1/2	9	350 psi	2.97
	K	1	—	2-1/2 to 4	12	75 psi	2.93
Reinforced Masonry	PM	1	1	—	6	2500 psi	3.10
	PL	1	—	1/4 to 1/2	4	2500 psi	3.41

Note: The total aggregate should be between 2.25 to 3 times the sum of the cement and lime used.

The labor cost to mix the mortar is included in the labor cost on brickwork.

Machine mixing is usually specified on jobs of any size. There is a large price saving over hand mixing and mortar is more uniform.

There are two types of mortar color used. Prices in Section 04060 are for the inert additive type with about 100 lbs. per M brick as the typical quantity required. These colors are also available in smaller batch size bags (1 lb. to 15 lb.) which can be placed directly into the mixer without measuring. The other type is premixed and replaces the masonry cement with ranges in price from $5 to $15 per 70 lb. bag. Dark green color has the highest cost.

R04060-200 Miscellaneous Mortar (material only)

Quantities	Glass Block Mortar		Gypsum Cement Mortar	
White Portland cement at $17.75 per bag	7 bags	$124.25		
Gypsum cement at $11.40 per 80 lb. bag			11.25 bags	$128.25
Lime at $5.60 per 50 lb. bag	280 lbs.	31.36		
Sand at $17.50 per C.Y.*	1 C.Y.	17.50	1 C.Y.	17.50
Mixing machine and fuel		7.75		7.75
Total per C.Y.		$180.86		$153.50
Approximate Total per C.F.		$ 6.70		$ 5.69

* Includes 10 mile haul

Masonry | R040 | Mortar & Masonry Accessories

R04080-500 Masonry Reinforcing

Horizontal joint reinforcing helps prevent wall cracks where wall movement may occur and in many locations is required by code. Horizontal joint reinforcing is generally not considered to be structural reinforcing and an unreinforced wall may still contain joint reinforcing.

Reinforcing strips come in 10' and 12' lengths and in truss and ladder shapes, with and without drips. Field labor runs between 2.7 to 5.3 hours per 1000 L.F. for wall thicknesses up to 12".

The wire meets ASTM A82 for cold drawn steel wire and the typical size is 9 ga. sides and ties with 3/16" diameter also available. Typical finish is mill galvanized with zinc coating at .10 oz. per S.F. Class I (.40 oz. per S.F.) and Class III (.80 oz per S.F.) are also available, as is hot dipped galvanizing at 1.50 oz. per S.F.

Masonry | R042 | Unit Masonry

R04210-050 Brick Chimneys

Quantities	16" x 16"		20" x 20"		20" x 24"		20" x 32"	
Brick at $300 per M	28 brick	$ 8.40	37 brick	$11.10	42 brick	$12.60	51 brick	$15.30
Type M mortar at $3.70 per C.F.	.5 C.F.	1.89	.6 C.F.	2.26	1.0 C.F.	3.77	1.3 C.F.	4.90
Flue tile (square)(per foot)	8" x 8"	3.00	12" x 12"	5.70	2 @ 8" x 12"	7.54	2 @ 12" x 12"	11.40
Install tile & brick, crew D-1	.055 day	15.71	.073 day	20.85	.083 day	23.70	.10 day	28.56
Total per L.F. high		$29.00		$39.91		$47.61		$60.16

Material costs include 3% waste for brick and 25% waste for mortar.

Labor costs are bare costs and do not include contractor's O&P.

Labor for chimney brick using D-1 crew is 31 hours per thousand brick or about $530 per thousand brick. An 8" x 12" flue takes 33 brick and two 8" x 8" flues take 37 brick.

R04210-100 Economy in Bricklaying

Have adequate supervision. Be sure bricklayers are always supplied with materials so there is no waiting. Place best bricklayers at corners and openings.

Use only screened sand for mortar. Otherwise, labor time will be wasted picking out pebbles. Use seamless metal tubs for mortar as they do not leak or catch the trowel. Locate stack and mortar for easy wheeling.

Have brick delivered for stacking. This makes for faster handling, reduces chipping and breakage, and requires less storage space. Many dealers will deliver select common in 2' x 3' x 4' pallets or face brick packaged. This affords quick handling with a crane or forklift and easy tonging in units of ten, which reduces waste.

Use wider bricks for one wythe wall construction. Keep scaffolding away from wall to allow mortar to fall clear and not stain wall.

On large jobs develop specialized crews for each type of masonry unit.

Consider designing for prefabricated panel construction on high rise projects.

Avoid excessive corners or openings. Each opening adds about 50% to labor cost for area of opening.

Bolting stone panels and using window frames as stops reduces labor costs and speeds up erection.

Masonry | R042 | Unit Masonry

R04210-120 Common and Face Brick Prices

Prices are based on truckload lot purchases for Common Brick and carload lots for Face Brick. Prices are per M, (thousand), brick.

	Material			Installation				Total			
	Brick per M Delivered		Mortar	Common in 8" Wall		Face Brick, 4" Veneer		Common in 8" Wall		Face Brick, 4" Veneer	
City	Face	3/8" Joint	Bare Costs	Incl. O & P	Bare Costs	Incl. O & P	Bare Costs	Incl. O & P	Bare Costs	Incl. O & P	
Atlanta	$190	$250	$38.10	$278	$ 471	$334	$ 565	$512	$ 728	$ 623	$ 883
Baltimore	235	280	for 8" Wall	320	542	384	651	600	850	704	1,002
Boston	305	480	and	573	970	687	1,164	925	1,357	1,213	1,742
Buffalo	260	350	$31.40	491	832	590	998	797	1,168	981	1,429
Chicago	260	380	for 4" Wall	520	881	624	1,057	826	1,218	1,047	1,522
Cincinnati	230	330		384	650	461	781	659	953	832	1,189
Cleveland	235	325		464	785	556	942	744	1,093	923	1,345
Columbus	235	350		379	641	454	769	659	949	846	1,200
Dallas	200	275		273	463	328	555	517	731	643	901
Denver	210	310		318	538	381	645	572	818	732	1,031
Detroit	235	275		487	825	585	990	768	1,133	900	1,336
Houston	250	275		284	481	341	578	580	806	656	924
Indianapolis	255	265		388	657	466	788	689	988	770	1,123
Kansas City	260	300		416	704	499	845	722	1,041	840	1,220
Los Angeles	250	325		483	818	580	982	779	1,144	946	1,385
Memphis	190	235		273	463	328	555	507	720	601	856
Milwaukee	280	365		458	776	550	931	785	1,135	957	1,379
Minneapolis	255	400		473	801	568	961	774	1,132	1,011	1,449
Nashville	185	245		267	452	321	543	496	704	604	855
New Orleans	210	380		236	400	283	480	491	680	706	945
New York City	300	355		649	1,098	778	1,318	996	1,480	1,176	1,755
Philadelphia	245	345		516	873	619	1,048	806	1,193	1,006	1,473
Phoenix	340	350		296	501	355	601	684	928	747	1,032
Pittsburgh	210	275		420	711	504	854	675	991	819	1,200
St. Louis	235	265		249	422	299	506	529	730	603	841
San Antonio	260	325		429	726	514	871	735	1,062	880	1,274
San Diego	285	360		562	951	674	1,141	893	1,316	1,076	1,584
San Francisco	350	435		470	795	563	954	868	1,233	1,043	1,481
Seattle	340	390		473	801	568	961	861	1,228	1,001	1,438
Washington, D.C.	210	240		330	558	396	670	584	838	674	976
Average	$235	$305		$410	$ 690	$490	$ 823	$700	$1,010	$ 850	$1,226

Common building brick manufactured according to ASTM C62 and facing brick manufactured according to ASTM C216 are the two standard bricks available for general building use. Building brick is made in three grades; SW, where high resistance to damage caused by cyclic freezing is required; MW, where moderate resistance to cyclic freezing is needed; and NW, where little resistance to cyclic freezing is needed. Facing brick is made in only the two grades SW and MW. Additionally, facing brick is available in three types; FBS, for general use; FBX, for general use where a higher degree of precision and lower permissible variation in size than FBS is needed; and FBA, for general use to produce characteristic architectural effects resulting from non-uniformity in size and texture of the units.

In figuring above installation costs, a D-8 Crew (with a daily output of 1.5 M) was used for the 4" veneer. A D-8 Crew (with a daily output of 1.8 M) was used for the 8" solid wall.

In figuring the total cost including overhead and profit, an allowance of 10% was added to the sum of the cost of the brick and mortar. Also, 3% breakage was included for both the bare costs and the costs with overhead and profit. If bricks are delivered palletized with 280 to 300 per pallet, or packaged, allow only 1-1/2% for breakage. Then add $10 per M to the cost of brick and deduct two hours helper time. The net result is a savings of $30 to $40 per M in place. Packaged or palletized delivery is practical when a job is big enough to have a crane or other equipment available to handle a package of brick. This is so on all industrial work but not always true on small commercial buildings.

There are many types of red face brick. The prices above are for the most usual type used in commercial, apartment house or industrial construction. If it is possible to obtain the price of the actual brick to be used, it should be done and substituted in the table. The use of buff and gray face is increasing, and there is a continuing trend to the Norman, Roman, Jumbo and SCR brick.

See R04210-500 for brick quantities per S.F. and mortar quantities per M brick. (Average prices for the various sizes are listed in Division 4)

Common red clay brick for backup is not used that often. Concrete block is the most usual backup material with occasional use of sand lime or cement brick. Sand lime cost about $15 per M less than red clay and cement brick are about $5 per M less than red clay. These figures may be substituted in the common brick breakdown for the cost of these items in place, as labor is about the same. Occasionally common brick is being used in solid walls for strength and as a fire stop.

Brick panels built on the ground floor and then crane erected to the upper floors have proven to be economical. This allows the work to be done under cover and without scaffolding.

Masonry | R042 | Unit Masonry

R04210-180 Brick in Place

Table below is for common bond with 3/8" concave joints and includes 3% waste for brick and 25% waste for mortar. Crew costs are bare costs.

Item	8" Common Brick Wall 8" x 2-2/3" x 4"		Select Common Face 8" x 2-2/3" x 4"		Red Face Brick 8" x 2-2/3" x 4"	
1030 brick delivered	$250 per M	$257.50	$300 per M	$309.00	$325 per M	$334.75
Type N mortar @ $2.99 per C.F.	12.5 C.F.	38.13	10.3 C.F.	31.42	10.3 C.F.	31.42
Installation using indicated crew	Crew D-8 @ .556 Days	406.55	Crew D-8 @ .667 Days	487.71	Crew D-8 @ .667 Days	487.71
Total per M in place		$702.18		$828.13		$853.88
Total per S.F. of wall	13.5 bricks/S.F.	$ 9.48	6.75 bricks/S.F.	$ 5.59	6.75 bricks/S.F.	$ 5.76

R04210-185 Reinforced Brick in Walls

Table below is for common bond with 3/8" concave joints and includes 3% waste. Standard 8" x 2-2/3" x 4" bricks.

Item	8" to 9" Thick Wall		16" to 17" Thick Wall	
1030 brick (select common)	$305 per M	$309.00	$305 per M	$309.00
Type PL mortar @ $3.41 per C.F.	12.5 C.F.	42.63	13.9 C.F.	47.40
Reinforcing bars delivered	50 lb. @ .27 per lb.	13.00	50 lb. @ .27 per lb.	13.00
Installation incl. reinforcing using indicated crew	Crew D-8 @ .517 Days	378.03	Crew D-8 @ .513 Days	375.11
Total per M in place		$742.66		$744.50
Total per S.F. of wall	8" wall, 13.5 brick/S.F.	$ 10.03	16" wall, 27.0 brick/S.F.	$ 20.10

Masonry — R042 Unit Masonry

R04210-500 Brick, Block & Mortar Quantities

Type Brick	Nominal Size (incl. mortar) L x H x W	Modular Coursing	Number of Brick per S.F.	C.F. of Mortar per M Bricks, Waste Included 3/8" Joint	C.F. of Mortar per M Bricks, Waste Included 1/2" Joint
Standard	8 x 2-2/3 x 4	3C=8"	6.75	10.3	12.9
Economy	8 x 4 x 4	1C=4"	4.50	11.4	14.6
Engineer	8 x 3-1/5 x 4	5C=16"	5.63	10.6	13.6
Fire	9 x 2-1/2 x 4-1/2	2C=5"	6.40	550 # Fireclay	—
Jumbo	12 x 4 x 6 or 8	1C=4"	3.00	23.8	30.8
Norman	12 x 2-2/3 x 4	3C=8"	4.50	14.0	17.9
Norwegian	12 x 3-1/5 x 4	5C=16"	3.75	14.6	18.6
Roman	12 x 2 x 4	2C=4"	6.00	13.4	17.0
SCR	12 x 2-2/3 x 6	3C=8"	4.50	21.8	28.0
Utility	12 x 4 x 4	1C=4"	3.00	15.4	19.6

For Other Bonds Standard Size — Add to S.F. Quantities in Table to Left

Bond Type	Description	Factor
Common	full header every fifth course	+20%
	full header every sixth course	+16.7%
English	full header every second course	+50%
Flemish	alternate headers every course	+33.3%
	every sixth course	+5.6%
Header = W x H exposed		+100%
Rowlock = H x W exposed		+100%
Rowlock stretcher = L x W exposed		+33.3%
Soldier = H x L exposed		—
Sailor = W x L exposed		-33.3%

Concrete Blocks Nominal Size	Approximate Weight per S.F. Standard	Approximate Weight per S.F. Lightweight	Blocks per 100 S.F.	Mortar per M block Partitions	Mortar per M block Back up
2" x 8" x 16"	20 PSF	15 PSF	113	16 C.F.	36 C.F.
4"	30	20		31	51
6"	42	30		46	66
8"	55	38		62	82
10"	70	47		77	97
12"	85	55		92	112

R04210-550 Brick Veneer in Place

Table below is for running bond with 3/8" concave joints and includes 3% waste for brick and 25% waste for mortar.

Item	Buff Face Brick 8" x 2-2/3" x 4"		Red Norman Face Brick 12" x 2-2/3" x 4"		Buff Roman Face Brick 12" x 2" x 4"	
1030 brick delivered	$390 per M	$401.70	$675 per M	$ 695.25	$700 per M	$ 721.00
Type N mortar @ $3.05 per C.F.	10.3 C.F.	31.42	14.0 C.F.	42.70	13.4 C.F.	40.87
Installation using indicated crew	Crew D-8 @ .667 days	487.71	Crew D-8 @ .690 days	504.53	Crew D-8 @ .667 days	487.71
Total per M in place		$920.83		$1,242.48		$1,249.58
Total per S.F. of wall	6.75 brick/S.F.	$ 6.22	4.50 brick/S.F.	$ 5.59	6.00 brick/S.F.	$ 7.50

Masonry — R042 Unit Masonry

R04220-200 Concrete Block

8" x 16" block, sand aggregate blocks with 3/8" joints for partitions.

City	Material Per Block, Delivered 8" Thick	Material Per Block, Delivered 4" Thick	113 Block, Delivered 8" Thick	113 Block, Delivered 4" Thick	City	Material Per Block, Delivered 8" Thick	Material Per Block, Delivered 4" Thick	113 Block, Delivered 8" Thick	113 Block, Delivered 4" Thick
Atlanta	$.74	$1.00	$83.62	$113.00	Memphis	.68	.94	$76.84	$106.22
Baltimore	.65	.85	73.45	96.05	Milwaukee	.76	1.10	85.88	124.30
Boston	.60	.85	67.80	96.05	Minneapolis	.75	1.01	84.75	114.13
Buffalo	.70	1.15	79.10	129.95	Nashville	.77	1.02	87.01	115.26
Chicago	.60	.92	67.80	103.96	New Orleans	.82	1.15	92.66	129.95
Cincinnati	.60	.85	67.80	96.05	New York City	.77	1.03	87.01	116.39
Cleveland	.66	.93	74.58	105.09	Philadelphia	.70	.80	79.10	90.40
Columbus	.69	.83	77.97	93.79	Phoenix	.62	.86	70.06	97.18
Dallas	.61	1.02	68.93	115.26	Pittsburgh	.72	.92	81.36	103.96
Denver	.73	1.05	82.49	118.65	St. Louis	.76	1.10	85.88	124.30
Detroit	.77	1.05	87.01	118.65	San Antonio	.75	.96	84.75	108.48
Houston	.75	.97	84.75	109.61	San Diego	.63	.86	71.19	97.18
Indianapolis	.70	1.00	79.10	113.00	San Francisco	.88	1.45	99.44	163.85
Kansas City	.72	1.20	81.36	135.60	Seattle	.88	1.27	99.44	143.51
Los Angeles	.81	.95	91.53	107.35	Washington D.C.	.66	1.07	74.58	120.91
					Average	$.72	$1.01	$80.91	$113.60

The cost of sand aggregate block is based on the delivered price of 113 blocks, which is the equivalent of a 100 S.F. wall area. Mortar for a 100 S.F. wall area will average $10.50 for 4" thick block and $21.10 for 8" thick block.

Cost for 100 S.F. of 8" x 16" Concrete Block Partitions to Four Stories High, Tooled Joint One Side

8" x 16" Sand Aggregate	4" Thick Block		8" Thick Block		12" Thick Block	
113 block delivered	$.72 Ea.	$ 79.10	$1.01 Ea.	$113.00	$1.60 Ea.	$175.15
Type N mortar at $3.05 per C.F.	3.5 C.F.	10.36	7.0 C.F.	20.72	10.4 C.F.	30.78
Installation Crew (Bare Costs)	D-2 @ 9.32 L.H.	164.22	D-2 @ 10.68 L.H.	188.18	D-6 @ 14.70 L.H.	303.41
Total per 100 S.F.		$253.68		$321.90		$509.34
Add for filling cores solid	6.7 C.F.	$ 85.58	25.8 C.F.	$183.19	42.2 C.F.	$251.22

Cost for 100 S.F. of 8" x 16" Concrete Block Backup with Trowel Cut Joints

8" x 16" Sand Aggregate	4" Thick Block		8" Thick Block		12" Thick Block	
113 block delivered	$.72 Ea.	$ 79.10	$1.01 Ea.	$113.00	$1.60 Ea.	$175.15
Type N mortar at $3.05 per C.F.	5.8 C.F.	17.17	9.3 C.F.	27.53	12.7 C.F.	37.59
Installation Crew (Bare Costs)	D-2 @ 9.20 L.H.	162.10	D-2 @ 10.10 L.H.	177.96	D-6 @ 13.20 L.H.	272.45
Total per 100 S.F.		$258.37		$318.49		$485.19

Special block: corner, jamb and head block are same price as ordinary block of same size. Tabulated on the next page are national average prices per block. Labor on specials is about the same as equal sized regular block. Bond beam and 16" high lintel blocks cost 30% more than regular units of equal size. Lintel blocks are 8" long and 8" or 16" high. Costs in individual cities may be factored from the table above.

Use of motorized mortar spreader box will speed construction of continuous walls.

Masonry | R042 | Unit Masonry

R04270-700 Glass Block

Nominal Size (Including Mortar)	Cost of Blocks Each, Zone 1 Contractor Prices							
	Truckload or Carload				Less Than Truckload or Carload			
	Regular	Thinline	Essex	Solar Reflective	Regular	Thinline	Essex	Solar Reflective
6" x 6"	$ 3.25	$ 2.10	—	$ 7.00	$ 3.50	$ 2.35	—	$ 8.00
8" x 8"	4.10	2.35	4.70	$ 9.00	4.25	2.75	4.85	10.00
Solid 8" x 8"	22.00	—	—	—	24.00	—	—	—
4" x 8"	2.75	2.50	—	—	3.00	3.00	—	—
12" x 12"	9.25	—	—	—	11.00	—	—	—

Size	Per 100 S.F.		Per 1000 Block					
	No. of Block	Mortar " Joint	Asphalt Emulsion	Caulk	Expansion Joint	Panel Anchors	Wall Mesh	
6" x 6"	410 ea.	5.0 C.F.	.17 gal.	1.5 gal.	80 L.F.	20 ea.	500 L.F.	
8" x 8"	230	3.6	.33	2.8	140	36	670	
12" x 12"	102	2.3	.67	6.0	312	80	1000	
Approximate quantity per 100 S.F.			.07 gal.	0.6 gal.	32 L.F.	9 ea.	51,	68, 102 L.F.

Accessories required for installation — Wall ties: run galvanized double steel mesh full length of joint at $.32 per L.F.— fiberglass expansion joints at sides and top at $.40 per L.F.— silicone caulking one gallon at $50.00 does 95 L.F. asphalt emulsion one gallon does 600 L.F. at $4.50 per gallon in one gallon lots. If blocks are not set in wall chase use 2'-0" long wall anchors at $1.50 each at 2'-0" O.C.

Quantities	Cost per 100 S.F. (Truckload or Carload Price)					
	6" x 6" Block		8" x 8" Block		12" x 12" Block	
Block, delivered	410 @ $3.25	$1,332.50	230 @ $4.10	$ 943.00	102 @ $9.25	$ 943.50
Mortar @ $6.30 per C.F.	5.0 C.F.	31.50	3.6 C.F.	22.68	2.3 C.F.	14.49
Ties, expansion joints, etc.		92.37		92.82		102.00
Crew D-8 @	.690 days	504.53	.465 days	340.01	.417 days	304.91
Total per 100 S.F.		$1,960.90		$1,398.51		$1,364.90

Masonry | R044 | Stone

R04430-500 Ashlar Veneer

Quantities			Low Price Stone		High Price Stone	
4" Stone, random ashlar,	2.5	tons average	$275 per ton	$ 687.50	$375 per ton	$ 937.50
Type N mortar @	$3.05	C.F.	20 C.F.	61.00	20 C.F.	61.00
50 - 1" x 1/4" x 6" galv. stone anchors @	$1.10	each		55.00		55.00
Using crew D-8 @	$731.20	per day	.714 days	522.00	.833 days	609.09
Total per 100 S.F.				$1,325.50		$1,662.59

Stone coverage varies from 30 S.F. to 50 S.F. per ton. Mortar quantities include 1" backing bed.

Masonry | R049 | Restoration & Cleaning

R04930-100 Cleaning Face Brick

On smooth brick a person can clean 70 S.F. an hour; on rough brick 50 S.F. per hour. Use one gallon muriatic acid to 20 gallons of water for 1000 S.F. Do not use acid solution until wall is at least seven days old, but a mild soap solution may be used after two days. Commercial cleaners cost from $9 to $12 per gallon.

Time has been allowed for clean-up in brick prices.

Metals | R051 | Structural Metal Framing

R05120-220 Steel Estimating Quantities

One estimate on erection is that a crane can handle 35 to 60 pieces per day. Say average is 45. With usual sizes of beams, girders, and columns, this would amount to about 20 tons. Allowing $1,187 per day rental for the crane including fuel and operator, this would amount to about $59 per ton for crane charge. The type of connection greatly affects the speed of erection. Moment connections for continuous design slow down production and increase erection costs.

Short open web bar joists can be set at the rate of 75 to 80 per day, with 50 per day being the average for setting long span joists.

After main members are calculated, add the following for usual allowances: base plates 2% to 3%; column splices 4% to 5%; and miscellaneous details 4% to 5%, for a total of 10% to 13% in addition to main members.

Ratio of column to beam tonnage varies depending on type of steels used, typical spans, story heights and live loads.

It is more economical to keep the column size constant and to vary the strength of the column by using high strength steels. This also saves floor space. Buildings have recently gone as high as ten stories with 8" high strength columns. For light columns under W8X31 lb. sections, concrete filled steel columns are economical.

High strength steels may be used in columns and beams to save floor space and to meet head room requirements. High strength steels in some sizes sometimes require long lead times.

Round, square and rectangular columns, both plain and concrete filled, are readily available and save floor area, but are higher in cost per pound than rolled columns. For high unbraced columns, tube columns may be less expensive.

Below are average minimum figures for the weights of the structural steel frame for different types of buildings using A36 steel, rolled shapes and simple joints. For economy in domes, rise to span ratio = .13. Open web joist framing systems will reduce weights by 10% to 40%. Composite design can reduce steel weight by up to 25% but additional concrete floor slab thickness may be required. Continuous design can reduce the weights up to 20%. There are many building codes with different live load requirements and different structural requirements, such as hurricane and earthquake loadings which can alter the figures.

*See R13128-310 for Domes and Thin Shell Structures.

Structural Steel Weights per S.F. of Floor Area

Type of Building	No. of Stories	Avg. Spans	L.L. #/S.F.	Lbs. Per S.F.	Type of Building	No. of Stories	Avg. Spans	L.L. #/S.F.	Lbs. Per S.F.
Steel Frame Mfg.	1	20'x20' 30'x30' 40'x40'	40	8 13 18	Apartments	2-8 9-25	20'x20'	40	8 14
Parking garage	4	Various	80	8.5	Office	to 10 20 30 over 50	Various	80	10 18 26 35
Domes (Schwedler)*	1	200' 300'	30	10 15					

R05120-250 Subpurlins

Bulb tee and truss tee subpurlins are structural members designed to support and reinforce a variety of roof deck systems such as precast cement fiber roof deck tiles, monolithic roof deck systems, and gypsum or lightweight concrete over formboard. Other uses include interstitial service ceiling systems, wall panel systems, and joist anchoring in bond beams.

The table below shows bare costs for material only for subpurlins spaced at 32-5/8" O.C., purchased in lots of 6,000 to 10,000 L.F. (see Division 05120-140 for pricing on a square foot basis). Maximum span is based on a 3-span condition with a total allowable vertical load of 55 psf.

	Bulb Tees, Painted			Truss Tees, Painted							
Type	Wt. per L.F.	Cost per L.F.	Max. Span	Size	Wt. per L.F.	Cost per L.F.	Max. Span	Size	Wt. per L.F.	Cost per L.F.	Max. Span
112	1.48 #	$ .85	5'-6"	2"	1.01 #	$1.32	5'-9"	3"	1.06 #	$1.37	7'-3"
158	1.68	.86	6'-4"		1.17	1.38	6'-0"				
168	1.87	.91	7'-8"		1.75	1.61	7'-6"		1.90	1.69	9'-0"
178	2.15	1.00	8'-9"	2-1/2"	1.06	1.35	6'-9"	3-1/2"	1.06	1.39	7'-9"
218	3.19	1.26	10'-2"		1.31	1.44	6'-9"		1.31	1.46	8'-6"
228	3.87	1.60	12'-0"		1.90	1.67	8'-9"		1.90	1.71	10'-9"

Wood & Plastics — R061 Rough Carpentry

R06100-010 Thirty City Lumber Prices

Prices for boards are for #2 or better or sterling, whichever is in best supply. Dimension lumber is "Standard or Better" either Southern Yellow Pine (S.Y.P.), Spruce-Pine-Fir (S.P.F.), Hem-Fir (H.F.) or Douglas Fir (D.F.). The species of lumber used in a geographic area is listed by city. Plyform is 3/4" BB oil sealed fir or S.Y.P. whichever prevails locally, 3/4" CDX is S.Y.P. or Fir.

These are prices at the time of publication and should be checked against the current market price. Relative differences between cities will stay approximately constant.

City	Species	Contractor Purchases per M.B.F. S4S Dimensions 2"x4"	2"x6"	2"x8"	2"x10"	2"x12"	4"x4"	Boards 1"x6"	1"x12"	Contractor Purchases per M.S.F. 3/4" Ext. Plyform	3/4" Thick CDX T&G
Atlanta	S.P.F.	$603	$577	$644	$819	$915	$824	$757	$1,289	$900	$804
Baltimore	S.P.F.	596	528	642	703	681	943	1000	1500	746	789
Boston	S.P.F.	600	568	630	790	830	1042	1920	1550	1032	831
Buffalo	H.F.	513	553	579	642	650	755	1280	1610	635	515
Chicago	S.P.F.	608	586	614	748	685	690	936	1543	779	774
Cincinnati	S.P.F.	599	683	691	795	882	863	1200	1446	824	904
Cleveland	S.P.F.	673	631	650	892	893	787	984	1910	1030	837
Columbus	S.P.F.	566	722	696	894	908	744	1419	1947	1186	775
Dallas	S.P.F.	771	656	609	690	731	668	728	1029	718	718
Denver	H.F.	631	631	646	766	658	958	771	1131	1196	900
Detroit	S.P.F.	550	589	614	851	886	594	1147	1859	804	889
Houston	S.Y.P.	528	536	536	592	641	802	707	855	985	703
Indianapolis	S.Y.F.	633	615	705	795	835	888	917	1177	825	840
Kansas City	D.F.	590	570	570	645	670	864	1050	1300	785	830
Los Angeles	D.F.	525	536	625	642	654	696	671	1385	690	790
Memphis	S.P.F.	698	654	693	834	770	1068	909	1314	954	1045
Milwaukee	S.P.F.	785	750	862	900	995	842	850	1237	828	859
Minneapolis	S.P.F.	559	545	536	675	754	674	1197	1409	1138	753
Nashville	S.P.F.	609	625	642	804	753	864	831	1543	910	754
New Orleans	S.Y.P.	577	577	561	612	570	732	591	1547	870	717
New York City	S.P.F.	539	522	605	666	577	1044	800	1101	1064	796
Philadelphia	H.F.	576	561	599	704	676	804	1223	1468	749	796
Phoenix	D.F.	614	658	715	838	768	1134	1178	1455	1308	924
Pittsburgh	H.F.	584	624	679	749	780	799	1300	1490	780	830
St. Louis	S.P.F.	488	470	412	438	506	858	910	654	742	691
San Antonio	S.Y.P.	527	531	618	662	737	616	743	712	753	802
San Diego	D.F.	522	522	550	618	588	768	1059	1178	1284	904
San Francisco	D.F.	490	490	487	630	625	840	760	1355	906	937
Seattle	S.P.F.	638	627	673	794	801	817	666	910	800	792
Washington, DC	S.P.F.	567	548	586	752	716	908	819	1210	1126	820
Average		$592	$590	$622	$731	$738	$830	$977	$1,337	$912	$811

To convert square feet of surface to board feet, 4% waste included.

S4S Size	Multiply S.F. by	T & G Size	Multiply S.F. by	Flooring Size	Multiply S.F. by
1 x 4	1.18	1 x 4	1.27	25/32" x 2-1/4"	1.37
1 x 6	1.13	1 x 6	1.18	25/32" x 3-1/4"	1.29
1 x 8	1.11	1 x 8	1.14	15/32" x 1-1/2"	1.54
1 x 10	1.09	2 x 6	2.36	1" x 3"	1.28
				1" x 4"	1.24

Wood & Plastics — R061 Rough Carpentry

R06110-030 Lumber Product Material Prices

The price of forest products fluctuates widely from location to location and from season to season depending upon economic conditions. The table below indicates National Average material prices in effect Jan. 1, 2000. The table shows relative differences between various sizes, grades and species. These percentage differentials remain fairly constant even though lumber prices in general may change significantly during the year.

Availability of certain items depends upon geographic location and must be checked prior to firm price bidding.

National Average Contractor Price, Quantity Purchase

Dimension Lumber, S4S, #2 & Better, KD

Framing Lumber per MBF	Species	2"x4"	2"x6"	2"x8"	2"x10"	2"x12"
	Douglas Fir	$548	$555	$589	$681	$661
	Spruce	614	604	635	767	780
	Southern Yellow Pine	544	548	571	622	649
	Hem-Fir	576	592	625	715	691

Heavy Timbers, Fir

3" x 4" thru 3" x 12"	$1,164
4" x 4" thru 4" x 12"	1,326
6" x 6" thru 6" x 12"	2,000
8" x 8" thru 8" x 12"	2,055
10" x 10" and 10" x 12"	2,086

S4S "D" Quality or Clear, KD

Boards per MBF *See also Cedar Siding	Species	1"x4"	1"x6"	1"x8"	1"x10"	1"x12"
	Sugar Pine	$1,491	$1,645	$1,757	$2,079	$2,506
	Idaho Pine	1,288	1,645	1,624	1,834	2,394
	Engleman Spruce	1,260	1,722	1,750	1,841	2,275
	So. Yellow Pine	1,275	1,498	1,589	1,505	1,695
	Ponderosa Pine	1,300	1,624	1,645	1,869	2,394
	Redwood, CVG	3,940	3,800	3,750	3,770	5,940

S4S # 2 & Better or Sterling, KD

Species	1"x4"	1"x6"	1"x8"	1"x10"	1"x12"
Sugar Pine	$602	$805	$861	$987	$1,288
Idaho Pine	1,029	1,036	1,022	1,043	1,253
Engleman Spruce	630	756	847	966	1,295
So. Yellow Pine	504	728	790	616	833
Ponderosa Pine	553	756	812	945	1,218

Flooring per MSF

1" x 4" Vertical grain, Fir "B" & better	$2,400
2-1/4" x 25/32", Oak, clear	3,200
Select	3,050
#1 common	2,700
Oak, prefinished, standard & better	3,750
Standard	3,200
2-1/4" x 25/32" Maple, select	$2,970
#2 & better	2,700
2-1/4" x 33/32" Maple, #2 & better	3,132
3-1/4" x 33/32" Maple, #2 & better	2,916
Parquet, unfinished, 5/16", minimum	2,700
Maximum	5,000

Siding per MBF

Clapboard, Cedar, beveled	
1/2" x 6" thru 1/2" x 8", clear	$1,344
"A" grade	1,246
"B" grade	933
3/4" x 10" "clear"	2,592
"A" grade	2,334
Redwood, beveled	
1/2" x 6" thru 1/2" x 8", vertical grain, clear	1,907
3/4" x 10" vertical grain, clear	2,889
*Rough sawn, Cedar, T&G, "A" grade 1" x 4"	$2,434
1" x 6"	3,201
"STK" grade, 1" x 6"	1,614
Board, "STK" grade, 1" x 8"	1,561
Board & Batten 1" x 12"	1,561
Cedar channel siding 1" x 8", #3 & better	$1,092
Factory stained	1,292
White Pine siding, T&G, rough sawn	$637
Factory stained	837

Shingles per CSF

Red Cedar	
5X—16" long #1 regular	$151
#2	170
18" long perfections #1	184
#2	128
Resquared & Rebutted #1	116
Handsplit shakes, resawn	
24" long, 1/2" to 3/4"	145
18" long, 1/2" to 3/4"	125
White Cedar shingles	
16" long, extra grade (East Coast)	$125
Clear, 1st grade (East Coast)	115
Fire retardant Red Cedar	
5X — 16" long	$251
18" long perfections	284
Handsplit & resawn	
24" long, 1/2" to 3/4"	245
18" long, 1/2" to 3/4"	225

Wood & Plastics — R061 Rough Carpentry

R06160-020 Plywood

There are two types of plywood used in construction: interior, which is moisture resistant but not waterproofed, and exterior, which is waterproofed.

The grade of the exterior surface of the plywood sheets is designated by the first letter: A, for smooth surface with patches allowed; B, for solid surface with patches and plugs allowed; C, which may be surface plugged or may have knot holes up to 1" wide; and D, which is used only for interior type plywood and may have knot holes up to 2-1/2" wide. "Structural Grade" is specifically designed for engineered applications such as box beams. All CC & DD grades have roof and floor spans marked on them.

Underlayment grade plywood runs from 1/4" to 1-1/4" thick. Thicknesses 5/8" and over have optional tongue and groove joints which eliminates the need for blocking the edges. Underlayment 19/32" and over may be referred to as Sturd-i-Floor.

The price of plywood can fluctuate widely due to geographic and economic conditions. When one or two local prices are known, the relative prices for other types and sizes may be found by direct factoring of the prices in the table below.

Typical uses for various plywood grades are as follows:

AA-AD Interior — cupboards, shelving, paneling, furniture
BB Plyform — concrete form plywood
CDX — wall and roof sheathing
Structural — box beams, girders, stressed skin panels
AA-AC Exterior — fences, signs, siding, soffits, etc.
Underlayment — base for resilient floor coverings
Overlaid HDO — high density for concrete forms & highway signs
Overlaid MDO — medium density for painting, siding, soffits & signs
303 Siding — exterior siding, textured, striated, embossed, etc.

Grade	Type	4'x8'	Type	4'x8'	4'x10'
		National Average Price in Lots of 10 MSF, per MSF-January 2000			
Sanded Grade	1/4" Interior AD	$ 609	1/4" Exterior AC	$ 616	$ 696
	3/8"	693	3/8"	707	727
	1/2"	805	1/2"	819	871
	5/8"	959	5/8"	937	1,045
	3/4"	1,050	3/4"	1,064	1,239
	1"	1,221	1"	1,380	1,427
	1-1/4"	1,464	Exterior AA, add	155	160
	Interior AA, add	155	Exterior AB, add	130	135
			CD Structural 1	Underlayment	
Unsanded Grade 4' x 8' Sheets	5/16" CDX	$ 400	5/16", 4'x8' sheets $365	3/8", 4'x8' sheets	$ 572
	3/8"	405	3/8" 459	1/2"	733
	1/2"	531	1/2" 647	5/8"T&G	756
	5/8"	666	5/8" 777	3/4"T&G	1,211
	3/4"	806	3/4" 889	1-1/8" 2-4-1T&G	1,897
	3/4" T&G	831			
Form Plywood	5/8" Exterior, oiled BB, plyform	$ 966	5/8" HDO (overlay 2 sides)		$2,148
	3/4" Exterior, oiled BB, plyform	1,036	3/4" HDO (overlay 2 sides)		2,214
Overlaid 4'x8' Sheets	Overlay 2 Sides MDO 3/8" thick	$1,212	Overlay 1 Side MDO 3/8" thick		$1,029
	1/2"	1,418	1/2"		1,186
	5/8"	1,600	5/8"		1,332
	3/4"	1,830	3/4"		1,562
303 Siding	Fir, rough sawn, natural finish, 3/8" thick	$ 707	Texture 1-11	5/8" thick, Fir	$1,093
	Redwood	1,890		Redwood	1,890
	Cedar	1,890		Cedar	1,537
	Southern Yellow Pine	539		Southern Yellow	850
Waferboard/O.S.B.	1/4" sheathing	$ 298	19/32" T&G		$ 407
	7/16" sheathing	435	23/32" T&G		594

For 2 MSF to 10 MSF, add 10%. For less than 2 MSF, add 15%.

Wood & Plastics — R061 Rough Carpentry

R06170-100 Wood Roof Trusses

Loading figures represent live load. An additional load of 10 psf on the top chord and 10 psf on the bottom chord is included in the truss design. Spacing is 24″ O.C.

Span in Feet	Cost per Truss for Different Live Loads and Roof Pitches					
	Flat	4 in 12 Pitch		5 in 12 Pitch		8 in 12 Pitch
	40 psf	30 psf	40 psf	30 psf	40 psf	30 psf
20	$ 66	$ 48	$ 50	$ 48	$ 52	$ 58
22	73	53	55	53	57	63
24	79	58	60	58	62	69
26	86	62	64	63	67	75
28	92	67	69	68	72	81
30	99	72	74	73	77	86
32	123	86	88	87	91	101
34	131	92	94	93	97	107
36	139	97	99	99	103	113
38	146	103	105	104	108	120
40	154	108	110	110	114	126

Wood & Plastics — R064 Architectural Woodwork

R06430-100 Wood Stair, Residential

One Flight with 8′-6″ Story Height, 3′-6″ Wide Oak Treads Open One Side, Built in Place				
Item	Quantity	Unit Cost	Bare Costs	Costs Incl. Subs O & P
Treads 10-1/2″ x 1-1/16″ thick	11 Ea.	$ 27.50	$ 302.50	$ 332.80
Landing tread nosing, rabbeted	1 Ea.	12.50	12.50	13.80
Risers 3/4″ thick	12 Ea.	12.40	148.80	163.60
Single end starting step (range $175 to $210)	1 Ea.	193.00	193.00	212.40
Balusters (range $4.50 to $25.00)	22 Ea.	7.75	170.50	187.60
Newels, starting & landing (range $40 to $120)	2 Ea.	73.00	146.00	160.60
Rail starter (range $39 to $190)	1 Ea.	67.00	67.00	73.80
Handrail (range $4.50 to $10.30)	26 L.F.	7.00	182.00	200.20
Cove trim	50 L.F.	.78	39.00	43.00
Rough stringers three - 2 x 12's, 14' long	84 B.F.	.74	62.16	68.40
Carpenter's installation: Bare Cost Incl Subs O & P	36 Hrs.	$ 19.70 $ 33.75	$ 709.20	$1,215.00
	Total per Flight		$2,032.66	$2,671.20

Add for rail return on second floor and for varnishing or other finish. Adjoining walls or landings must be figured separately.

Thermal & Moist. Protection | R071 | Waterproofing & Dampproofing

R07110-010   1/2" Pargeting (rough dampproofing plaster)

1:2-1/2 Mix, 4.5 C.F. Covers 100 S.F., 2 Coats, Waste Included	Regular Portland Cement		Waterproofed Portland Cement	
1.7 lbs. integral waterproofing admixture			$.65 per lb.	$ 1.11
1.7 bags Portland cement	$7.10 per bag	$ 12.07	$7.10 per bag	12.07
4.25 C.F. sand @ $18.70 per C.Y.		2.94		2.94
Labor mix, apply, crew D-1 @ $285.60 per day	.40 days	114.24	.40 days	114.24
Total Bare Cost per 100 S.F.		$129.25		$130.36

Thermal & Moist. Protection | R073 | Shingles & Roofing Tiles

R07310-020   Roof Slate

16", 18" and 20" are standard lengths and slate usually comes in random widths. For standard 3/16" thickness use 1-1/2" copper nails. Allow for 3% breakage.

Quantities per Square	Unfading Vermont Colored	Weathering Sea Green	Buckingham, Virginia Black
Slate delivered (incl. punching)	$385.00	$286.00	$470.00
# 30 Felt, 2.5 lbs. copper nails	14.49	14.49	14.49
Slate roofer 4.6 hrs. @ $17.10 per hr.	78.66	78.66	78.66
Total Bare Cost per Square	$478.15	$379.15	$563.15

Thermal & Moist. Protection | R075 | Membrane Roofing

R07510-020   Built-Up Roofing

Asphalt is available in kegs of 100 lbs. each; coal tar pitch in 560 lb. kegs. Prepared roofing felts are available in a wide range of sizes, weights & characteristics. However, the most commonly used are #15 (432 S.F. per roll, 13 lbs. per square) and #30 (216 S.F. per roll, 27 lbs. per square).

Inter-ply bitumen varies from 24 lbs. per sq. (asphalt) to 30 lbs. per sq. (coal tar) per ply, ± 25%. Flood coat bitumen also varies from 60 lbs. per sq. (asphalt) to 75 lbs. per sq. (coal tar), ± 25%. Expendable equipment (mops, brooms, screeds, etc.) runs about 16% of the bitumen cost. For new, inexperienced crews this factor may be much higher.

Rigid insulation board is typically applied in two layers. The first is mechanically attached to nailable decks or spot or solid mopped to non-nailable decks; the second layer is then spot or solid mopped to the first layer. Membrane application follows the insulation, except in protected membrane roofs, where the membrane goes down first and the insulation on top, followed with ballast (stone or concrete pavers). Insulation and related labor costs are NOT included in Div. 07510-300; they can be found in Div. 07510-700.

Prices shown are bare costs only.

4-Ply Felt on Insulated Deck	Asphalt/Glass Fiber Felt		Coal Tar/Organic Felt	
4 ply #15 felt (1/4 roll per ply)	$3.70 per ply	$ 14.80	$7.50 per ply	$ 30.00
156 lbs. hot asphalt/195 lbs. hot tar	$260 per ton	20.24	$550 per ton	53.63
500 lbs. roofing stone (1/4" to 1/2")	$21.50 per ton	5.38	$21.50 per ton	5.38
Installation, crew G-1, @ $1,320.20 per day	.050 days	66.01	.048 days	63.37
Total Bare Cost per Square in-place		$106.43		$152.38

Thermal & Moist. Protection | R075 | Membrane Roofing

R07550-030 Modified Bitumen Roofing

The cost of modified bitumen roofing is highly dependent on the type of installation that is planned. Installation is based on the type of modifier used in the bitumen. The two most popular modifiers are atactic polypropylene (APP) and styrene butadiene styrene (SBS). The modifiers are added to heated bitumen during the manufacturing process to change its characteristics. A polyethylene, polyester or fiberglass reinforcing sheet is then sandwiched between layers of this bitumen. When completed, the result is a pre-assembled, built-up roof that has increased elasticity and weatherablility. Some manufacturers include a surfacing material such as ceramic or mineral granules, metal particles or sand.

The preferred method of adhering SBS-modified bitumen roofing to the substrate is with hot-mopped asphalt (much the same as built-up roofing). This installation method requires a tar kettle/pot to heat the asphalt, as well as the labor, tools and equipment necessary to distribute and spread the hot asphalt.

The alternative method for applying APP and SBS modified bitumen is as follows. A skilled installer uses a torch to melt a small pool of bitumen off the membrane. This pool must form across the entire roll for proper adhesion. The installer must unroll the roofing at a pace slow enough to melt the bitumen, but fast enough to prevent damage to the rest of the membrane.

Modified bitumen roofing provides the advantages of both built-up and single-ply roofing. Labor costs are reduced over those of built-up roofing because only a single ply is necessary. The elasticity of single-ply roofing is attained with the reinforcing sheet and polymer modifiers. Modifieds have some self-healing characteristics and because of their multi-layer construction, they offer the reliability and safety of built-up roofing.

Doors & Windows | R081 | Metal Doors & Frames

R08110-100 Hollow Metal Frames

Table below lists material prices only.

Frame for Opening Size	16 Ga. Frames		16 Ga. UL Frames		16 Ga. Drywall Frames		16 Ga. UL Drywall Frames	
	6-3/4" Deep	8-3/4" Deep	6-3/4" Deep	8-3/4" Deep	7-1/8" Deep	8-1/4" Deep	7-1/8" Deep	8-1/4" Deep
2'-0" x 6'-8"	$ 69	$ 84	$ 78	$ 93	$ 83	$ 89	$ 92	$ 98
2'-6" x 6'-8"	69	84	78	93	83	89	92	98
3'-0" x 7'-0"	75	85	84	94	84	90	93	99
3'-6" x 7'-0"	80	87	89	96	88	94	97	103
4'-0" x 7'-0"	85	97	94	104	89	94	98	103
6'-0" x 7'-0"	90	101	108	126	99	106	117	124
8'-0" x 8'-0"	107	124	116	134	119	126	137	144

For welded frames, add $26.00.
For galvanized frames, add $22.00.

R08110-120 Hollow Metal Doors

Table below lists material prices only, not including hardware or labor.

Door Thickness and Size		Full Flush Doors, 18 Ga.				Flush Fire Doors			
		Hollow Metal		Composite Core		Hollow Metal "B"		Hollow Metal Class "A"	
		Plain	Glazed	Plain	Glazed	20 Ga.	18 Ga.	16 Ga.	18 Ga.
1-3/4"	3'-0" x 6'-8"	$189	$246	$214	$271	$179	$205	$255	$217
	3'-0" x 7'-0"	199	255	224	280	192	211	265	222
	3'-6" x 7'-0"	227	300	252	325	215	244	285	258
	3'-0" x 8'-0"	247	330	272	355	249	269	310	283
	4'-0" x 8'-0"	285	380	310	405	320	316	349	321
1-3/8"	2'-0" x 6'-8"	146	203	—	—	160	—	—	—
	2'-6" x 6'-8"	155	212	—	—	170	—	—	—
	3'-0" x 7'-0"	165	222	—	—	180	—	—	—

*Indicates 20 gauge doors.

R08110-130 Tin Clad Fire Door

6' x 7' Opening, A Label	Double Sliding Door		Double Swing Door	
Doors, 3 Ply		$ 510.00		$ 510.00
Frame, 8" channel (11.5 lb. per L.F.), installed	275 lb.	250.25	275 lb.	250.25
Track, hangers, hardware	13.75 L.F.	693.00	Straps and hinges	326.00
2 Carpenters @ $19.70 per hour each	16 hrs. total	315.20	16 hrs. total	315.20
Complete in place		$1,768.45		$1,401.45

Doors & Windows | R082 | Wood & Plastic Doors

R08210-100 Wood Doors

Table below, except where indicated, lists price per door only, not including frame, hardware or labor. For pre-hung exterior door units up to 3' x 7', add $125 per door for wood frame and hardware for types not listed under pre-hung. Pricing is for ten or more doors. Doors are factory trimmed for butts and locksets.

Door Thickness and Size		Flush Type Doors (Interior)					Architectural (1-3/4" Ext., 1-3/8" Int.)					
		Hollow Core		Solid Particle Core			Pine		Fir		Pre-hung	
		Lauan	Birch	Lauan	Birch	Oak	Panel	Glazed	Panel	Glazed	Pine Panel	Flush Birch S. C.
1-3/4"	2'-6" x 6'-8"	$36	$56	$ 73	$ 91 *	$ 97	$289	$304	$242	$257	$394	$250
	3'-0" x 6'-8"	41	64	79	96 *	106	311	326	247	262	417	260
	3'-0" x 7'-0"	52	71	97	102 *	113	354	369	260	270	464	274
	3'-6" x 7'-0"	63	82	104	125 *	137	—	—	—	—	—	—
	4'-0" x 7'-0"	69	97	112	136 *	147	—	—	—	—	—	—
1-3/8"	2'-0" x 6'-8"	33	47	69	76 *	86	110	—	140	—	177	140
	2'-6" x 6'-8"	35	49	73	83 *	91	124	139	159	—	190	147
	3'-0" x 6'-8"	40	51	77	94 *	98	148	163	182	—	220	157
	3'-0" x 7'-0"	46	56	82	99 *	106	166	—	206	—	—	—

*Add to the above for the following birch face door types:

Solid wood stave core, add $25
3/4 hour label door, add $61

1 hour label door, add $72
1-1/2 hour label door, add $82

8' high door, add 31%

Doors & Windows | R085 | Metal Windows

R08550-200 Replacement Windows

Replacement windows are typically measured per United Inch.

United Inches are calculated by rounding the width and height of the window opening up to the nearest inch, then adding the two figures.

The labor cost for replacement windows includes removal of sash, existing sash balance or weights, parting bead where necessary and installation of new window.

Debris hauling and dump fees are not included.

Doors & Windows — R088 Glazing

R08810-010 Glazing Labor

Glass sizes are estimated by the "united inch" (height + width). Table below shows the number of lights glazed in an eight hour period by the crew size indicated, for glass up to 1/4" thick. Square or nearly square lights are more economical on a S.F. basis. Long slender lights will have a high S.F. installation cost. For insulated glass reduce production by 33%. For 1/2" float glass reduce production by 50%. Production time for glazing with two glaziers per day averages: 1/4" float glass 120 S.F.; 1/2" float glass 55 S.F.; 1/2" insulated glass 95 S.F.; 3/4" insulated glass 75 S.F.

Glazing Method	United Inches per Light							
	40"	60"	80"	100"	135"	165"	200"	240"
Number of Men in Crew	1	1	1	1	2	3	3	4
Industrial sash, putty	60	45	24	15	18	—	—	—
With stops, putty bed	50	36	21	12	16	8	4	3
Wood stops, rubber	40	27	15	9	11	6	3	2
Metal stops, rubber	30	24	14	9	9	6	3	2
Structural glass	10	7	4	3	—	—	—	—
Corrugated glass	12	9	7	4	4	4	3	—
Storefronts	16	15	13	11	7	6	4	4
Skylights, putty glass	60	36	21	12	16	—	—	—
Thiokol set	15	15	11	9	9	6	3	2
Vinyl set, snap on	18	18	13	12	12	7	5	4
Maximum area per light	2.8 S.F.	6.3 S.F.	11.1 S.F.	17.4 S.F.	31.6 S.F.	47 S.F.	69 S.F.	100 S.F.
Daily Bare Crew Cost	$154.80	$154.80	$154.80	$154.80	$309.60	$464.40	$464.40	$619.20

R08810-100 Commercial Window Glass

Description		1/2" Insulated Glass		1/4" Plate Glass	
Glass per S.F.		$10.00	R Value	$3.75	R Value
Allow for breakage 10%		1.00	3.00	.38	.93
Glazing compound 1/2 lb. per S.F. @	$1.10 per lb.	.55	Average	.55	Average
Labor, 2 glaziers @	$19.35 per hour each (75 S.F./day)	4.13		4.13	
Total in place, per S.F.		$15.68		$8.81	

Finishes — R091 Metal Support Assemblies

R09110-610 Studs, Joists and Track

Bare material prices per 1000 L.F., for galvanized studs, joists and tracks. Panhead framing screws, 7/16" long are $13.00 per thousand; #6 x 1-5/8" long drywall screws are $10.00 per thousand.

Non-load bearing, 20 ga. stud and track are primarily used for curtain wall. (C.W.)

	Non-Load Bearing				Load Bearing 1-5/8" Flange—Light Gauge Structural					
	25 Ga.		20 Ga. (C.W.)		18 Ga.		16 Ga.		14 Ga.	
Size	Stud	Track	Stud	Track	Stud	Track	Stud	Track	Stud	Track
1-5/8"	$155	$150	$267	$266						
2-1/2"	170	163	285	289	$527	$441	$646	$549	—	—
3-5/8"	197	195	308	318	586	501	729	632	$980	—
4"	238	237	384	390	647	554	796	692	1,050	—
6"	315	310	447	458	785	705	969	870	1,331	—
8"					1,019	925	1,281	1,175	1,635	$1,516

	Non-Load Bearing				Load Bearing—Extra Wide Flange — 2" Flange					
	25 Ga.		22 Ga.		18 Ga.		16 Ga.		14 Ga.		12 Ga.
Size	C-H Stud	J-Track	C-H Stud	J-Track	Joist	Track	Joist	Track	Joist	Track	Joist	Track
2-1/2"	$647	$497	$788	$750	—	—	—	—	—	—	—	—
4"	799	745	866	709	—	—	—	—	—	—	—	—
6"					$912	$705	$1,201	$870	$1,537	$1,162	—	—
8"					1,152	925	1,434	1,175	1,826	1,516	—	—
10"					1,359	1,222	1,707	1,467	2,195	1,873	$3,231	$2,867
12"					1,660	1,430	1,944	1,729	2,512	2,247	3,669	3,382

Finishes

R092 Plaster & Gypsum Board

R09205-050 Gypsum Lath

Item	Nailed to Wood Studs		Clipped to Steel Studs	
	Quantities and Bare Cost	Incl. O & P	Quantities and Bare Cost	Incl. O & P
105 S.Y. 3/8" gypsum lath	$3.73 per S.Y.	$391.65	$430.82	$3.73 per S.Y.	$391.65	$430.82
Fasteners, nails & clips	6 lb. @ $1.20/lb.	7.20	7.92	600 @ $.07 ea.	42.00	46.20
Lather @ $19.20 and $31.45 per hr.	9.4 hours	180.48	295.63	10.7 hours	205.44	336.52
Total per 100 S.Y. in place		$579.33	$734.37		$639.09	$813.54

Regular lath comes in 16" x 32" and 16" x 48" sheets. Firestop gypsum base comes in 4' x 8' sheets. For nailing use 1-1/8" No. 13 ga. flathead blued nails.

R09205-060 Metal Lath

Painted Diamond Expanded	Nailed to Wood Studs		Screwed to Steel Studs	
	Quantity and Bare Cost	Incl. O & P	Quantity and Bare Cost	Incl. O & P
105 S.Y. 3.4 lb. lath	$2.31 per S.Y.	$242.55	$266.81	$2.31 per S.Y.	$242.55	$266.81
Fasteners, corner beads, etc.		4.38	4.82		9.88	10.87
Lather @ $19.20 and $31.45 per hr.	10.0 hrs.	192.00	314.50	10.7 hrs.	205.44	336.52
Total per 100 S.Y. in place		$438.93	$586.13		$457.87	$614.20

Material prices of other types of lath in Div. 09200 may be substituted in the table above to arrive at their costs in place. For nailing on studs use a 1" roofing nail with 7/16" head. For nailing on ceiling use 1-1/2" No. 11 ga. barbed roofers nail with 7/16" head for holding power.

R09210-105 Gypsum Plaster

Quantities for 100 S.Y.	2 Coat, 5/8" Thick		3 Coat, 3/4" Thick		
	Base	Finish	Scratch	Brown	Finish
	1:3 Mix	2:1 Mix	1:2 Mix	1:3 Mix	2:1 Mix
Gypsum plaster	1300 lb.		1350 lb.	650 lb.	
Sand	1.75 C.Y.		1.16 C.Y.	.84 C.Y.	
Finish hydrated lime		340 lb.			340 lb.
Gauging plaster		170 lb.			170 lb.

Total, in Place for 100 S.Y. on Walls				2 Coat, 5/8" Thick			3 Coat, 3/4" Thick		
				Quantities	Bare Cost	Incl. O & P	Quantities	Bare Cost	Incl. O & P
Gypsum plaster @	$13.02	per 80	lb. bag	1300 lb.	$211.58	$232.74	2000 lb.	$325.50	$358.05
Finish hydrated lime @	$5.60	per 50	lb. bag	340 lb.	38.08	41.89	340 lb.	38.08	41.89
Gauging plaster @	$19.02	per 100	lb. bag	170 lb.	32.33	35.56	170 lb.	32.33	35.56
Sand @	$17.25	per C.Y.		1.7 C.Y.	29.33	32.26	2.0 C.Y.	34.50	37.95
J-1 crew @	$17.45	& $29.18	per L.H.	34.8 L.H.	607.26	1,015.46	42.0 L.H.	732.90	1,225.56
Cleaning, staging, handling, patching				3.6 L.H.	62.82	105.05	4.0 L.H.	69.80	116.72
Total per 100 S.Y. in place					$981.40	$1,462.96		$1,233.11	$1,815.73

Finishes | R092 | Plaster & Gypsum Board

R09210-115 Vermiculite or Perlite Plaster

Proportions: Over lath, scratch coat and brown coat,
100# gypsum plaster to 2 C.F. aggregate; over masonry,
100# gypsum plaster to 3 C.F. aggregate.

Quantities for 100 S.Y.			2 Coat, 5/8" Thick			3 Coat, 3/4" Thick		
			Quantities	Bare Cost	Incl. O & P	Quantities	Bare Cost	Incl. O & P
Gypsum plaster @	$13.02 per 80	lb. bag	1250 lb.	$ 203.44	$ 223.78	2250 lb.	$ 366.19	$ 402.81
Vermiculite or perlite @	$13.03 per	bag	7.8 bags	101.63	111.79	11.3 bags	147.24	161.96
Finish hydrated lime @	$ 5.60 per 50	lb. bag	340 lb.	38.08	41.89	340 lb.	38.08	41.89
Gauging plaster @	$19.02 per 100	lb. bag	170 lb.	32.33	35.56	170 lb.	32.33	35.56
J-1 crew @	$18.81 & $30.68	per L.H.	40.0 L.H.	752.50	1,227.15	50.0 L.H.	940.63	1,533.94
Cleaning, staging, handling, patching			3.6 L.H.	67.73	110.44	4.0 L.H.	75.25	122.72
Total per 100 S.Y. in place				$1,195.71	$1,750.61		$1,599.72	$2,298.88

R09220-300 Stucco

Quantities for 100 S.Y. 3 Coats, 1" Thick			On Wood Frame			On Masonry		
			Quantities	Bare Cost	Incl. O & P	Quantities	Bare Cost	Incl. O & P
Portland cement @	$ 7.10 per bag		29 bags	$ 205.90	$ 226.49	21 bags	$149.10	$164.01
Sand @	$15.85 per C.Y.		2.6 C.Y.	41.21	45.33	2 C.Y.	31.70	34.87
Hydrated lime @	$ 5.60 per 50 lb. bag		180 lb.	20.16	22.18	120 lb.	13.44	14.78
Painted stucco mesh @	$ 2.32 per S.Y., 3.6#		105 S.Y.	243.60	267.96	—	—	—
Furring nails and jute fiber				13.84	15.22		—	—
Mix and install with crew indicated			J-2 @ .74 days	670.51	1,094.28	J-1 @ 0.5 days	376.25	613.58
Cleaning, staging, handling, patching			.10 days	85.16	143.63	.10 days	85.16	143.63
Total per 100 S.Y. in place				$1,280.38	$1,815.09		$655.65	$970.87

Finishes | R094 | Terrazzo

R09400-100 Terrazzo Floor

5000 S.F. 5/8" Terrazzo Topping Quantities for 100 S.F.		Bonded to Concrete 1-1/8" Bed, 1:4 Mix			Not Bonded 2-1/8" Bed and 1/4" Sand		
		Quantities	Bare Cost	Incl. O & P	Quantities	Bare Cost	Incl. O & P
Portland cement @	$7.10 per bag	6 bags	$ 42.60	$ 46.86	8 bags	$ 56.80	$ 62.48
Sand @	$15.85 per C.Y.	10 C.F.	5.87	6.46	20 C.F.	11.74	12.91
Divider strips for 4' square panels, zinc, 14 ga.		55 L.F.	50.60	55.66	55 L.F.	50.60	55.66
Terrazzo fill, domestic aggregates		600 lb.	180.00	198.00	600 lb.	180.00	198.00
15 lb. tarred felt			—	—		2.00	2.20
Mesh 2 x 2 #14 galvanized			—	—		14.75	16.23
Crew J-3 @	$25.09 & $36.62 per day	12.30 L.H.	308.61	450.43	13.90 L.H.	348.75	509.02
Total per 100 S.F. in place		Bonded	$587.68	$757.41	Not Bonded	$664.64	$856.50

2' x 2' panels require 1 L.F. divider strip per S.F.; 6' x 6' panels need .33 L.F. per S.F.

Finishes — R096 Flooring

R09658-600 Resilient Flooring and Base

Description	12" x 12" x 3/32" V.C. Tile			4" x 1/8" Vinyl Base		
	Quantities	Bare Cost	Incl. O & P	Quantities	Bare Cost	Incl. O & P
Vinyl composition, tile, standard line	100 S.F.	$112.00	$123.20	100 L.F.	$ 63.00	$ 69.30
Vinyl cement @ $10.95 per gallon	0.8 gallon	8.76	9.64	0.5 gallon	5.48	6.03
Tile layer @ $19.20 and $31.10 per L.H.	1.5 L.H.	28.80	46.65	2.7 L.H.	51.84	83.97
Total per 100 S.F. in place		$149.56	$179.49	100 L.F.	$120.32	$159.30

Finishes — R097 Wall Finishes

R09700-700 Wall Covering

Quantities for 100 S.F.	Medium Price Paper			Expensive Paper		
	Quantities	Bare Cost	Incl. O & P	Quantities	Bare Cost	Incl. O & P
Paper @ $30.00 and $57.00 per double roll	1.6 dbl. rolls	$48.00	$ 52.80	1.6 dbl. rolls	$ 91.20	$100.32
Wall sizing @ $19.60 per gallon	.25 gallon	4.90	5.39	.25 gallon	4.90	5.39
Vinyl wall paste @ $ 9.00 per gallon	0.4 gallon	3.60	3.96	0.4 gallon	3.60	3.96
Apply sizing @ $17.95 and $29.95 per hour	0.3 hour	5.39	8.99	0.3 hour	5.39	8.99
Apply paper @ $17.95 and $29.95 per hour	1.2 hours	21.54	35.94	1.5 hours	26.93	44.93
Total including waste allowance per 100 S.F.		$83.43	$107.08		$132.02	$163.59

This is equivalent to about $66.95 and $102.25 per double roll complete in place. Most wallpapers now come in double rolls only. To remove old paper allow 1.3 hours per 100 S.F.

Finishes | R099 | Paints & Coatings

R09910-100 Paint

Material prices per gallon in 5 gallon lots, up to 25 gallons. For 100 gallons, deduct 10%.

Exterior, Alkyd (oil base)	
Flat	$23.00
Gloss	22.25
Primer	22.00

Exterior, Latex (water base)	
Acrylic stain	19.00
Gloss enamel	23.00
Flat	19.50
Primer	21.75
Semi-gloss	22.50

Interior, Alkyd (oil base)	
Enamel undercoater	20.75
Flat	18.50
Gloss	23.00
Primer sealer	17.25
Semi-gloss	22.25

Interior, Latex (water base)	
Enamel undercoater	15.50
Flat	15.75
Floor and deck	21.00
Gloss	22.50
Primer sealer	18.00
Semi-gloss	21.50

Masonry, Exterior	
Alkali resistant primer	$19.25
Block filler, epoxy	20.25
Block filler, latex	10.00
Latex, flat or semi-gloss	17.00

Masonry, Interior	
Alkali resistant primer	15.75
Block filler, epoxy	23.25
Block filler, latex	10.25
Floor, alkyd	21.00
Floor, latex	21.25
Latex, flat acrylic	11.25
Latex, flat emulsion	11.00
Latex, sealer	12.25
Latex, semi-gloss	18.75

Varnish and Stain	
Alkyd clear	21.25
Polyurethane, clear	21.25
Primer sealer	15.00
Semi-transparent stain	15.75
Solid color stain	15.75

Metal Coatings	
Galvanized	24.50
High heat	48.50
Machinery enamel, alkyd	25.00
Normal heat	25.50
Rust inhibitor ferrous metal	$23.50
Zinc chromate	20.00

Heavy Duty Coatings	
Acrylic urethane	51.50
Chlorinated rubber	29.75
Coal tar epoxy	29.25
Metal pretreatment (polyvinyl butyral)	23.50
Polyamide epoxy finish	31.50
Polyamide epoxy primer	31.50
Silicone alkyd	32.00
2 component solvent based acrylic epoxy	59.50
2 component solvent based polyester epoxy	88.50
Vinyl	27.00
Zinc rich primer	94.50

Special Coatings/Miscellaneous	
Aluminum	22.75
Dry fall out, flat	12.25
Fire retardant, intumescent	36.50
Linseed oil	10.25
Shellac	18.25
Swimming pool, epoxy or urethane base	36.00
Swimming pool, rubber base	27.75
Texture paint	11.25
Turpentine	10.75
Water repellent 5% silicone	17.75

R09910-220 Painting

Item	Coat	One Gallon Covers			In 8 Hours a Laborer Covers			Labor-Hours per 100 S.F.		
		Brush	Roller	Spray	Brush	Roller	Spray	Brush	Roller	Spray
Paint wood siding	prime	250 S.F.	225 S.F.	290 S.F.	1150 S.F.	1300 S.F.	2275 S.F.	.695	.615	.351
	others	270	250	290	1300	1625	2600	.615	.492	.307
Paint exterior trim	prime	400	—	—	650	—	—	1.230	—	—
	1st	475	—	—	800	—	—	1.000	—	—
	2nd	520	—	—	975	—	—	.820	—	—
Paint shingle siding	prime	270	255	300	650	975	1950	1.230	.820	.410
	others	360	340	380	800	1150	2275	1.000	.695	.351
Stain shingle siding	1st	180	170	200	750	1125	2250	1.068	.711	.355
	2nd	270	250	290	900	1325	2600	.888	.603	.307
Paint brick masonry	prime	180	135	160	750	800	1800	1.066	1.000	.444
	1st	270	225	290	815	975	2275	.981	.820	.351
	2nd	340	305	360	815	1150	2925	.981	.695	.273
Paint interior plaster or drywall	prime	400	380	495	1150	2000	3250	.695	.400	.246
	others	450	425	495	1300	2300	4000	.615	.347	.200
Paint interior doors and windows	prime	400	—	—	650	—	—	1.230	—	—
	1st	425	—	—	800	—	—	1.000	—	—
	2nd	450	—	—	975	—	—	.820	—	—

Special Construction — R131 | Pre-Engin. Struct. & Aquatic Facil.

R13128-210 Pre-engineered Steel Buildings

These buildings are manufactured by many companies and normally erected by franchised dealers throughout the U.S. The four basic types are: Rigid Frames, Truss type, Post and Beam and the Sloped Beam type. Most popular roof slope is low pitch of 1″ in 12″. The minimum economical area of these buildings is about 3000 S.F. of floor area. Bay sizes are usually 20′ to 24′ but can go as high as 30′ with heavier girts and purlins. Eave heights are usually 12′ to 24′ with 18′ to 20′ most typical.

Material prices shown in Division 13128-700 are bare costs for the building shell only and do not include floors, foundations, interior finishes or utilities. Costs assume at least three bays of 24′ each, a 1″ in 12″ roof slope, and they are based on 30 psf roof load and 20 psf wind load and no unusual requirements. Costs include the structural frame, 26 ga. colored steel roofing, 26 ga. colored steel siding, fasteners, closures, trim and flashing but no allowance for doors, windows, skylights, gutters or downspouts. Very large projects would generally cost less for materials than the prices shown.

Conditions at the site, weather, shape and size of the building, and labor availability will affect the erection cost of the building.

R13128-520 Swimming Pools

Pool prices given per square foot of surface area include pool structure, filter and chlorination equipment where required, pumps, related piping, diving boards, ladders, maintenance kit, skimmer and vacuum system. Decks and electrical service to equipment are not included.

Residential in-ground pool construction can be divided into two categories: vinyl lined and gunite. Vinyl lined pool walls are constructed of different materials including wood, concrete, plastic or metal. The bottom is often graded with sand over which the vinyl liner is installed. Costs are generally in the $16 to $23 per S.F. range. Vermiculite or soil cement bottoms may be substituted for an added cost.

Gunite pool construction is used both in residential and municipal installations. These structures are steel reinforced for strength and finished with a white cement limestone plaster. Residential costs run from $26 to $38 per S.F. surface. Municipal costs vary from $50 to $75 because plumbing codes require more expensive materials, chlorination equipment and higher filtration rates.

Municipal pools greater than 1,800 S.F. require gutter systems to control waves. This gutter may be formed into the concrete wall. Often a vinyl, stainless steel gutter or gutter and wall system is specified, which will raise the pool cost an additional $51 to $150 per L.F. of gutter installed and up to $290 per L.F. if a gutter and wall system is installed.

Competition pools usually require tile bottoms and sides with contrasting lane striping. Add $12 per S.F. of wall or bottom to be tiled.

Special Construction — R132 | Hazardous Material Remediation

R13281-120 Asbestos Removal Process

Asbestos removal is accomplished by a specialty contractor who understands the federal and state regulations regarding the handling and disposal of the material. The process of asbestos removal is divided into many individual steps. An accurate estimate can be calculated only after all the steps have been priced.

The steps are generally as follows:

1. Obtain an asbestos abatement plan from an industrial hygienist.
2. Monitor the air quality in and around the removal area and along the path of travel between the removal area and transport area. This establishes the background contamination.
3. Construct a two part decontamination chamber at entrance to removal area.
4. Install a HEPA filter to create a negative pressure in the removal area.
5. Install wall, floor and ceiling protection as required by the plan, usually 2 layers of fireproof 6 mil polyethylene.
6. Industrial hygienist visually inspects work area to verify compliance with plan.
7. Provide temporary supports for conduit and piping affected by the removal process.
8. Proceed with asbestos removal and bagging process. Monitor air quality as described in Step #2. Discontinue operations when contaminate levels exceed applicable standards.
9. Document the legal disposal of materials in accordance with EPA standards.
10. Thoroughly clean removal area including all ledges, crevices and surfaces.
11. Post abatement inspection by industrial hygienist to verify plan compliance.
12. Provide a certificate from a licensed industrial hygienist attesting that contaminate levels are within acceptable standards before returning area to regular use.

Mechanical — R151 Plumbing

R15100-420 Plumbing Fixture Installation Time

Item	Rough-In	Set	Total Hours	Item	Rough-In	Set	Total Hours
Bathtub	5	5	10	Shower head only	2	1	3
Bathtub and shower, cast iron	6	6	12	Shower drain	3	1	4
Fire hose reel and cabinet	4	2	6	Shower stall, slate		15	15
Floor drain to 4 inch diameter	3	1	4	Slop sink	5	3	8
Grease trap, single, cast iron	5	3	8	Test 6 fixtures			14
Kitchen gas range		4	4	Urinal, wall	6	2	8
Kitchen sink, single	4	4	8	Urinal, pedestal or floor	6	4	10
Kitchen sink, double	6	6	12	Water closet and tank	4	3	7
Laundry tubs	4	2	6	Water closet and tank, wall hung	5	3	8
Lavatory wall hung	5	3	8	Water heater, 45 gals. gas, automatic	5	2	7
Lavatory pedestal	5	3	8	Water heaters, 65 gals. gas, automatic	5	2	7
Shower and stall	6	4	10	Water heaters, electric, plumbing only	4	2	6

Fixture prices in front of book are based on the cost per fixture set in place. The rough-in cost, which must be added for each fixture, includes carrier, if required, some supply, waste and vent pipe connecting fittings and stops. The lengths of rough-in pipe are nominal runs which would connect to the larger runs and stacks. The supply runs and DWV runs and stacks must be accounted for in separate entries. In the eastern half of the United States it is common for the plumber to carry these to a point 5' outside the building.

Mechanical — R155 Heating

R15500-050 Factor for Determining Heat Loss for Various Types of Buildings

General: While the most accurate estimates of heating requirements would naturally be based on detailed information about the building being considered, it is possible to arrive at a reasonable approximation using the following procedure:

1. Calculate the cubic volume of the room or building.
2. Select the appropriate factor from Table 1 below. Note that the factors apply only to inside temperatures listed in the first column and to 0°F outside temperature.
3. If the building has bad north and west exposures, multiply the heat loss factor by 1.1.
4. If the outside design temperature is other than 0°F, multiply the factor from Table 1 by the factor from Table 2.
5. Multiply the cubic volume by the factor selected from Table 1. This will give the estimated BTUH heat loss which must be made up to maintain inside temperature.

Table 1 — Building Type	Conditions	Qualifications	Loss Factor*
Factories & Industrial Plants General Office Areas at 70°F	One Story	Skylight in Roof	6.2
		No Skylight in Roof	5.7
	Multiple Story	Two Story	4.6
		Three Story	4.3
		Four Story	4.1
		Five Story	3.9
		Six Story	3.6
	All Walls Exposed	Flat Roof	6.9
		Heated Space Above	5.2
	One Long Warm Common Wall	Flat Roof	6.3
		Heated Space Above	4.7
	Warm Common Walls on Both Long Sides	Flat Roof	5.8
		Heated Space Above	4.1
Warehouses at 60°F	All Walls Exposed	Skylights in Roof	5.5
		No Skylight in Roof	5.1
		Heated Space Above	4.0
	One Long Warm Common Wall	Skylight in Roof	5.0
		No Skylight in Roof	4.9
		Heated Space Above	3.4
	Warm Common Walls on Both Long Sides	Skylight in Roof	4.7
		No Skylight in Roof	4.4
		Heated Space Above	3.0

*Note: This table tends to be conservative particularly for new buildings designed for minimum energy consumption.

Table 2 — Outside Design Temperature Correction Factor (for Degrees Fahrenheit)									
Outside Design Temperature	50	40	30	20	10	0	-10	-20	-30
Correction Factor	.29	.43	.57	.72	.86	1.00	1.14	1.28	1.43

Mechanical — R157 Air Conditioning & Ventilation

R15700-020 Air Conditioning Requirements

BTU's per hour per S.F. of floor area and S.F. per ton of air conditioning.

Type of Building	BTU per S.F.	S.F. per Ton	Type of Building	BTU per S.F.	S.F. per Ton	Type of Building	BTU per S.F.	S.F. per Ton
Apartments, Individual	26	450	Dormitory, Rooms	40	300	Libraries	50	240
Corridors	22	550	Corridors	30	400	Low Rise Office, Exterior	38	320
Auditoriums & Theaters	40	300/18*	Dress Shops	43	280	Interior	33	360
Banks	50	240	Drug Stores	80	150	Medical Centers	28	425
Barber Shops	48	250	Factories	40	300	Motels	28	425
Bars & Taverns	133	90	High Rise Office—Ext. Rms.	46	263	Office (small suite)	43	280
Beauty Parlors	66	180	Interior Rooms	37	325	Post Office, Individual Office	42	285
Bowling Alleys	68	175	Hospitals, Core	43	280	Central Area	46	260
Churches	36	330/20*	Perimeter	46	260	Residences	20	600
Cocktail Lounges	68	175	Hotel, Guest Rooms	44	275	Restaurants	60	200
Computer Rooms	141	85	Corridors	30	400	Schools & Colleges	46	260
Dental Offices	52	230	Public Spaces	55	220	Shoe Stores	55	220
Dept. Stores, Basement	34	350	Industrial Plants, Offices	38	320	Shop'g. Ctrs., Supermarkets	34	350
Main Floor	40	300	General Offices	34	350	Retail Stores	48	250
Upper Floor	30	400	Plant Areas	40	300	Specialty	60	200

*Persons per ton
12,000 BTU = 1 ton of air conditioning

Mechanical — R158 Air Distribution

R15810-050 Ductwork

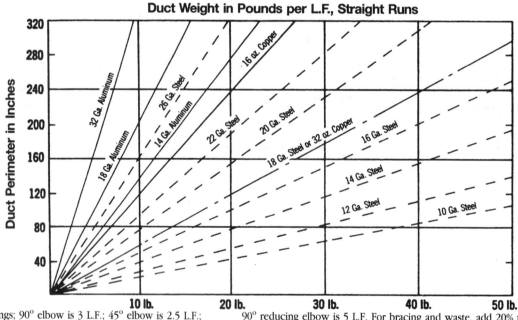

Duct Weight in Pounds per L.F., Straight Runs

Add to the above for fittings; 90° elbow is 3 L.F.; 45° elbow is 2.5 L.F.; offset is 4 L.F.; transition offset is 6 L.F.; square-to-round transition is 4 L.F.; 90° reducing elbow is 5 L.F. For bracing and waste, add 20% to aluminum and copper, 15% to steel.

Electrical — R161 Wiring Methods

R16132-210 Conductors in Conduit

Table below lists maximum number of conductors for various sized conduit using THW, TW or THWN insulations.

Copper Wire Size	1/2"			3/4"			1"			1-1/4"			1-1/2"			2"			2-1/2"			3"			3-1/2"			4"		
	TW	THW	THWN	TW	THW	THWN	TW	THW	THWN	TW	THW	THWN	TW	THW	THWN	TW	THW	THWN	TW	THW	THWN	THW	THWN	THW	THWN		THW	THWN		
#14	9	6	13	15	10	24	25	16	39	44	29	69	60	40	94	99	65	154	142	93		143		192						
#12	7	4	10	12	8	18	19	13	29	35	24	51	47	32	70	78	53	114	111	76	164	117		157						
#10	5	4	6	9	6	11	15	11	18	26	19	32	36	26	44	60	43	73	85	61	104	95	160	127			163			
#8	2	1	3	4	3	5	7	5	9	12	10	16	17	13	22	28	22	36	40	32	51	49	79	66	106		85	136		
#6		1	1		2	4		4	6		7	11		10	15		16	26		23	37	36	57	48	76		62	98		
#4		1	1		1	2		3	4		5	7		7	9		12	16		17	22	27	35	36	47		47	60		
#3		1	1		1	1		2	3		4	6		6	8		10	13		15	19	23	29	31	39		40	51		
#2		1	1		1	1		2	3		4	5		5	7		9	11		13	16	20	25	27	33		34	43		
#1					1	1		1	1		3	3		4	5		6	8		9	12	14	18	19	25		25	32		
1/0					1	1		1	1		2	3		3	4		5	7		8	10	12	15	16	21		21	27		
2/0					1	1		1	1		1	2		3	3		5	6		7	8	10	13	14	17		18	22		
3/0					1	1		1	1		1	1		2	3		4	5		6	7	9	11	12	14		15	18		
4/0						1		1	1		1	1		1	2		3	4		5	6	7	9	10	12		13	15		
250 kcmil								1	1		1	1		1	1		2	3		4	4	6	7	8	10		10	12		
300								1	1		1	1		1	1		2	3		3	4	5	6	7	8		9	11		
350									1		1	1		1	1		1	2		3	3	4	5	6	7		8	9		
400											1	1		1	1		1	1		2	3	4	5	5	6		7	8		
500											1	1		1	1		1	1		1	2	3	4	4	5		6	7		
600												1		1	1		1	1		1	1	3	3	4	4		5	5		
700														1	1		1	1		1	1	2	3	3	4		4	5		
750														1	1		1	1		1	1	2	2	3	3		4	4		

Crews

Crew No.	Bare Costs Hr.	Bare Costs Daily	Incl. Subs O & P Hr.	Incl. Subs O & P Daily	Cost Per Labor-Hour Bare Costs	Cost Per Labor-Hour Incl. O&P
Crew A-1						
1 Building Laborer	$14.45	$115.60	$24.75	$198.00	$14.45	$24.75
1 Gas Eng. Power Tool		70.00		77.00	8.75	9.63
8 L.H., Daily Totals		$185.60		$275.00	$23.20	$34.38
Crew A-1A						
1 Skilled Worker	$19.75	$158.00	$33.90	$271.20	$19.75	$33.90
1 Shot Blaster, 20"		142.00		156.20	17.75	19.53
8 L.H., Daily Totals		$300.00		$427.40	$37.50	$53.43
Crew A-2						
2 Laborers	$14.45	$231.20	$24.75	$396.00	$14.92	$25.33
1 Truck Driver (light)	15.85	126.80	26.50	212.00		
1 Light Truck, 1.5 Ton		172.95		190.25	7.21	7.93
24 L.H., Daily Totals		$530.95		$798.25	$22.13	$33.26
Crew A-2A						
2 Laborers	$14.45	$231.20	$24.75	$396.00	$14.92	$25.33
1 Truck Driver (light)	15.85	126.80	26.50	212.00		
1 Light Truck, 1.5 ton		172.95		190.25		
1 Concrete Saw		110.00		121.00	11.79	12.97
24 L.H., Daily Totals		$640.95		$919.25	$26.71	$38.30
Crew A-3						
1 Truck Driver (heavy)	$16.20	$129.60	$27.05	$216.40	$16.20	$27.05
1 Dump Truck, 12 Ton		365.30		401.85	45.66	50.23
8 L.H., Daily Totals		$494.90		$618.25	$61.86	$77.28
Crew A-3A						
1 Truck Driver (light)	$15.85	$126.80	$26.50	$212.00	$15.85	$26.50
1 Pickup truck (4x4)		144.65		159.10	18.08	19.89
8 L.H., Daily Totals		$271.45		$371.10	$33.93	$46.39
Crew A-3B						
1 Equip. Oper. (medium)	$19.90	$159.20	$33.00	$264.00	$18.05	$30.02
1 Truck Driver (heavy)	16.20	129.60	27.05	216.40		
1 Dump Truck, 16 Ton		447.25		492.00		
1 F.E. Loader, 3 C.Y.		400.80		440.90	53.00	58.30
16 L.H., Daily Totals		$1136.85		$1413.30	$71.05	$88.32
Crew A-3C						
1 Equip. Oper. (light)	$19.10	$152.80	$31.70	$253.60	$19.10	$31.70
1 Wheeled Skid Steer Loader		225.80		248.40	28.23	31.05
8 L.H., Daily Totals		$378.60		$502.00	$47.33	$62.75
Crew A-4						
2 Carpenters	$19.70	$315.20	$33.75	$540.00	$19.12	$32.48
1 Painter, Ordinary	17.95	143.60	29.95	239.60		
24 L.H., Daily Totals		$458.80		$779.60	$19.12	$32.48
Crew A-5						
2 Laborers	$14.45	$231.20	$24.75	$396.00	$14.61	$24.94
.25 Truck Driver (light)	15.85	31.70	26.50	53.00		
.25 Light Truck, 1.5 Ton		43.24		47.55	2.40	2.64
18 L.H., Daily Totals		$306.14		$496.55	$17.01	$27.58
Crew A-6						
1 Chief Of Party	$22.10	$176.80	$38.15	$305.20	$20.93	$36.03
1 Instrument Man	19.75	158.00	33.90	271.20		
16 L.H., Daily Totals		$334.80		$576.40	$20.93	$36.03
Crew A-7						
1 Chief Of Party	$22.10	$176.80	$38.15	$305.20	$20.40	$34.72
1 Instrument Man	19.75	158.00	33.90	271.20		
1 Rodman/Chainman	19.35	154.80	32.10	256.80		
24 L.H., Daily Totals		$489.60		$833.20	$20.40	$34.72
Crew A-8						
1 Chief Of Party	$22.10	$176.80	$38.15	$305.20	$20.14	$34.06
1 Instrument Man	19.75	158.00	33.90	271.20		
2 Rodmen/Chainmen	19.35	309.60	32.10	513.60		
32 L.H., Daily Totals		$644.40		$1090.00	$20.14	$34.06
Crew A-9						
1 Asbestos Foreman	$21.00	$168.00	$36.75	$294.00	$20.56	$36.01
7 Asbestos Workers	20.50	1148.00	35.90	2010.40		
64 L.H., Daily Totals		$1316.00		$2304.40	$20.56	$36.01
Crew A-10						
1 Asbestos Foreman	$21.00	$168.00	$36.75	$294.00	$20.56	$36.01
7 Asbestos Workers	20.50	1148.00	35.90	2010.40		
64 L.H., Daily Totals		$1316.00		$2304.40	$20.56	$36.01
Crew A-10A						
1 Asbestos Foreman	$21.00	$168.00	$36.75	$294.00	$20.67	$36.18
2 Asbestos Workers	20.50	328.00	35.90	574.40		
24 L.H., Daily Totals		$496.00		$868.40	$20.67	$36.18
Crew A-10B						
1 Asbestos Foreman	$21.00	$168.00	$36.75	$294.00	$20.63	$36.11
3 Asbestos Workers	20.50	492.00	35.90	861.60		
32 L.H., Daily Totals		$660.00		$1155.60	$20.63	$36.11
Crew A-10C						
3 Asbestos Workers	$20.50	$492.00	$35.90	$861.60	$20.50	$35.90
1 Flatbed truck		172.95		190.25	7.21	7.93
24 L.H., Daily Totals		$664.95		$1051.85	$27.71	$43.83
Crew A-10D						
2 Asbestos Workers	$20.50	$328.00	$35.90	$574.40	$19.66	$33.56
1 Equip. Oper. (crane)	20.65	165.20	34.25	274.00		
1 Equip. Oper. Oiler	17.00	136.00	28.20	225.60		
1 Hydraulic crane, 33 ton		694.95		764.45	21.72	23.89
32 L.H., Daily Totals		$1324.15		$1838.45	$41.38	$57.45
Crew A-11						
1 Asbestos Foreman	$21.00	$168.00	$36.75	$294.00	$20.56	$36.01
7 Asbestos Workers	20.50	1148.00	35.90	2010.40		
2 Chipping Hammers		33.90		37.30	.53	.58
64 L.H., Daily Totals		$1349.90		$2341.70	$21.09	$36.59
Crew A-12						
1 Asbestos Foreman	$21.00	$168.00	$36.75	$294.00	$20.56	$36.01
7 Asbestos Workers	20.50	1148.00	35.90	2010.40		
1 Large Prod. Vac. Loader		529.80		582.80	8.28	9.11
64 L.H., Daily Totals		$1845.80		$2887.20	$28.84	$45.12
Crew A-13						
1 Equip. Oper. (light)	$19.10	$152.80	$31.70	$253.60	$19.10	$31.70
1 Large Prod. Vac. Loader		529.80		582.80	66.23	72.85
8 L.H., Daily Totals		$682.60		$836.40	$85.33	$104.55

Crews

Crew No.		Bare Costs	Incl. Subs O & P		Cost Per Labor-Hour	
Crew B-1	Hr.	Daily	Hr.	Daily	Bare Costs	Incl. O&P
1 Labor Foreman (outside)	$16.45	$131.60	$28.20	$225.60	$15.12	$25.90
2 Laborers	14.45	231.20	24.75	396.00		
24 L.H., Daily Totals		$362.80		$621.60	$15.12	$25.90
Crew B-2	Hr.	Daily	Hr.	Daily	Bare Costs	Incl. O&P
1 Labor Foreman (outside)	$16.45	$131.60	$28.20	$225.60	$14.85	$25.44
4 Laborers	14.45	462.40	24.75	792.00		
40 L.H., Daily Totals		$594.00		$1017.60	$14.85	$25.44
Crew B-3	Hr.	Daily	Hr.	Daily	Bare Costs	Incl. O&P
1 Labor Foreman (outside)	$16.45	$131.60	$28.20	$225.60	$16.27	$27.47
2 Laborers	14.45	231.20	24.75	396.00		
1 Equip. Oper. (med.)	19.90	159.20	33.00	264.00		
2 Truck Drivers (heavy)	16.20	259.20	27.05	432.80		
1 F.E. Loader, T.M., 2.5 C.Y.		784.00		862.40		
2 Dump Trucks, 16 Ton		894.50		983.95	34.97	38.47
48 L.H., Daily Totals		$2459.70		$3164.75	$51.24	$65.94
Crew B-3A	Hr.	Daily	Hr.	Daily	Bare Costs	Incl. O&P
4 Laborers	$14.45	$462.40	$24.75	$792.00	$15.54	$26.40
1 Equip. Oper. (med.)	19.90	159.20	33.00	264.00		
1 Hyd. Excavator, 1.5 C.Y.		713.20		784.50	17.83	19.61
40 L.H., Daily Totals		$1334.80		$1840.50	$33.37	$46.01
Crew B-3B	Hr.	Daily	Hr.	Daily	Bare Costs	Incl. O&P
2 Laborers	$14.45	$231.20	$24.75	$396.00	$16.25	$27.39
1 Equip. Oper. (med.)	19.90	159.20	33.00	264.00		
1 Truck Drivers (heavy)	16.20	129.60	27.05	216.40		
1 Backhoe Loader, 80 H.P.		273.15		300.45		
1 Dump Trucks, 16 Ton		447.25		492.00	22.51	24.76
32 L.H., Daily Totals		$1240.40		$1668.85	$38.76	$52.15
Crew B-3C	Hr.	Daily	Hr.	Daily	Bare Costs	Incl. O&P
3 Laborers	$14.45	$346.80	$24.75	$594.00	$15.81	$26.81
1 Equip. Oper. (med.)	19.90	159.20	33.00	264.00		
1 F.E. Loader, 3.75 C.Y.		542.40		596.65	16.95	18.65
32 L.H., Daily Totals		$1048.40		$1454.65	$32.76	$45.46
Crew B-4	Hr.	Daily	Hr.	Daily	Bare Costs	Incl. O&P
1 Labor Foreman (outside)	$16.45	$131.60	$28.20	$225.60	$15.08	$25.71
4 Laborers	14.45	462.40	24.75	792.00		
1 Truck Driver (heavy)	16.20	129.60	27.05	216.40		
1 Tractor, 4 x 2, 195 H.P.		325.60		358.15		
1 Platform Trailer		152.00		167.20	9.95	10.95
48 L.H., Daily Totals		$1201.20		$1759.35	$25.03	$36.66
Crew B-5	Hr.	Daily	Hr.	Daily	Bare Costs	Incl. O&P
1 Labor Foreman (outside)	$16.45	$131.60	$28.20	$225.60	$15.94	$27.09
3 Laborers	14.45	346.80	24.75	594.00		
1 Equip. Oper. (med.)	19.90	159.20	33.00	264.00		
1 Air Compr., 250 C.F.M.		127.00		139.70		
2 Air Tools & Accessories		39.00		42.90		
2-50 Ft. Air Hoses, 1.5" Dia.		14.40		15.85		
1 F.E. Loader, T.M., 2.5 C.Y.		784.00		862.40	24.11	26.52
40 L.H., Daily Totals		$1602.00		$2144.45	$40.05	$53.61
Crew B-5A	Hr.	Daily	Hr.	Daily	Bare Costs	Incl. O&P
1 Foreman	$16.45	$131.60	$28.20	$225.60	$16.20	$27.38
6 Laborers	14.45	693.60	24.75	1188.00		
2 Equip. Oper. (med.)	19.90	318.40	33.00	528.00		
1 Equip. Oper. (light)	19.10	152.80	31.70	253.60		
2 Truck Drivers (heavy)	16.20	259.20	27.05	432.80		
1 Air Compr. 365 C.F.M.		174.40		191.85		
2 Pavement Breakers		39.00		42.90		
8 Air Hoses w/Coup.,1"		41.60		45.75		
2 Dump Trucks, 12 Ton		730.60		803.65	10.27	11.29
96 L.H., Daily Totals		$2541.20		$3712.15	$26.47	$38.67
Crew B-5B	Hr.	Daily	Hr.	Daily	Bare Costs	Incl. O&P
1 Powderman	$19.75	$158.00	$33.90	$271.20	$18.02	$30.18
2 Equip. Oper. (med.)	19.90	318.40	33.00	528.00		
3 Truck Drivers (heavy)	16.20	388.80	27.05	649.20		
1 F.E. Ldr. 2-1/2 CY		400.80		440.90		
3 Dump Trucks, 16 Ton		1341.75		1475.95		
1 Air Compr. 365 C.F.M.		174.40		191.85	39.94	43.93
48 L.H., Daily Totals		$2782.15		$3557.10	$57.96	$74.11
Crew B-5C	Hr.	Daily	Hr.	Daily	Bare Costs	Incl. O&P
3 Laborers	$14.45	$346.80	$24.75	$594.00	$16.66	$27.98
1 Equip. Oper. (medium)	19.90	159.20	33.00	264.00		
2 Truck Drivers (heavy)	16.20	259.20	27.05	432.80		
1 Equip. Oper. (crane)	20.65	165.20	34.25	274.00		
1 Equip. Oper. Oiler	17.00	136.00	28.20	225.60		
2 Dump Truck, 16 Ton		894.50		983.95		
1 F.E. Lder, 3.75 C.Y.		1035.00		1138.50		
1 Hyd. Crane, 25 Ton		563.90		620.30	38.96	42.86
64 L.H., Daily Totals		$3559.80		$4533.15	$55.62	$70.84
Crew B-6	Hr.	Daily	Hr.	Daily	Bare Costs	Incl. O&P
2 Laborers	$14.45	$231.20	$24.75	$396.00	$16.00	$27.07
1 Equip. Oper. (light)	19.10	152.80	31.70	253.60		
1 Backhoe Loader, 48 H.P.		203.00		223.30	8.46	9.30
24 L.H., Daily Totals		$587.00		$872.90	$24.46	$36.37
Crew B-7	Hr.	Daily	Hr.	Daily	Bare Costs	Incl. O&P
1 Labor Foreman (outside)	$16.45	$131.60	$28.20	$225.60	$15.69	$26.70
4 Laborers	14.45	462.40	24.75	792.00		
1 Equip. Oper. (med.)	19.90	159.20	33.00	264.00		
1 Chipping Machine		211.05		232.15		
1 F.E. Loader, T.M., 2.5 C.Y.		784.00		862.40		
2 Chain Saws, 36"		98.40		108.25	22.78	25.06
48 L.H., Daily Totals		$1846.65		$2484.40	$38.47	$51.76
Crew B-7A	Hr.	Daily	Hr.	Daily	Bare Costs	Incl. O&P
2 Laborers	$14.45	$231.20	$24.75	$396.00	$16.00	$27.07
1 Equip. Oper. (light)	19.10	152.80	31.70	253.60		
1 Rake w/Tractor		219.05		240.95		
2 Chain Saws, 18"		56.80		62.50	11.49	12.64
24 L.H., Daily Totals		$659.85		$953.05	$27.49	$39.71

Crews

Crew No.	Bare Costs		Incl. Subs O & P		Cost Per Labor-Hour	
Crew B-8	Hr.	Daily	Hr.	Daily	Bare Costs	Incl. O&P
1 Labor Foreman (outside)	$16.45	$131.60	$28.20	$225.60	$16.79	$28.26
2 Laborers	14.45	231.20	24.75	396.00		
2 Equip. Oper. (med.)	19.90	318.40	33.00	528.00		
2 Truck Drivers (heavy)	16.20	259.20	27.05	432.80		
1 Hyd. Crane, 25 Ton		559.20		615.10		
1 F.E. Loader, T.M., 2.5 C.Y.		784.00		862.40		
2 Dump Trucks, 16 Ton		894.50		983.95	39.96	43.95
56 L.H., Daily Totals		$3178.10		$4043.85	$56.75	$72.21
Crew B-9	Hr.	Daily	Hr.	Daily	Bare Costs	Incl. O&P
1 Labor Foreman (outside)	$16.45	$131.60	$28.20	$225.60	$14.85	$25.44
4 Laborers	14.45	462.40	24.75	792.00		
1 Air Compr., 250 C.F.M.		127.00		139.70		
2 Air Tools & Accessories		39.00		42.90		
2-50 Ft. Air Hoses, 1.5" Dia.		14.40		15.85	4.51	4.96
40 L.H., Daily Totals		$774.40		$1216.05	$19.36	$30.40
Crew B-9A	Hr.	Daily	Hr.	Daily	Bare Costs	Incl. O&P
2 Laborers	$14.45	$231.20	$24.75	$396.00	$15.03	$25.52
1 Truck Driver (heavy)	16.20	129.60	27.05	216.40		
1 Water Tanker		205.90		226.50		
1 Tractor		325.60		358.15		
2-50 Ft. Disch. Hoses		12.80		14.10	22.68	24.95
24 L.H., Daily Totals		$905.10		$1211.15	$37.71	$50.47
Crew B-9B	Hr.	Daily	Hr.	Daily	Bare Costs	Incl. O&P
2 Laborers	$14.45	$231.20	$24.75	$396.00	$15.03	$25.52
1 Truck Driver (heavy)	16.20	129.60	27.05	216.40		
2-50 Ft. Disch. Hoses		12.80		14.10		
1 Water Tanker		205.90		226.50		
1 Tractor		325.60		358.15		
1 Pressure Washer		60.40		66.45	25.20	27.72
24 L.H., Daily Totals		$965.50		$1277.60	$40.23	$53.24
Crew B-9C	Hr.	Daily	Hr.	Daily	Bare Costs	Incl. O&P
1 Labor Foreman (outside)	$16.45	$131.60	$28.20	$225.60	$14.85	$25.44
4 Laborers	14.45	462.40	24.75	792.00		
1 Air Compr., 250 C.F.M.		127.00		139.70		
2-50 Ft. Air Hoses, 1.5" Dia.		14.40		15.85		
2 Breaker, Pavement, 60 lb.		39.00		42.90	4.51	4.96
40 L.H., Daily Totals		$774.40		$1216.05	$19.36	$30.40
Crew B-10	Hr.	Daily	Hr.	Daily	Bare Costs	Incl. O&P
1 Equip. Oper. (med.)	$19.90	$159.20	$33.00	$264.00	$19.90	$33.00
8 L.H., Daily Totals		$159.20		$264.00	$19.90	$33.00
Crew B-10A	Hr.	Daily	Hr.	Daily	Bare Costs	Incl. O&P
1 Equip. Oper. (med.)	$19.90	$159.20	$33.00	$264.00	$19.90	$33.00
1 Roll. Compact., 2K Lbs.		94.25		103.70	11.78	12.96
8 L.H., Daily Totals		$253.45		$367.70	$31.68	$45.96
Crew B-10B	Hr.	Daily	Hr.	Daily	Bare Costs	Incl. O&P
1 Equip. Oper. (med.)	$19.90	$159.20	$33.00	$264.00	$19.90	$33.00
1 Dozer, 200 H.P.		806.80		887.50	100.85	110.94
8 L.H., Daily Totals		$966.00		$1151.50	$120.75	$143.94

Crew No.	Bare Costs		Incl. Subs O & P		Cost Per Labor-Hour	
Crew B-10C	Hr.	Daily	Hr.	Daily	Bare Costs	Incl. O&P
1 Equip. Oper. (med.)	$19.90	$159.20	$33.00	$264.00	$19.90	$33.00
1 Dozer, 200 H.P.		806.80		887.50		
1 Vibratory Roller, Towed		110.00		121.00	114.60	126.06
8 L.H., Daily Totals		$1076.00		$1272.50	$134.50	$159.06
Crew B-10D	Hr.	Daily	Hr.	Daily	Bare Costs	Incl. O&P
1 Equip. Oper. (med.)	$19.90	$159.20	$33.00	$264.00	$19.90	$33.00
1 Dozer, 200 H.P		806.80		887.50		
1 Sheepsft. Roller, Towed		126.40		139.05	116.65	128.32
8 L.H., Daily Totals		$1092.40		$1290.55	$136.55	$161.32
Crew B-10E	Hr.	Daily	Hr.	Daily	Bare Costs	Incl. O&P
1 Equip. Oper. (med.)	$19.90	$159.20	$33.00	$264.00	$19.90	$33.00
1 Tandem Roller, 5 Ton		133.20		146.50	16.65	18.32
8 L.H., Daily Totals		$292.40		$410.50	$36.55	$51.32
Crew B-10F	Hr.	Daily	Hr.	Daily	Bare Costs	Incl. O&P
1 Equip. Oper. (med.)	$19.90	$159.20	$33.00	$264.00	$19.90	$33.00
1 Tandem Roller, 10 Ton		224.00		246.40	28.00	30.80
8 L.H., Daily Totals		$383.20		$510.40	$47.90	$63.80
Crew B-10G	Hr.	Daily	Hr.	Daily	Bare Costs	Incl. O&P
1 Equip. Oper. (med.)	$19.90	$159.20	$33.00	$264.00	$19.90	$33.00
1 Sheepsft. Roll., 130 H.P.		567.60		624.35	70.95	78.05
8 L.H., Daily Totals		$726.80		$888.35	$90.85	$111.05
Crew B-10H	Hr.	Daily	Hr.	Daily	Bare Costs	Incl. O&P
1 Equip. Oper. (med.)	$19.90	$159.20	$33.00	$264.00	$19.90	$33.00
1 Diaphr. Water Pump, 2"		29.40		32.35		
1-20 Ft. Suction Hose, 2"		6.50		7.15		
2-50 Ft. Disch. Hoses, 2"		10.80		11.90	5.84	6.42
8 L.H., Daily Totals		$205.90		$315.40	$25.74	$39.42
Crew B-10I	Hr.	Daily	Hr.	Daily	Bare Costs	Incl. O&P
1 Equip. Oper. (med.)	$19.90	$159.20	$33.00	$264.00	$19.90	$33.00
1 Diaphr. Water Pump, 4"		65.65		72.20		
1-20 Ft. Suction Hose, 4"		12.50		13.75		
2-50 Ft. Disch. Hoses, 4"		17.00		18.70	11.89	13.08
8 L.H., Daily Totals		$254.35		$368.65	$31.79	$46.08
Crew B-10J	Hr.	Daily	Hr.	Daily	Bare Costs	Incl. O&P
1 Equip. Oper. (med.)	$19.90	$159.20	$33.00	$264.00	$19.90	$33.00
1 Centr. Water Pump, 3"		38.00		41.80		
1-20 Ft. Suction Hose, 3"		9.50		10.45		
2-50 Ft. Disch. Hoses, 3"		12.80		14.10	7.54	8.29
8 L.H., Daily Totals		$219.50		$330.35	$27.44	$41.29
Crew B-10K	Hr.	Daily	Hr.	Daily	Bare Costs	Incl. O&P
1 Equip. Oper. (med.)	$19.90	$159.20	$33.00	$264.00	$19.90	$33.00
1 Centr. Water Pump, 6"		157.20		172.90		
1-20 Ft. Suction Hose, 6"		22.50		24.75		
2-50 Ft. Disch. Hoses, 6"		41.10		45.20	27.60	30.36
8 L.H., Daily Totals		$380.00		$506.85	$47.50	$63.36
Crew B-10L	Hr.	Daily	Hr.	Daily	Bare Costs	Incl. O&P
1 Equip. Oper. (med.)	$19.90	$159.20	$33.00	$264.00	$19.90	$33.00
1 Dozer, 75 H.P.		279.60		307.55	34.95	38.45
8 L.H., Daily Totals		$438.80		$571.55	$54.85	$71.45

Crews

Crew No.	Bare Costs		Incl. Subs O & P		Cost Per Labor-Hour	
Crew B-10M	Hr.	Daily	Hr.	Daily	Bare Costs	Incl. O&P
1 Equip. Oper. (med.)	$19.90	$159.20	$33.00	$264.00	$19.90	$33.00
1 Dozer, 300 H.P.		1108.00		1218.80	138.50	152.35
8 L.H., Daily Totals		$1267.20		$1482.80	$158.40	$185.35
Crew B-10N	Hr.	Daily	Hr.	Daily	Bare Costs	Incl. O&P
1 Equip. Oper. (med.)	$19.90	$159.20	$33.00	$264.00	$19.90	$33.00
1 F.E. Loader, T.M., 1.5 C.Y.		329.95		362.95	41.24	45.37
8 L.H., Daily Totals		$489.15		$626.95	$61.14	$78.37
Crew B-10O	Hr.	Daily	Hr.	Daily	Bare Costs	Incl. O&P
1 Equip. Oper. (med.)	$19.90	$159.20	$33.00	$264.00	$19.90	$33.00
1 F.E. Loader, T.M., 2.25 C.Y.		450.00		495.00	56.25	61.88
8 L.H., Daily Totals		$609.20		$759.00	$76.15	$94.88
Crew B-10P	Hr.	Daily	Hr.	Daily	Bare Costs	Incl. O&P
1 Equip. Oper. (med.)	$19.90	$159.20	$33.00	$264.00	$19.90	$33.00
1 F.E. Loader, T.M., 2.5 C.Y.		784.00		862.40	98.00	107.80
8 L.H., Daily Totals		$943.20		$1126.40	$117.90	$140.80
Crew B-10Q	Hr.	Daily	Hr.	Daily	Bare Costs	Incl. O&P
1 Equip. Oper. (med.)	$19.90	$159.20	$33.00	$264.00	$19.90	$33.00
1 F.E. Loader, T.M., 5 C.Y.		1035.00		1138.50	129.38	142.31
8 L.H., Daily Totals		$1194.20		$1402.50	$149.28	$175.31
Crew B-10R	Hr.	Daily	Hr.	Daily	Bare Costs	Incl. O&P
1 Equip. Oper. (med.)	$19.90	$159.20	$33.00	$264.00	$19.90	$33.00
1 F.E. Loader, W.M., 1 C.Y.		233.60		256.95	29.20	32.12
8 L.H., Daily Totals		$392.80		$520.95	$49.10	$65.12
Crew B-10S	Hr.	Daily	Hr.	Daily	Bare Costs	Incl. O&P
1 Equip. Oper. (med.)	$19.90	$159.20	$33.00	$264.00	$19.90	$33.00
1 F.E. Loader, W.M., 1.5 C.Y.		303.00		333.30	37.88	41.66
8 L.H., Daily Totals		$462.20		$597.30	$57.78	$74.66
Crew B-10T	Hr.	Daily	Hr.	Daily	Bare Costs	Incl. O&P
1 Equip. Oper. (med.)	$19.90	$159.20	$33.00	$264.00	$19.90	$33.00
1 F.E. Ldr, W.M.,2.5CY		400.80		440.90	50.10	55.11
8 L.H., Daily Totals		$560.00		$704.90	$70.00	$88.11
Crew B-10U	Hr.	Daily	Hr.	Daily	Bare Costs	Incl. O&P
1 Equip. Oper. (med.)	$19.90	$159.20	$33.00	$264.00	$19.90	$33.00
1 F.E. Loader, W.M., 5.5 C.Y.		861.20		947.30	107.65	118.42
8 L.H., Daily Totals		$1020.40		$1211.30	$127.55	$151.42
Crew B-10V	Hr.	Daily	Hr.	Daily	Bare Costs	Incl. O&P
1 Equip. Oper. (med.)	$19.90	$159.20	$33.00	$264.00	$19.90	$33.00
1 Dozer, 700 H.P.		2752.00		3027.20	344.00	378.40
8 L.H., Daily Totals		$2911.20		$3291.20	$363.90	$411.40
Crew B-10W	Hr.	Daily	Hr.	Daily	Bare Costs	Incl. O&P
1 Equip. Oper. (med.)	$19.90	$159.20	$33.00	$264.00	$19.90	$33.00
1 Dozer, 105 H.P.		393.00		432.30	49.13	54.04
8 L.H., Daily Totals		$552.20		$696.30	$69.03	$87.04
Crew B-10X	Hr.	Daily	Hr.	Daily	Bare Costs	Incl. O&P
1 Equip. Oper. (med.)	$19.90	$159.20	$33.00	$264.00	$19.90	$33.00
1 Dozer, 410 H.P.		1397.00		1536.70	174.63	192.09
8 L.H., Daily Totals		$1556.20		$1800.70	$194.53	$225.09

Crew No.	Bare Costs		Incl. Subs O & P		Cost Per Labor-Hour	
Crew B-10Y	Hr.	Daily	Hr.	Daily	Bare Costs	Incl. O&P
1 Equip. Oper. (med.)	$19.90	$159.20	$33.00	$264.00	$19.90	$33.00
1 Vibratory Drum Roller		372.00		409.20	46.50	51.15
8 L.H., Daily Totals		$531.20		$673.20	$66.40	$84.15
Crew B-11	Hr.	Daily	Hr.	Daily	Bare Costs	Incl. O&P
1 Equipment Oper. (med.)	$19.90	$159.20	$33.00	$264.00	$17.17	$28.88
1 Laborer	14.45	115.60	24.75	198.00		
16 L.H., Daily Totals		$274.80		$462.00	$17.17	$28.88
Crew B-11A	Hr.	Daily	Hr.	Daily	Bare Costs	Incl. O&P
1 Equipment Oper. (med.)	$19.90	$159.20	$33.00	$264.00	$17.17	$28.88
1 Laborer	14.45	115.60	24.75	198.00		
1 Dozer, 200 H.P.		806.80		887.50	50.43	55.47
16 L.H., Daily Totals		$1081.60		$1349.50	$67.60	$84.35
Crew B-11B	Hr.	Daily	Hr.	Daily	Bare Costs	Incl. O&P
1 Equipment Oper. (med.)	$19.90	$159.20	$33.00	$264.00	$17.17	$28.88
1 Laborer	14.45	115.60	24.75	198.00		
1 Dozer, 200 H.P.		806.80		887.50		
1 Air Powered Tamper		15.80		17.40		
1 Air Compr. 365 C.F.M.		174.40		191.85		
2-50 Ft. Air Hoses, 1.5" Dia.		14.40		15.85	63.21	69.53
16 L.H., Daily Totals		$1286.20		$1574.60	$80.38	$98.41
Crew B-11C	Hr.	Daily	Hr.	Daily	Bare Costs	Incl. O&P
1 Equipment Oper. (med.)	$19.90	$159.20	$33.00	$264.00	$17.17	$28.88
1 Laborer	14.45	115.60	24.75	198.00		
1 Backhoe Loader, 48 H.P.		203.00		223.30	12.69	13.96
16 L.H., Daily Totals		$477.80		$685.30	$29.86	$42.84
Crew B-11K	Hr.	Daily	Hr.	Daily	Bare Costs	Incl. O&P
1 Equipment Oper. (med.)	$19.90	$159.20	$33.00	$264.00	$17.17	$28.88
1 Laborer	14.45	115.60	24.75	198.00		
1 Trencher, 8' D., 16" W.		540.90		595.00	33.81	37.19
16 L.H., Daily Totals		$815.70		$1057.00	$50.98	$66.07
Crew B-11L	Hr.	Daily	Hr.	Daily	Bare Costs	Incl. O&P
1 Equipment Oper. (med.)	$19.90	$159.20	$33.00	$264.00	$17.17	$28.88
1 Laborer	14.45	115.60	24.75	198.00		
1 Grader, 30,000 Lbs.		516.00		567.60	32.25	35.48
16 L.H., Daily Totals		$790.80		$1029.60	$49.42	$64.36
Crew B-11M	Hr.	Daily	Hr.	Daily	Bare Costs	Incl. O&P
1 Equipment Oper. (med.)	$19.90	$159.20	$33.00	$264.00	$17.17	$28.88
1 Laborer	14.45	115.60	24.75	198.00		
1 Backhoe Loader, 80 H.P.		273.15		300.45	17.07	18.78
16 L.H., Daily Totals		$547.95		$762.45	$34.24	$47.66
Crew B-12	Hr.	Daily	Hr.	Daily	Bare Costs	Incl. O&P
1 Equip. Oper. (crane)	$20.65	$165.20	$34.25	$274.00	$20.65	$34.25
8 L.H., Daily Totals		$165.20		$274.00	$20.65	$34.25
Crew B-12A	Hr.	Daily	Hr.	Daily	Bare Costs	Incl. O&P
1 Equip. Oper. (crane)	$20.65	$165.20	$34.25	$274.00	$20.65	$34.25
1 Hyd. Excavator, 1 C.Y.		538.00		591.80	67.25	73.98
8 L.H., Daily Totals		$703.20		$865.80	$87.90	$108.23

Crews

Crew No.	Bare Costs		Incl. Subs O & P		Cost Per Labor-Hour	
Crew B-12B	Hr.	Daily	Hr.	Daily	Bare Costs	Incl. O&P
1 Equip. Oper. (crane)	$20.65	$165.20	$34.25	$274.00	$20.65	$34.25
1 Hyd. Excavator, 1.5 C.Y.		713.20		784.50	89.15	98.07
8 L.H., Daily Totals		$878.40		$1058.50	$109.80	$132.32
Crew B-12C	Hr.	Daily	Hr.	Daily	Bare Costs	Incl. O&P
1 Equip. Oper. (crane)	$20.65	$165.20	$34.25	$274.00	$20.65	$34.25
1 Hyd. Excavator, 2 C.Y.		1017.00		1118.70	127.13	139.84
8 L.H., Daily Totals		$1182.20		$1392.70	$147.78	$174.09
Crew B-12D	Hr.	Daily	Hr.	Daily	Bare Costs	Incl. O&P
1 Equip. Oper. (crane)	$20.65	$165.20	$34.25	$274.00	$20.65	$34.25
1 Hyd. Excavator, 3.5 C.Y.		2126.00		2338.60	265.75	292.33
8 L.H., Daily Totals		$2291.20		$2612.60	$286.40	$326.58
Crew B-12E	Hr.	Daily	Hr.	Daily	Bare Costs	Incl. O&P
1 Equip. Oper. (crane)	$20.65	$165.20	$34.25	$274.00	$20.65	$34.25
1 Hyd. Excavator, .5 C.Y.		342.80		377.10	42.85	47.14
8 L.H., Daily Totals		$508.00		$651.10	$63.50	$81.39
Crew B-12F	Hr.	Daily	Hr.	Daily	Bare Costs	Incl. O&P
1 Equip. Oper. (crane)	$20.65	$165.20	$34.25	$274.00	$20.65	$34.25
1 Hyd. Excavator, .75 C.Y.		455.80		501.40	56.98	62.67
8 L.H., Daily Totals		$621.00		$775.40	$77.63	$96.92
Crew B-12G	Hr.	Daily	Hr.	Daily	Bare Costs	Incl. O&P
1 Equip. Oper. (crane)	$20.65	$165.20	$34.25	$274.00	$20.65	$34.25
1 Power Shovel, .5 C.Y.		477.00		524.70		
1 Clamshell Bucket, .5 C.Y		48.80		53.70	65.73	72.30
8 L.H., Daily Totals		$691.00		$852.40	$86.38	$106.55
Crew B-12H	Hr.	Daily	Hr.	Daily	Bare Costs	Incl. O&P
1 Equip. Oper. (crane)	$20.65	$165.20	$34.25	$274.00	$20.65	$34.25
1 Power Shovel, 1 C.Y.		516.80		568.50		
1 Clamshell Bucket, 1 C.Y.		70.00		77.00	73.35	80.69
8 L.H., Daily Totals		$752.00		$919.50	$94.00	$114.94
Crew B-12I	Hr.	Daily	Hr.	Daily	Bare Costs	Incl. O&P
1 Equip. Oper. (crane)	$20.65	$165.20	$34.25	$274.00	$20.65	$34.25
1 Power Shovel, .75 C.Y.		492.60		541.85		
1 Dragline Bucket, .75 C.Y.		33.20		36.50	65.73	72.30
8 L.H., Daily Totals		$691.00		$852.35	$86.38	$106.55
Crew B-12J	Hr.	Daily	Hr.	Daily	Bare Costs	Incl. O&P
1 Equip. Oper. (crane)	$20.65	$165.20	$34.25	$274.00	$20.65	$34.25
1 Gradall, 3 Ton, .5 C.Y.		618.20		680.00	77.28	85.00
8 L.H., Daily Totals		$783.40		$954.00	$97.93	$119.25
Crew B-12K	Hr.	Daily	Hr.	Daily	Bare Costs	Incl. O&P
1 Equip. Oper. (crane)	$20.65	$165.20	$34.25	$274.00	$20.65	$34.25
1 Gradall, 3 Ton, 1 C.Y.		817.85		899.65	102.23	112.45
8 L.H., Daily Totals		$983.05		$1173.65	$122.88	$146.70
Crew B-12L	Hr.	Daily	Hr.	Daily	Bare Costs	Incl. O&P
1 Equip. Oper. (crane)	$20.65	$165.20	$34.25	$274.00	$20.65	$34.25
1 Power Shovel, .5 C.Y.		477.00		524.70		
1 F.E. Attachment, .5 C.Y.		56.20		61.80	66.65	73.32
8 L.H., Daily Totals		$698.40		$860.50	$87.30	$107.57

Crew No.	Bare Costs		Incl. Subs O & P		Cost Per Labor-Hour	
Crew B-12M	Hr.	Daily	Hr.	Daily	Bare Costs	Incl. O&P
1 Equip. Oper. (crane)	$20.65	$165.20	$34.25	$274.00	$20.65	$34.25
1 Power Shovel, .75 C.Y.		492.60		541.85		
1 F.E. Attachment, .75 C.Y.		101.20		111.30	74.23	81.65
8 L.H., Daily Totals		$759.00		$927.15	$94.88	$115.90
Crew B-12N	Hr.	Daily	Hr.	Daily	Bare Costs	Incl. O&P
1 Equip. Oper. (crane)	$20.65	$165.20	$34.25	$274.00	$20.65	$34.25
1 Power Shovel, 1 C.Y.		516.80		568.50		
1 F.E. Attachment, 1 C.Y.		133.20		146.50	81.25	89.38
8 L.H., Daily Totals		$815.20		$989.00	$101.90	$123.63
Crew B-12O	Hr.	Daily	Hr.	Daily	Bare Costs	Incl. O&P
1 Equip. Oper. (crane)	$20.65	$165.20	$34.25	$274.00	$20.65	$34.25
1 Power Shovel, 1.5 C.Y.		714.30		785.75		
1 F.E. Attachment, 1.5 C.Y.		161.00		177.10	109.41	120.35
8 L.H., Daily Totals		$1040.50		$1236.85	$130.06	$154.60
Crew B-12P	Hr.	Daily	Hr.	Daily	Bare Costs	Incl. O&P
1 Equip. Oper. (crane)	$20.65	$165.20	$34.25	$274.00	$20.65	$34.25
1 Crawler Crane, 40 Ton		714.30		785.75		
1 Dragline Bucket, 1.5 C.Y.		48.95		53.85	95.41	104.95
8 L.H., Daily Totals		$928.45		$1113.60	$116.06	$139.20
Crew B-12Q	Hr.	Daily	Hr.	Daily	Bare Costs	Incl. O&P
1 Equip. Oper. (crane)	$20.65	$165.20	$34.25	$274.00	$20.65	$34.25
1 Hyd. Excavator, 5/8 C.Y.		411.60		452.75	51.45	56.60
8 L.H., Daily Totals		$576.80		$726.75	$72.10	$90.85
Crew B-12R	Hr.	Daily	Hr.	Daily	Bare Costs	Incl. O&P
1 Equip. Oper. (crane)	$20.65	$165.20	$34.25	$274.00	$20.65	$34.25
1 Hyd. Excavator, 1.5 C.Y.		713.20		784.50	89.15	98.07
8 L.H., Daily Totals		$878.40		$1058.50	$109.80	$132.32
Crew B-12S	Hr.	Daily	Hr.	Daily	Bare Costs	Incl. O&P
1 Equip. Oper. (crane)	$20.65	$165.20	$34.25	$274.00	$20.65	$34.25
1 Hyd. Excavator, 2.5 C.Y.		1673.00		1840.30	209.13	230.04
8 L.H., Daily Totals		$1838.20		$2114.30	$229.78	$264.29
Crew B-12T	Hr.	Daily	Hr.	Daily	Bare Costs	Incl. O&P
1 Equip. Oper. (crane)	$20.65	$165.20	$34.25	$274.00	$20.65	$34.25
1 Crawler Crane, 75 Ton		939.20		1033.10		
1 F.E. Attachment, 3 C.Y.		300.40		330.45	154.95	170.45
8 L.H., Daily Totals		$1404.80		$1637.55	$175.60	$204.70
Crew B-12V	Hr.	Daily	Hr.	Daily	Bare Costs	Incl. O&P
1 Equip. Oper. (crane)	$20.65	$165.20	$34.25	$274.00	$20.65	$34.25
1 Crawler Crane, 75 Ton		939.20		1033.10		
1 Dragline Bucket, 3 C.Y.		81.60		89.75	127.60	140.36
8 L.H., Daily Totals		$1186.00		$1396.85	$148.25	$174.61
Crew B-13	Hr.	Daily	Hr.	Daily	Bare Costs	Incl. O&P
1 Labor Foreman (outside)	$16.45	$131.60	$28.20	$225.60	$15.82	$26.91
4 Laborers	14.45	462.40	24.75	792.00		
1 Equip. Oper. (crane)	20.65	165.20	34.25	274.00		
1 Hyd. Crane, 25 Ton		559.20		615.10	11.65	12.82
48 L.H., Daily Totals		$1318.40		$1906.70	$27.47	$39.73

Crews

Crew No.	Bare Costs		Incl. Subs O & P		Cost Per Labor-Hour	
Crew B-13A	Hr.	Daily	Hr.	Daily	Bare Costs	Incl. O&P
1 Foreman	$16.45	$131.60	$28.20	$225.60	$16.79	$28.26
2 Laborers	14.45	231.20	24.75	396.00		
2 Equipment Operator	19.90	318.40	33.00	528.00		
2 Truck Drivers (heavy)	16.20	259.20	27.05	432.80		
1 Crane, 75 Ton		939.20		1033.10		
1 F.E. Lder, 3.75 C.Y.		1035.00		1138.50		
2 Dump Trucks, 12 Ton		730.60		803.65	48.30	53.13
56 L.H., Daily Totals		$3645.20		$4557.65	$65.09	$81.39
Crew B-13B	Hr.	Daily	Hr.	Daily	Bare Costs	Incl. O&P
1 Labor Foreman (outside)	$16.45	$131.60	$28.20	$225.60	$15.99	$27.09
4 Laborers	14.45	462.40	24.75	792.00		
1 Equip. Oper. (crane)	20.65	165.20	34.25	274.00		
1 Equip. Oper. Oiler	17.00	136.00	28.20	225.60		
1 Hyd. Crane, 55 Ton		808.20		889.00	14.43	15.88
56 L.H., Daily Totals		$1703.40		$2406.20	$30.42	$42.97
Crew B-13C	Hr.	Daily	Hr.	Daily	Bare Costs	Incl. O&P
1 Labor Foreman (outside)	$16.45	$131.60	$28.20	$225.60	$15.99	$27.09
4 Laborers	14.45	462.40	24.75	792.00		
1 Equip. Oper. (crane)	20.65	165.20	34.25	274.00		
1 Equip. Oper. Oiler	17.00	136.00	28.20	225.60		
1 Crawler Crane, 100 Ton		1139.00		1252.90	20.34	22.37
56 L.H., Daily Totals		$2034.20		$2770.10	$36.33	$49.46
Crew B-14	Hr.	Daily	Hr.	Daily	Bare Costs	Incl. O&P
1 Labor Foreman (outside)	$16.45	$131.60	$28.20	$225.60	$15.56	$26.48
4 Laborers	14.45	462.40	24.75	792.00		
1 Equip. Oper. (light)	19.10	152.80	31.70	253.60		
1 Backhoe Loader, 48 H.P.		203.00		223.30	4.23	4.65
48 L.H., Daily Totals		$949.80		$1494.50	$19.79	$31.13
Crew B-15	Hr.	Daily	Hr.	Daily	Bare Costs	Incl. O&P
1 Equipment Oper. (med)	$19.90	$159.20	$33.00	$264.00	$17.01	$28.42
.5 Laborer	14.45	57.80	24.75	99.00		
2 Truck Drivers (heavy)	16.20	259.20	27.05	432.80		
2 Dump Trucks, 16 Ton		894.50		983.95		
1 Dozer, 200 H.P.		806.80		887.50	60.76	66.84
28 L.H., Daily Totals		$2177.50		$2667.25	$77.77	$95.26
Crew B-16	Hr.	Daily	Hr.	Daily	Bare Costs	Incl. O&P
1 Labor Foreman (outside)	$16.45	$131.60	$28.20	$225.60	$15.39	$26.19
2 Laborers	14.45	231.20	24.75	396.00		
1 Truck Driver (heavy)	16.20	129.60	27.05	216.40		
1 Dump Truck, 16 Ton		447.25		492.00	13.98	15.37
32 L.H., Daily Totals		$939.65		$1330.00	$29.37	$41.56
Crew B-17	Hr.	Daily	Hr.	Daily	Bare Costs	Incl. O&P
2 Laborers	$14.45	$231.20	$24.75	$396.00	$16.05	$27.06
1 Equip. Oper. (light)	19.10	152.80	31.70	253.60		
1 Truck Driver (heavy)	16.20	129.60	27.05	216.40		
1 Backhoe Loader, 48 H.P.		203.00		223.30		
1 Dump Truck, 12 Ton		365.30		401.85	17.76	19.54
32 L.H., Daily Totals		$1081.90		$1491.15	$33.81	$46.60
Crew B-18	Hr.	Daily	Hr.	Daily	Bare Costs	Incl. O&P
1 Labor Foreman (outside)	$16.45	$131.60	$28.20	$225.60	$15.12	$25.90
2 Laborers	14.45	231.20	24.75	396.00		
1 Vibrating Compactor		60.60		66.65	2.53	2.78
24 L.H., Daily Totals		$423.40		$688.25	$17.65	$28.68

Crew No.	Bare Costs		Incl. Subs O & P		Cost Per Labor-Hour	
Crew B-19	Hr.	Daily	Hr.	Daily	Bare Costs	Incl. O&P
1 Pile Driver Foreman	$21.45	$171.60	$40.00	$320.00	$19.19	$34.86
4 Pile Drivers	19.45	622.40	36.25	1160.00		
1 Equip. Oper. (crane)	20.65	165.20	34.25	274.00		
1 Building Laborer	14.45	115.60	24.75	198.00		
1 Crane, 40 Ton & Access.		714.30		785.75		
60 L.F. Leads, 15K Ft. Lbs.		264.00		290.40		
1 Hammer, 15K Ft. Lbs.		311.40		342.55		
1 Air Compr., 600 C.F.M.		278.00		305.80		
2-50 Ft. Air Hoses, 3" Dia.		33.80		37.20	28.60	31.46
56 L.H., Daily Totals		$2676.30		$3713.70	$47.79	$66.32
Crew B-19A	Hr.	Daily	Hr.	Daily	Bare Costs	Incl. O&P
1 Pile Driver Foreman	$21.45	$171.60	$40.00	$320.00	$19.69	$35.21
4 Pile Drivers	19.45	622.40	36.25	1160.00		
2 Equip. Oper. (crane)	20.65	330.40	34.25	548.00		
1 Equip. Oper. Oiler	17.00	136.00	28.20	225.60		
1 Crawler Crane, 75 Ton		939.20		1033.10		
60 L.F. Leads, 25K Ft. Lbs.		315.00		346.50		
1 Air Compressor, 750 CFM		300.80		330.90		
4-50 Ft. Air Hose, 3" Dia.		67.60		74.35	25.35	27.89
64 L.H., Daily Totals		$2883.00		$4038.45	$45.04	$63.10
Crew B-20	Hr.	Daily	Hr.	Daily	Bare Costs	Incl. O&P
1 Labor Foreman (out)	$16.45	$131.60	$28.20	$225.60	$15.12	$25.90
2 Laborer	14.45	231.20	24.75	396.00		
24 L.H., Daily Totals		$362.80		$621.60	$15.12	$25.90
Crew B-20A	Hr.	Daily	Hr.	Daily	Bare Costs	Incl. O&P
1 Labor Foreman	$16.45	$131.60	$28.20	$225.60	$17.60	$29.56
1 Laborer	14.45	115.60	24.75	198.00		
1 Plumber	21.95	175.60	36.30	290.40		
1 Plumber Apprentice	17.55	140.40	29.00	232.00		
32 L.H., Daily Totals		$563.20		$946.00	$17.60	$29.56
Crew B-21	Hr.	Daily	Hr.	Daily	Bare Costs	Incl. O&P
1 Labor Foreman (out)	$16.45	$131.60	$28.20	$225.60	$15.91	$27.09
2 Laborer	14.45	231.20	24.75	396.00		
.5 Equip. Oper. (crane)	20.65	82.60	34.25	137.00		
.5 S.P. Crane, 5 Ton		138.23		152.05	4.94	5.43
28 L.H., Daily Totals		$583.63		$910.65	$20.85	$32.52
Crew B-21A	Hr.	Daily	Hr.	Daily	Bare Costs	Incl. O&P
1 Labor Foreman	$16.45	$131.60	$28.20	$225.60	$18.21	$30.50
1 Laborer	14.45	115.60	24.75	198.00		
1 Plumber	21.95	175.60	36.30	290.40		
1 Plumber Apprentice	17.55	140.40	29.00	232.00		
1 Equip. Oper. (crane)	20.65	165.20	34.25	274.00		
1 S.P. Crane, 12 Ton		401.35		441.50	10.03	11.04
40 L.H., Daily Totals		$1129.75		$1661.50	$28.24	$41.54
Crew B-22	Hr.	Daily	Hr.	Daily	Bare Costs	Incl. O&P
1 Labor Foreman (out)	$16.45	$131.60	$28.20	$225.60	$16.22	$27.57
2 Laborer	14.45	231.20	24.75	396.00		
.75 Equip. Oper. (crane)	20.65	123.90	34.25	205.50		
.75 S.P. Crane, 5 Ton		207.34		228.05	6.91	7.60
30 L.H., Daily Totals		$694.04		$1055.15	$23.13	$35.17

Crews

Crew No.	Bare Costs		Incl. Subs O & P		Cost Per Labor-Hour	
Crew B-22A	Hr.	Daily	Hr.	Daily	Bare Costs	Incl. O&P
1 Labor Foreman (out)	$16.45	$131.60	$28.20	$225.60	$16.97	$28.90
1 Skilled Worker	19.75	158.00	33.90	271.20		
2 Laborers	14.45	231.20	24.75	396.00		
.75 Equipment Oper. (crane)	20.65	123.90	34.25	205.50		
.75 Crane, 5 Ton		207.34		228.05		
1 Generator, 5 KW		46.90		51.60		
1 Butt Fusion Machine		188.00		206.80	11.64	12.80
38 L.H., Daily Totals		$1086.94		$1584.75	$28.61	$41.70
Crew B-22B	Hr.	Daily	Hr.	Daily	Bare Costs	Incl. O&P
1 Skilled Worker	$19.75	$158.00	$33.90	$271.20	$17.10	$29.33
1 Laborer	14.45	115.60	24.75	198.00		
1 Electro Fusion Machine		86.00		94.60	5.38	5.91
16 L.H., Daily Totals		$359.60		$563.80	$22.48	$35.24
Crew B-23	Hr.	Daily	Hr.	Daily	Bare Costs	Incl. O&P
1 Labor Foreman (outside)	$16.45	$131.60	$28.20	$225.60	$14.85	$25.44
4 Laborers	14.45	462.40	24.75	792.00		
1 Drill Rig, Wells		1796.00		1975.60		
1 Light Truck, 3 Ton		180.45		198.50	49.41	54.35
40 L.H., Daily Totals		$2570.45		$3191.70	$64.26	$79.79
Crew B-23A	Hr.	Daily	Hr.	Daily	Bare Costs	Incl. O&P
1 Labor Foreman (outside)	$16.45	$131.60	$28.20	$225.60	$16.93	$28.65
1 Laborers	14.45	115.60	24.75	198.00		
1 Equip. Operator, medium	19.90	159.20	33.00	264.00		
1 Drill Rig, Wells		1796.00		1975.60		
1 Pickup Truck, 3/4 Ton		130.20		143.20	80.26	88.28
24 L.H., Daily Totals		$2332.60		$2806.40	$97.19	$116.93
Crew B-23B	Hr.	Daily	Hr.	Daily	Bare Costs	Incl. O&P
1 Labor Foreman (outside)	$16.45	$131.60	$28.20	$225.60	$16.93	$28.65
1 Laborer	14.45	115.60	24.75	198.00		
1 Equip. Operator, medium	19.90	159.20	33.00	264.00		
1 Drill Rig, Wells		1796.00		1975.60		
1 Pickup Truck, 3/4 Ton		130.20		143.20		
1 Pump, Cntfgl, 6"		157.20		172.90	86.81	95.49
24 L.H., Daily Totals		$2489.80		$2979.30	$103.74	$124.14
Crew B-24	Hr.	Daily	Hr.	Daily	Bare Costs	Incl. O&P
1 Cement Finisher	$18.90	$151.20	$30.80	$246.40	$17.68	$29.77
1 Laborer	14.45	115.60	24.75	198.00		
1 Carpenter	19.70	157.60	33.75	270.00		
24 L.H., Daily Totals		$424.40		$714.40	$17.68	$29.77
Crew B-25	Hr.	Daily	Hr.	Daily	Bare Costs	Incl. O&P
1 Labor Foreman	$16.45	$131.60	$28.20	$225.60	$16.12	$27.31
7 Laborers	14.45	809.20	24.75	1386.00		
3 Equip. Oper. (med.)	19.90	477.60	33.00	792.00		
1 Asphalt Paver, 130 H.P.		1309.00		1439.90		
1 Tandem Roller, 10 Ton		224.00		246.40		
1 Roller, Pneumatic Wheel		240.80		264.90	20.16	22.17
88 L.H., Daily Totals		$3192.20		$4354.80	$36.28	$49.48

Crew No.	Bare Costs		Incl. Subs O & P		Cost Per Labor-Hour	
Crew B-25B	Hr.	Daily	Hr.	Daily	Bare Costs	Incl. O&P
1 Labor Foreman	$16.45	$131.60	$28.20	$225.60	$16.43	$27.79
7 Laborers	14.45	809.20	24.75	1386.00		
4 Equip. Oper. (medium)	19.90	636.80	33.00	1056.00		
1 Asphalt Paver, 130 H.P.		1309.00		1439.90		
2 Rollers, Steel Wheel		448.00		492.80		
1 Roller, Pneumatic Wheel		240.80		264.90	20.81	22.89
96 L.H., Daily Totals		$3575.40		$4865.20	$37.24	$50.68
Crew B-26	Hr.	Daily	Hr.	Daily	Bare Costs	Incl. O&P
1 Labor Foreman (outside)	$16.45	$131.60	$28.20	$225.60	$16.63	$28.43
6 Laborers	14.45	693.60	24.75	1188.00		
2 Equip. Oper. (med.)	19.90	318.40	33.00	528.00		
1 Rodman (reinf.)	21.10	168.80	39.25	314.00		
1 Cement Finisher	18.90	151.20	30.80	246.40		
1 Grader, 30,000 Lbs.		516.00		567.60		
1 Paving Mach. & Equip.		1324.00		1456.40	20.91	23.00
88 L.H., Daily Totals		$3303.60		$4526.00	$37.54	$51.43
Crew B-27	Hr.	Daily	Hr.	Daily	Bare Costs	Incl. O&P
1 Labor Foreman (outside)	$16.45	$131.60	$28.20	$225.60	$14.95	$25.61
3 Laborers	14.45	346.80	24.75	594.00		
1 Berm Machine		73.00		80.30	2.28	2.51
32 L.H., Daily Totals		$551.40		$899.90	$17.23	$28.12
Crew B-28	Hr.	Daily	Hr.	Daily	Bare Costs	Incl. O&P
2 Carpenters	$19.70	$315.20	$33.75	$540.00	$17.95	$30.75
1 Laborer	14.45	115.60	24.75	198.00		
24 L.H., Daily Totals		$430.80		$738.00	$17.95	$30.75
Crew B-29	Hr.	Daily	Hr.	Daily	Bare Costs	Incl. O&P
1 Labor Foreman (outside)	$16.45	$131.60	$28.20	$225.60	$15.82	$26.91
4 Laborers	14.45	462.40	24.75	792.00		
1 Equip. Oper. (crane)	20.65	165.20	34.25	274.00		
1 Gradall, 3 Ton, 1/2 C.Y.		618.20		680.00	12.88	14.17
48 L.H., Daily Totals		$1377.40		$1971.60	$28.70	$41.08
Crew B-30	Hr.	Daily	Hr.	Daily	Bare Costs	Incl. O&P
1 Equip. Oper. (med.)	$19.90	$159.20	$33.00	$264.00	$17.43	$29.03
2 Truck Drivers (heavy)	16.20	259.20	27.05	432.80		
1 Hyd. Excavator, 1.5 C.Y.		713.20		784.50		
2 Dump Trucks, 16 Ton		894.50		983.95	66.99	73.69
24 L.H., Daily Totals		$2026.10		$2465.25	$84.42	$102.72
Crew B-31	Hr.	Daily	Hr.	Daily	Bare Costs	Incl. O&P
1 Labor Foreman (outside)	$16.45	$131.60	$28.20	$225.60	$14.85	$25.44
4 Laborers	14.45	462.40	24.75	792.00		
1 Air Compr., 250 C.F.M.		127.00		139.70		
1 Sheeting Driver		12.20		13.40		
2-50 Ft. Air Hoses, 1.5" Dia.		14.40		15.85	3.84	4.22
40 L.H., Daily Totals		$747.60		$1186.55	$18.69	$29.66
Crew B-32	Hr.	Daily	Hr.	Daily	Bare Costs	Incl. O&P
1 Laborer	$14.45	$115.60	$24.75	$198.00	$18.54	$30.94
3 Equip. Oper. (med.)	19.90	477.60	33.00	792.00		
1 Grader, 30,000 Lbs.		516.00		567.60		
1 Tandem Roller, 10 Ton		224.00		246.40		
1 Dozer, 200 H.P.		806.80		887.50	48.34	53.17
32 L.H., Daily Totals		$2140.00		$2691.50	$66.88	$84.11

Crews

Crew No.	Bare Costs		Incl. Subs O & P		Cost Per Labor-Hour	
Crew B-32A	Hr.	Daily	Hr.	Daily	Bare Costs	Incl. O&P
1 Laborer	$14.45	$115.60	$24.75	$198.00	$18.08	$30.25
2 Equip. Oper. (medium)	19.90	318.40	33.00	528.00		
1 Grader, 30,000 Lbs.		516.00		567.60		
1 Roller, Vibratory, 29,000 Lbs.		432.40		475.65	39.52	43.47
24 L.H., Daily Totals		$1382.40		$1769.25	$57.60	$73.72
Crew B-32B	Hr.	Daily	Hr.	Daily	Bare Costs	Incl. O&P
1 Laborer	$14.45	$115.60	$24.75	$198.00	$18.08	$30.25
2 Equip. Oper. (medium)	19.90	318.40	33.00	528.00		
1 Dozer, 200 H.P.		806.80		887.50		
1 Roller, Vibratory, 29,000 Lbs.		432.40		475.65	51.63	56.80
24 L.H., Daily Totals		$1673.20		$2089.15	$69.71	$87.05
Crew B-32C	Hr.	Daily	Hr.	Daily	Bare Costs	Incl. O&P
1 Labor Foreman	$16.45	$131.60	$28.20	$225.60	$17.51	$29.45
2 Laborers	14.45	231.20	24.75	396.00		
3 Equip. Oper. (medium)	19.90	477.60	33.00	792.00		
1 Grader, 30,000 Lbs.		516.00		567.60		
1 Roller, Steel Wheel		224.00		246.40		
1 Dozer, 200 H.P.		806.80		887.50	32.23	35.45
48 L.H., Daily Totals		$2387.20		$3115.10	$49.74	$64.90
Crew B-33	Hr.	Daily	Hr.	Daily	Bare Costs	Incl. O&P
1 Equip. Oper. (med.)	$19.90	$159.20	$33.00	$264.00	$19.90	$33.00
.25 Equip. Oper. (med.)	19.90	39.80	33.00	66.00		
10 L.H., Daily Totals		$199.00		$330.00	$19.90	$33.00
Crew B-33A	Hr.	Daily	Hr.	Daily	Bare Costs	Incl. O&P
1 Equip. Oper. (med.)	$19.90	$159.20	$33.00	$264.00	$19.90	$33.00
.25 Equip. Oper. (med.)	19.90	39.80	33.00	66.00		
1 Scraper, Towed, 7 C.Y.		81.00		89.10		
1.25 Dozer, 300 H.P.		1385.00		1523.50	146.60	161.26
10 L.H., Daily Totals		$1665.00		$1942.60	$166.50	$194.26
Crew B-33B	Hr.	Daily	Hr.	Daily	Bare Costs	Incl. O&P
1 Equip. Oper. (med.)	$19.90	$159.20	$33.00	$264.00	$19.90	$33.00
.25 Equip. Oper. (med.)	19.90	39.80	33.00	66.00		
1 Scraper, Towed, 10 C.Y.		204.80		225.30		
1.25 Dozer, 300 H.P.		1385.00		1523.50	158.98	174.88
10 L.H., Daily Totals		$1788.80		$2078.80	$178.88	$207.88
Crew B-33C	Hr.	Daily	Hr.	Daily	Bare Costs	Incl. O&P
1 Equip. Oper. (med.)	$19.90	$159.20	$33.00	$264.00	$19.90	$33.00
.25 Equip. Oper. (med.)	19.90	39.80	33.00	66.00		
1 Scraper, Towed, 12 C.Y.		204.80		225.30		
1.25 Dozer, 300 H.P.		1385.00		1523.50	158.98	174.88
10 L.H., Daily Totals		$1788.80		$2078.80	$178.88	$207.88
Crew B-33D	Hr.	Daily	Hr.	Daily	Bare Costs	Incl. O&P
1 Equip. Oper. (med.)	$19.90	$159.20	$33.00	$264.00	$19.90	$33.00
.25 Equip. Oper. (med.)	19.90	39.80	33.00	66.00		
1 S.P. Scraper, 14 C.Y.		1626.00		1788.60		
.25 Dozer, 300 H.P.		277.00		304.70	190.30	209.33
10 L.H., Daily Totals		$2102.00		$2423.30	$210.20	$242.33

Crew No.	Bare Costs		Incl. Subs O & P		Cost Per Labor-Hour	
Crew B-33E	Hr.	Daily	Hr.	Daily	Bare Costs	Incl. O&P
1 Equip. Oper. (med.)	$19.90	$159.20	$33.00	$264.00	$19.90	$33.00
.25 Equip. Oper. (med.)	19.90	39.80	33.00	66.00		
1 S.P. Scraper, 24 C.Y.		1913.00		2104.30		
.25 Dozer, 300 H.P.		277.00		304.70	219.00	240.90
10 L.H., Daily Totals		$2389.00		$2739.00	$238.90	$273.90
Crew B-33F	Hr.	Daily	Hr.	Daily	Bare Costs	Incl. O&P
1 Equip. Oper. (med.)	$19.90	$159.20	$33.00	$264.00	$19.90	$33.00
.25 Equip. Oper. (med.)	19.90	39.80	33.00	66.00		
1 Elev. Scraper, 11 C.Y.		698.55		768.40		
.25 Dozer, 300 H.P.		277.00		304.70	97.56	107.31
10 L.H., Daily Totals		$1174.55		$1403.10	$117.46	$140.31
Crew B-33G	Hr.	Daily	Hr.	Daily	Bare Costs	Incl. O&P
1 Equip. Oper. (med.)	$19.90	$159.20	$33.00	$264.00	$19.90	$33.00
.25 Equip. Oper. (med.)	19.90	39.80	33.00	66.00		
1 Elev. Scraper, 20 C.Y.		984.80		1083.30		
.25 Dozer, 300 H.P.		277.00		304.70	126.18	138.80
10 L.H., Daily Totals		$1460.80		$1718.00	$146.08	$171.80
Crew B-34A	Hr.	Daily	Hr.	Daily	Bare Costs	Incl. O&P
1 Truck Driver (heavy)	$16.20	$129.60	$27.05	$216.40	$16.20	$27.05
1 Dump Truck, 12 Ton		365.30		401.85	45.66	50.23
8 L.H., Daily Totals		$494.90		$618.25	$61.86	$77.28
Crew B-34B	Hr.	Daily	Hr.	Daily	Bare Costs	Incl. O&P
1 Truck Driver (heavy)	$16.20	$129.60	$27.05	$216.40	$16.20	$27.05
1 Dump Truck, 16 Ton		447.25		492.00	55.91	61.50
8 L.H., Daily Totals		$576.85		$708.40	$72.11	$88.55
Crew B-34C	Hr.	Daily	Hr.	Daily	Bare Costs	Incl. O&P
1 Truck Driver (heavy)	$16.20	$129.60	$27.05	$216.40	$16.20	$27.05
1 Truck Tractor, 40 Ton		427.60		470.35		
1 Dump Trailer, 16.5 C.Y.		133.60		146.95	70.15	77.17
8 L.H., Daily Totals		$690.80		$833.70	$86.35	$104.22
Crew B-34D	Hr.	Daily	Hr.	Daily	Bare Costs	Incl. O&P
1 Truck Driver (heavy)	$16.20	$129.60	$27.05	$216.40	$16.20	$27.05
1 Truck Tractor, 40 Ton		427.60		470.35		
1 Dump Trailer, 20 C.Y.		136.10		149.70	70.46	77.51
8 L.H., Daily Totals		$693.30		$836.45	$86.66	$104.56
Crew B-34E	Hr.	Daily	Hr.	Daily	Bare Costs	Incl. O&P
1 Truck Driver (heavy)	$16.20	$129.60	$27.05	$216.40	$16.20	$27.05
1 Truck, Off Hwy., 25 Ton		700.00		770.00	87.50	96.25
8 L.H., Daily Totals		$829.60		$986.40	$103.70	$123.30
Crew B-34F	Hr.	Daily	Hr.	Daily	Bare Costs	Incl. O&P
1 Truck Driver (heavy)	$16.20	$129.60	$27.05	$216.40	$16.20	$27.05
1 Truck, Off Hwy., 22 C.Y.		1058.00		1163.80	132.25	145.48
8 L.H., Daily Totals		$1187.60		$1380.20	$148.45	$172.53
Crew B-34G	Hr.	Daily	Hr.	Daily	Bare Costs	Incl. O&P
1 Truck Driver (heavy)	$16.20	$129.60	$27.05	$216.40	$16.20	$27.05
1 Truck, Off Hwy., 34 C.Y.		1405.00		1545.50	175.63	193.19
8 L.H., Daily Totals		$1534.60		$1761.90	$191.83	$220.24

Crews

Crew No.	Bare Costs		Incl. Subs O & P		Cost Per Labor-Hour	
Crew B-34H	**Hr.**	**Daily**	**Hr.**	**Daily**	**Bare Costs**	**Incl. O&P**
1 Truck Driver (heavy)	$16.20	$129.60	$27.05	$216.40	$16.20	$27.05
1 Truck, Off Hwy., 42 C.Y.		1599.00		1758.90	199.88	219.86
8 L.H., Daily Totals		$1728.60		$1975.30	$216.08	$246.91
Crew B-34J	**Hr.**	**Daily**	**Hr.**	**Daily**	**Bare Costs**	**Incl. O&P**
1 Truck Driver (heavy)	$16.20	$129.60	$27.05	$216.40	$16.20	$27.05
1 Truck, Off Hwy., 60 C.Y.		2240.00		2464.00	280.00	308.00
8 L.H., Daily Totals		$2369.60		$2680.40	$296.20	$335.05
Crew B-34K	**Hr.**	**Daily**	**Hr.**	**Daily**	**Bare Costs**	**Incl. O&P**
1 Truck Driver (heavy)	$16.20	$129.60	$27.05	$216.40	$16.20	$27.05
1 Truck Tractor, 240 H.P.		531.20		584.30		
1 Low Bed Trailer		384.75		423.25	114.49	125.94
8 L.H., Daily Totals		$1045.55		$1223.95	$130.69	$152.99
Crew B-35	**Hr.**	**Daily**	**Hr.**	**Daily**	**Bare Costs**	**Incl. O&P**
1 Laborer Foreman (out)	$16.45	$131.60	$28.20	$225.60	$18.65	$31.48
1 Skilled Worker	19.75	158.00	33.90	271.20		
1 Welder (plumber)	21.95	175.60	36.30	290.40		
1 Laborer	14.45	115.60	24.75	198.00		
1 Equip. Oper. (crane)	20.65	165.20	34.25	274.00		
1 Electric Welding Mach.		48.80		53.70		
1 Hyd. Excavator, .75 C.Y.		455.80		501.40	12.62	13.88
40 L.H., Daily Totals		$1250.60		$1814.30	$31.27	$45.36
Crew B-35A	**Hr.**	**Daily**	**Hr.**	**Daily**	**Bare Costs**	**Incl. O&P**
1 Laborer Foreman (out)	$16.45	$131.60	$28.20	$225.60	$17.81	$30.05
2 Laborers	14.45	231.20	24.75	396.00		
1 Skilled Worker	19.75	158.00	33.90	271.20		
1 Welder (plumber)	21.95	175.60	36.30	290.40		
1 Equip. Oper. (crane)	20.65	165.20	34.25	274.00		
1 Equip. Oper. Oiler	17.00	136.00	28.20	225.60		
1 Welder, 300 amp		84.00		92.40		
1 Crane, 75 Ton		939.20		1033.10	18.27	20.10
56 L.H., Daily Totals		$2020.80		$2808.30	$36.08	$50.15
Crew B-36	**Hr.**	**Daily**	**Hr.**	**Daily**	**Bare Costs**	**Incl. O&P**
1 Labor Foreman (outside)	$16.45	$131.60	$28.20	$225.60	$17.03	$28.74
2 Laborers	14.45	231.20	24.75	396.00		
2 Equip. Oper. (med.)	19.90	318.40	33.00	528.00		
1 Dozer, 200 H.P.		806.80		887.50		
1 Aggregate Spreader		73.80		81.20		
1 Tandem Roller, 10 Ton		224.00		246.40	27.62	30.38
40 L.H., Daily Totals		$1785.80		$2364.70	$44.65	$59.12
Crew B-36A	**Hr.**	**Daily**	**Hr.**	**Daily**	**Bare Costs**	**Incl. O&P**
1 Labor Foreman (outside)	$16.45	$131.60	$28.20	$225.60	$17.85	$29.96
2 Laborers	14.45	231.20	24.75	396.00		
4 Equip. Oper. (med.)	19.90	636.80	33.00	1056.00		
1 Dozer, 200 H.P.		806.80		887.50		
1 Aggregate Spreader		73.80		81.20		
1 Roller, Steel Wheel		224.00		246.40		
1 Roller, Pneumatic Wheel		240.80		264.90	24.03	26.43
56 L.H., Daily Totals		$2345.00		$3157.60	$41.88	$56.39

Crew No.	Bare Costs		Incl. Subs O & P		Cost Per Labor-Hour	
Crew B-36B	**Hr.**	**Daily**	**Hr.**	**Daily**	**Bare Costs**	**Incl. O&P**
1 Labor Foreman (outside)	$16.45	$131.60	$28.20	$225.60	$17.64	$29.59
2 Laborers	14.45	231.20	24.75	396.00		
4 Equip. Oper. (medium)	19.90	636.80	33.00	1056.00		
1 Truck Driver, Heavy	16.20	129.60	27.05	216.40		
1 Grader, 30,000 Lbs.		516.00		567.60		
1 F.E. Loader, crl, 1.5 C.Y.		366.85		403.55		
1 Dozer, 300 H.P.		1108.00		1218.80		
1 Roller, Vibratory		432.40		475.65		
1 Truck, Tractor, 240 H.P.		531.20		584.30		
1 Water Tanker, 5000 Gal.		205.90		226.50	49.38	54.32
64 L.H., Daily Totals		$4289.55		$5370.40	$67.02	$83.91
Crew B-37	**Hr.**	**Daily**	**Hr.**	**Daily**	**Bare Costs**	**Incl. O&P**
1 Labor Foreman (outside)	$16.45	$131.60	$28.20	$225.60	$15.56	$26.48
4 Laborers	14.45	462.40	24.75	792.00		
1 Equip. Oper. (light)	19.10	152.80	31.70	253.60		
1 Tandem Roller, 5 Ton		133.20		146.50	2.78	3.05
48 L.H., Daily Totals		$880.00		$1417.70	$18.34	$29.53
Crew B-38	**Hr.**	**Daily**	**Hr.**	**Daily**	**Bare Costs**	**Incl. O&P**
2 Laborers	$14.45	$231.20	$24.75	$396.00	$16.00	$27.07
1 Equip. Oper. (light)	19.10	152.80	31.70	253.60		
1 Backhoe Loader, 48 H.P.		203.00		223.30		
1 Hyd. Hammer, (1200 lb)		213.60		234.95	17.36	19.09
24 L.H., Daily Totals		$800.60		$1107.85	$33.36	$46.16
Crew B-39	**Hr.**	**Daily**	**Hr.**	**Daily**	**Bare Costs**	**Incl. O&P**
1 Labor Foreman (outside)	$16.45	$131.60	$28.20	$225.60	$14.78	$25.32
5 Laborers	14.45	578.00	24.75	990.00		
1 Air Compr., 250 C.F.M.		127.00		139.70		
2 Air Tools & Accessories		39.00		42.90		
2-50 Ft. Air Hoses, 1.5" Dia.		14.40		15.85	3.76	4.13
48 L.H., Daily Totals		$890.00		$1414.05	$18.54	$29.45
Crew B-40	**Hr.**	**Daily**	**Hr.**	**Daily**	**Bare Costs**	**Incl. O&P**
1 Pile Driver Foreman (out)	$21.45	$171.60	$40.00	$320.00	$19.19	$34.86
4 Pile Drivers	19.45	622.40	36.25	1160.00		
1 Building Laborer	14.45	115.60	24.75	198.00		
1 Equip. Oper. (crane)	20.65	165.20	34.25	274.00		
1 Crane, 40 Ton		714.30		785.75		
1 Vibratory Hammer & Gen.		1218.00		1339.80	34.51	37.96
56 L.H., Daily Totals		$3007.10		$4077.55	$53.70	$72.82
Crew B-41	**Hr.**	**Daily**	**Hr.**	**Daily**	**Bare Costs**	**Incl. O&P**
1 Labor Foreman (outside)	$16.45	$131.60	$28.20	$225.60	$15.21	$25.97
4 Laborers	14.45	462.40	24.75	792.00		
.25 Equip. Oper. (crane)	20.65	41.30	34.25	68.50		
.25 Equip. Oper. Oiler	17.00	34.00	28.20	56.40		
.25 Crawler Crane, 40 Ton		178.57		196.45	4.06	4.46
44 L.H., Daily Totals		$847.87		$1338.95	$19.27	$30.43
Crew B-42	**Hr.**	**Daily**	**Hr.**	**Daily**	**Bare Costs**	**Incl. O&P**
1 Labor Foreman (outside)	$16.45	$131.60	$28.20	$225.60	$16.69	$28.25
4 Laborers	14.45	462.40	24.75	792.00		
1 Equip. Oper. (crane)	20.65	165.20	34.25	274.00		
1 Welder	21.95	175.60	36.30	290.40		
1 Hyd. Crane, 25 Ton		559.20		615.10		
1 Gas Welding Machine		84.00		92.40		
1 Horz. Boring Csg. Mch.		498.00		547.80	20.38	22.42
56 L.H., Daily Totals		$2076.00		$2837.30	$37.07	$50.67

Crews

Crew B-43

Crew No.	Hr.	Daily	Hr.	Daily	Bare Costs	Incl. O&P
1 Labor Foreman (outside)	$16.45	$131.60	$28.20	$225.60	$14.85	$25.44
4 Laborers	14.45	462.40	24.75	792.00		
1 Drill Rig & Augers		1796.00		1975.60	44.90	49.39
40 L.H., Daily Totals		$2390.00		$2993.20	$59.75	$74.83

Crew B-44

Crew No.	Hr.	Daily	Hr.	Daily	Bare Costs	Incl. O&P
1 Pile Driver Foreman	$21.45	$171.60	$40.00	$320.00	$18.60	$33.59
4 Pile Drivers	19.45	622.40	36.25	1160.00		
1 Equip. Oper. (crane)	20.65	165.20	34.25	274.00		
2 Laborer	14.45	231.20	24.75	396.00		
1 Crane, 40 Ton, & Access.		714.30		785.75		
45 L.F. Leads, 15K Ft. Lbs.		198.00		217.80	14.25	15.68
64 L.H., Daily Totals		$2102.70		$3153.55	$32.85	$49.27

Crew B-45

Crew No.	Hr.	Daily	Hr.	Daily	Bare Costs	Incl. O&P
1 Building Laborer	$14.45	$115.60	$24.75	$198.00	$15.33	$25.90
1 Truck Driver (heavy)	16.20	129.60	27.05	216.40		
1 Dist. Tank Truck, 3K Gal.		374.95		412.45	23.43	25.78
16 L.H., Daily Totals		$620.15		$826.85	$38.76	$51.68

Crew B-46

Crew No.	Hr.	Daily	Hr.	Daily	Bare Costs	Incl. O&P
1 Pile Driver Foreman	$21.45	$171.60	$40.00	$320.00	$17.28	$31.13
2 Pile Drivers	19.45	311.20	36.25	580.00		
3 Laborers	14.45	346.80	24.75	594.00		
1 Chain Saw, 36" Long		49.20		54.10	1.03	1.13
48 L.H., Daily Totals		$878.80		$1548.10	$18.31	$32.26

Crew B-47

Crew No.	Hr.	Daily	Hr.	Daily	Bare Costs	Incl. O&P
1 Blast Foreman	$16.45	$131.60	$28.20	$225.60	$15.45	$26.48
1 Driller	14.45	115.60	24.75	198.00		
1 Crawler Type Drill, 4"		312.55		343.80		
1 Air Compr., 600 C.F.M.		278.00		305.80		
2-50 Ft. Air Hoses, 3" Dia.		33.80		37.20	39.02	42.92
16 L.H., Daily Totals		$871.55		$1110.40	$54.47	$69.40

Crew B-47A

Crew No.	Hr.	Daily	Hr.	Daily	Bare Costs	Incl. O&P
1 Drilling Foreman	$16.45	$131.60	$28.20	$225.60	$18.03	$30.22
1 Equip. Oper. (heavy)	20.65	165.20	34.25	274.00		
1 Oiler	17.00	136.00	28.20	225.60		
1 Quarry Drill		486.05		534.65	20.25	22.28
24 L.H., Daily Totals		$918.85		$1259.85	$38.28	$52.50

Crew B-47C

Crew No.	Hr.	Daily	Hr.	Daily	Bare Costs	Incl. O&P
1 Laborer	$14.45	$115.60	$24.75	$198.00	$16.77	$28.23
1 Equip. Oper. (light)	19.10	152.80	31.70	253.60		
1 Air Compressor, 750 CFM		300.80		330.90		
2-50' Air Hose, 3"		33.80		37.20		
1 Air Track Drill, 4"		312.55		343.80	40.45	44.49
16 L.H., Daily Totals		$915.55		$1163.50	$57.22	$72.72

Crew B-47E

Crew No.	Hr.	Daily	Hr.	Daily	Bare Costs	Incl. O&P
1 Laborer Forman	$16.45	$131.60	$28.20	$225.60	$14.95	$25.61
3 Laborers	14.45	346.80	24.75	594.00		
1 Truck, Flatbed, 3 ton		180.45		198.50	5.64	6.20
32 L.H., Daily Totals		$658.85		$1018.10	$20.59	$31.81

Crew B-48

Crew No.	Hr.	Daily	Hr.	Daily	Bare Costs	Incl. O&P
1 Labor Foreman (outside)	$16.45	$131.60	$28.20	$225.60	$15.82	$26.91
4 Laborers	14.45	462.40	24.75	792.00		
1 Equip. Oper. (crane)	20.65	165.20	34.25	274.00		
1 Centr. Water Pump, 6"		157.20		172.90		
1-20 Ft. Suction Hose, 6"		22.50		24.75		
1-50 Ft. Disch. Hose, 6"		20.55		22.60		
1 Drill Rig & Augers		1796.00		1975.60	41.59	45.75
48 L.H., Daily Totals		$2755.45		$3487.45	$57.41	$72.66

Crew B-49

Crew No.	Hr.	Daily	Hr.	Daily	Bare Costs	Incl. O&P
1 Labor Foreman (outside)	$16.45	$131.60	$28.20	$225.60	$16.47	$28.74
5 Laborers	14.45	578.00	24.75	990.00		
1 Equip. Oper. (crane)	20.65	165.20	34.25	274.00		
2 Pile Drivers	19.45	311.20	36.25	580.00		
1 Hyd. Crane, 25 Ton		559.20		615.10		
1 Centr. Water Pump, 6"		157.20		172.90		
1-20 Ft. Suction Hose, 6"		22.50		24.75		
1-50 Ft. Disch. Hose, 6"		20.55		22.60		
1 Drill Rig & Augers		1796.00		1975.60	35.49	39.04
72 L.H., Daily Totals		$3741.45		$4880.55	$51.96	$67.78

Crew B-50

Crew No.	Hr.	Daily	Hr.	Daily	Bare Costs	Incl. O&P
1 Pile Driver Foremen	$21.45	$171.60	$40.00	$320.00	$17.77	$31.96
6 Pile Drivers	19.45	933.60	36.25	1740.00		
1 Equip. Oper. (crane)	20.65	165.20	34.25	274.00		
5 Laborers	14.45	578.00	24.75	990.00		
1 Crane, 40 Ton		714.30		785.75		
60 L.F. Leads, 15K Ft. Lbs.		264.00		290.40		
1 Hammer, 15K Ft. Lbs.		311.40		342.55		
1 Air Compr., 600 C.F.M.		278.00		305.80		
2-50 Ft. Air Hoses, 3" Dia.		33.80		37.20		
1 Chain Saw, 36" Long		49.20		54.10	15.87	17.46
104 L.H., Daily Totals		$3499.10		$5139.80	$33.64	$49.42

Crew B-51

Crew No.	Hr.	Daily	Hr.	Daily	Bare Costs	Incl. O&P
1 Labor Foreman (outside)	$16.45	$131.60	$28.20	$225.60	$15.02	$25.62
4 Laborers	14.45	462.40	24.75	792.00		
1 Truck Driver (light)	15.85	126.80	26.50	212.00		
1 Light Truck, 1.5 Ton		172.95		190.25	3.60	3.96
48 L.H., Daily Totals		$893.75		$1419.85	$18.62	$29.58

Crew B-52

Crew No.	Hr.	Daily	Hr.	Daily	Bare Costs	Incl. O&P
1 Labor Foreman	$16.45	$131.60	$28.20	$225.60	$16.35	$28.15
1 Carpenter	19.70	157.60	33.75	270.00		
4 Laborers	14.45	462.40	24.75	792.00		
.5 Rodman (reinf.)	21.10	84.40	39.25	157.00		
.5 Equip. Oper. (med.)	19.90	79.60	33.00	132.00		
.5 F.E. Ldr., T.M., 2.5 C.Y.		392.00		431.20	7.00	7.70
56 L.H., Daily Totals		$1307.60		$2007.80	$23.35	$35.85

Crew B-53

Crew No.	Hr.	Daily	Hr.	Daily	Bare Costs	Incl. O&P
1 Building Laborer	$14.45	$115.60	$24.75	$198.00	$14.45	$24.75
1 Trencher, Chain, 12 H.P.		102.50		112.75	12.81	14.09
8 L.H., Daily Totals		$218.10		$310.75	$27.26	$38.84

Crew B-54

Crew No.	Hr.	Daily	Hr.	Daily	Bare Costs	Incl. O&P
1 Equip. Oper. (light)	$19.10	$152.80	$31.70	$253.60	$19.10	$31.70
1 Trencher, Chain, 40 H.P.		210.90		232.00	26.36	29.00
8 L.H., Daily Totals		$363.70		$485.60	$45.46	$60.70

Crews

Crew No.	Bare Costs		Incl. Subs O & P		Cost Per Labor-Hour	
Crew B-54A	Hr.	Daily	Hr.	Daily	Bare Costs	Incl. O&P
.17 Labor Foreman (outside)	$16.45	$22.37	$28.20	$38.35	$19.40	$32.30
1 Equipment Operator (med.)	19.90	159.20	33.00	264.00		
1 Wheel Trencher, 67HP		409.35		450.30	43.73	48.11
9.36 L.H., Daily Totals		$590.92		$752.65	$63.13	$80.41
Crew B-54B	Hr.	Daily	Hr.	Daily	Bare Costs	Incl. O&P
.25 Labor Foreman (outside)	$16.45	$32.90	$28.20	$56.40	$19.21	$32.04
1 Equipment Operator (med.)	19.90	159.20	33.00	264.00		
1 Wheel Trencher, 150HP		565.20		621.70	56.52	62.17
10 L.H., Daily Totals		$757.30		$942.10	$75.73	$94.21
Crew B-55	Hr.	Daily	Hr.	Daily	Bare Costs	Incl. O&P
1 Laborers	$14.45	$115.60	$24.75	$198.00	$15.15	$25.63
1 Truck Driver (light)	15.85	126.80	26.50	212.00		
1 Auger, 4" to 36" Dia		483.00		531.30		
1 Flatbed 3 Ton Truck		180.45		198.50	41.47	45.61
16 L.H., Daily Totals		$905.85		$1139.80	$56.62	$71.24
Crew B-56	Hr.	Daily	Hr.	Daily	Bare Costs	Incl. O&P
2 Laborer	$14.45	$231.20	$24.75	$396.00	$14.45	$24.75
1 Crawler Type Drill, 4"		312.55		343.80		
1 Air Compr., 600 C.F.M.		278.00		305.80		
1-50 Ft. Air Hose, 3" Dia.		16.90		18.60	37.97	41.76
16 L.H., Daily Totals		$838.65		$1064.20	$52.42	$66.51
Crew B-57	Hr.	Daily	Hr.	Daily	Bare Costs	Incl. O&P
1 Labor Foreman (outside)	$16.45	$131.60	$28.20	$225.60	$16.09	$27.34
3 Laborers	14.45	346.80	24.75	594.00		
1 Equip. Oper. (crane)	20.65	165.20	34.25	274.00		
1 Barge, 400 Ton		453.65		499.00		
1 Power Shovel, 1 C.Y.		516.80		568.50		
1 Clamshell Bucket, 1 C.Y.		70.00		77.00		
1 Centr. Water Pump, 6"		157.20		172.90		
1-20 Ft. Suction Hose, 6"		22.50		24.75		
20-50 Ft. Disch. Hoses, 6"		411.00		452.10	40.78	44.86
40 L.H., Daily Totals		$2274.75		$2887.85	$56.87	$72.20
Crew B-58	Hr.	Daily	Hr.	Daily	Bare Costs	Incl. O&P
2 Laborers	$14.45	$231.20	$24.75	$396.00	$16.00	$27.07
1 Equip. Oper. (light)	19.10	152.80	31.70	253.60		
1 Backhoe Loader, 48 H.P.		203.00		223.30		
1 Small Helicopter, w/pilot		3540.00		3894.00	155.96	171.55
24 L.H., Daily Totals		$4127.00		$4766.90	$171.96	$198.62
Crew B-59	Hr.	Daily	Hr.	Daily	Bare Costs	Incl. O&P
1 Truck Driver (heavy)	$16.20	$129.60	$27.05	$216.40	$16.20	$27.05
1 Truck, 30 Ton		325.60		358.15		
1 Water tank, 5000 Gal.		205.90		226.50	66.44	73.08
8 L.H., Daily Totals		$661.10		$801.05	$82.64	$100.13
Crew B-60	Hr.	Daily	Hr.	Daily	Bare Costs	Incl. O&P
1 Labor Foreman (outside)	$16.45	$131.60	$28.20	$225.60	$16.59	$28.07
3 Laborers	14.45	346.80	24.75	594.00		
1 Equip. Oper. (crane)	20.65	165.20	34.25	274.00		
1 Equip. Oper. (light)	19.10	152.80	31.70	253.60		
1 Crawler Crane, 40 Ton		714.30		785.75		
45 L.F. Leads, 15K Ft. Lbs.		198.00		217.80		
1 Backhoe Loader, 48 H.P.		203.00		223.30	23.24	25.56
48 L.H., Daily Totals		$1911.70		$2574.05	$39.83	$53.63

Crew No.	Bare Costs		Incl. Subs O & P		Cost Per Labor-Hour	
Crew B-61	Hr.	Daily	Hr.	Daily	Bare Costs	Incl. O&P
1 Labor Foreman (outside)	$16.45	$131.60	$28.20	$225.60	$14.85	$25.44
4 Laborers	14.45	462.40	24.75	792.00		
1 Cement Mixer, 2 C.Y.		222.00		244.20		
1 Air Compr., 160 C.F.M.		98.40		108.25	8.01	8.81
40 L.H., Daily Totals		$914.40		$1370.05	$22.86	$34.25
Crew B-62	Hr.	Daily	Hr.	Daily	Bare Costs	Incl. O&P
2 Laborers	$14.45	$231.20	$24.75	$396.00	$16.00	$27.07
1 Equip. Oper. (light)	19.10	152.80	31.70	253.60		
1 Loader, Skid Steer		116.40		128.05	4.85	5.34
24 L.H., Daily Totals		$500.40		$777.65	$20.85	$32.41
Crew B-63	Hr.	Daily	Hr.	Daily	Bare Costs	Incl. O&P
5 Laborers	$14.45	$578.00	$24.75	$990.00	$14.45	$24.75
1 Loader, Skid Steer		116.40		128.05	2.91	3.20
40 L.H., Daily Totals		$694.40		$1118.05	$17.36	$27.95
Crew B-64	Hr.	Daily	Hr.	Daily	Bare Costs	Incl. O&P
1 Laborer	$14.45	$115.60	$24.75	$198.00	$15.15	$25.63
1 Truck Driver (light)	15.85	126.80	26.50	212.00		
1 Power Mulcher (small)		115.80		127.40		
1 Light Truck, 1.5 Ton		172.95		190.25	18.05	19.85
16 L.H., Daily Totals		$531.15		$727.65	$33.20	$45.48
Crew B-65	Hr.	Daily	Hr.	Daily	Bare Costs	Incl. O&P
1 Laborer	$14.45	$115.60	$24.75	$198.00	$15.15	$25.63
1 Truck Driver (light)	15.85	126.80	26.50	212.00		
1 Power Mulcher (large)		283.85		312.25		
1 Light Truck, 1.5 Ton		172.95		190.25	28.55	31.41
16 L.H., Daily Totals		$699.20		$912.50	$43.70	$57.04
Crew B-66	Hr.	Daily	Hr.	Daily	Bare Costs	Incl. O&P
1 Equip. Oper. (light)	$19.10	$152.80	$31.70	$253.60	$19.10	$31.70
1 Backhoe Ldr. w/Attchmt.		187.20		205.90	23.40	25.74
8 L.H., Daily Totals		$340.00		$459.50	$42.50	$57.44
Crew B-67	Hr.	Daily	Hr.	Daily	Bare Costs	Incl. O&P
1 Millwright	$20.60	$164.80	$33.60	$268.80	$19.85	$32.65
1 Equip. Oper. (light)	19.10	152.80	31.70	253.60		
1 Forklift		171.20		188.30	10.70	11.77
16 L.H., Daily Totals		$488.80		$710.70	$30.55	$44.42
Crew B-68	Hr.	Daily	Hr.	Daily	Bare Costs	Incl. O&P
2 Millwrights	$20.60	$329.60	$33.60	$537.60	$20.10	$32.97
1 Equip. Oper. (light)	19.10	152.80	31.70	253.60		
1 Forklift		171.20		188.30	7.13	7.85
24 L.H., Daily Totals		$653.60		$979.50	$27.23	$40.82
Crew B-69	Hr.	Daily	Hr.	Daily	Bare Costs	Incl. O&P
1 Labor Foreman (outside)	$16.45	$131.60	$28.20	$225.60	$16.24	$27.48
3 Laborers	14.45	346.80	24.75	594.00		
1 Equip Oper. (crane)	20.65	165.20	34.25	274.00		
1 Equip Oper. Oiler	17.00	136.00	28.20	225.60		
1 Truck Crane, 80 Ton		1098.00		1207.80	22.88	25.16
48 L.H., Daily Totals		$1877.60		$2527.00	$39.12	$52.64

Crews

Crew No.	Bare Costs		Incl. Subs O & P		Cost Per Labor-Hour	
Crew B-69A	Hr.	Daily	Hr.	Daily	Bare Costs	Incl. O&P
1 Labor Foreman	$16.45	$131.60	$28.20	$225.60	$16.43	$27.71
3 Laborers	14.45	346.80	24.75	594.00		
1 Equip. Oper. (medium)	19.90	159.20	33.00	264.00		
1 Concrete Finisher	18.90	151.20	30.80	246.40		
1 Curb Paver		454.00		499.40	9.46	10.40
48 L.H., Daily Totals		$1242.80		$1829.40	$25.89	$38.11
Crew B-69B	Hr.	Daily	Hr.	Daily	Bare Costs	Incl. O&P
1 Labor Foreman	$16.45	$131.60	$28.20	$225.60	$16.43	$27.71
3 Laborers	14.45	346.80	24.75	594.00		
1 Equip. Oper. (medium)	19.90	159.20	33.00	264.00		
1 Cement Finisher	18.90	151.20	30.80	246.40		
1 Curb/Gutter Paver		911.85		1003.05	19.00	20.90
48 L.H., Daily Totals		$1700.65		$2333.05	$35.43	$48.61
Crew B-70	Hr.	Daily	Hr.	Daily	Bare Costs	Incl. O&P
1 Labor Foreman (outside)	$16.45	$131.60	$28.20	$225.60	$17.07	$28.78
3 Laborers	14.45	346.80	24.75	594.00		
3 Equip. Oper. (med.)	19.90	477.60	33.00	792.00		
1 Motor Grader, 30,000 Lb.		516.00		567.60		
1 Grader Attach., Ripper		64.00		70.40		
1 Road Sweeper, S.P.		218.40		240.25		
1 F.E. Loader, 1-3/4 C.Y.		303.00		333.30	19.67	21.63
56 L.H., Daily Totals		$2057.40		$2823.15	$36.74	$50.41
Crew B-71	Hr.	Daily	Hr.	Daily	Bare Costs	Incl. O&P
1 Labor Foreman (outside)	$16.45	$131.60	$28.20	$225.60	$17.07	$28.78
3 Laborers	14.45	346.80	24.75	594.00		
3 Equip. Oper. (med.)	19.90	477.60	33.00	792.00		
1 Pvmt. Profiler, 450 H.P.		3442.00		3786.20		
1 Road Sweeper, S.P.		218.40		240.25		
1 F.E. Loader, 1-3/4 C.Y.		303.00		333.30	70.78	77.85
56 L.H., Daily Totals		$4919.40		$5971.35	$87.85	$106.63
Crew B-72	Hr.	Daily	Hr.	Daily	Bare Costs	Incl. O&P
1 Labor Foreman (outside)	$16.45	$131.60	$28.20	$225.60	$17.42	$29.31
3 Laborers	14.45	346.80	24.75	594.00		
4 Equip. Oper. (med.)	19.90	636.80	33.00	1056.00		
1 Pvmt. Profiler, 450 H.P.		3442.00		3786.20		
1 Hammermill, 250 H.P.		1177.00		1294.70		
1 Windrow Loader		929.85		1022.85		
1 Mix Paver 165 H.P.		1440.00		1584.00		
1 Roller, Pneu. Tire, 12 T.		240.80		264.90	112.96	124.26
64 L.H., Daily Totals		$8344.85		$9828.25	$130.38	$153.57
Crew B-73	Hr.	Daily	Hr.	Daily	Bare Costs	Incl. O&P
1 Labor Foreman (outside)	$16.45	$131.60	$28.20	$225.60	$18.11	$30.34
2 Laborers	14.45	231.20	24.75	396.00		
5 Equip. Oper. (med.)	19.90	796.00	33.00	1320.00		
1 Road Mixer, 310 H.P.		1008.00		1108.80		
1 Roller, Tandem, 12 Ton		224.00		246.40		
1 Hammermill, 250 H.P.		1177.00		1294.70		
1 Motor Grader, 30,000 Lb.		516.00		567.60		
.5 F.E. Loader, 1-3/4 C.Y.		151.50		166.65		
.5 Truck, 30 Ton		162.80		179.10		
.5 Water Tank 5000 Gal.		102.95		113.25	52.22	57.44
64 L.H., Daily Totals		$4501.05		$5618.10	$70.33	$87.78
Crew B-74	Hr.	Daily	Hr.	Daily	Bare Costs	Incl. O&P
1 Labor Foreman (outside)	$16.45	$131.60	$28.20	$225.60	$17.86	$29.88
1 Laborer	14.45	115.60	24.75	198.00		
4 Equip. Oper. (med.)	19.90	636.80	33.00	1056.00		
2 Truck Drivers (heavy)	16.20	259.20	27.05	432.80		
1 Motor Grader, 30,000 Lb.		516.00		567.60		
1 Grader Attach., Ripper		64.00		70.40		
2 Stabilizers, 310 H.P.		1394.10		1533.50		
1 Flatbed Truck, 3 Ton		180.45		198.50		
1 Chem. Spreader, Towed		101.20		111.30		
1 Vibr. Roller, 29,000 Lb.		432.40		475.65		
1 Water Tank 5000 Gal.		205.90		226.50		
1 Truck, 30 Ton		325.60		358.15	50.31	55.34
64 L.H., Daily Totals		$4362.85		$5454.00	$68.17	$85.22
Crew B-75	Hr.	Daily	Hr.	Daily	Bare Costs	Incl. O&P
1 Labor Foreman (outside)	$16.45	$131.60	$28.20	$225.60	$18.10	$30.29
1 Laborer	14.45	115.60	24.75	198.00		
4 Equip. Oper. (med.)	19.90	636.80	33.00	1056.00		
1 Truck Driver (heavy)	16.20	129.60	27.05	216.40		
1 Motor Grader, 30,000 Lb.		516.00		567.60		
1 Grader Attach., Ripper		64.00		70.40		
2 Stabilizers, 310 H.P.		1394.10		1533.50		
1 Dist. Truck, 3000 Gal.		374.95		412.45		
1 Vibr. Roller, 29,000 Lb.		432.40		475.65	49.67	54.64
56 L.H., Daily Totals		$3795.05		$4755.60	$67.77	$84.93
Crew B-76	Hr.	Daily	Hr.	Daily	Bare Costs	Incl. O&P
1 Dock Builder Foreman	$21.45	$171.60	$40.00	$320.00	$19.67	$35.33
5 Dock Builders	19.45	778.00	36.25	1450.00		
2 Equip. Oper. (crane)	20.65	330.40	34.25	548.00		
1 Equip. Oper. Oiler	17.00	136.00	28.20	225.60		
1 Crawler Crane, 50 Ton		850.80		935.90		
1 Barge, 400 Ton		453.65		499.00		
1 Hammer, 15K Ft. Lbs.		311.40		342.55		
60 L.F. Leads, 15K Ft. Lbs.		264.00		290.40		
1 Air Compr., 600 C.F.M.		278.00		305.80		
2-50 Ft. Air Hoses, 3" Dia.		33.80		37.20	30.44	33.48
72 L.H., Daily Totals		$3607.65		$4954.45	$50.11	$68.81
Crew B-77	Hr.	Daily	Hr.	Daily	Bare Costs	Incl. O&P
1 Labor Foreman	$16.45	$131.60	$28.20	$225.60	$15.13	$25.79
3 Laborers	14.45	346.80	24.75	594.00		
1 Truck Driver (light)	15.85	126.80	26.50	212.00		
1 Crack Cleaner, 25 H.P.		76.95		84.65		
1 Crack Filler, Trailer Mtd.		145.20		159.70		
1 Flatbed Truck, 3 Ton		180.45		198.50	10.07	11.07
40 L.H., Daily Totals		$1007.80		$1474.45	$25.20	$36.86
Crew B-78	Hr.	Daily	Hr.	Daily	Bare Costs	Incl. O&P
1 Labor Foreman	$16.45	$131.60	$28.20	$225.60	$14.85	$25.44
4 Laborers	14.45	462.40	24.75	792.00		
1 Paint Striper, S.P.		209.00		229.90		
1 Flatbed Truck, 3 Ton		180.45		198.50		
1 Pickup Truck, 3/4 Ton		130.20		143.20	12.99	14.29
40 L.H., Daily Totals		$1113.65		$1589.20	$27.84	$39.73

Crews

Crew No.	Bare Costs		Incl. Subs O & P		Cost Per Labor-Hour	
Crew B-79	Hr.	Daily	Hr.	Daily	Bare Costs	Incl. O&P
1 Labor Foreman	$16.45	$131.60	$28.20	$225.60	$14.95	$25.61
3 Laborers	14.45	346.80	24.75	594.00		
1 Thermo. Striper, T.M.		245.05		269.55		
1 Flatbed Truck, 3 Ton		180.45		198.50		
2 Pickup Truck, 3/4 Ton		260.40		286.45	21.43	23.58
32 L.H., Daily Totals		$1164.30		$1574.10	$36.38	$49.19
Crew B-80	Hr.	Daily	Hr.	Daily	Bare Costs	Incl. O&P
1 Labor Foreman	$16.45	$131.60	$28.20	$225.60	$15.12	$25.90
2 Laborer	14.45	231.20	24.75	396.00		
1 Flatbed Truck, 3 Ton		180.45		198.50		
1 Fence Post Auger, TM		370.00		407.00	22.94	25.23
24 L.H., Daily Totals		$913.25		$1227.10	$38.06	$51.13
Crew B-80A	Hr.	Daily	Hr.	Daily	Bare Costs	Incl. O&P
3 Laborers	$14.45	$346.80	$24.75	$594.00	$14.45	$24.75
1 Flatbed Truck, 3 Ton		180.45		198.50	7.52	8.27
24 L.H., Daily Totals		$527.25		$792.50	$21.97	$33.02
Crew B-80B	Hr.	Daily	Hr.	Daily	Bare Costs	Incl. O&P
3 Laborers	$14.45	$346.80	$24.75	$594.00	$15.61	$26.49
1 Equip. Oper. (light)	19.10	152.80	31.70	253.60		
1 Crane, flatbed mnt		283.20		311.50	8.85	9.74
32 L.H., Daily Totals		$782.80		$1159.10	$24.46	$36.23
Crew B-81	Hr.	Daily	Hr.	Daily	Bare Costs	Incl. O&P
1 Laborer	$14.45	$115.60	$24.75	$198.00	$15.33	$25.90
1 Truck Driver (heavy)	16.20	129.60	27.05	216.40		
1 Hydromulcher, T.M.		301.75		331.95		
1 Tractor Truck, 4x2		325.60		358.15	39.21	43.13
16 L.H., Daily Totals		$872.55		$1104.50	$54.54	$69.03
Crew B-82	Hr.	Daily	Hr.	Daily	Bare Costs	Incl. O&P
1 Laborer	$14.45	$115.60	$24.75	$198.00	$16.77	$28.23
1 Equip. Oper. (light)	19.10	152.80	31.70	253.60		
1 Horiz. Borer, 6 H.P.		49.80		54.80	3.11	3.42
16 L.H., Daily Totals		$318.20		$506.40	$19.88	$31.65
Crew B-83	Hr.	Daily	Hr.	Daily	Bare Costs	Incl. O&P
1 Tugboat Captain	$19.90	$159.20	$33.00	$264.00	$17.17	$28.88
1 Tugboat Hand	14.45	115.60	24.75	198.00		
1 Tugboat, 250 H.P.		444.80		489.30	27.80	30.58
16 L.H., Daily Totals		$719.60		$951.30	$44.97	$59.46
Crew B-84	Hr.	Daily	Hr.	Daily	Bare Costs	Incl. O&P
1 Equip. Oper. (med.)	$19.90	$159.20	$33.00	$264.00	$19.90	$33.00
1 Rotary Mower/Tractor		209.50		230.45	26.19	28.81
8 L.H., Daily Totals		$368.70		$494.45	$46.09	$61.81
Crew B-85	Hr.	Daily	Hr.	Daily	Bare Costs	Incl. O&P
3 Laborers	$14.45	$346.80	$24.75	$594.00	$15.89	$26.86
1 Equip. Oper. (med.)	19.90	159.20	33.00	264.00		
1 Truck Driver (heavy)	16.20	129.60	27.05	216.40		
1 Aerial Lift Truck, 80'		569.60		626.55		
1 Brush Chipper, 130 H.P.		211.05		232.15		
1 Pruning Saw, Rotary		22.25		24.50	20.07	22.08
40 L.H., Daily Totals		$1438.50		$1957.60	$35.96	$48.94

Crew No.	Bare Costs		Incl. Subs O & P		Cost Per Labor-Hour	
Crew B-86	Hr.	Daily	Hr.	Daily	Bare Costs	Incl. O&P
1 Equip. Oper. (med.)	$19.90	$159.20	$33.00	$264.00	$19.90	$33.00
1 Stump Chipper, S.P.		165.60		182.15	20.70	22.77
8 L.H., Daily Totals		$324.80		$446.15	$40.60	$55.77
Crew B-86A	Hr.	Daily	Hr.	Daily	Bare Costs	Incl. O&P
1 Equip. Oper. (medium)	$19.90	$159.20	$33.00	$264.00	$19.90	$33.00
1 Grader, 30,000 Lbs.		516.00		567.60	64.50	70.95
8 L.H., Daily Totals		$675.20		$831.60	$84.40	$103.95
Crew B-86B	Hr.	Daily	Hr.	Daily	Bare Costs	Incl. O&P
1 Equip. Oper. (medium)	$19.90	$159.20	$33.00	$264.00	$19.90	$33.00
1 Dozer, 200 H.P.		806.80		887.50	100.85	110.94
8 L.H., Daily Totals		$966.00		$1151.50	$120.75	$143.94
Crew B-87	Hr.	Daily	Hr.	Daily	Bare Costs	Incl. O&P
1 Laborer	$14.45	$115.60	$24.75	$198.00	$18.81	$31.35
4 Equip. Oper. (med.)	19.90	636.80	33.00	1056.00		
2 Feller Bunchers, 50 H.P.		868.80		955.70		
1 Log Chipper, 22" Tree		2017.00		2218.70		
1 Dozer, 105 H.P.		279.60		307.55		
1 Chainsaw, Gas, 36" Long		49.20		54.10	80.37	88.40
40 L.H., Daily Totals		$3967.00		$4790.05	$99.18	$119.75
Crew B-88	Hr.	Daily	Hr.	Daily	Bare Costs	Incl. O&P
1 Laborer	$14.45	$115.60	$24.75	$198.00	$19.12	$31.82
6 Equip. Oper. (med.)	19.90	955.20	33.00	1584.00		
2 Feller Bunchers, 50 H.P.		868.80		955.70		
1 Log Chipper, 22" Tree		2017.00		2218.70		
2 Log Skidders, 50 H.P.		784.00		862.40		
1 Dozer, 105 H.P.		279.60		307.55		
1 Chainsaw, Gas, 36" Long		49.20		54.10	71.40	78.54
56 L.H., Daily Totals		$5069.40		$6180.45	$90.52	$110.36
Crew B-89	Hr.	Daily	Hr.	Daily	Bare Costs	Incl. O&P
1 Skilled Worker	$19.75	$158.00	$33.90	$271.20	$17.10	$29.33
1 Building Laborer	14.45	115.60	24.75	198.00		
1 Cutting Machine		48.00		52.80	3.00	3.30
16 L.H., Daily Totals		$321.60		$522.00	$20.10	$32.63
Crew B-89A	Hr.	Daily	Hr.	Daily	Bare Costs	Incl. O&P
1 Skilled Worker	$19.75	$158.00	$33.90	$271.20	$17.10	$29.33
1 Laborer	14.45	115.60	24.75	198.00		
1 Core Drill (large)		60.60		66.65	3.79	4.17
16 L.H., Daily Totals		$334.20		$535.85	$20.89	$33.50
Crew B-89B	Hr.	Daily	Hr.	Daily	Bare Costs	Incl. O&P
1 Equip. Oper. (light)	$19.10	$152.80	$31.70	$253.60	$17.48	$29.10
1 Truck Driver, Light	15.85	126.80	26.50	212.00		
1 Wall Saw, Hydraulic, 10 H.P.		133.20		146.50		
1 Generator, Diesel, 100 KW		207.05		227.75		
1 Water Tank, 65 Gal.		13.00		14.30		
1 Flatbed Truck, 3 Ton		180.45		198.50	33.36	36.69
16 L.H., Daily Totals		$813.30		$1052.65	$50.84	$65.79

Crews

Crew No.	Bare Costs		Incl. Subs O & P		Cost Per Labor-Hour	
Crew B-90	Hr.	Daily	Hr.	Daily	Bare Costs	Incl. O&P
1 Labor Foreman (outside)	$16.45	$131.60	$28.20	$225.60	$16.30	$27.49
3 Laborers	14.45	346.80	24.75	594.00		
2 Equip. Oper. (light)	19.10	305.60	31.70	507.20		
2 Truck Drivers (heavy)	16.20	259.20	27.05	432.80		
1 Road Mixer, 310 H.P.		1008.00		1108.80		
1 Dist. Truck, 2000 Gal.		350.30		385.35	21.22	23.35
64 L.H., Daily Totals		$2401.50		$3253.75	$37.52	$50.84
Crew B-90A	Hr.	Daily	Hr.	Daily	Bare Costs	Incl. O&P
1 Labor Foreman	$16.45	$131.60	$28.20	$225.60	$17.85	$29.96
2 Laborers	14.45	231.20	24.75	396.00		
4 Equip. Oper. (medium)	19.90	636.80	33.00	1056.00		
2 Graders, 30,000 Lbs.		1032.00		1135.20		
1 Roller, Steel Wheel		224.00		246.40		
1 Roller, Pneumatic Wheel		240.80		264.90	26.73	29.40
56 L.H., Daily Totals		$2496.40		$3324.10	$44.58	$59.36
Crew B-90B	Hr.	Daily	Hr.	Daily	Bare Costs	Incl. O&P
1 Labor Foreman	$16.45	$131.60	$28.20	$225.60	$17.51	$29.45
2 Laborers	14.45	231.20	24.75	396.00		
3 Equip. Oper. (medium)	19.90	477.60	33.00	792.00		
1 Roller, Steel Wheel		224.00		246.40		
1 Roller, Pneumatic Wheel		240.80		264.90		
1 Road Mixer, 310 H.P.		1008.00		1108.80	30.68	33.75
48 L.H., Daily Totals		$2313.20		$3033.70	$48.19	$63.20
Crew B-91	Hr.	Daily	Hr.	Daily	Bare Costs	Incl. O&P
1 Labor Foreman (outside)	$16.45	$131.60	$28.20	$225.60	$17.64	$29.59
2 Laborers	14.45	231.20	24.75	396.00		
4 Equip. Oper. (med.)	19.90	636.80	33.00	1056.00		
1 Truck Driver (heavy)	16.20	129.60	27.05	216.40		
1 Dist. Truck, 3000 Gal.		374.95		412.45		
1 Aggreg. Spreader, S.P.		640.40		704.45		
1 Roller, Pneu. Tire, 12 Ton		240.80		264.90		
1 Roller, Steel, 10 Ton		224.00		246.40	23.13	25.44
64 L.H., Daily Totals		$2609.35		$3522.20	$40.77	$55.03
Crew B-92	Hr.	Daily	Hr.	Daily	Bare Costs	Incl. O&P
1 Labor Foreman (outside)	$16.45	$131.60	$28.20	$225.60	$14.95	$25.61
3 Laborers	14.45	346.80	24.75	594.00		
1 Crack Cleaner, 25 H.P.		76.95		84.65		
1 Air Compressor		78.30		86.15		
1 Tar Kettle, T.M.		23.20		25.50		
1 Flatbed Truck, 3 Ton		180.45		198.50	11.22	12.34
32 L.H., Daily Totals		$837.30		$1214.40	$26.17	$37.95
Crew B-93	Hr.	Daily	Hr.	Daily	Bare Costs	Incl. O&P
1 Equip. Oper. (med.)	$19.90	$159.20	$33.00	$264.00	$19.90	$33.00
1 Feller Buncher, 50 H.P.		434.40		477.85	54.30	59.73
8 L.H., Daily Totals		$593.60		$741.85	$74.20	$92.73
Crew B-94A	Hr.	Daily	Hr.	Daily	Bare Costs	Incl. O&P
1 Laborer	$14.45	$115.60	$24.75	$198.00	$14.45	$24.75
1 Diaph. Water Pump, 2"		29.40		32.35		
1-20 Ft. Suction Hose, 2"		6.50		7.15		
2-50 Ft. Disch. Hoses, 2"		10.80		11.90	5.84	6.42
8 L.H., Daily Totals		$162.30		$249.40	$20.29	$31.17

Crew No.	Bare Costs		Incl. Subs O & P		Cost Per Labor-Hour	
Crew B-94B	Hr.	Daily	Hr.	Daily	Bare Costs	Incl. O&P
1 Laborer	$14.45	$115.60	$24.75	$198.00	$14.45	$24.75
1 Diaph. Water Pump, 4"		65.65		72.20		
1-20 Ft. Suction Hose, 4"		12.50		13.75		
2-50 Ft. Disch. Hoses, 4"		17.00		18.70	11.89	13.08
8 L.H., Daily Totals		$210.75		$302.65	$26.34	$37.83
Crew B-94C	Hr.	Daily	Hr.	Daily	Bare Costs	Incl. O&P
1 Laborer	$14.45	$115.60	$24.75	$198.00	$14.45	$24.75
1 Centr. Water Pump, 3"		38.00		41.80		
1-20 Ft. Suction Hose, 3"		9.50		10.45		
2-50 Ft. Disch. Hoses, 3"		12.80		14.10	7.54	8.29
8 L.H., Daily Totals		$175.90		$264.35	$21.99	$33.04
Crew B-94D	Hr.	Daily	Hr.	Daily	Bare Costs	Incl. O&P
1 Laborer	$14.45	$115.60	$24.75	$198.00	$14.45	$24.75
1 Centr. Water Pump, 6"		157.20		172.90		
1-20 Ft. Suction Hose, 6"		22.50		24.75		
2-50 Ft. Disch. Hoses, 6"		41.10		45.20	27.60	30.36
8 L.H., Daily Totals		$336.40		$440.85	$42.05	$55.11
Crew B-95	Hr.	Daily	Hr.	Daily	Bare Costs	Incl. O&P
1 Equip. Oper. (crane)	$20.65	$165.20	$34.25	$274.00	$17.55	$29.50
1 Laborer	14.45	115.60	24.75	198.00		
16 L.H., Daily Totals		$280.80		$472.00	$17.55	$29.50
Crew B-95A	Hr.	Daily	Hr.	Daily	Bare Costs	Incl. O&P
1 Equip. Oper. (crane)	$20.65	$165.20	$34.25	$274.00	$17.55	$29.50
1 Laborer	14.45	115.60	24.75	198.00		
1 Hyd. Excavator, 5/8 C.Y.		411.60		452.75	25.73	28.30
16 L.H., Daily Totals		$692.40		$924.75	$43.28	$57.80
Crew B-95B	Hr.	Daily	Hr.	Daily	Bare Costs	Incl. O&P
1 Equip. Oper. (crane)	$20.65	$165.20	$34.25	$274.00	$17.55	$29.50
1 Laborer	14.45	115.60	24.75	198.00		
1 Hyd. Excavator, 1.5 C.Y.		713.20		784.50	44.58	49.03
16 L.H., Daily Totals		$994.00		$1256.50	$62.13	$78.53
Crew B-95C	Hr.	Daily	Hr.	Daily	Bare Costs	Incl. O&P
1 Equip. Oper. (crane)	$20.65	$165.20	$34.25	$274.00	$17.55	$29.50
1 Laborer	14.45	115.60	24.75	198.00		
1 Hyd. Excavator, 2.5 C.Y.		1673.00		1840.30	104.56	115.02
16 L.H., Daily Totals		$1953.80		$2312.30	$122.11	$144.52
Crew C-1	Hr.	Daily	Hr.	Daily	Bare Costs	Incl. O&P
2 Carpenters	$19.70	$315.20	$33.75	$540.00	$17.16	$29.40
1 Carpenter Helper	14.80	118.40	25.35	202.80		
1 Laborer	14.45	115.60	24.75	198.00		
32 L.H., Daily Totals		$549.20		$940.80	$17.16	$29.40
Crew C-2	Hr.	Daily	Hr.	Daily	Bare Costs	Incl. O&P
1 Carpenter Foreman (out)	$21.70	$173.60	$37.20	$297.60	$17.52	$30.03
2 Carpenters	19.70	315.20	33.75	540.00		
2 Carpenter Helpers	14.80	236.80	25.35	405.60		
1 Laborer	14.45	115.60	24.75	198.00		
48 L.H., Daily Totals		$841.20		$1441.20	$17.52	$30.03

Crews

Crew No.	Bare Costs		Incl. Subs O & P		Cost Per Labor-Hour	

Crew C-2A	Hr.	Daily	Hr.	Daily	Bare Costs	Incl. O&P
1 Carpenter Foreman (out)	$21.70	$173.60	$37.20	$297.60	$19.02	$32.33
3 Carpenters	19.70	472.80	33.75	810.00		
1 Cement Finisher	18.90	151.20	30.80	246.40		
1 Laborer	14.45	115.60	24.75	198.00		
48 L.H., Daily Totals		$913.20		$1552.00	$19.02	$32.33

Crew C-3	Hr.	Daily	Hr.	Daily	Bare Costs	Incl. O&P
1 Rodman Foreman	$23.10	$184.80	$43.00	$344.00	$18.61	$33.34
3 Rodmen (reinf.)	21.10	506.40	39.25	942.00		
1 Equip. Oper. (light)	19.10	152.80	31.70	253.60		
3 Laborers	14.45	346.80	24.75	594.00		
3 Stressing Equipment		78.00		85.80		
.5 Grouting Equipment		125.00		137.50	3.17	3.49
64 L.H., Daily Totals		$1393.80		$2356.90	$21.78	$36.83

Crew C-4	Hr.	Daily	Hr.	Daily	Bare Costs	Incl. O&P
1 Rodman Foreman	$23.10	$184.80	$43.00	$344.00	$19.94	$36.56
2 Rodmen (reinf.)	21.10	337.60	39.25	628.00		
1 Building Laborer	14.45	115.60	24.75	198.00		
3 Stressing Equipment		78.00		85.80	2.44	2.68
32 L.H., Daily Totals		$716.00		$1255.80	$22.38	$39.24

Crew C-5	Hr.	Daily	Hr.	Daily	Bare Costs	Incl. O&P
1 Rodman Foreman	$23.10	$184.80	$43.00	$344.00	$19.14	$34.21
2 Rodmen (reinf.)	21.10	337.60	39.25	628.00		
1 Equip. Oper. (crane)	20.65	165.20	34.25	274.00		
2 Building Laborers	14.45	231.20	24.75	396.00		
1 Hyd. Crane, 25 Ton		559.20		615.10	11.65	12.82
48 L.H., Daily Totals		$1478.00		$2257.10	$30.79	$47.03

Crew C-6	Hr.	Daily	Hr.	Daily	Bare Costs	Incl. O&P
1 Labor Foreman (outside)	$16.45	$131.60	$28.20	$225.60	$15.53	$26.33
4 Laborers	14.45	462.40	24.75	792.00		
1 Cement Finisher	18.90	151.20	30.80	246.40		
2 Gas Engine Vibrators		76.00		83.60	1.58	1.74
48 L.H., Daily Totals		$821.20		$1347.60	$17.11	$28.07

Crew C-7	Hr.	Daily	Hr.	Daily	Bare Costs	Incl. O&P
1 Labor Foreman (outside)	$16.45	$131.60	$28.20	$225.60	$16.06	$27.11
5 Laborers	14.45	578.00	24.75	990.00		
1 Cement Finisher	18.90	151.20	30.80	246.40		
1 Equip. Oper. (med.)	19.90	159.20	33.00	264.00		
1 Equip. Oper. (oiler)	17.00	136.00	28.20	225.60		
2 Gas Engine Vibrators		76.00		83.60		
1 Concrete Bucket, 1 C.Y.		27.20		29.90		
1 Hyd. Crane, 55 Ton		808.20		889.00	12.66	13.92
72 L.H., Daily Totals		$2067.40		$2954.10	$28.72	$41.03

Crew C-8	Hr.	Daily	Hr.	Daily	Bare Costs	Incl. O&P
1 Labor Foreman (outside)	$16.45	$131.60	$28.20	$225.60	$16.79	$28.15
3 Laborers	14.45	346.80	24.75	594.00		
2 Cement Finishers	18.90	302.40	30.80	492.80		
1 Equip. Oper. (med.)	19.90	159.20	33.00	264.00		
1 Concrete Pump (small)		656.00		721.60	11.71	12.89
56 L.H., Daily Totals		$1596.00		$2298.00	$28.50	$41.04

Crew C-8A	Hr.	Daily	Hr.	Daily	Bare Costs	Incl. O&P
1 Labor Foreman (outside)	$16.45	$131.60	$28.20	$225.60	$16.27	$27.34
3 Laborers	14.45	346.80	24.75	594.00		
2 Cement Finishers	18.90	302.40	30.80	492.80		
48 L.H., Daily Totals		$780.80		$1312.40	$16.27	$27.34

Crew C-8B	Hr.	Daily	Hr.	Daily	Bare Costs	Incl. O&P
1 Labor Foreman (outside)	$16.45	$131.60	$28.20	$225.60	$15.94	$27.09
3 Laborers	14.45	346.80	24.75	594.00		
1 Equipment Operator	19.90	159.20	33.00	264.00		
1 Vibrating Screed		49.20		54.10		
1 Vibratory Roller		432.40		475.65		
1 Dozer, 200 HP		806.80		887.50	32.21	35.43
40 L.H., Daily Totals		$1926.00		$2500.85	$48.15	$62.52

Crew C-8C	Hr.	Daily	Hr.	Daily	Bare Costs	Incl. O&P
1 Labor Foreman (outside)	$16.45	$131.60	$28.20	$225.60	$16.43	$27.71
3 Laborers	14.45	346.80	24.75	594.00		
1 Cement Finisher	18.90	151.20	30.80	246.40		
1 Equipment Operator (med.)	19.90	159.20	33.00	264.00		
1 Shotcrete Rig, 12 CY/hr		346.00		380.60	7.21	7.93
48 L.H., Daily Totals		$1134.80		$1710.60	$23.64	$35.64

Crew C-8D	Hr.	Daily	Hr.	Daily	Bare Costs	Incl. O&P
1 Labor Foreman (outside)	$16.45	$131.60	$28.20	$225.60	$17.23	$28.86
1 Laborers	14.45	115.60	24.75	198.00		
1 Cement Finisher	18.90	151.20	30.80	246.40		
1 Equipment Operator (light)	19.10	152.80	31.70	253.60		
1 Compressor, 250 CFM		127.00		139.70		
2 Hoses, 1", 50'		10.40		11.45	4.29	4.72
32 L.H., Daily Totals		$688.60		$1074.75	$21.52	$33.58

Crew C-8E	Hr.	Daily	Hr.	Daily	Bare Costs	Incl. O&P
1 Labor Foreman (outside)	$16.45	$131.60	$28.20	$225.60	$17.23	$28.86
1 Laborers	14.45	115.60	24.75	198.00		
1 Cement Finisher	18.90	151.20	30.80	246.40		
1 Equipment Operator (light)	19.10	152.80	31.70	253.60		
1 Compressor, 250 CFM		127.00		139.70		
2 Hoses, 1", 50'		10.40		11.45		
1 Concrete Pump (small)		656.00		721.60	24.79	27.27
32 L.H., Daily Totals		$1344.60		$1796.35	$42.02	$56.13

Crew C-10	Hr.	Daily	Hr.	Daily	Bare Costs	Incl. O&P
1 Laborer	$14.45	$115.60	$24.75	$198.00	$17.42	$28.78
2 Cement Finishers	18.90	302.40	30.80	492.80		
24 L.H., Daily Totals		$418.00		$690.80	$17.42	$28.78

Crew C-11	Hr.	Daily	Hr.	Daily	Bare Costs	Incl. O&P
1 Skilled Worker Foreman	$21.75	$174.00	$37.30	$298.40	$20.16	$34.44
5 Skilled Worker	19.75	790.00	33.90	1356.00		
1 Equip. Oper. (crane)	20.65	165.20	34.25	274.00		
1 Truck Crane, 150 Ton		1563.00		1719.30	27.91	30.70
56 L.H., Daily Totals		$2692.20		$3647.70	$48.07	$65.14

Crew C-12	Hr.	Daily	Hr.	Daily	Bare Costs	Incl. O&P
1 Carpenter Foreman (out)	$21.70	$173.60	$37.20	$297.60	$19.32	$32.91
3 Carpenters	19.70	472.80	33.75	810.00		
1 Laborer	14.45	115.60	24.75	198.00		
1 Equip. Oper. (crane)	20.65	165.20	34.25	274.00		
1 Hyd. Crane, 12 Ton		453.45		498.80	9.45	10.39
48 L.H., Daily Totals		$1380.65		$2078.40	$28.77	$43.30

Crew C-13	Hr.	Daily	Hr.	Daily	Bare Costs	Incl. O&P
2 Struc. Steel Worker	$21.25	$340.00	$41.90	$670.40	$20.73	$39.18
1 Carpenter	19.70	157.60	33.75	270.00		
1 Gas Welding Machine		84.00		92.40	3.50	3.85
24 L.H., Daily Totals		$581.60		$1032.80	$24.23	$43.03

Crews

Crew No.	Bare Costs		Incl. Subs O & P		Cost Per Labor-Hour	
Crew C-14	**Hr.**	**Daily**	**Hr.**	**Daily**	**Bare Costs**	**Incl. O&P**
1 Carpenter Foreman (out)	$21.70	$173.60	$37.20	$297.60	$17.56	$30.19
3 Carpenters	19.70	472.80	33.75	810.00		
2 Carpenter Helpers	14.80	236.80	25.35	405.60		
4 Laborers	14.45	462.40	24.75	792.00		
2 Rodmen (reinf.)	21.10	337.60	39.25	628.00		
2 Rodman Helpers	14.80	236.80	25.35	405.60		
2 Cement Finishers	18.90	302.40	30.80	492.80		
1 Equip. Oper. (crane)	20.65	165.20	34.25	274.00		
1 Crane, 80 Ton, & Tools		1098.00		1207.80	8.07	8.88
136 L.H., Daily Totals		$3485.60		$5313.40	$25.63	$39.07
Crew C-14A	**Hr.**	**Daily**	**Hr.**	**Daily**	**Bare Costs**	**Incl. O&P**
1 Carpenter Foreman (out)	$21.70	$173.60	$37.20	$297.60	$19.56	$33.90
16 Carpenters	19.70	2521.60	33.75	4320.00		
4 Rodmen (reinf.)	21.10	675.20	39.25	1256.00		
2 Laborers	14.45	231.20	24.75	396.00		
1 Cement Finisher	18.90	151.20	30.80	246.40		
1 Equip. Oper. (med)	19.90	159.20	33.00	264.00		
1 Gas Engine Vibrator		38.00		41.80		
1 Concrete Pump (small)		656.00		721.60	3.47	3.82
200 L.H., Daily Totals		$4606.00		$7543.40	$23.03	$37.72
Crew C-14B	**Hr.**	**Daily**	**Hr.**	**Daily**	**Bare Costs**	**Incl. O&P**
1 Carpenter Foreman (out)	$21.70	$173.60	$37.20	$297.60	$19.53	$33.78
16 Carpenters	19.70	2521.60	33.75	4320.00		
4 Rodmen (reinf.)	21.10	675.20	39.25	1256.00		
2 Laborers	14.45	231.20	24.75	396.00		
2 Cement Finishers	18.90	302.40	30.80	492.80		
1 Equip. Oper. (med)	19.90	159.20	33.00	264.00		
1 Gas Engine Vibrator		38.00		41.80		
1 Concrete Pump (small)		656.00		721.60	3.34	3.67
208 L.H., Daily Totals		$4757.20		$7789.80	$22.87	$37.45
Crew C-14C	**Hr.**	**Daily**	**Hr.**	**Daily**	**Bare Costs**	**Incl. O&P**
1 Carpenter Foreman (out)	$21.70	$173.60	$37.20	$297.60	$18.49	$32.00
6 Carpenters	19.70	945.60	33.75	1620.00		
2 Rodmen (reinf)	21.10	337.60	39.25	628.00		
4 Laborers	14.45	462.40	24.75	792.00		
1 Cement Finisher	18.90	151.20	30.80	246.40		
1 Gas Engine Vibrator		38.00		41.80	.34	.37
112 L.H., Daily Totals		$2108.40		$3625.80	$18.83	$32.37
Crew C-14D	**Hr.**	**Daily**	**Hr.**	**Daily**	**Bare Costs**	**Incl. O&P**
1 Carpenter Foreman (out)	$21.70	$173.60	$37.20	$297.60	$19.45	$33.46
18 Carpenters	19.70	2836.80	33.75	4860.00		
2 Rodmen (reinf.)	21.10	337.60	39.25	628.00		
2 Laborers	14.45	231.20	24.75	396.00		
1 Cement Finisher	18.90	151.20	30.80	246.40		
1 Equip. Oper. (med.)	19.90	159.20	33.00	264.00		
1 Gas Engine Vibrator		38.00		41.80		
1 Concrete Pump (small)		656.00		721.60	3.47	3.82
200 L.H., Daily Totals		$4583.60		$7455.40	$22.92	$37.28
Crew C-14E	**Hr.**	**Daily**	**Hr.**	**Daily**	**Bare Costs**	**Incl. O&P**
1 Carpenter Foreman (out)	$21.70	$173.60	$37.20	$297.60	$18.89	$33.34
2 Carpenters	19.70	315.20	33.75	540.00		
4 Rodmen (reinf.)	21.10	675.20	39.25	1256.00		
3 Laborers	14.45	346.80	24.75	594.00		
1 Cement Finisher	18.90	151.20	30.80	246.40		
1 Gas Engine Vibrator		38.00		41.80	.43	.48
88 L.H., Daily Totals		$1700.00		$2975.80	$19.32	$33.82

Crew No.	Bare Costs		Incl. Subs O & P		Cost Per Labor-Hour	
Crew C-14F	**Hr.**	**Daily**	**Hr.**	**Daily**	**Bare Costs**	**Incl. O&P**
1 Laborer Foreman (out)	$16.45	$131.60	$28.20	$225.60	$17.64	$29.17
2 Laborers	14.45	231.20	24.75	396.00		
6 Cement Finishers	18.90	907.20	30.80	1478.40		
1 Gas Engine Vibrator		38.00		41.80	.53	.58
72 L.H., Daily Totals		$1308.00		$2141.80	$18.17	$29.75
Crew C-14G	**Hr.**	**Daily**	**Hr.**	**Daily**	**Bare Costs**	**Incl. O&P**
1 Laborer Foreman (out)	$16.45	$131.60	$28.20	$225.60	$17.28	$28.70
2 Laborers	14.45	231.20	24.75	396.00		
4 Cement Finishers	18.90	604.80	30.80	985.60		
1 Gas Engine Vibrator		38.00		41.80	.68	.75
56 L.H., Daily Totals		$1005.60		$1649.00	$17.96	$29.45
Crew C-14H	**Hr.**	**Daily**	**Hr.**	**Daily**	**Bare Costs**	**Incl. O&P**
1 Carpenter Foreman (out)	$21.70	$173.60	$37.20	$297.60	$19.26	$33.25
2 Carpenters	19.70	315.20	33.75	540.00		
1 Rodman (reinf.)	21.10	168.80	39.25	314.00		
1 Laborer	14.45	115.60	24.75	198.00		
1 Cement Finisher	18.90	151.20	30.80	246.40		
1 Gas Engine Vibrator		38.00		41.80	.79	.87
48 L.H., Daily Totals		$962.40		$1637.80	$20.05	$34.12
Crew C-15	**Hr.**	**Daily**	**Hr.**	**Daily**	**Bare Costs**	**Incl. O&P**
1 Carpenter Foreman (out)	$21.70	$173.60	$37.20	$297.60	$18.15	$31.09
2 Carpenters	19.70	315.20	33.75	540.00		
3 Laborers	14.45	346.80	24.75	594.00		
2 Cement Finishers	18.90	302.40	30.80	492.80		
1 Rodman (reinf.)	21.10	168.80	39.25	314.00		
72 L.H., Daily Totals		$1306.80		$2238.40	$18.15	$31.09
Crew C-16	**Hr.**	**Daily**	**Hr.**	**Daily**	**Bare Costs**	**Incl. O&P**
1 Labor Foreman (outside)	$16.45	$131.60	$28.20	$225.60	$17.74	$30.62
3 Laborers	14.45	346.80	24.75	594.00		
2 Cement Finishers	18.90	302.40	30.80	492.80		
1 Equip. Oper. (med.)	19.90	159.20	33.00	264.00		
2 Rodmen (reinf.)	21.10	337.60	39.25	628.00		
1 Concrete Pump (small)		656.00		721.60	9.11	10.02
72 L.H., Daily Totals		$1933.60		$2926.00	$26.85	$40.64
Crew C-17	**Hr.**	**Daily**	**Hr.**	**Daily**	**Bare Costs**	**Incl. O&P**
2 Skilled Worker Foremen	$21.75	$348.00	$37.30	$596.80	$20.15	$34.58
8 Skilled Workers	19.75	1264.00	33.90	2169.60		
80 L.H., Daily Totals		$1612.00		$2766.40	$20.15	$34.58
Crew C-17A	**Hr.**	**Daily**	**Hr.**	**Daily**	**Bare Costs**	**Incl. O&P**
2 Skilled Worker Foremen	$21.75	$348.00	$37.30	$596.80	$20.16	$34.58
8 Skilled Workers	19.75	1264.00	33.90	2169.60		
.125 Equip. Oper. (crane)	20.65	20.65	34.25	34.25		
.125 Crane, 80 Ton, & Tools		137.25		151.00	1.69	1.86
81 L.H., Daily Totals		$1769.90		$2951.65	$21.85	$36.44
Crew C-17B	**Hr.**	**Daily**	**Hr.**	**Daily**	**Bare Costs**	**Incl. O&P**
2 Skilled Worker Foremen	$21.75	$348.00	$37.30	$596.80	$20.16	$34.57
8 Skilled Workers	19.75	1264.00	33.90	2169.60		
.25 Equip. Oper. (crane)	20.65	41.30	34.25	68.50		
.25 Crane, 80 Ton, & Tools		274.50		301.95		
.25 Walk Behind Power Tools		13.25		14.60	3.51	3.86
82 L.H., Daily Totals		$1941.05		$3151.45	$23.67	$38.43

Crews

Crew No.	Bare Costs		Incl. Subs O & P		Cost Per Labor-Hour	
Crew C-17C	Hr.	Daily	Hr.	Daily	Bare Costs	Incl. O&P
2 Skilled Worker Foremen	$21.75	$348.00	$37.30	$596.80	$20.17	$34.57
8 Skilled Workers	19.75	1264.00	33.90	2169.60		
.375 Equip. Oper. (crane)	20.65	61.95	34.25	102.75		
.375 Crane, 80 Ton & Tools		411.75		452.95	4.96	5.46
83 L.H., Daily Totals		$2085.70		$3322.10	$25.13	$40.03
Crew C-17D	Hr.	Daily	Hr.	Daily	Bare Costs	Incl. O&P
2 Skilled Worker Foremen	$21.75	$348.00	$37.30	$596.80	$20.17	$34.56
8 Skilled Workers	19.75	1264.00	33.90	2169.60		
.5 Equip. Oper. (crane)	20.65	82.60	34.25	137.00		
.5 Crane, 80 Ton & Tools		549.00		603.90	6.54	7.19
84 L.H., Daily Totals		$2243.60		$3507.30	$26.71	$41.75
Crew C-17E	Hr.	Daily	Hr.	Daily	Bare Costs	Incl. O&P
2 Skilled Worker Foremen	$21.75	$348.00	$37.30	$596.80	$20.15	$34.58
8 Skilled Workers	19.75	1264.00	33.90	2169.60		
1 Hyd. Jack with Rods		60.50		66.55	.76	.83
80 L.H., Daily Totals		$1672.50		$2832.95	$20.91	$35.41
Crew C-18	Hr.	Daily	Hr.	Daily	Bare Costs	Incl. O&P
.125 Labor Foreman (out)	$16.45	$16.45	$28.20	$28.20	$14.67	$25.13
1 Laborer	14.45	115.60	24.75	198.00		
1 Concrete Cart, 10 C.F.		52.00		57.20	5.78	6.36
9 L.H., Daily Totals		$184.05		$283.40	$20.45	$31.49
Crew C-19	Hr.	Daily	Hr.	Daily	Bare Costs	Incl. O&P
.125 Labor Foreman (out)	$16.45	$16.45	$28.20	$28.20	$14.67	$25.13
1 Laborer	14.45	115.60	24.75	198.00		
1 Concrete Cart, 18 C.F.		83.20		91.50	9.24	10.17
9 L.H., Daily Totals		$215.25		$317.70	$23.91	$35.30
Crew C-20	Hr.	Daily	Hr.	Daily	Bare Costs	Incl. O&P
1 Labor Foreman (outside)	$16.45	$131.60	$28.20	$225.60	$15.94	$26.97
5 Laborers	14.45	578.00	24.75	990.00		
1 Cement Finisher	18.90	151.20	30.80	246.40		
1 Equip. Oper. (med.)	19.90	159.20	33.00	264.00		
2 Gas Engine Vibrators		76.00		83.60		
1 Concrete Pump (small)		656.00		721.60	11.44	12.58
64 L.H., Daily Totals		$1752.00		$2531.20	$27.38	$39.55
Crew C-21	Hr.	Daily	Hr.	Daily	Bare Costs	Incl. O&P
1 Labor Foreman (outside)	$16.45	$131.60	$28.20	$225.60	$15.94	$26.97
5 Laborers	14.45	578.00	24.75	990.00		
1 Cement Finisher	18.90	151.20	30.80	246.40		
1 Equip. Oper. (med.)	19.90	159.20	33.00	264.00		
2 Gas Engine Vibrators		76.00		83.60		
1 Concrete Conveyer		182.00		200.20	4.03	4.43
64 L.H., Daily Totals		$1278.00		$2009.80	$19.97	$31.40
Crew C-22	Hr.	Daily	Hr.	Daily	Bare Costs	Incl. O&P
1 Rodman Foreman	$23.10	$184.80	$43.00	$344.00	$21.37	$39.58
4 Rodmen (reinf.)	21.10	675.20	39.25	1256.00		
.125 Equip. Oper. (crane)	20.65	20.65	34.25	34.25		
.125 Equip. Oper. Oiler	17.00	17.00	28.20	28.20		
.125 Hyd. Crane, 25 Ton		69.90		76.90	1.66	1.83
42 L.H., Daily Totals		$967.55		$1739.35	$23.03	$41.41

Crew No.	Bare Costs		Incl. Subs O & P		Cost Per Labor-Hour	
Crew C-23	Hr.	Daily	Hr.	Daily	Bare Costs	Incl. O&P
2 Skilled Worker Foremen	$21.75	$348.00	$37.30	$596.80	$19.97	$34.05
6 Skilled Workers	19.75	948.00	33.90	1627.20		
1 Equip. Oper. (crane)	20.65	165.20	34.25	274.00		
1 Equip. Oper. Oiler	17.00	136.00	28.20	225.60		
1 Crane, 90 Ton		1022.00		1124.20	12.78	14.05
80 L.H., Daily Totals		$2619.20		$3847.80	$32.75	$48.10
Crew C-24	Hr.	Daily	Hr.	Daily	Bare Costs	Incl. O&P
2 Skilled Worker Foremen	$21.75	$348.00	$37.30	$596.80	$19.97	$34.05
6 Skilled Workers	19.75	948.00	33.90	1627.20		
1 Equip. Oper. (crane)	20.65	165.20	34.25	274.00		
1 Equip. Oper. Oiler	17.00	136.00	28.20	225.60		
1 Truck Crane, 150 Ton		1563.00		1719.30	19.54	21.49
80 L.H., Daily Totals		$3160.20		$4442.90	$39.51	$55.54
Crew C-25	Hr.	Daily	Hr.	Daily	Bare Costs	Incl. O&P
2 Rodmen (reinf.)	$21.10	$337.60	$39.25	$628.00	$16.93	$31.50
2 Rodmen Helpers	12.75	204.00	23.75	380.00		
32 L.H., Daily Totals		$541.60		$1008.00	$16.93	$31.50
Crew D-1	Hr.	Daily	Hr.	Daily	Bare Costs	Incl. O&P
1 Bricklayer	$20.00	$160.00	$33.85	$270.80	$17.85	$30.23
1 Bricklayer Helper	15.70	125.60	26.60	212.80		
16 L.H., Daily Totals		$285.60		$483.60	$17.85	$30.23
Crew D-2	Hr.	Daily	Hr.	Daily	Bare Costs	Incl. O&P
3 Bricklayers	$20.00	$480.00	$33.85	$812.40	$18.28	$30.95
2 Bricklayer Helpers	15.70	251.20	26.60	425.60		
40 L.H., Daily Totals		$731.20		$1238.00	$18.28	$30.95
Crew D-3	Hr.	Daily	Hr.	Daily	Bare Costs	Incl. O&P
3 Bricklayers	$20.00	$480.00	$33.85	$812.40	$18.35	$31.08
2 Bricklayer Helpers	15.70	251.20	26.60	425.60		
.25 Carpenter	19.70	39.40	33.75	67.50		
42 L.H., Daily Totals		$770.60		$1305.50	$18.35	$31.08
Crew D-4	Hr.	Daily	Hr.	Daily	Bare Costs	Incl. O&P
1 Bricklayer	$20.00	$160.00	$33.85	$270.80	$16.31	$27.68
3 Bricklayer Helpers	15.70	376.80	26.60	638.40		
1 Building Laborer	14.45	115.60	24.75	198.00		
1 Grout Pump, 50 C.F./hr		67.20		73.90		
1 Hoses & Hopper		31.60		34.75		
1 Accessories		10.55		11.60	2.73	3.01
40 L.H., Daily Totals		$761.75		$1227.45	$19.04	$30.69
Crew D-5	Hr.	Daily	Hr.	Daily	Bare Costs	Incl. O&P
1 Block Mason Helper	$15.70	$125.60	$26.60	$212.80	$15.70	$26.60
8 L.H., Daily Totals		$125.60		$212.80	$15.70	$26.60
Crew D-6	Hr.	Daily	Hr.	Daily	Bare Costs	Incl. O&P
3 Bricklayers	$20.00	$480.00	$33.85	$812.40	$17.85	$30.23
3 Bricklayer Helpers	15.70	376.80	26.60	638.40		
48 L.H., Daily Totals		$856.80		$1450.80	$17.85	$30.23
Crew D-7	Hr.	Daily	Hr.	Daily	Bare Costs	Incl. O&P
1 Tile Layer	$19.20	$153.60	$31.10	$248.80	$17.33	$28.08
1 Tile Layer Helper	15.45	123.60	25.05	200.40		
16 L.H., Daily Totals		$277.20		$449.20	$17.33	$28.08

Crews

Crew No.	Bare Costs		Incl. Subs O & P		Cost Per Labor-Hour	
Crew D-8	Hr.	Daily	Hr.	Daily	Bare Costs	Incl. O&P
3 Bricklayers	$20.00	$480.00	$33.85	$812.40	$18.28	$30.95
2 Bricklayer Helpers	15.70	251.20	26.60	425.60		
40 L.H., Daily Totals		$731.20		$1238.00	$18.28	$30.95
Crew D-9	Hr.	Daily	Hr.	Daily	Bare Costs	Incl. O&P
3 Block Masons	$20.00	$480.00	$33.85	$812.40	$17.85	$30.23
3 Block Mason Helpers	15.70	376.80	26.60	638.40		
48 L.H., Daily Totals		$856.80		$1450.80	$17.85	$30.23
Crew D-10	Hr.	Daily	Hr.	Daily	Bare Costs	Incl. O&P
1 Stone Mason Foreman	$21.50	$172.00	$36.40	$291.20	$18.61	$31.37
1 Stone Mason	19.50	156.00	33.00	264.00		
2 Stone Mason Helpers	15.70	251.20	26.60	425.60		
1 Equip. Oper. (crane)	20.65	165.20	34.25	274.00		
1 Truck Crane, 12.5 Ton		401.35		441.50	10.03	11.04
40 L.H., Daily Totals		$1145.75		$1696.30	$28.64	$42.41
Crew D-11	Hr.	Daily	Hr.	Daily	Bare Costs	Incl. O&P
2 Stone Masons	$19.50	$312.00	$33.00	$528.00	$18.23	$30.87
1 Stone Mason Helper	15.70	125.60	26.60	212.80		
24 L.H., Daily Totals		$437.60		$740.80	$18.23	$30.87
Crew D-12	Hr.	Daily	Hr.	Daily	Bare Costs	Incl. O&P
2 Stone Masons	$19.50	$312.00	$33.00	$528.00	$17.60	$29.80
2 Stone Mason Helpers	15.70	251.20	26.60	425.60		
32 L.H., Daily Totals		$563.20		$953.60	$17.60	$29.80
Crew D-13	Hr.	Daily	Hr.	Daily	Bare Costs	Incl. O&P
1 Stone Mason Foreman	$21.50	$172.00	$36.40	$291.20	$18.76	$31.64
2 Stone Masons	19.50	312.00	33.00	528.00		
2 Stone Mason Helpers	15.70	251.20	26.60	425.60		
1 Equip. Oper. (crane)	20.65	165.20	34.25	274.00		
1 Truck Crane, 12.5 Ton		401.35		441.50	8.36	9.20
48 L.H., Daily Totals		$1301.75		$1960.30	$27.12	$40.84
Crew E-1	Hr.	Daily	Hr.	Daily	Bare Costs	Incl. O&P
2 Struc. Steel Workers	$21.25	$340.00	$41.90	$670.40	$21.25	$41.90
1 Gas Welding Machine		84.00		92.40	5.25	5.78
16 L.H., Daily Totals		$424.00		$762.80	$26.50	$47.68
Crew E-2	Hr.	Daily	Hr.	Daily	Bare Costs	Incl. O&P
1 Struc. Steel Foreman	$23.25	$186.00	$45.85	$366.80	$21.48	$41.28
4 Struc. Steel Workers	21.25	680.00	41.90	1340.80		
1 Equip. Oper. (crane)	20.65	165.20	34.25	274.00		
1 Crane, 90 Ton		1022.00		1124.20	21.29	23.42
48 L.H., Daily Totals		$2053.20		$3105.80	$42.77	$64.70
Crew E-3	Hr.	Daily	Hr.	Daily	Bare Costs	Incl. O&P
1 Struc. Steel Foreman	$23.25	$186.00	$45.85	$366.80	$21.92	$43.22
2 Struc. Steel Worker	21.25	340.00	41.90	670.40		
1 Gas Welding Machine		84.00		92.40	3.50	3.85
24 L.H., Daily Totals		$610.00		$1129.60	$25.42	$47.07
Crew E-4	Hr.	Daily	Hr.	Daily	Bare Costs	Incl. O&P
1 Struc. Steel Foreman	$23.25	$186.00	$45.85	$366.80	$21.75	$42.89
3 Struc. Steel Workers	21.25	510.00	41.90	1005.60		
1 Gas Welding Machine		84.00		92.40	2.63	2.89
32 L.H., Daily Totals		$780.00		$1464.80	$24.38	$45.78

Crew No.	Bare Costs		Incl. Subs O & P		Cost Per Labor-Hour	
Crew E-5	Hr.	Daily	Hr.	Daily	Bare Costs	Incl. O&P
1 Struc. Steel Foremen	$23.25	$186.00	$45.85	$366.80	$21.41	$41.49
7 Struc. Steel Workers	21.25	1190.00	41.90	2346.40		
1 Equip. Oper. (crane)	20.65	165.20	34.25	274.00		
1 Crane, 90 Ton		1022.00		1124.20		
1 Gas Welding Machine		84.00		92.40	15.36	16.90
72 L.H., Daily Totals		$2647.20		$4203.80	$36.77	$58.39
Crew E-6	Hr.	Daily	Hr.	Daily	Bare Costs	Incl. O&P
1 Struc. Steel Foreman	$23.25	$186.00	$45.85	$366.80	$21.20	$40.97
12 Struc. Steel Workers	21.25	2040.00	41.90	4022.40		
1 Equip. Oper. (crane)	20.65	165.20	34.25	274.00		
1 Equip. Oper. (light)	19.10	152.80	31.70	253.60		
1 Crane, 90 Ton		1022.00		1124.20		
1 Gas Welding Machine		84.00		92.40		
1 Air Compr., 160 C.F.M.		98.40		108.25		
2 Impact Wrenches		55.60		61.15	10.50	11.55
120 L.H., Daily Totals		$3804.00		$6302.80	$31.70	$52.52
Crew E-7	Hr.	Daily	Hr.	Daily	Bare Costs	Incl. O&P
1 Struc. Steel Foreman	$23.25	$186.00	$45.85	$366.80	$21.41	$41.49
7 Struc. Steel Workers	21.25	1190.00	41.90	2346.40		
1 Equip. Oper. (crane)	20.65	165.20	34.25	274.00		
1 Crane, 90 Ton		1022.00		1124.20		
2 Gas Welding Machines		168.00		184.80	16.53	18.18
72 L.H., Daily Totals		$2731.20		$4296.20	$37.94	$59.67
Crew E-8	Hr.	Daily	Hr.	Daily	Bare Costs	Incl. O&P
1 Struc. Steel Foreman	$23.25	$186.00	$45.85	$366.80	$21.38	$41.56
9 Struc. Steel Workers	21.25	1530.00	41.90	3016.80		
1 Equip. Oper. (crane)	20.65	165.20	34.25	274.00		
1 Crane, 90 Ton		1022.00		1124.20		
4 Gas Welding Machines		336.00		369.60	15.43	16.98
88 L.H., Daily Totals		$3239.20		$5151.40	$36.81	$58.54
Crew E-9	Hr.	Daily	Hr.	Daily	Bare Costs	Incl. O&P
2 Struc. Steel Foremen	$23.25	$372.00	$45.85	$733.60	$21.19	$40.67
5 Struc. Steel Workers	21.25	850.00	41.90	1676.00		
1 Welder Foreman	23.25	186.00	45.85	366.80		
5 Welders	21.25	850.00	41.90	1676.00		
1 Equip. Oper. (crane)	20.65	165.20	34.25	274.00		
1 Equip. Oper. Oiler	17.00	136.00	28.20	225.60		
1 Equip. Oper. (light)	19.10	152.80	31.70	253.60		
1 Crane, 90 Ton		1022.00		1124.20		
5 Gas Welding Machines		420.00		462.00	11.27	12.39
128 L.H., Daily Totals		$4154.00		$6791.80	$32.46	$53.06
Crew E-10	Hr.	Daily	Hr.	Daily	Bare Costs	Incl. O&P
1 Struc. Steel Foreman	$23.25	$186.00	$45.85	$366.80	$21.92	$43.22
2 Struc. Steel Workers	21.25	340.00	41.90	670.40		
4 Gas Welding Machines		336.00		369.60		
1 Truck, 3 Ton		180.45		198.50	21.52	23.67
24 L.H., Daily Totals		$1042.45		$1605.30	$43.44	$66.89
Crew E-11	Hr.	Daily	Hr.	Daily	Bare Costs	Incl. O&P
2 Painters, Struc. Steel	$18.65	$298.40	$37.80	$604.80	$17.25	$33.45
1 Building Laborer	14.45	115.60	24.75	198.00		
1 Air Compressor 250 C.F.M.		127.00		139.70		
1 Sand Blaster		31.60		34.75		
1 Sand Blasting Accessories		10.55		11.60	7.05	7.75
24 L.H., Daily Totals		$583.15		$988.85	$24.30	$41.20

Crews

Crew No.	Bare Costs		Incl. Subs O & P		Cost Per Labor-Hour	
Crew E-12	**Hr.**	**Daily**	**Hr.**	**Daily**	**Bare Costs**	**Incl. O&P**
1 Welder Foreman	$23.25	$186.00	$45.85	$366.80	$21.18	$38.78
1 Equip. Oper. (light)	19.10	152.80	31.70	253.60		
1 Gas Welding Machine		84.00		92.40	5.25	5.78
16 L.H., Daily Totals		$422.80		$712.80	$26.43	$44.56
Crew E-13	**Hr.**	**Daily**	**Hr.**	**Daily**	**Bare Costs**	**Incl. O&P**
1 Welder Foreman	$23.25	$186.00	$45.85	$366.80	$21.87	$41.13
.5 Equip. Oper. (light)	19.10	76.40	31.70	126.80		
1 Gas Welding Machine		84.00		92.40	7.00	7.70
12 L.H., Daily Totals		$346.40		$586.00	$28.87	$48.83
Crew E-14	**Hr.**	**Daily**	**Hr.**	**Daily**	**Bare Costs**	**Incl. O&P**
1 Struc. Steel Worker	$21.25	$170.00	$41.90	$335.20	$21.25	$41.90
1 Gas Welding Machine		84.00		92.40	10.50	11.55
8 L.H., Daily Totals		$254.00		$427.60	$31.75	$53.45
Crew E-16	**Hr.**	**Daily**	**Hr.**	**Daily**	**Bare Costs**	**Incl. O&P**
1 Welder Foreman	$23.25	$186.00	$45.85	$366.80	$22.25	$43.88
1 Welder	21.25	170.00	41.90	335.20		
1 Gas Welding Machine		84.00		92.40	5.25	5.78
16 L.H., Daily Totals		$440.00		$794.40	$27.50	$49.66
Crew E-17	**Hr.**	**Daily**	**Hr.**	**Daily**	**Bare Costs**	**Incl. O&P**
1 Structural Steel Foreman	$23.25	$186.00	$45.85	$366.80	$22.25	$43.88
1 Structural Steel Worker	21.25	170.00	41.90	335.20		
1 Power Tool		2.40		2.65	.15	.17
16 L.H., Daily Totals		$358.40		$704.65	$22.40	$44.05
Crew E-18	**Hr.**	**Daily**	**Hr.**	**Daily**	**Bare Costs**	**Incl. O&P**
1 Structural Steel Foreman	$23.25	$186.00	$45.85	$366.80	$21.38	$40.91
3 Structural Steel Workers	21.25	510.00	41.90	1005.60		
1 Equipment Operator (med.)	19.90	159.20	33.00	264.00		
1 Crane, 20 Ton		532.55		585.80	13.31	14.65
40 L.H., Daily Totals		$1387.75		$2222.20	$34.69	$55.56
Crew E-19	**Hr.**	**Daily**	**Hr.**	**Daily**	**Bare Costs**	**Incl. O&P**
1 Structural Steel Worker	$21.25	$170.00	$41.90	$335.20	$21.20	$39.82
1 Structural Steel Foreman	23.25	186.00	45.85	366.80		
1 Equip. Oper. (light)	19.10	152.80	31.70	253.60		
1 Power Tool		2.40		2.65		
1 Crane, 20 ton		532.55		585.80	22.29	24.52
24 L.H., Daily Totals		$1043.75		$1544.05	$43.49	$64.34
Crew E-20	**Hr.**	**Daily**	**Hr.**	**Daily**	**Bare Costs**	**Incl. O&P**
1 Structural Steel Foreman	$23.25	$186.00	$45.85	$366.80	$20.89	$39.73
5 Structural Steel Workers	21.25	850.00	41.90	1676.00		
1 Equip. Oper. (crane)	20.65	165.20	34.25	274.00		
1 Oiler	17.00	136.00	28.20	225.60		
1 Power Tool		2.40		2.65		
1 Crane, 40 ton		704.80		775.30	11.05	12.16
64 L.H., Daily Totals		$2044.40		$3320.35	$31.94	$51.89
Crew E-22	**Hr.**	**Daily**	**Hr.**	**Daily**	**Bare Costs**	**Incl. O&P**
1 Skilled Worker Foreman	$21.75	$174.00	$37.30	$298.40	$20.42	$35.03
2 Skilled Worker	19.75	316.00	33.90	542.40		
24 L.H., Daily Totals		$490.00		$840.80	$20.42	$35.03

Crew No.	Bare Costs		Incl. Subs O & P		Cost Per Labor-Hour	
Crew E-24	**Hr.**	**Daily**	**Hr.**	**Daily**	**Bare Costs**	**Incl. O&P**
3 Structural Steel Worker	$21.25	$510.00	$41.90	$1005.60	$20.91	$39.68
1 Equipment Operator (medium)	19.90	159.20	33.00	264.00		
1-25 ton crane		559.20		615.10	17.48	19.22
32 L.H., Daily Totals		$1228.40		$1884.70	$38.39	$58.90
Crew F-3	**Hr.**	**Daily**	**Hr.**	**Daily**	**Bare Costs**	**Incl. O&P**
2 Carpenters	$19.70	$315.20	$33.75	$540.00	$17.93	$30.49
2 Carpenter Helpers	14.80	236.80	25.35	405.60		
1 Equip. Oper. (crane)	20.65	165.20	34.25	274.00		
1 Hyd. Crane, 12 Ton		453.45		498.80	11.34	12.47
40 L.H., Daily Totals		$1170.65		$1718.40	$29.27	$42.96
Crew F-4	**Hr.**	**Daily**	**Hr.**	**Daily**	**Bare Costs**	**Incl. O&P**
2 Carpenters	$19.70	$315.20	$33.75	$540.00	$17.93	$30.49
2 Carpenter Helpers	14.80	236.80	25.35	405.60		
1 Equip. Oper. (crane)	20.65	165.20	34.25	274.00		
1 Hyd. Crane, 55 Ton		808.20		889.00	20.21	22.23
40 L.H., Daily Totals		$1525.40		$2108.60	$38.14	$52.72
Crew F-5	**Hr.**	**Daily**	**Hr.**	**Daily**	**Bare Costs**	**Incl. O&P**
2 Carpenters	$19.70	$315.20	$33.75	$540.00	$17.25	$29.55
2 Carpenter Helpers	14.80	236.80	25.35	405.60		
32 L.H., Daily Totals		$552.00		$945.60	$17.25	$29.55
Crew F-6	**Hr.**	**Daily**	**Hr.**	**Daily**	**Bare Costs**	**Incl. O&P**
2 Carpenters	$19.70	$315.20	$33.75	$540.00	$17.79	$30.25
2 Building Laborers	14.45	231.20	24.75	396.00		
1 Equip. Oper. (crane)	20.65	165.20	34.25	274.00		
1 Hyd. Crane, 12 Ton		453.45		498.80	11.34	12.47
40 L.H., Daily Totals		$1165.05		$1708.80	$29.13	$42.72
Crew F-7	**Hr.**	**Daily**	**Hr.**	**Daily**	**Bare Costs**	**Incl. O&P**
2 Carpenters	$19.70	$315.20	$33.75	$540.00	$17.08	$29.25
2 Building Laborers	14.45	231.20	24.75	396.00		
32 L.H., Daily Totals		$546.40		$936.00	$17.08	$29.25
Crew G-1	**Hr.**	**Daily**	**Hr.**	**Daily**	**Bare Costs**	**Incl. O&P**
1 Roofer Foreman	$19.05	$152.40	$35.45	$283.60	$16.11	$29.99
4 Roofers, Composition	17.05	545.60	31.75	1016.00		
2 Roofer Helpers	12.75	204.00	23.75	380.00		
1 Application Equipment		171.60		188.75		
1 Tar Kettle/Pot		49.00		53.90		
1 Crew Truck		197.60		217.35	7.47	8.21
56 L.H., Daily Totals		$1320.20		$2139.60	$23.58	$38.20
Crew G-2	**Hr.**	**Daily**	**Hr.**	**Daily**	**Bare Costs**	**Incl. O&P**
1 Plasterer	$18.65	$149.20	$31.20	$249.60	$16.25	$27.37
1 Plasterer Helper	15.65	125.20	26.15	209.20		
1 Building Laborer	14.45	115.60	24.75	198.00		
1 Grouting Equipment		250.00		275.00	10.42	11.46
24 L.H., Daily Totals		$640.00		$931.80	$26.67	$38.83
Crew G-3	**Hr.**	**Daily**	**Hr.**	**Daily**	**Bare Costs**	**Incl. O&P**
2 Sheet Metal Workers	$21.50	$344.00	$36.30	$580.80	$17.98	$30.52
2 Building Laborers	14.45	231.20	24.75	396.00		
32 L.H., Daily Totals		$575.20		$976.80	$17.98	$30.52

Crews

Crew No.	Bare Costs		Incl. Subs O & P		Cost Per Labor-Hour	
Crew G-4	Hr.	Daily	Hr.	Daily	Bare Costs	Incl. O&P
1 Labor Foreman (outside)	$16.45	$131.60	$28.20	$225.60	$15.12	$25.90
2 Building Laborers	14.45	231.20	24.75	396.00		
1 Light Truck, 1.5 Ton		172.95		190.25		
1 Air Compr., 160 C.F.M.		98.40		108.25	11.31	12.44
24 L.H., Daily Totals		$634.15		$920.10	$26.43	$38.34
Crew G-5	Hr.	Daily	Hr.	Daily	Bare Costs	Incl. O&P
1 Roofer Foreman	$19.05	$152.40	$35.45	$283.60	$15.73	$29.29
2 Roofers, Composition	17.05	272.80	31.75	508.00		
2 Roofer Helpers	12.75	204.00	23.75	380.00		
1 Application Equipment		171.60		188.75	4.29	4.72
40 L.H., Daily Totals		$800.80		$1360.35	$20.02	$34.01
Crew G-6A	Hr.	Daily	Hr.	Daily	Bare Costs	Incl. O&P
2 Roofers Composition	$17.05	$272.80	$31.75	$508.00	$17.05	$31.75
1 Small Compressor		20.95		23.05		
2 Pneumatic Nailers		37.90		41.70	3.68	4.05
16 L.H., Daily Totals		$331.65		$572.75	$20.73	$35.80
Crew G-7	Hr.	Daily	Hr.	Daily	Bare Costs	Incl. O&P
1 Carpenter	$19.70	$157.60	$33.75	$270.00	$19.70	$33.75
1 Small Compressor		20.95		23.05		
1 Pneumatic Nailer		18.95		20.85	4.99	5.49
8 L.H., Daily Totals		$197.50		$313.90	$24.69	$39.24
Crew H-1	Hr.	Daily	Hr.	Daily	Bare Costs	Incl. O&P
2 Glaziers	$19.35	$309.60	$32.10	$513.60	$20.30	$37.00
2 Struc. Steel Workers	21.25	340.00	41.90	670.40		
32 L.H., Daily Totals		$649.60		$1184.00	$20.30	$37.00
Crew H-2	Hr.	Daily	Hr.	Daily	Bare Costs	Incl. O&P
2 Glaziers	$19.35	$309.60	$32.10	$513.60	$17.72	$29.65
1 Building Laborer	14.45	115.60	24.75	198.00		
24 L.H., Daily Totals		$425.20		$711.60	$17.72	$29.65
Crew H-3	Hr.	Daily	Hr.	Daily	Bare Costs	Incl. O&P
1 Glazier	$19.35	$154.80	$32.10	$256.80	$17.08	$28.73
1 Helper	14.80	118.40	25.35	202.80		
16 L.H., Daily Totals		$273.20		$459.60	$17.08	$28.73
Crew J-1	Hr.	Daily	Hr.	Daily	Bare Costs	Incl. O&P
3 Plasterers	$18.65	$447.60	$31.20	$748.80	$17.45	$29.18
2 Plasterer Helpers	15.65	250.40	26.15	418.40		
1 Mixing Machine, 6 C.F.		54.50		59.95	1.36	1.50
40 L.H., Daily Totals		$752.50		$1227.15	$18.81	$30.68
Crew J-2	Hr.	Daily	Hr.	Daily	Bare Costs	Incl. O&P
3 Plasterers	$18.65	$447.60	$31.20	$748.80	$17.74	$29.56
2 Plasterer Helpers	15.65	250.40	26.15	418.40		
1 Lather	19.20	153.60	31.45	251.60		
1 Mixing Machine, 6 C.F.		54.50		59.95	1.14	1.25
48 L.H., Daily Totals		$906.10		$1478.75	$18.88	$30.81
Crew J-3	Hr.	Daily	Hr.	Daily	Bare Costs	Incl. O&P
1 Terrazzo Worker	$19.10	$152.80	$30.95	$247.60	$17.33	$28.08
1 Terrazzo Helper	15.55	124.40	25.20	201.60		
1 Terrazzo Grinder, Electric		50.00		55.00		
1 Terrazzo Mixer		74.20		81.60	7.76	8.54
16 L.H., Daily Totals		$401.40		$585.80	$25.09	$36.62

Crew No.	Bare Costs		Incl. Subs O & P		Cost Per Labor-Hour	
Crew J-4	Hr.	Daily	Hr.	Daily	Bare Costs	Incl. O&P
1 Tile Layer	$19.20	$153.60	$31.10	$248.80	$17.33	$28.08
1 Tile Layer Helper	15.45	123.60	25.05	200.40		
16 L.H., Daily Totals		$277.20		$449.20	$17.33	$28.08
Crew K-1	Hr.	Daily	Hr.	Daily	Bare Costs	Incl. O&P
1 Carpenter	$19.70	$157.60	$33.75	$270.00	$17.78	$30.13
1 Truck Driver (light)	15.85	126.80	26.50	212.00		
1 Truck w/Power Equip.		180.45		198.50	11.28	12.41
16 L.H., Daily Totals		$464.85		$680.50	$29.06	$42.54
Crew K-2	Hr.	Daily	Hr.	Daily	Bare Costs	Incl. O&P
1 Struc. Steel Foreman	$23.25	$186.00	$45.85	$366.80	$20.12	$38.08
1 Struc. Steel Worker	21.25	170.00	41.90	335.20		
1 Truck Driver (light)	15.85	126.80	26.50	212.00		
1 Truck w/Power Equip.		180.45		198.50	7.52	8.27
24 L.H., Daily Totals		$663.25		$1112.50	$27.64	$46.35
Crew L-1	Hr.	Daily	Hr.	Daily	Bare Costs	Incl. O&P
.25 Electrician	$22.10	$44.20	$36.15	$72.30	$21.98	$36.27
1 Plumber	21.95	175.60	36.30	290.40		
10 L.H., Daily Totals		$219.80		$362.70	$21.98	$36.27
Crew L-2	Hr.	Daily	Hr.	Daily	Bare Costs	Incl. O&P
1 Carpenter	$19.70	$157.60	$33.75	$270.00	$17.25	$29.55
1 Carpenter Helper	14.80	118.40	25.35	202.80		
16 L.H., Daily Totals		$276.00		$472.80	$17.25	$29.55
Crew L-3	Hr.	Daily	Hr.	Daily	Bare Costs	Incl. O&P
1 Carpenter	$19.70	$157.60	$33.75	$270.00	$20.18	$34.23
.25 Electrician	22.10	44.20	36.15	72.30		
10 L.H., Daily Totals		$201.80		$342.30	$20.18	$34.23
Crew L-3A	Hr.	Daily	Hr.	Daily	Bare Costs	Incl. O&P
1 Carpenter Foreman (outside)	$21.70	$173.60	$37.20	$297.60	$21.63	$36.90
.5 Sheet Metal Worker	21.50	86.00	36.30	145.20		
12 L.H., Daily Totals		$259.60		$442.80	$21.63	$36.90
Crew L-4	Hr.	Daily	Hr.	Daily	Bare Costs	Incl. O&P
1 Skilled Workers	$19.75	$158.00	$33.90	$271.20	$17.27	$29.63
1 Helper	14.80	118.40	25.35	202.80		
16 L.H., Daily Totals		$276.40		$474.00	$17.27	$29.63
Crew L-5	Hr.	Daily	Hr.	Daily	Bare Costs	Incl. O&P
1 Struc. Steel Foreman	$23.25	$186.00	$45.85	$366.80	$21.45	$41.37
5 Struc. Steel Workers	21.25	850.00	41.90	1676.00		
1 Equip. Oper. (crane)	20.65	165.20	34.25	274.00		
1 Hyd. Crane, 25 Ton		559.20		615.10	9.99	10.98
56 L.H., Daily Totals		$1760.40		$2931.90	$31.44	$52.35
Crew L-5A	Hr.	Daily	Hr.	Daily	Bare Costs	Incl. O&P
1 Structural Steel Foreman	$23.25	$186.00	$45.85	$366.80	$21.60	$40.98
2 Structural Steel Workers	21.25	340.00	41.90	670.40		
1 Equip. Oper. (crane)	20.65	165.20	34.25	274.00		
1 Crane, SP, 25 Ton		563.90		620.30	17.62	19.38
32 L.H., Daily Totals		$1255.10		$1931.50	$39.22	$60.36

Crews

Crew No.	Bare Costs		Incl. Subs O & P		Cost Per Labor-Hour	
Crew L-6	Hr.	Daily	Hr.	Daily	Bare Costs	Incl. O&P
1 Plumber	$21.95	$175.60	$36.30	$290.40	$22.00	$36.25
.5 Electrician	22.10	88.40	36.15	144.60		
12 L.H., Daily Totals		$264.00		$435.00	$22.00	$36.25
Crew L-7	Hr.	Daily	Hr.	Daily	Bare Costs	Incl. O&P
1 Carpenters	$19.70	$157.60	$33.75	$270.00	$16.87	$28.77
2 Carpenter Helpers	14.80	236.80	25.35	405.60		
.25 Electrician	22.10	44.20	36.15	72.30		
26 L.H., Daily Totals		$438.60		$747.90	$16.87	$28.77
Crew L-8	Hr.	Daily	Hr.	Daily	Bare Costs	Incl. O&P
1 Carpenters	$19.70	$157.60	$33.75	$270.00	$18.19	$30.90
1 Carpenter Helper	14.80	118.40	25.35	202.80		
.5 Plumber	21.95	87.80	36.30	145.20		
20 L.H., Daily Totals		$363.80		$618.00	$18.19	$30.90
Crew L-9	Hr.	Daily	Hr.	Daily	Bare Costs	Incl. O&P
1 Skilled Worker Foreman	$21.75	$174.00	$37.30	$298.40	$18.26	$31.11
1 Skilled Worker	19.75	158.00	33.90	271.20		
2 Helpers	14.80	236.80	25.35	405.60		
.5 Electrician	22.10	88.40	36.15	144.60		
36 L.H., Daily Totals		$657.20		$1119.80	$18.26	$31.11
Crew L-10	Hr.	Daily	Hr.	Daily	Bare Costs	Incl. O&P
1 Structural Steel Foreman	$23.25	$186.00	$45.85	$366.80	$21.72	$40.67
1 Structural Steel Worker	21.25	170.00	41.90	335.20		
1 Equip. Oper. (crane)	20.65	165.20	34.25	274.00		
1 Hyd. Crane, 12 Ton		453.45		498.80	18.89	20.78
24 L.H., Daily Totals		$974.65		$1474.80	$40.61	$61.45
Crew M-1	Hr.	Daily	Hr.	Daily	Bare Costs	Incl. O&P
3 Elevator Constructors	$22.90	$549.60	$37.80	$907.20	$21.75	$35.90
1 Elevator Apprentice	18.30	146.40	30.20	241.60		
5 Hand Tools		95.00		104.50	2.97	3.27
32 L.H., Daily Totals		$791.00		$1253.30	$24.72	$39.17
Crew M-3	Hr.	Daily	Hr.	Daily	Bare Costs	Incl. O&P
1 Electrician Foreman (out)	$24.10	$192.80	$39.40	$315.20	$19.94	$33.04
1 Common Laborer	14.45	115.60	24.75	198.00		
.25 Equipment Operator, Medium	19.90	39.80	33.00	66.00		
1 Elevator Constructor	22.90	183.20	37.80	302.40		
1 Elevator Apprentice	18.30	146.40	30.20	241.60		
.25 Crane, SP, 4 x 4, 20 ton		118.83		130.70	3.49	3.84
34 L.H., Daily Totals		$796.63		$1253.90	$23.43	$36.88
Crew M-4	Hr.	Daily	Hr.	Daily	Bare Costs	Incl. O&P
1 Electrician Foreman (out)	$24.10	$192.80	$39.40	$315.20	$19.81	$32.84
1 Common Laborer	14.45	115.60	24.75	198.00		
.25 Equipment Operator, Crane	20.65	41.30	34.25	68.50		
.25 Equipment Operator, Oiler	17.00	34.00	28.20	56.40		
1 Elevator Constructor	22.90	183.20	37.80	302.40		
1 Elevator Apprentice	18.30	146.40	30.20	241.60		
.25 Crane, hyd, SP, 4WD, 40 ton		195.95		215.55	5.44	5.99
36 L.H., Daily Totals		$909.25		$1397.65	$25.25	$38.83
Crew Q-1	Hr.	Daily	Hr.	Daily	Bare Costs	Incl. O&P
1 Plumber	$21.95	$175.60	$36.30	$290.40	$19.75	$32.65
1 Plumber Apprentice	17.55	140.40	29.00	232.00		
16 L.H., Daily Totals		$316.00		$522.40	$19.75	$32.65

Crew No.	Bare Costs		Incl. Subs O & P		Cost Per Labor-Hour	
Crew Q-1C	Hr.	Daily	Hr.	Daily	Bare Costs	Incl. O&P
1 Plumber	$21.95	$175.60	$36.30	$290.40	$19.80	$32.77
1 Plumber Apprentice	17.55	140.40	29.00	232.00		
1 Equip. Oper. (medium)	19.90	159.20	33.00	264.00		
1 Trencher, Chain		540.90		595.00	22.54	24.79
24 L.H., Daily Totals		$1016.10		$1381.40	$42.34	$57.56
Crew Q-2	Hr.	Daily	Hr.	Daily	Bare Costs	Incl. O&P
1 Plumber	$21.95	$175.60	$36.30	$290.40	$19.02	$31.43
2 Plumber Apprentices	17.55	280.80	29.00	464.00		
24 L.H., Daily Totals		$456.40		$754.40	$19.02	$31.43
Crew Q-3	Hr.	Daily	Hr.	Daily	Bare Costs	Incl. O&P
2 Plumbers	$21.95	$351.20	$36.30	$580.80	$19.75	$32.65
2 Plumber Apprentices	17.55	280.80	29.00	464.00		
32 L.H., Daily Totals		$632.00		$1044.80	$19.75	$32.65
Crew Q-4	Hr.	Daily	Hr.	Daily	Bare Costs	Incl. O&P
2 Plumbers	$21.95	$351.20	$36.30	$580.80	$20.85	$34.47
1 Welder (plumber)	21.95	175.60	36.30	290.40		
1 Plumber Apprentice	17.55	140.40	29.00	232.00		
1 Electric Welding Mach.		48.80		53.70	1.53	1.68
32 L.H., Daily Totals		$716.00		$1156.90	$22.38	$36.15
Crew Q-5	Hr.	Daily	Hr.	Daily	Bare Costs	Incl. O&P
1 Steamfitter	$22.10	$176.80	$36.55	$292.40	$19.90	$32.90
1 Steamfitter Apprentice	17.70	141.60	29.25	234.00		
16 L.H., Daily Totals		$318.40		$526.40	$19.90	$32.90
Crew Q-6	Hr.	Daily	Hr.	Daily	Bare Costs	Incl. O&P
1 Steamfitters	$22.10	$176.80	$36.55	$292.40	$19.17	$31.68
2 Steamfitter Apprentices	17.70	283.20	29.25	468.00		
24 L.H., Daily Totals		$460.00		$760.40	$19.17	$31.68
Crew Q-7	Hr.	Daily	Hr.	Daily	Bare Costs	Incl. O&P
2 Steamfitters	$22.10	$353.60	$36.55	$584.80	$19.90	$32.90
2 Steamfitter Apprentices	17.70	283.20	29.25	468.00		
32 L.H., Daily Totals		$636.80		$1052.80	$19.90	$32.90
Crew Q-8	Hr.	Daily	Hr.	Daily	Bare Costs	Incl. O&P
2 Steamfitters	$22.10	$353.60	$36.55	$584.80	$21.00	$34.72
1 Welder (steamfitter)	22.10	176.80	36.55	292.40		
1 Steamfitter Apprentice	17.70	141.60	29.25	234.00		
1 Electric Welding Mach.		48.80		53.70	1.53	1.68
32 L.H., Daily Totals		$720.80		$1164.90	$22.53	$36.40
Crew Q-9	Hr.	Daily	Hr.	Daily	Bare Costs	Incl. O&P
1 Sheet Metal Worker	$21.50	$172.00	$36.30	$290.40	$19.35	$32.67
1 Sheet Metal Apprentice	17.20	137.60	29.05	232.40		
16 L.H., Daily Totals		$309.60		$522.80	$19.35	$32.67
Crew Q-10	Hr.	Daily	Hr.	Daily	Bare Costs	Incl. O&P
2 Sheet Metal Workers	$21.50	$344.00	$36.30	$580.80	$20.07	$33.88
1 Sheet Metal Apprentice	17.20	137.60	29.05	232.40		
24 L.H., Daily Totals		$481.60		$813.20	$20.07	$33.88
Crew Q-11	Hr.	Daily	Hr.	Daily	Bare Costs	Incl. O&P
2 Sheet Metal Workers	$21.50	$344.00	$36.30	$580.80	$19.35	$32.67
2 Sheet Metal Apprentices	17.20	275.20	29.05	464.80		
32 L.H., Daily Totals		$619.20		$1045.60	$19.35	$32.67

Crews

Crew No.	Bare Costs		Incl. Subs O & P		Cost Per Labor-Hour	
Crew Q-12	Hr.	Daily	Hr.	Daily	Bare Costs	Incl. O&P
1 Sprinkler Installer	$22.05	$176.40	$36.50	$292.00	$19.85	$32.85
1 Sprinkler Apprentice	17.65	141.20	29.20	233.60		
16 L.H., Daily Totals		$317.60		$525.60	$19.85	$32.85
Crew Q-13	Hr.	Daily	Hr.	Daily	Bare Costs	Incl. O&P
2 Sprinkler Installers	$22.05	$352.80	$36.50	$584.00	$19.85	$32.85
2 Sprinkler Apprentices	17.65	282.40	29.20	467.20		
32 L.H., Daily Totals		$635.20		$1051.20	$19.85	$32.85
Crew Q-14	Hr.	Daily	Hr.	Daily	Bare Costs	Incl. O&P
1 Asbestos Worker	$20.50	$164.00	$35.90	$287.20	$18.45	$32.30
1 Asbestos Apprentice	16.40	131.20	28.70	229.60		
16 L.H., Daily Totals		$295.20		$516.80	$18.45	$32.30
Crew Q-15	Hr.	Daily	Hr.	Daily	Bare Costs	Incl. O&P
1 Plumber	$21.95	$175.60	$36.30	$290.40	$19.75	$32.65
1 Plumber Apprentice	17.55	140.40	29.00	232.00		
1 Electric Welding Mach.		48.80		53.70	3.05	3.36
16 L.H., Daily Totals		$364.80		$576.10	$22.80	$36.01
Crew Q-16	Hr.	Daily	Hr.	Daily	Bare Costs	Incl. O&P
2 Plumbers	$21.95	$351.20	$36.30	$580.80	$20.48	$33.87
1 Plumber Apprentice	17.55	140.40	29.00	232.00		
1 Electric Welding Mach.		48.80		53.70	2.03	2.24
24 L.H., Daily Totals		$540.40		$866.50	$22.51	$36.11
Crew Q-17	Hr.	Daily	Hr.	Daily	Bare Costs	Incl. O&P
1 Steamfitter	$22.10	$176.80	$36.55	$292.40	$19.90	$32.90
1 Steamfitter Apprentice	17.70	141.60	29.25	234.00		
1 Electric Welding Mach.		48.80		53.70	3.05	3.36
16 L.H., Daily Totals		$367.20		$580.10	$22.95	$36.26
Crew Q-17A	Hr.	Daily	Hr.	Daily	Bare Costs	Incl. O&P
1 Steamfitter	$22.10	$176.80	$36.55	$292.40	$20.15	$33.35
1 Steamfitter Apprentice	17.70	141.60	29.25	234.00		
1 Equip. Oper. (crane)	20.65	165.20	34.25	274.00		
1 Truck Crane, 12 Ton		453.45		498.80		
1 Electric Welding Mach.		48.80		53.70	20.93	23.02
24 L.H., Daily Totals		$985.85		$1352.90	$41.08	$56.37
Crew Q-18	Hr.	Daily	Hr.	Daily	Bare Costs	Incl. O&P
2 Steamfitters	$22.10	$353.60	$36.55	$584.80	$20.63	$34.12
1 Steamfitter Apprentice	17.70	141.60	29.25	234.00		
1 Electric Welding Mach.		48.80		53.70	2.03	2.24
24 L.H., Daily Totals		$544.00		$872.50	$22.66	$36.36
Crew Q-19	Hr.	Daily	Hr.	Daily	Bare Costs	Incl. O&P
1 Steamfitter	$22.10	$176.80	$36.55	$292.40	$20.63	$33.98
1 Steamfitter Apprentice	17.70	141.60	29.25	234.00		
1 Electrician	22.10	176.80	36.15	289.20		
24 L.H., Daily Totals		$495.20		$815.60	$20.63	$33.98
Crew Q-20	Hr.	Daily	Hr.	Daily	Bare Costs	Incl. O&P
1 Sheet Metal Worker	$21.50	$172.00	$36.30	$290.40	$19.90	$33.37
1 Sheet Metal Apprentice	17.20	137.60	29.05	232.40		
.5 Electrician	22.10	88.40	36.15	144.60		
20 L.H., Daily Totals		$398.00		$667.40	$19.90	$33.37

Crew No.	Bare Costs		Incl. Subs O & P		Cost Per Labor-Hour	
Crew Q-21	Hr.	Daily	Hr.	Daily	Bare Costs	Incl. O&P
2 Steamfitters	$22.10	$353.60	$36.55	$584.80	$21.00	$34.63
1 Steamfitter Apprentice	17.70	141.60	29.25	234.00		
1 Electrician	22.10	176.80	36.15	289.20		
32 L.H., Daily Totals		$672.00		$1108.00	$21.00	$34.63
Crew Q-22	Hr.	Daily	Hr.	Daily	Bare Costs	Incl. O&P
1 Plumber	$21.95	$175.60	$36.30	$290.40	$19.75	$32.65
1 Plumber Apprentice	17.55	140.40	29.00	232.00		
1 Truck Crane, 12 Ton		453.45		498.80	28.34	31.17
16 L.H., Daily Totals		$769.45		$1021.20	$48.09	$63.82
Crew Q-22A	Hr.	Daily	Hr.	Daily	Bare Costs	Incl. O&P
1 Plumber	$21.95	$175.60	$36.30	$290.40	$18.65	$31.08
1 Plumber Apprentice	17.55	140.40	29.00	232.00		
1 Laborer	14.45	115.60	24.75	198.00		
1 Equip. Oper. (crane)	20.65	165.20	34.25	274.00		
1 Truck Crane, 12 Ton		453.45		498.80	14.17	15.59
32 L.H., Daily Totals		$1050.25		$1493.20	$32.82	$46.67
Crew Q-23	Hr.	Daily	Hr.	Daily	Bare Costs	Incl. O&P
1 Plumber Foreman	$23.95	$191.60	$39.60	$316.80	$21.93	$36.30
1 Plumber	21.95	175.60	36.30	290.40		
1 Equip. Oper. (medium)	19.90	159.20	33.00	264.00		
1 Power Tools		2.40		2.65		
1 Crane, 20 Ton		532.55		585.80	22.29	24.52
24 L.H., Daily Totals		$1061.35		$1459.65	$44.22	$60.82
Crew R-1	Hr.	Daily	Hr.	Daily	Bare Costs	Incl. O&P
1 Electrician Foreman	$22.60	$180.80	$36.95	$295.60	$19.75	$32.68
3 Electricians	22.10	530.40	36.15	867.60		
2 Helpers	14.80	236.80	25.35	405.60		
48 L.H., Daily Totals		$948.00		$1568.80	$19.75	$32.68
Crew R-1A	Hr.	Daily	Hr.	Daily	Bare Costs	Incl. O&P
1 Electrician	$22.10	$176.80	$36.15	$289.20	$18.45	$30.75
1 Helper	14.80	118.40	25.35	202.80		
16 L.H., Daily Totals		$295.20		$492.00	$18.45	$30.75
Crew R-2	Hr.	Daily	Hr.	Daily	Bare Costs	Incl. O&P
1 Electrician Foreman	$22.60	$180.80	$36.95	$295.60	$19.88	$32.91
3 Electricians	22.10	530.40	36.15	867.60		
2 Helpers	14.80	236.80	25.35	405.60		
1 Equip. Oper. (crane)	20.65	165.20	34.25	274.00		
1 S.P. Crane, 5 Ton		276.45		304.10	4.94	5.43
56 L.H., Daily Totals		$1389.65		$2146.90	$24.82	$38.34
Crew R-3	Hr.	Daily	Hr.	Daily	Bare Costs	Incl. O&P
1 Electrician Foreman	$22.60	$180.80	$36.95	$295.60	$22.01	$36.09
1 Electrician	22.10	176.80	36.15	289.20		
.5 Equip. Oper. (crane)	20.65	82.60	34.25	137.00		
.5 S.P. Crane, 5 Ton		138.23		152.05	6.91	7.60
20 L.H., Daily Totals		$578.43		$873.85	$28.92	$43.69
Crew R-4	Hr.	Daily	Hr.	Daily	Bare Costs	Incl. O&P
1 Struc. Steel Foreman	$23.25	$186.00	$45.85	$366.80	$21.82	$41.54
3 Struc. Steel Workers	21.25	510.00	41.90	1005.60		
1 Electrician	22.10	176.80	36.15	289.20		
1 Gas Welding Machine		84.00		92.40	2.10	2.31
40 L.H., Daily Totals		$956.80		$1754.00	$23.92	$43.85

Crews

Crew No.	Bare Costs		Incl. Subs O & P		Cost Per Labor-Hour	
Crew R-5	**Hr.**	**Daily**	**Hr.**	**Daily**	**Bare Costs**	**Incl. O&P**
1 Electrician Foreman	$22.60	$180.80	$36.95	$295.60	$19.49	$32.30
4 Electrician Linemen	22.10	707.20	36.15	1156.80		
2 Electrician Operators	22.10	353.60	36.15	578.40		
4 Electrician Groundmen	14.80	473.60	25.35	811.20		
1 Crew Truck		197.60		217.35		
1 Tool Van		214.40		235.85		
1 Pickup Truck, 3/4 Ton		130.20		143.20		
.2 Crane, 55 Ton		161.64		177.80		
.2 Crane, 12 Ton		90.69		99.75		
.2 Auger, Truck Mtd.		359.20		395.10		
1 Tractor w/Winch		298.80		328.70	16.51	18.16
88 L.H., Daily Totals		$3167.73		$4439.75	$36.00	$50.46
Crew R-6	**Hr.**	**Daily**	**Hr.**	**Daily**	**Bare Costs**	**Incl. O&P**
1 Electrician Foreman	$22.60	$180.80	$36.95	$295.60	$19.49	$32.30
4 Electrician Linemen	22.10	707.20	36.15	1156.80		
2 Electrician Operators	22.10	353.60	36.15	578.40		
4 Electrician Groundmen	14.80	473.60	25.35	811.20		
1 Crew Truck		197.60		217.35		
1 Tool Van		214.40		235.85		
1 Pickup Truck, 3/4 Ton		130.20		143.20		
.2 Crane, 55 Ton		161.64		177.80		
.2 Crane, 12 Ton		90.69		99.75		
.2 Auger, Truck Mtd.		359.20		395.10		
1 Tractor w/Winch		298.80		328.70		
3 Cable Trailers		422.40		464.65		
.5 Tensioning Rig		141.20		155.30		
.5 Cable Pulling Rig		819.50		901.45	32.22	35.45
88 L.H., Daily Totals		$4550.83		$5961.15	$51.71	$67.75
Crew R-7	**Hr.**	**Daily**	**Hr.**	**Daily**	**Bare Costs**	**Incl. O&P**
1 Electrician Foreman	$22.60	$180.80	$36.95	$295.60	$16.10	$27.28
5 Electrician Groundmen	14.80	592.00	25.35	1014.00		
1 Crew Truck		197.60		217.35	4.12	4.53
48 L.H., Daily Totals		$970.40		$1526.95	$20.22	$31.81
Crew R-8	**Hr.**	**Daily**	**Hr.**	**Daily**	**Bare Costs**	**Incl. O&P**
1 Electrician Foreman	$22.60	$180.80	$36.95	$295.60	$19.75	$32.68
3 Electrician Linemen	22.10	530.40	36.15	867.60		
2 Electrician Groundmen	14.80	236.80	25.35	405.60		
1 Pickup Truck, 3/4 Ton		130.20		143.20		
1 Crew Truck		197.60		217.35	6.83	7.51
48 L.H., Daily Totals		$1275.80		$1929.35	$26.58	$40.19
Crew R-9	**Hr.**	**Daily**	**Hr.**	**Daily**	**Bare Costs**	**Incl. O&P**
1 Electrician Foreman	$22.60	$180.80	$36.95	$295.60	$18.51	$30.85
1 Electrician Lineman	22.10	176.80	36.15	289.20		
2 Electrician Operators	22.10	353.60	36.15	578.40		
4 Electrician Groundmen	14.80	473.60	25.35	811.20		
1 Pickup Truck, 3/4 Ton		130.20		143.20		
1 Crew Truck		197.60		217.35	5.12	5.63
64 L.H., Daily Totals		$1512.60		$2334.95	$23.63	$36.48
Crew R-10	**Hr.**	**Daily**	**Hr.**	**Daily**	**Bare Costs**	**Incl. O&P**
1 Electrician Foreman	$22.60	$180.80	$36.95	$295.60	$20.97	$34.48
4 Electrician Linemen	22.10	707.20	36.15	1156.80		
1 Electrician Groundman	14.80	118.40	25.35	202.80		
1 Crew Truck		197.60		217.35		
3 Tram Cars		541.20		595.70	15.39	16.93
48 L.H., Daily Totals		$1745.20		$2467.85	$36.36	$51.41

Crew No.	Bare Costs		Incl. Subs O & P		Cost Per Labor-Hour	
Crew R-11	**Hr.**	**Daily**	**Hr.**	**Daily**	**Bare Costs**	**Incl. O&P**
1 Electrician Foreman	$22.60	$180.80	$36.95	$295.60	$20.04	$33.09
4 Electricians	22.10	707.20	36.15	1156.80		
1 Helper	14.80	118.40	25.35	202.80		
1 Common Laborer	14.45	115.60	24.75	198.00		
1 Crew Truck		197.60		217.35		
1 Crane, 12 Ton		453.45		498.80	11.63	12.79
56 L.H., Daily Totals		$1773.05		$2569.35	$31.67	$45.88
Crew R-12	**Hr.**	**Daily**	**Hr.**	**Daily**	**Bare Costs**	**Incl. O&P**
1 Carpenter Foreman	$20.20	$161.60	$34.60	$276.80	$18.00	$31.23
4 Carpenters	19.70	630.40	33.75	1080.00		
4 Common Laborers	14.45	462.40	24.75	792.00		
1 Equip. Oper. (med.)	19.90	159.20	33.00	264.00		
1 Steel Worker	21.25	170.00	41.90	335.20		
1 Dozer, 200 H.P.		806.80		887.50		
1 Pickup Truck, 3/4 Ton		130.20		143.20	10.65	11.71
88 L.H., Daily Totals		$2520.60		$3778.70	$28.65	$42.94
Crew R-15	**Hr.**	**Daily**	**Hr.**	**Daily**	**Bare Costs**	**Incl. O&P**
1 Electrician Foreman	$22.60	$180.80	$36.95	$295.60	$21.68	$35.54
4 Electricians	22.10	707.20	36.15	1156.80		
1 Equipment Operator	19.10	152.80	31.70	253.60		
1 Aerial Lift Truck		251.35		276.50	5.24	5.76
48 L.H., Daily Totals		$1292.15		$1982.50	$26.92	$41.30
Crew R-18	**Hr.**	**Daily**	**Hr.**	**Daily**	**Bare Costs**	**Incl. O&P**
.25 Electrician Foreman	$22.60	$45.20	$36.95	$73.90	$17.65	$29.57
1 Electrician	22.10	176.80	36.15	289.20		
2 Helpers	14.80	236.80	25.35	405.60		
26 L.H., Daily Totals		$458.80		$768.70	$17.65	$29.57
Crew R-19	**Hr.**	**Daily**	**Hr.**	**Daily**	**Bare Costs**	**Incl. O&P**
.5 Electrician Foreman	$22.60	$90.40	$36.95	$147.80	$22.20	$36.31
2 Electricians	22.10	353.60	36.15	578.40		
20 L.H., Daily Totals		$444.00		$726.20	$22.20	$36.31
Crew R-21	**Hr.**	**Daily**	**Hr.**	**Daily**	**Bare Costs**	**Incl. O&P**
1 Electrician Foreman	$22.60	$180.80	$36.95	$295.60	$22.17	$36.27
3 Electricians	22.10	530.40	36.15	867.60		
.1 Equip. Oper. (med.)	19.90	15.92	33.00	26.40		
.1 Hyd. Crane 25 Ton		56.39		62.05	1.72	1.89
32. L.H., Daily Totals		$783.51		$1251.65	$23.89	$38.16
Crew R-22	**Hr.**	**Daily**	**Hr.**	**Daily**	**Bare Costs**	**Incl. O&P**
.66 Electrician Foreman	$22.60	$119.33	$36.95	$195.10	$19.04	$31.63
2 Helpers	14.80	236.80	25.35	405.60		
2 Electricians	22.10	353.60	36.15	578.40		
37.28 L.H., Daily Totals		$709.73		$1179.10	$19.04	$31.63
Crew R-30	**Hr.**	**Daily**	**Hr.**	**Daily**	**Bare Costs**	**Incl. O&P**
.25 Electricians	$24.10	$48.20	$39.40	$78.80	$17.55	$29.38
1 Electricians	22.10	176.80	36.15	289.20		
2 Laborers, (Semi-Skilled)	14.45	231.20	24.75	396.00		
26 L.H., Daily Totals		$456.20		$764.00	$17.55	$29.38

Historical Cost Indexes

The following tables are the revised Historical Cost Indexes based on a 30-city national average with a base of 100 on January 1, 1993.

The indexes may be used to:
1. Estimate and compare construction costs for different years in the same city.
2. Estimate and compare construction costs in different cities for the same year.
3. Estimate and compare construction costs in different cities for different years.
4. Compare construction trends in any city with the national average.

EXAMPLES

1. Estimate and compare construction costs for different years in the same city.

    A. To estimate the construction cost of a building in Lexington, KY in 1970, knowing that it cost $900,000 in 2000.

    Index Lexington, KY in 1970 = 26.9
    Index Lexington, KY in 2000 = 100.9

    $$\frac{\text{Index 1970}}{\text{Index 2000}} \times \text{Cost 2000} = \text{Cost 1970}$$

    $$\frac{26.9}{100.9} \times \$900,000 = \$240,000$$

    Construction Cost in Lexington, KY in 1970 = $240,000

    B. To estimate the current construction cost of a building in Boston, MA that was built in 1978 for $900,000.

    Index Boston, MA in 1978 = 54.0
    Index Boston, MA in 2000 = 138.1

    $$\frac{\text{Index 2000}}{\text{Index 1978}} \times \text{Cost 1978} = \text{Cost 2000}$$

    $$\frac{138.1}{54.0} \times \$900,000 = \$2,301,500$$

    Construction Cost in Boston in 2000 = $2,301,500

2. Estimate and compare construction costs in different cities for the same year.

    To compare the construction cost of a building in Topeka, KS in 2000 with the known cost of $800,000 in Baltimore, MD in 2000.

    Index Topeka, KS in 2000 = 101.0
    Index Baltimore, MD in 2000 = 108.1

    $$\frac{\text{Index Topeka}}{\text{Index Baltimore}} \times \text{Cost Baltimore} = \text{Cost Topeka}$$

    $$\frac{101.0}{108.1} \times \$800,000 = \$747,500$$

    Construction Cost in Topeka in 2000 = $747,500

3. Estimate and compare construction costs in different cities for different years.

    To compare the construction cost of a building in Detroit, MI in 2000 with the known construction cost of $4,000,000 for the same building in San Francisco, CA in 1978.

    Index Detroit, MI in 2000 = 124.7
    Index San Francisco, CA in 1978 = 62.6

    $$\frac{\text{Index Detroit 1999}}{\text{Index San Francisco 1978}} \times \text{Cost San Francisco 1978} = \text{Cost Detroit 2000}$$

    $$\frac{124.7}{62.6} \times \$4,000,000 = \$7,968,000$$

    Construction Cost in Detroit in 2000 = $7,968,000

4. Compare construction trends in any city with the national average.

    To compare the construction cost in Las Vegas, NV from 1975 to 2000 with the increase in the National Average during the same time period.

    Index Las Vegas, NV for 1975 = 42.8 For 2000 = 124.3
    Index 30 City Average for 1975 = 43.7 For 2000 = 119.6

    A. National Average increase From 1975 to 2000 = $\frac{\text{Index — 30 City 2000}}{\text{Index — 30 City 1975}}$

    $$= \frac{119.6}{43.7}$$

    National Average increase From 1975 to 2000 = 2.74 or 274%

    B. increase for Las Vegas, NV From 1975 to 2000 = $\frac{\text{Index Las Vegas, NV 2000}}{\text{Index Las Vegas, NV 1975}}$

    $$= \frac{124.3}{42.8}$$

    Las Vegas increase 1975 — 2000 = 2.90 or 290%

    Conclusion: Construction costs in Las Vegas are higher than National average costs and increased at a greater rate from 1975 to 2000 than the National Average.

Historical Cost Indexes

Year	National 30 City Average	Alabama					Alaska	Arizona		Arkansas		California				
		Birmingham	Huntsville	Mobile	Montgomery	Tuscaloosa	Anchorage	Phoenix	Tuscon	Fort Smith	Little Rock	Anaheim	Bakersfield	Fresno	Los Angeles	Oxnard
Jan 2000	119.6E	103.8E	99.9E	99.7E	94.3E	94.2E	148.1E	106.0E	105.5E	95.2E	96.4E	129.1E	125.0E	128.5E	130.1E	129.8E
1999	116.6	101.2	97.4	97.7	92.5	92.8	145.9	105.5	103.3	93.1	94.4	127.9	123.6	126.9	128.7	128.2
1998	113.6	96.2	94.0	94.8	90.3	89.8	143.8	102.1	101.0	90.6	91.4	125.2	120.3	123.9	125.8	124.7
1997	111.5	94.6	92.4	93.3	88.8	88.3	142.0	101.8	100.7	89.3	90.1	124.0	119.1	122.3	124.6	123.5
1996	108.9	90.9	91.4	92.3	87.5	86.5	140.4	98.3	97.4	86.5	86.7	121.7	117.2	120.2	122.4	121.5
1995	105.6	87.8	88.0	88.8	84.5	83.5	138.0	96.1	95.5	85.0	85.6	120.1	115.7	117.5	120.9	120.0
1994	103.0	85.9	86.1	86.8	82.6	81.7	134.0	93.7	93.1	82.4	83.9	118.1	113.6	114.5	119.0	117.5
1993	100.0	82.8	82.7	86.1	82.1	78.7	132.0	90.9	90.9	80.9	82.4	115.0	111.3	112.9	115.6	115.4
1992	97.9	81.7	81.5	84.9	80.9	77.6	128.6	88.8	89.4	79.7	81.2	113.5	108.1	110.3	113.7	113.7
1991	95.7	80.5	78.9	83.9	79.8	76.6	127.4	88.1	88.5	78.3	80.0	111.0	105.8	108.2	110.9	111.4
1990	93.2	79.4	77.6	82.7	78.4	75.2	125.8	86.4	87.0	77.1	77.9	107.7	102.7	103.3	107.5	107.4
1989	91.0	77.5	76.1	81.1	76.8	73.7	123.4	85.3	85.6	75.6	76.4	105.6	101.0	101.7	105.3	105.4
1988	88.5	75.7	74.7	79.7	75.4	72.3	121.4	84.4	84.3	74.2	75.0	102.7	97.6	99.2	102.6	102.3
1987	85.7	74.0	73.8	77.9	74.5	71.7	119.3	79.7	80.6	72.2	72.8	100.9	96.1	98.0	100.0	101.1
1986	83.7	72.8	71.7	76.7	72.3	69.4	117.3	78.9	78.8	70.7	72.3	99.2	94.2	93.8	97.8	100.4
1985	81.8	71.1	70.5	72.7	70.9	67.9	116.0	78.1	77.7	69.6	71.0	95.4	92.2	92.6	94.6	96.6
1984	80.6	69.7	69.0	75.2	69.4	66.3	113.8	79.4	79.2	69.1	70.3	93.4	89.8	90.6	92.1	94.0
1983	78.2	67.4	68.8	72.9	69.4	65.5	104.5	77.5	77.9	66.8	68.7	90.9	88.8	88.3	89.7	91.8
1982	72.1	63.5	63.1	67.3	64.4	61.8	96.1	72.3	71.7	62.2	64.6	83.1	83.0	82.5	80.9	83.4
1981	66.1	60.0	58.1	62.2	61.3	58.1	91.5	69.0	67.5	57.7	60.2	75.7	76.5	75.2	74.8	77.1
1980	60.7	55.2	54.1	56.8	56.8	54.0	91.4	63.7	62.4	53.1	55.8	68.7	69.4	68.7	67.4	69.9
1979	54.9	49.9	48.9	52.5	50.8	48.7	82.5	55.9	56.5	47.8	49.5	62.9	61.9	62.5	61.7	62.9
1978	51.3	46.7	45.5	48.8	46.9	45.0	75.3	51.9	52.3	44.8	45.4	57.4	57.1	57.7	57.0	58.6
1977	47.9	43.1	43.2	45.1	42.4	40.5	70.6	48.7	48.5	41.6	42.2	53.2	52.8	53.2	53.7	53.4
1975	43.7	40.0	40.9	41.8	40.1	37.8	57.3	44.5	45.2	39.0	38.7	47.6	46.6	47.7	48.3	47.0
1970	27.8	24.1	25.1	25.8	25.4	24.2	43.0	27.2	28.5	24.3	22.3	31.0	30.8	31.1	29.0	31.0
1965	21.5	19.6	19.3	19.6	19.5	18.7	34.9	21.8	22.0	18.7	18.5	23.9	23.7	24.0	22.7	23.9
1960	19.5	17.9	17.5	17.8	17.7	16.9	31.7	19.9	20.0	17.0	16.8	21.7	21.5	21.8	20.6	21.7
1955	16.3	14.8	14.7	14.9	14.9	14.2	26.6	16.7	16.7	14.3	14.5	18.2	18.1	18.3	17.3	18.2
1950	13.5	12.2	12.1	12.3	12.3	11.7	21.9	13.8	13.8	11.8	11.6	15.1	15.0	15.1	14.3	15.1
1945	8.6	7.8	7.7	7.8	7.8	7.5	14.0	8.8	8.8	7.5	7.4	9.6	9.5	9.6	9.1	9.6
1940	6.6	6.0	6.0	6.1	6.0	5.8	10.8	6.8	6.8	5.8	5.7	7.4	7.4	7.4	7.0	7.4

Year	National 30 City Average	California							Colorado			Connecticut				
		Riverside	Sacramento	San Diego	San Francisco	Santa Barbara	Stockton	Vallejo	Colorado Springs	Denver	Pueblo	Bridge-Port	Bristol	Hartford	New Britain	New Haven
Jan 2000	119.6E	127.8E	131.3E	126.3E	146.7E	128.5E	130.2E	137.1E	108.6E	111.9E	108.6E	123.3E	123.1E	123.5E	122.7E	123.0E
1999	116.6	126.5	129.4	124.9	145.1	127.2	127.9	135.3	107.0	109.1	107.1	121.1	121.3	121.2	121.1	121.3
1998	113.6	123.7	125.9	121.3	141.9	123.7	124.7	132.5	103.3	106.5	103.7	119.1	119.4	120.0	119.7	120.0
1997	111.5	122.6	124.7	120.3	139.2	122.4	123.3	130.6	101.1	104.4	102.0	119.2	119.5	119.9	119.7	120.0
1996	108.9	120.6	122.4	118.4	136.8	120.5	121.4	128.1	98.5	101.4	99.7	117.4	117.6	117.9	117.8	118.1
1995	105.6	119.2	119.5	115.4	133.8	119.0	119.0	122.5	96.1	98.9	96.8	116.0	116.5	116.9	116.3	116.5
1994	103.0	116.6	114.8	113.4	131.4	116.4	115.5	119.7	94.2	95.9	94.6	114.3	115.0	115.3	114.7	114.8
1993	100.0	114.3	112.5	111.3	129.6	114.3	115.2	117.5	92.1	93.8	92.6	108.7	107.8	108.4	106.3	106.1
1992	97.9	111.8	110.8	109.5	127.9	112.2	112.7	115.2	90.6	91.5	90.9	106.7	106.0	106.6	104.5	104.4
1991	95.7	109.4	108.5	107.7	125.8	110.0	111.0	113.4	88.9	90.4	89.4	97.6	97.3	98.0	97.3	97.9
1990	93.2	107.0	104.9	105.6	121.8	106.4	105.4	111.4	86.9	88.8	88.8	96.3	95.9	96.6	95.9	96.5
1989	91.0	105.2	103.1	103.6	119.2	104.6	103.7	109.8	85.6	88.7	87.4	94.5	94.2	94.9	94.2	94.6
1988	88.5	102.6	100.5	101.0	115.2	101.4	101.2	107.2	83.8	87.0	85.6	92.8	92.6	93.3	92.6	93.0
1987	85.7	100.3	98.7	99.0	111.0	99.3	99.6	103.2	81.6	84.8	83.8	92.4	93.0	93.3	93.0	92.6
1986	83.7	98.6	94.4	96.9	108.0	97.0	96.1	99.1	81.9	83.1	82.7	88.5	88.7	89.4	88.6	88.8
1985	81.8	94.8	92.2	94.2	106.2	93.2	93.7	97.0	79.6	81.0	80.4	86.1	86.2	87.2	86.1	86.3
1984	80.6	91.8	93.8	92.0	102.8	90.6	91.9	95.6	79.8	84.1	80.5	83.5	83.4	84.4	83.0	83.2
1983	78.2	89.4	92.0	90.3	101.0	88.8	91.0	93.0	77.3	80.5	77.4	79.6	79.9	81.1	79.5	79.3
1982	72.1	82.1	84.9	83.8	91.9	81.8	84.9	85.7	71.2	73.4	75.0	71.9	71.2	73.2	71.4	72.0
1981	66.1	75.5	77.7	74.5	82.8	75.9	77.4	78.5	65.5	66.3	65.4	67.0	66.3	67.3	66.3	67.0
1980	60.7	68.7	71.3	68.1	75.2	71.1	71.2	71.9	60.7	60.9	59.5	61.5	60.7	61.9	60.7	61.4
1979	54.9	62.2	64.0	61.6	67.5	65.0	64.2	65.2	54.9	56.0	54.0	56.0	55.3	56.4	55.4	55.8
1978	51.3	57.7	60.0	58.3	62.6	59.2	59.5	59.9	51.1	52.0	50.6	52.5	50.9	51.9	51.8	52.0
1977	47.9	53.3	55.7	53.4	58.1	53.0	54.8	54.2	47.2	48.3	47.0	49.3	48.6	48.6	48.3	49.0
1975	43.7	47.3	49.1	47.7	49.8	46.1	47.6	46.5	43.1	42.7	42.5	45.2	45.5	46.0	45.2	46.0
1970	27.8	31.0	32.1	30.7	31.6	31.3	31.8	31.8	27.6	26.1	27.3	29.2	28.2	29.6	28.2	29.3
1965	21.5	23.9	24.8	24.0	23.7	24.1	24.5	24.5	21.3	20.9	21.0	22.4	21.7	22.6	21.7	23.2
1960	19.5	21.7	22.5	21.7	21.5	21.9	22.3	22.2	19.3	19.0	19.1	20.0	19.8	20.0	19.8	20.0
1955	16.3	18.2	18.9	18.2	18.0	18.4	18.7	18.6	16.2	15.9	16.0	16.8	16.6	16.8	16.6	16.8
1950	13.5	15.0	15.6	15.0	14.9	15.2	15.4	15.4	13.4	13.2	13.2	13.9	13.7	13.8	13.7	13.9
1945	8.6	9.6	10.0	9.6	9.5	9.7	9.9	9.8	8.5	8.4	8.4	8.8	8.7	8.8	8.7	8.9
1940	6.6	7.4	7.7	7.4	7.3	7.5	7.6	7.6	6.6	6.5	6.5	6.8	6.8	6.8	6.8	6.8

Historical Cost Indexes

Year	National 30 City Average	Connecticut Norwalk	Connecticut Stamford	Connecticut Waterbury	Delaware Wilmington	D.C. Washington	Florida Fort Lauderdale	Florida Jacksonville	Florida Miami	Florida Orlando	Florida Tallahassee	Florida Tampa	Georgia Albany	Georgia Atlanta	Georgia Columbus	Georgia Macon
Jan 2000	119.6E	122.0E	126.8E	124.1E	118.1E	114.1E	102.9E	99.2E	102.2E	101.5E	94.1E	99.7E	94.9E	105.9E	94.4E	97.4E
1999	116.6	120.0	122.0	121.2	116.6	111.5	101.4	98.0	100.8	100.1	92.5	97.9	93.5	102.9	92.8	95.8
1998	113.6	118.8	120.8	120.1	112.3	109.6	99.4	96.2	99.0	98.4	90.9	96.3	91.4	100.8	90.4	93.3
1997	111.5	118.9	120.9	120.2	110.7	106.4	98.6	95.4	98.2	97.2	89.9	95.5	89.9	98.5	88.7	91.6
1996	108.9	117.0	119.0	118.4	108.6	105.4	95.9	92.6	95.9	95.1	88.0	93.6	86.6	94.3	83.9	88.3
1995	105.6	115.5	117.9	117.3	106.1	102.3	94.0	90.9	93.7	93.5	86.5	92.2	85.0	92.0	82.6	86.8
1994	103.0	113.9	116.4	115.8	105.0	99.6	92.2	88.9	91.8	91.5	84.5	90.2	82.3	89.6	80.6	83.7
1993	100.0	108.8	110.6	104.8	101.5	96.3	87.4	86.1	87.1	88.5	82.1	87.7	79.5	85.7	77.8	80.9
1992	97.9	107.2	109.0	103.1	100.3	94.7	85.7	84.0	85.3	87.1	80.8	86.2	78.2	84.3	76.5	79.6
1991	95.7	100.6	103.2	96.5	94.5	92.9	85.1	82.8	85.2	85.5	79.7	86.3	76.3	82.6	75.4	78.4
1990	93.2	96.3	98.9	95.1	92.5	90.4	83.9	81.1	84.0	82.9	78.4	85.0	75.0	80.4	74.0	76.9
1989	91.0	94.4	97.0	93.4	89.5	87.5	82.4	79.7	82.5	81.6	76.9	83.5	73.4	78.6	72.4	75.3
1988	88.5	92.3	94.9	91.8	87.7	85.0	80.9	78.0	81.0	80.1	75.4	81.9	71.8	76.9	70.8	73.6
1987	85.7	92.0	92.7	92.5	85.1	82.1	78.8	76.0	77.9	76.6	74.4	79.4	72.2	73.6	70.3	72.3
1986	83.7	88.2	89.7	87.8	83.8	80.8	78.9	75.0	79.9	76.2	72.3	79.1	68.5	72.0	67.5	69.8
1985	81.8	85.3	86.8	85.6	81.1	78.8	76.7	73.4	78.3	73.9	70.6	77.3	66.9	70.3	66.1	68.1
1984	80.6	82.6	83.7	82.5	79.7	79.1	73.8	72.6	75.6	73.0	69.6	76.5	65.9	68.6	65.2	67.5
1983	78.2	78.5	79.8	78.7	76.3	76.0	71.5	70.4	72.9	71.1	67.6	73.6	65.6	68.6	64.7	65.3
1982	72.1	71.7	72.0	71.8	69.1	69.4	65.4	65.2	65.7	66.0	62.8	67.7	60.1	61.9	59.3	60.0
1981	66.1	66.2	66.2	67.3	63.4	64.9	60.3	61.0	60.7	62.0	58.6	62.1	56.2	58.3	55.7	56.1
1980	60.7	60.7	60.9	62.3	58.5	59.6	55.3	55.8	56.5	56.7	53.5	57.2	51.7	54.0	51.1	51.3
1979	54.9	55.2	55.6	57.0	53.6	53.9	50.0	51.7	50.8	51.8	48.9	51.8	46.5	48.3	46.1	46.4
1978	51.3	51.5	51.6	52.9	50.0	51.5	47.0	47.7	47.6	48.0	45.4	48.4	42.7	45.0	42.5	42.4
1977	47.9	48.1	48.3	49.0	47.1	48.4	43.7	43.3	45.7	44.9	42.0	45.9	41.6	42.1	37.7	40.8
1975	43.7	44.7	45.0	46.3	42.9	43.7	42.1	40.3	43.2	41.5	38.1	41.3	37.5	38.4	36.2	36.5
1970	27.8	28.1	28.2	28.9	27.0	26.3	25.7	22.8	27.0	26.2	24.5	24.2	23.8	25.2	22.8	23.4
1965	21.5	21.6	21.7	22.2	20.9	21.8	19.8	17.4	19.3	20.2	18.9	18.6	18.3	19.8	17.6	18.0
1960	19.5	19.7	19.7	20.2	18.9	19.4	18.0	15.8	17.6	18.3	17.2	16.9	16.7	17.1	16.0	16.4
1955	16.3	16.5	16.5	17.0	15.9	16.3	15.1	13.2	14.7	15.4	14.4	14.1	14.0	14.4	13.4	13.7
1950	13.5	13.6	13.7	14.0	13.1	13.4	12.5	11.0	12.2	12.7	11.9	11.7	11.5	11.9	11.0	11.3
1945	8.6	8.7	8.7	8.9	8.4	8.6	7.9	7.0	7.8	8.1	7.6	7.5	7.4	7.6	7.0	7.2
1940	6.6	6.7	6.7	6.9	6.4	6.6	6.1	5.4	6.0	6.2	5.8	5.7	5.7	5.8	5.5	5.6

Year	National 30 City Average	Georgia Savannah	Hawaii Honolulu	Idaho Boise	Idaho Pocatello	Illinois Chicago	Illinois Decatur	Illinois Joliet	Illinois Peoria	Illinois Rockford	Illinois Springfield	Indiana Anderson	Indiana Evansville	Indiana Fort Wayne	Indiana Gary	Indiana Indianapolis
Jan 2000	119.6E	98.0E	145.6E	112.3E	111.6E	131.7E	115.0E	124.5E	119.4E	122.3E	116.0E	109.9E	111.9E	108.5E	118.0E	112.6E
1999	116.6	96.0	143.0	110.2	109.6	129.6	113.0	122.6	116.4	120.7	113.8	107.1	109.3	106.8	112.8	110.6
1998	113.6	93.7	140.4	107.4	107.0	125.2	110.1	119.8	113.8	115.5	111.1	105.0	107.2	104.5	111.1	108.0
1997	111.5	92.0	139.8	104.6	104.7	121.3	107.8	117.5	111.5	113.1	108.9	101.9	104.4	101.8	110.3	105.2
1996	108.9	88.6	134.5	102.2	102.1	118.8	106.6	116.2	109.3	111.5	106.5	100.0	102.1	99.9	107.5	102.7
1995	105.6	87.4	130.3	99.5	98.2	114.2	98.5	110.5	102.3	103.6	98.1	96.4	97.2	95.0	100.7	100.1
1994	103.0	85.3	124.0	94.8	95.0	111.3	97.3	108.9	100.9	102.2	97.0	93.6	95.8	93.6	99.1	97.1
1993	100.0	82.0	122.0	92.2	92.1	107.6	95.6	106.8	98.9	99.6	95.2	91.2	94.3	91.5	96.7	93.9
1992	97.9	80.8	120.0	91.0	91.0	104.3	94.4	104.2	97.3	98.2	94.0	89.5	92.9	89.9	95.0	91.5
1991	95.7	79.5	106.1	89.5	89.4	100.9	92.3	100.0	95.9	95.8	91.5	87.8	91.4	88.3	93.3	89.1
1990	93.2	77.9	104.7	88.2	88.1	98.4	90.9	98.4	93.7	94.0	90.1	84.6	89.3	83.4	88.4	87.1
1989	91.0	76.0	102.8	86.6	86.5	93.7	89.4	92.8	91.6	92.1	88.6	82.3	87.8	81.7	86.5	85.1
1988	88.5	74.5	101.1	83.9	83.7	90.6	87.5	89.8	88.5	87.9	86.8	80.8	85.5	80.0	84.6	83.1
1987	85.7	72.0	99.1	81.4	81.5	86.6	84.4	86.7	86.3	85.2	85.0	78.8	82.4	78.1	81.7	80.4
1986	83.7	70.8	97.5	80.6	80.3	84.4	83.5	85.3	85.1	84.5	83.4	77.0	80.9	76.5	79.6	78.6
1985	81.8	68.9	94.7	78.0	78.0	82.4	81.9	83.4	83.7	83.0	81.5	75.2	79.8	75.0	77.8	77.1
1984	80.6	67.9	90.7	76.6	76.8	80.2	80.4	82.2	83.6	80.6	80.1	73.5	77.4	73.5	77.3	75.9
1983	78.2	66.3	87.2	76.0	75.8	79.0	78.8	80.6	81.5	78.9	78.8	70.9	75.0	71.2	75.2	74.0
1982	72.1	61.1	79.3	71.0	70.1	75.0	72.7	74.4	75.3	72.6	72.7	66.4	69.8	67.1	70.5	68.2
1981	66.1	57.1	73.4	65.4	64.8	68.6	67.2	68.8	70.3	67.3	67.0	61.5	64.3	61.5	65.3	62.4
1980	60.7	52.2	68.9	60.3	59.5	62.8	62.3	63.4	64.5	61.6	61.1	56.5	59.0	56.7	59.8	57.9
1979	54.9	47.2	63.0	54.4	53.7	56.5	56.0	57.0	58.4	54.8	55.6	50.8	53.3	51.2	54.0	51.9
1978	51.3	43.6	58.9	50.0	49.9	52.9	51.9	52.7	53.8	50.4	51.5	46.7	49.1	46.9	49.4	48.7
1977	47.9	40.2	52.0	45.1	45.2	49.0	47.7	49.1	49.4	46.8	47.5	43.2	45.7	43.0	46.7	45.4
1975	43.7	36.9	44.6	40.8	40.5	45.7	43.1	44.5	44.7	42.7	42.4	39.5	41.7	39.9	41.9	40.6
1970	27.8	21.0	30.4	26.7	26.6	29.1	28.0	28.6	29.0	27.5	27.6	25.4	26.4	25.5	27.1	26.2
1965	21.5	16.4	21.8	20.6	20.5	22.7	21.5	22.1	22.4	21.2	21.3	19.5	20.4	19.7	20.8	20.7
1960	19.5	14.9	19.8	18.7	18.6	20.2	19.6	20.0	20.3	19.2	19.3	17.7	18.7	17.9	18.9	18.4
1955	16.3	12.5	16.6	15.7	15.6	16.9	16.4	16.8	17.0	16.1	16.2	14.9	15.7	15.0	15.9	15.5
1950	13.5	10.3	13.7	13.0	12.9	14.0	13.6	13.9	14.0	13.3	13.4	12.3	12.9	12.4	13.1	12.8
1945	8.6	6.6	8.8	8.3	8.2	8.9	8.6	8.9	9.0	8.5	8.6	7.8	8.3	7.9	8.4	8.1
1940	6.6	5.1	6.8	6.4	6.3	6.9	6.7	6.8	6.9	6.5	6.6	6.0	6.4	6.1	6.5	6.3

Historical Cost Indexes

Year	National 30 City Average	Indiana Muncie	Indiana South Bend	Indiana Terre Haute	Iowa Cedar Rapids	Iowa Davenport	Iowa Des Moines	Iowa Sioux City	Iowa Waterloo	Kansas Topeka	Kansas Wichita	Kentucky Lexington	Kentucky Louisville	Louisiana Baton Rouge	Louisiana Lake Charles	Louisiana New Orleans
Jan 2000	119.6E	109.1E	107.0E	111.2E	108.7E	111.4E	108.5E	98.9E	98.4E	101.0E	101.2E	100.9E	108.1E	97.8E	99.2E	101.5E
1999	116.6	105.7	103.7	107.9	104.1	109.2	107.6	96.7	96.4	98.7	99.3	99.7	106.6	96.1	97.6	99.5
1998	113.6	103.6	102.3	106.0	102.5	106.8	103.8	95.1	94.8	97.4	97.4	97.4	101.5	94.7	96.1	97.7
1997	111.5	101.3	101.2	104.6	101.2	105.6	102.5	93.8	93.5	96.3	96.2	96.1	99.9	93.3	96.8	96.2
1996	108.9	99.4	99.1	102.1	99.1	102.0	99.1	90.9	90.7	93.6	93.8	94.4	98.3	91.5	94.9	94.3
1995	105.6	95.6	94.6	97.2	95.4	96.6	94.4	88.2	88.9	91.1	91.0	91.6	94.6	89.6	92.7	91.6
1994	103.0	93.2	93.0	95.8	93.0	92.3	92.4	85.9	86.8	89.4	88.1	89.6	92.2	87.7	89.9	89.5
1993	100.0	91.0	90.7	94.4	91.1	90.5	90.7	84.1	85.2	87.4	86.2	87.3	89.4	86.4	88.5	87.8
1992	97.9	89.4	89.3	93.1	89.8	89.3	89.4	82.9	83.8	86.2	85.0	85.8	88.0	85.2	87.3	86.6
1991	95.7	87.7	87.5	91.7	88.5	88.0	88.4	81.8	81.6	84.4	83.9	84.1	84.4	83.2	85.3	85.8
1990	93.2	83.9	85.1	89.1	87.1	86.5	86.9	80.4	80.3	83.2	82.6	82.8	82.6	82.0	84.1	84.5
1989	91.0	81.9	83.5	86.7	85.3	84.9	85.3	79.0	78.8	81.6	81.2	81.3	80.1	80.6	82.7	83.4
1988	88.5	80.4	82.1	84.9	83.4	83.1	83.6	76.8	77.2	80.1	79.5	79.4	78.2	79.0	81.1	82.0
1987	85.7	78.5	79.6	83.0	82.4	81.1	81.2	75.4	75.6	78.9	78.4	77.3	76.3	77.1	78.0	81.3
1986	83.7	76.9	77.5	81.1	79.5	80.0	79.8	73.8	73.7	76.9	76.2	76.5	75.3	76.1	78.1	79.2
1985	81.8	75.2	76.0	79.2	75.9	77.6	77.1	72.1	72.3	75.0	74.7	75.5	74.5	74.9	78.5	78.2
1984	80.6	73.6	75.1	78.0	80.3	77.8	76.6	74.4	73.7	75.1	74.6	75.6	73.7	77.3	78.3	77.5
1983	78.2	71.2	74.5	76.1	79.6	77.0	75.5	74.6	73.0	73.2	72.4	74.9	74.1	75.1	76.6	74.7
1982	72.1	66.0	68.3	69.9	73.5	71.7	71.1	70.1	67.7	68.4	66.1	69.4	69.2	69.3	70.5	69.2
1981	66.1	61.0	63.1	65.1	68.6	66.7	67.1	65.5	63.5	63.0	62.2	64.7	65.1	64.0	64.8	62.7
1980	60.7	56.1	58.3	59.7	62.7	59.6	61.8	59.3	57.7	58.9	58.0	59.3	59.8	59.1	60.0	57.2
1979	54.9	50.7	52.6	54.0	56.6	54.1	55.6	53.6	52.0	53.5	52.9	53.2	54.5	53.5	54.5	52.6
1978	51.3	46.8	48.2	49.9	52.3	50.0	51.1	49.2	48.1	50.0	48.9	49.2	50.0	49.2	50.2	48.5
1977	47.9	43.4	44.7	46.4	48.1	45.7	47.6	45.3	44.2	46.2	45.8	45.8	46.1	45.2	45.2	46.0
1975	43.7	39.2	40.3	41.9	43.1	41.0	43.1	41.0	40.0	42.0	43.1	42.7	42.5	40.4	40.6	41.5
1970	27.8	25.3	26.2	27.0	28.2	26.9	27.6	26.6	25.9	27.0	25.5	26.9	25.9	25.0	26.7	27.2
1965	21.5	19.5	20.2	20.9	21.7	20.8	21.7	20.5	20.0	20.8	19.6	20.7	20.3	19.4	20.6	20.4
1960	19.5	17.8	18.3	18.9	19.7	18.8	19.5	18.6	18.2	18.9	17.8	18.8	18.4	17.6	18.7	18.5
1955	16.3	14.9	15.4	15.9	16.6	15.8	16.3	15.6	15.2	15.8	15.0	15.8	15.4	14.8	15.7	15.6
1950	13.5	12.3	12.7	13.1	13.7	13.1	13.5	12.9	12.6	13.1	12.4	13.0	12.8	12.2	13.0	12.8
1945	8.6	7.8	8.1	8.4	8.7	8.3	8.6	8.2	8.0	8.3	7.9	8.3	8.1	7.8	8.3	8.2
1940	6.6	6.0	6.3	6.5	6.7	6.3	6.6	6.3	6.2	6.4	6.1	6.4	6.3	6.0	6.4	6.3

Year	National 30 City Average	Louisiana Shreveport	Maine Lewiston	Maine Portland	Maryland Baltimore	Massachusetts Boston	Massachusetts Brockton	Massachusetts Fall River	Massachusetts Lawrence	Massachusetts Lowell	Massachusetts New Bedford	Massachusetts Pittsfield	Massachusetts Springfield	Massachusetts Worcester	Michigan Ann Arbor	Michigan Dearborn
Jan 2000	119.6E	96.0E	106.0E	105.7E	108.1E	138.1E	129.3E	128.4E	129.6E	130.1E	128.3E	117.0E	121.4E	127.0E	123.0E	125.2E
1999	116.6	94.8	105.0	104.7	106.4	136.2	126.7	126.3	127.3	127.4	126.2	115.0	118.8	123.8	117.5	122.7
1998	113.6	92.0	102.8	102.5	104.1	132.8	125.0	124.7	125.0	125.3	124.6	114.2	117.1	122.6	116.0	119.7
1997	111.5	90.1	101.7	101.4	102.2	132.1	124.4	124.7	125.1	125.6	124.7	114.6	117.4	122.9	113.5	117.9
1996	108.9	88.5	99.5	99.2	99.6	128.7	121.6	122.2	121.7	122.1	122.2	112.7	115.0	120.2	112.9	116.2
1995	105.6	86.7	96.8	96.5	96.1	128.6	119.6	117.7	120.5	119.6	117.2	110.7	112.7	114.8	106.1	110.5
1994	103.0	85.0	95.2	94.9	94.4	124.9	114.2	113.7	116.8	116.9	111.6	109.2	111.1	112.2	105.0	109.0
1993	100.0	83.6	93.0	93.0	93.1	121.1	111.7	111.5	114.9	114.3	109.4	107.1	108.9	110.3	102.8	105.8
1992	97.9	82.3	91.7	91.7	90.9	118.0	110.0	109.8	110.9	110.2	107.7	105.7	107.2	108.6	101.5	103.2
1991	95.7	81.6	89.8	89.9	89.1	115.2	105.9	104.7	107.3	105.5	105.1	100.6	103.2	105.7	95.1	98.3
1990	93.2	80.3	88.5	88.6	85.6	110.9	103.7	102.9	105.2	102.8	102.6	98.7	101.4	103.2	93.2	96.5
1989	91.0	78.8	86.7	86.7	83.5	107.2	100.4	99.7	102.9	99.7	99.8	94.7	96.4	99.1	89.6	94.6
1988	88.5	77.0	84.0	84.1	81.2	102.5	96.7	96.4	97.7	96.6	96.4	91.9	92.7	94.4	85.8	91.4
1987	85.7	74.5	81.3	81.4	78.6	97.3	94.5	94.8	94.9	94.8	95.0	89.4	90.5	92.2	87.4	88.7
1986	83.7	74.1	79.0	79.0	75.7	95.1	91.2	90.8	92.5	90.8	90.7	87.1	87.8	89.8	81.7	85.3
1985	81.8	73.6	76.7	77.0	72.7	92.8	88.8	88.7	89.4	88.2	88.6	85.0	85.6	86.7	80.1	82.7
1984	80.6	73.3	75.0	75.1	72.4	88.1	85.5	85.8	86.8	84.1	84.8	82.8	83.5	83.6	78.9	81.1
1983	78.2	72.2	73.1	72.9	70.6	84.0	82.5	82.2	82.7	80.9	81.3	79.5	80.6	81.3	77.7	79.4
1982	72.1	67.0	67.4	67.4	64.7	76.9	75.7	75.0	74.5	73.6	74.2	72.3	73.7	73.1	73.7	75.8
1981	66.1	62.5	62.2	63.1	59.0	67.8	68.7	69.2	68.4	67.1	68.7	66.3	67.0	66.7	68.6	70.1
1980	60.7	58.7	57.3	58.5	53.6	64.0	63.7	64.1	63.4	62.7	63.1	61.8	62.0	62.3	62.9	64.0
1979	54.9	53.1	52.3	52.6	48.2	57.9	58.0	57.5	57.3	57.0	57.4	56.2	55.9	56.0	57.0	57.8
1978	51.3	49.2	48.6	48.8	45.9	54.0	53.4	53.3	53.5	53.1	53.2	52.6	52.3	52.2	52.5	54.1
1977	47.9	44.8	46.2	45.2	44.4	49.3	50.4	50.1	50.5	49.5	49.9	49.2	50.0	49.4	48.6	48.6
1975	43.7	40.5	41.9	42.1	39.8	46.6	45.8	45.7	46.2	45.7	46.1	45.5	45.8	46.0	44.1	44.9
1970	27.8	26.4	26.5	25.8	25.1	29.2	29.1	29.2	29.1	28.9	29.0	28.6	28.5	28.7	28.5	28.9
1965	21.5	20.3	20.4	19.4	20.2	23.0	22.5	22.5	22.5	22.2	22.4	22.0	22.4	22.1	22.0	22.3
1960	19.5	18.5	18.6	17.6	17.5	20.5	20.4	20.4	20.4	20.2	20.3	20.0	20.1	20.1	20.1	20.2
1955	16.3	15.5	15.6	14.7	14.7	17.2	17.1	17.1	17.1	16.9	17.1	16.8	16.9	16.8	16.7	16.9
1950	13.5	12.8	12.9	12.2	12.1	14.2	14.1	14.1	14.1	14.0	14.1	13.8	14.0	13.9	13.8	14.0
1945	8.6	8.1	8.2	7.8	7.7	9.1	9.0	9.0	9.0	8.9	9.0	8.9	8.9	8.9	8.8	8.9
1940	6.6	6.3	6.3	6.0	6.0	7.0	6.9	7.0	7.0	6.9	6.9	6.8	6.9	6.8	6.8	6.9

Historical Cost Indexes

Year	National 30 City Average	Michigan						Minnesota			Mississippi		Missouri			
		Detroit	Flint	Grand Rapids	Kala-mazoo	Lansing	Sagi-naw	Duluth	Minne-apolis	Roches-ter	Biloxi	Jackson	Kansas City	St. Joseph	St. Louis	Spring-field
Jan 2000	119.6E	124.7E	116.9E	102.5E	111.0E	116.7E	114.5E	123.9E	130.6E	119.3E	97.0E	93.1E	116.8E	107.3E	122.2E	102.3E
1999	116.6	122.6	113.7	100.9	106.1	111.5	110.1	120.3	126.5	117.1	95.7	91.8	114.9	106.1	119.8	101.1
1998	113.6	119.5	112.3	99.7	104.9	110.1	108.7	117.7	124.6	115.5	92.1	89.5	108.2	104.2	115.9	98.9
1997	111.5	117.6	110.8	98.9	104.2	108.7	107.5	115.9	121.9	114.1	90.9	88.3	106.4	102.4	113.2	97.3
1996	108.9	116.0	109.7	93.9	103.5	107.8	106.9	115.0	120.4	113.5	87.1	85.4	103.2	99.9	110.1	94.6
1995	105.6	110.1	104.1	91.1	96.4	98.3	102.3	100.3	111.9	102.5	84.2	83.5	99.7	96.1	106.3	89.5
1994	103.0	108.5	102.9	89.7	95.0	97.1	101.1	99.1	109.3	101.1	82.4	81.7	97.3	92.8	103.2	88.2
1993	100.0	105.4	100.7	87.7	92.7	95.3	98.8	99.8	106.7	99.9	80.0	79.4	94.5	90.9	99.9	86.2
1992	97.9	102.9	99.4	86.4	91.4	94.0	97.4	98.5	105.3	98.6	78.8	78.2	92.9	89.5	98.6	84.9
1991	95.7	97.6	93.6	85.3	90.0	92.1	90.3	96.0	102.9	97.3	77.6	76.9	90.9	88.1	96.4	83.9
1990	93.2	96.0	91.7	84.0	87.4	90.2	88.9	94.3	100.5	95.6	76.4	75.6	89.5	86.8	94.0	82.6
1989	91.0	94.0	90.2	82.4	85.8	88.6	87.4	92.5	97.4	93.7	75.0	74.1	87.3	85.1	91.8	80.9
1988	88.5	90.8	87.8	80.7	84.1	86.3	85.6	90.5	94.6	91.9	73.5	72.9	84.9	82.9	89.0	79.2
1987	85.7	87.8	85.4	79.7	82.6	85.2	84.5	88.5	92.0	90.6	72.4	71.3	81.7	82.7	85.5	78.2
1986	83.7	84.3	82.8	77.5	80.9	82.7	81.5	87.2	89.7	88.4	70.6	69.9	79.9	79.4	83.5	75.6
1985	81.8	81.6	80.7	75.8	79.4	80.0	80.2	85.2	87.9	86.5	69.4	68.5	78.2	77.3	80.8	73.2
1984	80.6	79.4	79.5	74.3	78.7	79.0	79.9	84.9	86.9	86.0	68.6	67.5	77.7	76.7	77.7	72.7
1983	78.2	77.0	77.4	74.1	76.2	77.1	78.0	81.6	81.4	82.6	68.0	67.3	76.2	76.2	76.2	71.6
1982	72.1	74.8	73.5	69.9	72.3	73.2	73.7	74.6	75.2	75.6	63.1	62.6	70.2	70.3	69.2	66.5
1981	66.1	68.6	68.2	64.6	67.1	67.5	68.5	68.8	69.6	69.8	58.7	59.1	63.5	65.7	64.5	61.4
1980	60.7	62.5	63.1	59.8	61.7	60.3	63.3	63.4	64.1	64.8	54.3	54.3	59.1	60.4	59.7	56.7
1979	54.9	56.2	56.4	53.7	55.5	55.8	56.8	59.2	58.5	58.7	49.3	49.3	54.4	54.6	54.9	51.3
1978	51.3	52.6	52.0	49.1	51.7	51.4	53.2	53.1	54.1	53.3	45.0	45.8	50.1	51.4	51.1	47.9
1977	47.9	48.7	48.4	44.9	48.4	47.4	49.1	49.5	50.0	48.9	41.2	42.3	46.8	48.8	48.2	44.5
1975	43.7	45.8	44.9	41.8	43.7	44.1	44.8	45.2	46.0	45.3	37.9	37.7	42.7	45.4	44.8	41.1
1970	27.8	29.7	28.5	26.7	28.1	27.8	28.7	28.9	29.5	28.9	24.4	21.3	25.3	28.3	28.4	26.0
1965	21.5	22.1	21.9	20.6	21.7	21.5	22.2	22.3	23.4	22.3	18.8	16.4	20.6	21.8	21.8	20.0
1960	19.5	20.1	20.0	18.7	19.7	19.5	20.1	20.3	20.5	20.2	17.1	14.9	19.0	19.8	19.5	18.2
1955	16.3	16.8	16.7	15.7	16.5	16.3	16.9	17.0	17.2	17.0	14.3	12.5	16.0	16.6	16.3	15.2
1950	13.5	13.9	13.8	13.0	13.6	13.5	13.9	14.0	14.2	14.0	11.8	10.3	13.2	13.7	13.5	12.6
1945	8.6	8.9	8.8	8.3	8.7	8.6	8.9	8.9	9.0	8.9	7.6	6.6	8.4	8.8	8.6	8.0
1940	6.6	6.8	6.8	6.4	6.7	6.6	6.9	6.9	7.0	6.9	5.8	5.1	6.5	6.7	6.7	6.2

Year	National 30 City Average	Montana		Nebraska		Nevada		New Hamphire		New Jersey				NM	NY
		Billings	Great Falls	Lincoln	Omaha	Las Vegas	Reno	Man-chester	Nashua	Camden	Jersey City	Newark	Pater-son	Trenton	Albu-querque	Albany
Jan 2000	119.6E	113.2E	114.0E	98.7E	107.0E	124.3E	118.2E	111.0E	111.0E	128.8E	131.0E	133.1E	133.0E	131.0E	108.5E	115.6E
1999	116.6	112.1	112.7	96.6	104.8	121.9	114.4	109.6	109.6	125.3	128.8	131.6	129.5	129.6	106.7	114.6
1998	113.6	109.7	109.1	94.9	101.2	118.1	111.4	110.1	110.0	124.3	127.7	128.9	128.5	127.9	103.8	113.0
1997	111.5	107.4	107.6	93.3	99.5	114.6	109.8	108.6	108.6	121.9	125.1	126.4	126.4	125.2	100.8	110.0
1996	108.9	108.0	107.5	91.4	97.4	111.8	108.9	106.5	106.4	119.1	122.5	122.5	123.8	121.8	98.6	108.3
1995	105.6	104.7	104.9	85.6	93.4	108.5	105.0	100.9	100.8	107.0	112.2	111.9	112.1	111.2	96.3	103.6
1994	103.0	100.2	100.7	84.3	91.0	105.3	102.2	97.8	97.7	105.4	110.6	110.1	110.6	108.2	93.4	102.4
1993	100.0	97.9	97.8	82.3	88.7	102.8	99.9	95.4	95.4	103.7	109.0	108.2	109.0	106.4	90.1	99.8
1992	97.9	95.5	96.4	81.1	87.4	99.4	98.4	90.4	90.4	102.0	107.5	107.0	107.8	103.9	87.5	98.5
1991	95.7	94.2	95.1	80.1	86.4	97.5	95.8	87.8	87.8	95.0	98.5	95.6	100.0	96.6	86.2	96.1
1990	93.2	92.9	93.8	78.8	85.0	96.3	94.5	86.3	86.3	93.0	93.5	93.8	97.2	94.5	84.9	93.2
1989	91.0	91.3	92.2	77.3	83.6	94.8	92.3	84.8	84.8	90.4	91.4	91.8	95.5	91.9	83.3	88.4
1988	88.5	89.5	90.4	75.8	82.0	93.0	90.5	83.2	83.2	87.7	89.5	89.7	92.5	89.5	81.7	86.8
1987	85.7	85.4	87.3	75.3	79.8	91.5	88.3	82.4	82.5	86.1	88.3	88.7	90.3	87.9	78.1	84.5
1986	83.7	86.2	87.3	72.6	78.8	90.1	87.2	79.7	79.7	84.6	86.1	86.7	87.8	86.1	78.9	81.7
1985	81.8	83.9	84.3	71.5	77.5	87.6	85.0	78.1	78.1	81.3	83.3	83.5	84.5	82.8	76.7	79.5
1984	80.6	83.2	82.9	70.9	77.6	85.9	83.3	76.4	75.5	78.2	80.3	80.1	81.3	78.5	75.9	77.3
1983	78.2	80.3	80.0	72.1	77.3	83.0	81.9	73.0	72.2	74.6	76.5	76.7	76.5	74.2	73.1	74.2
1982	72.1	74.3	74.1	67.8	72.5	76.9	75.8	67.0	66.7	67.9	69.8	70.4	70.1	69.1	67.1	69.2
1981	66.1	69.5	70.3	64.1	68.2	70.2	68.5	61.5	61.1	62.5	65.0	65.1	64.8	63.4	63.3	64.4
1980	60.7	63.9	64.6	58.5	63.5	64.6	63.0	56.6	56.0	58.6	60.6	60.1	60.0	58.9	59.0	59.5
1979	54.9	57.5	58.9	52.9	56.2	58.4	56.7	51.2	50.8	53.5	55.6	55.3	54.7	54.0	54.3	54.2
1978	51.3	53.4	54.0	49.3	52.2	53.9	51.9	47.5	47.1	50.1	51.7	51.4	51.7	50.8	49.8	50.6
1977	47.9	48.9	49.1	44.6	48.0	49.7	48.1	43.9	43.6	47.2	47.4	47.2	47.7	47.0	45.1	47.7
1975	43.7	43.1	43.8	40.9	43.2	42.8	41.9	41.3	40.8	42.3	43.4	44.2	43.9	43.8	40.3	43.9
1970	27.8	28.5	28.9	26.4	26.8	29.4	28.0	26.2	25.6	27.2	27.8	29.0	27.8	27.4	26.4	28.3
1965	21.5	22.0	22.3	20.3	20.6	22.4	21.6	20.6	19.7	20.9	21.4	23.8	21.4	21.3	20.6	22.3
1960	19.5	20.0	20.3	18.5	18.7	20.2	19.6	18.0	17.9	19.0	19.4	19.4	19.4	19.2	18.5	19.3
1955	16.3	16.7	17.0	15.5	15.7	16.9	16.4	15.1	15.0	16.0	16.3	16.3	16.3	16.1	15.6	16.2
1950	13.5	13.9	14.0	12.8	13.0	14.0	13.6	12.5	12.4	13.2	13.5	13.5	13.5	13.3	12.9	13.4
1945	8.6	8.8	9.0	8.1	8.3	8.9	8.7	8.0	7.9	8.4	8.6	8.6	8.6	8.5	8.2	8.5
1940	6.6	6.8	6.9	6.3	6.4	6.9	6.7	6.2	6.1	6.5	6.6	6.6	6.6	6.6	6.3	6.6

Historical Cost Indexes

Year	National 30 City Average	New York								North Carolina					N. Dakota	Ohio
		Bing-hamton	Buffalo	New York	Roches-ter	Schen-ectady	Syracuse	Utica	Yonkers	Charlotte	Durham	Greens-boro	Raleigh	Winston-Salem	Fargo	Akron
Jan 2000	119.6E	111.4E	122.0E	158.3E	118.6E	116.1E	115.1E	111.4E	142.6E	90.4E	91.4E	91.5E	91.7E	91.1E	98.1E	117.8E
1999	116.6	108.6	120.2	155.9	116.8	114.7	113.7	108.5	140.6	89.3	90.2	90.3	90.5	90.1	96.8	116.0
1998	113.6	109.0	119.1	154.4	117.2	114.2	113.6	108.6	141.4	88.2	89.1	89.2	89.4	89.0	95.2	113.1
1997	111.5	107.2	115.7	150.3	115.2	111.2	110.6	107.0	138.8	86.8	87.7	87.8	87.9	87.6	93.5	110.6
1996	108.9	105.5	114.0	148.0	113.3	109.5	108.2	105.2	137.2	85.0	85.9	86.0	86.1	85.8	91.7	107.9
1995	105.6	99.3	110.1	140.7	106.6	104.6	104.0	97.7	129.4	81.8	82.6	82.6	82.7	82.6	88.4	103.4
1994	103.0	98.1	107.2	137.1	105.2	103.5	102.3	96.5	128.1	80.3	81.1	81.0	81.2	81.1	87.0	102.2
1993	100.0	95.7	102.2	133.3	102.0	100.8	99.6	94.1	126.3	78.2	78.9	78.9	78.9	78.8	85.8	100.3
1992	97.9	93.7	100.1	128.2	99.3	99.3	98.1	90.5	123.9	77.1	77.7	77.8	77.8	77.6	83.1	98.0
1991	95.7	89.5	96.8	124.4	96.0	96.7	95.5	88.7	121.5	76.0	76.6	76.8	76.7	76.6	82.6	96.8
1990	93.2	87.0	94.1	118.1	94.6	93.9	91.0	85.6	111.4	74.8	75.4	75.5	75.5	75.3	81.3	94.6
1989	91.0	85.4	91.8	114.8	92.6	89.6	88.4	84.1	102.1	73.3	74.0	74.0	74.0	73.8	79.7	92.9
1988	88.5	83.6	89.4	106.5	87.9	87.7	86.5	82.4	99.9	71.5	72.2	72.2	72.2	72.0	78.2	91.1
1987	85.7	82.4	85.8	102.6	85.8	85.3	85.1	82.1	98.9	70.6	71.2	71.2	71.1	71.0	77.4	90.1
1986	83.7	80.2	84.4	99.8	83.7	82.4	83.3	79.2	96.6	68.4	69.1	69.1	69.1	68.9	75.4	87.7
1985	81.8	77.5	83.2	94.9	81.6	80.1	81.0	77.0	92.8	66.9	67.6	67.7	67.6	67.5	73.4	86.8
1984	80.6	75.4	81.3	91.1	80.6	78.5	79.0	76.1	89.5	66.3	66.6	66.8	66.2	66.0	72.2	82.8
1983	78.2	73.2	78.2	86.5	77.1	75.3	76.5	74.5	85.4	64.6	65.1	65.7	64.9	64.6	70.5	79.4
1982	72.1	67.5	70.3	78.3	71.3	69.8	70.2	68.4	77.8	58.7	60.0	60.5	59.5	59.4	66.1	73.2
1981	66.1	62.6	65.2	71.6	65.6	64.7	64.6	63.5	71.0	55.3	56.6	56.7	55.9	56.3	61.7	67.7
1980	60.7	58.0	60.6	66.0	60.5	60.3	61.6	58.5	65.9	51.1	52.2	52.5	51.7	50.8	57.4	62.3
1979	54.9	52.8	56.0	60.2	55.0	54.9	56.2	53.2	60.0	45.9	46.9	47.1	46.4	45.9	52.2	56.3
1978	51.3	50.0	52.6	56.4	51.7	50.8	52.2	49.1	54.7	41.9	43.0	42.9	42.4	41.9	47.8	52.1
1977	47.9	46.5	49.0	52.9	48.3	48.5	49.1	46.0	51.8	40.2	41.5	40.7	40.0	39.6	44.2	48.5
1975	43.7	42.5	45.1	49.5	44.8	43.7	44.8	42.7	47.1	36.1	37.0	37.0	37.3	36.1	39.3	44.7
1970	27.8	27.0	28.9	32.8	29.2	27.8	28.5	26.9	30.0	20.9	23.7	23.6	23.4	23.0	25.8	28.3
1965	21.5	20.8	22.2	25.5	22.8	21.4	21.9	20.7	23.1	16.0	18.2	18.2	18.0	17.7	19.9	21.8
1960	19.5	18.9	19.9	21.5	19.7	19.5	19.9	18.8	21.0	14.4	16.6	16.6	16.4	16.1	18.1	19.9
1955	16.3	15.8	16.7	18.1	16.5	16.3	16.7	15.8	17.6	12.1	13.9	13.9	13.7	13.5	15.2	16.6
1950	13.5	13.1	13.8	14.9	13.6	13.5	13.8	13.0	14.5	10.0	11.5	11.5	11.4	11.2	12.5	13.7
1945	8.6	8.3	8.8	9.5	8.7	8.6	8.8	8.3	9.3	6.4	7.3	7.3	7.2	7.1	8.0	8.8
1940	6.6	6.4	6.8	7.4	6.7	6.7	6.8	6.4	7.2	4.9	5.6	5.6	5.6	5.5	6.2	6.8

Year	National 30 City Average	Ohio									Oklahoma			Oregon		PA
		Canton	Cincin-nati	Cleve-land	Colum-bus	Dayton	Lorain	Spring-field	Toledo	Youngs-town	Lawton	Oklahoma City	Tulsa	Eugene	Port-land	Allen-town
Jan 2000	119.6E	113.1E	110.0E	121.5E	111.9E	109.1E	115.2E	109.1E	115.7E	114.2E	99.0E	99.2E	97.6E	124.7E	125.8E	119.8E
1999	116.6	110.6	107.9	118.7	109.5	107.1	112.0	106.4	113.6	111.9	97.5	97.4	96.2	120.9	124.3	117.8
1998	113.6	108.5	105.0	114.8	106.8	104.5	109.4	104.1	111.2	109.0	94.3	94.5	94.5	120.0	122.2	115.7
1997	111.5	106.4	102.9	112.9	104.7	102.6	107.4	102.2	108.5	107.2	92.9	93.2	93.0	118.1	119.6	113.9
1996	108.9	103.2	100.2	110.1	101.1	98.8	103.8	98.4	105.1	104.1	90.9	91.3	91.2	114.4	116.1	112.0
1995	105.6	98.8	97.1	106.4	99.1	94.6	97.1	92.0	100.6	100.1	85.3	88.0	89.0	112.2	114.3	108.3
1994	103.0	97.7	95.0	104.8	95.4	93.0	96.0	90.1	99.3	98.9	84.0	86.5	87.6	107.6	109.2	106.4
1993	100.0	95.9	92.3	101.9	93.9	90.6	93.9	87.7	98.3	97.2	81.3	83.9	84.8	107.2	108.8	103.6
1992	97.9	94.5	90.6	98.7	92.6	89.2	92.2	86.4	97.1	96.0	80.1	82.7	83.4	101.1	102.6	101.6
1991	95.7	93.4	88.6	97.2	90.6	87.7	91.1	84.9	94.2	92.7	78.3	80.7	81.4	99.5	101.1	98.9
1990	93.2	92.1	86.7	95.1	88.1	85.9	89.4	83.4	92.9	91.8	77.1	79.9	80.0	98.0	99.4	95.0
1989	91.0	90.3	84.6	93.6	85.8	82.6	87.9	81.0	91.4	88.4	75.8	78.5	78.4	95.2	97.0	92.4
1988	88.5	89.1	82.9	92.1	83.9	81.0	85.2	79.5	89.6	86.9	74.3	77.0	76.8	93.5	95.3	89.8
1987	85.7	87.7	80.9	89.6	81.5	79.5	86.9	78.0	86.1	85.4	72.6	74.4	74.7	91.4	91.9	86.9
1986	83.7	84.9	78.8	87.3	79.7	78.0	81.6	76.1	84.6	83.8	71.6	74.1	73.9	90.0	91.7	84.7
1985	81.8	83.7	78.2	86.2	77.7	76.3	80.7	74.3	84.7	82.7	71.2	73.9	73.8	88.7	90.1	81.9
1984	80.6	81.3	77.4	82.5	76.1	75.9	79.5	73.9	83.5	80.3	70.6	72.9	72.8	89.5	89.3	78.4
1983	78.2	76.9	74.9	77.9	74.5	71.6	77.2	71.9	80.4	77.3	69.1	71.5	72.2	89.0	89.2	75.0
1982	72.1	70.8	70.1	71.6	68.6	65.2	70.8	66.5	74.6	71.7	63.5	65.2	66.2	82.3	83.6	68.2
1981	66.1	65.8	65.3	66.0	63.9	60.6	65.7	61.3	69.4	67.2	57.0	60.3	61.5	74.2	74.5	62.7
1980	60.7	60.6	59.8	61.0	58.6	56.6	60.8	56.2	64.0	61.5	52.2	55.5	57.2	68.9	68.4	58.7
1979	54.9	55.2	54.3	54.8	52.7	51.2	55.2	51.0	57.9	56.0	47.7	49.8	51.4	61.9	60.6	54.1
1978	51.3	51.1	50.4	51.3	49.3	47.2	50.3	46.9	54.0	52.2	44.5	46.3	48.0	57.2	56.0	50.4
1977	47.9	47.7	46.9	48.2	45.3	44.3	46.7	42.9	50.2	48.7	40.6	43.0	44.8	51.5	51.7	46.7
1975	43.7	44.0	43.7	44.6	42.2	40.5	42.7	40.0	44.9	44.5	38.6	38.6	39.2	44.6	44.9	42.3
1970	27.8	27.8	28.2	29.6	26.7	27.0	27.5	25.4	29.1	29.6	24.1	23.4	24.1	30.4	30.0	27.3
1965	21.5	21.5	20.7	21.1	20.2	20.0	21.1	19.6	21.3	21.0	18.5	17.8	20.6	23.4	22.6	21.0
1960	19.5	19.5	19.3	19.6	18.7	18.1	19.2	17.8	19.4	19.1	16.9	16.1	18.1	21.2	21.1	19.1
1955	16.3	16.3	16.2	16.5	15.7	15.2	16.1	14.9	16.2	16.0	14.1	13.5	15.1	17.8	17.7	16.0
1950	13.5	13.5	13.3	13.6	13.0	12.5	13.3	12.3	13.4	13.2	11.7	11.2	12.5	14.7	14.6	13.2
1945	8.6	8.6	8.5	8.7	8.3	8.0	8.5	7.8	8.6	8.4	7.4	7.1	8.0	9.4	9.3	8.4
1940	6.6	6.7	6.5	6.7	6.4	6.2	6.5	6.1	6.6	6.5	5.7	5.5	6.1	7.2	7.2	6.5

Historical Cost Indexes

Year	National 30 City Average	Pennsylvania Erie	Pennsylvania Harris-burg	Pennsylvania Phila-delphia	Pennsylvania Pitts-burgh	Pennsylvania Reading	Pennsylvania Scranton	RI Provi-dence	South Carolina Charles-ton	South Carolina Colum-bia	South Dakota Rapid City	South Dakota Sioux Falls	South Dakota Chatta-nooga	Tennessee Knox-ville	Tennessee Memphis	Tennessee Nash-ville
Jan 2000	119.6E	115.1E	114.2E	132.7E	121.3E	116.3E	118.0E	123.2E	90.3E	89.4E	94.0E	97.4E	97.1E	95.2E	100.9E	100.3E
1999	116.6	113.8	112.8	129.8	119.6	114.8	116.3	121.6	89.0	88.1	92.4	95.8	95.9	93.4	99.7	98.6
1998	113.6	109.8	110.5	126.6	117.1	112.5	113.7	120.3	88.0	87.2	90.7	93.6	94.9	92.4	97.2	96.8
1997	111.5	108.5	108.8	123.3	113.9	110.8	112.3	118.9	86.5	85.6	89.3	92.1	93.6	90.8	96.1	94.9
1996	108.9	106.8	105.7	120.3	110.8	108.7	109.9	117.2	84.9	84.1	87.2	90.0	91.6	88.8	94.0	91.9
1995	105.6	99.8	100.6	117.1	106.3	103.6	103.8	111.1	82.7	82.2	84.0	84.7	89.2	86.0	91.2	87.6
1994	103.0	98.2	99.4	115.2	103.7	102.2	102.6	109.6	81.2	80.6	82.7	83.1	87.4	84.1	89.0	84.8
1993	100.0	94.9	96.6	107.4	99.0	98.6	99.8	108.2	78.4	77.9	81.2	81.9	85.1	82.3	86.8	81.9
1992	97.9	92.1	94.8	105.3	96.5	96.7	97.8	106.9	77.1	76.8	80.0	80.7	83.6	77.7	85.4	80.7
1991	95.7	90.5	92.5	101.7	93.2	94.1	94.8	96.1	75.0	75.8	78.7	79.7	81.9	76.6	83.0	79.1
1990	93.2	88.5	89.5	98.5	91.2	90.8	91.3	94.1	73.7	74.5	77.2	78.4	79.9	75.1	81.3	77.1
1989	91.0	86.6	86.5	94.2	89.4	87.9	88.8	92.4	72.1	72.8	75.7	77.0	78.5	73.7	81.2	74.8
1988	88.5	84.9	84.0	89.8	87.7	85.3	86.4	89.1	70.5	71.3	74.2	75.5	77.0	72.1	79.7	73.1
1987	85.7	82.6	81.8	86.8	85.6	82.9	83.3	86.3	70.2	70.2	73.6	74.9	74.4	70.2	77.5	71.0
1986	83.7	81.5	79.6	84.9	83.6	79.7	81.3	85.2	67.4	68.1	71.0	72.3	74.2	69.2	75.3	68.1
1985	81.8	79.6	77.2	82.2	81.5	77.7	80.0	83.0	65.9	66.9	69.7	71.2	72.5	67.7	74.3	66.7
1984	80.6	78.3	75.4	79.0	78.9	75.9	78.8	80.3	64.3	65.5	68.7	70.4	71.4	66.4	73.3	66.3
1983	78.2	76.3	72.2	74.4	75.5	74.0	75.0	76.3	65.3	65.2	67.7	69.7	69.3	64.1	72.8	65.1
1982	72.1	71.0	66.3	69.0	70.9	67.7	68.4	70.0	59.9	60.0	64.4	67.6	64.5	59.8	66.5	62.2
1981	66.1	64.6	61.7	63.4	66.3	62.5	62.3	64.3	56.3	55.9	59.7	63.6	59.9	55.9	60.3	57.3
1980	60.7	59.4	56.9	58.7	61.3	57.7	58.4	59.2	50.6	51.1	54.8	57.1	55.0	51.6	55.9	53.1
1979	54.9	54.0	52.6	54.2	55.6	53.8	53.5	53.3	45.7	44.4	49.7	51.5	49.8	46.4	51.0	47.0
1978	51.3	50.2	49.6	51.3	51.4	49.7	50.0	50.2	42.1	40.7	46.0	47.4	45.9	42.8	47.4	43.9
1977	47.9	46.6	45.2	48.1	48.3	46.5	46.9	47.0	39.2	39.4	42.5	44.3	42.5	40.2	45.0	40.7
1975	43.7	43.4	42.2	44.5	44.5	43.6	41.9	42.7	35.0	36.4	37.7	39.6	39.1	37.3	40.7	37.0
1970	27.8	27.5	25.8	27.5	28.7	27.1	27.0	27.3	22.8	22.9	24.8	25.8	24.9	22.0	23.0	22.8
1965	21.5	21.1	20.1	21.7	22.4	20.9	20.8	21.9	17.6	17.6	19.1	19.9	19.2	17.1	18.3	17.4
1960	19.5	19.1	18.6	19.4	19.7	19.0	18.9	19.1	16.0	16.0	17.3	18.1	17.4	15.5	16.6	15.8
1955	16.3	16.1	15.6	16.3	16.5	15.9	15.9	16.0	13.4	13.4	14.5	15.1	14.6	13.0	13.9	13.3
1950	13.5	13.3	12.9	13.5	13.6	13.2	13.1	13.2	11.1	11.1	12.0	12.5	12.1	10.7	11.5	10.9
1945	8.6	8.5	8.2	8.6	8.7	8.4	8.3	8.4	7.1	7.1	7.7	8.0	7.7	6.8	7.3	7.0
1940	6.6	6.5	6.3	6.6	6.7	6.4	6.5	6.5	5.4	5.5	5.9	6.2	6.0	5.3	5.6	5.4

Year	National 30 City Average	Texas Abilene	Texas Amarillo	Texas Austin	Texas Beaumont	Texas Corpus Christi	Texas Dallas	Texas El Paso	Texas Fort Worth	Texas Houston	Texas Lubbock	Texas Odessa	Texas San Antonio	Texas Waco	Texas Wichita Falls	Utah Ogden
Jan 2000	119.6E	93.3E	95.9E	97.3E	101.0E	95.2E	103.0E	92.0E	99.3E	106.1E	97.5E	93.5E	99.7E	96.4E	96.9E	104.7E
1999	116.6	91.8	94.5	96.0	99.7	94.0	101.0	90.7	97.6	104.6	96.0	92.1	98.0	94.8	95.5	103.3
1998	113.6	89.8	92.7	94.2	97.9	91.8	97.9	88.4	94.5	101.3	93.3	90.1	94.8	92.7	92.9	98.5
1997	111.5	88.4	91.3	92.8	96.8	90.3	96.1	87.0	93.3	100.1	91.9	88.8	93.4	91.4	91.5	96.1
1996	108.9	86.8	89.6	90.9	95.3	88.5	94.1	86.7	91.5	97.9	90.2	87.1	92.3	89.7	89.9	94.1
1995	105.6	85.2	87.4	89.3	93.7	87.4	91.4	85.2	89.5	95.9	88.4	85.6	88.9	86.4	86.8	92.2
1994	103.0	83.3	85.5	87.0	91.8	84.6	89.6	82.2	87.4	93.4	87.0	83.9	87.0	84.9	85.4	89.4
1993	100.0	81.4	83.4	84.9	90.0	82.8	87.8	80.2	85.5	91.1	84.8	81.9	85.0	83.0	83.5	87.1
1992	97.9	80.3	82.3	83.8	88.9	81.6	86.2	79.0	84.1	89.8	83.6	80.7	83.9	81.8	82.4	85.1
1991	95.7	79.2	81.1	82.8	87.8	80.6	85.9	78.0	83.3	87.9	82.7	79.8	83.3	80.6	81.5	84.1
1990	93.2	78.0	80.1	81.3	86.5	79.3	84.5	76.7	82.1	85.4	81.5	78.6	80.7	79.6	80.3	83.4
1989	91.0	76.6	78.7	79.7	85.1	77.8	82.5	75.2	80.6	84.0	80.0	77.0	78.7	78.1	78.7	81.9
1988	88.5	75.0	77.2	78.3	83.9	76.3	81.1	73.6	79.8	82.5	78.4	75.5	77.1	76.5	77.4	80.4
1987	85.7	75.9	76.4	76.0	81.3	75.6	79.9	72.2	77.7	81.3	77.3	74.6	75.7	75.2	77.4	79.1
1986	83.7	72.5	74.1	75.3	80.5	73.5	78.9	70.7	76.7	80.3	75.4	72.4	74.3	72.8	74.5	77.3
1985	81.8	71.1	72.5	74.5	79.3	72.3	77.6	69.4	75.1	79.6	74.0	71.2	73.9	71.7	73.3	75.2
1984	80.6	70.5	71.6	72.8	78.7	71.6	79.8	68.3	77.1	80.7	72.7	69.8	73.1	71.3	72.4	74.9
1983	78.2	68.3	70.6	71.0	75.0	70.2	77.8	66.3	74.9	79.8	71.0	69.0	71.8	69.4	70.1	76.1
1982	72.1	62.7	65.5	64.9	65.7	64.7	70.7	62.6	68.2	72.1	65.6	62.5	66.0	64.3	64.8	69.7
1981	66.1	57.9	60.1	59.3	62.4	59.6	63.1	58.6	61.0	65.1	61.0	58.3	60.5	59.2	60.0	65.3
1980	60.7	53.4	55.2	54.5	57.6	54.5	57.9	53.1	57.0	59.4	55.6	57.2	55.0	54.9	55.4	62.2
1979	54.9	48.6	49.8	49.0	52.3	48.7	51.5	48.5	51.4	53.3	50.3	51.1	49.2	49.7	49.4	53.3
1978	51.3	44.7	45.8	46.2	48.0	44.9	48.2	45.3	47.5	50.2	46.5	43.9	46.0	46.4	47.1	49.6
1977	47.9	42.9	42.5	43.1	43.9	42.2	44.9	41.7	44.7	47.6	42.9	41.5	42.5	42.1	41.1	45.4
1975	43.7	37.6	39.0	39.0	39.6	38.1	40.7	38.0	40.4	41.2	38.9	37.9	39.0	38.6	38.0	40.0
1970	27.8	24.5	24.9	24.9	25.7	24.5	25.5	23.7	25.9	25.4	25.1	24.6	23.3	24.8	24.5	26.8
1965	21.5	18.9	19.2	19.2	19.9	18.9	19.9	19.0	19.9	20.0	19.4	19.0	18.5	19.2	18.9	20.6
1960	19.5	17.1	17.4	17.4	18.1	17.1	18.2	17.0	18.1	18.2	17.6	17.3	16.8	17.4	17.2	18.8
1955	16.3	14.4	14.6	14.6	15.1	14.4	15.3	14.3	15.2	15.2	14.8	14.5	14.1	14.6	14.4	15.7
1950	13.5	11.9	12.1	12.1	12.5	11.9	12.6	11.8	12.5	12.6	12.2	12.0	11.6	12.1	11.9	13.0
1945	8.6	7.6	7.7	7.7	8.0	7.6	8.0	7.5	8.0	8.0	7.8	7.6	7.4	7.7	7.6	8.3
1940	6.6	5.9	5.9	5.9	6.1	5.8	6.2	5.8	6.2	6.2	6.0	5.9	5.7	5.9	5.8	6.4

Historical Cost Indexes

Year	National 30 City Average	Utah Salt Lake City	Vermont Burlington	Vermont Rutland	Virginia Alexandria	Virginia Newport News	Virginia Norfolk	Virginia Richmond	Virginia Roanoke	Washington Seattle	Washington Spokane	Washington Tacoma	West Virginia Charleston	West Virginia Huntington	Wisconsin Green Bay	Wisconsin Kenosha
Jan 2000	119.6E	105.9E	99.3E	98.8E	107.9E	96.9E	98.0E	100.6E	91.0E	125.2E	118.1E	123.5E	111.9E	114.7E	114.7E	119.1E
1999	116.6	104.5	98.2	97.7	106.1	95.6	96.5	98.8	89.8	123.3	116.7	121.6	110.6	113.4	112.1	115.8
1998	113.6	99.5	97.8	97.3	104.1	93.7	93.9	97.0	88.3	119.4	114.3	118.3	106.7	109.0	109.5	112.9
1997	111.5	97.2	96.6	96.3	101.2	91.6	91.7	92.9	86.9	118.1	111.7	117.2	105.3	107.7	105.6	109.1
1996	108.9	94.9	95.1	94.8	99.7	90.2	90.4	91.6	85.5	115.2	109.2	114.3	103.1	104.8	103.8	106.4
1995	105.6	93.1	91.1	90.8	96.3	86.0	86.4	87.8	82.8	113.7	107.4	112.8	95.8	97.2	97.6	97.9
1994	103.0	90.2	89.5	89.3	93.9	84.6	84.8	86.3	81.4	109.9	104.0	108.3	94.3	95.3	96.3	96.2
1993	100.0	87.9	87.6	87.6	91.6	82.9	83.0	84.3	79.5	107.3	103.9	106.7	92.6	93.5	94.0	94.3
1992	97.9	86.0	86.1	86.1	90.1	81.0	81.6	82.0	78.3	105.1	101.4	103.7	91.4	92.3	92.0	92.1
1991	95.7	84.9	84.2	84.2	88.2	77.6	77.9	79.8	77.3	102.2	100.0	102.2	89.7	88.6	88.6	89.8
1990	93.2	84.3	83.0	82.9	86.1	76.3	76.7	77.6	76.1	100.1	98.5	100.5	86.1	86.8	86.7	87.8
1989	91.0	82.8	81.4	81.3	83.4	74.9	75.2	76.0	74.2	96.1	96.8	98.4	84.5	84.8	84.3	85.6
1988	88.5	81.3	79.7	79.7	81.0	73.4	73.7	74.1	72.3	94.2	95.0	96.6	82.8	83.1	82.0	83.0
1987	85.7	79.8	79.0	79.0	77.9	71.0	71.7	72.7	70.2	91.9	92.4	92.9	81.1	81.3	80.2	81.7
1986	83.7	78.1	76.4	76.4	76.8	70.3	70.5	71.0	67.8	90.5	91.9	92.8	79.9	80.2	77.8	79.3
1985	81.8	75.9	74.8	74.9	75.1	68.7	68.8	69.5	67.2	88.3	89.0	91.2	77.7	77.7	76.7	77.4
1984	80.6	75.2	73.7	73.7	75.0	67.9	68.3	68.8	66.2	89.2	88.1	90.0	75.4	75.8	75.0	75.1
1983	78.2	74.5	70.9	71.5	72.9	65.9	66.5	68.9	66.0	87.6	86.9	88.8	73.8	74.4	73.4	73.2
1982	72.1	68.1	64.9	65.4	67.0	60.6	60.9	63.7	61.0	80.7	80.6	81.7	67.6	68.4	67.5	69.1
1981	66.1	61.9	59.3	59.7	61.6	56.0	56.8	58.2	55.6	75.7	72.9	73.8	62.6	63.3	63.6	63.5
1980	60.7	57.0	55.3	58.3	57.3	52.5	52.4	54.3	51.3	67.9	66.3	66.7	57.7	58.3	58.6	58.3
1979	54.9	52.5	49.8	51.2	51.7	47.6	47.6	49.1	46.8	61.1	60.1	60.0	51.8	52.5	52.7	53.2
1978	51.3	49.0	46.1	47.2	48.9	44.0	44.9	46.0	43.6	56.7	55.4	55.6	48.2	48.5	48.7	49.5
1977	47.9	45.0	44.5	46.3	44.6	41.2	41.5	42.5	41.2	51.8	51.1	50.7	44.7	43.2	44.8	45.5
1975	43.7	40.1	41.8	43.9	41.7	37.2	36.9	37.1	37.1	44.9	44.4	44.5	41.0	40.0	40.9	40.5
1970	27.8	26.1	25.4	26.8	26.2	23.9	21.5	22.0	23.7	28.8	29.3	29.6	26.1	25.8	26.4	26.5
1965	21.5	20.0	19.8	20.6	20.2	18.4	17.1	17.2	18.3	22.4	22.5	22.8	20.1	19.9	20.3	20.4
1960	19.5	18.4	18.0	18.8	18.4	16.7	15.4	15.6	16.6	20.4	20.8	20.8	18.3	18.1	18.4	18.6
1955	16.3	15.4	15.1	15.7	15.4	14.0	12.9	13.1	13.9	17.1	17.4	17.4	15.4	15.2	15.5	15.6
1950	13.5	12.7	12.4	13.0	12.7	11.6	10.7	10.8	11.5	14.1	14.4	14.4	12.7	12.5	12.8	12.9
1945	8.6	8.1	7.9	8.3	8.1	7.4	6.8	6.9	7.3	9.0	9.2	9.2	8.1	8.0	8.1	8.2
1940	6.6	6.3	6.1	6.4	6.2	5.7	5.3	5.3	5.7	7.0	7.1	7.1	6.2	6.2	6.3	6.3

Year	National 30 City Average	Wisconsin Madison	Wisconsin Milwaukee	Wisconsin Racine	Wyoming Cheyenne	Canada Calgary	Canada Edmonton	Canada Hamilton	Canada London	Canada Montreal	Canada Ottawa	Canada Quebec	Canada Toronto	Canada Vancouver	Canada Winnipeg
Jan 2000	119.6E	117.0E	119.7E	118.8E	98.3E	116.3E	116.2E	130.1E	127.6E	123.0E	128.9E	124.3E	133.5E	128.7E	116.2E
1999	116.6	115.9	117.4	115.5	96.9	115.3	115.2	128.1	125.6	120.8	126.8	121.9	131.2	127.1	115.2
1998	113.6	110.8	113.4	112.7	95.4	112.5	112.4	126.3	123.9	119.0	124.7	119.6	128.5	123.8	113.7
1997	111.5	106.1	110.1	109.3	93.2	110.7	110.6	124.4	121.9	114.6	122.8	115.4	125.7	121.9	111.4
1996	108.9	104.4	107.1	106.5	91.1	109.1	109.0	122.6	120.2	112.9	121.0	113.6	123.9	119.0	109.7
1995	105.6	96.5	103.9	97.8	87.6	107.4	107.4	119.9	117.5	110.8	118.2	111.5	121.6	116.2	107.6
1994	103.0	94.5	100.6	96.1	85.4	106.7	106.6	116.5	114.2	109.5	115.0	110.2	117.8	115.1	105.5
1993	100.0	91.3	96.7	93.8	82.9	104.7	104.5	113.7	111.9	106.9	112.0	107.0	114.9	109.2	102.8
1992	97.9	89.2	93.9	91.7	81.7	103.4	103.2	112.4	110.7	104.2	110.8	103.6	113.7	108.0	101.6
1991	95.7	86.2	91.6	89.3	80.3	102.1	102.0	108.2	106.7	101.8	106.9	100.4	109.0	106.8	98.6
1990	93.2	84.3	88.9	87.3	79.1	98.0	97.1	103.8	101.4	99.0	102.7	96.8	104.6	103.2	95.1
1989	91.0	81.8	86.4	85.0	77.8	95.6	94.7	98.2	96.5	94.9	98.5	92.5	98.5	97.6	92.9
1988	88.5	79.9	84.1	82.4	76.3	93.9	93.0	95.4	93.2	90.6	94.0	88.3	94.8	95.9	90.0
1987	85.7	78.1	81.1	81.1	76.7	90.2	89.3	89.7	90.1	87.2	89.6	85.0	90.4	94.2	87.1
1986	83.7	76.1	78.9	78.6	73.4	91.2	90.1	88.9	87.9	84.8	87.7	82.4	89.3	93.2	85.3
1985	81.8	74.3	77.4	77.0	72.3	90.2	89.1	85.8	84.9	82.4	83.9	79.9	86.5	89.8	83.4
1984	80.6	72.5	76.3	74.7	73.8	89.6	88.1	85.1	84.1	81.6	82.9	80.0	85.2	89.2	82.5
1983	78.2	71.9	74.0	73.0	73.8	84.6	82.8	80.1	79.1	76.4	77.6	75.8	79.9	83.3	77.9
1982	72.1	65.9	70.5	68.6	68.3	76.2	74.1	73.7	71.8	70.2	72.3	69.6	71.8	76.4	70.7
1981	66.1	61.4	64.9	63.3	62.7	70.6	68.1	68.8	67.1	64.6	65.8	64.3	67.5	70.3	65.4
1980	60.7	56.8	58.8	58.1	56.9	64.9	63.3	63.7	61.9	59.2	60.9	59.3	60.9	65.0	61.7
1979	54.9	51.4	53.0	52.8	51.2	58.9	58.5	58.1	56.6	54.4	55.5	53.9	56.0	59.2	56.1
1978	51.3	47.3	49.9	49.2	47.5	54.9	54.4	54.0	52.6	50.4	51.7	49.7	51.9	55.1	51.7
1977	47.9	44.0	46.9	45.7	44.8	49.9	49.9	48.8	48.2	43.7	48.0	43.0	48.2	50.2	47.8
1975	43.7	40.7	43.3	40.7	40.6	42.2	41.6	42.9	41.6	39.7	41.5	39.0	42.2	42.4	39.2
1970	27.8	26.5	29.4	26.5	26.0	28.9	28.6	28.5	27.8	25.6	27.6	26.0	25.6	26.0	23.1
1965	21.5	20.6	21.8	20.4	20.0	22.3	22.0	22.0	21.4	18.7	21.2	20.1	19.4	20.5	17.5
1960	19.5	18.1	19.0	18.6	18.2	20.2	20.0	20.0	19.5	17.0	19.3	18.2	17.6	18.6	15.8
1955	16.3	15.2	15.9	15.6	15.2	17.0	16.8	16.7	16.3	14.3	16.2	15.3	14.8	15.5	13.3
1950	13.5	12.5	13.2	12.9	12.6	14.0	13.9	13.8	13.5	11.8	13.4	12.6	12.2	12.8	10.9
1945	8.6	8.0	8.4	8.2	8.0	9.0	8.8	8.8	8.6	7.5	8.5	8.0	7.8	8.2	7.0
1940	6.6	6.2	6.5	6.3	6.2	6.9	6.8	6.8	6.6	5.8	6.6	6.2	6.0	6.3	5.4

Location Factors

Costs shown in *Means cost data publications* are based on National Averages for materials and installation. To adjust these costs to a specific location, simply multiply the base cost by the factor for that city. The data is arranged alphabetically by state and postal zip code numbers. For a city not listed, use the factor for a nearby city with similar economic characteristics.

STATE/ZIP	CITY	Residential	Commercial
ALABAMA			
350-352	Birmingham	.86	.87
354	Tuscaloosa	.82	.80
355	Jasper	.75	.76
356	Decatur	.81	.82
357-358	Huntsville	.83	.84
359	Gadsden	.81	.82
360-361	Montgomery	.81	.79
362	Anniston	.74	.75
363	Dothan	.81	.79
364	Evergreen	.81	.79
365-366	Mobile	.83	.84
367	Selma	.80	.78
368	Phenix City	.81	.79
369	Butler	.80	.78
ALASKA			
995-996	Anchorage	1.26	1.25
997	Fairbanks	1.26	1.25
998	Juneau	1.26	1.25
999	Ketchikan	1.31	1.30
ARIZONA			
850,853	Phoenix	.93	.90
852	Mesa/Tempe	.88	.85
855	Globe	.90	.87
856-857	Tucson	.91	.88
859	Show Low	.92	.88
860	Flagstaff	.94	.90
863	Prescott	.91	.87
864	Kingman	.91	.87
865	Chambers	.90	.86
ARKANSAS			
716	Pine Bluff	.80	.80
717	Camden	.71	.71
718	Texarkana	.75	.74
719	Hot Springs	.70	.70
720-722	Little Rock	.81	.81
723	West Memphis	.80	.80
724	Jonesboro	.80	.80
725	Batesville	.77	.77
726	Harrison	.77	.77
727	Fayetteville	.70	.67
728	Russellville	.79	.76
729	Fort Smith	.83	.80
CALIFORNIA			
900-902	Los Angeles	1.10	1.10
903-905	Inglewood	1.07	1.07
906-908	Long Beach	1.08	1.08
910-912	Pasadena	1.07	1.07
913-916	Van Nuys	1.09	1.09
917-918	Alhambra	1.09	1.09
919-921	San Diego	1.11	1.07
922	Palm Springs	1.10	1.07
923-924	San Bernardino	1.10	1.06
925	Riverside	1.12	1.08
926-927	Santa Ana	1.10	1.07
928	Anaheim	1.11	1.09
930	Oxnard	1.14	1.09
931	Santa Barbara	1.11	1.08
932-933	Bakersfield	1.11	1.06
934	San Luis Obispo	1.22	1.09
935	Mojave	1.10	1.06
936-938	Fresno	1.12	1.08
939	Salinas	1.12	1.12
940-941	San Francisco	1.21	1.24
942,956-958	Sacramento	1.11	1.10
943	Palo Alto	1.14	1.17
944	San Mateo	1.16	1.18
945	Vallejo	1.13	1.15
946	Oakland	1.16	1.19
947	Berkeley	1.16	1.18
948	Richmond	1.14	1.17
949	San Rafael	1.24	1.18
950	Santa Cruz	1.16	1.14
951	San Jose	1.21	1.19
952	Stockton	1.14	1.10
953	Modesto	1.14	1.10
CALIFORNIA (CONT'D)			
954	Santa Rosa	1.14	1.17
955	Eureka	1.10	1.09
959	Marysville	1.11	1.10
960	Redding	1.11	1.10
961	Susanville	1.11	1.10
COLORADO			
800-802	Denver	.98	.94
803	Boulder	.89	.85
804	Golden	.96	.92
805	Fort Collins	.98	.92
806	Greeley	.91	.85
807	Fort Morgan	.97	.91
808-809	Colorado Springs	.94	.91
810	Pueblo	.94	.92
811	Alamosa	.90	.88
812	Salida	.90	.88
813	Durango	.88	.86
814	Montrose	.86	.84
815	Grand Junction	.91	.86
816	Glenwood Springs	.96	.91
CONNECTICUT			
060	New Britain	1.03	1.04
061	Hartford	1.03	1.04
062	Willimantic	1.03	1.04
063	New London	1.03	1.02
064	Meriden	1.03	1.04
065	New Haven	1.03	1.04
066	Bridgeport	1.01	1.04
067	Waterbury	1.04	1.04
068	Norwalk	.99	1.03
069	Stamford	1.03	1.07
D.C.			
200-205	Washington	.94	.96
DELAWARE			
197	Newark	.99	1.00
198	Wilmington	.98	.99
199	Dover	.99	1.00
FLORIDA			
320,322	Jacksonville	.85	.84
321	Daytona Beach	.88	.87
323	Tallahassee	.77	.79
324	Panama City	.71	.73
325	Pensacola	.86	.84
326,344	Gainesville	.85	.82
327-328,347	Orlando	.88	.86
329	Melbourne	.89	.88
330-332,340	Miami	.84	.86
333	Fort Lauderdale	.85	.87
334,349	West Palm Beach	.86	.83
335-336,346	Tampa	.82	.84
337	St. Petersburg	.82	.84
338	Lakeland	.81	.83
339,341	Fort Myers	.81	.81
342	Sarasota	.80	.82
GEORGIA			
300-303,399	Atlanta	.84	.89
304	Statesboro	.66	.68
305	Gainesville	.70	.74
306	Athens	.77	.81
307	Dalton	.69	.68
308-309	Augusta	.78	.79
310-312	Macon	.82	.82
313-314	Savannah	.81	.82
315	Waycross	.75	.75
316	Valdosta	.77	.77
317	Albany	.78	.80
318-319	Columbus	.79	.79
HAWAII			
967	Hilo	1.26	1.22
968	Honolulu	1.26	1.22

Location Factors

STATE/ZIP	CITY	Residential	Commercial
STATES & POSS.			
969	Guam	.85	.82
IDAHO			
832	Pocatello	.95	.94
833	Twin Falls	.81	.80
834	Idaho Falls	.85	.84
835	Lewiston	1.11	1.03
836-837	Boise	.95	.94
838	Coeur d'Alene	.98	.91
ILLINOIS			
600-603	North Suburban	1.08	1.07
604	Joliet	1.06	1.05
605	South Suburban	1.08	1.07
606	Chicago	1.12	1.11
609	Kankakee	.99	.99
610-611	Rockford	1.04	1.03
612	Rock Island	1.05	.97
613	La Salle	1.06	.98
614	Galesburg	1.05	.98
615-616	Peoria	1.07	1.01
617	Bloomington	1.04	.99
618-619	Champaign	1.03	1.00
620-622	East St. Louis	.99	.99
623	Quincy	.97	.95
624	Effingham	1.00	.97
625	Decatur	1.00	.97
626-627	Springfield	1.01	.98
628	Centralia	.98	.98
629	Carbondale	.96	.96
INDIANA			
460	Anderson	.94	.93
461-462	Indianapolis	.96	.95
463-464	Gary	1.02	1.00
465-466	South Bend	.92	.90
467-468	Fort Wayne	.91	.92
469	Kokomo	.91	.90
470	Lawrenceburg	.91	.88
471	New Albany	.92	.88
472	Columbus	.95	.92
473	Muncie	.93	.92
474	Bloomington	.95	.92
475	Washington	.93	.93
476-477	Evansville	.94	.94
478	Terre Haute	.95	.94
479	Lafayette	.90	.90
IOWA			
500-503,509	Des Moines	.96	.92
504	Mason City	.86	.80
505	Fort Dodge	.83	.78
506-507	Waterloo	.88	.83
508	Creston	.90	.85
510-511	Sioux City	.89	.83
512	Sibley	.80	.78
513	Spencer	.81	.79
514	Carroll	.85	.81
515	Council Bluffs	.94	.88
516	Shenandoah	.82	.77
520	Dubuque	.97	.87
521	Decorah	.90	.80
522-524	Cedar Rapids	1.00	.92
525	Ottumwa	.94	.86
526	Burlington	.92	.86
527-528	Davenport	.96	.94
KANSAS			
660-662	Kansas City	.96	.95
664-666	Topeka	.86	.85
667	Fort Scott	.86	.84
668	Emporia	.82	.81
669	Belleville	.88	.82
670-672	Wichita	.88	.85
673	Independence	.82	.80
674	Salina	.85	.81
675	Hutchinson	.80	.77
676	Hays	.86	.82
677	Colby	.86	.82
678	Dodge City	.86	.83
679	Liberal	.79	.76
KENTUCKY			
400-402	Louisville	.94	.91
403-405	Lexington	.88	.85

STATE/ZIP	CITY	Residential	Commercial
KENTUCKY (CONT'D)			
406	Frankfort	.94	.87
407-409	Corbin	.79	.74
410	Covington	.96	.93
411-412	Ashland	.95	.96
413-414	Campton	.78	.74
415-416	Pikeville	.82	.83
417-418	Hazard	.77	.74
420	Paducah	.95	.90
421-422	Bowling Green	.94	.89
423	Owensboro	.91	.89
424	Henderson	.94	.92
425-426	Somerset	.76	.73
427	Elizabethtown	.92	.88
LOUISIANA			
700-701	New Orleans	.86	.85
703	Thibodaux	.85	.85
704	Hammond	.83	.82
705	Lafayette	.85	.82
706	Lake Charles	.83	.83
707-708	Baton Rouge	.83	.82
710-711	Shreveport	.81	.81
712	Monroe	.79	.79
713-714	Alexandria	.78	.78
MAINE			
039	Kittery	.78	.79
040-041	Portland	.87	.89
042	Lewiston	.88	.89
043	Augusta	.78	.78
044	Bangor	.92	.92
045	Bath	.78	.78
046	Machias	.83	.83
047	Houlton	.80	.80
048	Rockland	.83	.83
049	Waterville	.79	.78
MARYLAND			
206	Waldorf	.89	.89
207-208	College Park	.90	.90
209	Silver Spring	.90	.90
210-212	Baltimore	.91	.91
214	Annapolis	.88	.89
215	Cumberland	.88	.89
216	Easton	.74	.75
217	Hagerstown	.90	.88
218	Salisbury	.78	.79
219	Elkton	.84	.85
MASSACHUSETTS			
010-011	Springfield	1.05	1.03
012	Pittsfield	.99	.99
013	Greenfield	1.02	1.00
014	Fitchburg	1.09	1.05
015-016	Worcester	1.11	1.07
017	Framingham	1.07	1.08
018	Lowell	1.10	1.10
019	Lawrence	1.10	1.10
020-022, 024	Boston	1.16	1.17
023	Brockton	1.07	1.09
025	Buzzards Bay	1.04	1.06
026	Hyannis	1.06	1.07
027	New Bedford	1.07	1.08
MICHIGAN			
480,483	Royal Oak	1.03	1.02
481	Ann Arbor	1.05	1.04
482	Detroit	1.06	1.05
484-485	Flint	.98	.99
486	Saginaw	.96	.97
487	Bay City	.96	.97
488-489	Lansing	1.02	.99
490	Battle Creek	.99	.93
491	Kalamazoo	1.00	.94
492	Jackson	1.00	.97
493,495	Grand Rapids	.90	.87
494	Muskegan	.96	.93
496	Traverse City	.89	.86
497	Gaylord	.89	.90
498-499	Iron mountain	.98	.95
MINNESOTA			
550-551	Saint Paul	1.11	1.09
553-555	Minneapolis	1.14	1.10

Location Factors

STATE/ZIP	CITY	Residential	Commercial
556-558	Duluth	1.03	1.05
559	Rochester	1.04	1.01
560	Mankato	1.01	1.00
561	Windom	.90	.89
562	Willmar	.93	.92
563	St. Cloud	1.09	1.02
564	Brainerd	1.06	.99
565	Detroit Lakes	.88	.95
566	Bemidji	.91	.98
567	Thief River Falls	.87	.94
MISSISSIPPI			
386	Clarksdale	.71	.67
387	Greenville	.82	.78
388	Tupelo	.72	.73
389	Greenwood	.73	.69
390-392	Jackson	.82	.78
393	Meridian	.77	.76
394	Laurel	.73	.70
395	Biloxi	.86	.82
396	Mccomb	.69	.68
397	Columbus	.72	.73
MISSOURI			
630-631	St. Louis	1.00	1.03
633	Bowling Green	.92	.94
634	Hannibal	1.00	.93
635	Kirksville	.85	.89
636	Flat River	.94	.97
637	Cape Girardeau	.93	.96
638	Sikeston	.89	.91
639	Poplar Bluff	.89	.91
640-641	Kansas City	1.01	.98
644-645	St. Joseph	.86	.90
646	Chillicothe	.80	.83
647	Harrisonville	.93	.91
648	Joplin	.83	.85
650-651	Jefferson City	.96	.90
652	Columbia	.94	.88
653	Sedalia	.94	.88
654-655	Rolla	.87	.82
656-658	Springfield	.84	.86
MONTANA			
590-591	Billings	.98	.96
592	Wolf Point	.98	.96
593	Miles City	.97	.95
594	Great Falls	.97	.96
595	Havre	.96	.95
596	Helena	.96	.95
597	Butte	.94	.93
598	Missoula	.95	.94
599	Kalispell	.94	.93
NEBRASKA			
680-681	Omaha	.91	.90
683-685	Lincoln	.88	.83
686	Columbus	.75	.74
687	Norfolk	.85	.84
688	Grand Island	.88	.84
689	Hastings	.83	.79
690	Mccook	.77	.74
691	North Platte	.84	.80
692	Valentine	.78	.75
693	Alliance	.76	.72
NEVADA			
889-891	Las Vegas	1.06	1.05
893	Ely	.96	.97
894-895	Reno	.95	1.00
897	Carson City	.97	.99
898	Elko	.93	.96
NEW HAMPSHIRE			
030	Nashua	.93	.94
031	Manchester	.93	.94
032-033	Concord	.92	.93
034	Keene	.79	.80
035	Littleton	.81	.82
036	Charleston	.77	.78
037	Claremont	.77	.78
038	Portsmouth	.92	.91

STATE/ZIP	CITY	Residential	Commercial
NEW JERSEY			
070-071	Newark	1.14	1.12
072	Elizabeth	1.09	1.07
073	Jersey City	1.12	1.11
074-075	Paterson	1.12	1.12
076	Hackensack	1.10	1.10
077	Long Branch	1.11	1.09
078	Dover	1.12	1.10
079	Summit	1.08	1.06
080,083	Vineland	1.11	1.07
081	Camden	1.12	1.09
082,084	Atlantic City	1.11	1.09
085-086	Trenton	1.13	1.11
087	Point Pleasant	1.11	1.09
088-089	New Brunswick	1.12	1.10
NEW MEXICO			
870-872	Albuquerque	.89	.91
873	Gallup	.89	.91
874	Farmington	.89	.91
875	Santa Fe	.89	.91
877	Las Vegas	.88	.90
878	Socorro	.87	.89
879	Truth/Consequences	.87	.87
880	Las Cruces	.85	.85
881	Clovis	.90	.90
882	Roswell	.90	.90
883	Carrizozo	.91	.91
884	Tucumcari	.90	.90
NEW YORK			
100-102	New York	1.34	1.34
103	Staten Island	1.28	1.28
104	Bronx	1.28	1.28
105	Mount Vernon	1.18	1.18
106	White Plains	1.17	1.17
107	Yonkers	1.20	1.20
108	New Rochelle	1.18	1.18
109	Suffern	1.12	1.12
110	Queens	1.29	1.29
111	Long Island City	1.30	1.30
112	Brooklyn	1.30	1.30
113	Flushing	1.30	1.30
114	Jamaica	1.29	1.29
115,117,118	Hicksville	1.27	1.27
116	Far Rockaway	1.30	1.30
119	Riverhead	1.27	1.27
120-122	Albany	.98	.98
123	Schenectady	.98	.98
124	Kingston	1.11	1.09
125-126	Poughkeepsie	1.13	1.12
127	Monticello	1.09	1.08
128	Glens Falls	.95	.93
129	Plattsburgh	.96	.94
130-132	Syracuse	.99	.97
133-135	Utica	.91	.94
136	Watertown	.94	.97
137-139	Binghamton	.94	.94
140-142	Buffalo	1.07	1.03
143	Niagara Falls	1.06	1.02
144-146	Rochester	.99	1.00
147	Jamestown	.97	.94
148-149	Elmira	.95	.93
NORTH CAROLINA			
270,272-274	Greensboro	.76	.77
271	Winston-Salem	.76	.77
275-276	Raleigh	.77	.77
277	Durham	.76	.77
278	Rocky Mount	.69	.69
279	Elizabeth City	.71	.71
280	Gastonia	.75	.76
281-282	Charlotte	.75	.76
283	Fayetteville	.76	.76
284	Wilmington	.74	.76
285	Kinston	.68	.68
286	Hickory	.67	.68
287-288	Asheville	.74	.76
289	Murphy	.67	.68
NORTH DAKOTA			
580-581	Fargo	.78	.83
582	Grand Forks	.78	.82
583	Devils Lake	.78	.82
584	Jamestown	.78	.82
585	Bismarck	.81	.85

Location Factors

STATE/ZIP	CITY	Residential	Commercial
NORTH DAKOTA (CONT'D)			
586	Dickinson	.79	.83
587	Minot	.81	.85
588	Williston	.78	.82
OHIO			
430-432	Columbus	.96	.94
433	Marion	.91	.92
434-436	Toledo	.99	.98
437-438	Zanesville	.91	.90
439	Steubenville	.97	.97
440	Lorain	1.03	.97
441	Cleveland	1.09	1.02
442-443	Akron	1.00	.99
444-445	Youngstown	.99	.96
446-447	Canton	.96	.95
448-449	Mansfield	.95	.94
450	Hamilton	.98	.92
451-452	Cincinnati	.99	.93
453-454	Dayton	.93	.92
455	Springfield	.94	.92
456	Chillicothe	1.01	.95
457	Athens	.88	.87
458	Lima	.95	.94
OKLAHOMA			
730-731	Oklahoma City	.82	.84
734	Ardmore	.83	.82
735	Lawton	.84	.83
736	Clinton	.80	.82
737	Enid	.83	.82
738	Woodward	.82	.81
739	Guymon	.70	.69
740-741	Tulsa	.85	.82
743	Miami	.85	.83
744	Muskogee	.76	.74
745	Mcalester	.76	.78
746	Ponca City	.82	.81
747	Durant	.79	.81
748	Shawnee	.79	.81
749	Poteau	.85	.81
OREGON			
970-972	Portland	1.08	1.06
973	Salem	1.07	1.06
974	Eugene	1.06	1.05
975	Medford	1.06	1.05
976	Klamath Falls	1.07	1.06
977	Bend	1.07	1.06
978	Pendleton	1.04	1.02
979	Vale	.99	.97
PENNSYLVANIA			
150-152	Pittsburgh	1.04	1.02
153	Washington	1.03	1.01
154	Uniontown	1.02	1.00
155	Bedford	1.04	.97
156	Greensburg	1.03	1.01
157	Indiana	1.06	.99
158	Dubois	1.04	.97
159	Johnstown	1.05	.98
160	Butler	1.00	.98
161	New Castle	1.00	.97
162	Kittanning	1.03	1.00
163	Oil City	.90	.95
164-165	Erie	.98	.97
166	Altoona	1.05	.96
167	Bradford	.99	.97
168	State College	.98	.98
169	Wellsboro	.94	.95
170-171	Harrisburg	.97	.96
172	Chambersburg	.96	.95
173-174	York	.97	.95
175-176	Lancaster	.96	.94
177	Williamsport	.92	.92
178	Sunbury	.96	.95
179	Pottsville	.96	.95
180	Lehigh Valley	1.02	1.01
181	Allentown	1.02	1.01
182	Hazleton	.97	.96
183	Stroudsburg	.96	.95
184-185	Scranton	.97	1.00
186-187	Wilkes-Barre	.94	.97
188	Montrose	.94	.97
189	Doylestown	.94	1.05

STATE/ZIP	CITY	Residential	Commercial
PENNSYLVANIA (CONT'D)			
190-191	Philadelphia	1.14	1.12
193	Westchester	1.07	1.05
194	Norristown	1.09	1.07
195-196	Reading	.97	.98
RHODE ISLAND			
028	Newport	1.02	1.04
029	Providence	1.02	1.04
SOUTH CAROLINA			
290-292	Columbia	.72	.75
293	Spartanburg	.72	.75
294	Charleston	.74	.76
295	Florence	.71	.73
296	Greenville	.72	.75
297	Rock Hill	.65	.68
298	Aiken	.66	.68
299	Beaufort	.69	.71
SOUTH DAKOTA			
570-571	Sioux Falls	.88	.82
572	Watertown	.85	.79
573	Mitchell	.84	.79
574	Aberdeen	.86	.80
575	Pierre	.86	.80
576	Mobridge	.86	.79
577	Rapid City	.85	.79
TENNESSEE			
370-372	Nashville	.84	.84
373-374	Chattanooga	.83	.82
375,380-381	Memphis	.85	.85
376	Johnson City	.80	.79
377-379	Knoxville	.80	.80
382	Mckenzie	.69	.69
383	Jackson	.68	.75
384	Columbia	.76	.76
385	Cookeville	.69	.69
TEXAS			
750	Mckinney	.90	.83
751	Waxahackie	.83	.83
752-753	Dallas	.91	.87
754	Greenville	.80	.75
755	Texarkana	.90	.79
756	Longview	.85	.75
757	Tyler	.92	.81
758	Palestine	.75	.75
759	Lufkin	.79	.79
760-761	Fort Worth	.85	.84
762	Denton	.88	.80
763	Wichita Falls	.82	.82
764	Eastland	.76	.75
765	Temple	.78	.77
766-767	Waco	.82	.81
768	Brownwood	.74	.73
769	San Angelo	.80	.77
770-772	Houston	.88	.89
773	Huntsville	.75	.75
774	Wharton	.77	.78
775	Galveston	.86	.87
776-777	Beaumont	.83	.85
778	Bryan	.82	.83
779	Victoria	.81	.81
780	Laredo	.79	.79
781-782	San Antonio	.83	.84
783-784	Corpus Christi	.81	.80
785	Mc Allen	.80	.78
786-787	Austin	.79	.82
788	Del Rio	.70	.70
789	Giddings	.75	.74
790-791	Amarillo	.81	.81
792	Childress	.76	.79
793-794	Lubbock	.80	.82
795-796	Abilene	.78	.78
797	Midland	.80	.81
798-799,885	El Paso	.78	.77
UTAH			
840-841	Salt Lake City	.90	.89
842,844	Ogden	.90	.88
843	Logan	.91	.89
845	Price	.82	.81
846-847	Provo	.90	.89

Location Factors

STATE/ZIP	CITY	Residential	Commercial
VERMONT			
050	White River Jct.	.73	.72
051	Bellows Falls	.74	.73
052	Bennington	.70	.69
053	Brattleboro	.74	.73
054	Burlington	.83	.84
056	Montpelier	.81	.82
057	Rutland	.84	.83
058	St. Johnsbury	.74	.75
059	Guildhall	.73	.74
VIRGINIA			
220-221	Fairfax	.89	.90
222	Arlington	.90	.90
223	Alexandria	.90	.91
224-225	Fredericksburg	.85	.85
226	Winchester	.79	.80
227	Culpeper	.80	.80
228	Harrisonburg	.76	.77
229	Charlottesville	.84	.82
230-232	Richmond	.87	.85
233-235	Norfolk	.82	.82
236	Newport News	.83	.82
237	Portsmouth	.81	.81
238	Petersburg	.86	.84
239	Farmville	.75	.74
240-241	Roanoke	.78	.77
242	Bristol	.80	.75
243	Pulaski	.71	.70
244	Staunton	.78	.76
245	Lynchburg	.81	.77
246	Grundy	.70	.70
WASHINGTON			
980-981,987	Seattle	1.01	1.06
982	Everett	.98	1.04
983-984	Tacoma	1.06	1.04
985	Olympia	1.06	1.04
986	Vancouver	1.11	1.05
988	Wenatchee	.94	.98
989	Yakima	1.04	1.02
990-992	Spokane	1.01	1.00
993	Richland	1.02	1.01
994	Clarkston	1.01	1.00
WEST VIRGINIA			
247-248	Bluefield	.88	.88
249	Lewisburg	.90	.90
250-253	Charleston	.94	.94
254	Martinsburg	.77	.77
255-257	Huntington	.95	.97
258-259	Beckley	.92	.92
260	Wheeling	.94	.96
261	Parkersburg	.94	.96
262	Buckhannon	.97	.94
263-264	Clarksburg	.98	.95
265	Morgantown	.98	.95
266	Gassaway	.94	.94
267	Romney	.91	.91
268	Petersburg	.95	.92
WISCONSIN			
530,532	Milwaukee	1.02	1.01
531	Kenosha	1.02	1.01
534	Racine	1.05	1.00
535	Beloit	1.00	.98
537	Madison	1.01	.99
538	Lancaster	.93	.91
539	Portage	.99	.96
540	New Richmond	1.02	.94
541-543	Green Bay	1.00	.97
544	Wausau	.97	.93
545	Rhinelander	.97	.93
546	La Crosse	.98	.95
547	Eau Claire	1.04	.96
548	Superior	1.03	.97
549	Oshkosh	.97	.94
WYOMING			
820	Cheyenne	.88	.83
821	Yellowstone Nat. Pk.	.82	.79
822	Wheatland	.84	.80
823	Rawlins	.83	.79
824	Worland	.80	.77
825	Riverton	.84	.80
826	Casper	.87	.83

STATE/ZIP	CITY	Residential	Commercial
WYOMING (CONT'D)			
827	Newcastle	.81	.77
828	Sheridan	.85	.82
829-831	Rock Springs	.84	.80
CANADIAN FACTORS (reflect Canadian currency)			
ALBERTA			
	Calgary	1.01	.98
	Edmonton	1.01	.98
	Fort McMurray	1.00	.97
	Lethbridge	1.00	.97
	Lloydminster	1.00	.97
	Medicine Hat	1.00	.97
	Red Deer	1.00	.97
BRITISH COLUMBIA			
	Kamloops	1.04	1.05
	Prince George	1.06	1.07
	Vancouver	1.07	1.08
	Victoria	1.06	1.07
MANITOBA			
	Brandon	.99	.98
	Portage la Prairie	.99	.98
	Winnipeg	.99	.98
NEW BRUNSWICK			
	Bathurst	.94	.92
	Dalhousie	.94	.92
	Fredericton	.97	.95
	Moncton	.94	.92
	Newcastle	.94	.92
	Saint John	.97	.95
NEWFOUNDLAND			
	Corner Brook	.96	.95
	St. John's	.96	.95
NORTHWEST TERRITORIES			
	Yellowknife	.93	.92
NOVA SCOTIA			
	Dartmouth	.98	.97
	Halifax	.98	.97
	New Glasgow	.98	.97
	Sydney	.96	.95
	Yarmouth	.98	.97
ONTARIO			
	Barrie	1.10	1.08
	Brantford	1.12	1.10
	Cornwall	1.10	1.08
	Hamilton	1.14	1.10
	Kingston	1.10	1.08
	Kitchener	1.06	1.04
	London	1.10	1.08
	North Bay	1.09	1.07
	Oshawa	1.10	1.08
	Ottawa	1.11	1.09
	Owen Sound	1.10	1.08
	Peterborough	1.10	1.08
	Sarnia	1.12	1.10
	St. Catharines	1.06	1.04
	Sudbury	1.05	1.03
	Thunder Bay	1.06	1.04
	Toronto	1.13	1.12
	Windsor	1.07	1.05
PRINCE EDWARD ISLAND			
	Charlottetown	.93	.91
	Summerside	.93	.92
QUEBEC			
	Cap-de-la-Madeleine	1.05	1.04
	Charlesbourg	1.05	1.04
	Chicoutimi	1.04	1.03
	Gatineau	1.04	1.03
	Laval	1.05	1.04
	Montreal	1.11	1.04
	Quebec	1.13	1.05
	Sherbrooke	1.05	1.04
	Trois Rivieres	1.05	1.04

Location Factors

STATE/ZIP	CITY	Residential	Commercial
SASKATCHEWAN			
	Moose Jaw	.93	.93
	Prince Albert	.93	.93
	Regina	.93	.93
	Saskatoon	.93	.93
YUKON			
	Whitehorse	.93	.92

Abbreviations

A	Area Square Feet; Ampere	Cab.	Cabinet	d.f.u.	Drainage Fixture Units
ABS	Acrylonitrile Butadiene Stryrene; Asbestos Bonded Steel	Cair.	Air Tool Laborer	D.H.	Double Hung
A.C.	Alternating Current; Air-Conditioning; Asbestos Cement; Plywood Grade A & C	Calc	Calculated	DHW	Domestic Hot Water
		Cap.	Capacity	Diag.	Diagonal
		Carp.	Carpenter	Diam.	Diameter
		C.B.	Circuit Breaker	Distrib.	Distribution
A.C.I.	American Concrete Institute	C.C.A.	Chromate Copper Arsenate	Dk.	Deck
AD	Plywood, Grade A & D	C.C.F.	Hundred Cubic Feet	D.L.	Dead Load; Diesel
Addit.	Additional	cd	Candela	DLH	Deep Long Span Bar Joist
Adj.	Adjustable	cd/sf	Candela per Square Foot	Do.	Ditto
af	Audio-frequency	CD	Grade of Plywood Face & Back	Dp.	Depth
A.G.A.	American Gas Association	CDX	Plywood, Grade C & D, exterior glue	D.P.S.T.	Double Pole, Single Throw
Agg.	Aggregate			Dr.	Driver
A.H.	Ampere Hours	Cefi.	Cement Finisher	Drink.	Drinking
A hr.	Ampere-hour	Cem.	Cement	D.S.	Double Strength
A.H.U.	Air Handling Unit	CF	Hundred Feet	D.S.A.	Double Strength A Grade
A.I.A.	American Institute of Architects	C.F.	Cubic Feet	D.S.B.	Double Strength B Grade
AIC	Ampere Interrupting Capacity	CFM	Cubic Feet per Minute	Dty.	Duty
Allow.	Allowance	c.g.	Center of Gravity	DWV	Drain Waste Vent
alt.	Altitude	CHW	Chilled Water; Commercial Hot Water	DX	Deluxe White, Direct Expansion
Alum.	Aluminum			dyn	Dyne
a.m.	Ante Meridiem	C.I.	Cast Iron	e	Eccentricity
Amp.	Ampere	C.I.P.	Cast in Place	E	Equipment Only; East
Anod.	Anodized	Circ.	Circuit	Ea.	Each
Approx.	Approximate	C.L.	Carload Lot	E.B.	Encased Burial
Apt.	Apartment	Clab.	Common Laborer	Econ.	Economy
Asb.	Asbestos	C.L.F.	Hundred Linear Feet	EDP	Electronic Data Processing
A.S.B.C.	American Standard Building Code	CLF	Current Limiting Fuse	EIFS	Exterior Insulation Finish System
Asbe.	Asbestos Worker	CLP	Cross Linked Polyethylene	E.D.R.	Equiv. Direct Radiation
A.S.H.R.A.E.	American Society of Heating, Refrig. & AC Engineers	cm	Centimeter	Eq.	Equation
		CMP	Corr. Metal Pipe	Elec.	Electrician; Electrical
A.S.M.E.	American Society of Mechanical Engineers	C.M.U.	Concrete Masonry Unit	Elev.	Elevator; Elevating
		CN	Change Notice	EMT	Electrical Metallic Conduit; Thin Wall Conduit
A.S.T.M.	American Society for Testing and Materials	Col.	Column		
		CO_2	Carbon Dioxide	Eng.	Engine, Engineered
Attchmt.	Attachment	Comb.	Combination	EPDM	Ethylene Propylene Diene Monomer
Avg.	Average	Compr.	Compressor		
A.W.G.	American Wire Gauge	Conc.	Concrete	EPS	Expanded Polystyrene
AWWA	American Water Works Assoc.	Cont.	Continuous; Continued	Eqhv.	Equip. Oper., Heavy
Bbl.	Barrel	Corr.	Corrugated	Eqlt.	Equip. Oper., Light
B. & B.	Grade B and Better; Balled & Burlapped	Cos	Cosine	Eqmd.	Equip. Oper., Medium
		Cot	Cotangent	Eqmm.	Equip. Oper., Master Mechanic
B. & S.	Bell and Spigot	Cov.	Cover	Eqol.	Equip. Oper., Oilers
B. & W.	Black and White	C/P	Cedar on Paneling	Equip.	Equipment
b.c.c.	Body-centered Cubic	CPA	Control Point Adjustment	ERW	Electric Resistance Welded
B.C.Y.	Bank Cubic Yards	Cplg.	Coupling	E.S.	Energy Saver
BE	Bevel End	C.P.M.	Critical Path Method	Est.	Estimated
B.F.	Board Feet	CPVC	Chlorinated Polyvinyl Chloride	esu	Electrostatic Units
Bg. cem.	Bag of Cement	C.Pr.	Hundred Pair	E.W.	Each Way
BHP	Boiler Horsepower; Brake Horsepower	CRC	Cold Rolled Channel	EWT	Entering Water Temperature
		Creos.	Creosote	Excav.	Excavation
B.I.	Black Iron	Crpt.	Carpet & Linoleum Layer	Exp.	Expansion, Exposure
Bit.; Bitum.	Bituminous	CRT	Cathode-ray Tube	Ext.	Exterior
Bk.	Backed	CS	Carbon Steel, Constant Shear Bar Joist	Extru.	Extrusion
Bkrs.	Breakers			f.	Fiber stress
Bldg.	Building	Csc	Cosecant	F	Fahrenheit; Female; Fill
Blk.	Block	C.S.F.	Hundred Square Feet	Fab.	Fabricated
Bm.	Beam	CSI	Construction Specifications Institute	FBGS	Fiberglass
Boil.	Boilermaker			F.C.	Footcandles
B.P.M.	Blows per Minute	C.T.	Current Transformer	f.c.c.	Face-centered Cubic
BR	Bedroom	CTS	Copper Tube Size	f'c.	Compressive Stress in Concrete; Extreme Compressive Stress
Brg.	Bearing	Cu	Copper, Cubic		
Brhe.	Bricklayer Helper	Cu. Ft.	Cubic Foot	F.E.	Front End
Bric.	Bricklayer	cw	Continuous Wave	FEP	Fluorinated Ethylene Propylene (Teflon)
Brk.	Brick	C.W.	Cool White; Cold Water		
Brng.	Bearing	Cwt.	100 Pounds	F.G.	Flat Grain
Brs.	Brass	C.W.X.	Cool White Deluxe	F.H.A.	Federal Housing Administration
Brz.	Bronze	C.Y.	Cubic Yard (27 cubic feet)	Fig.	Figure
Bsn.	Basin	C.Y./Hr.	Cubic Yard per Hour	Fin.	Finished
Btr.	Better	Cyl.	Cylinder	Fixt.	Fixture
BTU	British Thermal Unit	d	Penny (nail size)	Fl. Oz.	Fluid Ounces
BTUH	BTU per Hour	D	Deep; Depth; Discharge	Flr.	Floor
B.U.R.	Built-up Roofing	Dis.;Disch.	Discharge	F.M.	Frequency Modulation; Factory Mutual
BX	Interlocked Armored Cable	Db.	Decibel		
c	Conductivity, Copper Sweat	Dbl.	Double	Fmg.	Framing
C	Hundred; Centigrade	DC	Direct Current	Fndtn.	Foundation
C/C	Center to Center, Cedar on Cedar	DDC	Direct Digital Control	Fori.	Foreman, Inside
		Demob.	Demobilization	Foro.	Foreman, Outside

Abbreviations

Fount.	Fountain	J.I.C.	Joint Industrial Council	M.C.P.	Motor Circuit Protector
FPM	Feet per Minute	K	Thousand; Thousand Pounds;	MD	Medium Duty
FPT	Female Pipe Thread		Heavy Wall Copper Tubing, Kelvin	M.D.O.	Medium Density Overlaid
Fr.	Frame	K.A.H.	Thousand Amp. Hours	Med.	Medium
F.R.	Fire Rating	KCMIL	Thousand Circular Mils	MF	Thousand Feet
FRK	Foil Reinforced Kraft	KD	Knock Down	M.F.B.M.	Thousand Feet Board Measure
FRP	Fiberglass Reinforced Plastic	K.D.A.T.	Kiln Dried After Treatment	Mfg.	Manufacturing
FS	Forged Steel	kg	Kilogram	Mfrs.	Manufacturers
FSC	Cast Body; Cast Switch Box	kG	Kilogauss	mg	Milligram
Ft.	Foot; Feet	kgf	Kilogram Force	MGD	Million Gallons per Day
Ftng.	Fitting	kHz	Kilohertz	MGPH	Thousand Gallons per Hour
Ftg.	Footing	Kip.	1000 Pounds	MH, M.H.	Manhole; Metal Halide; Man-Hour
Ft. Lb.	Foot Pound	KJ	Kiljoule	MHz	Megahertz
Furn.	Furniture	K.L.	Effective Length Factor	Mi.	Mile
FVNR	Full Voltage Non-Reversing	K.L.F.	Kips per Linear Foot	MI	Malleable Iron; Mineral Insulated
FXM	Female by Male	Km	Kilometer	mm	Millimeter
Fy.	Minimum Yield Stress of Steel	K.S.F.	Kips per Square Foot	Mill.	Millwright
g	Gram	K.S.I.	Kips per Square Inch	Min., min.	Minimum, minute
G	Gauss	kV	Kilovolt	Misc.	Miscellaneous
Ga.	Gauge	kVA	Kilovolt Ampere	ml	Milliliter, Mainline
Gal.	Gallon	K.V.A.R.	Kilovar (Reactance)	M.L.F.	Thousand Linear Feet
Gal./Min.	Gallon per Minute	KW	Kilowatt	Mo.	Month
Galv.	Galvanized	KWh	Kilowatt-hour	Mobil.	Mobilization
Gen.	General	L	Labor Only; Length; Long;	Mog.	Mogul Base
G.F.I.	Ground Fault Interrupter		Medium Wall Copper Tubing	MPH	Miles per Hour
Glaz.	Glazier	Lab.	Labor	MPT	Male Pipe Thread
GPD	Gallons per Day	lat	Latitude	MRT	Mile Round Trip
GPH	Gallons per Hour	Lath.	Lather	ms	Millisecond
GPM	Gallons per Minute	Lav.	Lavatory	M.S.F.	Thousand Square Feet
GR	Grade	lb.; #	Pound	Mstz.	Mosaic & Terrazzo Worker
Gran.	Granular	L.B.	Load Bearing; L Conduit Body	M.S.Y.	Thousand Square Yards
Grnd.	Ground	L. & E.	Labor & Equipment	Mtd.	Mounted
H	High; High Strength Bar Joist;	lb./hr.	Pounds per Hour	Mthe.	Mosaic & Terrazzo Helper
	Henry	lb./L.F.	Pounds per Linear Foot	Mtng.	Mounting
H.C.	High Capacity	lbf/sq.in.	Pound-force per Square Inch	Mult.	Multi; Multiply
H.D.	Heavy Duty; High Density	L.C.L.	Less than Carload Lot	M.V.A.	Million Volt Amperes
H.D.O.	High Density Overlaid	Ld.	Load	M.V.A.R.	Million Volt Amperes Reactance
Hdr.	Header	LE	Lead Equivalent	MV	Megavolt
Hdwe.	Hardware	LED	Light Emitting Diode	MW	Megawatt
Help.	Helpers Average	L.F.	Linear Foot	MXM	Male by Male
HEPA	High Efficiency Particulate Air	Lg.	Long; Length; Large	MYD	Thousand Yards
	Filter	L & H	Light and Heat	N	Natural; North
Hg	Mercury	LH	Long Span Bar Joist	nA	Nanoampere
HIC	High Interrupting Capacity	L.H.	Labor Hours	NA	Not Available; Not Applicable
HM	Hollow Metal	L.L.	Live Load	N.B.C.	National Building Code
H.O.	High Output	L.L.D.	Lamp Lumen Depreciation	NC	Normally Closed
Horiz.	Horizontal	L-O-L	Lateralolet	N.E.M.A.	National Electrical Manufacturers
H.P.	Horsepower; High Pressure	lm	Lumen		Assoc.
H.P.F.	High Power Factor	lm/sf	Lumen per Square Foot	NEHB	Bolted Circuit Breaker to 600V.
Hr.	Hour	lm/W	Lumen per Watt	N.L.B.	Non-Load-Bearing
Hrs./Day	Hours per Day	L.O.A.	Length Over All	NM	Non-Metallic Cable
HSC	High Short Circuit	log	Logarithm	nm	Nanometer
Ht.	Height	L.P.	Liquefied Petroleum; Low Pressure	No.	Number
Htg.	Heating	L.P.F.	Low Power Factor	NO	Normally Open
Htrs.	Heaters	LR	Long Radius	N.O.C.	Not Otherwise Classified
HVAC	Heating, Ventilation & Air-	L.S.	Lump Sum	Nose.	Nosing
	Conditioning	Lt.	Light	N.P.T.	National Pipe Thread
Hvy.	Heavy	Lt. Ga.	Light Gauge	NQOD	Combination Plug-on/Bolt on
HW	Hot Water	L.T.L.	Less than Truckload Lot		Circuit Breaker to 240V.
Hyd.;Hydr.	Hydraulic	Lt. Wt.	Lightweight	N.R.C.	Noise Reduction Coefficient
Hz.	Hertz (cycles)	L.V.	Low Voltage	N.R.S.	Non Rising Stem
I.	Moment of Inertia	M	Thousand; Material; Male;	ns	Nanosecond
I.C.	Interrupting Capacity		Light Wall Copper Tubing	nW	Nanowatt
ID	Inside Diameter	M²CA	Meters Squared Contact Area	OB	Opposing Blade
I.D.	Inside Dimension; Identification	m/hr; M.H.	Man-hour	OC	On Center
I.F.	Inside Frosted	mA	Milliampere	OD	Outside Diameter
I.M.C.	Intermediate Metal Conduit	Mach.	Machine	O.D.	Outside Dimension
In.	Inch	Mag. Str.	Magnetic Starter	ODS	Overhead Distribution System
Incan.	Incandescent	Maint.	Maintenance	O.G.	Ogee
Incl.	Included; Including	Marb.	Marble Setter	O.H.	Overhead
Int.	Interior	Mat; Mat'l.	Material	O & P	Overhead and Profit
Inst.	Installation	Max.	Maximum	Oper.	Operator
Insul.	Insulation/Insulated	MBF	Thousand Board Feet	Opng.	Opening
I.P.	Iron Pipe	MBH	Thousand BTU's per hr.	Orna.	Ornamental
I.P.S.	Iron Pipe Size	MC	Metal Clad Cable	OSB	Oriented Strand Board
I.P.T.	Iron Pipe Threaded	M.C.F.	Thousand Cubic Feet	O. S. & Y.	Outside Screw and Yoke
I.W.	Indirect Waste	M.C.F.M.	Thousand Cubic Feet per Minute	Ovhd.	Overhead
J	Joule	M.C.M.	Thousand Circular Mils	OWG	Oil, Water or Gas

Abbreviations

Oz.	Ounce	S.	Suction; Single Entrance; South	THHN	Nylon Jacketed Wire
P.	Pole; Applied Load; Projection	SCFM	Standard Cubic Feet per Minute	THW.	Insulated Strand Wire
p.	Page	Scaf.	Scaffold	THWN;	Nylon Jacketed Wire
Pape.	Paperhanger	Sch.; Sched.	Schedule	T.L.	Truckload
P.A.P.R.	Powered Air Purifying Respirator	S.C.R.	Modular Brick	T.M.	Track Mounted
PAR	Weatherproof Reflector	S.D.	Sound Deadening	Tot.	Total
Pc., Pcs.	Piece, Pieces	S.D.R.	Standard Dimension Ratio	T-O-L	Threadolet
P.C.	Portland Cement; Power Connector	S.E.	Surfaced Edge	T.S.	Trigger Start
P.C.F.	Pounds per Cubic Foot	Sel.	Select	Tr.	Trade
P.C.M.	Phase Contract Microscopy	S.E.R.;	Service Entrance Cable	Transf.	Transformer
P.E.	Professional Engineer; Porcelain Enamel; Polyethylene; Plain End	S.E.U.	Service Entrance Cable	Trhv.	Truck Driver, Heavy
		S.F.	Square Foot	Trlr	Trailer
		S.F.C.A.	Square Foot Contact Area	Trlt.	Truck Driver, Light
Perf.	Perforated	S.F.G.	Square Foot of Ground	TV	Television
Ph.	Phase	S.F. Hor.	Square Foot Horizontal	T.W.	Thermoplastic Water Resistant Wire
P.I.	Pressure Injected	S.F.R.	Square Feet of Radiation		
Pile.	Pile Driver	S.F. Shlf.	Square Foot of Shelf	UCI	Uniform Construction Index
Pkg.	Package	S4S	Surface 4 Sides	UF	Underground Feeder
Pl.	Plate	Shee.	Sheet Metal Worker	UGND	Underground Feeder
Plah.	Plasterer Helper	Sin.	Sine	U.H.F.	Ultra High Frequency
Plas.	Plasterer	Skwk.	Skilled Worker	U.L.	Underwriters Laboratory
Pluh.	Plumbers Helper	SL	Saran Lined	Unfin.	Unfinished
Plum.	Plumber	S.L.	Slimline	URD	Underground Residential Distribution
Ply.	Plywood	Sldr.	Solder		
p.m.	Post Meridiem	SLH	Super Long Span Bar Joist	US	United States
Pntd.	Painted	S.N.	Solid Neutral	USP	United States Primed
Pord.	Painter, Ordinary	S-O-L	Socketolet	UTP	Unshielded Twisted Pair
pp	Pages	sp	Standpipe	V	Volt
PP; PPL	Polypropylene	S.P.	Static Pressure; Single Pole; Self-Propelled	V.A.	Volt Amperes
P.P.M.	Parts per Million			V.C.T.	Vinyl Composition Tile
Pr.	Pair	Spri.	Sprinkler Installer	VAV	Variable Air Volume
P.E.S.B.	Pre-engineered Steel Building	spwg	Static Pressure Water Gauge	VC	Veneer Core
Prefab.	Prefabricated	Sq.	Square; 100 Square Feet	Vent.	Ventilation
Prefin.	Prefinished	S.P.D.T.	Single Pole, Double Throw	Vert.	Vertical
Prop.	Propelled	SPF	Spruce Pine Fir	V.F.	Vinyl Faced
PSF; psf	Pounds per Square Foot	S.P.S.T.	Single Pole, Single Throw	V.G.	Vertical Grain
PSI; psi	Pounds per Square Inch	SPT	Standard Pipe Thread	V.H.F.	Very High Frequency
PSIG	Pounds per Square Inch Gauge	Sq. Hd.	Square Head	VHO	Very High Output
PSP	Plastic Sewer Pipe	Sq. In.	Square Inch	Vib.	Vibrating
Pspr.	Painter, Spray	S.S.	Single Strength; Stainless Steel	V.L.F.	Vertical Linear Foot
Psst.	Painter, Structural Steel	S.S.B.	Single Strength B Grade	Vol.	Volume
P.T.	Potential Transformer	sst	Stainless Steel	VRP	Vinyl Reinforced Polyester
P. & T.	Pressure & Temperature	Sswk.	Structural Steel Worker	W	Wire; Watt; Wide; West
Ptd.	Painted	Sswl.	Structural Steel Welder	w/	With
Ptns.	Partitions	St.; Stl.	Steel	W.C.	Water Column; Water Closet
Pu	Ultimate Load	S.T.C.	Sound Transmission Coefficient	W.F.	Wide Flange
PVC	Polyvinyl Chloride	Std.	Standard	W.G.	Water Gauge
Pvmt.	Pavement	STK	Select Tight Knot	Wldg.	Welding
Pwr.	Power	STP	Standard Temperature & Pressure	W. Mile	Wire Mile
Q	Quantity Heat Flow	Stpi.	Steamfitter, Pipefitter	W-O-L	Weldolet
Quan.; Qty.	Quantity	Str.	Strength; Starter; Straight	W.R.	Water Resistant
Q.C.	Quick Coupling	Strd.	Stranded	Wrck.	Wrecker
r	Radius of Gyration	Struct.	Structural	W.S.P.	Water, Steam, Petroleum
R	Resistance	Sty.	Story	WT., Wt.	Weight
R.C.P.	Reinforced Concrete Pipe	Subj.	Subject	WWF	Welded Wire Fabric
Rect.	Rectangle	Subs.	Subcontractors	XFER	Transfer
Reg.	Regular	Surf.	Surface	XFMR	Transformer
Reinf.	Reinforced	Sw.	Switch	XHD	Extra Heavy Duty
Req'd.	Required	Swbd.	Switchboard	XHHW; XLPE	Cross-Linked Polyethylene Wire Insulation
Res.	Resistant	S.Y.	Square Yard		
Resi.	Residential	Syn.	Synthetic	XLP	Cross-linked Polyethylene
Rgh.	Rough	S.Y.P.	Southern Yellow Pine	Y	Wye
RGS	Rigid Galvanized Steel	Sys.	System	yd	Yard
R.H.W.	Rubber, Heat & Water Resistant; Residential Hot Water	t.	Thickness	yr	Year
		T	Temperature; Ton	Δ	Delta
rms	Root Mean Square	Tan	Tangent	%	Percent
Rnd.	Round	T.C.	Terra Cotta	~	Approximately
Rodm.	Rodman	T & C	Threaded and Coupled	Ø	Phase
Rofc.	Roofer, Composition	T.D.	Temperature Difference	@	At
Rofp.	Roofer, Precast	T.E.M.	Transmission Electron Microscopy	#	Pound; Number
Rohe.	Roofer Helpers (Composition)	TFE	Tetrafluoroethylene (Teflon)	<	Less Than
Rots.	Roofer, Tile & Slate	T. & G.	Tongue & Groove; Tar & Gravel	>	Greater Than
R.O.W.	Right of Way				
RPM	Revolutions per Minute	Th.; Thk.	Thick		
R.R.	Direct Burial Feeder Conduit	Thn.	Thin		
R.S.	Rapid Start	Thrded	Threaded		
Rsr	Riser	Tilf.	Tile Layer, Floor		
RT	Round Trip	Tilh.	Tile Layer, Helper		

Index

A

Abandon catch basin	287
Abatement asbestos	468
ABC extinguisher	448
Abrasive floor tile	419
tread	419
ABS DWV pipe	486
Absorption testing	274
water chillers	502
Access door & panels	396
door basement	312
door duct	507
door floor	396
door roof	384
Accessories bath	451
bathroom	418, 451
door	410
door and window	410
drywall	418
duct	507
plaster	415
roof	384
Accessory drainage	384
masonry	314
Acid proof floor	422
Acoustic ceiling board	420
Acoustical and plaster ceiling	187, 188, 189
ceiling	188, 420, 421
door	397, 450
enclosure	261
folding partition	450
panel	427, 450
partition	450
sealant	387, 418
treatment	427
wallboard	418
Acrylic ceiling	421
sign	447
wall coating	436
wallcovering	426
wood block	423
Act. tools & fast. powder	326
Adhesive cement	424
roof	378
wallpaper	426
Adobe brick	318
masonry	318
Aerial lift	279
Aggregate exposed	311
stone	303
testing	274
Air cleaner electronic	509
compressor	279
compressor control system	472
compressor sprinkler system	474
conditioner central station	503
conditioner cooling & heating	503
conditioner direct expansion	504
conditioner fan coil	504
conditioner packaged terminal	503
conditioner portable	503
conditioner receptacle	523
conditioner removal	480
conditioner rooftop	503
conditioner self-contained	504
conditioner thru-wall	503
conditioner window	503
conditioner wiring	525
conditioning	578
conditioning computer	503
conditioning fan	508
conditioning ventilating	472, 503, 507-509
cooled belt drive condenser	502
curtain	262
extractor	507
filter	509
filter roll type	509
filtration	468
handling unit	503
hose	280
return grille	509
sampling	468
spade	280
supply pneumatic control	472
supply register	509
supported storage tank cover	261
supported structure	261
tool	280
vent roof	385
Air-conditioner air cooled	229, 235, 237
chilled water	227, 229
direct expansion	231, 233, 235, 237
fan coil	227, 229, 237
self-contained	233, 235
single zone	231
split system	237
water cooled	233
Air-conditioning air cooled	227
Airless sprayer	469
Alarm burglar	473
detection	473
fire	474
residential	524
valve sprinkler	475
Aluminum and glass door	158
ceiling tile	421
chain link	301
column	362
coping	317, 318
diffuser perforated	508
door	157-159, 397-399
door frame	398
downspout	383
drip edge	383
ductwork	506
edging	305
entrance	398
flagpole	447
flashing	381, 492
foil	369, 371
folding	180
gravel stop	168, 384
grille	509
gutter	172, 384
ladder	335
louver	444
nail	339
reglet	384
rivet	326
rolling grill	158
roof	374
sash	400
service entrance cable	514
sheet metal	383
shingle	372
siding	154, 375
siding accessories	375
siding paint	433
sliding door	395
stair	335
storefront	399
storm door	395
tile	373, 420
tube frame	412
window	161, 400
window awning	162
window casement	161
window demolition	292
window jalousie	162
window picture	162
window projecting	161
window single hung	161
window sliding	161
wire	517
Anchor bolt	314
brick	314
buck	314
chemical	325
expansion	325
framing	340
hollow wall	325
joist	340
masonry	314
nailing	325
partition	314
rafter	340
rigid	314
screw	325
sill	340
steel	314
stone	314
wall	324
Anechoic chamber	261
Antenna system	536
T.V.	536
wire TV	515
Apartment call system	536
Appliance	255, 457
plumbing	498, 499
residential	457, 524
Apron wood	353
Arch laminated	351
radial	351
Architectural equipment	255-258, 457
fee	272, 545
woodwork	356
Area clean-up	469
Armored cable	513
Arrester lightning	464
Asbestos abatement	468
demolition	468
disposal	470
felt	378
removal	467
removal process	575
Ash receiver	452
Ashlar veneer	561
Asphalt base sheet	378
block	300
block floor	300
coating	368
curb	299
distributor	280
felt	377, 378
flashing	381
flood coat	378
paper	371
paver	281
primer	380, 424
sheathing	349
shingle	372
sidewalk	299
Asphaltic emulsion	303
paving	547
Astragal	410
molding	354
Atomizer water	469
Attic stair	458
Audio masking system	261
Audiometric room	261
Auditorium chair	258
Auger	277
hole	286
Automatic fire suppression	475
opener	407
transfer switch	530
washing machine	458
Automotive equipment	255
Awning	448
window	401

B

Backfill	293, 294
compaction	548
planting pit	304
trench	294
Backflow preventer	490
Backhoe	294
excavation	294
trenching	264
Backsplash countertop	338
Backup block	316
Baffle roof	385
Bag disposable	469
glove	467
Baked enamel door	390
enamel frame	392
Balanced door	157-159, 399
Ball wrecking	282
Baluster	361
Balustrade painting	433
Band joist framing	333
molding	353
Bank air-conditioning	227, 229, 231, 233, 235, 237
equipment	255
Bankrun gravel	286
Baptistry	255
Bar grab	451
towel	452
Zee	418
Barber equipment	255
Bark mulch	302
mulch redwood	303
Barrel	280
Barricade	280
Barrier separation	468
Base cabinet	356, 357
carpet	425
ceramic tile	418
column	340
course	298
cove	423
gravel	298
molding	352
quarry tile	419
resilient	423
road	298
sheet	378
sink	357
stone	298
terrazzo	420
vanity	360
wood	352, 353, 362, 363
Baseboard demolition	291
heat	505
register	509
Basement stair	312
Basic meter device	526
Basketweave fence	301
Bath accessories	451
communal	495
steam	464
whirlpool	495
Bathroom	495
accessories	418, 451
accessory curtain rod	254

Index

exhaust fan . . . 508
faucet . . . 494
fixture . . . 495, 496
Bathtub enclosure . . . 444
removal . . . 480
Batt insulation . . . 369
Battery inverter . . . 527
light . . . 535
Bay window . . . 160
Bead casing . . . 418
corner . . . 415
parting . . . 354
Beam & girder framing . . . 341
bondcrete . . . 416
ceiling . . . 361
drywall . . . 416, 417
fireproofing . . . 385
hanger . . . 340
laminated . . . 350, 351
mantel . . . 362
plaster . . . 416
steel . . . 327
wood . . . 341, 342
Bed molding . . . 353
Bedding brick . . . 300
pipe . . . 294
Beech tread . . . 361
Belgian block . . . 299
Bell & spigot pipe . . . 481
Bench park . . . 302
players . . . 302
Berm pavement . . . 299
Bevel siding . . . 377
Birch door . . . 392
molding . . . 354
paneling . . . 355
stair . . . 361
wood frame . . . 394
Bituminous block . . . 299
coating . . . 368
paver . . . 281
paving . . . 265, 547
Blanket insulation . . . 369, 480
sound attenuation . . . 427
Blaster shot . . . 282
Bleacher . . . 257
Blind . . . 258, 259
exterior . . . 363
venetian . . . 460
window . . . 460
Block asphalt . . . 300
asphalt floor . . . 299
backup . . . 316
belgian . . . 299
bituminous . . . 299
Block, brick and mortar . . . 559
Block concrete . . . 315-317, 321
concrete bond beam . . . 316
concrete exterior . . . 317
decorative concrete . . . 316, 317
filler . . . 436
floor . . . 423
glass . . . 318, 561
grooved . . . 316
ground face . . . 316
hexagonal face . . . 317
insulation . . . 314
lightweight . . . 321
lintel . . . 317
partition . . . 175, 321
profile . . . 316, 317
reflective . . . 318
scored . . . 317
scored split face . . . 317
slump . . . 316
split rib . . . 316
wall glass . . . 152
wall removal . . . 287
Blocking . . . 341
carpentry . . . 341
wood . . . 344
Blockout slab . . . 308
Blown in cellulose . . . 368
in fiberglass . . . 368
in insulation . . . 368
Blueboard . . . 416
Bluegrass sod . . . 303
Bluestone sidewalk . . . 299
sill . . . 318
step . . . 299
Board & batten fence . . . 301
& batten siding . . . 377
bulletin . . . 442
ceiling . . . 420
directory . . . 447
fence . . . 301
gypsum . . . 416
insulation . . . 369
paneling . . . 356
sheathing . . . 349
valance . . . 357
verge . . . 353
Boiler . . . 499
coal-fired hydronic . . . 222
demolition . . . 480
electric . . . 499
electric steam . . . 499
fire tube . . . 500
fossil fuel hydronic . . . 222
gas fired . . . 499
gas pulse . . . 499
general . . . 480
hot water . . . 499, 500
hot-water and steam . . . 219
hydronic fossil fuel . . . 221, 222
oil fired . . . 499
oil-fired hydronic . . . 221, 222
scotch marine . . . 500
steam . . . 499, 500
water tube . . . 500
Bollard . . . 327
light . . . 534
Bolt . . . 324
anchor . . . 314
expansion . . . 325
steel . . . 324, 340
toggle . . . 325
Bolt-on circuit-breaker . . . 532
Bond performance . . . 544, 545
Bondcrete . . . 416
Bookcase . . . 356, 357
Boom lift . . . 279
Boost transformer buck . . . 527
Booth for attendant . . . 257
telephone . . . 451
Boring . . . 286
cased . . . 286
Borrow . . . 286
Bow window . . . 402
Bowling alley . . . 260
alley air-conditioning . 227, 229, 231
233, 235, 237
Bowstring truss . . . 351
Box . . . 520
distribution . . . 296
mail . . . 449
out . . . 308
outlet cover . . . 521
pull . . . 521
stair . . . 360
termination . . . 536
Boxed headers & beams . . . 329, 333
Boxes & wiring device . . . 520, 521
Bracing . . . 341
let-in . . . 340
metal joists . . . 332
stud walls . . . 328
Brass hinge . . . 408
screw . . . 339
Brazed connection . . . 513
Breaker circuit . . . 530
Brick . . . 558
adobe . . . 318
anchor . . . 314
bedding . . . 300
Brick, block and mortar . . . 559
Brick cart . . . 280
chimney simulated . . . 446
chimneys . . . 556
cleaning . . . 322
demolition . . . 289, 290
edging . . . 305
face . . . 315
forklift . . . 280
molding . . . 353
paving . . . 300
removal . . . 287
shelf . . . 308
sidewalk . . . 300
sill . . . 318
step . . . 299
veneer . . . 315, 559
veneer demolition . . . 291
wash . . . 322
Bricklaying . . . 556
Bridging . . . 341
metal joists . . . 332
steel . . . 341
stud walls . . . 329
wood . . . 341
Broiler without oven . . . 256
Bronze butterfly valve . . . 488
plaque . . . 447
push-pull plate . . . 409
valve . . . 488
Broom cabinet . . . 357
finish concrete . . . 311
Brown coat . . . 420
Brownstone . . . 320
Brush clearing . . . 293
cutter . . . 278
Buck anchor . . . 314
boost transformer . . . 527
rough . . . 344
Bucket concrete . . . 277
crane . . . 278
Buggy concrete . . . 277, 311
Builder's risk insurance . . . 538
Building air-conditioning . . . 227, 229,
231 , 233, 235, 237
demolition . . . 286
directory . . . 536
greenhouse . . . 465
hardware . . . 407
insulation . . . 368
paper . . . 371
permit . . . 272
prefabricated . . . 465
slab . . . 102
sprinkler . . . 475
Built-in range . . . 457
shower . . . 496
Built-up roof . . . 377, 378
roofing . . . 166, 567
Bulk bank measure excavating . . 294
Bulkhead door . . . 396
formwork . . . 308
Bulldozer . . . 278, 279, 293
Bulletin board . . . 255, 442
Bullnose block . . . 317
Bumper dock . . . 454
door . . . 408
wall . . . 408, 445
Burglar alarm . . . 473
Burial cable direct . . . 514
Burlap curing . . . 311
rubbing . . . 311
Bush hammer . . . 311
hammer concrete . . . 311
Butt fusion machine . . . 280
Butterfly valve bronze . . . 488
Butyl caulking . . . 387
flashing . . . 382
Buzzer system . . . 536
BX cable . . . 513

C

Cabinet base . . . 356, 357
broom . . . 357
casework . . . 357
corner base . . . 357
corner wall . . . 357
current transformer . . . 521
demolition . . . 291
door . . . 358, 359
dormitory . . . 259
electrical . . . 521
fire equipment . . . 448
hardboard . . . 356
hardware . . . 359
hinge . . . 359
hotel . . . 452
household . . . 259
kitchen . . . 356
medicine . . . 452
oven . . . 357
school . . . 259
shower . . . 443
stain . . . 428
strip . . . 536
transformer . . . 521
varnish . . . 428
wall . . . 357
Cable armored . . . 513
BX . . . 513
electric . . . 513
heating . . . 505
jack . . . 283
MC . . . 513
sheathed nonmetallic . . . 514
sheathed romex . . . 514
tray . . . 517
tray alum cover . . . 517
tray galv cover . . . 517
tray ladder type . . . 517
undercarpet system . . . 515
underground feeder . . . 514
Cafe door . . . 393
Call system apartment . . . 536
system emergency . . . 536
Canopy . . . 254, 448
entrance . . . 448
framing . . . 346
Cant roof . . . 345
Cantilever retaining wall . . . 302
Canvas awning . . . 448
Cap post . . . 340
service entrance . . . 515
Carbon dioxide exting. . . . 448, 475
Carpentry finish . . . 352
rough . . . 341
Carpet . . . 424

INDEX

Index

base	425
felt pad	425
floor	424
nylon	425
padding	425
removal	289
stair	425
urethane pad	425
wool	425
Carrier ceiling	415
channel	414
fixture	493
fixture support	493
Cart brick	280
concrete	277, 311
Car-wash water-heater	498
Case work	357
Cased boring	286
Casement window	400, 402
Casework cabinet	357
custom	356
ground	348
painting	428
stain	428
varnish	428
Casing bead	418
wood	353
Cast in place concrete	310
in place terrazzo	420
iron bench	302
iron drain	491
iron fitting	482
iron pipe	481
iron pipe fitting	482, 485
iron stair	335
iron trap	492
Catch basin	298
basin and manhole	264
basin precast	298
basin removal	287
door	359
Caulking	387
lead	482
oakum	482
polyurethane	387
sealant	387
Cavity wall insulation	368
wall reinforcing	314
Cedar closet	356
drop siding	154, 155
fence	301
paneling	356
post	362
roof deck	351
roof plank	349
shingle	372
siding	377
stair	360
Ceiling acoustical	188, 420, 421
acoustical and plaster	188
beam	361
board	420
board acoustic	420
board fiberglass	420
bondcrete	416
carrier	415
demolition	288
diffuser	508
drill	324
drywall	417
eggcrate	421
fan	508
framing	342
furring	348, 414, 415
heater	457
insulation	368
lath	415
luminous	421, 535
molding	353
painting	435, 436
plaster	415, 416
plaster and acoustical	188, 189
stair	458
support	326
suspended	415, 421
tile	421
Cellar door	396
Cellular deck concrete	312
Cellulose blown in	368
insulation	368
Cement adhesive	424
flashing	380
gypsum	314
masonry	314
masonry unit	316, 317, 321
mortar	419, 555
parging	368
Central station air handling unit	503
vacuum	454
Centrifugal liquid chiller	502
pump	281
Ceramic coating	437
mulch	303
tile	418, 419
tile countertop	338
tile demolition	289
tile floor	418
Chain hoist	283
hoist door	397
link fence	301
link fence paint	430
link fence removal	287
saw	282
trencher	279
Chair molding	354
rail demolition	291
Chalkboard	255, 442
freestanding	442
liquid chalk	442
wall hung	442
Chamber decontamination	468
Channel carrier	414
frame	390
furring	414, 418
siding	377
steel	415
Charges disposal	470
Check swing valve	489
Checkout counter	256
Chemical anchor	325
dry extinguisher	448
Chiller water	502
Chimney	321
accessories	446
brick	321
demolition	290
flue	320
foundation	310
metal	445
simulated brick	446
vent	501
China cabinet	356
Chipper brush	278
Chipping hammer	280
Chips wood	303
Chlorination system	466
Church equipment	255, 258
Chute linen or refuse	254
mail	254, 449
rubbish	292
CI roof drain system	198
C.I.P. concrete	310
C.I.P. square column	104
Circuit-breaker	530
bolt-on	532
Circular saw	282
Circulating pump	493
Cladding	375
Clamp water pipe ground	513
Clamshell bucket	278
Clapboard painting	433
Classroom heating & cooling	505
Clay masonry	315
roofing tile	373
tile	373
tile coping	318
Cleaner steam	282
Cleaning brick	322
face brick	562
masonry	322
up	284
Cleanout tee	491
Clean-up area	469
Clear & grub	293
Clearing brush	293
site	293
Climbing jack	283
Clip plywood	340
Clock timer	524
Closer door	408
Closet cedar	356
pole	354
rod	356
water	496
Clothes dryer residential	457
CMU	315
Coal tar pitch	378, 380
tar roofing	166
Coat brown	420
glaze	436
scratch	420
Coating bituminous	368
ceramic	437
epoxy	437
flood	379
glazed	436
roof	380
rubber	368
silicone	368
spray	368
trowel	368
wall	436
water repellent	368
waterproofing	368
Coiling door & grilles	396
Coin laundry water heater	499
Cold applied roofing	379
roofing	379
Coldformed framing	328
joists	332
metal joists	333
Collected stone	320
Collection box lobby	449
Collector solar energy system	471
Colonial door	392
wood frame	394
Column	327
base	340
bondcrete	416
brick	321
demolition	290
drywall	417
fireproof	385
footing	98
lally	327
laminated wood	352
lath	415
pipe	327
plaster	415, 416
removal	290
square tied	105
steel	107
wood	110, 342, 362
Combination device	522
storm door	392
Comfort station	261
Commercial air-conditioning	227, 229, 231, 233, 235
dish washer	256
door	390, 396
folding partition	450
gutting	290
water heater	498
Common brick	315
nail	339
Communal bath	495
Communication system	535
Compact fill	294
Compaction	294
backfill	548
soil	293
Compactor	255
earth	278
residential	457
Compartments shower	443
toilet	442
Compensation workers'	273
Component sound system	535
sprinkler system	474
Composite door	390
insulation	371
Composition flooring	424
Compressive strength	274
Compressor air	279
tank mounted	472
Computer air conditioning	503
grade voltage regulator	528
power cond transformer	529
regulator transformer	529
transformer	529
Concrete beam	111
block	315-317, 321, 560
block back-up	315
block bond beam	316
block decorative	316, 317
block demolition	288
block exterior	317
block foundation	317
block grout	314
block insulation	368
block painting	436
block partition	175
broom finish	311
bucket	277
buggy	311
bush hammer	311
cart	277, 311
cast in place	310
cellular deck	312
C.I.P.	310
C.I.P. wall	130
conveyer	277
coping	317
curb	299
curing	311
cutout	288
cylinder	274
darby finish	311
demolition	288
drill	324
elevated slab	310
finish	298, 311
float	277
float finish	311

Index

footing . . . 98, 310
formwork . . . 308
foundation . . . 310
furring . . . 348
hand trowel finish . . . 311
lightweight . . . 554
lintel . . . 312
mixer . . . 277
monolithic finish . . . 311
panel tilt up . . . 135
paver . . . 281
paving . . . 298
pipe . . . 296, 297, 550
placing . . . 310, 551
precast . . . 312
precast flat . . . 131, 134
precast fluted window . . . 131, 132, 134
precast mullion . . . 131, 134
precast ribbed . . . 131, 132, 134
precast specialties . . . 132
precast unit price . . . 131, 134
protection . . . 308
pump . . . 277
ready mix . . . 310
reinforcement . . . 309
removal . . . 287
restoration . . . 312
retaining wall . . . 302
roof deck . . . 312
saw . . . 277
septic tank . . . 296
shingle . . . 373
sidewalk . . . 299
sill . . . 318
slab . . . 310
slab multispan joist . . . 113
spreader . . . 281
stair . . . 310
stamping . . . 311
structural . . . 310
testing . . . 274
tile . . . 373
tilt up . . . 553
trowel . . . 277
utility vault . . . 286
vibrator . . . 277
wall . . . 311
wheeling . . . 311
Condenser . . . 502
air cooled belt drive . . . 502
Conditioner portable air . . . 503
Conditioning ventilating air . . . 503
Conductive floor . . . 423
Conductor . . . 513
& grounding . . . 514, 515, 517
wire . . . 516
Conductors . . . 579
Conduit & fitting flexible . . . 526
electrical . . . 518
fitting . . . 519
high installation . . . 518
in slab . . . 518
in slab PVC . . . 518
in trench electrical . . . 518
in trench steel . . . 518
intermediate . . . 518
intermediate steel . . . 518
PVC . . . 518
rigid in slab . . . 518
rigid steel . . . 518
Confessional . . . 256
Connection brazed . . . 513
motor . . . 526
standpipe . . . 474
Connector joist . . . 340
stud . . . 326
timber . . . 340
Constant voltage transformer . . . 529
Construction aids . . . 274
management fee . . . 272
Contactor lighting . . . 530, 531
Contractor equipment . . . 277
overhead . . . 274
pump . . . 281
Control & motor starter . . . 532
board . . . 254
center motor . . . 532
component . . . 473
hand-off-automatic . . . 531
radiator supply . . . 490
station . . . 531
station stop-start . . . 531
system cooling . . . 472
system electronic . . . 472
system split system . . . 472
system VAV box . . . 472
system ventilator . . . 472
tower . . . 260
valve heating . . . 490
Conveyor . . . 277
Cooking equipment . . . 457
range . . . 457
Cooler beverage . . . 256
water . . . 495
Cooling & heating A/C . . . 503
control system . . . 472
tower . . . 502
towers fiberglass . . . 502
towers stainless . . . 502
Coping . . . 318, 319
aluminum . . . 317, 318
clay tile . . . 318
concrete . . . 317
precast . . . 132
removal . . . 291
terra cotta . . . 318
Copper cable . . . 513
downspout . . . 383
drum trap . . . 492
DWV tubing . . . 483
fitting . . . 483
flashing . . . 169, 381, 492
gravel stop . . . 168
gutter . . . 172
pipe . . . 483
reglet . . . 384
rivet . . . 326
roof . . . 381
roof edge . . . 168
wall covering . . . 425
wire . . . 516, 517
Core drill . . . 274, 277
testing . . . 274
Cork floor . . . 423
tile . . . 423
wall tile . . . 425
Corner base cabinet . . . 357
bead . . . 415
wall cabinet . . . 357
Cornice . . . 318
drain . . . 491
molding . . . 353
painting . . . 433
Corridor air-conditioning . . . 227, 229, 231 , 233, 235, 237
Corrugated metal pipe . . . 297
roof tile . . . 373
siding . . . 375, 376
Counter door . . . 396
flashing . . . 384
top . . . 338
top demolition . . . 291
window post office . . . 254
Countertop backsplash . . . 338
sink . . . 496
Course base . . . 298
Cove base . . . 423
base ceramic tile . . . 418
base terrazzo . . . 420
molding . . . 353
molding scotia . . . 353
Cover cable tray alum . . . 517
cable tray galv . . . 517
ground . . . 303
outlet box . . . 521
pool . . . 466
Covering and paint wall . . . 183, 184
and tile floor . . . 185
CPVC pipe . . . 486
Crane bucket . . . 278
crawler . . . 282
hydraulic . . . 283
Crawler crane . . . 282
Crew survey . . . 272
Crown molding . . . 353
Crushed stone sidewalk . . . 299
Cubicle shower . . . 496
toilet . . . 442
Cup sink . . . 202
Cupola . . . 446
Curb . . . 299
asphalt . . . 299
concrete . . . 299
edging . . . 299
formwork . . . 308
granite . . . 299, 319
inlet . . . 299
precast . . . 299
removal . . . 287
roof . . . 345
terrazzo . . . 420
Curing concrete . . . 311
paper . . . 371
Current transformer cabinet . . . 521
Curtain damper fire . . . 507
rod . . . 451
wall . . . 412
Curved stair . . . 361
Custom casework . . . 356
Cut stone trim . . . 320
Cutout . . . 288
counter . . . 338
slab . . . 288
Cutter brush . . . 278
Cutting block . . . 338
torch . . . 282
Cylinder concrete . . . 274
lockset . . . 409

D

Damper . . . 507
multi-blade . . . 507
Darby finish concrete . . . 311
Darkroom . . . 261
Deadbolt . . . 408
Deciduous shrub . . . 304
tree . . . 304
Deciduous tree . . . 305
Deck and joist on bearing wall . . . 124
concrete cellular . . . 312
metal . . . 328
roof . . . 328, 349
steel . . . 328
wood . . . 349, 351
Decontamination chamber . . . 468
enclosure . . . 468
equipment . . . 469
Decorative block . . . 316
Decorator device . . . 522
switch . . . 522
Deep freeze . . . 457
Dehumidifier . . . 457
Delivery charge . . . 293
Demolition . . . 286, 287, 289, 291, 480
asbestos . . . 468
baseboard . . . 291
boiler . . . 480
brick . . . 289, 290
building . . . 286
cabinet . . . 291
ceiling . . . 288
ceramic tile . . . 289
chimney . . . 290
column . . . 290
concrete . . . 288
concrete block . . . 288
door . . . 289
ductwork . . . 480
electric . . . 512
enclosure . . . 469
fireplace . . . 291
flooring . . . 289
framing . . . 290
glass . . . 292
granite . . . 291
gutter . . . 291
hammer . . . 278
house . . . 286
HVAC . . . 480
joist . . . 290
masonry . . . 288, 290
metal stud . . . 292
millwork . . . 291
paneling . . . 291
partition . . . 292
pavement . . . 287
plaster . . . 288, 292
plumbing . . . 480
plywood . . . 288, 291, 292
post . . . 290
rafter . . . 290
railing . . . 291
roofing . . . 291, 468
selective . . . 288
siding . . . 291
site . . . 286, 287
tile . . . 288
truss . . . 290
walls and partitions . . . 292
window . . . 292
wood . . . 288
Demountable partition . . . 449
Dental equipment . . . 256
metal interceptor . . . 491
Department store A/C . . . 229, 231
Derrick . . . 283
guyed . . . 283
stiffleg . . . 283
Detection alarm . . . 473
system . . . 473
Detector infra-red . . . 473
motion . . . 473
smoke . . . 473
Detention equipment . . . 256
Device & box wiring . . . 520
combination . . . 522
decorator . . . 522
panic . . . 409
receptacle . . . 523
residential . . . 521
wiring . . . 525

INDEX

Index

Diaphragm pump 281	lauan 182	Drill ceiling 324	**E**
Diesel generator 527	louvered pine 182	concrete 324	
Diffuser ceiling 508	mall front 157, 158	core 274, 277	Earth compactor 278
linear 508	mall front panel 182	drywall 325	vibrator 294
opposed blade damper 508	metal 182, 390, 397	hammer 280	Earthwork 293
perforated aluminum 508	mirror 411	plaster 325	equipment 277, 293
rectangular 508	molding 354	rig 286	Edge drip 383
steel 509	moulded 393	steel 280	form 308
T-bar mount 509	opener 397, 407	track 280	Edging 305
Dimmer switch 522, 525	overhead 157-159, 397	wall 324	aluminum 305
Direct burial cable 514	overhead commercial 397	wood 340	curb 299
Directory 447	panel 393	Drinking fountain 495	Eggcrate ceiling 188, 421
board 254, 447	paneled pine 182	fountain support 493	Ejector pump 497
building 536	partition 449	Drip edge 383	Elbow downspout 384
Disappearing stairs 458	passage 392	edge aluminum 383	Electric appliance 457
stairway 254	patio 157, 158	Drive pin 326	baseboard heater 505
Discharge hose 281	plastic 392	Driver sheeting 280	boiler 499
Dishwasher 457	prehung 157, 158, 392	Driveway 299	cable 513
water-heater 498	pull 409	removal 287	demolition 512
Dispenser napkin 451	release 536	Drop pan ceiling 421	dumbwaiter 478
soap 451	removal 289	Drug store air-conditioning . 227, 229,	fixture 533-535
toilet tissue 451, 452	residential 394	231, 233, 235, 237	generator 280
towel 451, 452	revolving 157, 158	Drum trap 492	generator set 527
Dispenser towel 254	rolling 157-159, 396	trap copper 492	heater 457
Dispersion nozzle 475	rolling overhead 182	Dry cleaner 256	heating 505
Display case 254	roof 385	pipe sprinkler head 475	heating boiler 219, 220
Disposable bag 469	rough buck 344	transformer 528	lamp 535
Disposal 287	sauna 464	wall partition 177	log 446
asbestos 470	sectional 397	Dryer clothes 457	metallic tubing 518
charges 470	shower 444	receptacle 523	pool heater 500
field 296	sill 354, 394, 410	vent 458	service 530
garbage 457	sliding 182, 395	Drywall 416	stair 458
Distribution box 296	special 396, 397	accessories 418	switch 525, 530
Distributor asphalt 280	stain 429	ceiling 187, 188	unit heater 506
Divider strip terrazzo 420	stainless steel and glass 157	column 417	utility 297
Diving board 260, 466	steel 157, 158, 390, 397	cutout 288	water heater 498
Dock bumper 256, 454	stop 408, 412	demolition 292	wire 516
Dome 261	storm 157, 158, 395	drill 325	Electrical cabinet 521
Door & grilles coiling 396	swing 182	frame 391	conduit 518
& panels access 396	switch burglar alarm 473	gypsum 416	nonmetallic tubing 519
accessories 410	threshold 394	nail 339	site work 297
acoustical 397, 450	trim vinyl 376	painting 435	Electronic air cleaner 509
aluminum 157-159, 398, 399	varnish 430	partition 176	control system 472
automatic entrance 399	weatherstrip 409	prefinished 417	Elevated conduit 518
balanced 157, 399	weatherstrip garage 410	screw 418	slab concrete 310
bell residential 523	wood 157, 158, 182, 392	Duck tarpaulin 276	Elevator 478
bulkhead 396	Doors & windows interior paint . 429	Duct access door 507	passenger 192
bumper 408	Dormer gable 348	accessories 507	Embossed print door 393
buzzer system 536	Double acting door 397	flexible insulated 506	Emergency call system 536
cabinet 358, 359	hung window 400, 402, 403	flexible noninsulated 506	lighting 534
cafe 393	oven 255	furnace 500	Employer liability 273
catch 359	Dowels reinforcing 309	HVAC 506	EMT 518
chain hoist 397	Downspout 172, 383, 384	insulation 480	Emulsion adhesive 424
closer 408	aluminum 383	liner 481	asphaltic 303
commercial 182, 390	copper 383	silencer 508	pavement 300
composite 390	elbow 384	underground 297	sprayer 280
counter 396	lead coated copper 383	Ductile iron pipe cement lined . 295	Encapsulation 470
demolition 289	painting 432	Ductwork 506, 578	pipe 470
entrance 398, 399	steel 383	aluminum 506	Enclosure bathtub 444
fire 157, 182, 390	strainer 383	demolition 480	decontamination 468
flexible 397	Dozer 278, 293	fabric coated flexible 506	demolition 469
floor 396	Dragline bucket 278	fabricated 506	shower 444
frame 390, 391, 394	Drain 491	galvanized 506	swimming pool 465
frame interior 394	cast iron 491	rectangular 506	telephone 451
garage 397	floor 491	rigid 506	ENT 519
glass 395, 397, 399	Drainage accessory 384	Dumbwaiter 478	Entrance aluminum 398
glass and aluminum 157, 158	pipe 296	electric 478	and storefront 398
handle 359	site 298	manual 478	cable service 514
hardware 407	trap 492	Dump charge 292	canopy 448
hollow metal 182	Drains roof 491	truck 279	door 392, 398, 399
industrial 396	Drapery 259	Dumpster 292	frame 394
interior 182	hardware 460	Duplex receptacle 525	lock 408
kalamein 182	rings 460	Duranodic gravel stop 168	screen 443
kick plate 408	Drawer 360	Dust partition 292	sliding 399
knocker 410	track 359	DWV pipe ABS 486	weather cap 519
labeled 390	wood 359	tubing copper 483	EPDM roof 379

624

Index

Epoxy coating 437
  floor 424
  grout 419
  terrazzo 424
  wall coating 436
Equipment 277
  earthwork 277, 293
  fire 214, 473
  insurance 273
  loading dock 454
  rental 277, 282
Erosion control 295
Estimate electrical heating .. 505
Estimating 538
Evergreen shrub 303, 304
  tree 303
Excavating bulk bank measure . 294
  equipment 550
Excavation 293, 294, 549
  and backfill 100
  backhoe 294
  hand 293, 294
  hauling 294
  planting pit 304
  septic tank 296
  structural 294
  trench 264, 294
Excavator rental 277
Exhaust hood 457
  vent 509
Exhauster roof 508
Exit and emergency lighting .. 534
  light 535
Expansion anchor 325
  bolt 325
  joint 415
  shield 325
  tank 489
  tank steel 489
Expense office 273
Exposed aggregate 311
  aggregate coating 437
Exterior blind 363
  concrete block 317
  door frame 394
  lighting fixture 534
  molding 353
  plaster 416
  pre-hung door 392
  wood frame 394
  wood siding 154, 155
Extinguisher abc 448
  carbon dioxide 448
  chemical dry 448
  fire 448
  standard 448
Extractor air 507

F

Fabric flashing 382
  welded wire 309
Fabricated ductwork 506
Fabrication metal 335
Fabrications plastic 364
Face brick 315, 557
  brick cleaning 562
Facing stone 319
Factory air-conditioning 227, 229, 231, 235, 237
Fan 508
  air conditioning 508
  bathroom exhaust 508
  ceiling 508
  coil air conditioning 504

paddle 524
propeller exhaust 508
residential 524
roof 508
utility set 508
ventilation 524
wiring 524
Fascia aluminum 375
  board 344
  board demolition 290
  metal 375
  vinyl 376
  wood 353
Fastener timber 339
  wood 339
Fastenings metal 324
Faucet & fitting 494
  bathroom 494
Fee architectural 272
Feeder cable underground 514
Felt 378
  asphalt 377, 378
  carpet pad 425
  tarred 378
Fence and gate 300
  auger, truck mounted 277
  board 301
  board & batten 301
  cedar 301
  chain link 301
  metal 301
  open rail 301
  picket paint 430
  removal 287
  wire 300
  wood 301
Fiber cement shingle 372
  cement siding 375
  optic cable 514
  optics 513
  reinforcing 309
Fiberboard insulation 170, 370
Fiberglass tank 466
Fiberglass blown in 368
  ceiling board 420
  cooling towers 502
  cupola 446
  door 397
  insulation 368, 369
  panel 375, 427
  planter 305
  shower stall 443
  tank 466
  wall lamination 437
  wool 368
Fiberglass/urethane 170
Field assembly panel siding .. 153
  disposal 296
  office expense 274
Fieldstone 320
Fill 293, 294
  gravel 286, 294
Filler block 436
Film polyethylene 303
  transport 257
Filter air 509
  iron removal 498
  mechanical media 509
  swimming pool 466
  water 497
Filtration air 468
  equipment 465
Fin tube radiation 505
Fine grade 303
Finish carpentry 352
  concrete 298, 311

floor 423, 436
hardware 407
nail 339
wall 311
Finisher floor 277
Finishes wall 425
Fir column 362
  floor 422
  molding 354
  roof deck 351
  roof plank 349
Fire alarm 474
  call pullbox 473
  damage repair 437
  damper curtain type 507
  door 157, 390
  door frame 391
  equipment 473
  equipment cabinet 448
  escape 254
  escape disappearing 458
  extinguisher 448
  extinguisher portable 448
  extinguishing 208-211, 213-215
  horn 473
  hose 473
  hose adapter 473
  hose nozzle 474
  hose storage cabinet 448
  hose valve 474
  hydrant 474
  protection 448
  protection system 473
  resistant ceiling 188, 189
  resistant drywall 417
  retardant lumber 338
  retardant plywood 338
  suppression automatic 475
  tube boiler 500
  valve 474
Firebrick 321
Fireplace box 321
  built-in 446
  demolition 291
  free standing 445
  mantel 362
  masonry 321
  prefabricated 254, 445
Fireproofing 385
  plaster 385
  spray 385
  steel column 111
Firestop gypsum 415
  wood 343
Firestopping 385
Fitting cast iron 482
  cast iron pipe 483
  conduit 519
  copper 483
  grooved joint pipe 485
  plastic 486
  plastic pipe 487
  PVC 486
  steel 485
Fixture bathroom 495, 496
  carrier 493
  electric 533
  exterior mercury vapor 534
  fluorescent 533
  incandescent 534
  interior light 533
  mercury vapor 534
  metal halide 534
  mirror light 534
  plumbing 491-493
  removal 480

residential 523
support handicap 493
wire 515
Flagging 300, 422
  slate 300
Flagpole 254, 446
  aluminum 447
Flashing 169, 381
  aluminum 381, 492
  asphalt 381
  butyl 382
  cement 380
  copper 381, 492
  counter 384
  fabric 382
  lead coated 382
  masonry 381
  membrane 378
  PVC 382
  stainless 382
  valley 372
  vent 492
Flat precast concrete wall . 131, 134
Flatbed truck 279
Flexible conduit & fitting ... 526
  door 397
  ductwork fabric coated 506
  insulated duct 506
  transparent curtain 261
Float concrete 277
  finish concrete 311
  glass 410
Floater equipment 273
Floating floor 423
  pin 408
Flood coating 379
Floodlight 280, 281
Floor 422
  acid proof 422
  asphalt block 300
  brick 422
  carpet 185, 186, 424
  ceramic tile 419
  composition 185, 186
  concrete topping 186
  conductive 423
  cork 423
  door 396
  drain 491
  epoxy 424
  finish 436
  finisher 277
  flagging 300
  framing removal 289
  framing steel joist 119
  insulation 368
  marble 319
  mat 259
  multi span joist slab 113
  nail 339
  oak 422
  paint 431
  parquet 422
  plank 343
  plywood 350
  polyethylene 423
  quarry tile 419
  register 509
  removal 287
  resilient 185, 186
  rubber 424
  sander 282
  stain 431
  subfloor 350
  terrazzo 420, 572
  tile and covering 185

625

Index

tile terrazzo	420
underlayment	185, 350
varnish	432
vinyl	424
wood	185, 186, 422
Flooring	422
composition	424
demolition	289
Flue chimney	320
chimney metal	501
liner	320, 321
prefab metal	501
tile	320
Fluorescent fixture	533
lamp	535
Flush tube framing	412
Fluted window precast concrete	131
FM200 fire extinguisher	475
Foam glass insulation	369
insulation	369
Foil aluminum	369, 371
back insulation	370
Folding accordion partition	450
door shower	444
partition	450
Food warmer counter	256
Footing concrete	310
reinforcing	309
removal	287
spread	98, 308
Forklift brick	280
Form edge	308
polystyrene	308
slab	328
Formbloc	316
Formwork bulkhead	308
concrete	308
curb	308
plywood	308
structural	308
wall	308
Fossil fuel hydronic heating	221
Foundation backfill	100
chimney	310
concrete	310
concrete block	317
excavation	100
mat	310
removal	286
wall	317
Fountain drinking	495
wash	494
Frame baked enamel	392
door	390, 391, 394
drywall	391
entrance	394
fire door	391
labeled	391
metal	391
metal butt	392
steel	391
welded	391
window	404
wood	357, 394
Framing anchor	340
band joist	333
beam & girder	341
canopy	346
coldformed	328
demolition	290
heavy	342
joists	333
laminated	351
metal	326
metal joists	333
metal partitions	330
metal studs	330
partition	414
removal	289
roof rafters	344
sleepers	346
steel	327
steel joist	119
timber	342
tube	412
wall	346
window wall	412
wood	341, 342
Freestanding chalkboard	442
Freeze deep	457
Front end loader	294
shovel	278
Frozen food chest	257
FRP panel	426
Fryer with submerger	256
Fuel tank	466
Full vision door	399
vision glass	411
Furnace duct	500
gas	501
gas fired	500
hot air	500
oil fired	500
wall	501
Furnishing	259
cabinet	259
Furnishings site	302
Furring and lathing	414
ceiling	348, 414, 415
channel	414, 418
metal	414
steel	414
wall	348, 415
Fusible link closer	408
Fusion machine butt	280

G

Gable dormer	348
Galvanized ductwork	506
gravel stop	168
roof	374
roof edge	168
steel reglet	384
Garage cost	260
door	397
door weatherstrip	410
Garbage disposal	457
disposer	255
Garden house	261
Gas fired boiler	499
fired furnace	500
fired infra-red heater	505
fired pool heater	500
fired space heater	501
furnace	501
generator	527
heat air conditioner	503
heating boiler	219, 220
log	446
pipe	296
pulse combustion	499
vent	501
water heater	498
Gasoline generator	527
Gate metal	301
valve	489
General contractor's overhead	539
Generator construction	280
diesel	527
emergency	527
gas	527
gasoline	527
set	527
Geodesic	261
Girder wood	341, 343
Glass	410, 411
and aluminum door	157, 158
bead	412
bead molding	354
block	318, 561
bulletin board	442
demolition	292
door	395, 397-399
door shower	444
fiber felt	166
float	410
full vision	411
heat reflective	411
lined water heater	458
low emissivity	411
masonry	318
mirror	411, 451
panel	163
shower stall	444
tempered	410, 411
tile	411
tinted	410
window	411
Glaze coat	436
Glazed ceramic tile	418
coating	436
wall coating	436
Glazing	410
labor	570
Glove bag	467
Glued laminated	350, 351
laminated construction	351
Grab bar	254, 451
Gradall	278
Grade fine	303
Grader motorized	278
Grading	293
Grandstand	260
Granite	319
building	319
chips	303
curb	299, 319
demolition	291
indian	299
paver	319
sidewalk	300
veneer wall	143
Grass cloth wallpaper	426
lawn	303
seed	303
sprinkler	300
Grating	335
Gravel bankrun	286
base	298
fill	286, 294
pea	303
stop	172, 384
stop aluminum	168
stop duranodic	168
Gravity retaining wall	302
Grease interceptor	491
Greenhouse	261, 465
Grid spike	340
Grille air return	509
aluminum	509
decorative wood	362
painting	432
roll up	396
top coiling	396
window	404
Grinder terrazzo	277
Grooved block	316
joint pipe	485
Ground	348
clamp water pipe	513
cover	303
face block	316
rod	512, 513
wire	513
Grounding	512
& conductor	514, 515, 517
wire brazed	513
Group wash fountain	494
Grout	314
concrete block	314
epoxy	419
tile	418
wall	314
Guard gutter	384
wall	445
Guards wall & corner	445
Gutter aluminum	384
demolition	291
guard	384
strainer	384
vinyl	384
wood	384
Gutting	290
Guyed derrick	283
Guying tree	305
Gypsum block demolition	288
board	416
board accessories	418
board ceiling	188, 189
cement	314
drywall	416
fabric wallcovering	426
firestop	415
lath	415, 571
lath nail	339
partition	449
plaster	415, 571
plaster ceiling	188, 189
plaster partition	179
sheathing	349
weatherproof	349

H

Hair interceptor	491
Half round molding	354
Hammer bush	311
chipping	280
demolition	278
drill	280
hydraulic	281
Hand carved door	393
clearing	293
excavation	293, 294
hole	286
rail	336
split shake	373
trowel finish concrete	311
Handicap fixture support	493
lever	408
opener	407
ramp	310
shower	496
tub shower	495
water cooler	495
Handle door	359
Handling rubbish	292
unit air	503
Hand-off-automatic control	531
Handrail	445
and railing	335

626

Index

wood 354, 360
Hangar 260, 261
Hanger beam 340
   joist 340
Hardboard cabinet 356
   paneling 355
   siding 377
   soffit 355
   tempered 355
   underlayment 350
Hardware 407
   cabinet 359
   door 407
   drapery 460
   finish 407
   ooor 407
   rough 340
   window 407
Hardwood floor 422
   grille 362
Hasp 408
Hatch roof 384
Haul earth 100
Hauling excavation 294
   truck 295
Hay 302
Head sprinkler 475
Header wood 346
Headers & beams boxed 329, 333
Hearth 321
Heat baseboard 505
   loss 577
   pump residential 525
   radiant 505
   reflective glass 411
Heater contractor 281
   electric 457
   electric wall 506
   floor mounted space 501
   infra-red 505
   infra-red quartz 506
   sauna 464
   swimming pool 500
   terminal 505
   unit 501, 505
   warm air 501
   water 458, 498
Heating 480, 481, 499-501
   & cooling classroom 505
   cable 505
   control valve 490
   electric 505
   estimate electrical 505
   fin-tube 222
   hot air 501, 505
   hot-water and steam 219, 220
   hydronic 499
   hydronic fossil fuel 222
   oil-fired hydronic 221
   panel radiant 506
   terminal unit-heater 221
Heavy framing 342
Hemlock column 362
Hex nut 324
   nuts steel 324
Hexagonal face block 317
High bay lighting 534
   build coating 437
   installation conduit 518
   intensity discharge lamp 535
   intensity discharge lighting .. 534
   rise glazing 411
   strength block 316
Highway paver 300
Hinge brass 408
   cabinet 359

residential 408
   steel 408
Hip rafter 345
Hoist 255
   and tower 283
   contractor 283
   lift equipment 282
Holdown 340
Hole drilling 324
Hollow concrete block partition . 174
   metal 391
   metal door 390
   metal frames 568
Honed block 317
Hood range 457
Hook robe 451
Hook-up electric service 242
Horizontal aluminum siding 375
   auger 277
   vinyl siding 376
Horn fire 473
Hose adapter fire 473
   air 280
   discharge 281
   fire 473
   nozzle 474
   rack 474
   reel 474
   suction 281
   water 281
Hospital door hardware 407
   furniture 258
   tip pin 408
   whirlpool 495
Hot air furnace 500
   air heating 501, 505
   tub 495
   water boiler 499, 500
   water heating 499, 505
Hotel cabinet 452
   furnishing 258
   lockset 409
Hot-water and steam boiler 219
House demolition 286
   telephone 536
Housewrap 371
HTRW 467
Humidifier 457
   room 504
Humus peat 302
HVAC demolition 480
   duct 506
Hydrant fire 474
Hydraulic crane 283
   hammer 281
   jack 283

I

Ice skating rink 260
Icemaker 457
Impact wrench 280
In insulation blown 368
   slab conduit 518
Incandescent fixture 534
Incinerator 257
Incubator 256
Indian granite 299
Industrial door 396
   window 400
Infra-red detector 473
   heater 505
   heater gas fired 505
   quartz heater 506
Inlet curb 299

Insecticide 295
Insulation 170, 368, 369, 480
   batt 369
   blanket 369, 480
   board 369
   building 368
   cavity wall 368
   cellulose 368
   composite 371
   duct 480
   fiberglass 368, 369
   foam 369
   foam glass 369
   insert 314
   isocyanurate 369
   masonry 368
   mineral fiber 370
   polystyrene 368, 369
   removal 467
   roof 370
   roof deck 370
   shingle 372
   vapor barrier 371
   vermiculite 368
   wall 314, 369
Insurance 273, 538
   builder risk 273
   equipment 273
   public liability 273
Integrated ceiling 260, 261
Interceptor 491
   grease 491
   metal recovery 491
Intercom 535
Interior door frame 394
   finish 183, 184
   light fixture 533
   pre-hung door 392
   wood frame 394
Intermediate conduit 518
Interval timer 522
Intrusion system 473
Inverter battery 527
Investigation subsurface 286
Iron body valve 489
   removal filter 498
   wrought 336
Ironer commercial 256
Ironspot brick 422
Irrigation system 300
Isocyanurate insulation 369
Isolation transformer 529

J

Jack cable 283
   hydraulic 283
   mud 277
   rafter 345
Jackhammer 280
Jalousie 401
Joint expansion 415
   reinforcing 314
   sealer 387
Joist anchor 340
   connector 340
   demolition 290
   hanger 340
   metal 328
   sister 343
   steel 328
   steel on beam 119, 123
   steel on beam and wall .. 118, 122
   steel on girder and wall 125
   web stiffeners 334

wood 343, 351
   wood pitched 127
Joists cold formed 332
   framing 333
Jute mesh 295

K

Kalamein door 392
Kennel fence 300
Kettle/pot tar 282
Key station on pedestal 257
Keyway 308
Kick plate 408
   plate door 408
Kiln dried lumber 338
Kiosk 260
Kitchen appliance 457
   cabinet 356
   equipment 256, 454
   sink 496
   sink faucet 494
Knocker door 410

L

Lab equip. 258
   grade voltage regulator 529
Labeled door 390
   frame 391
Laboratory equipment 256, 258
Ladder 281, 335
   building 335
   steel 335
   swimming pool 466
Lag screw 325
Lally column 327
Laminated beam 350, 351
   construction glued 351
   countertop 338
   countertop plastic 338
   epoxy & fiberglass 437
   framing 351
   glued 350
   roof deck 350
   veneer members 352
   wood 350, 351
Lamp fluorescent 535
   high intensity discharge 535
   metal halide 535
   post 336
Lampholder 526
Lamphouse 257
Landfill fees 292
Landing newel 361
Landscape light 534
   surface 423
Laser 281
Latch set 409
Latex caulking 387
   underlayment 424
Lath gypsum 415, 571
   metal 415, 571
Lattice molding 354
Laundry equipment 256
   faucet 494
   sink 495
   tray 495
   water heater 499
   water heater coin 499
Lavatory 495
   faucet 494

Index

removal 480
support 493
vanity top 495
wall hung 496
Lawn grass 303
Lazy susan 357
Leaching pit 296
Lead caulking 482
   coated copper downspout ... 383
   coated downspout 383
   coated flashing 382
   flashing 382
   paint encapsulation 471
   paint removal 470
   roof 381
Lean-to type greenhouse 465
Let-in bracing 340
Letter slot post office 254
Lever handicap 408
Liability employer 273
   insurance 539
Library air-conditioning . 227, 229, 231
                           233, 235
   equipment 256, 258
Lift aerial 279
Light bollard 534
   exit 535
   fixture interior 533
   landscape 534
   portable 280
   stand 469
   support 327
   switch 248
   tower 281
   track 535
   weight block partition ... 175
Lighting 534, 535
   contactor 530, 531
   emergency 534
   exit and emergency 534
   fixture exterior 534
   high intensity discharge . 534
   incandescent 534
   mercury vapor 534
   metal halide 534
   outdoor 280
   outlet 523
   residential 523
   strip 533
   track 535
Lightning arrester 464
   protection 464
   suppressor 521
Lightweight block 321
   concrete 554
Limestone 303, 319
   coping 318
   veneer wall 142
Linear diffuser 508
Linen wallcovering 426
Liner duct 481
   flue 320
Lintel 312, 319, 327
   block 317
   concrete 312
   precast 312
Liquid chiller centrifugal 502
Load center indoor 531
   center plug-in breaker . 531, 532
   center rainproof 531
Loadbearing studs 328
Loadcenter residential 521
Loader front end 294
   tractor 279
   vacuum 469
Loading dock equipment 454

Loam 293
Lobby collection box 449
Lock entrance 408
Locker 254
   metal 447
   steel 447
   wire mesh 447
Locking receptacle 523
Lockset cylinder 409
   hotel 409
Log electric 446
   gas 446
Louver 444
   aluminum 444
   midget 445
   redwood 362
   ventilation 362
   wall 445
   wood 362
Louvered blind 363
   door 392
Low-voltage silicon rectifier . 525
   switching 525
   switchplate 525
   transformer 525
Lube equipment 255
Lumber 563, 564
   core paneling 355
   kiln dried 338
   treatment 338
Luminous ceiling 421, 535
   panel 421

M

Machine trowel finish 311
   welding 282
Magnetic motor starter 532
Mahogany door 393
Mail box 254, 449
   box call system 536
   chute 449
Main office expense 273
Mall front 399
   front door 157, 158
Management fee construction ... 272
Manhole precast 298
   removal 287
Mantel beam 362
   fireplace 362
Manual dumbwaiter 478
Maple countertop 338
Marble 319
   base 319
   chips 303
   coping 318
   countertop 338
   floor 319
   screen 443
   sill 318
   sink 202
   synthetic 422
   tile 422
Masonry accessory 314
   adobe 318
   anchor 314
   brick 321, 422
   cement 314
   clay 315
   cleaning 322
   cornice 318
   demolition 288, 290
   fireplace 321
   flashing 381
   furring 348

glass 318
insulation 368
nail 339
painting 436
partition 174
point 322
reinforcing 314, 556
removal 287, 290
saw 282
sill 318
step 299
toothing 288
unit 315
wall 302, 316
wall tie 314
Mat foundation 310
Material hoist 283
MC cable 513
Mechanical media filter 509
Medical center air-conditioner ... 238
   center air-conditioning .. 228, 230,
                     232, 234, 236
   equipment 256
Medicine cabinet 254, 452
Membrane curing 311
   flashing 378
   roofing 377
Mercury vapor exterior fixture ... 534
   vapor fixture 534
Metal butt frame 392
   chimney 260, 445
   deck 328
   deck & joist on bearing wall .. 124
   door 390, 397, 398
   fabrication 335
   fascia 375
   fastenings 324
   fence 301
   flue chimney 501
   frame 391
   framing 326
   furring 414
   gate 301
   halide fixture 534
   halide lamp 535
   halide lighting 534
   hollow 391
   joist 328
   joists bracing 332
   joists bridging 332
   joists coldformed 333
   joists framing 333
   lath 415, 571
   locker 447
   molding 415
   ornamental 336
   partitions framing 330
   pipe 295
   recovery interceptor 491
   roof 374
   sash 400
   screen 400
   sheet 381
   shelf 450
   siding panel 153
   siding support 112
   stair 335
   stud 414
   stud demolition 292
   stud partitions 330
   stud walls 330
   studs 330
   studs framing 330
   support assemblies 414
   threshold 410
   tile 419

toilet partition 443
truss 327
window 400, 401
Metallic coating 99
   foil 369
Meter device basic 526
   socket 526
   water supply 490
   water supply domestic 490
Microphone wire 515
Microtunneling 295
Microwave oven 457
Mill construction 342
Millwork 352, 357
   demolition 291
Mineral fiber ceiling 421
   fiber insulation 370
   roof 380
   wool blown in 368
Mirror 254, 411, 451
   ceiling board 421
   door 411
   glass 411
   light fixture 534
   wall 411
Mix planting pit 304
Mixer concrete 277
   mortar 277, 281
   plaster 281
Mixing valve 494
   valve shower 496
Mobilization 293
Modified bitumen roofing 568
Modular flexible wirnig system .. 515
   office 180
   office system 449
Modulating damper motorized .. 507
Module tub-shower 495
Moil point 280
Molding 353
   base 352
   brick 353
   ceiling 353
   chair 354
   cornice 353
   exterior 353
   hardboard 355
   metal 415
   pine 353
   trim 354
   wood 361
   wood transition 423
Monel rivet 326
Monitor support 326
   voltage 530
Monolithic finish concrete 311
Monument survey 272
Mop roof 379
Mortar 555
Mortar, brick and block 559
Mortar cement 419
   mixer 277, 281
   thinset 419
Moss peat 303
Motion detector 473
Motor connection 526
   control center 532
   generator 280
   starter 532
   starter & control 532
   starter enclosed & heated . 532
   starter magnetic 532
   starter w/fused switch ... 532
   support 327
Motorized modulating damper .. 507
Moulded door 393

Index

Mounting board plywood 349
Movable louver blind 460
   office partition 449
Movie equipment 257, 258
   screen 454
Moving shrub 302
   tree 302
Mud jack 277
Mulch 302
   bark 302
   ceramic 303
   stone 303
Mullion precast concrete .. 131, 134
   vertical 412
Multi-blade damper 507
Multi-channel rack enclosure ... 514
Muntin window 404
Music practice room 261

N

Nail 339
Nailer 281
   pneumatic 280
   steel 343
   wood 343
Napkin dispenser 451
Neoprene flashing 169, 382
Newel 361
Night depository 255
No hub pipe 482
Non-removable pin 408
Nozzle dispersion 475
   fire hose 474
   fog 474
   playpipe 474
Nut hex 324
Nylon carpet 424-427

O

Oak door frame 394
   floor 422
   molding 354
   paneling 355
   stair tread 361
   threshold 394
Oakum caulking 482
Off highway truck 279
Office air-conditioner 238
   air-conditioning . 228, 230, 232, 234, 236
   expense 273
   furniture 258
   partition 449
   partition movable 449
   system modular 449
Oil fired boiler 499
   fired furnace 500
   heater temporary 281
   interceptor 491
   water heater 498
Onyx 303
Ooor hardware 407
Open rail fence 301
Opener automatic 407
   door 397, 407
   handicap 407
Operating cost 277
   cost equipment 277
Organic felts 166
Ornamental metal 336
   railing 336
OSHA testing 469

Outdoor lighting 280
Outlet box cover 521
   box plastic 521
   box steel 520
   lighting 523
Oven 457
   cabinet 357
   microwave 457
Overhaul 292
Overhead and profit 543
   commercial door 397
   contractor 274, 539
   door 157-159, 397

P

P trap 492
   trap running 492
Packaging waste 470
Padding carpet 425
Paddle fan 524
Paint 574
   aluminum siding 433
   and covering wall 183, 184
   cabinets 183, 184
   casework 183, 184
   ceiling 183, 184
   chain link fence 430
   doors & windows interior .. 429
   encapsulation lead 471
   fence picket 430
   floor 431
   floor concrete 431
   floor wood 431
   miscellaneous metal .. 183, 184
   removal 470
   siding 433
   structural steel 183, 184
   trim, exterior 434
   walls masonry, exterior ... 435
   woodwork 183, 184
Painting 574
   balustrade 433
   casework 428
   ceiling 435, 436
   clapboard 433
   concrete block 436
   cornice 433
   decking 431
   downspout 432
   drywall 435
   grille 432
   masonry 436
   pipe 432
   plaster 435
   siding 433
   sprayer 281
   steel siding 433
   stucco 433
   swimming pool 465
   trim 432
   truss 433
   wall 436
   window 430
Paints & coatings 428
Palladian windows 403
Pan form stair 335
   shower 382
Panel acoustical 427, 450
   concrete tilt up 135
   divider 450
   door 393
   fiberglass 375
   FRP 426
   luminous 421

   portable 450
   radiant heat 506
   spandrel glass 163
Panelboard 533
   w/circuit-breaker 532
Paneling 355
   board 356
   cutout 289
   demolition 291
   hardboard 355
   plywood 355
   wood 355
Panic device 409
   device door hardware 408
Paper building 371
   curing 311
   sheathing 371
Paperhanging 425
Paperholder 452
Pargeting 567
Parging cement 368
Park bench 302
Parking equipment 257
Parquet floor 422
Particle board siding 377
   board underlayment 350
Parting bead 354
Partition 449, 450
   accordion 180
   acoustical 180, 450
   anchor 314
   block 321
   commercial 180
   demolition 292
   demountable 449
   door 449
   dust 292
   folding 180, 450
   folding leaf 450
   framing 414
   gypsum 449
   gypsum board 178
   handicap 181
   hospital 255
   industrial 180
   marble 181
   metal 181
   movable 180
   movable office 449
   office 449
   plasterboard 178
   plastic laminate 181
   porcelain enamel 181
   shower 443
   shower stall 254
   sound board 178
   steel 449
   steel folding 180
   support 326, 327
   toilet 181, 442
   urinal 181
   vinyl folding 180
   wood folding 180
   wood frame 344
Partitions demountable 449
Passage door 392
   lock set 409
Patch core hole 274
   roof 380
Patching concrete floor 312
Patient wall system 258
Patio door 157-159, 395
Pavement 299
   berm 299
   demolition 287
   emulsion 300

Paver bituminous 281
   concrete 281
   floor 422
   highway 300
Paving asphaltic 547
   brick 300
   concrete 298
Pea gravel 303
Peastone 303
Peat humus 302
   moss 303
Pedestal access floor 262
Pedestrian traffic control 258
Pegboard 355
Perforated aluminum pipe 297
   ceiling 421
   pipe 297
Performance bond 544, 545
Perlite insulation 369
   plaster 187-189, 416, 572
   plaster ceiling 188, 189
   sprayed 437
Permit building 272
Personal respirator 469
Phone booth 451
PIB roof 379
Pickup truck 282
Picture window 400, 403
Pier brick 321
Pilaster toilet partition 443
   wood column 362
Pile sod 304
Pin floating 408
   non-removable 408
Pine door frame 394
   fireplace mantel 362
   floor 423
   molding 353
   roof deck 350
   shelving 356
   siding 377
   stair 361
   stair tread 360
Pipe & fittings 483, 486, 487, 489
   bedding 294
   cast iron 481
   column 327
   concrete 296, 297, 550
   copper 483
   corrugated metal 297
   covering 481
   covering fiberglass 481
   CPVC 486
   drainage 296
   DWV ABS 486
   elevated installation 480
   encapsulation 470
   fitting cast iron 482, 483, 485
   fitting copper 483
   fitting DWV 487
   fitting grooved joint ... 485
   fitting no hub 482
   fitting plastic 486-488
   fitting soil 482
Pipe, FRP plastic 486
Pipe gas 296
   grooved joint 485
   insulation removal 467
   no hub 482
   painting 432
   perforated aluminum ... 297
   plastic 486
   polyethylene 296
   PVC 296
   rail 335, 336
   removal 287, 480

Index

sewage	296
shock absorber	491
single hub	482
soil	482
stainless	336
steel	297, 485
subdrainage	297
vitrified clay	296, 297
weld joint	485
Piping subdrainage	297
Pit leaching	296
Pitch coal tar	378
emulsion tar	300
pocket	384
pockets	384
Pivoted window	400
Placing concrete	310, 551
Plain tube framing	412
Plank floor	343
precast	114
roof	349
sheathing	349
Plant bed preparation	304
screening	278
Planter	305
fiberglass	305
Planting	302
Plaque	255, 447
Plaster	414
accessories	415
and acoustical ceiling	187
beam	416
ceiling	187, 416
column	415
cutout	289
demolition	288, 292
drill	325
ground	348
gypsum	415
mixer	281
painting	435
perlite	416
thincoat	416
wall	416
Plasterboard	187, 188, 416, 417
ceiling	188, 189
Plastic door	392
fabrications	364
faced hardboard	355
fitting	486
laminated countertop	338
outlet box	521
pipe	486
pipe fitting	487, 488
pipe, FRP	486
skylight	406
trap	492
Plate shear	340
wall switch	526
Platform trailer	282
Plating zinc	339
Players bench	302
Plug in circuit-breaker	532
Plugmold raceway	519
Plumbing	490
appliance	498, 499
demolition	480
fixture	491-493, 576
fixtures removal	480
Plywood	565
clip	340
demolition	288, 291, 292
floor	350
formwork	308
joist	351
mounting board	349

paneling	355
sheathing, roof and walls	349
shelving	356
siding	377
sign	447
soffit	355
subfloor	350
treatment	338
underlayment	350
Pneumatic control system	472
nailer	280
tube system	255
Pocket door frame	394
pitch	384
Pockets pitch	384
Point masonry	322
moil	280
Pole closet	354
Police connect panel	473
Polycarbonate panel	163
Polyethylene film	303
floor	423
pipe	296
pool cover	466
septic tank	296
tarpaulin	276
Polypropylene shower	496
trap	492
Polystyrene blind	363
ceiling	421
ceiling panel	421
insulation	368, 369
Polysulfide caulking	387
Polyurethane caulking	387
varnish	437
Polyvinyl chloride roof	379
Pool accessory	466
cover	466
cover polyethylene	466
heater electric	500
heater gas fired	500
swimming	465
Porcelain enamel sink	202
tile	419
Porch molding	353
Portable air compressor	279
booth	260
fire extinguisher	448
heater	280
light	280
panel	450
Post cap	340
cedar	362
demolition	290
lamp	336
wood	342
Postal specialty	449
Powder act. tools & fast.	326
Power conditioner transformer	527, 529
supply uninterruptible	527
system undercarpet	515
wiring	530
Precast beam L shaped	111
beam T shaped	111
catch basin	298
concrete	312
concrete flat	131, 134
concrete fluted window	131, 134
concrete mullion	131, 134
concrete ribbed	131, 132, 134
concrete specialties	132, 133
concrete unit price	131, 134
conrete	552
coping	318
curb	299

lintel	312
manhole	298
plank	114
receptor	444
sill	318
stair	312
terrazzo	420
wall	552
window sills	133
Pre-engineered steel buildings	575
structure	465
Prefabricated building	465
fireplace	445
Prefinished drywall	417
floor	423
shelving	356
Preformed roof panel	374
roofing & siding	374
Prehung door	157-159, 392
Preparation plant bed	304
surface	437, 439
Pressure booster pump	497
valve relief	489
wash	438
Pretreatment termite	295
Primer asphalt	380, 424
Profile block	316, 317
Projected window	400, 401
Projection screen	254, 454
Projector mechanism	257
Propeller exhaust fan	508
Property line survey	272
Protection fire	448
lightning	464
slope	295
termite	295
worker	469
P&T relief valve	489
Pull box	521
door	359
Pulse boiler gas	499
Pump	497
circulating	493
concrete	277
diaphragm	281
gasoline	255
pressure booster	497
rental	281
sewage ejector	497
staging	274
submersible	281
sump	458, 497
trash	281
water	281, 296
Purlin roof	343
Push button lock	409
Push-pull plate	409
Puttying	437
PVC conduit	518
conduit in slab	518
fitting	486
flashing	382
gravel stop	384
pipe	295, 296, 486
siding	376
waterstop	309

Q

Quarry tile	419
Quarter round molding	354
Quartz	303
heater	506
Quoins	319

R

Rabbeted cedar bev. siding	154, 155
Raceway	515, 517, 518
plugmold	519
surface	519
wiremold	519
wireway	520
Rack hose	474
Radial arch	351
Radiant heat	505
heating panel	506
Radiation fin tube	505
Radiator supply control	490
supply valve	490
thermostat control sys.	472
Radio tower	260
Rafter	344
anchor	340
demolition	290
Rail crash	445
hand	336
pipe	335, 336
trolley	445
wall	336
Railing aluminum	335
demolition	291
ornamental	336
steel	336
wood	354, 360, 361
Railroad tie	305
tie step	299
Ramp handicap	310
Ranch plank floor	423
Range cooking	457
hood	457
receptacle	523, 525
restaurant	256
Ready mix concrete	310, 551
Receptacle air conditioner	523
device	523
dryer	523
duplex	525
locking	523
range	523, 525
telephone	523
television	523
waste	452
weatherproof	523
Receptor precast	444
shower	444
terrazzo	444
Reciprocating water chiller	502
Rectangular diffuser	508
ductwork	506
Rectifier low-voltage silicon	525
Red cedar shingle siding	154
Redwood bark mulch	303
cupola	446
louver	362
paneling	356
shiplap siding	154
siding	377
trim	354
Refinish floor	423
Reflective block	318
insulation	369
Refrigerant removal	470
Refrigeration residential	458
Refrigerator	255
walk in	262
Register air supply	509
baseboard	509
Reglet	384
aluminum	384
galvanized steel	384

Index

Regulator automatic voltage 528
   voltage 528, 529
Reinforced brick 558
   PVC roof 379
Reinforcement concrete 309
Reinforcing dowels 309
   fiber 309
   footing 309
   joint 314
   masonry 314
   steel 309
   wall 309
Release door 536
Removal air conditioner 480
   asbestos 467
   block wall 287
   catch basin 287
   chain link fence 287
   concrete 287
   curb 287
   driveway 287
   fence 287
   floor 287
   foundation 286
   insulation 467
   lavatory 480
   masonry 287
   paint 470
   pipe 287, 480
   pipe insulation 467
   plumbing fixtures 480
   refrigerant 470
   shingle 292
   sidewalk 287
   sod 304
   stone 287
   tree 293
   VAT 468
   water closet 480
   window 293
Rendering 272
Rental equipment 277, 282
   equipment rate 274, 277
   generator 280
Repair fire damage 437
Repellent water 436
Replacement windows 405, 569
Resaturant roof 380
Residential alarm 524
   appliance 457, 524
   device 521
   door 394
   door bell 523
   elevator 478
   fan 524
   fixture 523
   folding partition 450
   greenhouse 465
   gutting 290
   heat pump 525
   hinge 408
   lighting 523
   loadcenter 521
   lock 409
   refrigeration 458
   service 521
   smoke detector 524
   stair 361
   storm door 395
   switch 521
   water heater 498, 525
   wiring 521, 524
Resilient floor 423
   flooring 573
Respirator 469
   personal 469

Resquared shingle 372
Restaurant air-conditioner 238
   air-conditioning . 228, 230, 232, 234
                                , 236
   furniture 258, 259
Restoration concrete 312
   window 470
Retaining wall 301, 302
Retarder vapor 371
Revolving door 157-159
Ribbed precast concrete 131, 132, 134
   waterstop 309
Ridge board 345
   cap 372
   shingle 373
   shingle clay 373
   shingle slate 372
   vent 445
Rig drill 286
Rigid anchor 314
   conduit in-trench 518
   in slab conduit 518
   insulation 369
Ring split 340
   toothed 340
Rings drapery 460
Riser standpipe 213
   wood stair 361
Rivet 326
   tool 326
Road base 298
Robe hook 451
Rod closet 356
   curtain 451
   ground 512, 513
Roll roof 377
   roofing 380
   type air filter 509
   up grille 182, 396
Roller earth 278
Rolling door 157-159, 396, 399
   grille 157
   overhead door 182
   service door 396
Romex copper 514
Roof accessories 384
   adhesive 378
   aluminum 374
   baffle 385
   beam 351
   built-up 377, 378
   cant 345, 378
   clay tile 373
   coating 380
   copper 381
   CSPE 379
   deck 328, 349
   deck and joist 124
   deck concrete 312
   deck insulation 370
   deck laminated 350
   deck wood 351
   drain system 198
   drains 491
   EPDM 379
   exhauster 508
   fill 554
   framing removal 289
   gravel stop 172
   hatch 384
   insulation 370
   joist steel 122, 123, 125, 126
   joist wood 127
   lead 381
   metal 374

   mineral 380
   modified bitumen 380
   mop 379
   nail 339
   panel preformed 374
   patch 380
   PIB 379
   polyvinyl chloride 379
   purlin 343
   rafter 345
   rafters framing 344
   reinforced PVC 379
   resaturant 380
   roll 377
   sheathing 349
   shingle 372
   single-ply 379
   skylight 406
   slate 372, 567
   specialties, prefab 383
   steel 374
   tile 373
   truss 327, 343, 351
   ventilator 385
   zinc 381
Roofing & siding preformed 374
   cold 379
   demolition 291, 468
   membrane 377
   modified bitumen 380
   roll 380
   sheet metal 381
   single ply 379
Rooftop air conditioner 503
   air-conditioner 231
   single zone 231
   space heater 501
Room humidifier 504
Rosewood door 393
Rosin paper 371
Rotary hammer drill 280
Rough buck 344
   carpentry 341
   hardware 340
   stone wall 320
Rough-in sink countertop 496
   sink raised deck 496
   sink service floor 496
   tub 495
Round diffuser 509
Rubber base 423
   coating 368
   floor 424
   floor tile 424
   sheet 424
   threshold 410
   tile 423
   tired roller 278
Rubbish chute 292
   handling 292

S

Safe 257
Safety switch 530
Salamander 281
Sales tax 272, 544
Salt treatment lumber 338
Sampling air 468
Sand fill 286
Sandblasting equipment 281
Sander floor 282
Sanding 437
   floor 423
Sandstone 320

   flagging 300
Sandwich panel siding 153
   panel skylight 407
Sanitary base cove 418
Sash aluminum 400
   metal 400
   security 400
   steel 400
   wood 404
Sauna 257, 464
   door 464
Saw 282
   chain 282
   circular 282
   concrete 277
   masonry 282
Sawn cedar batten siding 154
Scaffold steel tubular 275
Scaffolding 546
   tubular 274
Scale 254
School air-conditioner 238
   air-conditioning . 228, 230, 232, 234
                                , 236
   door hardware 407
   equipment 257, 258
Scoreboard 257
Scored block 317
   split face block 317
Scotch marine boiler 500
Scrape after damage 437
Scraper 278
Scratch coat 420
Screen entrance 443
   fence 301
   metal 400
   molding 353
   projection 454
   security 400
   sight 450
   urinal 443
   window 400
Screening plant 278
Screw anchor 325
   brass 339
   drywall 418
   lag 325
   sheet metal 339
   steel 339
   wood 339
Screws lag 326
Seal pavement 300
Sealant 368
   acoustical 418
   caulking 387
Sealcoat 300
Sealer joint 387
Seating 257
   lecture hall 258
Sectional door 397
Security gate 254
   sash 400
   screen 400
Seeding 303, 550
Selective demolition 288
Self propelled crane 283
Self-contained air conditioner ... 504
Separation barrier 468
Septic system 296
   tank 296
   tank concrete 296
Service electric 530
   entrance cable aluminum 514
   entrance cap 515
   residential 521
   sink 496

Index

sink faucet	494
Sewage ejector pump	497
pipe	296
Shade	259
Shake wood	373
Shear plate	340
wall	349
Sheathed nonmetallic cable	514
romex cable	514
Sheathing	349
asphalt	349
gypsum	349
paper	371
roof	349
Sheathing, roof & walls plyw.	349
Sheet base	378
floor	424
metal	381
metal aluminum	383
metal edge	168
metal roofing	381
metal screw	339
Sheeting driver	280
Sheetrock	417
Shelf brick	308
metal	450
Shellac door	430
Shelter	260
Shelving	356
metal	254
pine	356
plywood	356
prefinished	356
storage	450
wood	356
Shield expansion	325
undercarpet	516
Shielded sound wire	515
Shielding lead	262
lead X-ray	260
Shingle	372
aluminum	372
asphalt	372
concrete	373
removal	292
ridge	373
roof	372
stain	433
strip	372
wood	372
Shock absorber	491
absorber pipe	491
absorbing door	397
Shooting range	260
Shot blaster	282
Shovel front	278
Shower arm	496
built-in	496
by-pass valve	494
compartments	443
cubicle	496
door	444
enclosure	444
glass door	444
handicap	496
pan	382
partition	443
polypropylene	496
receptor	444
stall	496
surround	444
Shredder	257
Shrub broadleaf evergreen	304
deciduous	304
evergreen	303, 304
moving	302
Shutter	460
Sidelight	394
Sidewalk	299, 422
asphalt	299
brick	300
concrete	299
removal	287
Siding aluminum	154, 375
aluminum panel	153
bevel	377
cedar	377
cedar drop	154
demolition	291
fiber cement	375
fiberglass	375
field assembly panel	153
hardboard	377
metal panel	153
nail	339
paint	433
painting	433
plywood	377
PVC	376
rabbetted cedar bev	154
red cedar shingle	154
redwood	377
redwood shiplap	154
removal	292
sandwich panel	153
sawn cedar batten	154
stain	433
steel	376
steel panel	153
support metal	112
vertical T. and G. redwood	154
vinyl	376
wood	376
wood exterior	154, 155
wood product	377
Sign	447
exit	254
Silencer duct	508
Silicone caulking	387
coating	368
water repellent	368
Sill	319, 345
anchor	340
door	354, 394, 410
masonry	318
precast	318
quarry tile	419
Sillcock	494
Silo	260
Silt fence	295
Single hub pipe	482
hung window	400, 401
oven	255
ply roofing	379
zone rooftop unit	503
Single-ply roof	379
Sink	496
base	357
countertop	496
countertop rough-in	496
kitchen	496
laundry	495
raised deck rough-in	496
removal	480
service	496
service floor rough-in	496
support	493
waste treatment	492
Siren	473
Sister joist	343
Site clearing	293
demolition	286, 287
drainage	298
furnishings	302
improvement	300, 302
preparation	286
work electrical	297
Skirtboard	361
Skylight	171, 406
removal	292
roof	406
Slab blockout	308
concrete	310
cutout	288
form	328
multispan joist C.I.P.	113
on grade	308
on grade removal	288
precast	312
textured	310
Slate	320
flagging	300
removal	292
roof	372
shingle	372
sidewalk	300
sill	319
stair	320
tile	422
Slatwall	426
Sleeper	346
Sleepers framing	346
Sliding alum. & glass panels	159
door	395
electric door	159
entrance	399
glass door	395
mirror	452
panel door	399
patio door	159
window	401, 404
Slop sink	495, 496
Slope protection	295
Slotted pipe	296
Slump block	316
Small tools	276
Smoke detector	473
hatch	171
vent chimney	501
Soap dispenser	451
holder	452
Socket meter	526
Sod	303
Sodium low pressure fixture	534
Soffit	346
aluminum	375
drywall	417
plaster	416
plywood	355
vinyl	376
wood	355
Softener water	458
Soil compaction	293
compactor	278
pipe	482
tamping	293
treatment	295
Solar energy	471
Solid concrete block partition	174
core door	157-159
wood door	393
Sound absorbing panel	261
attenuation	427
system	257
system component	535
wire shielded	515
Spa bath	495
Space heater floor mounted	501
heater rental	281
heater rooftop	501
Spade air	280
Spandrel glass panel	163
Spanish roof tile	373
Special construction	260-262
door	396, 397
systems	473, 506
wires & fittings	515
Specialties & accessories, roof	383
piping HVAC	490
precast concrete	132
Specialty item	254
Spike grid	340
Spiral stair	335, 361
Split rib block	316
ring	340
system control system	472
Sport court	260
Spray coating	368
fireproofing	385
substrate	469
Sprayer airless	469
emulsion	280
Spread footing	98, 308, 310
Spreader concrete	281
Sprinkler dry pipe system	210, 211
grass	300
head	475
system component	474
wet pipe system	208
Square tied column	105
Stacked bond block	317
Stage equipment	258
Staging aids	276
pump	274
Stain cabinet	428
casework	428
door	429
floor	431
lumber	352
shingle	433
siding	377, 433
truss	433
Stainless cooling towers	502
flashing	382
pipe	336
reglet	384
screen	443
steel and glass door	157
steel bolt	324
steel gravel stop	384
steel gutter	172
steel hinge	408
steel rivet	326
steel shelf	452
steel sink	201, 202
steel storefront	399
Stair	361
basement	312, 360
carpet	425
ceiling	458
climber	478
concrete	310
electric	458
metal	335
pan form	335
precast	312
removal	290
residential	361
slate	320
spiral	361
stringer	344
tread tile	419
tread wood	361
wood	360, 361

632

Index

Stairs disappearing 458
   fire escape 335
Stairway door hardware 408
Stairwork and handrails 360
Stall shower 496
   toilet 443
   type urinal 494
   urinal 494
Stamping concrete 311
   texture 311
Standard extinguisher 448
Standpipe connection 474
   equipment 214
Starter board & switch 532
   motor 532
Starting newel 361
Station control 531
   scrub 256
Stationary ventilator mushroom .. 509
Steam and hot-water boiler . 219, 220
   bath 464
   bath residential 464
   boiler 499, 500
   boiler electric 499
   cleaner 282
   pressure valve 490
Steel anchor 314
   bin wall 301
   bolt 324, 340
   bridging 341
   channel 415
   column 108
   conduit in slab 518
   conduit in trench 518
   conduit intermediate 518
   conduit rigid 518
   deck 328
   diffuser 509
   door 157, 158, 390, 396, 397
   downspout 383
   drill 280
   edging 305
   estimating 562
   expansion tank 489
   fitting 485
   flashing 382
   frame 391
   framing 327
   furring 414
   galv. gutter 172
   gravel stop 384
   hex nuts 324
   hinge 408
   joist 328
   joist on beam 119, 123
   joist on beam and wall ... 118, 122
   joist on girder and wall 125
   ladder 335
   lintel 327
   locker 447
   nailer 343
   partition 449
   pipe 297, 485
   pipe fitting 485
   reinforcing 309
   roof 374
   roof drain system 198
   sash 400
   scaffold tubular 275
   screw 339
   siding 376
   sink 201, 202
   structural 326, 327
   stud 414
   testing 274
   window 160-162, 400

window casement 161
window double hung 160
window picture 161
window projected 161
window security 161
Steeple 256
   tip pin 408
Step bluestone 299
   brick 299
   masonry 299
   railroad tie 299
   stone 319
Sterilizer 256
Stiffleg derrick 283
Stockade fence 301
Stone 318
   aggregate 303
   anchor 314
   base 298
   collected 320
   faced wall 142
   fill 286
   floor 320
   ground cover 303
   mulch 303
   paver 300, 319
   removal 287
   step 319
   threshold 318
   trim cut 320
   veneer wall 142
   wall 302
Stool cap 354
   window 319, 320
Stop door 354
   gravel 384
Stop-start control station 531
Storage shelving 450
   tank 466
Store air-conditioning . 227, 233, 235, 237
Storefront aluminum 399
Storm door 157-159, 395
   drainage system 198
   window 405
Strainer downspout 383
   gutter 384
   roof 383
   wire 383
   Y type iron body 490
Stranded wire 517
Strap tie 340
Straw 302
Street sign 254
Strength compressive 274
Stringer stair 344
Strip cabinet 536
   floor 422
   footing 310
   lighting 533
   shingle 372
Structural concrete 310
   excavation 294
   formwork 308
   steel 326, 327
Structure pre-engineered 465
Stucco 416, 572
   painting 433
Stud connector 326
   demolition 290
   metal 414
   partition 344
   partitions metal 330
   steel 414
   wall 176, 177, 344, 414
   wall wood 154, 155

walls bracing 328
walls bridging 329
walls metal 330
Studs 570
   loadbearing 328
   metal 330
Studwall 176
Subcontractor O & P 273
Subdrainage pipe 297
   piping 297
   system 297
Subfloor 350
   plywood 350
   wood 350
Submersible pump 281, 296
Subpurlins 562
Subsurface investigation 286
Suction hose 281
Sump pump 458, 497
Supermarket air-conditioning ... 227, 230, 232, 233, 235, 237
Support carrier fixture 493
   ceiling 326
   drinking fountain 493
   lavatory 493
   light 327
   monitor 326
   motor 327
   partition 326, 327
   sink 493
   urinal 494
   water closet 494
   water cooler 494
Suppressor lightning 521
   transformer 530
Surface landscape 423
   preparation 437, 439
   raceway 519
Surfacing 300
Surfactant 469
Surgery table 257
Surround shower 444
   tub 444
Survey crew 272
   monument 272
   property line 272
   topographic 272
Suspended ceiling 415, 421
Suspension mounted heater 501
   system 187-189
   system ceiling 188, 189, 414
Swimming pool ... 260, 262, 465, 466
   pool enclosure 262, 465
   pool heater 500
   pool ladder 466
   pools 575
Swing check valve 488
Switch box 520
   box plastic 521
   decorator 522
   dimmer 522, 525
   electric 525, 530
   general duty 530
   residential 521
   safety 530
Switching low-voltage 525
Switchplate low-voltage 525
Synthetic marble 422
System antenna 536
   boiler fossil fuel 221
   boiler oil fired 222
   buzzer 536
   control 472
   dry pipe sprinkler 210, 211
   fin-tube radiation 222
   fire sprinkler 208-211

FM200 fire suppression 215
heating hydronic 221, 222
hydronic fossil fuel 221
irrigation 300
light pole 246
packaged water-heater 498
riser dry standpipe 213
septic 296
sprinkler 474
standpipe riser dry 213
subdrainage 297
T.V. 536
undercarpet data processing .. 516
undercarpet power 515
undercarpet telephone 516
VAV box control 472
VHF 536
wall exterior brick 144
wet pipe sprinkler 208, 209

T

Table top 338
Tamper 280
Tamping soil 293
Tandem roller 278
Tank expansion 489
   fibergalss 466
   oil & gas 489
   septic 296
   storage 466
Tar kettle/pot 282
   paper 303
   pitch emulsion 300
   roof 380
Tarpaulin 276
   duck 276
   polyethylene 276
Tarred felt 378
Tavern air-conditioning . 227, 229, 231, 233, 235, 237
Tax 272, 545
   sales 272
   social security 272
   unemployment 272
T-bar mount diffuser 509
Teak floor 423
   molding 353, 354
   paneling 355
Tee cleanout 491
   pipe 485
Telephone booth 451
   enclosure 451
   house 536
   receptacle 523
   undercarpet 516
   wire 515
Television receptacle 523
Teller automated 255
   window 255
Temperature relief valve 489
Tempered glass 410, 411
   hardboard 355
Tempering valve 489
Temporary oil heater 281
Tennis court fence 301
Tension structure 262
Terminal A/C packaged 503
   heater 505
Termination box 536
Termite pretreatment 295
   protection 295
Terne coated flashing 382
Terra cotta coping 318
   cotta demolition 288

633

Index

Terrazzo 420
base 420
epoxy 424
floor 420, 572
precast 420
receptor 444
wainscot 420
Testing 274
OSHA 469
Texture 1-11 siding 154
stamping 311
Textured slab 310
Thermostat 473
control sys. radiator 472
wire 515, 525
Thincoat plaster 416
Thinset ceramic tile 418
mortar 419
Threshold 410
door 394
stone 318, 320
wood 354
Thru-wall air conditioner 503
Tie rafter 345
railroad 305
strap 340
wall 314
Tile 418, 424
aluminum 373
and covering floor 185, 186
ceiling 188, 189, 421
ceramic 418, 419
clay 373
concrete 373
cork 423
cork wall 425
demolition 288
flue 320
glass 411
grout 418
marble 422
metal 419
porcelain 419
quarry 419
roof 373
rubber 423
slate 422
stainless steel 419
stair tread 419
vinyl 424
wall 418
window sill 419
Tilt up concrete 553
up concrete panel 135
Timber connector 340
fastener 339
framing 342
laminated 351
Timer clock 524
interval 522
Tin clad door 393
clad fire door 568
Tinted glass 410
Toggle bolt 325
Toilet accessory 451
bowl 496
compartments 442
door hardware 408
partition 442
stall 443
tissue dispenser 254, 451, 452
trailer 282
Tool air 280
rivet 326
Tools small 276
Toothed ring 340
Toothing masonry 288
Top coiling grille 396
demolition counter 291
dressing 293
table 338
Topographic survey 272
Topping epoxy 424
Torch cutting 282
Towel bar 254, 452
dispenser 451, 452
Tower cooling 502
light 281
Track drawer 359
drill 280
light 535
lighting 535
traverse 460
Tractor 279, 294
loader 279
truck 282
Traffic detector 257
Trailer platform 282
toilet 282
truck 279, 282
Transceiver 514
Transfer switch automatic 530
Transformer 527
cabinet 521
computer 529
constant voltage 529
dry type 528
high-voltage 528
isolation 529
low-voltage 525
power conditioner 529
silicon filled 528
suppressor 530
UPS 527
Transient voltage suppressor 530
Transit 282
Transition molding wood 423
Transom lite frame 392
Transplanting 302
Trap cast iron 492
drainage 492
grease 491
P 492
plastic 492
polypropylene 492
Trash pump 281
Traverse 460
track 460
Travertine 320
Tray cable 517
laundry 495
Tread abrasive 419
stone 320
wood 361
Treatment acoustical 427
lumber 338
plywood 338
wood 338
Tree 304
deciduous 304
Tree deciduous 305
Tree evergreen 303
guying 305
moving 302
removal 293
Trench backfill 264, 294
box 282
excavation 294
utility 294
Trencher 279
chain 279
Trim exterior 353
Trim, exterior paint 434
Trim molding 354
painting 432
redwood 354
tile 418
wood 353
Trolley rail 445
Trowel coating 368
concrete 277
Truck dump 279
flatbed 279
hauling 295
loading 294
mounted crane 283
off highway 279
pickup 282
rental 279
tractor 282
trailer 279, 282
vacuum 282
Truss bowstring 351
demolition 290
flat wood 351
metal 327
painting 433
plate 340
roof 327, 343, 351
stain 433
varnish 433
Tub enclosure 254
hot 495
redwood 495
rough-in 495
shower handicap 495
surround 444
Tube framing 412
Tubing copper 483
electric metallic 518
Tub-shower module 495
Tubular scaffolding 274
steel joist 351
Tumbler holder 452
Tunneling 295
Turned column 362
Turnstile 255
TV antenna 536
antenna wire 515
system 536

U

Undercarpet power system 515
shield 516
telephone system 516
Undereave vent 445
Underground duct 297
Underlayment 350
hardboard 350
latex 424
Unemployment tax 272
Uninterruptible power supply 527
Unit heater 501, 505
heater electric 506
masonry 315
price precast concrete 134
Urethane wall coating 437
Urinal 494
screen 443
stall 494
stall type 494
support 494
wall hung 494
Utility electric 297
set fan 508
site work 297
trench 294
vault 286

V

Vacuum central 454
cleaning 454
loader 469
truck 282
Valance board 357
Valley flashing 372
rafter 345
Valve 488, 489
balancing & shut off 490
bronze 488
bronze butterfly 488
check swing 489
fire 474
fire hose 474
gate 489
heating control 490
hot water radiator 490
iron body 489
mixing 494
radiator supply 490
relief pressure 489
shower by-pass 494
shower mixing 496
sprinkler alarm 475
steam pressure 490
swing check 488
tempering 489
water pressure 489
Vanity base 360
top lavatory 202, 495
Vapor barrier 371
barrier sheathing 349
retarder 371
Variable volume damper 507
Varnish cabinet 428
casework 428
door 430
floor 432, 436
polyurethane 437
truss 433
VAT removal 468
Vault utility 286
VCT removal 289
Veneer ashlar 561
brick 315, 559
core paneling 355
granite 319
members laminated 352
removal 291
wall stone 142
Venetian blind 460
Vent chimney 501
chimney all fuel 501
dryer 458
exhaust 509
flashing 492
metal chimney 501
ridge 445
ridge strip 445
Ventilating air conditioning 472, 507, 508, 509
Ventilation fan 524
louver 362
Ventilator control system 472
mushroom stationary 509
roof 385
Verge board 353
Vermiculite insulation 368
Vertical aluminum siding 375
T. and G. redwood siding 154, 155

INDEX

634

Index

vinyl siding 376
VHF system 536
Vibrator concrete 277
earth 278, 294
Vibratory equipment 278
roller 278
Vinyl blind 363
composition floor 424
door trim 376
downspout 383
faced wallboard 417
fascia 376
floor 424
gutter 384
siding 376
siding accessories 376
soffit 376
tile 424
wall coating 437
wallpaper 426
window trim 376
Vitreous china lavatory 202
Vitreous-china service sink 202
Vitrified clay pipe 296, 297
Vocational shop equipment 257
Voltage monitor 530
regulator 528, 529
regulator computer grade 528
regulator lab grade 529
regulator std grade 529
Volume control damper 507

W

Wainscot ceramic tile 418
molding 354
quarry tile 419
terrazzo 420
Walk 299
Wall & corner guards 445
anchor 324
bumper 408, 445
cabinet 357
canopy 448
cast in place 99
ceramic tile 419
coating 436
concrete 311
covering 573
curtain 412
cutout 288
drill 324
drywall 416
finish 311
Wall finishes 425
Wall formwork 308
foundation 317
framing 346
framing removal 289
furnace 501
furring 348, 415
grout 314
guard 445
heater 457
heater electric 506
hung lavatory 496
hung urinal 494
hydrant 474
insulation 314, 368, 369
lath 415
louver 445
masonry 302, 316
mirror 411
painting 436
paneling 355
plaster 415, 416
precast concrete 131, 134, 552
rail 336
reinforcing 309
removal 287
retaining 301, 302
shear 349
sheathing 349
siding 377
steel bin 301
stone faced metal stud 142
stone faced wood stud 142
structural support 112
stucco 416
stud 344, 414
switch plate 526
tie 314
tie masonry 314
tile 418
tile cork 425
tilt up panel concrete 135
wood stud 154, 155
Wallboard acoustical 418
Wallcovering 425
acrylic 426
gypsum fabric 426
Wallguard 445
Wallpaper 425, 426
grass cloth 426
vinyl 426
wall 183
Walls and partitions demolition .. 292
Walnut door frame 394
floor 423
Wardrobe unit 259
wood 358
Warm air heater 501
Wash bowl 495
brick 322
fountain 494
fountain group 494
Washer 340
commercial coin operated 256
residential 256, 458
Washing machine automatic 458
Waste handling 257
packaging 470
receptacle 452
treatment sink 492
Water atomizer 469
chiller 502
chillers absorption 502
closet 496
closet removal 480
closet support 494
cooler 495
cooler handicap 495
cooler support 494
distribution system 295
filter 497
filter dirt and rust 497
heater 458, 498, 525
heater commercial 498
heater electric 498
heater gas 498
heater laundry 499
heater oil 498
heater removal 480
heater residential 498
heating hot 499, 505
hose 281
pipe ground clamp 513
pressure relief valve 489
pressure valve 489
pump 281, 296, 458, 493
purification 497
repellent 436
repellent coating 368
repellent silicone 368
softener 458
supply domestic meter 490
supply meter 490
tank sprayer 282
tempering valve 489
trailer 282
treatment potable 497
tube boiler 500
well 296
Water-heater 498
car-wash 498
dishwasher 498
system packaged 498
Waterproofing 368
coating 368
Waterstop 309
PVC 309
ribbed 309
Weather cap entrance 519
Weatherproof receptacle 523
Weatherstrip 396, 409
and seals 409
door 409
Web stiffeners joist 334
Weld joint pipe 485
Welded frame 391
wire fabric 309
Welding machine 282
Well 296
water 296
Wheelbarrow 282
Whirlpool bath 495
Window 400, 407
air conditioner 503
aluminum 400
awning 401
blind 363, 460
casement 400, 402
casing 353
demolition 292
double hung 400, 402, 403
frame 404
glass 411
grille 404
hardware 407
industrial 400
metal 400, 401
muntin 404
painting 430
picture 400, 403
pivoted 400
projected 400
removal 292, 293
restoration 470
screen 400
sill 318
sill marble 319
sill tile 419
sills precast 133
sliding 404
steel 400
stool 319, 320
storm 405
trim set 354
trim vinyl 376
wall framing 412
wood 401-404
Winter protection 308
Wire aluminum 517
copper 516, 517
electric 516
fence 300
fixture 515
ground 513
mesh 416
mesh locker 447
strainer 383
telephone 515
thermostat 525
THW 516
THWN-THHN 517
T.V. antenna 515
Wiremold raceway 519
raceway, non-metallic 519
Wireway raceway 520
Wiring air conditioner 525
device 525
device & box 520
fan 524
methods 513
power 530
residential 521, 524
Wood base 352, 362, 363
beam 341, 342
blind 363, 460
block floor 422
block floor demolition 289
blocking 344
bridging 341
canopy 346
casing 353
chips 303
column 342, 362
cupola 446
deck 349, 351
demolition 288
door 157, 158, 392, 569
drawer 359
fascia 353
fastener 339
fence 301
fiber sheathing 349
fiber soffit 355
fiber subfloor 350
fiber underlayment 350
floor 422, 423
floor demolition 289
folding panel 258
folding partition 450
frame 357, 394
framing 341
furring 348
girder 341
gutter 172, 384
handrail 354
joist 343, 351
joist pitched 127
laminated 350, 351
louver 362
molding 361
nailer 343
panel door 393
paneling 355
partition 344
planter 305
product siding 377
railing 360, 361
roof deck 349
roof deck demolition 291
roof trusses 566
sash 404
screw 339
shake 373
sheathing 349
shelving 356
shingle 168, 372
sidewalk 299
siding 376
siding demolition 292

INDEX

635

Index

siding exterior 154, 155
sill 346
soffit 355
stair 360, 361, 566
storm door 392
stud wall 155
subfloor 350
threshold 354
tread 361
treatment 338
trim 353
truss 351
veneer wallpaper 426
wardrobe 358
window 401-404
window awning 160
window casement 160
window demolition 293
window double hung 160
window picture 160
window sliding 160
Woodburning stove 254
Woodwork architectural .. 356
Wool fiberglass 368
Worker protection 469
Workers' compensation . 273, 540-542
Wrecking ball 282
Wrench impact 280
Wrought iron 336

X

X-ray 257
   unit 256

Y

Y type iron body strainer 490
Yellow pine floor 423

Z

Z bar suspension 414
Zee bar 418
Zinc divider strip 420
   plating 339
   roof 381
   terrazzo strip 420
   weatherstrip 409

Division Notes

	CREW	DAILY OUTPUT	LABOR-HOURS	UNIT	2000 BARE COSTS				TOTAL INCL O&P
					MAT.	LABOR	EQUIP.	TOTAL	

Division Notes

	CREW	DAILY OUTPUT	LABOR-HOURS	UNIT	2000 BARE COSTS				TOTAL INCL O&P
					MAT.	LABOR	EQUIP.	TOTAL	

Division Notes

	CREW	DAILY OUTPUT	LABOR-HOURS	UNIT	2000 BARE COSTS				TOTAL INCL O&P
					MAT.	LABOR	EQUIP.	TOTAL	

Division Notes

		CREW	DAILY OUTPUT	LABOR-HOURS	UNIT	2000 BARE COSTS				TOTAL INCL O&P
						MAT.	LABOR	EQUIP.	TOTAL	

Division Notes

Notes

Notes

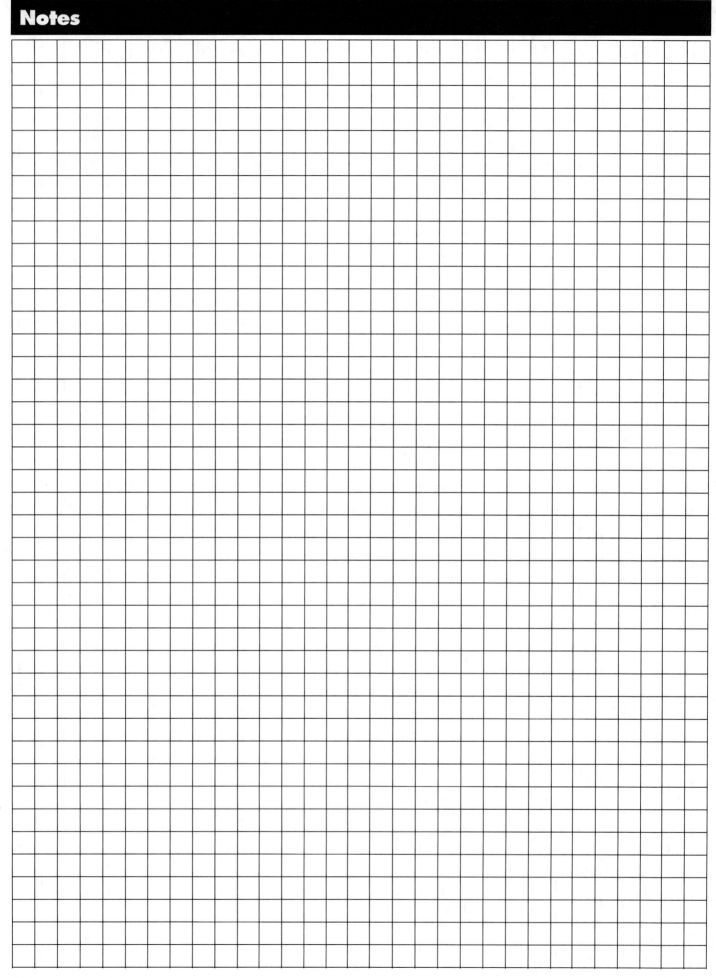

Notes

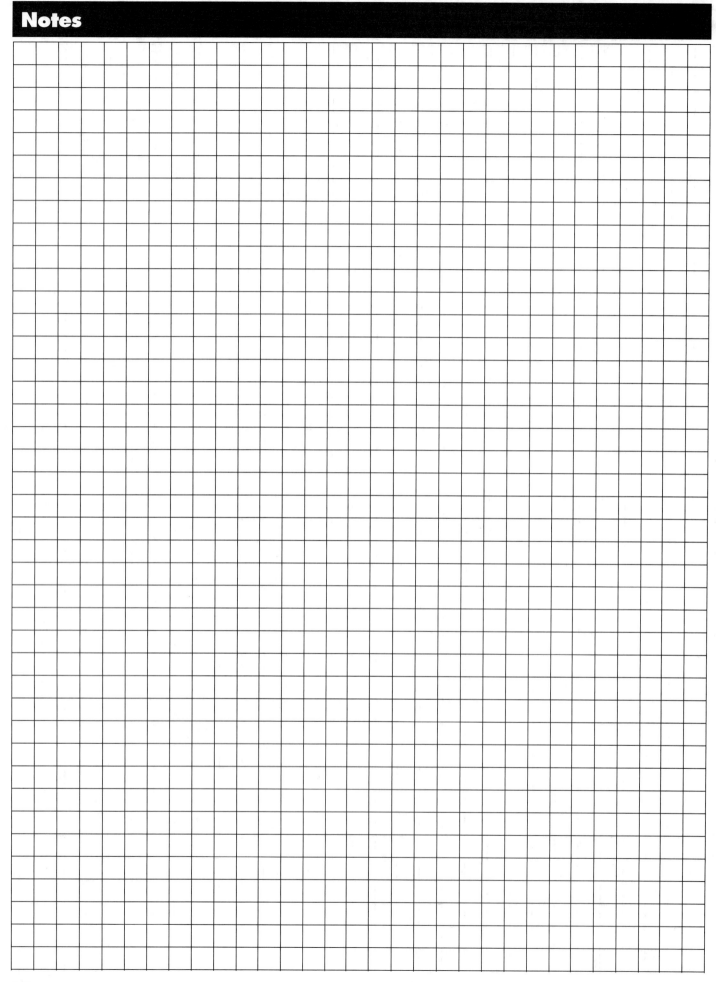

CMD Group...

R.S. Means Company, Inc., a CMD Group company, the leading provider of construction cost data in North America, supplies comprehensive construction cost guides, related technical publications and educational services.

CMD Group, a leading worldwide provider of proprietary construction information, is comprised of three synergistic product groups crafted to be the complete resource for reliable, timely and actionable construction market data. In North America, CMD Group encompasses:
- Architects' First Source
- Construction Market Data (CMD)
- Associated Construction Publications
- Manufacturer's Survey Associates (MSA)
- R.S. Means
- CMD Canada
- BIMSA/Mexico
- Worldwide, CMD Group includes Byggfakta Scandinavia (Denmark, Estonia, Finland, Norway and Sweden) and Cordell Building Information Services (Australia).

First Source for Products, available in print, on the Means CostWorks CD, and on the Internet, is a comprehensive product information source. In alliance with The Construction Specifications Institute (CSI) and Thomas Register, Architects' First Source also produces CSI's SPEC-DATA® and MANU-SPEC® as well as CADBlocks.℠ Together, these products offer commercial building product information for the building team at each stage of the construction process.

CMD Exchange, a single electronic community where all members of the construction industry can communicate, collaborate, and conduct business, better, faster, easier.

Construction Market Data provides complete, accurate and timely project information through all stages of construction. Construction Market Data supplies industry data through productive leads, project reports, contact lists, market penetration analysis and sales evaluation reports. Any of these products can pinpoint a county, look at a state, or cover the country. Data is delivered via paper, e-mail or the Internet.

Construction Market Data Canada serves the Canadian construction market with reliable and comprehensive information services that cover all facets of construction. Core services include: Buildcore Product Source, a preliminary product selection tool available in print and on the Internet; CMD Building Reports, a national construction project lead service; CanaData, statistical and forecasting information; Daily Commercial News, a construction newspaper reporting on news and projects in Ontario; and Journal of Commerce, reporting news in British Columbia and Alberta.

Manufacturers' Survey Associates is a quantity survey and specification service whose experienced estimating staff examines project documents throughout the bidding process and distributes edited information to its clients. Material estimates, edited plans and specifications and distribution of addenda form the heart of the Manufacturers' Survey Associates product line, which is available in print and on CD-ROM.

Associated Construction Publications, founded in 1930, are a group of 14 regional magazines focused on highway and heavy construction and cost data. With one of the world's largest editorial staffs dedicated to construction coverage, the magazine group focuses on local and regional news of projects in all phases of construction, precise project volumes, plus material and labor costs. The magazines have a cumulative ABC and BPA audited circulation of 115,000, an estimated pass-along readership of one-half million, and more than 25,000 pages of annual advertising.

Clark Reports is the premier provider of industrial construction project data, noted for providing earlier, more complete project information in the planning and pre-planning stages for all segments of industrial and institutional construction.

Byggfakta Scandinavia AB, founded in 1936, is the parent company for the leaders of customized construction market data for Denmark, Estonia, Finland, Norway and Sweden. Each company fully covers the local construction market and provides information across several platforms including subscription, ad-hoc basis, electronically and on paper.

Cordell Building Information Services, with its complete range of project and cost and estimating services, is Australia's specialist in the construction information industry. Cordell provides in-depth and historical information on all aspects of construction projects and estimation, including several customized reports, construction and sales leads, and detailed cost information among others.

For more information, please visit our website at www.cmdg.com

CMD Group Corporate Offices
30 Technology Parkway South #100
Norcross, GA 30092-2912
(770) 417-4000
(770) 417-4002 (fax)
www.cmdg.com

Means Project Cost Report

By filling out and returning the Project Description, you can receive a discount of $20.00 off any one of the Means products advertised in the following pages. The cost information required includes all items marked (✔) except those where no costs occurred. The sum of all major items should equal the Total Project Cost.

$20.00 Discount per product for each report you submit.

DISCOUNT PRODUCTS AVAILABLE—FOR U.S. CUSTOMERS ONLY—STRICTLY CONFIDENTIAL

Project Description (No remodeling projects, please.)

- ✔ Type Building _____
- ✔ Location _____
-    Capacity _____
- ✔ Frame _____
- ✔ Exterior _____
- ✔ Basement: full ☐ partial ☐ none ☐ crawl ☐
- ✔ Height in Stories _____
- ✔ Total Floor Area _____
-    Ground Floor Area _____
- ✔ Volume in C.F. _____
-    % Air Conditioned _____ Tons _____
-    Comments _____

-    Owner _____
-    Architect _____
-    General Contractor _____
- ✔ Bid Date _____
-    Typical Bay Size _____
- ✔ Labor Force: _____ % Union _____ % Non-Union
- ✔ Project Description (Circle one number in each line)
  - 1. Economy 2. Average 3. Custom 4. Luxury
  - 1. Square   2. Rectangular 3. Irregular 4. Very Irregular

	✔	**Total Project Cost**			$
A	✔	**General Conditions**			$
B	✔	**Site Work**			$
BS		Site Clearing & Improvement			
BE		Excavation	(	C.Y.	)
BF		Caissons & Piling	(	L.F.	)
BU		Site Utilities			
BP		Roads & Walks Exterior Paving	(	S.Y.	)
C	✔	**Concrete**			$
C		Cast in Place	(	C.Y.	)
CP		Precast	(	S.F.	)
D	✔	**Masonry**			$
DB		Brick	(	M	)
DC		Block	(	M	)
DT		Tile	(	S.F.	)
DS		Stone	(	S.F.	)
E	✔	**Metals**			$
ES		Structural Steel	(	Tons	)
EM		Misc. & Ornamental Metals			
F	✔	**Wood & Plastics**			$
FR		Rough Carpentry	(	MBF	)
FF		Finish Carpentry			
FM		Architectural Millwork			
G	✔	**Thermal & Moisture Protection**			$
GW		Waterproofing-Dampproofing	(	S.F.	)
GN		Insulation	(	S.F.	)
GR		Roofing & Flashing	(	S.F.	)
GM		Metal Siding/Curtain Wall	(	S.F.	)
H	✔	**Doors and Windows**			$
HD		Doors	(	Ea.	)
HW		Windows	(	S.F.	)
HH		Finish Hardware			
HG		Glass & Glazing	(	S.F.	)
HS		Storefronts	(	S.F.	)

J	✔	**Finishes**			$
JL		Lath & Plaster	(	S.Y.	)
JD		Drywall	(	S.F.	)
JM		Tile & Marble	(	S.F.	)
JT		Terrazzo	(	S.F.	)
JA		Acoustical Treatment	(	S.F.	)
JC		Carpet	(	S.Y.	)
JF		Hard Surface Flooring	(	S.F.	)
JP		Painting & Wall Covering	(	S.F.	)
K	✔	**Specialties**			$
KB		Bathroom Partitions & Access.	(	S.F.	)
KF		Other Partitions	(	S.F.	)
KL		Lockers	(	Ea.	)
L	✔	**Equipment**			$
LK		Kitchen			
LS		School			
LO		Other			
M	✔	**Furnishings**			$
MW		Window Treatment			
MS		Seating	(	Ea.	)
N	✔	**Special Construction**			$
NA		Acoustical	(	S.F.	)
NB		Prefab. Bldgs.	(	S.F.	)
NO		Other			
P	✔	**Conveying Systems**			$
PE		Elevators	(	Ea.	)
PS		Escalators	(	Ea.	)
PM		Material Handling			
Q	✔	**Mechanical**			$
QP		Plumbing	(No. of fixtures		)
QS		Fire Protection (Sprinklers)			
QF		Fire Protection (Hose Standpipes)			
QB		Heating, Ventilating & A.C.			
QH		Heating & Ventilating	(BTU Output		)
QA		Air Conditioning	(	Tons	)
R	✔	**Electrical**			$
RL		Lighting	(	S.F.	)
RP		Power Service			
RD		Power Distribution			
RA		Alarms			
RG		Special Systems			
S	✔	**Mech./Elec. Combined**			$

Product Name _____

Product Number _____

Your Name _____

Title _____

Company _____
   ☐ Company
   ☐ Home   Street Address _____

City, State, Zip _____

☐ Please send _____ forms.

Please specify the Means product you wish to receive. Complete the address information as requested and return this form with your check (product cost less $20.00) to address below.

R.S. Means Company, Inc.,
Square Foot Costs Department
100 Construction Plaza, P.O. Box 800
Kingston, MA 02364-9988

R.S. Means Company, Inc... a tradition of excellence in Construction Cost Information and Services since 1942.

For more information visit Means Web Site at www.rsmeans.com

Table of Contents
Annual Cost Guides, Page 2
Reference Books, Page 6
Seminars, Page 11
Consulting Services, Page 13
Electronic Data, Page 14
New Titles, Page 15
Order Form, Page 16

Book Selection Guide

The following table provides definitive information on the content of each cost data publication. The number of lines of data provided in each unit price or assemblies division, as well as the number of reference tables and crews is listed for each book. The presence of other elements such as an historical cost index, city cost indexes, square foot models or cross-referenced index is also indicated. You can use the table to help select the Means' book that has the quantity and type of information you most need in your work.

Unit Cost Divisions	Building Construction Costs	Mechanical	Electrical	Repair & Remodel.	Square Foot	Site Work Landsc.	Assemblies	Interior	Concrete Masonry	Open Shop	Heavy Construc.	Residential	Light Commercial	Facil. Construc.	Plumbing	Western Construction Costs
1	1108	587	588	780		1035		488	1000	1098	1075	430	625	1524	670	1095
2	3126	1460	506	2292		9120		1150	1636	3339	5584	1152	1217	5043	1819	3100
3	1516	90	68	764		1320		193	1935	1489	1492	263	208	1356	43	1493
4	875	28	0	675		770		642	1193	857	677	320	405	1143	0	853
5	1743	176	173	829		780		837	681	1710	1054	634	661	1733	95	1726
6	1396	95	82	1297		482		1317	334	1364	590	1515	1454	1393	59	1749
7	1365	178	91	1331		492		545	454	1366	353	803	1037	1359	188	1367
8	1898	60	0	1900		335		1818	723	1878	9	1131	1180	2030	0	1898
9	1610	48	0	1466		184		1706	322	1556	113	1250	1350	1805	48	1603
10	978	58	32	551		204		823	197	982	0	253	454	979	251	978
11	1157	434	212	621		152		915	35	1051	96	117	239	1220	369	1050
12	352	0	0	48		226		1536	29	343	0	71	68	1537	0	343
13	1384	1217	493	582		443		1071	106	1355	302	230	1331	965	1080	1335
14	364	43	0	261		36		331	0	365	39	9	18	363	17	363
15	2518	14708	764	2249		1802		1471	77	2518	2088	1065	1549	12986	10913	2547
16	1517	801	10627	1195		1098		1289	60	1529	911	703	1286	10207	708	1453
17	459	369	459	0		0		0	0	460	0	0	0	459	369	459
Totals	23366	20352	14095	16841		18479		16132	8782	23260	14383	9946	13082	46102	16629	23412

Assembly Divisions	Building Construction Costs	Mechanical	Electrical	Repair & Remodel.	Square Foot	Site Work Landsc.	Assemblies	Interior	Concrete Masonry	Open Shop	Heavy Construc.	Residential	Light Commercial	Facil. Construc.	Plumbing	Western Construction Costs
1		0	0	202	131	738	775	0	689		752	844	131	89	0	
2		0	0	40	33	35	48	9	49		0	589	32	0	0	
3		0	0	443	1114	0	3064	228	1245		0	1546	806	174	0	
4		0	0	714	1353	0	3172	255	1273		0	2046	1171	26	0	
5		0	0	288	227	0	459	0	0		0	1082	223	25	0	
6		0	0	1006	845	0	1295	1723	152		0	1082	738	330	0	
7		0	0	42	89	0	179	159	0		0	723	33	71	0	
8		2257	149	998	1691	0	2720	926	0		0	1694	1191	1114	2084	
9		0	1346	355	357	0	1283	306	0		0	241	358	319	0	
10		0	0	0	0	0	0	0	0		0	0	0	0	0	
11		0	0	365	465	0	724	153	0		0	0	466	201	0	
12		501	160	539	84	2392	699	0	698		541	0	84	119	850	
Totals		2758	1655	4992	6389	3165	14418	3750	4106		1293	9847	5233	2468	2934	

Reference Section	Building Construction Costs	Mechanical	Electrical	Repair & Remodel.	Square Foot	Site Work Landsc.	Assemblies	Interior	Concrete Masonry	Open Shop	Heavy Construc.	Residential	Light Commercial	Facil. Construc.	Plumbing	Western Construction Costs
Tables	150	46	84	66	4	84	223	59	83	146	55	49	70	86	49	148
Models					102							32	43			
Crews	408	408	408	389		408		408	408	391	408	391	391	389	408	408
City Cost Indexes	yes	yes	yes	yes	yes	yes	yes	yes	yes	yes	yes	yes	yes	yes	yes	yes
Historical Cost Indexes	yes	yes	yes	yes	yes	yes	yes	yes	yes	yes	yes	no	yes	yes	yes	yes
Index	yes	yes	yes	yes	no	yes	yes	yes	yes	yes	yes	yes	yes	yes	yes	yes

Annual Cost Guides

For more information visit Means Web Site at www.rsmeans.com

Means Building Construction Cost Data 2000

Available in Both Softbound and Looseleaf Editions

The "Bible" of the industry comes in the standard softcover edition or the looseleaf edition.

Many customers enjoy the convenience and flexibility of the looseleaf binder, which increases the usefulness of *Means Building Construction Cost Data 2000* by making it easy to add and remove pages. You can insert your own cost information pages, so everything is in one place. Copying pages for faxing is easier also. Whichever edition you prefer, softbound or the convenient looseleaf edition, you'll get the *DesignBuildIntelligence* newsletter at no extra cost.

$85.95 per copy, Softbound
Catalog No. 60010

$109.95 per copy, Looseleaf
Catalog No. 61010

Means Building Construction Cost Data 2000

Offers you unchallenged unit price reliability in an easy-to-use arrangement. Whether used for complete, finished estimates or for periodic checks, it supplies more cost facts better and faster than any comparable source. Over 23,000 unit prices for 2000. The City Cost Indexes cover over 930 areas, for indexing to any project location in North America. Order and get the *DesignBuildIntelligence* newsletter sent to you FREE. You'll have year-long access to the Means Estimating **Hotline** FREE with your subscription. Expert assistance when using Means data is just a phone call away.

$85.95 per copy
Over 650 pages, illustrated, available Oct. 1999
Catalog No. 60010

Means Building Construction Cost Data 2000

Metric Version

The Federal Government has stated that all federal construction projects must now use metric documentation. The *Metric Version* of *Means Building Construction Cost Data 2000* is presented in metric measurements covering all construction areas. Don't miss out on these billion dollar opportunities. Make the switch to metric today.

$89.95 per copy
Over 650 pages, illustrated, available Nov. 1999
Catalog No. 63010

For more information
visit Means Web Site
at www.rsmeans.com

Annual Cost Guides

Means Mechanical Cost Data 2000

• HVAC • Controls

Total unit and systems price guidance for mechanical construction . . . materials, parts, fittings, and complete labor cost information. Includes prices for piping, heating, air conditioning, ventilation, and all related construction.

Plus new 2000 unit costs for:

- Over 2500 installed HVAC/controls assemblies
- "On Site" Location Factors for over 930 cities and towns in the U.S. and Canada
- Crews, labor and equipment

$89.95 per copy
Over 600 pages, illustrated, available Oct. 1999
Catalog No. 60020

Means Plumbing Cost Data 2000

Comprehensive unit prices and assemblies for plumbing, irrigation systems, commercial and residential fire protection, point-of-use water heaters, and the latest approved materials. This publication and its companion, *Means Mechanical Cost Data*, provide full-range cost estimating coverage for all the mechanical trades.

$89.95 per copy
Over 500 pages, illustrated, available Oct. 1999
Catalog No. 60210

Means Electrical Cost Data 2000

Pricing information for every part of electrical cost planning: More than 15,000 unit and systems costs with design tables; clear specifications and drawings; engineering guides and illustrated estimating procedures; complete labor-hour and materials costs for better scheduling and procurement; the latest electrical products and construction methods.

- A Variety of Special Electrical Systems including Cathodic Protection
- Costs for maintenance, demolition, HVAC/ mechanical, specialties, equipment, and more

$89.95 per copy
Over 450 pages, illustrated, available Oct. 1999
Catalog No. 60030

Means Electrical Change Order Cost Data 2000

You are provided with electrical unit prices exclusively for pricing change orders—based on the recent, direct experience of contractors and suppliers. Analyze and check your own change order estimates against the experience others have had doing the same work. It also covers productivity analysis and change order cost justifications. With useful information for calculating the effects of change orders and dealing with their administration.

$89.95 per copy
Over 450 pages, available Oct. 1999
Catalog No. 60230

Means Facilities Maintenance & Repair Cost Data 2000

Published in a looseleaf format, *Means Facilities Maintenance & Repair Cost Data* gives you a complete system to manage and plan your facility repair and maintenance costs and budget efficiently. Guidelines for auditing a facility and developing an annual maintenance plan. Budgeting is included, along with reference tables on cost and management and information on frequency and productivity of maintenance operations.

The only nationally recognized source of maintenance and repair costs. Developed in cooperation with the Army Corps of Engineers.

$199.95 per copy
Over 600 pages, illustrated, available Dec. 1999
Catalog No. 60300

Means Square Foot Costs 2000

It's Accurate and Easy To Use!

- **Updated 2000 price information,** based on nationwide figures from suppliers, estimators, labor experts and contractors.
- "How-to-Use" Sections, with **clear examples** of commercial, residential, industrial, and institutional structures.
- Realistic graphics, offering true-to-life illustrations of building projects.
- Extensive information on using square foot cost data, including **sample estimates** and **alternate pricing methods.**

$99.95 per copy
Over 450 pages, illustrated, available Nov. 1999
Catalog No. 60050

Annual Cost Guides

For more information visit Means Web Site at www.rsmeans.com

Means Repair & Remodeling Cost Data 2000
Commercial/Residential

You can use this valuable tool to estimate commercial and residential renovation and remodeling.

Includes: New costs for hundreds of unique methods, materials and conditions that only come up in repair and remodeling. PLUS:
- Unit costs for over 16,000 construction components
- Installed costs for over 90 assemblies
- Costs for 300+ construction crews
- Over 930 "On Site" localization factors for the U.S. and Canada.

$79.95 per copy
Over 600 pages, illustrated, available Oct. 1999
Catalog No. 60040

Means Facilities Construction Cost Data 2000

For the maintenance and construction of commercial, industrial, municipal, and institutional properties. Costs are shown for new and remodeling construction and are broken down into materials, labor, equipment, overhead, and profit. Special emphasis is given to sections on mechanical, electrical, furnishings, site work, building maintenance, finish work, and demolition. More than 45,000 unit costs plus assemblies and reference sections are included.

$209.95 per copy
Over 1150 pages, illustrated, available Nov. 1999
Catalog No. 60200

Means Residential Cost Data 2000

Contains square foot costs for 30 basic home models with the look of today—plus hundreds of custom additions and modifications you can quote right off the page. With costs for the 100 residential systems you're most likely to use in the year ahead. Complete with blank estimating forms, sample estimates and step-by-step instructions.

Means Light Commercial Cost Data 2000

Specifically addresses the light commercial market, which is an increasingly specialized niche in the industry. Aids you, the owner/designer/contractor, in preparing all types of estimates, from budgets to detailed bids. Includes new advances in methods and materials. Assemblies section allows you to evaluate alternatives in the early stages of design/planning.

Over 13,000 unit costs for 2000 ensure you have the prices you need... when you need them.

$74.95 per copy
Over 550 pages, illustrated, available Dec. 1999
Catalog No. 60170

$76.95 per copy
Over 600 pages, illustrated, available Dec. 1999
Catalog No. 60180

Means Assemblies Cost Data 2000

Means Assemblies Cost Data 2000 takes the guesswork out of preliminary or conceptual estimates. Now you don't have to try to calculate the assembled cost by working up individual components costs. We've done all the work for you.

Presents detailed illustrations, descriptions, specifications and costs for every conceivable building assembly—240 types in all—arranged in the easy-to-use UniFormat system. Each illustrated "assembled" cost includes a complete grouping of materials and associated installation costs including the installing contractor's overhead and profit.

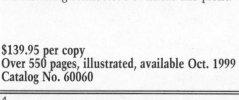

Means Site Work & Landscape Cost Data 2000

Means Site Work & Landscape Cost Data 2000 is organized to assist you in all your estimating needs. Hundreds of fact-filled pages help you make accurate cost estimates efficiently.

Updated for 2000!
- Demolition features—including ceilings, doors, electrical, flooring, HVAC, millwork, plumbing, roofing, walls and windows
- State-of-the-art segmental retaining walls
- Flywheel trenching costs and details
- Updated Wells section
- Thousands of landscape materials, flowers, shrubs and trees

$139.95 per copy
Over 550 pages, illustrated, available Oct. 1999
Catalog No. 60060

$89.95 per copy
Over 600 pages, illustrated, available Nov. 1999
Catalog No. 60280

For more information
visit Means Web Site
at www.rsmeans.com

Annual Cost Guides

Means Open Shop Building Construction Cost Data 2000

The latest costs for accurate budgeting and estimating of new commercial and residential construction... renovation work... change orders... cost engineering. *Means Open Shop BCCD* will assist you to...
- Develop benchmark prices for change orders
- Plug gaps in preliminary estimates, budgets
- Estimate complex projects
- Substantiate invoices on contracts
- Price ADA-related renovations

$89.95 per copy
Over 650 pages, illustrated, available Oct. 1999
Catalog No. 60150

Means Heavy Construction Cost Data 2000

A comprehensive guide to heavy construction costs. Includes costs for highly specialized projects such as tunnels, dams, highways, airports, and waterways. Information on different labor rates, equipment, and material costs is included. Has unit price costs, systems costs, and numerous reference tables for costs and design. Valuable not only to contractors and civil engineers, but also to government agencies and city/town engineers.

$89.95 per copy
Over 450 pages, illustrated, available Nov. 1999
Catalog No. 60160

Means Building Construction Cost Data 2000
Western Edition

This regional edition provides more precise cost information for western North America. Labor rates are based on union rates from 13 western states and western Canada. Included are western practices and materials not found in our national edition: tilt-up concrete walls, glu-lam structural systems, specialized timber construction, seismic restraints, landscape and irrigation systems.

$89.95 per copy
Over 600 pages, illustrated, available Dec. 1999
Catalog No. 60220

Means Heavy Construction Cost Data 2000
Metric Version

Make sure you have the Means industry standard metric costs for the federal, state, municipal and private marketplace. With thousands of up-to-date metric unit prices in tables by CSI standard divisions. Supplies you with assemblies costs using the metric standard for reliable cost projections in the design stage of your project. Helps you determine sizes, material amounts, and has tips for handling metric estimates.

$89.95 per copy
Over 450 pages, illustrated, available Nov. 1999
Catalog No. 63160

Means Construction Cost Indexes 2000

Who knows what 2000 holds? What materials and labor costs will change unexpectedly? By how much?
- Breakdowns for 305 major cities.
- National averages for 30 key cities.
- Expanded five major city indexes.
- Historical construction cost indexes.

$198.00 per year/$49.50 individual quarters
Catalog No. 60140

Means Interior Cost Data 2000

Provides you with prices and guidance needed to make accurate interior work estimates. Contains costs on materials, equipment, hardware, custom installations, furnishings, labor costs . . . every cost factor for new and remodel commercial and industrial interior construction, including updated information on office furnishings, plus more than 50 reference tables. For contractors, facility managers, owners.

$89.95 per copy
Over 550 pages, illustrated, available Oct. 1999
Catalog No. 60090

Means Concrete & Masonry Cost Data 2000

Provides you with cost facts for virtually all concrete/masonry estimating needs, from complicated formwork to various sizes and face finishes of brick and block, all in great detail. The comprehensive unit cost section contains more than 8,500 selected entries. Also contains an assemblies cost section, and a detailed reference section which supplements the cost data.

$79.95 per copy
Over 450 pages, illustrated, available Nov. 1999
Catalog No. 60110

Means Labor Rates for the Construction Industry 2000

Complete information for estimating labor costs, making comparisons and negotiating wage rates by trade for over 300 cities (United States and Canada). With 46 construction trades listed by local union number in each city, and historical wage rates included for comparison. No similar book is available through the trade.

Each city chart lists the county and is alphabetically arranged with handy visual flip tabs for quick reference.

$189.95 per copy
Over 300 pages, available Dec. 1999
Catalog No. 60120

Reference Books

For more information visit Means Web Site at www.rsmeans.com

Project Scheduling & Management for Construction
New Revised Edition
By David R. Pierce, Jr.

A comprehensive, yet easy-to-follow guide to construction project scheduling and control—from vital project management principles through the latest scheduling, tracking, and controlling techniques. The author is a leading authority on scheduling with years of field and teaching experience at leading academic institutions. Spend a few hours with this book and come away with a solid understanding of this essential management topic.

$64.95 per copy
Over 250 pages, illustrated, Hardcover
Catalog No. 67247A

Means Landscape Estimating Methods
New 3rd Edition
By Sylvia H. Fee

This revised edition offers expert guidance for preparing accurate estimates for new landscape construction and grounds maintenance. Includes a complete project estimate featuring the latest equipment and methods, and **two totally new chapters—Life Cycle Costing and Landscape Maintenance Estimating.**

$62.95 per copy
Over 300 pages, illustrated, Hardcover
Catalog No. 67295A

Total Productive Facilities Management
By Richard W. Sievert, Jr.

A New Operational Standard for Facilities Management.

The TPFM program incorporates the best of today's cost and project management, quality, and value engineering principles. The book includes:

- Realistic ways to collect and use benchmarking data.
- Value Engineering to find economical answers to the organization's needs.
- Selecting the best scheduling, control, and contracting methods for construction projects.

$79.95 per copy
Over 270 pages, illustrated, Hardcover
Catalog No. 67321

Cyberplaces: The Internet Guide for Architects, Engineers & Contractors
By Paul Doherty

Internet applications for business and project management. Includes Book, CD-ROM and Web Site.

The CD-ROM offers built-in links to web sites, tours of captured sites, FREE browser software and a document workshop, plus a test for Continuing Education Credits. **The Web Site** keeps the book current with updates on new technologies, interactive workshops, links to new tools and sites, and reports from top firms.

$59.95 per copy
Over 700 pages, illustrated, Softcover
Catalog No. 67317

Value Engineering: Practical Applications
...For Design, Construction, Maintenance & Operations
By Alphonse Dell'Isola, P.E., leading authority on VE in construction

A tool for immediate application—for engineers, architects, facility managers, owners and contractors. Includes: Making the Case for VE—The Management Briefing, Integrating VE into Planning and Budgeting, Conducting Life Cycle Costing, Integrating VE into the Design Process, Using VE Methodology in Design Review and Consultant Selection, Case Studies, VE Workbook, and a Life Cycle Costing program on disk.

$79.95 per copy
Over 450 pages, illustrated, Hardcover
Catalog No. 67319

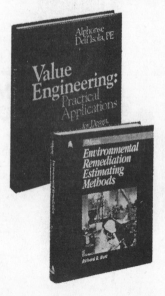

Means Environmental Remediation Estimating Methods
By Richard R. Rast

The first-ever guide to estimating any size environmental remediation project... anywhere in the country.

Field-tested guidelines for estimating 50 standard remediation technologies. This resource will help you: prepare preliminary budgets, develop detailed estimates, compare costs, select solutions, estimate liability, review quotes, and negotiate settlements.

A valuable support tool for *Means Environmental Remediation Unit Price* and *Assemblies* books.

$99.95 per copy
Over 600 pages, illustrated, Hardcover
Catalog No. 64777

For more information visit Means Web Site at www.rsmeans.com

Reference Books

Cost Planning & Estimating for Facilities Maintenance

In this unique book, a team of facilities management authorities shares their expertise at:
- Evaluating and budgeting maintenance operations
- Maintaining & repairing key building components
- Applying *Means Facilities Maintenance & Repair Cost Data* to your estimating

With the special maintenance requirements of the 10 major building types.

$82.95 per copy
Over 475 pages, Hardcover
Catalog No. 67314

Facilities Planning & Relocation

By David D. Owen

An A–Z working guide—complete with the checklists, schematic diagrams and questionnaires to ensure the success of every office relocation. Complete with a step-by-step manual, 100-page technical reference section, over 50 reproducible forms in hard copy and on computer diskettes.
Winner of the International Facility Managers Assoc. "Distinguished Author of the Year" award.

$109.95 per copy, Textbook-384 pages,
Forms Binder-146 pages, illustrated, Hardcover
Catalog No. 67301

Means Facilities Maintenance Standards

Unique features of this one-of-a-kind working guide for facilities maintenance

A working encyclopedia that points the way to solutions to every kind of maintenance and repair dilemma. With a labor-hours section to provide productivity figures for over 180 maintenance tasks. Included are ready-to-use forms, checklists, worksheets and comparisons, as well as analysis of materials systems and remedies for deterioration and wear.

$159.95 per copy, 600 pages, 205 tables, checklists and diagrams, Hardcover
Catalog No. 67246

HVAC: Design Criteria, Options, Selection
Expanded Second Edition

By William H. Rowe III, AIA, PE

Includes Indoor Air Quality, CFC Removal, Energy Efficient Systems and Special Systems by Building Type. Helps you solve a wide range of HVAC system design and selection problems effectively and economically. Gives you clear explanations of the latest ASHRAE standards.

$84.95 per copy
Over 600 pages, illustrated, Hardcover
Catalog No. 67306

The Facilities Manager's Reference

By Harvey H. Kaiser, PhD

The tasks and tools the facility manager needs to accomplish the organization's objectives, and develop individual and staff skills. Includes Facilities and Property Management, Administrative Control, Planning and Operations, Support Services, and a complete building audit with forms and instructions, widely used by facilities managers nationwide.

$86.95 per copy, over 250 pages,
with prototype forms and graphics, Hardcover
Catalog No. 67264

Facilities Maintenance Management

By Gregory H. Magee, PE

Now you can get successful management methods and techniques for all aspects of facilities maintenance. This comprehensive reference explains and demonstrates successful management techniques for all aspects of maintenance, repair and improvements for buildings, machinery, equipment and grounds. Plus, guidance for outsourcing and managing internal staffs.

$86.95 per copy
Over 280 pages with illustrations, Hardcover
Catalog No. 67249

Understanding Building Automation Systems

- Direct Digital Control
- Security/Access Control
- Energy Management
- Life Safety
- Lighting

By Reinhold A. Carlson, PE & Robert Di Giandomenico

The authors, leading authorities on the design and installation of these systems, describe the major building systems in both an overview with estimating and selection criteria, and in system configuration-level detail.

Now $39.98 per copy, limited quantity
Over 200 pages, illustrated, Hardcover
Catalog No. 67284

The ADA in Practice

By Deborah S. Kearney, PhD

Helps you meet and budget for the requirements of the Americans with Disabilities Act. Shows how to do the job right by understanding what the law requires, allows, and enforces. Includes a "buyers guide" for 70 ADA-compliant products.

*Winner of IFMA's "Distinguished Author of the Year" award.

See page 9 for another ADA resource.

$72.95 per copy
Over 600 pages, illustrated, Softcover
Catalog No. 67147A

Reference Books

For more information visit Means Web Site at www.rsmeans.com

Basics for Builders: Plan Reading & Material Takeoff
By Wayne J. DelPico

For Residential and Light Commercial Construction

A valuable tool for understanding plans and specs, and accurately calculating material quantities. Step-by-step instructions and takeoff procedures based on a full set of working drawings.

$35.95 per copy
Over 420 pages, Softcover
Catalog No. 67307

Means Illustrated Construction Dictionary Unabridged Edition

Written in contractor's language, the information is adaptable for report writing, specifications or just intelligent discussion. Contains over 13,000 construction terms, words, phrases, acronyms and abbreviations, slang, regional terminology, and hundreds of illustrations.

$99.95 per copy
Over 700 pages, Hardcover
Catalog No. 67292

Superintending for Contractors:
How to Bring Jobs in On-time, On-Budget
By Paul J. Cook

This book examines the complex role of the superintendent/field project manager, and provides guidelines for the efficient organization of this job. Includes administration of contracts, change orders, purchase orders, and more.

$35.95 per copy
Over 220 pages, illustrated, Softcover
Catalog No. 67233

Means Forms for Contractors
For general and specialty contractors

Means editors have created and collected the most needed forms as requested by contractors of various-sized firms and specialties. This book covers all project phases. Includes sample correspondence and personnel administration.
Full-size forms on durable paper for photocopying or reprinting.

$79.95 per copy
Over 400 pages, three-ring binder, more than 80 forms
Catalog No. 67288

Risk Management for Building Professionals

By Thomas E. Papageorge, RA

Outlines a master plan for phasing a risk management system into each company function. Guides the manager in writing a detailed Risk Management Plan for each project. Contains sample procedures manual.

Now $29.98 per copy, limited quantity
Over 200 pages, illustrated, Hardcover
Catalog No. 67254

Means Heavy Construction Handbook

Informed guidance for planning, estimating and performing today's heavy construction projects. Provides expert advice on every aspect of heavy construction work including hazardous waste remediation and estimating. To assist planning, estimating, performing, or overseeing work.

$74.95 per copy
Over 430 pages, illustrated, Hardcover
Catalog No. 67148

Estimating for Contractors:
How to Make Estimates that Win Jobs
By Paul J. Cook

Estimating for Contractors is a reference that will be used over and over, whether to check a specific estimating procedure, or to take a complete course in estimating.

$35.95 per copy
Over 225 pages, illustrated, Softcover
Catalog No. 67160

HVAC Systems Evaluation
By Harold R. Colen, PE

You get direct comparisons of how each type of system works, with the relative costs of installation, operation, maintenance and applications by type of building. With requirements for hooking up electrical power to HVAC components. Contains experienced advice for repairing operational problems in existing HVAC systems, ductwork, fans, cooling coils, and much more!

$84.95 per copy
Over 500 pages, illustrated, Hardcover
Catalog No. 67281

Quantity Takeoff for Contractors:
How to Get Accurate Material Counts
By Paul J. Cook

Contractors who are new to material takeoffs or want to be sure they are using the best techniques will find helpful information in this book, organized by CSI MasterFormat division.

Now $17.98 per copy, limited quantity
Over 250 pages, illustrated, Softcover
Catalog No. 67262

Roofing: Design Criteria, Options, Selection

By R.D. Herbert, III

This book is required reading for those who specify, install or have to maintain roofing systems. It covers all types of roofing technology and systems. You'll get the facts needed to intelligently evaluate and select both traditional and new roofing systems.

Now $31.48 per copy
Over 225 pages with illustrations, Hardcover
Catalog No. 67253

For more information visit Means Web Site at www.rsmeans.com

Reference Books

Means Estimating Handbook

This comprehensive reference covers a full spectrum of technical data for estimating, with information on sizing, productivity, equipment requirements, codes, design standards and engineering factors.
Means Estimating Handbook will help you: evaluate architectural plans and specifications, prepare accurate quantity takeoffs, prepare estimates from conceptual to detailed, and evaluate change orders.

$99.95 per copy
Over 900 pages, Hardcover
Catalog No. 67276

Means Repair and Remodeling Estimating Third Edition

By Edward B. Wetherill & R.S. Means

Focuses on the unique problems of estimating renovations of existing structures. It helps you determine the true costs of remodeling through careful evaluation of architectural details and a site visit.
New section on disaster restoration costs.

$69.95 per copy
Over 450 pages, illustrated, Hardcover
Catalog No. 67265A

Successful Estimating Methods:
From Concept to Bid
By John D. Bledsoe, PhD, PE

A highly practical, all-in-one guide to the tips and practices of today's successful estimator. Presents techniques for all types of estimates, *and* advanced topics such as life cycle cost analysis, value engineering, and automated estimating.
Estimate spreadsheets available at Means Web site.

$64.95 per copy
Over 300 pages, illustrated, Hardcover
Catalog No. 67287

Means Productivity Standards for Construction

Expanded Edition *(Formerly Man-Hour Standards)*

Here is the working encyclopedia of labor productivity information for construction professionals, with labor requirements for thousands of construction functions in CSI MasterFormat.
Completely updated, with over 3,000 new work items.

$159.95 per copy
Over 800 pages, Hardcover
Catalog No. 67236A

Means Electrical Estimating Methods Second Edition

Expanded version includes sample estimates and cost information in keeping with the latest version of the CSI MasterFormat. Contains new coverage of Fiber Optic and Uninterruptible Power Supply electrical systems, broken down by components and explained in detail. A practical companion to *Means Electrical Cost Data*.

$64.95 per copy
Over 325 pages, Hardcover
Catalog No. 67230A

Means Mechanical Estimating Methods Second Edition

This guide assists you in making a review of plans, specs and bid packages with suggestions for takeoff procedures, listings, substitutions and pre-bid scheduling. Includes suggestions for budgeting labor and equipment usage. Compares materials and construction methods to allow you to select the best option.

$64.95 per copy
Over 350 pages, illustrated, Hardcover
Catalog No. 67294

Means Scheduling Manual
Third Edition
By F. William Horsley

Fast, convenient expertise for keeping your scheduling skills right in step with today's cost-conscious times. Covers bar charts, PERT, precedence and CPM scheduling methods. Now updated to include computer applications.

$64.95 per copy
Over 200 pages, spiral-bound, Softcover
Catalog No. 67291

Maintenance Management Audit

This annual audit program is essential for every organization in need of a proper assessment of its maintenance operation. The forms in this easy-to-use workbook allow managers to identify and correct problems and enhance productivity. Includes electronic forms on disk.

Now $32.48 per copy, limited quantity
125 pages, spiral bound, illustrated, Hardcover
Catalog No. 67299

Means Square Foot Estimating Methods Second Edition

By Billy J. Cox and F. William Horsley

Proven techniques for conceptual and design-stage cost planning. Steps you through the square foot cost process, demonstrating faster, better ways to relate the design to the budget. Now updated to the latest version of UniFormat.

$69.95 per copy
Over 300 pages, illustrated, Hardcover
Catalog No. 67145A

Means ADA Compliance Pricing Guide

Gives you detailed cost estimates for budgeting modification projects, including estimates for each of 260 alternates. You get the 75 most commonly needed modifications for ADA compliance, each an assembly estimate with detailed cost breakdown.

$72.95 per copy
Over 350 pages, illustrated, Softcover
Catalog No. 67310

Reference Books

For more information visit Means Web Site at www.rsmeans.com

Means Unit Price Estimating Methods
2nd Edition
$59.95 per copy
Catalog No. 67303

Legal Reference for Design & Construction
By Charles R. Heuer, Esq., AIA
Now $54.98 per copy, limited quantity
Catalog No. 67266

Means Plumbing Estimating Methods
New 2nd Edition
By Joseph J. Galeno & Sheldon T. Greene
$59.95 per copy
Catalog No. 67283

Structural Steel Estimating
By S. Paul Bunea, PhD
Now $39.98 per copy, limited quantity
Catalog No. 67241

Business Management for Contractors
How to Make Profits in Today's Market
By Paul J. Cook
Now $17.98 per copy, limited quantity
Catalog No. 67250

Understanding Legal Aspects of Design/Build
By Timothy R. Twomey, Esq., AIA
$79.95 per copy
Catalog No. 67259

Construction Paperwork
An Efficient Management System
By J. Edward Grimes
Now $26.48 per copy, limited quantity
Catalog No. 67268

Contractor's Business Handbook
By Michael S. Milliner
Now $21.48 per copy, limited quantity
Catalog No. 67255

Basics for Builders: How to Survive and Prosper in Construction
By Thomas N. Frisby
$34.95 per copy
Catalog No. 67273

Successful Interior Projects Through Effective Contract Documents
By Joel Downey & Patricia K. Gilbert
Now $34.98 per copy, limited quantity
Catalog No. 67313

The Building Professional's Guide to Contract Documents
By Waller S. Poage, AIA, CSI, CCS
$64.95 per copy
Catalog No. 67261

Illustrated Construction Dictionary, Condensed
$59.95 per copy
Catalog No. 67282

Managing Construction Purchasing
By John G. McConville, CCC, CPE
Now $31.48 per copy, limited quantity
Catalog No. 67302

Hazardous Material & Hazardous Waste
By Francis J. Hopcroft, PE, David L. Vitale, M. Ed., & Donald L. Anglehart, Esq.
Now $44.98 per copy, limited quantity
Catalog No. 67258

Fundamentals of the Construction Process
By Kweku K. Bentil, AIC
Now $34.98 per copy, limited quantity
Catalog No. 67260

Construction Delays
By Theodore J. Trauner, Jr., PE, PP
$59.95 per copy
Catalog No. 67278

Basics for Builders: Framing & Rough Carpentry
By Scot Simpson
$24.95 per copy
Catalog No. 67298

Interior Home Improvement Costs
6th Edition
$19.95 per copy
Catalog No. 67308B

Exterior Home Improvement Costs
6th Edition
$19.95 per copy
Catalog No. 67309B

Concrete Repair and Maintenance Illustrated
By Peter H. Emmons
$69.95 per copy
Catalog No. 67146

How to Estimate with Metric Units
Now $24.98 per copy, limited quantity
Catalog No. 67304

For more information
visit Means Web Site
at www.rsmeans.com

Seminars

Developing Facility Assessment Programs

This two-day program concentrates on the management process required for planning, conducting, and documenting the physical condition and functional adequacy of buildings and other facilities. Regular facilities condition inspections are one of the facility department's most important duties. However, gathering reliable data hinges on the design of the overall program used to identify and gauge deferred maintenance requirements. Knowing where to look . . . and reporting results effectively are the keys. This seminar is designed to give the facility executive essential steps for conducting facilities inspection programs.

Inspection Program Requirements • Where is the deficiency? • What is the nature of the problem? • How can it be remedied? • How much will it cost in labor, equipment and materials? • When should it be accomplished? • Who is best suited to do the work?

Note: Because of its management focus, this course will not address trade practices and procedures.

Repair and Remodeling Estimating

Repair and remodeling work is becoming increasingly competitive as more professionals enter the market. Recycling existing buildings can pose difficult estimating problems. Labor costs, energy use concerns, building codes, and the limitations of working with an existing structure place enormous importance on the development of accurate estimates. Using the exclusive techniques associated with Means' widely acclaimed **Repair & Remodeling Cost Data**, this seminar sorts out and discusses solutions to the problems of building alteration estimating. Attendees will receive two intensive days of eye-opening methods for handling virtually every kind of repair and remodeling situation . . . from demolition and removal to final restoration.

Mechanical and Electrical Estimating

This seminar is tailored to fit the needs of those seeking to develop or improve their skills and to have a better understanding of how mechanical and electrical estimates are prepared during the conceptual, planning, budgeting and bidding stages. Learn how to avoid costly omissions and overlaps between these two interrelated specialties by preparing complete and thorough cost estimates for both trades. Featured are order of magnitude, assemblies, and unit price estimating. In combination with the use of **Means Mechanical Cost Data**, **Means Plumbing Cost Data** and **Means Electrical Cost Data**, this seminar will ensure more accurate and complete Mechanical/Electrical estimates for both unit price and preliminary estimating procedures.

Unit Price Estimating

This seminar shows how today's advanced estimating techniques and cost information sources can be used to develop more reliable unit price estimates for projects of any size. It demonstrates how to organize data, use plans efficiently, and avoid embarrassing errors by using better methods of checking.

You'll get down-to-earth help and easy-to-apply guidance for:
- making maximum use of construction cost information sources
- organizing estimating procedures in order to save time and reduce mistakes
- sorting out and identifying unusual job requirements to improve estimating accuracy.

Square Foot Cost Estimating

Learn how to make better preliminary estimates with a limited amount of budget and design information. You will benefit from examples of a wide range of systems estimates with specifications limited to building use requirements, budget, building codes, and type of building. And yet, with minimal information, you will obtain a remarkable degree of accuracy.

Workshop sessions will provide you with model square foot estimating problems and other skill-building exercises. The exclusive Means building assemblies square foot cost approach shows how to make very reliable estimates using "bare bones" budget and design information.

Scheduling and Project Management

This seminar helps you successfully establish project priorities, develop realistic schedules, and apply today's advanced management techniques to your construction projects. Hands-on exercises familiarize participants with network approaches such as the Critical Path Method. Special emphasis is placed on cost control, including use of computer-based systems. Through this seminar you'll perfect your scheduling and management skills, ensuring completion of your projects *on time* and *within budget*. Includes hands-on application of **Means Scheduling Manual** and **Means Building Construction Cost Data**.

Facilities Maintenance and Repair Estimating

With our Facilities Maintenance and Repair Estimating seminar, you'll learn how to plan, budget, and estimate the cost of ongoing and preventive maintenance and repair for all your buildings and grounds. Based on R.S. Means' groundbreaking cost estimating book, this two-day seminar will show you how to decide to either contract out or retain maintenance and repair work in-house. In addition, you'll learn how to prepare budgets and schedules that help cut down on unplanned and costly emergency repair projects. Facilities Maintenance and Repair Estimating crystallizes what facilities professionals have learned over the years, but never had time to organize or document. This program covers a variety of maintenance and repair projects, from underground storage tank removal, roof repair and maintenance, exterior wall renovations, and energy source conversions, to service upgrades and estimating energy-saving alternatives.

Managing Facilities Construction and Maintenance

In you're involved in new facility construction, renovation or maintenance projects and are concerned about getting quality work done on time and on or below budget, in Means' seminar **Managing Facilities Construction and Maintenance** you'll learn management techniques needed to effectively plan, organize, control and get the most out of your limited facilities resources.

Learn how to develop budgets, reduce expenditures and check productivity. With the knowledge gained in this course, you'll be better prepared to successfully sell accurate project budgets, timing and manpower needs to senior management . . . plus understand how to evaluate the impact of today's facility decisions on tomorrow's budget.

Call 1-800-448-8182 for more information

Seminars

For more information visit Means Web Site at www.rsmeans.com

2000 Means Seminar Schedule

Location	Dates
Las Vegas, NV	March 13 - 16
Dallas, TX	April 10-13
Washington, DC	April 17-20
Denver, CO	May 22-25
San Francisco, CA	June 12-15
Cape Cod, MA	September 11-14
Washington, DC	September 25-28
San Diego, CA	October TBD
Atlantic City, NJ	November TBD
Orlando, FL	November 13-16

Registration Information

Register Early... Save up to $150! Register 30 days before the start date of a seminar and save $150 off your total fee. *Note: This discount can be applied only once per order.*

How to Register Register by phone today! Means toll-free number for making reservations is: **1-800-448-8182**.

Individual Seminar Registration Fee $875 To register by mail, complete the registration form and return with your full fee to: Seminar Division, R.S. Means Company, Inc., 63 Smiths Lane, Kingston, MA 02364.

Federal Government Pricing All Federal Government employees save 25% off regular seminar price. Other promotional discounts cannot be combined with Federal Government discount.

Team Discount Program Two to four seminar registrations: $760 per person—Five or more seminar registrations: $710 per person—Ten or more seminar registrations: Call for pricing.

Consecutive Seminar Offer One individual signing up for two separate courses at the same location during the designated time period pays only $1,400. You get the second course for only $525 (**a 40% discount**). Payment must be received at least ten days prior to seminar dates to confirm attendance.

Refunds Cancellations will be accepted up to ten days prior to the seminar start. There are no refunds for cancellations postmarked later than ten working days prior to the first day of the seminar. A $150 processing fee will be charged for all cancellations. Written notice or telegram is required for all cancellations. Substitutions can be made at any time before the session starts. **No-shows are subject to the full seminar fee.**

AACE Approved Courses The R.S. Means Construction Estimating and Management Seminars described and offered to you here have each been approved for 14 hours (1.4 recertification credits) of credit by the AACE International Certification Board toward meeting the continuing education requirements for re-certification as a Certified Cost Engineer/Certified Cost Consultant.

AIA Continuing Education R.S. Means is registered with the AIA Continuing Education System (AIA/CES) and is committed to developing quality learning activities in accordance with the CES criteria. R.S. Means seminars meet the AIA/CES criteria for Quality Level 2. AIA members will receive (28) learning units (LUs) for each two day R.S. Means Course.

Daily Course Schedule The first day of each seminar session begins at 8:30 A.M. and ends at 4:30 P.M. The second day is 8:00 A.M.–4:00 P.M. Participants are urged to bring a hand-held calculator since many actual problems will be worked out in each session.

Continental Breakfast Your registration includes the cost of a continental breakfast, a morning coffee break, and an afternoon break. These informal segments will allow you to discuss topics of mutual interest with other members of the seminar. (You are free to make your own lunch and dinner arrangements.)

Hotel/Transportation Arrangements R.S. Means has arranged to hold a block of rooms at each hotel hosting a seminar. To take advantage of special group rates when making your reservation be sure to mention that you are attending the Means Seminar. You are of course free to stay at the lodging place of your choice. (**Hotel reservations and transportation arrangements should be made directly by seminar attendees.**)

Important Class sizes are limited, so please register as soon as possible.

Registration Form

Call 1-800-448-8182 to register or FAX 1-800-632-6732

Please register the following people for the Means Construction Seminars as shown here. Full payment or deposit is enclosed, and we understand that we must make our own hotel reservations if overnight stays are necessary.

☐ Full payment of $ _____ enclosed.
☐ Bill me

Name of Registrant(s)
(To appear on certificate of completion)

P.O. #: _____
GOVERNMENT AGENCIES MUST SUPPLY PURCHASE ORDER NUMBER

Firm Name _____
Address _____
City/State/Zip _____
Telephone No. _____ Fax No. _____
E-Mail Address _____
Charge our registration(s) to: ☐ MasterCard ☐ VISA ☐ American Express ☐ Discover
Account No. _____ Exp. Date _____
Cardholder's Signature _____
Seminar Name   City   Dates

Please mail check to: R.S. MEANS COMPANY, INC., 63 Smiths Lane, P.O. Box 800, Kingston, MA 02364 USA

Consulting Services Group

Solutions For Your Construction and Facilities Cost Management Problems

We are leaders in the cost engineering field, supporting the unique construction and facilities management costing challenges of clients from the Federal Government, Fortune 1000 Corporations, Building Product Manufacturers, and some of the World's Largest Design and Construction Firms.

Developers of the Dept. of Defense Tri-Services Estimating Database

Recipient of SEARS 1997 Chairman's Award for Innovation

Research...

- **Custom Database Development**—Means expertise in construction cost engineering and database management can be put to work creating customized cost databases and applications.
- **Data Licensing & Integration**—To enhance applications dealing with construction, any segment of Means vast database can be licensed for use, and harnessed in a format compatible with a previously developed proprietary system.
- **Cost Modeling**—Pre-built custom cost models provide organizations that expend countless hours estimating repetitive work with a systematic time-saving estimating solution.
- **Database Auditing & Maintenance**—For clients with in-house data, Means can help organize it, and by linking it with Means database, fill in any gaps that exist and provide necessary updates to maintain current and relevant proprietary cost data.

Estimating...

- **Estimating Service**—Means expertise is available to perform construction cost estimates, as well as to develop baseline schedules and establish management plans for projects of all sizes and types. Conceptual, budget and detailed estimates are available.
- **Benchmarking**—Means can run baseline estimates on existing project estimates. Gauging estimating accuracy and identifying inefficiencies improves the success ratio, precision and productivity of estimates.
- **Project Feasibility Studies**—The Consulting Services Group can assist in the review and clarification of the most sound and practical construction approach in terms of time, cost and use.
- **Litigation Support**—Means is available to provide opinions of value and to supply expert interpretations or testimony in the resolution of construction cost claims, litigation and mediation.

Training...

- **Core Curriculum**—Means educational programs, delivered on-site, are designed to sharpen professional skills and to maximize effective use of cost estimating and management tools. On-site training cuts down on travel expenses and time away from the office.
- **Custom Curriculum**—Means can custom-tailor courses to meet the specific needs and requirements of clients. The goal is to simultaneously boost skills and broaden cost estimating and management knowledge while focusing on applications that bring immediate benefits to unique operations, challenges, or markets.
- **Staff Assessments and Development Programs**—In addition to custom curricula, Means can work with a client's Human Resources Department or with individual operating units to create programs consistent with long-term employee development objectives.

MEMBER:

www.eas.asu.edu/joc/

For more information and a copy of our capabilities brochure, please call 1-800-448-8182 and ask for the Consulting Services Group, or reach us at www.rsmeans.com

MeansData™

CONSTRUCTION COSTS FOR SOFTWARE APPLICATIONS
Your construction estimating software is only as good as your cost data.

Software Integration

A proven construction cost database is a mandatory part of any estimating package. We have linked MeansData™ directly into the industry's leading software applications. The following list of software providers can offer you MeansData™ as an added feature for their estimating systems. Visit them on-line at **www.rsmeans.com/demo/** for more information and free demos. Or call their numbers listed below.

ACT
Applied Computer Technologies
Facility Management Software
919-851-7172

AEPCO, Inc.
301-670-4642

ArenaSoft ESTIMATING
888-370-8806

ASSETWORKS, Inc.
Facility Management Software
800-659-9001

BSD
Building Systems Design, Inc.
888-BSD-SOFT

CDCI
Construction Data Controls, Inc.
800-285-3929

CMS
Computerized Micro Solutions
800-255-7407

CONAC GROUP
604-273-3463

CONSTRUCTIVE COMPUTING, Inc.
800-456-2113

ESTIMATING SYSTEMS, Inc.
800-967-8572

G2 Estimator
A Div. of Valli Info. Syst., Inc.
800-627-3283

GEAC COMMERCIAL SYSTEMS, Inc.
800-554-9865

GRANTLUN CORPORATION
602-897-7750

HCI SYSTEMS, Inc.
800-750-4424

IQ BENECO
801-565-1122

MC^2
Management Computer Controls
800-225-5622

PRISM COMPUTER CORPORATION
Facility Management Software
800-774-7622

PYXIS TECHNOLOGIES
888-841-0004

QUEST SOLUTIONS, Inc.
800-452-2342

RICHARDSON ENGINEERING SERVICES, Inc.
602-497-2062

SANDERS SOFTWARE, Inc.
800-280-9760

STN, Inc.
Workline Maintenance Systems
800-321-1969

TIMBERLINE SOFTWARE CORP.
800-628-6583

TMA SYSTEMS, Inc.
Facility Management Software
800-862-1130

US COST, Inc.
800-955-1385

VERTIGRAPH, Inc.
800-989-4243

WENDLWARE
714-895-7222

WINESTIMATOR, Inc.
800-950-2374

DemoSource™ One-stop shopping for the latest cost estimating software for just **$19.95.** This evaluation tool includes product literature and demo diskettes for ten or more estimating systems, all of which link to MeansData™. **Call 1-800-334-3509 to order.**

**FOR MORE INFORMATION ON ELECTRONIC PRODUCTS CALL
1-800-448-8182 OR FAX 1-800-632-6732.**

MeansData™ is a registered trademark of R.S. Means Co., Inc., *A CMD Group* Company.

For more information
visit Means Web Site
at www.rsmeans.com

New Titles

Facilities Operations & Engineering Reference
An all-in-one technical reference for planning & managing facility projects & solving day-to-day operations problems

Ten major sections address:

- **Management Skills**
- **Economics** (budgeting/cost control, financial analysis, VE, etc.)
- **Civil Engineering & Construction Practices**
- **Maintenance** (detailed staffing guidance and job descriptions, CMMS, planning/scheduling, training, work orders, preventive/predictive maintenance)
- **Energy Efficiencies** (optimizing energy use, including heating, cooling, lighting, and water)
- **HVAC**
- **Mechanical Engineering**
- **Instrumentation & Controls**
- **Electrical Engineering**
- **Environmental, Health & Safety**

Produced jointly by R.S. Means and
the Association for Facilities Engineering.

$99.95 per copy
Over 700 pages, Illustrated, Hardcover
Catalog No. 67318

Planning & Managing Interior Projects New Second Edition
by Carol E. Farren, CFM

Revised edition addresses the major changes in technology and business that have affected all aspects of interiors projects. Guides you through every step in managing interior design and construction for commercial endeavors, from initial client meetings to post-project administration. Corporate downsizing, company mergers, and advances in information technology have revolutionized the workplace. This book addresses all of these changes, showing how to plan effectively for current and future organizational needs. Includes new sections on:

- Evaluating space requirements and selecting the right site
- Alternative work models (telecommuting, hoteling, etc.)
- Records and document management
- Telecommunications and data issues
- Working with consultants
- Environmental considerations

$69.95 per copy
Over 400 pages, Illustrated, Hardcover
Catalog No. 67245A

Residential & Light Commercial Construction Standards
A Unique Collection of Industry Standards That Define Quality in Construction
For Contractors & Subcontractors, Owners, Developers, Architects & Engineers, Attorneys & Insurance Personnel

Compiled from the nation's major building codes, and from scores of publications and reports from professional institutes and other authorities, this one-of-a-kind resource enables you to:

- Set a standard for subcontractors and employees
- Protect yourself against defect claims
- Substantiate your own claim for inferior workmanship
- Resolve disputes
- Overview installation methods
- Answer client questions with an authoritative reference

$59.95 per copy
Over 500 pages, Illustrated, Softcover
Catalog No. 67322

ORDER TOLL FREE 1-800-334-3509
OR FAX 1-800-632-6732.

2000 Order Form

Qty.	Book No.	COST ESTIMATING BOOKS	Unit Price	Total
	60060	Assemblies Cost Data 2000	$139.95	
	60010	Building Construction Cost Data 2000	85.95	
	61010	Building Const. Cost Data–Looseleaf Ed. 2000	109.95	
	63010	Building Const. Cost Data–Metric Version 2000	89.95	
	60220	Building Const. Cost Data–Western Ed. 2000	89.95	
	60110	Concrete & Masonry Cost Data 2000	79.95	
	50140	Construction Cost Indexes 2000	198.00	
	60140A	Construction Cost Index–January 2000	49.50	
	60140B	Construction Cost Index–April 2000	49.50	
	60140C	Construction Cost Index–July 2000	49.50	
	60140D	Construction Cost Index–October 2000	49.50	
	60310	Contr. Pricing Guide: Framing/Carpentry 2000	36.95	
	60330	Contr. Pricing Guide: Resid. Detailed 2000	36.95	
	60320	Contr. Pricing Guide: Resid. Sq. Ft. 2000	39.95	
	64020	ECHOS Assemblies Cost Book 2000	149.95	
	64010	ECHOS Unit Cost Book 2000	99.95	
	54000	ECHOS (Combo set of both books)	214.95	
	60230	Electrical Change Order Cost Data 2000	89.95	
	60030	Electrical Cost Data 2000	89.95	
	60200	Facilities Construction Cost Data 2000	209.95	
	60300	Facilities Maintenance & Repair Cost Data 2000	199.95	
	60160	Heavy Construction Cost Data 2000	89.95	
	63160	Heavy Const. Cost Data–Metric Version 2000	89.95	
	60090	Interior Cost Data 2000	89.95	
	60120	Labor Rates for the Const. Industry 2000	189.95	
	60180	Light Commercial Cost Data 2000	76.95	
	60020	Mechanical Cost Data 2000	89.95	
	60150	Open Shop Building Const. Cost Data 2000	89.95	
	60210	Plumbing Cost Data 2000	89.95	
	60040	Repair and Remodeling Cost Data 2000	79.95	
	60170	Residential Cost Data 2000	74.95	
	60280	Site Work & Landscape Cost Data 2000	89.95	
	60050	Square Foot Costs 2000	99.95	
		REFERENCE BOOKS		
	67147A	ADA in Practice	72.95	
	67310	ADA Pricing Guide	72.95	
	67298	Basics for Builders: Framing & Rough Carpentry	24.95	
	67273	Basics for Builders: How to Survive and Prosper	34.95	
	67307	Basics for Builders: Plan Reading & Takeoff	35.95	
	67261	Building Prof. Guide to Contract Documents	64.95	
	67312	Building Spec Homes Profitably	29.95	
	67250	Business Management for Contractors	17.98	
	67146	Concrete Repair & Maintenance Illustrated	69.95	
	67278	Construction Delays	59.95	
	67268	Construction Paperwork	26.48	
	67255	Contractor's Business Handbook	21.48	
	67314	Cost Planning & Est. for Facil. Maint.	82.95	
	67317	Cyberplaces: The Internet Guide for A/E/C	59.95	
	67230A	Electrical Estimating Methods–2nd Ed.	64.95	
	64777	Environmental Remediation Est. Methods	99.95	
	67160	Estimating for Contractors	35.95	
	67276	Estimating Handbook	99.95	
	67249	Facilities Maintenance Management	86.95	

Qty.	Book No.	REFERENCE BOOKS (Con't)	Unit Price	Total
	67246	Facilities Maintenance Standards	$159.95	
	67264	Facilities Manager's Reference	86.95	
	67318	Facilities Operations & Engineering Reference	99.95	
	67301	Facilities Planning & Relocation	109.95	
	67231	Forms for Building Const. Profess.	94.95	
	67288	Forms for Contractors	79.95	
	67260	Fundamentals of the Construction Process	34.98	
	67258	Hazardous Material & Hazardous Waste	44.98	
	67148	Heavy Construction Handbook	74.95	
	67308B	Home Improvement Costs–Interior Projects	19.95	
	67309B	Home Improvement Costs–Exterior Projects	19.95	
	67304	How to Estimate with Metric Units	24.98	
	67306	HVAC: Design Criteria, Options, Select.–2nd Ed.	84.95	
	67281	HVAC Systems Evaluation	84.95	
	67282	Illustrated Construction Dictionary, Condensed	59.95	
	67292	Illustrated Construction Dictionary, Unabridged	99.95	
	67295A	Landscape Estimating–3rd Ed.	62.95	
	67266	Legal Reference for Design & Construction	54.98	
	67299	Maintenance Management Audit	32.48	
	67302	Managing Construction Purchasing	31.48	
	67294	Mechanical Estimating–2nd Ed.	64.95	
	67245A	Planning and Managing Interior Projects, 2nd Ed.	69.95	
	67283A	Plumbing Estimating Methods, 2nd Ed.	59.95	
	67236A	Productivity Standards for Constr.–3rd Ed.	159.95	
	67247A	Project Scheduling & Management for Constr.	64.95	
	67262	Quantity Takeoff for Contractors	17.98	
	67265A	Repair & Remodeling Estimating–3rd Ed.	69.95	
	67322	Residential & Light Commercial Const. Stds.	59.95	
	67254	Risk Management for Building Professionals	29.98	
	67253	Roofing: Design Criteria, Options, Selection	31.48	
	67291	Scheduling Manual–3rd Ed.	64.95	
	67145A	Square Foot Estimating Methods–2nd Ed.	69.95	
	67241	Structural Steel Estimating	39.98	
	67287	Successful Estimating Methods	64.95	
	67313	Successful Interior Projects	34.98	
	67233	Superintending for Contractors	35.95	
	67321	Total Productive Facilities Management	79.95	
	67284	Understanding Building Automation Systems	39.98	
	67259	Understanding Legal Aspects of Design/Build	79.95	
	67303	Unit Price Estimating Methods–2nd Ed.	59.95	
	67319	Value Engineering: Practical Applications	79.95	

MA residents add 5% state sales tax

Shipping & Handling**

Total (U.S. Funds)*

Prices are subject to change and are for U.S. delivery only. *Canadian customers may call for current prices. **Shipping & handling charges: Add 7% of total order for check and credit card payments. Add 9% of total order for invoiced orders.

Send Order To: ADDV-1001

Name (Please Print) _____

Company _____

☐ Company
☐ Home   Address _____

_____

City/State/Zip _____

Phone # _____ P.O. # _____

Mail To: R.S. Means Company, Inc., P.O. Box 800, Kingston, MA 02364-0800   (Must accompany all orders being billed)